Modern Energy Technology

Vol. I

Research and Education Association.
11

MODERN ENERGY TECHNOLOGY

- NUCLEAR
- COAL
- PETROLEUM
- SOLAR
- GEOTHERMAL
- FUEL CELLS
- OIL SHALE
- TAR SANDS
- ORGANIC WASTES

VOL. I

Staff of Research and Education Association

Research and Education Association
342 Madison Avenue
New York, N.Y. 10017

MODERN ENERGY TECHNOLOGY

Printed in the United States of America

Library of Congress Catalog Card Number 75-13406

International Standard Book Number 0-87891-506-0

PREFACE

The need to supplement our energy sources available from currently known oil reserves in the world, is now well accepted by the industrialized nations. The need for supplementary energy sources became particularly apparent after the Mideast war of 1973 when oil producing nations applied a strangle-hold to many of the oil consuming nations through boycotts and price demands. The industrialized oil consuming nations are particularly susceptible to political and economic pressure because many of them have little or no oil reserves of their own and must import virtually all their requirements. The problem became especially acute for these nations after that war when the oil producing nations united to force up the price of oil. The resultant effect produced a sharp inflationary surge in the economies of the industrialized nations who were already plagued by their past inflationary problems. Due to the large increase in the cost of energy, the prices of almost all manufactured products increased. This, in turn, led to a decrease in the demand for manufactured goods. As a result, economic recession became superimposed on a difficult-to-manage inflation. The inflation in prices of consumer goods coupled with recessions in industrial activities and hence employment, brought several nations to the brink of financial bankruptcy with an accompanying large deficit in the trade balance.

The industrialized nations are generally in agreement that to avoid political and economic pressures that may have disastrous effects to the extent of provoking another war, it is essential to find and develop alternate energy sources to add to known oil reserves. The main problem, is to select the energy sources which are most promising from the viewpoint of cost, development time, environmental pollution effects, and availability in quantities to meet future demands. An intensive development program is required to bring about the desired results.

One apparent alternate possibility is to search for new or unknown oil sources either on-shore or off-shore. Any large findings of new oil sources could sharply change the political and economic relationships prevailing among the oil producing and consuming nations, and bring about marked reductions in oil prices. This relief, however, would be short-lived, since all oil reserves will eventually become depleted. U. S. Geological data indicates that the U. S. for example, has only a 12-to-15 year supply of measured and estimated oil reserves, plus a 30-60 year possible supply in still-untapped reservoirs on the mainland, and on underwater continental shelves. It is therefore essential to consider sources other than oil as well.

Among the most important other sources are nuclear, coal, natural gas, oil shale, solar, geothermal, hydrogen fuel cells, and organic wastes.

The nuclear source has particular advantages because it is available in sufficient quantity to meet future demand, and the technology has already been developed to the extent that the required number of power plants can be built. Fear of the effects of nuclear wastes, radiation, and plant accidents has delayed the construction and installation of an adequate number of nuclear plants in the U. S. To this date, however, such fears are unfounded because techniques are available to cope safely with the presence and effects of nuclear power plants. The risks to life and health from nuclear plants can be readily reduced to a level far below that encountered by society in the operation of automobiles, for example.

It is expected that fears of nuclear power plants will be overcome and nuclear installations will be accepted. If and when a major proportion of the U. S. energy demand is supplied from nuclear sources, a practical arrangement would be to have a few large central installations rather than a large number of local installations. The few central installations can be closely controlled from the viewpoint of safeguarding nuclear materials against theft and misuse, and of safety in plant operation. Satisfactory plant operation requires skilled personnel, which is as yet, in limited supply. It is therefore advantageous to aim for few central installations in lieu of attempting to spread nuclear plants among numerous local installations. Such central installations, furthermore, can be equipped economically with complex and sophisticated devices to protect health and safeguard against theft of nuclear materials. Central installations can be located so that they will provide for efficiency in power transmission to local communities. This will, in turn, result in minimum cost to the consumer.

The chief objection to the use of coal is that it is accompanied by the release of pollutants to the atmosphere. However, the U. S. has a large coal reserve (estimated to last several hundred years) and these may be an important interim source of energy until other sources are sufficiently developed.

Gasification and liquefaction of coal are under current development to convert this energy source into a product that can be more conveniently transported and utilized. These processes remain to be developed, however, so that they may be economically practical.

Natural gas can be recovered from the same source from which oil is obtained. However, the measured and estimated reserves point to a supply of only 16-to-22 years. Potential reserves may provide an additional 40-to-80 year supply.

Oil shale is available in the U. S. in large quantity, but the economical recovery of oil from this source is far off.

Solar energy meets the requirements of adequate supply and it will not pollute the environment. It will require, however, many years of development until this source is practical to use on a large scale. The same is true for geothermal energy, except that the latter may emit pollutants into the environment.

The world's energy demands by 1985 will probably not be met by any single source. Stationary power plants may be operated, for example, from nuclear, coal, solar, or geothermal sources, and leave the more limited oil for use in transportation.

There is urgency in achieving a solution to the energy problem. We have little time. To develop energy sources that meet the requirements of future power demands, price, and environmental regulations, represents a herculean task estimated to cost over half a trillion dollars over the next ten years. This expenditure, however, is a small price for the U. S. to pay to remain independent of foreign political demands. In addition, the search for the energy solution will open large new areas of employment and opportunities for industry. The rewards are there for those who can meet the challenge.

In preparing the contents of the two volumes of this publication, a massive amount of information available from government agencies, universities, and industry, was screened, analyzed, and reduced into a form believed to be most useful to the reader. Contributions are gratefully acknowledged from the many persons and organizations referenced throughout the two volumes. Particularly significant contributions are thankfully included from the staffs of the organizations listed below.

Max Fogiel, Ph. D. , Editor

CONTRIBUTORS TO "MODERN ENERGY TECHNOLOGY"

ACF Industries... Aerojet-General Nucleonics... Aircraft Nuclear Propulsion... Alco Products... Allis-Chalmers... AMF Atomics.. American Association of Petroleum Geologists ...American Chemical Society... American Gas Association... American Geophysical Union.. American Institute of Chemical Engineers... American Institute of Mining... American Mining Congress.. American National Standards Institute... American Nuclear Society... American Petroleum Institute.. American Society of Mechanical Engineers.. American Standard, Inc... Appalachian Regional Commission .. Argonne National Laboratory.. Associated Universities, Inc... Atomic Power Development Associates... Atomics International.. AVCO... Babcode & Wilcox... Battelle Memorial Institute... Bendix... Bettis Atomic Power Laboratory... Blaw-Knox... Broken Hill Proprietary Co., LTD.. Brookhaven National Laboratory.. Bureau of Land Management... Bureau of Mines... Bureau of Solid Waste Management... Burns & Roe... California... Institute of Technology... Carpco Research and Engineering ... Charles River Associates... Colorado School of Mines ...Combustion Engineering. .Consolidated Edison.. Consultants Bureau .. Cook Electronic... Council on Environmental Quality... Curtiss-Wright.. Daystrom, Inc.. Dept. of Defense... Dept. of Environmental Resources .. Dept. of Interior... Detroit Edison... Dow Chemical.. E. I. Du Pont.. Ebasco Service... Edward Federal Water Pollution Control Admin...Foster Associates ... Foster Wheeler ... General Dynamics Corp... General Electric... General Motors ... General Nuclear Engineering... Georgia Institute of Technology... Gould-National Batteries ... Graphics Management... Gulf General Atomic ... Gulf Radiation Technology... H. K. Ferguson ... Hanford Engineering Development Laboratory ... Hittman Associates... Hughes Aircraft... Idaho Nuclear... Ideal Electric and Manufacturing ... Independent Petroleum Assoc. of America... Ingalls Shipbuilding... Institute of Electrical and Electronics Engineers... Institute of Gas Technology ... Institute of Mechanical Engineers.. Instrument Society of America... International Atomic Agency ... International District Heating Association... International Federation for Information Processing Congress... International Solar Energy Conference... Internuclear... Intersociety Energy Conversion Engineering Conference ... Interstate Oil Compact Commission... Intertechnology ... Iowa State Univ. of Science and Technology... Japan Atomic Energy Research Institute... Joint Committee on Atomic Energy... Kaiser Engineers... Kaman Aircraft... Kentucky Division of Reclamation... Knolls Atomic Power Laboratory... Lawrence Radiation Laboratory... Lockheed Aircraft... Los Alamos Scientific Laboratory... Manhattan Engineer District... Mare Island Naval Shipyard... Marine Technology Society... Marton Marietta... Maxon Construction ... Metallurgist and Petroleum Engineers... Mid-Continent Oil and Gas Association... Milletron... Mitre... Molten-Salt Group... Montana Agricultural Experimental Station...Morgantown Coal Research Center ... NASA ... National Academy of Engineering ... National Academy of Sciences... National Bureau of Standards... National Coal Association... National Petroleum Council... Nattional Reactor Testing Station. National Research Associates... National Research Council.. National Science Foundation... Naval Research Laboratory... Nevada Test Site... N. Y. Shipbuilding... Newport News Shipbuilding... North American Aviation... North Carolina State University ... Northern States Power Co... Nuclear Rocket Dev. Station... Nuclear Systems Associates... Oak Ridge National Laboratories... Oil Subcommittee of National Petroleum Council ...Organization for Economic Cooperation and Development... Pacific Northwest Laboratory ... Pa. Dept. of Environmental Resources ... Phillips Petroleum Co... Pipeline Industry... Pittsburgh Energy Research Center... Portsmouth Naval Shipyard... San Francisco Bay Naval Shipyard... Smithsonian Institution ... Society Petroleum Engineering... Solar Energy Society... Tennessee Valley Authority... Textron... Transatlantic American Nuclear Society... Tripartite Conference... Tyco Laboratories... United Aircraft... United Nuclear ... United Nuclear Industries... U. S. A. Standards Institute... U. S. Bureau of the Census... U. S. Dept. of Commerce.. U. S. Dept. of Health, Education and Welfare... U. S. Dept. of Interior... U. S. Dept. of Labor... U. S. Geological Survey... U. S. Naval Hydrographic Office... University of California at Los Angeles ... Univ. of Minnesota ... Univ. of Montana... Univ. of Tennessee and Oak Ridge... Univ. of Tenn. Space Institute... Univ. of Texas at Austin... Univ. of Tokyo... University of Utah... Univ. of Washington... Univ. of Wyoming ... W. Virginia Surface Mining and Reclamation Association... W. Virginia Univ.. Western Electric... Westinghouse Atomic Power Divisions... World Energy Conference... World Meteorological Organization... Wyoming Geological Association

CONTENTS

SECT. 4 PROCESS INSTRUMENTATION 696

SECT. 5 NEUTRON-FLUX SIGNAL CONDITIONING 747

SECT. 6 TRANSFER-FUNCTION MEASUREMENT SYSTEMS 772

SECT. 7 CONTROL-ROD DRIVES AND INDICATING SYSTEMS 798

SECT. 10 INSTALLATION OF INSTRUMENTATION SYSTEMS 852

PART I

REVIEW OF THE ENERGY FIELD

THE POTENTIAL FOR ENERGY CONSERVATION

ENERGY PROGRAM PLANNING

CHAPTER 1

A SUMMARY REVIEW OF THE ENERGY FIELD

The use of energy in various fuel cycles is discussed below in terms of conventional market mechanisms and under the assumptions that there would be no major change in American life styles, reduction in economic growth or significant shifts in energy technology. It also is assumed that the quantities of energy currently being used generally result in the lowest immediate overall cost of the materials or services involved. Energy costs are viewed as being fairly steady. If energy costs were to rise by a factor of 10, while other costs remained constant, the character of energy consumption would undoubtedly be modified.

The application of some of these assumptions can be illustrated in the following case examples:

(i) The size of the wire which an electric utility uses to transmit electricity from its transformer to an individual vendor is that which provides electricity at lowest cost. Making the diameter of the line larger or using a material with less electrical resistance would reduce energy losses but at increased capital costs. If energy costs were to rise ten times while line costs remained constant, the utility would be expected to install lines of larger diameter, or lines made of other materials—at least in new homes—to reduce line losses and thus minimize overall costs.

(ii) If oil prices were to rise, a number of secondary recovery techniques, already discovered for oil recovery but not now economically attractive, would become feasible and probably be widely used. Recoveries might be increased dramatically.

BUILDING THE RESOURCE BASE

Policies for research and regulation aimed at improving recovery rates would serve to expand our resource base. In a recent Symposium on Coal and Public Policies[1] it was noted that under present working conditions for the development of fossil fuels our overall energy system is probably not more than 10 percent efficient. Although recoveries of energy in the gas industry are high (sometimes up to 90 to 95 percent) the average recovery for petroleum in the United States is approximately 30 percent and in the range of 50 percent for coal.

While opportunities exist for increasing the recoveries of all fossil fuels—and several types of nuclear fuels—the number of remaining sites where hydroelectric power can be generated is too small to be of any major significance in the total energy picture of the future. Geothermal resources that are now being used are quite limited. If we must depend on present technology for using the heat in the earth's interior, geothermally-generated energy will most likely remain of importance in only a few local areas. With the development of new technology geothermal sources may eventually supply a larger portion of future energy demands.

Domestic coal reserves are very large and sufficient quantities have been identified to satisfy anticipated demands far into the future. It is now generally accepted that there are hundreds of years of supply even if coal is used at greatly increased rates. Should massive amounts be required for the conversion of coal to gas and coal to oil, as well as for electricity generation, then the supply would be exhausted at a much faster pace than presently foreseen. Even under these circumstances the U.S. supply of coal should last well into the next century.

The time-span for nuclear fuels is somewhat different. At or near present prices proved reserves to supply the light water reactors that are either now in operation, under construction or planned should be ample until 1990 to 1995. A report to the Secretary of Interior in June of 1970 [2] stressed the need for conserving existing nuclear reserves as follows:

> The projected operation of light-water reactors with existing known resources will result in uranium ore demands which may cause uranium costs to increase unless more uranium is found and a commercial breeder is developed on a timely basis. However, because of the high energy content of uranium, power costs in a light-water reactor would increase only 0.3 mill per kilowatt hour if the present $8 per pound price level of uranium were to double. Exploration is continuing and more uranium will undoubtedly be found, but more efficient use of existing reserves would be very desirable. Breeder reactors will increase the fuel potential of any given amount of natural uranium by a factor of over 30. In addition, an increase in uranium price to $100 per pound would increase breeder reactor power costs by only 0.2 mill per kilowatt hour.

Unless breeder reactors are in commercial operation by the end of the century, uranium reserves, mineable at low cost, will be in short supply in which case a large exploration program may be needed to discover additional low cost supplies. Should such an exploration effort be relatively unsuccessful the costs of generating electricity from nuclear fuels would increase.

At present, the reserve situations for natural gas and oil are nearly identical. Previously, oil reserve to consumption ratios had remained relatively constant for a number of years (at about 12). In recent years they have dropped to under 10—about the lowest level considered satisfactory for efficient operation.

Since 1948 the United States has been a net importer of petroleum and in 1970 imports of all types of petroleum products reached nearly 25 percent of domestic demand. Although the gas and oil reserve situation is now very similar this has only come about in the last few years. Until about 1960 the gas reserve to production ratio had been as high as 20 but starting in 1968 (excluding the Alaska discoveries) more gas has been consumed each year than discovered. This, together with the high growth rate for gas, had reduced the reserve to production ratio to about 11 in 1971. In view of the new concern about environmental pollution, even greater pressures will be placed on gas supplies since this is the cleanest of the fossil fuels.

In a report prepared by the National Petroleum Council,[3] domestic oil and gas supplies were discussed as follows:

Oil

Domestic production of conventional crude oil, excluding the Alaskan North Slope, was projected to increase slightly in the next 1 to 3 years, but to decline thereafter. By 1975, crude oil production (excluding the North Slope) would be 8.55 million B/D compared to 9.1 million B/D in 1970, By 1985 the corresponding figure would be 7.9 million B/D. Thus, were it not for the North Slope, domestic crude oil production would be lower in 1985 than in 1970 by 1.2 million B/D, a substantial percentage of the starting figure. With condensates and gas liquids also lower, the domestic trend would have turned down steeply.

Therefore, in order to meet growing demands for petroleum liquids, imports would have to increase more than fourfold by 1985, reaching a rate of 14.8 million B/D in that year. Assuming the availability of foreign supply, oil imports would then account for 57 percent of total petroleum supplies and would represent 25 percent of total energy consumption. Most of the imports would have to originate in the Eastern Hemisphere because of the limited potential for increased imports from Western Hemisphere sources.

Gas

Under current regulatory policies and Federal leasing policies, however, the supplies of domestic natural gas (excluding North Slope) could be expected to fall from 21.82 TCF in 1970 to 13 TCF in 1985. By this time, another 1.50 TCF would be contributed by the Alaskan North Slope.

The potential for improving the domestic resource base for both oil and gas is considerable. For example, it has been estimated that the ultimate reserves of oil in the lower 48 States could be between 575 billion and 2,400 billion barrels. Present proved reserves are 36.5 billion barrels.[4] The ultimate natural gas resources have been estimated to be between 1,250 trillion and 2,175 trillion cubic feet. Present proved reserves are estimated at about 275 trillion cubic feet.

The oil and gas industries have repeatedly argued that these new resources would be located if incentives for drilling become more adequate. A recent study released by the National Petroleum Council estimated that 55 percent of the recoverable oil and 66 percent of the recoverable gas are still in the ground. Drilling rates have declined steadily since the mid-1950's. Now slightly more than half as many wells are drilled each year as were drilled in 1956 (which was the peak year).

To discover additional oil onshore it will become necessary to look for deeper formations, previously overlooked shallower formations, and new targets in lean or marginal areas. Most of the easy-to-find structural traps have probably been discovered but the more difficult-to-find "subtle" traps cannot be detected by the instruments or methods now available.

Although the methods used to identify favorable target locations are better than they were in the 1920's and 1930's, exploration for oil and gas still involves very high risks. Significant research opportunities exist for developing new technology and methods that would increase the finding ratio and greatly reduce costs.

As shown on table 1-1 the U.S. petroleum reserves represent less than 20 percent of the world total of discovered reserves. If we can continue to rely on overseas supplies (as we have since 1948) then our own resource base for oil is not as critically important. If, on the other hand, reliance on overseas resources is disqualified as a primary source for policy reasons of security, possible price manipulations by the few countries with large reserves, or balance of payments problems, then an entirely different set of policy options must be considered for building reserves.

Table 1-1 —WORLD CRUDE OIL-IN-PLACE DISCOVERABLE AND DISCOVERED

[In billions of barrels]

	Existing [1]	Discoverable [2]	Discovered	
			January 1, 1962 [3]	January 1, 1966 [4]
United States	1,600	1,000	346	386
Canada, Mexico, Central America, and Caribbean	500	300	50	77
South America (including Venezuela)	800	500	214	238
Total Western Hemisphere	2,900	1,800	610	701
Europe	500	300	21	26
Africa	1,800	1,100	56	139
Middle East (excluding Turkey)	1,400	900	793	928
Southern Asia	200	100	7	8
U.S.S.R., China, Mongolia	2,900	1,800	90	122
Indonesia, Australia, et cetera	300	200	21	25
Total Eastern Hemisphere	7,100	4,400	988	1,248
Total world	10,000	6,200	1,598	1,949

[1] Total oil-in-place (Hendricks, 1965).
[2] Total discoverable oil-in-place (Hendricks, 1965).
[3] Original oil-in-place in known reservoirs as at Januray 1, 1962 (Torrey, Moore and Weber, 1963).
[4] Hypothetical calculation of original oil-in-place in reservoirs as known at January 1, 1966 (D.C. Ion, 1967).

Source: U.S. Energy Outlook 1971–85, National Petroleum Council, vol. II.

Until recently U.S. reliance on imports of natural gas and petroleum followed different patterns: While we imported some gas from Canada and Mexico (less than 4 percent of consumption) foreign reserves of natural gas were of little interest. Now, however, in view of the present shortage of natural gas in the United States, all possible sources of supply are of interest. Because the technology for moving natural gas as a liquid (LNG) has been shown to be technically feasible the world reserves of gas are theoretically potentially available to the United States at least to supply peak demands. U.S. proved gas reserves represent about 20 percent of the world reserves. The other major known deposits outside of North America are in the Middle East, North Africa, and the U.S.S.R. Plans for bringing North African gas to the eastern seaboard of the United States are well advanced and may provide some relief for the anticipated natural gas shortage.

Manufacture of synthetic natural gas from coal, oil shale, and petroleum fractions is presently under active experimentation. Plants using petroleum as a feedstock are under construction. If coal or oil shale are the raw materials there are no foreign policy implications. If petroleum fractions are to be used the same issues as noted earlier with respect to foreign sources of petroleum supply apply here also.

PRODUCTION OF FUELS

Coal.—In the production of coal by underground methods, the generally accepted recovery rate is 50 percent. In a recent study it was found that the average recovery rate was 57 percent. In some mines recoveries approach 90 percent but only under the special circumstances where reserves are limited and the coal is of high-grade metallurgical quality and high recovery rates are economical. In the few long-wall mining operations (a mining method extensively used in Europe) the average recovery rates are significantly higher than for the U.S. method of room and pillar mining.

Opportunities for increasing recovery rates through technological improvements were discussed at a recent Cornell workshop on energy and the environment:

> Coal mining recoveries can be significantly improved through adoption of currently available technology. The bulk of underground mining operations in the United States is conducted with the classic room and pillar technique which requires the preservation of large areas of the scene as a supporting network. In Europe over the past 15 years the long-wall technology has developed which has permitted recoveries in the range of 80–90 percent of the coal initially in place. This technology has been developed on a relatively hit-or-miss basis in Europe, and should be amenable to a systematic research attack to reduce the cost and improve the effectiveness of long-wall technology.

Since longwall operations contribute to increased occupational safety (a subject of growing legislative importance), it may be expected that the number of mines using this method will increase. However, a major finding of the U.S. Bureau of Mines study shows that many mines still operate at very low recovery rates—in one instance only 29 percent.

Strip mining recoveries are assumed to average about 80 percent (this does not include contour stripping). Some stripping operations are thought to be even higher but when all of the coal that might be recovered (outcrop coal that has a low heating value), the losses at the boundaries of the property, and coal that must be left under streams and rivers) is included, 80 percent appears to be a rounded average for most strip mines.

Considering the current large reserves of coal, it is unlikely that major changes would be made in mining techniques to increase recovery efficiency, unless new economic or social forces are imposed on the industry.[14]

The latest year for which published data are available on the energy consumption for producing, transporting the coal to the tipple and for upgrading at the coal preparation plants is 1963. In that year nearly 464 million tons of coal were produced, about 305 million from underground mines and the balance from strip and auger operations.[15]

Table 1-2 shows that in 1963 96.4 x 10^{12} B.t.u. which was only 0.8 percent of the energy produced in mining was used to extract, transport to the surface, and prepare the coal (including drying of wet coal).

Uranium.—Because the uranium present in deposits, unlike coal, is always found in relatively dilute quantities mixed with large volumes of other rocks and minerals, large amounts of unwanted material must be removed during the mining operation. For this reason, and since the low grade part of the deposit may be deliberately not mined, it is difficult to evaluate the percentage of uranium re-

Table 1-2

Quantities:	*Estimated B.t.u.'s* $10^{(12)}$
633,000 tons coal	15. 9
1,993 barrels of distillate oil	11. 5
223,000 barrels of residual oil	1. 3
982 million cubic feet of natural gas	. 9
Other fuels ($2,660,000) estimate	10. 0
Electric energy 5,047 billion kilowatt hours at 30 percent efficiency	56. 8
Total	96. 4

covery. Like coal, however, the mining may be done by either surface or underground methods, the choice depending on such factors as how close the deposit is to the surface, the type of rock and soil that is associated with the uranium ore, and the size, shape and grade of the deposit.

The Bureau of Mines estimated [16] that the energy consumed in mining the very low grade uranium deposits in Chattanooga shales, even under these unfavorable circumstances, is insignificant.

Petroleum.—Average recovery of petroleum from a reservoir is now about 30 percent, including secondary and tertiary recovery. The recovery percentage has been increasing on the average about 0.35 percentage points per year and, in the best situations, recoveries as high as 80 percent are possible. The degree of recovery of oil in place, unlike coal, is based on the physical properties of the reservoir (permeability porosity, initial gas pressure and rock structure) and on the physical properties of the oil (viscosity and chemical constitution.) Recovery can vary from as much as 5 percent to 80 percent.

A recent report prepared by a Subcommittee of the National Petroleum Council [17] discussed the source of increased recovery efficiency as follows:

> The cumulative recovery efficiency trends shown for each region and major geographical areas also tell an interesting story. For the United States as a whole (excluding North Slope) the recovery efficiency was increasing by approximately 0.35 percentage points per year or growing at a compound rate of 1.32 percent per year during the 14-year historical period. During the 15-year forecast period, it is projected to grow at a compound rate of 1.35 percent per year (increase 0.45 percentage points per year). This is partly due to the increase in offshore activity where the fields discovered have a higher initial primary recovery efficiency than has been experienced in the onshore fields as a whole. However, the increase in secondary and tertiary activity level provides most of the increased recovery efficiency.

Secondary methods for increasing the amount of oil recovered from reservoirs assist in increasing the amount of oil produced. For example, as the natural reservoir pressure decreases during production, the amount of oil extracted eventually is reduced to uneconomic rates of production. By supplementing the natural reservoir pressure, reducing the oil viscosity, increasing the reservoir permeability, or changing the forces in the reservoir the amount of oil that can be recovered is increased. Methods for increasing petroleum recovery by such methods are generally either water flooding, injection of solvent or carbon dioxide, hydraulic or explosive fractioning, or the use of wetting agents as flood water additives.

Tertiary recovery methods following water flooding are also being used. This usually involves some thermal methods or the injection of miscible fluids, polymers, emulsions, and so forth, so that the residual oil can be removed economically in greater quantities. Such methods are under study and may be used more widely in the future.

Lowering the viscosity of oil to increase its flow characteristics by heat can be accomplished in several ways. One method is by the injection of hot steam that will increase the temperature of the oil; however, there are limitations to the use of this method in many locations. In-situ combustion or "fireflooding," in which part of the oil is burned, has much greater potential and, while as much as 5 to 15 percent of the oil is consumed in providing the heat, most of the oil is recovered. The injection of hot water has also had limited use.

A number of research problems must be resolved if recovery efficiencies are to be improved as a means of significantly extending our resource base. Research needs include an improved understanding of (1) porosity and permeability in oil reservoirs, (2) how fluids flow through the pore network, (3) the internal structure of the reservoir bodies, and (4) the nature of surface forces in the reservoir. New information on the various methods of stimulating both secondary and tertiary recovery would also be useful. Since the addition of solvents to improve the flow of oil and underground combustion techniques are not well understood, further studies in these subjects would be extremely useful; neither of these techniques can be used routinely but both offer significant recovery improvement possibilities. In a recent listing of study opportunities, Interior Secretary Morton [18] included "development of improved oil recovery techniques such as miscible fluid or thermal methods" among much needed R. and D.

There are institutional problems associated with the efficient recovery of oil. Because of the physical nature of oil and gas deposits, conservation can only be achieved if an entire reservoir is treated using the best engineering practices. The courts, however, have ruled that oil or gas (or both) belong to the party who brings the resource to the surface—the "law of capture." Since multiple ownership of a reservoir often occurs, this has led to an excessive number of wells being drilled.

A report prepared by Arthur D. Little, Inc.,[19] stressed the need for such regulations:

> Resource conservation regulation is largely concerned with oil and gas and originated in the major producing States. These regulations have developed over a number of years in parallel ways as between States but usually differing in specifics. One group of regulations might be classed as those to insure good engineering practice and concern themselves with such factors as well spacing, drilling regulations, salt water disposal, waste of gas produced along with oil, and related problems * * * Texas, and its adjoining states of Oklahoma, Louisiana, and New Mexico, developed market demand proration regulations. These controlled well spacing, assigned allowable maximum producing rates to individual wells, and then, based on estimates of the market demand for oil, prorated the State demand for oil to the individual well.

Other institutional barriers affect the opportunity to achieve maximum conservation at the state level. The A. D. Little report adds:

> Unitization is a related problem to proration in the area of

correlative rights. In large concession areas outside the United States, individual operators usually will control the development of the entire reservoir without regard to the conflicting interests of a multiplicity of surface owners. The operator can then locate wells and institute secondary recovery according to engineering economic criteria rather than lease boundaries. Unitization—the agreement of the owners in a field to pool their interests under one operator for joint development—permits development along similar lines to individual ownership. Compulsory unitization statutes are law in a number of States, but have not passed in others, such as Texas and California, largely because independent producers and leaseholders fear the power of major oil companies in the unit to operate to their disadvantage. The absence of unitization tends to make new field drilling move along lease lines rather than reservoir lines and may impede the ability of operators to justify secondary recovery programs.

The Department of the Interior commented on the unitization practice as follows:[20]

> Such practices have prevented premature abandonment in times of depressed markets, and have contributed materially to the ultimate recovery of energy resources.

Natural Gas.—Unlike the situation for crude oil (where recoveries are normally low) there is little room to improve natural gas recoveries in normal reservoirs since 80 percent recovery is now generally attained. There are large numbers of reservoirs identified, however, that contain considerable gas resources with too low a permeability to be produced economically. Since the volume of gas contained is very large (the Bureau of Mines estimates 317 trillion cubic feet, compared with proved reserves currently being produced of 275 trillion cubic feet)[21] the development of methods to increase the permeability to economic levels should be possible. Conventional methods, such as hydraulic fracturing or the use of conventional explosives, appear to be uneconomical. while the use of nuclear explosive devices to increase the flow rates to economic levels offers a more feasible prospect. The work to date on this method has been summarized in an AEC report:[22]

> Two nuclear-explosive gas-stimulation experiments have been conducted, and a substantial quantity of gas has been produced with this new technology. In this report, an economic projection of commercial production after the successful completion of a research and development program has been made for the nuclear-explosive stimulation of a well in the Green River Basin in Wyoming. The internal rate of return on investment varies from 8 to 26 percent, depending on the properties of the reservoir, with an assumed wellhead price for gas of 30 cents per thousand cubic feet.

However, the application of this method is opposed by environmentalists who argue that it is unknown how long it will take the radioactivity of the gas, caused by the detonation of the nuclear device, to reach levels that are sufficiently low to permit safe distribution. There is also concern about long term groundwater contamination resulting from the nuclear explosion.

Natural gas also occurs in coal seams. While the quantities sometimes are large, by the time they are diluted so that safe mining of the coal is possible, it is generally uneconomical to consider the sale of the gas. While gas from coal seams is sold in some European countries, gas prices are significantly higher than in the United States. Moreover, the difference in methods of mining allow methane to be collected at much higher concentrations. Methods of collecting this gas at high concentrations in the United States would not only permit its sale in conventional natural gas markets but would reduce the risks of explosions in mines. Research on methods to achieve these combined advantages are now underway.

UPGRADING OF FUEL RESOURCES

Coal.—During the mining of coal some impurities—including rock and other undesirable strata—are removed along with the coal substance. Generally the large pieces can be separated easily from the coal substance. In addition to these impurities, a certain amount of other unwanted material is brought to the surface along with the coal.

In the United States about 78 percent of all mined coal is subjected to washing or "coal preparation" for the removal of unwanted material. During the washing operation [23] about 75 percent of the raw coal is recovered as clean coal. The balance contains impurities found in the raw coal and a certain amount of clean coal that is displaced with the rejected material. The amount of coal that is lost in this operation depends on a large number of factors: (1) The nature of the coal seam; (2) the types of impurities; (3) the mechanical adequacy of the coal washing equipment; (4) the efficiency with which the equipment is operated; and (5) the value of the coal that is being sold. In general, losses of coal to the waste stream in modern well-operated plants are in the range of 3 percent to 5 percent.

Uranium.—When the ore has been transported to the ore processing plant and crushed to separate the uranium oxide from the extraneous wastes, large volumes of additional waste products are released. In most modern mills at least 95 percent recovery of the uranium in the ores is accomplished. Two additional processing steps are used to refine the impure uranium oxide concentrate and to convert the product into a fuel that can be used in nuclear reactors. Losses in these final two steps are also low.

The energy costs in upgrading uranium oxide are very small. At an estimated cost of 1 cent per kilowatt-hour, the direct use of energy to recovery per pound of uranium oxide (the fuel used in nuclear reactors) from a small copper leaching solution plant is 0.1 percent of the energy that will eventually be produced by the uranium oxide. Energy costs are also expected to be negligible for recovery of uranium from phosphate rock in the low-grade Florida phosphates and from Chattanooga shales.

Petroleum.—The efficiency of use of crude feedstocks at refineries has changed between 1945 and 1960, partly as a result of improved refinery technology and partly as a result of modifications that have been made in the product mix as the demand for different petroleum fractions varied over time.

There have been large increases in the percentage of distillate fuels and some increase in the percentage of the barrel that appears as lubricating oils and other products. Both of these products have increased at the expense of the amount of residual oil produced. These shifts in product yields required major changes in refinery technology including new techniques such as catalytic cracking, alkylation, coking, hydrocracking and hydrogen treating. These changes increased the complexity of the refinery operation and made more difficult any attempt to increase efficiency in the use of the crude feedstock at the refinery. However, the trend toward fewer refineries of larger size has helped to contribute to economies of scale not only in the economics of the process but in the energy units consumed per barrel of product.

The efficiency of energy use during the refining operation was calculated from the quantity of fuels consumed in 1970. In 1970 12.5 percent of the petroleum heating value in the crude run was required in the refinery while making marketable products from crude oil.

In 1970 for each B.t.u. of crude oil fed to the refinery 95 percent appeared in the final products. This high figure results from the purchased power and fuel which is not included in the B.t.u.'s in the feed stock going into the refinery, some of which appears as B.t.u.'s in the final products.

Natural Gas.—About 85 percent of the natural gas marketed in 1968 was processed before use to remove water, hydrogen sulfide, carbon dioxide and other impurities. From this gas 550 million barrels of natural gas liquids were removed and sold at a value of $1.1 billion. In 1968, 867 million cubic feet of helium were also extracted (valued at $29 million) and 684,000 long tons of sulfur (valued at $26 million) were removed from the gas before injection into the gas pipeline distribution system. The degree of treatment of the gas before use varies widely from field to field but in some instances it is economically attractive and in others technically necessary to remove impurities before use in order to have the gas meet pipeline quality standards.

The energy required for gas treatment varies widely. In some cases no energy is needed for gas treatment before use, since it is free of natural gas liquids or impurities; in others, the energy used generally creates a salable byproduct with the energy increment appearing in the price of that byproduct. It is only rarely that gas is treated solely to remove an impurity in order to meet pipeline standards. The amount of energy used at natural gas processing plants is probably less than 1 percent of that in the gas sold. Since total field use (pumping, drilling, extraction loss and plant fuel) represents only 3.5% of gas produced.

TRANSPORTATION

Transportation represents a large share of the total average price of delivered energy. For coal in 1969 the average value added by rail transportation was 39 percent of the delivered cost; for petroleum crudes shipped to refineries, transportation represented 50 to 60 percent of the delivered costs; and for natural gas delivered to the ultimate consumer, transportation represented 66 percent of the total cost.

Coal.—Railroads moved 71 percent, waterways nearly 13 percent and trucks 12 percent of the coal mined in the United States. The small balance was used at the mines. Assuming the B.t.u. required per ton mile for railroads is 680,[26] with an average length of haul of about 250 miles, the 397.8 million tons of coal moved by rail in 1969 would have required 68 x 10^{12} B.t.u. Since this tonnage of coal contained 1 x 10^{16} B.t.u., the amount of energy used in transporting coal by rail was 0.7 percent of the heating value of the coal. For waterways the percentage used to transport would be slightly less (0.6 percent) and for trucks somewhat more (2.7 percent) of the fuel value.

Uranium.—Uranium mining is largely confined to the four States of Colorado, New Mexico, Utah, and Wyoming. Other States produce only 6 percent of the balance of the uranium ore that must be shipped to mills for upgrading. Since most of the milling operations are located in the States where the ore is processed and relatively small quantities of ore are needed (compared to the large tonnages for coal) the energy costs for shipping of uranium are negligible. Even less important in terms of energy consumption is the cost of transporting the enriched ores to enrichment plants and the shipping of these products to fabrication and reprocessing plants. While the distances under these circumstances are much greater, the quantities are measured in pounds rather than tons and energy transportation costs are too small to be a factor in overall energy use.

Petroleum.—In 1968 pipelines were used to deliver nearly 76 percent of the crude oil received at refineries, tankers and barges carried 23 percent and the remaining 1 percent was delivered by tank cars and tank trucks. The energy used to move the pipeline oil represented about 0.7 percent of the energy in the oil moved to the refinery.

Tanker transportation from foreign sources represented only about 8 percent (35 percent of 23 percent) of the crude oil transported; although these are long distance hauls, the total contribution to energy use in transportation of oil is small, particularly as tanker sizes have grown larger. In this connection a 60,000-ton tanker moving products (or crude) from Texas to the east coast consumes about 1.2 percent of the fuel that it carries.

Pipelines move 45 percent of the product from the refinery to distribution points and it is estimated that they consume 0.7 percent of the energy in the fuel moved. Barges move about 25 percent of the products and their use of energy would be less than the 1.2 percent for the much longer tanker hauls noted above. Trucks move the other 40 percent for local distribution and if one assumes an average of 100 miles for delivery in trucks with 12,000 gallons capacity the energy used would be only 0.2 percent of that contained in the fuel. Thus all energy transportation requirements for petroleum are in the range of about 1.0 percent or less of the total energy in the product.

The energy costs of bringing the petroleum products to the final user will obviously vary greatly with the type of product, the nature and size of the consuming units and other factors. Uses which require the refined product to be delivered by truck will have the highest B.t.u. per ton/mile. If it is assumed that all the household and commercial sector fuels and all the transportation sector fuels are delivered in this manner, the average length of haul must be estimated to determine if this portion of fuels transportation is important. A 7,800-gallon truck delivering gasoline to a gasoline service station uses about 50 gallons of gasoline to deliver about 30,000 gallons—or about 0.2 percent of the energy delivered. In the case of the household and commercial sectors trucks which carry 1,500-2,600 gallons are used to deliver about 7,000 gallons while consuming 28 gallons of fuel. Thus about 0.4 percent of the fuel is consumed in delivery.

By contrast, in delivering fuels to industrial customers (which is mostly residual oil and still gas) it is believed that pipelines and large trucks are used more widely than other forms of transportation. One would anticipate that the distances involved would be greater;

otherwise different transport modes would be used. If, as a first approximation, it is assumed that that portion delivered by pipeline (which is five times as efficient as truck movement in B.t.u.'s per ton/mile) is estimated to be balanced by an increase in the average distance that fuels must be moved to industrial users (an unproved but probably reasonable first approximation) then the energy used in the distribution of petroleum products overall would be about 0.2 percent of their delivered heat value. That portion delivered in 6,000 gallon trucks would on the average use 50 gallons of oil to deliver 24,000 gallons—again about 0.2 percent of the energy in the fuel delivered.

Natural gas.—Virtually all natural gas is delivered by pipelines from the producing wells to the natural gas processing plant. From there it is delivered by pipeline transmission companies to distribution facilities that are also virtually all pipelined to their final destination. In 1969 about 2.7 percent of the natural gas was reported to have been used for pipeline fuel. This value includes the amount used in gathering the gas from the wells (generally short distances with small diameter pipe and thus high pressure drops) and from the wells to the utility distribution systems (long distance but large diameter pipes with small pressure drops).

The energy used in distributing the gas within each utility system will vary with the concentration of the customers, the size of the pipelines originally installed and still in use, and other factors that are peculiar to each gas utility, but since most of the gas is normally delivered to the gas utility at a pressure sufficiently high to distribute to the ultimate consumer, the 2.7 percent reported above has this energy already included. Estimating energy losses in the distribution system is difficult (the difference between what the utility buys and what is metered to the customers includes losses such as broken mains, leaking connections, poor meters, etc.) For a local utility in the Washington D.C. area the difference is reported to be 4 percent which is designated as "unaccounted for." For the United States as a whole the difference between the gas produced at the wells and that which cannot be accounted for is between 3 and 4 percent, a value similar to that is used in transporting the gas which gives a reasonable check on the local utility's experience. A portion could also be lost in leakage from pipelines all the way from the wellhead to the domestic consumer. Since the cost of reducing this waste may be so much greater than any benefits, more efficient methods will probably not be sought until the price of gas goes very much higher than it is today.

Electricity.—The best opportunities to make significant savings in the transportation of energy are in the form of electricity. While the line losses are a function of a number of engineering variables, the two most important factors are the distance of transmission and the voltage used. In alternating current (a.c.) transmission systems the lines used to move the electricity either directly to a large user or transmission to substations utilize voltages that start at about 50 kilovolts but have reached voltages as high as 765 kilovolts; plans are underway to test the feasibility of a 1,500 kilovolts AC overhead line. The trends are toward ever higher voltages (see table 1-3) because this reduces line losses.

Figure 1-1 indicates, for example, that, in transporting 1,000 megawatts of power for 100 miles, even at 765 kilovolts, losses run about 0.4 percent and increase to 1 percent at 500 kilovolts. Because of these factors, the utility industry has greatly intensified its efforts

Table 1-3—TRANSMISSION LINE MILEAGES IN U.S., 230 KILOVOLT AND ABOVE [1]

	230 kv.	287 kv.	345 kv.	500 kv.	765 kv.	±400 kv.(d.c.)	Total
1940	2,327	647					2,974
1950	7,383	791					8,174
1960	18,701	1,024	2,641	13			22,379
1970	40,600	1,020	15,180	7,220	500	850	65,370
1980	59,560	870	32,670	20,180	3,540	1,670	118,490
1990	67,180	560	47,450	33,400	8,940	1,670	159,200

[1] By 1990 there may be significant applications of a.c. voltages higher than 765 kv. and more extensive use of HVDC than that shown in the table.

Source: The 1970 Federal Power Survey: Federal Power Commission.

Figure 1-1
Resistance Losses
Per 100 Circuit Miles of Transmission

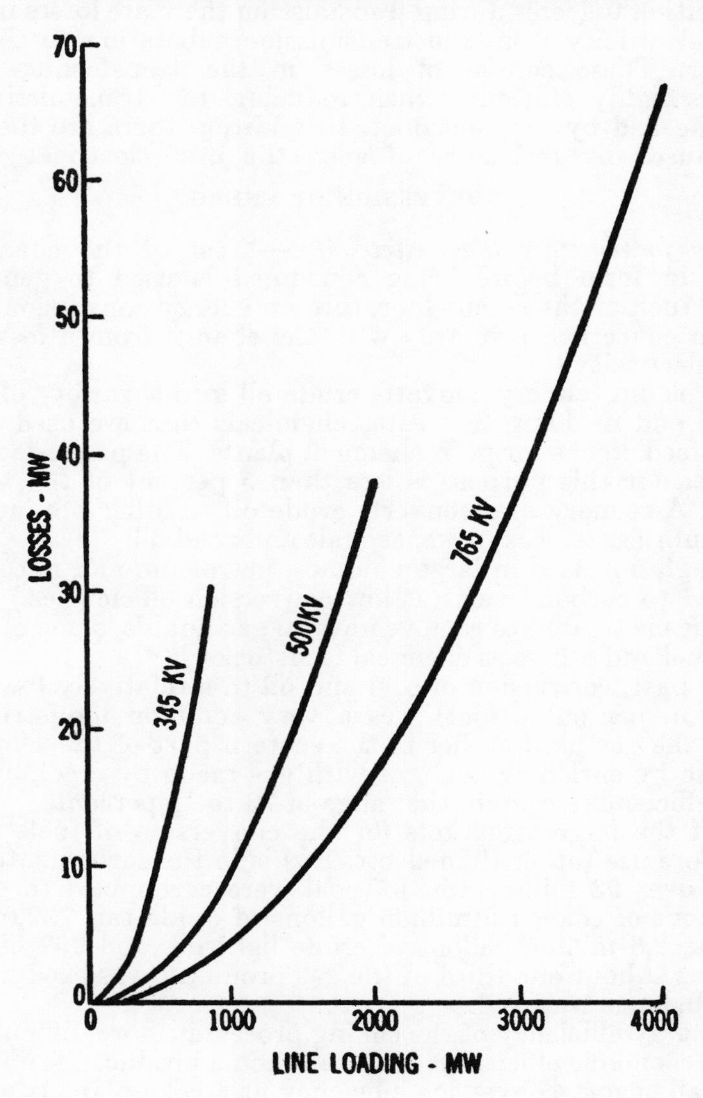

not only to increase the voltages of transmission but to vastly increase the number of interconnections between transmission lines so a particular kilowatt that is generated can be moved the shortest possible distance.

An alternative to a.c. transmission is to move the electricity in the form of direct current (d.c.). The economic advantages of the d.c. line do not appear until large blocks of power are to be moved long distances. D.c. also has advantages for underwater transmission, even over short distances, and may be more economical under certain circumstances for underground lines in heavily developed metropolitan areas. The energy losses of comparable a.c. and d.c. lines are difficult to estimate since so many other factors influence the line design for moving comparable blocks of electrical energy over a given distance. Line losses (unlike a.c. current) are small for d.c. current but the losses of converting the a.c. current generated at the plant to d.c. and the losses of reversing this process at the point of delivery are relatively high.

In addition to losses during transmission there are losses in distributing the electricity from the transmission substation to the ultimate consumer. These consist of losses in the transformers (although they are highly efficient) when reducing the transmission voltage to that needed by the consumer. In addition there are the usual line losses caused by resistance of even the best electrical conductors.

CONVERSION OF ENERGY

Energy forms other than electricity.—Most of the energy that is changed in form before being consumed is used to generate electricity. Much of the recent literature on energy conversion has therefore been concerned primarily with the change from a fossil or fissile fuel to electricity.

A petroleum refinery converts crude oil into a variety of other fuel products and in doing so creates chemicals that are used directly or serve as feed stocks for petrochemical plants. The percentage of petreleum used for this purpose is less than 5 percent of the total but is growing. A refinery also converts crude oil to other nonfuel products such as lubricants, wax, coke, asphalt and road oil.

Although not used in large volumes, petroleum and natural gas are converted to carbon black (at low conversion efficiencies.) Moreover, natural gas is treated to remove natural gas liquids, some of which end up as a fuel and others as chemical feedstocks.

In the past, conversion of coal and oil to a relatively low B.t.u. gas (550 B.t.u. per cubic foot) was a very common industrial process. Most of the gas used earlier in the eastern part of the United States was made by enriching coal gas with gas made by cracking oil. Conversion efficiencies were in the range of 76 to 78 percent.

One of the largest markets for the conversion of fuels to another form before use (other than electricity) is in the carbonization of coal. In 1969 over 93 million tons of coal were carbonized to produce 69 million tons of coke, 7.6 million gallons of crude tar, 732,000 tons of ammonia, 2.6 million gallons of crude light oil, and 962 billion cubic feet of gas. About one-third of the gas produced was used in the plant but the balance was sold.

The energy efficiency of the coking process is more difficult to assess than the economic efficiency. If the ammonia produced is not included, the overall energy conversion efficiency at a coke plant is probably in the range of 93 to 95 percent.

Conversion to electricity.—By far the largest percentage of electric energy produced in the United States is by the combustion of a fossil fuel or the use of fissile fuels to produce steam at elevated temperature and pressure. This steam is used to drive a steam turbine which generates electricity.

A recent report said: [28]

> The most dramatic increase in fuel-conversion efficiency in this century has been achieved by the electric power industry. In 1900 less than 5 percent of the energy in the fuel was converted to electricity. Today the average efficiency is around 33 percent. The increase has been achieved largely by increasing the temperature of the steam entering the turbines that turn the electric generators and by building larger generating units. In 1910 the typical inlet temperature was 500° F. The latest units take steam superheated to 1000°.

The most modern fossil fuel plants can achieve a conversion efficiency of from 38 to 40 percent.—The lower value is generally being designed into newer plants because it produces a lower overall cost for electricity. The efficiency of steam turbine electric generating plants is limited by the thermodynamics of the cycle. If higher steam pressures and temperatures were used, the efficiency would be increased; but with the present state of metallurgy this cannot be achieved at low costs. Although the best new fossil plants are designed for this range of 38 to 40 percent efficiency, the average for the country as a whole in 1971 was about 30 percent efficiency.

Powerplants using coal have a somewhat higher energy conversion efficiency that those using oil; and those using oil have a better fuel efficiency than those using natural gas. However, both the capital costs of these plants and adverse environmental effects are in the opposite direction of production efficiency.

The efficiency of light water electric generating plants using enriched uranium is considerably lower—the average being in the 30 to 32 percentage range. (Gas cooled high temperature nuclear reactors have an efficiency of about 40 percent.) Not only do these plants create additional thermal pollution problems but unless breeder reactors are developed questions must be raised concerning the availability of low cost uranium resources. At present steam-electric, fossil fuel and nuclear stations contribute 82 percent of the electricity produced in the United States. A small amount (about 0.4 percent) is generated by internal combustion engines.

Hydroelectric power, which now produces about 17 percent of our total electric demand, is highly efficient as a user of the potential energy stored in water. Turbines of over 90 percent efficiency are available but the number of sites for new plants is limited, thus the amount of electricity generated at hydroelectric plants will continue to occupy a declining share of total generation. Hydroelectric pumped storage installations (where water can be pumped back to the reservoir using off-peak power) are increasing in number. This is by far the lowest cost approach for storing electric energy, yet pumped storage actually reduces overall energy efficiency (although it is economically attractive) because it requires about 3 kilowatt-hours of off-peak power to raise the water to the level where it will produce 2 kilowatt-hours of peak power.

As gas turbines (manufactured basically for airplane use) are developed to operate at higher temperatures and for the longer periods of on-stream time demanded by central electric generating stations, the efficiency of these units can be expected to go up sharply (turbines operating in airplanes are about 26 percent efficient at 1700° F. while industrial turbines would be about 32 percent efficient for the same temperature); combined steam-turbine, gas-turbine plants will then begin to be attractive both economically and from an efficiency standpoint.

Since electric energy consumption has been growing at 2½ times the rate of all energy use, methods to increase its efficiency of production are imperative if our resources are to be extended and reductions achieved in the adverse enviromental effects created by electrical generation.

Numerous opportunities exist for generating electricity more efficiently and for using more constructively the chemical byproducts that have created environmental pollution problems.

Recent developments in conversion technology.—Following World War II long-distance pipelines brought low-cost natural gas into the East and gas production from coal and oil virtually disappeared. The developing shortage of natural gas has now reversed this trend; numerous announcements have recently been made for the construction of large plants to convert petroleum fractions and coal to a high B.t.u. gas (1,000 B.t.u. per cubic foot) as a substitute for natural gas.

ENERGY STORAGE

The storage of coal, uranium, natural gas, and oil present no particularly difficult energy efficiency problems. Storage is practiced largely as the least costly method of meeting peak demands or as a means of assuring energy supply should there be some unexpected interruptions in the normal schedule. When manufactured gas was the principal product on the market, it was stored in metal gas holders—either at elevated pressures or near atmospheric pressure. As natural gas became available the use of these holders declined and more and more gas was stored in depleted oil or gas wells, aquifers and salt domes.

Oil is generally stored in metal containers above ground.

The economic cost of stockpiling these fuels varies with the resource and the particular situation at any given location. Although there is some energy consumed in stockpiling, the energy implications are relatively insignificant. For gas some of the energy losses are the pressure drop loss in moving the gas into abandoned wells; for oil the energy losses are in the form of electricity used for pumping the oil in and out of storage; for coal and uranium, losses occur in bulldozing and compacting the piles in a way to prevent environmental damage. (In the case of coal there are also losses due to spontaneous combustion of the coal pile.)

Electricity storage differs sharply from the storage methods used for the fuels from which it is generated. Moreover, electric generation is a capital intensive industry and since the peak-to-average-demand is much greater than for the original fuels used to generate it, the capacity built to meet these sharp peak demands is economically very unattractive so that efficient storage methods are important.

The simple storage methods that are available for the other fuels do not exist for electricity. The only widely used method is pumped storage which has an efficiency of about 67 percent. Batteries are used for storage of small quantities of elctricity at low voltages but they are extremely expensive and the efficiency of use is about

70 percent. Table 1-4 gives some information on the characteristics of certain battery types.

Table 1-4.—CHARACTERISTICS OF COMMON STORAGE-BATTERY SYSTEMS[1]

Battery type [2]	Volts per cell: Open-circuit voltage	Volts per cell: Average closed-circuit voltage	Number of cells	Cycle life	Energy (watt-hours per pound)	Density (watt-hours per cubic inch)	Initial cost [3]	Cost per kilowatt-hour [4]
Lead-acid:								
Automotive:								
Regular	2.1	1.9	6	150	12	1.3	$0.50	$0.33
Bus and diesel	2.1	1.9	6	300	12	1.3	60	.22
Industrial:								
Flat plate	2.1	1.9	6	1,500	11	1.2	90	.09
Iron clad	2.1	1.9	6	2,000	12	1.3	90	.08
Nickel-iron (E line)	1.5	1.2	10	2,500	13	1.1	190	.08
Nickel-cadmium:								
Pocket type	1.3	1.2	10	1,000	8	.4	320	.31
Sintered plate type	1.3	1.2	10	2,000	13	1.0	400	.21
Silver-cadmium	1.3	1.1	11	400	35	3.5	1,200	2.50
Silver-zinc	1.8	1.5	8	100	55	4.5	1,000	8.40

[2] 12-volt, 100-ampere-hour size, at 6-hour rate, deep-cycle service.
[3] User cost calculated for 100-ampere equivalent.
[4] Including amortization over an average cycle life and charging current.

In the Electric Research Council (ERC) Report on R. & D.[29] the development of bulk electric storage batteries was given priority No. 2 (very important) with benefits expected in the more efficient use of installed generating equipment and the elimination of highly polluting and inefficient steam and gas turbines for peaking power. This kind of technology, however, is still in its infancy.

Another proposal that has been made for storage of electricity is the use of compressed air. The advantages of compressed air over pumped storage are as follows: [30]

> [It] can be located almost anywhere that firm rock can be found at depth, uses only about one-tenth of the land surface that equivalent pumped-water storage does, and uses the stored energy more effectively; but it can be utilized only in combination with gas turbine generators. At present air energy storage requires equipment development and, therefore, there are uncertainties in equipment cost estimates. There are also large uncertainties in excavation cost estimates and in possible environmental effects.

The Oak Ridge report states that effectiveness of storage of power in compressed gases is 79 percent as compared to 67 percent for pumped storage.

Another proposal for energy storage is to generate hydrogen by electrolysis of water and to store the hydrogen either as a gas or liquid for use as a nonpolluting fuel for electricity generation.

The 1972 NSF-Cornell study on energy and environment [31] raised the following questions with regard to hydrogen:

> The panel was intrigued by this long-range prospect and urged that studies aimed at assessing the implication of this long-range role of H_2 be supported.
>
> In the shorter run hydrogen may play a key role as a medium of energy transport and storage. Here the question

is economic. Can H_2 be manufactured electrolytically at sufficiently low cost? If it can, would its uses give to an electricity-dominated energy system a flexibility that is inherent in a chemical energy system but not in an electrical energy system? These are fundamental questions which deserve some attention and support.

UTILIZATION OF ENERGY

There are several background observations to be made concerning the factors influencing energy use that are common to all major market sectors (household, industrial, transportation):

Increasing use of solid state circuitry, reduction of friction losses and use of superconducting motors are methods that are applicable to some degree in all three market sectors.

Table 1-5 —EFFICIENCIES FOR RESIDENTIAL ENERGY USES

	Energy efficiency [1]				Economic efficiency gas-electric [2]
	Gas (percent)	Electricity (percent)		Gas-electric	
		Utilization	Ultimate [3]		
Space heating	60	100	28	2.1	3.6
Cooking	40	80	22	1.8	3.0
Water heating	62	91	25	2.5	4.1
Clothes drying	50	50	14	3.6	6.1

[1] Energy efficiency data from Consumer BuHetin annual (1972) and Gas Engineers handbook (1966).
[2] Econonic efficiency is the ratio of useful energy delivered per unit price for gas relative to electricity. Base prices used are for 1969: 0.000601 cent B.t.u. for gas and 0.000613 cent/B.t.u. for electricity.
[3] Assumes 27.8 percent efficiency (for 1970) of generation, transmission, and distribution for electricity.

Statistics on end use utilization of energy are collected in a variety of ways by a number of different organizations. The most internally consistent set of data (and these have deficiencies) are those collected by the U.S. Bureau of Mines. Using the Bureau's method of classifying end uses, the categories selected are household and commercial, industrial, transportation and electric utility generation.

Household and Commercial.—Consumption in this sector represented 35.3 percent of all the energy demand[35] in the United States in 1970. The major energy uses are space heating and cooling, heating of hot water for washing and cleaning, drying of clothes, and the electricity needed for operating a wide variety of devices ranging from electric toothbrushes to electric home heating. Some household and commercial uses provide competition between the fuel forms: space heating and cooling can be provided by coal, oil, gas and electricity (which is produced by any of these primary fuel sources and from nuclear fuels). For other commercial and household uses the competition is less complex but there is still ample competition between electricity and gas for clothes dryers and hot water heaters, and increasing competition for space heating and cooling.

The decline in the use of coal for space heating and hot water production can be expected to continue but the competition between oil, gas and electricity for these markets is continuing. Electric heating is energy intensive (all other factors remaining constant) because of losses in generation and transmission (see previous discussion) but it is clean, versatile and highly efficient at the point of use. Strong promotional efforts have resulted in the construction of "all electric homes" in a number of locations but these efforts have generally been most successful where electricity has been available at low cost (e.g., TVA area) and for new homes where adequate insulation has reduced heat losses to acceptable levels.

The number of electrically heated dwellings for selected years is shown in table 1-6 with projections to the year 1990:

Table 1-6 —ELECTRICALLY HEATED DWELLINGS IN THE UNITED STATES, 1964–90

Year:	Millions
1964	2.2
1966	2.7
1968	3.4
1970	4.2
1980	12.5
1990	24.0

Source: 1970 Federal Power Survey.

Most electric heating projects have used the "resistance type" in which the heat is provided by deliberately designing the resistance of the wire to convert electricity to heat. (The same effect is defined as "line losses" during transmission and distribution.) By the use of heat pumps it would be possible to increase the efficiency of electricity use for heating by some factor. For example, the amount of electricity saved when using a heat pump and supplemental electric heat over using resistance heating alone was estimated to be 60 percent in New York City and 50 percent in Buffalo and Albany. However, heat pumps are much more expensive to install, have not generally been reliable and have required high maintenance costs. They also require a "heat sink"—either to water or to the ground. As a result of these factors, heat pumps have not become widely used. Since no new technology is required and large energy savings are possible, as energy prices increase the use of heat pumps may become more widespread.

Natural gas is a clean fuel and, from the consumer's standpoint, requires less attention (no fuel tank, no fuel deliveries). Where natural gas lines have been available it has been the preferred fuel for new home construction; 32 percent of all natural gas went into the household and commercial market in 1968 and served over 40 million customers. The efficiency of use of gas depends not only on the original design of the burner (which can be quite simple) but on how well it is maintained. A well-designed burner properly adjusted should give between 70 and 80 percent efficiency. Very little more could be done to increase this efficiency since the losses are mostly to the stack and are nearly impossible to reduce. However, there is growing evidence that, because of the gas burners' simple design (which requires little maintenance and therefore adjustment), many gas burners are operating at much lower efficiencies—perhaps in the 60 to 70 percent range or even lower. Neither the consumer nor the distributor would be aware of this reduction in efficiency; moreover, the gas companies have had little incentive to maintain burner efficiency in order to sell less gas.

Distillate oil supplies most of the rest of the space heating market since the number of homes using coal is now relatively small. Oil burners tend to be less efficient than gas burners (although new ones are said to have a 70 to 80 percent efficiency) and to require more maintenance. However, before the burner needs to be repaired it can be operating at efficiencies much below the original design value. Hottel and Howard [36] suggest that in oil burners the on-off feature as well as other factors can reduce efficiency to as low as 50 to 60 percent and in oil-fired water heaters to as low as 30 percent.

In the hot water heating and clothes drying markets the same factors operate that affect the choice between electricity and gas for home heating: the relative price of the appliances; the relative cost of gas and electricity in a given area; the availability of gas; and the importance to the consumer of the somewhat cleaner operation with electricity. Efficiencies for electric [37] and gas dryers are about the same (much of the loss is to the stack) depending on design, amount of insulation, et cetera. For electric immersion hot water heaters efficiencies [38] are in the range of 80 percent (the balance is lost due to radiation). For gas hot water heaters it would be somewhat lower. In addition to the 20 percent radiation loss there would be a stack loss of about 30 percent.

In recent years (before the gas shortage) gas air conditioning in large commercial establishments and apartments had been making inroads into the central electric air conditioning market despite high initial cost and size. Gas air conditioning is said to be less troublesome and the promotional rates set by gas utility companies to encourage the summer use of gas (to maintain high load factors on pipelines) were reported to allow gas air conditioning to compete economically with central electric air conditioning despite the much lower efficiency of energy use. For large consumers, if an engine driven gas air conditioner is used (not many are because of high operating and maintenance costs with present units) a gas air conditioner is only slightly less efficient than an electric home central air conditioner with an average efficiency. If an absorption cycle is used (the types used in all residences and even in most large installations) the efficiency is about half that of the engine driven type.

Room electric air conditioners are an excellent example of the wide variations in energy consumption per unit of useful work that exists among equipment currently being sold. Figure 1-2 is a tabulation of the efficiencies of room air conditioners of different sizes for two different voltages. For small sizes a variation in energy use of as much as a factor of two can be found. In fact, some manufacturers advertise a low capital cost, high energy consuming unit as well as a high capital cost, low energy consuming model. As the public becomes educated to overall cost considerations and as energy prices increase there should be a shift toward more purchases of the more efficient units. If this does not occur the use of more efficient units can be achieved by requiring that manufacturers meet specified standards.

Table 1-7 is a listing of the wattage rating and estimated annual kw.-hr. consumption of various electric appliances. There is no indication on the table as to the efficiency of energy use of each of these appliances. Detailed data are lacking for most of these appliances and large variations can be expected for some appliances among different manufacturers. Nevertheless, where the electricity is used for electric motors, one can assume a fairly high efficiency of use of electricity since motor design and manufacture have become highly standardized. This would include the motors on dehumidifiers, dishwashers, fans, and oil burners. However, for many of the large energy consuming devices (whether using efficient motors or not) the design of the appliance can be such that the motor must operate for greatly different periods of time and require different amounts of energy. In some instances (refrigerators, freezers, etc.) the type and quality of the insulation alone could have a large impact on yearly electricity consumption.

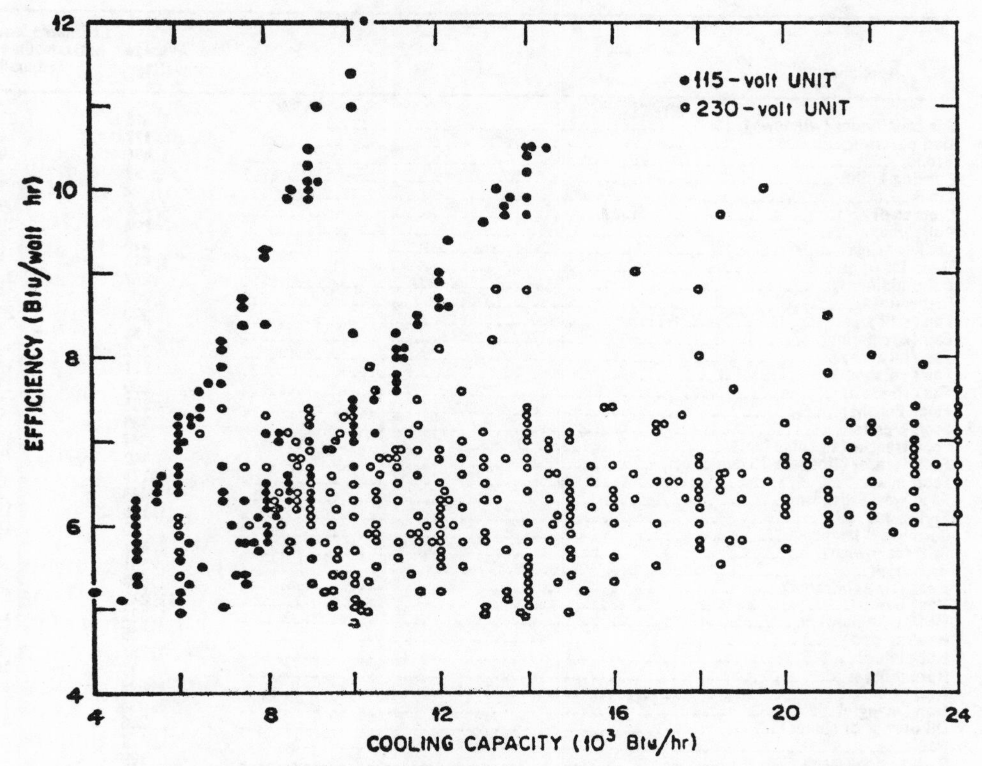

Figure 1-2 —Efficiency of room air conditioners as a function of unit size.[1]

Reduced demand in the residential and commercial sector can be achieved in a number of ways. The President's clean energy message of June 4, 1971, suggested that additional insulation would be required for all homes with FHA guaranteed mortgages. A simple administrative order was all that was needed to reduce energy consumption for space heating and cooling for all new homes covered by the FHA.

Ways other than installing insulation to reduce heat losses from residential and commercial installations were contained in a recent advertisement by Owens/Corning Fiberglas. These were:

1. Weatherstrip and caulk around all doors and windows. (Leaking air could waste 15 to 30 percent of heat consumed.)
2. Install storm windows or insulating glass. (These can reduce heat losses through windows by 50 percent.)
3. Maintain the heating plant in good condition.
4. Close windows and draperies at night.
5. Stop heat loss to the attic.
6. Turn off lights and TV's, stop faucet leaks and lower the thermostat. Hot water faucet leaks can add considerably to the cost of energy, and lowering the thermostat 4° F. can save 3 percent of energy costs.

House design and location of a particular piece of property can also have an impact on energy use. For example, where possible, placing a home in a location to protect it from cold Northwest winds in the winter and under trees to avoid the direct rays of summer sun can be

Table 1-7 —APPROXIMATE WATTAGE RATING AND ESTIMATED ANNUAL KILOWATT-HOUR CONSUMPTION OF ELECTRICAL APPLIANCES UNDER NORMAL USE—1969

Appliance	Average wattage	Estimated kw.-hrs. consumed annually
Air conditioner (window) [1]	1,566	1,389
Bed covering	177	147
Broiler	1,436	100
Carving knife	92	8
Clock	2	17
Clothes dryer [1]	4,856	993
Coffeemaker	894	106
Cooker (eggs)	516	14
Deep fat fryer	1,448	83
Dehumidifier [1]	257	377
Dishwater [1]	1,201	363
Fan (attic)	370	291
Fan (circulating)	88	43
Fan (furnace) [1]	292	394
Fan (rollabout)	171	138
Fan (window)	200	170
Floor Polisher	305	15
Food blender	386	15
Food freezer (15 cubic feet) [1]	341	1,195
Food freezer (frostless 15 cubic feet) [1]	440	1,761
Food mixer	127	13
Food waste disposer	445	30
Frying pan	1,196	186
Germicidal lamp	20	141
Grill (sandwich)	1,161	33
Hair dryer	381	14
Heat lamp (infrared)	250	13
Heat pump [1]	11,848	16,003
Heater (radiant)	1,322	176
Heating pad	65	10
Hot plate	1,257	90
Humidifier	117	163
Iron (hand)	1,088	144
Iron (mangle)	1,494	158
Oil burner or stoker [1]	266	410
Radio	71	86
Radio-phonograph	109	109
Range [1]	12,207	1,175
Refrigerator (12 cubic feet) [1]	241	728
Refrigerator (frostless 12 cubic feet [1]	321	1,217
Refrigerator-freezer (14 cubic feet) [1]	326	1,137
Refrigerator-freezer (frostless 14 cubic feet) [1]	615	1,829
Roaster	1,333	205
Sewing machine	75	11
Shaver	14	18
Sun lamp	279	16
Television (B. & W.) [1]	237	362
Television (color) [1]	332	502
Toaster	1,146	39
Toothbrush	7	5
Vacuum cleaner	630	46
Vibrator	40	2
Waffle iron	1,116	22
Washing machine (automatic)	512	103
Washing machine (nonautomatic)	286	76
Water heater (standard) [1]	2,475	4,219
Water heater (quick recovery) [1]	4,474	4,811
Water pump	460	231

[1] Large consumers, 300 kw.-hrs. annually or more.

Source: Edison Electric Institute.

useful in conserving fuels. Home design offers numerous opportunities to keep radiation losses to the minimum.

A publication by the Office of Consumer Affairs [39] suggested 11 ways to reduce domestic fuel consumption but covered essentially the same suggestions mentioned above.

Except for the insulation requirement, most of the above methods suggested for reducing energy consumption will result in only relatively small savings, even if people can be persuaded that it is necessary and desirable to try to take these energy saving steps.

Other methods, such as the "save-a-watt" advertising campaign of Consolidated Edison Co., of New York appear to have reduced demand to a limited degree.

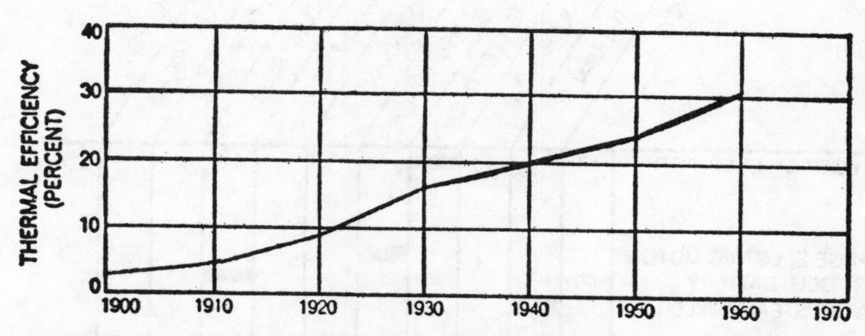

EFFICIENCY OF FUEL-BURNING POWER PLANTS in the U.S. increased nearly tenfold from 3.6 percent in 1900 to 32.5 percent last year. The increase was made possible by raising the operating temperature of steam turbines and increasing the size of generating units.

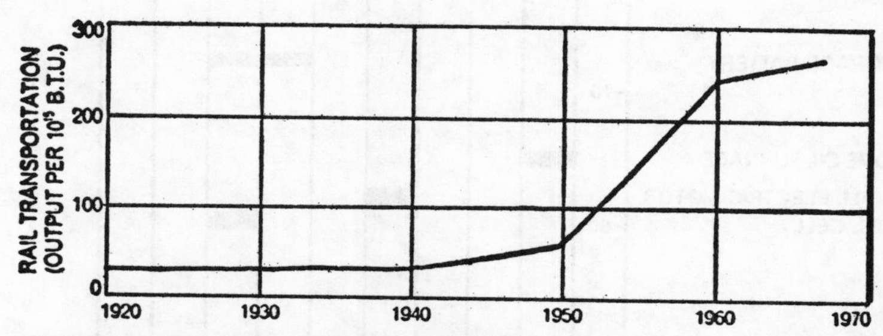

EFFICIENCY OF RAILROAD LOCOMOTIVES can be inferred from the energy needed by U.S. railroads to produce a unit of "transportation output." The big leap in the 1950's reflects the nearly complete replacement of steam locomotives by diesel-electric units.

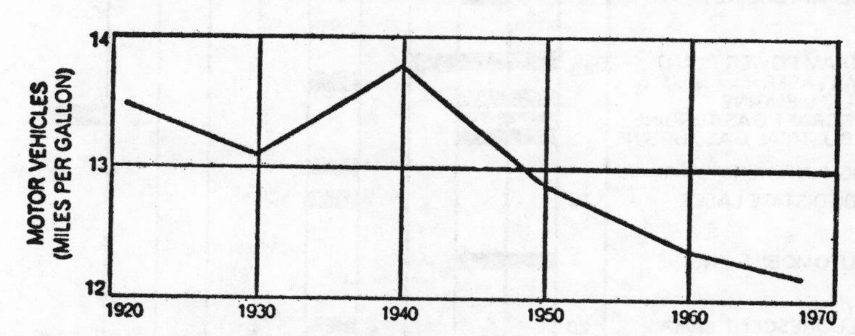

EFFICIENCY OF AUTOMOBILE ENGINES is reflected imperfectly by miles per gallon of fuel because of the increasing weight and speed of motor vehicles. Manufacturers say that the thermal efficiency of the 1920 engine was about 22 percent; today it is about 25 percent.

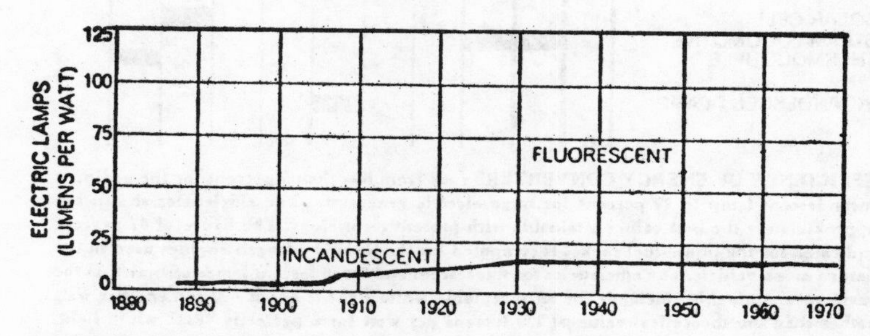

EFFICIENCY OF ELECTRIC LAMPS depends on the quality of light one regards as acceptable. Theoretical efficiency for perfectly flat white light is 220 lumens per watt. By enriching the light slightly in mid-spectrum one could obtain about 400 lumens per watt. Thus present fluorescent lamps can be said to have an efficiency of either 36 percent or 20.

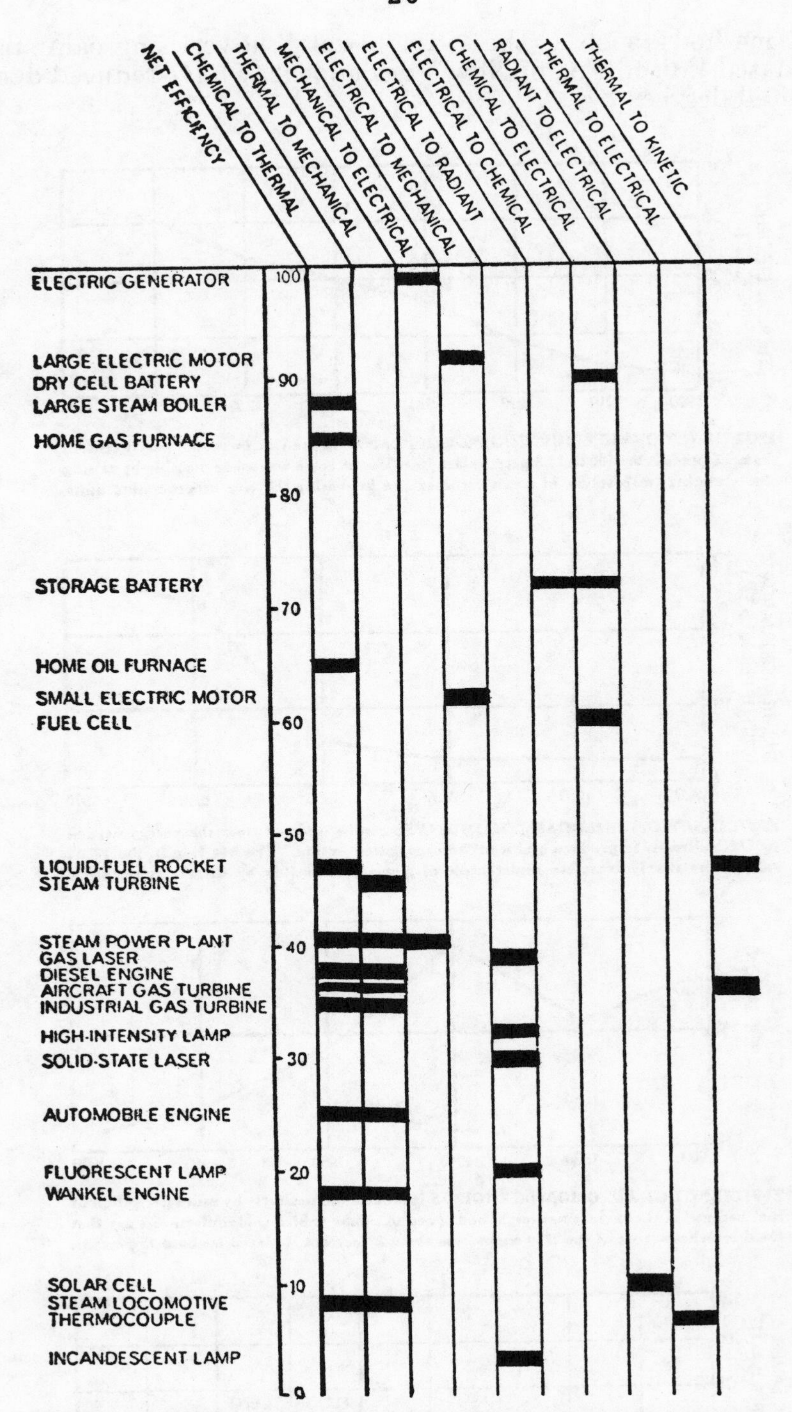

EFFICIENCY OF ENERGY CONVERTERS runs from less than 5 percent for the ordinary incandescent lamp to 99 percent for large electric generators. The efficiencies shown are approximately the best values attainable with present technology. The figure of 47 percent indicated for the liquid-fuel rocket is computed for the liquid-hydrogen engines used in the Saturn moon vehicle. The efficiencies for fluorescent and incandescent lamps assume that the maximum attainable efficiency for an acceptable white light is about 400 lumens per watt rather than the theoretical value of 220 lumens per watt for a perfectly "flat" white light.

Nevertheless many opportunities exist for making what may be substantial savings in some energy sectors. A number of these are discussed by Fred Dubin [40] in a recent article. He said:

> In a recent study by the Stanford Research Institute for the U.S. Office of Science and Technology, it was indicated that residential and commercial buildings consumed 20 percent of the Nation's energy for space heating; 4 percent for water heating; 2½ percent for air conditioning, which is rapidly increasing; and 2.4 percent for illumination; or a total of almost 29 percent of all energy in only residential and commercial buildings.

He suggested a number of ways in which this could be done without changing life style: examine to determine what lighting levels are really needed and whether rooms should be heated when they are continuously used or only used infrequently; use energy at one level for refrigeration but recycle the rejected heat for producing hot water.

Mr. Dubin listed a number of ways in which energy is lost from buildings or for which there are opportunities for savings. These are:

> 1. Heat losses and cooling loads due to transmission through walls, ceilings, and floors are greater with lightweight panel construction lack of adequate insulation, and are affected by color, orientation, shape and angle of incidence to sun's rays.
>
> 2. Solar transmission is of course greatest through glazing but may also be a significant factor on flat roofs; 60 square feet of glazing on the western exposure of a building will require a ton of refrigeration for cooling and consume a kw per hour.
>
> 3. Outdoor air (we used to call it "fresh air") for ventilation and infiltration are tremendous sources of energy consumption both summer and winter. The use of better filtration and odor control devices permits the reduction of outdoor air.
>
> 4. Chimneys, fire places, and combustion devices, such as in the kitchen, laboratory, and laundry exhaust hoods.
>
> 5. Individual small buildings, as opposed to high-density residential development, increase energy consumption because the skin-area-to-building-volume ratio is very great.
>
> 6. Single-purpose buildings with short occupancy and lack of diversity.
>
> 7. Lighting levels at unrealistically high intensities rather than efficient lighting design. A light bulb is a very inefficient energy converter, only 10 percent to 14 percent of the energy consumed results in useful lighting, while the rest goes into heat.
>
> 8. Sloppy calculations, excess safety factors, failure to account for people and appliance loads result in oversized equipment and inefficient operation.
>
> 9. The stack effect in buildings introduces outdoor air when building design is inadequate to prevent it.

10. The lack of energy storage systems results in a loss of energy which otherwise could be stored and used.

11. Improper zoning within the buildings and ventilation of unoccupied areas.

12. Air exhausted from buildings contains heat energy in the winter and cooling energy in the summer; with proper heat recovery equipment, this lost energy could be transferred to incoming outdoor air.

The article concludes that "the systems mentioned above and others elsewhere through the country can whittle away at the 29 percent of the national energy consumption in residential and commercial buildings" but that other methods can be used to reduce demand such as codes to limit energy use, better building design aimed at reducing energy requirements, the use of megastructures which lessen energy use and total energy systems.

In a report to the National Science Foundation, prepared by Inter-technology Corp.[41] the following comment is made concerning domestic energy usage:

> Domestic energy consumption is about 22 percent of the U.S. total. Little R & D has been done toward effecting: (1) reduction of total use, (2) more efficient utilization, (3) waste heat use and reuse, (4) economically preventable or reducible losses, (5) the on-site employment of prime movers whose waste heat would be a usable resource rather than a headache, and (6) a substantive investigation into what can only be called a profligate waste of energy, largely sensible heat.

The Federal General Services Administration (GSA) has announced plans to build a 100,000 sq. ft. Federal office building designed to test new designs and equipment to reduce energy consumption in office buildings. GSA has been monitoring 400 existing buildings to develop base data for future office construction. In addition, several New York State buildings with "heat recovery" systems have already been built and operated. In the buildings the excess heat in the interior area of the buildings, instead of being discharged to the atmosphere, is used to heat the peripheral areas which are heat deficient. This reduces both the cooling and heating loads significantly.

Other evidence that energy conservation in commercial and residential buildings is receiving attention is the number of conferences and forums that are being conducted to educate architects and engineers in methods of reducing energy consumption.

Industrial sector: The industrial sector is the largest consumer of energy, using 40.7 percent, including electricity, in 1968. Because of the myriads of uses that energy finds in the thousands of industrial operations, it is impossible to attempt to generalize on how efficiently energy is being consumed at present. Table 1-8 shows the total cost of purchases of fuels and electricity by industry groups. For the less intensive energy industries where the consumption is largely for space heating and cooling, lighting and the operation of electric motors the same methods that have been suggested for conservation in the residential and commercial sectors would apply. Obviously, even in these less intensive energy consuming industries, there may be opportunities for significant energy saving by redesign of the special machinery and equipment used in a given industry.

But the energy intensive industries are relatively few and, as shown in table 1-8, these are primary metals (using 21.5 percent of total industrial demand), chemicals and allied products, (15.4 percent) food and kindred products (8.5 percent) stone, clay, and glass processing, (8.3 percent) and paper and allied products (7.4 percent).

Table 1-8 —THE COST OF FUELS AND ELECTRICITY USED BY INDUSTRY GROUPS, PER CENT OF TOTAL, AND PER CENT OF VALUE ADDED AND VALUE OF SHIPMENTS,

SIC No.		Total cost of purchased fuels and electricity as percent of total	Cost of purchased fuels and electricity, as percent of—	
			Value added	Value of shipments
20	Food & kindred products	8.5	2.5	0.8
21	Tobacco manufacturers	.2	.7	.3
22	Textile mill products	3.7	3.5	1.4
23	Apparel and other textile	1.2	.9	.4
24	Lumber and wood products	2.7	4.2	1.9
25	Furniture and fixtures	.8	1.4	.8
26	Paper and allied products	7.4	5.9	2.8
27	Printing and publishing	1.6	.9	.6
28	Chemicals and allied products	15.4	5.0	2.8
29	Petroleum and coal products	5.8	8.3	2.1
30	Rubber and plastic products, nec	2.2	2.6	1.4
31	Leather and leather products	.5	1.3	.7
32	Stone_ clay and glass products	8.3	7.6	4.4
33	Primary metals	21.5	8.3	3.6
34	Fabricated metal products	4.5	2.0	1.0
35	Machinery, except electrical	4.8	1.3	.8
36	Elec. equipment and supplies	3.7	1.2	.7
37	Transportation equipment	4.9	1.4	.6
38	Instruments and related	.7	.9	.6
39 & 19	Miscellaneous manufacturing, ordinance and accessories	1.5	1.2	.7
	Total manufacturing	100.0	3.0	1.4

In a study at Oak Ridge National Laboratory [42] the energy requirements for the production of metals was compared in terms of equivalent coal energy. The data for five metals produced from a variety of ores or other raw materials are shown in table 1-9. Energy requirements for magnesium and iron should remain relatively constant in the future. Poorer grade bauxite ores to produce aluminum do not require much additional energy but when clays and anorthosite are used energy requirements are increased appreciably. Recycling of copper scrap will use much less energy than producing copper from ores. As copper and titanium ore of poorer grade must be used, energy consumption will increase appreciably. Recycling of titanium also requires much less energy than producing it from ores.

A more detailed analysis of the primary metal market, based on census data for 1963, indicates that the steel industry used 58 percent of the energy consumed by this sector, of which 51 percent was in the blast furnace and steel mills. Since that time coke rates have declined steadily in blast furnace operations and the opportunities for further reduction in energy use are limited without development of new technology of steelmaking. With present processes it is necessary that enough coke be present to provide heat to melt the ore, the amount of carbon monoxide needed to reduce the iron oxide to metallic iron, and the amount of coke required to give the required physical strength to the blast furnace charge. In recent testimony before a House subcommittee, Jaske [43] stated that there are "projections of excellent progress in reductions per unit heat (energy) requirements for the

Table 1-9 -SUMMARY OF THE ENERGY REQUIREMENTS FOR THE PRODUCTION AND RECYCLING OF METALS

Metal	Ore or main source	Equivalent coal energy (kwh./ton metal)
Magnesium	Sea water	90,930
Aluminum	Bauxite (50 to 30 percent Al_2O_3)	51,470–59,540
	Clays	65,972
	Anorthosite	72,360
Iron	High grade hematite	3,180
	Magnetic taconites	3,560
	Iron laterites	5,180
Copper	1 percent sulfide ore	13,530
	0.3 percent sulfide ore	24,760
	98 percent Cu scrap recycle	590
	Impure Cu scrap recycle	1,560
Titanium	High grade rutile	126,280
	Ilmenite bearing minerals, e.g., sands, rocks, etc	150,120–157,080
	High grade Ti soils	206,750
	Ti scrap recycle	39,000

production of steel using the basic oxygen process and other related new reduction processes." Estimates based on the plots shown in that testimony show possible savings of as much as 25 percent by the year 2000 in the steel industry.

The other major energy consuming product of the primary metals industry is in the production of primary aluminum. Of the 22 percent of the energy used by the nonferrous industries—in primary metal production in 1963—aluminum used 16 percent. Electricity requirements for aluminum production—the major energy form used—have reached the level of better than 35 percent overall efficiency—for the process used—and if new aluminum producing processes are not devoloped further decreases in energy use per unit of aluminum produced can only occur slowly although efficiencies as high as 45 percent may be attainable.

Another large user of energy in the industrial sector is chemicals and allied products. This is a more diverse sector than that of primary metals and it is more difficult to determine whether energy savings are possible. Eighty percent of the energy used in this category went into the manufacture of a variety of basic chemicals. Of this, 21 percent went into manufacture of organic chemicals—not included with this category are such materials as fibers, plastic, rubber, drugs, and so forth—and 34 percent into inorganic chemicals not otherwise classified, these others include cleaning and toilet goods, paints and varnishes, agricultural chemicals, and such other chemicals materials as explosives, glue, printing ink, and so forth. Because of the diversity of products in this category it is not possible to identify individual examples of where energy savings are possible but, even with low-cost energy, one would not expect it to be wasted indiscriminately in large amounts by industries that are highly competitive and energy intensive. Rather, only small savings would be expected unless there were major changes in technology or substitutes or synthetics were developed.

A more detailed breakdown of energy consumption in the individual industries is needed to see where and how much energy savings can be achieved. Until this information is developed the only prudent guess would be that energy savings (using existing technology) would be nominal.

In a newly issued book [44] the following comments were made about this sector:

MATERIALS AND RECYCLING

A substantial part of the consumption of energy in general and electricity in particular is consumed in the extraction and processing of metals, paper and glass. * * * Among them they account for more than 10 percent of U.S. energy use and almost 15 percent of electricity use. All of these materials are recyclable, yet today only a small fraction of consumption comes from recycling. Usually, recycling uses less energy per pound of product than obtaining new material by the usual processes. Certain other costs, however, are higher for recycling; otherwise profit-conscious industry would be recycling more and mining, logging and so on less. Therefore a substantial increase in the price of energy would have three desirable effects: First, since materials would become more expensive, waste and planned obsolescence would be discouraged. Second, this change would reduce per capita consumption of energy. Finally, since the cost of recycling would increase proportionately less than the cost of the alternatives, industry would be forced to rely more heavily on recycling of raw materials.

The disproportionate energy consumption associated with the production of aluminum deserves special attention. Five times as much energy is necessary to produce a pound of aluminum as to produce a pound of steel. Thus, for example, even though twice as many cans are made from a pound of aluminum as from a pound of steel, the energy requirement per can is two and a half times greater when aluminum is used. The consumption of aluminum in cans and packaging, which now accounts for about 10 percent of all aluminum use, is doubling every 7 years. Surely this massive proliferation of disposable and nondegradable containers is a negative contribution to our standard of living rather than a positive one, exclusive of the waste of energy it represents.

One additional possibility for reducing domestic industrial consumption would be the potential for manufacturing energy intensive products in areas abroad where energy is low cost or being wasted. Underdeveloped hydroelectric generating sites could be used for producing the electricity to manufacture aluminum, electrochemical and other products using large blocks of electricity. The natural gas currently being "flared" in the Middle East and elsewhere could be used to make hydrogen for ammonia manufacture or for carbon black. Low-cost crude oils of the Middle East could be used to manufacture petrochemicals. The overall economics of such developments will depend on such other factors as transportation costs to markets, construction and operating costs abroad compared to those in the United States, and the relative costs of energy in the United States and in other countries.

Transportation: The transportation sector consumed 24 percent (the small amount of electricity consumed by this sector has not been included) of all the energy produced in the United States in 1968. Except for small quantities of natural gas, petroleum products dominated this market (95 percent)—53 percent of the refined products of crude oil were used in the transportation sector.

Gasoline is consumed in highway transport, aviation, mechanized farming and power boating. Light diesel oils are used for truck and bus fuel, locomotive fuel, vessel use and by the military. In all of these applications the petroleum product is consumed either in an internal combustion engine (gasoline or diesel) or in a gas turbine (aviation uses). The efficiency in use for highway transportation varies widely during start-up, idling, acceleration, and deceleration. Efficiency of use is also heavily influenced by engine design, the size of the engine for a given car, car and truck size and numerous other variables. The spark ignited engine has a theoretical efficiency of over 30 percent at a compression ratio of 10 to 1, but cars actually operate in the range of about 18 percent—near the practical limit of efficiency considering driving patterns, losses during start-up, deceleration, idling, and so forth. Diesel engines, particularly large stationary ones, have a somewhat higher average efficiency—in the range of 35 percent, but in practice diesel engines have a wide range of efficiencies and for similar reasons.

Opportunities to increase efficiencies are not very great without new arrangements for land use or mass transportation, new laws restricting horsepower per unit of car weight or completely new approaches to the uses of energy for transportation. The reasons are that, although even average efficiencies appear low, there are thermodynamic limits to what can be achieved with present engine designs and fuels. Obviously improvements can and should be made in those parts of the cycle where efficiencies are lowest. Moreover, provisions of the recent Clean Air Acts have resulted in designs that will bring about some fuel savings through the use of the gasoline that formerly evaporated from the gas tank and through the new equipment required for using the energy units contained in crankcase blowby.

Obviously changing transportation modes can result in significant energy savings as shown by tables 1-10, 1-11, and 1-12.

The use of buses for urban mass transportation would result in a saving of more than 75 percent in B.t.u.'s per passenger-mile over the use of automobiles (table 1-11). In intercity passenger transportation, buses are the most efficient method of transporting passengers while the airplane is the least efficient, requiring nearly nine times as much energy per passenger-mile as a bus (table 1-10). For intercity freight pipelines, waterways, and railroads are much more efficient than trucks and from 55 to 80 times more efficient than airplanes (table 1-12). As energy prices rise, or new laws are enacted to abate pollution or restrict the use of energy intensive devices, a shift toward more energy-efficient methods of moving people and freight can be expected to occur.

There are also other potentials for reducing the total amount of passenger-miles traveled. Much of the airline travel (an energy intensive method of transportation) is done in connection with the conduct of business. Through the use of advanced communication systems, such as multiplex TV conferences and real time transmission of letters and other business papers, at least a portion of business travel could be reduced. A significant reduction in passenger-miles could also be achieved through land use and community planning. Distances to work and to shopping could be reduced and, with proper planning, mass transportation would be more convenient and consequently more widely used.

There are also technological opportunities for reducing energy used in the transportation sector Entirely new types of internal

Table 1-10 —ENERGY-EFFICIENCY FOR INTERCITY PASSENGER TRAFFIC

	Passenger-miles per gallon	B.t.u. per passenger-mile
1. Buses	125	1,090
2. Railroads	80	1,700
3. Automobiles	32	4,250
4. Airplanes	14	9,700

Table 1-11 —ENERGY-EFFICIENCY FOR URBAN PASSENGER TRAFFIC

	Vehicle-miles per gallon	Passengers, per vehicle	Passenger-miles per gallon	B.t.u. per Passenger-mile
1. Bicycles			765.0	180
2. Walking			450.0	300
3. Buses	5.35	20.6	110.0	1,240
4. Automobiles	14.15	1.9	26.9	5,060

Table 1-12 —ENERGY-EFFICIENCY FOR INTERCITY FREIGHT TRANSPORT

	Ton-miles per gallon	B.t.u. per ton-mile
1, Pipelines	300.0	450
2. Waterways	250.0	540
3. Railroads	200.0	680
4. Trucks	58.0	2,340
5. Airways	3.7	37,000

conbustion engines, new engines with higher efficiencies, external combustion engines, fuel cells, et cetera, might also be developed. Increased efficiency might be achieved by improved gas turbines capable of operating at higher temperatures than can be used at present. Storage of the energy that is being wasted during deceleration and idling in fly wheels or other storage devices could also improve overall energy utilization.

SUMMARY

In addition to opportunities of conserving the resource base in all phases of the energy cycle, there are large quantities of the less abundant fuels still to be found. Improved geological and other methods of finding these deposits, along with whatever incentives are required to stimulate further exploration and development of these resources, could improve the resource base appreciably. Very large additions to recoverable reserves could be obtained if the present low-recovery efficiencies for already discovered oil (30 percent) and coal (57 percent) could be improved. In this connection, a number of research and development opportunities exist that require new funding and field testing.

In the upgrading and transportation of fuel resources the best opportunity to reduce energy use is at petroleum refineries where between 10 and 15 percent of the energy is required to operate the refinery. For electricity significant energy savings might be achieved by reducing transmission losses. Very large increases in the efficiency of converting fuels to electricity might be possible through the development of new methods of generating electricity that would increase the present conversion efficiency from 40 percent in the best plants (average about 30 percent to 50 to 55 percent).

In the utilization of energy in the residential and commercial sector the largest consumption is for space heating and cooling, hot water heating, cooking, clothes drying, lighting and refrigeration. The best opportunities for energy conservation are reduction in the losses occuring in the heating and cooling of buildings. Improved building design, more insulation, more care in the use of lights and televisions, realistic lighting levels and better use of the heat energy in the air already in the buildings are some of the means proposed to reduce energy demands.

It is difficult to identify opportunities for increasing the efficiency of energy use in the industrial sector, because of the wide diversity of end uses. Of the energy intensive industries, new methods of steelmaking promise reductions in energy consumption of up to 25 percent and similar reductions might be possible in improved cell design for the manufacture of aluminum. Lower grade ores that will probably be used in the future will increase energy consumption, but recycling of some metals might permit significant reductions in energy requirements. Production of energy intensive products abroad in areas where there are large amounts of low-cost energy, or where energy is being wasted, would reduce U.S. energy requirements.

The major use of fuels in the transportation sector is for automobiles and trucks; here there is little that can be done to improve practical efficiency under current conditions. Reduction in the use of energy in this sector will have to come by changing the modes of transportation, greater use of mass transport, restrictions imposed by law, or reducing the amount of travel through better land use and community planning, or more advanced communications methods that will decrease the necessity for travel in connection with business operations.

Systems Approaches to Energy Conservation

The objective of this section is to describe recently applied or experimental methods of energy and electricity development that can help to reduce total demand or utilize new resources. Such methods include: (i) utilizing an increased percentage of the original energy input in a series of systems; (ii) achieving greater efficiencies through the use of larger units; (iii) developing solar energy as a means of storing energy for both space heating and cooling; (iv) utilizing the heat in waste products; and (v) producing useful fuels from organic wastes.

In testimony before the House subcommittee [1] Jaske described one such systems approach using the neighborhood fuel cell to generate electricity combined with other improvements to reduce residential demand as follows:

> New systems such as the electrically interconnected, neighborhood fuel cell concept with full thermal energy recovery for water heating and other domestic uses deserve more attention than they are currently receiving. Such systems could, in conjunction with better insulation, improved lighting efficiency, and more efficient air conditioning equipment using regenerative recovery of energy in ventilation air, offer a means to provide improved sophistication in home

living with total energy use reduction by as much as 75 percent over present uncoordinated systems. Needed fuel gas supplies are much easier to provide and gas service corridors require an order of magnitude less space than electrical transmission lines for equivalent energy transmission rates.

Most of these methods of conserving energy have so far received little attention in the United States, for fuel costs have generally been low and the use of added fuel to reduce capital investment has been less costly than to increase efficiency. The technology is still in early development stages for some of the systems described; for others limitations are apparent in overall economics rather than engineering aspects.

DISTRICT HEATING

District heating, which utilizes a central energy plant to produce space heating and cooling for a number of commercial and industrial establishments, universities and groups of houses, has been applied for many years both in the United States and abroad.[2]

When used in combination with a plant generating electricity, steam is discharged at some point in a steam turbine; the steam is then used for space heating and cooling. A district heating system allows significant savings in heat units, particularly when combined with the generation of electricity. There are heat losses during distribution of the steam to the various customers as well as additional operation and maintenance costs; thus district heating is largely used in areas which are densely populated.[3]

The increased efficiency of energy in this method results from the use of low temperature steam for heating and cooling, thus permitting the electricity to be generated with 50 percent less fuel. Although less electricity is produced the low pressure steam is used constructively for heating and cooling. Moreover, boiler efficiencies may be as much as 6 percent higher because additional equipment for saving energy units can be justified economically.

Producing steam in this way in large plants under carefully controlled conditions (compared to using a number of smaller plants) also has environmental benefits. Since the large plants are normally operated at higher efficiencies, thermal pollution is reduced and air pollutants are generally discharged from higher stacks, reducing their effects at ambient level.

THE TOTAL ENERGY SYSTEM (GROUP TO ADVANCE TOTAL ENERGY [GATE])

For a number of years the gas industry has promoted the use of onsite generation of electricity, using natural gas in either gas turbines or gas engines with the waste heat applied in space heating and cooling. Small conventional devices are used during peak periods. This method is particularly applicable in shopping centers, large apartment houses, small factories, motels, hospitals, and schools. It could also be used in large single-unit homes located in densely populated areas. By 1968 approximately 300 installations were in operation; in the absence of a gas shortage, many more could be constructed.

In 1968 the dollar savings in power costs for this kind of system were estimated to have covered the cost of the initial installation in 7 years. Where the waste heat can also be used as process heat[4] the dollar savings are even greater. A recent estimate indicates, for example, that an investment of $300,500 in a total energy system at a brick plant saved the company involved $3,100 in monthly utility bills.[5]

The GATE program has been sold on the basis of lower overall cost. Equally important, however, is the fact that it can raise the efficiency of use from 25 to 30 percent for small electricity generating stations to 65 to 85 percent for total energy installations.

In terms of electric utility plants, the installations are small—from 1,500 to 7,000 kw—but by taking advantage of all the possible energy savings, considerable economic advantages are possible. For example, fluorescent light operates most efficiently (by a factor of 2) at 400 cycles rather than the 60 cycles delivered by the electric utility systems. A GATE installation can be designed to generate electricity at 400 cycles without increased costs and at greatly reduced electric utility use.

In a recent publication prepared for the National Science Foundation [6] the advantages of the total energy system were described as follows:

> The concept of onsite generation, from a basic fuel, of the electrical energy locally required is very attractive, because the waste heat (65 percent) is onsite available at very desirable temperatures (especially in contrast to central station electrical plants). This waste heat can be used for heating, cooling, water recycling, and other purposes. The trend toward this concept is growing despite bitter and active opposition from electric utilities.
>
> Reliability, voltage and frequency regulation for special purposes (computers, etc.), and economy are attractive attributes of total energy. Considerable engineering R. & D. opportunities exist in total energy.
>
> The Nation should consider whether it would be desirable to foster total energy officially as a policy of first choice, for residential, commercial and industrial applications. Investment returns are attractive to the builder or owner, without subsidy. Convenience and reliability are improved, and the national interest is served—all while the operating costs are reduced.

Although a number of onsite installations have been made, there are reported operation and maintenance difficulties that still need to be resolved. Unlike central station plants, skilled operators are not generally used or available.

Recent studies have been made of the use of the fuel cell (with its electric high conversion efficiency of up to 70 percent) in place of the gas engine or gas turbine for onsite generation. This system would be particularly useful in situations where the electric load is high in comparison to the heating and cooling loads.

In another of the systems studied the fuel cells would be used in small neighborhood power stations to which the fuel would be piped (natural gas or hydrogen). Not only could the electricity be generated more efficiently, and with less air and water pollution per unit generated, but overall transmission and distribution losses would be reduced.

THE DEPARTMENT OF HOUSING AND URBAN DEVELOPMENT PROGRAM

The Department of Housing and Urban Development (HUD), in seeking methods to reduce both capital and operating costs in the development of public housing, recently looked into the potentials for using several modifications of the total energy concept as a means of promoting modern energy utilization techniques in housing.

Under an HUD contract, Hittman Associates were asked to examine the components of "total energy balance" in a home, in an effort to achieve optimum use of energy.[7] As an example, they learned that 14 percent of energy consumed could be saved by operating an air conditioner using either inside or outside air depending upon which had the best physical properties to reduce electric demand.[8]

The study was designed to include the energy demands in the building, producing the materials of its construction and construction techniques as alternative means of reducing overall energy use in housing. The study is still underway but important energy savings appear possible.

APPLICATIONS COMBINING LAND USE PLANNING AND ENERGY USE

The lower electrical generation efficiencies of nuclear lightwater reactors (those presently being used) result in a 40- to 50-percent increase of thermal pollution over fossil fuel plants. In a series of recent papers,[9] S. Beall, Jr., proposed a nuclear power complex which would provide electricity for a city designed to use total electric power for integrated industrial, commercial, and household uses. Low-temperature warm water discharged from the condensor would supply heat for greenhouses, other agricultural uses, and for heating and cooling residences and commercial and industrial establishments.

Mr. Beall sketched an imaginary new city with a climate similar to Philadelphia's in which projected energy requirements would be 500,000 kilowatts and two-thirds of the residents would live in three-story apartment buildings near the center of town; schools, libraries, and shopping centers would be located within walking distance. The powerplant would be operated to provide 300° F. water for distribution (with a 150° to 200° F. returning). Each apartment and office building would have its own unit for heating and cooling. Table 1-13 gives a summary of how the heat would be utilized.

Table 1-13 —SUMMARY OF URBAN HEAT USES

Item	Capacity, MW(th)
Capacity of heat source	2,268
Annual average thermal power	2,041
Annual average net electrical power	463
Annual average internal power consumption	29
Annual average rate of district heat production	457
Peak summer district heating load (300–202° F.)	1,144
Peak winter district heating load (300–148° F.)	1,088
Minimum district heat load	250
Industrial steam load	368
Sewage distillation steam at 32 p.s.i.g	90
Peak greenhouse heating	243
Annual average greenhouse heating	33
Annual average heat to condensers[1]	601
Maximum heat to condensers[1]	898

[1] This heat to be disposed of by the greenhouses, etc. of the foodgrowing complex, leaving no heat to be discarded to urface waters.

Mr. Beall summarized the advantages of this approach as follows:

> Comparing these numbers to an equivalent electric-only powerplant for the same city, we found that we could prevent

the burning of over 3 million barrels of oil [10] each year with its accompanying air pollution, and—by using heat that would otherwise have been thrown away—reduce thermal pollution of the biosphere (from this station) by 40 percent.

SOLAR ENERGY USE FOR SPACE HEATING AND COOLING

The use of solar energy as a fuel saver and supplement to conventional methods of heating and cooling has been receiving research attention under studies sponsored by the National Science Foundation. A collector of solar heat is installed in a home behind a transparent wall and the heat is stored in water or in some other material to be used at some future time for heating or cooling. Savings of up to 50 percent or higher in household fuel use may be accomplished.

Although conventional fuels are conserved, this approach will not be economically attractive in many locations in the United States unless the equipment can be manufactured under mass production conditions. On the other hand, it is possible by this capture of solar energy to store coolness and use it during periods of peak electrical demands for air conditioning.[11]

Commerical applications of solar systems which store coolness have been made where extreme peak to average demands ratios are high—for example in churches which are little used during long periods of time but have a high cooling demand for short periods.

UTILIZATION OF SOLID WASTES FOR ENERGY PURPOSES

In many European countries (and to a limited extent in the United States and Canada) solid wastes have been burned to produce steam and electric energy. Most of the plants using waste as fuel have been constructed in instances where the costs of both solid waste disposal and fuels are high.[12]

The general trend toward higher fuel prices and waste disposal costs in industrial nations has created additional interest in these types of plants. For example, in Vienna, Austria, a district heating plant has been designed to provide 200 megawatts of power with refuse contributing 25 to 30 percent of the heat required.[13] Calculations have recently been made for an incineration plant in France which would use the heat for district heating by industry and for electricity generation.[14]

Studies of refuse disposal from commercial and industrial wastes in the United States have been conducted by the U.S. Bureau of Mines, Department of the Interior, using carbonization, hydrogeneration and incineration. As an example, the typical wet municipal refuse from Altoona, Pa. (from which glass and metals have been separated), was shown to have a heating value (even though wet with garbage) of 4,827 B.t.u./pound, (about one-third the heating value of a high value bituminous coal).

To date the following results have been obtained by pyrolysis [15] (the process is about 83 percent efficient for each ton of wet refuse used): For municipal refuse 810,000 B.t.u. are recovered per ton of waste while, for industrial refuse, 1,690,000 B.t.u. per ton are recovered. Residential, commercial, and industrial waste in the United States is estimated to be about 350 million tons per year. If this quantity could be pyrolyzed, about 17 million tons of coal-equivalent fuel value could be recovered, helping to solve the waste disposal problem and also leading to the production of other useful byproducts such as ammonium sulfate.

Pyrolysis of the 1 million tons of tires scrapped annually could also produce useful heat. The chief advantage in this particular case, however, is the disposal of wastes and the production of valuable chemicals. When heated to 1,650° F., 15.9 million B.t.u. per ton of tires are found in the residue along with 8.7 million B.t.u. in the gas that is produced. (Approximately 24.6 times 10^{12} B.t.u. would be produced from 1 million tons of tires, equivalent to 1 million tons of coal.)

As noted earlier, incineration has been used as the most economical method for disposal of solid wastes. For various reasons it has not generally been worthwhile to attempt to capture the heating value of the disposed wastes, but there is now increasing interest in the United States (as there has been historically in European countries) to use these wastes for generating electricity or steam for process use. It has been estimated that the heat in industrial, commercial, and residential wastes from which solids have been removed is equivalent to 160 million tons of coal.

One difficulty is that the low heating value of refuse restricts the top steam temperature attainable; thus, if an incinerator plant produces electricity from the steam, it does so at low efficiencies. Several interesting experiments have been announced by the Environmental Protection Agency in which refuse is to be mixed with coal in such quantities to produce only a minor reduction in electric generation efficiency (if any) and still to obtain the benefits of the heat units in the refuse.

This and other methods being investigated by EPA were described in an article in "Ecology": [16]

> The Environmental Protection Agency is backing a system to turn garbage into electricity. Combustion Power Co., Inc., of Palo Alto, Calif., has designed an electric power generating refuse burner, the CPU-400, that consists of a gas turbine costing $1.8 million and sold by General Electric, Pratt-Whitney and Westinghouse. The operation is somewhat tricky. The turbine blades are set in motion and compress air which gets hot enough to melt sand in a chamber. After being ground up by a shredder, manufactured by Eidel International, Albuquerque, the garbage is fed onto the boiling sand.
>
> The garbage has already been cleaned of metal, glass and rocks in the shredding process, and the remaining organic material will now burn at a predictable rate. the extremely high heat breaks down sulfur dioxide and hydrochloric acid normally released by the plastics in the garbage.
>
> The CPU-400 has a capacity of 400 tons a day, the refuse of 160,000 people. The heating value of dry garbage is 6,300 B.t.u.'s per pound. (Coal has a B.t.u. range of less than 6,000 to 14,000 per pound.) Converting garbage to electricity via the turbine should yield 15,000 kilowatts of power.
>
> Two other gadgets supported by the EPA are the Torrax system being tried in Erie County, N.Y., built by Carborundum, which consumes garbage without oxygen and generates fuel gas therefrom. The other is a system in St. Louis by the Union Electric Power Co. which simply burns garbage instead of coal in the utility's furnace to make steam.

Another experiment in using industrial waste as a powerplant fuel is being conducted at Fort Wayne, Ind. [17] Industrial waste has a less variable composition than residential waste and this makes it more attractive as a fuel supplement for powerplants. The electric generating plant at Fort Wayne uses sized coal and if a low-cost low-sulfur fuel were available the useful life of the plant would be prolonged.

A recycling company already extracts the paper and paperboard from industrial and municipal works and by further processing has devised a method to separate and make into "cubettes" the below-grade fibrous materials along with small quantities of wood, leather, and other carbonaceous materials. Two combustion tests using a 3 to 1 ratio of coal to "cubettes" have been encouraging. No operating difficulties were encountered. The potential for the Fort Wayne area alone is for the production of 120 tons per day of coal equivalent or about the quantity that could be used by two to three 40,000 kilowatt stoker fired plants.

Although the use of the heating values in refuse has received relatively little attention in the United States to date, this situation can be expected to change. As population increases and our economy grows, an increasing amount of solid wastes will be generated, and the cost of their disposal (already starting to increase because of exploding interest in environmental matters) will grow dramatically. Compounded with the inevitable increase in the real price of all fuels, this situation can be expected to encourage the economical use of such refuse in many locations.

UTILIZATION OF REFUSE AND OTHER WASTES TO PRODUCE LIQUID AND GASEOUS FUELS

Urban refuse, cellulosic wastes, wood waste, and sewage sludge have been successfully converted to heavy oil in a laboratory experiment by heating the refuse under pressure with carbon monoxide and steam. Conversions of the dry organic matter have been about 40 percent—yielding the equivalent of two barrels of oil per ton of dry ash-free waste. Such wastes are low in sulfur—providing a superior oil product.

Although this method is still in the preliminary stages of experimentation, the Bureau of Mines stated in a recent report: [18]

> Cellulosic materials, all other carbohydrates, wood wastes (largely cellulose and lignin), urban wastes (mostly cellulose plus other carbohydrates, proteins, fats, and small amounts of other organic materials), sewage sludge, agricultural wastes, and bovine manure can be converted to oil with carbon monoxide and water. Some plastics depolymerize and dissolve in the product oil; some remain as part of the unconverted residue. But since the plastic content of urban refuse is relatively small, the presence of these materials is expected to have only a minor influence on oil composition and yield.
>
> A significant part of the energy demand of the Nation can be obtained on a renewable basis by converting nearly every kind of organic solid waste to a low-sulfur oil by treatment under pressure with carbon monoxide and water. Methods for lowering carbon monoxide consumption and for operating at lower pressures have been found; these offer the potential of low processing costs for converting cellulosic wastes to oil. While the effects of temperature, pres-

sure, and water on this process have been explored, more work is required to find optimum conditions for the conversion. A continuous unit has operated successfully, and preliminary results have been obtained on the conversion of sucrose to oil.

In a recent article it was proposed [19] that organic and urban waste be converted by anaerobic action (the reaction of microorganisms in the absence of air) into methane which could be distributed (50 to 80%) and sludge (50 to 20%) which could be disposed of by incineration or for fertilizer. The anaerobic action occurs spontaneously but, in commercial production of methane by this method the natural process would have to be speeded up. The method is used in the initial step in the treatment of sewage but no attempt has been made to maximize the amount of methane or to use mixtures of solid waste (garbage and paper products) and sewage. It is thought that the mixture would be a better raw material for methane production than sewage or solid wastes alone.

The potential for methane from this source is large since 1.5 billion tons of solid waste produced each year could generate as much as 30 trillion cubic feet of gas—an amount larger than current gas consumption in the United States.

Capital costs for this conversion may be high since it requires 7 days of digestion to complete the reactions. Plants would have to be located to minimize collection costs as this is the most expensive part of the overall system. If located at a feedlot with 100,000 cattle, the 150,000 tons of dry organic waste generated per year would yield 3 billion cubic feet of methane. If the feedlot were in Kansas City, this gas would be worth (at wholesale) about $500,000. However, insufficient process and engineering data are available on which to estimate the economics of such a plant.

SUMMARY

Several approaches for using a larger amount of the energy in the original fuel as part of a total energy system have been developed. Some already are in commercial use, others have been tested and indicate considerable promise. In view of the anticipated rise in the price of fuels, apparent shortages of supply, and growing concern over ways to reduce the adverse environmental impacts of energy, increased interest can be expected for the further development of these types of systems. Some of the methods under experimental use will involve additional extensive research and development, while others simply need improvement in existing technology or the overcoming of inertia accompanying the introduction of new technology. The great promise that these various systems hold suggests that they deserve careful study to bring about either their rejection or wider commercial use.

THE ENVIRONMENTAL FACTOR

Except in a few cases in the above discussion where recent innovations have been described, the impacts of new environmental standards on energy use and efficiency have not so far been considered. Current writings on this subject generally reflect the position that additional energy required to meet new standards for cleaning up the environment will range from relatively insignificant amounts to as high as 25 percent more energy than presently consumed.

Actually, a major research effort has not yet been mounted to determine the overall implications of environmental cleanup needs in terms of energy demand. If the public demands both secondary and tertiary treatment of all water wastes, recycling of everything that can be recycled, restoring all disturbed land to its original or an improved condition, and selecting the least environmentally damaging route for electric transmission lines, then, in most cases, these objectives can only be achieved by using more energy rather than less. To illustrate, a recent environmental impact statement on the leasing of uranium deposits noted three new environmental requirements that would consume additional energy (assuming no other changes were made or substitutes used):

> All solid or liquid waste shall be disposed of by using accepted State and Federal disposal methods and following the State and Federal laws, regulations, and standards.
>
> Access to the withdrawal site is across forest land over old, poorly drained roads. All roads used for access should be properly drained with culverts and/or water bars and reseeded to stabilize them. If road construction becomes necessary the standards will be determined by the Forest Service.
>
> All disturbed areas must be returned as nearly as practicable to their original condition, or to a condition to be agreed upon by both the lessee and the surface manager as to the satisfactory standards for such reclamation. This reclamation should be accomplished as soon as practicable after the damage has occurred.

The following discussion examines four industrial areas in which environmental concern will affect energy demand. Two of the examples are from industries which account for nearly 48 percent of total energy consumed. The third example is the recycling of cans and bottles. The fourth involves recycling of aluminum.

POWERPLANTS

Basic assumptions must be made in calculating the increased heat inputs for a 1,000-megawatt new electric generating plant needed to satisfy new environmental standards: In the base case there would be no particulate clean up, no sulfur oxide removal and, once through-cooling of condenser water (assuming that the resulting thermal pollution will be acceptable). In the alternate case particulates would have to be removed by electrostatic precipitators with 99-percent efficiency, 85-percent removal of sulfur oxides would be achieved by wet scrubbing methods and a mechanical draft cooling tower would be used to reduce thermal pollution to acceptable levels. While there would be a range in both capital and operating costs (which will differ for each location depending on the nature and quality of the coal ash and the sulfur content of the coal as well as the meteorological conditions where the cooling tower is used) an average of capital costs and energy operating costs for large plants is assumed to be:[22]

	No controls	Environmental controls
Capital costs (dollar per killowatt):		
SO_2 removal (scrubber)	0	40–60
Electrostatic precipitator	0	8–12
Thermal pollution control (by mechanical draft tower)	0	10–12
Total		54–84
Energy operating costs—percent of electricity generated:		
SO_2 removal (2-stage scrubber)	0	.8
Electrostatic precipitator	0	.2
Mechanical draft tower	0	.8
Total		[1] 1.6

[1] 480 B.t.u. per kilowatt-hour of energy input.

SO^2 removal will also require 120 B.t.u./kilowatt-hour for reheating the cooled gases. No increase in energy has been assigned to nitrogen oxide removal or disposal of any solid wastes that might be created.

In this illustration the total energy units increased by 600 B.t.u. (480 plus 120) over the heat input to the plant without environmental controls. The most modern plants have a total heat consumption of 9,500 B.t.u. per kilowatt-hour; thus, with environmental controls, an additional 6.5 percent of the plant B.t.u. input would be required.

Additional energy would also be expended to manufacture the equipment used to control pollution, but the exact amount is difficult to estimate; however, first order approximation will be attempted here.

It is assumed that of the total cost of the added equipment about one-half is in the cost of fabricated metal vessels made of steel. Primary metals consume 21.5 percent of the total industrial energy and steel accounts for 58 percent. With industrial energy representing 39.8 percent of all energy, about 5 percent of all energy was used in steel production. All steel products were valued at $55,152 times 10^6 in 1969, so that the B.t.u. used per dollar in producing steel products can be estimated at 60,000 B.t.u. for each dollar of finished product.

With the assumption that one-half the average additional capital investment of control equipment added to the plant was the fabricated metals (with an average life of 35 years), an additional 0.6 percent of energy would be required to construct the control equipment. Although some energy use would also be associated with the other one-half of dollars of capital expenditure, none have been included in this estimate. The gross total increase in energy used for this assumed case is about 7.1 percent.

TRANSPORTATION BY AUTOS AND TRUCKS

The following tables (1-14, 1-15, and 1-16) indicate the increased use of fuel that is to be expected for light vehicles and heavy-duty trucks.

In discussing these tables the report made the following comment on light vehicles:

> . . . The increases in operating costs are due to increased fuel consumption. For light duty vehicles, annual fuel consumption is based on a reference of 760 gallons per vehicle. The baseline for fuel consumption comparisons is model year 1973, and an additional 7 percent beginning model year 1975.

Thus, by 1975 an additional 15 percent of energy input (fuel consumption) to the automobile will be required to meet environmental standards. The energy consumed in maintaining the environmentally acceptable vehicle must also be added. If the 90,000 B.t.u./dollar of GNP[25] (the average value in 1970) is used as a basis for estimating additional energy consumption by 1975 this would add an additional 1,200,000 B.t.u./year, which would increase energy consumption for the environmentally acceptable car by an additional 1.2 percent.[26] Therefore, by 1975 for the same size car and motor overall energy consumption could be increased by nearly 16.2 percent (before including the energy required to build the control equipment).[27]

The report made the following comment on heavy duty trucks:

> For heavy duty gasoline vehicles, increases are based on annual averages of 1,380 gallons of fuel with a 15-percent increase in consumption beginning model year 1975.

Using the same method as above for assigning energy costs maintenance would add another 2.8 percent to the 15 percent, making a total increase in energy use of nearly 18 percent for heavy duty vehicles.

Increased costs of manufacture of equipment to meet environmental standards are shown in table 1-15. An increased investment per car or truck of $250 has been assumed with a 10-year life of the vehicle.

Table 1-14 ANNUALIZED UNIT COST INCREASES FOR LIGHT DUTY VEHICLES

[In dollars]

Cost type	Model year— 1968	1969	1970	1971	1972	1973	1974	1975	1976	1977
Increased fuel use				15.90	15.90	21.50	21.50	40.60	40.60	40.60
Maintenance	6.10	6.10	6.10	12.70	12.70	15.50	15.50	[1] 50.10	[1] 60.40	[1] 60.40
Maintenance offsets								[1] −36.30	[1] −36.30	[1] −36.30
Total	6.10	6.10	6.10	28.60	28.60	37.00	37.00	54.40	64.70	64.70

[1] Offsets for reduced requirements for present type tune-ups and exhaust system maintenance.

Table 1-15 —ANNUALIZED UNIT COST INCREASES FOR HEAVY DUTY TRUCKS [1]

[In dollars]

Cost type	Model year— 1968–69	1970–72	1973	1974	1975	1976	1977
Gasoline engines:							
Increased fuel use [1]	None				68.40	68.40	68.40
Maintenance	None		9.90	9.90	31.70	31.70	31.70
Total operating and maintenance penalties	None		9.90	9.90	100.10	100.10	100.10
Annualized control investment costs [2]	None	1.80	8.20	8.20	48.40	48.40	48.40
Total annualized cost increase	None	1.80	18.10	18.10	148.50	148.50	148.50
Diesel engines:							
Increased fuel use [3]	None		None	None	222.00	222.00	222.00
Annualized control investment costs [2]	None		None	None	200.00	200.00	200.00
Total annualized cost increase	None		None	None	422.00	422.00	422.00

[1] Based on averaged of 1,380 gallons of fuel per year as baseline, fuel at 33 cents per gallon.
[2] Based on 5 years engine life, annualized straightline basis.
[3] Based on average of 10,660 gallons of fuel per year as baseline, fuel at 26 cents per gallon.

Table 1-16 NATIONAL COSTS FOR MOBILE SOURCE COMPLIANCE

[In millions of dollars]

Cost type (increases)	Fiscal year— 1968	1969	1970	1971	1972	1973	1974	1975	1976	1977	Totals, 1968 through 1977
Light-duty vehicles:											
New investment	40.6	52.9	114.6	267.5	369.5	809.5	954.0	2,528.0	4,061.0	4,463.0	13,660.6
Assembly line testing [1]					2.5	34.8	35.4	36.3	36.9	37.8	183.7
Annual operating and maintenance	41.5	97.2	154.4	379.3	662.6	1,016.1	1,394.7	1,916.8	2,581.6	3,267.3	11,511.5
Total	82.1	150.1	269.0	646.8	1,034.6	1,860.4	2,384.1	4,481.1	6,679.5	7,768.1	25,355.8
Heavy-duty gasoline vehicles:											
New investment			4.2	5.4	5.7	22.9	28.6	147.1	185.2	195.0	594.1
Annual operating and maintenance						5.3	12.5	72.6	153.4	236.6	480.4
Total			4.2	5.4	5.7	28.2	41.1	219.7	338.6	431.6	1,074.5
Gasoline price [2] increases due to lead removal								10.7	22.3	34.8	67.8
Total, all gasoline vehicles	82.1	150.1	275.2	652.2	1,040.3	1,888.6	2,425.2	4,711.5	7,040.4	8,234.5	26,498.1
Heavy-duty diesel vehicles:											
New investment								68.0	92.0	101.0	261.0
Annual operating								15.1	31.5	49.5	96.1
Total								83.1	123.5	150.5	357.1
All vehicles, total	82.1	150.1	273.2	652.2	1,040.3	1,888.6	2,425.2	4,794.6	7,163.9	8,385.0	26,855.2

[1] Assumes inspection program (partially implemented 1972) with 3 percent of production tested beginning 1973, average cost $3 per vehicle produced (includes foreign and domestic production for U.S. consumption).

[2] Considers changes in demand patterns and fuel penalties as a result of controls as well as added costs of producing gasoline.

Again, using the methodology previously applied to translate dollars of GNP to energy units by 1975, the following additional energy costs (over those shown previously) would be:

	Percent increase due to new capital investment
Light-duty vehicles	2. 2
Heavy-duty vehicles (gasoline)	2. 2
Heavy-duty vehicles (diesel)	2. 2

Thus the total for a light gasoline vehicle is in the order of 18.4 (15+1.2+2.2) percent increase in energy inputs.

ENERGY FOR BOTTLES AND CANS

In a recent article, Bruce M. Hannon [28] pointed out that the share of the market for soft drinks and beer in returnable containers declined between 1958 and 1966. Soft drinks in returnable bottles declined from 98 percent to 80 percent and beer from 55 percent to 35 percent. This decline is expected to continue in the future.

The objective of Mr. Hannon's article was to examine the claims that recycling of materials would not only conserve resources, but would reduce economic and energy costs. Indirect energy commitments (such as those required to make the machines, delivery trucks, paper, and packaging) were neglected although they were estimated to be about 0.4 to 0.5 percent of the energy expended by the container industry.

The energy costs in B.t.u. for 1 gallon of soft drinks expended in returnable and throwaway bottles are compared in table 1-17.

Table 1-17 —ENERGY EXPENDED (IN B.T.U.'s) FOR 1 GALLON OF SOFT DRINK IN 16-OUNCE RETURNABLE OR THROW-AWAY BOTTLES

Operation	Returnable (8 fills)	Throwaway
Raw material acquisition	990	5, 195
Transportation of raw materials	124	650
Manufacture of container	7, 738	40, 624
Manufacture of cap (crown)	1, 935	1, 935
Transportation to bottler	361	1, 895
Bottling	6, 100	6, 100
Transportation to retailer	1, 880	1, 235
Retailer and consumer		
Waste collection	89	468
Separation, sorting, return for processing, 30 percent recycle	1, 102	5, 782
Total energy expended in B.t.u.'s per gallon:		
Recycled	19, 970	62, 035
Not recycled	19, 220	58, 100

Source: "Bottles, Cans and Energy", op. cit.

These energy costs have been calculated in terms of B.t.u. inputs, for example, estimating the B.t.u. inputs to the electric generating plant rather than the B.t.u.'s in the kilowatt hours used.

The table makes it clear that the least intensive energy consumer is the returnable bottle—by a factor of nearly three. However, as the table indicates, for glass bottles—using the assumption above—recycling of the throwaway bottles requires slightly more energy—(7 percent)—than if the bottles are not recycled.

A similar energy analysis for 1 gallon of soft drink in 12-ounce cans is shown on table 1-18, where a throwaway metal container is compared to returnable glass system:

Table 1-18. *Energy expended (in B.t.u.'s) for 1 gallon of soft drink in 12-ounce cans*

OPERATION:	*B.t.u.'s per gallon*
Mining (2.5 lbs. of ore per lb. of finished steel)	1,570
Transportation of ore (1,000 miles by barge)	560
Manufacture of finished steel from ore	27,600
Aluminum lid (11.9 percent of total can weight; 4.7 times the unit steel energy)	12,040
Transportation of finished steel 392 miles	230
Manufacture of cans (4 percent waste)	3,040
Transportation to bottler (300 miles average)	190
Transportation to retailer	6,400
Retailer and consumer	
Waste collection	110
Total energy for can container system [1]	51,830
Total energy for 12 oz. returnable glass system	17,820
Ratio of total energy expended by can container system to that expended by 12 oz. returnable glass	2.91

[1] The all-aluminum can system consumes 33 percent more energy than the bimetal (steel and aluminum) can system.

Source: Bottles, Cans, and Energy, *op. cit.*

A bimetal throwaway system uses nearly three times as much energy as a returnable glass container. If an all-aluminum nonreturnable can were used it would require nearly four times as much energy as the returnable glass system.

Table 1-19 compares the energy ratios for returnable glass—in one case returnable plastic—with various throwaway systems. Paper, although nearly twice as energy intensive as returnable glass containers, is by far the best from an energy-consuming standpoint.

Table 1-19 —ENERGY RATIOS FOR VARIOUS BEVERAGE CONTAINER SYSTEMS

[The energy per unit beverage expended by a throwaway container system divided by the energy per unit beverage expended by a returnable container system [1]]

Container type					
Throwaway	Returnable	Quantity	Beverage	Returnable fills	Energy ratios
Glass	Glass	16 ounces	Soft drink	15	4.4
Can	"	12 ounce	"	15	2.9
Glass	"	"	Beer	19	3.4
Can	"	"	"	19	3.8
Paper	"	½ gallon	Milk	33	1.8
Plastic [2]	Plastic	"	"	50	2.4

[1] Without remelting loop (discarded bottles and cans are not returned for remanufacture)
[2] High-density polyethylene.

Source: Bottles, Cans, and Energy, Op. cit.

By recalculation of the data in this table, the energy use comparison of each of the throwaway options can be estimated. This information is shown on Table 1-20 with allowance made for the different throwaway containers—assuming a straight line relationship between container size and energy consumed:

Table 1-20

Throwaway	Quantity	Beverage	Energy ratio
Glass	16 ounce	Soft drink	2.4
Can	12 ounce	"	2.1
Glass	"	Beer	2.5
Can	"	"	2.8
Paper	½ gallon	Milk	[1] 1.0
Plastic	"	"	1.4

[1] Assigned value of 1.

Source: Bottles, cans, and energy, op. cit.

While paper and plastic remain the least energy intensive, there is a slight advantage for glass (from an energy standpoint) over a metal container in a throwaway system.

RECYCLING ALUMINUM

Very large amounts of the energy used in producing aluminum might be saved if more of it were recycled. The present status of recovery of aluminum from secondary sources has been described in a Department of Interior publication[29] as follows:

> The recovery of secondary aluminum averages slightly over 11 percent of the total aluminum supply shown in figure 1. The scrap from which secondary aluminum is recovered is designated as either "new scrap" or "old scrap." New scrap is generated from (1) fabricating operations, such as alloying, stamping, forging, extruding, machining, and casting and (2) rejected semifabricated and manufactured items. Old scrap is aluminum that has been used in end products and is collected for metal recovery after the products are worn out or discarded. New scrap consumption and metal recovery data are based on quantities treated outside of the generating plant and do not include "run-around" scrap consumed by the generating company.
>
> New scrap is the source of nearly 81 percent of the secondary aluminum recovered. The relatively small recovery from old scrap reflects the fact that the aluminum industry is a growing industry. The tonnage of aluminum in use, estimated at about 40 million tons, represents a growing reservoir of old scrap for the future.
>
> Consumers of aluminum scrap may be divided into the following four major categories: (1) Secondary smelters; (2) primary producers and fabricators; (3) foundries and miscellaneous manufacturers; and (4) chemical plants. The secondary smelters accounted for about 75 percent of the scrap consumed from 1959 through 1968. Much of this was consumed in the production of castings.

The utilization of "old scrap" is only 19 percent of the total and reflects the economic problems associated with collecting from a variety of sources large enough volumes of aluminum products which are similar alloys—that is, aluminum mixed with known amounts and kinds of other metals. Separation before melting is done by secondary recovery aluminum industry to meet many of the specifications for its reuse. As the number of different alloys has increased to supply new markets which require different alloys to best meet a particular end use, the difficulty, cost of collecting, and even identifying alloys and of treating the various types to get a satisfactory product has grown more complex.

On the other hand, if it were decided in advance that the primary product was going to be recycled, it should be possible to arrange a system that would permit the easy identification and more rational collection of the aluminum products to be discarded. A good example of a system that might easily be devised is aluminum engine blocks in automobiles. In practice the very diverse number of uses for aluminum might make this possible for only a small part of the aluminum produced. Nevertheless, were it accomplished, the energy comparisons between primary production and secondary recovery would be as follows:

Per pound of aluminum

	B.t.u.
Primary production, 7 kilowatt-hours (39 percent efficiency electric generating plant)	62,000
Secondary recovery (approximately)	5,000

The energy efficiency of secondary recovery is about 12 times that of primary production. No attempt has been made to compare the energy requirements of the other parts of these two systems.

SUMMARY

Generally environmental enhancement requires that some additional energy be used; this, in turn, causes more pollution. No attempt has been made to quantify all the additional energy demands of environmental control efforts but calculations have been made for two important energy consuming sectors (powerplants and the automobile) which together consume nearly 50 percent of all energy. For powerplants an increase of about 7 percent in the fuels used will be required to remove particulates and sulfur dioxide and to control thermal pollution. An additional 18.4 percent in energy use will be needed to meet 1975 auto emission standards.

Two examples of the energy used for recycling of solid wastes show the following: When bottles are recycled somewhat larger quantities of energy (7 percent) are consumed than when new bottles are manufactured and not reused. The recycling of aluminum, on the other hand, would use only one-twelfth the energy now consumed if it were produced by electrolysis from the aluminum ore.

New Technology and Energy Sources

In the discussions above, a number of significant opportunities for conservation and energy savings through research and development have been identified. Examples include methods for improving coal and oil recoveries, for reducing petroleum consumption in mobile units by changing the nature of the transportation system, and for conserving electric energy used for residential, commercial, and industrial purposes.

This section will focus on research designed to: (1) build the resource base (e.g., SO_2 reduction and breeder reactors); (2) convert a large energy resource into more useful or more environmentally acceptable forms for the consumer (e.g., coal to gas or oil, oil shale to petroleum); (3) improve the efficiency of conversion processes; and (4) bring into commercial use entirely new energy resources that are not now widely used. The topics are covered in approximately the order in which they may be expected to become commercially useful.

BUILDING THE RESOURCE BASE

Sulfur Oxide Controls.—The enactment of laws in some States (e.g., New Jersey) with respect to sulfur oxide emissions essentially disqualifies coal from being used in electric powerplants. Under the criteria determined by the Federal Clean Air Act, many other States have already, or will soon set limits for the sulfur content of coals at a level low enough to disqualify coal deposits in their natural (or conventionally upgraded) state as a fuel for power generation. Since coal is by far the most abundant indigenous fossil fuel resource (80 percent of total), methods to insure its use in electric power generation should be devised while still meeting the new environmental standards.

There are several alternative methods to achieve this objective: (1) the coal can be gasified to a low B.t.u. fuel from which the sulfur can be extracted by known methods or, if higher efficiency of energy use is important, by methods still to be developed;[1] (2) coal can be liquified to form a low sulfur fuel oil suitable for boiler fuel;[1] or (3) the sulfur oxides can be removed from the stack gases after combustion of high sulfur fuels (either coal or residual oil). All of these methods will increase energy consumption but will enable the vast coal deposits to be used in an environmentally acceptable manner.

The most advanced of these methods is removal of sulfur oxides from stack gases. Experimental work on this approach is more than 40 years old and a wide variety of processes have been tested experimentally. Some suggested methods are based only on paper studies; others have been tested at the bench scale and pilot plant level. Limestone scrubbing processes have been tested on a relatively large scale for several years in two plants (about 125 megawatt capacity) in the United States. Experiments in limestone scrubbing have also been conducted abroad.

However, no single process for stack gas scrubbing has so far been commercially proven. At least 20 large-scale installations have been announced and three processes are being tested during 1972 with Federal support to demonstrate their feasibility at powerplants. Results have not yet been obtained but by early 1973 the status of technical, engineering, and economic feasibility of these processes will have been determined. The President's energy message also called for three additional processes to be demonstrated over the next few years.

Breeder Reactor.—The commercial development of a breeder reactor would extend the life of present, economically useful uranium resources by a factor of at least 50 to 70 and would also bring thorium into the energy resource base by using alternative technologies. In addition, high-cost uranium not now considered a part of the resource base would then be economically attractive. A technically and economically successful breeder would essentially guarantee a fuel for electricity generation for hundreds of years.

In the President's energy message [2] the significance of the breeder was noted as follows:

> Our best hope today for meeting the Nation's growing demand for economical clean energy lies with the fast breeder reactor. Because of its highly efficient use of nuclear fuel, the breeder reactor could extend the life of our natural uranium fuel supply from decades to centuries, with far less impact on the environment than the powerplants which are operating today.

The breeder reactor is able to operate at about 40 percent efficiency and produce more fuel (fissionable material) than was originally in the fertile fuel fed to the reactor. It can do this because it changes the fuels that were not fissionable but present in the reactor to ones that are fissionable.

There are several variations of the breeder reactor that can be developed but progress in the United States has been slow. Much of the research undertaken was to prove the feasibility of using liquid metal coolants (sodium) for the reactor. In addition to the liquid metal fast breeder reactor (LMFBR) another breeder reactor looks attractive—the gas cooled fast breeder reactor (GCFBR). This reactor would use helium as the coolant at pressure 50 to 75 times atmospheric.

The GCFBR, if developed, would apply knowledge gained from the operation of the high-temperature nonbreeding gas-cooled reactors. In addition to breeding fuels, the efficiency of generation would also be in the range of 40 percent, thus reducing thermal pollution problems associated with the present nonbreeder reactors (which have efficiencies of 30–32 percent). The difficulties and advantages of this method were summarized by Dr. Manson Benedict [3] as follows:

> The major drawbacks of the gas-cooled fast reactor stem from the relatively poor heat-removal characteristics of helium gas, compared with liquid sodium, and the most serious consequence is the difficulty of preventing overheating of the fuel if in some way the helium loses its pressure.
>
> The GCFR is, nevertheless, an attractive concept because it promises high breeding ratio, high thermal efficiency, low capital cost, and freedom from the problems of handling sodium that complicate the liquid metal breeder. Because of its advantages over the liquid metal breeder, the gas-cooled fast breeder should receive substantially increased development funding.

The LMFBR is receiving the most attention throughout the world and is the approach that the AEC has decided to use to develop breeder reactors. A number of LMFBR's have been built abroad using the excellent heat removal characteristics of liquid sodium. Although use of sodium avoids the problems of overheating of the GCFBR it poses other engineering drawbacks but new technology has been reported largely to have overcome them. One problem, for example, is the reported tendency of metals to swell after prolonged heavy neutron discharge. The solution to swelling will most likely be found by using various exotic alloys and stainless steel.[4]

The AEC has supported work by three companies on testing the components of an LMFBR and requested each of these firms to submit a proposal to build a demonstration reactor in the 300 to 500 MW range costing about $500 million.

During the early part of 1972 the efforts of the AEC to bring LMFBR technology to commercial realization were concentrated on making a "cooperative arrangement between the nuclear and electrical industries for the design, development, construction and operation of the first U.S. LMFBR demonstration plant." [5]

As a result of a series of meetings, it was agreed that a joint industry-Government program would be conducted by Commonwealth Edison, TVA and the AEC. In making the announcement,[6] the AEC estimated that the project would cost $500 million, with the utility industry providing nearly one-half of the funds. The demonstration reactor would be located at a TVA site although the exact location, plant size and other characteristics were still to be determined.

Manufacture of synthetics from coal, oil shale and tar sands.—The processes to be described under this category generally use more energy than they supply but would provide fuel that might otherwise be untapped in the form that the consumer desires. While more energy rather than less would be used based on the original resource, it is clear that the consumer would prefer to utilize the energy in a more convenient and more environmentally acceptable form, even at higher economic and energy costs.

Coal gasification to high B.t.u. gas.—Numerous processes have been investigated to produce a synthetic gas to supplement natural gas which is in short supply. Only one of these (the Lurgi process) has reached the commercial stage. It is based on the use of coals with special characteristics and has not been used in the United States. Other processes have been investigated on a much smaller scale and considerable development work is necessary before they can be brought to a commercial stage.

In testifying in hearings before the House Interior and Insular Affairs Committee, the representative of the AGA said: [7]

> Throughout 1970, the industry considered various possibilities and alternatives for accelerating the coal gasification research effort to help solve the gas supply problem. This led to the formal recommendation to the Department of Interior in January of 1971, which subsequently became the basis for the Interior-AGA joint program which seeks to evaluate the known coal gasification processes through the pilot plant stage. The program as it is presently constituted would take 4 years at an annual expenditure level of $30 million ($20 million from Government and $10 million from industry) to pilot the principal processes and select those elements from each which offer the greatest promise of success in demonstration and ultimately commercial application.
>
> Coal gasification, at present, constitutes the most promising single technological advancement in the field of developing substitute natural gas supplies.
>
> Inherent in such a program is the basic understanding that at this time no single coal gasification process in its totality constitutes the best method for commercial application.

These programs are designed to discover how to produce gas from coal at the lowest cost per million B.t.u. All of them will produce less energy in the gas than were present in the fuels to manufacture them. Most of the processes can be expected to be in the range of 65 percent efficiency of conversion. This means that we will have consumed 35 percent of the inherent energy in the original coal in the conversion process.

The imminent status of these developments is shown by the announcement in a recent paper presented at a meeting of the American Institute of Chemical Engineers: [8]

> Recently, El Paso Natural Gas Co. announced that it will construct a plant in the Four Corners area of the United States to produce 250 million cubic feet of gas per day at a plant investment of about $250 million. This investment represents a profound new step in modern fuels-supply technology.

High B.t.u. gas from petroleum products.—The shortage of natural gas combined with the anticipated ability to import petroleum products from abroad under conventional marketing conditions has resulted in the announcement of the construction of numerous plants to produce gas from various petroleum fractions. These first plants will use either existing and proven Japanese or British technology but their number is not expected to become large although about 1 billion cu. ft. per day of production has already been announced. Although this is a quick and easy way to overcome local gas shortages, the economic costs (compared to those associated with coal gasification) are expected to be higher and would result in even a greater dependence on oil imports than is now anticipated. From an energy efficiency standpoint they will be 90 percent efficient compared to only 65 percent for coal gasification.

Coal liquification to various petroleum products.—During World War II the Germans developed two methods (Fischer Tropsch and coal hydrogenation) to produce oil from coal in very large scale plants. While the economics were not attractive, they proved that conventional petroleum fractions could be produced from coal in large scale plants. Conversion efficiencies were in the range of 56 percent for the Fischer Tropsch process and 38 percent for the coal hydrogenation process. At the experimental plant of the Bureau of Mines at Louisiana, Missouri, constructed after World War II, the calculated efficiency for the coal hydrogenation plant was 55 percent. As in the case of coal gasification, an energy penalty is paid for producing the more convenient fuel form.

In two new processes being supported by the Office of Coal Research the efficiency of conversion in the FMC process (coal to some gas, oil and large amounts of char) is 85 percent efficient and for the de-ashed coal process the efficiency is about 75 percent.

There has been somewhat less interest and activity in coal to oil research than there has been in converting coal to gas. Nevertheless, some research has been carried out by both industrial firms and the Federal Government in an attempt to improve on the German processes and to make them suitable for American coals.

The direction of this research has recently changed from making a crude oil from which conventional petroleum fractions could be produced to making a low-grade, low-sulfur oil from coal which could meet the new sulfur oxide standards and be used as a boiler fuel. Little information is available either on the economic or energy costs involved because the research is relatively new and data have not been reported. However, because the product contains less hydrogen than for those processes previously tested, the overall energy efficiency should be somewhat higher.

Production of oil shale.—The production of shale oil from which conventional petroleum products can be made has attracted the attention of numerous investigators for 50 years or more. While numerous processes have been tested on a demonstration scale in the United States (and commercial production carried out in numerous foreign countries for many years) shale oil has not been able to compete economically with domestic petroleum. As with all other conversion processes, there are energy losses at each step.

If room and pillar mining is used (as has been proposed for the mining of the richest, most easily available deposits) about 75 percent of the resource would be recovered during mining—a much better percentage than the average 30 percent in recovery of oil in place.

Methods have been designed (but not tested) to exploit the very thick deep seams that would yield much higher recoveries—in the order of 90 or more percent. During crushing and subsequent retorting, the percentage recovery of the energy in the shale could vary [9] (depending on the process selected) from 90 to 100 percent.[10] Upgrading the crude shale oil to make it suitable for a recovery feedstock would further reduce the amount of energy recovered, and final upgrading in a conventional refinery would reduce the yield about 13 percent.

The present situation with respect to shale oil development in the United States has been described as follows:[11]

> A renewed retorting development program, which would involve the Bureau's plant under a lease agreement with Development Engineering, Inc., of Denver is projected. The retort proposed by this company is similar in many regards to the gas combustion retort, but was developed for commercial use in lime calcining.

An alternative method of exploiting oil shale is to retort it in place—i.e., an in situ operation that avoids mining the shale. Processes for accomplishing in situ retorting have been tested by both industry and the Bureau of Mines but neither the economic cost nor energy efficiencies are yet known.

The President's energy message [12] called for "a leasing program to develop our vast oil shale resources" and the Department of Interior responded by issuing drilling permits on a number of federally owned oil shale deposits. Presumably leases will be offered after the results of the drilling have been evaluated. However, as the above quote indicates, there have been no greatly intensified mining or retorting research programs undertaken either by the private or public sector. Such programs will be needed if shale oil is to be produced at the lower costs needed to become competitive with crude petroleum.

Tar Sands.—One commercial tar sand plant producing 45,000 barrels of oil a day is in operation in Canada and a second license has been granted. Surface mining of the tar sands is practiced in the one commercial operation and, as in the case of uranium, it is difficult to give meaningful estimates of the percentage recovery, since only the higher grade materials are wanted. Of the higher grade tar sands that are suitable for processing 80 to 90 percent or more should be recovered during surface mining.

During extraction of the oil from the sands recoveries of 90–95 percent are achieved. Subsequent refining to finished petroleum products is estimated to reduce the original content about 13 percent.

Although tar sand resources in the United States are known to exist, the number of locations where economically attractive deposits occur are thought to be relatively small but some work is now underway to get additional information.

Although in situ processing of tar sands has not been practiced commercially, there are several experiments being conducted that will test methods such as injection of hot fluids (steam, water, natural gas, or inert gas) in-situ combustion and solvent extraction. The economic and energy costs associated with these methods have yet to be determined.

IMPROVED METHODS OF GENERATING ELECTRICITY

The current status of the development of new methods of generating electricity was recently summarized by AEC Commissioner James Ramey:[13]

> There has been considerable discussion in the press and elsewhere regarding the potential virtues of the more novel

sources of electric power, including among others, magnetohydrodynamics (MHD), harnessing the Gulf Stream, solar energy, and tidal power. Various proponents have intimated that these sources are just around the corner. This is just not so. The overwhelming consensus of the experts who have studied this matter is that, while limited uses of these other sources may be possible, it is extremely doubtful that we will see their large-scale employment in this century.

In spite of this rather negative report, there is very active interest in finding ways to increase the current level of 38 to 40 percent efficiency inherent in the steam turbine cycle.

Gas turbine-steam turbine cycles.—Gas turbines used in modern airplanes operate at temperatures of 1600° F. or above. Much longer periods of use would be required if turbines were used in utility plants. The efficiency of gas turbine-steam turbine cycles, reflecting three generations of development (corresponding to what might be accomplished in the three decades of 1970's, 1980's, and 1990's), is shown in table 1-21 below: [14]

Table 1-21 —PROPOSED COGAS POWER SYSTEMS

Generation	I	II	III
Number of gas turbines	3	2	2
Turbine inlet temperature, °F	2200	2800	3100
Compressor pressure ratio	8	12	20
Percent of airflow bled for cooling	4.7	8.5	9.0
Turbine exhaust temperature, °F	1297	1514	1485
Compressor-turbine overall length (feet)	33	27	26
Single steam turbine, of size (megawatts)	431	381	312
Stack temperature, °F	314	219	241
System efficiency [1]	47.0	54.5	57.7
Total capital cost (millions)	$109.3	$94.0	$89.3

[1] Electric generator losses and auxiliary power requirements not included. Multiply by 0.96.

This table indicates that with adequate R. & D. efficiencies of the combined cycle could reach about 55 percent in 1990–2000.

The scope of ongoing research effort was outlined in a paper prepared for a recent coal symposium as follows: [15]

Such organizations as United Aircraft, Institute of Gas Technology, and Westinghouse Electric Corp. have proposed research in the combined cycle. The United Aircraft system would consist of a gasifier to convert coal to a clean fuel gas with a low B.t.u. content. Air would be compressed, mixed with this fuel in a combustor, and then exhausted through a compressor drive and power turbine to an unfired heat recovery boiler. Steam from the boiler would be expanded to a turbine and then condensed. Approximately two-thirds of the power would be from the gas turbine, with the remaining power produced by the steam turbine. United says the principal attractions of the system, in addition to the possibility of removing all possible pollutants at the lowest possible cost, would be according to their report a much lower capital cost; a bus bar power cost approximately 10 percent less than for conventional 1980 fossil-fuel steam station plants, a much higher overall efficiency. United says even the first generation of plants would be more efficient than conventional units, and later plants, with the inlet temperatures of the turbines as high as 3,000° F., would perform far better.

For the long term the reserves of oil and gas are too small to be used for central station power generation. A low cost method must be devised to produce a clean gas from coal that can be used in gas turbines. Processes for making low B.t.u. gas from coal were commonplace before natural gas replaced coal gas. However, it is doubtful that these methods would be used again since it should be possible to devise lower cost methods using new technology. Many of the processes that have been proposed for making high B.t.u. gas are also thought to be capable of making a low B.t.u. gas at a lower cost—by substituting air for oxygen and by eliminating the final methanization step. According to Hottel and Howard [16] the thermal efficiency of the gas plant will be about 77 percent; thus, the values shown in table 1-20 must be reduced by this amount to calculate the overall fuel to electricity efficiency—gasification efficiency times the conversion efficiency to electricity in the combined cycle.

Dual fluid cycles.—The use of two fluids is a method to increase electric generating efficiency of the steam turbine cycle. In topping cycles a fluid is used which can be heated to higher temperatures at lower pressures than steam, thus, avoiding the pressure and temperature limitations now existing for steam boilers. Some years ago a few plants were constructed which used mercury as the topping fluid but the costs and hazards made these plants unattractive. More recently Oak Ridge National Laboratories [17] has concluded that potassium could be used as the topping material:

> About 200,000 hours of operation of boiling potassium systems, including over 15,000 hours of potassium vapor turbine operation under the space power plant program, suggest that a potassium vapor topping cycle, with a turbine inlet temperature of ~1500° F. merits consideration. A design study has been carried out to indicate the size, cost, and development problems of the new types of equipment required. The results indicate that a potassium vapor cycle superimposed on a conventional 1,050° F. steam cycle would give an overall thermal efficiency of about 54 percent as compared to only 40 percent from a conventional steam cycle. Thus, the proposed system would have a fuel consumption only 75 percent and a heat rejection rate only 50 percent that of a conventional plant.

Other proposals suggest the use of carbon dioxide and helium (in place of steam) as the working fluid as a means of avoiding the need for two working fluids, while obtaining higher temperatures and greater efficiencies than steam permits.

In bottoming cycles the overall plant efficiency is increased by reducing the final temperature at which heat is rejected from the plant, using a second fluid at the lower temperature end of the system. Freon, ammonia, and other fluids have been suggested for this purpose but little experimental work has been conducted. The opportunities for increased energy efficiency using bottoming cycles are less than for topping cycles.

Magnetohydrodynamics.—This method of electricity generation and its potential for increasing energy efficiencies have been described as follows: [18]

> MHD is potentially capable of serving as a high-tempera-

ture "topping" device to be operated in series with a steam turbine and generator in producing electricity. Some time ago MHD was being advanced as the energy converter of the future. In such a device the fuel is burned at a high temperature, and the gaseous products of combustion are made electrically conducting by the injection of a "seed" material, such as potassium carbonate. The electrically conducting gas travels at high velocity through a magnetic field, and in the process creates a flow of direct current. If the MHD technology can be developed, it should be possible to design fossil-fuel powerplants with an efficiency of 45 to 50 percent.

Advanced MHD cycles are said to be capable of efficiencies approaching 60 percent.

Experimental work on MHD has been underway for more than 20 years both here and abroad. A plant with 25 megawatts of MHD capacity out of total plant capacity of 75 megawatts is now being tested using natural gas in the U.S.S.R. Data from these tests are not available.

Small scale research on various aspects of the engineering problems associated with MHD is still being conducted in the United States. However, the additional engineering and operating problems that MHD will encounter when coal is used directly as the fuel—which for the long term is the only adequate indigenous resource in the United States for fossil fuel power generation—are not well understood. Extensive research on the effects of coal combustion products on generator performance and life and on stack gas cleaning is still needed.

Fuel cells.—Fuel cells are a means of converting chemical energy into electrical energy. A Department of Interior Study [19] stated that:

> A fuel cell is a device that converts into electrical energy the chemical energy released from the reaction between a fuel and an oxidant via an electrolytic medium. As a result of the direct conversion, efficiencies up to 75 percent are deemed practical expectations.

There was active interest in fuel cell research between 1900 and 1930 but the development of the internal combustion engine reduced interest until the launching of the current space program.

As with most of the other new techniques that have been suggested, the fuel cells now available require a clean fuel—preferably hydrogen, or hydrogen made from a clean fuel such as methane. Fuel cells, as indicated above, have a very high conversion efficiency. Continued research, therefore, appears justified. Although fuel cells may not be feasible for use in single family residences, they may be very useful for multiple unit dwellings, shopping centers, and electric utility substations. These potential uses of fuel cells were discussed in the Federal Power Commission's 1970 National Power Survey: [20]

> A strong effort is being made to develop a fuel cell for residential and commercial service. In 1967, a team of 23 natural gas utilities undertook a $20 million, 3-year research program to develop a natural gas fuel cell. The team (TARGET—Team to Advance Research for Gas Energy Transformation, Inc.) awarded the contract for the first phase of the program to the Pratt & Whitney Division of United Aircraft Corp. By 1971 there were 32 TARGET members

working in conjunction with Pratt & Whitney and the Institute of Gas Technology. They were at the mid-point of a 9-year research program under which $41,500,000 were to be spent by the end of 1972. One of the early results of this research is the world's first natural gas fuel cell home which was dedicated in Farmington, Conn., on May 19, 1971. The research program ultimately will include 37 test installations and a wide variety of residential, commercial, and industrial applications.

Fuel cells are also being considered for use in substations, where they would provide baseload or peaking power to the existing electric power system. This would reduce the need for central station power and transmission lines and consequently reduce the effect of power generation on the environment. The fuel cell holds potential for future use in the transportation field, where batteries are presently used, and in the electrochemical industries where low-voltage direct current power is required.

Although the future of the fuel cell is uncertain, it could have a bright future. A successful fuel cell module of 10 to 15 kilowatts, envisioned for single family residences, could be coupled in banks to serve larger users. Looking to the larger applications, the fuel cell, because of its modular capability and the possibility of tailoring power output to a specific customer's needs, could find broad application in industrial, commercial, and apartment complexes. Since many of the fuel cell applications now being actively considered would not be dependent on electric transmission and distribution networks, successful conclusion of the current experiments would permit significant future environmental advantages. The fuel cell is not, however, expected to replace central station power generation.

Other novel methods of electricity generation:

(1) *Thermionic Converters*

In thermionic generation heat is applied to a material and the electrons emitted as a result of the higher temperature flow to a receiver. This flow of electrons permits an external load to be connected and electricity is thus generated and used.

Thermionic generation of electricity is possible with any heat source but little effort has been made to apply this principle to central station power development. The report of the above-noted National Power Survey stated: [21]

> Units have operated satisfactorily at relatively low power levels and efficiencies for thousands of hours, but many problems hinder economic application in the commercial power market. Present reports show conversion efficiencies of 15 to 25 percent and power densities as high as 50 watts/cm^2 but these for the most part are laboratory accomplishments. Maximum obtainable efficiency is limited to the Carnot efficiency of an ideal heat engine operating between inlet and exhaust temperatures which correspond to the emitter and collector temperatures.

(2) Thermoelectric Generation

The National Power Survey described thermoelectric generation as follows:

The thermoelectric generator is a device which converts heat energy directly into low-voltage direct-current electricity. It makes use of the 100-year-old "Seebeck principle" that a voltage difference is produced between the ends of two joined dissimilar conductors when heat is applied. With advancements in the technology of semiconductor materials it became possible to produce usable generating units. Usable amounts of electric power are produced by connecting several generating units into thermopiles for use as a single generator. It is also possible, by varying the materials used for thermoelectric conductors, to operate the generators in different segments of a wide range of operation temperatures. * * *

With possible efficiencies in generator technology in the order of 30 percent, it appears that overall efficiencies of thermoelectric systems cannot exceed 10 to 15 percent. Efforts are being made to increase the internal efficiency of the process but the rate of progress has tapered off.

(3) Electrogasdynamics

A simple description of electrogasdynamics and the current status of research were outlined in the National Power Survey as follows:

The electrogasdynamic generator is a device which, utilizing a moving gas, converts heat energy to electricity by transporting charged particles "uphill" against an electric field. The electric field opposes the flow of a gas containing charged particles. In the process of overcoming this opposition, the electrical potential of the charged particle is increased. The device converts the kinetic energy of the moving gas to high voltage dc electricity by providing an external circuit between a charged particle emitter and collector. * * *

A pilot plant was constructed at the Foster Wheeler Research Laboratories in Carteret, N.J. Tests on the pilot model uncovered major technical problems and led to termination of the project on September 30, 1968. The original design proved unacceptable because the precipitation of charged particles on the walls of the "slender channel" gas duct resulted in unacceptable energy losses. An evaluation of the Gourdine project made by three scientific entities for the Office of Coal Research indicates that the EGD generator as envisioned by Gourdine Systems, Inc., requires some major breakthroughs before a practical unit can be considered possible.

DEVELOPMENT OF NEW ENERGY SOURCES FOR COMMERCIAL USES

Fusion.—Scientists are now attempting to unlock the fusion of certain kinds of hydrogen atoms, deuterium, and tritium for the purpose of releasing useful energy. Deuterium, a virtually unlimited source of fusion fuel, occurs in recoverable amounts in the oceans. Tritium occurs in nature in extremely small amounts but could be produced from lithium. A practicable nuclear fusion reaction has yet to be demonstrated in a laboratory; if the process can be successfully demonstrated it would provide a virtually unlimited source of fuel at low cost and with environmental effects much smaller than fossil fuels or uranium.

The report entitled "R. & D. and National Progress (1964)" identified cost and safety as two major advantages of fusion: [22]

> It has been estimated that deuterium in the oceans could provide 20 billion years' energy, at current consumption rates. Even for present extraction methods, deuterium's cost per unit of fuel value is only a small fraction of the cost of conventional fuels.
>
> Inherent safety. Unlike fission powerplants, there can be no explosion or runaway chain reaction. The fusion process is analogous to chemical burning, rather than to fission, and the energy content of the fuel in the reaction chamber at any instant is relatively small. Further, the reaction products and most radioactivity induced in the surrounding vessel are short-lived, and the products of the "complete" fusion process are nonradioactive nuclei only.

If fusion does prove practicable, the fuel likely to be used first would be a combination of deuterium and tritium. The availability of lithium from which tritium is made would not be a limiting factor. For example, a recent analysis of lithium supplies [23] indicates that, even with the most conservative possible assumption as to known and inferred reserves on dry land, the supplies are estimated to contain energy totaling some 50,000 times the energy used in the United States in 1970 to generate electricity. The lithium content of the sea would come to more than 1 billion times this amount.

The idea of fusion predates World War II. Since 1951, the Atomic Energy Commission as well as the U.S.S.R. have systematically funded research to gain fuller understanding of fusion necessary to demonstrate and apply a controllable process for the production of useful energy. Through continued improvements in technologies for creating magnetic fields and through increased experimental knowledge of behavior of plasmas [24] of hydrogen atoms under conditions approaching those postulated as necessary for fusion, it should eventually be possible to demonstrate a proof of principle for fusion.

According to the chief scientist of the AEC's fusion program, the general requirements for achieving useful power from controlled thermonuclear reactions are.[25]

> 1. To heat a dilute gas of fusion fuel to a temperature of hundreds of millions of degrees where it is in the plasma state.
> 2. To contain it free from any contact with material walls or from contamination by impurities long enough for a significant fraction of the fuel to react; and finally,
> 3. To extract that energy released and convert it to a useful form.

Because no known material could contain the reaction under the temperatures required, powerful magnetic fields will have to be used. Much of the current research and development is concentrated on building large experimental machines that can contain the desired reaction and heat the fusion fuel to the necessary temperatures.

At the moment, no fusion machine has demonstrated a useful fusion reaction although there are more than 200 planned or operating machines in 14 countries of the world. Nor is there any certainty that such a controlled process can in fact be demonstrated. It seems unlikely that a demonstration fusion powerplant will be in operation much before the end of this century.

In 1970 the late AEC Commissioner Theos J. Thompson, a noted nuclear scientist, spoke of a return to the spirit of qualified optimism which is absolutely essential for finding the right key to unlock the promise of fusion power. Sounding a note of caution against over-optimism, he said.[26]

> One of the problems which we face today is the danger of misleading Congress and the general public in regard to the time scale within which fusion can be developed. A real milestone will occur when a fusion process is demonstrated by which more energy is produced than is consumed. * * * Some of my optimistic fusion scientist friends believe that this goal will be reached within the next decade. Others are more pessimistic, and say by the end of the century. And even today, some believe it may never happen. I can report to you that among thoughtful scientists directly working in this field there is a steadily growing confidence in the fusion community's ability to pick that still stubborn lock.

In a recent assessment of the prospects of fusion for the lay public, Dr. Edward Teller wrote as follows:[27]

> What are the prospects? If one is optimistic, one might hope that within a few years we shall have an apparatus that produces more electricity than it consumes. This would mean a proof of feasibility in principle, but not yet in practice. The real question is whether energy produced in this manner would be competitive.
>
> From the point of view of the fuel, the prospects are excellent. Deuterium, the primary fuel, can be extracted from seawater. It is cheap and for all practical purposes limitless. The other fuel—tritium—is generated as needed by letting neutrons from the reaction react with a "blanket" of lithium surrounding the reaction chamber.
>
> From the viewpoint of engineering and capital investment, no early success can be expected. Controlled fusion will require much more delicate and sophisticated operation than we encounter in today's nuclear reactors. It would be most surprising if economic feasibility * * * could be established before the last decade of our century.
>
> But if we do succeed, there is a big advantage to be gained: Fusion reactors are apt to be safe and clean. There is a good chance that controlled fusion will become the standard source—one that will be plentiful and consistent with all the requirements of a clean environment.

Scientific feasibility must be demonstrated and this may take another 10 years or more. Estimates of cost range from $600 to $1,200 million to reach this point. The engineering feasibility of fusion reactors would then have to be proven. This might be possible by the year 1990, assuming a demonstration of scientific feasibility by 1980 and also rapid resolution of a number of anticipated engineering problems. Estimates for the cost of development range from at least $1,200 to $2,500 million.[28] Assuming technological practicability is demonstrated and that each megawatt of installed generating capacity costs $500,000, then each fusion plant of 1,000 megawatts output would require an additional captital investment of $500 million. The benefits would be great—unlimited electric power at low costs for thousands of years.

Geothermal Energy.—An attractive but still unutilized resource in most parts of the United States is geothermal energy. Demonstrated technology exists to build and operate geothermal powerplants with electrical outputs in the range of several hundred megawatts (which is about one-fifth to one-third the size of the large conventional and nuclear steam electric powerplants now being built.) Geothermal energy in California is being exploited rapidly and exploration throughout the West is accelerating. Geothermal energy is seen as providing a relatively low-polluting energy source that may capture a sizable portion of the new electricity generation market in some areas.

Geothermal energy is the natural heat of the earth. Temperatures in the earth rise with increasing depth. At the base of the continental crust (15 to 30 miles deep), temperatures range from 200 degrees C. to 4,500 degrees C. (390 degrees F. to 8,150 degrees F.). But most of the earth's heat is far too deeply buried to be tapped by man, even under the most optimistic assumptions of technological development.

In nature geothermal heat is sometimes found concentrated in restricted volumes; economically significant concentrations of geothermal energy (elevated temperatures in a range of 40 degrees C. to 380 degrees C.) are found in permeable rocks.

There are three types of geothermal systems: dry steam, hot water, and hot dry rock. The Cornell Workshop[31] described these as follows:

> (1) Dry steam exists in the ground in many places in the world where temperatures are high and pressures low. These vapor-dominated systems usually have a temperature range from 350 to 550° F. and pressures of several hundred pounds per square inch.
>
> (2) Hot water areas are much more abundant than dry steam and occur where geothermal water pressures are greater than the bubble point. Temperatures can range from tepid to in excess of 680° F. Pressures are hydrostatic or nearly so.
>
> (3) The problem with hot dry rock systems is finding areas where sufficient hot rock lies at economic drilling depths, fracturing the rock, circulating a heat exchange fluid, and withdrawing the heat to drive a generating system. The lowest hot rock temperature that seems usable at present is about 350° F. The hottest temperature anticipated for sustained recovery may be about 1,000° F., but no practical upper limit has yet been identified. Obviously, the hotter rocks are more attractive as energy sources.

(1) *Dry steam.*—Of dry steam, the workshop wrote:[32]

> There have been several dry steam fields developed for power in the world. Major dry steam fields are under development in Japan, Italy, and the United States. The Geysers field in northern California may become one of the largest dry steam fields in the world. At the end of 1971 it was producing a capacity of 192 megawatts and contracts for sale of 600 megawatts by 1976 had been negotiated. * * * The geo-

physical estimates of the field extent submitted to the State of California suggest 25,000 megawatts of sustained power potential for this field which has multiple productive zones. There has been insufficient production history of the deeper zones to detect the field decline rate, but it appears to have a life in excess of 30 years, especially if supported with water injection to maintain reservoir pressure.

Another dry steam field exists at Yellowstone and two others are suspected, one in California and one in New Mexico. * * * One estimate for the U.S. dry steam potential is 100,000 megawatts, with a lifetime of at least 20 years.

(2) *Hot water systems.*—Of hot water systems the workshop said: [33]

> Hot water systems are estimated to be about 20 times as numerous as dry steam systems and also may be of larger size. Over a thousand have been identified in the United States on the basis of hot springs. One of the large ones occurs in the Imperial Valley of California where a conservative estimate of the power potential is 30,000 megawatts. The total power potential in the United States from this source is not known, but is estimated to be between 10^6 and 10^7 megawatts. The life of the resource is also not known but studies in the Imperial Valley suggest 100 to 300 years.

Insufficient experience is presently available on actual operation of hot water systems to permit a definitive appraisal of either the economics or the environmental problems associated with the reinjection of water into the ground that is characteristic of this process.

(3) *Hot dry rock systems.*—Of hot, dry rock systems, the workshop said: [34]

> Even though current geothermal developments center on dry steam and hot water systems, almost all of the enormous heat content of the earth's interior is present in dry rock. * * * It appears that geothermal energy alone is capable of meeting all American power requirements for several centuries if the hot dry rocks resource proves to be practical. This provides a major incentive for placing a high priority on appraisal of this energy source and its technology. There has been no test of the hot rock concept to date, and all projections for these systems are based on theoretical studies. These are being carried out by the Los Alamos Scientific Laboratory with the support of AEC.

Production of electricity from geothermal energy is attractive environmentally because no solid atmospheric pollutants are emitted and radiation hazard is not involved. But geothermal generation has some environmental effects, including noise during drilling and testing, escape of volatile, noxious chemicals and disposal of waste cooling waters. Highly mineralized water wastes from either hot water or vapor-dominated systems can pollute streams or ground water. Consequently, Federal and State regulations require reinjecting such objectionable wastes back into a deep reservoir. Thermal pollution is also a problem, particularly in hot-water systems but it can be solved in part by reinjection of unwanted warm water and residual steam condensate.

The 1970 FPC report [35] discussed the potentials of geothermal energy as follows:

> Geothermal enthusiasts foresee future installations involving deep drilling through the earth's mantle (20 to 30 miles) making it possible to tap energy sources almost anywhere on earth. They also envision producing high-pressure steam by the injection and recirculation of water through huge subterranean hot cavities or reservoirs created by underground nuclear explosions. This procedure is also expected to provide a slurry of metal ores suitable for additional processing and marketing. Experiments in recirculation techniques have already been initiated. Overall, however, geothermal generation is not expected to become a significant factor in central station generation in most sections of the country.

Four conclusions of the Cornell workshop [36] about geothermal energy provide a good synopsis of its present status which is much more optimistic about the role of geothermal energy than earlier studies. The workshop concluded that:

> 1. This energy source is developing rapidly in California because of its substantial economic advantages.
> 2. Federal support for a national geothermal appraisal and inventory is badly needed to assist long-range planning.
> 3. A rapid, major investment should be made to see if the hot dry rock geothermal investment can be demonstrated to be technically and commercially feasible.
> 4. Further careful studies should be made of the environmental factors associated with geothermal developments. These include waste heat, noxious fumes, and earth subsidence.

The view generally supported the position of Professor Rex of the University of California Riverside Geothermal Resources Program as expressed in a recent article on geothermal steam.[37]

The Geothermal Steam Act of 1970 (Public Law 91-581), authorizes the Secretary of the Interior to issue leases for the development and utilization of geothermal steam and associated geothermal resources on various public lands. The Geological Survey is active in identifying known geothermal deposits so that leasing can be accelerated on these lands. This renewed emphasis by government in geothermal energy was reflected in the President's energy message [38] which promised development of a geothermal leasing program beginning this fall.

Solar Energy.—Small, direct uses of solar energy have been made over the years, but none on a scale which would sustain a substantial new industry. Scientists, engineers and innovators are now beginning to think beyond small scale uses to the large scale generation of electricity or to solar production of hydrogen and oxygen for use as clean fuels.

Solar energy has the advantages of being readily available everywhere and free of pollution. Characteristics which adversely affect the practicality and cost of its use are its high variability (because of hourly, daily, and seasonal changes and irregular variations of cloudiness) and its low intensity in comparison with the energy released by combustion and other processes.

Ten years ago the Joint Committee on Atomic Energy in hearings on the frontiers for atomic energy research looked into the solar potentials. The Joint Committee heard testimony that sufficient economy could be achieved, probably within a decade, to place domestic solar heating on a competitive basis with conventional fuels wherever the conditions of substantial heating loads, ample sunshine, and comparatively expensive fuels existed. According to witness Dr. George O. G. Lof, then associated with Resources for the Future, Inc.:[39]

> Domestic solar energy utilization will probably have its first application and greatest economic attractiveness in areas where both heating and cooling are needed. The first installations will be expensive, but costs should decrease after mass production methods have been established.

Present technology for several small-scale applications of solar energy include heating and cooling of buildings, heating of water, drying of materials, providing high temperature heat energy for experiments, supplying electricity for space equipment, or for low power equipment are all technically feasible. Until recently high initial costs and the competition of cheap alternative sources of energy have discouraged emergence of an industry to produce, market and service solar energy generators and equipment. The most important potential near-term use in the United States would be full or supplementary heating and cooling of houses, schools, shopping centers, and other structures that have large exposed roof areas which could be used for solar collecting apparatus. While a rapid deployment of solar energy for heating and cooling of enclosed spaces (or "space conditioning") cannot be expected to reduce appreciably the present use of fossil fuel and electricity, the widespread use of supplementary space conditioning powered by solar energy could slow down the rate of growth in demand for fuels and for electricity.

Many potential options exist for use of solar energy. Briefly, solar energy may be: (1) intercepted in space as it approaches the earth, converted there into electricity, and beamed to the ground as microwaves for conversion into electricity; (2) it may be absorbed by collectors at the surface to produce electricity directly; (3) it may be absorbed by collectors to provide heat either for direct use in heating or refrigeration, or to generate electricity; finally, (4) it may be absorbed by cultivated plant life, ranging from algae to trees, to produce food or fuel. Additionally, some indirect effects of solar energy upon nature may be tapped, notably wind and temperature differentials in sea water.

Probably the most spectacular proposal for large scale use of solar energy is to put solar cells in a large space array, convert the incoming energy into microwaves which would be beamed down to earth, collected there on a large receiving antenna and converted into electricity. The Cornell workshop on energy and the environment concluded that beneficial application of this approach is remote.[40]

The workshop assessed the space absorption and conversion approaches as follows:

> In our opinion, the requirement of tenfold reductions in (1) weight/kilowatt of the space array, (2) cost/unit weight to orbit, and (3) component cost, provide a very large cumulative barrier to success and make success in this area much less likely than success of a ground-based system. Additionally, the main environmental hazard appears to be

associated with transmitting power in the 1,000 to 3,000 MHZ band to large arrays on earth at power densities of 100 watts per square meter. Experts disagree on the biological effects of long-term exposure to this power density, with implications as follows. A lower level would be clearly uneconomic. Thus, a lower permitted exposure would force increasingly severe beam control. Time of beneficial installation is likely to be remote.

The most common approach to using solar energy is to absorb the incoming radiation energy at the surface of the earth as a means of generating electricity or heat, which in small applications can be used to heat or cool buildings and to dry products.

Several schemes for collecting and converting solar energy on a large scale can be considered. The first involves the use of flat plates of photovoltaic cells [41] having an efficiency of about 10 percent.

A recent method was proposed by Aden B. Meinel and Marjorie P. Meinel of the University of Arizona utilizes the hothouse effect by means of selective coatings on transparent pipes carrying a molten mixture of sodium and potassium which would be heated by solar energy to a temperature well above the boiling point of water. By means of a heat exchanger, this heat is retained at a constant temperature in a large insulated chamber filled with a mixture of sodium and potassium chlorides that is large enough to store enough heat for at least 1 day's collection. Heat extracted from this chamber would raise steam to operate a conventional steam electric powerplant.

The Cornell Workshop assessed use of direct thermal cycles as meriting active exploration and anticipated that suitable technology could be made available at somewhat uncertain cost in the 1980's: [42]

> The second scheme uses direct thermal cycles (heating a low-pressure transfer fluid), tilting arrays, and probably some crude focusing of sunlight. Estimates of presently proposed schemes, using elements that absorb visible sunlight efficiently and reradiate infrared (thermal) energy poorly, are that they now cost too much by a factor of about 10. This is the nearest that any such scheme seems to practical realization; bearing in mind the magnitude of possible benefits, an outstanding factor of about only 10, with reasonable chance for reductions, we see sufficient cause for active exploration. In both these earth-based schemes the local environmental difficulties, except for foregone land availability at 50 to 200 km^2 per 1,000 megawatts, seem minimal. The system could be made available, at somewhat uncertain cost, in the 1980's.
>
> Waste heat disposal from all these solar schemes can be arranged to cause virtually no incremental heating of the earth. Electric transmission problems are exacerbated because of the necessity of siting in low cloud (desert) regions.

A second method would use the idea of trapping solar energy by growing algae or other plants, which could be used as food or burned for energy.

The Cornell Workshop viewed this scheme as unresolved, but pointed out that the possibilities should be explored in a coherent solar energy program. Present conversion efficiencies, in practice, run about 1 to 2 percent in the algae chlorella, which is unacceptably low. The workshop noted rumors about higher efficiencies with such plants as water hyacinths or pineapples, but reported it had no data and expressed concern about the ecological unsoundness of a one-crop system.

Power generation using the temperature difference between warm surface sea water and the colder water at greater depths has been considered for some time. Principles of thermodynamics affirm that where a temperature difference exists, power generation should be possible but costs may be too high to make its use economically competitive with alternative solar methods of generating electricity.

One drawback of solar energy is its obvious intermittent nature. A means to store energy is needed so that the users of the output of a solar powerplant do not lose their power during hours of darkness One proposed solution has been to use solar energy to generate hydrogen from water. The hydrogen could be collected, stored, and distributed as needed, with the supply replenished during daylight hours by electrolysis of water using the electricity from solar energy. A variation of this idea is to concentrate the radiation energy from the sun by means of cheap plastic lenses to cause thermal dissociation of water. Proponents of these approaches note that hydrogen at a cost competitive with natural gas would be an almost ideal fuel. It would be usable not only for electric power generation but for direct heating. In liquified form, it might well be usable for transportation.[43]

Several recent reviews of the potential for large scale use of solar energy have been published. An MIT study concluded:[44]

> With 25 percent of the U.S. energy consumption converted to electricity at an average efficiency of 32 percent and on the assumption that solar energy is convertible to process heat at 30 percent and to electrical energy at 5 percent * * * one can readily calculate that the 48-State total U.S. land area receives as solar energy only 150 times our present energy needs. A 1,000 megawatt (24-hour average) powerplant operating in a 1,400 B.t.u./day solar climate with an efficiency of 5 percent would require 37 square miles of ground coverage, compared with a few hundred acres for a nuclear or fossil-fuel plant. Major replacement of present power sources by solar energy has poor prospects of success. The proposal that instead of covering the earth with solar energy collectors we put up giant satellites to intercept solar energy, convert it there to microwave radiation, beam it to the earth in a dilute enough beam not to make a death ray of it, reconvert it to electrical energy and then transmit it to our cities appears even less attractive in its long-range possibilities. It puts in series at least four steps each one of which is far beyond our present capability except at prohibitive cost.
>
> * * * Domestic hot water from the sun is economically significant in many areas today, solar house heating in some, and its prospects are improving; solar distillation to produce fresh water from saline water is economic in areas of extremely high fossil fuel cost (certainly not in the U.S. mainland); solar electric power from photovoltaic cells is significant in space research where the laws of terrestrial economics are inapplicable, and it has some chance of becoming much cheaper. There are certainly enough of these areas to justify a vigorous research program, but a major effect on the national energy picture is not to be expected.

The MIT study concluded that little additional development effort is needed with respect to solar heating and air conditioning:[45]

> * * * Any improvements in ruggedness, reliability, efficiency and cheapness of roof collectors will of course hasten the day of acceptance of solar houses. It must be remembered, however, that a constraint is put on the appearance and orientation of the house. Furthermore, solar heat is more expensive the smaller the installation, with 10 cents to 50 cents per million B.t.u. added to the cost of solar heat as the house size drops from 25,000 B.t.u./DD to 15,000 B.t.u./DD.[46] For these reasons, and particularly because many householders still install heating on a lowest cost basis, solar heating cannot be expected to have a large influence on the U.S. energy scene for many years. There are parts of the world, however, where it could become more important soon.

The 1970 Federal Power Commission survey concluded:[47]

> The efficiencies and cost-weight characteristics of existing solar conversion devices place them in an unfavorable economic position today. However, recent discoveries of organic compounds possessing semiconductor and photovoltaic properties, together with theoretical and experimental data, indicate the organic compounds may provide conversion devices with high efficiency and low cost-weight ratios. Successful development of such devices would make the possibility of orbital power stations more nearly feasible. In this concept, solar stations orbiting the earth would collect solar energy from the sun's rays and convert the solar energy to electric energy for electric microwave transmission to earth. The design problems include a number of separate areas such as orbital characteristics, conversion devices, transmission facilities, and the reception of power on earth.
>
> At this time, the development of orbital solar collection stations for central station power generation appears to be within the projected capabilites of system engineering and would not require the discovery of any new physical principles. However, the necessary development efforts, the possible excessive costs of the resultant technology, and unknowns such as reliability and harmful effects on the environment would preclude application of such systems in the foreseeable future.

The National Academy of Science concluded in a report to AID that, for developing countries:[48]

> Solar energy is an energy resource widely available in areas with energy needs in development countries.
>
> Useful applications of solar energy are now being made.
>
> Other applications are in various stages of experimental development.
>
> Solar energy is a resource that has the capability to meet energy needs substantially beyond the applications now being made, and this potential can be realized only with further research and development.

Windpower.—Until the last 100 years windpower played an important role in providing energy to mankind. As noted in Vansant's book:[49]

> Wind energy is ubiquitous and free, and much use was made of windmills in 19th-century rural America, mainly for applications that could tolerate the intermittent nature of the power source, such as pumping water. Although the wind source is inexhaustible and free, however, the converter (windmill) proved to be cumbersome, costly, and ineffective in comparison with the alternatives of gasoline engines and electric motors. When these alternatives were not available, windmills did their jobs satisfactorily, until they were made obsolete by new power sources. This illustration is not intended to intimate that wind energy should necessarily be written off as a power source but simply that having a source of energy is not enough. A practical and affordable converter must also be available to enable men to apply that energy.

Because of inherent limitations windpower has received little serious attention in recent years. The last serious attempt to harness windpower was made in the early 1940's when a 1.2 megawatt aerogenerator was built and operated for a period; it finally failed because of a broken rotor blade. The chief disadvantage, in addition to the small scale to which it is limited, is the interruptible nature of its operation.

Tidal Power.—The total tidal power of the earth has been estimated at 1.4 billion kilowatts, of which 1.1 billion kilowatts would be captured in bays and estuaries.[50] Although it appears that the total tidal energy could be a significant world energy input it is doubtful that much of it will be harnessed for either economic or environmental reasons. In the United States, for example, only the Northeast and Alaska have tidal ranges sufficient to be of interest and even for these locations a turbine would have to be developed to operate economically under low hydrostatic heads.

Ocean Currents.—The potential energy of ocean currents is real and the technology for converting this energy resource is probably available. However, the velocities are relatively small and the efficiency of conversion to human use is low.[51] The main barrier is that the capital investment needed is prohibitive relative to the value of the energy output. Energy must cost at least an order of magnitude greater than today's value of $1 per million B.t.u. before such a method of obtaining energy would be considered.

SUMMARY

New technology must eventually be developed and applied to help provide the increasing amounts of clean energy required by the United States and world population and economic growth. The methods discussed in this section still require extensive research and development to bring about commercial use. For the United States, methods must be found which permit the use of very large deposits of high sulfur coals in an acceptable manner. Several approaches are possible: (1) stack gas scrubbing, (2) low-B.t.u., low-sulfur gas from coal and (3) a low-sulfur oil from coal.

In order to extend our nuclear fuel resource base, the breeder reactor must be developed. Methods also should be developed to supplement our limited supplies of natural gas and crude oil by production from coal, oil shale, and tar sands.

The conversion of fuels to electricity rejects 60 to 70 percent of the energy in the fuel as unusable heat. Methods to increase this efficiency are particularly valuable in extending the resource base and reducing pollution per unit of electricity generated. A variety of potential methods for achieving better use of fuels such as the gas-turbine/steam-turbine cycle, dual fluid cycles, MHD, and fuel cells have been compared with respect to the research needs to make them commercial.

Renewable clean energy sources, such as fusion, geothermal, and solar energy, may be needed by the end of the century.

REFERENCES

The publications referred to in chapter 1 may be found in "Technological Processes for Conservation of Energy" by the Committee on Interior and Insular Affairs-U. S. Senate

The following Executive Summary has been excerpted from "PROJECT INDEPENDENCE"

BACKGROUND

-- Domestic energy demand has been growing at 4-5 percent per year.

-- The U.S. was self-sufficient in energy through about 1950, but our situation has deteriorated rapidly since then.

- Coal production is still at 1940's levels,
- Crude oil production has been declining since 1970,
- Natural gas consumption has been exceeding new discoveries since 1968.

-- Our dependence on foreign oil has grown to 35 percent of domestic petroleum consumption in 1973 (see Figure 1-3).

-- The world oil market is dominated by several Middle East countries.

- They have 60 percent of world reserves (see Figure 1-4).
- They produce 70 percent of world oil exports (see Figure 1-5).

-- The 1973 embargo demonstrated our domestic vulnerability to insecure imports.

- The embargo affected 14 percent of U.S. petroleum consumption.
- Its economic impact was a $10-20 billion drop in GNP.
- 500,000 additional people were unemployed at its peak.

Figure 1-3

U.S. ENERGY PRODUCTION AND CONSUMPTION 1947-1973

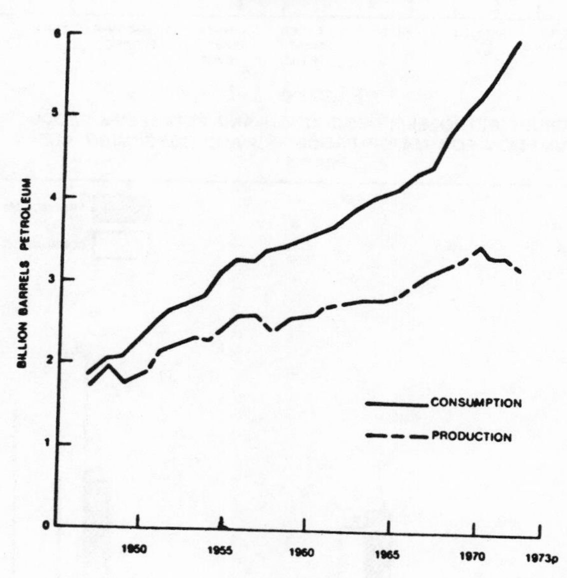

WORLD OIL ASSESSMENT

-- The world oil price will largely determine U.S. energy prices and, in turn, affect both United States supply possibilities and rate of energy growth.

-- World oil prices are highly uncertain and could decline to about $7 per barrel (FOB U.S.) and might fall somewhat lower.

- World supply/demand can be brought into balance at $7, but would require significant OPEC production cutbacks from the expected doubling of their capacity by 1985.

- OPEC has already cut back production 10 percent in four months to eliminate the estimated 2-3 million barrel per day (MMBD) world surplus.

-- Major OPEC cutbacks would be required to sustain $11 world oil prices.

- In the short term, prices can be supported by moderate production cutbacks.
- Much of the expected increase over 1973 OPEC production levels must be foregone by 1985 to support $11 prices.
- Decisions by major oil exporters will be more political than economic because greater revenues are not needed by the key suppliers to support their economic growth.

Figure 1-4

1973 CRUDE PETROLEUM RESERVES FOR MAJOR PRODUCING AREAS

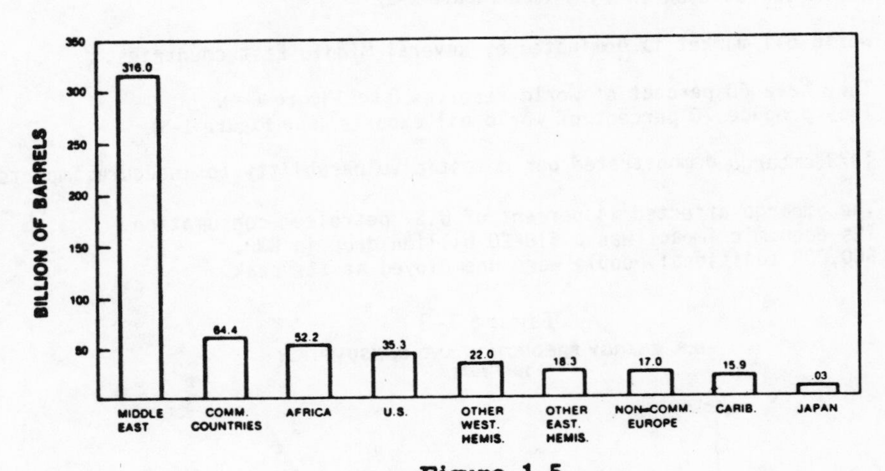

Figure 1-5

1973 CRUDE PETROLEUM PRODUCTION AND PETROLEUM PRODUCT CONSUMPTION FOR MAJOR PRODUCING AND CONSUMING AREAS [MB/D]

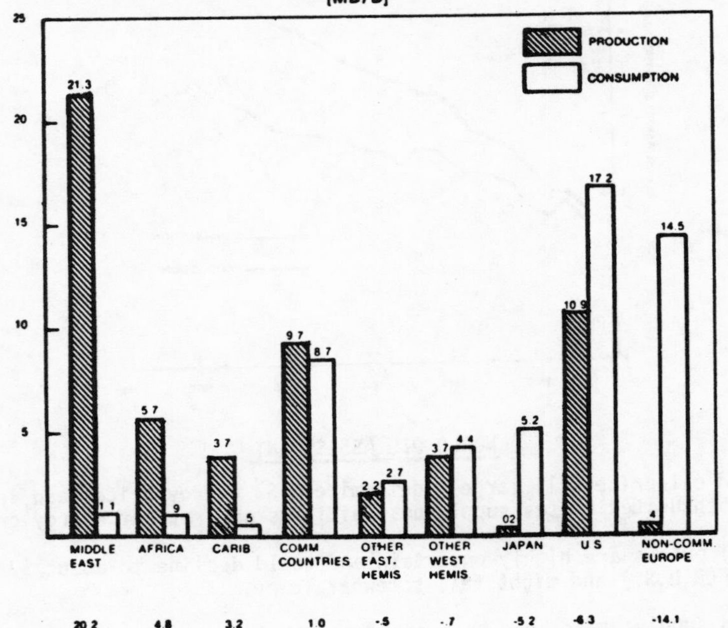

-- Foreign sources of oil have a significant probability of being insecure in the 1974-1985 time frame.

-- The resolution of pressing international financial, economic and political problems will ultimately determine world oil prices and security of supply.

- The study contrasts differences in the United States' situation based on a $7 and on an $11 world oil price.
- The study also estimates potential levels of world oil disruptions and their impact on the U.S.

DOMESTIC ENERGY THROUGH 1985: THE BASE CASE

-- If major policy initiatives are not implemented, the U.S. energy picture will be substantially different from pre-1974 trends, and is described below.

Energy Demand

-- At $11 world oil prices, domestic energy demand will grow at substantially lower rates than it has in the past.

- Total demand will grow at a rate of 2.7 percent per year between 1972 and 1985, compared to 4-5 percent during 1960-1970,
- 1985 demand will be about 103 quadrillion Btu's (quads) as contrasted with most other forecasts in the 115-125 quads range,
- Electric demand will also be below its recent high growth rates,
- Petroleum demand will be about constant between 1974 and 1977 and only grow at about 1-2 percent per year thereafter.

-- At $7 prices, total energy demand will grow at 3.2 percent through 1985, and petroleum consumption will be about 5 million barrels per day (MMBD) higher than at $11 levels by 1985.

Energy Supply

-- Petroleum production is severely constrained in the short run and greatly affected by world oil prices in the long run (Figure 1-6).

- Between 1974 and 1977, there is little that can prevent domestic production from declining or at best remaining constant.
- By 1985, at $7 world oil prices, production will rise to 8.9 MMBD from the current 8.6 MMBD. "Lower 48" production will decline from 8.2 MMBD to 4.2 MMBD, but is offset by Alaskan and Outer Continental Shelf (OCS) production.
- If oil prices remain at $11, production could reach 12.8 million barrels per day by 1985. This further increase comes mainly from the use of more expensive secondary and tertiary recovery in the lower 48 States.

-- Coal production will increase significantly, but is limited by lack of markets.

- By 1985, coal use will be between 1.0 and 1.1 billion tons per year depending on world oil prices.
- Production could be expanded greatly by 1985, but lower electric growth, increasing nuclear capacity and environmental restrictions limit this increase.

-- Potential increases in natural gas production are limited, but continued regulation could result in significant declines.

- Continued regulation at today's price will reduce production to 15.2 TCF by 1985, or 38 percent below the deregulated case.
- With deregulation of gas, production will rise from 22.4 trillion cubic feet (TCF) in 1972 to 24.6 TCF by 1985. Alaska production will be 1.6 TCF of this total.

-- Nuclear power is expected to grow from 4.5 percent to 30 percent of total electric power generation.

Figure 1-6
EFFECTS OF $7 VS $11 FOREIGN OIL

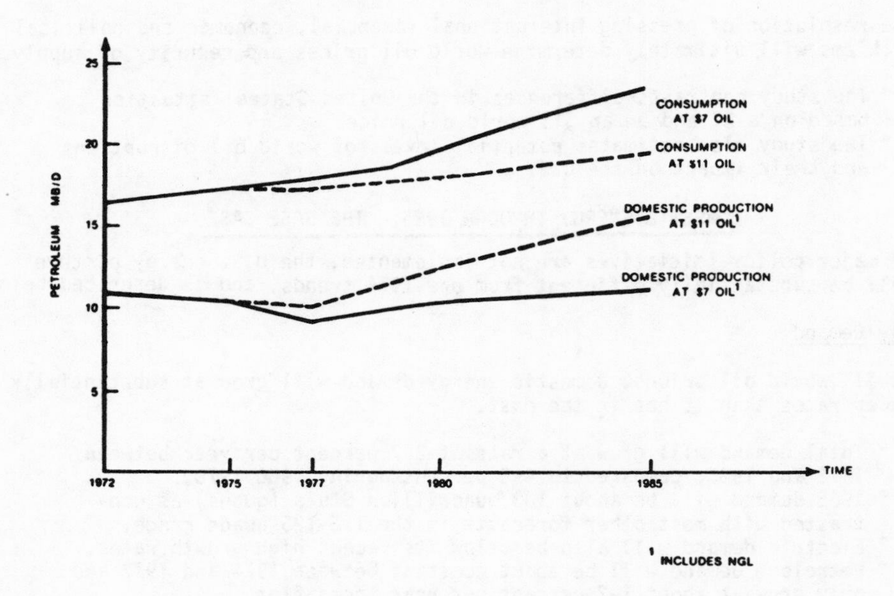

- This forecast is lower than many others due to continued schedule deferments, construction delays and operating problems.

-- Synthetic fuels will not play a major role between now and 1985.

- At $7 they are marginally economic.
- At $11 they are economic, but given first commercial operation in the late 1970's, their contribution by 1985 is small.
- Research and development (R&D) on these technologies is important if they are to replace a growing liquid and gaseous fuels gap which may develop after 1985.

-- Shale oil could reach 250,000 B/D by 1985 at $11 prices, but would be lower if expectations for $7 prices prevail.

-- Geothermal, solar and other advanced technologies are large potential sources, but will not contribute to our energy requirements until after 1985.

- R&D is needed so that these important sources, which can have less environmental impact than current sources and are renewable (do not deplete existing reserves), can be useful beyond 1985.

Constraints and Barriers

-- Even achieving the Base Case will require actions to alleviate potentially serious barriers.

- Rather than stimulating coal use, current Clean Air Act requirements could, by mid-1975, preclude 225 million tons of coal now used in utilities.
- The financial situation of the electric utility industry is particularly critical, and inadequate rates of return will not only reduce their internal funds, but hamper their ability to attract debt or equity financing.
- Current manpower, equipment, and materials shortages are likely to persist in the short-term and inhibit production increases.
- Continued problems with growth in the nuclear industry are possible, unless reliability problems, future shortages of enrichment capacity, and the waste disposal problem are resolved.

Oil Imports and Domestic Vulnerability

-- Oil imports will remain level or rise in the next few years, no matter what long-term actions we take.

-- Our domestic vulnerability to future disruptions is dependent on world oil prices.

- At $7 oil and no new domestic policy actions, imports will reach 12.3 MMBD in 1985, of which 6.2 MMBD are susceptible to disruption. A one year embargo could cost the economy $205 billion.
- At $11, imports will decline to 3.3 MMBD by 1985, and only 1.2 MMBD are susceptible to disruption, at a cost of $40 billion for a one year embargo.

Economic and Environmental Assessment

-- Higher energy prices are likely in any event, but $11 world oil will magnify these price trends and have several major effects.

- $11 oil prices, as opposed to $7, will reduce U.S. economic growth from 3.7 percent to 3.2 percent.
- Dollar outflows for petroleum imports will be higher in the near term for $11 than for $7, but by 1980 the situation will be reversed.
- At $11 oil prices, large regional price disparities exist with eastern oil-dependent regions at the high end of the spectrum.
- At $7 prices, these disparities narrow and the Northeast is no longer the highest cost region.
- Because energy costs as a percentage of total consumption are higher for lower income groups, higher energy costs will impact the poor more heavily.

-- Energy production through 1985 will have mixed environmental impacts.

- Most sources of water pollution should be below 1972 levels, due to the Federal water pollution standards.
- Emission controls will lessen the air pollution impact of increased energy use, but some regions will still be affected significantly.
- Surface mining will continue to increase and problems of secondary economic development in the West and Alaska are likely.

ALTERNATIVE ENERGY STRATEGIES

-- U.S. options to reduce vulnerability fall into three distinct categories. While each have significant impact, a national energy policy will probably combine elements from each.

Accelerating Domestic Supply

-- Federal policies to lease the Atlantic OCS, reopen the Pacific OCS and tap the Naval Petroleum Reserves can dramatically increase domestic oil production.

- At $7 prices, domestic production by 1985 could rise from 8.9 MMBD to 12.8 MMBD.
- At $11 prices, production could reach as high as 17 MMBD, although less is needed to achieve zero imports (Figure 1-7).

-- Shale oil production could reach one MMBD in 1985.

- Prices close to $11 would be needed for economic viability.
- Potential water and environmental constraints would have to be overcome.

-- Accelerating nuclear power plant construction does not reduce imports much; in general, it replaces new coal-fired power plants.

-- Accelerating synthetic fuel production would require by-passing key research steps and may not be cost-effective or practical in the 1985 time frame.

-- Accelerating domestic energy production could be inhibited by several key constraints:

- In the short-term, many shortages of materials, equipment, and labor will persist.
- By 1985, however, most critical shortages will be overcome sufficiently to meet the requirements of the accelerated supply scenario.
- Availability of drilling rigs and fixed and mobile platforms will be a major constraint in reaching the projected oil levels.
- Financial and regulatory problems in the utility and railroad industries could hamper their ability to purchase needed facilities and equipment.
- Water availability will be a problem in selected regions by 1985.

Figure 1-7

EFFECT OF DOMESTIC PRODUCTION ON PETROLEUM IMPORTS

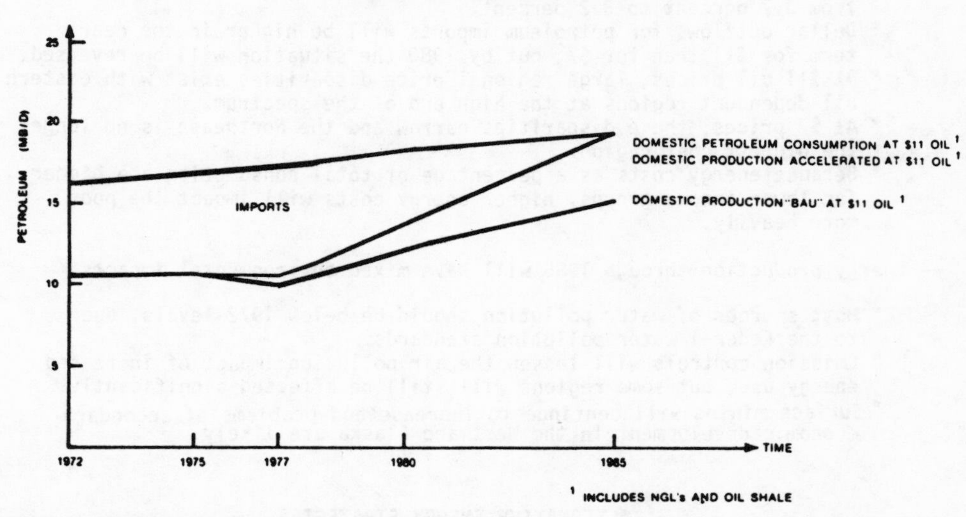

Energy Conservation and Demand Management

-- Energy conservation actions can reduce demand growth to about 2.0 percent per year between 1972 and 1985.

- To achieve reductions beyond those induced by price could require new standards on products and buildings, and/or subsidies and incentives.
- Major actions could include standards for more efficient new autos, incentives to reduce miles traveled, incentives for improved thermal efficiency in existing homes and offices and minimum thermal standards for new homes and offices.
- Petroleum demand could be reduced by 2.2 MMBD by 1985 (Figure 1-8).
- Electricity consumption could be reduced from 12.3 quads to about 11.0 quads in 1985, compared with 5.4 in 1972.

-- Demand management can further reduce dependence on limited oil and gas supplies by actions that involve switching from petroleum and natural gas consumption to coal or coal-fired electric power.

- Switching existing power plants and industrial users, prohibiting new oil or gas-fired power plants and encouraging electric space heating is most important at lower oil prices, and can substitute 400 million tons of coal per year for 2.5 MMBD of petroleum and 2.5 TCF per year of natural gas.

- Implementation may be limited by environmental restrictions and financial inability of the electric utility industry to support a large electrification strategy.
- Electrification to increase coal use in the pre-1985 period must be weighed against the possibility of increasing coal use by liquefication and gasification in the post-1985 period.

Figure 1-8

CONSERVATION MANAGEMENT

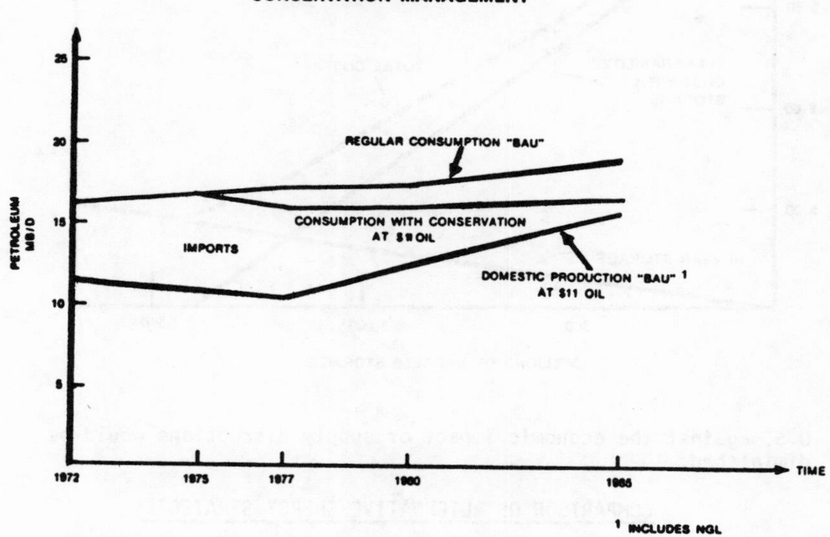

Emergency Programs

-- Standby conservation or curtailment measures can reduce vulnerability.

- Depending on the level of demand in 1985, curtailment measures in response to an embargo can cut consumption by 1-3 MMBD.
- At higher world oil prices curtailment is less effective because there is less "fat" in energy consumption.
- They involve almost no cost when not needed and relatively small administrative costs and some economic impact when implemented.
- They can be instituted in 60-90 days.

-- Emergency storage is cost-effective in reducing the impact of an embargo (Figure 1-9).

- Storage to insure against a one MMBD cutoff for one year would cost $6.3 billion over ten years.
- A one MMBD interruption of oil supply for one year during that period could cost the economy $30-40 billion.
- This cost effectiveness holds for any level of insecure imports, and applies if there is a one-in-five chance of one disruption in ten years.

-- The International Energy Program (IEP) developed in Brussels will foster consumer nation cooperation and reduce the United States' economic impact of a supply disruption during the next several years.

- It can reduce the likelihood of an import disruption.
- It includes a formula for allocating shortages which avoids excessive bidding and divisive scramble for oil by the participants during the most vulnerable period of the next few years. If, in the 1980's, the United States achieves the low import levels which are possible at high oil prices and by pursuing aggressive strategies of accelerating supply and conservation, the IEP would still act as a deterrent to an interruption, but its utility in protecting the

Figure 1-9
COSTS OF VULNERABILITY AS A FUNCTION OF STOCKPILE SIZE

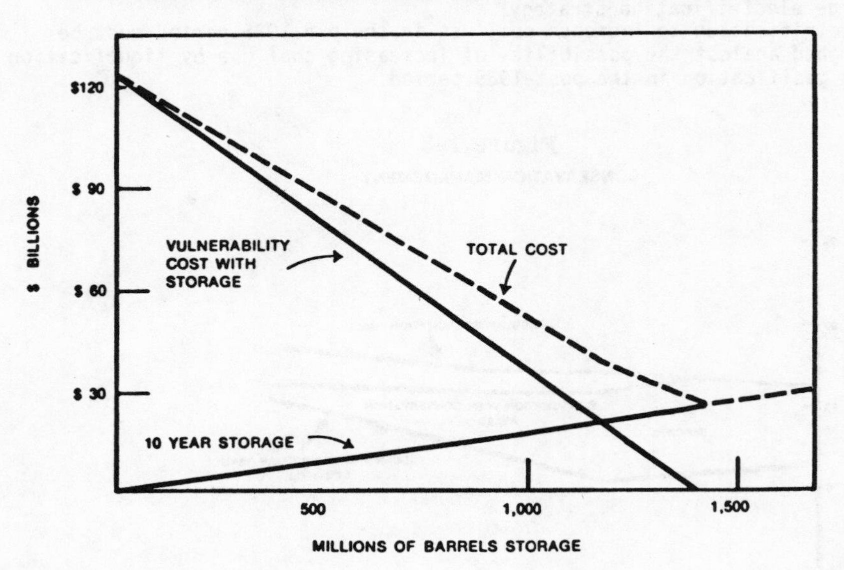

U.S. against the economic impact of supply disruptions would be diminished.

COMPARISON OF ALTERNATIVE ENERGY STRATEGIES

Import Vulnerability

-- Domestic supply and demand actions can greatly reduce U.S. vulnerability to import disruptions by 1985.

  • At $7 per barrel oil with all supply and demand actions implemented, 3 MMBD could still be subject to cutoff.
  • At $11, either all the demand actions or only a portion of the supply strategy would completely eliminate our vulnerability

-- Domestic supply and demand strategies are cheaper in economic terms than imported oil or any other emergency option.

  • At either $7 or $11, they have a lower present resource cost than imports, and reduce insecure imports.

-- After domestic actions, standby demand curtailment is most effective in reducing vulnerability.

-- Demand curtailment and storage can be designed to buffer against large levels of insecure imports.

Economic and Regional Impacts

-- Accelerating domestic supply or reducing demand will mean lower energy costs for the Nation and, hence, higher economic growth.

-- Reduced energy costs will benefit lower income groups.

-- Increased domestic supply may result in wider regional price disparities than if no action is taken.

-- The economy can absorb the increased financial costs of reducing vulnerability.

-- Both supply and demand actions will have economic impacts, regionally and in key sectors of the economy.

Environmental Impacts

-- The conservation strategy has the lowest environmental impact.

-- A demand management strategy which substitutes coal for oil and gas will result in the greatest increase in environmental impact over the base case.

-- The accelerated supply strategy has mixed environmental impacts.

- Air pollution is lower due to more nuclear plants and increased oil and gas production.
- Solid waste is up dramatically due to increased shale oil production.
- Many virgin resource areas will be disturbed for the first time.

MAJOR UNCERTAINTIES

-- The degree to which price will dampen demand

- If demand is much less sensitive to price than is assumed, we will be much more vulnerable in 1977 and in 1985 at higher world oil prices.
- Mandatory energy conservation measures or diversification and acceleration of supply hedge against this uncertainty.

-- The ultimate production potential of frontier oil areas

- Literally all the new oil production forecast comes from frontier areas in Alaska and the Atlantic OCS, or from improved tertiary recovery techniques.
- If the frontier areas do not prove productive:

  - 1985 domestic production could decline to 5 MMBD at $7 oil prices;
  - at $11 and with accelerated supply actions, total production could still not exceed about 11 MMBD.

- Synthetic fuels, switching from petroleum and gas to coal and mandatory conservation may be necessary if frontier areas are not lucrative.

-- Time required to implement domestic measures

- While lead times were taken into account, other factors could delay achievement.
- Federal inaction or local opposition.
- Materials and equipment constraints.
- Delays in private investment decisions due to price uncertainty.

POLICY IMPLICATIONS

-- Although $11 world oil prices make achievement of self-sufficiency easier, the United States is still better off economically with lower world oil prices. The implementation of a limited number of major supply or demand actions could make us self-sufficient. By 1985, we could be at zero imports at $11, and down to 5.6 MMBD of imports at $7 prices.

- Not all of these actions may be warranted, but they indicate we have significant flexibility when one considers:

  - some projected imports in 1985 are from secure sources.
  - some insecure imports can be insured against.
  - not all of the supply and demand actions must be implemented to achieve the desired result.

-- While we cannot delay all action, we can pick from those that make the most economic, environmental and regional sense.

-- Accelerating domestic supply, while economic, has some important drawbacks:

- It will adversely affect environmentally clean areas.
- It requires massive regional development in areas which may not benefit from or need increased supply,
- It is a gamble on as yet unproved reserves of oil and gas.
- It may well be constrained by key materials and equipment shortages.

-- Implementing a conservation strategy has positive environmental effects and alleviates constraint problems, but:

- It requires intervention and regulation in previously free market areas.
- It results in increased nonmarket costs due to more limited individual choice and changed lifestyles.

-- While cost effective, there are several important ramifications to a storage program.

- It will take a few years to implement and our vulnerability will be greatest during that period.
- It requires more imports now, which will act to sustain cartel prices. in the near term.
- We could suffer major capital losses -- $4 billion for each one billion barrels stored if the world oil price drops from $11 to $7.

-- Our actions to increase domestic self-sufficiency could have an appreciable impact on world oil price.

- U.S. reduction in imports can make even $7 hard for OPEC to maintain.
- World oil price reductions could jeopardize domestic energy investments and could require price guarantees or other supports.

-- Any domestic energy policy must be designed to resolve uncertainties and minimize the risk of not anticipating world oil prices correctly.

- Policy programs should include actions to reduce domestic uncertainty, such as exploring the frontier areas.
- Policies may be needed to avoid or defer major investments or actions, if they involve significant costs of being wrong, until world uncertainty is reduced.
- A flexible and dynamic approach must be balanced against the need for a stable long term policy which encourages domestic energy investment.

PROJECT INDEPENDENCE OVERVIEW

This section summarizes the Federal Energy Administration's (FEA) Project Independence Report. A background section summarizes domestic and world energy trends through 1973, leading up to the Arab oil embargo, and discusses its effects. The following section defines energy independence and summarizes the objectives and approach of the blueprint study. The major results of the study are then summarized, and a final section discusses the policy implications of the work to date.

Background

Domestic Energy Trends. Domestic energy demand has been growing rapidly since the turn of the century -- keeping pace with a growing population, increasing industrialization and greater affluence. For the past 10 years U.S. demand for energy has grown at an annual rate of between 4 and 5 percent. Today U.S. per capita energy consumption is eight times the average of the rest of the world.

Through 1950, the United States was totally self-sufficient; it easily met its growing thirst for energy with cheap and abundant domestic fuels -- coal, oil, and gas, and hydroelectric power. But by 1960, imports of crude oil and petroleum products accounted for 15 percent of total domestic petroleum consumption, and by 1973 imports had jumped to 35 percent. This growing dependence was the result of increasing demand, cheap imports, and a steadily deteriorating domestic supply situation.

- Coal production has not increased appreciably since 1943.
- Exploration for crude oil peaked in 1956; domestic crude oil production leveled off in 1970 and has been declining ever since.
- Since 1968, we have been consuming natural gas faster than we discover it.
- Nuclear power, the dream of the 1960's, has been plagued by technical and regulatory difficulties.

The U.S. now relies on oil for 46 percent of its energy and on natural gas for 32 percent, while coal, which provides only 17 percent of our consumption, represents over 95 percent of our total domestic energy resources. As contrasted with the more than 800 years of coal supply available at the current rate of consumption, we have only about ten years of proven oil and gas reserves.

The Arab Embargo. Our growing dependence on foreign oil was considered a blessing by many -- cheap foreign oil meant lower consumer prices and a more competitive domestic industry in world markets. At the same time as our dependence on foreign oil was growing, the world oil market came to be dominated by a few Middle East countries with massive reserves and production costs of literally pennies per barrel. The disadvantages of this dependency were at least hinted at when an oil embargo was imposed in 1967 as a result of international tensions in the Middle East. However, its short duration resulted in literally no domestic economic impact. It was not until some Middle East countries cut back production and exports 4-5 MMB/D and imposed a total embargo on crude oil shipments to the U.S. and other countries in the winter of 1973 that the full impact of this dependence became apparent.

Imposed in October 1973, the embargo was slow to take effect, but by January 1974, imports were down by 2.7 million barrels per day, reducing total petroleum supply 14 percent below expected consumption. Simultaneously, prices jumped to unheard of levels, rising from about $3 per barrel in September 1973 to over $11 per barrel by January 1974. Although massive unemployment, blackouts and other major disruptions were avoided, the embargo still had an appreciable impact. In addition to the inconvenience of long gasoline lines, it is estimated that:

- GNP dropped by $10 to $20 billion during the embargo.
- Unemployment caused by the embargo amounted to five hundred thousand workers.
- Of the 9.8 percent increase in consumer prices, a third was due to higher world oil prices.

In addition to its domestic impact, the embargo and the rapid increase in the price of oil had major international implications -- for industrialized and less developed countries alike. But in spite of its highly negative impact, the embargo could have occurred at a much worse time. It is true that a few years earlier, an embargo would have had no appreciable domestic effect. On the other hand, if the embargo had been delayed until several years in the future, when petroleum imports would have amounted to 50 percent of domestic consumption, there would have been no effective means of cushioning the domestic economy from the impact. The embargo made obvious the need to reevaluate our domestic and international energy policies and to fashion a new energy program to hold our vulnerability to acceptable levels.

Energy Independence and Other Goals for a Long Term Energy Strategy

In response to the embargo and the realization of the potential costs of being increasingly dependent on foreign sources of energy, the President established the goal of energy independence for the United States by 1980. There was considerable divergence of opinion, however, regarding the meaning of energy independence, and there are likely to be even greater differences as to how the goal should be achieved once it is ultimately defined. To some, energy independence is a situation in which the United States receives no energy through imports, i.e., it produces all of its energy domestically. To others, independence is a situation in which the United States does import to meet some of its energy requirements, but only up to a point of "acceptable" political and economic vulnerability. The definition of independence and the criteria for evaluating it are central to the choice of a U.S. energy strategy.

If the objective of a long-term energy policy were simply to reduce imports, then the strategy -- although not necessarily the means for implementing it -- would, indeed, be very simple to set forth. Unfortunately, the reduction of imports, if it means the substitution of more expensive domestic energy sources, could very easily be accompanied by much higher domestic energy prices, inflation, a drop in real gross national product (GNP), supply risks, and a number of other undesirable effects, such as environmental degradation and depletion of reserves. Rather than reducing imports per se, our objective should be to reduce our vulnerability to disruptions of imports for this reason, "independence" is better than "self-sufficiency" as a description of our objective.

One way that an energy strategy can reduce our dependence on imports is by reducing imports through offsetting increases in domestic supply and/or reductions in demand. Other steps that could be taken instead of, or in conjunction with, a reduction of imports would be to build emergency supplies to reduce the amount of actual interruption, or to develop standby demand curtailment and allocation programs to minimize the impact of an interruption.

Any policy which reduces our dependence on imports and economic vulnerability to supply disruptions will have other effects that must be considered. These are discussed below.

The Domestic Economic Impact of the Strategy. The overall economic impact of an energy strategy can be measured through the standard indicators -- real growth in the GNP, the rate of inflation, and the unemployment rate. The effect on the balance of payments, and any extraordinary impact concentrated on particular localities, economic sectors or income groups, must also be measured.

The Environmental Impact of the Strategy. The quality of life is measured in more than economic terms, and environmental quality is a key element of such an assessment. In addition to air and water pollution, the impact on land use and recreation for each energy strategy must also be measured. This is usually a very difficult evaluation to perform, because the trade-offs are incommensurate. How, for instance, should one value the preservation of a remote, pristine Alaska wilderness area versus the development of offshore oil near major recreational areas on the northeast coast?

The Degree of Federal Intervention Required to Implement the Strategy. An otherwise acceptable strategy might involve an intolerable or unfeasible amount of Federal intervention. Under this criterion are included financial intervention in the form of taxes or subsidies, and new energy-related regulations. The analysis must also include institutional intervention into environmental, health and safety matters, and licensing and regulatory barriers.

The Effect of the Strategy on World Oil Prices. If domestic strategies can significantly affect U.S. imports, then such strategies may also affect the world supply/demand balance. Hence, they must be weighed in terms of their international ramifications and, consequently, their impact on U. S. import price expectations.

Alternative Energy Strategies for the United States

Because of the almost limitless number of alternate individual policy actions, FEA has focused on four broad alternatives to contrast the effects of very different national energy policies. These strategies are intended to illustrate as clearly as possible the very different energy futures available to the U.S. For this reason, they are designed to overlap very little. Practically speaking, any final Project Independence program would almost surely be a mixed strategy taking elements from each.

The four broad strategies which were evaluated are:

- A Base Case, in which existing policies continue and only limited new actions are considered.
- An Accelerated Supply Strategy, in which the Federal Government takes a number of key actions to increase the domestic supply of energy. The principal actions are :
  - Greatly accelerating Federal leasing of the Outer Continental shelf.
  - Opening the Naval Petroleum Reserves to commercial exploration and development.
  - Removing some regulatory delays in nuclear power development.
- A Conservation Strategy, which would reduce demand for petroleum by actions such as:
  - Setting minimum mileage standards for new automobiles.
  - Provide incentives and standards to increase residential insulation and energy-use efficiency.
- An Emergency Preparedness Strategy in which reliance is placed on a standby emergency curtailment program and on a storage program to provide emergency reserves in case of an import supply disruption.

A Special Case -- the International Energy Program (IEP). The U.S. as a member of the 12-nation Energy Coordinating Group, has recently approved the IEP agreement. This agreement has the following principal provisions:

- A formula for allocating supply shortfalls among participating countries to minimize the effects of an embargo.
- Stockpiling requirements for all participants in proportion to their levels of imports.
- Development of emergency conservation measures amounting to 7 to 10 percent of consumption to mitigate the effects of an embargo.

If the international agreement is achieved, it will, of course, apply in combination with whatever other domestic strategy is selected. Therefore, its effects will be evaluated in this report.

Study Approach

The previous sections discussed the alternative energy strategies and evaluation criteria for the Project Independence Blueprint. This section describes the overall study approach, but does not give the actual methodology, i.e., the assumptions, data, and equations used in the various models.

FEA's study approach was shaped by three underlying considerations: the importance of making explicit the dependence of supply, demand and policy alternatives on price; the need to consider domestic supply, demand and constraints on a regional rather than a national basis, and the desirability of structuring the overall energy system in one cohesive, analytical framework.

With respect to price, the impacts of recent price increases only point up the necessity of understanding the impact of alternative prices on energy growth, production possibilities and policy alternatives.

As far as regionality is concerned, the simple fact is that the United States is not homogenous and does not demand or supply fuels equally by region. In fact, many of our most important policy questions are dominated by regional considerations, such as how to transport and utilize our oil resources in Alaska or the question of water availability in the West.

Calculations were made indicating how much production could be achieved for each of the sources of energy under different world oil prices and under two alternative assumptions -- Business-as-Usual and Accelerated Development. In each case, account was taken of the lead times associated with increasing production from each of the sources. Separate projections of supply as a function of price were made for oil, natural gas, coal, nuclear, synthetic fuels, shale oil, solar, and geothermal energy. As an adjunct to the supply projections, requirements for scarce resources - capital, manpower, equipment and materials, transportation, and water - were also estimated. The energy costs included the costs of meeting current or expected environmental standards.

Estimates were made of the costs at which key facilities - refineries, natural gas plants, and electric utilities - could be built and their leadtimes. These facility costs, when coupled with the costs of primary fuels, resulted in projections of the availability and price of electricity and of petroleum products. Similarly, transportation facility lead times and costs for each fuel were estimated between each region where fuels would be produced and each region where fuels would be consumed.

At the same time that energy production levels, costs and lead times were projected, the demand for each energy product (including electricity) for each region, as a function of price, was developed. In addition to reductions in demand induced by higher prices, the impact of specific conservation measures was also forecast.

These estimates constitute the basic data required for the analysis forecasts of supply and demand for each source of energy, each price, and each region; the costs of facilities and transportation; and the requirements of each type of energy for scarce resources. Then all of the supply, demand, facility and transportation data were combined to provide estimates of the price of each fuel and the quantity of each which would be consumed; at the same time the scarce resources required to produce the fuels are estimated, to see if any critical shortages occur.

The energy sector of the economy is analyzed in this way for the years 1977, 1980, 1985. This description is sufficient to yield important information on such factors as the consumption of each fuel, imports of petroleum, regional supply gaps and price variations, shortages of scarce resources, etc. While this study uses the most sophisticated set of models and analyses yet applied to energy forecasting, it represents only the best estimate of what is still a highly uncertain situation.
It does, however, provide a rigorous analysis and is especially useful in evaluating changes in the energy forecast due to policy actions.

The energy forecast for the alternative strategies are evaluated in terms of their environmental and economic impact. The impact of an embargo on the U.S. economy is then estimated for each strategy and level of imports. The last step in the analysis is to assess the international impact of each U.S. strategy, based chiefly on the level of imports that the U.S. would require, and its effect on world-wide demand, and hence, on oil price.

Energy Through 1985 -- What Happens If We Continue with our Present Energy Posture?

Before alternative energy policies are contemplated, there must be a careful evaluation of the situation through 1985 if current policies continue. This section first summarizes the world oil situation and then summarizes its implications for domestic energy trends through 1985.

World Oil Through 1985. The world oil price is determined by a complex set of political and economic considerations. Two important factors are (1) the world supply and demand for oil, and (2) the political cohesion required of the petroleum exporting nations to maintain prices at higher-than- market levels.

The Middle East countries currently possess 58 percent of the world's total proven petroleum reserves. In 1973, they accounted for 37 percent of the world's production, but only 2 percent of the consumption. Of the 29 million barrels of oil exported each day, 70 percent come from the countries of the Middle East. They dominate the world export market now, and will probably continue to set future world oil prices.

Because the behavior of the embargo countries is uncertain, any assessment of future world oil prices must weigh the conflicting forces of market economics and the politics of cartel behavior. To assess the probable situation, world demand for petroleum through 1985 was forecast at prices of $4, $7, and $11 per barrel. Taking transportation into account, these prices are roughly consistent with Persian Gulf prices of $3, $6 and $9 per barrel. All prices are given in constant mid-1973 dollars.

Using these prices, Table 1-22 below summarizes the import needs of the major oil consuming nations of the world in 1985.

Table 1-22

Estimated Imports by Region, as a Function of Price Per Barrel at the U.S. 1985 (millions of barrels per day)

	Price Per Barrel			For Comparison
	$4	$7	$11	1973 Imports
U. S.	21.4	12.4	3.3	6.3
Western Europe	22.9	15.8	12.3	14.1
Japan	11.9	9.8	8.6	5.2
Other	7.6	5.5	4.1	3.8
TOTAL	63.8	43.5	28.3	29.4

As expected, the world demand for oil imports in 1985 at $4 per barrel is more than double that of $11 per barrel. Countries like the U.S. and Canada show the greatest sensitivity to price due to their domestic supply potential at higher prices. On the other hand Japan, which has no domestic resources, can only reduce imports slightly at higher prices.

Table 1-23 shows that the Middle East countries can almost double their production if called for by their economic and political objectives.

Table 1-23

Estimated Export Capacity of Middle East Countries (Millions of Barrels per Day)

	1973 Actual Export Production	1985 Maximum Potential Export Capacity
Saudi Arabia	7.4	19.3
Other Arab	8.4	18.2
Iran	5.5	7.2
Other	8.1	8.4
TOTAL	29.4	53.1

The comparison of 1985 world import demand and 1985 world export supply capacity is shown in Table 1-24. It shows a balance between potential supply capacity and demand somewhere between $4 and $7 per barrel U.S. price. Also indicated is the production capability of OPEC countries which cannot absorb all of the revenues generated at alternative world oil prices. Thus, if the OPEC countries which do not need all the revenues generated forego 22 percent of their growth in production capacity, a price of about $7 per barrel could be maintained.

Table 1-24
World Energy Supply and Demand, 1985
(Millions of Barrels per Day)

	At $4	At $7	At $11
Potential 1985 capacity	53.1	53.1 MMBD	53.1 MMBD
Estimated 1985 demand	58.9	43.5 MMBD	28.3 MMBD
Production to be foregone	*	9.6 MMBD	24.8 MMBD
Production in OPEC countries not absorbing maximum revenues	*	43.3 MMBD	48.7 MMBD
Percent foregone production required of these countries, 1985	*	22%	51%

*Not applicable; at $4, world import demand will exceed OPEC's capability to supply it.

This 22 percent cutback would still allow them to increase production by 50 percent over present levels. To maintain an $11 price, they would have to keep production at their present levels. The difference in price more than offsets the reduced production; that is, at $11 and reduced production, more revenues are generated. The question in the long run is whether OPEC has the cohesion or desire to forego some or all of its growth. In the short run, they have already demonstrated their capability to reduce production and maintain higher prices. During the last several months when there was an estimated surplus of roughly 2-3 MMBD in the world market. OPEC has successfully controlled production to maintain the current price.

This argument suggests that in the Base Case, by 1985, a $7 price is likely, a price between $4 and $7 is possible, but unlikely, and action by OPEC to maintain a price close to $11 is possible although this will be affected by politics rather than economics. Therefore, throughout this report each strategy is evaluated at two price levels, $7 and $11 per barrel. The higher price brings us closer to self sufficiency with a minimum of new Government action, while the lower price can provide a large degree of independence provided that relatively strong Federal action is taken. The two views of the problem provide a range within which major policy decisions can influence our energy future.

Domestic Supply and Demand. FEA's 1985 forecast is largely dependent on three key assumptions: (1) the rate and sectoral trends in economic growth; (2) the physical availability, economics and technical aspects of future energy supply; and (3) the energy policies in effect which directly shape or constrain the energy sectors.

FEA's forecast of energy supply and demand starts with a forecast of domestic economic activity. Table 1-25 shows the key parameters in the forecast.

This forecast also necessitated several key assumptions about future energy policies:

- Regulation of natural gas prices will not discourage production which would otherwise be economically justified, or deregulation of new natural gas occurs.

- Allocation and price controls will be removed by the end of 1975.
- Current Clean Air Act provisions will be modified to alleviate the clean fuels deficits.
- Current tax laws on depletion allowances, profits, etc., will remain unchanged.
- Natural gas imports from Canada will be available at $1.20 per thousand cubic feet (MCF) up to 2.1 trillion cubic feet (TCF) annually and liquified natural gas imports will be available at $2.00 per MCF.
- The Trans-Alaska Pipeline will be completed on schedule.
- No major Pacific, Atlantic or Gulf of Alaska OCS leasing is slated.

Table 1-25

Parameters of the Economic Forecast for 1985

	1973 Actual	1985 Forecast	Average Annual Growth Rate
Population	210 million	236 million	0.96%
GNP 1/	$.84 trillion	$1.28 trillion	3.5%

1/ 1958 constant dollars

Energy Demand Through 1985: Table 1-26 shows the FEA forecasts of total U. S. energy demand at $4, $7, and $11 per barrel throughout the time period.

Table 1-26

U. S. Total Energy Demand in the Base Case as a Function of Price (In Quadrillion Btu's per Year) 1/

		Total Demand	
	1972	1985 Estimated	
Price of Imports	Actual	Quads	% Growth
$4	72.1	118.3	3.8
$7	72.1	109.6	3.2
$11	72.1	102.9	2.7

1/ Three units of energy used in this report are quadrillion Btu's per year (Quads), millions of barrels of oil per day (MMBD), and trillions of cubic feet of natural gas per year (TCF). These are related by the formulas 1MMBD=2 quad; 1TCF=1 quad.

FEA's forecast of energy demand through 1985 is lower than most forecasts published heretofore, if $11 prices are assumed, as indicated in Table 1-27.

Table 1-27

Comparison of FEA Projection of Energy Demand With Other Projections, 1985 (quadrillion Btu's per year)

	Projected Demand 1985	Compound Annual Growth Rate 1972-1985
FEA @ $4 per barrel	118.3	3.8%
FEA @ $11 per barrel	102.9	2.7%
Ford Foundation, Energy Policy Project	115.0	3.6%
Dupree-West	116.6	3.7%
National Petroleum Council	124.9	4.2%

At higher oil prices, the decline in energy growth rates is greatest in those sectors which are most dependent on petroleum and least able to switch to other fuels. Table 1-28 compares the FEA base cases with historical growth rates by sector.

Table 1-28
Comparison of Sectoral Growth Rates

Sector	1972 Consumption Quads	1985 Forecast $7	1985 Forecast $11	Compound Growth Rates 1960-1972	Compound Growth Rates 1972-1985 $7	Compound Growth Rates 1972-1985 $11
Household and Commercial	18.0	25.9	25.1	3.9	2.8	2.6
Industrial	22.9	30.4	29.0	3.1	2.2	1.8
Transportation	18.1	24.5	22.0	3.9	2.3	1.5
TOTAL	59.0	80.8	76.1	3.6	2.4	2.0

Energy Supply Through 1985. Energy supply changes between 1974 and 1985 will be the result of a complex set of factors, including:

- Changing patterns of demand which may preclude certain fuel uses or create new demands for particular fuels.
- Relative costs of producing and transporting each fuel to demand centers.
- Trends in physical availability and production techniques which can change relative prices over time.

The FEA forecast is based on an analysis of these factors and the development of detailed price and quantity supply curves for each major energy producing region.

With $11 world oil the supply of domestic energy in the Base Case is projected to grow at a faster rate than domestic energy consumption, from 60.4 quads in 1972 to about 96.4 quads in 1985. A gap of 6.5 quads is projected to be filled by petroleum imports, as shown in Table 1-29.

Table 1-29
Energy Consumption in 1985
(in quads)

	@ $7 Oil	@ $11 Oil	1972 For Comparison
Domestic Energy Supply	84.3	96.4	60.4
Petroleum Imports	24.8	6.5	11.7
Total Energy Consumption	109.1	102.9	72.1

By contrast, at $7 oil, consumption increases, domestic production is lower and imports are 24.8 quads. This is more than triple the $11 case, and equals 23 percent of total consumption. The high level of imports at $7 prices in the Base Case forecast is due to:

- Much higher costs of domestic petroleum production than previously estimated.
- Limited expansion of nuclear power.
- Little contribution by 1985 from new technologies such as oil shale and synthetic fuels.
- Only moderate increases in coal production, due to limited opportunities for the use of coal.

Domestic coal supply is not expected to be a constraint. Coal production from our massive reserves can be expanded without significant increases in production costs. Coal, however, is not suitable for meeting many forms of demand and is projected to contribute only moderately to expanded domestic energy production, rising from 650 million tons in 1974 to about 1,085 million tons in 1985, assuming $11 oil prices. At $7 world oil, utilities and industry consume more oil, and coal reaches only 945 million tons by 1985. These figures include exports, which were 70 million tons in 1974 and are forecasted to rise to at least 85 million tons by 1985.

The petroleum findings of the FEA projections are significantly different from other forecasts and deserve detailed discussion. Exploration and drilling costs have increased rapidly, while reserves of the most economical sources of petroleum have rapidly declined. This situation makes it very difficult to maintain or expand the current level of petroleum production -- even at high petroleum prices.

Table 1-30

Crude Production (Million Barrels Per Day)

	1974	1985 @ $7 Imports	1985 @ $11 Imports
Lower 48 States and Gulf of Mexico	8.2	4.2	7.4
North Slope		2.5	2.5
Other Alaska	0.2	0.1	.5
Atlantic and Pacific OCS	0.1	2.1	2.1
Other	—	.3	.6
Total	8.5	9.2	13.1

The most important observation is that production from the 1974 proven reserves is expected to decrease from 8.2 MMBD now to 4.2 MMB/D in 1985 at $7 prices. Even with oil prices of $7 and the opening up of vast new areas such as Alaska, total domestic production would remain approximately constant. At $11 prices, production from already discovered oil reserves would be roughly stable, but this is only through extensive application of expensive secondary and tertiary recovery, in part using techniques not yet fully developed. Production increases at $11 come largely from new discoveries envisioned on the OCS as well as production of known but yet undeveloped Alaskan reserves.

FEA's estimate for domestic gas production is significantly lower than other forecasts. Assuming $11 oil and gas prices of about $1.00 at the city gate, FEA projects domestic production for 1985 at 24.6 TCF, as contrasted with 22.1 TCF in 1972. At the $7 oil import price, decreases in domestic petroleum production levels lead to a drop of 1.3 TCF from associated gas production. If current regulated prices for new gas continue, at $.42 per MCF at the wellhead, domestic production is expected to decline to 15.2 TCF by 1985.

Domestic oil and gas supplies after 1985-1990 may show a return to declining domestic petroleum and natural gas production. Higher prices now would lead to an ever more rapid decline in production eventually.

Total electric generation is expected to reach 12.3 quads of distributed power in 1985, more than double 1972 levels, but substantially less than most other forecasts. In 1985 nuclear power is forecast to provide 30 percent of all electricity generation, compared to 4.5 percent today. This forecast is much lower than that of many other estimates; in part due to the delays experienced during the last few years in getting nuclear plants into commercial operation. It is very unlikely that nuclear power can increase any faster, although base load nuclear power plants are less expensive than fossil fuel plants on a total cost basis. The projection assumed that only those nuclear plants already planned would be available before 1985. More would be built if construction, licensing and delivery delays, material unavailability and other problems could be overcome. If these problems persist, even less nuclear power and a corresponding increase in fossil fuels, particularly coal, are likely. Little contribution is expected by 1985 from new technologies under current government R&D policies. The first commercial plant producing oil from oil shale is not expected to be in operation before the late 1970's, as will many of the more advanced commercial synthetic plants. Summarizing this discussion, the composition of the projected domestic energy supply in 1985 is given in Table 1-31.

Table 1-31
Composition of U.S. Domestic
Energy Production, 1985
(quadrillion Btu's)

	1972	1985 @ $7 Per Barrel	1985 @ $11 Per Barrel
Oil	22.4	23.1	31.3
Gas	22.1	23.9	24.6
Coal	12.5	19.9	22.9
Nuclear	0.6	12.5	12.5
Other	2.9	4.9	5.1
Total Production	60.4	84.3	96.4

The Impact of the Base Case

Given this base case forecast, it is now possible to evaluate its impact and desirability.

Oil Imports and Vulnerability. Table 1-32 shows the domestic petroleum situation, level of imports, and the loss which would result from an interruption in imports for $7 and $11 world oil prices without IEP. At $7 prices, imports are almost quadruple those at $11 prices due to increasing consumption and less economic domestic production. The estimated GNP cost of an embargo is found by estimating the level of insecure imports and the expected length of an interruption. At $11 per barrel for foreign crude, a one year embargo could cost the economy $40 billion. For the $7 world oil case our imports from insecure sources are substantially higher and the economic impact of a one year disruption could be about $205 billion.

Table 1-32
Imports and Vulnerability in 1985

	Price of World Oil	
	$7	$11
Imports	12.4 MMBD	3.3 MMBD
Insecure Imports	6.2 MMBD	1.2 MMBD
Economic Impact of a One Year Interruption	$205 Billion	$40 Billion

This table highlights the levels of imports, and the high cost to the U.S. from an embargo if we continue our current energy policies and oil prices drop. It also shows the importance of price -- at $11 world prices we would hardly have an import problem.

Impact on the Domestic Economy. Table 1-33 shows the rates of inflation and growth in the real GNP, and the outflows for imported oil at the two world oil prices. The impacts on total balance of payments have not been estimated, because of the difficulties of estimating world investment flows, exports to Middle East countries, etc.

Table 1-33
Economic Impact of Base Case
1985

Indicator	$7	$11
Domestic energy cost	$1.30/Million Btu	$1.40/Million Btu
Inflation rate (1973-1990, CPI)	6.2%	6.4%
Average annual growth in GNP (1973-1985)	3.7%	3.2%
Annual Outflow of funds from U.S. in payment for oil	$31.7 billion	$13.2 billion

If the world price of crude remains at $11/bbl, price stability will be more difficult to achieve. Should the price of crude drop to $7/bbl this would reduce the inflation rate by about two-tenths of a point per year over the next decade. Lower oil prices could also affect economic growth and purchasing power. GNP would grow at 3.7 percent during the 1973-1985 period rather than at 3.2 percent if oil dropped from $11 to $7 per barrel.

With $11 crude prices, rising incomes are more than sufficient to offset rising prices so that the purchasing power of the average person increases over 1973. Real consumption would rise significantly should the world price drop from $11 to $7.

If the world price of crude were to drop to $7 the immediate result is to lower the dollar outflow during the next few years, but in the longer run the effect is to increase the dollar outflow over what it would have been if the price of crude had remained at $11. In 1977, outflows would be $23.7 and $26.5 billion for $7 and $11 world oil respectively. But by 1985, outflow for oil imports will be $13.2 billion at $11, or 58 percent below outflows at $7 prices. However, since the U.S. will have the largest relatively stable capital market in the world, a return flow of OPEC investment funds could offset any trade outflow. Balance of payments considerations will therefore not be of overriding importance to energy policy decisions, once the reflow of funds gets under way. It should be further emphasized that while high oil prices may reduce outflows of dollars, higher oil prices also mean billions of dollars in extra costs for energy which we would avoid at lower prices.

The recent upward shift in energy costs has had a large detrimental effect on the poor. They were hit both by large rises in the prices of energy products, which require a disproportionate share of their budgets and by a lowering of the share of labor income, their major source, due to higher unemployment. At $11 world oil prices these economic burdens are substantially greater than they are at $7 prices.

In terms of the price of energy products the hardest hit groups are now, and will continue to be, those who live in the regions with the highest energy prices, such as the Northeast.

Environmental Assessment. The environmental impact from the "Business As Usual" forecast are determined by three factors:

- Increasing aggregate level of domestic energy production and consumption
- Changing energy mix
- Levels of environmental controls

While total energy production and use rises from 72 quads in 1972 to 103-109 quads in 1985 (depending on world oil prices), the environmental impacts are mixed and in some cases lower than 1972, due to the dramatic increases in the use of pollution control devices mandated by Federal air and water pollution control legislation now being implemented. The following table compares total emissions for selected categories in 1972 and 1985, under different world oil price assumptions. Air emissions at $11 oil are generally higher than at $7 oil despite reduced demand, because coal production and use increases.

Table 1-34
Environmental Comparisons 1972-1985

Pollutant Category	1972 Level	1985 Level $7	1985 Level $11
Water Pollution:			
Dissolved solids (tons/day)	37,000	5,200	5,800
Suspended solids (tons/day)	7,600	240	300
Thermal discharges (billion Btu/day)	19,500	24,500	24,000
Air Pollution:			
Particulates (tons/day)	1,800	2,200	2,300
Nitrogen Oxides (tons/day)	38,000	41,800	46,700
Sulfur Oxides (tons/day)	58,900	47,100	53,700
Hydrocarbons (tons/day)	33,200	18,800	18,800
Carbon Monoxide (tons/day)	7,900	1,000	1,400

Potential Constraints to Base Case Forecast

Even achieving the increases in domestic production implied by the Base Case forecast could be inhibited by several key barriers and constraints which will be particularly important in the next few years.

The increased coal use forecast is predicated on resolution of the "clean fuels deficit" problem. Current State regulatory standards for sulfur emissions, promulgated under the Clean Air Act are more stringent than required to meet national ambient air quality standards and could preclude the use of about 225 million tons of coal now being burned by utilities. Unless this problem is resolved, along with the longer term-issues of allowable stack gas cleaning technology, increased coal use is unlikely.

Of equal importance to the coal industry and the economy as a whole is the financial viability of the utility industry. Current rates of return appear inadequate to finance needed investments and, if the financial situation does not improve, the industry may be unable to attract needed debt or equity capital. This has already caused new plant cancellations and construction stretchouts and these threaten to continue.

Finally, the energy industries have been plagued in recent months by shortages of key manpower skills, equipment, and materials. By 1985, none of these items will be in short supply, but during the next one to three years shortages could constrain energy growth. Clear directions in energy policy are needed to stimulate investment, not only in energy, but in the capital goods industries which are critical to the energy industry's expansion.

Degree of Federal Intervention. Current energy policies involve little Federal intervention beyond a sizable R&D budget of about $2.3 billion per year. All conservation programs implied in the Base Case are voluntary. Stimulation of increased energy supply does not go beyond a gradual increase in leasing and continuation of current tax policies. Also, significantly less intervention is expected in price controls for oil and natural gas. However, significant regulation is expected to continue with respect to environmental controls, health and safety, utility industry operations, etc.

Energy Demand and Supply Under Alternative Energy Strategies

Implementation of any of these alternative strategies can dramatically affect domestic energy supply and demand by 1985. The net impacts are summarized below for the cases of $7 and $11 imported oil.

Table 1-35
U.S. Energy Supply, Demand and Imports in 1985

At $7 World Oil Prices

	U.S. Energy Demand (quads)	Domestic Energy Supply (quads)	Imports[1/] Oil (MMB/D)	Imports[1/] Natural Gas (TCF)
Base Case w/ and w/o Emergency Programs	109.1	84.2	12.4	0
Accelerated Supply	109.6	92.6	8.5	0
Conservation	99.2	79.6	9.8	0
Accelerated Supply plus Conservation	99.7	88.5	5.6	0

At $11 World Oil Prices

	U.S. Energy Demand (quads)	Domestic Energy Supply (quads)	Imports[1/] Oil (MMB/D)	Imports[1/] Natural Gas (TCF)
Base Case w/ and w/o Emergency Programs	102.9	96.3	3.3	0
Accelerated Supply	104.2	104.2	0	0
Conservation	94.2	91.8	1.2	0
Accelerated Supply plus Conservation	96.3	96.3	0	0

1/ 1 MMB/D of oil = 2 quads; 1 TCF of natural gas = 1 quad.

Although this table does not address vulnerability directly, it leads to several significant conclusions:

- At $11 oil, the import problem is not very serious, and an Accelerated Supply program can eliminate imports entirely.
- Even at $7 oil, supply and conservation measures can bring imports down to levels below 1974.
- Especially at $7, the strategies can produce a large reduction in U.S. demand for world petroleum, which could be expected to put downward pressure on world oil prices.

Comparative Evaluation of Oil Strategies

Oil Import Vulnerability. Table 1-36 shows the effect of each of the strategies in reducing the vulnerability of the U.S. to supply disruptions.

These are the basic building blocks of the strategy with respect to supply security. At $7, the insecure imports total 6.2 MMBD. An accelerated supply program can reduce that vulnerability by 1.7 MMBD. long-term conservation can reduce it by 1.0 MMBD, and together they can accomplish a long-term reduction of 3.2 MMBD. If no long-term conservation program is undertaken, emergency demand curtailment can reduce vulnerability by 2.5 MMBD. However, if long-term conservation is implemented, then the possible savings from emergency demand curtailment is reduced to 1.5 MMBD. Hence, long-term conservation, accelerated domestic supply and demand curtailment can reduce vulnerability to 1.5 MMBD.

At $11, insecure imports are only 1.2 MMBD. These imports can be completely eliminated by an accelerated supply program, or completely offset by emergency demand curtailment in case of an interruption. Long-term conservation measures can reduce imports by 1.0 MMBD and the remainder can still be easily offset by emergency demand reductions if necessary.

Emergency storage could be built to eliminate vulnerability to any level of insecure imports resulting from any other combination of supply and conservation actions at either price.

Table 1-36
Effect of Strategies on Imports, 1985

	$7 (MMBD)	$11 (MMBD)
Total Oil Imports, Base Case	12.4	3.3
Insecure Imports, Base Case	6.2	1.2
Reduction of Insecure Imports through Accelerated Supply	1.7	1.2
Reduction of Insecure Imports through Conservation	1.0	1.0
Reduction of Insecure Imports through Accelerated Supply and Conservation	3.2	1.2
Reduction of Demand for Insecure Imports by Demand Curtailment (without long-term Conservation)	2.5	1.2
Reduction of Demand for Insecure Imports by Demand Curtailment (with long term Conservation)	1.5	1.2
Reduction of Vulnerability to Insecure Imports by Storage	Any Level	

The International Energy Program (IEP), includes a formula for allocating shortages which avoids excessive bidding and divisive scramble for oil by the participants during the most vulnerable period of the next few years. If, in the 1980's, the United States achieves the low import levels which are possible at high oil prices and by pursuing aggressive strategies of accelerating supply and conservation, the IEP would still act as a deterrent to an interruption, but its utility in protecting the U.S. against the economic impact of supply disruptions would be diminished.

Table 1-37 summarizes the evaluation of the strategies in terms of reducing the economic costs of a supply disruption.

The most striking observation is that at $11 world oil prices our direct vulnerability is minimal and any of our strategies effectively reduces it to zero, although our allies will still be dependent on imports. The emergency storage strategy makes no effect to reduce imports; instead, it uses a one billion barrel emergency storage plus an emergency conservation program to cover any expected economic damage from an import disruption. The conservation strategy reduces the economic loss substantially without eliminating it.

At $7 world oil prices, when the Base Case volume of imports is much greater, none of the policy strategies by itself would eliminate our economic vulnerability, although the implementation of all of the strategies would eliminate it.

It should be reemphasized that regardless of price, all of the strategies greatly reduce the vulnerability to interruption, as they were designed to, compared to the base case.

<u>Economic Impact of the Strategies</u>. The alternative domestic strategies have different economic impacts. One of the key parameters is the average resource cost shown in Table 1-38 .

In general accelerating domestic supply reduces average resource costs as imported oil is replaced with equally cheap or less expensive domestic oil. The effect is most pronounced at $11 world oil prices where domestic supply increases reduce U.S. resource costs by almost 10 percent. Conservation and combined strategies also reduce domestic costs. However, their impact are overstated because consumer investment costs for conservation, such as for home insulation, are not included in the calculations.

As indicated in Table 1-39 changes in inflation and economic growth for the Base Case and Accelerated Supply Case are negligible.

Table 1-37

Economic Costs of an Oil Import Interruption

	At $7 Imported Oil		At $11 Imported Oil	
	1985 Oil Imports (Millions of Barrels/Day)	Loss in GNP From a 12 Month Import Interruption ($ Billions)	1985 Oil Imports (Millions of Barrels/Day)	Loss in GNP From a 12 Month Import Interruption ($ Billions)
Base Case	12.4	205	3.3	40
Accelerated Supply	8.5	149	0	0
Conservation	9.8	172	1.2	0
Supply plus Conservation	5.6	99	0	0
Emergency Storage Program[1/]	12.4	36	3.3	0

1/ Standby demand curtailment plus a 1 billion barrel storage program.

Table 1-38

Resource Cost of Alternative Energy Strategies
1985

	$7 Oil				$11 Oil			
	BAU	Acc. Supply	Consv.	Acc Supply and Consv.	BAU	Acc. Supply	Consv.	Acc Supply and Consv.
Resource Cost ($/MMBtu)	$1.31	1.24	1.23	1.17	$1.40	1.25	1.28	1.14

Table 1-39

Economic Impact of Selected Energy Strategies at $11 Prices

	Average Annual Rates: 1973-85 Inflation	GNP Growth
Base Case	6.4	3.2
Accelerated Supply	6.3	3.2

Trade outflows for oil imports increase to $30 billion annually without domestic policy actions. However, these estimates do not include the effects of capital inflows or changes elsewhere in the balance of payments accounts. The domestic strategies which minimize imports also minimize the balance of payments deficits, as can be seen in Table 1-40. At $11 oil, the supply strategies eliminate imports and eliminate any contribution of oil to the balance of trade outflows. At $7 prices, domestic strategies are also effective, but trade outflows persist. While this measure is not adequate for assessing the economic desirability of a particular strategy, it does indicate the effect of the strategies on alleviating potential pressure on the dollar and reducing the net transfer of wealth out of the U.S.

The four alternative strategies have very different domestic economic impacts. In terms of national economic growth, both the accelerated supply and conservation strategies have positive influence because they repre-

Table 1-40
Balance of Payments Impact of Energy Strategies

	1985			
	At $7 Imports		At $11 Imports	
	Imports (MMBD)	Annual Outflows for Oil ($ Billions)	Imports (MMBD)	Annual Outflow for Oil ($ Billions)
Base Case	12.4	$31.7	3.3	$13.2
Accelerated Supply	8.5	21.8	0	0
Conservation	9.8	25.0	1.2	4.8
Accelerated Supply plus conservation	5.6	14.3	0	0

sent actions which have lower resource costs than continuing to use either $7 or $11 imported oil. Therefore, they result in lower priced domestic energy. However, this is where the similarity ends. Accelerating domestic supply will cause significant regional growth--primarily in the West and Alaska as new resources are tapped. Similarly, effects of growth--and inflationary trends--will be focused in the oil industry and related supplying industries. However, due to water constraints in several regions, energy industry growth will mean less growth in other sectors. Therefore, the net effect on employment and regional earnings may not be positive in all cases.

By contrast, the energy conservation strategy will stimulate growth in insulation and similar industries, but, by increasing auto prices and reducing vehicle miles traveled, it may result in changes in the auto recreational, and leisure industries. An important question with respect to these impacts is whether they will cause actual industrial and regional dislocations or just change the paths of future growth.

The emergency program has little direct economic effect -- except to the extent it requires production of storage facilities. However, the aggregate economic effect of such a program might not be small if purchases of oil on the world market for the storage system help maintain current high prices, or drive them even higher.

In general, the domestic strategies do not adversely affect lower income groups because energy costs do not rise. However, the Accelerated Supply and Conservation strategies do increase regional energy price disparities and low income groups in the higher energy-cost regions, such as the Northeast, will suffer disproportionately.

Environmental Comparisons. Table 1-41 compares the environmental impacts of the alternative strategies. The energy strategies postulated imply significantly different impacts from the Base Case scenario. The Base Case assumes moderate growth in energy use, but environmental impacts in many cases are lower due to the installation of better pollution control equipment.

The Accelerated Supply case has several impacts that differ from the Base Case, including:

- Reduced air pollution impact from coal due to increasing nuclear power and oil development
- Reduction of oil spills, etc., because OCS drilling results in fewer spills than imports by tanker.
- More solid waste due to increased production of oil shale.
- Significant disruption of virgin frontier areas in the West for coal and in Alaska and the Atlantic and Pacific OCS for oil.

As expected, the conservation strategy is most beneficial environmentally.

when consumption is reduced by nine quads, almost all categories of pollution are reduced as compared with the Base Case.

Table 1-41
Environmental Impacts of Energy Strategies
(Selected Indicators)

		1985 at $11 Oil Imports Alternate Energy Strategies		
Air Pollution	1972	Base Case	Accelerated Supply	Conservation
Particulates (tons/day)	1,800	2,200	2,300	1,800
NO_X (tons/day)	30,000	46,800	43,000	38,400
SO_X (tons/day)	58,900	53,700	48,800	41,500
Water Pollution				
Disolved Solids (tons/day)	37,000	5,800	5,500	5,000
Suspended solids	7,600	300	260	210
Solid Waste				
1000 tons/day	900	1,100	2,300	900
Land Disruption 1,000 acres	19,800	26,700	21,800	17,900

Government Intervention. Most of the alternative energy policies would involve changes in Federal intervention, beyond the relatively restricted Federal involvement of current energy policies. The emergency protection strategy obviously requires the least additional Federal intervention: enough funds to set up a storage system and some regulatory powers to set up a standby allocation and conservation program. The accelerated supply strategy requires stepped up Federal leasing, reduced regulatory impediments and possible price guarantees. Further, allocation or other Federal actions to make available other resources, such as tubular goods, may be required. The Energy Conservation Strategy would require increased Federal regulation of new products and subsidies for certain energy saving investments.

Impact on the International Price of Oil. Table 1-24 showed that potential supply at $7 a barrel would exceed demand in 1985 even in the base case; the figures are 53.1 vs. 41.3 MMBD. Political action by OPEC could keep prices at $7 or more. Further reductions in world demand through U.S. action could put increased downward pressure on world oil prices. The Accelerated Supply Strategy would reduce U.S. imports by 5 MMBD at the $7 price; the combined Supply and Conservation Strategy would lead to a reduction of 7 MMBD in imports. Added to the potential supply excess of 12 MMBD, these U.S. strategies might drive prices even lower.

The supply strategy is, however, particularly sensitive to the price of imports. Most of the oil components of the supply acceleration--the offshore drilling and the Naval Petroleum Reserve in Alaska -- will produce oil at approximately $6-6.50 per barrel; if prices drop below this level, these production investments will not be economic. Furthermore, the risk of lower prices may inhibit investors.

Energy Balance Beyond 1985. The key element in the energy balance beyond 1985 is the rate at which the new technology program is brought in. Gasification of coal, solar heating and cooling, and geothermal energy sources appear promising at the higher prices of oil which are likely to prevail. While an R&D program is not a major component of pre-1985 strategies, it will play a large and growing role in the post-1985 period.

The second observation is that economic sources of oil will

to be exhausted in the late 1980's. Higher prices -- $11 rather than $7 -- could lead to the opening of otherwise uneconomic sources (e.g., shale oil), but they will nonetheless lead to faster depletion of existing conventional oil reserves. The conservation strategy, by reducing the country's high consumption rate of fuels, would leave us in a slightly better position beyond 1985. Under both the Base Case and the Accelerated Supply Strategy imports could begin to increase rapidly between 1990 and 2000. Demand will be growing at 2.5 percent per year and domestic oil supply will be declining. Therefore, new technologies to convert coal to oil and gas must be on the rise to offset these trends.

Finally, continued and growing use of nuclear power in the post-1985 time frame will require substantial expansion of our uranium reserve base and extraction industry, greatly increased fuel enrichment capacity and the ultimate resolution of the waste disposal problem.

Major Uncertainties

The results of this study and its implications for national energy policy could be affected by these uncertainties.

1. The degree to which price will dampen demand: If demand is much less sensitive to price than assumed in the report, the U.S. will be much more vulnerable then we thought in 1977 and 1985 at the $11 price. One way to protect against this uncertainty is through mandatory energy conservation measures, which assure lower levels of growth then do high prices or by diversifying and accelerating domestic supply.

2. The ultimate production potential of frontier oil areas: Literally all the new oil production is forecast from frontier areas in Alaska, from the Atlantic OCS or from improved tertiary recovery techniques. Each of these requires further exploration or new technology. If these areas do not prove productive, domestic production will decline to less than 5 MMBD (instead of 9 MMBD) at $7 oil prices. Even at $11,or in the accelerated supply actions, total oil production could not exceed about 11 MMBD. If new oil production did not occur, then synthetic fuels, switching from petroleum and gas to coal, and mandatory conservation may be necessities.

3. The schedule on which new energy production can be delivered: Even with existing technology, delays could occur in investment decisions, construction, delivery of equipment, or licensing and regulation. Energy industry evaluation of expected world oil prices or uncertainties about domestic energy policy could all lead to additional delays and reduced domestic energy supplies.

Policy Implications

While the Blueprint does not present policy recommendations, there are a number of important policy implications which can be drawn from the results which are essential to the formulation of a rational and effective energy policy.

World Oil Prices and Domestic Policy. The domestic energy situation with current policies and the impact of the major energy strategies is dramatically affected by world oil prices. Yet, the prevailing price, is and likely will remain, uncertain for months, if not years. There are several important implications:

- While higher prices make achievement of domestic self-sufficiency much easier, the United States and the world will be better off with lower prices. Our economic growth and world financial stability more than offset any advantage to the U.S. from higher world oil prices.

- Many more actions--supply, demand and emergency measures--will be needed to reach the same level of vulnerability if $7, rather than $11, world oil prices prevail. Hence, with some likelihood of lower prices, as well as unexpected delays and problems in implementing any major programs, and risk aversion by industry, more programs may be needed to achieve a given level of imports than the forecasts might indicate.

- Because world prices are uncertain, domestic energy policy must balance the need to resolve uncertainties and stimulate needed, long-term energy investments. Unfortunately many programs to resolve price uncertainty also imply cumbersome, inefficient and inflexible programs such as tariffs or subsidies. A difficult tradeoff is involved.

- A domestic energy policy must be flexible and dynamic in dealing with uncertain world oil prices. New energy policies must weigh potential gains against the "costs of being wrong." Policies which cannot adjust to changing world oil prices might involve unacceptable domestic costs.

- Finally, our domestic policies are likely to influence world oil prices in a way which can thwart our domestic programs. Programs to increase domestic supplies at relatively high world oil prices will cut our imports and put downward pressure on the world oil price. This in turn may make some domestic investment uneconomic. Conversely, taking no domestic actions only increases the probability that high world oil prices will continue.

<u>Zero Imports and Domestic Vulnerability.</u> The preceding analysis leads to several major conclusions about the desirability of zero imports and our capability to reach that goal.

Over the next few years, it will be difficult to reduce or even maintain our current level of imports. Hence, our vulnerability may well increase. By 1985, however, just the opposite will be true. The United States has a tremendous range of alternatives which can reduce or eliminate our domestic vulnerability to import disruptions. While higher oil prices make reductions in vulnerability easier to achieve, even at $7 there are numerous programs which, if implemented, would reverse preembargo trends.

While zero imports is achievable, it is simply not warranted economically or politically. Some imports are from secure sources. Others are from insecure sources but they can be insured against through emergency demand curtailment measures or standby storage. In any event, zero imports is not synonymous with eliminating domestic vulnerability to import disruptions. Vulnerability should be a measure of an energy policy, not the level of imports per se. Further decreasing vulnerability can involve significant economic and environmental tradeoffs. These should be weighed concurrently.

It is clear from the analysis that at high world oil prices import vulnerability will almost take care of itself, but at lower prices our situation will worsen without new policy initiatives. While the United States does not need to implement every possible action to increase supply, reduce demand or implement emergency standby programs, some action will be necessary.

<u>Implications of Domestic Energy Strategies</u>. Each of the domestic strategies evaluated in the Blueprint will require major new initiatives and each has an important set of implications, uncertainties and drawbacks.

- Accelerating domestic supply is possible without major subsidies or guarantees, provided expectations for world oil prices are between $7 and $11. Because increased supply must rely on existing technologies through 1985, a supply strategy generally involves some additional environmental impacts, although much less than proportional to the growth in energy. These impacts either involve increased development in previously undisturbed areas, such as Alaska, the Atlantic OCS or the West, or increased problems associated with much greater coal use. Difficult environment/energy tradeoffs must be made. As important as the environmental trade-off is the question of regional development. At present, there is much debate over the advantages and disadvantages of increasing national energy supply from regions which generally do not need the energy for themselves. Resolution of the regional/national question is central to increasing domestic supplies. With respect to increased oil production, most increases must come from as yet unproven reserves. Resolving the uncertainty with respect to these areas is important in terms of gaining sufficient lead times to pursue other courses of action if it should become necessary to do so.

- Reducing the rate of growth in energy demand has several appealing aspects, including its positive environmental impact and the avoidance of potential constraints or bottlenecks, such as limited water supply, and materials and equipment shortages. It is also clear that some level of conservation beyond the price induced level can be achieved without significant economic impact. Conversely, there are conservation programs which could reduce economic growth, industrial output and our standard of living. To achieve conservation savings greater than those induced by price alone will require additional Federal intervention in the marketplace. This intervention, and the nonmarket costs of conservation, such as more limited individual choices and changed lifestyles, must be weighed against the positive environmental and energy impacts of a conservation strategy.

- Emergency storage programs appear cost effective in reducing the economic impact of future oil supply disruptions. However, this benefit must be weighed against several import limitations. It may take several years to design and implement a major storage program, yet our vulnerability is highest now and storage may be of little value five to ten years from now. Secondly, building storage will require greater imports now. Purchases in the world market for storage will tend to sustain higher prices in the shortrun and put additional strains on the international financial system. Finally, if we purchase storage now to avoid the risk of a large economic loss from an embargo, we also risk a possibly large capital loss if the oil we store drops dramatically in price.

CHAPTER 2

THE POTENTIAL FOR ENERGY CONSERVATION

A considerable number of projections, both private and governmental, are in substantial agreement as to energy demand for the next 15-20 years. There is a greater variation in projections as to how the demand can be met by supply--the principal differences having to do with timing of availability of nuclear power expansion, the extent to which coal will contribute and the extent to which domestic gas will contribute as well as associated environmental impacts. The latter source depends on the price of well head gas and the related uncertainty about the supply elasticity of gas.

There is substantial agreement that the shortfall in domestic fuel supply will be provided almost entirely by imported oil. While imported liquid natural gas (LNG) and synthetic natural gas (SNG) made from imported petroleum will contribute to the energy fuel supply, that contribution will be minor compared to total energy demand.

One current creditable estimate [1/] of required foreign oil in million b/d is:

1971	1975	1980	1985
3.7	6.5-8.4	9.2-11.6	13.8-16.5

The estimates of energy consumption [2/] in equivalent barrels of oil in million b/d are:

1971	1975	1980	1985
32.6	37.9	45.4	55.1

Studies are underway examining how this estimated demand for foreign oil can be reduced through increasing currently estimated U.S. supply of fuels. The same ultimate objective as sought by increasing supply can be achieved by measures to reduce demand--energy conservation. In addition, such a program should generally, although not always, contribute both to environmental objectives and to consumer objectives.

Energy conservation measures can reduce estimated energy demand in 1980 by as much as the equivalent of 7.3 million b/d of oil at an estimated annual value of $10.7 billion[3/]. (This estimate is to be compared with projected 1980 oil imports of 9.2-11.6 million b/d.)

The most significant additional energy conservation measures are the installation of improved insulation in both new and old homes and the use of more efficient air conditioners; a shift of intercity freight from trucks to rail, intercity passengers from air to rail and bus, and urban passengers from automobiles to motorized mass transit; and the introduction of more efficient industrial processes and equipment. By 1980, all suggested measures could reduce demand by the equivalent of 2.4 million b/d in the residential/commercial sector, 2.3 million b/d in the transportation sector, and 2.6 million b/d in the industrial sector.

The following charts, Figs. 2-1 to 2-3 summarize those conservation measures which could have high payoffs over the short-term (1972-1975), mid-term (1976-1980), and long-term (beyond 1980) either in terms of large BTU savings or in terms of public awareness and support of energy conservation. The economic, environmental, and socio-political impact of these measures will vary, as will the likelihood of their being implemented. Designations High, Medium, Low to describe these impacts are of course subjective but hopefully reflect a general consensus of opinion. The title "High Payoff Conservation Measures" does not necessarily indicate high payoff in terms of BTU savings but also high payoff in the sense of feasibility, public awareness and the potential for implementation.

As Figure 2-4 illustrates, the suggested measures can have a significant effect on future energy consumption. While the conservation measures proposed in this report will not, taken alone, eliminate the need for increased oil imports, they can substantially reduce this need. For example, if the long-term measures were adopted, the currently estimated required oil imports could be cut in half by 1990. [4/] Energy conservation can also delay the attainment of a given level of consumption and thus postpone requirements for increases in imports, thereby giving more time for desirable adjustments in energy supply which might contribute further to closing the currently estimated "energy gap." For example, projected 1980 petroleum consumption in the absence of conservation measures could be delayed until 1982 with the suggested short-term measures, until 1986 with the suggested mid-term measures, or until 1991 with the long-term measures.

Fig. 2-1.

HIGH PAYOFF CONSERVATION MEASURES

Short-Term (1972 - 1975)

Conservation Measures	Direct Payoff			Other Effects			Likelihood of Implementation	Lead Agencies
	Annual Savings Quadrillion BTU by 1975	Public		Economic Impact	Benefit to Environment	Social/ Political Impact		
		Awareness	Acceptability					
Transportation*								
1. Accelerate improvement of motorized mass transit including measures to improve traffic flow.	0.4	Low	Medium	Medium - High	High	High	Medium	DOT
2. Improve automobile energy efficiency through use of low loss tires, improved engine tuning.	0.9	High	High	Medium - High	High	High	Medium	DOT
3. Inject energy issue into appropriate national programs (environmental, health, urban reform, etc.). Initiate special conservation programs.	-	High	High	-	Medium	Low	High	Executive Office
4. Promulgate energy efficiency standards for transportation.	-	High	High	-	Medium	Low	Medium	DOT
Residential/Commercial								
1. Tax incentives for adding insulation and storm windows in existing homes.	0.1	High	High	Medium	High	High	Medium	IRS, HUD, Executive Office
2. Educational program to encourage good energy conservation practices in the home.	0.05	High	High	Low	High	High	High	OCA

* Transportation savings based on projections developed by Hirst, Energy Consumption of Transportation in the U. S., Oak Ridge National Laboratory ORNL-NSF-EP-15 March 1972. (Modified to reflect changes in energy efficiency.)

Fig. 2-1 (continued)

HIGH PAYOFF CONSERVATION MEASURES
Short-Term (1972-1975)

Conservation Measures	Direct Payoff			Other Effects			Likelihood of Implementation	Lead Agencies
	Annual Savings Quadrillion BTU by 1975	Public		Economic Impact	Benefit to Environment	Social/ Political Impact		
		Awareness	Acceptability					
Industry								
1. Raise energy prices by tax and/or regulation.	1.6	Medium	Low	Low	High	Medium	Medium	Treasury, FPC Executive Office
2. Increase recycling and reuse of materials and products.	0.2	High	High	Medium	High	Medium	High	EPA, Bureau of Mines
Electric Utilities*								
1. Alleviate construction delays. **	0.5	Low	Medium	Low	Medium	Low	Medium	FPC, AEC, EPA, Department of Labor, Executive Office
2. Smooth out daily demand cycle by shifting some loads to off-peak hours.	0.5	Medium	Low	Low	Medium	Medium	Low	FPC
Environment								
1. Review regulations and programs with objective of meeting environmental standards while using least energy and avoiding scarce fuels	-	-	-	-	-	-	-	-

* The savings in the electric utility sector have already been assumed in the based projections of energy consumption in this report. (See Section II.)

** New plants can be brought on line faster to replace older, less efficient plants. (See Appendix F for further explanation.)

Fig. 2-2

HIGH PAYOFF CONSERVATION MEASURES
Mid-Term (1976 - 1980)

Conservation Measures	Direct Payoff			Other Effects				
	Annual Savings Quadrillion BTU by 1980	Public						
		Awareness	Acceptability	Economic Impact	Benefit to Environment	Social/ Political Impact	Likelihood of Implementation	Lead Agencies
Transportation *								
1. Expand intercity surface transportation service.	0.9	Low	Medium	Medium-High	High	High	Medium	DOT
2. Expand high speed and motorized mass transit service. Implement feeder service (e.g., "dial-a-bus").	0.3	Low	Medium	Medium-High	High	High	Medium	DOT
3. Improve freight handling systems through freight consolidation and containerization.	0.9	Low	High	Low	Medium	Low	Medium	DOT
4. Emphasize transportation issue in urban development (pedestrian oriented clusters).	0.2	High	High	Medium-High	High	High	Low	DOT, HUD, Executive Office
5. Improve automobile energy efficiency through improved engines and drive trains, improved traffic flow, use of low loss tires and improved engine tuning.	0.3	High	Medium	High	High	Medium	Medium	DOT
Residential/Commercial								
1. Further upgrade FHA minimum property standard for new single and multi-family dwellings to require more insulation.	0.9	Medium	Medium	Medium	High	Medium	Medium	FHA

* Transportation savings based on projections developed by Hirst, Energy Consumption of Transportation in the U. S., Oak Ridge National Laboratory ORNL-NSF-EP-15 March 1972. (Modified to reflect changes in energy efficiency.)

Fig. 2-2 (continued)

HIGH PAYOFF CONSERVATION MEASURES
Mid-Term (1976 - 1980)

Conservation Measures	Direct Payoff			Other Effects			Likelihood of Implementation	Lead Agencies
	Annual Savings Quadrillion BTU by 1980	Public		Economic Impact	Benefit to Environment	Social/ Political Impact		
		Awareness	Acceptability					
Residential/Commercial (Continued)								
2. Increase price of fuel by tax levied at the point of production.	Large	High	Low	High	High		Medium	IRS
3. Utilize above revenues for R&D to increase efficiency of energy utilization in the residential/commercial sector.	Large	Medium	High	Low	High	High	Medium	HUD, GSA, DOC, EPA, Executive Office
4. Establish minimum efficiency standards for furnaces, air conditioners, and appliances.	0.6	Low	High	Medium	Medium	High	High	GSA, DOC, HUD, Executive Office
5. Require energy consumption of all appliances to be stated on nameplate, price tag, and in any advertisement that quotes a price.	0.6	High	High	Low	Medium	Medium	High	DOC
Industry								
1. Raise energy prices by tax and/or regulation.	3.8	Medium	Low	Low	High	Medium	Medium	FPC

Fig. 2-2 (continued)

HIGH PAYOFF CONSERVATION MEASURES
Mid-Term (1976 - 1980)

Conservation Measures	Direct Payoff			Other Effects			Likelihood of Implementation	Lead Agencies
	Annual Savings Quadrillion BTU by 1980	Public		Economic Impact	Benefit to Environment	Social/ Political Impact		
		Awareness	Acceptability					
Industry (Continued)								
2. Increase recycling and reuse of materials and products.	0.4	High	High	Medium	High	Medium	High	EPA, Bureau of Mines, Executive Office
Electric Utilities *								
1. Alleviate construction delays. **	0.5	Low	Medium	Low	Medium	Low	Medium	FPC, AEC, EPA, Department of Labor, Exec. Off., FPC
2. Smooth out daily demand cycle by shifting some loads to off-peak hours.	0.5	Medium	Low	Low	Medium	Medium	Low	
3. Increase research and development efforts.	Large	Medium	High	Low	High	High	High	FPC, AEC, NSF, OST, EPA, Electric Research Council (Industry Council), Exec. Office
Environment								
1. Review regulations and programs with objective of meeting environmental standards while using least energy and avoiding scarce fuels.	-	-	-	-	-	-	-	-

* The savings in the electric utility sector have already been assumed in the basic projections of energy consumption in this report. (See Section II.)

** New plants can be brought on line faster to replace older, less efficient plants. (See Appendix F for further explanation.)

Fig. 2-3

HIGH PAYOFF CONSERVATION MEASURES
Long-Term (beyond 1980)

Conservation Measures	Direct Payoff			Other Effects			Likelihood of Implementation	Lead Agencies
	Annual Savings Quadrillion BTU by 1990	Public		Economic Impact	Benefit to Environment	Social/ Political Impact		
		Awareness	Acceptability					
Transportation*								
1. New freight handling systems.	2.0	Low	High	Medium	High	High	Medium	DOT
2. New mass transit systems.	2.0	Low	Medium	Medium-High	High	High	Medium	DOT
3. Improved urban design.	2.0	Low	High	High	High	High	Medium	DOT, HUD, Executive Office
4. New engines (hybrid, non-petroleum).	1.0	Low	High	Low-Medium	High	High	High	DOT
Residential/Commercial								
1. Develop non-fuel energy sources (solar and wind energy).	Large	High	High	High	High	High	Medium	AEC, NASA, DOC, Executive Office
Industry								
1. Raise energy prices by tax and/or regulation	5.5	Medium	Low	Low	High	Medium	Medium	Treasury/FPC, Executive Office
2. Increase recycling and reuse of materials and products.	1.1	High	High	Medium	High	Medium	High	EPA, Bureau of Mines, Executive Office
Electric Utilities**								
1. Increase research and development efforts	Large	Medium	High	Low	High	High	High	FPC, AEC, NSF, OST, EPA, Electric Research Council (Industry Council, Executive Office)
Environment								
1. Review regulations and programs with objective of meeting environmental standards while using least energy and avoiding scarce fuels	-	-	-	-	-	-	-	-

* Transportation savings based on projections developed by Hirst, Energy Consumption of Transportation in the U. S. Oak Ridge National Laboratory ORNL-NSF-EP-15 March 1972. (Modified to reflect changes in energy efficiency.)

** The savings in the electric utility sector have already been assumed in the basic projections of energy consumption in this report. (See Section II.)

Figure 2-4

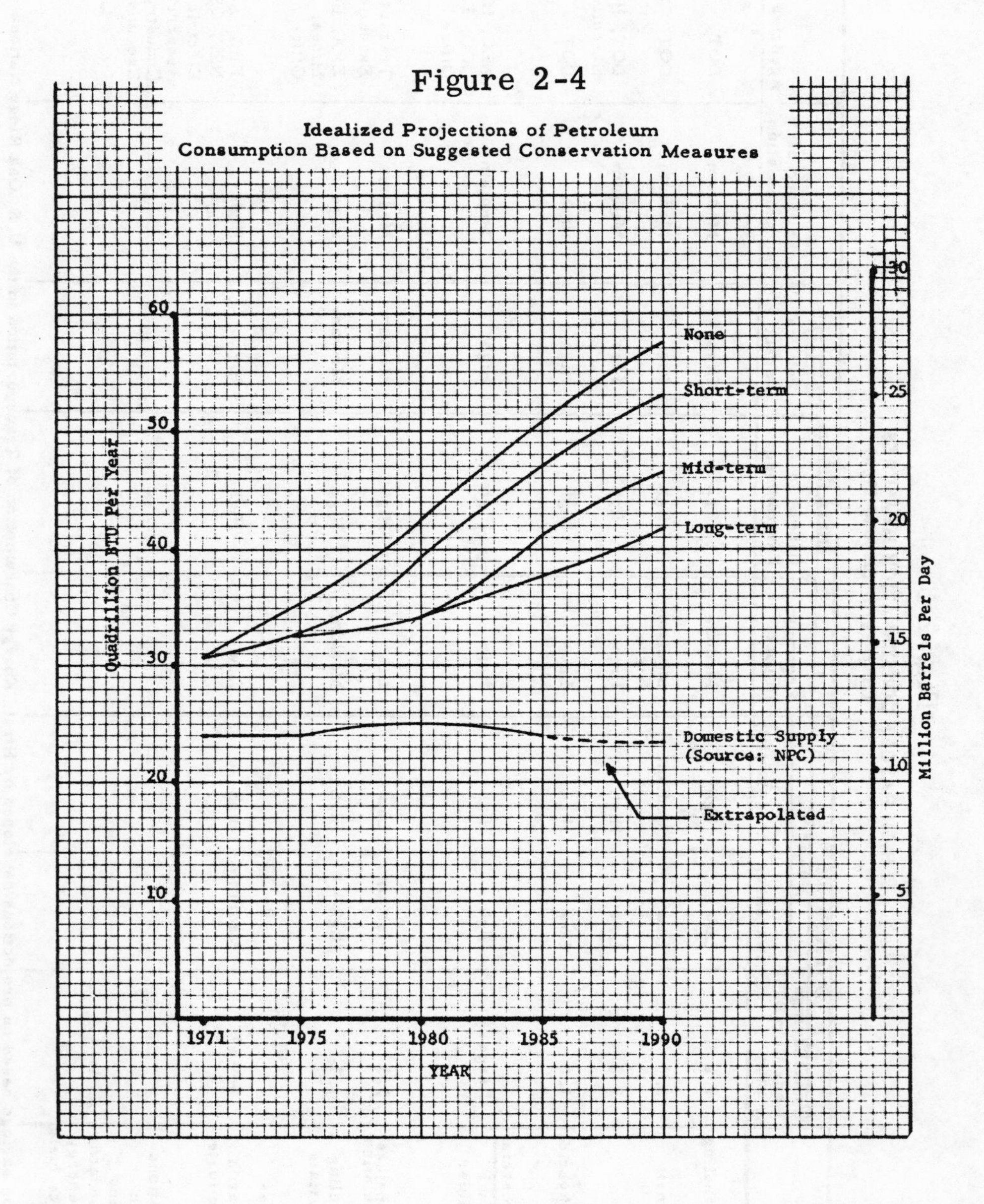
Idealized Projections of Petroleum
Consumption Based on Suggested Conservation Measures
Quadrillion BTU Per Year
60
50
40
30
20
10
None
Short-term
Mid-term
Long-term
Domestic Supply
(Source: NPC)
Extrapolated
1971
1975
1980
1985
1990
YEAR
Million Barrels Per Day
30
25
20
15
10
5

PATTERNS OF ENERGY SUPPLY AND DEMAND

Table 2-1 presents U. S. energy supply and demand by source and consuming sector for 1971 and projections in five-year intervals from 1975 to 1990. The projections were made on a "most probable" basis by the Department of the Interior. 1/ The major projected change between now and 1990 is that nuclear power will significantly increase its contribution to energy sources, largely at the expense of natural gas.

Total gross energy input is projected to increase by 39 percent from 1971 to 1980 (an average compounded increase of 3. 8 percent per year) and by 46 percent from 1980 to 1990 (an average increase of 3. 9 percent per year). At present, fossil fuels account for 95 percent, nuclear power for 0. 6 percent and hydropower for 4 percent of gross energy inputs. The projected percentages for 1980 are 89 percent, 7 percent, and 4 percent respectively. The differences are accounted for by fossil fuel growth rates which are slightly smaller than average and by a nuclear power growth rate (37 percent per year) which is much larger than average. For 1990, the projected percentages are 80 percent for fossil fuels, 16 percent for nuclear power and 3 percent for hydropower. The differences are accounted for by a natural gas growth rate (1. 3 percent per year) which is smaller than average and a nuclear power growth rate (13 percent per year) which is much larger than average.

The major projected change between now and 1990 in the consuming sectors is a tripling in the energy devoted to electrical generation. 2/ At present, industrial use accounts for 29 percent, household and commercial use for 21 percent and transportation and electrical generation for 25 percent each of total energy consumption. By 1980, this breakdown is expected to change only slightly, the major change being an increase to 31 percent of total energy consumption for

Table 2-1

United States total gross consumption of energy resources by major sources and consuming sectors, 1971 and projected to 1975, 1980, 1985, and 1990 on a "Most Probable" basis

Year	Consuming Sector	Coal 1/ million short tons	Petroleum 2/ million barrels	Natural Gas billion cubic feet	Nuclear power million kilowatt-hours	Hydropower million kilowatt-hours
1971	Household and Commercial--	14.6	1,149.6	7,125.0		
	Industrial----------------	164.3	982.0	10,125.0		
	Transportation------------	0.3	3,004.9	800.0		
	Electrical Generation-----	331.6	386.9	4,000.0	37,899	266,320
	Synthetic Gas-------------	-	-	-	-	-
	Total---------------	510.8	5,523.4	22,050.0	37,899	266,320
1975	Household and Commercial--	12	1,221	8,400.0		
	Industrial----------------	169	1,186	11,387.0		
	Transportation------------	-	3,360	989.3		
	Electrical Generation-----	384	573	3,685.7	240,150	350,000
	Synthetic gas-------------	-	-	-	-	-
	Total---------------	565	6,340	24,462.0	240,150	350,000
1980	Household and Commercial--	11	1,356	9,195		
	Industrial----------------	175	1,383	12,124		
	Transportation------------	-	3,992	1,358		
	Electrical Generation-----	460	800	3,492	630,300	420,000
	Synthetic gas-------------	19	84	-		
	Total---------------	665	7,615	26,169	630,300	420,000
1985	Household and Commercial--	4	1,546	9,758		
	Industrial----------------	190	1,663	12,842		
	Transportation------------		4,739	1,591		
	Electrical Generation-----	613	1,064	3,346	1,130,000	469,600
	Synthetic Gas-------------	86	128	-	-	-
	Total---------------	893	9,140	27,537	1,130,000	469,600
1990	Household and Commercial---	-	1,680	10,200	-	-
	Industrial----------------	236	1,950	14,300	-	-
	Transportation------------	-	5,660	1,740	-	-
	Electrical Generation------	661	1,010	3,100	2,430,000	540,000
	Synthetic Gas-------------	133	120	-	-	-
	Total----------------	1,030	10,420	29,340	2,430,000	540,000

1/ Includes anthracite, bituminous, and lignite coals.
2/ Petroleum products refined and processed from crude oil, including still gas, liquefied refinery gas, and natural gas liquids.

Source: Department of Interior

electrical generations. (This represents an average compounded increase of 6.2 percent per year.) This trend is expected to continue from 1980 to 1990, with electrical generation consumption increasing to 38 percent of total energy consumption (an average increase of 5.9 percent per year). Transportation is expected to be the second fastest growing sector, with projected increases of 35 percent from 1971 to 1980 (an average of 3.4 percent per year) and 41 percent from 1980 to 1990 (an average of 3.5 percent per year). For the entire time period 1971-1990, industrial use is expected to increase by 53 percent (2.3 percent per year) and household and commercial use by 41 percent (1.8 percent per year). Synthetic gas is not expected to be a major factor, with a projected use of only 3 percent by 1990.

ENERGY CONSERVATION AND ENVIRONMENTAL POLLUTION

In nearly all cases the extraction of fuel from the earth and of energy from fuels has significant environmental impact. Current and future mining practices (both underground and surface) should seek to limit environmental damages to surface areas and water resources that occurred in some past mining activities. Disposal of heat from nuclear powered generating plants needs to be accomplished in such a way as to avoid thermal pollution of rivers and lakes.

Federal and state pollution control regulations are likely to result in the additional use of energy and increase the demand for environmentally clean fuels. For example, the combined effect of air quality control plans required of the states under the Clean Air Act might require importing as much as an additional 1 billion barrels of low sulfur oil for use mainly in electric power generation in 1975. The Clean Air Act also requires a major reduction in automobile pollution which, when fully implemented in 1976, will introduce a fuel penalty of 5 percent or more for each technique needed to control different emissions. A fully equipped car (1976 or later) will experience a (15-30) percent fuel penalty (over a comparable 1967 automobile) with the best estimate being 15 percent at this time.

Seen in the context of an increased standard of living (new home appliances, vacation homes, air conditioning, travel) the extra fuel seems inexpensive. From a conservation point of view, however, we must begin somewhere to conserve scarce fuel supplies.

The close interdependence of these issues points to the need to develop programs which will meet the objectives of both pollution control and energy conservation. These programs should consist of measures to (1) make pollution control systems more energy efficient and (2) encourage greater energy conservation by choosing applications which will both reduce energy consumption and pollution.

Efficient Pollution Control Systems

To meet environmental quality standards industry invested some $9.3 billion in pollution control in 1970. This annual level is expected to double by 1975 (Figure 2-5). As a percentage of total annual capital expenditures, this ranged up to 10 percent for the iron and steel industries, 2.6 percent for transportation, and 3.8 percent for electric utilities (Table 2-2). However, some of these programs may have the effect of increasing energy consumption unnecessarily. It is not clear that all these investments produce energy efficient pollution control systems. For firms whose energy costs are significant with respect to profits, efficient pollution controls probably have been installed. For other firms, Federal standards may have to be devised.

Relating traditional profit incentives to efficient pollution control will be difficult until the principle of pollution control itself has

become common practice. The capital investment and the institutional policies (e.g., environmental protection and licensing policies) take substantial time to implement on a regional or a national basis.

Figure 2-5
GROWTH IN ANNUAL POLLUTION CONTROL COSTS
1970 and 1975
(BILLIONS OF DOLLARS)

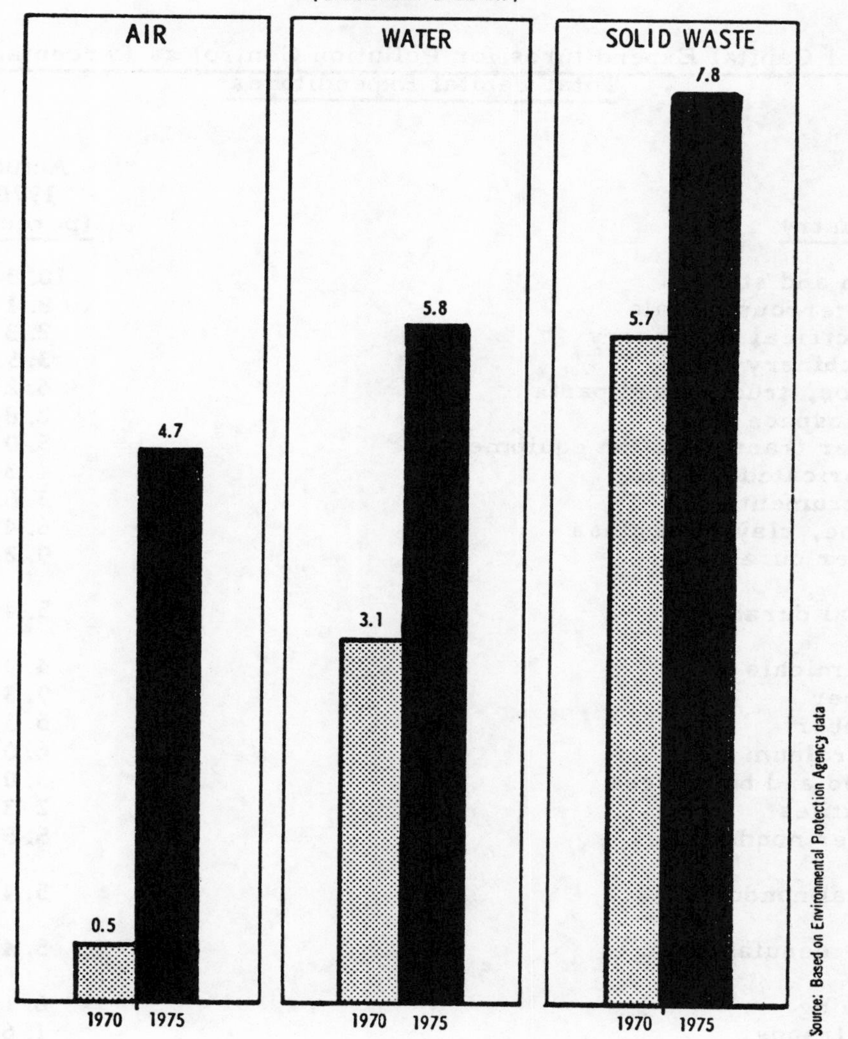

In some cases energy conservation is the fortuitous result of investment in other programs. For example urban mass transit systems, which are promoted to shorten commuter time and reduce highway costs, also produce several times less pollutants per passenger

mile than the automobile and reduce energy consumption as well. But the multiple benefits promised by urban mass transit will remain limited pending the availability of sufficient capital to build and modernize.

Table 2-2

Total Capital Expenditures for Pollution Control as Percentage of Total Capital Expenditures

Industry	Actual 1970 (percent)
Iron and steel	10.3
Nonferrous metals	8.1
Electrical machinery	2.3
Machinery	3.5
Autos, trucks, and parts	4.2
Aerospace	2.8
Other transportation equipment	5.0
Fabricated metals	4.3
Instruments	3.6
Stone, clay, and glass	6.4
Other durables	9.2
Total durables	5.4
Chemicals	4.9
Paper	9.3
Rubber	5.3
Petroleum	6.0
Food and beverages	3.0
Textiles	2.3
Other nondurables	5.5
Total nondurables	5.4
All manufacturing	5.4
Mining	6.1
Railroads	1.6
Airlines	0.7
Other transportation	0.3
Communications	0
Electric utilities	3.8
Gas utilities	4.4
Commercial	0.6
All business	3.1

Energy Conservation Research

There is presently projected an inadequate supply of fuels of all kinds to meet the aggregate demand which the States' plans for pollution control will generate. The Clean Air Act provides for the setting of national primary (health) and secondary (welfare) ambient air quality standards. The fifty States have submitted implementation plans (SIP's), which if adopted and carried out, would lead to deficits summarized in Table 2-3.

Research on energy conservation is underway (EPA) to refine 5 and 10 year projections of demand. Best estimates now are that the additional need created by the SIP's for low sulfur imported oil will be greater than 150 million and less than 1,070 million barrels in 1975.

The two air pollutants which mainly are the subject of air pollution controls are sulfur oxides and nitrogen oxides. About 47 percent of NO_x emissions are caused by transportation sources. It has been estimated that the application of pollution control devices to automobiles will result in a fuel penalty of 15-30 percent in 1976 model cars. The impact on the fuels market of NO_x control builds up to about 2.5 percent of 1980 U. S. annual petroleum consumption or 200 million barrels of petroleum.

The primary means of meeting SO_x regulations rests with control at stationary sources because 75 percent of U. S. totals of SO_x come from Industry, Commercial/Residential, and Utility emissions. Low sulfur fuels are the primary means of controlling SO_x because proven technologies for SO_x removal are not yet available.

The State Implementation Plans have been analyzed for their impact on fuel supplies in Table 2-3. The impact of the fuel

Table 2-3

SUMMARY OF AVAILABILITY AND REQUIREMENTS STATIONARY SOURCES

	1970	1975					1977				
	CONSUMPTION	REQUIRED AT SIP	AVAILABLE AT SIP[1]	DEFICIT AT SIP	AVAILABLE AT ALL SULFUR LEVELS	HIGH SULFUR SURPLUS (ABOVE SIP)	REQUIRED AT SIP	AVAILABLE AT SIP[1]	DEFICIT AT SIP	AVAILABLE AT ALL SULFUR LEVELS	HIGH SULFUR SURPLUS (ABOVE SIP)
COAL (10^6 tons)	424	592[2]	268[3]	324	586[3]	318	615	290	325	675	385
OIL (10^6 barrels)	1,248	1,439	1,048	391	1,726	678	1,505	1,144	361	1,945	801
GAS[4] (10^{12} ft^3)	21.4	25.2	24.8	0.4	24.8	0[5]	25.7	24.3	1.4	24.3	0[5]

[1]Includes reasonable blending and transfer.

[2]Excludes power plant retirements estimated at 7,000 Mw of installed capacity.

[3]Includes 25 x 10^6 tons currently under development, in eastern slope Rocky Mountain states.

[4]Assumes no switching to gas from other fuels (see discussion in Section III, Approach).

[5]Sulfur content in gas is negligible.

deficits (at SIP) can be greatly reduced by tailoring the SIP's in selected regions of the U. S.

Imported low-sulfur oil is projected over a series of strategies for the years 1975 and 1977, (additional annual projections are being developed), ranging from 128 million barrels to 1,070 million barrels.

Conclusion

In view of the need for time (e. g., five years) to achieve widespread acceptance and implementation of pollution control practices and the lack of clearly defined incentives for earlier action (e. g., availability of cheap capital to industry, to utilities) to achieve energy conservation in pollution control, the best near term measure is careful planning for the future.

All proposed actions which have an effect on energy consumption, including environmental studies and programs, should entail consideration of their energy costs (and benefits). An explicit treatment of this issue should provide decision-makers (business, government, the public) with the information necessary to make the tradeoffs which we all must get accustomed to thinking about and acting on. For example, urban mass transit, in practice, would reduce automobile consumption of scarce fuel as well as reduce air pollution where it is greatest--in the cities. The costs and benefits of urban mass transit should be reassessed for leverage on pollution control.

Environmental programs do not offer attractive possibilities for significant energy savings in the next five years. They do possess a mechanism for public examination of the effect on energy consumption of energy conservation measures. They should include that examination, particularly where program analyses show a potential increase in energy consumption and/or a particular impact on the more scarce fuels such as gas.

In summary, even as environmental considerations now enter into every energy program, the national interest requires that broad environmental programs be centrally concerned with energy efficiency.

REFERENCES

The publications referred to in chapter 2 may be found in "The Potential for Energy Conservation-A Staff Study" by the Executive Office of the President-Office of Emergency Preparedness

Excerpt From "PROJECT INDEPENDENCE"

DOMESTIC ENERGY SUPPLY
SUMMARY

To assess the domestic energy situation between now and 1985, we have to analyze the effects of existing policies upon future supplies and identify alternative actions to increase domestic production. FEA's evaluation of the cost and levels of potential production was undertaken by nine separate interagency fuel task forces which evaluated:

1. Oil
2. Natural Gas
3. Coal
4. Nuclear
5. Shale Oil
6. Synthetic Fuels
7. Solar
8. Geothermal
9. Facilities

The final reports of these individual task forces are available separately from the Government Printing Office.

Methodology and Assumptions

The fuel task forces developed estimates of potential production levels for each fuel, as a function of price. These figures were prepared for 1977, 1980, and 1985, taking account of production leadtimes and institutional factors which could affect the rate of growth. Data were compiled for the major producing regions of the country, which are different for each fuel, as well as for the Nation as a whole. Thus, the task forces analyzed nine coal regions, twelve oil and gas regions, and nine electric reliability councils. The regional approach highlights differing production costs, potential recoverable reserves, finding rates, transportation capacities and costs, and provides a basis for analysis of environmental impacts and production constraints.

Energy production is largely determined by an array of Government policies and by economic conditions. The fuel task forces evaluated potential porduction under a Business As Usual (BAU) case and Accelerated Development (AD) case. The Business As Usual strategy assumes, to the extent possible, a continuation of existing policies and no new actions to stimulate supply or to remove barriers that limit production. It assumes there will be no changes in current tax policies; phasing out of current price and allocation; programs during 1975; implementation of current environmental regulations; and continuation of the $11 billion energy research and development program. The Accelerated Development strategy assumed the implementation of incentives or other programs to stimulate supply and the relaxation of selected key barriers that inhibit production. It should be noted that many of the options considered under the AD strategy would require Congressional approval.

The major differences between the BAU and AD strategies are with regard to degree of Government intervention, rate of leasing, regulatory controls, and relaxation of institutional barriers. A comparison of major BAU and AD assumptions is shown in Table 2-4 . More detailed descriptions of assumptions are contained in the Task Force reports.

Table 2-4.
Comparison of BAU and AD Assumptions

Energy Source	Business As Usual Assumptions	Accelerated Development Assumptions
Oil	Moderate OCS leasing program (1-3 million acres per year); Prudhoe Bay developed with one pipeline	Accelerated OCS leasing program, including Atlantic and Gulf of Alaska; expanded Alaskan program assuming additional pipeline and authority to develop Naval Petroleum Reserve No. 4
Natural Gas	Phased deregulation of new natural gas; LNG facilities in Alaska	Deregulation of new natural gas; additional gas pipelines in Alaska; gas produced in tight formations
Coal	Some Federal coal land leasing; phased implementation of Clean Air Act with installation of effective stack gas control equipment; moderate strip mining legislation	Same as BAU with additional leasing and larger new mines
Nuclear	No change in licensing or regulations; added enrichment and reprocessing capability	Streamlined siting and licensing to reduce leadtimes; increased reliability; additional uranium availability; material allocation
Synthetic Fuels	No change from current policies	Streamlined licensing and siting; financial incentives; increased water availability
Shale Oil	No change from existing policies	Additional leasing of Federal lands; modification of Colorado air quality standards; financial incentives; increased water availability
Geothermal	Continued R&D and Federal leasing programs	Leasing of Federal lands; streamlined licensing and regulatory procedures; financial incentives
Solar	Continued R&D program	Additional R&D expenditures and financial incentives

Reserves

The United States has abundant coal reserves. At current levels of usage, our coal reserves could last about 800 years, well beyond the availability of other fossil fuels (see Table 2-5 for a comparison of reserves for different fuels). Although about 90 percent of the strip mineable coal can be recovered, only about half of the underground coal reserves can be recovered with current mining technology. Coal production has traditionally been concentrated in Appalachia, but most of the Nation's reserves are in the Midwest and Northern Great Plains.

Oil and gas reserves are considerably more limited. Even with extensions, revision, and discovery of new pools in known fields, proven reserves are less than 46 billion barrels, or less than 10 years of domestic supply (includes Prudhoe Bay, but not Naval Petroleum Reserve No. 4). The greatest potential for new discoveries lies in the offshore areas (OCS) and in Alaska. Many of the areas of greatest promise are publicly owned and to date have not been explored. For example, of the 70,000 square miles on the North Slope of Alaska that are potentially favorably-producing areas all but 27,000 square miles are reserved in the Arctic National Wildlife Range and Naval Petroleum Reserve No. 4 (NPR-4). Natural gas reserves are even more concentrated in Alaska, which has about one-third of the Nation's gas potential. The high oil and gas potential in these largely unexplored areas underscores the uncertainty of future levels of production.

Table 2-5
Proven Reserves

Source	Fuel Units	Quadrillion Btu's	Years Left at 1972 Consumption Levels
Coal			
high sulfur (more than 1%)	273 billion tons	6908	
low sulfur (less than 1%)	160 billion tons	3838	
TOTAL	433 billion tons	10746	823
Oil			
lower 48 (crude)	30 billion barrels	176	
natural gas liquids	6 billion barrels	37	
Alaska	10 billion barrels	59	
TOTAL	46 billion barrels	272	8
Gas			
lower 48	218 TCF	225	
Alaska	32 TCF	32	
TOTAL		257	11
Shale	20-170 billion barrels	116-986	3-28
Tar Sands	29 billion barrels	168	28

In addition to the oil in conventional fields, there are a large number of shallow oil fields containing oil-saturated sand reserviors; deposits, primarily in Utah, of oil-impregnated rocks, known as tar sands; shale oil; and heavy crude oil reserves, mainly in California. The ability to produce oil from these areas is heavily dependent upon the price of world oil and the environmental concerns of local residents.

Major Findings

Actual production levels will be affected by demand for various fuels; interfuel substitution; the price of world oil and access to imports; availability of materials, equipment, water, capital, manpower, and transportation; and Government actions. The fuel task forces assumed there were no constraints to production and did not take account of expected demand. These potential levels of production are combined with the FEA demand forecast and transportation cost estimates to develop a least cost way of producing and delivering energy to demand centers.

Under the Business As Usual strategy, and if world oil is $11 a barrel. oil production could increase to 15 million barrels per day (MMBD). Under Accelerated Development at $11 oil, production could increase to 20 MMBD The AD case assumes large-scale development in NPR-4; the Atlantic, Alaskan, and Pacific OCS; and a more extensive leasing program. Although the potential exists for 20 MMBD production, actual production levels will probably be lower. There is considerable uncertainty regarding NPR-4, Alaska, and the Atlantic OCS, and it is unlikely that all institutional and equipment barriers could be overcome.

The potential for coal development is virtually unlimited under accelerated conditions if no equipment, manpower, or demand constraints are assumed. Coal production could be over 2 billion tons per year in 1985 under these assumptions, although demand limitations are likely to keep production to about 1 billion

Table 2-6
Maximum 1985 Production Levels Under BAU and AD
(at $11 Oil)

Source	BAU Potential	AD Potential
Oil	15.0 million barrels/day	20.0 million barrels/day
Natural Gas	23.4 trillion cubic feet/year	29.3 trillion cubic feet/year
Coal	1.1 billion tons/year	2.1 billion tons/year
Nuclear	234 million kilowatts	275 million kilowatts
Coal Gasification	0.5 trillion cubic feet/year	1.0 trillion cubic feet/year
Coal Liquefaction	-0-	500,000 barrels/day
Shale Oil	250,000 barrels/day	1.0 million barrels/day
Geothermal	6000 megawatts	15,000 megawatts
Solar Heating & Cooling	0.3 quadrillion Btu's	0.6 quadrillion Btu's
Solar Electricity	41 million MWh/yr	151 million MWh/yr

tons in 1985. Natural gas production can be increased through increasing availability. Synthetic fuels, geothermal, solar, and shale oil levels are considerably higher under AD, although still relatively small contributors to total energy production before 1985. Nuclear power also shows large increases under AD as regulatory delays are reduced through streamlined siting and licensing practices. (See Table 2-6 for BAU and AD production levels.)

The variation in world oil price has a significant effect on potential production levels of source energy sources. Obviously, oil is very sensitive to imported prices. Production under AD conditions could be 20 MMBD in 1985 at $11 oil, but would be 16.9 MMBD at $7 oil. The major reason for these differences is that some sources that are economic at the higher prices are not economic when $7 a barrel imports are available. These sources, such as secondary and tertiary recovery, some Alaskan oil, and some heavy crude oil are produced at greater rates in the AD case. At $4 oil, domestic production would decline considerably, as most of the Alaskan production would not be economic.

The variation of world oil prices has virtually no effect on the supply of coal. The supply curve for coal is basically inelastic in the $7-11 oil price range. In the long term, after any immediate constraints have been relieved, coal can be produced in greater quantities than it can be consumed. The supply of nuclear power is relatively inelastic between $7 and $11 oil, as its availability depends more upon capital requirements, regulatory and delivery delays, perceptions about safety, and electricity demand, than on the price of oil. Nuclear power has the cheapest life cycle cost for base load electric lower generation and, along with coal, could substitute for some petroleum demand. Natural gas is generally cheaper to produce and transport than oil and is less sensitive to varying oil prices.

Considerations for Policy Development

The Accelerated Supply Case reflects substantial increases in the rate of production of several energy sources by 1985, including nuclear power and oil from Alaska and the Outer Continental Shelf. The increases are premised on Government action (Federal, state and local) and industry initiatives that are unlikely to take place unless a number of policy problems can be solved, most of which involve reforms in the regulatory process or the reconcilation of conflicting economic, environmental and social interests. This portion of the chapter highlights several of the most important policy problems.

Energy Facilities. A pervasive problem affecting attainment of the supply increases in the Accelerated Supply Case is the complexity and inconsistency of siting, licensing and regulatory procedures. Leadtimes for new facilities have increased beyond the normal engineering and construction delays. It takes eight or nine years to start up a nuclear plant from the time the decision is made to build the plant.

CHAPTER 3

ENERGY PROGRAM PLANNING

ADVANCED TECHNOLOGY NEEDS FOR FUEL RECOVERY AND UTILIZATION

Introduction

Within a relatively short time and in an unexpected manner, several circumstances are coincidentally occurring that are increasing the uncertainty of meeting future energy requirements in the United States. Much of the dilemma centers around the need for substantially increased environmental purification and protection. The following factors are intended to briefly summarize the situation:

(1) Clean air, particularly with respect to sulfur dioxide content, is the primary incentive for using low sulfur fuels. Use of gas and low sulfur fuel oils has begun to accelerate. Meanwhile, part of the abundant coal reserves has been limited in use because of high sulfur contents and the lack of economic desulfurization technology.

(2) The inability to obtain approval for new plant sites is causing delays in refinery planning and construction.

(3) Environmental concerns regarding the land and the sea are delaying the use of domestic energy resources.

(4) Oil production rate in the lower 48 states is projected to begin declining within the decade. Recently it has become clear that forecasts of the supply of natural gas and associated liquids have been overestimated.

(5) The need for substantially greater quantities of imported oils, combined with the difficulty in obtaining plant sites and oil import allocations, is forcing the export of important refinery capacity to foreign countries.

(6) Progress in the development of technology to use U.S. energy resources such as coal, oil shale, and uranium (breeder reactor and U-238 isotope conversion), may not occur in time to allow commercialization in quantities required to avert a serious energy shortage.

These events are producing a great deal of uncertainty and a need for careful planning that shows comprehension of the entire social and economic structure of the United States. The future security of the United States will depend on the rapid implementation of new energy developments and the continued innovative search for new ideas.

The U.S. Energy Situation, 1970-1980

Before examining the U.S. energy balance, a brief review of energy sources and their interrelationships will be useful in obtaining an overall view.

Petroleum

The United States is faced with an increasing shortage of domestic petroleum. Estimates of known oil for 1971 in the United States (including the North Slope of Alaska) may be categorized as follows:

	Billions of Barrels	Percent of Total
Cumulative past production	96	22.6%
Recoverable reserves	38	9.0
Potentially recoverable oil	61	14.4
Unrecoverable oil	230	54.0
Total oil discovered	425	100.0%

Crude oil recovery efficiency has increased over the last two decades despite decreasing crude oil prices, as shown in Figure3-1.

The lower 48 states' production is expected to begin a decline within this decade. Alaskan oil will provide an important addition to domestic production, but it will not be enough. U.S. imports of foreign oil are expected to rise as shown below:

	1970		1980	
	Thousands of B/cd	Percent	Thousands of B/cd	Percent
Domestic production	11,312	76.8%	12,780	50.5%
Imports	3,418	23.2	12,520	49.5
Total supply	14,730	100.0%	25,300	100.0%

Refinery processing schemes are expected to change considerably. The removal of lead from gasoline will require the production of higher clear octane number gasolines. Refineries are expected to be producing considerably greater amounts of low-sulfur fuel oils to balance the energy shortfall caused by insufficient gas. The gas oil portion of fuel oil will command a premium value, because it is relatively easy to desulfurize compared with the residuum. Future refineries also will be more complex because of the requirements for octane improvement and fuel oil desulfurization.

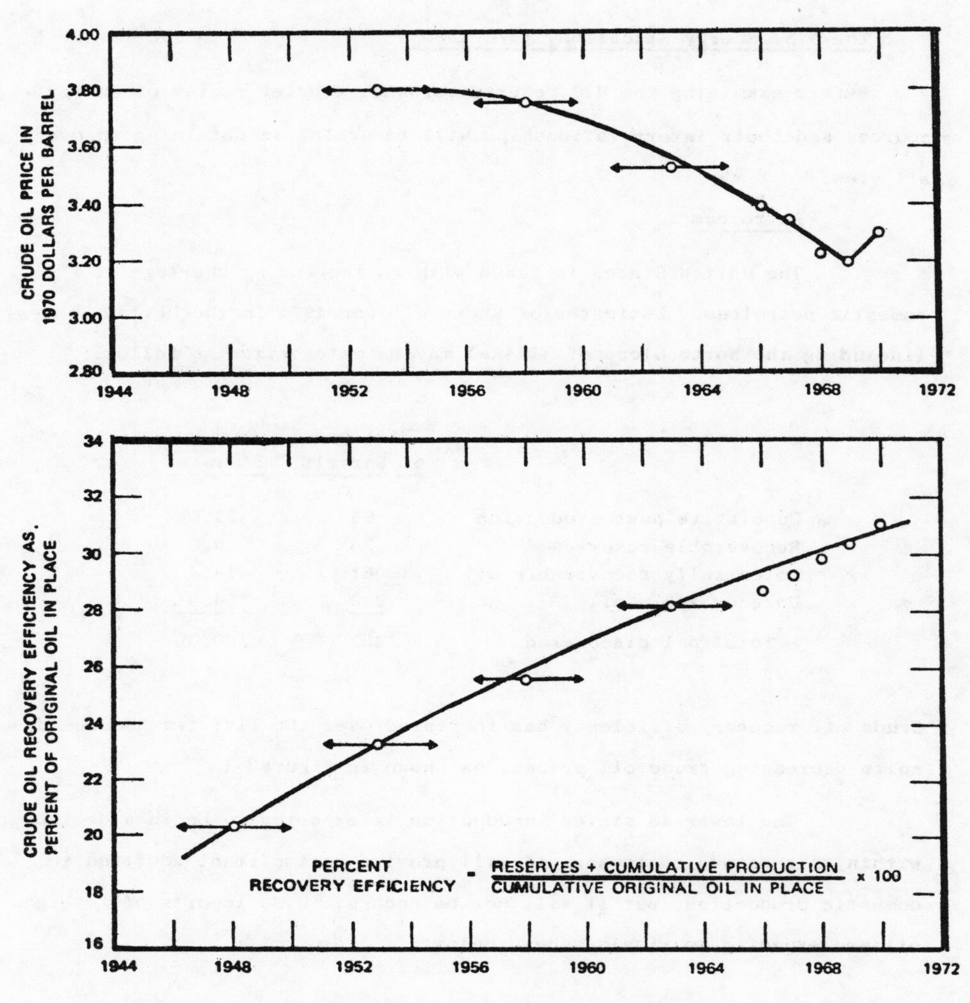

FIGURE 3-1 HISTORICAL PATTERN OF U.S. CRUDE OIL RECOVERY EFFICIENCY AND PRICE

Gas

At existing price levels, the supply of natural gas in the United States has become limited. The recent regulation of prices by the Federal Power Commission below parity with other low-sulfur fuels is stimulating an excessive demand for gas. Furthermore, the lower prices for gas have reduced the incentive for exploration and development of new gas reserves. The dependence on gas and the shortage of supply are forcing the conversion of oil to gas. It is expected that by 1980 about 700,000 barrels per calendar day of liquid hydrocarbons (primarily naphtha) will be converted to substitute natural gas. The natural gas suppliers and gas transmission companies have already announced several planned installations. These companies will be bidding for light hydrocarbon feedstocks in direct competition with the petrochemical manufacturers. The utility companies, for the most part, may be able to outbid the petrochemical companies, because the utility has considerably less risk in an FPC-approved project.

Supplemental supplies of gas are expected from Alaska, Canada, and via liquified natural gas (LNG) imports. Additionally, by 1980, coal gasification is expected to begin a relatively long period of gas production.

Coal

The United States has abundant potentially usable reserves of coal. These are optimistically estimated at 800 billion tons, or the equivalent, if liquefied, of 2,500 billion barrels of synthetic crude oil. These figures leave little doubt that coal will become the dominant future source of oil and gas. However, during this decade the direct use of coal as an energy source is expected to remain nearly constant in the range of 6,500 to 7,000 thousands of barrels per calendar day (B/CD) of oil heating value equivalent. The use of coal in the form of substitute natural gas (SNG) is expected to emerge by the end of this decade. By 1980 only 0.4 trillion cubic feet per year of SNG is anticipated. After 1980, the trend of coal gasification should accelerate.

Oil Shale

The deposits of oil shale in the United States are a potential source of future energy supply. However, commercialization in significant quantities is not expected during the coming decade. Estimates of potential oil from shale vary widely, depending on the concentration of Kerogen (hydrocarbon source) in the various deposits. Deposits of oil shale with a concentration of 35 gallons per ton or greater are estimated to have a potential for production of 75 billion barrels of oil.

Athabasca Tar Sands

These tar sands deposits in Alberta are estimated to have the enormous potential equivalent of 600 to 1,000 billion barrels of synthetic crude oil. This is more than the world's entire proven reserves of conventional crude oil. The exploitation of this energy source has already been commercialized by Great Canadian Oil Sands, Ltd. (GCOS) at the rate of 45,000 barrels per stream day. However, a manufacturing facility of this nature is extremely capital intensive and requires greater than average risks.

Nuclear Energy

Projections of U.S. energy supply have assumed an important contribution from nuclear energy sources. It is anticipated that the power generated from nuclear plants in 1980 will be equivalent to having converted 6 million barrels per calendar day of oil, gas, or coal at 31 percent efficiency into electrical energy. If these plants cannot be installed, the major alternative to meeting the forecasted energy demand will be the importation of greater amounts of oil, primarily from the Eastern Hemisphere.

Petrochemicals

The major problem that petrochemical manufacturers are facing is the limited supply of low-cost hydrocarbon feedstock. Cheap foreign naphtha is expected to become less available because of the demand for this material for gasification in the United States. One proposed SNG plant in the United States has already contracted for 75,000 barrels per day of imported naphtha at 8.3 cents per gallon, which is about 30 percent higher than foreign naphtha has been in recent years.

The limitation of natural gas supply will also limit the production of natural gas liquids. The production of hydrocarbon chemicals has been based on the use of low-cost natural gas liquids and equivalent refinery streams. It is clear that future hydrocarbon feedstocks for petrochemicals will emanate from petroleum. It is probable that the gas oil will be used increasingly for the production of olefins despite the increased value of gas oil in low-sulfur fuel oil. This is especially true as long as gasoline remains leaded. The removal of lead will liberate naphtha-type hydrocarbons that may become petrochemical feedstock. Most of this naphtha will remain in gasoline even at low lead levels. Most of the low octane naphtha will not become available until lead is eliminated. Sources of petrochemical feedstocks are estimated below.

	1970		1980	
	Thousands of B/cd	Percent	Thousands of B/cd	Percent
Gas (oil equivalent)	242	25.2%	395	20.9%
Natural gas liquids and equivalent refinery streams	620	64.4	592	31.4
Imported naphtha	100	10.4	200	10.6
Other liquid hydrocarbons			700	37.1
Total	962	100.0%	1,887	100.0%

Thus, petrochemical feedstocks will become less available at past prices because of (1) the growing shortage of natural gas and an attendant upward pressure on gas and LPG prices, making the alternative use of natural gas liquids for LPG more attractive; (2) the increasing use of naphtha for gasification; and (3) the increasing necessity of using gas oil in low-sulfur fuel oil to make up the energy shortfall.

In an effort to simplify the complex mixture of energy supply types and their innumerable uses, energy balances for 1970 and 1980 were made by converting all energy forms to acceptable oil heating value equivalents. The oil figures, however, were not modified. Assumed energy equivalence factors are shown below.

Gas	1,035 Btu/scf
Oil	5.85 million Btu/bbl
Coal	25.2 million Btu/st
Power generation efficiency	
1970	30.3%
1980	31.0%
Hydroelectric and nuclear	9,500 Btu/kWh

U. S. Energy Supply

Table 3-1 shows that petroleum supplied almost one-half the energy needs in 1970. Domestic oil accounted for the bulk of the oil supply. Residual oil from the Caribbean and Western Hemisphere crude oil primarily from Canada and Venezuela accounted for a major portion of the imported oil. Eastern Hemisphere crude oil imports primarily from the Middle East and Africa were only 2.5 percent of the total petroleum supply or about 10.6 percent of total oil imports. The U.S. strategy for imported oil clearly was to minimize dependence on geographically distant and politically unreliable sources.

Domestic production of gas of 22 trillion cubic feet in 1970 (equivalent in heating value to over 10 million barrels per calendar day of oil) supplied about one-third of total energy requirements. Additionally, 0.8 trillion cubic feet were obtained from Canada in 1970.

The demand for energy and hydrocarbons for chemicals may be conveniently divided into several basic categories. This allows one to quantitatively comprehend the magnitude of energy consumed in these sectors as shown below.

Demand Sector	**Percent of Total 1970 Energy Supply**
Residential and commercial	**19.3%**
Industrial plants	**27.4**
Transportation	**24.7**
Electric generation	**20.4**
Hydroelectric (supply)	**1.2**
Nuclear (supply)	**0.1**
Chemicals	**2.9**
Other petroleum products	**4.0**
Total	**100.0%**

These figures leave little doubt that the U.S. society is truly industrialized and strongly dependent on automotive transportation. The unparalleled success of achieving a highly industrialized economy and the accompanying increases in standard of living in the United States

may be attributed in part to the abundance of domestic energy sources. However, the United States is now facing a situation that indicates a much greater dependence on foreign energy supplies.

Table 3-1

U.S. ENERGY SUPPLY

	1970 Thousands of B/cd*	1970 Percentage of Total	1980 Thousands of B/cd*	1980 Percentage of Total
Coal				
Solid	6,508	19.2%	6,700	12.9%
Gas			276	0.5
Subtotal	6,508	19.2%	6,976	13.4%
Gas				
Domestic	10,638	31.3	10,186	19.7
Imports	402	1.2	1,350	2.6
Subtotal	11,040	32.5%	11,536	22.3%
Oil				
Domestic	11,312	33.2	12,780	24.7
Western hemisphere crude imports	959	2.8	2,700	5.2
Eastern hemisphere crude imports	365	1.1	6,418	12.4
Light products imports	566	1.7	1,102	2.1
Residual oil imports	1,528	4.5	2,300	4.4
Total imports	3,413	10.1	12,520	24.2
Refinery gains	324	1.0	555	1.1
Subtotal	15,054	44.3%	25,855	50.0%
Hydroelectric	1,250	3.7	1,370	2.6
Nuclear	100	0.3	6,050	11.7
Total energy supply	33,952	100.0%	51,787	100.0%

* Oil or oil heating value equivalent.

The U.S. energy supply picture is essentially complete when coal is considered. This source accounted for about 20 percent of total supply and was used primarily for the generation of electricity and as fuel for industrial operations.

Energy Consumption and Gross National Product

One may question the basic premise that requires the satisfaction of total energy demands in the United States. In view of the predicted increasing dependence on foreign sources of energy and the questionable availability of these supplies on a long term basis, it is imperative that the United States detect and reduce nonproductive uses of energy. However, over the last 40 years, it is clear that increased energy consumption was productive as measured by the Gross National Product (GNP) and as shown in Figure 3-4 . .An extrapolation of the trend shown to

Fig. 3-2

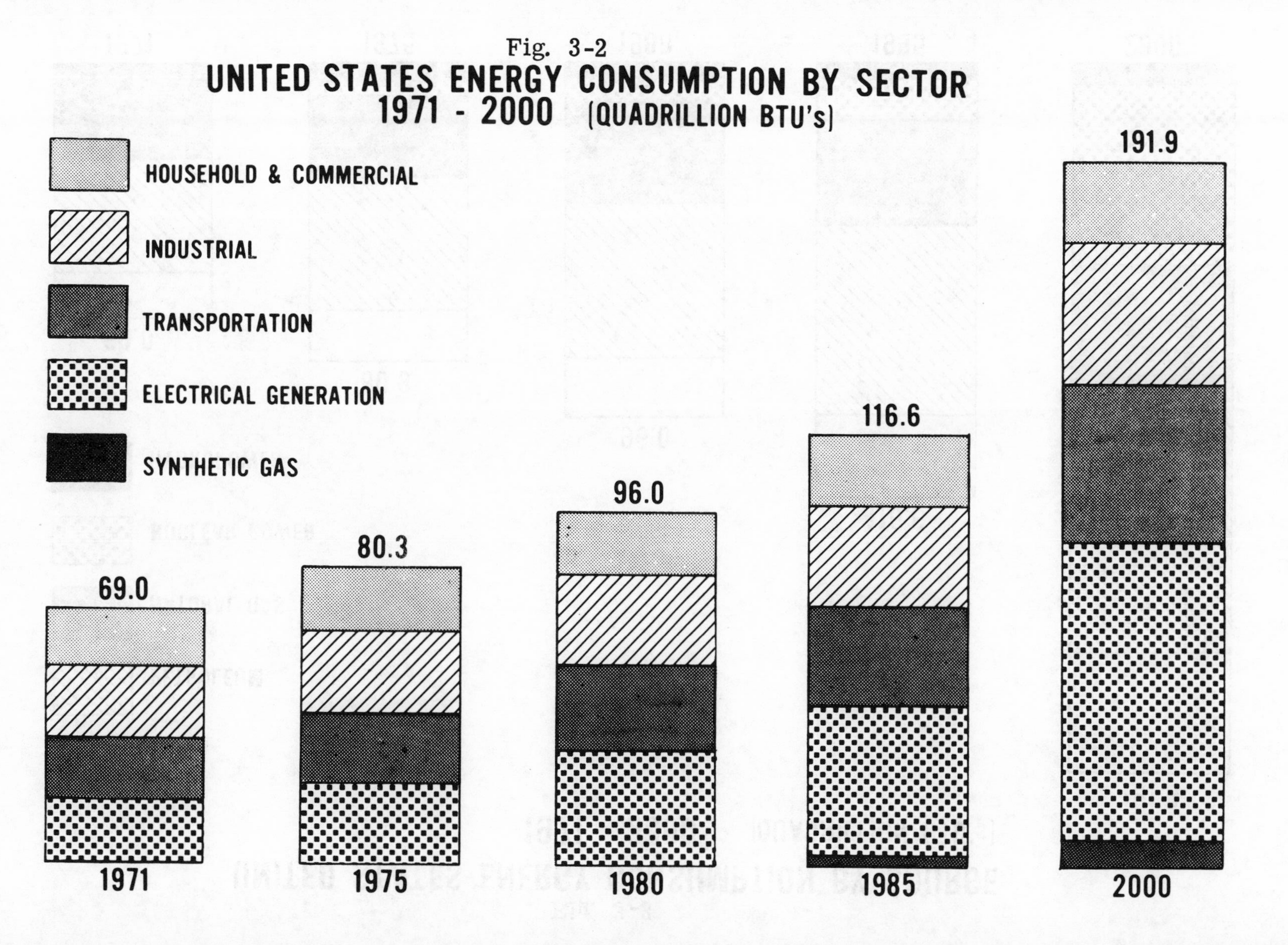
UNITED STATES ENERGY CONSUMPTION BY SECTOR
1971 - 2000 (QUADRILLION BTU's)
HOUSEHOLD & COMMERCIAL
INDUSTRIAL
TRANSPORTATION
ELECTRICAL GENERATION
SYNTHETIC GAS
69.0
80.3
96.0
116.6
191.9
1971
1975
1980
1985
2000

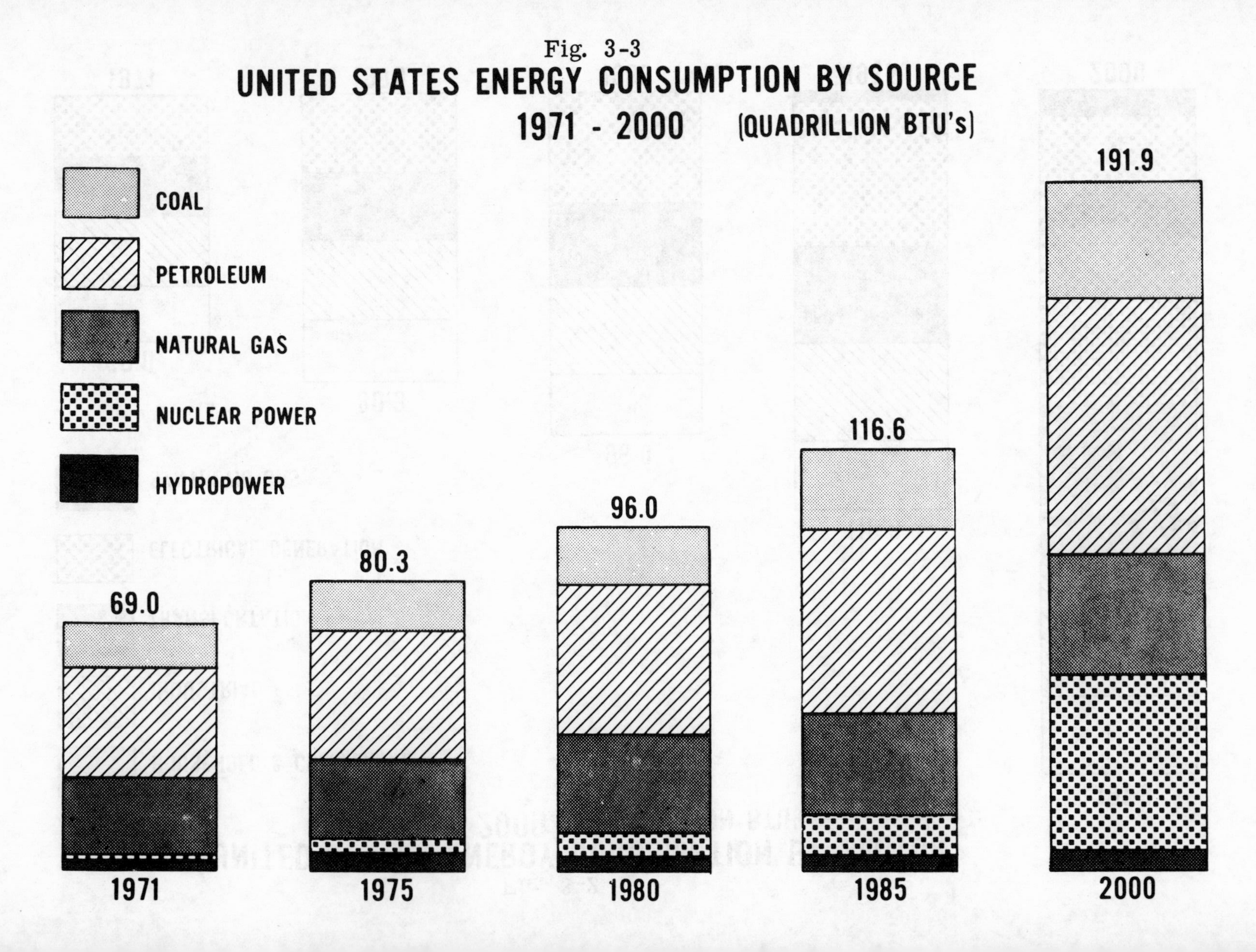

Fig. 3-3
UNITED STATES ENERGY CONSUMPTION BY SOURCE
1971 - 2000 (QUADRILLION BTU's)
COAL
PETROLEUM
NATURAL GAS
NUCLEAR POWER
HYDROPOWER
69.0
80.3
96.0
116.6
191.9
1971
1975
1980
1985
2000

zero GNP appears to have a positive intercept of energy consumption, which may be an amount of energy that is nonproductive, if, in fact, the intercept is real.

On the basis of the extremely good correlation between energy consumption and GNP, it is reasonable to conclude that the GNP will increase only if greater quantities of energy are supplied and consumed within the framework of the U.S. society as it now exists.

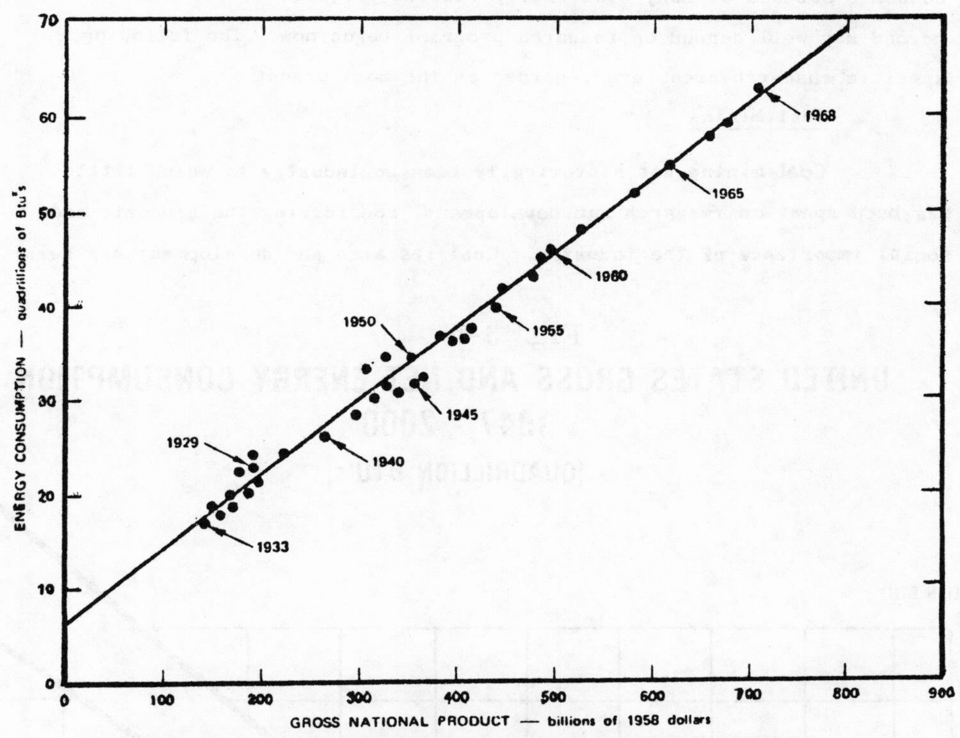

FIGURE 3-4 U.S. ENERGY CONSUMPTION AND GROSS NATIONAL PRODUCT.

Technology Status for Future Resource Recovery and Utilization

Introduction

From the preceeding discussion, it is clear that the need exists for research and development for utilization of fuels and their recovery from the earth. The United States is at a critical crossroads in its industrial development pattern. No longer is the United States self-sufficient in resources; in fact, the pattern of resource recovery and utilization grows more difficult each year. Much research is being conducted by energy-associated industry, but it is becoming increasingly clear that the time and money expenditures associated with such activity are not sufficient to bring research efforts to fruition fast enough to avoid major problems with fuel supply.

In view of the seriousness of the U.S. energy crisis, massive support is needed for research and development programs in the fields of fuel production, processing, power generation, energy transportation, and energy utilization. The energy crisis in the near term (until about 1980) can be eased only by large expenditures for existing development programs. It should be recognized, however, that little fundamental research has been done in many areas pertaining to energy. The ultimate technical and economic success of many U.S. energy-related projects in the 1980s and beyond may well depend on research programs begun now. The following specific research areas are regarded as the most urgent.

Coal Mining

Coal mining has historically been an industry in which little has been spent on research and development, considering the economic and social importance of the industry. Coal research and development has been

Fig. 3-5

UNITED STATES GROSS AND NET ENERGY CONSUMPTION 1947 - 2000

(QUADRILLION BTU's)

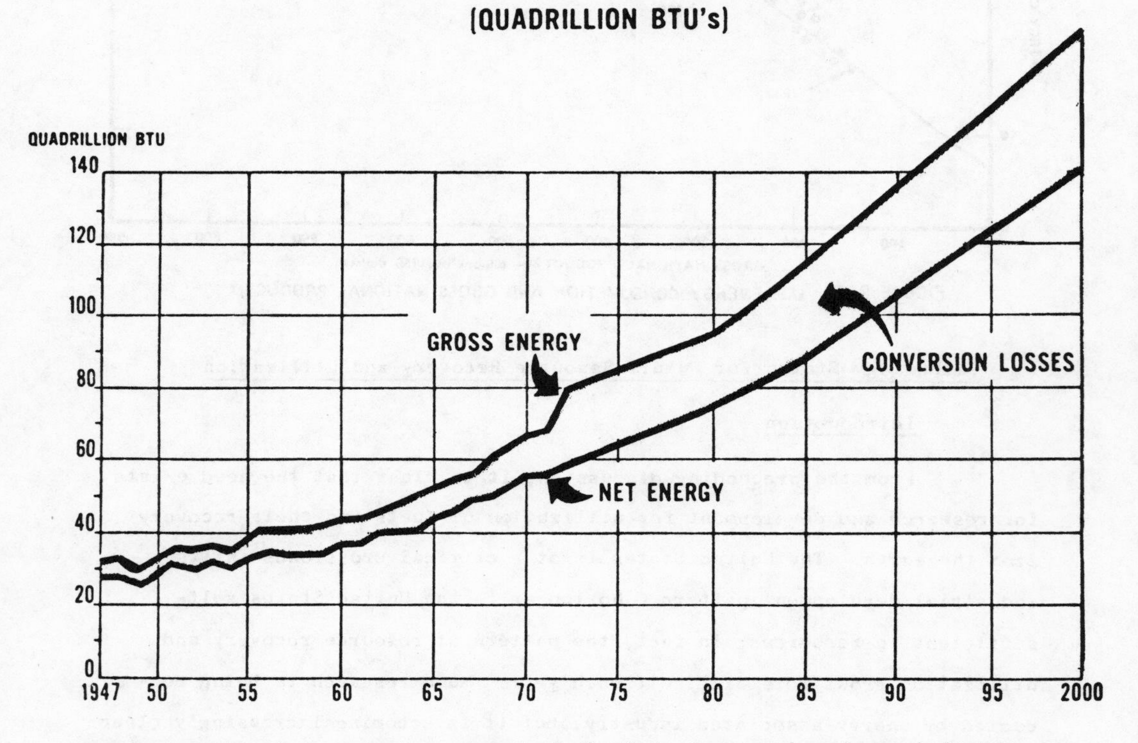

funded primarily by the Department of the Interior, coal and gas industry associations, and a few petroleum and chemical companies. Moreover, the work done on coal research and development pales in significance when compared with that for petroleum exploration and refining.

Fig. 3-6

GROSS ENERGY CONSUMPTION PER DOLLAR OF 1958 GROSS NATIONAL PRODUCT 1947 - 2000 (THOUSANDS OF BTU's)

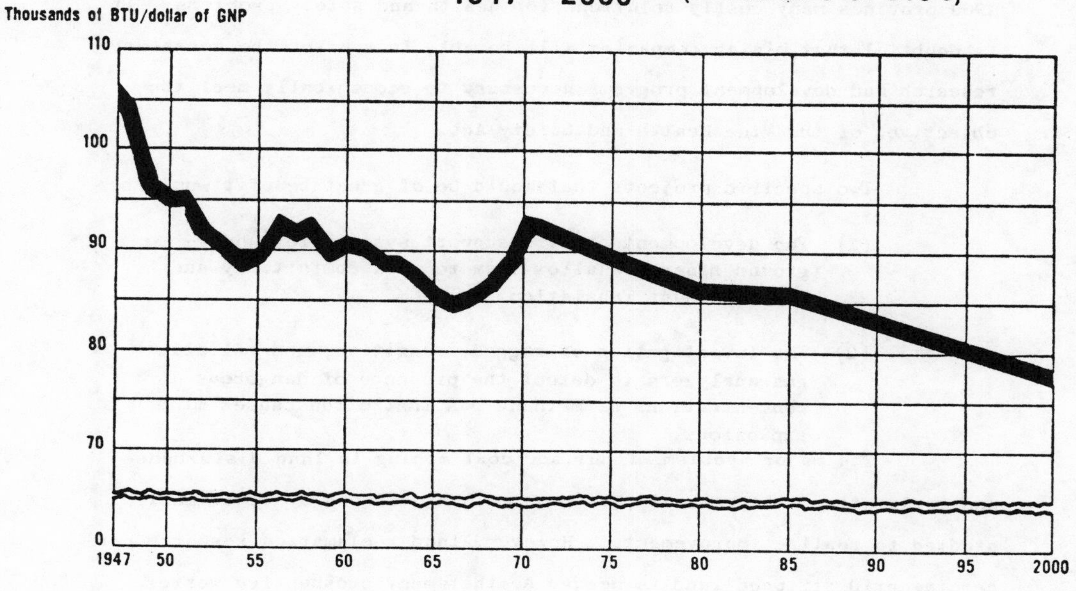

Fig. 3-7

UNITED STATES NET AND GROSS ENERGY INPUTS PER CAPITA 1947 - 2000 (MILLIONS OF BTU's)

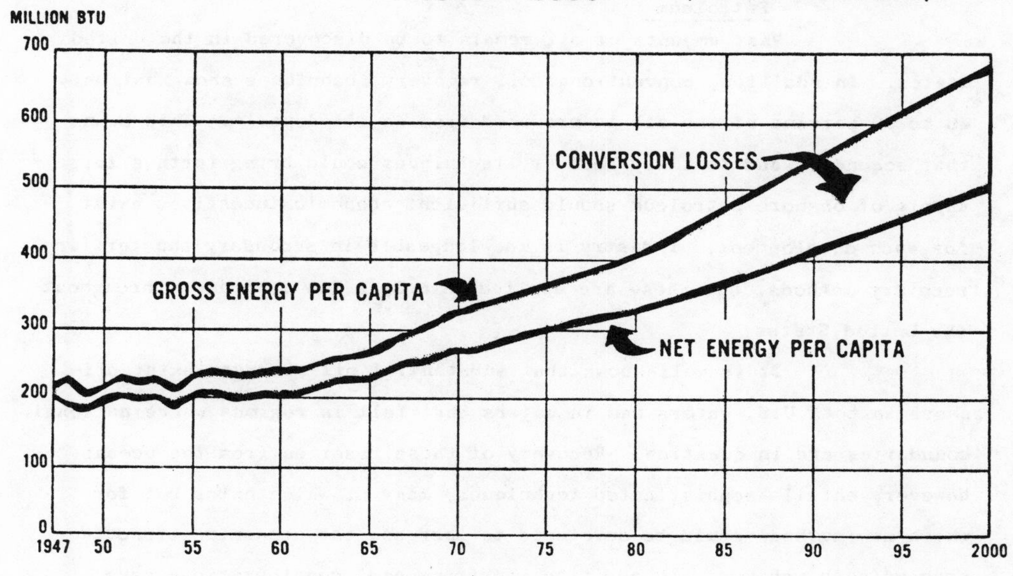

Coal mining is a labor intensive industry. The work in an underground mine is dangerous, uncomfortable, and unhealthy. As such, mining does not attract a sufficient number of trainable personnel. The health and safety record of underground coal mining is poor in spite of postwar technological improvements. The Mine Health and Safety Act of 1969 provides many costly solutions for health and safety problems. It is doubtful that mining companies will be able to sponsor the necessary research and development programs necessary to economically meet the objectives of the Mine Health and Safety Act.

Two specific projects that would be of great benefit would be:

(1) The development of life support systems for underground miners to allow them to work comfortably and to avoid dust inhalation.

(2) The investigation of rugged, sensitive, and reliable gas analyzers to detect the presence of dangerous concentrations of methane gas that often causes mine explosions.

The major problem of surface coal mining is land disturbance and reclamation. In general, this problem has not been systematically studied to realize improvements. However, land reclamation research concerning arid stripped land is needed so that many prospective western coal fields may be mined. Cheap soil stabilization and revegetation methods should be studied.

Petroleum and Gas Recovery

Petroleum

Vast amounts of oil remain to be discovered in the United States. In addition, conventional oil recovery techniques show that only 40 to 50 percent of the oil is produced from an oil deposit. This means that secondary and tertiary recovery techniques could bring forth a large supply of onshore petroleum should sufficient economic incentives exist for such development. Industry is knowledgeable in secondary and tertiary recovery methods, but these are applied in only a few locations throughout the United States.

It is well-known that substantial oil reserves exist offshore in both U.S. waters and in waters that fall in regions where national boundaries are in question. Recovery of these reserves from the oceans, however, entails sophisticated techniques, many of which exist but for which new methods could be developed to increase the economic attractiveness of such schemes. In addition, environmental considerations have held up development of these resources to a large extent. Should such delays persist, the United States will become increasingly dependent on imported petroleum.

The U.S. petroleum companies have considerable knowledge in offshore recovery techniques. However, there appears to be incentives to develop methods to recover oil by installing recovery facilities directly on the ocean floor itself.

The recovery of petroleum-type liquids from coal and oil shale merits serious consideration. The United States has potentially more than 100 times the crude oil reserves in coal compared with conventional petroleum. As mentioned earlier, our oil shale reserves are estimated at 75 billion barrels for shale containing more than 30 gallons per ton of liquids. Lower grade shales increase these reserves by several orders of magnitude.

Both the government and some petroleum companies are conducting research on methods for producing synthetic crude oil from coal. A number of processes are at the pilot plant stage of development--most with funding by the Office of Coal Research of the Department of Interior.

The size of a synthetic crude oil plant from the coal industry may well be limited by the ability of the United States to recover its coal reserves for such an operation. For a 100,000 barrel per day synthetic crude oil plant, roughly 35,000 to 50,000 tons per day of coal are needed. The United States is not used to handling such vast amounts of solid materials, and this may well limit the development of a significant resource base of petroleum from coal.

The principal technological problems for converting coal to oil lie in two areas: development of hydrotreating catalysts for all stages of processing and development of materials that withstand the high pressures and high temperatures and can handle the abrasive characteristics of coal/oil mixtures.

In addition to these two problem areas, methods to produce cheap hydrogen are needed for any coal-upgrading process scheme. In some processes, since oxygen requirements are large, an improvement in oxygen production economics would be of significant value.

In an environmentally concerned society, any fuel that has the potential of emitting the oxides of sulfur or nitrogen to the atmosphere would be closely regulated. Although crude oil derived from coal can be desulfurized, at the sulfur level competitive with conventional crude oil, large amounts of nitrogen remain. For example, a 0.3 percent sulfur coal-derived crude oil contains nearly 1.0 percent nitrogen; a 0.3 percent sulfur conventional petroleum low-sulfur oil contains virtually no nitrogen. Therefore, the coal liquid must be denitrogenated to be competitive--at a serious cost disadvantage. Processes for removal of nitrogen from coal-derived crude oil must ultimately be developed for effective marketing of these materials.

A shale oil industry should be considered to be in the national interest, since it provides a petroleum back-up reserve of over 75 billion barrels of oil--nearly three times the reported Alaskan resource. This size is estimated from deposits that are believed to be marginally attractive at today's prices. Should the United States go after all the oil shale deposits for oil recovery, the reserves would be significantly larger than this. The process development problems associated with an oil shale industry are primarily those associated with environmental protection--disposal of the spent shale, control of nitrogen oxide emissions from burning shale oil, control of nitrogen and other odorants from a shale retort, and, the most important, mining the oil shale without damaging the Colorado and Utah countrysides. Retort development has been funded principally by industry, although some of the earliest and most extensive work was done under sponsorship of the Bureau of Mines. Most of the Bureau's activities today are devoted to in-situ recovery of shale oil--a very fertile area for research.

Gas

As pointed out earlier, the U.S. gas supply is on a serious decline at a time when demand is increasing because of its attractiveness as a nonpolluting fuel. Since much of the problem of gas supply and demand is due to overregulation, it is imperative that policies be set to provide the maximum incentives to find this premium fuel. Currently, the low wellhead price for natural gas has discouraged producers to seek new supplies in unconventional regions.

A higher price for natural gas would enable producers to look for additional gas supplies in regions of the Unites States that have remained virtually untapped. The two principal areas in this regard are offshore natural gas and gas in deep deposits. The former requires technology virtually the same as for offshore petroleum recovery. The latter requires technical and economic advances in deep well drilling, including low-cost materials of construction for the drills themselves.

Industry and the government are examining the problems associated with the production of substitute natural gas (SNG) from coal or oil at a relatively slow pace compared with the need for supply.

Most problems associated with coal gasification processes are related more to process development rather than to specific technicalities. There is considerable difficulty, however, in obtaining a coal handling system that will enable solids to be fed continuously into a coal gasification reactor operating at 1,000 pounds per square inch pressure. One process uses a slurry feed system and the problem of handling coal-oil slurry at the pressure is formidable.

One area that merits further examination is that in which explosive devices are used to stimulate gas recovery from tight gas sands deposits. One technique that uses nuclear devices has met with marginal success. However, the rapidly expanding need for gas does prompt a definitive estimate or development in this area.

The same holds true for in-situ gasification of coal. A recent Bureau of Mines report shows this to be a development area that is riddled with technical as well as economic questions.

Electric Power Generation

Electric power generation currently accounts for about 25 percent of U.S. energy use. It is predicted that this percentage will increase to 37 percent in 1985. Today's fossil-fueled power plants operate at a maximum electrical generation efficiency of about 36 percent. Advanced power cycles have been proposed that could increase electrical generation efficiency to well over 50 percent.

Advanced power cycles systems depend on high temperature gas turbines combined with lower temperature steam turbines. The principal technological limitation on developing such systems is the temperature limitation on the turbine blade material. Currently, turbine blades may be designed to withstand a temperature of about 1,800°F. Advanced power cycles with an efficiency of about 33 percent could be operated within this temperature limitation. An efficiency of about 58 percent could be achieved with an advanced power cycle if the inlet temperature could be increased to 3,000°F. Continued large-scale research is needed in high temperature turbine blade materials and blade design including internal or transpiration cooling.

Other methods of improving electric generation efficiency include fuel cells and magnetohydrodynamics (MHD). Neither method is as far along in development as are advanced power cycles.

Small-scale fuel cells to supply electricity to individual home or apartment houses, however, are under current development. The increased overall efficiency of power utilization (no transmission line losses) make small-scale fuel cells a worthwhile subject of further research and development.

Energy Transportation

Much of the energy supply problem can be attributed to the fact that fossil fuel supplies are located far from the energy markets. Large oil and gas discoveries are being made in remote Arctic regions. Gas associated with oil in remote areas often must be flared because it is not economical to transport it to market. The largest part of undeveloped U.S. coal reserves, especially low-sulfur coal, lies in the mountains far from the eastern and Pacific Coast markets. In the case of electricity, most new power plants are being located away from metropolitan areas for environmental and esthetic reasons.

The well-publicized problems of obtaining approval for the Alaskan oil pipeline in Alaska points out the problem of Arctic oil and gas transportation. Publicly funded research is needed on the Arctic environment in addition to the large sums spent by private industry. With discovery of gas and oil on the Canadian Arctic Islands, yet more difficult transportation problems loom. Proposals made so far include undersea (and under ice) pipelines, LNG submarines, dirigible LNG tankers, and icebreaker tankers. A vast amount of work is needed in cooperation with Canada on the entire Arctic environment, including Arctic Ocean ice packs, permafrost, Arctic meteorology, glaciology, and biology. Special emphasis needs to be placed on research on effects and prevention of oil spills on Arctic land and seas.

The conventional mode of coal transportation is by rail or, where applicable, by barge. A promising new method of coal transportation is that of slurry pipelining. Coal slurry pipelining is now commercial with a 273 mile, 15,000 ton per day pipeline in operation in Arizona. Nevertheless, support for fundamental research is needed on the flow characteristics of slurry particles of various types of coal. Additional development work is also needed on auxiliary systems for coal grinding, slurry dewatering, and coal drying.

Long distance electrical transmission at the present time results in large power losses. Developmental programs are under way on reducing transmission losses by use by cryogenic temperatures for transmission cables to reduce electrical resistance. The use of liquid nitrogen (-320°F) or liquid hydrogen (-423°F) would reduce transmission losses considerably. In the case of liquid hydrogen temperatures, there are good prospects of developing superconducting materials whose use would result in nil transmission losses. A higher level of research in the field of cryogenic conductors and superconductors should be funded. New low resistance materials may also be used for generator windings, greatly reducing the size and cost of large generators.

Energy Utilization

The possibilities for saving energy are numerous. Many savings particularly in industrial uses, will be made as a result of simple economics. Others will be possible only by educating the general public as to ways of saving fuel and power--and, consequently money. Still other savings require technological advances, including those previously discussed for electrical generation and transmission. Finally, future energy savings require social acceptance of limits to population growth and the growth of per capita energy usage.

The category of savings of energy of primary interest in this study is the one that depends on technological advances. Two important end uses of energy are home heating (and cooling) and automobile propulsion. Both end uses have inelastic supply curves, i.e., increasing the price does not affect the demand. However, the design of the dwelling or automobile does have a substantial effect on the energy usage.

Improved insulation, storm doors and windows, and awnings can greatly reduce home heating and cooling requirements. New research funds could be profitably used to study methods for partial solar heating of houses by storing heat in hot water during sunny periods. A more controversial proposal would be to study cheap tamper-proof systems for remotely controlling individual thermostats during periods of energy shortages or environmental crises.

Fuel conservation in the automobile is a more complex problem involving both quantity and quality of the fuel consumed. The most obvious method of fuel conservation is to have fewer automobile miles driven. Additional mass transit systems should be developed, particularly for commuting to and from work in densely populated urban areas. The preference shown by the commuter for the convenience of the automobile should not be ignored. For example, studies should be made of "mini-rapid transit," a scheme whereby publicly-owned and maintained vehicles (car or small buses) are driven to and from work by car-pooling commuters.

Fuel economy per mile driven can be improved by reducing engine size or increasing the compression ratio, requiring higher octane gasoline. New engine development may be needed to economically meet impending federal environmental standards. The Wankel engine is already commercial but in need of improvement. Types of engines for which exploratory development money should be available include electric, hybrid electric, and steam-powered (external combustion). These new engines might be more practical for powering "minirapid transit" vehicles than for private automobiles.

ENGINEERING PROBLEMS IN COAL GASIFICATION

The developing coal gasification industry presents a number of technical problems. These may be classified as both technical problems and problems associated with process development.

Figure 3-7 is a general process scheme of a pipeline gas (methane) from coal plant. The drawing shows the major operations and chemical reactions entailed in forming CH_4 from coal or lignite.

One of the most important reactions in the coal gasifier is the "devolatilization reaction."

$$\text{Coal volatile matter} \xrightarrow{1100\text{-}1500^{\circ}F} CH_4 + C. \qquad (1)$$

This reaction is exothermic, and it is important to take full advantage of exothermic reactions to reduce the required heat input to the gasification system. High pressure favors reaction (1) since it prevents decomposition of CH_4 to C and H_2.

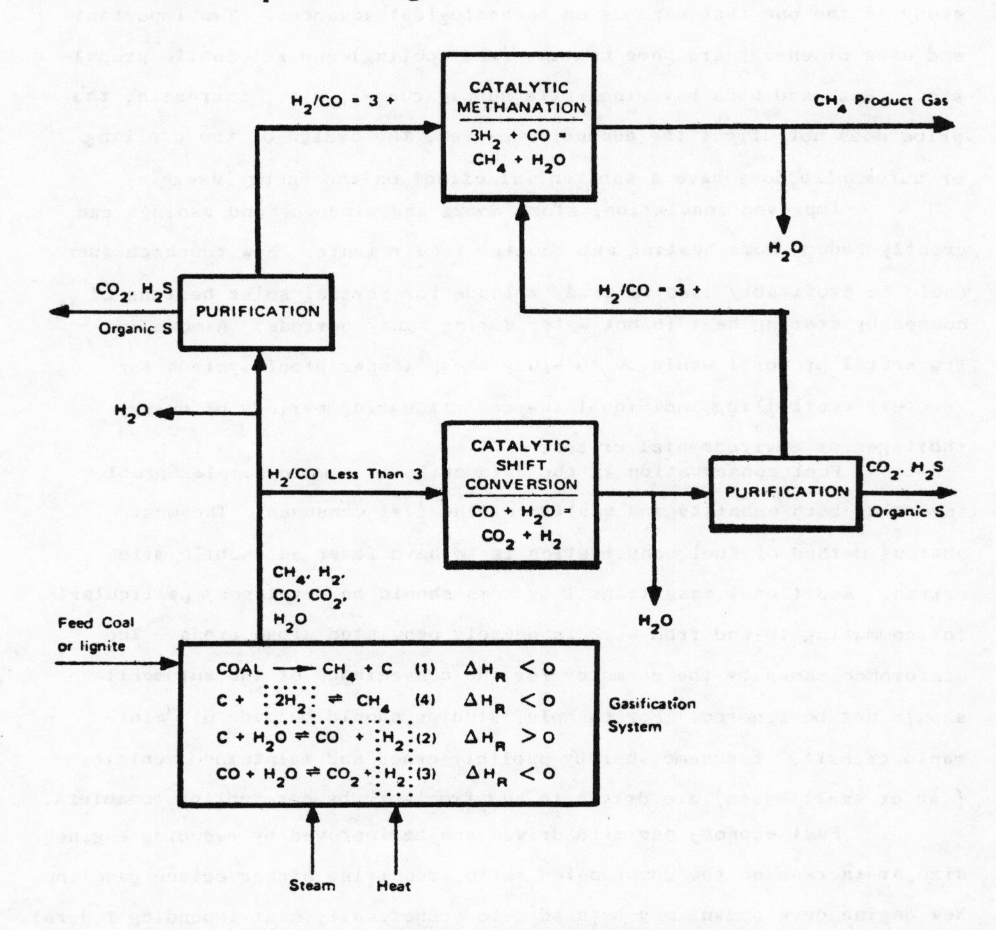

FIGURE 3-7 METHANE FROM COAL GASIFICATION
GENERAL PROCESS SCHEME

After devolatilization, large quantities of carbon remain as char. To produce methane from the char, a supplemental source of hydrogen is required. This hydrogen is generally produced in situ by the strongly endothermic steam-carbon reaction:

$$C + H_2O \overset{>1700^{\circ}F}{\rightleftharpoons} CO + H_2 \qquad (2)$$

This reaction results in a large and expensive indirect heat input to any gasification system. In terms of reaction rate, reaction (2) is the slowest of the gasifier reactions. Generally, high pressure increases the rate of this reaction.

Another important reaction in any gasification scheme is the water-gas shift reaction:

$$CO + H_2O \rightleftharpoons H_2 + CO_2 \quad (3)$$

This reaction is only mildly exothermic, thus it is not significant as a source of heat for the steam-carbon reaction zone. The major importance of the shift reaction is its ability to generate additional hydrogen within the gasifier for promoting the "hydrogasification" reaction:

$$C + 2H_2 \xrightleftharpoons{1700^\circ F} CH_4 \quad (4)$$

Reaction (4) is highly exothermic, and good yields of methane can be obtained at 1700°F, a temperature high enough for adequate maintenance of the steam-carbon reaction rate. Thus, if methane can be formed via hydrogasification in a steam-carbon reaction zone, the exothermic reaction heat from such methane formation can supply heat to the steam-carbon reaction. High pressure favors the formation of CH_4 by reaction (4).

The shift reaction can be used externally from the gasification system to adjust the H_2 to CO ratio to 3, which is the required ratio for catalytically producing methane from H_2 and CO:

$$3H_2 + CO \xrightarrow[700^\circ F]{Catalyst} CH_4 + H_2O \quad (5)$$

Reaction (5) is strongly exothermic, but since it is carried out at low temperature in a separate reactor, it cannot be used to supply steam-carbon reaction heat. High pressure increases the rate and methane yield from reaction (5).

Before catalytic methanation of the gas, H_2S, organic sulfur and surplus CO_2 must be removed. High pressure generally results in lowest cost acid gas purification processes.

Thus, from the above discussion, operation of the gasifier at high pressure has several advantages:

(1) It favors the yield of methane in the gasifier either from the devolatilization or hydrogasification reactions--a high yield of methane has large economic significance, since it results in a lower required input of expensive indirect heat.

(2) It lowers the costs of acid gas purification and catalytic methanation.

Operation at high pressure is one of the two major differences between the U.S. processes and the Lurgi process. Most U.S. processes are programmed to operate at pressures of 1000 psi or higher, whereas the Lurgi gasifier operates at about 350 psi. The Lurgi gasifier was built to produce low Btu town gas or chemical synthesis gas. In these applications methane formation is desirably suppressed. This is a major reason that the U.S. processes are expected to yield methane at lower costs.

The other major advantage of the U.S. processes is in the gasification reactor throughput. The Lurgi gasifier is a relatively low-rate fixed bed system, whereas the U.S. processes, for the most part, are based on relatively high throughput fluidized bed or entrained systems. For example, to produce 250 million scf sd of gas, about 30 Lurgi gasifiers are required as opposed to only three in the Bigas entrained process as it is now conceived.

However, the potential for economic gain in the U.S. processes presents several problems and areas for creative innovation. The first problem faced in operating a gasifier at high pressure is that of coal feeding and withdrawal of solids residues. The most common method for feeding or withdrawing solids at high pressure is the lock hopper. A lock hopper system that might be considered for delivering coal against a back pressure of 1000 psi is shown in Figure 3-8. In this system, coal at ambient pressure is delivered from feed hopper M-303 to lock hopper V-301 during its fill cycle. V-301 is then pressured by compressed gas from holder V-304 during its pressure cycle. When at pressure, the contents of V-301 are dumped into continuous feed hopper V-302, which must have enough holdup capacity to allow V-301 to complete its cycles. After dumping coal, V-301 is depressured by releasing gas to V-304.

Although lock hoppers have operated successfully on Lurgi gasifiers at 300 to 400 psi pressure, the question remains about how long a vessel such as V-301 can stand up under the stress of repeated cycling from 1000 psi to ambient pressure. For example, using a total fill-to-fill cycle time of 30 minutes, V-301 would be subject to a 1000 psi pressure change about 17,000 times a year. This problem might be overcome by staging V-301 to take, say, 500 psi increments, but now three lock hoppers are needed instead of two. Even with only two, the volume of lock hoppers is usually a major fraction of the volume of gasifiers, and in one process (the Bigas process), they actually would be larger than the gasifiers. Another defect of lock hoppers is the large quantity of energy needed to pressure them to 1000 psi.

Another obvious way to get coal into a 1000 psi gasifier is to slurry it in. Since water (or steam) is a reactant in the gasifier, this seems like an appropriate thing to do. Unfortunately, however, the economics of a coal gasification system suffer greatly with water used as a slurry medium. This is principally because to vaporize water within a gasifier requires very great quantities of costly indirect heat (~ $1.50 MM Btu). This cost is compounded, because the heat used to vaporize water within the gasifier generally cannot be entirely used outside the gasifier, since gasifier waste heat from dry coal fed gasifiers is now used for at least a portion of gasifier steam generation.

The adverse economics of a water slurry system can be offset by using a low latent heat organic liquid. The Institute of Gas technology is in fact now trying organic liquids--specifically benzene and toluene. A

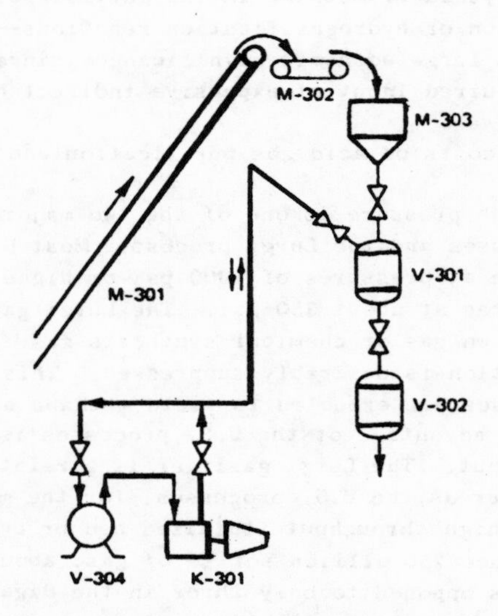

Fig. 3-8 LOCK HOPPER FEED SYSTEM

diagram of its proposed slurry system is shown in Figure 3-9. The most significant operation in this system is the vaporization of the slurry oil in a fluidized bed at 625°F by hot gas rising from the bottom of the gasifier. The use of an aromatic oil in this service is relatively satisfactory from the standpoint of slurry oil makeup costs, since aromatics are quite refractory at the conditions of the bed. However, even if oil losses are held to 1% of the circulation rate, the cost of slurry oil makeup can amount to about 3 cents per MSCF of product gas.

Another difficulty with aromatic oils in this service is their tendency to dissolve organic matter from coals even in small quantities. If this happens, the dissolved materials, after being released from solution in the slurry drying bed, could have characteristics quite different from other solid components in the bed, resulting in loss of fluidization control, fines carry-over, or even bed agglomeration. The tendency of aromatic oils to dissolve coal components might also vary somewhat from coal to coal, or even on an hourly basis when coal from a single source is processed.

Even if a slurry feed system were efficiently operable, such a system would be somewhat cumbersome and would require several additional control points in a process.

In an attempt to simplify and lower the cost of high pressure coal feeding, Bituminous Coal Research Inc. (BCR) conceived the idea of a "piston coal feeder." In this concept, which is shown in Figure 3-10, ground coal is first delivered to auxiliary feed vessel B. The coal is then transferred with low pressure inert gas into high pressure cylinder A. Cylinder A is then pressured with product gas, and, finally, the coal and gas are pushed into the gasifier by a hydraulically driven piston.

Fig. 3-11 is a drawing of the gasification section of the BCR BIGAS process. In this system, coal ground to about 200 mesh size is entrained by steam into the top stage (R-1) of a two-stage gasifier. The coal is

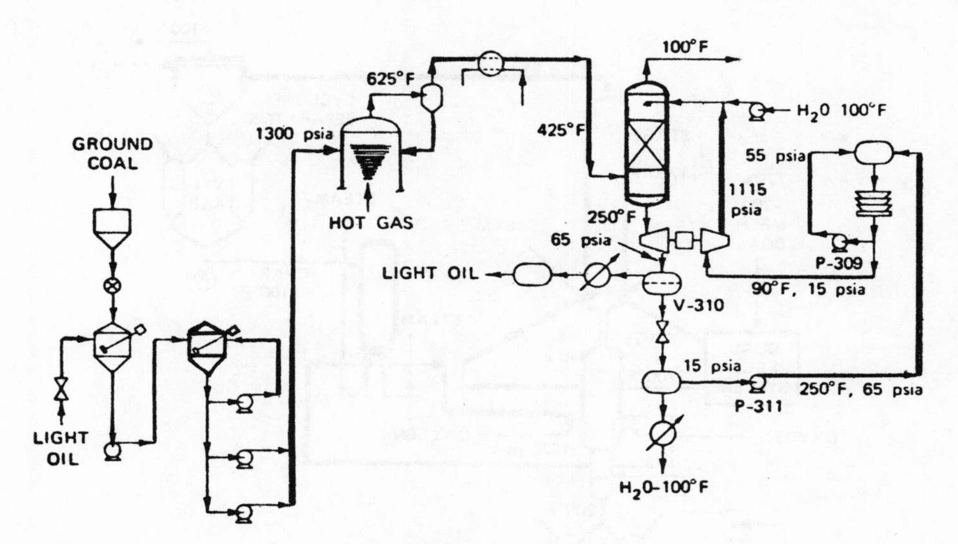

Fig. 3-9 SLURRY FEED SYSTEM

then swept upward by hot (2700°F) gas (CO_2, H_2, and CO) rising from the bottom stage (R-2). During a 6-second residence time in the top stage, the coal is devolatilized to CH_4, a small amount of devolatilized char is hydrogasified to CH_4, and additional char reacts with steam to form additional CO and H_2. Char that does not react is separated from the gasifier offgas in separator M-1 and returned to R-2 through lock hoppers

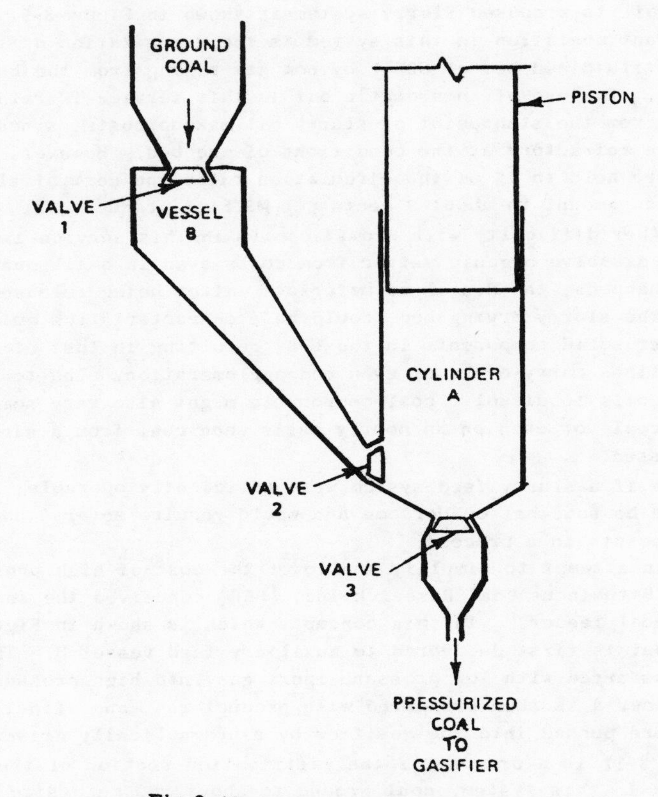

Fig. 3-10 CONCEPTUAL PISTON COAL FEEDER

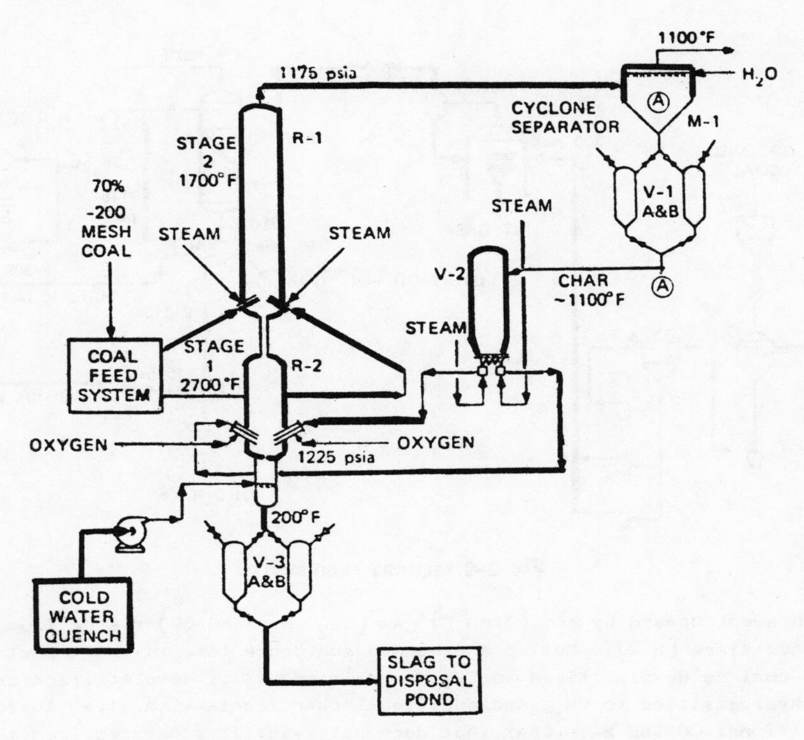

Fig. 3-11 BIGAS GASIFICATION SYSTEM

V-1 and feed hopper V-2. This recycled char is fed with steam into R-2, where it is contacted with oxygen at a controlled temperature of 2700°F. The char is reacted to CO_2, H_2, and CO, and these gases flow to R-1, completing the gasification circuit. Residence time in R-2 is about two seconds. The ash in the coal is slagged, quenched with water, and the ash-water slurry is let down in pressure through cyclically operated lock hoppers V-3.

If the costs of the facilities shown in **Fig. 3-11** are separated into a cost for coal feeding, char recycling, and slag discharge and a cost for the gasifiers, the cost for solids handling would be about five times the cost of the gasifiers. This multiple would not apply for other processes because the Bigas process has a relatively high gasifier throughput; but the multiple is presented to show that, in financial terms, there is substantial room for improvement in solids handling methods.

Referring again to **Fig. 3-11**, the circled points A represent locations at which recycled char is essentially sulfur free. A system for delivering a portion of this char directly to the burner of a fired heater would be of significant economic benefit for two reasons:

- Combustion of this char would replace combustion of as received coal, and thus SO_2 scrubbing of combustion offgas would not be required.

- More as received coal would pass through the gasifier, resulting in relatively more CH_4 being made from coal volatile matter at a unit cost substantially less than the cost of an equivalent quantity of CH_4 made from char.

These economic benefits of char withdrawal also apply to the Bureau of Mines Synthane process.

Various flow schemes for processing gas leaving a gasification system to product CH_4 are presented in **Fig. 3-12**.

Scheme A is a typical scheme now proposed for coal gasification. Shift conversion is proposed over standard iron-chromia shift catalysts. There now are serious questions about the life and mechanical strength of these catalysts in this service. The H_2S content of the shift feed gas will be several orders of magnitude higher than that encountered in current commercial usage of iron-chromia catalysts. Also, CO partial pressure will be higher than normal. The high H_2S content will affect the life of these catalysts adversely, and it is suspected that the high CO partial pressure will lead to serious carbon deposition. BASF of Germany has tested a sulfided Co/Mo catalyst under conditions almost as severe as those encountered in coal gasification. Although this catalyst appears to have merit in this application, it is only one catalyst, and there may be prospects for additional catalyst developments in this service.

Scheme B in **Fig. 3-12** represents a process sequence felt to be adequate from a demonstrated commercial standpoint. In this scheme, low temperature copper-zinc based shift conversion catalysts can be used. But these catalysts are poisoned by H_2S, and as a result, an additional purification step and an additional reheat step are required when compared with scheme A. This illustrates the reason for the incentive to develop scheme A.

If a scheme such as C could be developed, then no heat removal or reheat would be required in the shift, purification, or methanation train. However, this scheme would be restrictive, since purification would be required at only one temperature.

The most valuable improvement of this train is represented by scheme

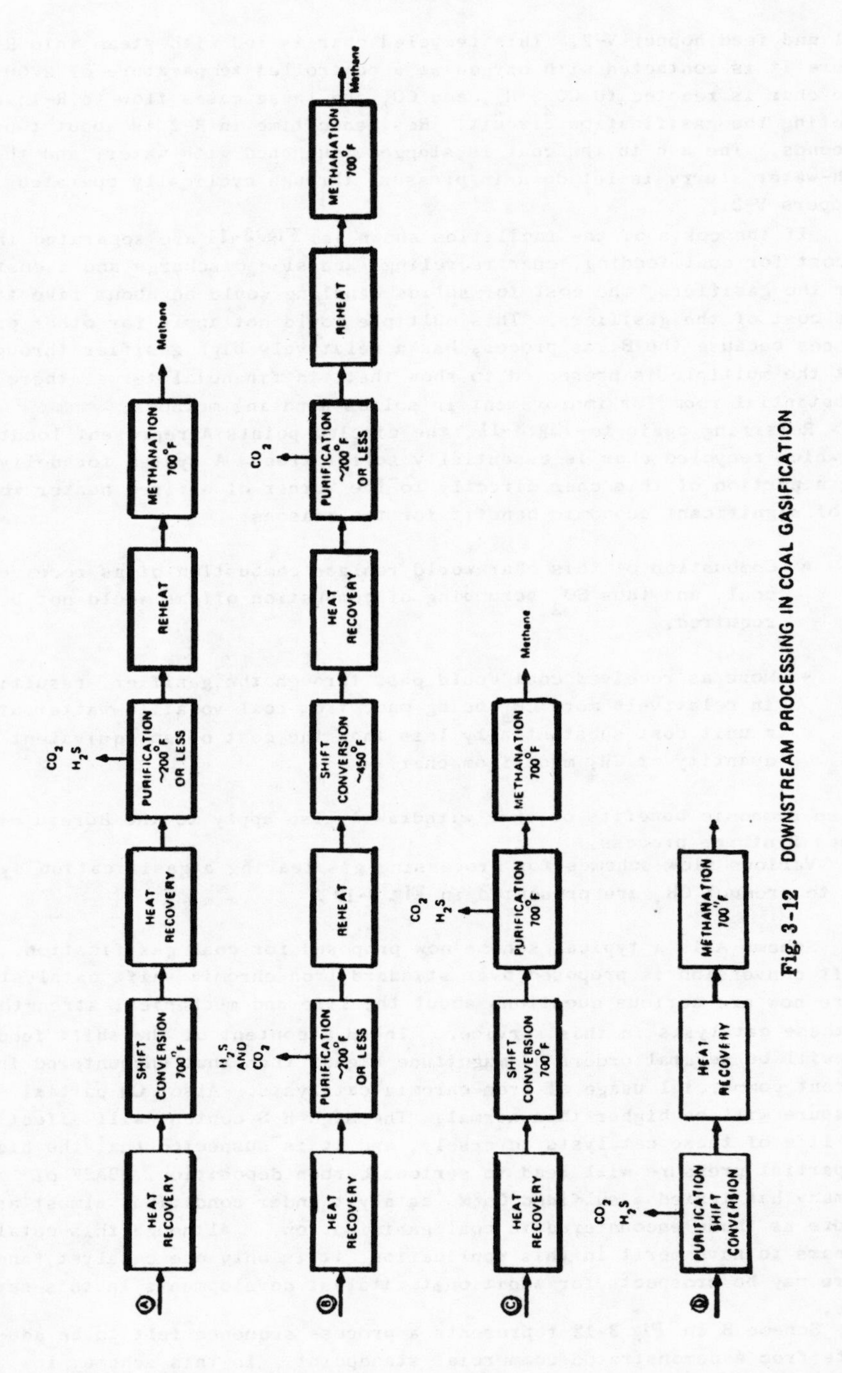

Fig. 3-12 DOWNSTREAM PROCESSING IN COAL GASIFICATION

D, where purification and shift conversion are assumed to occur simultaneously at close to gasifier temperature. One way to accomplish this is to use technology similar to the "CO_2 acceptor" process being developed by Consolidation Coal Co. In this process, CO_2 and H_2S react readily and almost completely with dolomite (or lime) at temperatures as high as 1600°F.

$$MgO \cdot CaO + CO_2 \rightarrow MgO \cdot CaCO_3$$

$$MgO \cdot CaO + H_2S \rightarrow MgO \cdot CaS + H_2O$$

The dolomite can be regenerated by heating in the presence of air:

$$MgO \cdot CaCO_3 \xrightarrow{\Delta} MgO \cdot CaO + CO_2$$

$$MgO \cdot CaS + 1\text{-}1/2\ O_2 \xrightarrow{\Delta} MgO \cdot CaO + SO_2$$

By adding a small amount of shift conversion catalyst to the circulating dolomite, the shift reaction could be driven strongly toward H_2 and CO_2 production, since CO_2 would be absorbed as it was formed. Thus a combined "hot" purification and shift conversion could be performed in one step.

Perhaps other systems such as the CO_2 acceptor process can be developed. Such developments would be of great value to coal gasification processes.

Even if "hot" purification systems could not be developed, it seems that opportunities still could exist in the development of new low temperature processes. Simply from the sheer number of acid gas purification processes now available, chances seem good that new processes could be developed--particularly for the specific conditions and requirements in coal gasification plants. Some of these conditions and requirements now are:

- High CO_2 partial pressures in feed gas (~ 300 psi).
- H_2S concentrations of over 1% in the feed gas.
- H_2S removal to less than 0.5 ppm (required to prevent poisoning of methanation catalysts).
- Selective removal of H_2S so that facilities for recovery of sulfur as elemental sulfur can be minimized in cost.
- Almost complete removal of organic sulfur (i.e., COS and CS_2) for protection of methanation catalyst.

The requirements for almost complete removal of sulfur compounds could change if a sulfur resistant methanation catalyst could be developed--another opportunity for development.

The last step in making CH_4 from coal is the catalytic methanation step:

$$3H_2 + CO \xrightarrow[700^\circ F]{\text{Catalyst}} CH_4 + H_2O + 94{,}000 \text{ Btu}$$

This step has not been performed commercially under conditions in a coal gasification plant--the most important variance from current commercial practice being the high mole percent CO in the methanator feed gas.

Current commercial practice is limited to hydrogen producing plants where toxic CO is methanated, but the methanation feed gas usually contains less than 1% CO. In this case, the heat of reaction is easily taken up by the product gas with only a slight rise in temperature, and equilibrium conversion is still attained. However, in a coal gasification plant, CO content typically might be 15 to 20 mole percent. In this case, reaction heat must be removed from the reactor to hold temperature down for good equilibrium conversion. Also, temperature within the reactor must be controlled closely to prevent hot spots and attendant carbon deposition.

The Institute of Gas Technology, the U.S. Bureau of Mines, and other

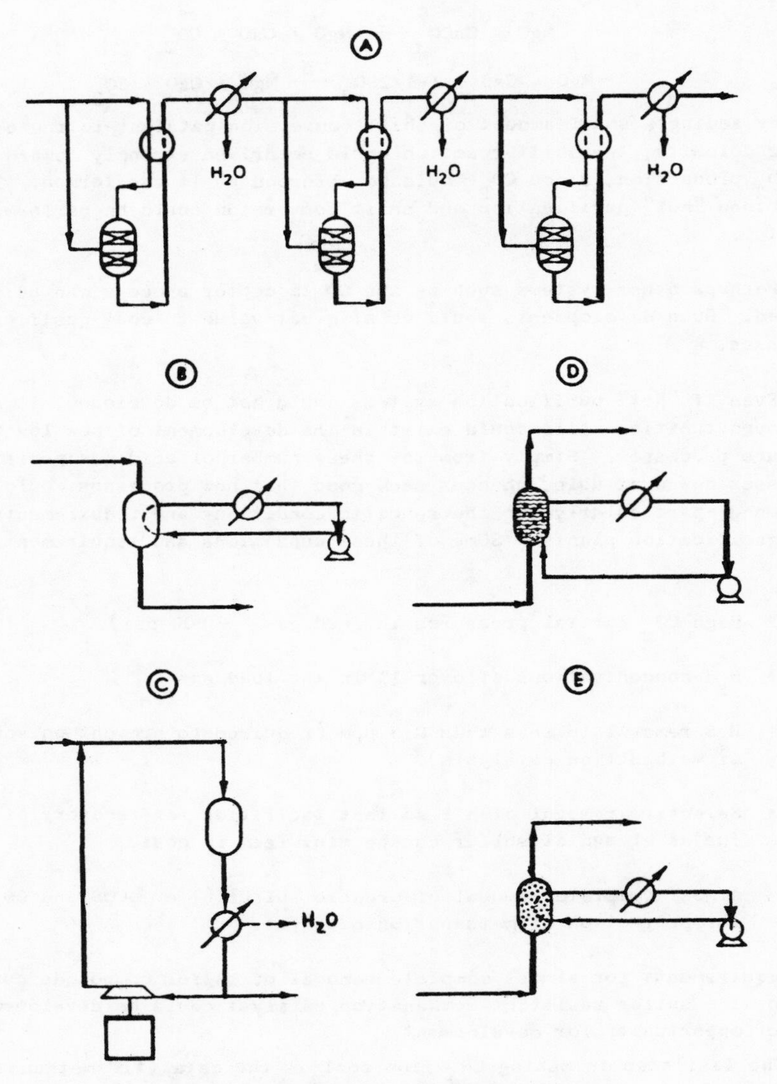

Fig. 3-13 CATALYTIC METHANATION SCHEMES

groups have conducted research into proposed methanation systems for this application for several years. However, at this time, optimized systems--from both cost and operability standpoints--have not been arrived at. Thus, this area is ripe for engineering opportunities.

Several schemes for controlling reaction heat during methanation are now being investigated. Some of these are shown in Fig. 3-13.

Scheme A is a "cold gas quench" system. In this system, a multistage fixed catalyst bed reactor system is used with cold feed gas taking up the heat of reaction between stages.

Scheme B is a "heat extraction" system in which the heat of reaction is taken up by a circulating fluid. The catalyst in this scheme can be in packed beds or flame sprayed on the exterior of the tubes through which the heat removal fluid is circulating.

Scheme C is a "hot gas recycle" system in which large quantities of product gas are recycled to the reactor to dilute the CO content of feed gas so the heat of reaction can be taken up in one pass through the reactor. Reaction heat is removed from the product gas before recycle. The recycle gas is maintained at a temperature of about 550°F to preheat feed gas to 500°F, thus the term "hot gas recycle." In this scheme, the catalyst can be in a fixed bed or it can be flame sprayed on internal surfaces.

A wet fluidized bed (or ebullated bed) reactor is used in scheme D. In this system, the catalyst is kept in a suspended state by the velocities of circulating liquid and feed gas. Reaction heat is removed from the circulating liquid external to the reactor.

A dry fluidized bed reactor is used in scheme E in which the catalyst is fluidized by feed gas, and heat is removed by fluid circulating through tubes immersed in the bed.

In addition to the numerous schemes being investigated for methanation, the catalyst can be nickel, cobalt, molybdenum, iron, ruthenium, or combinations of these on various supports. As previously mentioned, a sulfur resistant catalyst would be the most valuable. At this time, optimized catalyst and operating systems have not been proven. Thus, large opportunities exist for development in this area.

ADVANCED RESEARCH IN RAPID EXCAVATION FOR OIL SHALE RECOVERY

Statement of the Problem

The application of rapid excavation technology in routine operations remains to be demonstrated. The Committee on Rapid Excavation of the U.S. National Research Council estimated that the minimum market in the 1970-90 period will be $69 billion--about evenly divided between public works and mining. These estimates were made before the current national concern with protection and clean-up of the environment developed. That concern, with the heavy emphasis it portends on urban mass transportation and independent sewage and storm water disposal, will increase the public sector demand for tunneling under cities. At least equally important will be the application of rapid excavation techniques to encourage economic development of mineral and energy resources such as oil shale. Finally, rapid excavation technology could be used in construction of hardened military sites, storage facilities, and transportation or communication arteries at strategic sites.

The potential for rapid excavation technology for oil shale development is expecially promising. Oil shale is essentially a marlstone that contains organic matter (Kerogen) that can be recovered by heating and converted into a crude oil. However, because the Kerogen is only a small fraction of the total rock, large volumes of oil shale must be mined to supply conversion plants (a commonly used figure is 34 gallons of recovered shale oil per ton of rock). Thus, a relatively small plant supplying only 50,000 barrels per day of shale oil would require from 62,500 to 72,500 tons of raw shale per day; larger plants would require proportionally more shale. These large quantities of oil shale will probably have to be mined by underground methods because overburden is too thick for surface mining, even if environmental effects were accounted for.

Conventional underground mining methods for oil shale recovery require cyclical operations--preparation, drilling, blasting, removal of broken rock, and roof bolting. This would have to be done on a massive scale--rooms and pillars 60 feet square are envisioned in most treatments of oil shale mining. There is no doubt that cyclical mining methods can provide oil shale for conversion, since this has been demonstrated in pilot operations by the Bureau of Mines and private companies. It has not been shown that conventional mining is capable of providing a continuous, high volume supply of oil shale necessary for operation of a conversion plant. In this respect, the application of rapid excavation technology to oil shale development appears to show exceptional promise.

Objectives

The general objective of the program is to produce information, technical data, and demonstrated methodology that will enable assessment of the cost and opportunities for rapid excavation applications to oil shale recovery. Specific objectives are to focus on obtaining data on the properties of oil shale critical for design of excavation equipment and practices, on applying research results, and on anticipating and solving new problems and bottlenecks in the several related technologies that support rapid excavation operations.

Recommended Future Research Topics

A number of research topics could be attacked in appraisal of the potentials for application of rapid excavation technology to oil shale recovery. For preliminary discussion purposes, topics might be chosen from the following list.

- Oil shale weakening or cutting methods.
- Handling of mined shale.
- Advanced geological exploration.
- Improved grouting procedures for water control.
- Improved methods of analyzing the interrelationships

of excavation operations.

- Mechanical boring of noncircular underground openings for haulage and disposal.
- Treatment of processed oil shale and reclamation of mined lands.

Oil Shale Weakening or Cutting Methods

Improvements in the rapid excavation of oil shale will require improvements in materials used for bits and cutters to provide better wear-resistance and greater life expectancies and to develop new techniques to weaken the shale so that existing cutting materials can be used. Oil shale is a tough, resilient rock that will present excavation problems (especially in bit wear) unless it can be weakened to allow greater ease in excavation. Since the economy and efficiency of tungsten carbide alloys in cutting equipment are not likely to be rivaled by other materials at an early date, greatest promise for improvement in oil shale excavation appears to be in reducing the rock's resistance to breaking, either by physical alteration of rock or new methods of fragmentation.

Oil shale may be weakened by thermal and chemical treatments, and such treatments can reduce the work necessary to fracture laboratory-scale samples by 50 percent. It would be helpful if a means could be developed to combine rock treatment with partial fragmentation through controlled shock or by new techniques using explosives.

One approach would be to modify a rotary tunnel boring machine to treat and excavate the oil shale essentially in the same operation. A heated spray of an appropriate solution directed at the face would precede the excavating machine. A series of advance cutting tools would cut kerfs in the face and create additional surface area to facilitate action of the heated chemical spray on the oil shale. The main action of the boring machine would be a series of conventional cylindrical or cone-shaped cutter bits that would act on the weakened rocks between the kerfs. One problem with this approach would be the disposal of water used to spray the face; however, it is conceivable that recovery of this water could be used to facilitate the removal of mined shale as a slurry.

Handling of Mined Shale

Progress in rapid excavation for oil shale recovery will require improvements in material handling technology to remove mined shale at rates comparable with advance of the excavator. Cyclical loading and haulage systems for shale removal appear to be inadequate to achieve the required capacity for conversion operations, and continuous systems appear to have great advantages.

Conveyors for transportation of rock fragments are well known, and the feasibility of using pipelines to carry finely divided rock as slurries has been demonstrated in dredging, coal pipelines, and laboratory work. If a means could be found to implement a crusher and hydraulic system required for an efficient oil shale pipeline, it may be possible to achieve oil shale removal rates of from 25 to 50 cubic yards per minute. This should meet peak capacity requirements of 20 feet per hour of borings up to 30 feet in diameter (although this is only half the height of target shale areas, a two-level mining bench approach could be used).

One approach to the problem would be to develop a self-contained pipeline intake unit that would follow the excavation equipment in the mine. The pipeline intake unit would comprise a crusher, a hydraulic unit to mix the crushed oil shale with water introduced from the surface to the unit under pressure, and a high capacity pump to drive the resulting slurry through a pipe to a stockpile area for plant feed. Flexible pipe could be used for the interval immediately following the unit to facilitate advances. After a sufficient advance, the flexible pipe could be connected to more permanent pipes for intake and outflow.

The pipeline unit might be a companion piece to a mechanical boring machine, following it directly and receiving the discharge of oil shale fragments from the borer's conveyor. The pipeline unit could also be used in excavations resulting from more conventional means, if fed by a ramp-type loading machine and shuttle car similar to those used in underground coal mining.

Advanced Geological Exploration

Geological conditions are the main factor in determining the difficulty of rapid excavation. Advance determination of rock characteristics is important both in establishing excavation parameters before excavation and in defining geologic factors ahead of excavations in progress. This is especially critical for oil shale because of its horizontal and vertical variability.

Present geological and geophysical techniques for defining geological conditions were developed for exploration purposes, and they lack the precision required for engineering measurements associated with surveying for excavation dimensions or characteristics. If a means can be found to adapt existing geological techniques to the small targets presented by individual excavations, it should be possible to predict the geological conditions to be encountered in such operations.

An approach to the problem of exploring prospective excavations would be to combine programs of rapid core or full-hole drilling with systematic measurements by modified in-hole logging equipment. In addition to more conventional methods, investigation of acoustical characteristics of rocks at several frequencies may be capable of identifying geological structures to determine their amenability for rapid excavation technology. It probably will be necessary to correlate such data with those from other sources, since the complexity of acoustical returns will make interpretation difficult. Such possibilities of determining rock characteristics by measuring reflections from a radioactive source within a bore hole should be examined.

The problem of exploring geological conditions ahead of rapid excavations in progress could be approached by modifying existing boring equipment to permit small core drilling along the center line of, or in a directionally drilled hole roughly parallel to, the direction of advance of the main excavation. In addition to obtaining rock samples from the advance area for study, the pilot or side drill hole could serve as a means to carry out logging measurements to define geophysical factors along the route such as the presence of ground water or faulted or bad ground. Measurements could be performed by an instrument package incorporated into the boring machines to facilitate routine operations without requiring curtailment of operations.

Similar instrumentation could be mounted on a jumbo for use with normal operations in conventional excavations.

Improved Grouting Procedures for Water Control

Stability of underground openings is vital to the success of programs in rapid excavation, especially for large openings such as in oil shale mining. Accordingly, the rate of installation of ground support measures must keep pace with the advance of the excavation, and the measures must be compatible with the underground environment. Grouting to reduce water inflows and contribute to added strength or prevent subsidence is an important part of the overall problem of excavation stability, especially if grout can be applied quickly. Limited data for the

Piceance Creek oil shale area of Colorado suggest that water could present formidable problems in large scale mining operation.

Development of chemical grouting materials has shown that certain silicates, resins, and polymers can effectively reduce the porosity and permeability of several soils and rocks. If a means can be found to apply grout materials to moist rock, and if greater efficiency of application can be devised to permit lower grout costs, the stability of excavations could be improved.

One approach would be to continue research on new chemical grouts, the influence of moisture and soil character on grout setting processes, and the means of grout application. The ideal grout would be low in initial cost and applied in one operation. Polymeric materials whose viscosity can be varied through adjusting mixtures may be applicable to a number of situations, simplifying the equipment and processes required.

A complementary approach would be to develop mixing and application equipment that will apply the grout to the excavation in a single operation. The equipment should be sufficiently versatile to accomodate suspended-solid grouts, such as cement, in addition to the chemicals.

Improved Methods of Analyzing the Interrelationships of Excavation Operations

The rapid excavation process comprises a combination of several distinct operations. Each excavation project attempts to apply an optimum combination of separate components to achieve its objective. The result is a number of special cases and a lack of standardization which impairs the incentive to invest in development of compatible units of equipment that can be used widely.

Analysis of the capacity and output of separate operations (and the equipment used) in relation to the efficiency of the subsequent operations to determine the highest efficiency of the overall process can yield the information required to identify improvements in existing equipment and practices.

Mechanical Boring of Noncircular Tunnels for Haulage and Disposal

Oil shale operations will require underground haulage of mined shale to the conversion plant, as well as disposal of processed shale from which Kerogen has been recovered. This could require special excavations and present special problems.

Tunnels for transportation should in most cases be 0 or horseshoe shaped. Mechanical boring methods for hard rock produce a circular shape. This requires cutting about 20 percent more rock than is necessary for the needed cross section.

An approach to this would be to have reaming cutters mounted tangentially at the rear of a boring machine. These cutters would remove a crescent-shaped cross section to provide a horseshoe shape.

Treatment of Processed Oil Shale and Reclamation of Mined Lands

Oil shale mining and processing removes only a small portion of the total rock (less than 1 percent by weight). It is not a simple task to dispose of the remainder. The acts of mining, crushing, and processing reduce the size of oil shale fragments and increase its volume by about one-third; even if the mine openings could be completely refilled, considerable amounts of processed oil shale would have to be disposed of at the surface--closer to two-thirds of the total material handled. Processed shale will be finely divided (especially when mined by rapid excavation techniques) and will contain water soluble salts that could contaminate streams or ground water unless steps are taken to protect the disposal area. Control of erosion of disposal areas through revegetation also remains to be demonstrated, and species of plants and fertilizer/nutrient requirements to support vegetative growth on processed shale need to be determined.

One means of finding the solution to treatment of processed oil shale and reclamation of mined lands or disposal sites is to examine factors such as those briefly described above in connection with operations conducted or in progress at the vicinity of the Naval oil shale reserves in Colorado. Previous work in that area produced quantities of oil shale for testing; the demonstration projects should provide a basis for conducting carefully structured experiments to examine the potential environmental effects of oil shale development through rapid excavation so as to determine prospective solutions to be incorporated into the design of excavation operations.

MATERIALS REQUIREMENTS IN ADVANCED ENERGY CONVERSION SYSTEMS

Statement of the Problem

As the demand for energy grows and conventional resources dwindle, the need for new energy generation and conversion systems becomes more pressing. Immediate objectives must include the development of new techniques for the more efficient recovery of known fossil fuel resources, together with evolutionary developments of current conversion systems, to allow known lower-quality resources such as shale oil and high sulfur coal to be used. At the same time, the developmental activities related to nuclear power should be supplemented by development of entirely new or underutilized energy systems--thermonuclear, MHD, fuel cells, solar and geothermal power--to ensure the continuing adequate supply of clean energy.

Both objectives will rely for their success on the availability of materials able to withstand ever more rigorous operating conditions. In many cases suitable materials are not available today to meet the demands of the future, and in many areas our basic understanding of the behavior of materials is too fragmentary to allow the materials scientist to develop improved materials on other than an empirical basis. Failures occurring during early service experience with new materials, which are a probable consequence of this method of materials development; are not likely to be greeted with enthusiasm either by the power generating authority or by an increasingly vocal, environmentally conscious society. Thus, it is important to attempt to assess the materials requirements that must be satisfied if advanced energy generation and conversion techniques are to succeed and, by matching these requirements with the properties of existing materials, to determine the areas where additional research and development are most urgently needed.

State of the Art

Recent technoeconomic studies of the future market for advanced materials in the electric power, process, and aerospace industries conducted by SRI indicate present or potential materials problems in all the energy generation and conversion systems considered. There is a striking consistency in the nature of the materials problems identified irrespective of the system considered; thus, it is possible to classify the future materials requirements into four major areas that apply to a greater or lesser extent to all the advanced systems so far proposed.

Materials for High Temperature Service

Many of the advanced systems feature very high operating temperature mechanical properties. Conventional nickel-based superalloys have reached a stage of development where it is difficult to foresee further substantial improvement in properties. Development of high temperature alloys based on the refractory metals necessitates the simultaneous development of long life, protective coating materials to provide adequate oxidation and nitridation resistance. The development of some form of self-repairing coating appears necessary before refractory metal alloys can be used in long term high temperature application with any confidence.

The use of ceramic materials appears more immediately promising, although a more detailed understanding of the interrelation between mechanical properties and microstructure of these materials is needed, for example, to allow optimization of high temperature creep behavior while retaining acceptable thermal shock and impact resistance. Additional practical experience of the problems encountered when ceramics are used in applications for which prior experience relates only to metals also appears to be essential.

Materials for Long Term Service in Aggressive Environments

This area dominates all others in advanced energy conversion systems and at the same time is the area where our present knowledge and understanding of the important factors are most rudimentary and incomplete. Even the seemingly innocuous step of diluting natural gas with a few percent of air can under certain circumstances lead to unforeseen catastrophic corrosion problems in both copper and steel piping. The potential problems in advanced energy systems are usually much more obvious than this, but a major difficulty remains that of considering in advance every circumstance that might lead to an environment-related failure during a 30-year or longer service life.

Another difficulty is to maintain adequate environmental resistance while satisfying other materials requirements. An illustration in a different field of application is provided by the development of a titanium alloy suitable for the hull of a deep-diving undersea vehicle. The requirements for this application include moderate strength, high toughness, good weldability and formability, as well as resistance to corrosion and stress corrosion cracking in sea water. The weldability requirement necessitates the use of an all alpha or near-alpha composition, and the first alloy developed for the application was the near-alpha Ti-8Al-2Cb-1Ta. However, since this alloy was subsequently found to exhibit an ordering reaction that severely impaired its toughness, the aluminum content was reduced to 7 percent. The resultant Ti-7Al-2Cb-1Ta alloy was delivered in considerable quantity before its susceptibility to accelerated crack growth in seawater was discovered. This necessitated a further reduction in aluminum content to 6 percent, which resulted in the alloy being too weak for the intended application. A 1 percent Mo addition was found to increase the strength sufficiently without leading to seawater stress corrosion susceptibility, and the resultant alloy Ti-6Al-2Cb-1Ta-1Mo is currently a top contender for a role as a marine structural metal. Its development illustrates the typical problems encountered when a range of properties must be optimized simultaneously.

The range of possibly damaging environments encountered in advanced energy generation and conversion systems is very wide and therefore will be summarized briefly with a few illustrative examples. Aqueous environments are commonly encountered in cooling and heat transfer systems and can lead to stress corrosion and corrosion fatigue as well as general corrosion problems. A wide range of prior experience exists from conventional systems. In most cases, suitable materials already exist, provided close control of solution chemistry is feasible and the service stresses to which the component is subjected are well understood. However, they have not always been used in the past because their use commonly requires a cost premium.

Experience has tended to show that the use of the best material available is often cost-effective in the long term even allowing for higher initial costs. For example, the British electric power industry is now seriously investigating the use of titanium tubing in steam condensers, taking the view that the initial materials investment (which is increased by more than an order of magnitude) will be more than offset by the decreased downtime and repair costs over a 30-year lifetime. In hot water reactors, additional problems can result from superimposed effects of radiation; for example, radiolytic decomposition of water can lead to a potential hydrogen embrittlement situation.

A major barrier to fundamental investigations of corrosion behavior in hot aqueous solutions has been the absence of adequate basic research techniques. For example, it is necessary to develop reliable reference electrodes to make electrochemical corrosion measurements in high temperature, high pressure environments. This is an area where extensive additional work is clearly indicated.

Liquid metals are widely proposed for use as coolants in advanced systems, and metal vapor/liquid loops appear to offer potential efficiency improvements. In this area, little or no backlog of service experience exists, and basic knowledge of the potential containment problems is in its infancy. Liquid metal embrittlement is a well known metallurgical phenomenon and, in long term service, mass transport of alloy constituents by hot liquid metals may lead to major problems. We need to know much more about the behavior of both metals and ceramics in the presence of liquid metal vapors before these environments can be used with confidence. Localized corrosion, stress corrosion, and corrosion fatigue behavior must be investigated in detail. Once again, the basic techniques required for generating the needed data are only in the early development stages. We also lack the instrumentation for continuous monitoring of the chemistry of the liquid metal coolants that is required, for example, to determine mass transport rates of carbon in service.

Environment-related problems are also encountered in most types of fuel cells and to a lesser extend in more conventional energy sources as well. Gas and fuel oil transmission pipelines and distribution systems are particularly subject to problems and may become increasingly so as lower quality resources are developed and put on-line. Most problems arise because of specific combinations of small quantities of impurities (e.g., sulfur compounds plus amines plus oxygen in natural gas) which are more likely to be found in fuels derived from lower quality resources.

Improved Materials for Structural and Mechanical Applications

The needs of advanced energy systems in this area are not very different from those of other advanced systems. Materials are required with greater stiffness, strength, and toughness than those available today. Composite materials offer the potential of greater stiffness and strength and may find wide application. Uranium enrichment by the centrifuge technique may become more attractive if advanced composite materials can reliably attain their projected properties, and novel propulsion systems based on composite material flywheels can become significant as fossil fuel resources decline. Emphasis on failure safe design will encourage the development and use of materials with greater fracture resistance than those available today, under both static (fracture toughness) and dynamic (fatigue) loading. Improvements in nondestructive testing and inspection techniques will also be required, aimed at decreasing the minimum detectable flaw size and hence improving our ability to identify potential sources of catastrophic failure in critical structural members.

Improved Fabrication Techniques

As new materials become available, fabrication techniques must be improved to enable us to fabricate useful products from them. Changes in design practice will also be needed--for example, design principles developed for relatively isotropic homogeneous metals cannot be applied when highly anisotropic materials such as composites are used. Design experience in the use of relatively brittle materials such as ceramics is also needed for cases in which other considerations dictate their choice.

Although of secondary importance, the use of energy-conserving fabrication techniques such as powder metallurgy and superplastic forming should be encouraged. Moreover, resource conservation as well as energy conservation appears increasingly critical. Fabrication techniques that lead to much higher levels of materials utilization than those common today are needed, and new processes must be developed to effectively and economically recycle mineral values that are currently wasted.

PRODUCTION AND UTILIZATION OF NONHYDROCARBON CHEMICAL FUELS

Statement of the Problem

It is widely recognized that "premium" fossil fuels (natural gas and low sulfur oil) are already in short supply and that all fossil fuels (including coal, lignite, tar sands, and shale) will be in short supply within a few generations at projected growth rates for total energy consumption. The priority that one places on the need for alternative, nonfossil-based chemical fuels depends on the time scale of the scenario under consideration. For the shorter term of one or two decades, the major priority is likely to be placed on the need for transformation of the relatively more abundant fossil fuels (coal and lignite) into the less abundant premium forms. Nevertheless, it is appropriate to consider what research should be devoted to the long term total energy problem whose solution will require chemical fuels made from nonfossil fuel sources.

The current and projected increased emphasis on pollution abatement makes it desirable to develop and stimulate the use of new, inherently less polluting fuel types. Here it is important to assess not only the principal products of combustion for each fuel but also the products resulting from incomplete combustion, from combustion of fuel impurities, and from side reactions such as nitrogen oxide formation.

Any consideration of new fuel types must take into account questions of safety, reliability, and economy. In regard to the latter, it is necessary to develop fuel types that are not only cheap to produce but are also cheap to transport and store and can be efficiently utilized.

Most of the above needs are not specific to defense fuel requirements, but apply quite generally to the total national energy plan. Many other needs are more easily identified with military goals; e.g., the need for fuel types offering advantages in portability, in specific energy storage capacity, in high power density or specific impulse, in light weight or added buoyancy, or in the capability of the fuel to perform other useful functions such as structure cooling.

This section is concerned with a relatively limited category of fuels, namely those that are not primarily hydrocarbons (a criterion that excludes essentially all fossil fuels) and that are vigorous reducing agents rather than vigorous oxidizing agents (which excludes such oxidizers as fluorine, NO_2, HNO_3, and ozone). Some potential candidate fuels, like sulfur or H_2O_2, can act either as oxidizing or reducing agents but are excluded here because their main fuel potential is as oxidizing agents.

Thus, this section is limited to the following fuel types:

- Hydrogen
- Inorganic hydrogen carriers (ammonia, hydrazines, silanes, boranes)
- Partially oxygenated carbon compounds (CO and CH_3OH)
- Active metals (e.g., Li, Na, Al, Mg, Zn).

It may be noted that solid hydrogen carriers such as metal hydrides are considered to be a variation of hydrogen itself, rather than to be in the second category.

State of the Art

The following discussion will attempt to cover for each fuel type, the current status of methods of production, utilization, storage and transmission and to identify the principal future opportunities and any critical technical barriers that must be overcome.

Hydrogen

Hydrogen is now produced commercially primarily by catalytic steam reforming of natural gas. This will doubtless remain the most economic production method as long as natural gas (or synthetic natural gas made from other fossil fuels) remains available at low cost. Other available methods for hydrogen production include the catalytic partial oxidation of hydrocarbons, the steam-iron method, and the electrolysis of water. Water electrolysis is a likely first step in the preparation of deuterium, which is a fuel for thermonuclear reactors. Therefore, hydrogen will be a readily available by-product of thermonuclear power generation.

Hydrogen finds practically no present use as a conventional fuel except in mixture with CO and other gases. Liquid hydrogen is highly successful as a rocket fuel and as a fuel cell material for power generation in spacecraft. Its disadvantages include its relatively high cost and its relatively hazardous character, the latter because of its wide explosive limits and its tendency to diffuse through metals. Because of the above factors, hydrogen is used industrially mainly as a captive intermediate to make other chemicals such as ammonia and methanol.

Hydrogen fuel cell technology is well advanced technically but still economically uncompetitive. Hydrogen has been evaluated as a non-polluting fuel for internal combustion engines and has been found to be eminently suitable, if appropriate engine modifications are made to counteract its tendency for preignition.

In regard to storage and transmission, hydrogen is handled almost exclusively in one of two forms: as a cryogenic liquid or as a highly compressed gas. Both methods are relatively costly and hazardous. Pipeline transport of hydrogen has been used between nearby petrochemical plants but has not yet been demonstrated on the scale of natural gas pipelining.

In contrast to the relatively modest current utilization of hydrogen, it is widely regarded that tremendous opportunities exist for growth in its utilization. As Gregory[1] and Weinberg and Hammond[2] have pointed out, unlimited hydrogen from the sea could serve as the key to synthesis of many kinds of potentially scarce resources, including metals and portable fuels.

Although electrical energy is undeniably convenient and clean, it suffers from a number of technical disadvantages, foremost among which are its lack of storability (without prior conversion to another form) and its high transmission cost over long distances. The reversible hydrogen fuel cell offers a simple, efficient way to store off-peak electrical energy, thereby reducing the capital requirements for future nuclear electric power plants. (The same advantages would also apply to fusion powered or solar powered electric plants, when and if these become realities.)

In regard to transmission costs, underground electric power lines are reported to cost 10 to 40 times as much as common overhead lines of similar capacity. Since we are accustomed to moving large quantities of natural gas across the country in pipelines, the same approach can be applied to hydrogen. Even today, it should be possible to make and deliver energy in the form of hydrogen more cheaply nationwide than the average selling price of electricity.

Even before the projected depletion of fossil fuel resources makes the cost of natural gas exceed that of nuclear-electric-based hydrogen, the "hydrogen economy" will probably be justified on the basis of resource conservation and pollution abatement. Schoeppel[3] has pointed out that hydrogen is substitutable for any other mineral fuel currently available. Its use has already been demonstrated in reciprocating internal combustion engines and turbines, and its adaptation to external combustion engines would be even simpler. Its light weight and high energy density as a liquid give it desirable storage features of particular value in aircraft and space vehicles. Its ability to chemisorb reversibly with certain metals to form hydrides enables it to be conveniently stored at ambient pressures and temperatures, thereby potentially making it even less hazardous than gasoline.

According to Witcofski,[4] a distinguishing feature of the projected hypersonic transport (HST) airplane will be its use of liquid hydrogen fuel, which has 2-3/4 times the energy per pound of conventional JP fuel. This large energy density more than compensates for the reduction in aerodynamic efficiency ascribable to housing the low density fuel. In addition, the large heat sink capacity of liquid hydrogen (10 percent of the combustion energy) allows active cooling of the airframe, which can thus be made of conventional aluminum structures. It is significant to the future prospects for a hypersonic transport (speed above about Mach 3) that it may avoid or overcome some of the environmental problems so critical to the decision to halt development of the U.S. supersonic transport (SST).

The problems that must be solved before hydrogen fueled vehicles can be universally accepted are: (1) the design and construction of adequate hydrogen production, handling, and distribution systems, (2) the development of hardware necessary to store and utilize hydrogen efficiently, and (3) the overcoming of the generally held but undeserved public attitude that hydrogen is too hazardous.

Inorganic Hydrogen Carriers (ammonia, hydrazine, silanes, boranes)

Ammonia production technology, now very advanced and economic, is based on the high pressure catalytic reaction of hydrogen and nitrogen.

Hydrazine, in turn, is produced mainly from ammonia by reaction with sodium hypochlorite, followed by dehyrdation and distillation. Silanes are made by acidic solvolysis of electro-positive metal silicides such as Mg_2Si. Boranes other than diborane are produced by pyrolysis of diborane, whereas the latter is synthesized by reaction of a metal hydride with a boron halide.

Most of today's ammonia production goes either directly into agriculture or into nitric acid manufacture. Ammonia has been evaluated as a fuel for internal combustion engines but has not found favor because of its very low combustion rate and its potential for formation of nitrogen oxides. Ammonia fueled engines are also subject to emissions of considerable unburned ammonia, a pollutant in its own right. The technology of ammonia fuel cells is highly advanced.

Hydrazine and its methyl-substituted derivatives have high heats of combustion and high specific impulse, which make them excellent rocket fuels. They also have good potential as fuels for fuel cells, except for their high cost. They are all extremely toxic and are subject to a detonation hazard because of their instability.

Silanes and boranes are very hazardous, being toxic, hygroscopic, and pyrophoric. All have high heats of combustion, making them attractive as rocket propellants. However, interest in them has waned because of the difficulty of exhausting the condensed phase combustion products, the inadequacy of engines for utilizing their available energy, and their high cost.

Ammonia, hydrazines, silanes, and boranes are all easily and economically stored and handled as liquids, at temperatures much more readily attained than those for liquid hydrogen. Pipelining of liquid ammonia is soon to be put into commercial practice on a large scale.

<u>Partially Oxygenated Carbon Compounds (carbon monoxide and methanol)</u>

Carbon monoxide is produced (generally in admixture with hydrogen and other gases) by various modifications of partial combustion or controlled oxidation of fossil fuels, either with air or steam. Its high toxicity, low heating value, and low boiling point all tend to favor its role as an intermediate to other fuels, rather than as a fuel for general use (except in low Btu gas mixtures, which appear to have a future for power generation and other industrial applications).

Methanol is produced by the high pressure catalytic reaction of carbon monoxide and hydrogen in equipment similar to that used for ammonia synthesis. It is a clean-burning fuel that has been used successfully in internal combustion engines and in fuel cells. It is being seriously considered as a substitute for liquefied natural gas, because of its easier storage and transportability. Its vapors are toxic, and it has only about half the heating value of methane on a weight basis.

The real opportunity for large scale use of methanol may come when liquid fossil fuels ultimately become depleted. Then it may be necessary to develop a methanol synthesis route for which the carbon monoxide is derived by hydrogen reduction of atmospheric carbon dioxide, as suggested by Williams et al.[5]

Active Metals (e.g., Li, Na, Al, Mg, Zn)

With the exception of zinc, which can be liberated from its ores by roasting and carbon reduction, all the active metals are made by electrolysis of their salts.

The principal fuel use of these and other metals is as the anode material in high energy density batteries. Lead/acid, nickel/cadmium, and nickel/iron secondary batteries will probably continue to provide rechargeable energy sources for applications where their low energy density and high cost per unit of power are unimportant. However, the limitations of the present storage batteries have stimulated research in high energy density systems such as (1) secondary zinc/air cells, (2) organic electrolyte cells, (3) molten salt cells, and (4) refuelable batteries.

The military has used mechanically refuelable batteries for communications equipment,[6] based on zinc/air, aluminum/air, magnesium/air, iron/air, and cadmium/air cells. None of the existing cells meets the requirements projected for the ultimate nonfossil fuel society. The improvements that must be made to achieve energy densities of 100 watt-hours per pound and power densities of 75 to 100 watts per pound at acceptable costs include:

- Develop lower cost cathode catalyst
- Reduce irreversibility of the cathode
- Reduce heat generation at rated current.
- Develop improved removal of discharged anodes

Avoid adsorption of CO_2 by alkaline electrolyte

Present Activities and Organizations

Bobo[7] gives an extensive list of federal, state, and independent agencies and their fuel information categories. Only a few agencies, such as EPA, AEC, and NASA, regularly publish lists of active research contracts. One of the most convenient tabulations available is the (copyrighted) Market Intelligence Report, published by DMS Inc. That report includes sections dealing with "Combustion and Ignition," "Energy Storage," "Fuels," and "Rocket Propellants." The report lists the major Army, Navy, Air Force, and NASA sponsors for each designated research topic, and gives the current funding levels for each program. What is needed is a systematic analysis of these topics for their relevance to energy-related matters treated in this report.

Recommendations for Further Studies

The development of automatic dual-fuel internal combustion engines appears to warrant a high priority effort. It is known, for example, that the addition of about 2 percent hydrogen to ammonia or to conventional fuels can permit lean operation, with greatly reduced nitrogen oxide formation. What is really needed is a dual-fuel system that will meet all safety and logistic requirements. This will probably entail improved methods for reversible storage of hydrogen in the form of hydrides, plus methods for odorization of hydrogen, for leak detection.

The concept of storage of combustion heat by dissociation of hydrides, with recovery of heat during recharging of the hydrides, needs to be demonstrated on a practical scale.

In the area of fuel cells, an urgent need still exists to develop cheaper and more effective catalysts, and to develop ways to use existing expensive catalysts more efficiently.

None of the existing refuelable battery systems meets projected future requirements, and improvements should be sought in reducing heat generation, reducing cathode irreversibility, and reducing catalyst costs.

As a long term goal, perhaps the most important effort should be the development of large scale reversible hydrogen fuel cells to be used as peak electrical energy storage devices in conjunction with advanced generating systems, whether they be conventional nuclear, breeder nuclear, solar, geothermal, or fusion powered.

WASTE HEAT FROM ENERGY PRODUCTION PROCESSES

Statement of the Problem

A typical modern power plant using pulverized coal discharges in the range of 60 percent of the input energy to the atmosphere as waste heat. About 10 percent of this waste goes up the stack with combustion gases and 50 percent goes out with the condenser cooling water. Although nuclear power plants claim to be relatively clean, the waste heat burden from these plants is even higher than for fossil fuel-fired plants. Almost two-thirds of the heat released in the reactor has to be dumped into the environment. This thermal waste problem becomes of greater importance as our energy resources decline. The ecological implications of this low-level heat burden are little understood. However, the diseconomy of discarding two-thirds of our energy resources as low grade heat indicates its importance to the energy supply situation and emphasizes the need for research.

State of the Art

Statements have been made for a dozen years or more that the total energy concept would have to be used at all levels to conserve energy and reduce pollution levels. Action at management levels in both the power industry and government is clearly necessary.

Over the past years, a large number of intelligent suggestions have been made for the deposition of waste heat. The AEC and the Interior Department have studied for several years the feasibility of using the waste heat from nuclear plants for water desalting. However, no nuclear plant now in construction or being planned has such provision for the use of its waste heat. A pioneering effort in the utilization of waste heat is now under way in Oregon in preparation for the use of heated water, when the 1,000 MW Trojan nuclear plant comes on line, to provide frost protection and to extend the crop growing season. This demonstration project entails 170 acres in efforts to improve the productivity of a number of crops. The results of most other projects that supply heating water, air conditioning, and greenhouse heating have been small indeed compared with the total heat wasted.

One of the obvious approaches to end this waste is for the government to tighten up on licensing by requiring that total energy concepts be used in the development of future power plants. To prepare for such an eventuality requires that proven processes be available to the wasted energy.

A complex heat cycle was recently proposed by A. P. Fraas of the AEC to reduce thermal discharge by improving the thermodynamic efficiency of a power plant by raising the operating temperatures. This would be accomplished in a binary system using potassium as a working fluid in the primary boiler-turbine complex and supercritical steam in the secondary system. Such a cycle is claimed to be capable of a 54.8 percent efficiency compared with a maximum of 40 percent for present plants and would reject only half the heat of a conventional steam cycle of comparable capacity.

Yet there is a practical problem. The use of potassium, or for that matter sodium or cesium, as a working fluid in a turbine has never been attempted on the scale required to be practical. Not only are the thermodynamic and fluid properties not adequately defined, but the possibility exists that an insurmountably severe materials problem will exist not only in the boiler but also definitely in the turbine. Almost nothing is known of the interaction of such vapors with present materials or future materials of construction of turbines.

Another development aimed at improved thermal efficiency concerns the use of helium to operate a helium turbine in a closed cycle nuclear reactor loop. Such a closed cycle gas turbine is claimed to provide an efficiency as high as 48 percent with a side stream of process steam for industry. This is far above the 33 percent efficiency of water-modulated nuclear plants common in the United States.

The foregoing suggests the practicality of a confined power plant/industrial complex to utilize waste heat from the power plant.

Another energy combination to be considered is the power plant/oil refinery complex whereby the incoming crude is given a preliminary preheat by using it on the condenser side of the power turbine. This would not only conserve waste water discarded by the power plant but it would also reduce the use of bunker oil to preheat the crude oil before refining.

Present Activities and Organization

The AEC has been investigating the use of sodium as a reactor heat transfer fluid for a number of years. It is believed that the physical property investigations have not encompassed the pressure range required for turbine operations. Some of the material problems apparently have been solved, although such a reactor is not a commercial reality at this time. The influence of such a vapor on turbine blade material at high temperatures and pressures is almost wholly unknown.

Since the technology required for predicting thermochemical reactions for rocket combusion processes has been well established, advanced fuel cycles using hydrocarbons and coal with oxygen are for the most part predictable. Here again though, the time scale of development is limited by materials requirements. Most material problems in rocket technology have been solved by trial and error rocket firings, and these solutions for a few minutes of operation are undoubtedly not relevant to the year-round continuous operating required of power plants. It is highly likely that there will have to be a trade-off between the operational pressure and temperatures against the economic life of materials used for advanced boiler techniques such as those proposed here. NASA technology would appear to be the best starting point in this respect.

There has been no unified analysis of advanced power cycles as proposed here. Many of the proposals are susceptible to theoretical thermodynamic and economic analysis, which should qualify them in terms of priority for allocation of research and development funds.

U.S. Bureau of Mines has scheduled a pilot plant for a coal-gasification process utilizing a fluidized-bed gasification with oxygen and steam to yield methane, hydrogen, and carbon monoxide with a heating of 900 Btu per cubic foot. The possibility of using this gas for a power plant feed stream for reaction with oxygen is an alternative to be considered. Another is to complete the oxidation in the fluidized bed and generate steam for power turbine operation. Again, the point is to raise the efficiency of energy conversion by operating at higher temperatures and to lower the pollution potential of NO_x by not using air.

Recommendations for Further Studies

To better understand the implications of the waste heat problem, further studies may be made in several areas. They include:

- Economic and thermodynamic analysis of utilization of power plant waste heat for liquefaction plants and the use of binary generative cycles using high vapor pressure refrigerants.
- Feasibility studies of stock piling LNG in underground reservoirs. There has been some work done in this respect.
- Initiation of R&D in advanced power boiler cycles utilizing oxygen rather than air for the combustion of coal or synthetic gas.
- Study of the interaction of sodium or potassium vapor with advanced turbine materials.
- Feasibility study of a closed-loop energy cycle using an LNG/oxygen boiler for steam operation and an air liquefaction plant for waste heat utilization.

LIQUEFIED NATURAL GAS (LNG) USAGE

Statement of the Problem

Some energy consuming installations need independent utilities for reasons of security, economics, and/or environmental impact. Heat, electricity, water supply, and waste water or sewage treatment might be logically provided by an internal plant. LNG is an extremely likely source of energy for such installations because of its increasing availability and because of its minimal contribution to atmospheric pollution relative to other fossil fuels.

It currently costs about 12.2 cents per million Btu for LNG at the producing site and another 5.4 cents per million Btu to regasify this fuel before combustion at the using site. Bringing LNG from its liquid state at -160°C to a gas at 0°C requires about 200,000 calories per kilogram. In existing facilities the regasification cost and energy are not recovered.

Possible uses of the LNG to reduce energy costs include air separation, refrigeration at -15°C, and to enable the cooling condenser in a power plant to increase generation efficiency. One estimate of the power saved or recovered ranks these three uses about an order of magnitude apart (0.5, 0.04, and 0.003 kWh/nm^3 LNG, respectively). A joint venture of five Japanese companies selected air separation to use LNG coldness to produce liquid nitrogen and liquid oxygen.

The liquid nitrogen/liquid oxygen ratio can be varied somewhat by shifting the plant operating conditions. The liquid nitrogen is useful as an inerting blanket for certain materials and manufacturing processes and as a refrigerant. The liquid oxygen uses include rocket propellants, aviators' breathing supply, and water treatment. This last-named use is of most interest currently, because the oxygen is useful by itself or as an intermediate for ozone manufacture.

Oxygen and ozone are technically unique in waste water and potable water treatment in some instances where ammonia and phenol content are high and there is a possibility that chlorine disinfection forms toxic materials. When produced from oxygen instead of air, ozone can compete in cost with chlorine and hypochlorite disinfection, especially when new ozone generator designs are applied.

State of the Art

LNG Coldness Recovery

The aforementioned joint venture of five Japanese companies is constructing an air separation plant near Tokyo to use the coldness from LNG supplied at the rate of 8 tons per hour to manufacture:

Liquid oxygen - 7,000 Nm^3/hr
Liquid nitrogen - 3,050 Nm^3/hr
Liquid argon - 150 Nm^3/hr

Three other air separation plants are in various stages of planning and design in Japan.

Ozone Generator Designs

Welsbach, W. R. Grace, and AIRCO are designing and manufacturing ozone generators in the size range considered suitable for water treatment plants in the United States. Research in a number of U.S. and foreign companies is directed toward reducing the size, cost, and power consumption of the generators. Conspicuous by its absence is work on ozone generators capable of operation near the boiling point of liquid oxygen, even though one company believes a 50 percent conversion of oxygen to ozone can be achieved in a single pass at this temperature compared with the 3 to 5 percent yield at ambient temperature. The present generators are about 10 percent efficient in their utilization of electrical energy. The current published efforts are directed toward design of "resonant" generators that can use pulses of power more efficiently and toward higher frequency power supplies. There is no evidence of more sophisticated application of useful potentials by solid state switching techniques.

Ozone Use in Water Treatment

More than 1,000 cities in Europe ozonate potable water (including a 238 million gallon per day plant for the city of Paris). Philadelphia, Pennsylvania; Louisville, Kentucky; Washington, D.C., and Wyoming, Michigan, are operating plants to obtain economic and technical data on a useful scale.

Military Applications of Cryogenics

Liquid nitrogen is carried on some military aircraft as a refrigerant for infrared sensors. Liquid oxygen is used for breathing oxygen supply and as a rocket propellant for certain launch vehicles. (These vehicles could obtain improved performance from liquid oxygen/liquid ozone mixtures.)

Future uses being studied include liquid methane, liquid hydrogen, and liquid oxygen for fueling hypersonic aircraft where the cryogenics can be used effectively to relieve the heat burden on the aerodynamic surfaces.

Propulsion for such aircraft may have the capabilities for operation from jet fuel or hydrogen in the sensible atmosphere and hydrogen/oxygen in a rocket mode outside the atmosphere.

Some military magnetohydrodynamic applications require superconductivity obtainable with the cryogenics.

For all of these, the state of the art of cryogenic manufacturing and handling exists. The thermodynamic characterization is not complete for all such applications.

In addition to the intense commercial and industrial activities in air separation and ozone manufacture and use, the Navy and Air Force have performed significant research on liquid oxygen/liquid ozone as a rocket propellant.

The military, especially the Air Force, has a continuing need for cryogenics, including liquid nitrogen and liquid oxygen. The installations using cryogenics are remote and are likely to be self-contained for utilities. These installations have a need for potable and waste water treatment systems and chemicals, perhaps liquid oxygen and ozone.

The likely fuel for some such defense installations will be LNG, and its heat absorption capability of 200,000 calories per kilogram during regasification might be recovered in an air separation plant that could supply all or part of the liquid nitrogen and liquid oxygen requirements of the installation.

LIQUID FUEL TRANSPORTATION

Materials of construction, propulsion, and the critical technology for several new bulk fuel transportation systems and concepts are now available. In most instances, the critical design data are in hand.

The range of fuel transport systems of possible interest is illustrated by four selections:

- Air cushion vehicle
- Liquid cargo aircraft
- Cargo airship
- Cargo submarine

These selections could transport from 450 to 90,000 barrels of product at speeds from 4 to 400 knots through ranges of 100 to 6,000 miles. Projected costs (some quite optimistic and preliminary) range from 10.0 to 0.5 cents per ton-mile.

Air-cushion vehicles capable of transporting 75 tons have been designed and constructed with no significant technology problems. The liquid cargo aircraft design is based on well-established technology and design data. The cargo airship designs propose use of existing, well-characterized materials and propulsion, but they are of such size that structural response to dynamic loads and some questions of aerodynamics are not yet resolved. Cargo submarine designs are based on well-characterized (and, in some instances, very low cost) materials of construction, and the hydrodynamic data are available, but the structural-to-total weight ratio depends on a number of design features and is not easily predictable.

References

1. Weinberg, A. M. and P. Hammond, Am. Scientist, 58, 412-418, (July-Aug. 1970).

2. Gregory, D. P., "A Hydrogen Energy System," presented at 163rd ACS Meeting (Div. of Fuel Chem.), Boston, Mass., (April 1972).

3. Schoeppel, R. J., "Prospects for Hydrogen Fueled Vehicles," presented at 163rd ACS Meeting (see Ref. 2).

4. Witcofski, R. D., "Hydrogen Fueled Hypersonic Transports," presented at 163rd ACS Meeting (see Ref. 2).

5. Williams, K. R. and N. Van Lookeren Campagne, "Synthetic Fuels from Atmospheric Carbon Dioxide," presented at 163rd ACS Meeting (see Ref. 2).

6. Blurton, K. F. and H. G. Oswin, "Refuelable Batteries," presented at 163rd ACS Meeting (see Ref. 2).

7. Bobo, D. L., et.al., "A Survey of Fuel and Energy Information Sources," MITRE Corporation Report to APCO, Contract No. F19628-68-C-0365, (November 1970).

CRUDE OIL

Excerpt From "PROJECT INDEPENDENCE"

Background

The first commercial well in the United States was drilled at Titusville, Pennsylvania, in 1859. For the next 30 years the United States was virtually the world's only source of crude oil and refined products - the principal product being illuminating oil. By the turn of the century, United States oil production increased to nearly 60 million barrels annually.

By World War I, oil production had expanded into 16 states. Drilling and refining technology developed rapidly and many companies emerged in this period, as competition was aided by court actions to dissolve the Standard Oil trust. In the years between the World Wars, oil's growth continued, enhanced by expanding requirements for gasoline, and was interrupted only by the depression.

Fears of "running out of oil" were expressed in the early 1920's, and American companies were urged by the Government to develop oil production abroad to augment domestic supplies. These circumstances were reversed within a decade. When depression-limited demand coincided with the development of the east Texas field in 1930 and other large fields soon after, the problem became one of management of surplus oil production. As a result of these circumstances oil production was concentrated in large, vertically integrated companies, and States became regulators and conservers of crude oil production. State regulatory agencies began to control well spacing, to restrict production to maximum efficient rates to prevent reservoir damage, and to prorate well production on the basis of market demand.

At the close of World War II, capacity was barely sufficient to meet market demand. By the mid-1950's, spurred by strong demand and rising prices, these lags had been overcome. Both oil reserves and the capacity to produce oil far exceeded needs, and drilling activity began a decline that lasted a decade and a half. As a consequence, the rate of reserve additions slowed; and 10 years later new annual additions to both oil and gas reserves became insufficient to replace production, and production began to decline.

Production peaked in 1970, reserves have fallen each year since 1966, and drilling effort has only recently reversed its long-term downward trend. Only the discovery of the Prudhoe Bay field in the Alaskan North Slope was a major exception to this trend.

With foreign oil available at costs far below those of domestic production, the major international oil companies greatly expanded production in foreign areas. Concerned over the national security aspects of rising imports and a deteriorating domestic industry, the Federal Government encouraged voluntary import restrictions under 1955 and 1957 programs. When these did not stem the rising tide of imports, President Eisenhower, in 1959, invoked the national security provision of the Trade Agreement Extension Act to establish mandatory oil import quotas.

Although this import quota program provided the domestic petroleum industry a measure of price protection from foreign competition, other policies and market demand prorationing limited the capacity that was used. Indeed, the United States was dependent on foreign sources for 19 percent of its oil supplies in 1959; by 1970, this dependence had grown to 26 percent. Consumption of petroleum rose from 6.5 million barrels per day in 1950 to 14.7 million barrels per day in 1970, an average compounded growth rate of 4.2 percent (see Figures 3-14 and 3-15 for trends in petroleum consumption). A large part of the increased imports (more than 40 percent) was in residual fuel oil, which was effectively decontrolled in 1966.

As the domestic oil-producing industry approached capacity operations, periodic modifications were made in the import quota levels in order to provide enough oil to meet domestic needs. On April 18, 1973, the President suspended the mandatory oil-import control program, replacing it with a system of license fees that escalate over time. The license fee program is designed to support the long-term restoration of domestic capacity, particularly in refining, while providing for the near-term need for access to imported oil supplies.

In 1969, a major oil spill in the Santa Barbara Channel and another on the Gulf Coast directed attention to the environmental risks of offshore production. As a result, the Department of the Interior suspended the leasing of OCS acreage, pending development of operating procedures and regulations to minimize the potential for future significant environmental damage. The delays in OCS leasing have affected oil production, as about one-third of the recoverable oil remains to be discovered, and many of the most favorable prospects for oil and gas production are believed to exist in the Federally controlled OCS. A 10 million acre leasing program is now planned for 1975.

Methodology and Assumptions

Future supplies of oil will be determined by four fundamental factors:

- The amount of oil resources remaining to be found.
- Our success in finding the remaining supply.
- Our ability to recover (produce) what is found.
- The costs of the necessary exploration and production efforts.

Future oil production and prices were estimated for each of the 12 National Petroleum Council (NPC) regions. (See Figure 3-16). The model used for these calculations was an updated version of an NPC model developed to produce a report on production for the Department of the Interior. 1/ A manual procedure similar to that applied for the NPC regions was used to analyze the Alaskan North Slope (including NPR #4), NPR #1, military reservations in the Gulf of Mexico and California, tar sands, and heavy hydrocarbons.

Methods used to estimate oil supply at a range of prices, for both BAU and AD scenarios, can be briefly summarized. First, optimistic estimates of annual exploratory drilling footage were developed for each region. These footages were multiplied by projected finding rates (barrels of oil found per exploratory foot drilled) to estimate discoveries of oil-in-place. Estimated recovery factors were then applied to calculate the volume of oil recoverable by primary method. It was assumed that annual production from proven reserves would be a constant fraction of the remaining reserves. The total footage required to process and fully develop these reserves was calculated by applying appropriate ratios to the amount of exploratory footage drilled. All of these estimates varied by producing regions, taking into account the unique characteristics of each.

Increments to the proven reserves were added at 5- and 10-year intervals to allow for secondary recovery. The extent of the secondary recovery at each interval was estimated by taking into consideration the magnitude of the primary recovery, along with the ultimate recovery potential in each region. Tertiary recovery was similarly estimated, except that only one phase of tertiary recovery was included in the 15-year interval of projection. For new fields, the first phase of the tertiary recovery was assumed to occur 10 years after the initial discovery of oil.

It was assumed that exploration and development projects in a particular region would be undertaken only if economically viable. Economic viability was determined by using the discounted cash flow method to calculate minimum acceptable prices per barrel of oil needed to attain a specified rate of return. The following major assumptions were used:

- A 10-percent required rate of return was used.
- Cash bonuses and rentals on leases were treated as economic rents and excluded as items of cost.
- Production declines exponentially in accord with a depletion factor or decline rate.
- The life of primary projects is 30 years, of secondary projects is 25 years, and of tertiary projects is 20 years.

Figure 3-15

DOMESTIC DEMAND FOR PETROLEUM

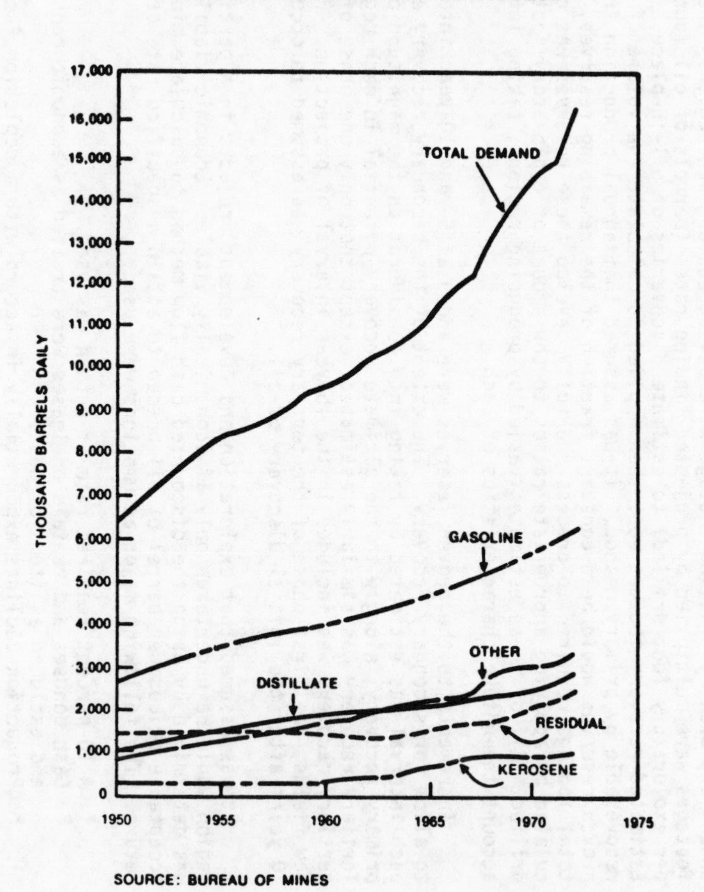

Figure 3-14

U.S. GROSS CONSUMPTION OF ENERGY RESOURCES

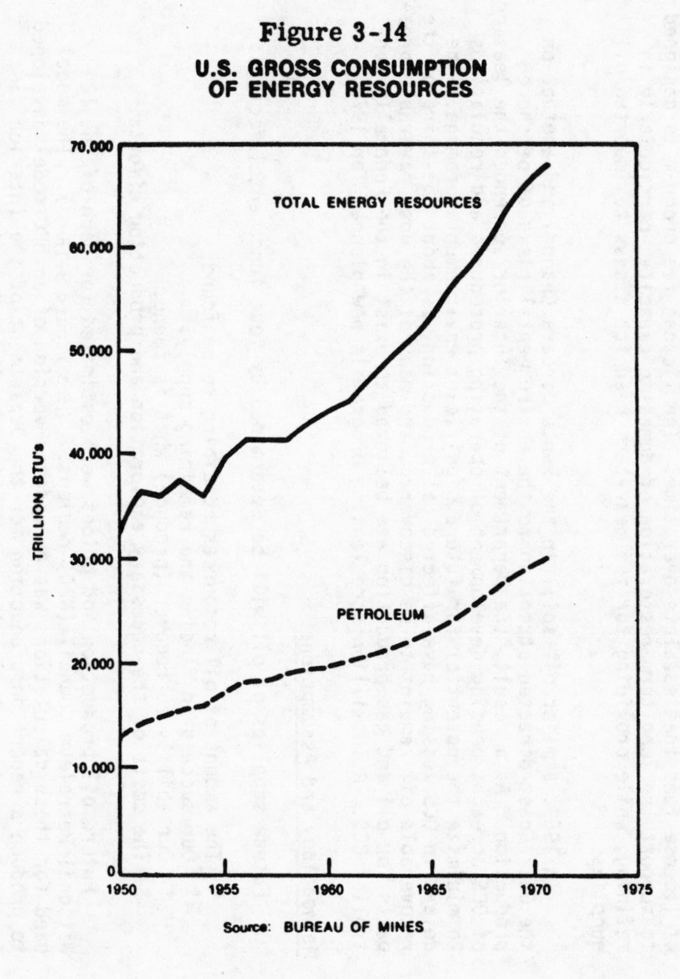

Investments, direct and indirect expenses, taxes, and deductions were estimated in relation to drilling footage and number of wells for primary projects, and to production for the secondary and tertiary projects.

Decisions on the feasibility of secondary projects depended on undertaking a primary project. If the required price per barrel of a secondary project was found to be lower than the required price of its related primary project, primary and secondary were evaluated jointly in calculating their minimum acceptable price. If the required price for secondary was greater than for its primary project, the secondary project was evaluated independently.

To obtain the regional supply estimates for each year and at various minimum acceptable prices, the yearly productions from all the primary, secondary, and tertiary projects found to be economically viable were aggregated.

The principal differences between the BAU and the AD scenarios are the assumptions that were made about Government policies on natural gas price regulation, OCS leasing, and Naval Petroleum Reserves. The BAU case assumed the economic regulation of natural gas prices where the prices are allowed to rise to clear the market. These prices were calculated in the supply-demand balancing analysis and the revenues from associated-dissolved gas were credited to the minimum acceptable price for oil.

Leasing of OCS lands was limited, in the BAU scenario, to levels consistent with the latest published Bureau of Land Management (BLM) schedules. Drilling activity consistent with these leasing levels and with recent regional drilling trends was projected at 63 million exploratory feet for the 1974 to 1988 period. Under the AD scenario, the availability of OCS lands did not limit drilling activity. Under these conditions, drilling was assumed to be limited by the availability of resource and economic extraction rates. Projected oil discoveries were constrained to 75 percent of the ultimate recoverable resources estimated by the NPC for each OCS region. This constraint resulted in a cumulative exploratory drilling assumption of 110 million feet in OCS areas for the 15-year drilling period, other than drilling in military reservations.

Under BAU conditions, royalties on offshore leases were assumed to remain at the existing one-sixth rate. Under the AD scenario, it was assumed that these royalties would be reduced to the statutory minimum of one-eighth.

No development of Federally owned petroleum reserves was assumed under BAU conditions. Under the AD scenario, however, it was assumed that development would occur in Naval Petroleum Reserve #1 (Elk Hills, California), Naval Petroleum Reserve #4 (North Slope, Alaska), and military reservations in the Gulf of Mexico and California OCS.

Advanced technology was assumed to increase substantially in the AD scenario over BAU levels, particularly in relation to enhanced recovery. Under AD conditions, the tertiary recovery factor was increased 33 percent over BAU assumptions. Finally, each scenario assumes that there will be no limitations on the availability of capital, manpower, materials or transportation.

Major Findings

The analysis was focused on developing estimates of production possibilities at various minimum acceptable prices. The following are the major task force findings:

1. Because of the long lead times required to bring new petroleum fields into production, domestic production of crude and natural gas liquids (NGL) will continue to decline for the next few years, regardless of higher prices or policies designed to encourage exploration. At minimum acceptable prices 1/ of $4 a barrel or less, production could continue to decline throughout the forecast period, even with new production from the OCS and Alaska. Even the development of NPR 1 (Elk Hills) and extensive OCS leasing could increase production by only about one million barrels.

Figure 3-16

NPC OIL REGIONS

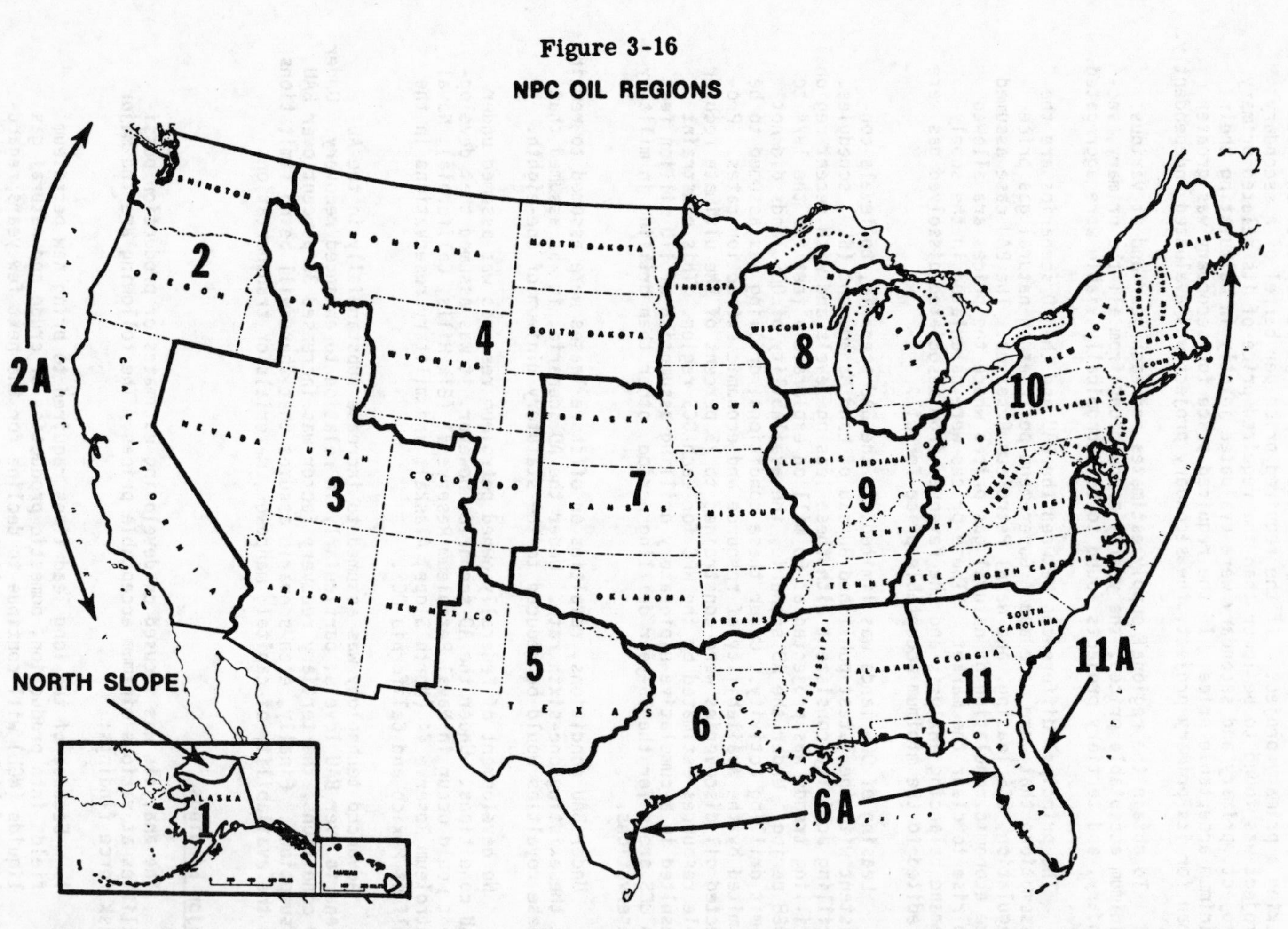

Source: NPC, FUTURE PETROLEUM PROVINCES OF THE UNITED STATES, A SUMMARY; JULY 1970 - WITH SLIGHT MODIFICATION.

Under both BAU and AD assumptions, minimum acceptable prices of $7 or higher would reverse the downward production trend. At $7 and $11 a barrel, respectively, production, under the BAU scenario, would increase to 11.1-12.2 million barrels a day by 1980, and to 11.9-15.0 million barrels a day by 1985, exceeding the all-time high production of 11.3 million barrels a day reached in 1970. (see Table 3-2 for crude oil and natural gas liquids estimates):

Table 3-2
Summation of Unconstrained Regional Production Possibilities for Crude Oil and Natural Gas Liquids
(Million barrels per day)

Minimum Acceptable Price Per Barrel	1974	1977	1980	1985
Business-As-Usual				
$ 4	10.5	9.0	9.8	9.8
7	10.5	9.5	11.1	11.9
11	10.5	9.9	12.2	15.0
Accelerated Development				
$ 4	10.5	9.7	11.1	11.6
7	10.5	10.2	12.9	16.9
11	10.5	10.3	13.5	20.0

2. At $11 per barrel oil, domestic onshore production would increase slightly under both BAU and AD assumptions. Almost half of the onshore production would be from new secondary and tertiary recovery, while conventional and new primary fields would decline considerably from 1974 levels.

The major new source of oil in these projections is Alaska. Under BAU assumptions, Alaska would provide 3 million barrels per day, mainly from the North Slope fields. If development of the Naval Petroleum Reserves is allowed, an additional 2 million barrels per day could be produced. The increased development in Alaska will shift the focus of United States oil production: Alaska could produce 20-25 percent of our oil by 1985, although it now accounts for less than 2 percent.

The 1985 projections also indicate substantial increases in lower 48 OCS production. Although production could increase about 1.2 million barrels per day under BAU assumptions (two-thirds of the increase from the Gulf of Mexico), OCS production could reach 4.3 million barrels per day (300 percent increase) under AD conditions. The major sources of this increase would be the offshore California and Atlantic fields. Considerable opposition to leasing in these areas could be expected (See Table 3-3 for a detailed description of potential production levels).

3. If production increases to 15-20 million barrels per day by 1985, this level of production could not be maintained indefinitely at these prices, as oil reserves at these prices would soon peak. Thus, in addition to potential constraints toward achieving these levels of production, the non-renewable nature of these resources should be considered.

4. At a minimum acceptable price of $7 a barrel, under the BAU scenario, almost 40 billion barrels of petroleum liquids would be produced from 1974 to 1985. This is almost equal to the 48 billion barrels of proved and indicated additional reserves of oil and natural gas liquids reported at the end of 1973 by the American Petroleum Institute. However, under AD conditions and at a price of $11 a barrel, cumulative production between 1974 and 1985 would be over 50 billion barrels. These production figures imply that huge additions to reserves would be needed in this time period. These additional reserves would about equal the most con-

servative estimates of undiscovered recoverable oil in the United States, although they would still be less than NPC estimates.

The uncertainties inherent in estimating future petroleum production (especially uncertainties having to do with the magnitude of undiscovered resources in as yet totally unexplored provinces and the finding rate per foot of exploratory drilling) are so great that numerical estimates of this type are highly speculative.

Table 3-3
Potential Rates of Domestic Oil Production
(Millions of barrels per day, at $11 oil)

Production Area	1974	BAU 1985	AD 1985
1. Onshore - Lower 48 States	8.9	9.1	9.9
• Conventional fields and new primary fields	6.9	3.4	3.5
• New secondary	---	2.4	2.4
• New tertiary	---	1.8	2.3
• Natural gas liquids	2.0	1.5	1.6
• Naval Petroleum Reserve #1	---	---	0.2
2. Alaska	0.2	3.0	5.3
• North Slope	---	2.5	2.5
• Southern Alaska (including OCS)	0.2	0.5	0.8
• Naval Petroleum Reserve #4	---	---	2.0
3. Lower 48 Outer Continental Shelf	1.4	2.6	4.3
• Gulf of Mexico	1.3	2.1	2.5
• California OCS	0.1	0.5	1.3
• Atlantic OCS	---	---	0.5
4. Heavy Crude Oil and Tar Sands	---	0.3	0.5
Total Potential Production	10.5	15.0	20.0

Sensitivity analyses show that, within a range of reasonable assumptions, different values regarding discount rates, financial costs, and finding rates could affect the quantities produced at $4, $7 and $11 per barrel in 1985 by 10 to 40 percent. Other assumptions about drilling costs, effective depletion rates, and co-product prices would affect production levels at these prices by as much as 15 percent.

Uncertainties regarding many of the factors used in the oil production model will be resolved only as additional exploration is undertaken. This is especially important in areas of high drilling and production costs, such as northern Alaska and deeper parts of the OCS, where even at high prices only giant fields might be economically feasible. Other uncertainties can be reduced through stabilized government policies affecting petroleum exploration and development.

In addition to the difficulties inherent in projecting the amount of future oil discoveries, a number of factors inherent in the model's assumption could lead to significantly different conclusions on costs and production levels, including:

1. Competition for available labor, material, and capital. Although the availability of labor, material, and capital has been assumed, oil must compete with other energy projects and with the production of non-energy goods and services for these resources.

2. Technology. The projected levels of crude oil supply will depend on utilization of secondary and tertiary recovery. Much of the technology needed

to achieve those production levels has not been applied commercially, though its principles are generally known. Depending on the technological success of oil recovery, different production rates and costs could result. These technological successes may require government support, responsiveness of private sector research to higher oil prices, and engineering achievements.

3. Access to resources. The Federal Government controls perhaps 40 percent of the remaining producible oil. The Government makes this acreage available for exploration and development with consideration of environmental and jurisdictional questions. The terms under which Government lands are made available affect the availability of capital, the rate at which lease areas are explored and produced, and the percent of oil-in-place ultimately recovered. State regulations, such as those relating to utilization and well spacing, may also influence production rates, recoveries, and costs.

4. Government price and fiscal policies. Government pricing and fiscal policies also affect production levels. Depletion allowances, tax rates, and fiscal uncertainty could constrain investments.

5. Environmental considerations. All extraction,manufacturing, and distribution processes affect the quality of the environment. The production of oil is no exception; any production level involves some direct and indirect environmental change. Oil spills can affect the marine environment and create aesthetic problems. The development of petroleum production also has social and economic implications, expecially in frontier areas. The Alaskan North Slope, with its fragile ecology and unpopulated areas, presents a particular set of problems. Its abundance of resources suggests that an extraordinary effort is needed to minimize environmental impact.

NATURAL GAS

Excerpt From "PROJECT INDEPENDENCE"

Natural gas is primarily methane, the most basic hydrocarbon. It is often found associated with oil in the same geologic formations, but is also found in geologic structures by itself. Its primary use is as a clean-burning fuel, but it is also used as a petrochemical feedstock.

Background

The first natural gas well was put into production in 1821 in Fredonia, New York. The discovery of oil in the U.S. in 1859 began a search that resulted in the discovery of large quantities of natural gas as well a supply for which there was no ready market at the time. Thus, the gas was flared as a part of the process of extracting oil from the ground. But once the possible uses and advantages of natural gas were discovered, it quickly replaced manufactured gas.

The first large-scale use of natural gas was in the manufacture of steel and glass in plants located in Pittsburgh. Initially, the use of gas was confined to areas near gas or oil fields, but the development of long-distance gas transmission systems in the 1930's broadened its market. During World War II, the war effort slowed down growth of gas pipeline and distribution systems. After the war, however, the availability of abundant supplies of natural gas--most of it found in the search for oil--and improved quality of pipe for high-pressure, long-distance delivery enabled the gas utility industry to expand rapidly and widely. Marketed gas production increased from four trillion cubic feet (TCF),in 1946,to eight TCF by 1952, and continued to grow at a 6.5 percent average annual rate in the 1950's and 1960's.

Natural gas now represents about one-third of the total energy consumed by the Nation and almost one-half of the non-transportation uses--an amount twice that supplied by either oil or coal. One-half of the gas is used for residential and commercial purposes, one-sixth for the generation of electricity. and one-third for industrial uses (See Figure 3-17 for natural gas utility sales trends).

Figure 3-17
GAS UTILITY INDUSTRY SALES BY CLASS OF SERVICE

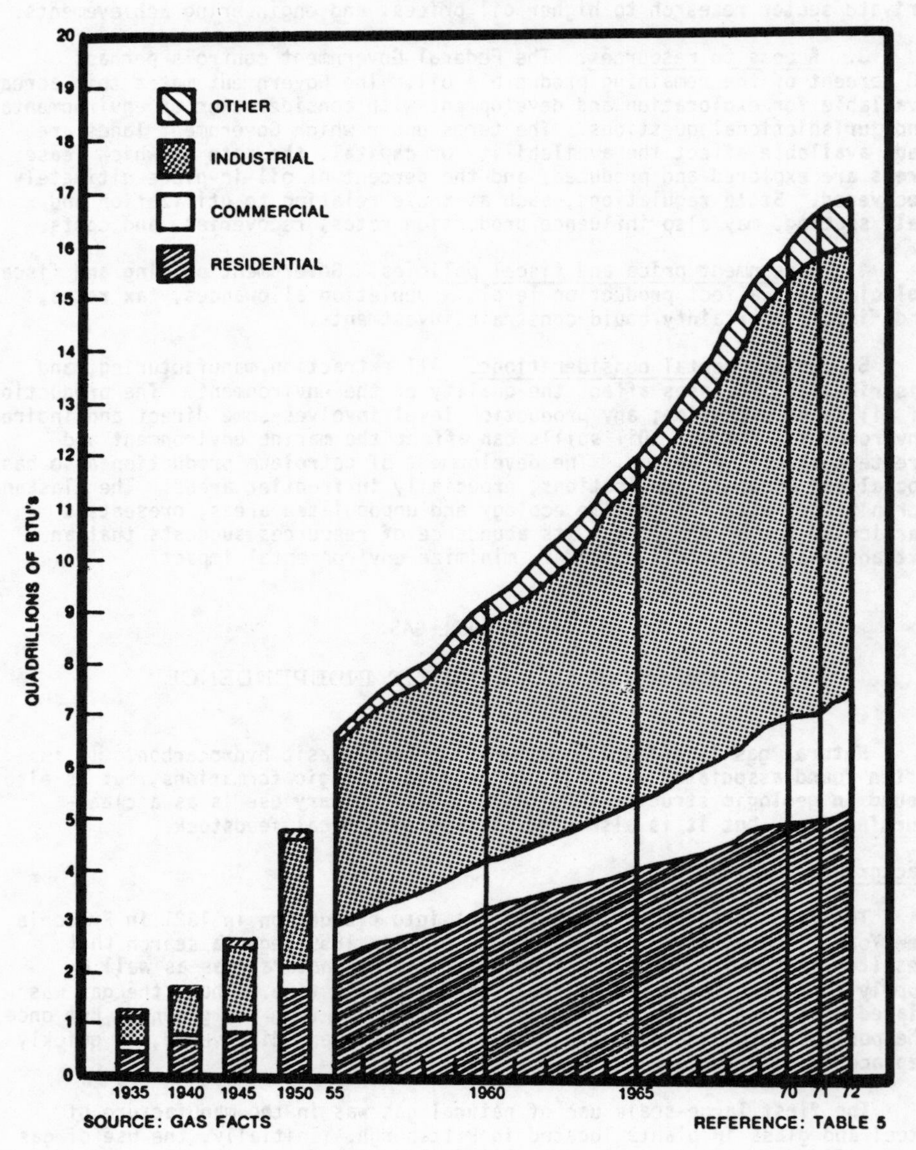

SOURCE: GAS FACTS REFERENCE: TABLE 5

In the 1970's, the demand for gas has exceeded its supply. Many gas distribution companies have found it necessary to deny gas service to new customers and to enforce contracts for interruptible gas sales. Additionally, the Federal Power Commission has set priorities on gas use.

The Natural Gas Act of 1938 gave the Federal Power Commission authority to regulate interstate pipelines and natural gas imports and exports. In 1954, in the landmark Phillips Petroleum case, the U.S. Supreme Court held that a firm which produces and gathers gas and sells it to a pipeline company is a natural gas company. As a result, the FPC began regulating the wellhead prices at which gas was sold in interstate commerce.

The approach for establishing producer's prices is based primarily on historical average industry costs. Drilling and exploration costs, on the one hand, have increased considerably in recent years; the cost per foot of a gas well increased 57 percent between 1961 and 1971. But the average price of gas, on the other hand, rose by only about 20 percent (Table 3-4 shows production and pricing trends). This price lag has impacted drilling and resulted in the erosion of gas reserves.

Proved gas reserves, the current estimated quantity of natural gas that can be reasonably recovered under existing economic and operating conditions, grew from 147 TCF in 1945 to a peak of 293 TCF in 1967. Since we are consuming 2 to 3 times as much natural gas as we are finding in the continental United States, proved reserves have declined from 1967,and were 250 TCF in 1973. Natural gas production grew from 4.8 TCF per year in 1945 to 22.7 TCF per year in 1971, but has now leveled off at between 22 and 23 TCF per year (See Figure 3-18).

Methodology and Assumptions

Future production possibilities and corresponding minimum acceptable prices[1/] were estimated for non-associated gas and natural gas liquids in each of the 12 regions defined by the National Petroleum Council (NPC). An adaptation of the NPC's natural gas supply computer program was utilized in the analysis.

There were several modifications made to this program,including development of a new section to calculate minimum acceptable price, using a discounted cash flow technique, and extensive updates and revisions to the data base through 1973 to reflect recent trends in critical variables. Some special sources of gas - Alaska, gas from tight formations, and gas occluded in coal seams - were not amenable to inclusion in the computer program and were therefore analyzed independently.

The detailed methodology used to estimate natural gas supplies is very similar to that used by the Oil Task Force. The most important assumptions common to both Business-As-Usual (BAU) and Accelerated Development (AD) scenarios are:

- A 10 percent after-tax rate of return on investment
- A depletion allowance of 22 percent
- Cash bonuses and rentals on leases are economic rents and therefore excluded as cost items

The third assumption is particularly important since it results in a definition of minimum acceptable prices different from that generally used in the industry; nevertheless, the assumption was made to facilitate analysis and provide consistency in comparisons with other energy sources.

The BAU scenario assumes changes in the regulatory environment and projected offshore leasing at levels consistent with current published Bureau of Land Management schedules. In the AD scenario, increased price incentives are assumed, and OCS areas are assumed available in earlier years. These assumptions were reflected in the analysis as follows:

- Drilling activity during the 1975-1978 period will increase at a 5.75 percent average annual rate under BAU conditions, and a 12.2 percent average rate under the AD scenario, although later rates of increase will be less under AD conditions.
- Offshore areas (California, Gulf of Mexico, and Atlantic) will account for roughly 20 percent of drilling activity by the mid-1980's under AD conditions, compared with 15 percent under the BAU scenario.
- Royalty rates were 1/6 offshore and 1/8 onshore under BAU; under AD conditions, they will be the statutory minimum of 1/8.
- Economic regulation of natural gas prices where prices are allowed to rise to clear market, or deregulation on new gas supplies.

Under the AD scenario, it was assumed reserves would be developed from

Figure 3-18
U.S. NATURAL GAS RESERVES

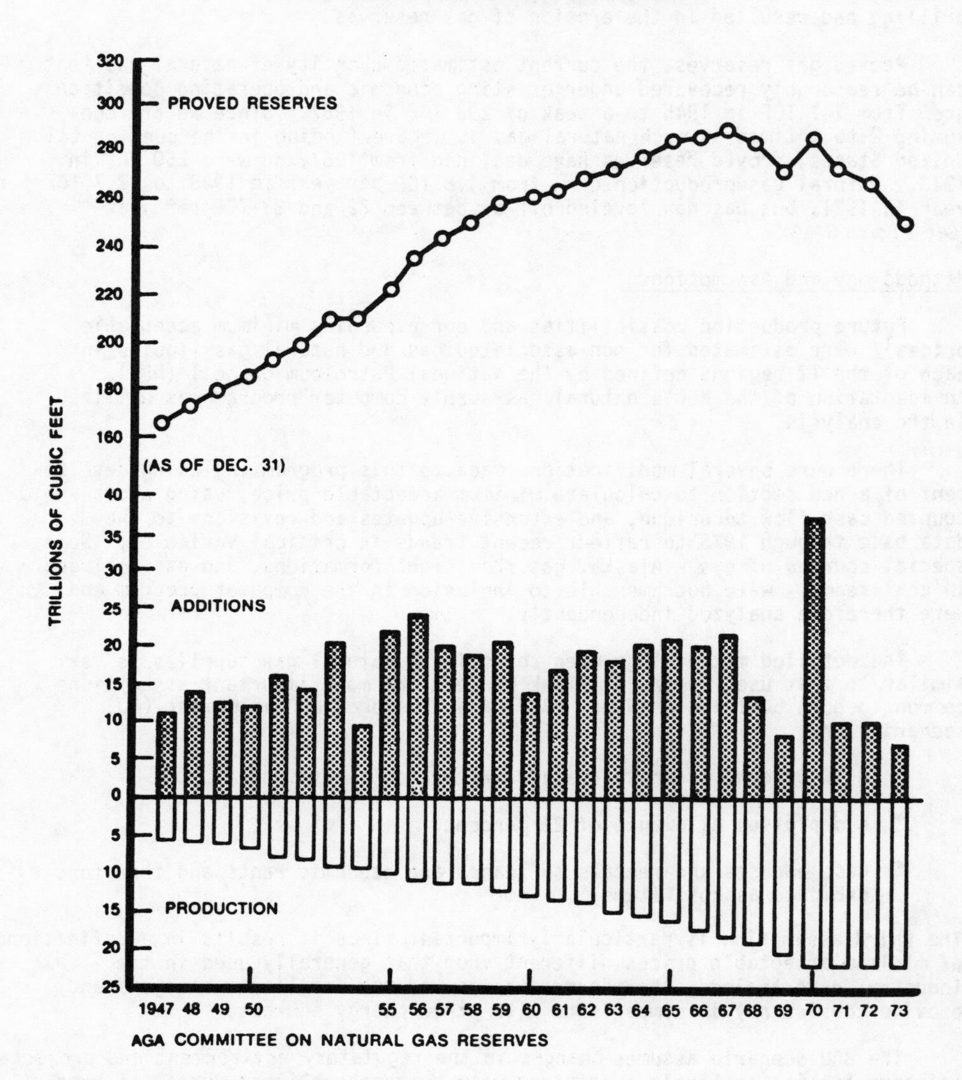

AGA COMMITTEE ON NATURAL GAS RESERVES

Naval Petroleum Reserve #4 (Alaska) for both non-associated and associated-dissolved gas, along with several minor onshore sources of associated-dissolved gas. R&D activities were assumed to result in recovery of non-associated gas from two minor special sources--tight reservoirs and gas--occluded in coal.

Major Findings

The projections of production possibilities hinge primarily on the projected success of the non-associated gas exploration effort. The major non-associated gas reserve additions are projected to occur Regions 6 and 6A in and around the Gulf of Mexico. These areas will also have fairly low acceptable selling prices. The Atlantic OCS could have large reserve increases by 1985 and could surpass Region 6 after 1985 under accelerated conditions (See Table 3-5 for regional additions to reserves). In both the BAU and AD scenarios, total annual findings peak late in the projection period and then begin to decline. This reflects projected drilling in both scenarios, and is indicative of the depletable nature of this resource. Newly found gas will come into production at higher than historical minimum prices as costs increase due to the expansion of exploration and drilling

Table 3-4
Marketed Production of Natural Gas and Average Wellhead Price
1945-1972

YEAR	MARKETED PRODUCTION: MILLIONS OF CUBIC FEET	MARKETED PRODUCTION: TRILLIONS OF BTU	AVERAGE WELLHEAD PRICE (CENTS PER MCF)
1945	4,049,002	4,481.7	4.9
1950	6,282,660	6,753.0	6.5
1951	7,457,359	8,016.7	7.3
1952	8,013,457	8,614.5	7.8
1953	8,396,916	9,026.7	9.2
1954	8,742,646	9,398.2	10.1
1955	9,405,351	10,110.4	10.4
1956	10,081,923	10,838.2	10.8
1957	10,680,258	11,481.0	11.3
1958	11,030,248	11,857.5	11.9
1959	12,046,115	12,919.5	12.9
1960	12,771,038	13,728.8	14.0
1961	13,254,025	14,248.1	15.1
1962	13,876,622	14,917.4	15.6
1963	14,746,663	15,862.7	15.8
1964	15,462,143	16,621.8	15.4
1965	16,039,753	17,242.7	15.6
1966	17,206,628	18,497.1	15.7
1967	18,171,326	19,534.2	16.0
1968	19,329,600	20,771.0	16.4
1969	20,698,240	22,250.6	16.7
1970	21,920,642	23,564.7	17.1
1971	22,493,017	24,180.0	18.2
1972	22,531,698	24,221.6	18.6

efforts in the face of generally declining findings rates (See Tables 3-6 and 3-7 for increments, at various minimum price intervals, of non-associated and associated gas, respectively).

The analyses lead to the following conclusions:

1. Because of the long lead-times required to bring natural gas production on stream, and because of anticipated declining finding rates, non-associated gas production from the lower 48 states should continue to decline until nearly 1980, regardless of price.

2. At a minimum acceptable price of $1.00 per MCF under BAU conditions, non-associated marketed production could increase from 16.7 TCF per year in 1974 to 18.1 TCF per year in 1985. The major sources of new gas would be in the offshore and onshore Gulf Coast region.

3. Under AD conditions, at a minimum acceptable price of $1.00 per MCF, marketed production could reach 21.3 TCF per year in 1985. Among the sources of further increases in non-associated gas production over the BAU case would be the Atlantic and California OCS.

4. Associated-dissolved gas production levels from the lower 48 states and southern Alaska OCS would depend significantly on oil prices. The 1974 production levels of 3.7 TCF per year would be reduced in 1977 at prices of $7 or less per barrel under both BAU and AD assumptions,

Table 3-5

SUMMARY OF NON-ASSOCIATED RESERVE ADDITION PROJECTIONS AND THEIR "MINIMUM ACCEPTABLE PRICES" LOWER 48 STATES 1/

NPC Region		1974		1977		1980		1985	
		Reserve Additions	"Price"	Reserve Additions	"Price"	Reserve Additions	"Price"	Reserve Additions	"Price"
2	BAU 2/	0.100	$0.60	0.156	$0.65	0.213	$0.66	0.278	$0.69
	ACC 3/	0.100	0.60	0.188	0.64	0.258	0.66	0.292	0.69
2A	BAU	0.0	--	0.105	0.69	0.129	0.71	0.277	0.80
	ACC	0.0	--	0.253	0.66	0.313	0.68	0.582	0.76
3	BAU	0.349	0.79	0.410	0.78	0.512	0.80	0.722	0.83
	ACC	0.349	0.79	0.494	0.78	0.611	0.80	0.728	0.83
4	BAU	0.407	0.35	0.530	0.48	0.621	0.51	0.840	0.58
	ACC	0.407	0.35	0.634	0.49	0.725	0.53	0.816	0.59
5	BAU	1.969	0.31	2.144	0.47	2.364	0.58	2.872	0.63
	ACC	1.969	0.31	2.534	0.48	2.772	0.60	2.742	0.67
6	BAU	3.992	0.43	4.245	0.54	4.412	0.61	4.428	0.86
	ACC	3.992	0.43	5.046	0.54	5.103	0.64	4.166	0.91
6A	BAU	3.753	0.29	5.938	0.35	7.195	0.44	6.774	0.71
	ACC	3.753	0.27	7.141	0.34	8.846	0.45	7.368	0.79
7	BAU	1.724	0.47	1.661	0.55	1.888	0.61	2.452	0.69
	ACC	1.724	0.47	1.978	0.56	2.206	0.62	2.424	0.70
8 & 9	BAU	0.049	0.77	0.037	1.04	0.034	1.23	0.036	1.81
	ACC	0.049	0.77	0.045	1.04	0.042	1.23	0.038	1.81
10	BAU	0.716	0.78	0.747	0.70	0.843	0.70	1.117	0.80
	ACC	0.716	0.78	0.901	0.70	0.976	0.73	1.101	0.81
11	BAU	0.0	--	0.003	5.78	0.007	5.80	0.010	5.79
	ACC	0.0	--	0.003	5.78	0.008	5.80	0.010	5.79
11A	BAU	0.0	--	0.0	--	0.064	0.89	1.847	0.92
	ACC	0.0	--	0.0	--	0.627	0.85	3.199	0.88
Sum of Additions:									
	BAU	13.059		15.976		18.282		21.653	
	ACC	13.059		19.217		22.487		23.466	

1/ Volumes in trillions of cubic feet, "prices" in cents per Mcf (constant 1973 dollars).
2/ Business as Usual Scenario.
3/ Accelerated Development Scenario.

but would increase in 1985. At $11 per barrel oil prices, associated-dissolved gas production would increase substantially over $7 levels.

5. Non-associated gas from both Alaskan regions and associated-dissolved gas from the North Slope could provide major quantities of new gas production. In 1974, this production amounts to only 0.1 TCF. per year. At oil prices of more than $7 per barrel, production under BAU conditions could reach 1.9 TCF per year in 1985, while production under AD conditions, with the development of NPR-4 and additional OCS leasing, could reach 3.6 TCF per year by 1985. The inclusion of transportation costs to the lower 48 states' markets would significantly affect prices.

6. Under the AD scenario, production of gas from tight formations would depend on successful development of recovery technology, but, if successful it could provide as much as 2.0 TCF per year in added gas production by 1985. The amount of gas recoverable from coal seams is forecast to be negligible.

7. If natural gas prices remain regulated at current levels, the outlook for increased gas supplies is not promising. At the current field price, wellhead production in 1985 could decline by over 6 TCF per year from 1974 levels (a decline of almost 30 percent). The share of natural gas in interstate markets would also be drastically reduced. The effects of price regulation predominantly impact non-associated gas.

Table 3-6
Total Non-Associated Natural Gas Production Possibilities
BAU[1/]

Price[2/]	1974	1977	1980	1985
@ 40¢ (or less)	16.522	15.222	13.337	9.483
@ 60¢ (or less)	16.670	15.847	16.028	16.655
@ 80¢ (or less)	16.670	16.073	16.389	18.139
@ $1.00 (or less)	16.670	16.075	16.394	18.152
@ $2.00 (or more)	16.670	16.075	16.400	18.172

AD[1/]

Price	1974	1977	1980	1985
@ $0.40 (or less)	16.552	15.284	13.652	9.100
@ $0.60 (or less)	16.670	16.029	17.781	19.260
@ $0.80 (or less)	16.670	16.265	18.096	21.344
@ $1.00 (or less)	16.670	16.267	18.103	21.348
@ $2.00 (or more)	16.670	16.267	18.110	21.371

1/ • Production projections are given for the lower 48 states, Alaska and for the natural gas from tight reservoirs.

• Production is given in trillion of cubic feet.

• AD = Accelerated Development

2/ Prices are given in cents per MCF, (in constant 1973 dollars)

Sensitivity analyses were performed to reflect uncertainties involved in estimating natural gas production. Finding rates were uniformly increased and decreased by 20 percent in these analyses, and discovery volumes differed from the BAU case by about 20 percent. Corresponding regional minimum acceptable prices were approximately 16-20 percent less with the higher finding rates and 24-28 percent higher with lower rates, indicating the considerable price sensitivity to finding rates.

In other sensitivity analyses, the after-tax rate of return on investment was set at 15 percent and 7.5 percent, resulting in price increases of 28-33 percent in the former case,and price decreases of 13 to 18 percent in the latter. Inclusion of lease bonus and rental costs increased prices by about 10 percent in onshore areas and by 36 to 265 percent, depending on the year and location in offshore areas, indicating the high degree of sensitivity of minimum acceptable prices.

Table 3-7

Total Associated - Dissolved Natural Gas Production Possibilities BAU1/

Minimum Acceptable Oil Price2/	1974	1977	1980	1985
$ 4.00	3.665	3.167	3.546	3.999
$ 7.00	3.665	3.365	4.003	5.824
$11.00	3.665	3.479	4.328	6.633

AD 1/

Minimum Acceptable Oil Price 2/	1974	1977	1980	1985
$ 4.00	3.665	3.327	3.803	5.190
$ 7.00	3.665	3.533	4.424	6.357
$11.00	3.655	3.539	4.553	7.978

1/ Production projections are given for the lower 48 states and Alaska.

- AD = Accelerated Development
- Production is given in trillion of cubic feet per year

2/ Minimum acceptable oil price is given in constant 1973 dollars per barrel, inasmuch as associated--dissolved natural gas is produced in conjunction with crude oil.

COAL

Excerpt From "PROJECT INDEPENDENCE"

History and Recent Trends

Background. The Nation's coal industry began in the 18th century with bituminous coal mined in Virginia and anthracite in Pennsylvania. Coal production increased steadily throughout the 19th century. Its uses included space heating, coal gas, steam generation, and as coke in steel production, and by the turn of this century, coal supplied 90 percent of the U.S. energy consumption.

However, during the first half of this century, coal consumption grew less rapidly than total energy consumption because more convenient and competitively priced domestic oil and natural gas became available, and new uses of oil (e.g., automobiles) expanded rapidly. By 1950, coal dropped to 38 percent of the Nation's energy consumption.

Figure 3-19

COAL'S DECLINING SHARE OF TOTAL UNITED STATES ENERGY CONSUMPTION

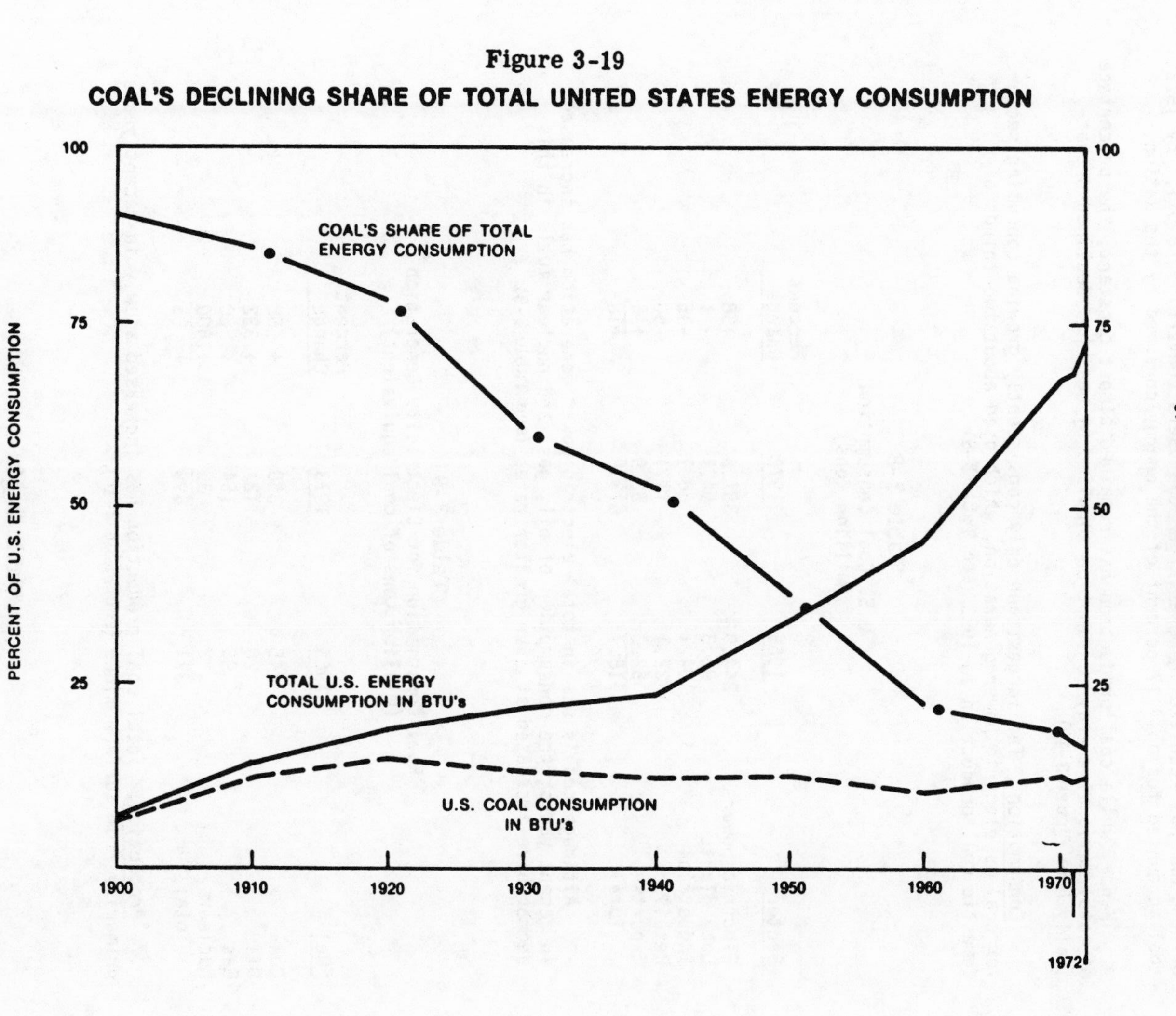

Since 1950, coal's declining role in the Nation's energy structure has been accelerated by Government actions. The stimulation of nuclear electric power reduced coal's role in generating electricity. The elimination, in 1966, of oil import quotas for residual oil on the East Coast resulted in many large coal users converting to cheaper and more convenient foreign oil. The implementation of the Clean Air Act during the 1970's has created significant uncertainties as to how much coal will be permitted to be burned, and has resulted in more large coal users converting to oil. By 1972. coal accounted for only 17 percent of the energy consumed by the Nation.

Hence, while coal production has remained almost constant, the percentage of total national energy consumption supplied by coal has declined dramatically (see Figure 3-19).

Consumption. The largest and only consistently growing consuming sector for coal is electric power generation, which used about two-thirds of domestic coal production in 1973 (see Table 3-8).

Table 3-8
U.S. Coal Consumption
(million tons)

Sector	1965	1973	Percent Change
Electric power	244.9	387.6	+58
Coke plants	95.3	94.1	- 1
Industrial	104.1	68.1	-35
Retail	22.1	11.1	-50
Exports	52.3	52.9	+ 1
Total	518.7	613.8	+18

Although coal's use in total electric power generation has increased in recent years, the consumption of oil, gas, and nuclear fuels in this sector have increased at even greater rates (see Table 3-9).

Table 3-9
Fuel Consumption for Electricity Generation
(million tons of coal equivalent)

Fuel	1965	1973	Percent Change
Coal	245	388	+ 58
Oil	28	121	+ 332
Gas	96	154	+ 60
Nuclear	2	36	+1800
Total	371	699	+ 88

Production. Total coal production has increased slowly in recent years, primarily from surface mines (see Table 3-10).

Table 3-10
U.S. Coal Production
(million tons)

Mine Type	1965	1973	Percent Change
Underground	338	300	-11
Surface	189	299	+58
Total	527	599	+14

Surface mining production has increased rapidly because surface mining costs have not increased as much as underground mining costs (see Table 3-11).

Table 3-11
Average Cost Per Ton FOB Mine
(in dollars)

Mine Type	1965	1973	Percent Change
Underground	4.93	10.30	+109
Surface	3.57	5.90	+ 46
Spread	1.36	4.40	+224

This cost differential is primarily due to productivity changes. Surface mining productivity has remained relatively constant at about 31 tons per man-day over the past five years, whereas underground productivity dropped nearly 30 percent from about 15.9 tons per man-day to 11 tons. This underground productivity decrease resulted in large part from compliance with the provision of the Mine Health and Safety Act of 1969, from an influx of untrained workers, and from difficulties in labor-management relations.

Historically, most coal production has occurred east of the Mississippi, but western coal production is growing rapidly. This is due, in large part, to the lower sulfur content of western coals. North Appalachian production has decreased due to its high sulfur content and to conversions from coal to oil in its northeastern and mid-Atlantic market area (see Table 3-12 for coal production by regions and Figure 3-20 for a map of coal regions).

Table 3-12
U.S. Coal Production
(million tons)

Regions	1965	1973	Percent Change
North Appalachia	191	177	- 7
South Appalachia	195	205	+ 5
Midwest	121	150	+ 24
Gulf	-	7	>1400
North Great Plains	6	32	+ 433
Rocky Mountain	13	24	+ 85
Pacific Coast	1	4	+ 300
Total	527	599	+ 14

Reserves. The Nation's coal reserves minable at current prices are enormous. At 1973 consumption levels, the Nation has enough coal reserves to last over 800 years (see Table 3-13 for estimated coal reserves).

Table 3-13
Coal Reserves

Region	Billion Tons	Quadrillion Btu's
North Appalachia	73.2	1922
South Appalachia	39.1	1052
Midwest	104.6	2492
Gulf	4.3	71
Northern Great Plains	175.4	3364
Rocky Mountains	23.7	568
Pacific Coast	13.6	262
Total	433.9	9731

On a tonnage basis about half these reserves are located east of the Mississippi and half west of the river. However, on a Btu-basis, over 55

Figure 3-20

PROJECT INDEPENDENCE COAL REGIONS

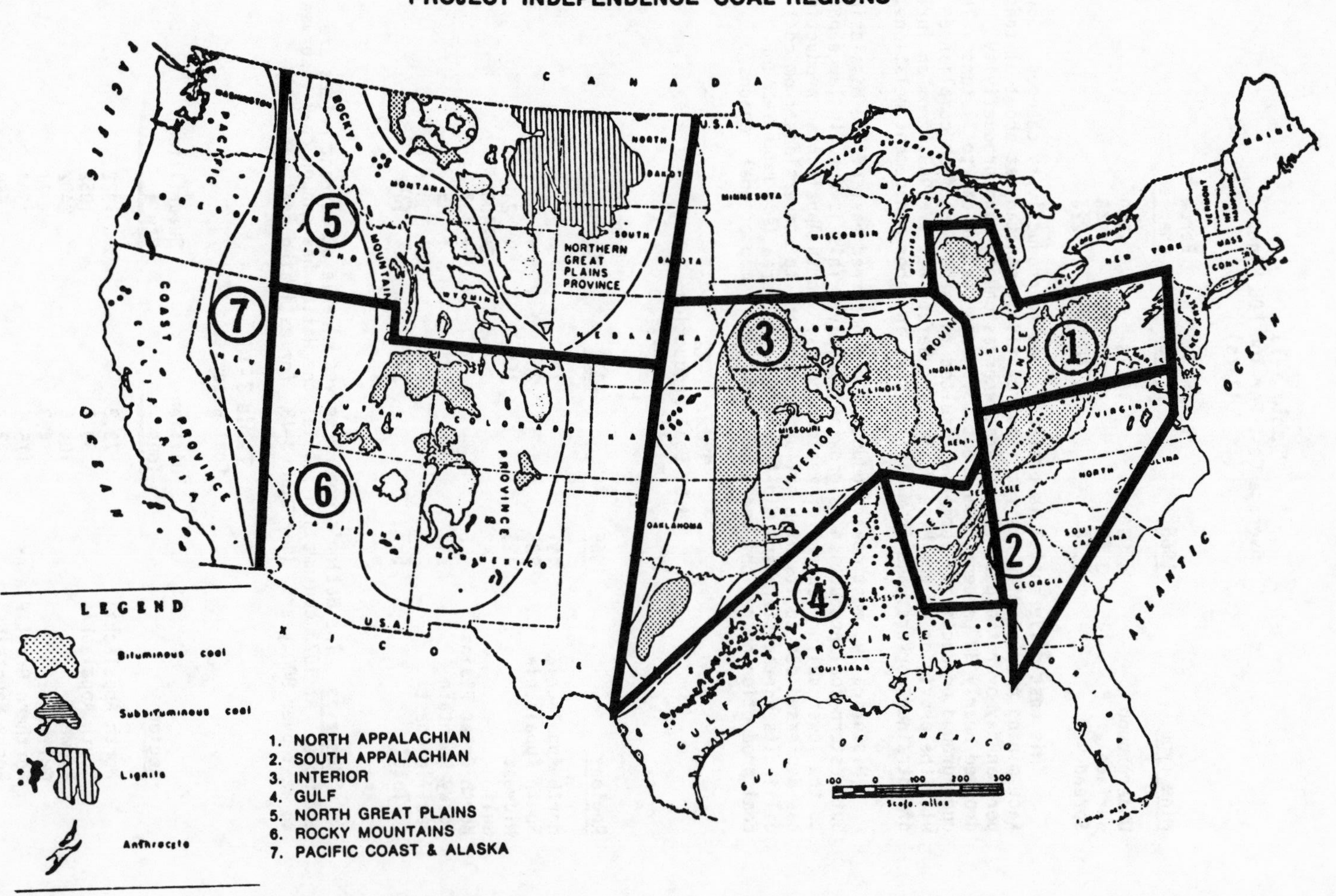

percent is located east of the river. This is because, on average, the Btu-content of western coal is less than eastern coal. Nearly all of the coal reserves in the East are privately owned, whereas most of the western coal reserves are owned by the Federal Government.

Only about 50% of underground and 90% of the surface minable coal can be physically recovered by conventional mining methods. Pillars of coal must generally be left in underground mines to support the roof, although modern practices often remove the pillars in deep mines where surface subsidence is not a problem. Certain surface-minable coal is inaccessible due to natural and man-made surface features. These recovery factors mean that the Nation's reserves would last about 500-600 years, rather than the 800 years mentioned above.

Approximately 60 percent of the Nation's coal reserves contain one percent or less sulfur by weight, and most of this is in the West. However, a smaller portion of reserves can meet the sulfur dioxide new source performance standard (NSPS) for large boilers (i.e., 1.2 pounds of SO_2 per million Btu's) established by the U.S. Environmental Protection Agency pursuant to the Clean Air Act Amendments of 1970. (See Table 3-14 for sulfur content comparisons by region.)

Table 3-14
Regional Mean Sulfur Content Comparisons

Region	Total Reserve Base* (billion tons)	Mean Sulfur Content (percent of weight)	Percent Sulfur Required to Meet SO_2 NSPS
Northern Appalachia	66.1	2.0	0.72
Southern Appalachia	38.9	1.0	0.72
Midwest	104.5	3.1	0.66
Gulf	4.3	1.0	0.42
Great Plains	175.4	0.5	0.51
Rocky Mountain	23.6	0.5	0.66
Pacific Coast	13.6	0.2	0.51
Total	426.4		

* Excludes anthracite coal

Nearly half of the low-sulfur eastern coal is used as coking coal at home and abroad. The combustion characteristics of these coals are unsuitable for many boilers presently in use. Much of these reserves is committed by long-term contracts to metallurgical and export markets. Nevertheless, the other half, which is still substantial, is available for use as steam coal.

Accordingly, while only about 21 percent of the coal currently mined is less than one percent sulfur by weight, there are vast reserves of low-sulfur coal that could be developed, primarily in the West and in southern Appalachia. Due to transportation costs, however, these coals could be more expensive in certain market areas.

Despite coal's abundance, there is currently a shortage of it, which is reflected by very high spot market coal prices. Coal production has not kept pace with coal demand. This situation has occurred for two reasons. First, demand has increased unexpectedly due to the Arab embargo, higher oil prices, and natural gas curtailments. Further, demand has recently been stimulated by attempts to build coal stockpiles to buffer against a threatened work stoppage.

Second, coal production has not increased as fast as it could have, because new coal mines require major initial capital outlays which must be recovered over 20 to 25 years. Hence, a coal producer must be reasonably certain that there will be a market for his coal for 20 to 25 years. But the market outlook for coal has been clouded by numerous uncertainties including strip mining legislation, western coal lands leasing policy, Clean

Air Act implementation, oil import policy, natural gas pricing policy, electricity demand forecasts, and nuclear capacity forecasts. Accordingly, coal producers have opened few mines in the 1970's, and the key related industries, such as railroads and mining equipment manufacturers, have delayed major investments. If these uncertainties are eliminated, the current coal shortage could be substantially alleviated in three to five years--the time required to open new mines. If the uncertainties are not cleared up, coal shortages could endure indefinitely.

Methodology

Two supply scenarios were developed as inputs to the Project Independence analysis. The Business-As-Usual (BAU) Scenario was developed based on recent trends; the Accelerated Development (AD) Scenario encompasses a number of institutional changes. A detailed discussion of assumptions is contained in the Coal Task Force Report.

First, production targets were established for each of seven coal-producing areas for a Business-As-Usual situation and for an Accelerated Development situation for 1977, 1980, 1985, and 1990. Using these production targets, it was then assumed that new mines would be opened annually to achieve the additional supply needed for each region in each year and to replace depleted production. A detailed discussion of assumptions and methodology follows below.

Assumptions Common to BAU and AD

It was assumed that a mine has a 20-year life. Wages and union welfare payments were considered as of November 12, 1973, and costs for materials and equipment are based on 1973 indices at 1974 dollars. Straight-line depreciation was used to determine annual depreciation by mine type; major items of equipment were depreciated over 20 years; other items were depreciated over four, six, or 10 years, depending on expected life. For new mines, the minimum acceptable selling price (i.e., the minimum price required to warrant the investment in a new mine) was computed to include a 15 percent rate of return, discounted after taxes,over the 20-year life of the mine. It was assumed that the 1972 ratio of 51:49 underground mine production to surface mine production would shift to a 40:60 ratio in 1990. Increased production of western coals to meet low-sulfur standards in eastern and midwestern markets, as well as for local demand, was assumed. Correspondingly, it was assumed that production from underground mines in high-sulfur coal areas of the East would decline over the years. Also, there would be some redistribution of eastern coals, from consuming areas, where they are not environmentally permissible,to areas where they could meet less stringent standards. Depletion of existing productive capacity was calculated at five percent of 1972 production for each year through 1980, and at three percent per year thereafter.

Additional assumptions applicable to both BAU and AD included:

- Coal production levels are based on long-term contract conditions.
- Estimated delivered cost of coal is the minimum acceptable selling price in 1973 dollars.
- Seventy-five percent of the new bituminous coal produced will undergo preparation. No lignite will require preparation.
- There will be 225 working days per year for deep mines and 250 working days for surface mines.

Assumptions Specific to BAU

Under BAU, it was assumed normal expansion in business activity would call for increased coal consumption - and therefore production - of 5.4 percent a year from 1972 through 1980, and 3.8 percent a year from 1980 through 1990. It was also assumed that pre-embargo public landleasing practices would be continued, and that mining capacity would not expand significantly in the earlier years because of the long lead-times required for mine development equipment and related requirements.

Assumptions Specific to AD

Production levels under AD conditions were set arbitrarily high to create an energy scenario that would not be constrained by available supply under even the most optimistic demand projection. It was also assumed that air pollution control regulations and oil prices would have no major adverse impacts on coal demand. Moreover, it was assumed that additional public lands would be leased, as needed, for new mines and expansion of existing mines; and synthetic fuels production would grow rapidly, beginning in 1980.

Major Findings

The coal industry has the capacity to satisfy almost any foreseeable demand for coal by 1985, at prices near 1972-1973 levels and considerably below current spot market levels. To the extent that investment decisions will have to be made in the immediate future to achieve long-range production goals, a sufficient return on investment and resolution of major uncertainties will be needed. Since unconstrained maximum production levels could be very high, Business-As-Usual and Accelerated Development scenarios were developed so that demand and resource constraints would limit supply. Under BAU, and certainly under AD, production potential exceeds most estimates of demand.

In the BAU case, total production could range from 755-1100 million tons per year from 1977-1985; while in Accelerated Development the range is 926-2063 million tons per year (see Table 3-15). Large increases could occur in Appalachian underground and northern Great Plains surface mining under these scenarios. The regional supply curves for coal are very different (see Figure 3-21 for BAU supply curves). All the supply curves, however, are relatively flat at reasonably low prices.

NUCLEAR FUELS

Excerpt From "PROJECT INDEPENDENCE"

Nuclear fuel production capability in the United States was initiated by the Manhattan Project, and spurred in the late 1940's by the demand for fissionable materials for nuclear weapons. Production capability was further increased in the 1950's as a result of the nuclear submarine program, and in the 1960's by the commercialization of nuclear electric power generation.

Early nuclear fuel production operations were sponsored by the United States Government, with some private involvement. After enactment of the Atomic Energy Act in 1954, private initiatives occurred in all phases of fuel production with the exception of enrichment and permanent disposal of high-level radioactive wastes. The Government continues to be responsible for regulation of nuclear operations, storage and permanent disposal of high-level radioactive wastes, and has a major role in the development of most nuclear fuels. It also remains the sole supplier of enrichment services for the country and most of the Western world, although private initiatives are under intensive study.

Fuel Cycle

The nuclear fuel cycle encompasses the exploration for reserves of uranium ore; mining the ore; milling and refining the ore to produce uranium concentrates (U_3O_8); production of uranium hexafloride (UF_6) from uranium concentrates to provide feed for isotopic enrichment; isotopic enrichment of UF_6 to attain reactor requirements; fabrication of nuclear reactor fuel (including converting UF_6 to uranium dioxide, pelletizing, encapsulating in rods, and assembling fuel elements); irradiation of nuclear fuel in the reactor; reprocessing irradiated fuel and converting

Figure 3-21

1985 BUSINESS-AS-USUAL REGIONAL SUPPLY CURVES

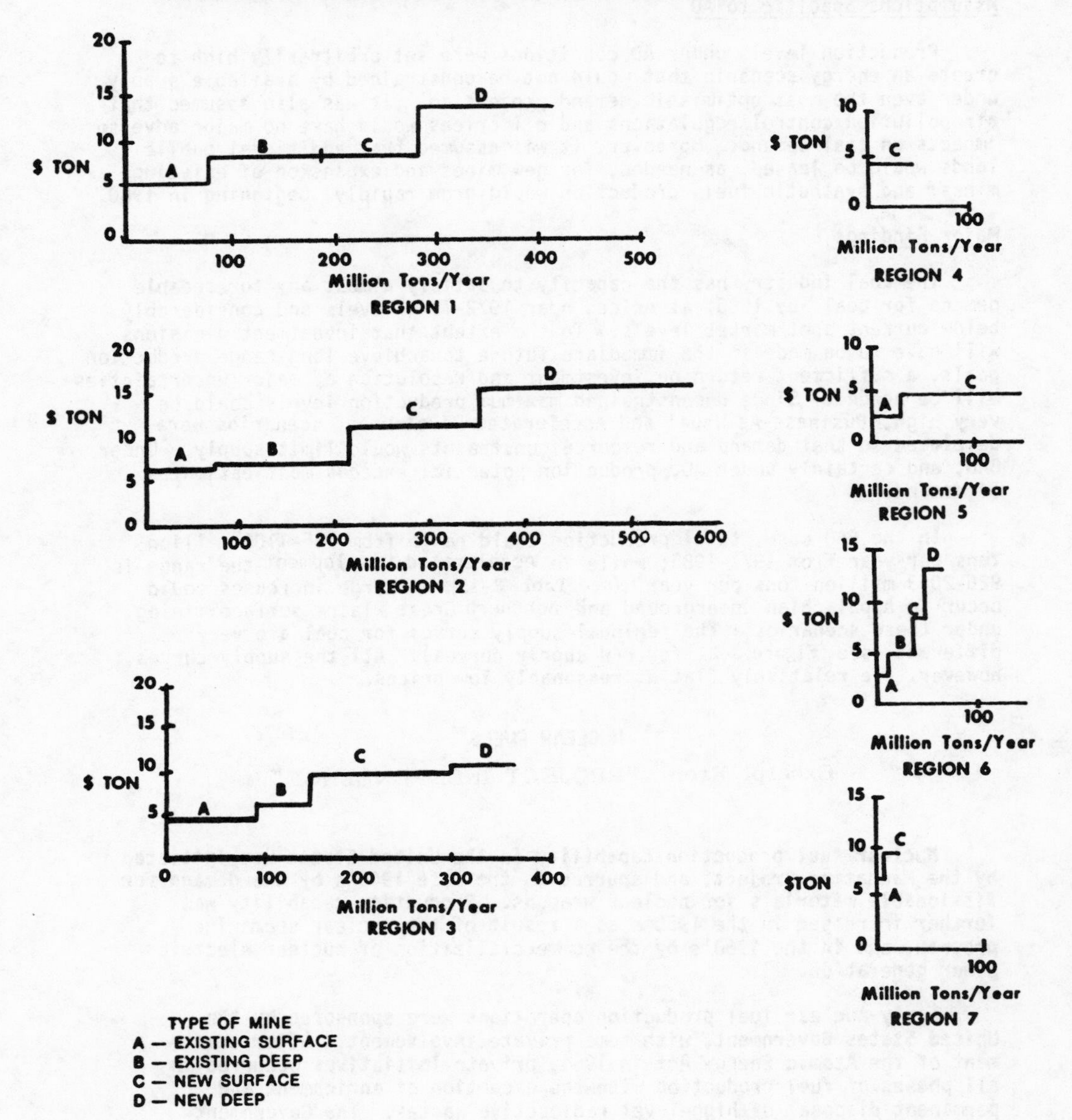

uranium to UF_6 (recycle through the gaseous diffusion plants) for re-enrichment; recovery of plutonium; radioactive waste management; and transportation of materials. (See Figure 3-22).

Fuel Cycle Requirements

The FEA analysis has estimated nuclear power plant capacity and related fuel cycle requirements as follows: (See Table 3-16):

Capital, manpower, and key equipment requirements are not major constraints on production in the long-run, although they may be serious short-term problems. Again, the energy policy issue with coal is not whether and how to stimulate its production, but rather whether and how to stimulate its consumption. An important factor constraining coal demand is the manner in which the Clean Air Act Amendments of 1970 will be implemented. If current emission regulations are enforced, about 225 million tons of coal could

Table 3-15
Coal Production Potential (million tons)

REGION	MINE TYPE	Business As Usual 1977	1980	1985	1990	Accelerated 1977	1980	1985	1990
Northern Appalachia									
	Surface	82.1	93.2	107.5	138.1	92.7	135.6	192.2	269.8
	Deep	118.7	123.4	135.9	145.8	128.4	177.4	267.0	352.2
	TOTAL	200.8	216.6	243.4	283.9	221.1	313.0	459.2	622.0
Southern Appalachia									
	Surface	92.6	111.0	137.2	164.6	132.5	166.6	245.1	359.3
	Deep	171.6	195.3	237.0	273.0	199.8	300.1	445.6	588.0
	TOTAL	264.2	306.3	374.2	437.6	332.3	466.7	690.7	947.3
Midwestern									
	Surface	118.8	133.7	159.1	187.9	139.1	196.5	287.7	416.3
	Deep	58.5	63.4	68.1	80.2	65.3	91.6	136.0	178.3
	TOTAL	177.3	197.1	227.2	268.1	204.4	288.1	423.7	594.6
Gulf									
	Surface	11.2	28.9	42.5	54.0	25.0	50.0	75.0	105.0
	Deep	-	-	-	-	-	-	-	-
	TOTAL	11.2	28.9	42.5	54.0	25.0	50.0	75.0	105.0
Northern Great Plains									
	Surface	64.4	99.6	152.0	184.5	98.5	185.9	302.6	380.6
	Deep	1.3	2.0	3.1	3.6	.7	.9	1.2	1.5
	TOTAL	65.7	101.6	155.1	188.1	99.2	186.8	303.8	382.1
Rocky Mountain									
	Surface	16.9	19.8	24.4	28.8	19.8	29.9	49.4	65.1
	Deep	10.8	12.3	14.7	17.3	12.3	18.1	27.2	38.8
	TOTAL	27.7	32.1	39.1	46.1	32.1	48.0	76.6	103.9
Pacific Coast									
	Surface	7.9	12.3	18.5	21.9	12.3	23.1	33.8	47.8
	Deep	.1	.1	.2	.2	.1	.2	.3	.4
	TOTAL	8.0	12.4	18.7	22.1	12.4	23.3	34.1	48.2
National									
	Surface	393.9	498.5	641.2	779.8	519.9	787.6	1185.8	1643.9
	Deep	361.0	396.5	459.0	520.1	406.6	588.3	877.3	1159.2
	TOTAL	754.9	895.0	1100.2	1299.9	926.5	1375.9	2063.1	2803.1

not be burned in 1975; if proposed non-deterioration regulations are implemented, there would be significant constraints on building new coal-fired powerplants; and if the current coal conversion legislation were not amended, there would likely be only limited conversions to coal. The alternatives available to the Federal Government to substantially increase coal consumption include amendments to the Clean Air Act and other coal conversion legislation.

Other important factors constraining coal demand are expectations about the price and the availability of oil and natural gas. Under one set of assumptions, coal and nuclear power would replace almost all the oil and natural gas being burned in the utility sector; under another set of plausible assumptions, they do not. These alternate assumptions about oil and gas price and supply could shift the demand for coal by a few hundred million tons per year. Similar effects (but of smaller magnitude) occur in other sectors. It would be possible for the Federal Government to ensure the direct or indirect substitution of coal for oil and gas (e.g., through the use of heavy taxes, or through regulation on coal conversions or the prohibition of oil and gas in new powerplants); however, such measures might generate significant costs and environmental impact.

A third important factor affecting coal demand is the assumption about the demand for electricity. Consumption of electricity is expected to grow more slowly than in the past, even after the substitution of electricity for oil and gas in certain sectors. Because of increasing prices, the extent to which such substitution takes place will depend on

Figure 3-22

NUCLEAR FUEL CYCLE FOR LWR's

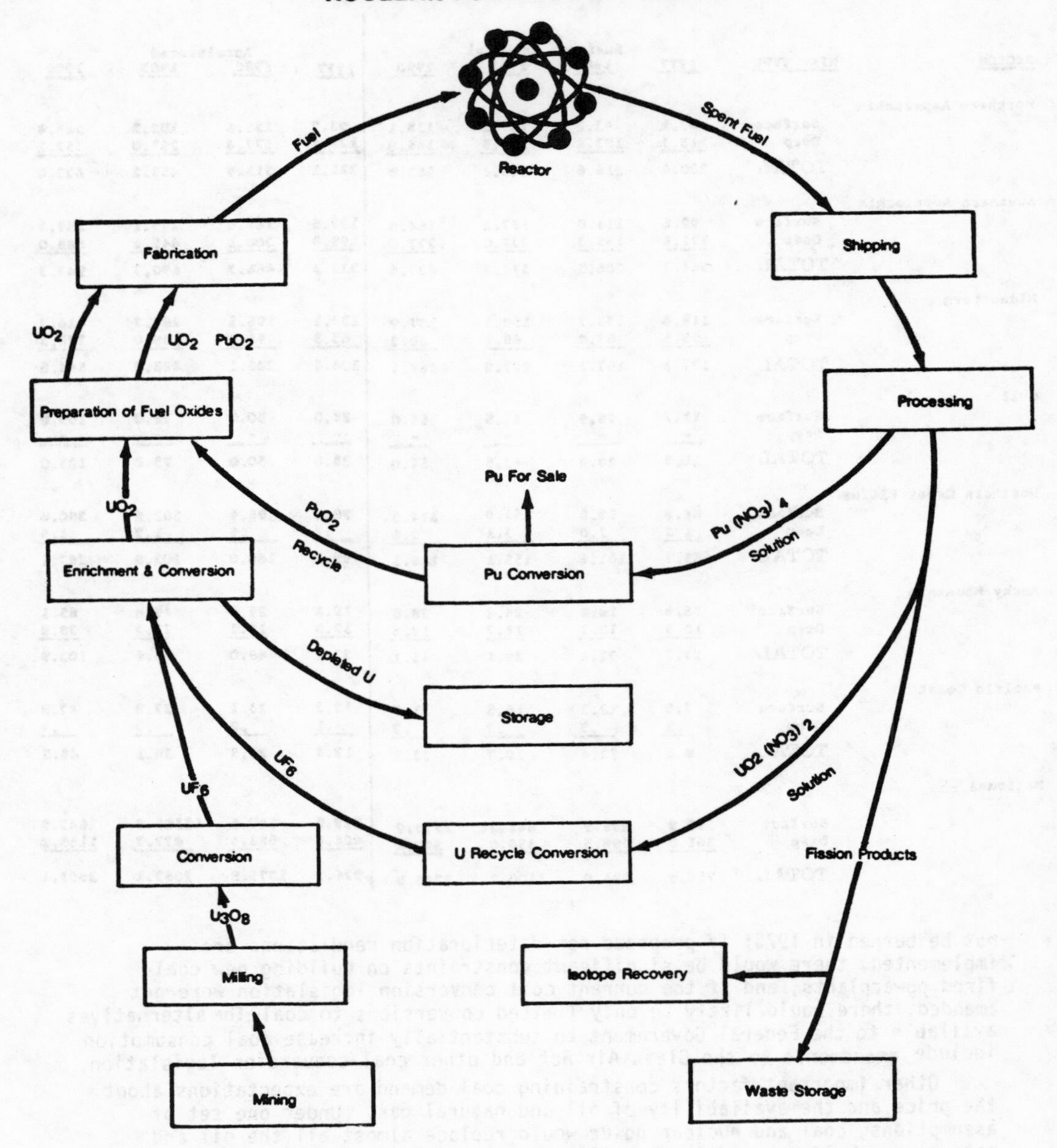

the relative prices of the energy sources and on attitudes and customs developed over many years. Further, if nuclear plant production slows down, coal-fired powerplants would increase.

The source of nuclear electric generating plant fuel is uranium ore. Uranium ores, like all minable natural resources, are depletable and of finite size. The Atomic Energy Commission estimates that the United States has 520,000 tons of uranium reserves producible at a cost of $15/lb or less and an additional 1,000,000 tons of potential resources producible at this cost.

At projected rates of exploration, mining and milling, the production capability of the industry will not be able to meet uranium fuel requirements for the Accelerated Case beyond 1985, unless the rate of exploration and mine/mill facility construction is accelerated in the near future.

Table 3-16
Fuel Cycle Requirements
(Accelerated Development Strategy)

	Requirements 1980	1985
Plant Capacity Nuclear (gigawatt)	93	240
Annual U_3O_8 1/	34,800	52,100
Cumulative U_3O_8 (cumulative tons from 1974)	141,500	351,900
Enrichment, Annual (SWU) 2/	19,900	32,700
Reprocessing, Annual (metric tons)	1,820	5,400

1/ U_3O_8 is uranium ore
2/ Separative Work Units, an arbitrary measure of the relationship of amount and assay of uranium feed to the enrichment process, the electric power used in separation, and the amount and assay of the enriched and depleted product and "tails." Tails are the depleted uranium which still contains some of the U_{235} isotope.

The assumptions inherent in this table include:

- Enrichment - Includes foreign demand based on commitments already made or being made to meet foreign requirements.
- Enrichment - Based on 0.2 percent tails assay through 1985, 0.3 percent thereafter.
- About 116,000 kilograms of separative work units are required to prepare enough uranium to support a typical light water reactor.

Since nuclear fuel requirements are directly related to nuclear power plants on line, the recent stretchout and cancellation of nuclear generating plant construction plans could have the effect of postponing projected 1985 capacity.

In order to meet the projected accelerated nuclear generating plant schedule, nuclear fuel production and uranium enrichment would have to more than double from 1980 to 1985.

There are a number of constraints to the potential expansion of nuclear power. These include:

- Uranium resources and exploratory activities to find them.
- Uranium mining and milling capacity.
- Uranium enrichment capacity.
- Spent fuel reprocessing capacity.
- Uncertainties about the schedule for on-line nuclear generating capacity.

Conditions which have constrained extensive exploratory efforts in the past include the relatively low price of uranium, the uncertain role of nuclear power in electric power generation, and insufficient planning for long-range nuclear fuel requirements. Possible Government actions to relieve these constraints could encompass financial incentives and greater access to exploration on public lands. Existing mining and milling capacity of 18,000 short tons of U_3O_8 per year, does not meet 1980 requirements. Fifteen to 20 new plants and an investment of almost $2 billion may be needed to meet demand.

Isotopic enrichment of uranium is necessary to provide fuel for light water moderated nuclear reactors. Enrichment results in a U-235 concentrate of 2 to 4 percent, as opposed to .7 percent in the natural state. The existing enrichment capacity (including authorized expansion of 27,000,000 SWU's per year) is sufficient to meet domestic and foreign requirements until

1983. New capacity additions of 8,750,000 SWU's per year may be required each 18 months; this capacity would cost about $1.5 billion. This capacity addition includes an allowance (approximately 30 percent) for meeting foreign requirements. This would represent a very large commitment by the private sector, or possibly the Government, with a long-term payback of investment. Research on centrifuge enrichment offers the potential of reducing this capital investment.

Recovering usable uranium by reprocessing "spent" nuclear fuel could reduce new uranium requirements by about 15 percent and enrichment service requirements by about 20 percent. While over 2,000 metric tons per year of reprocessing capacity is scheduled to be in service by 1977-78, this capacity is only adequate through 1980 and meets only half the 1985 requirements. Each new 1,500 metric ton per year reprocessing plant is estimated to cost about $200 million.

To achieve even the low estimates of nuclear growth would necessitate a reversal of recent trends in the ability of utilities to raise investment funds and in equipment delivery and construction schedules, as well as a reduction in licensing delays. Achievement of higher levels of nuclear power could require a national commitment of manpower and other resources. The long lead time required to achieve nuclear capacity additions severely limits the possibilty of increasing the number of nuclear plants which could become operational before the early 1980's.

Public acceptance of nuclear power is an important factor in overcoming the current problems constraining the use of nuclear power and the exploration and mining of uranium. Utility planning, site availability, licensing, schedules, and implementation of measures to shorten the construction period are all influenced by public acceptance. However, nuclear power generation has important advantages in being relatively insensitive to location of fuel sources because of low nuclear fuel transportation costs. Nuclear power may also be more environmentally acceptable than coal, providing that fears of nuclear accidents can be alleviated.

ELECTRICITY

Excerpt From "PROJECT INDEPENDENCE"

Background

Since electricity was first commercially produced in New York City in 1852, it has steadily become more important in the Nation's total energy picture. Electricity generation has doubled every ten years for the last 40 years. Since 1940, electricity has risen from 12 percent to 26 percent of total energy consumed. Electricity's rapid growth is attributable to its attractiveness relative to other energy sources. Electricity is a very flexible energy source. It is fully substitutable for other fuels in applications where stationary heat is required and most fuels can be used to generate electrical power. Until recently, the electrical utility industry experienced constant fuel costs and increasing fuel conversion efficiency and, as a consequence, electricity rates generally did not increase. However, recent increases in construction, fuel and operating costs have resulted in higher rates. Annual growth in electricity generation has been reduced from its historic level of about 7 percent to 0.6 percent this year, because of the mild weather, reduced business activity, the oil embargo, electricity rate increases, and conservation measures.

The installation of total generating capacity has been responsive to the growth in load. Hydroelectric power capacity has grown from 4 million kilowatts to 67 kilowatts in 1973, but its percentage of the total energy generation has shrunk from 30 percent to under 15 percent. Fossil fuel steam generation has grown from 70 percent to 80 percent of the total generation. In the last 15 years, nuclear powered steam has grown to almost 5 percent of the total capacity and is projected to provide a rapidly increasing portion

Table 3-17

Capacity - Electric Utility Industry
(Million Kilowatts at Year End)

	Total	Hydro	-Steam- Fossil	Nuclear	Internal Combustion
1920	13	4	9	0	
1925	21	6	15	0	.2
1930	32	8	23	0	.4
1935	34	9	24	0	.6
1940	40	11	28	0	.9
1945	50	15	34	0	1.1
1950	69	18	49	0	1.9
1955	114	25	87	0	2.4
1960	168	32	132	3	2.8
1965	236	44	188	1	3.4
1970	342	55	275	6	4.4
1973	432	67	311	21	2.6

of the electric energy generated (see Table 3-17 for distribution of electric capacity).

Coal, which traditionally accounted for 65 percent to 70 percent of all fossil fuels (oil, natural gas, and coal) used to generate electricity, declined to 54 percent in 1971-1972. As imported residual oil became more plentiful along the East Coast, the use of oil has increased to almost 20 percent of the total generation. Natural gas, a cleaner fuel, reached a peak of 30 percent of the generation in 1970, but fell back to 21 percent in 1973. (See Figure 3-23 for trends in fossil-fuel consumption.)

Nuclear power has become increasingly important in meeting the Nation's energy needs. In the last five years alone, nuclear power has increased from 1 percent to 4.5 percent of total electric energy generation. As compared to fossil fuels, nuclear fuels (uranium and thorium) are relatively inexpensive; however, nuclear-powered plants require larger capital investmemts. Setbacks in construction schedules due to engineering problems, poor quality control, equipment delays, lower-than-expected labor productivity, and licensing and regulatory delays have slowed nuclear growth. In 1973, only 21,000 megawatts were operational out of the 36,000 megawatts of nuclear capacity originally scheduled for completion. At present, it takes almost 10 years to bring some nuclear generating plants into production, compared to about 5 years for a fossil steam generator and 2-3 years for a gas turbine generator.

There are wide differences between regions in the types and amount of primary energy used in producing electricity (see Fig. 3-24). The East North Central and East South Central regions depend heavily (75 percent to 100 percent) on coal. The South Atlantic and West North Central area also depend principally (50 percent to 75 percent) on coal. The West South Central depends mostly on natural gas, the Pacific region on hydroelectric power, and New England on oil. The Mountain States region uses a variety of fuels, none of which account for as much as 50 percent of the total.

To minimize interruptions or voltage reductions in electrical service, utilities must be able to meet the peak demand, which may be twice the low demand point. There are major fluctuations in weekly and seasonal load curves with the greatest demand occurring in the summer months (see Figure 3-25). Planning generating capacity for varying load demand is divided into three general categories:

1. base loaded--that which is expected to operate at a high, and essentially constant load factor, generally on the order of 70 percent;

2. intermediate loaded--that which is expected to respond to

Figure 3-23

FOSSIL FUELS CONSUMED FOR ELECTRICITY GENERATION IN THE U.S.

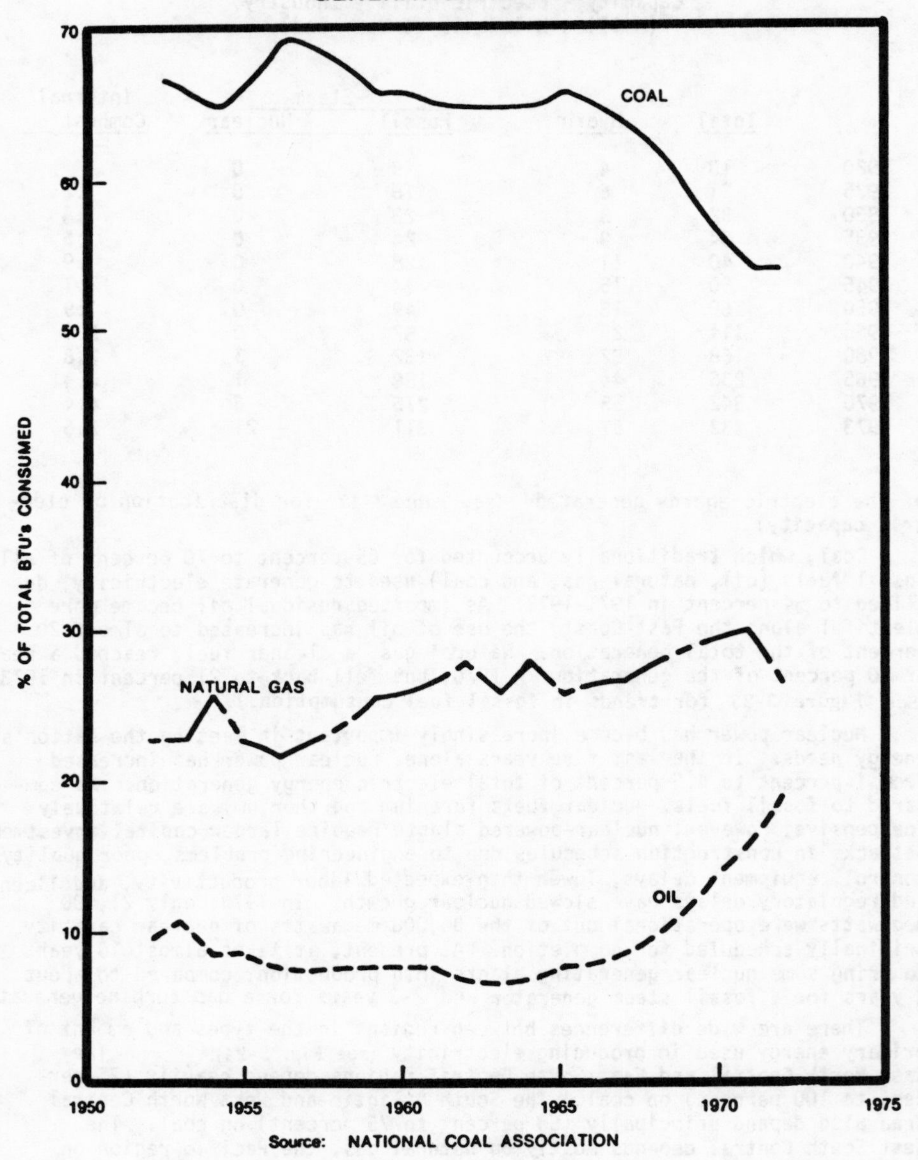

Source: NATIONAL COAL ASSOCIATION

a varying load demand and will operate with an annual plant capacity factor around 40 percent; and

3. peak loaded--that which can meet peak demands and shut down after the peak demand period ends and will operate with an annual plant capacity factor of about 10 percent.

To serve the base load, plants are selected which have the least fuel and operating cost (variable costs) and are generally nuclear plants, new, large, high efficiency coal-or oil-fired steam turbine plants, or hydroelectric plants where river flow is adequate to meet base load demand. New coal and nuclear plants are expected to cost $350-480 per kilowatt and power generation costs will be in the range of 14-22 mil/kilowatt hour.

Figure 3-24

REGIONAL TRENDS IN PRIMARY ENERGY CONSUMPTION IN ELECTRICITY

(1972)

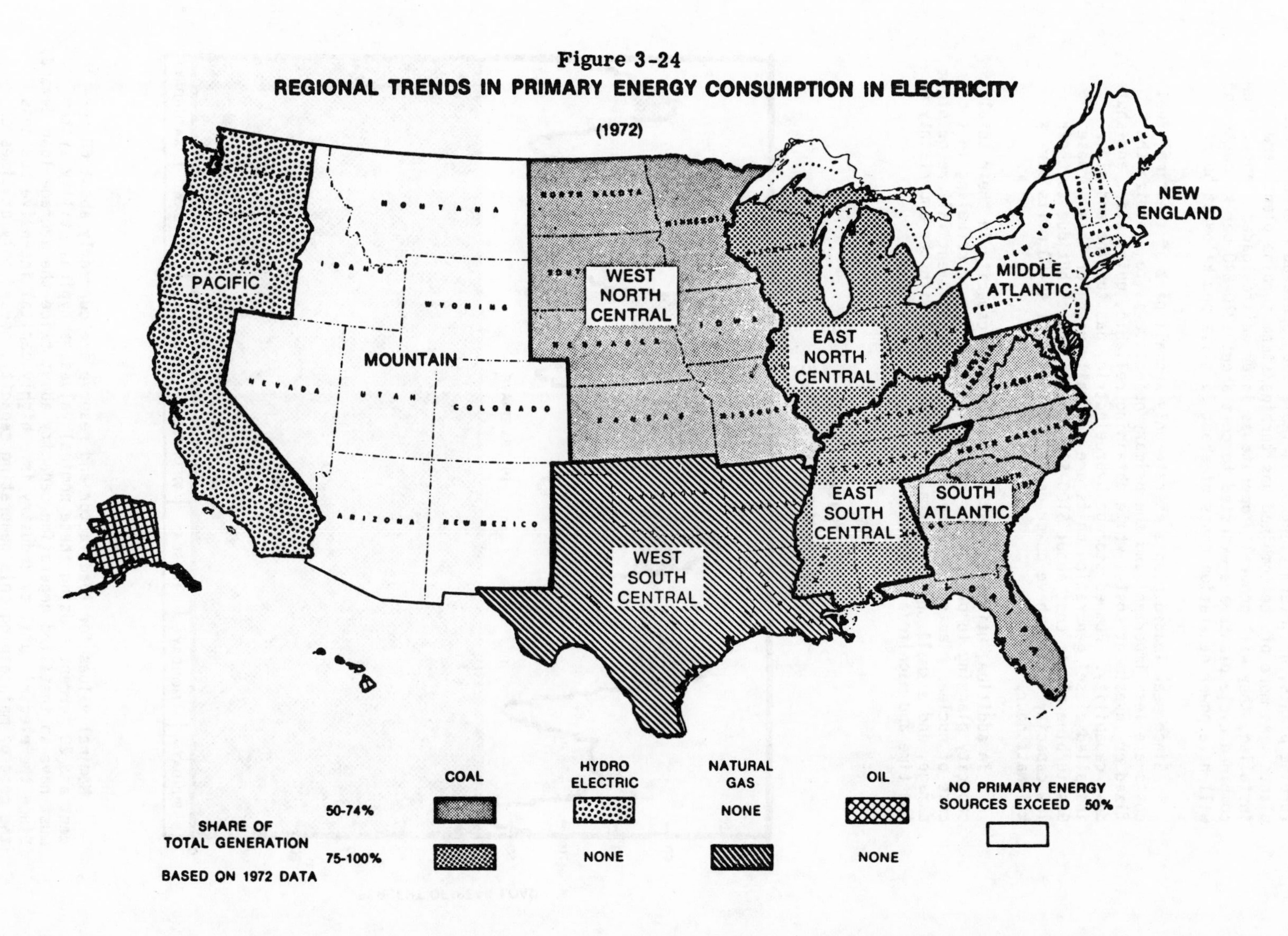

Intermediate loaded plants can be older, less efficient steam turbine plants, or intermittent or variable load operated hydroelectric plants. As more of the combined gas turbine/steam turbine plants are installed, they will generally operate as intermediate loaded plants. New combined cycle plants are expected to cost about $210-240 per kilowatt and will have power generation costs of about 25 mills per kilowatt hour.

Since peak loaded plants operate only a small part of the time, fuel costs are less important and the primary objective is to minimize the fixed or investment costs, while obtaining reliable, quick-start and shut-down capability. Pumped storage hydroelectric, gas turbine and older installed diesel generation units are generally used for this purpose. Gas turbine plants cost about $100-$120 per kilowatts but due to their low capacity factor, have power generation costs of 40-50 mills per kilowatt hour.

In addition, utilities must include reserve capacity in their total capacity planning to provide for routine maintenance or refueling in the case of nuclear plants, unplanned outages due to equipment failure or storm damage, and a small on-line extra capacity to deal with the uncertainty of the time and magnitude of peak demands.

Figure 3-25
WEEKLY LOAD CURVE

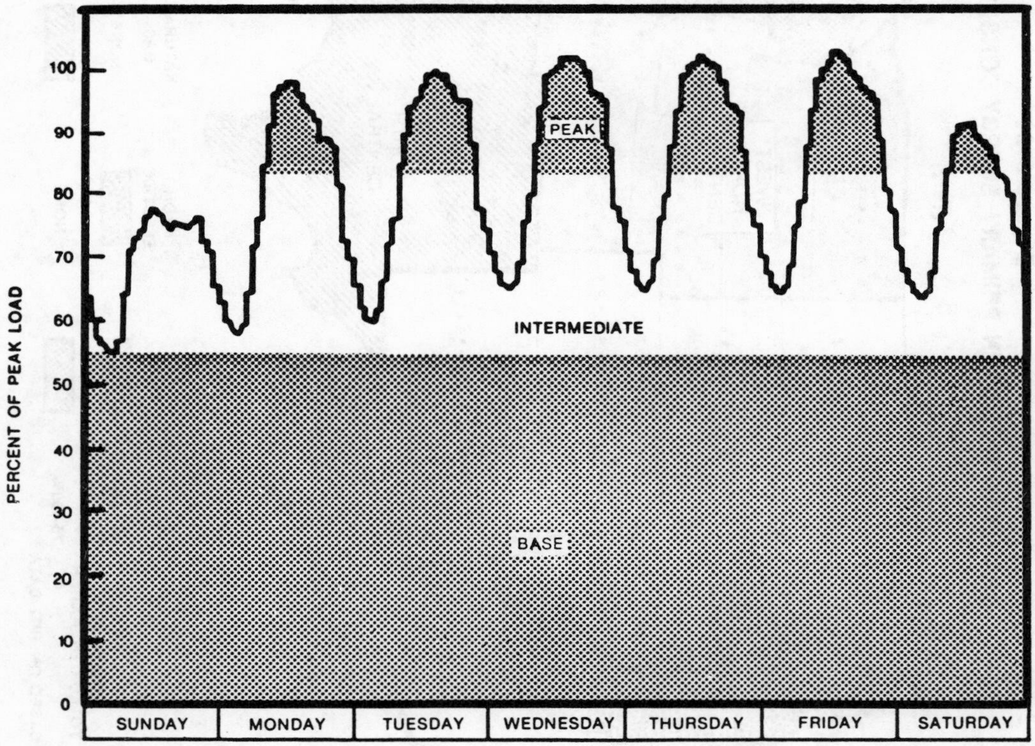

Nominal values for load factor and reserve are currently about 60 percent and 20 percent. Using these nominal values, a typical utility system must have an installed generation capacity about twice the average load demand. Since the electric utility industry is a highly capital intensive business, the cost of maintaining this generating capacity, needed only at times of high demand, increases electricity costs.

The electric power industry is made up of a number of utility systems, some owned by private companies (investor-owned utilities) some owned by the Federal Government or by other public bodies such as municipalities, states or public utility districts; and some owned by electric cooperatives; 250 investor-owned utilities account for 75 percent of the Nation's total generating capacity. The Federally-owned segment of the power industry accounts for 10 percent of the Nation's total generating capacity.

These utilities have joined into Reliability Councils that can plan the capability for and coordinate the dispatch and sharing of electrical loads over several States. (see Figure 3-26 for Reliability Councils). Such sharing has reduced the Nation's average reserve requirement from 50 percent over peak demand in the 1920's to about 20 percent today. When transmission voltage capacity was low, power plants had to be located near sources of use, i.e., major urban centers. As extremely high voltage transmission systems with lower losses of power have been developed, power plants can be located closer to fuel sources and away from congested urban demand centers.

An intricate system of Federal and State regulation has developed to control individual and collective rate structures and rates of return. The Federal Government controls licensing of certain plants, interstate sale of electricity, and registration of securities. Retail electricity rates and levels of intrastate sales are controlled by State public utility commissions.

Methodology

Projections of electricity generating facility requirements must account for the varying demand for electricity, price and availability of competing sources of energy and electricity generation equipment, and consideration of peak, intermediate, and base load generation requirements.

FEA evaluated plant requirements by estimating electricity demand by regions at various prices of competing fuels. The cost of stack gas emission control equipment and cooling towers to ease environmental impact were factored into cost estimates. Gross capacity additions needed to meet increased demand were calculated assuming historical ratios of capacity factor, reserve requirements and load variations; old coal, oil and gas-fired steam turbine plants were assumed to be retired at 3 percent of existing capacity per year. It was also assumed that a mix of new generation plants would be used to meet this higher demand, this mix being 75 percent base-load plants, 20 percent intermediate operating plants and 5 percent peak operation plants.

With this mix established, available nuclear capacity additions were applied to the base load requirement and available hydroelectric capacity additions to the intermediate and peak generation requirements. The remainder of the required capacity additions were met by fossil-fueled plants.

A variety of fossil-fueled generation plants is available to meet this demand, including:

coal-fired steam turbine plants,
oil-fired steam turbine plants,
oil-fired simple cycle gas turbine plants,
oil-fired combined cycle gas turbine plants,
coal-gas fired combined cycle gas turbine plants.

Since come new additions of natural gas-fired steam turbine plants are under construction but not yet operating, a limited number of such plants could also be selected. Selections of particular plants were based on least cost of generated power, and in proportion to the need for base, intermediate or peak operational mode requirements.

The principal differences between the Business-As-Usual and Accelerated Development estimates for electricity generation facility requirements occur in the projected growth of nuclear power. Nuclear generating capacity additions between 1973 and the beginning of 1985 are estimated to be 220,000 MWe under Accelerated Development and 184,000 MWe under Business-As-Usual development.

Figure 3-26

REGIONAL ELECTRIC RELIABILITY COUNCILS

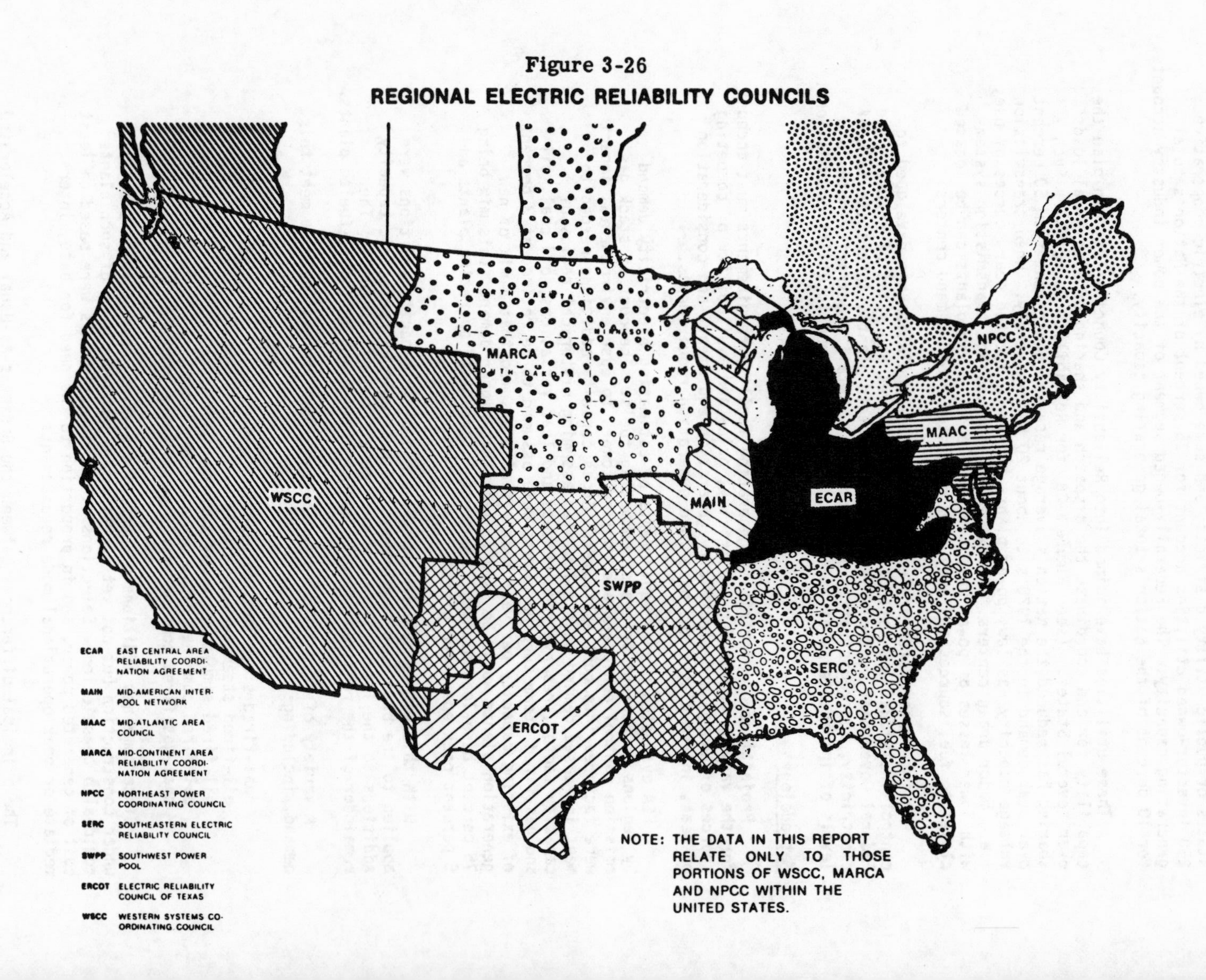
WSCC
MARCA
MAIN
ECAR
NPCC
MAAC
SWPP
SERC
ERCOT
ECAR EAST CENTRAL AREA RELIABILITY COORDINATION AGREEMENT
MAIN MID-AMERICAN INTER-POOL NETWORK
MAAC MID-ATLANTIC AREA COUNCIL
MARCA MID-CONTINENT AREA RELIABILITY COORDINATION AGREEMENT
NPCC NORTHEAST POWER COORDINATING COUNCIL
SERC SOUTHEASTERN ELECTRIC RELIABILITY COUNCIL
SWPP SOUTHWEST POWER POOL
ERCOT ELECTRIC RELIABILITY COUNCIL OF TEXAS
WSCC WESTERN SYSTEMS COORDINATING COUNCIL
NOTE: THE DATA IN THIS REPORT RELATE ONLY TO THOSE PORTIONS OF WSCC, MARCA AND NPCC WITHIN THE UNITED STATES.

Major Findings

The FEA has prepared two projections of future electricity demand--a Business-As-Usual Case, and an increased electrical case that entails greater Government participation in demand management. This latter strategy redistributes the aggregate consumption of energy to increase consumption of those fuels that can be produced domestically. In effect, this second projection calls for substitution of electricity for other energy end-use purposes which, in turn, would provide a demand basis for greater use of the Nation's coal and uranium resources. In addition, it calls for conversion of existing and planned oil and gas-fired electricity generating plants to the burning of coal, to the extent practicable. The results of these two projections are presented in Table 3-18 .

Table 3-18.

Electrical Capacity Projections

	Existing Capacity end-1973	1985 Projections(in gigawatts)1/ BAU $11/BBL.	2/ Demand Management
Total Electricity Capacity	424	922	1002
Growth Rate 1973-1985, %/yr.	--	6.3	7.4
Hydro Capacity GWe	65	100	100
Nuclear Capacity GWe	20	204	240 3/
Coal Capacity GWe	167	327	379
Oil Capacity GWe	78	81	64 4/
Gas Capacity GWe	61	48	48
Combustion Turbine GWe 5/	33	162	171

1/ Beginning of year projections (nuclear at end of year would be 234 and 275 for BAU and AD respectively.

2/ Without conservation.

3/ Accelerated nuclear construction schedules

4/ The demand management projection includes conversion of about 16,500 megawatts of existing oil-fired generation capacity to coal.

5/ These figures reflect projected increased market penetration of intermediate load combined cycle plants and continued use of gas turbine peaking plants.

Coal and nuclear power generally compete for base load generation capacity additions. The demand management projection would increase electricity demand so that coal use for electricity generation would increase substantially from the 1972 use of 330 million tons per year, and nulclear would increase from 20 MWe to 240 MWe.

The FEA projections of future electricity requirements are subject to a number of uncertainties. It is not yet clear whether recent reductions in electrical demand will continue. Future prices and availability of alternative fuels could reestablish the historical nuclear demand growth rate. If total electricity demand continues to grow less than peak demand, resulting in new lower load factors, increased generating capacity would be required. Recent downward trends in the efficiency of new large plants, unless corrected, would result in a need for a higher level of capability. If rate structures were adjusted to increase rates during peak periods (thus creating incentives to spread demand more evenly across the day), it is expected that the ratio of peak to average load would be reduced, along with the need for expensive peak capacity.

The electricity generation and delivery capacity projections have a major effect on the capital investments and other resource requirements of the energy industry. Each additional percentage increase in the rate of growth demand from 1974 to 1985 would require an increased investment of about $60 billion by 1985. The recent utility deferrals and cancellations of generating capacity additions (about 40,000 MW of nuclear plants and 31,000 MW of other plants by preliminary estimates) has already modified future plans. If cancellations of coal burning facilities continue, heavier reliance on oil imports and higher electric power costs would result. The inability of utilities to finance capacity additions could seriously curtail near-term commitments and long-term projections.

Hydroelectric power sites are limited and 80 percent of planned capacity installations up to 1983 are additions to operating generating facilities. The greatest potential for new hydroelectric development exists in Alaska. The Accelerated Development scenario represents a one-thousand percent increase above the planned development there, but is affected by long lead times. Electricity-intensive industry could locate in this region.

The public has become more aware of power plant siting issues. There has been greater concern over the health and safety effects of nuclear plant operations. The fear of large nuclear accidents, waste disposal leakages, sabotage, and thermal effluents is evident. Concerns with fossil fuel plants focus on the effects of harmful air emissions, thermal effluents and solid waste disposal.

Further tightening of constraints on power plant emissions could seriously limit the capability to reduce oil imports through increased use of coal. The uncertainty over the reliability of stack gas scrubbers and the health effects of sulfates will affect coal conversion and the ability to site new facilities.

In the period to 1985 the cost of generating and delivering electricity will continue to increase irrespective of inflation. The continual addition of new, higher cost generation and delivery systems will raise the average fixed costs per unit of capacity of utility systems. Strict environmental limitations result in further increased plant cost and reduced energy conversion efficiency of power generation plants. The higher cost of fossil fuels will increase production costs, and there are no near-term technology advances which can offset the effects of these cost increases.

Additional electricity generation constraints include site availability; competing land use; regulatory delays; construction delays due to management and engineering; short-term manpower and equipment availability; and reliability. Reliability is particularly important as average capacity in nuclear plants declined to about 60 percent. Increases in powerplant availability would result in fewer plants having to be built and would reduce capital requirements.

SHALE OIL

Excerpt From "PROJECT INDEPENDENCE"

Background

Oil shale is a laminated marlstone rock containing a tar-like organic material called kerogen. When heated to 450 to 600°C, kerogen undergoes pyrolysis (decomposition) to yield raw shale oil. The viscous petroleum-like material can be refined into a complete line of petroleum products by conventional refinery techniques. Oil shale yields from 10 to 40 gallons per ton (GPT) of processed shale. Commercial development requires a yield in the 25 to 35 gallons or more per ton range. Oil shale of at least 25 gallons per ton quality represents about 33 percent of known reserves. It is mined by either surface or deep mining methods similar to those used in large quarry operations. At present, there are two basic processes being developed for extracting the petroleum products from shale, retorting and in situ.

Commercial production of liquid fuels from oil shale began in France in 1838. The Scottish oil shale industry was inaugurated in 1850 and lasted for more than a century before production ceased because of the availability of lower cost substitute fuels. A small shale oil industry, using a somewhat different material, operated in Canada and the eastern United States in the 1860's, but disappeared following the discovery of oil in Pennsylvania. With concern over the adequacy of U.S. petroleum reserves immediately after World War I, there was renewed interest in western oil shale. However, by the end of the 1920's experimental projects were discontinued. Since World War II, experimental programs have been conducted to develop more efficient mining and extracting methods.

Until recently, nearly all of the research in mining techniques has been devoted to underground mining. The underground use of large, mobile equipment normally employed in surface operations was demonstrated by the Bureau of Mines in 1947. The Union Oil Company of California opened a similar mine in 1956 to supply shale for a retort demonstration. The Colony Development operation opened a large, underground mine in 1965 to supply shale for semicommercial retort operations and to demonstrate new mining equipment and techniques. Surface mining, employing techniques used in large, open-pit copper mines, has not been demonstrated, but is believed to be applicable to thick oil shale deposits which occur near the surface.

In addition to improved mining techniques, government and industry have sought to improve extracting methods. For example, several above-ground retorting systems have been tested on a pilot or demonstration scale, but none have yet been advanced to commercial sized equipment. Recent efforts have been devoted to developing the in situ retorting process, but there has been no satisfactory commercial demonstration of it to date.

Estimates of the oil shale resources in the Green River Formation of Colorado, Utah, and Wyoming have increased over the years as more information about them has become available. The Oil Shale Task Force analysis indicates that their resources are at least 1,800 billion barrels of oil, contained in a relatively small 25,000 square mile area. Of the known higher grade reserves, about 84 percent are located in Colorado, 10 percent in Utah, and 6 percent in Wyoming (see Figure 3-27). Colorado, the focus of industrial interest, has the smallest geographical area of oil shale deposits (3,000 square miles), but contains the richest, thickest, and best defined reserves.

The Federal Government is the largest owner of oil shale lands with 72 percent of the 11 million acres of possible commercial lands and nearly 80 percent of the oil shale resource. Several private companies own deposits on the peripheral portions of the Piceance Basin in Colorado, but development in these areas will probably depend on the rate and extent of public land leasing.

Leasing of Government oil shale lands was halted by Executive Action in 1930. On June 29, 1971, the Secretary of the Interior announced plans for a proposed prototype oil shale leasing program. This program involved a total of six tracts in Colorado, Utah, and Wyoming, but only the four tracts in Colorado and Utah have attracted interest. Development of this small portion of the United States oil shale resources could lead to production of about 250,000 barrels per day of shale oil. Additional leasing on a commercial scale is not anticipated until the environmental effects from initial development are examined.

Methodology, Assumptions, and Findings

The objective of the Interagency Oil Shale Task Force study was to assess the potential energy contribution to the Nation through oil shale development. Although it is not possible precisely to predict the level of production or the technologies that may be used, it is possible to

Figure 3-27

OIL SHALE AREAS COLORADO, UTAH, AND WYOMING

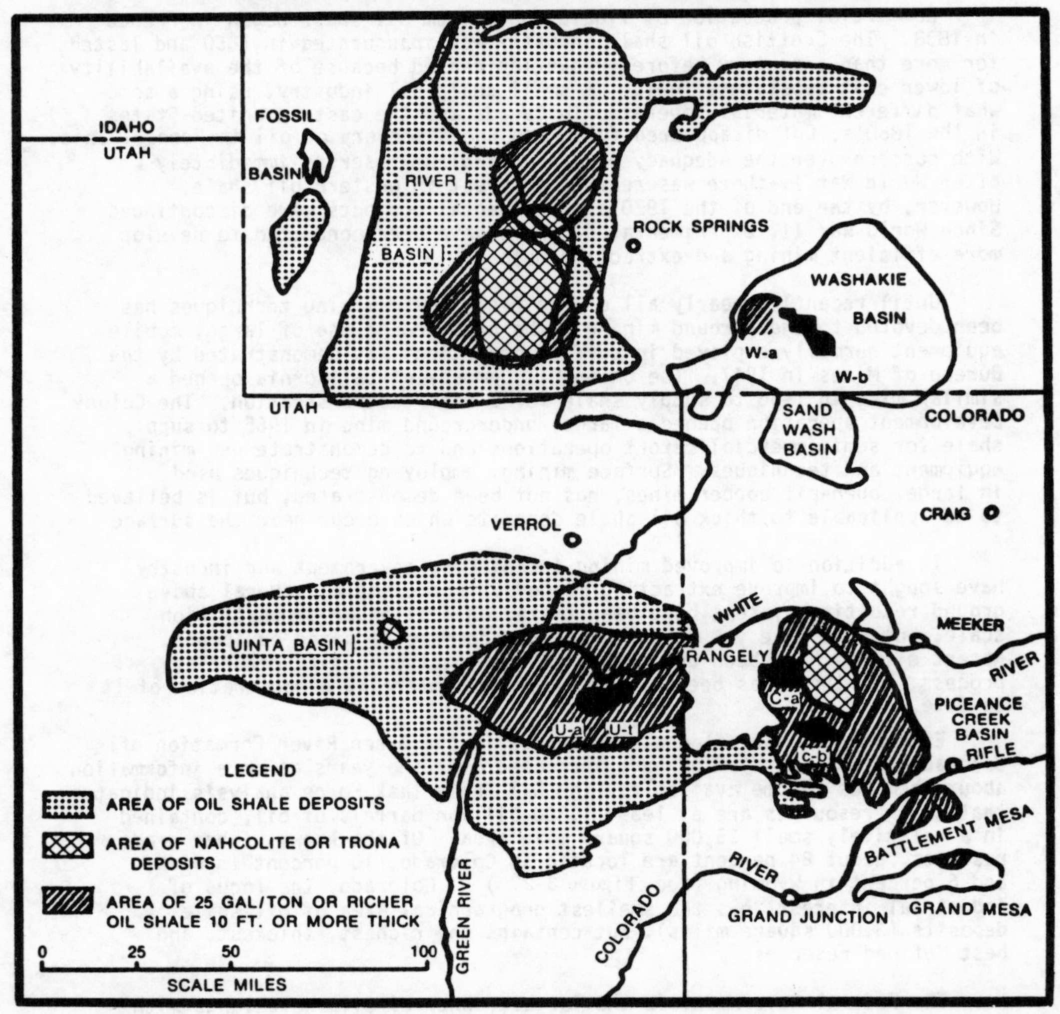

establish a range of possibilities. The range developed for this analysis reflects current judgment and plans already announted. Business-As-Usual (BAU) and Accelerated Development (AD) estimates were made for production and cost from 1976 to 1990.

The evaluations of production potential under various market conditions indicate that, if the world crude oil price were $4.00 per barrel, there would be no shale oil production, because the rate of return at this price would be less than 10 percent and therefore would not attract the necessary venture capital. Nevertheless, under the Business-As-Usual case and a world oil price of $7.00 per barrel, oil shale development is projected to reach 250,000 barrels per day by 1985 (see Table 3-19). This estimate has been established by the level of production that could be attained in Colorado under the State's air quality standard for sulfur dioxide. This standard would limit Colorado production to 200,000 barrels per day. Under BAU conditions, the minimum expected rate of return would be 15 percent. This schedule of production is expected to respond to normal economic incentives. Certain government actions and a stable world oil price of $11.00 per barrel, rather than $7.00 per barrel, would create the economic and institutional climate necessary to achieve higher levels of production. Under these conditions, production levels of one million barrels per day could be attained by 1985

Table 3-19
Shale Oil Production Potential

	World Oil Price, dollars/bbl	Shale Oil Production (thousand barrels per day)			
		1977	1980	1985	1990
Business-As-Usual	$7.00	0	50	250	450
Accelerated Development	$11.00	0	100	1,000	1,600
Expansion over Business-As-Usual		0	+50	+750	+1,250

Under the AD case and a world oil price of $11.00 per barrel the estimated return on investment would be about 20 percent. Higher crude oil prices would increase the return of capital, but probably would not stimulate higher production because of other constraints, such as lead times, environmental standards, and water availability.

The rate at which oil shale can be developed depends on a number of interrelated factors. Initially, it will depend on the availability of venture capital, which can expect to yield only a minimum acceptable rate of return. As the industry matures, however, profitability should increase, and factors other than venture capital will become constraints. (Estimates of capital and operating costs for various plant configurations, as well as material and labor requirements are shown in Table 3-20).

Table 3-20

Capital, Material and Labor Estimates for Oil Shale Production

	Undeground	Underground	Surface Mine	In situ	Modified In Situ	
	50,000 B/D	100,000 B/D	100,000 B/D	50,000 B/D	50,000 B/D	
	30 GPT	30 GPT	30 GPT	22 GPT	18GPT	25GPT
Capital Investment, Million dollars	280	520	600	380	310	280
Operating Cost(ex-depreciation), $ million/yr	45	80	80	140	110	90
Rate of Return (DCF)						
Shale Oil selling price - $8.35/bbl	14	16	15	-	11	15
Shale Oil selling price -$12.35/bbl	21	22	21	15	20	25
Total Engineering & Construction, Labor, man-years	2,400	4,200	4,200	3,200	3,800	3,800
Operating Labor, men	1,400	2,400	1,800	620	1,200	1,200
Steel Requirements, Thousand tons	70	130	130	410	65	65

Shale oil production will affect the environment. At production levels above 200,000 to 300,000 barrels per day, emissions from surface shale plants are expected to exceed the 1980 Colorado standards of 10 micrograms per cubic meter of sulfur dioxide in the Piceance Basin. Further, the quantities of water used and the plant effluents could affect local water acquifier pressures and the water quality. Shale development could concentrate industrial and population growth in a relatively unpopulated area, leading to serious questions concerning growth management.

Other factors limiting oil shale production are economics and land availability. Although shale oil is expected to be competitive at $7 a barrel oil, there is still uncertainty about the long-range world price of crude oil. If this price falls much below $7 per barrel, shale investments will not be made due to inadequate returns. This could suggest the need for incentives such as purchase agreements at a specified price and fiscal incentives such as depletion allowances, accelerated depreciation, and investment tax credits. Also, the Government could create a publicly owned or funded corporation to produce shale oil from public lands. Research to develop an <u>in situ</u> process could ease the environmental impacts from shale production. The Federal Government will also have to grant the access permits and take other steps necessary for any further shale oil development. Accelerated production will require leasing additional public land beginning about 1979.

The Government can stimulate levels of shale production in 1985 from up to one million barrels per day. Moving beyond the Business-As-Usual level will require a coordinated Federal program involving leasing, water availability, possible financial incentives, R&D and satisfying environmental criteria. State and local governments will be deeply involved in the decisionmaking and implementation process.

SYNTHETIC FUELS

Excerpt From "PROJECT INDEPENDENCE"

Background

Coal is the most abundant fossil fuel in the United States, but it is inconvenient to use, and the combustion of high-sulfur coal without clean-up of the resulting gases produces air pollution. For centuries people have tried to solve this problem by converting coal to synthetic liquid or gas fuels. Beginning in the late 19th century and continuing until after World War II in the United States, town gas and industrial producer gas was made from coal and sold commercially.

Coal conversion requires chemical processing at elevated temperature and pressure to remove the impurities and increase the hydrogen content of the products. The Lurgi process was developed in Germany in the 1930's to gasify non-coking coal with oxygen and steam. It produced a low-Btu gas that could be converted to electricity or used for industrial heating at or near the production site. Coal ash is removed during the gasification process. The Winkler process, which uses a fluidized bed reactor, was developed in Germany in the 1920s. Other processes include the Koppers-Topzek gasifier, developed in Germany in the 1930s; the Fischer-Tropsch process which also originated in Germany and is currently in use in South Africa, and several processes being developed in this country.

The United States undertook a major effort on coal synthetics in the late 1940's and early 1950's, but abandoned it because of unfavorable economics. Current synthetic fuel research in the Federal Government and under private sponsorship is focusing on increasing the spectrum of coals that can be used, reducing the relatively high capital investment and product costs (products costs might be cut as much as 20 - 30 percent), and increasing conversion efficiencies from 60-65 percent to 70 - 75 percent.

Several high-Btu processes have reached the pilot-plant level. The HYGAS, synthane, BI-GAS, and CO_2 Acceptor processes are being sponsored in part or completely by the Federal Government. High-Btu gas can be used in pipelines for shipment to demand regions.

Coal liquefaction technology is also being developed by Government and industry. A solvent refined coal pilot plant is now being built by the Government, and industry is active at the process development unit and laboratory stages. The liquid boiler feed and heavy liquid boiler fuel produced by liquefaction processes will be used as fuel oil. The naphtha produced can be utilized as a petrochemical feedstock.

Methodology and Assumptions

The potential production of synthetic fuel depends on the selection of the fuel product (liquid, high-Btu or low-Btu gas), the timely availability of new technology, and the cost of existing and new processes. To assess commercial feasibility, a competitive pricing formula was determined using the cost of coal and the synthetic fuel processing costs; a 15 percent discounted cash flow and 20-year plant life were assumed.

The processing costs are also strongly influenced by the variations in the price and quality of coal from various locations in the country and in the cost and availability of water. Synthetic fuel plant costs were determined for a range of coal costs for eastern and western locations (Table 3-21 summarizes major cost and resource differences).

Table 3-21
Economic Factors

Material/Element	Eastern Coal	Western Coal	Usage Ratio East/West
Coal ($/ton)	12.00	4.00	1/1.35
Process Water ($/Mgal)	0.25	0.40	1/0.85
Cooling Water ($/Mgal)	0.05	0.08	2/1
Electric Power ($/kwh)	0.015	0.015	1/1
Labor ($/hr)	6.50	6.50	1/1
Catalyst and Chemical	--	--	1/1
Coal Heat Content (Btu/lb)	11.500	8.500	--
Coal Water Content (%)	6.0	25.0	--
Coal Sulfur Content (%)	4.4	0.6	--

Additional major assumptions used in determining the price of synthetic fuels were that:

1. There would be no by-product credits from the sale of sulfur.

2. There would be no coal transportation costs.

3. Annual plant downtime for maintenance would be an average of 35 days.

Major Findings

There are no commercial synthetic plants currently in operation or under construction in the United States,and with several possible development routes to follow, the production potential could vary widely. At one extreme, the Government could discourage the construction of new technology plants and focus on the existing, but more expensive Lurgi gasification technology. In this case, due to economics, virtually no synthetic fuel production would occur by 1985. At the other extreme, the Government could subsidize the construction of plants, using existing technology and undertaking an accelerated program to develop advanced approaches. The major limitation to production in this case would be the lead times for demonstration and for construction.

The task force did not base its estimates on either of these extremes. It assumed that synthetic fuel capacity would be constructed using existing technology, primarily high-Btu Lurgi gasification plants built by gas utilities. The analysis also anticipates technological advances and demonstrations as a result of the Federal research and development program. The BAU approach assumes no change in the Government's pre-embargo policies, no relaxation or postponement of environmental standards, and no incentives or exceptional priority allowances. The Accelerated Case, on the other hand, assumes substantial relaxation or other modification of environmental standards, along with other incentives, priority allowances, accelerated research and faster construction.

Under the Accelerated Case, synthetic-liquid production could equal 500,000 barrels per day in 1985 (about 2-3 percent of petroleum consumption). Synthetic high-Btu gas production in 1985 under the accelerated case could be 1 TCF per year, which is also about 3 percent of gas consumption (See Figure 3-28). The largest potential for synthetic high-Btu gas will occur beyond the Project Independence timeframe.

Figure 3-28
GROWTH OF SYNTHETIC HIGH-BTU [PIPELINE] GAS ANNUAL PRODUCTION CAPACITY

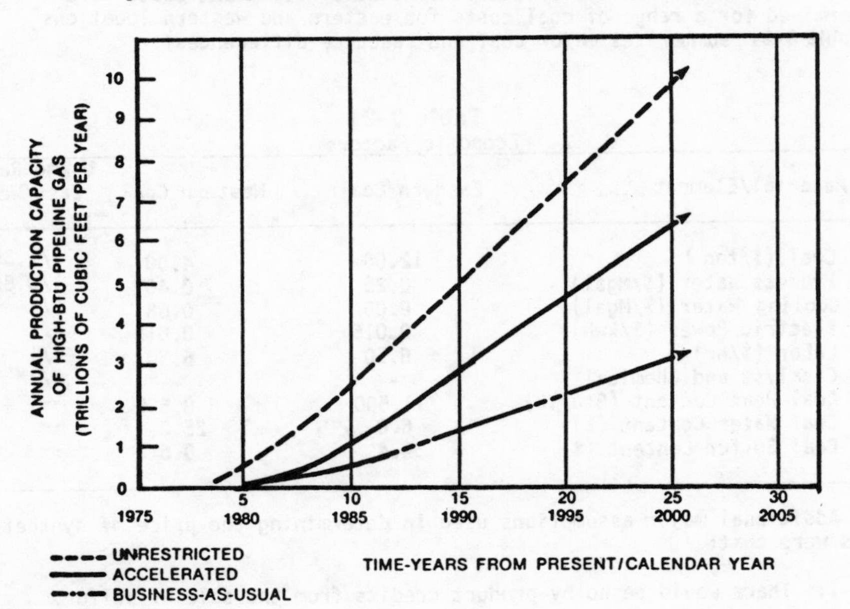

Although synthetic liquids are being produced in South Africa, technology availability and cost will limit the growth of synthetic liquid production in the United States. Production would probably not start until 1982, and even under the Accelerated Scenario only 500,000 barrels per day could be produced by 1985. There would be no BAU production in 1985. (See Figure 3-29 .)

The economical production of low-Btu (fuel) gas is dependent upon the development of new technology processes. No production of fuel gas is expected in the 1980's.

Although the synthetic fuels could only supply a small amount of our gas and oil requirements by 1985, the real potential for synthetic fuels will not be realized until the late 1980's or early 1990's.

Synthetic high-Btu gas costs of $2.00 or more per million Btu's make synthetic gas uncompetitive with natural gas if the latter is available and the price is significantly lower.

Figure 3-29

GROWTH OF SYNTHETIC LIQUID FUEL PRODUCTION CAPACITY

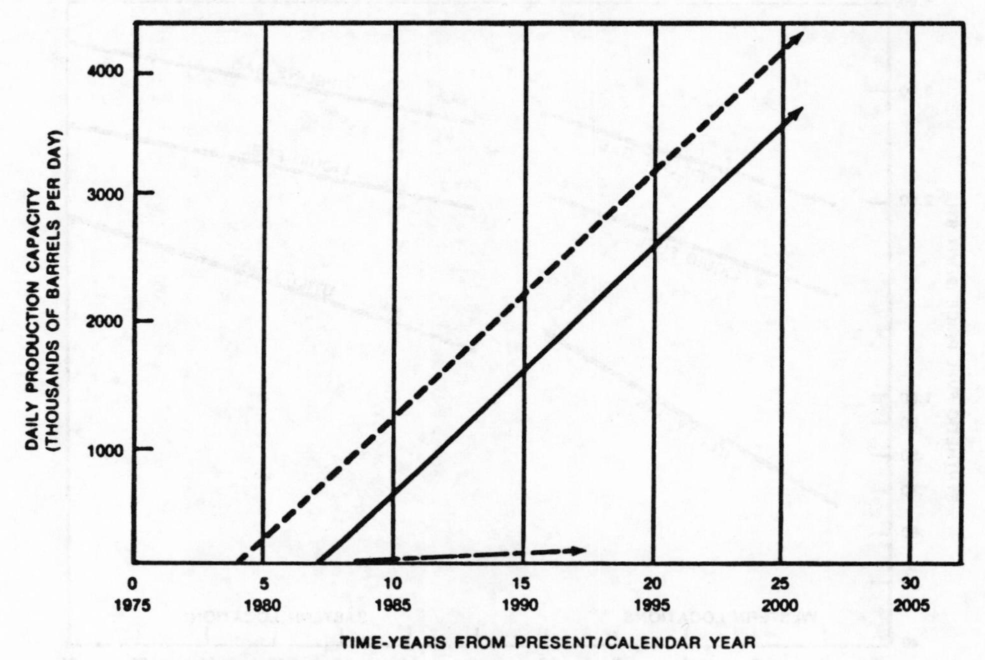

Figure 3-30 shows the cost for high-Btu (pipeline) gas, liquids, and low-Btu (utility) gas for a range of coal costs from eastern and western locations. Since synthetic fuel plants must be located where the raw materials, coal and water, are available, the transportation from the plants to the consuming regions is an additional cost.

However, at crude oil price levels of $11 per barrel, liquid fuels produced from coal costing less than $19 per ton are less expensive than natural petroleum products.

The major policy issues concerning synthetics from coal are whether existing technologies should be rapidly deployed; whether development of new, more efficient technologies should be accelerated; and whether the effort should favor particular products. A spectrum of liquids and gases, ranging from low-Btu gas and solvent-refined coal that are primarily power plant fuels to synthetic crude oil, chemical feedstocks, and pipeline quality gas could be produced. In addition, there are serious issues regarding who should assume the risk and pay the increased costs for the synthetic fuels.

SOLAR ENERGY

Parabolic collectors, similar to those being considered for solar thermal conversion systems, provided steam for irrigation pumping in Egypt in 1913. Current efforts are focused on building reliable, low cost systems that can be installed and used in a variety of ways.

Wind energy systems, for example, should be economically viable within a few years. If the aerodynamic technology developed over the last 30 years were applied, the system costs could be dramatically reduced and market applications would be greatly increased.

Figure 3-30

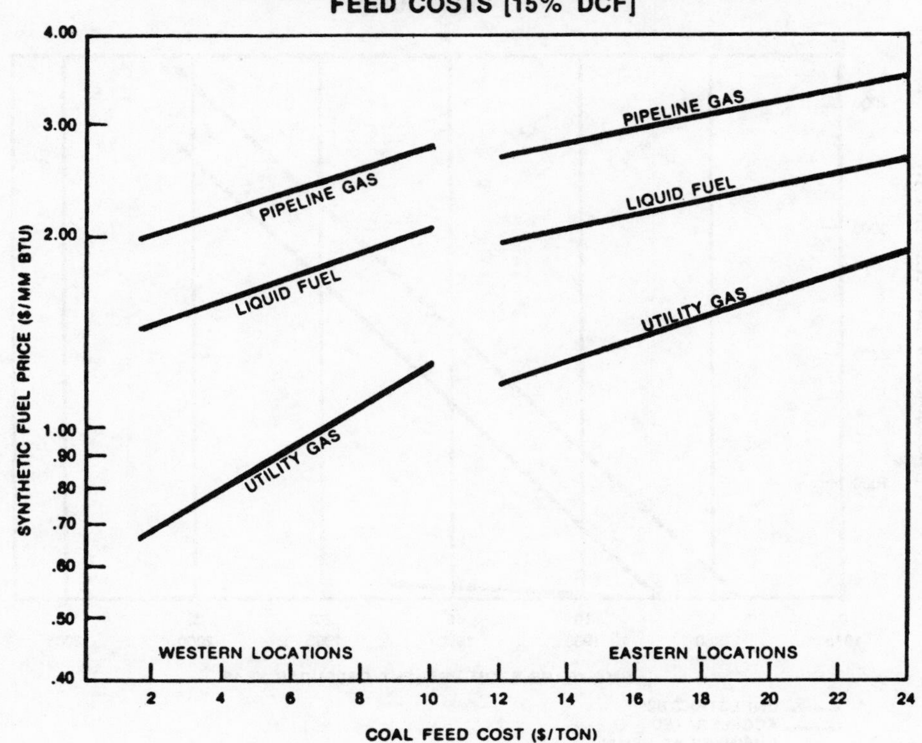

Photovoltaic systems (solar cells), are expected to have applications in two major market areas--the dispersed, on-site ("roof-top") uses and the small-to-moderate size, central-power-station facilities.

Bioconversion of fuels includes both near term and long term applications. Recycling of urban and agricultural wastes is now providing energy in some areas, while large biomass forms, both terrestrial and marine, are longer-range prospects.

The economic viability for these different applications will occur in different stages, but cost reductions will depend more on market volume than on technological breakthroughs. Indeed, economic viability for much of the dispersed market could be achieved by 1980 without any technological breakthroughs. However, considerable R&D would be required to achieve economic viability of photovoltaics for central power stations by the mid 1980's or early 1990's. And Government policy initiatives might be required to create a market large enough to sustain the high rates of production necessary to lower unit costs. A major government R&D program could result in a reduction of cost to $400 to $600 per Kwe (peak) by around 1980.

Solar thermal systems and ocean thermal systems require either substantial research or engineering development; and for most of these systems, low-cost storage will be needed.

Methodology and Assumptions

Both the Business-As-Usual (BAU) and the Accelerated Development (AD) cases were based on the successful completion of a $1 billion accelerated R&D program during the next five years. The alternative scenarios reflect different assumptions about the manner in which the new technologies would be implemented, but both envision a rapid implementation of the technologies as soon as they are demonstrated.

It was assumed that during the 1980's wind, photovoltaics, solar thermal and ocean thermal would only be used to generate electricity. The energy savings projected for the solar heating and cooling of buildings were based on only the heating and hot water requirements of projected new residential and small commercial buildings that otherwise would have selected space and water heating by electricity. Energy savings realized from solar cooling and from the direct use of other fuels (e.g., oil and gas) for heating and cooling would result in additional energy savings. In addition, savings projected for retrofitting were not included.

The cost to the consumer for using solar heating and cooling systems was assumed to equal the annualized cost ($/year) of the system, divided by the annual amount of solar energy Btu's/year utilized. It is a function of region (i.e., insolation, climate), system application, collector size and efficiency, mortgage rate, and time period, but does not include the cost for a conventional auxiliary system. The minimum system price at which production would take place in the United States is estimated to be $5 per square foot of collector area. In both the Accelerated Development and Business-As-Usual cases, market penetrations were assumed when the alternatives available to the consumer cost more than solar (i.e., at least $2.50/million Btu's on a national average basis).

The projected energy savings resulting from wind energy systems assumed a rapid rise in the production of windmills during the 1980's and a leveling off during the 1990's. Capital costs for wind systems were assumed to decrease from an estimated $500 - 1,000 per Kw today to approximately $250 per Kw by 1985. The cost of electricity is site specific and varies with wind velocity.

Capital and operating cost estimates for other solar energy technologies are highly speculative at this time because commercial systems have not yet been designed and mass produced. The task force considers its cost estimates to be conservative.

Major Findings

The two largest sources of solar energy through 1985--solar heating and cooling and wind electric systems--are not very dependent on new technology.

Solar heating and cooling, which is already being commercialized, is projected to be competitive with conventional systems at a cost of $2.50 per million Btu's of space heat. By 1985, solar energy could provide about 0.3 quadrillion Btu's of space heating under the BAU projections and with incentives (AD case), solar energy could provide 0.6 quadrillion Btu's of space heating. Solar heating can be thought of as a reduction in the demand for other conventional fuels for residential and commercial heating. The potential for solar electric and solar heating market penetration are indicated in Figures 3-31 and 3-32.

Wind energy systems offer the potential for a relatively inexpensive electrical source. Wind would be an intermittent source of electricity, and, taking account of the load factor, production would be measured in terms of annual electrical generation.

The acceptance of solar systems by consumers, which is difficult to predict, will clearly have a direct effect on production. User acceptance of new technology has traditionally been a long process. Some of the technologies, such as wind energy, may be aesthetically objectionable.

With the exception of bioconversion and ocean thermal conversion, all solar energy systems require some storage capability. Conventional backup systems must be utilized to the extent storage is inadequate during periods of no insolation or wind. There do not, however, appear to be any resource constraints that would impact upon the projected production levels. Normal growth in the materials industries should be able to sustain the proposed production levels.

Figure 3-31
SOLAR ELECTRIC POTENTIAL

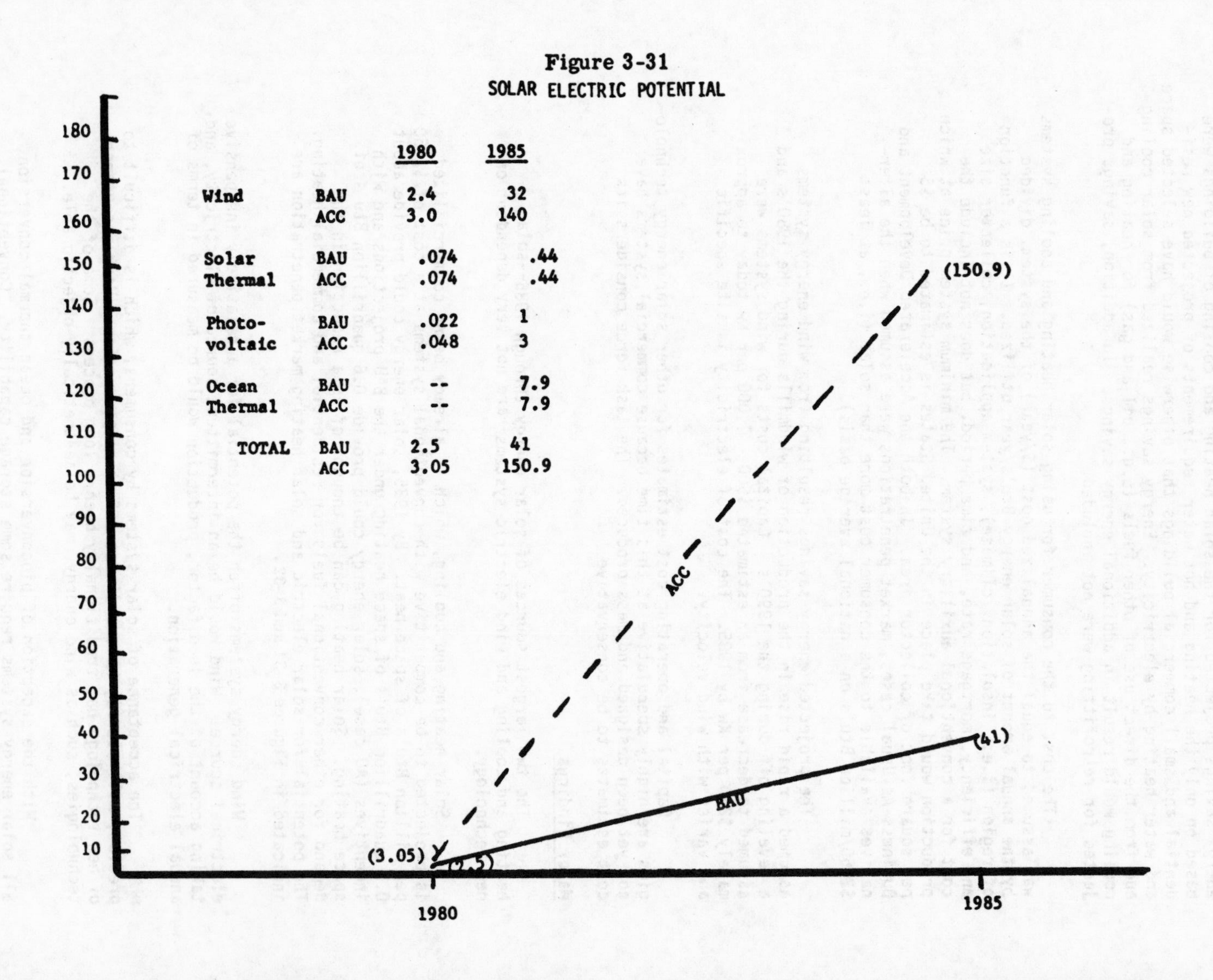

		1980	1985
Wind	BAU	2.4	32
	ACC	3.0	140
Solar Thermal	BAU	.074	.44
	ACC	.074	.44
Photo-voltaic	BAU	.022	1
	ACC	.048	3
Ocean Thermal	BAU	--	7.9
	ACC	--	7.9
TOTAL	BAU	2.5	41
	ACC	3.05	150.9

Figure 3-32
POTENTIAL FOR SOLAR HEATING AND COOLING OF BUILDINGS

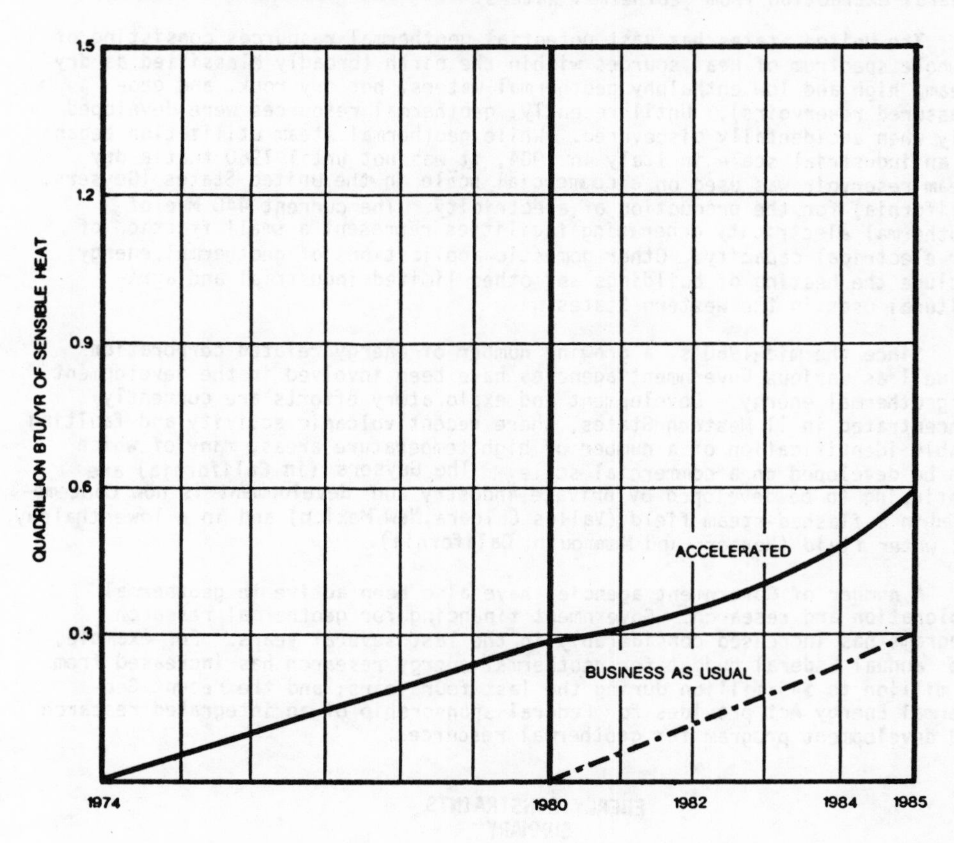

Full utilization of solar energy technologies could significantly reduce pollution levels. The environmental impacts of solar energy systems should be lower than those of conventional energy sources because solar systems are, by nature, essentially clean environmentally. That is, they do not produce any major health hazard such as radiation or pollution emissions. There are, however, some land use and aesthetic problems associated with solar collectors.

A wide range of financial and regulatory incentives are available to the Government to increase solar energy use. These include tax credits and deductions, investment tax credits, tax exemptions, accelerated depreciation on capital equipment, and guaranteed and insured loans. Other incentives that should be investigated include leasing of Federal lands for solar energy development, and guaranteed Government purchase of solar energy devices and components such as photovoltaic systems for use on Federal facilities and programs.

GEOTHERMAL ENERGY

Excerpt From "PROJECT INDEPENDENCE"

Background

Geothermal energy has only recently been recognized in the United States as a potentially important and domestically abundant source of energy. Besides generation of electric power, geothermal energy can be used for heating, cooling, and various household and industrial needs. (It should be noted that geothermal energy now has only minimal use in households in the United States.) Additional uses being developed are desalinization of water and mineral extraction from geothermal waters.

The United States has vast potential geothermal resources consisting of a whole spectrum of heat sources within the earth (broadly classified as dry steam, high and low enthalphy geothermal waters, hot dry rock, and geo-pressured reservoirs). Until recently, geothermal resources were developed only when accidentally discovered. While geothermal steam utilization began on an industrial scale in Italy in 1904, it was not until 1960 that a dry steam reservoir was used on a commercial scale in the United States (Geysers, California) for the production of electricity. The current 440 MWe of geothermal electricity generating facilities represent a small fraction of our electrical capacity. Other domestic applications of geothermal energy include the heating of buildings and other limited industrial and agri-cultural uses in the Western States.

Since the mid-1960's, a growing number of energy related corporations as well as various Government agencies have been involved in the development of geothermal energy. Development and exploratory efforts are currently concentrated in 11 Western States, where recent volcanic activity and faulting enable identification of a number of high temperature areas, many of which can be developed on a commercial scale. The Geysers (in California) are continuing to be developed by private industry and development is now contemp-lated in a flashed-steam field (Valles Caldera, New Mexico) and in a lowenthalphy, hot water fluid (Herber, and Mammouth, California).

A number of Government agencies have also been active in geothermal exploration and research. Government financing for geothermal research programs has increased considerably in the last several years. For example, the annual Federal budget for geothermal energy research has increased from $1 million to $45 million during the last four years; and the recent Geo-thermal Energy Act provides for Federal sponsorship of an integrated research and development program for geothermal resources.

ENERGY CONSTRAINTS SUMMARY

Excerpt From "PROJECT INDEPENDENCE"

Approach and Major Findings

Long-run energy supply forecasts have not generally taken into account requirements for materials, equipment, manpower, transportation, finance and water in projecting supply. Currently, energy industries are encountering very serious shortages of many resources. Orders for walking draglines used in strip-mining are backlogged until 1979; a shortage of shipyard personnel is limiting construction of off-shore drilling rigs; and lack of financing is causing utility companies to abandon expansion plans. An overall shortage of steel is already affecting the availability of steel plate for large equipment, and forgings and castings of small parts, for all energy industries.

An important element of the Project Independence Blueprint analysis is the determination of whether shortages of manpower, equipment, or other resources would constrain United States energy development in the period through 1985.

The approach used is straightforward -- requirements are compared to the capability to supply resources. The projected resource needs of the energy sector under alternative energy strategies are compared with the estimated availability of those resources in the same timeframe.

To perform the analysis, each major resource was assigned to a special "crosscut" task force. These task forces prepared detailed lists of materials, equipment, and manpower categories which could be in short supply.

They also specified how each fuel task force should estimate water and finance requirements. The fuel task forces then estimated the requirements for each of these categories for a typical coal mine, oil shale plant, refinery, or other operational unit.

Through this approach, resource requirements were projected for specified energy production levels in 1977, 1980, and 1985 under two basic energy strategies -- the Base Case (current energy policies) and the Accelerated Supply Strategy. Because production levels were greater under accelerated supply this became the key strategy for determining resource constraints, for if requirements could be met under accelerated supply they could also be fulfilled under a less intensive program of energy development.

The crosscut task forces projected the availability of each resource in 1977, 1980, and 1985, using the same economic forecast that was used to project energy demand; calculated how much of that resource had historically been used by the energy industry; and what would be a reasonable share for the energy sector in each of these years.

The transportation analysis provided the costs of moving different kinds of fuel by rail, barge and pipeline over different transportation links; the estimated capacities on each of these links; and comparisons of these capacities with projected requirements for each of the different energy strategies.

Since it takes resources to produce more resources, requirements for expansion of each category must also be examined. To expand transportation capacity, for example, requires capital, manpower, and steel. To expand steel production requires capital and manpower. In general, it was not possible to estimate these "secondary effects." But in several key areas, such as transportation, these requirements were carefully analyzed. Finally, a few special categories of resource requirements were investigated, such as the steel and tank building capacity required for an oil storage program, and materials, equipment, and financial requirements for certain energy conservation options.

The following are the major findings of the analysis:

1. The recent shortages of resources experienced by the energy industry are primarily due to recent dislocations in the American economy. The economy could not adapt quickly enough to price controls, surges in demand for domestic energy production as a result of rapid increases in foreign oil prices, an investment boom in coal production and oil exploration related to the same sharply higher energy prices, and new regulations for environmental protection and occupational health and safety.

2. The energy sector is not projected to increase significantly its share of business fixed investment by 1985, even under the accelerated domestic development strategies. Except for utilities, there should be no serious financial constraints.

3. In 1977, the United States' capacity to produce oil under the Accelerated Case, and the capacity to produce coal under the Base Case could be limited by shortages of certain categories of equipment and manpower.

4. In 1980, certain highly skilled categories of manpower which require long years of training may still be in short supply -- engineers, construction craftsmen, draftsmen and nuclear technicians.

5. By 1985, there will be no overall shortages of manpower or materials and equipment under both the Base Case and the Accelerated Supply Case, provided that clear energy policy direction is given so that capital and labor will be attracted to those equipment and skill classifications which require many years to develop.

6. In 1985, there may still be shortages of drilling rigs and fixed and mobile drilling platforms which could constrain or delay accelerated development of off-shore oil.

7. In 1985, there should not be shortages of transportation capacity for energy resources, provided that planning and Federal regulatory and other decisions are made soon.

8. By 1985, water will begin to be a serious constraint in the accelerated supply strategy for development of oil shale and synthetic fuels in the Upper Colorado Region. Water for energy will be in competition with agriculture in the northwestern part of the Missouri Region.

9. Any policy which seeks to stabilize planning and investment decisions will go a long way toward avoiding shortages and ensuring adequate resources for future energy development.

Tables 3-22 to 3-27 which follow summarize the key findings for each major resource. The sections which follow provide additional details on each resource. The appendix contains a complete listing of the resource categories investigated.

Findings On International Assessment

Excerpt From "PROJECT INDEPENDENCE"

1. Oil exporting countries will be faced with an increasingly stagnant international oil market if current oil prices prevail through 1980-1985. This is because little or no growth in world import demand is projected at current world oil prices. This could create pressure to reduce price and increase production substantially.

2. Both the individual and collective oil supply security of major oil importing countries can be enhanced immediately and materially by adopting a cohesive approach, such as that offered by the International Energy Program (IEP), to cope with major supply disruptions. The IEP will make it more difficult to target oil import shortfalls on specific importers. In the 1980-1985 period, the IEP can be complemented by building up supplemental United States petroleum stockpiles through the 1970's, to provide a high level of supply security at least cost for any level of United States import dependence likely to result, given the range of United States domestic initiatives and future world oil prices considered.

3. Substantial strategic petroleum stockpiles could contribute efficiently to future United States security even without an IEP. That security could be achieved at an average (annualized) cost of approximately $1 per barrel of import demand relying primarily on saltdome storage of crude in the Gulf Coast. Moreover stockpiles minimize any macro-economic impacts associated with oil import reductions. Acquisition of emergency stockpiles would result in a relatively modest increase in import prices (10 percent or less), concentrated substantially in the initial period of the buildup. This finding holds whether world oil prices continue at current levels or decline substantially through the buildup period.

4. Stockpile levels can easily be adjusted on a sequential basis as the world oil market evolves in 1975-1980 period. This would provide the United States with the flexibility needed to adapt its energy posture to the evolving international energy environment of 1980-1985.

5. There is a very limited potential for international supply diversification to materially improve United States supply security by 1980-1985 compared with that added by an IEP. An IEP agreement increases the incentive for participating countries to shift their collective import dependence to more secure supply sources, and reduce the market share vulnerable to interruption. The degree of diversification is nevertheless limited by the supply available

Table 3-22
Potentially-Constraining Materials and Equipment Categories/Accelerated Case
1975 - 1984

Items	Units	Estimated Availability to E-Sector 1975-1984	Estimated Consumption by E-Sector 1975-1984	Percentage of Consumption to Availability
Drilling Rigs Oil & Gas	item-years* workable	30,980	28,900	93%
Drilling Platforms Fixed	items	3,115	4,235	136%
Drilled Platforms Mobile	item-years *	5,154	7,099	138%
O.C. T.G.	10^6 tons	27.46	25.95	95%
Steel Products	10^6 tons	117.8	110.06	93%
Steel Pipe and Tubing	10^6 tons	27.15	22.67	83%
Walking Draglines	items	392	294	75%
Steam Turbine Generators Large	Gwe	525	435	85%

* Item-years represent the average number of the items available in a given year times the appropriate number of years.

Table 3-23
Critical Energy Sector Occupations

Operations and Maintenance

	1970 Energy Sector Direct Production Employment	Required Average Annual % Growth 1970-85 1/	Average Annual % Growth in Total Economy 1970-80
Electricians	18,081	2.9	2.3
Plumber/Pipefitters	7,477	4.6	3.0
Mine Operatives NEC	113,316	1.0	-1.9
Welder Flamecutters	9,355	2.3	2.8
Physicists	808	9.0	2.6
Nuclear Engineers	1,038	27.3	NA
Health Technicians	313	22.0	NA
Mettalurgical Engineers			
Power Station Operators	20,495	4.9	-0.6
Millwrights	426	12.5	1.2

1/ Source is United States Department of Labor.
Required to meet Accelerated Supply Scenario.

Table 3-24
Critical Energy Sector Occupations

Construction

	1970 Energy Sector Direct Production Employment	Required Average Annual % Growth 1970-85 1/	Average Annual % Growth in Total Economy 1970-80
Pipefitters/ Plumbers	15,398	21.6	1.9
Welders	5,782	16.3	3.7
Electricians	8,639	20.3	1.3
Boilermakers	1,764	23.2	-2.9
Millwrights	3,870	23.0	-4.7
Carpenters	2,760	19.5	-0.9
Electrical Engineers	1,038	17.8	4.7
Geologists	264	16.8	6.6
Engineers & Science Technicians, NEC	595	48.3	42.1
Draftsmen	3,237	10.0	10.0

1/ Source is United States Department of Labor. Required to meet Accelerated Supply Scenario.

Table 3-25
Energy Sector Financial Requirements

	1974 Energy Investment 1/			1975 to 1985 Accelerated Supply		
	$ Billions	%GNP	% Business Fixed Investment 2/	$ Billions	%GNP	% Business Fixed Investment 2/
Cumulative Investment Requirements	$30	2.3	21	$426	2.5	23

1/ Current energy investment is based on 1973-1974 data provided by the National Academy of Engineering.

2/ Energy sector capital requirements have historically averaged 23 percent of Business Fixed Investment. Energy sector includes coal mining, oil and gas extraction, oil refining, and electric power generation and distribution.

Table 3-26
Transportation

Transport Link Requiring Expansion	Cumulative Requirements for Capacity Expansion 1974-1985			
	(Capital) (Billions of Dollars)		Steel (Millions of Tons)	
	Base Case	Accelerated	Base Case	Accelerated
Natural Gas: by pipeline in Alaska and LNG tanker to west coast	3.9	6.4	5.8	8.1
Oil: by pipeline across Alaska; then by tanker to the west coast and east coast	5.4	6.9	3.1 (excluding tankers)	3.9
U.S. Oil Product Pipelines	1.5	2.2	2.8	4.1
Rail Expansion: • Locomotives • Hopper Cars • Track & Facilities	11.4	10.9	13.1	12.4
Waterway Improvements: • replaced and additional towboats and barges; • replaced locks, tankers	2.5	0.7	1.3	0.9
Total	24.7	27.1	26.1	29.4

from alternative sources, and the likely reallocation of available oil during an embargo.

6. Emergency demand conservation and supply allocation aré effective and efficient means of reducing the economic costs of an interruption in oil imports. These can reduce the level of oil consumption during a supply disruption, depending on the impacts of long term energy conservation measures, without imposing significant costs on the level of economic activity.

Research And Development

In presenting the conclusions of our analysis, we must make clear the essential uncertainty that underlies all of our projections, especially the long-term ones. It is not possible to predict what the energy demand or resources in the 21st century will be. We do not know what the population or economic growth rate will be at that time, and technological development is always somewhat uncertain. Thus, the basic R&D strategy cannot be considered a unique one: other strategies might be more appropriate if different underlying assumptions had been made with respect to the long-range future.

Nevertheless, we believe that there are certain elements of the R&D strategy that are largely independent of the long-range projections, and the strategies discussed in this chapter largely satisfy this criterion. The R&D strategies are realistic in that they take into account the long lead time required to bring new technologies into widespread use and are responsive to institutional forces.

Impacts Prior to 1985. The most critical technical problems in the Project Independence period involve increasing oil and gas supplies, using energy more efficiently and using available coal and uranium. Research and development can play a secondary, but necessary, role in overcoming these problems.

- Enhanced oil and gas recovery methods offer a large near term payoff in the terms of increased supply and recoverable resource base if the required research, development, and field testing is accelerated.

Table 3-27
Critical Water Regions

Millions of Acre Feet Per Year

Critical Water Regions	Surface Water Supplies (runoff)	Groundwater and Marine/Estuary Supplies (current use - 1970)	Total Water Supplies (Ground & Surface)	Total Consumptive Use as a Percent of Total Water Supplies-1985	Energy Related Consumptive Use as a Percent of Total Consumptive Use 1985
Upper Colorado	11.2 2/ (6.3)	.1	11.3 (6.4)	79.7	8.4
Lower Colorado	1.9 3/ (8.5)	5.0	6.9 (13.5)	34.1	1.1
Great Basin	2.8	4.6	7.4	51.4	1.4
Rio Grande	3.0	2.9	5.9	96.6	6.7
Missouri Basin	37.0	6.8	43.8	35.4	2.4

1/ The fresh surface water supplies used herein represent that amount of water originating from each region for (1) 50 percent of the total surface storage which existed as of January 1963 and (2) for a degree of certainty which can be assured 98 out of every 100 years. This material was derived from a paper prepared by the United States Geological Survey.

2/ The Colorado River Compact of 1922 required delivery of 75 million acre-feet of water in any 10-year period from the Upper Colorado River Basin to the Lower Colorado River Basin. Estimates of the water remaining for consumptive use in the Upper Basin range from 5.8 to 6.3 million acre-feet per year, depending upon assumptions used in interpretation of the Compact.

3/ The water available annually for consumptive use in the Lower Basin is increased by the amount released from the Upper Basin less 1.5 million acre-feet required to satisfy the U. S. - Mexico treaty obligations. This amount depends upon interpretations of the Colorado River Compact, could be as high as 8.5 million acre-feet per year.

- Conservation resulting from more efficient consumption of energy will depend primarily on widespread use of existing technology and improved design practices, but R& D aimed at improving consumption technologies and the process of implementing them is needed.

- Greater use of coal will require R&D to solve environmental problems related to its extraction and consumption. Similarly, growing use of nuclear power may be jeopardized unless R&D can resolve the remaining issues of safety and the fuel cycle.

Impact After 1985. The depletion of conventional oil and gas resources dominates the post-1985 period and leads to two fundamental observations:

1. Synthetics from coal and oil shale will be required after 1985 and their use will continue to grow rapidly.

-- Therefore, advanced technologies for producing coal synthetics and shale oil which maximize efficiency and minimize economic and environmental impacts should be developed to meet the post-1985 needs for liquids and gas.

2. The oil and gas shortfall is so large that major shifts in demand, most probably to electric power, as well as strong conservation measures will be needed. To achieve these post-1985 goals, new technologies for using energy more efficiently and for shifting to to electricity instead of oil and gas must be developed.

-- New energy sources not limited by conventional fossil fuel and uranium resources are needed. Technologies, such as the breeder, fusion and solar energy, will take decades to develop and introduce, but it is important to know that we will have one or more of these to support the shift in demand. R&D on these technologies must be pursued as a basis for charting the course of energy development.

PART II

NUCLEAR ENERGY

CHAPTER 4

NUCLEAR POWER FROM NOW TO YEAR 2000

A new forecast of the growth of nuclear power in the United States and the rest of the world has been prepared for use in AEC production planning and for other purposes. This forecast includes evaluation of domestic and foreign growth trends in nuclear power, foreign enrichment capabilities in the future, the timing and application of plutonium recycle technology, the role of the High Temperature Gas-Cooled Reactor (HTGR), and the introduction of the Liquid Metal-Cooled Fast-Breeder Reactor (LMFBR). The nuclear generating capacities in this forecast have been translated into demands for enrichment plant feed material, separative work, and other quantities based on the known and expected characteristics of nuclear reactors.

Three forecasts have been prepared both for the United States and for other non-Communist countries; a most likely, a high, and a low case. The forecasts, summarized in Table 4-1 are based on an evaluation of announced, or known, plants in the United States and in other countries and on extrapolations of trends in energy consumption and electricity generating capacity.

The total capacities and expected dates of operation of known plants have been used in deriving the most likely forecast. The high and low forecasts are based on estimates of the earliest and latest probable dates of commercial operation of these known plants. The forecasts of nuclear capacities to be installed in addition to these known plants are based on extrapolations through the year 2000 of trends in energy consumption and generating capacity addition patterns in the United States and in various other countries.

As shown in Table 4-1 the total nuclear generating capacity in the United States is forecast to be 132 GWe at the end of 1980, 280 GWe at the end of 1985, and 1,200 GWe at the end of 2000 in the most likely case. The total capacity in other non-Communist countries is forecast to be 141 GWe at the end of 1980, 303 GWe at the end of 1985, and 1,460 GWe at the end of 2000. In addition, the capacities in the Communist countries are forecast to be 20, 56, and 600 GWe at these points in time. World-wide total capacities are expected to be 293 GWe at the end of 1980, 639 GWe at the end of 1985, and 3,260 GWe at the end of 2000.

This forecast is slightly lower for the United States than the one presented a year ago, although still within the probable range suggested at the time. The reduction in the United States is due primarily to a general lengthening of nuclear project schedules and a slight reduction in the projected rate of growth of energy consumption. For other countries, the present forecast is slightly higher than the previous forecast. The total capacity of the United States and other countries is virtually the same in both forecasts. There is an increasing tendency in other nations, particularly Japan and the European Community, to turn to nuclear power to satisfy their needs for energy. This trend seems to be accelerating and may have been underestimated in previous forecasts.

Table 4-1

NUCLEAR ELECTRICAL CAPACITY IN GIGAWATTS

	End of Calendar Year					
	December 1972 Forecast				WASH-1139 (Rev. 1)[1]	
	1980	1985	1990	2000	1980	1985
United States						
Most Likely	132	280	508	1200	151	306
High	144	332	602	1500	166	344
Low	127	256	412	825	132	272
Foreign (excluding Communist Bloc)						
Most Likely	141	303	578	1460	124	276
High	153	358	704	1900		
Low	123	256	454	1035		
Communist Bloc						
Most Likely	20	56	146	600		

[1] Interpolated from fiscal year data.

The total separative work demands on enrichment plants operating at 0.30% ^{235}U tails assay corresponding to these capacity forecasts are shown in Table 4-2. The most likely cumulative separative work requirements, beginning in 1973,

Table 4-2

SEPARATIVE WORK REQUIREMENTS IN MILLIONS OF SEPARATIVE WORK UNITS

(All quantities on basis of enrichment plant tails assay of 0.30% ^{235}U)

	Calendar Year					
	December 1972 Forecast				WASH-1139 (Rev. 1)[1]	
	1980	1985	1990	2000	1980	1985
Annual						
United States						
Most Likely	15.3	30.0	52.8	74.3	17.1	32.3
High	17.4	35.8	62.5	92.3	18.9	36.7
Low	14.7	25.6	39.2	53.5	15.1	28.9
Foreign[2]						
Most Likely	13.8	27.9	47.7	84.6	13.1	26.0
High	15.2	34.3	60.6	115.1		
Low	11.7	23.0	35.4	59.5		
Cumulative from 1973						
United States						
Most Likely	72.4	190.8	405.5	1073.8	83.3	211.8
High	80.4	220.6	478.0	1292.0	92.8	237.7
Low	67.1	174.0	343.1	822.6	72.5	187.0
Foreign[2]						
Most Likely	63.4	172.3	371.0	1088.4	47.0	160.3
High	69.6	200.8	446.9	1375.3		
Low	55.5	146.2	300.3	803.7		

[1] Interpolated from fiscal year data.
[2] Excludes Communist Bloc nations.

in the United States are 72.4 million separative work units (SWU)[1] through 1980, 191 million SWU through 1985, and 1,074 million SWU through 2000. In foreign countries, Communist Bloc nations excluded, the cumulative demand through 1980 is expected to be 63.4 million SWU, 172 million through 1985, and 1,088 million through 2000. The cascade feed requirements corresponding to these separative work projections are shown in Table 4-3.

Recycle plutonium is assumed to be placed into reactors in 1977. One-fourth of the available plutonium is then recycled. This fraction is assumed to increase to one-half in 1978, to three-fourths in 1979, and to one in 1980. Recycle is expected to continue until the late 1980s, at which time the demand for plutonium as a breeder fuel will cause the recycle rate to decrease to zero over a short period. The result is that plutonium availability under these assumptions is not expected to limit the rate at which breeder reactors can be introduced. A similar pattern of plutonium recycle is expected in other countries, except that plutonium is not expected to be recycled in the United Kingdom nor in natural-uranium fueled reactors, located primarily in Canada.

DOMESTIC FORECAST

This forecast of the growth of nuclear power

Table 4-3

ENRCHMENT PLANT FEED REQUIREMENTS IN THOUSANDS OF METRIC TONS OF Uranium

(All quantities on basis of enrichment plant tails assay of 0.30% ^{235}U)

	Calendar Year					
	December 1972 Forecast				WASH-1139 (Rev. 1)[1]	
	1980	1985	1990	2000	1980	1985
Annual						
United States						
Most Likely	26.7	50.7	88.2	117.5	31.8	58.5
High	30.5	60.8	104.3	146.5	35.1	65.8
Low	25.8	42.9	64.2	84.5	28.1	51.7
Foreign[2]						
Most Likely	27.0	51.4	83.9	139.1	26.0	48.2
High	30.0	64.1	106.9	191.6		
Low	22.0	42.5	61.2	97.9		
Cumulative from 1973						
United States						
Most Likely	132.2	334.0	692.7	1775.1	155.8	390.1
High	146.8	387.3	817.5	2138.8	173.5	438.3
Low	122.7	303.6	582.0	1352.0	135.6	344.6
Foreign[2]						
Most Likely	127.1	330.9	686.3	1904.5	113.1	313.8
High	139.7	388.1	829.1	2416.1		
Low	110.4	279.1	551.2	1401.2		

[1] Interpolated from fiscal year data.
[2] Excludes Communist Bloc nations.

in the United States is based on an extrapolation of historic trends in energy consumption and generating capacity additions. Three variations, or cases, have been calculated—a most likely, a high, and a low case. These forecasts are shown in Fig. 4-1 and Table 4-4. The most likely case is recommended for use in planning studies. The high and low cases are intended to define ranges of the probable capacity to be installed and to be used in sensitivity analyses. By the end of 1980, the total nuclear electrical generating capacity installed in the United States is expected to be 132 GW, but will probably lie between 127 and 144 GW. Similarly, the total by the end of the year 1985 is expected to be 280 GW and to lie between 256 and 332 GW. By 1990, 508 GW are expected, with a range of from 412 to 602 GW; and by 2,000, 1,200 GW are expected, with the range extending from 825 to 1,500 GW.

The forecast is built up from two components. The first is the nuclear plants which are in operation, are under construction, or have been ordered. These plants are considered to be the total additions to nuclear generating capacity until about 1979. Some of these "known" plants contribute to the growth in the forecast until 1982. The other component is an extrapolation of the trend in these additions to the end of the century.

The forecast of nuclear capacity in the United States for the rest of the century has been derived from extrapolations of the trends in total energy consumption, patterns of use, and elec-

Fig. 4-1

INSTALLED NUCLEAR CAPACITY
U.S.

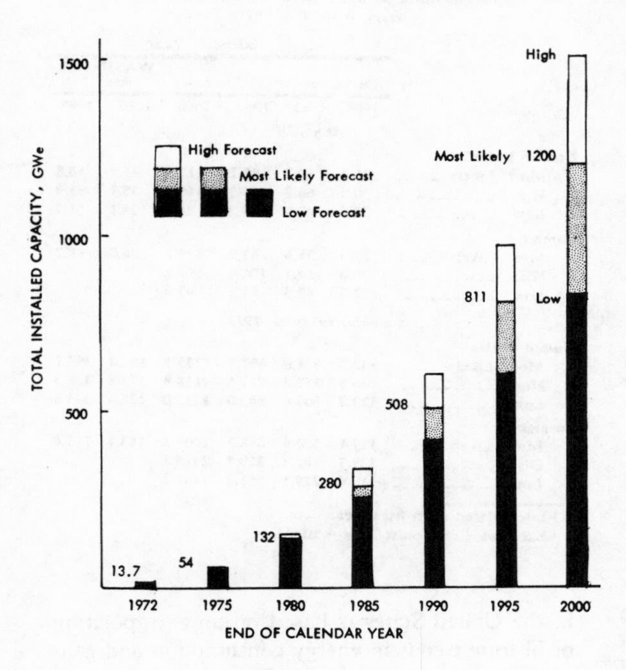

Table 4-4

FORECAST OF U.S. NUCLEAR POWER CAPACITY, GWe

Calendar Year	Most Likely Additions	Most Likely Cumulated	High Additions	High Cumulated	Low Additions	Low Cumulated
1973	15.2	28.9	17.5	31.2	8.0	21.7
1974	13.4	42.3	17.8	49.0	13.3	35.0
1975	11.9	54.2	7.9	56.9	17.1	52.1
1976	7.0	61.2	6.2	63.1	3.7	55.8
1977	8.1	69.3	12.0	75.1	8.8	64.6
1978	17.4	86.7	21.5	96.6	10.8	75.4
1979	16.6	103.3	20.2	116.8	22.3	97.7
1980	28.3	131.6	27.2	144.0	29.3	127.0
1981	25.4	157.0	28.	172.	24.	151.
1982	26.	183.	33.	205.	22.	173.
1983	28.	211.	37.	242.	26.	199.
1984	33.	244.	44.	286.	29.	228.
1985	36.	280.	46.	332.	28.	256.
1986	40.	320.	48.	380.	28.	284.
1987	44.	364.	51.	431.	31.	315.
1988	46.	410.	56.	487.	32.	347.
1989	47.	457.	55.	542.	33.	380.
1990	51.	508.	60.	602.	32.	412.
1995	67.	811.	82.	972.	41.	602.
2000	83.	1200.	122.	1500.	43.	825.

trical generating capacity additions. In 1960, just over 9 tonnes of coal equivalent energy were consumed per capita in the United States. In 1970, over 12 tonnes were consumed. This trend, which has existed since at least the 1920s, has been extrapolated to nearly 14 tonnes by 1980 and to 20 tonnes by 2000. The fraction of this energy used for the production of electricity has also shown a continuous increase, from 18 percent in 1960 to 24 percent in 1970. This fraction is projected to increase to 31 percent by 1980 and to 50 percent by 2000. The apparent load factor, defined as the total electricity generated in a year divided by the electricity which could have been generated in a year (8,760 hours) by the capacity available at the end of the year, has remained constant at about 50 percent for several years and is expected to remain at that level for the rest of the century. The heat rate, the energy consumed in production of electricity, has been, until recent years, steadily dropping—from 0.39 kg coal equiv/kWh (10,700 Btu/kWh) in 1960 to 0.37 (10,300) in 1970. This value is expected to drop to about 0.31 (8,500) for the most efficient plants in operation by 2000. If the energy consumed for electricity production is divided by the heat rate, the result is an estimate of the electricity production. A further division by the load factor and a conversion of units produces an estimate of the installed capacity. The total installed capacity in the United States was about 1 kW/capita in 1960 and about 1.7 in 1970. The method described above produces an estimate of 2.7 kW/capita in 1980 and 7 in 2000. These data can then be multiplied by the population of the United States, using the Bureau of Census Series D projection of population growth, to yield the total electrical generating capacity. This has increased from 175 GW in 1960 to 349 in 1970 and is expected to increase further to 630 GW in 1980 and to over 2000 GW in 2000. Nuclear generating capacity, which was about 3 percent of the total additions to all electrical generating capacity during the 1960s and will probably be about 44 percent of the additions during the 1970s, is forecast to average 72 percent of the additions during the 1980s and 81 percent during the 1990s. The detailed data used in this forecast are summarized in Table 4-5.

The high forecast of nuclear capacity was derived by the same method, but with use of the assumptions that population will increase according to the Census Bureau Series C projection, that the electrical share of total energy consumption will be 55 percent in the year 2000, and that nuclear additions will be about 90 percent of all additions to generating capacity after 1985. Similarly, the low forecast is based on assumptions of population Series E, an electrical share of 45 percent in 2000, and a market penetration by nuclear reactors of only 75 percent of the total

Table 4-5

FORECAST OF ENERGY CONSUMPTION AND GENERATING CAPACITY IN THE UNITED STATES

	1960	1970	1980	1990	2000
Energy Consumed, Metric Ton Coal Equivalent/Capita	9.04	12.07	13.70	16.82	20.15
Fraction for Electricity Generation	.18	.24	.31	.40	.50
Energy Consumed for Electricity Generation, Metric Ton Coal Equivalent/Capita	1.62	2.90	4.25	6.73	10.08
Apparent Load Factor	.51	.50	.51	.53	.53
Heat Rate, kg Coal equivalent/kWh	.39	.37	.35	.33	.31
Total Electric Generating Capacity/Capita, kw/Capita	.97	1.71	2.73	4.42	6.98
Total Electric Generating Capacity, GW	175.	349.	630.	1150.	2000.
Total Nuclear Generating Capacity, GW	.02	5.9	132.	508.	1200.

generating capacity additions after 1975.

This most likely forecast is slightly lower for the United States than the one presented a year ago, although it is still within the probable range suggested at that time, 132 to 166 at the end of 1980 and 272 to 344 at the end of 1985. The reduction is due primarily to two factors, a general lengthening of nuclear project schedules and a slight reduction in the rate of growth of energy consumption. Many events occur between ordering a reactor and beginning its operation, events whose impact on schedules have often been minimized in planning. These events include delays due to equipment delivery, labor stoppages, intervention, licensing, and other causes. Experience with these events has resulted in longer schedules for reactors now in the planning and construction stages, perhaps as much as a year longer, although the exact length of time is difficult to measure. The other factor reducing the forecast is a slight slowing of the rate of increase in energy consumption in the United States. During the 1960s, the nation increased its energy consumption at a rate exceeding the long-term rate of increase. This most likely forecast assumes that this rate of electricity growth will return to levels more consistent with the long-term rate. The penetration of nuclear reactors into the total market for generating capacity is similar for both this and the previous forecast.

The types of reactors which comprise the nuclear capacity in the United States for the remainder of the century are fast breeders, LMFBRs; high-temperature gas-cooled reactors, HTGRs; and light-water reactors, LWRs. The LMFBRs are expected to penetrate the nuclear market beginning in 1986 at the same percentage rate that LWRs penetrated the total thermal plant market beginning in 1967 (Fig. 4-2). Demonstration breeders are expected to become operational in 1980 and 1983. About 1.5 GW of LMFBR capacity is expected to be installed in 1986 and an additional 22 by 1990. The LMFBR capacity in 2000 is expected to be about

Fig. 4-2

NUCLEAR ADDITIONS AS A FRACTION OF TOTAL THERMAL ADDITIONS

U.S. - 1968-1983

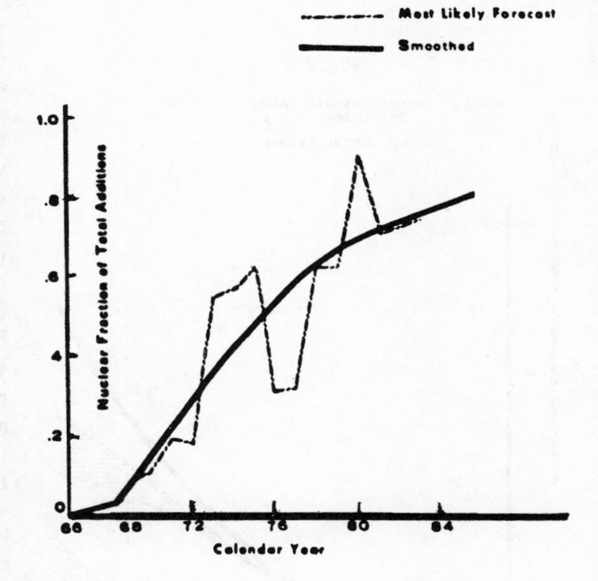

400 GW, approximately equal to the total electrical generating capacity in the United States today.

The HTGR is assumed to penetrate the non-breeder portion of the nuclear market to the extent of about 10 percent of additions around 1980 and to increase to 15 percent by 1985 and remain at that level for the rest of the century. Total HTGR capacity under this assumption will be 23 GW by 1985, 54 by 1990, and 100 by 2000. The remaining nuclear additions are expected to be LWRs, of which one-third are assumed to be boiling-water reactors, and two-thirds to be pressurized-water reactors.

Several additional sets of calculations were performed in the course of preparing this forecast using different assumptions about the expected reactor mix. One variation was based on the assumption that the LMFBR would not be introduced until 1990. Another was that the HTGR would capture a larger share of the nuclear market, about 20 percent by 1990 and 30 percent by 2000. The results of these cases are not reported here because the differences in separative work, feed, and other requirements from the base case described above are much smaller than the differences in these quantities between the most likely case and the high or low forecasts (Cf, sensitivity analyses, below). In other words, the uncertainty in the overall capacity forecast causes and feed requirements than the uncertainty in the reactor mix within a total capacity forecast. The range of reactor mixes resulting from the various assumptions used in these cases is shown in Fig. 4-3

Fig. 4-3

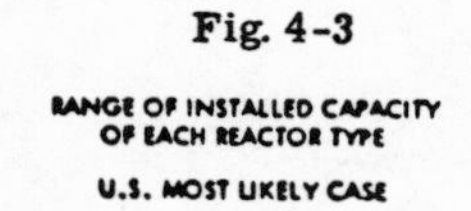

RANGE OF INSTALLED CAPACITY
OF EACH REACTOR TYPE
U.S. MOST LIKELY CASE

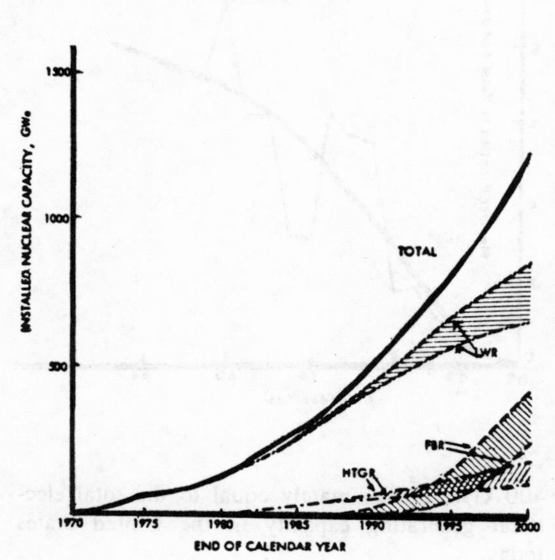

FOREIGN FORECAST

Three forecasts of foreign nuclear capacity have been prepared, a most likely, a high, and a low forecast. They are shown in Table 4-6 and Figure 4-4. By the end of 1980 the total nuclear capacity in other countries, Communist Bloc nations excluded, is expected to be 140 GW, 303 GW by 1985, 580 by 1990, and 1,460 by 2000. Here again, the high and low forecasts can be interpreted as probable ranges of the expected total nuclear capacity. In 1980, the range extends from 123 to 153 GW, from 256 to 358 in 1985, from 454 to 704 in 1990 and from 1035 to 1900 GW in the year 2000. For countries other than the U. S., the present forecast is slightly higher than the previous forecast. This is because of the increasing tendency of other nations, particularly Japan and the European Community, to turn to nuclear power to satisfy their needs for energy. This trend seems to be accelerating and may have been underestimated in previous forecasts.

The most likely forecast of foreign nuclear capacity has been derived by use of the methodology described above. This method has been applied to 18 major nations, to the Communist Bloc nations as a unit, and to the world as a whole, excluding the Peoples Republic of China. China has been excluded from all calculations because of the scarcity and unreliability of the pertinent data.

The LMFBR is expected to be introduced in other countries on nearly the same time schedule as in the United States. However, there are now existing or planned projects totalling over 5 GW which are likely to come into commercial operation before 1986. Following that year, the penetration of the LMFBR into the nuclear market is expected to occur at the same market-share rate as in the United States. Nearly 31 GW will be in operation by the end of 1990 and 415 GW by 2000. The HTGR is not expected to make a significant contribution before 1980, but is expected to capture 10 percent of the non-breeder market by 1990 and 15 percent by 2000. The total HTGR capacity will be 22 GW and 78 GW by these years in non-Communist countries.

Natural-uranium fueled reactors are assumed to penetrate the total foreign market at a rate equal to the projected nuclear additions in Canada. It is likely that the majority of the natural uranium reactors will be in Canada. The assumption as used is an attempt to allow for the probabilities, tending to offset each other that some natural-uranium fueled reactors will be built outside Canada and that not all the Canadian reactors will be of this type. The Advanced Thermal Reactor is expected to constitute a varying fraction of Japanese additions during the 1980s and early 1990s. These reactors have a fuel cycle which is based on the use of natural uranium and self-generated plutonium and are not included in the forecast of natural-uranium fueled reactors mentioned above. The Advanced Gas-cooled Reactor is assumed to penetrate the world market at a rate equal to the non-LMFBR additions in the United Kingdom. This assumption is based on a rationale similar to that used for the natural-

Table 4-6
FORECAST OF FOREIGN NUCLEAR CAPACITY, GWe

	Non-Communist						Communist	
	Most Likely		High		Low			
Calendar Year	Additions	Cumulated	Additions	Cumulated	Additions	Cumulated	Additions	Cumulated
1973	3.4	21.2	6.6	24.4	1.7	19.5	0.4	2.9
1974	8.8	30.0	7.0	31.4	6.1	25.6	1.4	4.3
1975	8.8	38.8	13.1	44.5	7.9	33.5	3.9	8.2
1976	20.1	58.9	16.1	60.6	14.0	47.5	1.9	10.1
1977	11.3	70.2	17.0	77.6	18.9	66.4	3.8	13.9
1978	26.2	96.4	21.1	98.7	15.9	82.3	1.0	14.9
1979	18.7	115.1	21.2	119.9	19.9	102.2	1.7	16.6
1980	25.4	140.5	33.4	153.3	20.4	122.6	2.9	19.5
1981	28.5	169.	33.7	187.	22.2	145.	5.	25.
1982	29.	198.	35.	222.	24.	169.	5.	30.
1983	33.	231.	41.	263.	26.	195.	7.	37.
1984	33.	264.	45.	308.	29.	224.	9.	46.
1985	39.	303.	50.	358.	32.	256.	10.	56.
1986	43.	346.	57.	415.	35.	291.	12.	68.
1987	49.	395.	64.	479.	37.	328.	16.	84.
1988	56.	451.	69.	548.	41.	369.	18.	102.
1989	61.	512.	75.	623.	42.	411.	21.	123.
1990	68.	580.	81.	704.	43.	454.	23.	146.
1995	86.	968.	112.	1182.	54.	700.	43.	318.
2000	110.	1460.	170.	1900.	78.	1035.	70.	600.

uranium reactors. Light-water reactors are expected to comprise the remainder of the nuclear capacity additions for the rest of the century. Pressurized-water and boiling-water reactors assumed to be built abroad in approximately equal numbers.

Fig. 4-4

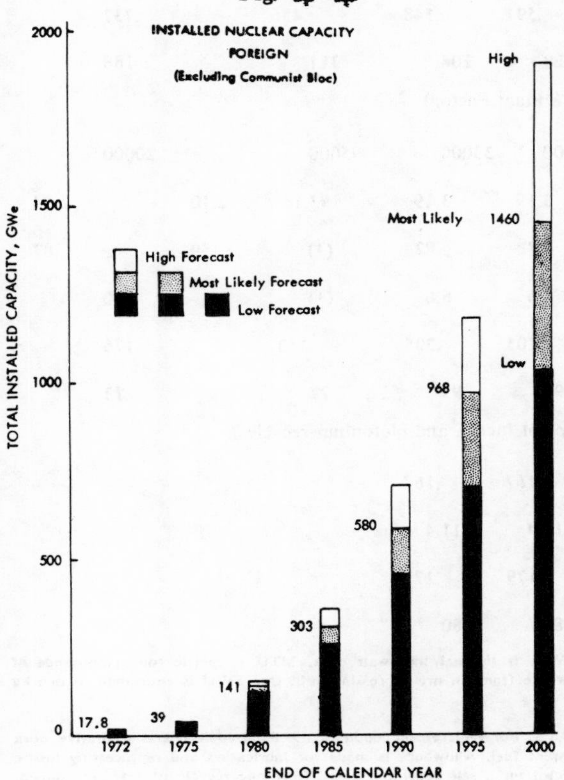

Additional cases were calculated for reactor mixes predicated on later LMFBR introduction and increased HTGR penetration. The range of reactor mixes used is shown in **Fig. 4-5.** The effects of various reactor mix assumptions on the overall requirements for separative work and feed materials are less than the uncertainties in the capacity forecast. Thus the results of these variation cases are not reported in detail here.

Fig. 4-5

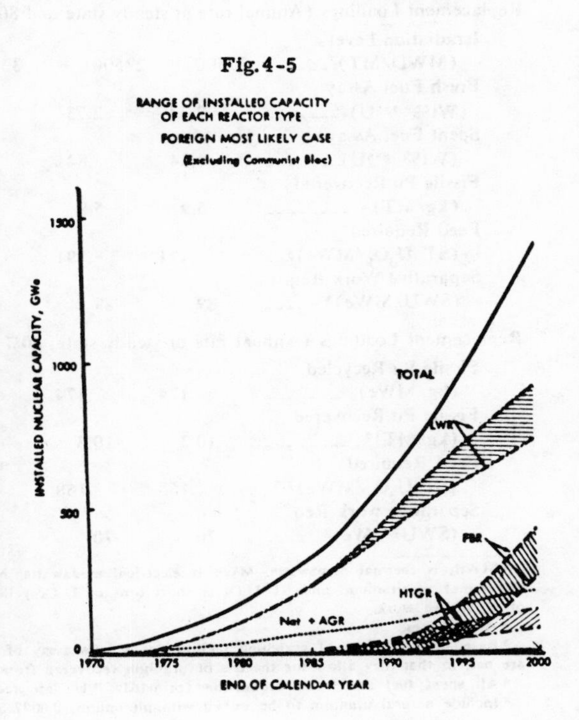

REACTOR AND FUEL CYCLE CHARACTERISTICS

For plants which will start operation before about 1977, enough specific information is available on design and expected operating characteristics to enable the fuel cycle to be represented separately for each reactor. For reactors which start at later times, generalized, or model, plant data have been used. The characteristics of these model plants are shown in Tables 4-7, 4-8. The models used to represent plants starting after about 1980 include some improvements which seem to be reasonable but which remain to be demonstrated. The characteristics given are for reactors in the 800 to 1,200 MWe size range. Although a number of units scheduled to go into operation in many countries are of much smaller size and have somewhat different characteristics, the ones shown for the larger plants have been used. The error introduced by this simplification is small because only a small fraction of the total additions to capacity is involved and the characteristics are not greatly different in most cases. Where reactor types to be constructed in a

Table 4-7

THERMAL REACTOR CHARACTERISTICS[1]

	BWR		PWR		HTGR	AGR	
	Thru 1980	After 1980	Thru 1980	After 1980		Inner Core	Outer Core
Thermal Efficiency (%)	34	34	33	33	39	42	
Specific Power (MWth/MT)	26	28	38	41	82	13	
Initial Core (Average)							
Irradiation Level	17000	17000	24000	24000	54500	13000	
Fresh Fuel Assay (Wt% ^{235}U)	2.03	2.03	2.63	2.63	93.15	1.49	1.78
Spent Fuel Assay (Wt% ^{235}U)	.86	.86	.85	.85	(4)	.75	1.00
Fissile Pu Recovered (kg/MT)[2]	4.8	4.8	5.8	5.8	(4)	2.5	
Feed Required (ST U_3O_8/MWe)[3]	.625	.580	.591	.548	.456	.737	
Separative Work Req. (SWU/MWe)[3]	200	185	224	208	311	188	
Replacement Loadings (Annual rate at steady state and 80% Plant Factor)							
Irradiation Level (MWD/MT)	27500	27500	33000	33000	95000	20000	
Fresh Fuel Assay (Wt% ^{235}U)	2.73	2.73	3.19	3.19	93.15	2.10	2.54
Spent Fuel Assay (Wt% ^{235}U)	.84	.84	.82	.82	(4)	.59	.87
Fissile Pu Recovered (kg/MT)[2]	5.9	5.9	6.6	6.6	(4)	4.0	
Feed Required (ST U_3O_8/MWe)[3]	.191	.191	.205	.205	.113	.176	
Separative Work Req. (SWU/MWe)[3]	89	89	99	99	77	73	
Replacement Loadings (Annual rate at steady state, 80% plant factor, and plutonium recycle.)							
Fissile Pu Recycled (kg/MWe)	.174	.174	.167	.167			
Fissile Pu Recovered (kg/MT)[2]	10.3	10.3	11.4	11.4			
Feed Required (ST U_3O_8/MWe)[3,5]	.168	.168	.179	.179			
Separative work Req. (SWU/MWe)[3]	70	70	80	80			

[1] MWth is thermal megawatts, MWe is electrical megawatts, MWDt is thermal megawatt days, MTU is metric tons (thousands of kilograms) of uranium, and ST U_3O_8 is short tons of U_3O_8 yellowcake from an ore processing mill. One SWU is equivalent to one kg of separative work.

[2] After losses.

[3] Based on operation of enriching facilities at a tails assay of 0.3%. For replacement loadings, the required feed and separative work are net, in that they allow for the use of uranium recovered from spent fuel. Allowance is made for fabrication and reprocessing losses.

[4] All spent fuel and fissile production (primarily ^{233}U) are recycled on a self generated basis. Only one recycle of ^{235}U is assumed.

[5] Include natural uranium to be spiked with plutonium; 0.0087 ST U_3O_8/MWe for BWR and 0.0067 for PWR.

Table 4-8

FAST REACTOR CHARACTERISTICS

	Reference	Advanced
Net Electrical Output (MWe)	1000	1000
Total Thermal Power[1] (MWth)	2400	2417
Core	2219	2081
Axial Blanket	107	195
Radial Blanket	74	141
Average Specific Power in Core (MWth/MT)[2]	116	155
Reactor Inventory[1]		
Core Uranium, kg	15460	10590
Core Fissile Plutonium, kg	2360	1690
Blanket Uranium, kg	28270	31110
Blanket Fissile Plutonium, kg	305	362
Burnup, MWD/MT Average[3]		
Core	67600	104059
Axial Blanket	4740	8725
Radial Blanket	7970	9051
Breeding Ratio	1.30	1.42
Compound Doubling Time (years)	13	7.8

[1] Mid-Cycle Equilibrium.
[2] Mid-Cycle Equilibrium Based on Mass Charged Initially ($^{238}U + ^{235}U + ^{239}U + ^{240}Pu + ^{241}Pu + ^{242}Pu$).
[3] End-of-cycle Equilibrium Based on Mass Charged Initially.

country are unknown, it is assumed in this forecast that additions to capacity will be divided according to the reactor mix schedule discussed above.

Nuclear plants are assumed to operate at the plant factors given below in percent.

Year of Commercial Operation	Year of Operation			
	1st	2nd	3rd thru 15th	16th and over
Thru 1975	50	70	80	dropping linearly at
1976-1980	60	75	80	2% per year to
1981-2000	70	80	80	50% in year 30

The plant factor is defined as the actual power produced during a period divided by the power that could have been produced had the reactor operated at full power level continuously throughout the period. Although a value of 80 percent is often used, a gradual buildup and later reduction has been assumed in these calculations. It has also been assumed that the fuel management practices of reactor operators will be such that the reactors will be refueled on an annual basis but that a varying fraction of the core will be replaced at each reload, the fraction depending on the plant factor for the preceding period. The data shown in **Tables 4-7, 4-8** are those that apply to operation at an 80 percent factor.

SEPARATIVE WORK AND FEED REQUIREMENTS

One of the most important uses of a forecast of nuclear power growth is in planning for the various supporting facilities and production requirements implied by the forecast. The separative work and feed requirements implied by these forecasts are shown in **Tables 4-9, 4-10** and in Figures **4-6, 4-7**. The reactor characteristics used to arrive at these separative work and feed requirements are those discussed above.

The annual separative work demand for the three domestic and foreign cases is shown in **Table 4-9**. The demand in 1980 in the United States is expected to be 15.3 million SWU[1], with a range of from 14.7 to 17.4 million SWU. In 1990, the demand will be about 53 million SWU with a range of from 39 to 62 million SWU. In 2000, the expected demand of 74 million SWU might range from 53 to 92 million SWU. The demand in foreign countries, Communist Bloc nations excluded, will be 14 million SWU in 1980 and range from less than 12 to over 15 million SWU. In 1990, the demand of 48 million might range from 35 to 61 million SWU. The 2000 expectation of 85 million SWU has a range extending from 60 to 115 million SWU. These demands are expressed in terms of amounts of enriched UF shipped from the enrichment plants and are based on the reference assumption that

[1] One SWU is equivalent to one kg of separative work.

Fig. 4-6

CUMULATIVE
WORLD-WIDE SEPARATIVE WORK DEMAND
(Excluding Communist Bloc)

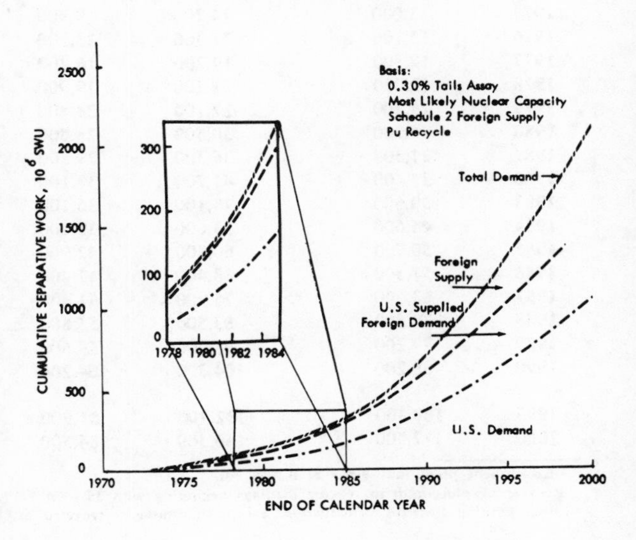

Table 4-9

TOTAL ANNUAL SEPARATIVE WORK DEMAND[1], 10^3 SWU/YR.

Calendar Year	United States			Foreign (excludes Communist Bloc)		
	Most Likely	High	Low	Most Likely	High	Low
1973	3,500	3,700	2,400	2,800	2,800	2,700
1974	4,300	5,200	4,200	3,700	4,900	2,800
1975	6,800	7,500	5,200	6,200	6,200	5,100
1976	8,100	9,200	7,200	5,900	6,700	6,600
1977	9,600	10,600	9,000	9,000	8,800	7,400
1978	10,700	11,800	11,100	9,600	11,100	8,800
1979	14,200	15,000	13,400	12,400	13,900	10,400
1980	15,300	17,400	14,700	13,800	15,200	11,700
1981	18,100	20,800	17,100	16,200	19,000	13,500
1982	20,600	24,100	19,400	19,000	22,500	15,800
1983	23,300	28,000	21,400	21,300	25,400	17,900
1984	26,400	31,500	23,400	24,600	30,000	20,500
1985	30,000	35,800	25,600	27,900	34,300	23,000
1986	34,100	40,700	28,500	31,800	39,300	26,000
1987	38,000	34,400	31,200	35,900	44,000	28,400
1988	42,300	50,200	33,900	39,700	48,900	31,100
1989	47,500	58,600	36,300	43,600	53,300	33,300
1990	52,800	62,500	39,200	47,700	60,600	35,400
1995	66,600	81,100	50,300	69,300	89,100	49,000
2000	74,300	92,300	53,500	84,600	115,100	59,500

[1] Enrichment plant tails assay at 0.30% ^{235}U.

Recycle of plutonium in reactor loadings beginning with 25% in 1977, 50% in 1978, 75% in 1979 and as much as possible thereafter, without limiting LMFBR construction, except that none is recycled in U. K. nor in natural-uranium fueled reactors.

Table 4-10

TOTAL ENRICHMENT PLANT FEED REQUIREMENTS,[1] MTU/YR. AS UF_6

Calendar Year	United States			Foreign (excludes Communist Bloc)		
	Most Likely	High	Low	Most Likely	High	Low
1973	7,000	7,600	4,900	5,800	5,700	5,800
1974	8,300	9,800	8,100	8,100	10,200	6,000
1975	13,000	14,200	9,900	12,800	13,000	10,600
1976	15,100	17,300	13,300	11,900	13,500	13,300
1977	17,500	19,200	16,400	18,400	17,400	14,900
1978	19,000	21,200	19,900	19,000	22,200	17,400
1979	25,700	27,100	24,400	24,200	27,500	20,300
1980	26,700	30,500	25,800	27,000	30,000	22,000
1981	31,500	36,300	29,800	30,900	36,800	25,500
1982	35,400	41,700	33,100	36,200	43,200	29,700
1983	39,600	48,100	36,100	39,800	48,300	33,300
1984	44,600	53,600	39,100	45,500	55,900	37,800
1985	50,700	60,800	42,900	51,400	64,100	42,500
1986	57,300	68,400	47,400	58,200	72,200	47,100
1987	63,500	75,900	51,400	65,000	79,800	50,700
1988	70,500	83,500	55,800	71,000	87,600	54,900
1989	79,200	98,200	59,600	77,400	94,500	58,100
1990	88,200	104,300	64,200	83,900	106,900	61,200
1995	108,300	132,100	81,900	118,600	153,600	83,300
2000	117,500	146,500	84,500	139,100	191,600	97,900

[1] Enrichment plant tails assay at 0.30% ^{235}U.

Recycle of plutonium in reactor loadings beginning with 25% in 1977, 50% in 1978, 75% in 1979 and as much as possible thereafter, without limiting LMFBR construction, except that none is recycled in U. K. nor in natural-uranium fueled reactors.

Fig. 4-7

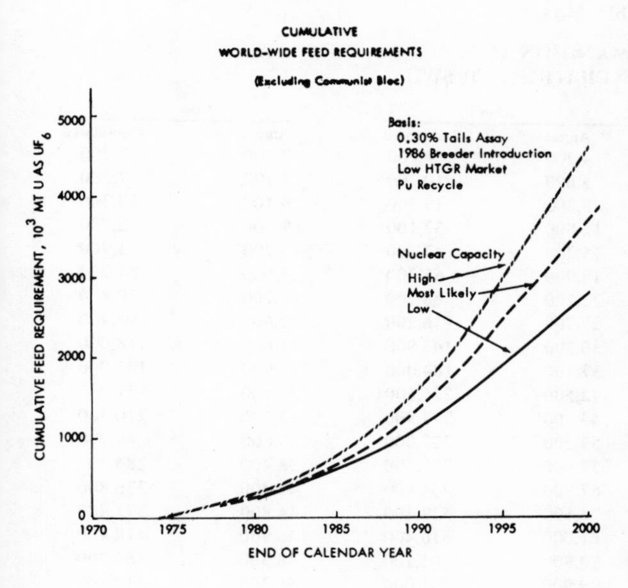

the operating tails assay of any enrichment plants will be 0.30% ^{235}U.

The feeds necessary to support these levels of enrichment activity are shown in **Table 4-10**. The 1980 requirements for the United States are 27, 26, and 31 thousand metric tons of uranium (MTU) and for foreign countries 27, 22, and 30 thousand MTU for the most likely, low, and high cases respectively. The 1990 requirements of 88 and 84 thousand MTU for the United States and foreign countries will range from 64 to 104 and 61 to 107 thousand MTU respectively. The requirements in the year 2000 are expected to be 118 thousand MTU in the United States and from 98 to 192 thousand MTU in foreign countries.

Fig. 4-8

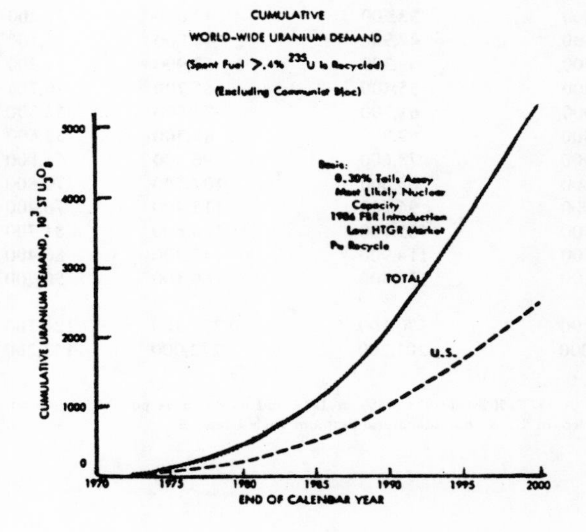

Although the requirements for feed and separative work mentioned above are expected to be needed in the times mentioned, it is unlikely that the total burden will be taken by facilities in the United States. It is assumed that the United States will supply between 40 and 60 percent of new foreign requirements for separative work during the 1980s and 1990s.

The total uranium demand for all reactors in the non-Communist world is shown in **Table 4-12** and in **Fig. 4-8**. About a million short tons of U_3O_8 will be required by 1985, and another million tons will be needed by 1990. The total requirements by the end of the century are expected to exceed 5 million tons of U_3O_8.

FOREIGN ENRICHMENT SUPPLY

Two estimates of the foreign enrichment supply capability are shown in **Table 4-13**. These estimates are based, in the early years, on the capability of existing enrichment plants in France and the United Kingdom. It is assumed, that beginning in the early 1980s, a few of the major nations will install enrichment capability which will enable them to supply all of their local growth in enrichment demand and to compete with other facilities in the world market. Two levels of success in this competition are assumed which result in the schedules shown in the table. The remainder of the world demand is assumed to be supplied by future enrichment facilities to be built in the United States.

FABRICATION, CONVERSION, AND REPROCESSING DEMANDS

The characteristics of reactors given above and the total capacity forecasts, together with the fuel cycle parameters also discussed, lead to an estimate of the expected demand for fuel fabrication, conversion from uranium to UF_6 for cascade feed, and chemical reprocessing of spent fuel. The results are given in **Table 4-14** for the United States and in **Table 4-15** for other countries. The calculations have been performed on a quarter-year basis. The LWR mixed-oxide data are plutonium recycle fuel and are considered to be natural uranium spiked with plutonium. The HTGR fissile and fertile demands are for highly enriched uranium and thorium. The cores for fast reactors are shown as FBR mixed oxides in the tables; the FBR blankets consist of depleted uranium.

Conversion to UF_6 is based on the assumption that the material to be fed to the cascades will

Table 4-11

TOTAL DEMAND ON U. S. ENRICHMENT FACILITIES[1], 10^3 SWU

Calendar Year	Most Likely		High		Low	
	Annual	Cumulated	Annual	Cumulated	Annual	Cumulated
1972	3,300	3,300	4,200	4,200	2,200	2,200
1973	5,600	8,900	5,800	10,200	4,400	7,200
1974	7,200	16,100	9,200	19,300	6,100	13,300
1975	12,000	28,100	12,800	32,100	9,400	22,700
1976	13,400	41,500	15,300	47,500	13,200	35,900
1977	18,100	59,600	18,900	66,300	15,900	51,700
1978	19,500	79,100	22,200	88,500	19,200	70,800
1979	25,400	104,500	27,700	116,200	22,600	93,400
1980	27,300	131,800	30,800	146,900	24,600	118,000
1981	31,600	163,400	37,100	184,000	27,900	145,900
1982	35,500	198,900	42,500	226,500	31,100	177,000
1983	38,500	237,400	47,400	273,900	33,300	210,300
1984	42,700	280,100	53,200	327,000	35,600	245,900
1985	47,600	327,700	59,900	386,900	38,400	284,300
1986	53,800	381,500	67,900	454,800	42,400	326,800
1987	59,100	440,600	74,600	529,400	44,800	371,400
1988	63,900	504,500	81,000	610,400	46,900	418,400
1989	70,000	574,500	90,800	701,100	48,500	466,900
1990	77,300	651,800	99,900	801,000	51,300	518,100
1995	102,300	1,125,300	136,600	1,413,800	65,700	817,700
2000	124,700	1,713,900	173,100	2,220,000	78,800	1,177,400

[1] Enrichment plant tails assay at 0.30% ^{235}U.
Recycle of plutonium in reactor loadings beginning with 25% in 1977, 50% in 1978, 75% in 1979 and as much as possible thereafter, without limiting LMFBR construction, except that none is recycled in U. K. nor in natural-uranium fueled reactors.
Foreign enrichment supply schedule 2 (see Appendix A).

Table 4-12

TOTAL URANIUM REQUIREMENTS,[1] ST U_3O_8/YR. IN YELLOW CAKE FROM MILL

Calendar Year	United States			Foreign (excluding Communist Bloc)		
	Most Likely	High	Low	Most Likely	High	Low
1973	10,000	11,300	8,400	12,400	13,800	11,100
1974	13,800	15,600	11,700	17,100	18,500	14,300
1975	18,200	20,400	15,100	19,700	20,900	19,000
1976	21,200	23,800	19,400	23,400	23,900	22,000
1977	24,000	26,600	23,900	28,600	30,100	24,800
1978	29,600	32,000	29,300	32,700	37,000	28,900
1979	34,500	38,000	33,000	38,500	42,600	32,200
1980	38,400	44,000	36,600	42,900	48,800	36,100
1981	44,200	51,400	41,500	49,300	57,900	41,200
1982	49,600	59,300	45,800	55,400	65,700	46,700
1983	55,900	67,400	50,000	61,900	74,600	52,300
1984	63,200	75,800	54,500	69,900	85,300	58,600
1985	71,500	85,500	59,900	78,600	96,600	65,000
1986	80,000	95,400	65,600	88,000	107,500	70,800
1987	88,400	105,100	71,000	96,900	118,300	76,400
1988	98,500	119,500	76,300	105,700	128,600	81,700
1989	109,500	131,600	81,600	114,900	142,300	86,400
1990	117,900	140,100	87,100	130,500	160,100	90,700
1995	142,600	173,700	100,100	170,100	221,300	121,100
2000	153,600	192,000	110,000	201,200	272,000	141,200

[1] Enrichment plant tails assay at 0.30% ^{235}U.
Recycle of plutonium in reactor loadings beginning with 25% in 1977, 50% in 1978, 75% in 1979 and as much as possible thereafter, without limiting LMFBR construction, except that none is recycled in U. K. nor in natural-uranium fueled reactors.

consist partly of natural uranium entering the fuel cycle and partly of recovered material from reactor discharges. The latter material is nearly all at enrichments ranging from 0.7 to 1.0 percent ^{235}U.

Reprocessing demands have similar definitions to fabrication demands and are assumed to occur in the third quarter following discharge from the reactor. Breeder cores and blankets are not separated because they are partially mixed in reprocessing and are similar in character at the reprocessing stage.

Table 4-13

FOREIGN SEPARATIVE WORK CAPABILITY

End of CY	Annual Supply, 10^3 SWU/yr. Schedule 1—High	Schedule 2—Low	End of CY	Annual Supply, 10^3 SWU/yr. Schedule 1—High	Schedule 2—Low
1972	400	400	1987	19200	14800
1973	450	450	1988	22500	18100
1974	500	500	1989	25300	21100
1975	600	500	1990	28000	23250
1976	800	500	1991	31800	25350
1977	950	500	1992	36400	28500
1978	1300	750	1993	41300	32000
1979	1900	1250	1994	47100	33600
1980	2700	1800	1995	50200	33600
1981	3900	2700	1996	50200	33800
1982	5600	4100	1997	51200	34100
1983	7700	6050	1998	52200	34200
1984	10300	8300	1999	52200	34200
1985	12800	10250	2000	52200	34200
1986	15600	12150			

Note: 1. Based on 0.3% tails assay.
2. Excludes Communist Bloc.

FUEL CYCLE PARAMETERS

The fuel cycle parameters which are used in this analysis are shown in **Table 4-16** For the typical enriched-uranium reactor, one quarter is required for conversion of refined U_3O_8 to UF_7, one quarter for enrichment, one for fabrication (two quarters for the fabrication of the larger quantities of first cores) and one for shipping and pre-loading inventories. An additional two quarters are allowed for pre-operational testing and startup for first core material.

When material is discharged from the reactor it is assumed to spend two quarter years in cooling and one in reprocessing. An additional quarter is allowed for converting the uranium to UF_6 and shipping it to an enrichment facility. Plutonium is considered available for fabrication three quarters after discharge. Because two quarters are required for fabrication, shipping and pre-loading inventories and because only self-generated plutonium recycle is assumed, the majority of the plutonium produced for recycle remains out of the reactor until the second annual reload following discharge.

For natural-uranium fueled reactors, the times assumed for the parts of the fuel cycle are similar to those assumed for enriched-uranium fueled reactors except that no time is allowed for the enrichment step. Uranium must thus be available from the refinery only 5 quarters before commercial operation. The plutonium from these reactors is assumed not to be recycled and the discharged material may be of sufficient value to warrant reprocessing only when fast reactor demands cause an increase in the value of the contained plutonium. Despite the uncertainty in this timing, the material from the natural-uranium fueled reactors is included in the tables as part of reprocessing loads after the normal cooling period. The plutonium thus appears in industry inventories as though it had been recovered at that time.

The material requirements shown throughout this report are based on the assumptions that 1 percent of the material fabricated is lost in processing. Where data are available, allowance is also made for the cold scrap generated during fabrication.. Losses of 1.3 percent for uranium and 1 percent for plutonium are assumed during the reprocessing of irradiated fuel.

It is assumed that the plutonium to be recycled will be mixed with natural uranium and fabricated into fuel elements and assemblies which are separate from those containing enriched uranium.

Table 4-14

FORECAST – UNITED STATES – MOST LIKELY – INCLUDES LOW HTGR, 86 LMFBR

PLUTONIUM RECYCLE

	FABRICATION LOAD IN METRIC TONS						CONVERSION	REPROCESSING LOAD IN METRIC TONS			
YEAR -CY-	LWR UO_2	LWR MIXED OXIDE	HTGR FISSILE	HTGR FERTILE	FBR MIXED OXIDE	FBR BLANKET	UF_6 -MT-	LWR MIXED OXIDE	HTGR FISSILE	HTGR FERTILE	FBR MIXED OXIDE
1972	750	0	1	11	0	0	3700	9	0	0	0
1973	1140	0	0	0	0	0	6600	92	0	0	0
1974	1350	0	0	2	0	0	8900	196	0	0	0
1975	1970	0	0	2	0	0	12400	810	0	2	0
1976	2700	34	0	6	0	0	13800	1190	0	2	0
1977	3300	179	2	33	0	0	15300	1610	0	2	0
1978	3300	400	3	62	0	0	19200	1880	0	2	0
1979	4600	400	3	61	10	5	23100	2100	0	4	0
1980	4600	430	7	121	5	1	25500	2400	0	11	0
1981	5300	480	9	148	7	2	29200	3000	1	22	0
1982	5900	610	11	188	19	8	33000	3600	1	32	5
1983	6800	830	15	250	9	1	37200	4300	2	52	5
1984	7400	1000	19	300	19	5	41800	4800	3	74	8
1985	8400	1060	23	350	56	23	48100	5400	4	101	15
1986	9600	1130	28	420	41	12	53700	6300	6	134	15
1987	10300	1080	32	460	180	75	59800	7100	7	172	21
1988	11400	1010	36	490	310	129	66600	8000	9	210	42
1989	12700	700	40	540	540	210	74300	9000	11	260	60
1990	14000	168	44	580	850	310	81900	9900	12	300	147
1995	17700	0	60	730	3300	1150	102600	14000	20	500	1710
2000	19700	0	70	830	6900	2500	110200	16800	26	640	5300

Table 4-15

FORECAST – FOREIGN – MOST LIKELY – INCLUDES LOW HTGR, 86 LMFBR

PLUTONIUM RECYCLE

	FABRICATION LOAD IN METRIC TONS						CONVERSION	REPROCESSING LOAD IN METRIC TONS			
YEAR -CY-	NAT & LWR UO_2	ATR & LWR MIXED OXIDE	HTGR FISSILE	HTGR FERTILE	FBR MIXED OXIDE	FBR BLANKET	UF_6 -MT-	LWR MIXED OXIDE	HTGR FISSILE	HTGR FERTILE	FBR MIXED OXIDE
1972	2500	0	0	0	16	8	5200	86	0	0	0
1973	3800	0	0	0	0	0	5800	370	0	0	0
1974	4200	0	0	0	8	1	10000	400	0	0	0
1975	5400	0	0	10	8	1	12400	640	0	0	9
1976	5200	17	0	0	8	1	15100	790	0	0	9
1977	6800	58	0	2	18	6	18600	1250	0	0	9
1978	7000	144	0	2	8	1	21300	1640	0	2	9
1979	8300	250	1	24	13	2	26600	2400	0	2	9
1980	9100	280	0	2	88	38	29600	3000	0	2	15
1981	10200	400	2	30	13	2	34900	3900	0	2	15
1982	11300	520	2	33	66	16	40500	4800	0	7	15
1983	12400	640	3	45	83	24	44800	5700	0	7	57
1984	13500	830	4	73	91	25	51900	6600	0	11	57
1985	15100	980	6	107	75	12	58600	7600	1	17	66
1986	16300	1150	8	138	139	38	66200	8600	1	23	84
1987	17900	1380	11	188	240	86	73100	9900	1	34	102
1988	18900	1650	15	230	420	156	80400	11100	2	54	102
1989	20500	1980	18	270	650	240	88000	12600	3	75	139
1990	21900	2200	21	310	970	340	98000	14200	4	101	230
1995	30100	1670	39	520	3500	1180	134900	22300	11	270	1900
2000	36600	2100	60	760	7100	2600	154600	29500	19	460	5500

This should result in lower fabrication costs than would be the case if plutonium were mixed throughout the fuel. To calculate the savings in feed and separative work to be gained from recycle, plutonium has been assumed to replace ^{235}U in reload fuel at a rate of 0.8 gram ^{235}U per gram of fissile plutonium used. The unburned plutonium plus the plutonium built up in the recycle fuel is assumed to total 58 percent of the plutonium orginally charged.

Table 4-16

FUEL CYCLE LEAD TIMES, QUARTER-YEARS

	FBR	Natural U Reactors	Enriched U Reactors
1. U_3O_8 procurement to enriched U withdrawal	2[a]	0	2
2. Enrichment	1[a]	0	1
3. a. Enriched U withdrawal for first cores to commercial operation	5[b]	5[d]	5
b. Enriched U withdrawal for reloads to charging	2[c]	2[e]	2
4. a. Fabrication—first cores	2	2	2
b. Fabrication—reloads	1	1	1
5. Discharge to reprocessing	2	2[f]	2
6. Discharge to return of spent fuel as enriched fuel to fabrication	4	4[f]	4
7. Discharge to return of Pu	3	3[e]	3

[a] Small amounts for initial load only.
[b] Also depleted U withdrawal for first cores to commercial operation.
[c] Depleted U withdrawal for reloads to charging.
[d] U_3O_8 procurement to commercial operation.
[e] U_3O_8 procurement to charging.
[f] Minimum time. Reprocessing may not occur until plutonium is needed for LMFBR inventories.

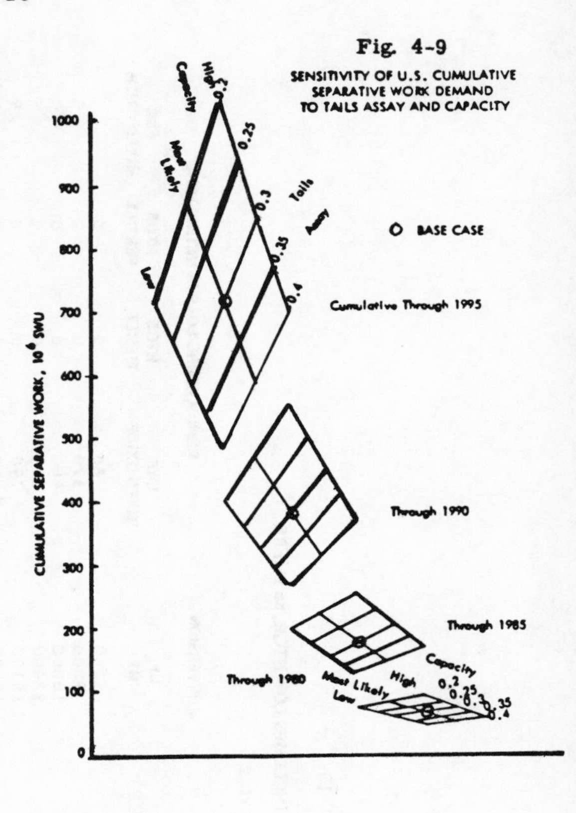

Fig. 4-9

SENSITIVITY OF U.S. CUMULATIVE SEPARATIVE WORK DEMAND TO TAILS ASSAY AND CAPACITY

SENSITIVITY ANALYSES

Several variations on the forecasts presented above were calculated to determine the sensitivity of various derived quantities to some of the assumptions underlying the basic forecast. These variations include enrichment tails assay, total installed nuclear capacity, mix of reactor types, and plutonium recycle. Separative work has been chosen as an example quantity for discussion of sensitivities.

The effect of changing the assumed enrichment tails assay and total capacity forecast is shown in **Fig. 4-9.** Operation of the enrichment facilities at various tails assays can almost exactly offset the effect of different forecasted capacities. Compared with most likely capacity forecast and an assay of 0.30 percent ^{235}U, virtually the same cumulative enrichment requirements result if the low forecast becomes an actuality and the cascades are operated at 0.2 percent tails or if the high capacity becomes actuality and the cascades are operated at 0.4 percent tails. This conclusion is unchanged at any point in time during the remainder of the century.

Fig. 4-10 shows the effect on the cumulative separative work demand of varying the mix of reactors actually built. The uncertainty in the capacity forecast produces an uncertainty in the cumulative separative work demand of over 20 percent, or about 150 million SWU by 1995. The differences in reactor mix produce an uncertainty in the demand for separative work of less than about 3 percent, or about 20 million SWU.

An additional variation that was investigated is plutonium recycle. The base case was based on the assumption that plutonium will be recycled on a self-generated basis beginning at relatively low rates in 1977 and will then slowly phase out in the late 1980s as breeder reactors begin to need the plutonium. An alternative assumption is that plutonium will not be recycled in thermal reactors. The effect of this assumption on the separative work demand is shown in **Fig. 4-11** . Plutonium recycle has the effect of delaying the cumulative demands by some 6 to 10 months.

Fig. 4-10

SENSITIVITY OF U.S. SEPARATIVE WORK DEMAND TO REACTOR MIX AND CAPACITY

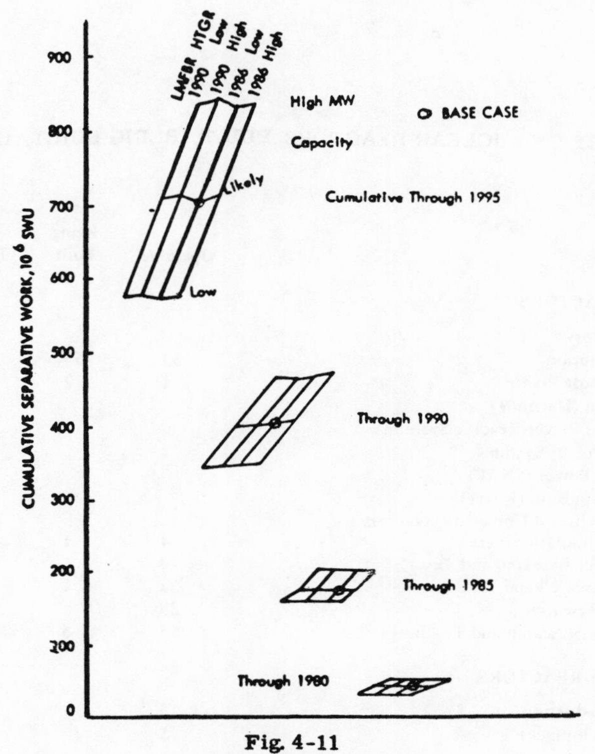

Fig. 4-11

WORLD-WIDE SEPARATIVE WORK REQUIREMENTS
(Excluding Communist Bloc)

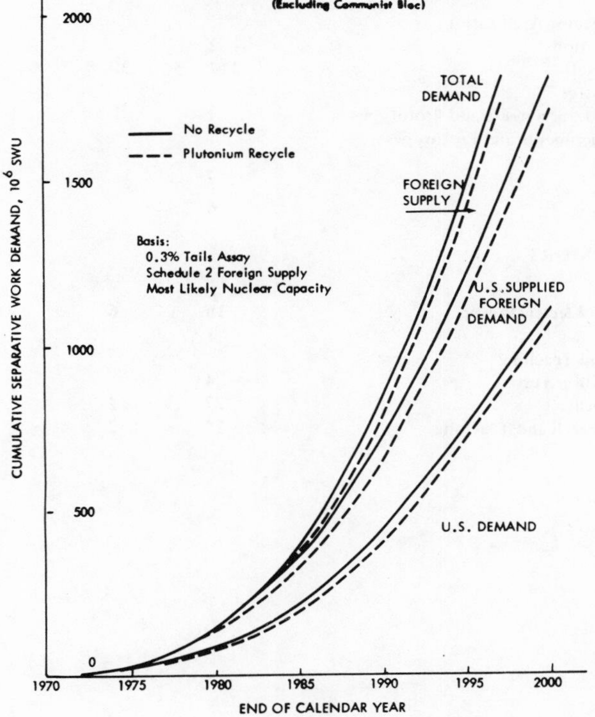

SUMMARY OF NUCLEAR REACTORS BUILT, BEING BUILT, OR PLANNED

	Operable	Being built	Planned	Shut down or dismantled
I. CIVILIAN REACTORS				
1. Power Reactors				
A. Central-Station	33	55	81	7
B. Dual-Purpose Plants	1	2		
C. Propulsion (Maritime)				1
2. Experimental Power-Reactor Systems				
A. Electric-Power Systems	1			25
B. Auxiliary Power (SNAP)				9
C. Space Propulsion (Rover)				21
3. Test, Research, and University Reactors				
A. General Irradiation Test	4	1		2
B. High-Power Research and Test	10			3
C. Safety Research and Test	2	1		9
D. General Research	28			30
E. University Research and Teaching	55	5	3	2
II. PRODUCTION REACTORS				
1. Materials Production	3			10
2. Process Development	5			
III. MILITARY REACTORS				
1. Defense Power-Reactor Applications				
A. Remote Installations	2			4
B. Propulsion (Naval)	117	30		6
2. Developmental Power				
A. Electric-Power Experiments and Prototypes	1			4
B. Propulsion Experiments and Prototypes	7			7
3. Test and Research				
A. Test	3			2
B. Research	6			3
IV. REACTORS FOR EXPORT				
1. Power Reactors				
A. Central-Station Electric Power	18	6	19	
B. Propulsion	1			
2. Test, Research, and Teaching				
A. General Irradiation Test	4			
B. General Research	27	2		1
C. University Research and Teaching	25	2		

CHAPTER 5

BREEDER REACTORS

A breeder is a reactor that produces more fuel than it consumes. (In the next section we will discuss the materials and processes involved in breeding.) If the breeder can be used in a nuclear power plant, it can provide the heat needed for the generation of electricity and simultaneously produce an excess of fissionable material that can be used to fuel other plants. The answer to this apparent contradiction is that breeder reactors do not give "something for nothing", but produce usable fissionable material from comparatively useless fertile material at a greater rate than the original fuel in the reactor is consumed in the fissioning process.

Fertile materials are abundant and relatively inexpensive, so by converting these materials into fissionable materials the problem of depleting our natural resources is shifted to the distant and very remote future. At the same time, the breeder reactor holds out enticing promises of reducing the cost of electric power generation and reducing its effect on the environment. These promises must be fulfilled before breeder reactors can be built on a large-scale commercial basis, and many people feel that these promises will soon be realized.

The Basic Processes

Breeder reactors are similar in many respects to the nuclear power reactors now in use or being built in many places throughout the world. Breeders, however, are unique in their ability to produce more fissionable material than they consume. To understand how this may be possible—and why other types of nuclear reactors do not "breed"—we must first look at some basic principles.

Fundamental to all nuclear reactors is the fission process. In this process, the impact of a neutron* on the nucleus of an atom of fissionable material can cause the nucleus to break apart, or fission.

Only a few isotopes† available in quantity are capable of sustaining the fission process. These isotopes are uranium-233 (^{233}U), uranium-235 (^{235}U), plutonium-239 (^{239}Pu), and plutonium-241 (^{241}Pu), and the term fissionable material refers to these four isotopes. All four undergo radioactive decay.‡ ^{235}U is the only one that exists in any quantity in nature; the other three, while perhaps present billions of years ago, have essentially decayed completely. Consequently, if we want ^{233}U, ^{239}Pu, or ^{241}Pu now, we must produce it artificially.

In order to create a fissionable material, we use a fertile material, which is a material that will join with or absorb a neutron and result in new fissionable material. The new fissionable material can produce more neutrons by fission and thus can continue the fissioning process. The fertile materials thorium-232, uranium-238, and plutonium-240 can produce fissionable ^{233}U, ^{239}Pu, and ^{241}Pu, respectively. Both ^{232}Th and ^{238}U exist in nature, while ^{240}Pu must be created artificially by neutron absorption in ^{239}Pu.

The absorption of a neutron by ^{232}Th or ^{238}U does not result directly in a new fissionable material, but first produces unstable, intermediate products. As indicated in the equations given below, the intermediate products quickly decay to produce the desired fissionable material.

$$(^{232}Th) + n^* \rightarrow (^{233}Th) \xrightarrow[\text{decay}]{\text{rapid}} (^{233}Pa) \xrightarrow[\text{decay}]{\text{rapid}} (^{233}U) \xrightarrow[\text{decay}]{\text{slow}}$$

$$(^{238}U) + n \rightarrow (^{239}U) \xrightarrow[\text{decay}]{\text{rapid}} (^{239}Np) \xrightarrow[\text{decay}]{\text{rapid}} (^{239}Pu) \xrightarrow[\text{decay}]{\text{slow}}$$

$$(^{240}Pu) + n \rightarrow (^{241}Pu) \xrightarrow[\text{decay}]{\text{slow}}$$

*The letter n represents a neutron. The decay of intermediate products results in each case above in the emission of a beta particle. A rapid decay means that the process will be essentially complete in hours or months, while a slow decay requires hundreds to hundreds of thousands of years.

As the equations show, ^{232}Th is converted to thorium-233, which decays to protactinium-233, which then decays to the desired fissionable material, ^{233}U. When ^{239}Pu is formed from ^{238}U, the intermediate products are uranium-239 and neptunium-239. Plutonium-241 is formed directly from ^{240}Pu.

We have discussed some of the processes involved in producing new fissionable material, and we will now consider another fundamental rule of nature—one concerning the number of neutrons produced in each fission. In Fig. 5-1 a single neutron striking the nucleus of an atom of fissionable material produces fission fragments and the release of free neutrons. The number of neutrons released varies, but is usually more than two. Since only one neutron is needed to continue the fission chain reaction, the other freed neutrons can be used to produce new fissionable material. These excess neutrons may be (1) absorbed by fertile material to produce new fissionable material, (2) absorbed nonproductively in structural material, fission fragments, control material, fuel, or the reactor coolant, or (3) escape from the reactor and be absorbed in the surrounding shielding material. In developing a breeder it therefore is necessary to design a reactor in which the necessary fissionable, fertile, structural, control, and coolant materials are present and in which as many of the

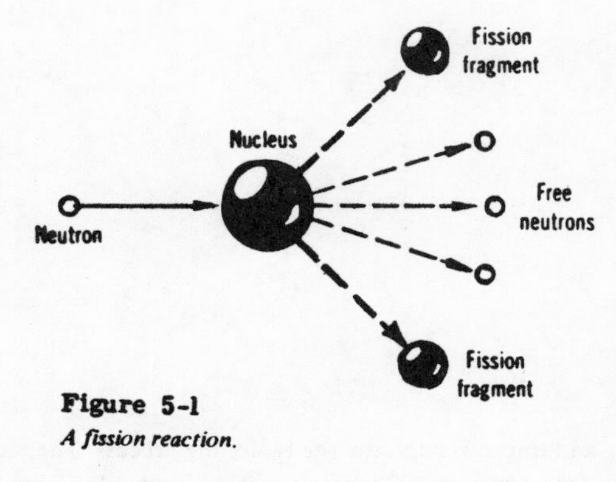

Figure 5-1
A fission reaction.

excess neutrons as possible are absorbed in fertile material. There are a number of different ways of doing this.

As stated earlier, the breeder reactor produces more fissionable material than it consumes. This means that for every neutron absorbed by an atom of fissionable material, more than one of the freed neutrons must be absorbed in an atom of fertile material to create more than one new atom of fissionable material. If only one atom of new fissionable material is produced per atom consumed, the amount of fuel (fissionable material) remains constant and breeding does not occur. If less than one new atom is produced, the amount of fuel decreases and the reactor is known as a "converter" or "burner".

Therefore in breeding the average number* of neutrons given off in the fission of a single atom must be greater than two—one to continue the fission process and at least fractionally more than another one to convert fertile material into new fissionable material. This means that breeding may be possible if more than two neutrons are produced, but if two or less are produced, breeding cannot occur regardless of reactor design.

Actually, considerably more than two neutrons per fission are necessary for breeding, because some will be lost through leakage and some lost by nonproductive absorption. The following table gives representative values for the number of neutrons produced per neutron absorbed by the

Average Number of Neutrons Produced* per Neutron Absorbed in Fissionable Material

Type of fissionable material	Thermal neutron absorbed	Fast neutron absorbed
Uranium-233	2.3	2.3
Uranium-235	2.1	2.0
Plutonium-239	1.9	2.4
Plutonium-241	2.1	2.7

*These are approximate values, and the actual value for any given reactor design may differ slightly depending on the details of the design.

different fissionable materials. The numbers indicate that breeding can probably be achieved with the least difficulty and with the greatest efficiency in a fast-neutron† reactor that uses plutonium-239 and plutonium-241 as the fissionable material. The table also indicates that breeding is unlikely, if not impossible, in thermal-neutron† reactors unless they use uranium-233 as the fissionable material.

At the present time, there are two basic breeder reactor materials "systems":

	Fissionable material used	Fertile material used	Fissionable material formed
(1)	Plutonium-239	Uranium-238	Plutonium-239
(2)	Uranium-233	Thorium-232	Uranium-233

Thermal-neutron or thermal reactors and fast-neutron or fast reactors are important classes of reactors. The neutrons produced in the fission process are generally traveling at very high speeds —about 30 or 40 million miles per hour — and thus are called fast neutrons. If certain materials called "moderators" (water or graphite, for example) are present, the fast neutrons collide with atoms of the moderator and lose speed with each collision.

After bouncing around from atom to atom, the neutrons will reach thermal equilibrium—a term which means that the neutrons have slowed to about the same speed as the moderator atoms. When this slowing down has occurred and thermal equilibrium has been reached, the neutrons are called thermal neutrons and are traveling at an average speed of only about 5000 to 10,000 miles per hour. Thus, the difference between thermal and fast neutrons is mainly a difference in their speed, which is related to the energy of the neutrons.

All reactors contain both thermal and fast neutrons, but the designers can select materials and arrangements that emphasize the thermal or the fast part of the mixture. A thermal reactor is one in which a moderator is present to slow down the neutrons, and the fissions are caused mainly by thermal neutrons. A fast reactor is one in which the amount of moderator material is made as small as possible, so that the fissions are caused principally by fast neutrons. All the very large nuclear power plants in the United States now use thermal reactors, and nearly all these reactors use ordinary water both as the moderator and as the reactor coolant (the fluid that removes the heat generated by the reactor).

The fast reactor is usually regarded as the most promising concept for breeding. In addition to the fact that fission by fast neutrons produces a relatively large number of free neutrons, the fast neutrons are not absorbed as readily as thermal neutrons by the structural material and reactor coolant. Furthermore, the fast neutrons can occasionally cause atoms of fertile material to undergo fission and thus create a few "bonus" neutrons. These bonus neutrons contribute appreciably to the amount of breeding that is possible with fast reactors.

Breeding ratio and doubling time are measures of the efficiency of a breeder. The breeding ratio is the number of atoms of fissionable material produced per atom of fissionable material consumed. Doubling time is the time required for a breeder reactor to produce as much fissionable material as the amount usually contained in its core plus the amount tied up in its fuel cycle (fabrication, reprocessing, etc.). Doubling time is a measure of the amount of breeding that is achieved in a given design: The more fissionable material produced during a given period by breeding, the shorter the doubling time. At present, researchers hope to achieve a doubling time of less than 10 years.

The production of excess fissionable material in the reactor is not the whole story, however. Periodically, fuel elements have to be removed from the reactor because (1) the fuel element containers do not last indefinitely and (2) the fuel must be chemically processed. This processing is necessary to recover the fissionable material and to clean the fuel. This cleaning removes the accumulated fission fragments, which increase the unproductive capture of neutrons. Once recovered, the fissionable material can be made into new fuel elements that are loaded back into the reactor (or another one) to continue the power-generating–breeding process. Excess fissionable material produced by breeding can be used in other reactors or stored until a sufficient amount is collected to provide fuel for a new reactor.

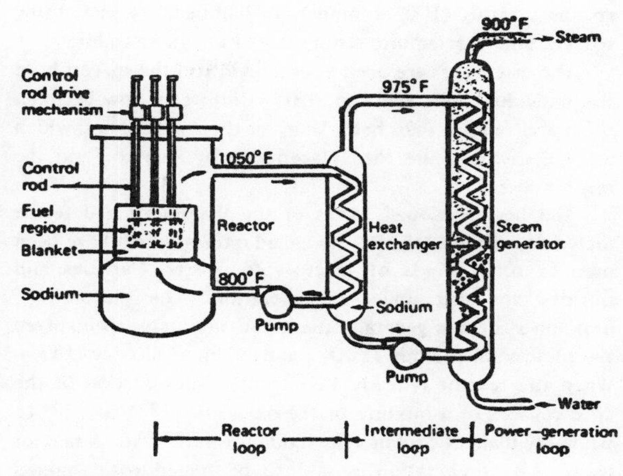

Fig. 5-2. *LMFBR nuclear steam supply.*

Kinds of Breeders

The reactors we will discuss are the liquid-metal-cooled breeder, the gas-cooled breeder, and the molten-salt breeder. There are other possibilities, and these are mentioned briefly.

Liquid Metal

This concept is properly called the liquid-metal, fast-breeder reactor (LMFBR). It is the highest priority civilian nuclear power activity of all the breeder type reactors. Indeed, President Nixon in his Energy Message to the U. S. Congress on June 4, 1971, stated that his program included "a commitment to complete the successful demonstration of the liquid-metal, fast-breeder reactor by 1980".

The basic arrangement of an LMFBR nuclear steam supply is shown in Fig. 5-2 (Each diagram that appears in this section shows steam and water pipes at the right side of the illustration. In a nuclear power plant, these pipes would be connected to the power-generating portion of the plant, which is not shown here.) Since the production of steam is an objective of all the breeders covered in this section, we will discuss the overall nuclear steam-supply system rather than just the reactors. The nuclear steam-supply system is the equivalent of the boiler in a fossil-fuel power plant. In all steam power plants, nuclear or fossil, electricity is generated in about the same way. Steam turns a turbine that drives a generator to produce electricity. With its energy almost gone, the steam leaves the turbine and is turned back into water in a condenser. The water from the condenser goes back to the steam-supply system, where it is formed into steam again. Although it isn't shown, we will call the power-generating portion of the plant the power-generation loop (Fig. 5-2).

As the figure shows, there are three basic loops in a power plant that uses an LMFBR. In addition to the power-generation loop, there are the reactor loop and the intermediate loop. In both the reactor loop and the intermediate loop, the fluid is liquid sodium. The liquid-metal, fast breeder operates with fast neutrons. One reason sodium is a good fast-reactor coolant is that it does not moderate, or slow down, neutrons as much as, say, water does. Sodium is also a very good heat-transfer agent and its boiling point is high. Consequently, it can operate at high temperatures without requiring a high pressure to prevent boiling. Now let's return to Fig. 5-2 to see what the temperatures in a typical LMFBR system might be.

In the reactor loop, the sodium is pumped to the reactor vessel at a temperature of about 800°F. In the vessel, the sodium flows through the reactor core, where its temperature is raised to about 1050°F. It leaves the reactor at this temperature and then passes through the heat exchanger, where it heats the sodium in the intermediate loop. This transfer of heat from the reactor loop to the intermediate loop reduces the temperature of the sodium in the reactor loop so that it is ready to be pumped back through the reactor to begin the cycle again and remove more heat.

The sodium in the intermediate loop is heated to about 975°F in the heat exchanger; it then passes through the steam generator and is returned to the heat exchanger. In the steam generator, the intermediate-loop sodium heats the water of the power-generation loop and produces steam at a temperature of 900°F.

The intermediate loop is a safety feature of the LMFBR. The loop separates the radioactive reactor loop from the power-generation loop; this eliminates the possibility of a single leak allowing radioactive material to enter the power-generation loop or resulting in a chemical reaction between the radioactive sodium of the reactor loop and the water or steam of the power-generation loop.

The core of a typical LMFBR consists of a central region of fissionable material (fuel) mixed with fertile material; the central region is surrounded by a blanket of fertile material. The fuel sustains the chain reaction and produces a surplus of neutrons that can be absorbed in the blanket. The absorptions in the blanket convert the fertile material into new fissionable material. The core consists of thousands of small vertical rods arranged so that the sodium coolant can flow along them and remove the heat that they generate. Since the reactor operates at very high temperatures, the rods must be made of special materials.

The fissionable and fertile materials are ceramics, which are able to withstand high temperatures. To protect the ceramics from possible damaging action by the flowing sodium, the fissionable and fertile materials are encased in thin tubes of stainless steel. The ceramic-filled tubes form rods that are about the same diameter as a pencil and are several feet long. Rods such as these are not very rigid, so groups of rods (100 or more) are bundled together, using spacers, and inserted into strong cases to form assemblies.

The assemblies are open at each end, and the spacers hold the individual rods apart so that sodium can flow between them and remove their heat. Many of these assemblies, which are quite strong, are then placed side by side to form the reactor core.

The best-developed forms of the fissionable and fertile materials used in LMFBR's are called oxides, which have been used in other kinds of reactors for years. Carbides and nitrides are also under consideration. The plutonium uranium system is generally used, and the materials involved are plutonium dioxide (PuO_2) and uranium dioxide (UO_2). When the reactor is ready to start, the fuel portion of the core consists of a mixture of the oxides of ^{239}Pu and ^{238}U, while the blanket contains the oxide form of ^{238}U. A reactor using this combination is said to be fueled with "mixed oxides", which is a reference to the PuO_2 - UO_2 portion of the core. The normal initial content of the mixed oxide is about 20% plutonium and 80% uranium-238; as this indicates, a small fraction of plutonium (the basic fissionable material) is sufficient to "drive" the reactor.

Space is provided in the core for another kind of assembly. It is called a control rod, and a number of these regulate the power level of the reactor. The control rods are made of materials that absorb neutrons, and they can be moved vertically in the channels provided for them. By changing the degree of insertion of the control rods in the core, either a smaller or a greater number of neutrons will be available to cause fissions or convert fertile material into fissionable material. Thus, the control rods do precisely what their name implies; they control, or regulate, the reaction rate between the neutrons and core materials.

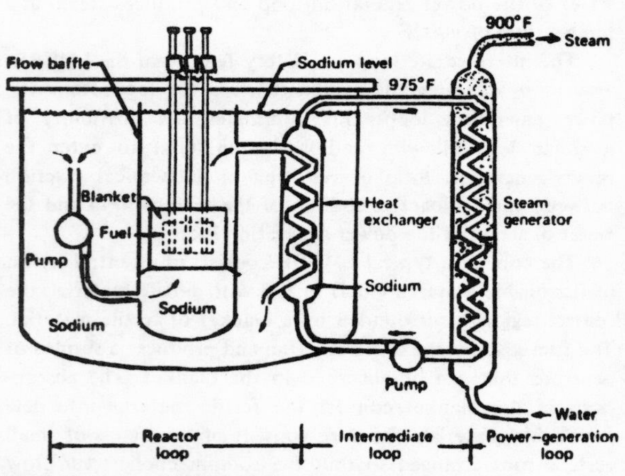

Fig. 5-3 *Pot-type LMFBR nuclear steam supply.*

We have discussed briefly one LMFBR plant, using in our example the materials that are specified most frequently. The physical arrangement we have described uses pipes to connect all major components of the nuclear steam-supply system. As you might have guessed from looking at Fig. 5-2 such a plant uses a loop-type system. Another type of system is pot-type system. This arrangement is illustrated in Fig. 5-3 which shows that the reactor loop is a self-contained unit in a relatively large vessel. There are advantages and disadvantages in both the loop and the pot concepts. For example, the pot-type system is more compact and its core can be cooled more easily if a pump fails, but it may be more difficult to repair.

Gas Cooled

The gas-cooled, fast-breeder reactor (GCFR) has been under study in the United States since the early 1960s and there are several incentives for developing it.

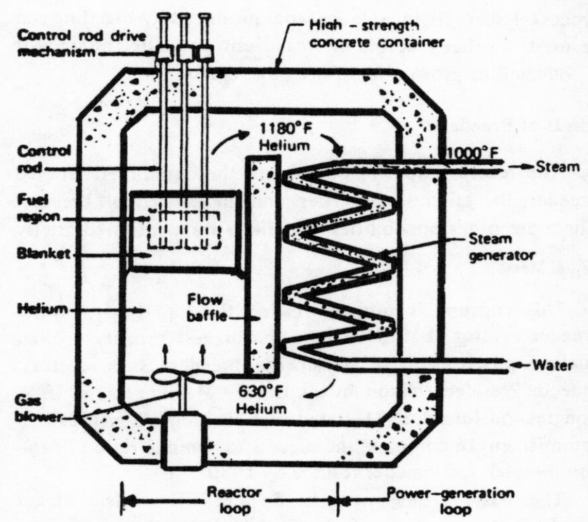

Fig. 5-4 *GCFR nuclear steam supply.*

Fig. 5-4 shows that the GCFR bears some resemblance to the pot-type LMFBR. The system is a closely coupled one in which the reactor cooling fluid is circulated within a single large vessel. Actually, the resemblance between the two kinds of breeders ends about there. The GCFR uses helium as the cooling fluid, although it (or any gas, for that matter) is not as good a heat-transfer agent as liquid sodium.

To qualify as a satisfactory cooling fluid, helium must be used at high pressures and must be passed through the reactor very rapidly. Under these high-pressure and high-velocity conditions the gas is reasonably effective in removing the heat produced by the nuclear processes in the core. The requirements of high coolant pressure and velocity affect adversely the cost of the GCFR because the vessel that contains the reactor must be very strong, and the gas blower must be powerful, but there are some compensating factors.

Since helium does not itself become radioactive or react chemically with water, as does sodium, the intermediate loop between the reactor loop and power-generation loop in the LMFBR is not needed in the GCFR. Thus, the inertness of the helium permits the use of the two-loop arrangement shown in Fig. 5-4. Heat is transferred from the reactor loop directly to the power-generation loop, so there is a reduction, in the number of principal components of the plant.

The general arrangement of the core in a gas-cooled, fast-breeder reactor is the same as that in a liquid-metal-cooled breeder. The fuel rods are probably a little larger in diameter for the GCFR, and there are more of them. The fuel materials under consideration are the same as those discussed for the LMFBR (oxides are used now, and carbides and nitrides may be used in the future), but the fraction of PuO_2 in the mixed-oxide fuel is less in the GCFR.

As you can see, there are similarities between the LMFBR and the GCFR.

Molten Salt

The molten-salt breeder reactor (MSBR) is a concept of American origin and evolution. When the United States started work on a nuclear-powered aircraft in 1947, the molten-salt reactor was among the systems considered for the power plant. Although the aircraft never became a reality, the molten-salt reactor did. The successful operation of an experimental molten-salt reactor, coupled with major advances in development that make possible a simpler plant, has led to the consideration of the MSBR, which is shown in Fig. 5-5.

The schematic drawing for the MSBR looks a lot like the one for the loop-type LMFBR; each shows a reactor loop, an intermediate loop, and, of course, a power-generation loop.

Salts are the key to the MSBR. Many salts are chemical compounds composed of a metal and an element such as chlorine, fluorine, or bromine. The best example of a salt is

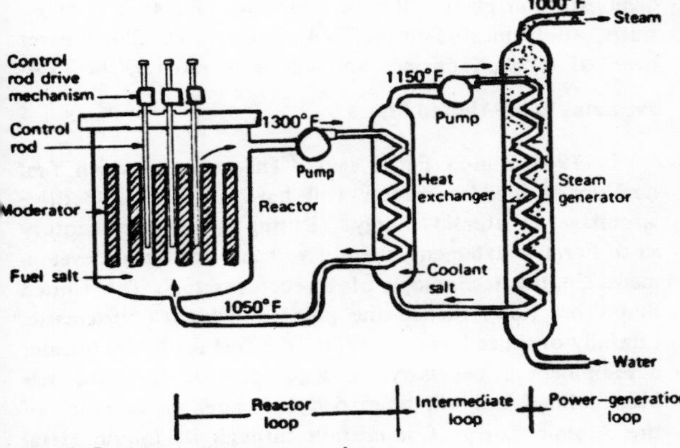

Fig. 5-5. *MSBR nuclear steam supply.*

common table salt, NaCl, in which sodium (Na) is the metal and chlorine (Cl) is the other element. Most people don't think of salts in the molten form, but the designers of the molten-salt reactor did and they developed an interesting plant.

The basic concept of the molten-salt breeder is one in which the fissionable material and the fertile material are circulated as a liquid through a region in which the fission process can occur. The fuel is a liquid called the fuel salt and consists of a combination of four fluorine-salt compounds. The metals used with fluorine to form the salts are uranium, thorium, lithium, and beryllium. The uranium and thorium are the fissionable and fertile materials used in the breeding process.* Since the uranium–thorium system operates on thermal neutrons, the MSBR is called a thermal breeder. Lithium and beryllium, which constitute most of the fuel salt, are used to dilute the fissionable and fertile materials to the proper concentrations.

The core region of the reactor has fixed, vertical "logs" of solid moderator material to slow down the neutrons. The solid moderator logs, which are graphite, are spaced apart so that the fuel salt can flow between them. When the fuel salt is in the core, the combined moderating effect of the salt itself and the graphite is such that the reactor can sustain the chain reaction. There is no possibility of a chain reaction outside the core (in the loop piping, for example), because the reaction requires a complex geometrical arrangement of materials.

The fuel salt circulates in the reactor loop during operation. As the salt flows through the core region, it momentarily becomes a part of the core and generates heat through the fission process that occurs in some of its constituents. When a particular volume of salt leaves the core region, the reaction in that volume stops; the salt then becomes the circulating fluid that transports the heat from the reactor to the heat exchanger.

In the heat exchanger, the energy (heat) produced in the reactor is transferred to the coolant salt of the intermediate loop for subsequent transfer to the water of the power-generation loop. The fluid in the intermediate loop is called the coolant salt because it, too, is a molten salt and because it removes heat from the fuel salt of the reactor loop. The mixture contains no fissionable or fertile materials, since there is no requirement for them in the intermediate loop. The coolant salt is a mixture of two salts that possess suitable qualities for the service intended and are relatively inexpensive.

The molten-salt breeder reactor holds promise as a system capable of producing low-cost power and, at the same time, using the abundant resources available for the uranium–thorium fuel cycle. Much development work is necessary, but, if the molten-salt concept proves commercially feasible, the United States will have developed a system that can play an important role in meeting future requirements for electric power.

Other Possibilities

Two other reactor concepts are being considered as possible breeders. The first of these is the light-water breeder, which has been under study in the U. S. for a number of years. The other is the steam-cooled breeder, which has been the subject of work in the U. S. and in other countries.

The light-water breeder is based in large part on the well-developed technology of commercial nuclear power plants that use pressurized-water reactors. The specification of the core for a light-water breeder (a thermal breeder, operating on the uranium–thorium cycle) represents a principal challenge to the reactor designers because the system must operate at highest neutron efficiency to breed. If the physics performance of the core can be shown to be satisfactory (from the breeding and economic points of view), most of the rest of the plant design can be based on an established foundation of experience.

Several groups in the United States have studied the steam-cooled breeder concept, but it appears that the most thorough investigation has been carried out by the West Germans. The studies indicate that the steam-cooled breeder is faced with a dilemma: At high steam pressures, the cost of the electric power produced is acceptably low but the reactor barely breeds; if the steam pressure is lowered, the breeding performance of the reactor improves but the cost of the electric power it produces increases. Research on the steam-cooled breeder has all but stopped.

Reactors in the United States

A number of reactors have been built in the United States for the purpose of demonstrating or developing the breeder concept. The world's first real breeder reactor was EBR-I, which had a core heat power rating of 1400 kilowatts and produced around 200 kw of electric power.* The fuel portion of the reactor core was quite small, and this is a characteristic of liquid-metal, fast-breeder reactors (LMFBR's). In EBR-I the fuel was in the usual form of small rods, and these rods occupied a space about the size of a regulation football.

The information and encouragement that scientists, engineers, and planners received from the success of EBR-I led to the design and construction of two larger fast-breeder reactors, the Experimental Breeder Reactor No. 2 (EBR-II) in Idaho and the Enrico Fermi Atomic Power Plant on the shore of Lake Erie in Michigan. Both these reactors are essentially pilot-plant LMFBR's, which were originally designed to demonstrate electric-power production with fast-reactor cores and to extend the store of knowledge that will ultimately make possible the construction of commercial nuclear power plants using breeder reactors.

EBR-II is designed to produce 62,500 kilowatts of heat. Its use has been oriented to irradiation testing of fuels and other materials for the fast breeder. The plant has operated satisfactorily for a number of years. Operations started in the late fall of 1961, and EBR-II began producing electric power in August 1964. Fuels and materials tests began in April 1965.

The Enrico Fermi Atomic Power Plant has a core heat rating of 200,000 kw. Although a series of operating problems delayed sustained power-demonstration runs, the Fermi plant has provided valuable insight into some of the problems that may face large, commercial breeder plants.

In the picturesque hills of Arkansas, a small fast reactor known as the Southwest Experimental Fast Oxide Reactor (SEFOR) operated until the middle of 1972. The specific purpose of this reactor was to provide more information regarding the dynamic performance of reactor cores that use the mixed plutonium–uranium oxide fuel referred to earlier in the chapter "Kinds of Breeders". SEFOR was a sodium-cooled reactor whose core was designed to produce 20,000 kw of heat. The plant did not generate electricity.

The U. S. Atomic Energy Commission has embarked on a comprehensive program aimed at developing the technology needed for the design, construction, and operation of fast-breeder, nuclear power plants. One of the most important items in this program is a large test reactor known as the Fast Flux Test Facility (FFTF). It is located near Hanford, Washington, and is scheduled for initial operation in the mid-1970s. A powerful facility from which meaningful test data can be obtained in a short time, the FFTF will have a core heat rating of 400,000 kw. The reactor is cooled by liquid sodium, and nine special areas are provided in the core for experiments. Although the FFTF reactor will be a fast-neutron reactor, it will not be a breeder. It will be used primarily to test fuels and other materials for future fast-breeder reactors.

The molten-salt reactor experiment at Oak Ridge National Laboratory in Tennessee began operation in 1965 and has furnished important data on the materials and systems required for a possible MSBR. Second, the design of a light-water breeder core is progressing, and such a core may be installed in an already existing nuclear power plant around the middle of the decade. It is appropriate that the plant selected for the breeder core is the Shippingport Atomic Power Station, which started operation in Pennsylvania in 1957 and was the first nuclear power plant to be operated commercially by an electric utility in the U. S.

The AEC has selected a proposal by the Commonwealth Edison Company of Chicago and the Tennessee Valley Authority to develop, own, and operate the first U. S. LMFBR (liquid-metal-cooled, fast-breeder reactor) demonstration plant. The project, which will cost about a half billion dollars, is a joint industry-government effort. The pledging of over $240,000,000 for this plant was an unprecedented cooperative endeavor involving all segments of the utility industry, including privately, publicly, and cooperatively owned companies. The AEC will contribute direct aid and services of around $250,000,000. The LMFBR demonstration plant will have a capacity of 300–500 megawatts, will be located on the TVA system on the Clinch River near Oak Ridge, Tennessee, and will be operated by the TVA.

Programs in the United States

In 1945 Enrico Fermi said, "The country which first develops a breeder reactor will have a great competitive advantage in atomic energy." During the quarter century since Fermi's statement, there have been significant developments in the technology of breeder reactors. The United States has now reached the point at which a systematic, carefully organized, and coordinated effort for future breeder development is necessary. A major part of this effort has been defined and will be carried out under the direction of the Atomic Energy Commission through its Liquid Metal Fast Breeder Reactor Program. Although there are several independent projects in the United States, sponsored generally by reactor manufacturers and electric power companies, the AEC's LMFBR Program is being aided in its implementation by the nuclear industry and the National Laboratories.*

The LMFBR Program Plan, published by the Atomic Energy Commission, covers a period of about 20 years and culminates in the 1980s with the introduction of commercial nuclear power plants using fast-breeder reactors. The stated objective of the plan is "... to achieve, through research and development, the technology necessary to design, construct, and safely, reliably, and economically operate a fast-breeder power plant in the utility environment: and to assure maximum development and use of a competitive, self-sustaining industrial capability in the program."

The LMFBR Program Plan divides the required development work into nine technical areas: (1) plant design, (2) components, (3) instrumentation and control, (4) sodium technology, (5) core design, (6) fuels and materials, (7) fuel recycle, (8) physics, and (9) safety. Implementation of the plan calls for the construction of a variety of test facilities as well as the use of facilities already in existence.

Foreign Developments

Many foreign countries have breeder-reactor programs. The programs vary widely in size, ranging from small, laboratory-type experiments of a specific nature to the construction of complete power plants. On the basis of their past experience and present activities, the United Kingdom, France, West Germany, and the USSR are considered the leading nations outside the United States in breeder-reactor development.

One of the similarities in the foreign programs is their strong emphasis on the design and construction of prototype power plants. The foreign programs will rely on the experience obtained with the medium-size prototype plants to provide the design bases for large commercial units. All the prototype plants are based on liquid-metal, fast-breeder reactors.

United Kingdom

As a part of their large nuclear-power program, the British have been engaged in the study and development of fast-breeder reactors since 1951.

In 1954 two experimental zero-power reactors were placed in operation at the Atomic Energy Research Establishment at Harwell. Their purpose was to provide basic physics information on fast-reactor cores and to verify that breeding could be a practical process.

On the basis of the information gained through analytical and experimental work, the United Kingdom Atomic Energy Authority (UKAEA) in 1955 began construction of a 60,000-kilowatt (heat) experimental power plant at Dounreay on the coast of Scotland.

The British decided fairly early to build an intermediate-size, prototype breeder reactor—that is, one considerably larger than the DFR, but not as large as commercial stations will be. In 1966 work began on the construction of the 250,000-kw (electric power) Prototype Fast Reactor (PFR), also at Dounreay.

France

In France, nuclear research is carried out largely by the Commissariat a l'Energie Atomique (C.E.A.), which corresponds to the U. S. Atomic Energy Commission. The C.E.A., which was created in 1945, has developed many nuclear reactors for research, production of materials for nuclear weapons, submarine propulsion, and electric-power generation. The nuclear power plants in France have generally been designed by the C.E.A., but they are operated by a nationalized electrical power organization.

The C.E.A. began fundamental research on liquid metals for nuclear power plant applications in 1953 at its Fontenay-aux-Roses Research Center. French work on fast breeders has continued and was given a substantial boost in 1962, when a contract was signed between the C.E.A. and Euratom for the design, construction, and operation of a fast reactor and associated facilities.

The focus of the C.E.A.–Euratom program is the sodium-cooled Rapsodie reactor. Rapsodie does not generate electric power, but its core was designed to produce 20,000 kilowatts of heat that is dissipated to the atmosphere through heat exchangers that operate on the same principle as automobile radiators.

Construction of Rapsodie began in 1962, and the reactor began operation in 1967. It has performed well and has given the French confidence in their capability to proceed with the next major step in a national breeder program: The construction of a 250,000-kilowatt (electric) prototype power plant. The plant is called Phenix and is scheduled to begin power operation in 1973 at Marcoule in southern France.

West Germany

The development of nuclear power plants in the Federal Republic of Germany got off to a late start. Following World War II, essentially no work was carried out in the nuclear field until 1955. Germany is unlike many countries in that it has no large national agency comparable to the U. S. Atomic Energy Commission. The Federal Government does give financial assistance to the individual German States for nuclear programs, but the States have considerable independence under the German system of government.

The Germans are making up for their late start in nuclear energy by establishing research centers, using their outstanding industrial capability, constructing many reactors, and undertaking joint programs with other countries and with Euratom.

Work on fast breeders has been carried out at Karlsruhe since 1960, with Euratom support of the breeder work starting in 1963. Several reactors are located at the Center, and among those having a direct bearing on the German fast-breeder program are SNEAK and KNK.

SNEAK is a zero-power facility that is used for physics research, while KNK is a power-producing nuclear plant. Although the sodium-cooled KNK is designed to operate initially as a thermal-neutron reactor, it will be converted to a fast reactor in the early 1970s. The Germans will continue to gain information and experience through participation in the SEFOR project in the United States.

A prototype sodium-cooled breeder, designated SNR, is scheduled to start power operation in the mid-1970s. SNR will have a power output of 300,000 kw (electric) and is a cooperative project between industry and government in Germany, Belgium, The Netherlands, and Luxembourg. The German role in the SNR project is the dominant one, but the Belgian and Dutch contributions are substantial.

USSR

The USSR has the most ambitious breeder-reactor construction program in the world. The size of the plants under construction (the largest has an electric power output of 600,000 kw) is impressive, but there have been significant delays in the schedules of some of the plants.

A low-power, experimental, fast-reactor core began operation at Obninsk, near Moscow, in 1955. This facility and others led the way for the construction of the BR-5 reactor, which began operation in 1959. The BR-5 core is rated at 5000 kw (heat), and this energy is dissipated to the atmosphere through heat exchangers.

On the basis of the experience gained from the BR-5, the Russians took a giant step and announced in 1964 that they would build a breeder plant for power production and water desalination near the shore of the Caspian Sea. The BN-350, as the plant is called, will produce 150,000 kw of electric power and enough process steam to desalt about 30,000,000 gallons of water per day. If the plant were used solely to produce electric power, it would have an output of 350,000 kw. The BN-350 is scheduled to begin operation in 1973.

The USSR is also constructing an advanced breeder, the BN-600, a 600,000-kw (electric) breeder power plant that is due to be completed in the mid-1970s. The Russian breeders of the future will probably be based on information obtained from the BN-600 and from a test reactor called the BOR-60.

CHAPTER 6

THE FISSION PROCESS

The fission process is usually described as the succession of three different phases. The compound nucleus first undergoes a long series of collective oscillations until one of them leads to the passing of the so-called saddle-point and followed very soon after by scission i.e. the splitting of the nucleus into two fragments of about equal mass. This is binary fission which occurs in most common cases. Other more complicated types of fragmentation can also happen but they are less frequent and will not be considered unless specifically mentioned.

Complete treatment of the fission process requires a dynamical study of the penetration of the multidimensional fission barrier. But, this is very complicated and only a few such attempts have been undertaken up to now. Most theoretical calculations are carried out on the assumption that the motion of the fissioning nucleus is slow enough for the fission process to be adiabatic. It is therefore of great importance to know the potential energy surface of the nucleus as a function of a set $\{s\}$ of deformation parameters. This potential energy cannot be obtained, at present, with microscopic calculations, taking into account the detailed structure of the nucleus. Fission barriers can nevertheless be calculated, in first approximation, from a macroscopic approach using the liquid drop model (LDM), the first one used to describe the fission process soon after its discovery.

But the LDM fission barrier is incapable of explaining several aspects of the fission process, for example : i) the variation of T_f°, the spontaneous fission half-life, from one nucleus to another ; ii) the existence of fission isomers ; iii) the phenomenon of intermediate structure in some subthreshold fission cross sections ; iv) the structures in some cross sections for near-threshold fission processes. In this section , we shall explain how fission barrier shapes, as calculated with the Strutinsky prescription, can provide an unified explanation of these aspects of the fission process and we shall show, in particular, what the various consequences are for the fission cross sections. An extensive list of references would be needed for the various topics presented here. Rather, we prefer to mention a few review papers [1,2,3,4,5] which contain the detailed relevant references.

FISSION BARRIER CALCULATIONS USING THE STRUTINSKY PRESCRIPTION

Improvements towards a better estimate of the fission barrier were first carried out by Myers and Swiatecki, taking into account the actual mass of the nucleus in its ground state. But the correction then applied to the LDM barrier was restricted to spherical or moderately deformed nuclei, for it was assumed, at that time, that the residual interactions would wash out shell structure effects at large deformations. In fact, as pointed out by Strutinsky, there is no reason to restrict the existence of shells to spherical or moderately deformed nuclei. The occurrence of shells in nuclei or, in other words, non-homogeneity in the single-particle level densities, is a consequence of finite nuclear size. In that sense, shells can also appear at large deformations and they are obtained in all calculations carried out with realistic potentials of arbitrary shape. The Strutinsky prescription assumes that the potential energy of any given nucleus as a function of deformation is given, <u>on the average</u>, by the LDM, and that the effect of shell structure is adequately reproduced by adding shell-energy corrections to the LDM energy. The total energy $E(\{s\})$ of a nucleus, at deformation $\{s\}$, can then be written as follows :

$$E_{(\{s\})} = E_{LDM}(\{s\}) + \sum_{N,P} \Delta E_{sh}(\{s\}), \quad (1)$$

where $E_{LDM}(\{s\})$ is the total energy at deformation $\{s\}$ as given by the liquid drop model ; $\Delta E_{sh}(\{s\})$ is the shell energy correction at deformation $\{s\}$ and $\sum_{N,P}$ simply means that the shell energy corrections must be applied for both neutrons and protons.

The shell-energy correction term $\Delta E_{sh}(\{s\})$, which is due to non-homogeneity in the single-particle energy level density, can be written as the difference between two terms :

$$\Delta E_{sh}(\{s\}) = 2\int_{-\infty}^{\lambda_0} E\,\mathring{g}(E,\{s\})\,dE - 2\int_{-\infty}^{\lambda} E\,\tilde{g}(E,\{s\})\,dE. \quad (2)$$

The first term is just the sum of the single-particle energies $E_{i(\{s\})}$ up to the Fermi energy λ_o. The factor 2, in front of the integral reflects the fact there are two nucleons (protons or neutrons) in every filled orbit. The density $\mathring{g}$ defined below :

$$\mathring{g}(E,\{s\}) = \sum_i \delta(E - E_i(\{s\})), \quad (3)$$

is the actual level density, as given by the shell model, which corresponds to a discrete set of levels i having energy $E_i(\{s\})$ at deformation $\{s\}$.

The second term in eq. (2) is very similar to the first one and gives the energy which would be obtained with a uniform density of levels $\tilde{g}$ (E, {s}) equal, on the average, to that of the nucleus under investigation. The density $\tilde{g}$ (E, {s}) is defined as follows :

$$\tilde{g}(E,\{s\}) = \mathring{g}(E,\{s\}) * \tilde{f}(E) \tag{4}$$

where the true level density $\mathring{g}$ (E, {s}) is folded into a suitably chosen weighting function $\tilde{f}(E)$.

Several types of potentials can be used to calculate the shell-energy corrections. One-centre potentials (Nilson, Saxon-Woods ...) were used exclusively in the early calculations of that type. Such potentials including high order odd and even deformation parameters are still widely employed for they prove to be valid for quite a wide range of deformations. For very large deformations, up to the scission point, double-centre potentials seem to be more realistic. Nevertheless, though the validity of the Strutinsky prescription is not formally established as yet, it is to date the only reliable method for calculating fission barriers in agreement with experimental results.

The shell-energy correction exhibits, as a function of deformation, an oscillatory behaviour which is illustrated in Fig.6-1. The effect of the shell-energy corrections on the shape of the fission barrier is of particular interest for actinide nuclei having neutron number in the vicinity of N = 146. For such nuclei, there is a negative shell-energy correction, at moderate deformation, which leads to a first minimum of the fission barrier corresponding to the ground state of the nucleus. In addition to this, there is also a second strong negative shell-energy correction at a deformation comparable to that of the LDM saddle-point. This results in a second minimum (see Fig.6-1) which plays an important role in the fission process at low energy. Even a third minimum is possible at very large deformations, near scission ; but it is not well pronounced and therefore is not very important for the fission mechanism.

The double-humped fission barrier provides the possibility for two types of states, called class I and class II, to exist with deformations comparable to those of the 1st and 2nd well respectively. The existence of class II states has important consequences for the understanding of several aspects of the fission process. We shall now examine and discuss these various aspects of fission.

SOME ASPECTS OF THE FISSION PROCESS FOR NUCLEI HAVING A DOUBLE-HUMPED FISSION BARRIER

Fission isomers

The double-humped barrier can explain immediately the existence of fission isomers. In the nuclear reactions where fission isomers have been observed, a compound nucleus is formed with an excitation which is usually above the two barriers. This compound nucleus then decays by neutron, proton or γ-ray emission. During the process of deexcitation, the nucleus can be caught in the second well of the double-humped fission barrier where subsequent γ-ray emission can eventually lead to the ground state or to one of the first excited states. These states in the second well are the fission isomers which have been observed with half-lives ranging from nanoseconds to seconds and for nuclei from uranium to berkelium. The γ-ray decay of these fission isomers to states lower in energy in the first well is strongly inhibited by the small penetrability of the inner barrier. On the other hand, the fission rate of decay is much faster for the fission isomers than for the ground state (in the first well). This is because the isomeric state is at a higher excitation energy and also because fission from this state occurs by tunneling through the outer barrier only instead of the whole barrier as in the case of the ground state.

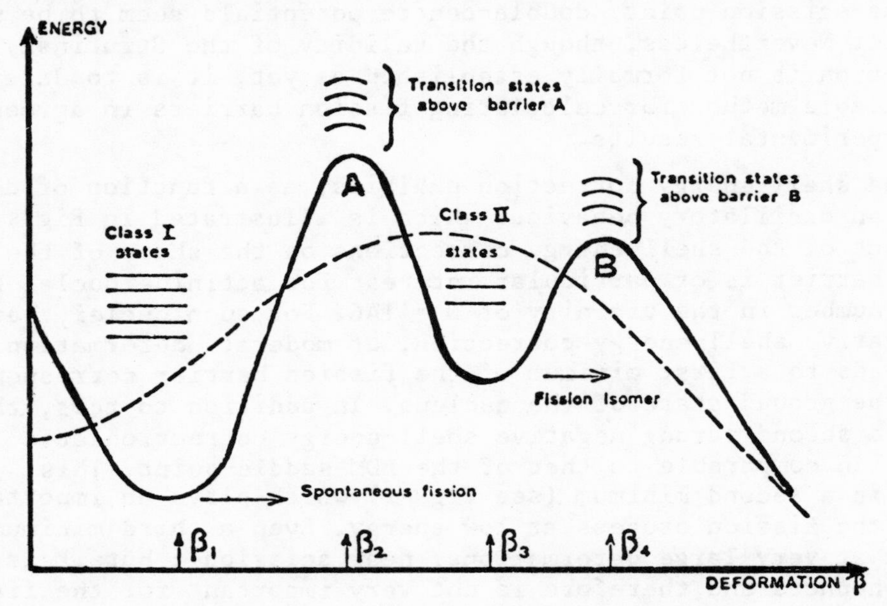

Fig. 6-1 - *Fission barrier (solid line) resulting from shell corrections to the LDM barrier (dashed line). The abscissa β gives only an indication of the magnitude of the deformation but does not specify the type of deformation. For certain classes of nuclei, for instance those having neutron number N in the vicinity of 146, the fission barrier presents two humps A and B, at deformations β_2 and β_4 respectively where the shell-energy corrections are positive. These two humps can be separated by a deep second well, at deformation β_3, where the shell-energy correction is negative. This kind of barrier shape has important consequences for the understanding of the fission process.*

Gross structure in some cross sections for near-threshold fission processes

The energy dependence of the cross sections for near-threshold fission processes is closely connected to the variation of the fission barrier penetrability with energy. For a single-humped parabolic barrier, such an energy variation is smooth. This is not the case for double-humped barriers for which the penetrability presents a sharp maximum each time the energy corresponds to that of a class-II vibrational level (in the second well) as is illustrated in Fig.6-2.

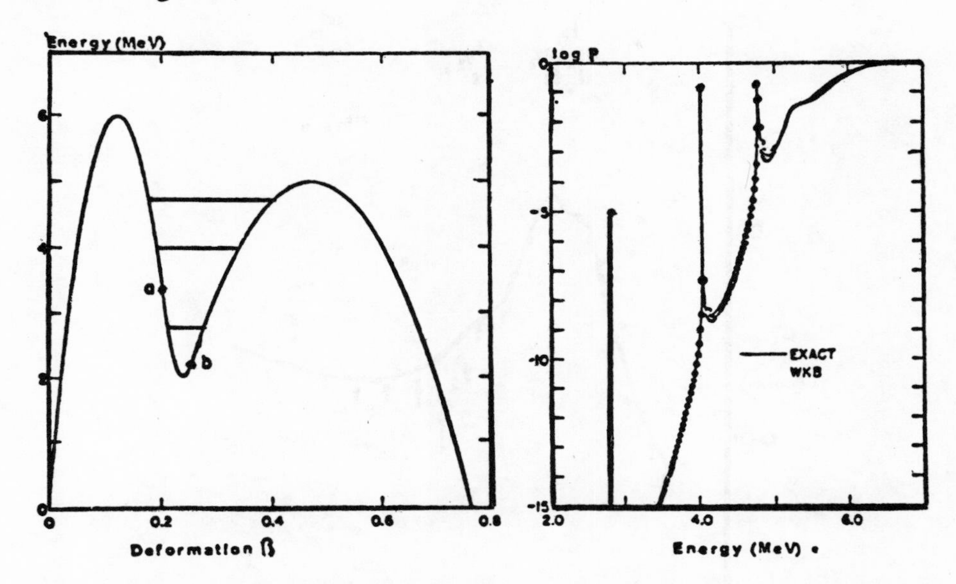

Fig. 6-2 - *Energy dependence of the penetrability P of a double-humped fission barrier. The asymmetric fission barrier is shown on the left and consists in three parabolas smoothly joined at connecting points a and b. The positions of the vibrational levels in the well are indicated by horizontal lines. The penetrability of this barrier is plotted on the right on a semilog scale. Results obtained from exact and WKB calculations are indicated. Note the sharp peaks in the penetrability each time the energy corresponds to that of a vibrational level.*

This behaviour of the penetrability can be responsible for gross structures in the nearthreshold (n,f) cross sections, in addition to other less pronounced structures due to the competition between the contributions of neutron and fission channels. One of the most striking examples of this phenomenon is the existence of a sharp peak observed at 720 keV in the ^{230}Th(n,f) cross section (Fig. 6-3).

Such a narrow peak cannot be interpreted in terms of the competition theory and is interpreted as being due to the first vibrational level in the shallow second well of the ^{231}Th fission barrier. Other examples can also be found in (d,pf) cross sections. One of them, the ^{239}Pu (d,pf) cross section, is plotted in Fig.6-4.There is a fairly large peak, at 5 MeV, below the neutron threshold, which is interpreted as being due to a class II vibrational level. But, in contrast to the ^{230}Th case, the second well is deeper for ^{239}Pu and the vibrational level responsible for the wide peak at 5 MeV is damped into the intrinsic states situated in the neighbourhood of 5 MeV. The peaks corresponding to these states are partially resolved, as can be seen in Fig. 6-4, due to the excellent resolution of the measurements.

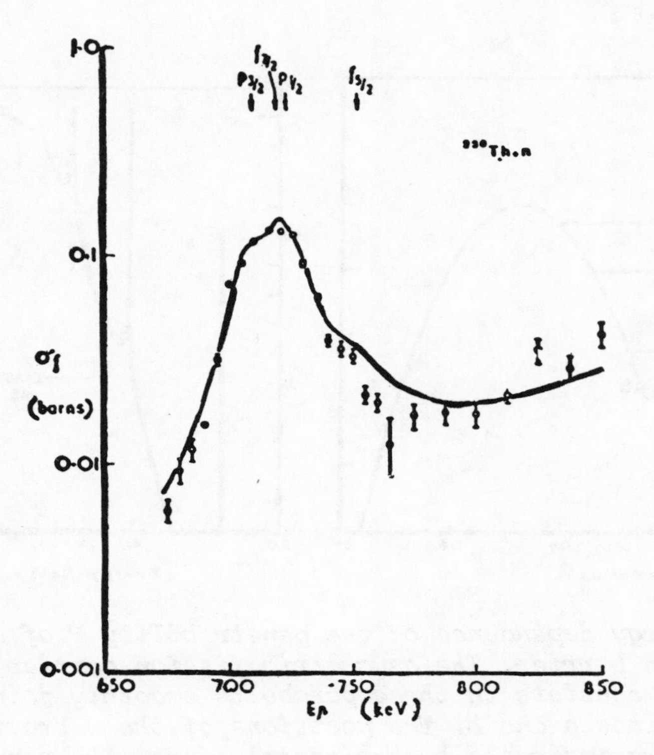

Fig. 6-3 - *The Harwell fission cross section of ^{230}Th measured between 600 keV and 850 keV with an energy resolution of 5 keV. The resonance at 715 keV is attributed to a vibrational level in the second well of the double-humped fission barrier and the solid line is a theoretical fit based on the rotational band structure indicated on the figure.*

Intermediate structure in sub-barrier fission cross sections

The double-humped fission barrier also provides the possibility for class II compound nucleus states to exist with deformations comparable to that of the second well. These states have properties similar to those of class-I states, but with a few differences due to their location in the second well of the barrier. For example, the spacing of class-II states is larger than for class-I states since, for a given total energy of the system the available intrinsic excitation energy of the levels is lower in the second well than in the first well. Also the fission width of a class-II state is larger since only the outer barrier has to be penetrated instead of the whole barrier for a class-I state. On the other hand, the neutron width of a class-II state is essentially zero because there is no overlap between the vibrational part of a class-II wave function and that of the target nucleus in its ground state coupled to an outgoing neutron wave.

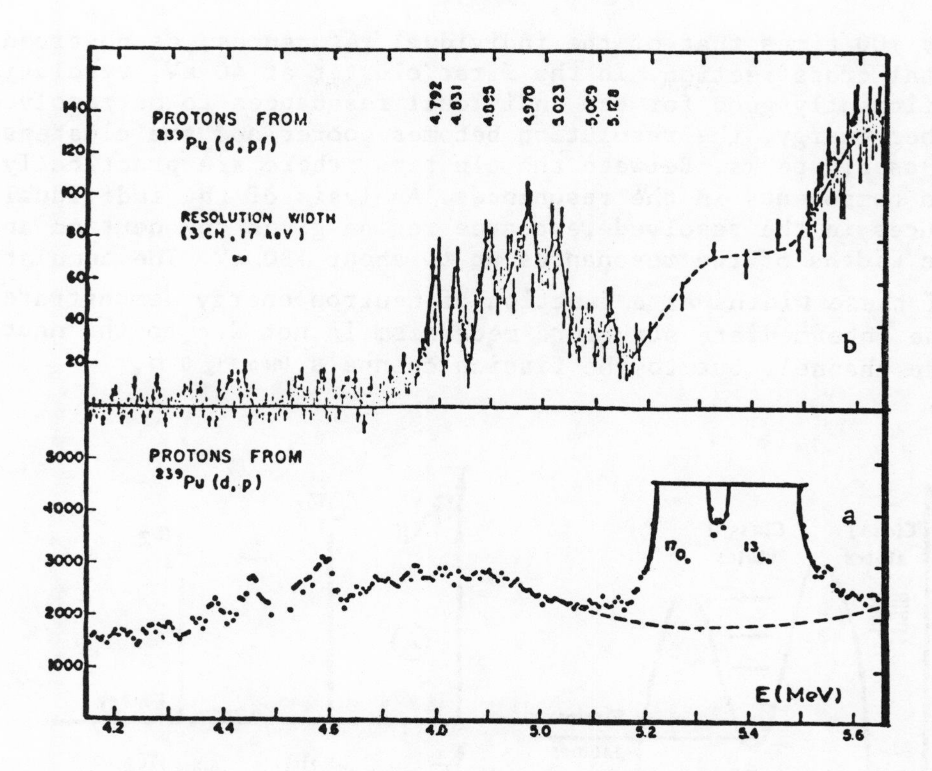

Fig.6-4 - *Direct experimental results about the* ^{239}*Pu (d,p) and* ^{239}*Pu (d,pf) reactions for excitation energies between 4.2 and 5.6. MeV. Diagrams a) and b) show the spectra of the proton singles and the protons in coincidence with fission, respectively. Structure appears clearly in the broad peak at 5 MeV (diagram b)). Error bars give statistical standard deviations.*

The existence of class-II states which have a large fission width and a large spacing is responsible for the phenomenon of intermediate structure in subthreshold fission cross sections. Below the fission threshold, i.e. for excitation energies below the top of the barrier, the inner barrier prevents strong mixing between class-I and class-II states. When the energy, spin and parity of class-I states match those of a class-II state, the coupling between the class-I and class-II states greatly enhances the fission components in the class-I states. Therefore, clusters of large fission resonances appear in the fission cross sections around the energies of the class-II states. This intermediate structure mechanism is illustrated in Fig. 6-5 where it is shown that the spacing D_{II} of the clusters, much larger than that of the class-I states, is actually the spacing of the class-II states having the required spin and parity. Such cases of intermediate structure have been observed for several non fissile nuclei : ^{234}U, ^{237}Np, ^{238}Pu, ^{240}Pu, ^{244}Cm. An illustration of this effect is shown in Fig.6-6 where the Saclay fission cross section of ^{237}Np is plotted as a function of neutron energy in a limited energy range. Clusters of large fission resonances appear at definite energies, for example a 40 eV with a spacing which is roughly 100 times that of the individual resonances, as observed in the total cross section. In the first cluster at 40 eV, resolution is sufficiently good for the individual resonances to be resolved. At higher energy, the resolution becomes poorer and the clusters appear as big peaks. Between the clusters, there are practically no fission components in the resonances. Analysis of the individual resonances in the resolved resonance region gives the neutron and fission widths of the resonances up to about 130 eV. The cumulative sums of these widths as a function of neutron energy demonstrate that the intermediate structure mechanism is not due to the neutron entrance channel, but to the fission channels (see Fig. 6-7).

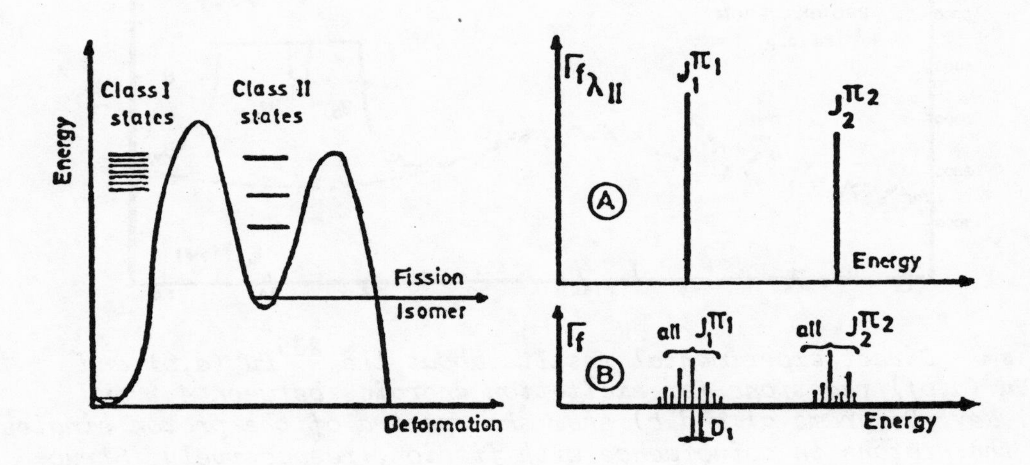

Fig. 6-5 - *Mechanism of intermediate structure in subthreshold fission cross sections. Clusters appear in the fission cross section when energy, spin and parity of a class-II state match those of the class-I resonances (at most two J^π values are possible). The fission widths are drawn at the energy of the respective levels for class-II states (diagram A) and for the observed resonances (diagram B).*

The type of coupling between class-I and class-II states depends on the relative heights of the inner and outer barriers. The weak coupling situation happens when the inner barrier is the higher. In that case, the width $\Gamma_{\lambda_{II}}$ of a cluster is the fission width $\Gamma_{f_{\lambda_{II}}}$ of the corresponding class-II state and the sum of the fission widths Γ_{f_λ} of the resonances in the cluster is the damping width of this class-II state.

$$\Gamma_{\lambda_{II}} = \Gamma_{f_{\lambda_{II}}} \quad , \tag{5}$$

$$\sum_\lambda \Gamma_{f_\lambda} = \Gamma_\downarrow \quad . \tag{6}$$

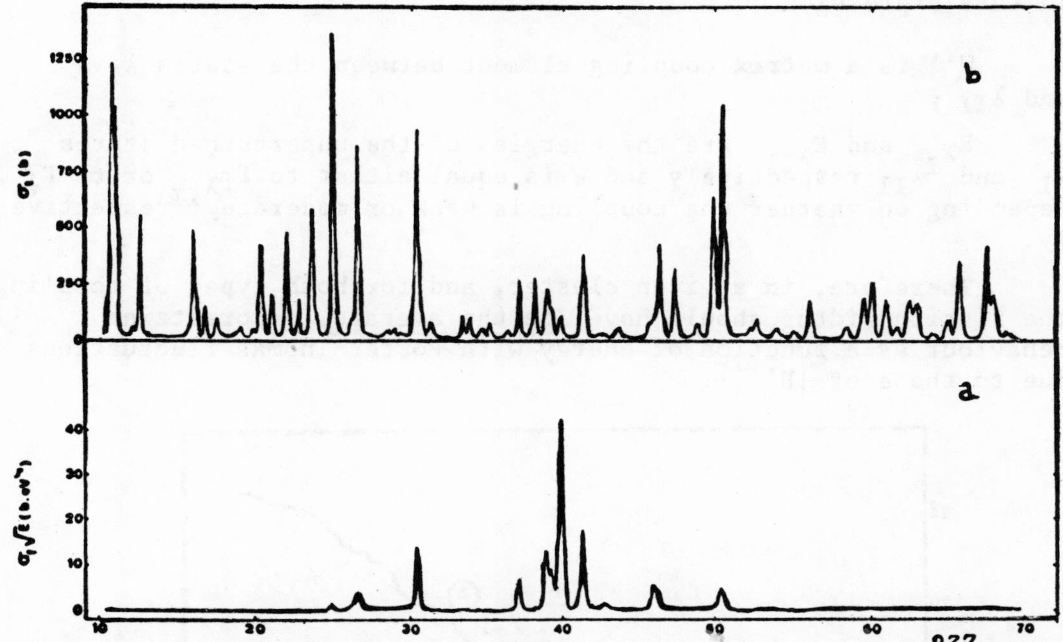

Fig.6-6 - *The Saclay total and fission cross sections of ^{237}Np are plotted as a function of neutron energy between 10 eV and 70 eV. A cluster of large fission resonances, about 10 eV wide, is clearly seen in the fission cross section at 40 eV, whereas practically no fission resonances are observed outside this fission cluster (curve a). This behaviour of the fission cross section is radically different from that of the total cross section (curve b) where resonances appear throughout the energy range, even in energy ranges where no fission resonances are observed.*

The moderate coupling situation happens when the outer barrier is the higher. The inner barrier, though lower in energy, is nevertheless high enough to prevent strong mixing between class-I and class-II states for such a mixing would wash out the intermediate structure effect. In that type of coupling, the following set of relations holds :

$$\Gamma_{\lambda_{II}} = \Gamma_{\dagger} , \quad (7)$$

$$\sum_{\lambda} \Gamma_{f_{\lambda}} = \Gamma_{f_{\lambda_{II}}} . \quad (8)$$

In a given cluster, the energy dependence of the fission widths $\Gamma_{f_{\lambda}}$ is given by the following relation :

$$\Gamma_{f_{\lambda}} = \Gamma_{f_{\lambda_{II}}} \frac{|H''|^2}{(E_{\lambda_I} - E_{\lambda_{II}})^2 + \frac{1}{4} W^2} . \quad (9)$$

In this expression :

H'' is a matrix coupling element between the states λ_I and λ_{II} ;

E_{λ_I} and $E_{\lambda_{II}}$ are the energies of the unperturbed states λ_I and λ_{II} respectively and W is equal either to $\Gamma_{f\lambda_{II}}$ or to $\Gamma_{\downarrow}$, depending on whether the coupling is weak or moderate respectively.

Therefore, in a given cluster, and for both types of coupling, the fission widths should have, on the average, a Lorentzian behaviour as a function of energy with Porter-Thomas fluctuations due to those of $|H''|^2$.

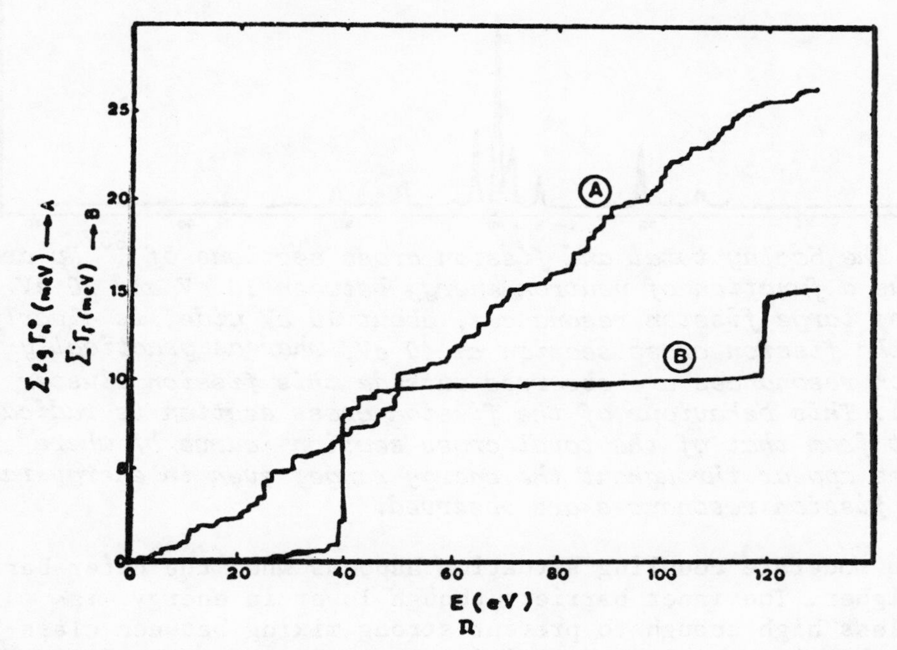

Fig. 6-7 - *The cumulative sums of the reduced neutron* ($2g\ \Gamma_n^0$) *and fission* (Γ_f) *widths, for the* ^{237}Np *resonances which could be analysed at low energy, are plotted as a function of neutron energy* E_n. *Histograms A and B correspond to reduced neutron widths and fission widths respectively. This picture demonstrates that the intermediate structure is not due to the neutron entrance channel but, to the fission exit channels.*

This seems to be the case for the ^{234}U resonances analysed in the first cluster at 700 eV (see Fig. 6-8) where the moderate coupling situation seems to apply according to other measurements (angular distribution of the fission fragments, for example). This seems to be also the case for the weak coupling situation for the ^{237}Np resonances situated in the first cluster at 40 eV (Fig. 6-9). But for ^{237}Np, the energy dependence of the fission widths is more complicated than for ^{234}U. This is because ^{237}Np is not an even-even nucleus ; therefore the "s" - wave ^{237}Np resonances have spin and parity $J^\pi = (I + 1/2)^\pi = 3^+$ or $J^\pi = (I - 1/2)^\pi = 2^+$ instead of $J^\pi = 1/2^+$ for those of ^{234}U. In the first cluster at 40 eV, supposed to be caused by one single class-II state, all the resonances enhanced in fission should have the same quantum numbers $J^\pi = 2^+$ or 3^+, those of this class-II state. The resonances having other J^π values should not be enhanced in fission and should have very small fission widths. These considerations explain the aspect of Fig. 6-9. The largest widths seem actually to be distributed, on the average, along a Lorentzian line with Porter-Thomas fluctuations. According to the intermediate structure mechanism described above, all the resonances corresponding to this category of fission widths should have the same J^π values (probably $(J,K)^\pi = (2,2)^+$, as determined from other measurements). The other resonances have small fission widths, with fluctuations, but distributed in a uniform manner as a function of neutron energy, and indeed so small than they can hardly be detected.

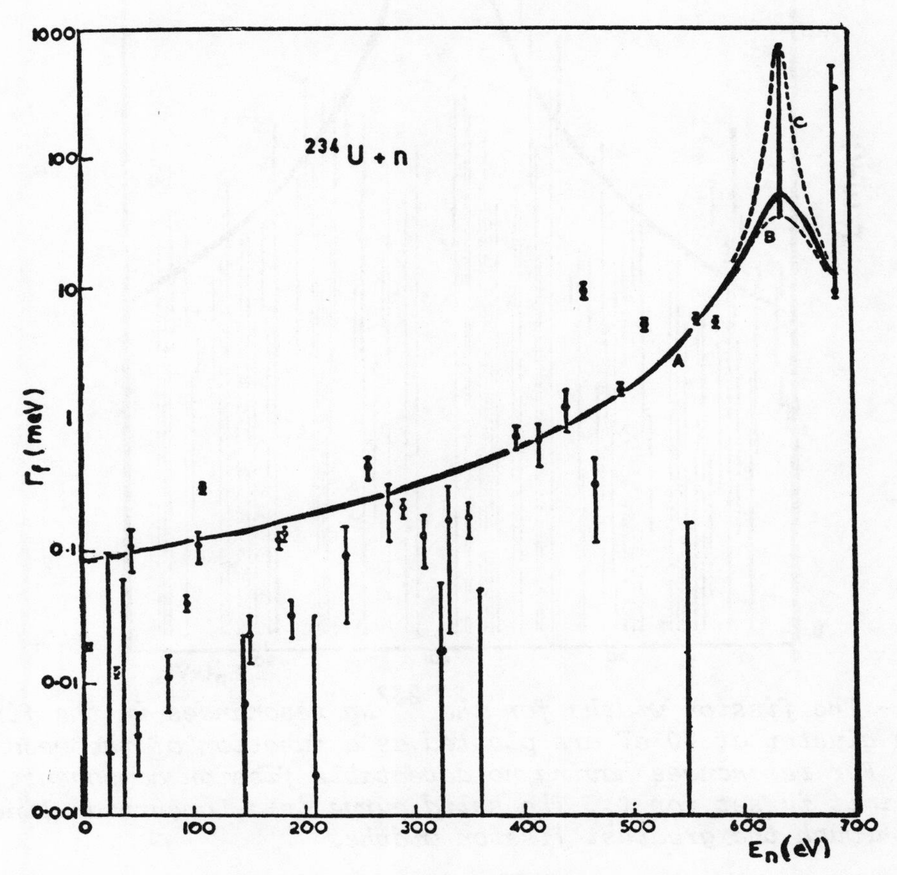

Fig. 6-8 - *The fission widths for the ^{234}U resonances in the first fission cluster at 700 eV are plotted as a function of the resonance energy. Curves A, B and C are Lorentzian lines. Curve A goes through the measured fission width at 638.4 eV whereas curves B and C go through points at one standard deviation from this datum.*

It is interesting to note that the intermediate structure phenomenon can explain the small thermal neutron fission cross sections measured for non fissile nuclei. These values are small just because, for thermal neutrons, the energy corresponds to a minimum in the intermediate structure.

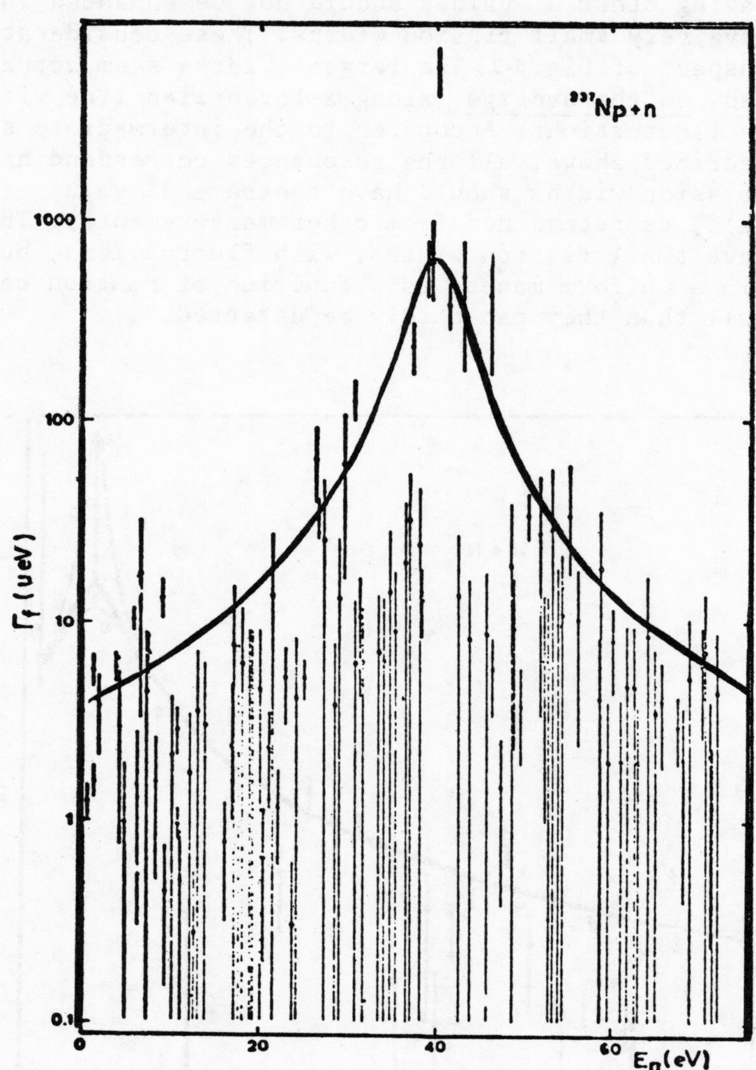

Fig.6-9 - *The fission widths for the* ^{237}*Np resonances in the first fission cluster at 40 eV are plotted as a function of the resonance energy. For resonances having no detectable fission component, an upper limit is set for* Γ_f. *The solid curve is a Lorentzian line drawn through the greatest fission widths.*

Intermediate structure in the fission cross sections of fissile nuclei

For fissile nuclei, the fission barrier can also present two humps but with heights below S_n , the neutron separation energy in the compound nucleus. This fission barrier corresponds to the deformation energy of the nucleus in its ground state. But according to the channel theory of Bohr, there is in fact one fission barrier for each of the fission exit channels i. For some of these channels, called j, fission can therefore occur below threshold and consequently, the corresponding cross section may present an intermediate structure effect. The total fission cross section σ_f can then be written :

$$\sigma_f = \sigma_f^{i.s.} + \sigma_f^N, \tag{10}$$

$$\sigma_f^{i.s} = \sum_j \sigma_f^j, \tag{11}$$

$$\sigma_f^N = \sum_{k \neq j} \sigma_f^k \tag{12}$$

In these expressions, $\sigma_f^{i.s.}$ corresponds to that part of the cross section where there can be intermediate structure and σ_f^N corresponds to a pure noise (without intermediate structure). Such an effect is illustrated in Fig. 6-10 .

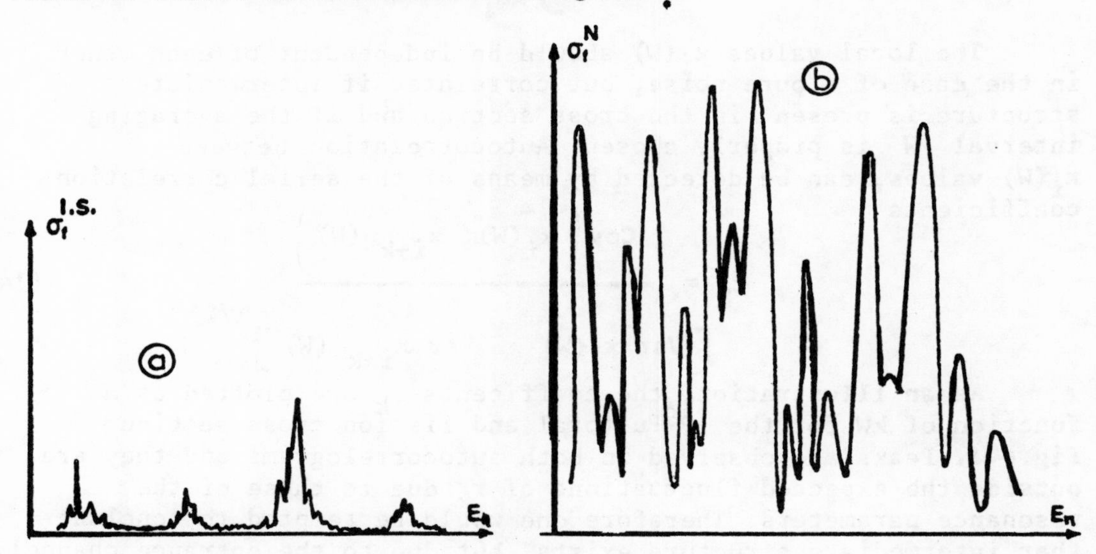

Fig. 6-10 - *Illustrative plots of two types of cross sections for fission through transition states having different excitation energies relative to the neutron separation energy S_n. In diagram a), the transition state is high enough in energy above S_n for the corresponding cross section $\sigma_f^{I.S.}$ to be small and to present an intermediate structure effect. This is in constrast to diagram b) for which the transition state is below S_n and where the corresponding cross section σ_f^N is large and is a pure noise, without intermediate structure.*

In general, for a fissile nucleus, it is very difficult to separate that part $\sigma_f^{i.s.}$ which presents an intermediate structure effect from the large fluctuations of the other part σ_f^N which tends to dominate since it corresponds to fission channels lower in energy. The intermediate structure effect will be greater and its detection less difficult if $\sigma_f^{i.s.}$ represents a significant fraction of σ_f. In this respect, the ^{239}Pu fission cross section is composed of $J^\pi = 0^+$ and $J^\pi = 1^+$ resonances ; but, in the 240 Pu compound nucleus, the main $J^\pi = 1^+$ fission exit channel is about 200 keV above S_n, therefore making it possible for intermediate structure to exist in the $J^\pi = 1^+$ fission cross section.

Several methods have been employed to detect the existence of intermediate structure in the presence of noise.

Autocorrelation techniques

These techniques, as developed first by Egelstaff, search for correlation between local average values $x_i(W)$ of the cross section $\sigma_f \sqrt{E}$.

$$x_i(W) = \frac{1}{W} \int_{E_1 + i W}^{E_1 + (i+1) W} \sigma_f \sqrt{E} \; dE . \tag{13}$$

The local values $x_i(W)$ should be independent of each other in the case of a pure noise, but correlated if intermediate structure is present in the cross section and if the averaging interval W is properly chosen. Autocorrelation between $x_i(W)$ values, can be detected by means of the serial correlation coefficients :

$$r_k = \frac{\mathrm{Cov}\left(x_i(W), \; x_{i+k}(W) \right)}{\left[\mathrm{Var}\, x_i(W) \quad \mathrm{Var}\, x_{i+k}(W) \right]^{1/2}} \tag{14}$$

As an illustration, the coefficents r_k are plotted as a function of kW for the ^{239}Pu total and fission cross sections in Fig.6-11. Peaks are observed in both autocorrelograms and they are outside the expected fluctuations of r_k due to those of the resonance parameters. Therefore one would be tempted to conclude that intermediate structure exists, but due to the entrance channel. In fact, as pointed out by Perez, the interpretation of such autocorrelograms is very delicate. This is so because the intermediate structure is due to class-II states which are not regularly spaced ; therefore the autocorrelogram should not present a periodical behaviour. Also, the sample of $x_i(W)$ values is of limited size and end effects can be the source, in the autocorrelograms, of spurious peaks not due to the effect of intermediate structure. Therefore this technique is difficult to use in its present stage of development.

Variance tests

This is perhaps the simplest statistical test since it consists in measuring the variations of the $x_i(W)$ values. More exactly, one calculates

$$H(W) = \frac{\text{Var } x_i(W)}{W. < x_i(W) >^2} . \tag{15}$$

According to Egelstaff, H(W) should take a constant value, independent of W, for a pure noise. In the case of ^{239}Pu, this constant value is calculated to be to about 6 ; but the H(W) values deduced from the measured cross section are far from being constant as can be seen in Fig.6-12. In fact, as pointed out by Perez, the behaviour of H(W) is much more similar to that obtained from a simulated cross section with built-in intermediate structure.

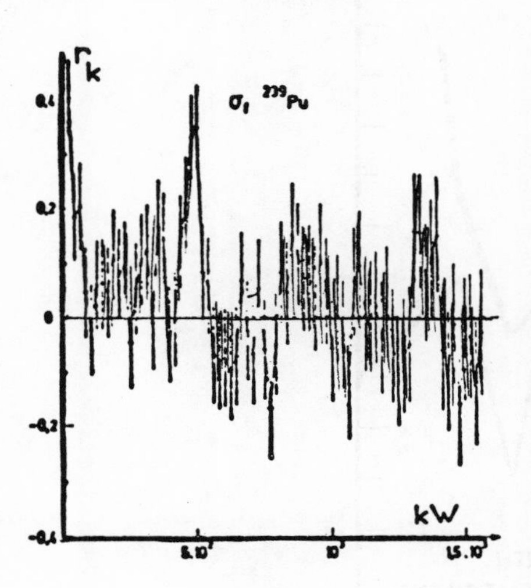

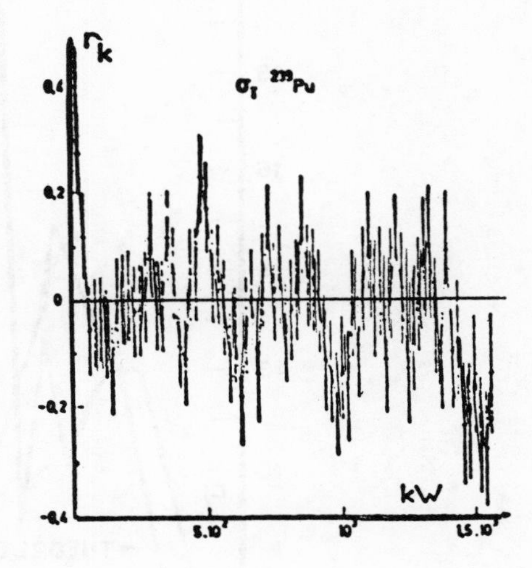

Fig.6-11 - *Comparison of the autocorrelograms $r_k = f(kW)$ for the Saclay total and fission cross sections of ^{239}Pu between 20 eV and 3020 eV. The averaging energy interval W for the two correlograms is equal to 20 eV.*

Distribution tests

These tests also reflect the variation properties of the local values of the cross section. But, rather than referring to the variance, as in b), the test is made on the distribution function of the local values $x_i(W)$. For a cross section presenting an intermediate structure effect, the fluctuations of $x_i(W)$ should be greater and therefore the distribution function should be wider than for a pure noise.

The distribution function for ^{239}Pu is not available yet, but preliminary calculations show that, for the measured cross sections, it is wider than expected from a pure noise.

Microscopic analysis of the cross sections

This method, though very complete, cannot be used over a wide energy range, due to the rapid deterioration of the experimental resolution as a function of neutron energy. Moreover, one has to separate the various channel components entering into the total fission width Γ_f. As in eq. (10), one can write :

$$\Gamma_f^J = \left(\Gamma_f^{i.s.}\right)_J + \left(\Gamma_f^N\right)_J \qquad (16)$$

This expression is valid for both spin states.

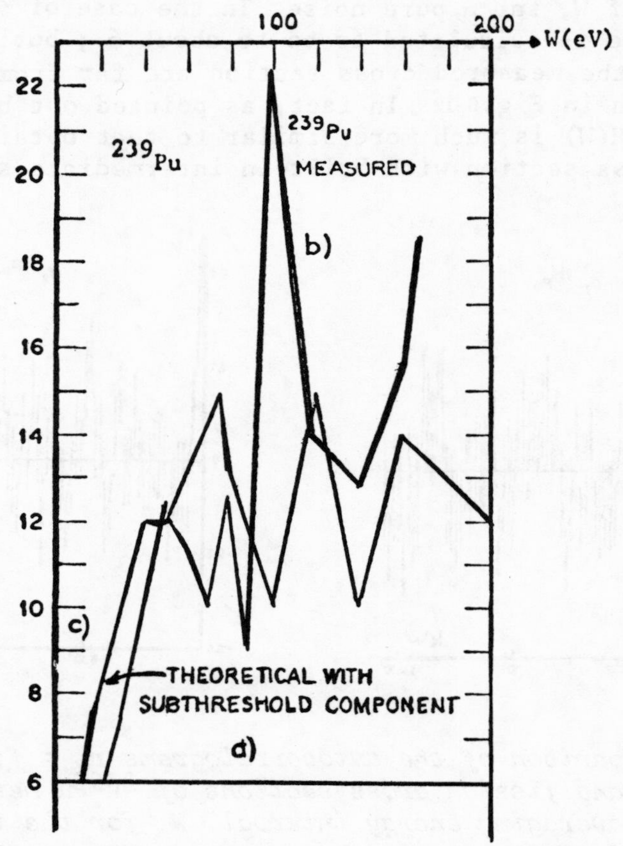

Fig. 6-12 - *Illustration of variance tests for detecting the presence of intermediate structure in fission cross sections. The ratio of the variance of cross section values averaged over several W intervals to the squared overall average value of the cross section is plotted as a function of the averaging interval W. The plots are given for :*

a) *the values obtained from the Egelstaff formula, using the known properties of the measured parameters for* ^{239}Pu *resonances;*

b) *the measured fission cross section of* ^{239}Pu*;*

c) *artificial fission cross sections of* ^{239}Pu *calculated with the Monte-Carlo technique. The cross section of* ^{239}Pu *contains an intermediate structure component.*

The width Γ_f^N should have fluctuations, but should remain constant, on the average, as a function of neutron energy, in contrast to the width $\Gamma_f^{l.s.}$ which, besides fluctuations, should exhibit a Lorentzian behaviour for energies near those of class-II states. Detection of this effect seems favorable for spin 1^+ resonances of ^{239}Pu, since most of the fission contribution comes from one exit channel which is partially closed. The 1^+ contribution to the fission cross section may thus present an effect of intermediate structure, which can be checked by the determination of the spins and the fission widths of the resonances. At present, these parameters have been measured up to 660 eV and the fission width $<\Gamma_f>_{1+}$, averaged over resonances in 110 eV intervals, is plotted in Fig. 6-13 as a function of the mean energy of the interval. The energy dependence of $<\Gamma_f>_{1+}$ shows rather extensive variations which are incompatible with a conventional statistical model. For example, the value of $<\Gamma_f>_{1+}$ between 550 eV and 660 eV is very small, as confirms the comparison of the fission width distributions for the intervals 550 eV - 660 eV and 0 - 660 eV. These energy variations of $<\Gamma_f>_{1+}$ are interpreted as being due to the coupling of the 1^+ resonances to broad 1^+ class-II doorway states ; one of them may have an energy of about 400 eV, where $<\Gamma_f>_{1+}$ takes the highest value.

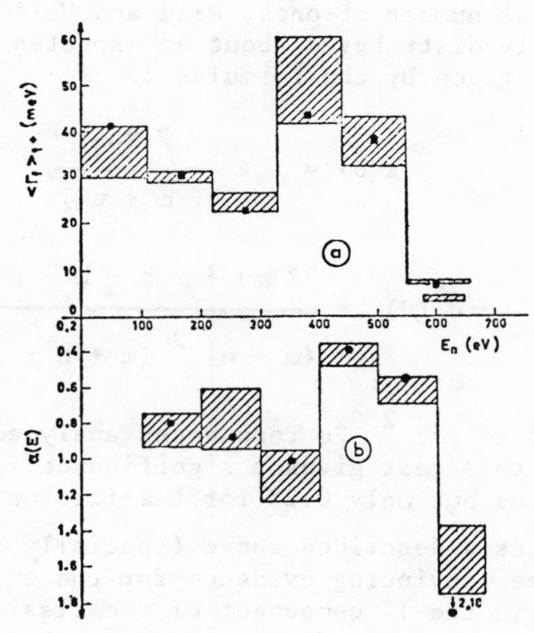

Fig. 6-13 - *Diagram a) shows the variation of the average value* $<\Gamma_f>_{1+}$ *of the fission widths of the* ^{239}Pu *neutron resonances having spin and parity* $J^\pi = 1^+$ *and situated in a given 110 eV neutron energy interval, as a function of the energy* E_n *of this interval. Hatched areas are between upper and lower limits of* $<\Gamma_f>_{1+}$ *as obtained directly from spin measurements. The black squares correspond to the* $<\Gamma_f>_{1+}$ *values. Note the extensive variations of* $<\Gamma_f>_{1+}$*, incompatible with those of conventional statistical models.*

- Diagram b) shows the variation of the ratio α of the capture to the fission cross sections of ^{239}Pu as a function of neutron energy. The hatched areas correspond to the direct measurements of α. Black circles correspond to values calculated from the resonances parameters. Note the correlation between the variations of α and those of $\langle\Gamma_f\rangle_{1+}$.

Use of the Wald and Wolfowitz distribution-free statistics

The application of these statistics to the detection of intermediate structure effects was suggested by James and is based on consideration of runs in a sequence. Let us consider, for example, a series of widths for the resonances situated in a given energy interval. These widths can be divided into two groups depending on whether they are greater or smaller than their mean value. These two groups can be marked as zeros and ones. After writing down the sequence of zeros and ones for the widths taken in the order of resonance energies, the Wald and Wolfowitz statistics U are found by counting the number of runs in the sequence. Here, a run is an unbroken series of zeros or ones. If m is the total number of zeros and n is the total number of ones, Wald and Wolfowitz have shown that U is normally distributed about an expected value $\varepsilon(U)$ with a variance $\sigma^2(U)$ given by the formulas :

$$\varepsilon(U) = 1 + \frac{2\,mn}{(m+n)}, \tag{17}$$

$$\sigma^2(U) = \frac{2\,mn\,(2mn - m - n)}{(m+n)^2\,(m+n-1)} \tag{18}$$

If applied to the ^{239}Pu resonances analysed below 660 eV, having $J^\pi = 1^+$, this test gives a significance level of 71.6% for the neutron widths but only 0.2% for the fission widths.

From the tests described above (specially tests d) and a)) there seems to be convincing evidence for the existence of intermediate structure in the 1^+ component of the fission cross section. This effect is quite important for reactor calculations, since it can explain the large fluctuation of α, the ratio of the capture to the fission cross section of ^{239}Pu. In particular, the high α value at 600 eV is due to the low $\langle\Gamma_f\rangle_{1+}$ value at this energy.

Other cases of intermediate structure probably exist for other fissile nuclei, such as ^{235}U ; but their detection, not possible at present by a detailed analysis of the microscopic cross sections, would require the use of autocorrelation techniques which need to be further developed.

SUMMARY

A new method for more accurate calculations of the fission barriers, taking into account the effect of shells at all deformations, has provided a better understanding of the fission process. Puzzling experimental results, such as fission isomers and intermediate structure, that remained previously unexplained with conventional models, could then be interpreted in terms of double-humped barrier shapes. This is quite important for the understanding and for the predictions of the fission cross sections. Near threshold, vibrational fission resonances can appear ; below threshold, intermediate structure can exist and influence the fission cross section for thermal neutrons ; the intermediate structure effect can also be present, though weaker. in the fission cross section of fissile nuclei.

REFERENCES

1) M.BRACK, Jens DAMGAARD, A.S. JENSEN, H.C. PAULI, V.M.STRUTINSKY and C.Y. WONG, Revs. Mod. Phys. 44, (1972), 320.

2) J.E. LYNN, Nuclear structure (I.A.E.A. Vienna, 1968)
Physics and Chemistry of Fission (I.A.E.A. Vienna,1969)

3) A. MICHAUDON,-Proceeding of the International Conference on Statistical Properties of Nuclei (Albany, August 1971),
-Advances in Nuclear Physics, Vol. 6 (in press).

4) S. BJØRNHOLM, Proceedings of the European Conference on Nuclear Physics, Aix-en-Provence (June 26 - July 1, 1972).

5) K. DIETRICH, Proceedings of the European Conference on Nuclear Physics, Aix-en-Provence (June 26 - July 1, 1972).

CHAPTER 7

FAST BREEDER REACTOR

As is true of reactor physics in general, we can look at fast breeder reactor physics from two complementary viewpoints. In one, we can look at the field from the point of view of the physicist associated with reactor design and operation. In the design phase he has to calculate such things as:

Critical size and composition,
Power distributions,
Reactivity coefficients,
Control rod worths,
Long term changes under burnup in reactivity, composition, and other reactor parameters,
and so forth.

His evaluation of the field is based strictly on consideration of how efficient ly, accurately, and reliably he can obtain his results.

Alternately, we may view the field by looking at the status of the technol ogy areas that have to be used in solving the various design problems and performing the required calculations. The technology areas include such items as:

Critical experiment facility, materials, and measurement technique capability,
Cross section data, including evaluated data files,
Computer codes related to such areas as cross section processing, multidimensional neutron diffusion, kinetics, fuel cycle, etc.

Although there are a very large number of types of design calculations, they are performed using a rather small set of calculational tools. Most of ou physics activities are thus aimed at developing and validating this set of tool This involves, among other things, establishing the data base by measurement and evaluation of the cross section data, perhaps with adjustment on the basis of integral data. Also required are the development of sufficiently powerful computer codes, commensurate with computer technology, validated both by numerical experiments and by comparison of experimental and calculated results to assure that the physical and mathematical approximations as well as the data base are sufficiently accurate. The appropriate experimental techniques must also be developed and the measurements on the appropriate integral systems performed.

It is evident that for any given design problem, a wide variety of technology areas or tools are required, and similarly, a reactor physics tool, onc developed, may generally be used for a variety of design problems. Because of this non-one to one correspondence between design problems and technology area it is often less than obvious what the impact of some improvement in reactor physics technology will be on actual fast reactors.

Among the design areas requiring physics studies are steady state performance, reactivity coefficients, control rods, fuel cycle, neutron irradiation, gamma heat deposition, shielding, kinetics, and accident analysis.

Interspersed with the more detailed look which we will take at some of these design areas, it may also be useful to take a similar brief look at the status of the major technology areas. These are primarily related to the three categories:

1. integral data - primarily from critical experiments - experiment and analysis
2. differential data - measurement, theory, and evaluation
3. theoretical methods - primarily implemented through computer codes.

Essential Features in Current Fast Reactor Design

In order to provide a framework for consideration of fast reactor physics problems, let us describe in a qualitative way the representative relevant features of a typical fast breeder reactor of current design. It does not correspond to any particular design. Arrangements are now in the formulative stage for a U.S. fast breeder demonstration reactor in the 300-500 MWe class. A reactor designer has not yet been selected. Various studies performed over the past few years, however, for both demonstration and full sized commercial systems and also the design of several European demonstration reactors all point to a set of reasonably well defined and generally accepted design features.

For a demonstration reactor sized system (∿400 MWe), the core would have a height and radius of ∿3 ft with axial and radial blankets ∿1-2 ft thick. The core and axial blankets would be made up of hexagonal subassemblies (∿4-6 in. across) containing bundles of fuel pins ∿1/4 in. diameter. The subassemblies comprising the radial blanket would have larger diameter pins. The subassembly boxes may have some space between them to allow for structural material swelling. The core has two enrichment zones. The zone volumes are approximately equal with an enrichment ratio of outer to inner zone of approximately 1.5, leading to a value of ∿1 for the power peak ratio of the center to inner part of the outer core.

Control, both for safety and shim, is achieved by the movement of poison rods (generally natural or enriched B_4C, or tantalum). The (∿10 to ∿40) control rods occupy normal subassembly positions and are arranged in a more or less regular configuration, perhaps located in several rings. The core volume fractions would typically be in the range of 0.35 fuel and fertile material, 0.45 sodium, and 0.20 structural and clad steel.

A typical core residence time might be several years, with partial reloading occurring on a six month or annual basis. The amount of shim control for burnup might be of the order of 3-4%, although this could vary considerably from design to design (depending very much on the core breeding ratio).

The amount of reactivity required for safety shutdown might be of the order of ∿5%. 2% or so will be required to go from cold-to-full-power operating conditions.

The main reactivity coefficients are those of Doppler and sodium void. They are primarily important in accident considerations. Other important reactivity coefficients are related to structural expansion, small motions related to core clamping, and possible bowing effects. Although the description is of a demonstration reactor and, except for size, also a full scale commercial power breeder (which would have a core diameter of ∿9 or 10 ft), it may also serve to catalogue most of the essential features of the FTR (except for the nickel reflector replacing the breeder uranium blanket).

Status of ZPR Facilities, Material, and Measurement Capability

The single most important observation one can make in considering the status of fast reactor physics is that generally there is a lack of large scale integral data. Only now are we beginning to obtain integral experimental data on the large dilute plutonium systems of interest. This consideration leads us naturally into the subject of our capability in the critical experiment program.

It may be worthwhile to recall some of the key issues and decisions of recent years in the integral data program. The decision to build ZPPR and to convert ZPR-6 and -9 to plutonium use provided facilities that potentially allow the study of plutonium systems up to commercial sizes.

These facilities have been described many times in the past. Very briefly, all three are two-half machines using plates (for the most part) as building blocks. When the halves are together, ZPPR is currently 10 ft × 10 ft × 8 ft and ZPR-6 and -9 are 8 ft × 8 ft × 8 ft. It is proposed to expand the ZPPR matrix to a size of 14 ft × 14 ft × 10 ft (ZPR-6 and -9 are expandable to a matrix size of 12 ft × 12 ft × 8 ft, but no such expansion is currently contemplated). With the ZPPR expansion, commercial sized systems can be studied on it, as well as many experiments of interest to the demonstration reactor related to behavior outside the blanket that are not possible with the current size matrix. Demonstration sized systems can now be studied on all three facilities, but measurements related to blanket phenomena would be seriously limited on ZPR-6 and -9.

A parallel decision was made several years back, involving an even larger capital investment, to acquire a full scale plutonium inventory. The related acquisitions, which involve over 3000 kg of plutonium (of several different types, but primarily metal plates, 1/4 in. thick, ∿2.5:1 U^{238}:Pu, 11.5% Pu^{240} isotopic content) permit the mockup of a full scale 1000 MWe commercial sized breeder reactor.

The major items comprising the ZPR material inventory are given below. As can be seen, the inventory includes a substantial amount (∿175 kg) of plutonium in metal form having a Pu^{240} content closer to that of systems of interest. It also includes a substantial inventory of mixed oxide rods (both of "normal" 11.5% Pu^{240} content and a smaller amount of "high" 26% Pu^{240} content).

One perennial question relates to the correlation of critical experiment results, obtained at room temperature with plate geometry, to reactor conditions at elevated temperatures with rod geometry. Such questions lead to considerations of the possible need for a full hot critical facility.

The decision, up to now at least, has been to suspend judgment on the basis that the need for a hot critical facility has not been shown. The program has proceeded on the assumption that the effects of temperature and geometry can be adequately studied within a zone in one of the ZPR facilities. This decision led to the acquisition of the mixed oxide rods and the sodium-filled calandria boxes to hold the rods. In order to study the effects of temperature, the Variable Temperature Rodded Zone (VTRZ) program was undertaken. Final decisions on a hot critical facility will await the completion of the VTRZ program, which is now getting ready to begin in-pile experiments.

A number of basic measurements are associated with the ZPR program.

1. Critical Configuration - Critical Mass

2. Reactivity Effects

   (a) Control Rod Worths - Shadowing & Tilting Effects, Subcriticality

MAJOR ZPR MATERIAL INVENTORY

	Material	Form / Composition	Isotopic composition	Amount
1.	BASIC Pu METAL,	1/4" PLATES, U 69.5 | Pu 28 | Mo 2.5 WEIGHT %,	239 86.1 | 240 11.5 | 241 1.7 %	∿3200 kg Pu
2.	Hi-240 Pu METAL	1/4" PLATES, FISSILE CONTENT SAME AS ABOVE,	239 67.3 | 240 25.9 | 241 5.3 %	∿ 175 kg Pu
3.	"SEFOR" Pu Metal,	1/4" PLATES, U:Pu ≈ 4:1		∿ 200 kg Pu
4.	Pu/Al METAL,	1/8" PLATES	239 95 | 240 4.5 | 241 0.5 %	∿ 200 kg Pu
5.	ENRICHED U PLATES,	VARIOUS THICKNESSES FROM 0.026" to 1/8",	93% ENRICHED	∿2100 kg U-235
6.	ENRICHED UO_2 PINS,	3/8" DIA. OF TWO ENRICHMENTS	(a) 16.4% U-235 (b) 46.5% U-235	317 kg U-235
7.	PuO_2|UO_2 PINS,	3/8" DIA. OF TWO Pu CONTENTS,	∿ 11.5% Pu-240 (a) 15% Pu → 150 kg Pu (b) 30% Pu → 130 kg Pu	280 kg Pu
8.	Hi-240 PuO_2|UO_2 PINS		239 67.5 | 240 25.9 | 241 5.3 %	27.5 kg Pu

9. 150 TONS DEPLETED U
   25 TONS U_3O_8
   8000 LITERS Na, 1/4" and 1/2" CANS

10. ALSO LARGE AMOUNTS OF

Fe_2O_3	SS	C
Boron	Al	Be
Tantalum	Ni	BeO

(b) Effects of Higher Plutonium Isotopes
(c) Rods vs. Plates - Heterogeneity Effects
(d) Doppler Coefficient - Mapping over Reactor
(e) Na Void Coefficient - Mapping over Reactor
(f) Worths of Various Materials - Small Samples & Large Changes
(g) Perturbation Theory Normalization Integral
(h) Heated Zone Effects - VTRZ

Power Distribution - Reaction Rate Distribution

(a) Fission Rates - U-235, U-238, Pu-239
(b) U-238 Capture and B-10(n,α) Rates
(c) Power Distributions near Interfaces and Control Rods
(d) In-cell Distributions - Rods vs. Plates
(e) γ-Ray Energy Deposition Particularly in Control Rods - Further development required
(f) Neutron Dosimetry for Radiation Damage Diagnostics - Further development required

Measurements Related to the Fuel Cycle

(a) Pu α
(b) Effects of Higher Plutonium Isotopes
(c) Effects of Fission Products - No current capability

Neutron Spectrum

(a) Proton Recoil
(b) Time-of-flight - Further development required.
(c) Spectral Indices - Reaction Ratios

Kinetics and Fluctuation Experiments

(a) Rossi α
(b) Noise Measurements
(c) Pulsed Neutron Measurements

Shielding Studies

Blanket Studies

Operational Experiments - ZPR Control Rod Calibrations

Most of the measurements given above have achieved a routine status. Even then, there is frequently continuing work required to further increase the accuracy. Because essentially all of the measurements are prone to possible unknown systemmatic error, it is a basic principle that if at all feasible at least two independent methods be developed for the measurement. Many of the measurements listed have attained that status. Included in the list are some measurements (e.g., fission products) that are hope rather than reality, and other measurements where clear discrepancies or anomalies are known to exist.

Kinds of Critical Experiments - Demonstration Reactor Benchmark Program

Although there has generally been agreement on the need for critical experiments, there has often been debate on the kind of experiment to be performed. More specifically, the debate usually centers on the issue of the choice of assembly configuration to be constructed and less frequently on the question of which measurements are to be made. Generally, the latter issue is largely determined by the current state of the art - one makes the measurements one can.

(The number or detail of the measurements may still be a subject of considerable discussion.)

The kinds of critical experiment that can be performed may be categorized in the following way.

(a) clean
(b) zone
(c) benchmark
(d) engineering mockup

(a) Clean experiments

By the usual definition, a clean critical experiment is a simple configuration, from the point of view of geometry or composition or both, which is designed to be effective in a diagnostic sense in the analysis of some integral parameters measured on it. It may achieve this by simplifying or eliminating computational uncertainties or by isolating the material or cross section of interest. Generally, it will not directly resemble any reactor of interest. Many such experiments have been performed in the fast reactor program in the past. The abstract case for such experiments is strong, and any well balanced program should probably have a component of such experiments. However, when balanced against other potential experiments, no clean experiments have, in fact, been selected over the past several years. Nor are any specifically planned now. It is not hard to imagine their inclusion in the program should, for example, a definitive clean experiment be proposed relative to say, the central worth discrepancy (discussed later).

(b) Zone experiments

Many of the integral parameters of interest are primarily composition dependent. In such cases, if a zone sufficiently large to establish the asymptotic spectrum is constructed, the required measurements can be made, often at a large savings of reactor materials and reactor time. A number of such experiments have been performed recently and are currently planned. Examples are the experiments on the high Pu^{240} fuel and on the mixed oxide rods. The inventories of these expensive materials acquired were in fact just those required for zone experiments. In a very general way, we expect that experiments requiring the full reactor configuration will tend to be performed on ZPPR and those that only require a zone may often be done on the smaller ZPR-6 or -9. This does not preclude the possibility of ZPPR zones, e.g., placing high Pu^{240} fuel or rods into various locations in ZPPR assemblies. We view zone experiments as playing a continuing important role in the program, but not everything can be measured in such experiments, and that, in fact, was the rationale for acquiring the full plutonium inventory.

(c) Benchmark experiments

These experiments have over the past several years formed the backbone of the critical experiment program. It was decided that the most useful set of experiments that could be performed at this stage of fast reactor development would be to undertake a series of benchmark experiments on systems that are realistically typical of reactors to be built. The measurements made on these systems and comparison with the various calculational techniques would provide the reactor designer with the necessary direct validation of his calculation techniques.

The benchmark program, aimed at demonstration reactor sized systems and established in consultation with industry, provides the required information prior to required engineering mockup studies. The assemblies studied are realistic, though slightly simplified, typical demonstration reactor configurations. The measurements focus on control rod effects, 2-zone core effects, distributed Doppler and sodium coefficients, blanket effects, local power distributions, effects of higher plutonium isotopes, and heterogeneity effects associated with plates vs. rods. The experiments have been mainly performed on ZPPR (Assemblies 2 and 3) with additional studies on ZPR-6. The VTRZ program is also viewed as being part of the Demonstration Reactor Benchmark Program.

(d) Engineering mockup critical (EMC) experiments

At the current stage of fast reactor technology, it is invariably assumed that at the appropriate time, corresponding to finalization of design, a mockup experiment will be performed in which the final design will be simulated as closely as feasible within the constraints of the critical facility and the material inventory available. We are currently in this phase for the FTR on ZPR-9. Previous experience tends to strongly indicate that such experiments are extremely useful to the reactor under consideration (in fact, after a number of engineering mockups for fast reactors later built, we have no instance of significant discrepancies between EMC predictions and actual reactor characteristics - of course, not all reactor parameters can be studied on critical experiments, most notably many related to the fuel cycle can not). However, they are not too useful for more general application, primarily because in the attempt to simulate all aspects of the design as closely as possible, the ability to isolate and validate specific calculational procedures is largely negated.

Steady State Performance - Power Distributions in Complex Geometry

In the steady state performance area, the problem of accurate power distribution, and particularly how they might be affected by measurement in a pin versus plate environment, remains a crucial problem. The power distribution relates directly to the specification of the hot spot. It can become a very complex phenomenon near interfaces or near control rods, and the difficulty is further increased because it must be obtained at the worst point during the cycle in which control rod movements, burnup and buildup of fuel further complicate the problem.

Critical experiment measurements pertaining to power distribuiton have been performed regularly on many assemblies but only recently have measurements been done on some of the configurations of control rods and core zones of interest to current fast reactor design. Such measurements to date are further limited in that they have been done in plate rather than pin geometry and clearly local power distributions are an area where the fine structure may be very significant. Such measurements in complex geometry using rods are planned in the Demonstration Reactor Benchmark program.

The experimental validation of calculation methods are particularly importan in this area because the procedures are subject to question with respect to the validity of the averging procedures used to obtain the multigroup constants. The need for transport rather than diffusion theory for calculation of detailed power distributions near sharply discontinuous, geometrically complex, boundaries also has not yet been adequately studied..

Central Worth Discrepancy

It has long been known, and observed on many assemblies at many laboratories, that when calculation and experiment are compared for small sample reactivity measurements in fast reactors, there is a systemmatic bias for the primary fissile and fertile isotopes, e.g., Pu^{239}, U^{235}, U^{238}, with the ratio for calculation to experiment (C/E) generally being well over unity. This systemmatic bias occurs in medium and larger sized assemblies fueled both with U^{235} and Pu^{239}. The bias is larger and more consistent in the Pu^{239} systems, generally with C/E $\sim$ 1.2-1.3. The bias in U^{235} systems tends to be somewhat less. The bias has a clear correlation with U^{238} content and with assembly size, but it should be noted that U^{238} content and size are themselves correlated. No other correlation, such as experimental configuration, measurement technique, calculation method, or data base used, tends to be decisive although all can have non-trivial effects.

It should be noted particularly that the cross sections used, and often much of the calculation procedure as well, are the same as those that give a value of k close to unity for the critical system and furthermore, that, all of the isotopes that are important in establishing the reactivity of the critical systems are ones that show the low experimental values for their small sample worth. It is this aspect that gives the problem its anomalous nature.

One other point should be noted. The discrepancy gets its name from the fact that small sample worth measurements are generally made at the center of the reactor. There is no conclusive evidence, however, that the bias is correlated to the position of measurement. In cases where the small sample worth has been mapped, the experimentally determined spatial distribution has tended to agree in shape with calculation.

The cause and cure of the central worth discrepancy is not known although there are a number of clues worthy of further study. The discrepancy may possibly gradually evaporate away as a result of many small corrections. It may due all or in part to wrong cross sections. The interpretation of the experiment in units of absolute reactivity involves relating a reactor period (the measured quantity, directly or indirectly) to absolute reactivity and this assumes knowledge of the delayed neutron parameters. A contributing factor to the discrepancy may thus be erroneous delayed neutron data (discussed in the next section) although it is hard to see how this could be the primary cause of the discrepancy. Until resolved, we are never quite sure how to handle the interpretation of reactivity measurements. Are we in fact getting the best agreement with theory for an arbitrary reactivity measurement when we get C/E = 1, or when we get the same value for C/E as we do for plutonium?

Delayed Neutron Parameters

Some recent measurements (Krick and Evans, and earlier Masters) have indicated that the delayed neutron yield of the isotopes important to the fast reactor (i.e., Pu^{239} and U^{238}) may be somewhat higher than previously believed (Keepin data). The importance of this is primarily related to the interpretation of reactivity measurements. Since the reactivity measurement gives the value in terms of dollars, the larger the value of the dollar, the larger the value of the experimentally deduced absolute reactivity. Hence a larger value of β reduces the central worth discrepancy bias. Better measurements of the delayed neutron parameters are now in progress. Some integral measurements from which β_{eff} can be deduced are also consistent with higher than the Keepin values of β (by $\sim$4%, which is about half the increase the recent measurement have given).

Another point worth bringing out is that in the past, quite frequently when the value of β_{eff} was calculated in a fast reactor system, it was done on the basis that ν is a function of E (which it is) but that β is constant. The latter is almost surely an erroneous approximation. The best experimental evidence (also consistent with theory) is that the delayed neutron yield per fission remains constant with energy (up to ~4 MeV, or in the range of interest) and thus as ν increases with energy, the value of β correspondingly decreases with energy. Furthermore, since the value of β, in the constant β approximation was taken from experiments on integral assemblies with much harder spectra than those in fast breeder reactors, β_{eff} was calculated too low. The correctly calculated β_{eff}, being higher, also helps (in a small way, ~3%) to reduce the central worth discrepancy.

Sodium Void Coefficient

Should all or part of a sodium cooled fast reactor become voided of sodium under accident conditions, the resulting reactivity changes could have a profound effect on the course of the accident. The essential point is that sodium removal in the central part of a (large) fast reactor adds reactivity. Sodium removal in the outer parts (axially and radially) of the core and in the blanket reduces reactivity. The reactivity history of the transient will thus crucially depend on the detailed space-time history of sodium voidage. It is not within the generally accepted scope of reactor physics to concern ourselves with the mechanisms of sodium voidage or, for that matter, the credibility of the event taking place at all. (Current safety analysis do consider such accidents, generally in the upper limit Design Basis accident category.) The voidage mechanisms are obviously very complex and may be related to boiling by heat transfer through the clad, to vaporization by injection of molten fuel into the sodium, to sodium expelled by suddenly-released fission gases from many failing pins during a transient, or to the passage through the core of an entrapped large gas bubble.

The physics problem consists of the accurate prediction of the reactivity change associated with any given voidage pattern. The problem is particularly difficult because of the great sensitivity of the coefficient to the details of the cross sections and because of the competition between the leakage and spectral components which are of opposite sign. For example, in a typical fast reactor spectrum (the center of ZPPR Assembly 2) going from ENDF Version 1 to Version 3 raised the calculated central worth of sodium by about 40%. In fact, C/E for Version 1 is $\approx$ 1.0 and thus for Version 3 it is $\approx$ 1.4. A value of C/E = 1 would seem better than one of 1.4, which would say that in this particular instance Version 1 is better than Version 3. But is it? The central worth of a plutonium sample has a calculated value of C/E $\approx$ 1.2 (for both Version 1 and 3), so that if we assume that we should get the value relative to plutonium correctly, then Version 1 and Version 3 are equally off in the sodium worth, one being low and the other high by about the same amount.

The above is an example of the kind of difficulty brought in by the central worth discrepancy. The key question will enter when we must assign an absolute value to the measured sodium coefficient. Should we or should we not normalize out the central worth discrepancy? Similar questions will arise with the Doppler coefficient, and other reactivity measurements. Probably the safest thing to do is to decide on a case by case basis which gives the more conservative value.

Another possibility arises from the consideration that quite frequently the absolute values of reactivity changes themselves are not the important quantities but rather, the degree to which the reactivities of various effects offset each other (e.g. Doppler vs sodium coefficient, or fuel burnup vs control rod movement) is important. The extent to which such consideration may be safely applied in reactor design is an area requiring further study, with particular emphasis on the validity of the conclusions under various assumptions as to the cause of the central worth discrepancy.

Because of the extreme sensitivity of the calculated sodium void worth to cross section data, particular reliance must be placed on the experimental determination. In principle, the measurement is straightforward. In practice great care must be used quite apart from the uncertainty in interpretation of the experimental result due to the central worth discrepancy. The measurement consists of substituting voided elements for the sodium-filled elements (generally small steel cans 1/4" or 1/2" thick) and measuring directly the resultant reactivity changes.

In complex geometries (e.g. the test loop and control rod configurations in the FTR mockup), the geometry must be accounted for rather closely both in the calculation and in the experiment or significantly erroneous results may ensue. (An error of 50% can result from modelling the FTR loop and control rod structure into annular rings. The sodium void coefficient at the same radial distance may vary by a factor of 3 depending upon whether it is at a control rod position or in the island between control rods).

Earlier measurements on the ZPR facilities showed that very different experimental results for the sodium coefficient were obtained for the same composition assembled with different drawer loading patterns. This led to concern that, at least for this parameter, a large heterogeneity effect may exist and that the results obtained in a plate configuration may be very different from the pin configuration of interest. Recent experiments directly comparing sodium void coefficients measured in pins and plates having identical average compositions show very little difference (∿10%) and of the order expected from theory. The small effect, compared to the large differences observed earlier with different plate configuration is mainly attributable to the fact that the earlier plate experiments had the fissile and fertile materials in totally separate plates whereas in the more recent experiments a large amount of the fertile material is in the fissile plates. The later plate configurations are thus intrinsically less heterogeneous.

The direct measurement of a large voided region is a very time consuming and laborious experiment since a large fraction of the total assembly has to be reloaded. For this reason, it is of great practical significance to establish the degree to which the reactivity of a large voided region may be established from a small sample mapping. The evidence so far is inconclusive. One obvious source of error is that the worth of sodium in a sodium filled environment is not expected to be the same as it is in a sodium voided environment. The degree of non-linearity in the sodium worth with sodium content in the environment is significant (∿30 or 40% difference between the two extremes, both by calculation and experiment). It is hoped that the small sample approach can be shown to be a reasonably quick and reliable, even if not extremely accurate, way of measuring the reactivity of a large voided region.

Doppler Effect

In a fast reactor transient, the Doppler coefficient is one of the major contributors to the safety of the system. It is effective both in terms of turning mild transients around and minimizing damage to the core or, in the case of very severe transients with violent disassembly, in very much reducing the destructive energy release. Its importance comes about primarily from the fact that in ceramic-fueled fast reactors, the Doppler effect is the only mechanism that yields an immediate negative feedback. (In a metal-fueled fast reactor excursion, a prompt negative feedback would also come from the axial expansion of the fuel).

Some years back, there was not even complete certainty as to the sign of the Doppler coefficient. As long as there was an even remote possibility that the Doppler coefficient might be positive and hence act as an autocatalytic agent in an excursion, the Doppler effect posed a most serious problem. Both experimental and theoretical work have long since totally removed that possibility. It is now definitely known that in any fast breeder reactor composition of interest, the negative U^{238} Doppler coefficient will totally dominate the fissile component of the Doppler effect (which may or may not be positive, but in any event is extremely small in magnitude compared to the U^{238} component).

The main current problem with the Doppler effect is in predicting its magnitude with reasonable and proven accuracy. This is particularly important since it plays a quantitative role in safety analyses. Generally, it is not very impor tant how large (negative) the Doppler coefficient is, providing it is known for sure that it is at least as large as some minimum value.

Small sample techniques have been perfected for measurement of the Doppler coefficient in critical experiments. Agreement between calculation and experiment has been reasonably good but a number of areas of uncertainty still remain. These primarily relate to such questions as extrapolation to high temperature, mapping of the Doppler coefficient over the full core which might be complex due to insertion of control rods, various enrichments, buildup and burnup of fuel, and presence of fission products. In the past several years, SEFOR full core experiments have also given extremely impressive verification and quantitative validation of the Doppler effect including the turning around of super prompt-critical excursions.

Neglecting the further uncertainty that arises from the central worth discrepancy, the estimated uncertainty of the Doppler coefficient as deduced from theory and small sample measurements is $\sim$25% with a C/E for U^{238} in a plutonium environment $\approx$ 1. (In a U^{235} fueled system, however, the agreement is not nearly so good with C/E $\approx$ 0.75 for the U^{238} Doppler coefficient.) The central worth discrepancy is generally ignored in this case, since not applying it as a normalization factor to the measured Doppler coefficient results in a lower and hence, more conservative value.

It should be noted that sodium voiding significantly reduces the Doppler effect (by close to 50%), with the factor being obtained by both calculation and experiment.

One major area of concern is the validity of the extrapolation of small sample results to full core Doppler effects. It will be one of the major purposes of the VTRZ program to answer this question since it will enable the Doppler coefficient to be measured in both cold and hot environments.

The experimental program utilizing the VTRZ will aim at confirmation of the reliability of the analytical techniques used for extrapolation of measurements of the Doppler and sodium voiding coefficients in plate-type criticals to the operating conditions of fast power reactors. It will concentrate on the effects of heterogeneity and elevated temperature. The VTRZ is currently being installed in ZPR-6.

In addition to the Doppler and sodium coefficients, there are other reactivity coefficients that may be of importance to fast reactor design. One area relates to accurate means of establishing the power coefficient, including such contributions as may arise from small motions involved in fuel bowing or effects associated with the core restraint mechanism. Such reactivity effects are small but important. The magnitude and sign are determined by the details of small complex motions. It is necessary to obtain reliable calculational procedures, which do not now exist, and, in addition, validate such calculational procedures with measurements. Mocking up such small displacements is a difficult area in the critical experiment program. Such difficulties have in fact manifested themselves in the FFTF critical experiments.

Control Rod Effects - Worths, Subcriticality Measurements

The major thrust of the ZPPR Assembly 3 program over the next year will relate to control rod effects. Problems to be studied include:

worths of individual rods or banks of rods fully or partially inserted;
mutual shadowing effects between rods;
the tilting of flux if rods are not inserted symmetrically;
maximum worth of a "stuck" rod;
power distributions in and near rods;
the effect of sodium voiding on rod worths (the reduced rod worth associated with the harder neutron spectrum in a voided system can be a major component to the positive sodium coefficient);
rod worths established both from subcriticality measurements and by returning the system to criticality by fuel spiking;
effect of rod size, geometry, and local fine structure.

Although there is some data available from previous critical experiments, the data is generally fragmentary. Appropriate experimental and calculational tools have not been validated.

The experiments to date generally show no clear bias to over-predict or under-predict rod worths. Values of C/E have generally ranged from slightly less than unity to about 1.2 (the latter about the value of the central worth discrepancy). Many of the calculations so far have been based on diffusion theory, although other studies have shown that transport theory will reduce the calculated worths by 10% or more.

Because of the lack of validated methods and relatively little experimental data, this area is represents one of the more pressing needs of the reactor designer. The near term experiments will focus on B_4C rods, although some tantalum experiments will also be performed.

The measurement of subcriticality, particularly for large values, presents a severe problem. Different measurement techniques have given inconsistent results. Measurements made by the rod-drop method, which are believed to be accurate, have been in agreement with one noise technique measurement and in disagreement with another. Until these discrepancies are resolved, or better understood, we will have significant uncertainties in this important area.

Fuel Cycle Effects - Breeding Gain, Plutonium Alpha,
Higher Plutonium Isotope Studies

The area in fast breeder reactor physics in which we have the greatest deficiencies of knowledge, is clearly the one related to fuel cycle neutronics. The primary reasons for this lack are (1) although there have been a number of fast reactors built and operated, there is as yet no experience of any reactor operating in a breeder mode - thus there is no backlog of realistic experience and results, (2) while critical experiments are an ideal way to study many aspects of breeder physics, they are not suitable for direct study of fuel cycle effects. (We should not sell them too short in this area, however, since they can be used to study many relevant effects indirectly.) (3) many of the crucial cross sections, e.g., fission and capture of the higher plutonium isotopes, fission products, are extremely difficult to measure.

With respect to relevant measurements that can be performed on critical facilities, further experimental studies with fuel containing large amounts of plutonium 240 and 241 are essential. Most of our measurements are on a fuel with plutonium isotopic concentrations not representative of the fuel to be used in actual reactors. It is therefore important that the differences and corrections be obtained reliably. The effects of the higher plutonium isotopes, fission product capture cross sections, and the plutonium capture-to-fission ratio all have significant or even decisive effects on the breeding gain and therefore play a crucial role in establishing the basic potential of different breeder systems. The uncertainty in the breeding gain, at least of the order of 0.1, is too large to make sufficiently reliable evaluations. Also, the required long term reactivity changes over the operating cycle have excessively large uncertainties.

Other Areas for Investigation in the ZPR Program

The following areas of ZPR activity have generally had only fragmentary or preliminary activity up to this point.

a. Neutron spectrometry by means other than proton recoil

In the measurement of the neutron spectrum it is important to validate the techniques used by comparing several different methods, since spectrometry measurements are vulnerable to systematic error. No single method has been developed that will satisfactorily cover the entire range of interest in fast reactors but there are ranges of overlap between various techniques. Time-of-flight, which is particularly useful at low energy, is currently being developed in ZPR-6. The significant range of overlap with the proton recoil measurements should provide the required assurance. Another important area in spectrometry relates to measurements at neutron energies of several MeV and above. This energy region is important to the characterization of radiation damage effects. NE 213 spectrum measurement devices are extremely promising in this energy range.

b. Radiation effects dosimetry

An important area is that of characterizing radiation damage phenomena. It is very useful to compare spectrum measurements by proven methods in critical facilities with other techniques, e.g., foil irradiations, which are then usable in actual reactor environments. From such measurements one can validate the procedures for unfolding the neutron spectrum from the responses of the various foils. It should be noted that it is not clear that the reactions that would be suitable in a high flux environment could be tested in the much lower critical experiment flux.

c. Gamma production, transport, and energy deposition

It is essential to validate the techniques used for measuring gamma heat deposition. The current methods, which are based on thermoluminiscent dosimetry (TLD), leave large uncertainties as to their validity.

d. Blanket studies

The most immediate problem is to obtain experimental information on the buildup of fuel in the blanket and the resulting effect on the system. In order to facilitate these studies, it is proposed to acquire an inventory of low enrichment mixed plutonium-uranium oxide rods.

e. Shielding

In addition to the ORNL shielding facilities, the critical facilities, particularly if the ZPPR matrix expansion takes place, provide a valuable tool for studying directly shielding phenomena effects considerably outside the blanket and reflector areas. Problems to be studied include the radiation field at instrument zones placed well outside the reflector. Such studies might include the effect of moderators or configurations selected to increase the sensitivity at the instrument zone. Generally, calculational methods must be validated for deep neutron transport, since the uncertainties with existing methods are large.

Cross Section Measurement Program - Status of Current Knowledge Relative to Fast Reactor Needs

An essential part of the fast breeder reactor physics program relates to cross section measurements, theory, and evaluation.

The productivity and capability of this activity were augmented as the ANL Fast Neutron Generator (FNG) and the ORNL ORELA facility became operational in the past few years. These facilities, along with the RPI linear accelerator, are the main facilities dedicated to the needs of the U. S. reactor development program.

These cross section measurement activities provide microscopic nuclear data for the reactor development program with top priority given to LMFBR requirements. The emphasis is on fast neutron induced phenomena and associated processes. It includes studies on fast neutron scattering, total, capture, and fission cross sections. It also includes studies on inelastic-gamma and gamma-production cross sections, fission properties (energetics, prompt and delayed fission neutron spectra and $\bar{\nu}$), particle emission cross sections, and standard cross sections and reactions. The ORELA facility focusses on low energy cross sections, and the FNG on higher energies.

The cross section evaluation functions are coordinated by the NNCSC at Brookhaven through CSEWG and result in the various versions of the ENDF files.

Although a vast amount of data has been accumulated, much of it of high precision, the status of cross section data relative to fast reactor needs is not good.

The main problem is that the most important cross sections, Pu^{239} capture and fission, U^{238} capture, U^{235} fission (important because other cross sections are measured relative to it), and perhaps U^{238} inelastic are all uncertain in important energy regions. The uncertainties are in the 10% range. Different measurers differ among each other to that extent.

Although existing evaluated data sets give excellent agreement with system criticality (to some extent by arbitrary adjustment), one has the uneasy feeling that many or all of the primary cross sections may yet have to undergo major readjustments, still yielding good criticality data but perhaps removing some current discrepancies in reaction rate ratios (and perhaps also helping with the central worth discrepancy). It should also be noted that our cross section data for the higher plutonium isotopes and fission products are extremely poor. The fission spectrum at low energies also may have important uncertainties and we have earlier discussed the implications of uncertainties in the delayed neutron data.

Integral data testing with ENDF Version 3 is in its early stages and some of these issues may be clearer after more of the testing has occurred.

It should be emphasized that the consequences of the uncertainties in cross sections manifest themselves particularly in reactor parameters that are not well suited for critical experiment studies, (e.g., burnup and fuel cycle effects). Thus, we have embarrassingly large uncertainties in the most fundamental quantity of the breeder, i.e., the breeding gain.

Cross Section Processing Codes - Obtaining Multigroup Constants by Averaging over Energy and Space

One of the major current areas of methods and code development is in cross section processing codes (i.e., averaging the differential data over appropriate energy and space distributions to obtain broad group constant). The most rigorous approach with respect to energy averaging is being taken by MC^2-2 (preliminary release expected at the end of this calendar year). If we include some of the MC^2-2 options that will follow within a reasonable time after the first release, then MC^2-2 probably represents the ultimate in rigor for the forseeable future insofar as the energy average is concerned.

It has become increasingly apparent, as the rigor of the energy averaging has increased, that the weak link is becoming progressively more the space average. MC^2-2 treats the space averaging for a ZPR like cell with some rigor (based on the CALHET, or if necessary, the RABBLE approach). It is not clear at this point whether a full space dependent MC^2 type code will eventually be required.

It should be noted that the running time of MC^2-2 may make it unsuitable for frequent routine use when the most rigorous options are employed. Its primary purpose is to serve as a standard against which less rigorous approximations may be tested. However, it will be possible to relax the rigor and, consequently, the running time.

A number of other less rigorous approaches, but perhaps adequate for most purposes, have been taken for the energy averaging, or space-energy averaging problems. These approximations largely rest on the fact that the main broad group cross section dependence on composition arises in the case of heavy element resonance cross sections. One family of such approaches is illustrated by ETOX-1DX. A composition independent intermediate library is used with a shielding factor (Bonderenko type approach) for the heavy elements, to obtain the heavy element intermediate group cross sections. An average over space, using one-dimensional diffusion theory, can then be used to collapse the intermediate group constants to broad group constants. A number of such codes exist and they are currently being consolidated into one code, combining the best features of each.

A somewhat different approach is used in the SDX option of MC^2-2. Here again an intermediate group composition independent library is formed. The heavy element resonance cross sections are derived for each composition using the rigorous MC^2-2 modules. An integral transport calculation may be performed at this point to obtain spatial self-shielding factors for unit cell homogenization. The resulting intermediate group, now composition-dependent, cross sections can then be collapsed over a one-dimensional spatial distribution to obtain broad group cross sections.

The extent to which such approaches are valid in obtaining broad group constants in regions having non-asymptotic fluxes must be further investigated, before final decisions are made on what, further developments are required in cross section processing codes.

Multi-dimensional (3D) Neutron Diffusion Design Codes

Because of the geometrical complexity of the fast breeder reactor, with interfaces and regions of different compositions, buildup and burnup of fuel, and insertion of control rods in various patterns, it is essential that effective means be developed for calculating the neutronic parameters in such configurations. It is a commonly accepted judgment, although certainly not obvious, that 3D codes will be required as design tools. It is also true that such codes already exist, e.g., 3DB. The real issue is whether codes can be developed that would be sufficiently fast running and reliable and capable of treating a reactor in sufficient detail, that they could be used as normal tools during design. The criteria would be particularly demanding if the 3D codes were used in conjunction with fuel management or space dependent kinetics problems, so that each calculation might require a series of neutronic solutions.

It is a presumption that direct solutions are unlikely to be sufficiently fast and therefore studies are being currently performed on the basis of approximate solutions, e.g., synthesis, finite elements. The optimum course is not yet established. Clearly the validity and efficiency of the selected approximation must be established in the systems of full complexity which provided the incentive for 3D solutions in the first place. Such validated approximate 3D codes are at least a few years away.

Developments in Space Dependent Kinetics

A major effort is currently directed toward the development of a two-dimensional space dependent kinetics multigroup code, FX2, using the factorization method. The model will be able to treat arbitrary fuel motion in addition to the basic capabilities required for safety analyses.

The formulation is capable of treating arbitrary fuel motions arising from fuel slumping or explosive disassembly, this is accomplished by mapping the moving reactor onto a fixed neutronics mesh grid. Other features of the code include mesh-cell dependent temperatures for interpolation of resonance cross sections and a method for reduction of the cost of space-time calculations by formation of reactivity tables which can be saved for later application to similar problems. The validity of the reactivity table approach is under question and studies on it are now being made. It may be replaced with an alternate approach or perhaps a more rigorous derivation of the reactivity table. All safety problems were considered carefully in the formulation of the code, including heat transfer, expansion, bowing, two-phase flow, fuel slumping, and high pressure disassembly. The code also will be capable of treating secondary burst problems when the physical models for the phenomena become available.

Studies are also being performed to establish under what conditions of fast reactor transients various space-dependent kinetics approximations (e.g., point kinetics, adiabatic, quasi-static, direct) are required. The picture is not yet completely clear, although it is definite that for some conditions the point kinetics is clearly inadequate. It is not clear that the adiabatic model may not always suffice. However, even if it does, the more rigorous quasi-static model may be preferable in that it is computationally no more difficult.

CHAPTER 8

FAST POWER BREEDERS

Parameters and Accuracy Required

Fast reactor physics is needed for accurate optimisation of the design and reliable operation of the station. Among the various problems brought about by fast power breeders, the most important neutronics topics are :

- **critical mass of the start-up and equilibrium cores. A total uncertainty (fuel fabrication included) of ± 1 % in Keff is desirable. This value corresponds to ± 2 % on critical enrichment or ± 5 % on critical mass after fabricating the pins (for PHENIX).**
- **power distributions in clean or equilibrium cores and blankets with control rods inserted. The maximal volumic power and integrated axial volumic power have to be known to ± 3 % and ± 2 % respectively.**
- **global breeding gain : ± 0.03 (absolute value)**
- **control rod worths (absorbers and diluents) an uncertainty of ± 5 % is acceptable.**

These predictions and target accuracies must be achieved for power reactors between 250 and 2000 MWe corresponding to an enrichment scale between 30 and 13 %, volumic fractions in the range 40 % for oxide fuel and sodium, and 20 % for stainless steel, Pu isotopic compositions up to 25 % and 15 % for Pu 240 and Pu 241 respectively.

As an example, the PHENIX start-up loading shown on figure 8-1(core midplane horizontal section) contains three enrichments :

Core 1 Pu E ≃ 18 %
Core 2 Pu E ≃ 25 %
Core 3 U 235 E ≃ 27 %

Modifications from previous detailed descriptions of PHENIX characteristics [3] concern mainly this third uranium core due to the delay on Pu availability. Plutonium and Uranium inventories are approximatively half ton each of fissile isotopes for this arrangment corresponding to nominal power (250 MWe). The transition between the start-up configuration and the equilibrium loading (two Pu cores) is obtained, as previously scheduled, by removing "diluents" (special stainless steel subassemblies).

EXPERIMENTAL PROGRAM

The main principles guiding the program of fast integral experiments carried out at MASURCA to achieve the design requests follow the ideas applied with success to thermal neutron physics :

a) Only systematic studies over the practical variation range of main fast reactor parameters can conduce to a physics understanding of fast neutron media : main sources of variation must be measured. Secondary effects and transpositions from experiments to the exact design conditions can then be calculated without large risks of errors.

b) Emphasis must be put on relative variation more than on absolute values to avoid systematic uncertainties.

c) First priority is given to fundamental mode neutron balance parameters in clean lattices, easy to calculate, and directly connected to the design parameters, without searching perfect simulation, too difficult to analyse and impossible to obtain.

d) Among the balance characteristics, the material buckling is considered the most important one ; firstly it gives a relation between production and absorption in the cell, independently of shape and surroundings, secondly it allows, with reflector savings, to define accurately the critical mass and the power distribution even in a two zone core.

e) Reaction rates (or spectral indices) in the fuel are necessary for a more detailed analysis of the balance.

f) Engineering problems must be studied, as far as possible, with the same policy, but only when the clean fundamental mode situation is well known : these problems can then be considered as perturbations of the reference neutron balance.

Fast reactor physics program at CEA proceeds from these lines and can be divided into four parts :

- clean lattices uranium and plutonium fuelled
- higher plutonium isotope effects
- burn-up parameters
- engineering problems as blanket effect, control rod worth , power distribution, sodium void effect, reactivity coefficient.

The MASURCA core RZ program, detailed in this section, is mainly devoted to the study of fundamental mode and design characteristic variations versus enrichment for lattices fuelled either with plutonium (cores Z2 - Z1 - Z3) or with uranium (cores R2 - R1 - R3) (Fig. 8-2). All these lattices have the same fuel-diluent volumic ratio (0.2) and the same diluent composition (45 % w/o Na in the cell - canned sodium metal square rods).

Metal fuel rods (1,27 cm diameter) are used both for uranium (Ni coated) and plutonium (stainless steel cladded). The oxygen is simulated by a square (1,27 cm) rod in iron oxide : going to the density of a power reactor fuel, the fuel volumic fraction in the cell is then 32 % (compared to 36 % for PHENIX). Plutonium contains 8 % Pu 240.

To vary the enrichment, depleted metal uranium rods (1,27 cm diameter) replace enriched uranium or plutonium rods . The whole variation range of power reactor enrichment is then covered with an equivalence uranium - plutonium and a relative value of fuel composition perfectly known ;

Z2	25.2 %	-	PHENIX outer core
Z1	18.3 %	-	PHENIX inner core and 1200 MWe outer core
Z3	11.9 %	-	1200 MWe inner core.

Besides the physics interest of this uranium - plutonium comparison , the Pu inventory (175 Kgs) lays down studies of Pu lattices in two-core assemblies with a reference uranium core that can be made critical with the U 235 inventory (700 Kgs).

Standard blankets are pure depleted uranium metal axially and pure depleted uranium oxide radially. Core heights are 60 cm or 90 cm. Arrangements of the assemblies in the facility were previously described [4] [5].

The core Z1 was built at SNEAK in 1970 with MASURCA fuel in the field of the close cooperation between G F K and C E A.

Space allows only a selection of the experiments to be discussed and compared with calculations. Those included are : material buckling and spectral indices for cell characteristics , critical mass, blanket and control rods parameters for engineering problems.

METHODS AND DATA

Cell calculations are performed with the most reliable methods to describe accurately all the neutron balance details in the fundamental mode. The cell code HETAIRE [6] calculates the spatially dependent resonant self-shielding and spatial fine structure in one or two-dimensional cells : it is based on first flight collision probabilities together with the subgroup method (including temperature dependence). The code uses 600 fine groups for elastic slowing down and transport cross-sections of light elements, 25 groups for all the other cross-sections. Average cross-sections over the cell are obtained in 25 groups for spatial calculations.

The whole reactor calculations are performed with usual multigroup codes in one or two-dimensional options and diffusion or transport approximations.

The data set corresponds to the CADARACHE Version 2 cross-section set, previously adjusted to fit integral experiments [7] .

NEUTRON BALANCE PARAMETERS

Experimental techniques - Two methods are used to measure material buckling of a fast lattice : flux mapping and progressive substitution method.

Material buckling determination in fast media by flux traverses must deal with harmonic effects coming from the outer zones and still large at the center of the test zone. From the previous detailed description of the method used [8] the major modification concerns the determination of the "experimental" adjoint flux traverse point by point by correlation between four measured fission rate traverses (U 238, Np 237, Pu 239, U 235) and five calculated ones (the same plus the calculated adjoint traverse). The curvature of this "experimental" adjoint flux traverse corresponds to the measured buckling. This solution leads to the same absolute value that the previous one but to a better accuracy in the range 0.1 to 0.15 10^{-4} cm^{-2} on material buckling.

Checks of the method used to eliminate harmonic effects were performed : modification of the height-diameter ratio of the core (R2 60 cm and 90 cm) to vary the harmonic influence on radial and axial directions, several axial blanket compositions to increase the harmonic effect (in the core R1, U metal was replaced by UO_2 - Na and by stainless steel - Na). All the experimental values are consistent in the quoted error bars.

The direct information deduced from progressive substitution method with a very good accuracy is the buckling difference between reference and substituted lattices [8] [9]. When sufficient amount of substituted fuel is available, cross check can be performed by flux mapping technique. Two new successful comparisons of these two methods were obtained on core Z2 (substitution Z2/R2) and core R2 half sodium voided (substitution R2 1/2 Na/R2). All these comparisons definitively bear out the quoted accuracies on the measured bucklings.

Comparison of calculations and experiments - The comparison between calculated results by the reference method, previously described, and experimental values, displayed on table 8-1 (ratio (meas. - cal.) / cal in %), shows a pretty good agreement for uranium cores and substantial differences for Pu cores. Fig. 8-3 presents the variation of calculated and measured values versus enrichment. It can be clearly seen that the slopes don't agree in the Pu case : the variation with energy of the difference production - absorption for Pu 239 must be corrected in the cross-section set. It is recalled that 1 % on B^2m corresponds to about 0.4 % in Keff in the 30 % enrichment range and 0.2 % in the 12 % enrichment range.

TABLE 8-1 - COMPARISON OF CALCULATED AND MEASURED B^2_m $(\frac{M-C}{C})$ %

ENRICH. %	URANIUM	ENRICH. %	PLUTONIUM
30.3	- 0.9 ± 0.5	25.2	- 2.4 ± 0.4
22.8	+ 0.9 ± 0.6	18.3	- 3.3 ± 0.8
15.3	- 1.3 ± 1.0	11.9	1.6 ± 2.1

LATTICE REACTION RATES

Experimental techniques - The goal is to measure the main reaction rates of the neutron balance in the exact place where these rates occur with smallest perturbation.

The standard method of measuring fission ratios $(\frac{F8}{F5}, \frac{F9}{F5})$ consists of the irradiation of fissile foils in the rods and in a dummy fission chamber located in a cavity in the test zone. each foil containing one isotope to high purity. Absolute values of these ratios are obtained by means of a separate measurement using calibrated fission chambers (flat or cylindrical) located in the same cavity. Calculated corrections are applied for self-shielding of foils (0.1 mm thick, 1.27cm diameter) in fission chambers and in rods. Harmonics effects are smaller than a few ‰. Total accuracy is estimated to ± 1.5 % on these ratios. The standard method of measuring the capture U 238 to fission U 235 rate ratio consists of the irradiation of foils in the rods and in a reference thermal column position at HARMONIE : hence it relies on the knowledge of the thermal cross-section ratio. Without errors on this thermal ratio, accuracy is estimated to about ± 1 % for uranium cores and ± 1.3 % for plutonium cases (additional correction for cladding caps).

Comparison of calculations and experiments - The discrepancies between calculated and experimental values for average-over-the cell reaction rate ratios versus enrichment are shown table 8-2 and Fig. 8-4 for F8/F5, table 8-3 and Fig. 8-5 for F9/F5, table 8-4 and Fig. 8-6 for C8/F5.

Systematic discrepancies on F8/F5 ratios can be explained by an increase of fission U 238 cross-section in the set, but a slope versus enrichment remains in both cores : a correction of the shape of fission U 235 cross-section with energy

TABLE 8-2 - COMPARISON OF CALCULATED AND MEASURED F8/F5 $(\frac{M-C}{C})$ %

ENRICH. %	URANIUM	ENRICH. %	PLUTONIUM
30.3	3.9 ± 1.6	25.2	3.8 ± 2
22.8	5.4 ± 1.1	18.3	4.0 ± 1.7
15.3	8.6 ± 1.2	11.9	8.9 ± 1.7

must be considered.

These measurements lend confidence to the average energy of Pu 239 fission spectrum used, higher by 6 % than the U 235 one : use the same fission spectrum for the two isotopes would increase the discrepancies in the three Pu cores by 6 %.

TABLE 8-3 - COMPARISON OF CALCULATED AND MEASURED F9/F5 $(\frac{M-C}{C})$ %

ENRICH. %	URANIUM	ENRICH. %	PLUTONIUM
30.3	-1.3 ± 2	25.2	-3 ± 1.5
22.8	2.4 ± 1.5	18.3	0.4 ± 1.5
15.3	1.7 ± 1.5	11.9	3.8 ± 1.5

Differences observed between measured and calculated slopes on the ratios F9/F5, mainly in Pu cores, indicate a need to correct the variation of this cross section ratio with energy in the cross-section set.

TABLE 8-4 - COMPARISON OF CALCULATED AND MEASURED C8/F5 $(\frac{M-C}{C})$ %

ENRICH. %	URANIUM	ENRICH. %	PLUTONIUM
30.3	1.5 ± 1.2	25.2	-1 ± 1.5
22.8	-0.1 ± 0.8	18.3	0.9 ± 1.2
15.3	1.6 ± 0.8	11.9	5.5 ± 1.5

The ratios C8/F5 are quite satisfactorily handled by the cross-section set, except for the softer Pu core that demonstrates a necessary revision of the low energy range cross-sections.

ENGINEERING PARAMETERS

It is not possible to detail all the studies performed in this field. In any case, the status of the art on these parameters does not progress as much as the cell studies, having started later on. General features of the measurements and typical examples will be described.

CRITICAL MASS

For all the previous assemblies, critical masses are calculated and compared to the experimental values in table 8-5 (deviations given in Keff). Corrections for cylindrical boundary, transport effect, shim rod have been applied. A fairly good agreement is obtained in all the cases on Keff inside a 0.5 % span.

TABLE 8-5 - COMPARISON OF CALCULATED AND MEASURED Keff $\left(\frac{MEA.-C}{C}\right)$ %

URANIUM		PLUTONIUM	
R2	+ 0.3 ± 0.06	Z2/R2	- 0.3 ± 0.06
R1	- 0.5 ± 0.06	Z1/R1	- 0.2 ± 0.06
R3/R2	0.0 ± 0.06	Z3/R2	+ 0.1 ± 0.06

However looking to the B^2_m discrepancies, it is concluded that this agreement results from compensating errors between cell calculations and blanket effects : for example, in the core R1, the calculated buckling is underpredicted by -0.9 % (that means - 0.4 % in Keff) and critical Keff overpredicted by 0.5 %. This conclusion demonstrates the interest of the RZ program to separate the sources of errors on critical enrichments and power distributions.

BLANKET EFFECTS

Major blankets effects studied are : reflector saving , power distribution (including γ heating) and Pu build-up.
Experimental values of reflector savings are directly deduced from the measured axial and radial buckling with an accuracy in the range 0.5 to 2 % depending of harmonic effects. Variations of reflector savings versus the enrichment of the core

for the same blanket are measured for two reference blankets : U metal axially, UO_2 radially. No systematic discrepancies on the variation with enrichment or fuel type (U or Pu) is observed in the quoted error bars. But differences between calculated and measured values, on an absolute scale, support the conclusions drawn from the buckling-critical mass comparison.

For the same core, variations of reflector savings with blanket composition are measured axially : UO_2-Na/U metal and stainless steel-Na/U metal comparisons performed on R1 indicate an overpredicted calculated variation for these two cases by about 4 % relative to the U metal absolute reflector savings (18.7 cm). This conclusion is supported by the Keff comparison : R1/UO_2-Na core is calculated overreactive by 0.3 % in Keff compared to the R1/U metal core.

Power distribution in clean blankets are obtained from fission U 235, U 238, Pu 239 and capture U 238 rate traverses measured by foils or fission chambers and from spectral indices measured at the core center. Accuracy on the power distribution normalised to 1 at the core center varies between 2 and 3 %. As an example, power distribution measured in the radial pure UO_2 blanket of the core R 2 is compared in fig.8-7 to the diffusion calculated distribution. The calculation overestimates power at the core-blanket interface by about 15 %. Transport approximation cannot explain the whole discrepancies. This general trend is obtained for all studied cases, such as the overestimation of the calculated capture U 238 blanket distribution (therefore of the Pu build-up) by about 5 %.

CONTROL RODS PARAMETERS

Control rod experiments are mainly devoted to reactivity worths of absorbers and diluents, power perturbations and γ heating into the rods.

The measurement of reactivity worth is performed by increasing the radius of the core to remain critical with the studied rod inserted in order to avoid large errors due to other techniques (like error an βeff for example). The experimental result is then firstly expressed in terms of core radius variation with the better accuracy and can be directly compared to values calculated in the same way. It is always possible to express later on the measurements in terms of Keff through the value in Keff of the peripherical cells. Depending on the rod worth, the accuracy of the first solution remains between $\pm$ 4 and $\pm$ 1 % and becomes $\pm$ 7 % for the second one.

First experiments deal with central fully-inserted rods : measurements of reactivity worth for a given material and increasing rod sizes, comparison of several materials for the same size. Natural Boron carbide, enriched Boron 10, Tantalum, Stainless Steel, Sodium rods are studied for rod size between 1.3 and

112 cm^2. Fig.8-8 shows the relative variation of reactivity worths (expressed in core radius variation) for the same volume of natural B4C versus the size of the rods (expressed in B 10 atomic density per cm). Normalisation is performed to the 7 cm^2 rod (N = 0.12 at B 10/cm). In both the cores R1 and R2, calculations underestimate the reactivity worth decrease with the rod size for the largest rods (- 5 $\pm$ 0.7 % in R1 for the 28 cm^2 rod : this rod corresponds to an antireactivity of - 1.8 % in Keff in this core).

A general overestimation of diluent (Na and stainless steel) calculated reactivity worths is found out(in the range - 15 %). Larger overprediction is obtained for Tantalum (- 20 %).
Other experiments concern partially inserted central B4C rods and full-inserted rods in symmetric radial positions.

Power distribution perturbations by rods are obtained by difference between radial power traverses in the core with the rod and in the clean core. In the two cases, fission rate traverses are measured with foils and fission chambers (U 235, U 238, Pu 239) together with fission rate ratios in a reference position. At a given point, power distribution perturbation $\frac{\Delta P}{P}(r)$ is expressed as the ratio of the difference between the perturbed power $P'(r)$ minus the unperturbed power P (r) over P (r). The core radius moving between the two cases, the power is normalised to the same radial integral value over the core. The accuracy on this ratio $\frac{\Delta P}{P}$ is estimated to $\pm$ 0.6 % . As an example, Fig.8-9 shows the radial power distribution perturbation $\frac{\Delta P}{P}(r)$ in core R1 versus the relative radius $\frac{R}{Ro}$ for a 28 cm^2 natural B4C rod. The agreement between calculation and experiment is not too bad, specially for the maximum power depression close to the rod.

CONSEQUENCES FOR FAST POWER REACTOR PROPERTIES

The final goal of the whole MASURCA RZ program is to use alltogether cell integral results to adjust cross-section set , to improve data and methods from systematic measurements of engineering parameters and then reach the design requests.This tremendous but exciting work cannot be described here. For example, a new version of the cross-section set must be operational at the end of this year.

However, the major improvements on the performance predictions of power breeders that are made by these studies can be identified, mainly in PHENIX case. They concern not only the modifications of the absolute values calculated by the standard method but also the uncertainties on these corrected predictions.

ENRICHMENT

Predictions of critical enrichments rely on several parameters. Looking to the PHENIX start-up loading (Fig. 8-1), one must deal with three zone cores, blankets, diluents, control rod holes, Pu isotopic contents, temperature and power coefficients, reactivity loss during build-up even for the start-up core : the program described previously allows to separate the problems and the RZ core results improve largely the predictions for the major items.

Imagine a theoritical model of fast power reactor with two zone cores in the RZ enrichment range, clean plutonium, U metal axial blanket and UO_2 radial blanket, without control rods and diluents at beginning of life. The RZ core results (B^2_m and critical mass) support the predicted critical enrichment calculated by the standard method to an accuracy of $\pm$0.5 % in Keff ($\pm$ 1 % in E). Even for the same model of the 3 zones start-up core of PHENIX (Fig. 8-1), same conclusions can be drawn, owing to the systematic U-Pu studied equivalence. Care must however be taken of the heterogeneity of the experimental lattices (estimated $\pm$ 0.1 % in Keff from systematic fine structure measurements) and of the knowledge of fuel composition (estimated $\pm$ 0.2 % in Keff for Pu). Finally the error on this theoretical model of PHENIX decreases from $\pm$ 1.9 % in Keff (estimated before these experiments) to $\pm$ .7 %.

Several differences remain between this theoretical model and the actual reactor :

• blankets : PHENIX upper and lower axial blankets consist respectively of stainless steel - sodium and UO_2 - sodium. The radial blanket composition is UO_2-Sodium. From the RZ results, the discrepancies between these blankets and the reference ones lead to a modification of - 0.4 % in Keff of the calculated value. Uncertainties on blanket effects vary from $\pm$ 1.0 in Keff (estimated before experiments) to $\pm$ 0.4 %.

Control rod holes and diluents : The improvements on diluent reactivity worth prediction do not interfere with the enrichment prediction : their positions can change in some amount. However the 6 control rods holes are definitively fixed. The RZ program shows an overprediction of the reactivity worth calculated value by 15 % corresponding to a correction of + 0.5 % in Keff for PHENIX. The accuracy is now estimated to $\pm$ 10 % for the corrected value (no estimation before the experiments).

To sum up, the RZ core results allow to correct the PHENIX predicted enrichment for the main sources of errors and decrease uncertainties for the major neutronics problems from about $\pm$ 2.5 % in Keff to $\pm$ 1.0 % in Keff. The total uncertainty for the start-up core of PHENIX or the equilibrium core (by adding other sources of errors like fuel composition for example) is now estimated to $\pm$ 1.5 % in Keff (on critical mass : $\pm$ 3 % by changing the enrichments or $\pm$ 7 % by varying the core radius in a two Pu zone core)

RELATIVE POWER DISTRIBUTION

From fission traverses and substitution experiments, the calculated volumic power distribution in the two or three zone clean model of PHENIX is considered to be accurate ± 3 % (integrated power for each subassembly). Volumic power at the upper part of the core is predicted to ± 10 %.

Power distribution in clean blankets is overestimated in the calculation by 15 % at the core blanket interface and by 5 % to 10 % for the whole blankets. The uncertainties is now estimated to 3 % after experiments.

Conclusion on the control rod power perturbations cannot be drawn so easily analyses of partially - inserted rods are just beginning . It can be tentatively concluded from full-inserted rod experiments that the power depression is correctly calculated for B4C rod to ± 2 % and underestimated for diluents by 2 ± 2 %.

BREEDING GAIN

All the RZ results support the low value of the breeding gain calculated by the standard method (GRG = 0.12 for PHENIX 2 Pu zone equilibrium core). Looking to the capture U 238 to fission Pu 239 rate ratio, it is clear that the calculated value are correct within an accuracy of ± 3 %. The α9 parameter is not directly measured in the RZ program but can be obtained from neutron balance analysis and from other experiments : K_∞ experiments performed in ERMINE, irradiation in RAPSODIE [2] and OSIRIS. A maximum error of ± 15 % can be estimated on the α average value from the whole analysis. Furthermore, the calculated Pu built-up in the blankets has been shown overpredicted by about 5 ± 2 %. Finally, all these experimental results support a maximum absolute error on the GRG in the range ± 0.05 for PHENIX or a 1200 MWe station. No satisfactory estimation of the error was possible before these experiments.

CONTROL RODS

The axial position of the control rod bank is not fixed up in PHENIX. So improvements on the absorber reactivity worth predictions do not interfere with enrichment prediction. The RZ results suggest an overpredicted calculated value of B4C absorber rod by about 10 %. The present accuracy is ± 10 % on the corrected value against ± 20 % estimated before these experiments. These conclusions are important for safety and operation reasons and the choice of the B10 enrichment.

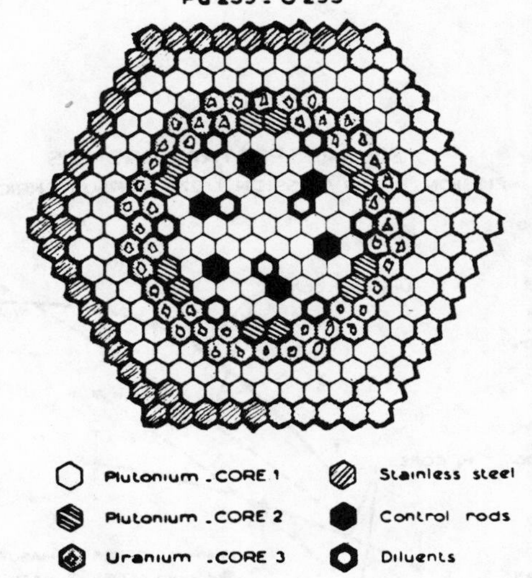

Fig. 8-1. PHENIX _ INITIAL CORE LOADING Pu 239 _ U 235

Fig. 8-2. MASURCA R-Z CELLS

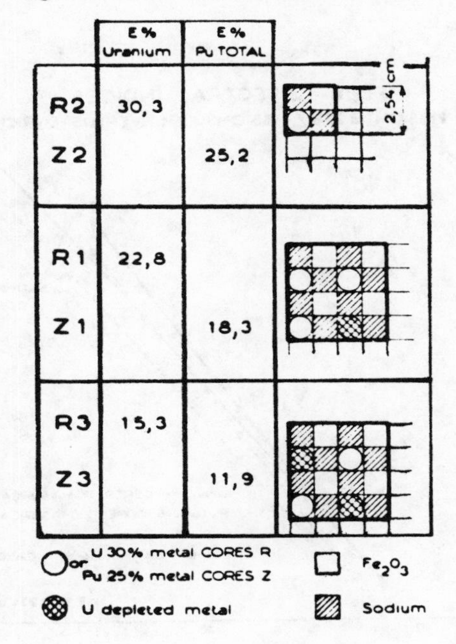

E % Uranium
E % Pu TOTAL
R2 30,3
Z2 25,2
R1 22,8
Z1 18,3
R3 15,3
Z3 11,9
2,54 cm
U 30% metal CORES R or Pu 25% metal CORES Z
U depleted metal
Fe2O3
Sodium

Fig. 8-3-MATERIAL BUCKLING VERSUS ENRICHMENT

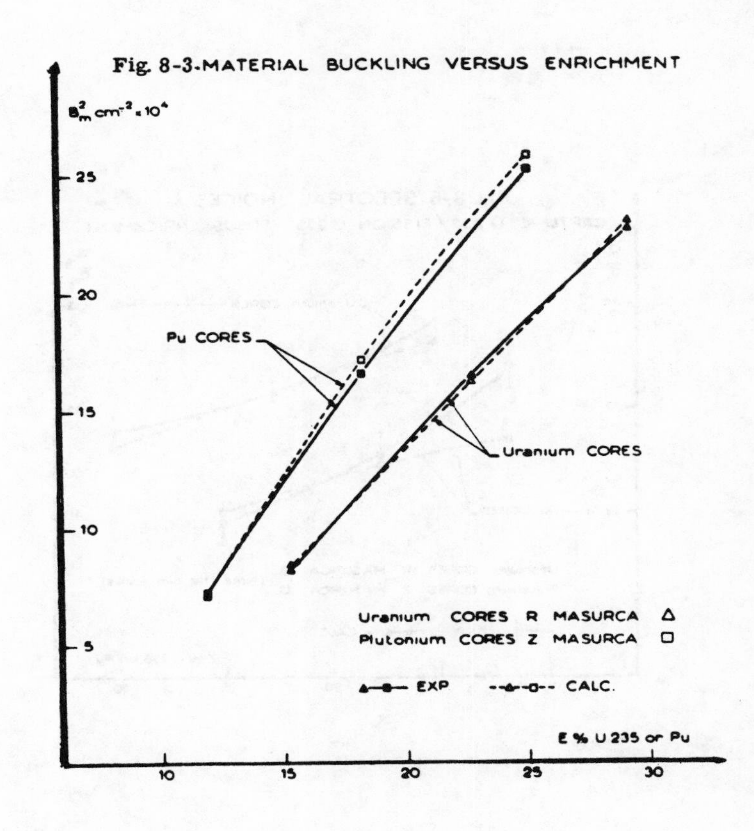

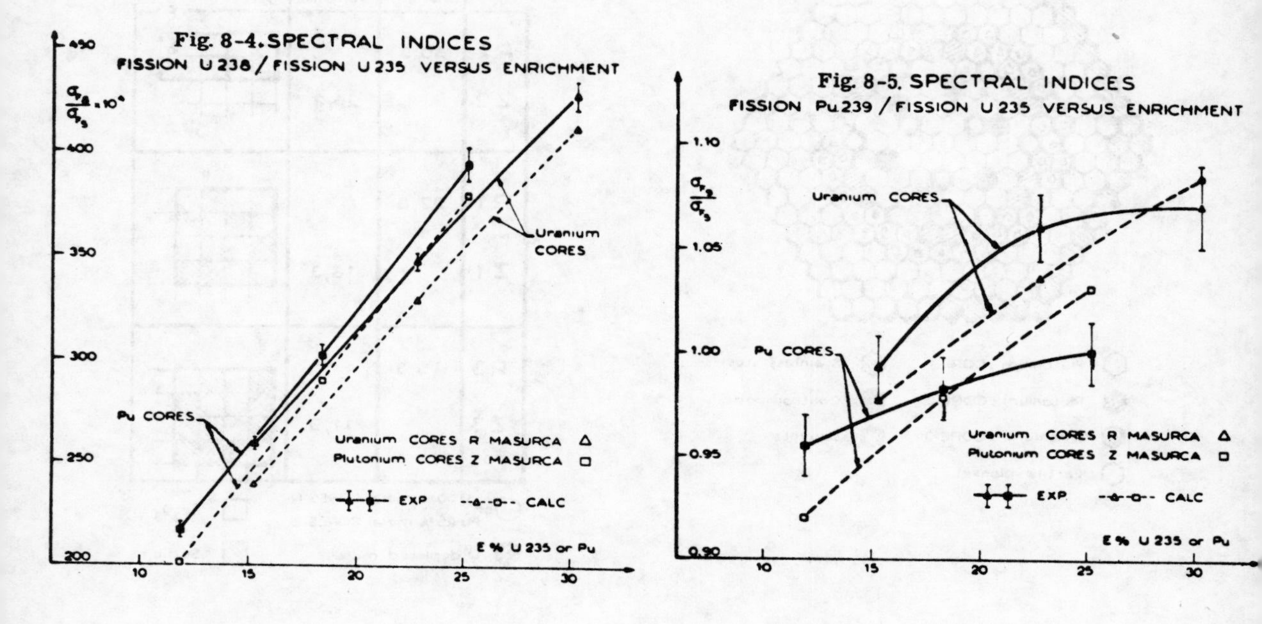

Fig. 8-4. SPECTRAL INDICES
FISSION U 238 / FISSION U 235 VERSUS ENRICHMENT
Uranium CORES
Pu CORES
Uranium CORES R MASURCA
Plutonium CORES Z MASURCA
EXP
CALC
E % U 235 or Pu
Fig. 8-5. SPECTRAL INDICES
FISSION Pu 239 / FISSION U 235 VERSUS ENRICHMENT
Uranium CORES
Pu CORES
Uranium CORES R MASURCA
Plutonium CORES Z MASURCA
EXP
CALC
E % U 235 or Pu

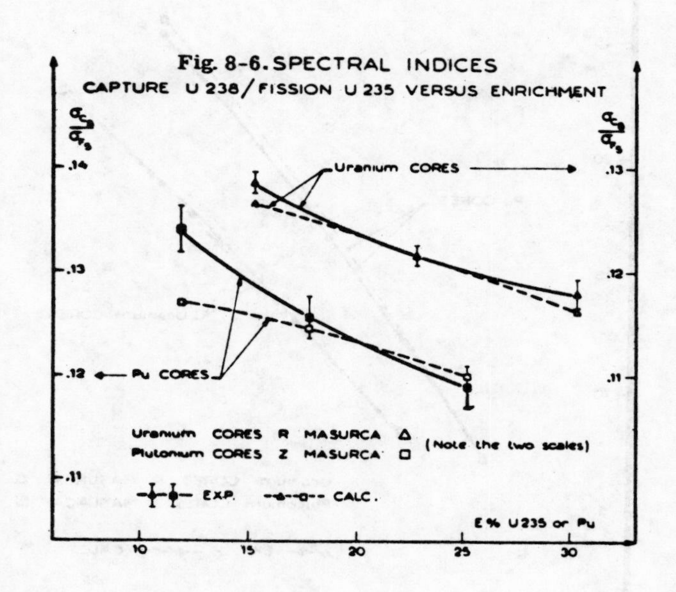

Fig. 8-6. SPECTRAL INDICES
CAPTURE U 238 / FISSION U 235 VERSUS ENRICHMENT
Uranium CORES
Pu CORES
Uranium CORES R MASURCA
Plutonium CORES Z MASURCA
(Note the two scales)
EXP.
CALC.
E % U 235 or Pu

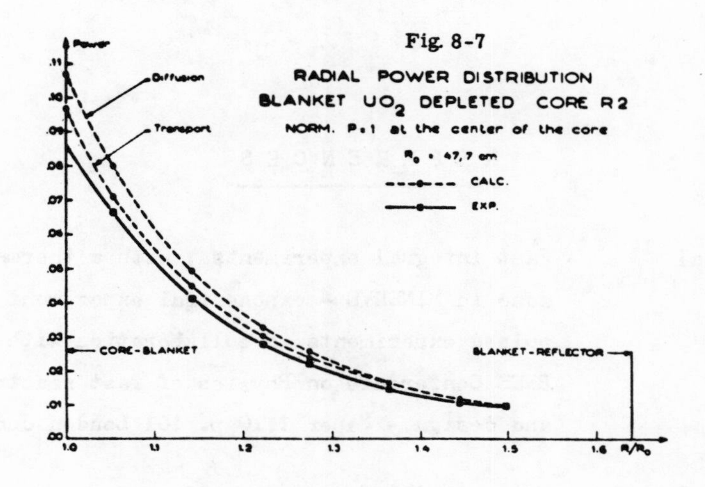

Fig. 8-7

RADIAL POWER DISTRIBUTION BLANKET UO_2 DEPLETED CORE R 2

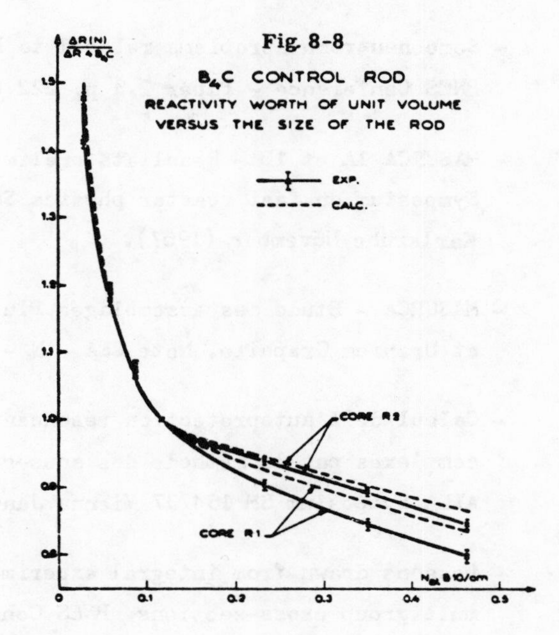

Fig. 8-8

B_4C CONTROL ROD

REACTIVITY WORTH OF UNIT VOLUME VERSUS THE SIZE OF THE ROD

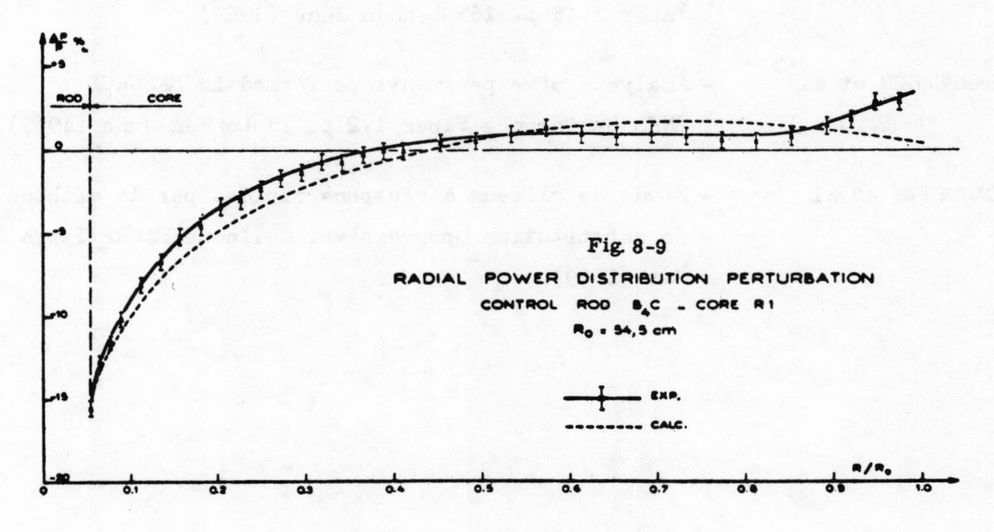

Fig. 8-9

RADIAL POWER DISTRIBUTION PERTURBATION

CONTROL ROD B_4C – CORE R 1

R E F E R E N C E S

[1] J. BOUCHARD et al - Fast integral experiments : with a thermal driver zone in MINERVE - exponential experiment with HARMONIE - pulsed experiments in collaboration with the GERMANS BNES Conference on Physics of fast reactor operation and design - Paper 1.10 p. 101 London June(1969).

[2] J. BOUCHARD et al - Paper to this meeting.

[3] P.P. CLAUZON et al - Some neutronics problems related to PHENIX BNES Conference - Paper 2.3 p. 222 London June (1969).

[4] J. KREMSER et al - MASURCA 1A et 1B - Resultats préliminaires Symposium on fast reactor physics SM 101/59 Vol.II p.3 Karlsruhe November (1967).

[5] O. TRETIAKOFF et al - MASURCA - Etude des assemblages Plutonium Graphite et Uranium Graphite. Note CEA - N - 1483 Septembre 1971.

[6] A. KHAIRALLAH et al - Calcul de l'autoprotection résonnante dans les cellules complexes par la méthode des sous-groupes. AIEA Symposium SM 154/37 Vienna January 1972.

[7] J.Y. BARRÉ et al - Lessons drawn from integral experiments on a set of multigroup cross-sections. BNES Conference Paper 1.15 p. 165 London June (1969).

[8] N. BARBERGER et al - Analysis of experiments performed in MASURCA BNES Conference Paper 1.2 p. 15 London June (1969).

[9] P. CAUMETTE et al - Etude de milieux à neutrons rapides par la méthode de substitution progressive. Colloque EACRP Ispra Mai (1969).

CHAPTER 9

FTR DESIGN PROBLEMS AND DEVELOPMENT AREAS

The Fast Flux Test Facility (FFTF) and its central component, the Fast Test Reactor (FTR), are integral parts of this nation's Liquid Metal Fast Breeder Reactor (LMFBR) program. The FFTF will provide the extensive irradiation test facilities required to develop fuels and materials for the fast breeders.

Development programs in support of the FFTF nuclear design were started quite early. The development of calculational tools for nuclear design and safety analyses was initiated in 1965. A critical experiment program on FTR configurations was started at ANL in 1966. An experimental shielding program was also initiated in 1966. Early shield experiments were carried out at the Shield Test and Irradiation Reactor (STIR) facility operated for the AEC by Atomics International (AI). More recent shield experiments are being done at ANL and at Oak Ridge National Laboratory (ORNL) as part of the overall LMFBR shield development effort.

In this section we will summarize the nuclear design features of the FTR and point out the more significant nuclear problem areas. A description of the nuclear development programs will be given and their impact on FFTF design decisions will be assessed. Finally, a brief discussion of the plans for nuclear startup and for the Irradiation Test Program will be presented.

FTR NUCLEAR DESIGN

In the early days of FTR design, the objective of obtaining high flux levels (maximizing watts/gms of fissile material) was a major nuclear design goal.[1,2] Other important design objectives called for easily accessible and ample test space and the utilization of core geometries and fuels prototypic of the proposed LMFBR's. The resolution of these design objectives has resulted in the present compact, oxide core consisting of 73 driver assemblies, 9 test positions, and 9 in-core control rods. At present, the core design is already quite well established and relatively little opportunity exists for adjustments. The main nuclear design problem, now, is to obtain accurate physics design information.

Summary Description of FTR Core Design

The reactor core consists of a hexagonal array of core assemblies arranged as shown in Figure 9-1. The initial core consists of 91 in-core positions surrounded by 9 fixed-shim or potentially moveable peripheral control rods and 99 reflector elements. All core and reflector units have the same lattice spacing of 4.715". The 9 test positions are arranged in a Y-shaped pattern compatible with three in-vessel fuel handling machines.

The first FTR core has an active height of 36" and a volume of 1034 liters. The design power of the core is 400 MW_t, with an expected total peak flux of 7×10^{15} n/cm^2-sec. The sodium coolant flow enters the reactor vessel via three 16" inlet nozzles. The coolant flows up through the assembly orifices and exits the reactor vessel through three 28" outlet nozzles. Coolant inlet temperature is 600°F and the average outlet temperature is 900°F.

Fuel assemblies are housed in hexagonally shaped ducts, 4.575" across flats and 12' long. A fuel assembly consists of a bundle of 217 fuel pins. Each pin is 0.230" OD with a clad wall thickness of 0.015". The cladding tube is made of 20% cold drawn type 316 steel. The pins are spaced by 0.056" spiral wire wrap. The fuel bearing portion of the pin consists of PuO_2/UO_2 fuel pellets stacked over a 36" length. Fuel enrichment is in the 20 to 25% range. The initial Pu composition is 86.5% Pu-239, 11.7% Pu-240, 1.6% Pu-241, and .2% Pu-242. Inconel reflector portions, 5.7" long, are located both above and below the fuel column. A 42" fission gas plenum is provided in the top of the fuel pin. The expected peak burnup is about 80 MWD/kg.

The reactor is controlled by means of 9 in-core B_4C rods and 9 peripheral fixed shim control rods. In the initial core design, 3 of the in-core rods are designated as safety rods. The remaining 6 in-core rods are used for shim control. The control margins are listed in Table 9-1.

The radial reflector consists of two kinds of reflector assemblies. The inner assemblies are supported by the core basket. The outer assemblies are shorter and are supported by the shield assembly. The active sections of the reflector consist of Inconel-600. The radial reflector is surrounded by the radial shield region made up of welded steel plates.

The core support structure locates the core components within the vessel. The core basket within the core support structure provides the support for the core and reflector assemblies. In-vessel storage is provided in three parts of the region between the core barrel and the vessel thermal liner. A core restraint mechanism governs the core configuration during operation.

Three instrument trees mounted in the reactor head position the core outlet instrumentation sensors above the driver fuel. They also provide secondary holddown for the core. Instrumentation monitors temperature and flow. A low level flux monitoring system determines the reactivity status of the reactor from full shutdown to very low power. The monitors are located in in-vessel thimbles and are retractable.

Some of the more important reactor parameters are listed as part of Figure 9-1.

Nuclear Design Status

Fuel Enrichment and Fuel Management

In the WARD study[4] two basic fuel management schemes for the FTR were analyzed. In both schemes it was assumed that no fuel elements were shuffled, rotated or reinserted. It was further assumed that the feed enrichment for all cases is equilibrium cycle feed. In one scheme - the "Maximum Burnup Refueling" scheme - the fuel assemblies with the highest burnup were discharged at the end of each cycle to maximize the burnup for the cycle. In the other scheme - the "Scatter Refueling" scheme - the discharge fuel is selected on a scattered basis throughout the core.

It was found that scatter refueling assures a higher cumulative discharge burnup and still achieved the same flux and power capability as the maximum burnup refueling scheme. (Average burnup for equilibrium operation: 49 MWD/kg, peak burnup: 80 MWD/kg.) Scatter refueling also provided a more uniform flux

LEGEND

- 73 - DRIVER FUEL ASSEMBLIES
  - -28 INNER (ROWS 1-4)
  - -45 OUTER (ROWS 5-6)
- OT: 4 - OPEN TEST ASSEMBLIES
- PROX OT INST: 1 - OPEN TEST ASSEMBLY WITH PROXIMITY INSTRUMENTATION
- GP CL: 2 - GENERAL PURPOSE CLOSED LOOPS
- SP CL: 2 - SPECIAL PURPOSE CLOSED LOOPS *
- SR: 3 - IN-CORE PRIMARY (SAFETY) RODS
- CR: 6 - IN-CORE SECONDARY (CONTROL) SHIM/SCRAM RODS
- PC: 9 - FIXED-SHIM OR POTENTIAL MOVABLE PERIPHERAL CONTROL RODS OR REFLECTORS
- 99 - REFLECTORS

* REACTOR CAPABILITY = 6 CLOSED LOOPS

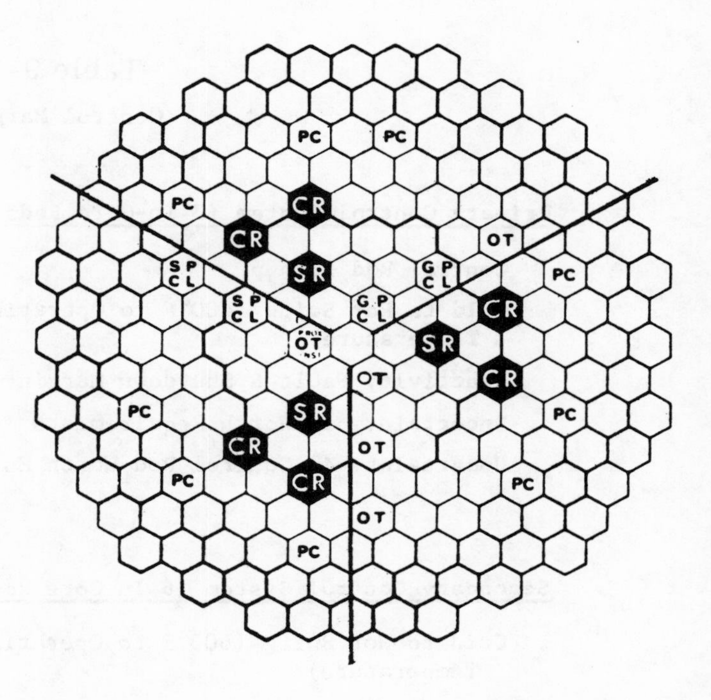

DESIGN VALUES OF ENGINEERING PARAMETERS

ITEM AND NUMBER SUPPLIED	VALUE EACH
REACTOR CORE	
POWER EXCLUDING CLOSED LOOPS	400 MWt
PEAK FLUX AT 400 MWt (n/cm^2 - sec)	7×10^{15}
DOPPLER CONSTANT (T dk/dt)	(-)0.005
LATTICE POSITION SPACING (91)	4.715 in.
FUEL DUCT (316 SS-20% CW)	
WALL THICKNESS	0.120 in.
OUTSIDE MEASUREMENT	4.575 in.
INITIAL CLOSED LOOPS (4)	2.3 MWt
COOLANT - MAX TEMP	1220°F
TEST DIAMETER	2.5 in.
REACTOR VESSEL (TYPE 304 SS)	
OVERALL HEIGHT	43 ft 1 in.
INLET NOZZLES (3) SIZE	16 in.
OUTLET NOZZLES (3) SIZE	28 in.
HEAD - DIAMETER (OD)	25 ft
HEAD - THICKNESS	49.81 in.
CONTAINMENT	
DIAMETER	135 ft
OVERALL HEIGHT	187 ft
DEPTH BELOW GRADE	78 ft
PRESSURE INT/EXT. (psig)	10/0.2
TEMPERATURE HIGH/LOW (°F)	250/(-)10
ALLOWABLE LEAK RATE (% PER DAY)	0.1
ELECTRICAL SYSTEM	
115 kV SUPPLY FEEDER (1)	
13.8 kV TRANSFORMER (1)	
13.8 kV STANDBY FEEDER (1)	3 MVA
480 V EMERGENCY GENERATOR (2)	1200 kW

ITEM AND NUMBER SUPPLIED	VALUE EACH
CLOSED LOOP EX-VESSEL MACHINE	
BORE DIAMETER (NOMINAL)	8 in.
HEAT REMOVAL CAPABILITY	10 kWt
PRIMARY HEAT TRANSPORT SYSTEM	
COOLING CIRCUITS (3)	133 MWt
HOT LEG-LOW PRESSURE PIPING OD	28 in.
-STRUCT. DESIGN TEMP AT 120 psig	1050° F
-VALVES (3) SIZE	28 in.
COLD LEG-PIPING SIZE OD	16 in.
-STRUCT. DESIGN TEMP AT 225 psig	830° F
-ISOLATION VALVES (3) SIZE	16 in.
-CHECK VALVES (3) SIZE	16 in.
PUMPS - (3) FLOW AT 500 ft Na HEAD	14,500 gpm
TEMPERATURE	1050° F
INTERMEDIATE HEAT EXCHANGERS (3)	133 MWt
STRUCT. DESIGN TEMP SHELL SIDE AT 225 psig	1050° F
STRUCT. DESIGN TEMP TUBE SIDE AT 250 psig	1050° F
SECONDARY HEAT TRANSPORT SYSTEM	
MAIN PIPING - SIZE OD	16 in.
BRANCH PIPING - SIZE OD	8 in.
PIPING IHX TO DHX STRUCT. DESIGN TEMP AT 250 psig	1000° F
PIPING DHX TO IHX STRUCT. DESIGN TEMP AT 250 psig	830° F
VALVES STRUCT. DESIGN TEMP AT 250 psig	1000° F
PUMPS - (3) FLOW AT 400 ft Na HEAD	14,500 gpm
STRUCT. DESIGN TEMP AT 250 psig	830° F
DHX UNITS (3) (4 MODULES/UNIT)	133 MWt
MODULES (12)	33 MWt
TUBE SIDE, STRUCT. DESIGN TEMP AT 250 psig	1000° F

Figure 9-1. FFTF Core Map

Table 9-1

Control Margins

Primary Control System (3 In-Core Rods)	% Δk/k
Control Rod Burnup	.05
Cold to Hot Swing (400°F to Operating Temperature)	1.30
Reactivity Fault & Shutdown Margin	1.20
Uncertainty in Worth Requirements	.35
Uncertainty in Control Rod Worth Estimates	.30
Total	3.20

Secondary Control System (6 In-Core Rods)	% Δk/k
Cold to Hot Swing (600°F to Operating Temperature)	.80
Reactivity Fault & Shutdown Margin	1.20
Fuel Burnup (15 MWD/kg)	2.60
Irradiation-induced Growth	.30
Operating Flexibility	.20
Uncertainty in Enrichment	.35
Uncertainty in Control Rod Worth Estimates	.55
Total	6.00

Peripheral Control System (9 Peripheral Rods)	% Δk/k
Test Allowance	1.00
Uncertainty in Enrichment	.50
Uncertainty in Initial k_{eff}	.50
Uncertainty in Control Rod Worth Estimates	.20
Total	2.20

to the core test positions. At present, scatter refueling is the accepted reference fuel management scheme for the FTR.

The effect of different enrichments on the core performance was investigated in detail. It was found that overenriching the feed does not improve the average discharge burnup for a fixed maximum allowable burnup. If enrichment is too high, additional shim control is necessary, resulting in non-optimum power distributions and a lowered power and flux potential. If the feed enrichment is too low, the performance is again non-optimum, resulting in a reduced cycle length. The penalties are more severe for underenrichment than for overenrichment

Fuel enrichments for FTR cores #1 and #2 have been set by WARD[5] with these guidelines in mind. The basic procedure used in setting the enrichment is as follows:

(1) A calculational method, cross section sets, and reactor modeling techniques are selected.

(2) The chosen calculational procedures are then tested on a series of critical experiments which closely simulate the FTR (the so-called Engineering Mockup Criticals, EMC) and a k_{eff} bias and uncertainty on this bias are established.

(3) The required excess reactivity margins for the FTR including uncertainty bands are determined.

(4) FTR enrichments are then calculated to be compatible with the k_{eff} requirements, the k_{eff} bias and the untainty bands.

Using the above approach, a total fissile Pu (Pu-239 + Pu-241) core loading of 551 kgs has been established. This implies a total Pu loading of 627 kg, and a total heavy metal loading of 2533 kg. The average fissile plutonium enrichment is therefore 551/2533 = 21.8%. The actual inner core zone enrichment is lower, and the outer core zone enrichment is higher to achieve a power flattened core.

Reactor calculations were carried out by means of two dimensional diffusion theory for a hexagonal reactor geometry. Axial buckling was determined from (R-Z) reactor models. The so-called FTR Set 300 cross sections[6] were employed. A test of these calculational schemes against a series of critical experiments indicated a reactivity bias of $k_{exp} - k_{calc} = .011 \pm .003$, i.e., for criticality, the calculated FTR core should have a $k_{eff} = .989$. Since the bias factors were derived from comparisons with plate type criticals, a further uncertainty of .002 $\Delta k/k$ was estimated due to heterogeneity effects.

The nominal excess reactivity requirements for the FTR were determined to be .0750 $\Delta k/k$. The uncertainty in this number was estimated to be $\pm .005$, and hence the enrichment was selected to yield a k_{eff} of 1.08 - after bias corrections, since possible overenrichment is considered less undesirable than possible underenrichment. A summary of the excess reactivity requirements is given in Table 9-2 below.

Table 9-2

Excess Reactivity Requirements

		$\Delta k/k$
1.	Fuel Burnup (100 day cycle)	.0272
2.	Equilibrium Feed Transition	.0313

3. Irradiation Induced Growth .0030
4. Loop Loading Variation .0075
5. Maneuverability .0010
6. Cold-to-Full Power Margin .0050

.0750

The uncertainty margin of .005 is still being assessed. Somewhat larger values up to ~.007 have been considered.

The control rod B_4C loadings are consistent with the indicated static control requirements.

Spatial Power Distributions

The nominal power lever for the FFTF is set at 400 MW_t. The spatial power distributions, specifically the peak-to-average power density values, enter directly into the determination of the reactor power level.

Spatial power distributions have been determined by WARD[7] and assessments of the uncertainties in the peak/average values have been made by comparison with EMC results.

Analyses of critical experiment results indicate that C/E values for radial peaking factors always exceed unity. Since the C/E ratio is consistently greater than unity, it has been accepted as a bias with a value of C/E = 1.03. To take calculational uncertainties and uncertainties in rod positioning into account, an uncertainty of ± 4% has been estimated for the radial peaking factor. For axial power distributions, a C/E bias of 1.02 has been determined from comparison with critical experiments. The uncertainty for this C/E value has been estimated at ± 2%.

The most recent radial and axial peaking factors for beginning of quasi equilibrium cycle conditions are tabulated below.

Table 9-3
Power Peaking Factors

	Calculated	Corrected for C/E
Radial	1.38	1.34 ± 4%
Axial	1.23	1.21 ± 2%

Power Coefficients

A knowledge of reactor power coefficients is essential for efficient and safe reactor operation. Power coefficients for the FTR have been estimated by WARD[8] based on fairly detailed calculations and comparison with applicable critical experiments.

The WARD calculations are based on the following startup mode:

(1) The reactor is refueled and startup occurs at a sodium temperature of 400°F.

(2) Isothermal reactor temperature of 600°F (or 800°F) is reached using pump power.

(3) Eighty percent coolant flow is established.

(4) The reactor reaches criticality at 600°F (or 800°F) isothermal conditions.

(5) Reactor power is increased to 80% at full 300°F ΔT across the reactor core.

(6) Reactor power and flow are increased to 100% retaining the 300°F ΔT margin.

The FTR power coefficients obtained by WARD are tabulated below. The power coefficients pertain to an incremental power change about the nominal 400 MW_t power level. They are all negative. Coefficients for both 600°F and the alternate 800°F inlet temperature conditions are shown.

Table 9-4

FTR Power Coefficients

Components	¢/MW at 600°F Inlet	¢/MW at 800°F Inlet
Doppler	.24	.22
Sodium Density	.02	.02
Radial Expansion	.20	.20
Axial Expansion	.10	.11
Total	.56	.55

From a safety point of view, the Doppler and sodium coefficient are of prime interest. The Doppler coefficient is an "inherent" coefficient which is essentially independent of mechanical design features. Its almost instantaneous response time and negative value makes it an extremely important safety feature of the FTR. The sodium coefficient is also of major interest in safety analyses. Even though the overall sodium coefficient value for FTR has been estimated to be very small and negative for FTR, the spatial dependence of this coefficient is quite pronounced and the central region FTR coefficient is positive.[9]

At present, the best nominal value for the beginning of life FTR Doppler constant, Tdk/dT, is -.0050 ± 20%. (Note: The Table 9-3 values are based on a previous estimate cf Tdk/dT - -.0047.) There are some indications that the nominal values may actually be somewhat more negative.

Because of FTR's relatively small size and resulting high enrichment, the Doppler constant is lower than what could be expected for larger oxide reactors. The presence of an inconel reflector accompanied by a softening of the edge spectrum tends to offset this trend somewhat. In Figure 9-2 reproduced from reference (11), the variation of Doppler constant as a function of core size for very simplified reactor geometries is illustrated. The 1000 liter, Ni reflected point represents an "FTR-like" core.

The FTR Doppler coefficient is a function of many reactor variables. Calculations indicate that as the core becomes depleted, fission products are

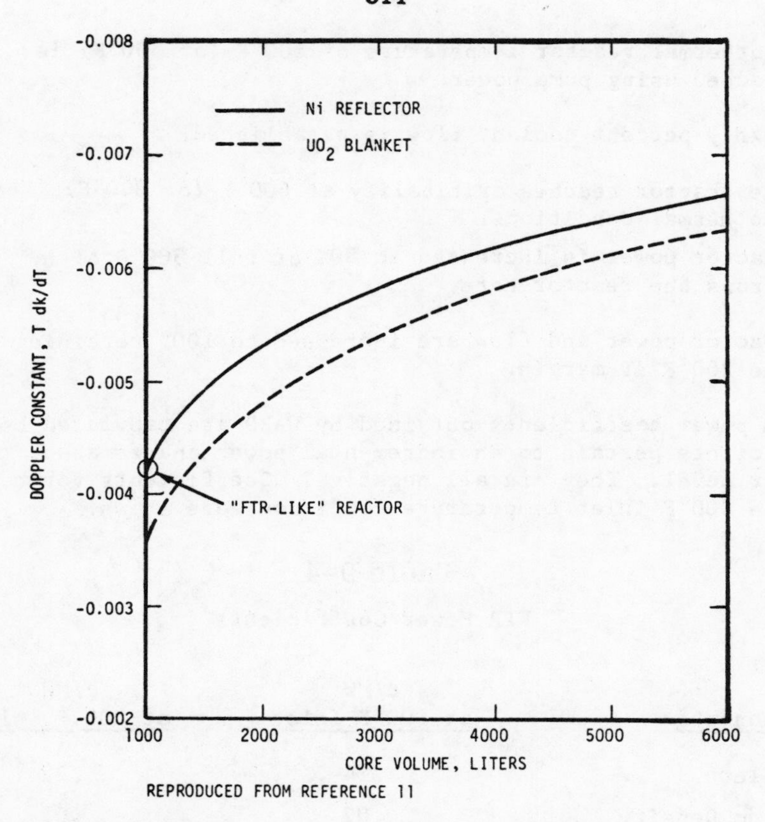

Figure 9-2. Variation in Doppler Constant with Core Volume and Reflector Material

accumulated, and control rods withdrawn, the Doppler coefficient becomes somewhat more negative. In a major hypothetical accident, the so-called "Na-out" coefficient is of interest. In this condition the FTR Doppler constant would be reduced by about 40%. Detailed accident analyses have been carried out which account for possible regional sodium ejections from the core. These analyses show that the reduction in the Doppler coefficient never gets below a value where energy releases would tend to increase sharply. Comprehensive operational and safety analyses utilizing the available power coefficient information have been performed for the FTR and the safety of the overall plant design has been satisfactorily established.

DEVELOPMENT OF ANALYTICAL PROGRAMS

The development of the analytical programs in support of FTR nuclear design was aimed at two main objectives:

- Development of programs which supplemented the existing array of calculational techniques and codes for fast reactors.
- Development of programs which improved - particularly in the area of speed and practical usage - existing fast reactor codes.

A system of codes was generated along these guidelines, compared and checked against existing methods and tied together into a useable design code package. This computer code system has now withstood the test of time for the last five years or so, and many of its subelements are being adopted by the reactor industry.

An outline of the HFRC (Hanford Fast Reactor Codes) package is shown in Figure 9-3. This code system basically has three branches covering core nuclear calculations, reactor accident calculations, and shield calculations. The inputs to the package are primarily physical reactor parameters and basic nuclear data. The output of the calculations is fed directly into core and shield design, and into overall reactor safety assessments. The nuclear data input to the design package is closely coupled with BNL's nuclear data files.

The details of some of the more important code modules are discussed below.

Cross Section Processing Routines

In the early days of FTR design, most of the nuclear data were derived from the so-called Russian ("ABN") Compilation.[12] The actual cross sections used were considerably modified. Modifications are documented in references (13) and (14).

More recently, the basic source of nuclear data for FTR design have been taken from the "Evaluated Nuclear Data File, Part B, ENDF/B." To transform these data into a form suitable for reactor calculations requires some method for generating multigroup cross sections. The self-shielding factor method (also referred to as "shielding factors" or "structure factors" method) for generating such group cross sections was developed in the early fifties at KAPL.[15,16] With the publication of the ABN cross sections in 1964, this method has now received wide recognition in the U.S. The codes ETOX/1DX[17,18,19] developed at PNL and HEDL utilize this structure method.

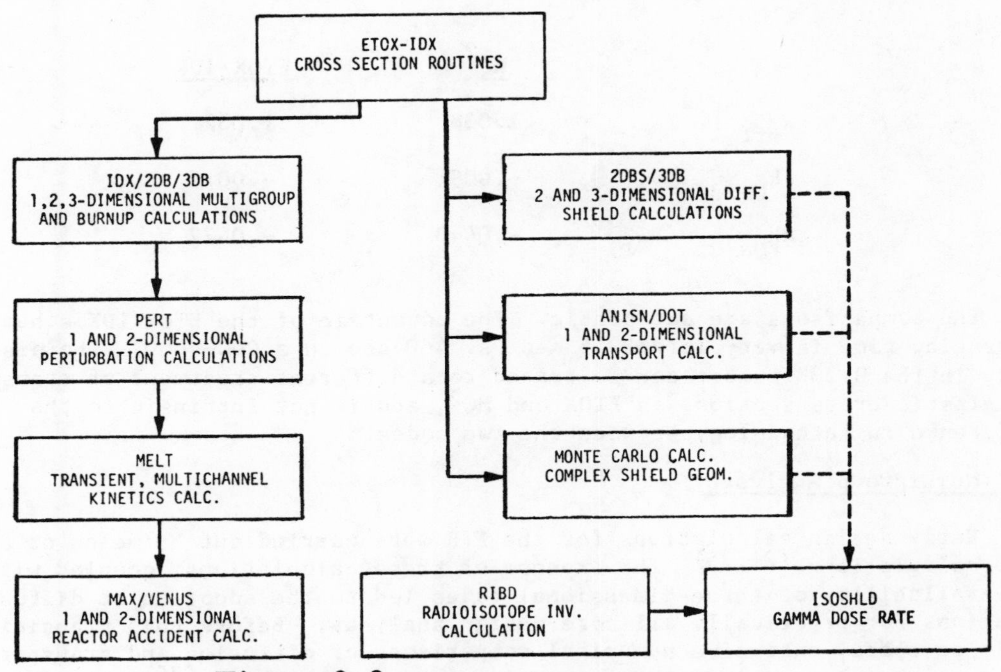

Figure 9-3. Hanford Fast Reactor Code Package

Shielding factors and infinitely dilute cross sections are generated for all isotopes of interest. The shielding factor, f, is defined by:

$$f = \frac{\bar{\sigma}}{\langle\sigma\rangle} ,$$

where: $\bar{\sigma}$ is the effective resonance self-shielded cross section, and

$\langle\sigma\rangle$ is the infinitely dilute cross section.

Shielding factors are computed for each energy group and nuclear process, for selected temperatures and background total cross section. The above calculations are performed in ETOX (ENDF/B to 1DX). Effective multigroup cross sections are then computed by the 1DX code by interpolating to the appropriate temperature, background cross section, group and nuclear process. 1DX also generates the usual one-dimensional diffusion information and can be used to iterate on the spectral flux profile.

The principal advantage of the ETOX/1DX method is the separation of the detailed resonance integration from the calculation of multigroup constants for a specific mix of isotopes. This results in greater calculational speed and ease of use. The major disadvantage of the method is that the flux weighting factors for a particular isotope do not reflect the energy dependence of the other cross sections in the mix.

The ETOX/1DX scheme has been compared with the more elaborate processing code MX^2.[20] For a typical fast reactor composition, some results[19] are shown in the table below.

Table 9-5

Comparison of ETOX/MC^2 Results

	MC^2	ETOX/1DX
k_{eff}	1.038	1.032
Δk_{U-238}	-.0091	-.0077
Δk_{Na}	-.0443	-.0472

The comparisons are acceptable. The advantage of the ETOX/1DX scheme in processing time is very large - 5 sec vs. 450 sec on a CDC-7600. The discrepancy in the U-238 number can be traced to a different treatment of transport and elastic cross sections in ETOX and MC^2, and is not intrinsic to the difference in methodology between the two codes.

Multigroup Analyses

Early design calculations for the FTR were carried out by means of S_n transport calculations.[21-25] The expense of these calculations, coupled with the non-availability of three-dimensional codes led to the adoption of diffusion calculations for practically all core design analyses. Before this transition was fully accepted, extensive numerical comparisons of diffusion and transport methods for selected fast reactor configurations were made.[26] The comparisons utilized the Los Alamos transport codes DTF-IV[27] and 2DF[28] and diffusion codes 1DX and 2DB.[29,30] Identical 5-group cross sections were used by each code in all calculations. Comparisons were made with respect to k_{eff}, reactivity coefficients, flux profiles and flux magnitudes far from the core source. The effects of mesh spacing, core geometry and volume on calculational results were also examined.

In the table below, comparisons of k_{eff} values for various one-dimensional, iron reflected cylinders with representative fast reactor core compositions are shown. Only isotropic scattering cross sections were used.

Table 9-6

k_{eff} Ratio as a Function of Quadrature Order for Cylindrical Reactors

Calculation	Computation Time (min) a.	k_{eff}/k_{eff} (S_{12}) 25 cm	50 cm	100 cm
DTF-IV, S_2	0.45	1.00465	1.00168	1.00043
S_4	0.98	1.00414	1.00140	1.00036
S_6	1.70	1.00104	1.00030	1.00008
S_8	2.65	1.00043	1.00014	1.00006
S_{12}	5.08	1.00000	1.00000	1.00000
1DX	0.02	0.98443	0.99633	0.99944

a. on a UNIVAC 1108

The transport k_{eff}'s are always larger than the diffusion k_{eff}'s. For a 50 cm radius (representative of the FTR) the k_{eff} bias is less than .5%. For a 50 cm sphere the bias would be close to 1% in k_{eff}. For large reactors (~100 cm in radius) the bias becomes negligible. The advantage in calculation time for 1-D geometries is enormous, i.e., a factor of about 5.08/.02 ~ 250 (!) for an S_{12}/diffusion comparison and about .98/.02 ~ 50 for an S_4/diffusion comparison. These comparisons also apply to uranium reflected reactors.

In Table 9-7, similar comparisons for 2-D calculations are shown. The comparison is for a cylindrical reactor, iron reflected radially and axially.

Table 9-7

Comparison of 2DB and 2DF

Parameter	2DB	2DF (S_2)	2DF (S_4)
k_{eff}	1.0810	1.0955	1.0888
Computation Time (Minutes)	2.1	12.7	30.0

---

*Reproduced from Reference (26).

To save computer time, the convergence criterion was less stringent than for the 1-D calculation and transport calculations were carried out only up to S_4. The 1% bias for the S_4 calculation is similar to the 1-D result. The saving in computation time is about a factor of 15.

Comparative transport/diffusion results for a central control rod (5 cm radius) in a 50 cm radius core are shown in Table 9-8.

Table 9-8

δk_{eff} as a Function of the Number of Mesh Intervals

Number of Mesh Intervals (rod, core, reflector)	δk 1DX	δk DTF-IV (S_4)	Ratio: 1DX / DTF-IV
30 (5, 15, 10)	-0.02199	-0.02166	1.015
60 (10, 30, 20)	-0.02211	-0.02161	1.023
120 (20, 60, 40)	-0.02216	-0.02160	1.025

The results indicate that the transport values are less sensitive to mesh spacing than the diffusion results. The bias in rod worth is about 2.5% for typical B_4C control rod compositions. As the B-10 content is increased, the transport diffusion discrepancy increases. The transport calculations generally show greater flux depressions near the rods than diffusion calculations. For design purposes, the discrepancies are small enough so that the use of simple correction factors is adequate. Similar conclusions also apply to peripherally located control rods.

Comparison of reactivity worth of Pu-239 and Na in central zones show remarkable agreement within a few percent.

For cell calculations, e.g., plate cells typical of the ZPR's, transport calculations are essential to obtain detailed spatial flux distributions. The reactivity calculations, however, are generally not sensitive to whether transport or diffusion calculations are carried out.

Gross flux distributions within the core and reflector regions can also be quite adequately handled via diffusion calculations. In fact, total flux distribution deviations are about a percent in the core regions and increase steadily to only about 10% to distances of 4.5 m (!) from the core center. For high energy fluxes (> .8 Mev), the discrepancies are, however, quite large at distances far from the core, e.g., 50% at 2.5 meters from the core center.

From the above discussion, it is evident that diffusion methods are quite adequate for fast reactor nuclear design calculations and that minor transport corrections can easily be made. Diffusion calculations also have considerable merit for fast reactor shielding analyses. The absence of hydrogenous media relaxes the dependence of the total flux on the high energy scattering treatment.

The diffusion codes which were developed for FTR nuclear analyses are 2DB and 3DB.[31,32] The perturbation code PERT[33,34] also finds frequent applicability. 2DB is designed explicitly for fast reactor analysis. Eigenvalues are

---

*Reproduced from Reference (26).

obtained by standard source-iteration techniques. Adjoint solutions are also generated. Criticality searches can be performed on buckling, time absorption, material concentrations and region dimensions. Criticality searches can be performed during burnup calculations to compensate for fuel depletion. Geometry options for the code are R-Z, R-θ, X-Y and a hexagonal grid. With the advent of fourth generation computers, three-dimensional analyses are becoming practical and this led to the development of 3DB. This code handles X-Y-Z, R-θ-Z and hex-Z geometries. It is quite similar to 2DB.

Shield Calculations

Shielding codes being utilized in the design analyses for FFTF include a number developed outside the Hanford Laboratory. The ORNL transport codes ANISN(35) and DOT(36) are used in addition to the especially adapted diffusion code, 2DBS(37) to generate spatial flux distributions. In 2DBS, flux convergence, particularly for points far from radiation sources, is accelerated. Additional output options, especially reaction and activation rate calculations, have been added. Monte Carlo codes developed at ORNL and BNW have also been employed where three dimensional analyses are required.

Fission product inventories and decay heat rates are treated employing the RIBD(38) code. A significant effort was expended to generate nuclear data libraries for RIBD that were appropriate for the FTR. Gamma dose rates for simple source geometries (cylinders, spheres, lines, parallelopipeds, points) are obtained using the point kernel code ISOSHLD.(39) The ISOSHLD photon source library(40) for fission products was revised and updated for FFTF project application. The more complicated distributed gamma sources that are associated with neutron capture are treated using transport codes such as ANISN and DOT.

The basic nuclear parameters for shield audit calculations are obtained from the ENDF/B files and processed by means of the ETOX code. A number of cross section studies have been made by ORNL for elements such as iron to seek the best possible evaluation. The design work by WARD employs a modified ABN set of cross sections.

Accident Calculations

A very extensive reactor safety program is presently under way at ANL. This program supplies both experimental and analytical results and techniques of direct applicability to LMFBR's and specifically to the FFTF.

To supplement this ANL program, some rather specific reactor kinetics codes have been developed at Hanford to treat the reactor conditions just prior to a hypothetical reactor disruption leading to melt-down and destruction of the reactor core. The Hanford codes are part of the series of MELT programs.(41,42,43)

MELT-II is a two-dimensional neutronics-heat transfer program which simulates core behavior up to the point of potential core disassembly. Transient fuel temperature distributions, radial and axial, can be calculated in ten specified locations in the core. This "multi-channel" feature of the MELT codes has now also been adopted in various ANL kinetics codes. Fuel movement and coolant voiding models are part of the present version of the program and the corresponding reactivity feedback coefficients are calculated. Doppler feedback is computed according to a spatial weighting scheme which takes into account the local spectrum hardening effects brought about by local sodium voiding. This particular refinement represents an important space-time correction of the underlying point-kinetics formalism of MELT.

The early versions of MELT were constructed in such a way as to feed directly into the one-dimensional hydrodynamic disassembly model, MAX.[44] The MAX code computes the energy release for large, hypothetical nuclear excursions. The MAX neutronics model utilizes the usual lumped parameter kinetics equations with tabular power and worth profiles. At the time of publication, an important new feature in MAX was the hydrodynamics model which consisted of many "moveable and compressible mesh points." ANL's later VENUS code[45] includes these features and has the additional advantage of being two dimensional. MELT-II and VENUS have now been coupled into a complete reactor accident analysis model.[46] Further improvements in MELT-II are presently under development as part of the MELT-III code package.

PLANS FOR NUCLEAR STARTUP AND THE IRRADIATION TEST PROGRAM

Nuclear Startup Tests

An active program of planning for the nuclear startup tests was started in 1971. An outline[80] of nuclear startup tests was prepared, estimates of startup schedules were made, and special equipment requirements were established.

The nuclear tests have been organized into three categories as follows:

(1) System Tests

(2) Nuclear Startup Tests, and

(3) Power Ascension and Full Power Tests.

In early plant layouts for the FFTF, provision was made for a so-called Nuclear Proof Test Facility (NPTF),[81-88] a zero-power critical assembly which closely resembled the FTR reactor. Most of the large U.S. thermal test reactors have this kind of prototypical critical facility. The major function of the NPTF was to contribute to safe, efficient and high-quality testing in the FTR. At this time, the NPTF facility has been deferred, and the present view is that the nuclear startup tests for the FTR would also include a series of tests which might be designated as core characterization tests. Some of these tests might have been done in the NPTF. The following tests would fall into this category:

(1) Determination of spatial dependence of selected material worths

(2) Determination of flux maps

(3) Determination of neutron energy spectrum maps

(4) Determination of flow and temperature coefficients

This series of four tests would complement similar measurements made previously in the critical facilities and would serve to provide adequate data for reactor operation in the absence of the NPTF.

Reactivity Monitoring and Anomaly Detection

A very important part of reactor operational procedures is the reactivity monitoring of the machine. Reactivity information provides a basic understanding of the reactor status to the operator and also serves an important safety function.

Reactivity is not measured directly, but is generally deduced from a variety of detector signals. The "Reactivity Monitoring and Anomaly System" (RMAAS) is a calculational system used to determine the reactivity status of the FTR.

RMAAS covers three classes of reactor operations:

- Reactor Shutdown
- Reactor Approach-to-Critical
- Reactor-at-Power

The shutdown category pertains to the FFTF reactivity status from -30$ to -1$. Reactor loading and reloading and other core changes take place in this reactivity domain.

The Approach-to-Critical category will provide reactor monitoring, both by means of off-line calculations as well as via an on-line computerized system.

When the reactor begins to generate power (1 kw - 1 MW range) and in the power range, reactor monitoring will be done on-line by means of a computer. At power, RMAAS will be used to determine the reactivity of the reactor, calculate the anomalous reactivity and provide output information to the operator.

On-line computer systems which perform reactivity calculations exist on many reactors. The FERMI system presently comes closest to the RMAAS system contemplated for the FFTF. FERMI's MDA[89] (Malfunction Detection Analyzer) system has been programmed to provide reactor operation messages in the anomaly range of 1¢ to 5¢.

In the shutdown reactivity range some problems currently exist. Subcritical measurements near one dollar subcritical have been made by various methods (inverse kinetics rod drop analyses, cross power spectral density noise analyses and polarity spectrum coherence neutron noise analysis) and systematic discrepancies have been uncovered. To measure reactivities in the -1$ to -30$ range depends largely on the ability to calculate the reactivity status of different shutdown reactor configurations in order to relate increased flux to reactivity. The procedure to be followed by RMAAS will rely on calculated correction factors to account for variations from the standard inverse multiplication rule. Checks of this procedure on the recent FTR-3 Critical[90] indicates that this approach has good potential.

It is expected that the RMAAS system will be operational in time for FFTF startup.

FFTF Irradiation Test Program Plan

Irradiation test plans are being developed to facilitate the orderly and timely utilization of the FFTF as the centralized test facility for the entire U.S. LMFBR program. These plans are closely coupled with nuclear design activities since prototypic LMFBR flux level intensity and spectral energy distributions are two of the prime characteristics of the irradiation program requirements. An initial reference test program has been developed for the irradiation testing of fuels and materials in the FFTF.[91] This program plan supersedes an earlier version issued in 1970.[92] In addition, a detailed "Users' Guide" has also been prepared[93] patterned after a similar document prepared for EBR-II.

The initial test plan addresses itself to a discussion of test requirements versus the FFTF capability of providing a suitable test environment. A system of generating test priorities is described. Examples of test environment requirements are burnup, transient behavior, fluence, neutron spectrum, flow rates and

ΔT, and coolant chemistry. In addition, specific mechanical and instrument requirements may be imposed by certain tests. In general, it is expected that some tests would be diverted to other facilities if these facilities can furnish a suitable test environment. The FFTF would be reserved, at least initially, for high priority tests which cannot be carried out in other facilities. In establishing the FTR testing priorities, the two criteria of "safe and effective operation of FTR" and "full and optimal utilization of the unique FTR testing capabilities" would be governing.

Fuels testing refers to irradiation tests for investigating part or all of the fuel-cladding-coolant system behavioral phenomena. Materials testing is to be taken in the context of non-fuel materials irradiation. The initial test schedule for the FFTF is shown in Table V-1. The position numbers in the table refer to the numbers shown on the FTR core map, Figure 9-4.

During the initial four cycles, the four instrumented open test positions and two of the closed loops are devoted to in-core performance measurements of FTR drivers and materials surveillance. The remaining two closed loops will be available initially for special purpose safety and general reactor technology tests. Early test results will not only be of direct benefit to FTR performance but also should be of direct benefit in the design and development of the LMFBR demonstration plants. The prototypicality of FFTF in this respect is shown in Table 9-10. A description of the FFTF closed loop capabilities is given in Table 9-11 and Figure 9-5.

SUMMARY

(1) The perpetual debate concerning the use of transport versus diffusion theory for core design of large (1000 liters and more) fast reactors can easily be decided in favor of diffusion theory. The flexibility gained by using the simpler diffusion approach can be exploited by utilizing more accurate, bona fide three-dimensional reactor calculational models with a well defined spatial and energy mesh. Minor transport corrections if necessary can always be superimposed on the diffusion results. The use of transport theory continues to have a legitimate place for detailed cell calculations.

(2) The use of transport theory for fast reactor shield design can and should be grossly augmented by diffusion calculations. Again, as in the case of core design, superior geometric modeling of the reactor made possible by the simpler diffusion approach is to be preferred over a more accurate description of reactor neutronics for an inferior geometric model.

(3) While it is true that calculational methods, computer capability, and nuclear cross section data have improved greatly in recent years, performance demands have certainly kept up with these improvements and sometimes have outpaced them. As a consequence, critical experiments as well as quasi-shield mockup experiments will continue to play a significant role in fast reactor design for some time to come. These experiments will serve as a means for further improvements in calcu-

lational techniques and nuclear parameters and also to obtain direct design information for a specific reactor design.

The technical complexities of FTR design have been considerable and real. The reasons for these complexities can be traced to two causes:

(1) The ultimate "testing function" of the FTR has resulted in an extreme interdependence of design features which simply does not arise on more conventional power reactors. It has been quite difficult to subdivide the entire design into separate systems which are reasonably independent and which interface with other systems in an easily defineable manner.

(2) The FTR is to serve both as a test bed for LMFBR fuels and materials irradiation and, at the same time, it is to be the vehicle for developing expertise and experience in the area of fast power reactor design and fabrication. These two design objectives often tend to conflict.

The general technical problems have spilled over into the nuclear design area to some extent. At this stage, the conflicting objectives have, however, been reasonably resolved and suitable compromises have been achieved.

REFERENCES

The publications referred to in chapter 9 may be found in "National Topic Meeting on New Developments in Reactor Physics and Shielding" by the U. S. Atomic Energy Commission Technical Information Center

Table 9-9

FFTF IRRADIATION TEST PROGRAM (INITIAL)
TEST POSITION LOADING VERSUS REACTOR OPERATING CYCLE

CORE ROW NO.		2			4			6		
POSITION NO.		1	2	3	4	5	6	7	8	9
CYCLE	TYPE	GPCL	OPEN TEST	OPEN TEST	GPCL	OPEN TEST	SPCL	OPEN TEST	OPEN TEST	SPCL
1974	1	FTR FUEL (7-37 PINS)	FTR DRIVER (217 PINS	MATERIALS TESTS (PROX. INST.) OR	FTR-TYPE DEMO. FUEL (7-19 PINS)	MATERIALS TESTS (SURVEIL-LANCE)	SPECIAL PURPOSE (1-19 PINS) OR MATERIALS	MATERIALS TESTS (GENERAL & SPECIAL)	FTR DRIVER (217 PINS) (WIRE WRAP OR GRIDDED)	SPECIAL PURPOSE (1-37 PINS)
	2									
1975	3			FTR DRIVER (217 PINS) (GRIDDED OR WIRE WRAP)		MATERIALS TESTS (GENERAL)				
	4									
1976	5	IMPROVED DEMO. FUEL	DEMO CORE FUEL (169 PINS)							
	6				LMFBR FUEL (7-19 PINS)				IMPROVED FTR DRIVER	
	7									
1977	8		IMPROVED DEMO CORE (127 PINS)					(ABSORBER TESTS)		
	9								LMFBR FUEL (127 PINS)	
1978	10									

CYCLE NUMBER

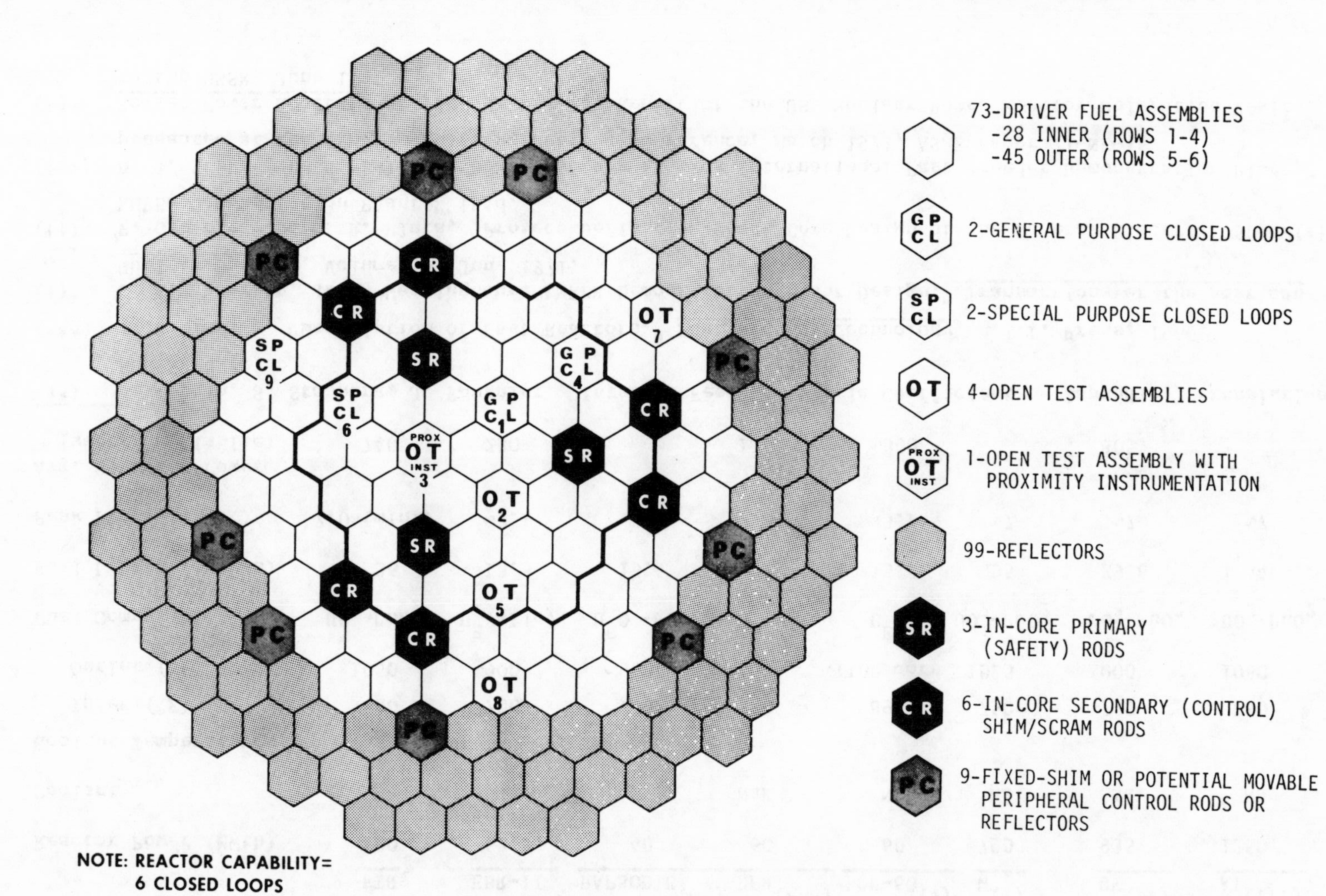

73-DRIVER FUEL ASSEMBLIES
-28 INNER (ROWS 1-4)
-45 OUTER (ROWS 5-6)
2-GENERAL PURPOSE CLOSED LOOPS
2-SPECIAL PURPOSE CLOSED LOOPS
4-OPEN TEST ASSEMBLIES
1-OPEN TEST ASSEMBLY WITH PROXIMITY INSTRUMENTATION
99-REFLECTORS
3-IN-CORE PRIMARY (SAFETY) RODS
6-IN-CORE SECONDARY (CONTROL) SHIM/SCRAM RODS
9-FIXED-SHIM OR POTENTIAL MOVABLE PERIPHERAL CONTROL RODS OR REFLECTORS
NOTE: REACTOR CAPABILITY= 6 CLOSED LOOPS

Figure 9-4. FTR Core Map with Test Positions

Table 9-10

COMPARISON OF FTR DESIGN WITH PROJECTED DEMO. PLANTS AND OTHER FAST IRRADIATION FACILITIES

						Demo Designs		
	FTR	EBR-II	RAPSODIE(**)	DFR(***)	BOR-60(#)	W(†)	GE(††)	AI(†††)
Reactor Power (MWth)	400	62.5	40	60	60	790	935	1250
Coolant	Na	Na	Na	NaK	Na	Na	Na	Na
Coolant Temperatures:								
Inlet (°F)	800	700	750	390	896	750	720	760
Outlet (°F)	1050	~900	~960	660	~1100 core	1025	1000	1060
Fuel Composition	UO_2-PuO_2	U^E-Alloy	U^EO_2-PuO_2	U^E-Alloy	U^EO_2	UO_2-PuO_2	UO_2-PuO_2	UO_2-PuO_2
Fuel Length (inches)	36	13.5	13.4	21	15.8	35	29.6	44
Peak Flux ($x10^{-15}$)	7.0 (min.)	~3.1	3	2.5	3.7	~7	~7	~7
Avg. Specific Power (watts/gm fissile)	740	230		170	~300		805	

(**) J. Martin, S. Stachura, J. Fournier, "Internal Feedback Static Coefficients - Rapsodie," translation EURFNR-837.

(***) J. G. Yevick, "Description of Fast Reactors," Fast Reactor Technology, M.I.T. Press, 1966.

(†) C. A. Anderson, Jr., "Westinghouse LMFBR Demonstration Plant Design," Transactions of the American Nuclear Society, Volume 14, June 1971.

(††) P. Greebler and P. R. Pluta, "Project Definition Phase Core Design of the General Electric 350-MW(e) LMFBR Demonstration Plant," ibid.

(†††) D. J. Stoker and R. Balent, "Design of the Atomics International Fast Breeder Demonstration Plant," presented at the ASME Nuclear Engineering Conference, March 1971, ASME Paper 71-NE-16.

(#) Soviet Power Reactors - 1970, WASH 1175, Report of the USA Nuclear Power Reactor Delegation Visit to the USSR, June 1970.

Table 9-11

MWt Closed Loop Testing Capabilities*

Testing Conditions	Values	Remarks
Maximum Na temperature from test section	1400°F	Achieved by internal by-passing and reducing total heat generation of test compatible with maximum flow capability.
Maximum Na hot leg temperature - primary piping	1200°F	Internals of CLIRA restricted to 1200°F above the mixing region.
Maximum Na cold leg temperature - primary piping	1000°	Includes both power and iso-thermal testing conditions.
Maximum Na flow rate	1.31×10^5 lb/hr	
Minimum Na flow rate	1.14×10^4 lb/hr	For normal operation control range. Control down to 5% of maximum required for decay cooling.
Maximum ΔT in primary	400°F	Corresponding test ΔT of 400°F requires bypass flow.
ΔP across test section	100 psi max.	
Minimum Na cold leg temperature - primary piping	520°F at 2.3 MWt	For lower temperature, accept lower power.
Maximum heat rejection capability of system	2.3 MWt	
Minimum cold leg temperature - secondary piping	420°F at 2.3 MWt system power	
Minimum test diameter in CLIRA	2.5 in.	Diameter of hole for test element.
Active length test element	3 ft.	Corresponds to active core.
Maximum test power	2.0 MWt	
Test train length	Capable of shortening to 12 ft.	For transfers out of containment.

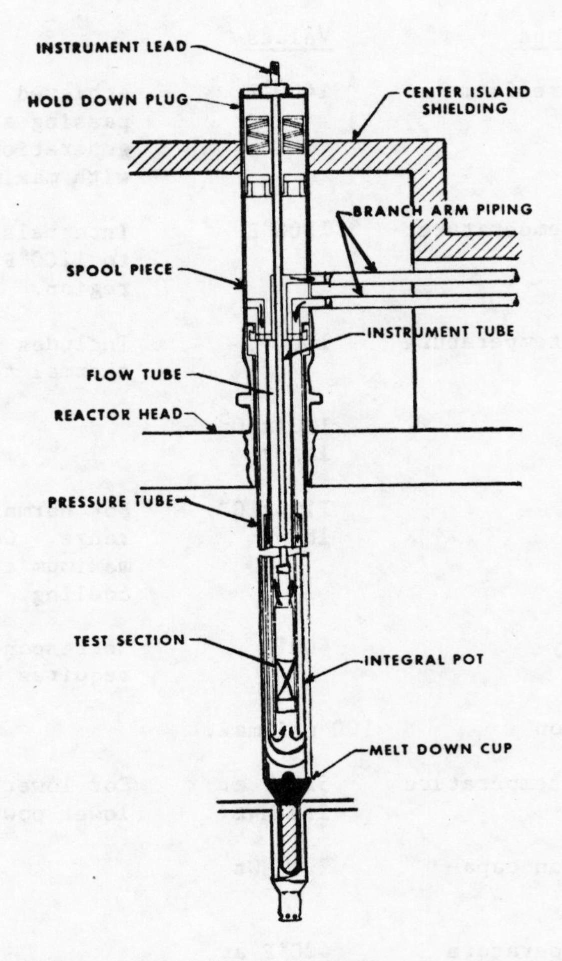

Figure 9-5. Closed Loop in Reactor Assembly

CHAPTER 10

FAST TEST REACTOR (FTR) SHIELDING

The purpose of this section is to present a summary of the calculational methods and relationship to the shielding design of the Fast Test Reactor (FTR). Emphasis is placed on the methods used to resolve the key design areas.

From a shielding viewpoint the FTR is unique in several aspects. Moderating materials such as graphite are not used within the vessel or around the reactor vessel for shielding purposes. The second unique aspect is a shielding design approach to minimize capital cost within a fixed reactor containment and equipment envelope. Shielding design has relied to a maximum extent on the envelope of the existing structures. As a result the order of magnitude error band typically used in shielding in the past has not been acceptable for the FTR design. The maximum utilization of benchmark and semiprototypic experiments coupled to the shielding analysis has been necessary. Both ANL FTR Critical Facility experiments and ORNL Tower Shielding Facility experiments have provided the confidence necessary in FTR shielding design analysis to ensure that the shielding design is adequate.

The key design areas which required development effort were the deep neutron penetration in sodium and steel, neutron streaming through annular gaps and steps, and streaming in the large reactor cavity[1]. These problem areas were uncertain enough that experimental work coupled to the analytical tools and methods was required to design the shielding. This development work was sponsored by the AEC under the LMFBR development program in shielding.

The experimental program which supported the FTR design yielded iron and sodium benchmarks for 2-1/2 to 15 feet of sodium and up to 36 inches of steel. In the case of neutron streaming both benchmark and prototypic gaps, annular slots and semiprototypic reactor cavity experiments were conducted and checked against cross sections, discrete ordinates transport theory and in some cases Monte Carlo techniques. By using exactly the same analytical tools to analyze the experimental data as was used in the design, reasonable correspondence with final results are expected. Extrapolations, of course, have had to be made due to physical size and the large number of penetrations which exist in the FTR design. Bias factors were applied only where necessary to correct the analytical design, fluences or doses.

With limited resources careful selection of experimental setups was required, in order that maximum design applications could be obtained from the experiments in a timely manner. Each experimental setup was analyzed with the DOT transport approximation. This same design tool and the same sets of cross sections were used to analyze the actual design configurations, thus cancelling the number of errors that typically are often made between the experimentalists.

By careful utilization of the experimental program coupled to the existing shielding analysis tools, a much closer shielding design for FTR has been achieved. Uncertainties are now predicted to be much smaller than originally were believed possible. This is not to say that inexpensive two and three dimensional transport shielding methods are not needed for future LMFBR design. Better tools are needed in shielding design analysis, particularly for the designer who faces day to day problems and design changes. Shielding costs in the future will be significantly reduced if the FTR design approach is used.

SHIELDING DESIGN DESCRIPTION

The status of the FTR shielding design and analysis is discussed. The experimental program has been completed and consisted of bulk penetration, streaming and large semiprototypic cavity experiments. The analysis of much of the experimental work has been completed with the DOT SN transport approximation. The design problems were all scoped with limits placed on them during the preliminary design phase of FTR which has been completed. The final design shielding analysis is nearing completion and will be completed over the next year. The final confirmation using high order SN transport approximations and some Monte Carlo work has not been completed. The first important area is in the complex reactor head. Manned access is desired on the head during power operation. Unlike FERMI, SEFOR and EBR-II, FTR has no graphite under the head for moderating material. The head itself in combination with additional steel and high nickel alloy materials beneath the head is utilized for shielding. Fig. 10-1 shows an elevation of the reactor cavity, vessel, fuel and head manned access area. The fission gas plenum constitutes a region which extends from about 6 inches to 48 inches above the FTR driver fuel. Three instrument trees and three in-vessel handling machines are major large stainless steel components in the sodium pool which penetrate the head. The vessel head consists of a combined thermal and nuclear radiation shield beneath a structural enclosure, the total of which is equivalent to 21 inches of steel. The reactor cavity is an annular nitrogen space 7 feet in width. The vessel support area is directly above the reactor cavity. The manned access area is directly above the head and vessel support area. This area is called the head compartment. This compartment contains parts of the components that penetrate the vessel head and peripheral areas and some of these parts require access during full power operation and all require access for maintenance and removal. Control rod drive lines penetrate the vessel head near its center and are connected to mechanisms above the head. Test loops also penetrate in this area. Although this section is devoted to neutron shielding, extensive gamma ray shielding is also required on the head because there is radioactive flowing sodium in the test loops. This sodium above, in addition to that sodium beneath, the reactor head must be shielded. Other penetrations into the head compartment area are sodium sample lines, low level flux monitors, fuel transfer ports, cover gas, core restraint mechanisms, ports for surveillance equipment, and power level instrumentation. All of these are considerations in determining the neutron environment in the head compartment in addition to the bulk neutron penetration through the head and the thermal radiation shield below.

One other major item which impacts significantly on the neutron dose in the head compartment area is the in-vessel stored fuel. The stored fuel is located in a trisector arrangement outside the core support structure and is well shielded from the active core by the radial shielding inside the core barrel. Nevertheless, fissioning in the stored fuel does result in a significant contribution to fast flux in the reactor cavity, as shown in Figure 10-2. This contribution must be considered in the peripheral parts of the head design, such as the in-vessel fuel handling machine and the support area around the head which includes the main concrete support ledge. Because of its significance in the shielding a major analysis effort has been expended in representing the stored fuel.

Figure 10-2 illustrates the impact of the stored fuel on the shielding design analysis. With stored fuel the total flux is about 4×10^8 nv just beneath the vessel support system. Without stored fuel the total flux is not significantly lower, but the fast flux is nearly 5 orders of magnitude higher. The same effect can be observed at the reactor beltline just outside the reactor vessel. This fast flux is the controlling aspect of the shielding design.

Figure 10-3 is an overall plan view showing the neutron fluxes in the reactor cavity and at other selected locations. The neutron flux is observed to be higher in the reactor cavity near the edge of the head than it is on the reactor centerline beneath the head. The spectrum, also, is significantly harder. The reason for this is that on the centerline there is a significant amount of sodium and steel for slowing down and attenuation. In the reactor cavity neutron scattering can occur over a height of approximately 50 feet of vessel, guard vessel and reactor cavity concrete wall. This neutron scattering makes the stored fuel contribution to neutron flux important even beneath the reactor vessel support ledge.

The first important area of design complexity was related to the desire to avoid graphite or moderating material under the head. Without moderating materials the age old iron window uncertainty problem at the 25 Kev level came back. The total neutron flux level under the head had been calculated in the early conceptual design stages to vary from approximately 5×10^7 to slightly over 10^{10}. This was another significant uncertainty which was largely dependent on the fission gas plenum and pool steel representation.

To resolve the neutron flux and pool steel problem, the error band was scoped by assuming the fission gas plenum to be a complete void and assuming it to be a homogeneous region. By coupling to the steel in the head and using S_N transport methods the neutron flux was found to vary by $\pm$ 1 decade around 1 mrem/hr. The iron windows nearly compensated in bulk penetration for spectrum differences and most of the apparent total flux difference below the head so it did not end up with 3 orders of magnitude uncertainty, but only one. The treatment of the iron in the Na pool below was found to contain a 10% uncertainty once the head area steel was coupled to the overall system problem. Detailed S_N analyses using an annular ring model and Monte Carlo analysis later confirmed the selection of a mid-point between the two fission gas plenum models. Thus a selection of the below-head shielding of the high nickel alloy Inconel and Stainless Steel was made. Today, confirmatory analysis using the S_N DOT transport approximation indicates that on bulk penetration we have a neutron dose of 0.5 mrem/hr within a factor of 3.

The second part of the head area dose design problem is the streaming through the major penetrations in the head. Fig. 10-4 illustrates many of the components that penetrate the head and their complexity in the head compartment. All the equipment which penetrates the head have annular gaps and voids which had to be mechanically offset to prevent excessive neutron and gamma ray streaming.

The In-Vessel Handling Machine (IVHM) is a typical example. The plug is about six feet in diameter, and the annular gap between it and the vessel head is 1/2 inch wide. On the basis of preliminary hand calculations of the required neutron attenuation, a two-step (offset) plug-IVHM design was selected. Fig. 10-5 shows the approach that was selected. The initial calculations using line of sight streaming formulas with offset reduction factors indicated that a two-offset, 1/16" gap clearance was necessary; however, 1/2-inch clearances were required in the mechanical design. Subsequent experimental efforts in the LMFBR program reduced the calculated dose rate by a factor of 70, thus allowing a more reasonable mechanical design. The reduction in streaming was a direct result of the annular slit series of stepped semiprototypic experiments done for FFTF at the Tower Shielding Facility (TSF).

The final design area of significant complexity to be discussed in this section is the reactor cavity source and its impact on the head support and surrounding vaults. These vaults include the main Heat Transport System Cells where the secondary sodium and structural materials must be protected from neutron activation.

The area in the reactor cavity at the head interface was a particularly difficult shielding problem from an analysis standpoint as well as from the design constraint standpoint. In a large area where there should be shielding there is none because of space allowance requirements for movement, inspection, insulation and maintenance. Referring to Fig. 10-3 the area of difficulty is between the reactor vessel and the main support. There are only 21 inches of carbon steel for shielding over a large annulus. Because of the hard spectrum and high total neutron flux, more neutron shielding is actually needed in this area than on the centerline of the reactor.

In order to fill in shielding around the head support area, the concrete containment support ledge was extended downward below the support area. In addition to the extended concrete support ledge, a collar of B_4C is placed on top of the guard vessel to fill in the neutron streaming paths to the vessel support area. The analysis of the neutron source with the stored fuel in place is a significantly complicated transport problem. With existing methods it is a multi-coupled S_N transport problem from the reactor core outward through the stored fuel, to the reactor cavity, upward in the reactor cavity and then through the head. In order to predict the streaming a high order angular quadrature must be used. Reference 10-2 describes this process.

The second part of the complex reactor cavity source problem is shown in Fig. 10-6 which shows the HTS pipeways in plan view. As can be seen there are a number of penetrations into these pipeways which contain the primary system valves. In addition there is a large shield (shadow shield) separating each pipeway from the corresponding main HTS cell which contains the secondary nonradioactive sodium. Although offsets exist between the reactor cavity penetrations and this shadow shield, there is still a significant neutron streaming path. The intermediate heat exchanger is directly on the other side of the shadow shield. This heat exchanger must be protected from a neutron flux that would result in significant activation of the secondary sodium. Again, coupled S_N transport approximation methods were used to calculate the neutron fluxes in the pipeway and the HTS cells. Specifications for the design were based on preliminary DOT ARD calculations. Much more detailed Monte Carlo analysis is planned by ORNL as a final confirmatory analysis. Confidence in the calculations is based on a recent Tower Shielding Experiment which mocked up the pipeway and its penetrations. Two mockups were made; one where the source-to-penetration angle was varied to determine the effect on streaming of source angular distribution and the second mocked up the pipeway with its staggered penetrations. This experiment was just completed at Oak Ridge.

Without the experimental program we would at this stage have little confidence in the analysis tools, cross sections, and methods that we have used in arriving at the basic design. The three series of experiments, the bulk penetration in sodium and iron, the annular slits and the semiprototypic mockup of the HTS pipeway have yielded valuable information in the area of deep penetration and neutron streaming analysis verification. From this series of experiments and with the shielding tools available, the three dimensional neutron transport problems in FTR have been and are continuing to be solved for the design of the shielding. The overall envelope is well established. Some supplementary shielding material may be required locally once the uncertainty analyses have been completed.

CONCLUSION

From a shielding viewpoint FFTF has to this point by use of key selected experiments and extensive analysis, designed closer to neutron dose rate/ fluence limits and activation of materials limits than previous sodium cooled reactors. Because of the LMFBR development program in shielding, significant progress beyond the one or two orders of magnitude uncertainty placed on shielding has been made. Significant progress has been made in the cross sections area, although better definition is still needed in certain areas as judged from the biases that we see. The LMFBR shielding program should continue to support future LMFBR reactors and, in particular develop simplier, rapid and more inexpensive multidimensional neutron transport tools. These tools are needed in order to enable the designer to make relatively quick analyses of many complex problems. The impact of these tools on capital costs of future LMFBR plants can be significant.

REFERENCES

1. J. L. Rathbun et al., "Shielding the Fast Flux Test Facility", ANS Transactions, Vol. 14, No. 2, p. 934, October 1971.
2. F. R. Mynatt and M. L. Gritzner, "Fast Reactor Shielding Methods Development", paper to be presented at National Topical Meeting, New Developments in Reactor Physics and Shielding, Kiamesha Lake, N.Y., September 12-15, 1972.
3. F. R. Mynatt, F. J. Muckenthaler and P. N. Stevens, "Development of Two-Dimensional Discrete Ordinates Transport Theory for Radiation Shielding", USAEC Report CTC INF-952 (1969).

REACTOR VESSEL ELEVATION

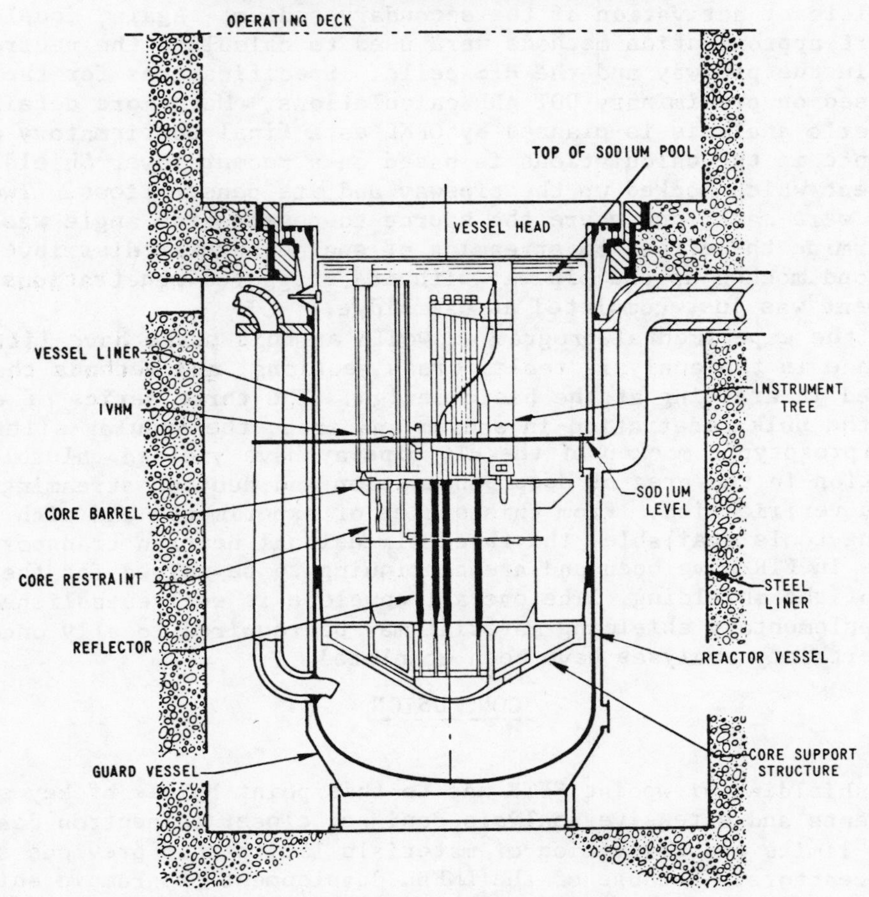

Figure 10-1. FFTF Reactor Vessel Elevation

IMPACT OF STORED FUEL ON NEUTRON FLUX

LOCATION - UNDER THE VESSEL SUPPORT STRAP LOCATION

FLUX	ϕ WITHOUT STORED FUEL	ϕ WITH STORED FUEL
>1 MeV	~40	2×10^6
>0.1 MeV	5×10^5	3×10^7
TOTAL	3×10^8	4×10^8

Figure 10-2. Impact of Stored Fuel on Neutron Flux

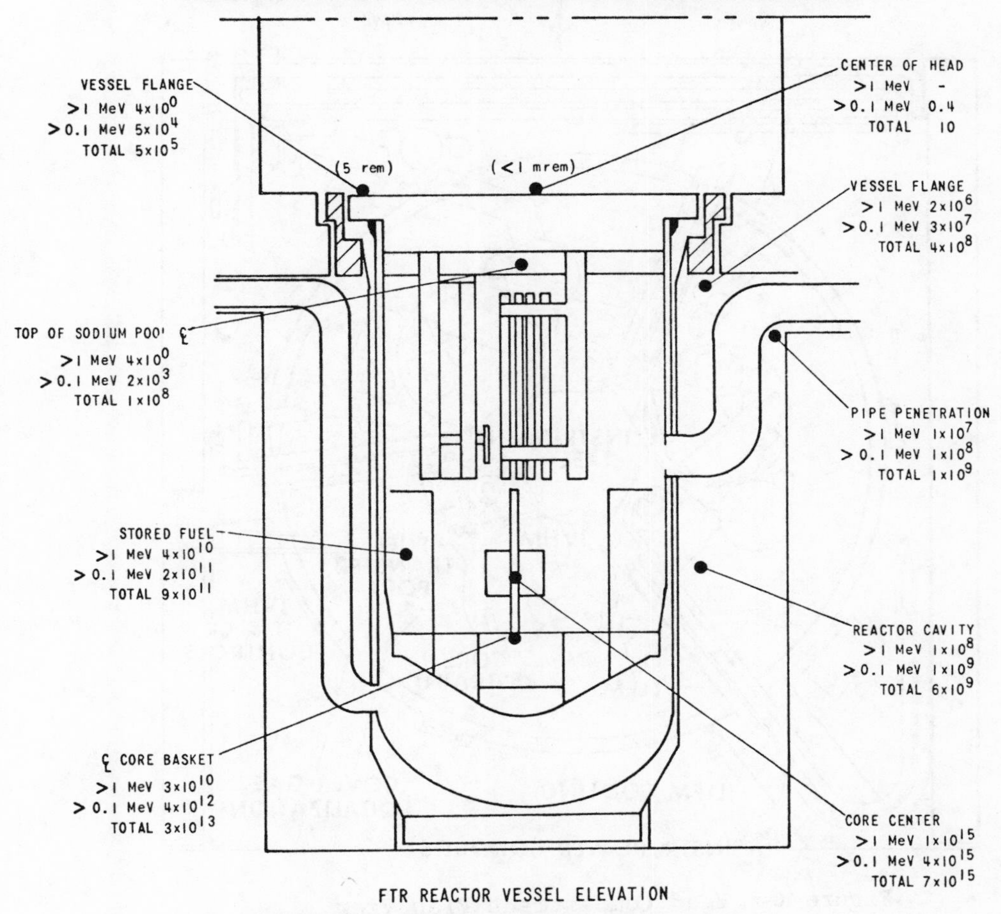

Figure 10-3. Neutron Flux Distribution

FAST REACTOR SHIELDING METHODS DEVELOPMENT

Advanced shielding methodology originated largely as a result of demanding shielding requirements in the aircraft nuclear propulsion program (ANP) and the space nuclear program (SNAP) and is now actively pursued by the military Defense Nuclear Agency and AEC space nuclear systems (SNS) and liquid metal fast breeder reactor (DRDT-LMFBR) programs. This methodology has developed rapidly since the mid 1960's when the Monte Carlo and discrete ordinates methods began to utilize the new super-computers. The basic characteristics of the advanced method are the flexible ability to solve widely different problems with the same data and technique and the ability to achieve an exact solution for any problem within the limits of the cross-section data or to economize for a particular problem type by appropriate adjustments to the method so as to not overcalculate the problem. By far the most important assumption is that almost any practical problem can be solved exactly, and, hence, that observed errors can be eventually traced to deficiencies in the cross sections or the transport calculational method, and that, when corrected, the improved cross sections or method is applicable to any problem of similar requirements.

HEAD COMPARTMENT ARRANGEMENT

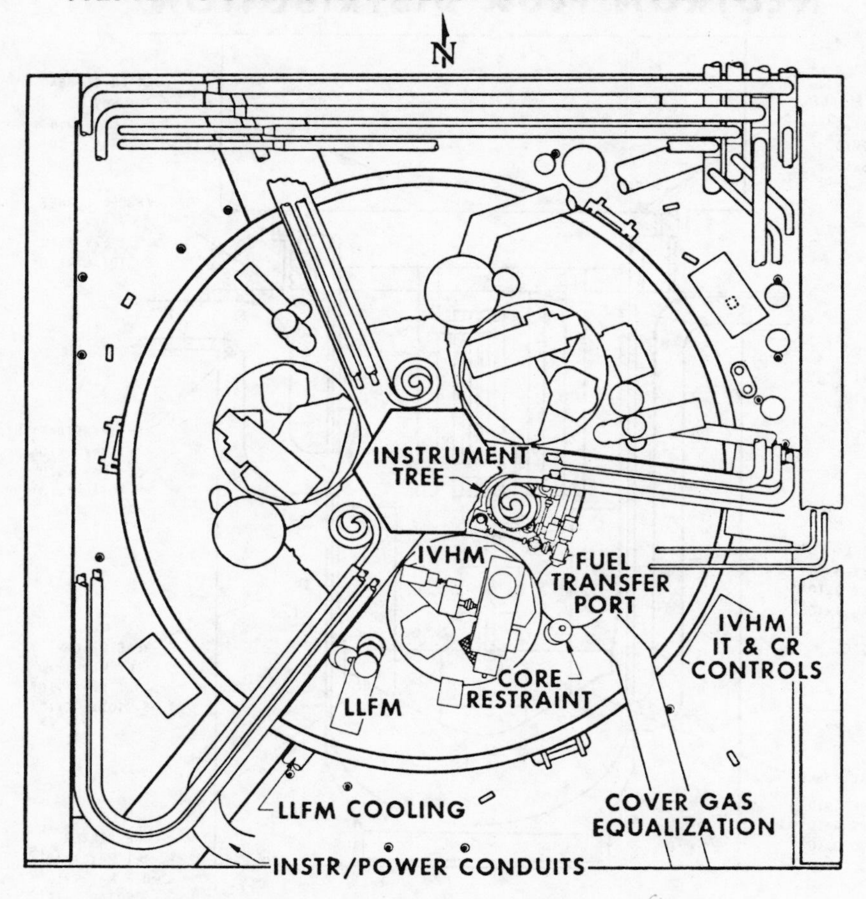

Figure 10-4. Head Compartment Plan View

The advanced shielding method must include five basic technologies:

(1) cross-section measurements
(2) cross-section evaluations
(3) cross-section processing
(4) integral experiments
(5) radiation transport calculations

(1) Cross-Section Measurements

For years the shielding community has needed improved measurements of the total cross sections, especially at the interference minima, and of secondary gamma-ray production cross sections for all neutron energies. Some feel that it is impossible to measure cross sections with sufficient accuracy for shielding.[2] This opinion largely derives from experiencing long delays in obtaining cross-section measurements adequate for shielding. Improved communication has led to significant progress as is demonstrated by improved total cross-section measurements for iron[4] and an efficient procedure for secondary gamma-ray production cross-section measurements.[5] While all cross sections cannot be measured to high precision in the available time, sensitivity analyses and integral experiments can be used to determine priorities and requirements for measurement of the most important cross sections.

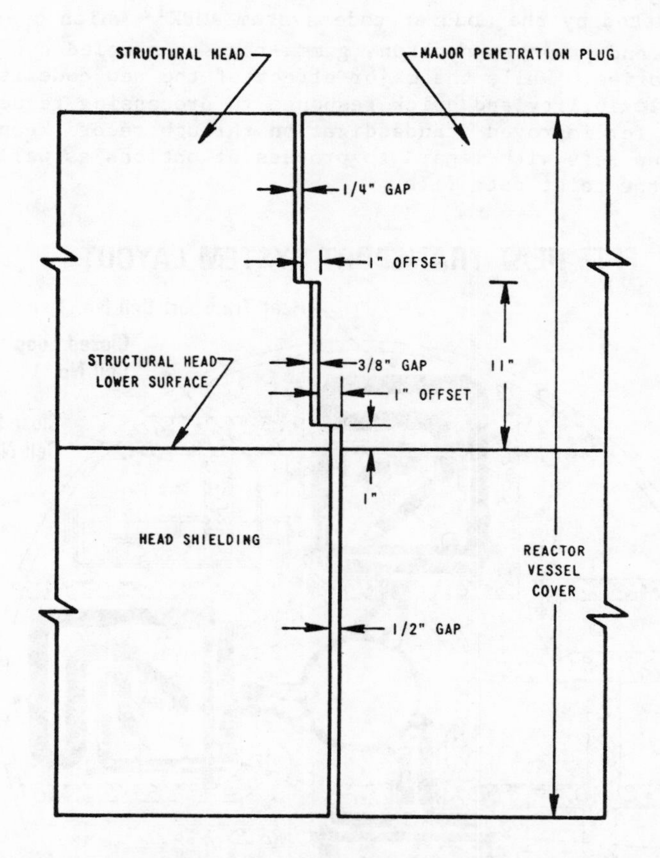

Figure 10-5. FTR Major Head Plug Penetration

(2) Cross-Section Evaluations

The evaluator must provide the calculator with a complete point cross-section set whether the appropriate data exists or not. He is the communication link between measurer and calculator, and he receives and participates in the conclusions of integral experiment analysis. It is important that the evaluator be continuously available in this role. As sensitivity analysis of integral experiments indicate possible cross-section errors, it is necessary for the evaluator to act on this information. Otherwise, an impasse is reached which can only be resolved by empirical adjustments or the proliferation of separate evaluations.

As an example of this process, within the last year analysis of an iron benchmark experiment[6] led to the remeasurements of the iron total cross section between 20 and 300 keV revealing factors of 2 errors in the interference minima[4] and within the last few months reanalysis of the iron broomstick experiment[7,8] has called for new total cross-section measurements between 1 and 4 MeV where 10% errors in the minima are suspected.

(3) Cross-Section Processing

The ENDF system format now includes sufficient flexibility for most shielding problems and the official version-III evaluations for many materials are expected to be useful for both shielding and core physics. In order to solve widely different problems, multigroup shielding calculations require the utmost in flexibility for group energy widths, scattering angular distribution expansions, and weighting functions. For several years, the most flexible neutron multigroup processing code was SUPERTOG[9] which had combined the virtues of CSP[10] and ETOG.[11] Secondary gamma-ray production cross sections were processed by a variety of codes such as POPOP4[12] since early ENDF evaluations had no data. These processing functions have now been inherited by the modular code system AMPX[12] which operates entirely from ENDF libraries and prepares neutron, gamma-ray, or coupled neutron-gamma-ray multigroup libraries. While the major effect of the new code is to provide greatly increased flexibility and quick response to processing requests, there is also opportunity for improved standardization through record keeping which identifies multigroup sets with regard to processing options as well as the mat and mod numbers of the point data file.

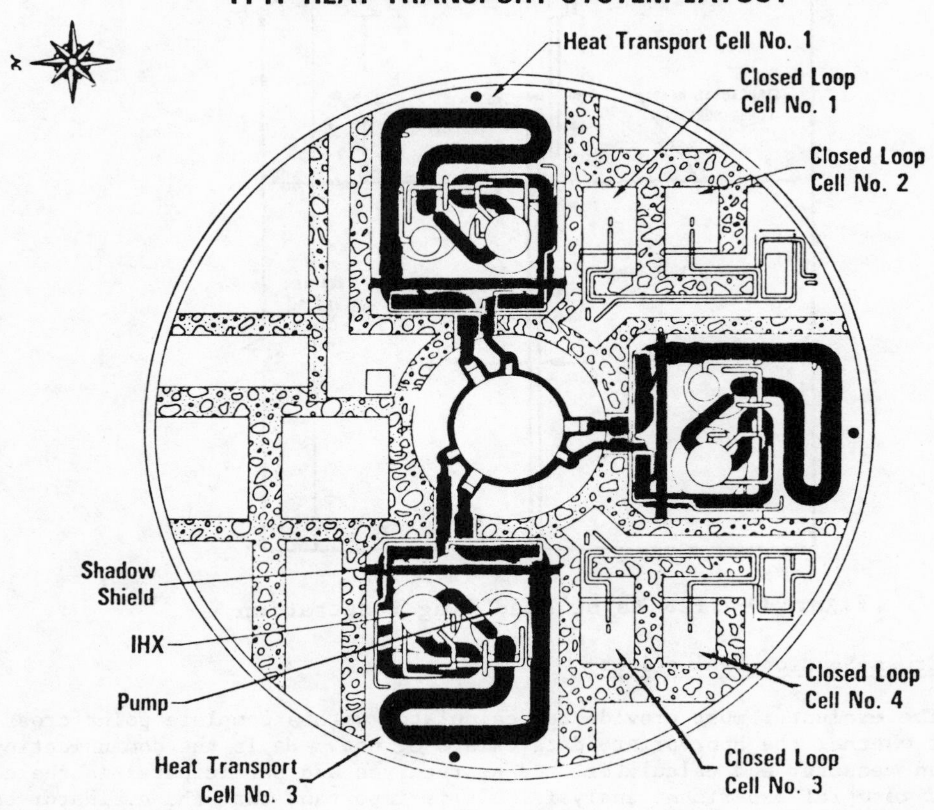

Figure 10-6 .FFTF Heat Transport System Layout

(4) Integral Experiments

In an empirical approach to shielding calculations, integral experiments serve the purpose of providing data for adjusting the computational algorithm. As such the experiments must closely resemble the design configuration and revised experiments are needed as the design changes even if essentially the same materials and sources are involved. In addition to the high cost of a continuous stream of experiments, experience has shown that shield designs change too quickly compared to the time needed for experiments, and one quickly is in the position of comparing apples and oranges.

In the advanced approach for shielding one can logically separate experiments for data testing and experiments for verifying basic methods or which serve as an expedient means of checking a complex calculational scheme having basic methods which are well known. A clear deficiency in this approach has been the need to determine if the data testing integral experiments are sensitive to the data in a way similar to the requirements of the design problem. Intuitive judgment and experience are most likely inadequate in this area and a sensitivity analysis approach[13] is being developed to fill the need.

In this approach, there are significant time delays in obtaining new cross-section measurements or in improving the calculational methods. When design decisions are required before the improvements are realized the present policy is to estimate the design problem uncertainty from the knowledge of the integral experiment.

The CSEWG shielding subcommittee has acknowledged several data testing experiments requiring only simple transport calculations[8,14,15] and one requiring a complex calculation.[16] Recent deep-penetration data testing experiments for iron[7] and sodium[17] are being fully documented. In all cases there is a clear need for sensitivity analyses when these experiments are used to evaluate uncertainties in a given design problem.

Integral experiments for methods verification ideally would involve only materials for which the cross-section data are adequately known. However, the lack of sufficient cross-section data and a fully developed sensitivity technology has led to a continuation of prototypical experiments for methods verification invariably involving both methods and data testing. In the analysis of such experiments the most exact methods are used and checked through intercomparison and data deficiencies are identified. When these are corrected through measurements and evaluations, the analysis should agree with experiment within experimental accuracy. Complex integral experiments are therefore regarded as data testing experiments where the appropriate sensitivity is incorporated by a prototypical configuration. In such experiments prototypicality must include the source and detector response as well as material configuration. A recent development is the technique for incorporating spectrum modifiers in such experiments to obtain prototype source spectra.[18,19] Methods verification is best performed through calculational intercomparisons, but this approach sometimes requires more time although less expense than experimental verification.
The CSEWG shielding subcommittee has acknowledged two methods testing experiments[20,21] while several others are being documented.[19,22-24] A peculiar example is the FTR lower axial shield prototype[19] where the streaming path was so complex and the need so urgent that an empirical streaming factor was sought. Although the conclusions obtained from evaluation of the experimental results were of significant value to the FTR design, the exercise demonstrated that the ambiguity in the empirical adjustment factor was large compared with the desired uncertainty.[18]

(5) Radiation Transport Calculations

As stated in the introduction, the key ingredient of the advanced shielding method is the ability to perform essentially exact numerical solutions of the linear Boltzmann transport equation for complex systems of practical interest. Although all facets of the shielding technology are required and are interdependent, this calculational capability is the crucial ingredient. Time and space do not allow a presentation here to substantiate this claim for the proof consists of a growing body of successful calculational comparisons and integral experiment analyses. A major aspect in achieving this area of confidence is the fact that the majority of practical problems are amenable to solution both by Monte Carlo and discrete ordinates methods in which the approximations are adjustable with ultimate accuracy limited only by computer capabilities. The utilization of general transport codes which do not have problem- or project-dependent limitations provide the methodology with flexibility. Similarly, for specific problem types time-saving approximations can be compared directly with precise solutions providing economy. This frees one from evaluating approximations by comparisons with integral experiments which invariably differ from the design problem.

At ORNL the most frequently used transport codes for reactor shield analysis are the multigroup MORSE[25] Monte Carlo and ANISN[26] and DOT-III[3] discrete ordinates codes. When non-multigroup "point" cross-section calculations are required, the 06R Monte Carlo code has been widely used in the past while for simple one-dimensional and two-dimensional problems point cross-section calculations with ANISN and DOT are being increasingly used.

A Plan For Economical Design Support Calculations

Overall economy in radiation transport calculations for LMFBR shield design can be achieved through a systematic approach utilizing the best features of each technique of the computational methods. An example of the possibilities is a procedure utilizing spacial coupling between discrete ordinates calculations and discrete ordinates and Monte Carlo calculations.

While limited to one- and two-dimensional geometries, multigroup discrete ordinates is the tool with greatest flexibility in the degree of approximation as one can vary the number of groups, angles, space points, and scattering expansion greatly reducing the cost in many cases while preserving the same accuracy. In addition to discrete ordinates, ANISN and several versions of DOT offer diffusion theory as an option by group and one can use diffusion theory mixed with discrete ordinates in the coupled calculations. Spacial coupling of DOT and MORSE calculations with the DOMINO[28] program offers a capability that makes solution of three-dimensional problems possible in a time frame compatible with shield design. This occurs when the deep penetration parts of the problem can be well approximated in two dimensions and the three-dimensional Monte Carlo part of the problem can be solved without extensive biasing.

Several discrete ordinates calculations can be advantageously coupled along spacial surfaces to solve a single problem. This procedure follows from the fact that one can take from a problem the angular flux at an interior surface, and, using this as an external boundary condition, delete the part of the problem on one side of the boundary and obtain identically the same result on the remaining side. The advantage is that, while the angular flux at the surface should include all scatterings on both sides of the surface, the entire problem need not be included. This allows one to bootstrap several small problem solutions to obtain the solution for a large problem.[29] Additional flexibility is gained by using an adjoint calculation for part of the problem and by the option to obtain an external surface angular flux from the first problem and use it as an internal boundary condition in the second problem.

The overlapped spacial coupling is used to advantage in three instances: (1) when a problem is too large for existing codes and computers and must be divided into subproblems, (2) when with existing codes one needs to change spacial mesh, group structure, or angular quadrature within a problem, and (3) when one can more efficiently perform parametric studies by isolating unchanging parts of the problem. In case (1) the total computer time used is greater than that which would be used to solve the entire problem in one calculation, but in cases (2) and (3) significant computer time can be saved.

The capabilities of forward and adjoint coupled calculations can be used to provide for inexpensive shield design studies. The procedure is to periodically determine baseline radiation environments by performing a forward two-dimensional discrete ordinates transport or transport diffusion calculation for the entire plant design to the extent where direct neutron and secondary gamma-ray radiation are important. At the same time, similar adjoint calculations are performed for each major criteria detector point (radiation damage response at core support and vessel points, various reaction rates in the radial cavity, and dose or activation responses in the head compartment and heat exchanger vaults). Although these baseline calculations are expensive they would likely be performed only four times a year or so. At any surface in the system the product of the forward angular flux and one of the adjoint angular fluxes gives the contribution per unit surface area to the detector represented by the adjoint. Integration over a surface effectively enclosing the core or the detector point gives the entire detector response.

These periodically updated forward and adjoint angular fluxes should be available for any purpose in core or shield design. Individual shield design studies then require calculations covering only the zone being designed. For example, the effects of materials or geometry changes in the radial shield would use the forward angular fluxes to define the reflector-shield or blanket-shield boundary condition and the adjoint fluxes at the outer surface as response functions. The radial shield design calculations could use discrete ordinates, diffusion, Monte Carlo, or any other method in any approximation. The calculation would be limited to the shield zone and the approximation could be checked against the existing discrete ordinates calculation for the unmodified design. The shield zone calculation could be one-, two-, or three-dimensional and effects of improvements to the baseline radial shield calculation could also be checked.

In addition to the procedure described above, the volume integration of the equivalent source[30] or adjoint difference[31] scheme can be used. In the adjoint difference approach the change in response at the ith detector is given by

$$\delta R_i = \int_{VOL} \Phi_P^*[\delta S + \delta\Sigma_s\phi - \delta\Sigma_T\phi]$$

where the symbols are from the operator form of the transport equation,

$$L\phi + \Sigma_T\phi = \Sigma_s\phi + S,$$

ϕ_P^* represents the perturbed system adjoint flux for the ith detector and $\delta\Sigma_s = \Sigma_S^P - \Sigma_s$, $\delta\Sigma_T = \Sigma_T^P - \Sigma_T$, and $\delta R_i = R_i^P - R_i$. The integration is only over the volume where the changes occur and if the unperturbed adjoint flux is used, the approximation is that of linear perturbation theory. Therefore, from the baseline forward and adjoint calculations, a linear perturbation approximation to the effects of radial shield composition changes on all the important detectors could be obtained directly by simple integration. These baseline forward and adjoint calculations also provide all the information for cross-section sensitivity analysis.

The procedure proposed here is less expensive than the present shield analysis for FTR. The baseline calculations are expensive but their frequency of occurrence is reduced. More importantly the angular fluxes are to be available projectwide and redundant calculations are limited to checking purposes. Of course the prime feature is the ability to minimize the calculational expenditures for design studies by utilizing the baseline calculations to their fullest extent.

The DOT-III[3] transport code and FACT[32] spacial coupling code utilize the spacial coupling procedure. Both forward-forward and forward-adjoint coupling have been performed in previous studies.[33,34] Some of the more recent calculations use many of the features of the procedure discussed above and the example presented here demonstrates the capability of the method.

An Example of Present Capability Analysis of the FTR Head Compartment Shielding Problem

Fig. 10-7 shows a somewhat simplified drawing of the FTR with many details, such as cavity shielding, omitted. After many studies it was determined that the dominant neutron transport path contributing toward neutron and secondary gamma-ray dose in the head compartment was transmission through the radial shield, induced fission in the stored fuel arrays, leakage into the radial cavity, and streaming through the complex vessel support system. In order to reduce the dose levels in the head compartment, the addition of a magnetite shield deck and B_4C shield ring in the radial cavity, and a borated polyethylene and steel shield floor in the head compartment are being studied.

Because of spectra differences the neutrons from fissions in the stored fuel dominate the dose rate above the cover by two orders of magnitude as compared with neutrons originating in the core. The fission distribution in the core was obtained from a diffusion mode DOT calculation and the total fission power in the stored fuel was determined from a three-dimensional coupled DOT-DOMINO-MORSE calculation.[35] The first-round analysis of the shielding problem utilized a four-step sequence of overlapped DOT-III cylindrical r-z calculations with fixed source distributions in the core and stored fuel. The stored fuel source was distributed uniformly over a homogenized annulus normalized to the total stored fuel power obtained from the three-dimensional calculation.

The first step in this sequence was a 50-neutron group S_6 P_1 13328 point DOT calculation for the reactor from the midpoint of the core extending out to a radius of over 550 cm and axially up to a height of 730 cm. This calculation could be used to define the source incident to the vessel support system; however,

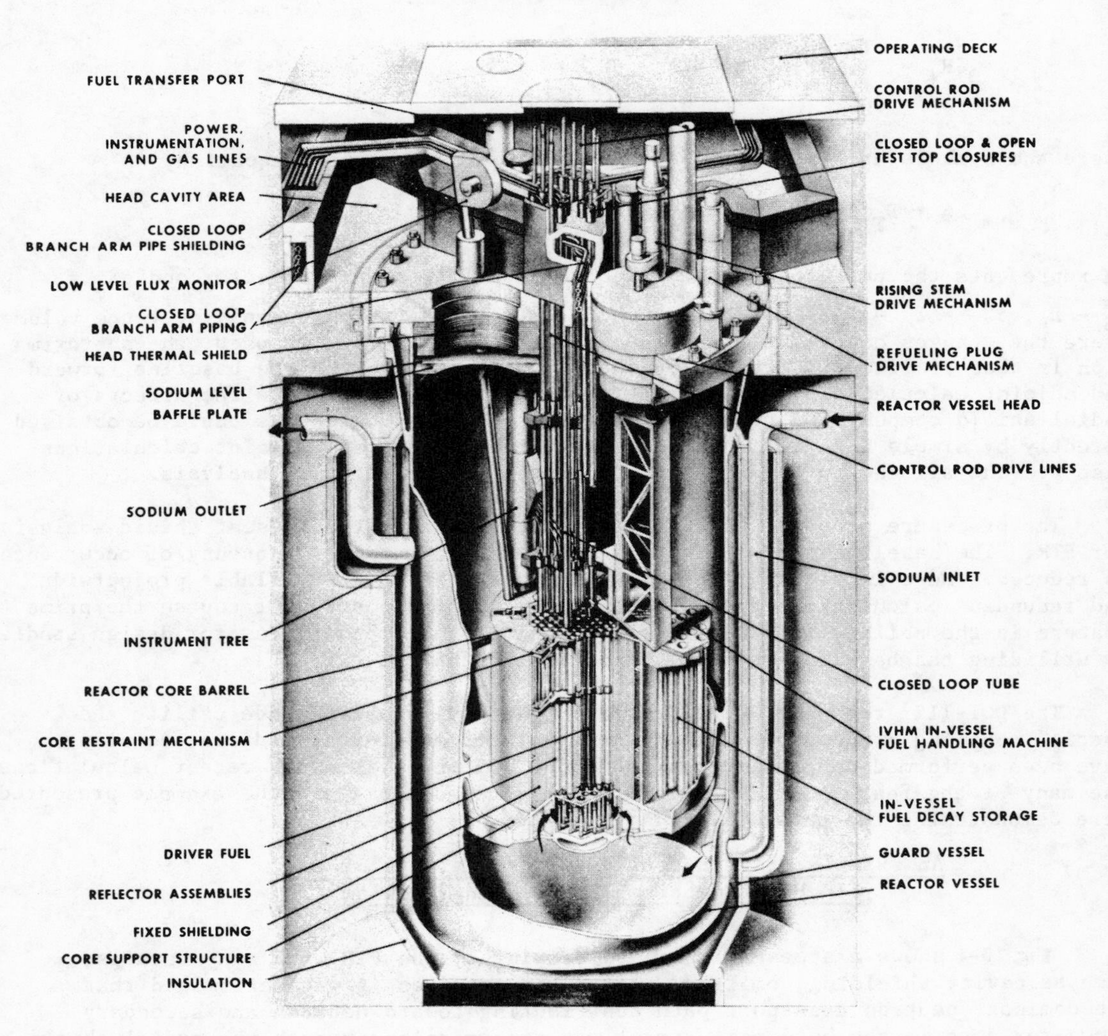

Fig. 10-7. A Pictorial Drawing of the FFTF Reactor.

it is assumed that the S_6 quadrature is inadequate for accurate determination of streaming in the cavity shield gaps, the geometric representation of the gap between the vessel and B_4C shield is poor, and gamma rays are not included.

A vertical boundary angular flux tape was obtained from step 1 at a radius of 295 cm which is at the outer surface of the baffle. The FACT code was then used to convert this boundary angular flux to an appropriate external left boundary flux for a 21-neutron group - 18 gamma-ray group S-166 P_1 2115 point calculation of the cavity. The purpose of this step 2 of the calculation is to obtain an adequate definition of streaming and include secondary gamma-ray transport in the cavity. The mesh in this step was different from that used in step 1, and separation of the cavity problem and shield deck problem was desired because additional shield deck configurations were to be studied. Step 2 produced a boundary flux tape for a horizontal boundary at an axial height of 500 cm.

Step 3 was a 21-neutron group - 18 gamma-ray group S-166 P_1 2880 point calculation and included the B_4C ring and magnetite deck in a zone extending radially from 295 to 555 cm and axially from 500 to 660 cm. The FACT code was used to prepare a bottom boundary flux condition from step 2 and a left boundary flux condition from step 1. Therefore, the calculation of this isolated zone incorporates the results of transport in zones below and to the left through the boundary conditions. A horizontal internal boundary angular flux tape was obtained at an axial height of 630 cm. The step 3 calculation extended to a height of approximately 660 cm, and it was assumed that the 30-cm overlap was sufficient to include the reflections from the materials above 630 cm.

The FACT code was again used to prepare a bottom boundary source for step 4 which was a 21-neutron group - 18 gamma-ray group S-166 P_1 4160 point calculation of the head region from the centerline to a radius of 555 cm and from an axial height of 630 cm to 886 cm. In this case the bottom boundary source from r = 0.0 to r = 295 cm was taken from step 1 results and the source from 295 cm to 555 cm was from step 3 results. This includes the thermal shield, cover, upper vessel flange, support arms, Z ring, support ledge, and the proposed head compartment shield floor consisting of 3 in. of borated polyethylene and 1 in. of steel.

Figs. 10-8 to 10-11 show computer plots of the geometries of steps 1-4 and the neutron isodose contours determined by the DOT calculations. Fig. 10-12 shows the secondary gamma-ray isodose contours for step 4. The portion of the step 4 calculation from the centerline to a radius of 295 cm was omitted from the plot to improve the enlargement of the complex section to the right of 295 cm.

Fig. 10-8 shows the shape of dose contours in the sodium pool and reactor cavity, while Fig. 10-9 shows that the dose contours in the S-166 cavity case are not greatly different from those in Fig. 10-8. Fig. 10-10 indicates that streaming is important in the inverted "L" gap between the B_4C shield and the magnetite shield and in the dogleg gap between the vessel flange and the B_4C shield. However, the design is reasonably well balanced inasmuch as additional shielding in this area would be circumvented by leakage outward from the vessel. The mission of the cavity shield is therefore to reduce the radiation incident on the head from cavity scattering to a level comparable with that from the sodium pool. The dimensions in this calculation are at operating temperatures since considerable movement takes place during startup.

The neutron dose contours in Fig. 10-11 show the importance of streaming in the gap between the vessel and the support arms. Dominant mechanical considerations excluded shielding from this area resulting in an imbalance in the shielding effectiveness. This calculation indicates that the maximum neutron dose at the top of the shield floor is 0.3 mrem/hr.

The gamma-ray dose contours in Fig. 10-12 show the effects of the large gap although in this case the gap between the Z ring and the support ledge is also an important streaming path. The calculation indicates that the maximum gamma-ray dose at the top of the shield floor is 0.43 mrem/hr.

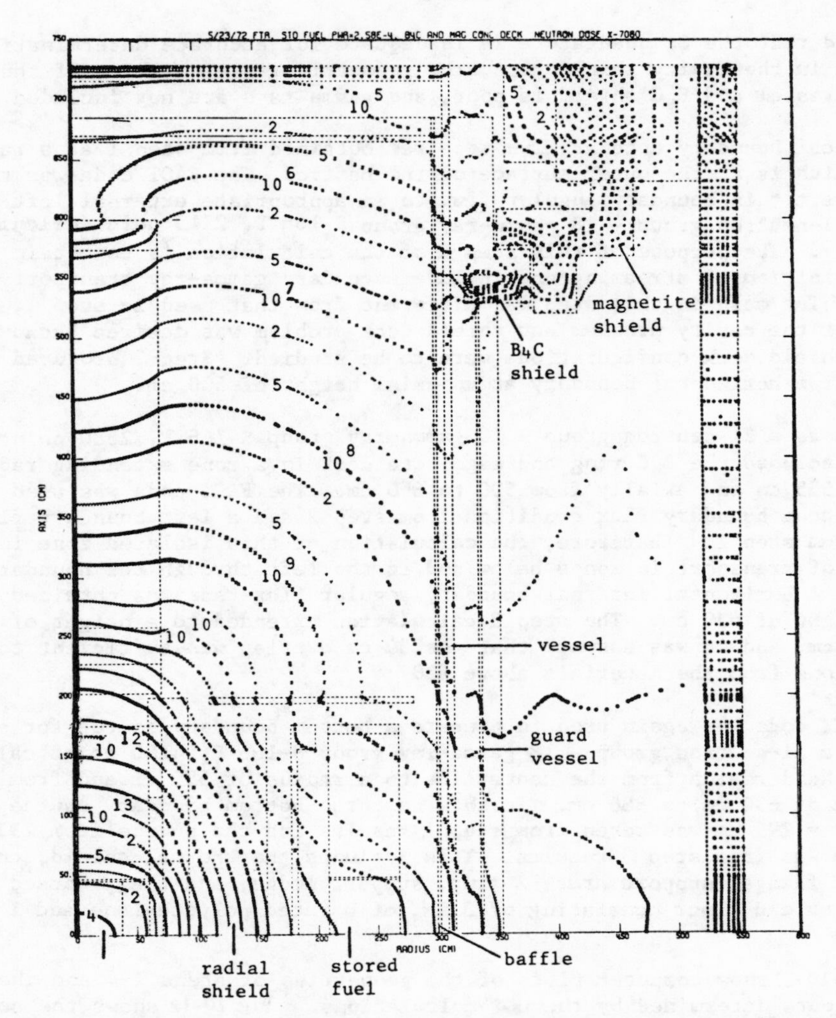

Fig. 10-8. Geometry and Neutron Dose Contours For Step 1 of the Calculatio
An S_6, P_1, 13328 Point R-Z Model of the Reactor.

In the four-step calculation described in the preceding, the coupling procedure was used between steps 1 and 2 to change group structure (thereby incorporatin gamma-ray groups), to change angular quadrature to allow use of a 166-angle set biased so as to provide a reasonably good calculation of streaming in annular slits, and to revise the mesh allocation reducing the running time and memory requirements for the S-166 cavity calculation. The group structure and angular quadrature were common to steps 2, 3, and 4. The separation of these steps was primarily used to reduce problem size and to reallocate the mesh in order to more accurately define the geometry.

A variation of the above procedure can be used to illustrate the capability for parametric design calculations. An adjoint calculation of step 4 may be performed to obtain the importance distribution (adjoint angular flux) at the 630-cm height surface. Given this tape and the top boundary angular flux tape from step 3, the FACT code performs the necessary integration giving the dose rate above the head compartment shield floor. A study of the effects of various clearance gaps in the B_4C and magnetite shield system then only involves the repetition of step 3 with its input boundary source tapes from steps 1 and 2, and the FACT code folding with the tape from adjoint step 4 to give the resulting changes in dose in the head compartment. The cost of the parametric study is reduced to a minimum by calculating only over the section being changed. Additionally, DOT need not be used in the step 3 calculations, but point kernel, Monte Carlo, and, if it were applicable, even diffusion theory could be used through appropriate coupling with the boundary conditions.

The three-dimensional aspects of this problem remain to be studied as the shield design progresses. The major concerns are penetrations of the shield floor in the head compartment and penetrations of the cavity shields by piping and other equipment. Coupling of intermediate results from these DOT calculations with MORSE is being used to investigate these problems.

Excluding these three-dimensional aspects of the problem, the uncertainty due to cross sections, multigroup approximations, P_ℓ truncation, and approximations inherent in the discrete ordinates method is estimated to be a factor of five. Since the sensitivity analysis tool is not yet in routine use to correlate integral experiments and design problems, this uncertainty estimate contains a large portion of intuitive judgment. Studies which contribute toward the verification of this calculation include integral experiments for neutron transport

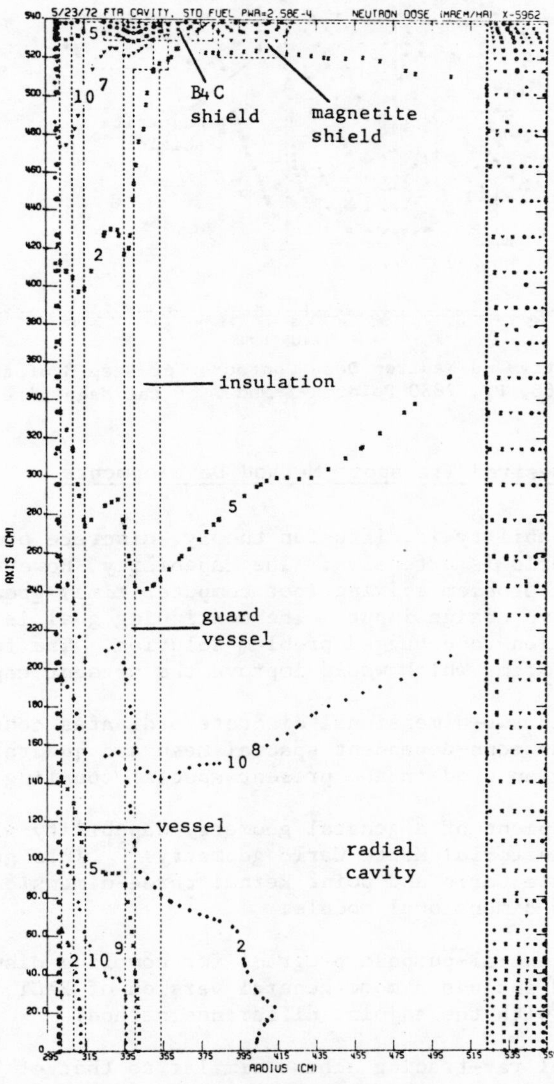

Fig. 10-9. Geometry and Neutron Dose Contours For Step 2 of the Calculation - A More Detailed S-166, P_1, 2115 Point R-Z Model of the Radial Cavity.

in iron,[6] stainless steel,[36] sodium,[17] neutron scattering in a concrete walled cavity,[24] and neutron streaming in offset annular slits[22] and coolant pipe chaseways.[23] Although the iron secondary gamma-ray production cross sections have been experimentally verified,[14,15,27] the neutron-to-gamma-dose ratio at the top of the cover is anomalously high, and a suitable integral experiment should be performed to verify the gamma-ray transport problem.

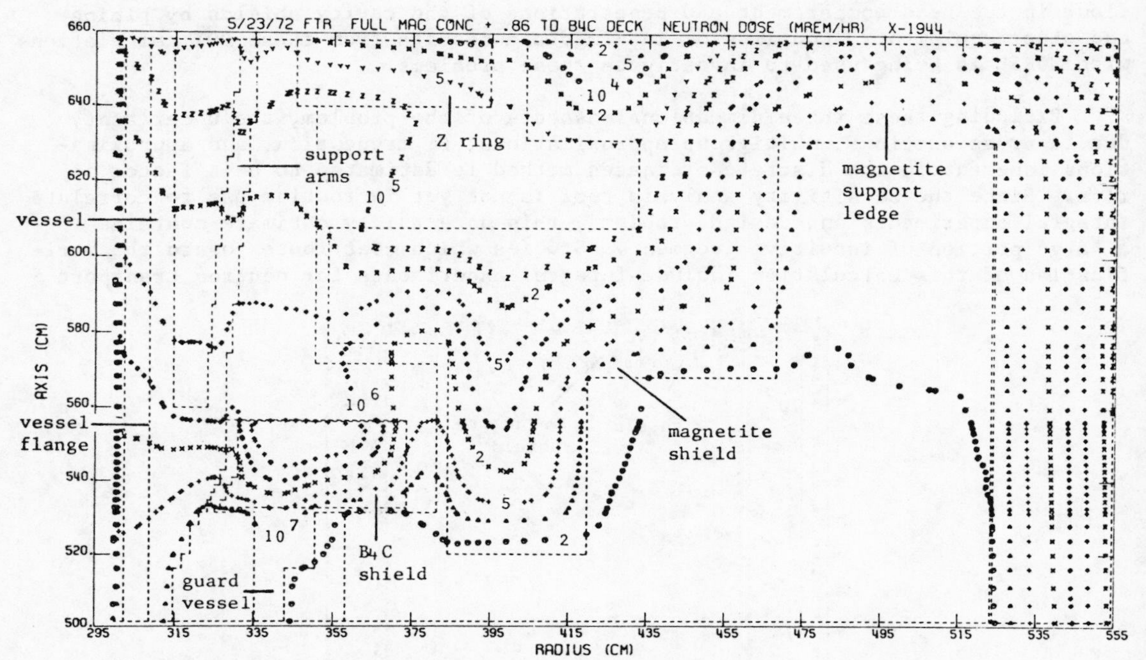

Fig. 10-10. Geometry and Neutron Dose Contours For Step 3 of the Calculation - A More Detailed S-166, P_1, 2880 Point R-Z Model of the Magnetite Concrete and B_4C Shields.

Desired Transport Method Developments

The present capability in diffusion theory, discrete ordinates transport theory, and Monte Carlo is extensive. The capability, however, is useless if the elapsed time for problem solving (not computer time) greatly exceeds the time required for effective design input. The continuing goal is to reduce the time from problem definition to debugged problem solution. The following is a brief list of development items which would improve the present capability:

(1) An improved two-dimensional discrete ordinates code having the capabilities of DOT-III plus zone-dependent spacial mesh and quadrature. This removes the inefficient overlap used in the present spacial coupling.

(2) The development of a general geometry capability with input similar to that of the combinatorial Monte Carlo geometry.[38] This geometric module should be used for both Monte Carlo and point kernel three-dimensional calculations and as the basis for two-dimensional models.

(3) Improved general-purpose programs for coupling discrete ordinates calculations. This would include a more general version of FACT,[32] and a general program for volume coupling with the adjoint difference method.

(4) An internal ray-tracing scheme similar to that of Vossebrecker[39] incorporated for use in diffusion theory and low-order discrete ordinates calculations.

(5) Increased capability for three-dimensional calculations to augment the present Monte Carlo and diffusion theory codes.

These improvements are needed but a higher priority is devoted to efforts for improving the means for utilizing the present capability in the most efficient way.

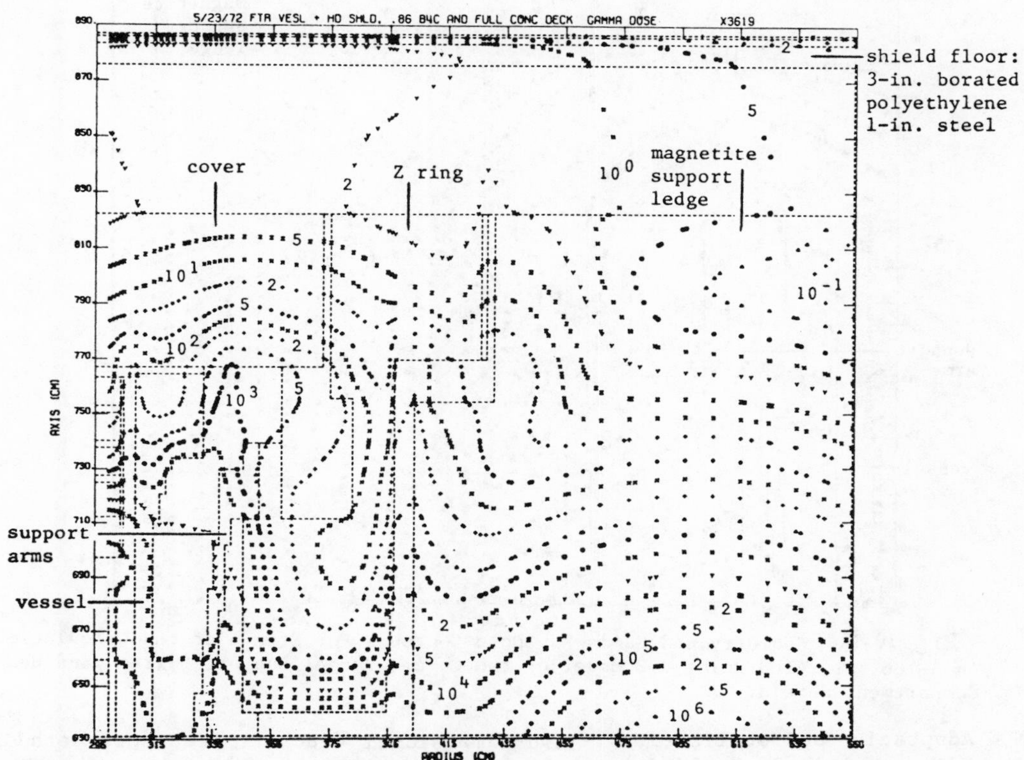

Fig.10-11. Geometry and Neutron Dose Contours For Step 4 of the Calculation - An S-166, P_1, 4160 Point R-Z Model of the Cover, Vessel Support System, and Head Compartment Shield.

Irradiation

EBR-II (Experimental Breeder Reactor No. 2) and the adjoining FEF (Fuel Examination Facility) were originally designed and operated to provide a small-plant demonstration of a sodium-cooled fast breeder reactor power plant. Following the successful demonstration of the plant, the need for an adequate fast reactor irradiation test and examination facility led to its operation for the testing of fast reactor fuel, structural, and poison materials. In this role, EBR-II has core loadings changed approximately one or more times each month for removal, insertion, and relocation of test-irradiation subassemblies. These core loadings are very heterogeneous and generally have a non-constant core radial periphery. Both for the characterization of the irradiation environments of individual test subassemblies and for the overall specification of the neutronics of the reactor system for operational and reactor modification, extensive reactor physics analyses are carried out.

The physics analyses of EBR-II loadings, operations, modifications, tests, and of environmental characteristics of each subassembly require extensive usage of current methods of analyses and also development and innovation of techniques. This is due to the heterogeneous complexity of the system, the frequent loading changes, and the user need for detailed environmental characterization. As such, the tests and analyses on EBR-II function also as indicators of the capabilities of current knowledge and methods of fast reactor physics in satisfying the LMFBR (Liquid Metal Fueled Breeder Reactor) program needs. They also point out areas where further development and data are required.

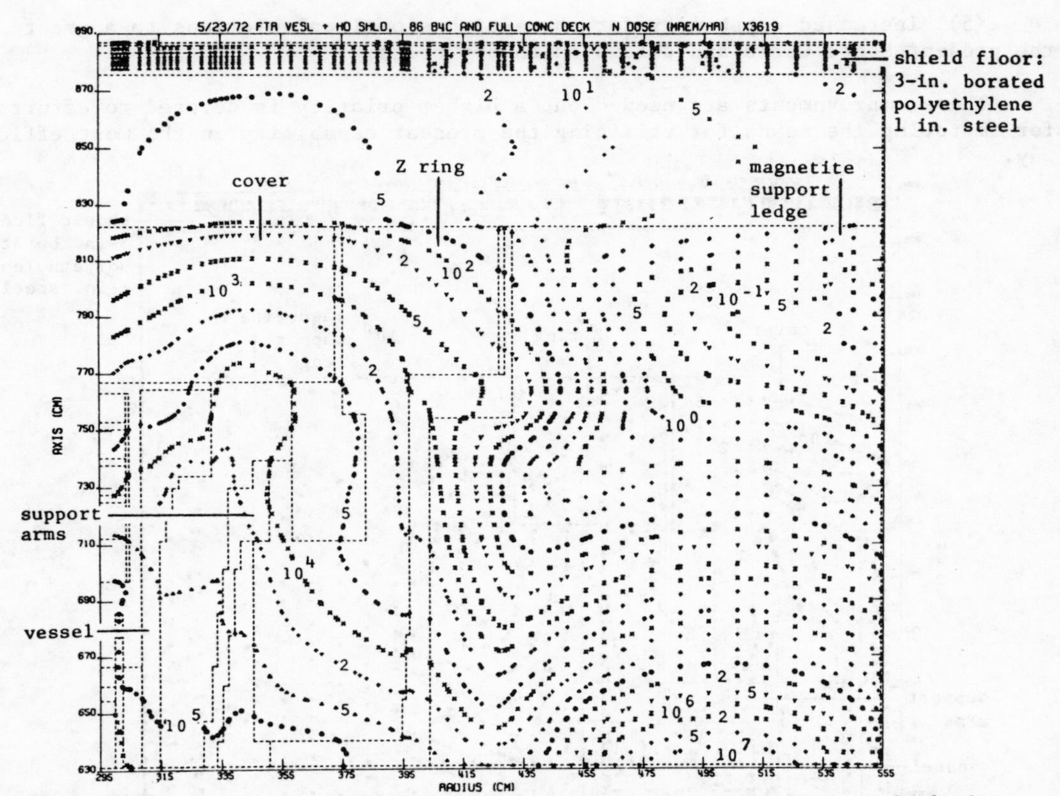

Fig. 10-12. Geometry and Gamma-Ray Dose Contours For Step 4 of the Calculation - An S-166, P_1, 4160 Point R-Z Model of the Cover, Vessel Support System, and Head Compartment Shield.

Adaptation and utilization of two-dimensional discrete-transport methods have been extensively applied to both neutronics and gamma[2] analyses of the heterogeneously-loaded EBR-II core loadings having both the former depleted-uranium-type blanket and the recently installed steel-rich reflector. Analytical studies for specific environmental tailoring of neutron and gamma levels for particular experimental subassemblies have been made.[3]

A burnup-history for the subassemblies in the EBR-II core and blanket regions has been underway, beginning with EBR-II Run No. 1.[4]

Criteria for choice of axial bucklings, especially in radial reflector regions containing light reflectors and the effects of adjusted reflector axial buckling upon radial traverses were made.[5]

Determination of neutron flux distribution and spectra in the EBR-II deep in neutron shields at instrument test thimbles has required R-θ geometry analyses which include detailed consideration of the radial variations in the effective axial bucklings of the radial blanket.[6]

Tests of space and energy distributions of neutrons and gamma flux in EBR-II type system have been made by calculational comparisons of critical masses, counter responses, TLD (thermoluminescent detectors), proton-recoil, and foil dosimetry measurements in mockups of EBR-II having blanketed, nickel, and steel reflected cores.[7-11]

Neutron dosimetry tests[12] by foil activations have been compared with analyses[13] for the capture to fission ratios of Pu and U isotopes at axial and radial positions throughout EBR-II. Analyses of a large number of spectrally-sensitive foils irradiated throughout the radially-blanketed EBR-II reactor are currently being evaluated.

Fine-group analyses of the spatial variation of neutron fluxes in EBR-II have been calculated both for comparison with measurements and to determine effect of spatially dependent cross section collapse upon the coarse-group collapsed spectra.[14] Effects of spatial cross section collapse for structural subassembly regions have also been studied.[15]

Physics aspects of using a higher burnup fuel pin (Mark-II) having a larger axial gas space have been under investigation.[16] The utilization of higher worth control rods having poison followers and the tritium production in the EBR-II reactor due to the B_4C contained in the high-worth control rods and in poison experimental subassemblies were investigated.[17,18]

The relative expansions of axially movable control rods relative to stationary core subassemblies have been calculated for the effect upon the power coefficient of reactivity. (As reactor power increases, the core subassemblies tend to thermally expand upward, whereas, control subassemblies supported at the top tend to thermally expand downward.)[19]

Detailed analysis of the effects of overcooling an EBR-II steel reflector upon the power coefficient have been made.[20]

Reactivity effects resulting from small outward or inward displacements of subassemblies have been calculated for EBR-II sized cores having both depleted-uranium type radial blanket and steel-rich radial reflector for use in conjunction with subassembly bowing analyses.[21]

The row-wise bowing configurations for both uranium-blanketed and steel-reflected EBR-II cores are under investigation using the BOW-V Code.[22] Currently, the bowing code together with the code CRASIB, which performs a creep and swelling analysis on subassembly cans, are being linked.[23]

Noise or signature analyses in fast reactor systems is currently being investigated for purpose of systems surveillance in EBR-II.[24] Neutronic-detectors, accelerometers, thermocouples, pressure-transducers, and flowmeters may be used as signature detectors.

Self-powered, neutron-gamma detectors for in-core use in EBR-II are being evaluated. These include experimental irradiations of detectors with rhodium, cobalt, and platinum as the emitter materials.[25]

Neutron Analyses for EBR-II Loadings

The methods utilized for neutronics analyses of EBR-II loadings are determined by the need for detailed predictions of both the overall global reaction rate intensities and shapes and of the detailed power, fission rates, and neutron flux[1] and gamma flux[2] levels for each specific subassembly. A representative core loading is shown in Fig.10-13. Analyses require delineation of the area of each individual hexagonal-shaped subassembly in XY-geometry as a rectangle of equal area and employing four mesh points per subassembly. In order to obtain sufficient accuracy of flux solutions through differing subassembly types, i.e., fueled, structural, and poisons discrete-ordinate, transport solutions using the S_4 approximations have been required. Local depressions (or increases) in flux that can occur in fueled experimental subassemblies as well as structural subassemblies depend primarily upon the amount of fissile and fertile isotopes contained relative to that in the adjacent or more dominant driver fuel.[3] For example, high energy (>2.23 MeV) flux depression or peaking factors have been calculated to range between ∿0.70 in a structural subassembly and ∿1.05 in a highly enriched experimental subassembly.

The XY-geometry transport calculations are made using the discrete-ordinate code, DOT. The weights of fissile, fertile, coolant, and structural components for each subassembly applicable to a given loading are used. Changes in isotopic compositions due to burnup in previous loading runs are accounted for. Calculated power (MWt) generated in individual subassemblies in the core configuration shown in Fig. 10-13 are given in Fig. 10-14.[26]

Auxiliary two-dimensional, RZ-geometry, analyses are carried out to obtain axial-variations, on a row-wise basis, and power normalizations. In these cases row-wise average compositions are used and details of axial geometry and compositions in pin tops, gaps, axial reflectors, etc. are included.

Scoping and operational calculations generally have utilized six neutron energy groups for economy. When required, a more detailed 29-group cross section set has been used. This set is based upon the ENDF/B (Version I) library. The set was derived by energy-collapse to 29-energy groups specifically for compositions representing core averages, blanket, axial gaps, and axial reflectors by the collapsing code, MC^2. The set comprised energies down to 0.68 eV in largely half-lethargy intervals.

The magnitudes of run-to-run variations of power density at various fiducial locations resulting from the run-to-run variations in the loading patterns are indicated in Fig. 10-15 where the per cent variations in ^{235}U fission density calculated from run to run at the twelve control rod positions located cylindrically in row 5 have been plotted.[27] These analyses were for thirty selected runs between runs 24 (December 1966) and 55A (March 1972) from the approximately 200 total EBR-II runs. Run-to-run variations in the spatial power distribution of EBR-II are of concern to reactor analysts as well as experimenters who use EBR-II as a facility for irradiating fuel and/or structural materials, since they both desire to minimize perturbations in irradiation conditions and to ideally maintain a stable reactor core environment. In general at the control rod positions the run-to-run variations were within ±5% for ^{235}U. Early run-to-run variations greater than ±5% are explained by the installation and subsequent removal of stainless steel reflector subassemblies in rows 7 and 8. Later significant variations often correspond to a change of subassembly in a particular position, a run of very short duration (0-100 MWd) due to changes in reactor operating conditions such as fission gas leak tests, and base power level changes, the latter two occurrences being accompanied by unscheduled loading changes. Lack of coincidence of the twelve series of plots in Fig. 10-15 indicate the presence of flux tilting. Effects of heterogeneity of loadings are also illustrated in Fig. 10-16, which is a contour map of the high energy flux for Run No. 32B.[1]

A burnup history of EBR-II, beginning with the first loading of the reactor and continuing to the present, is being calculated.[4] The programming effort, based on the REBUS[28] fuel-cycle modules of the ARC system and the DOT,[29] two-dimensional transport code, has been completed, and the series of computer runs for the history is underway. The calculations are being done in XY geometry, with 14-group cross sections, based on ENDF/B (Version I) data. Each subassembly, as well as a considerable portion of the blanket, is being studied separately through its entire lifetime in the reactor. The total is 352 subassemblies for each loading.

Gamma Analyses for EBR-II Loadings

Consideration of detailed distributions of gamma-energy depositions in the EBR-II loadings are necessary in order to obtain sufficiently accurate values of global and local power contours and levels.[2] Since ~14% of the energy released by fission and subsequent neutron capture is carried away by gamma rays, if it is assumed that all the energy is deposited at the location of the fission, the energy deposited in non-fueled regions such as structural subassemblies, and reflectors are not accounted for and, furthermore, the overall and local distributions throughout the system will be in error. The distribution of gamma energy deposition is flatter across a core than the fission rate distribution because of the large mean-free-path and the more forward scattering of gamma rays. The total power produced in a central subassembly is less than would be expected assuming all energy absorbed at location of fission. Power production throughout the core region is lowered also by the transport of gamma energy to reflector and blanket regions.

Gamma heating calculations are carried out in two-dimensional analyses using the discrete ordinates transport theory code DOT.[29] The gamma-production cross sections are obtained from the POPOP4 Code[30] which converts spectra of gamma rays from neutron-induced reactions to a required neutron-gamma energy-group structure. The gamma yield data used in POPOP4 comes from the POPOP4 library, which is a compendium of secondary gamma-ray yield and cross-section data for the various neutron-induced reactions. The gamma flux distribution that results from the computed gamma source is determined by performing a fixed-source gamma transport calculation utilizing gamma-ray cross sections obtained from the MUG Code.[31] Twenty energy groups of equal energy intervals have been used for the gamma transport analyses. The rate of heat production is then determined by multiplying the gamma fluxes by the gamma-energy-absorption cross sections.

Because total integrated gamma energy deposition relative to total fissions as obtained by way of the transport solutions do not necessarily result in the ratio required from fundamental data of reported gamma energy per fission corrections factors derived from an energy balance are applied.[2] This is accomplished by requiring the total transport-calculated gamma energy rate deposited in the reactor be consistent with basic data on the energy released per fission. Based on the total gamma energy resulting from fissions in ^{235}U, ^{238}U, and ^{239}Pu and the relative fission rates in a radially-blanketed EBR-II system the average gamma energy has been estimated to be 28.3 MeV/fission of a total of about 202.8 MeV/fission. Assuming negligible gamma leakage about 8.75 MWt of gamma heating occurs at 62.5 MWt operating power.

Calculated radial gamma-deposition distribution for a representative 91-subassembly size EBR-II core (not identical to that shown in Fig. 10-13 , surrounded radially by the blanket rich in depleted uranium, is shown in Fig. 10-17 for various axial levels in the system.[2] The core height is approximately 13.5 inches and the axial reflectors regions are of the steel and sodium compositions. The depression in row 2 is caused by the presence of a homogenized composition-mixture corresponding to three structural subassemblies diluting the fuel. The core-blanket interface is between rows 6 and 7. The discontinuity at the 55.8 cm axial fuel illustrates the change in gamma deposition between the steel-containing axial reflector and the depleted-uranium type blanket which surrounds it.

To obtain accurate relative values for gamma energy depositions for the individual subassemblies and the variations with core loadings, two-dimensional, XY-geometry gamma transport must be employed. Subassemblies may experience differing gamma heating rates depending not only on the composition of the subassembly but also on the adjacent subassemblies. These variation effects are shown on Fig. 10-18 where the gamma heating rates per gram of uranium are given at each subassembly position for a particular EBR-II loading pattern.[2]

Global loading pattern biases and neighboring subassembly effects can be utilized to specifically tailor the environments of particular experimental subassemblies. An experimental subassembly (XX03), for example, was designed for neutron irradiation at a controlled temperature. The temperature was to be obtained by gamma heating of the subassembly components. To obtain that temperature, the gamma flux in the subassembly had to produce $\sim$3.8 Watts per gram of Fe. Gamma transport analyses (normalized) showed that with the reactor operating at its design power of 62.5 MW(th), the gamma heating rate would be too high (5.6 Watts per gram in Fe). At this subassembly, transport theory analyses showed that it was possible to tailor the local reactor environment to accommodate XX03 by inserting subassemblies containing one-half as much fuel as driver subassemblies in the grid positions surrounding XX03 resulting in ~3.34 Watts per gram in Fe. The results of this study[3] showed the feasibility of modifying irradiation conditions significantly for a limited number of experiments.

Analyses of ZPR-3/EBR-II Critical Assemblies

A series of critical assemblies having EBR-II-type core compositions and radially reflected by depleted-uranium, nickel, and steel reflectors were constructed on the ZPR-3 critical facility.[7,8] The critical assemblies were directed toward obtaining experimental data relative to the effects of the varicus reflectors upon an EBR-II-type core and toward testing the reliability of cross sections and calculational techniques for such EBR-II type cores. These experiments and analyses assisted in the decision to replace the radial blanket by a steel-rich reflector and allowed calculational biases both for neutron reactions and gamma depositions in radial-axial directions to be obtained. In addition, experiments carried out in these criticals enable an improved understanding of the irradiation environment in EBR-II-type cores.

ZPR-3 Assembly 60 was an homogeneous approximation of EBR-II core loadings having a radial blanket rich in depleted uranium. The core represented a homogenization of typical numbers of uranium-fueled driver subassemblies, experimental fueled-subassemblies, experimental structural-subassemblies, and control-safety subassemblies. The assembly was axially reflected at top and bottom by sodium and stainless steel contained in gaps and reflectors. The nickel-rich-reflected ZPR-3 Assembly 61 had a similar though not identical core composition to compensate for the higher worth of the ~8.7-in.-thick nickel-rich radial reflector. This reflector was surrounded by ~8.7-in.-thick blanket of composition as in Assembly 60. The steel-rich reflected critical ZPR-3 Assembly 62 had the nickel-rich radial reflector replaced by a steel-rich reflector.

Two-dimensional, RZ geometry, S_4 neutron transport calculations resulted in calculated excess reactivities of about 1.7%, 2.2%, and 2.5%, respectively, for the contructed Assemblies 60, 61, and 62 at criticality.[7] These analyses utilized the 29-energy-group cross sections and included calculated corrections for the thin-plate heterogeneities and hexagonal core-outline.

Thermoluminescent dosimeter (TLD) dose mappings were performed in the EBR-II series of ZPR-3 critical assemblies.[9] The measured spatial distributions have been compared with neutronic calculations done with 29 neutron energy groups in the S_4P_0 approximation and gamma transport calculations with 20 energy groups in the S_8P_3 approximation. In Fig. 10-19 the comparison between the calculated and measured energy deposition in the radial direction for the uranium-blanketed EBR-II/ZPR-3 No. 60 Assembly is shown.[10] The calculated results are shown for deposition in iron because the TLD's were enclosed in steel sleeves. The calculations were done in RZ geometry with the hexagon-shaped core represented by a cylinder of equivalent volume. Analogous calculations and measurements have also been performed which show the effect of replacing the uranium blanket with steel and nickel reflectors on the radial and axial gamma heating rates.

Measurements of detailed neutron-flux spectra in the EBR-II-type critical assemblies and calculational comparisons have direct relevance to such EBR-II problems as dosimetry and material damage. Proton-recoil measurements were made at core center in core near to the core-reflector interface, and in the reflector of nickel-reflected and steel-reflected ZPR-3 Assemblies 61 and 62, respectively. The experimentally determined flux spectra at three spatial positions in ZPR-3 Assembly 62 are compared with calculated spectra in Fig. 10-20[8]. These calcuations were performed in one-diemnsional cylindrical geometry using the ANISN code in the diffusion-theory option. The 165-group cross-section set has a group width of 1/10 lethargy and was derived from the ENDF/B data files by energy-collapse specifically for the regional compositions of the assemblies by the MC^2 code. The complicated energy dependence, caused by the resonance scattering cross-section variations of many of the component materials, and its modification from core to reflector, is evident in both calculations and experiments.

Considerations in Analyses of Light Reflectors and Structural Subassemblies

In order to enhance the capability of EBR-II as a fast-neutron irradiation facility the substitution of a light radial-reflector rich in nickel or in steel has been considered. This change increases the reactivity of the system, allows for a larger power density if core size is kept constant or allows for a larger core size with less reduction in power density, gives a flatter power density, and results in more radial spectral variation. This enables a larger number of subassembly positions for experimental subassemblies of lower fuel worths, structurals, or poisons. Studies dealing with the installation of an inert radial reflector in EBR-II were made with regard to the choice of reflector material, size of reflector region, the change in the irradiation environment, the changes in plant operation, and the safety of the reactor. Nickel-rich and stainless-steel-rich reflectors were considered. Operating experience in EBR-II with each of these materials and neutronics analyses coupled with the measured data obtained in the ZPR-3 series of criticals led to the selection of three or four rows of stainless steel depending upon whether the core size is 6-row or 7-row. The former type blanket rich in depleted uranium surrounds this reflector.

Considerations of EBR-II type system radially reflected by light reflectors has led to analytical studies of effective axial bucklings and of spatially-dependent cross section effects.

Because of need for heterogeneous analyses of loading patterns and the non-circular core interface of EBR-II (and of the ZPR-III mockups) calculations including these details are carried out using two-dimensional discrete-ordinate transport calculations in XY-geometry. The use of low-absorption reflectors such as nickel or steel radially surrounding the EBR-II cores gives rise to the question of appropriate "effective" axial bucklings in such regions. (With cores reflected by the usual blanket rich in depleted uranium the identical axial buckling parameter has generally been used also in the blanket region.) Comparison k_{eff} calculations of one-dimensional cylindrical geometry with two-dimensional RZ geometry, were made.[5] Use of identical axial buckling (0.00235 cm^{-2}) in core and radial-reflector regions was compared with use of an adjusted reflector-region bucklings (0.00090 cm^{-2} and 0.00085 cm^{-1} respectively for the nickel and steel reflectors). The latter reflector bucklings were adjusted so that the k_{eff} of one-dimensional cylindrical calculations would be identical with corresponding values obtained by two-dimensional RZ-geometry. These studies were made using the compositions and dimensions of the EBR-II/ZPR-3 critical Assemblies 61 and 62, which are reflected radially by nickel-rich and steel-rich reflectors, respectively. Use of a reflector axial buckling identical with that for the core region leads to an error of about -3% in k_{eff} in these cases.

A related question is the shapes of calculated detector traverses by 1D and 2D analyses with the steel-reflector. Comparisons given in **Fig. 10-21** show that the radial-traverse shape for $^{235}U_f$ obtained from a one-dimensional analyses in which the k_{eff}-adjusted buckling in the reflector (described above) was used closely follows the axial-averaged (over an axial height equal to the core height) shape obtained from a corresponding RZ, two-dimensional analysis.[5] The shape differs from the core midplane shape of the RZ analyses. Also, use of identical buckling in the reflector as that in the core region results in large errors in shape. Since EBR-II studies usually are carried out in XY geometry in conjunction with axial-buckling parameters the calculation of midplane radial profiles of detector rates and foil activities must be corrected to the midplane shapes for comparison with measurements. Furthermore, the corrections will depend upon the spectral-sensitivity of the particular detecting isotope reaction.

Effects of resonance-scattering in neutronics analyses of nickel- and steel-reflected EBR-II-type systems have been studied[14] using fine-group solutions to generate spatially dependent, coarse-group cross sections to determine the effects such detailed considerations have upon resulting coarse-group spectra, and reactivities. One-dimensional, cylindrical-geometry, fine-energy-detailed neutron flux solutions having 371 equal-lethargy energy groups in the energy range 2.1 keV to 1.35 MeV were calculated using a composite of a library of angular scattering data and a discrete S_N transport code in the S_2 approximation. Coarse-group cross-section sets derived by fine-group cross-section collapse using the fine-group solutions corresponding to: (1) 63 mesh positions in the system; (2) to each of 11 subregions in the system; and (3) to 4 mesh positions (at center of each of the 4 regions), were utilized to calculate source-iterative (k_{eff} calculation) coarse-group solutions. Percent reactivity errors resulting from use of the 11-set and the 4-set approximations relative to the 63-set are about -0.05 and -0.16, respectively for the steel-reflected case. The corresponding percent reactivity error obtained assuming the simplification of central-point fine-flux solution as cross-section weighting spectrum for the entire core region together with the simplification of utilizing zero-buckling fine-flux solutions as cross-section weighting spectra for the steel-reflector region is -0.30. It was found that although the core to reflector leakages are sensitive to the number of spatial cross section sets used, coarse-group flux values, even at the core-reflector interface, do not differ significantly with the number of spatial cross section sets utilized.

A related study[15] has been made for light internal subassembly-sized regions. Steel-rich structural-test subassemblies, containing neither fissile nor fertile materials, are present in some locations in EBR-II core loadings. The neutron spectrum in a structural subassembly can be strongly influenced by the differing intragroup cross-section variations with energy of the structural subassembly composition and of the surrounding core. This can cause errors in calculated multigroup flux spectra in the energy range in which many of the structural and coolant materials of fast reactors contain resonance-scattering cross-section variations within a multigroup cross-section interval. Since normally the use of many fine-energy groups are prohibitive for most analyses, the possible errors incurred by utilizing weighted coarse-group cross sections are of interest.

In this study, fine-energy groups (~1/60 lethargy intervals) were again used to calculate fine-energy solutions, in the energy range of resonance scatterings, for the idealized situation wherein either a steel-rich or a nickel-rich subassembly is located at the center position of an EBR-II-type core. **Fig. 10-22** illustrates the fine spectral changes occurring between a central steel-rich region or a central nickel-rich region and a surrounding EBR-II core. Spatially-dependent coarse-group cross sections and spatially-independent coarse-group cross sections were derived. The resulting coarse-group spectra, for the case having a central steel subassembly are given in **Table 10-1**. The footnotes give the bases for the derivations of the collapsed cross section sets. Coarse-group spectra obtained using collapsed cross sections derived from the spatially dependent fine-resolution solutions result in much smaller errors than the simpler fundamental mode or the zero buckling approaches. This is especially evident at the lower energies where the effects of resonance scattering are more pronounced.

Analyses of Foil Activations

Use of the EBR-II for extensive irradiation tests of fuel, structural, and poison materials require that the irradiation environment for the tests in the EBR-II core be understood. This is required, for example, to assist in interpretation of tests on the various irradiation-induced phenomena; e.g., magnitude of the fuel and clad swelling with burnup. To gain confidence in the reliability of calculated irradiation environment, it is essential that comparison be made with measurements performed in some EBR-II core loadings. Use of neutron activated foils have been and are being employed in various EBR-II loadings. Furthermore, extensive analyses of spectrally-sensitive neutron-activated dosimetry foils exposed in the ZPR-3 critical mockups have also been made.

Table 10-1 Central Flux: Steel-Rich Central-Subassembly*

Energy Group	Lethargy at Group Bottom	Fine-Group Sum	Coarse-Group Calc / Fine-Group Calc: 63 Sets[a]	6 Sets[b]	FM[c]	$B^2 = 0$[d]
4	2.5	0.00249	1.00	1.00	1.00	0.98
5	3.0	0.00371	1.00	1.00	0.99	0.99
6	3.5	0.00379	1.00	1.00	1.02	0.98
7	4.0	0.00260	0.99	0.99	0.98	1.05
8	4.5	0.00189	0.98	0.98	1.05	1.01
9	5.0	0.00110	0.97	0.97	1.03	1.03
10	5.5	0.000613	0.99	0.98	0.97	1.04
11	6.0	0.000240	0.99	1.00	1.11	1.08
12	6.5	0.000148	0.99	0.99	0.90	1.14
13	7.0	0.000043	0.99	1.02	1.11	1.19
14	7.5	0.000012	0.99	1.04	1.17	1.50
15	8.5	0.000011	0.99	0.98	0.74	1.32

*Normalized to applied source of unity per centimeter of system height.

[a] Fifteen cross-section sets used in central-subassembly region; remaining sets used in core and blanket regions.

[b] One cross section set used in central-subassembly region; remaining sets used in core and blanket regions.

[c] Core fundamental-mode solution weighting of cross sections for central subassembly.

[d] Zero-buckling solution weighting of cross sections for central-subassembly region.

The utilization of an operating fast power reactor to obtain integral cross-section data is illustrated by the analyses[13] of the foil irradiations[12] carried out at numerous radial and axial positions throughout the EBR-II reactor to obtain the capture-to-fission ratios of ^{233}U, ^{235}U, ^{238}U, ^{239}Pu, ^{240}Pu, and ^{242}Pu. Since the EBR-II functions primarily as an irradiation facility for fast-fuel and structural subassembly testing, the core loadings and core-blanket peripheral shape can change from run to run. Calculational interpretation of integral data must take cognizance of the effects of subassembly variations at differing core locations. Analogously, calculational interpretation at axial positions allowed for such complexities as pin tops, gaps, etc. The effects of loading pattern variations on the values of alpha were estimated by calculation of the loading patterns for runs 7, 14, 17, and 20 whose loading patterns encompass the range of loading variations in the reactor sector of interest, during the foil irradiation runs 5 through 23. Although an attempt was made to retain a similar loading pattern in the sector of interest, changes did occur in the spectrum at some positions that especially affected the values of the fertile foils. For example, the calculated range of values for ^{240}Pu capture-to-fission ratios for these four runs at a blanket subassembly position (position 9B5) are 1.13, 1.07, 1.12, and 1.05. The measured value representing irradiation during runs 5 through 20 is 1.37. The values of capture-to-fission ratios calculated for the fissile isotopes were only slightly sensitive to the differing loading patterns in the sector of interest during these irradiation runs. Comparison values at various radial subassembly positions are shown in Fig. 10-22 for the fissile isotopes ^{233}U, ^{235}U, and ^{239}Pu for midplane core and radial blanket positions.

EBR-II has several irradiation facilities located in its neutron shields, some of which are used for proof testing of nuclear instruments. These test facilities called J and O thimbles are made of concentric tubes and in one thimble,

J_2, provision is made for heating or cooling to maintain a specified temperature. The inner neutron shield ≈33 cm thick, is located between the blanket and the reactor vessel and is composed of graphite-filled, stainless-steel cans, except in the arc regions of thimbles, where graphite is substituted by sodium. The outer neutron shield located between the reactor vessel and the sodium tank is ≈58 cm thick. The first part of this shield is graphite and the second part contains, besides the graphite, the J thimbles and a borated graphite region to shield the reactor components located further out in the sodium tank. The sodium tank contains the O thimbles. The calculation[6] of the irradiation environment in the thimbles is a deep-penetration neutron transport problem, since the distances from the neutron sources in the core and the blanket are quite large. The neutron spectrum varies from a fast spectrum in the core to an almost thermal spectrum in the vicinity of the thimbles. Thermal-group cross sections were calculated with the THERMOS Code, for use with the 29 group set below 0.69 eV. The θ direction variations of fluxes, obtained in S_4 approximation, at a radial location near the center of the J thimbles are shown in **Fig. 10-23**[6]. The effect of the inner shield "windows" in increasing the higher energy fluxes is clearly seen as well as the almost four orders of magnitude depression in the thermal flux caused by the absorption in the borated graphite region. Some foil irradiations were done in the J_2 and O_1 thimbles; the results of these calculations are in reasonable agreement with those measurements in the J_2 thimble, but not in the O_1 thimble.

During the series of EBR-II mockup criticals carried out in ZPR-3, spectrally-sensitive neutron-actuated dosimetry foils were irradiated and proton-recoil spectral measurements were made.[7,8] The foils were irradiated at core, interface, and reflector positions in the critical assemblies. Calculational comparisons[11] of these activations allow the foil methods applied to the EBR-II reactor to be evaluated from data derived from simpler loadings in critical assemblies. Relations between calculated and measured values can then be used in interpretations of EBR-II-irradiated dosimetry foils. In these cases packets containing 2- x 2-in. square thin foils of ^{235}U, ^{238}U, Au, Ni, and Al were positioned in the assembly drawers at right angles to the edges of the thin plates constituting the drawer composition. In this manner, the activations correspond to the approximately 2- x 2-in. area of the drawer representing an EBR-II subassembly. In addition, the activities in various vertical strips of the foils were measured to obtain effects of the intra-drawer loading patterns upon the activations to be obtained at core, interface, and reflector regions. The calculated relative activations for the foils are compared in **Table 10-2** for the blanketed assembly with measurements. Calculations include corrections for the spatial variations of the x-y geometry-deduced activations for deviations from correct midplane variations caused by use of axial buckling parameters, heterogeneity effects across the drawers in core, at interface, and reflector positions, and resonance self-shielding in the packet.

Table 10-2. Relative Foil Activations - ZPR-3 Assembly 60

Position		$^{235}U_f$	$^{238}U_{n,\gamma}$	$^{197}Au_{n,\gamma}$	$^{238}U_f$	$^{58}Ni_{n,p}$	$^{27}Al_{n,\alpha}$
Core	Calc.	1.000[a]	0.110	0.144	0.0705	0.0212	0.00106
	Calc./Meas.	1.00[a]	1.17	0.84	1.01	1.11	--
Core Near Interface	Calc.	0.636	0.0695	0.100	0.0358	0.0110	0.000055
	Calc./Meas.	1.01	1.00	0.83	0.94	1.12	--
Blanket	Calc.	0.325	0.0427	0.0659	0.00448	0.00136	0.0000078
	Calc./Meas.	1.01	1.00(5)	0.79	0.79(5)	0.95	--

[a]Calc. and meas. normalized to unity at core position for $^{235}U_f$ activity per ^{235}U atom.

Physics at Power

Many of the neutronic parameters that are intensively studied using analysis and zero power experimentation combine in a complicated manner to form integral response to determine the integral behavior of the system. In its irradiation test reactor mode of operation, the EBR-II has a new heterogeneous core configuration approximately once a month. From a configuration containing no in-core irradiation tests in 1964 and surrounded by a depleted uranium reflector, the current stainless steel reflected configuration contains more than fifty irradiation subassemblies which make up roughly half the core.

The understanding of the reactor behavior at power requires a monitoring of the integral responses on both long term and short-term bases. The former comprise principally the reactivity required to bring the reactor from zero to full power and the reactivity required to remain at power. The latter information is generally routinely obtained using rod oscillator and rod-drop methods as augmented by an analog reactivity meter and various programs for real-time digital data analysis. The latter include a post incident recall system. Monitoring of coolant temperatuer and flow information provides much insight on the unfolding of neutronic behavior.

From an inspection of the data summarized in **Table 10-3** it can be seen that all of the integral parameters of interest cited there have varied rather markedly throughout the history of operation.[32] While some of the changes can be attributed to major configurational alteration, lesser ones can be correlated with such simple variations as the use of recycled driver fuel vs. fresh (cold) driver fuel. Variations have been noted that affect reactor kinetics, having no bearing on safety, that have had operational implications. Variation in the long term effects have had implications on reactivity lifetime (e.g. length of an uninterrupted run) with a fixed control rod system. Non-linearities in feedback have been noted attributed to hydraulic design of the radial reflector. Such deduction has led to a new reflector design that more optimally utilized the fixed coolant flow capability.

The testing program using oscillator and rod drop methods has now proceeded to the point where, in the range of interest, the two methods are used with similar confidence. **Fig. 10-24** shows a comparison between a feedback function deduced from drop and oscillator rods on Run 49F.[33]

Table 10-3 also shows the deduced feedback coefficients from kinetic tests that can be attributed to effects with various time constants. These too may be seen to vary with run, from beginning to end of run and are frequently noted to be non-linear with power. The correlation of the deduced parameters with fundamental predictions gives confidence in understanding the operational behavior of the system. For example A_1, in **Table 10-3** can usually be correlated with prompt fuel expansion. Similar correlations may be made between A_2 on **Table 10-3** and the core-related coolant density coefficient. It is also noted that the zero time slope of the feedback curve following a good rod drop corresponds to $A_1 + A_2/2$. Hence the zero time slope is a very good practical indicator of prompt kinetics. **Fig. 10-25** shows a comparison between a measured and predicted feedback variation with time.[34]

A somewhat puzzling problem is that of the existence of a very small amplitude (normal plant monitoring equipment does not sense this) nominal 10 cps oscillation.[35] While the frequency correlates favorably with the fundamental vibrational frequency for a subassembly, of which there are over 600 in the reactor, the mechanism is not completely understood. Its amplitude has been found to correlate with control rod bank position.

Table 10-3

Tabulated Data on Rod Drop Curve-Fits for Various EBR-II Core Loadings

Power	Run No.	A_1	t_1	T_1	A_2	t_2	T_2	A_3	t_3	T_3	A_4	t_4	T_4	H(0), $ i=1,2	H(0), $	H(0), Ih	PRD/H(0)	$\frac{1}{\hat{}_{FB}}(0)$	Mark-IA Driver Fuel Subs, Row 1-7	Experimental Fueled Subs, Row 1-7	Control Rods and Struct'al Subs, Row 1-7	Rod Worth, $
20.0	48B	.022	.5		-.009	2.0	.05	.009	3.0		.023	5.0	9.6	.022	.045	14.0		.047	79	34	14	.0175
22.5	26B	.059	.2		+.054	.4		-.094	2.0	.2	.005	5.0	1.0	.113	.024	7.4		.384	63	16	48	.0417
30.0	26B	.079	.2		.070	.4	.05	-.120	2.0	.2	.005	5.0	1.0	.149	.034	10.5		.511	63	16	48	.0417
30.0	48B	.065	.5								.04	5.0		.065	.105	32.6		.138	79	34	14	.0175
40.0	48B	.12	.5					-.047	2.0		.044	5.0	6.5	.120	.117	36.3		.225	79	34	14	.0175
41.5	26B	.102	.2		.093	.4	.05	-.097	2.0	.2	.004	5.0	0.5	.195	.102	31.6		.656	63	16	48	.0417
41.5	29A	.08	.2		.064	.4	.06	-.09	3.0	1.0	.12	5.0	23.0	.144	.147	45.6		.554	62	20	45	.0495
41.5	29C	.098	.2		.083	.4	.06	-.067	2.0		.078	4.0	3.2	.181	.192	59.5		.680	62	21	44	.0408
45.0	26B	.107	.2		.098	.4		-.063	2.0	.2				.205	.142	44.0	.95	.749	63	16	48	.0417
50.0	35A	.18	.2		.042	.4		-.055	2.4		.03	2.5	8.0	.222	.197	61.1	.62	.996	78	33	16	.0194
50.0	37A (Start)	.11	.2		.146	.4		-.085	2.0		.012	6.0	6.0	.256	.183	56.7	.59	.875	82	28	17	.0215
50.0	37A (End)	.11	.2		.138	.4		-.085	2.0		.012	6.0	6.0	.248	.175	54.3		.855	82	28	17	.0215
50.0	38A	.11	.2		.12	.4		-.144	2.0		.045	6.0	6.0	.230	.131	40.6	.49	.786	100	12	15	.0165
50.0	38B	.11	.2		.131	.4		-.08	2.0					.241	.161	49.9	.55	.838	76	29	22	.0158
50.0	39B	.11	.2		.17	.4		-.085	2.0					.280	.195	60.5	.68	.933	76	32	19	.0195
50.0	40A	.11	.2		.16	.4		-.12	2.0		.03	4.0	1.2	.270	.180	55.8	.61	.898	77	35	15	.0170
50.0	42A	.08	.2	.09	.10	.6	.12	-.015	2.0					.180	.165	51.2	.57	.559	75	37	15	.0177
50.0	43A (Start)	.16	.2		.21	.4		-.17	2.0					.370	.200	62.0	.70	1.24	78	34	15	.0175
50.0	43A (End)	.08	.2	.06	.10	.6	.09	-.015	2.0					.180	.165	51.2		.559	78	34	15	.0175
50.0	44A (Start)	.12	.2	.09	.15	.4		-.1	2.0					.270	.170	52.7	.58	.925	79	34	14	.0176
50.0	44A (End)	.12	.2	.09	.12	.4	.12	-.09	2.0					.240	.150	46.5		.855	79	34	14	.0172
50.0	44B (Start)	.12	.2		.14	.4		-.11	2.0					.240	.130	40.3	.46	.895	83	30	14	.0176
50.0	44B (End)	.10	.2		.10	.4		-.03	2.0					.200	.170	52.7		.735	83	30	14	.0200
50.0	45A (Start)	.10	.2		.085	.4		+.03	2.0					.185	.215	66.7	.75	.728	89	23	15	.0195
50.0	45B (Start)	.07	.2		.09	.4		-.018	2.0					.160	.142	44.2	.50	.566	88	22	17	
50.0	46A (Start)	.10	.2		.048	.4		-.018	2.0					.148	.130	40.3	.47	.611	91	19	17	.0186
50.0	46A (End)	.08	.2		.01	.4		+.035	2.0		.037	5.4	5.0	.090	.162	50.2		.674	91	19	17	.0185
50.0	47A	.082	.2		.045	.4		+.038	2.0					.127	.165	51.2	.61	.504	87	25	15	.0180
50.0	47B	.11	.2		.10	.4		-.10	2.0					.210	.110	34.1	.41	.750	79	34	14	.0175
50.0	49F	.06	.2		.075	.4		-.02	2.0		.02	4.0		.135	.135	42.0	.51	.485	89	25	13	.020
56.0	46A (Start)	.085	.2		.0312	.4		.037	2.0		.034	5.4	5.0	.116	.187	58.0	.65	.528	91	19	17	
56.0	46A (End)	.085	.2		.0312	.4		.037	2.0					.116	.153	47.0		.522	91	19	17	.0186
56.0	47A (Start)	.10	.2		.075	.4		.015	2.0					.175	.190	59.0	.64	.695	87	25	15	.0185
56.0	47B (Start)	.13	.2		.11	.4		-.13	2.0		.07	4.0		.240	.180	56.0	.62	.878	79	34	14	.0175
58.0	49C	.06	.2		.08	.4		-.02	2.0		.08	4.0		.140	.200	62.0	.72	.510	74	39	14	.022
58.0	49F	.06	.2		.10	.4		-.02	2.0		.06	4.0		.160	.200	62.0	.65	.555	89	25	13	.020
58.0	51A (11" Bank)	.07	.2		.08	.4		-.02	2.0		.06	4.0		.150	.190	58.9	.66	.525				.0235
58.0	51A (14" Bank)	.05	.2		.06	.4		-.02	2.0		.05	4.0		.110	.140	43.4		.403				.0235
58.0	51A (End)	.045	.2		.07	.4		-.005	2.0		.04	4.0		.115	.150	46.5		.408				.0235
62.5	38A	.1375	.2		.15	.4		-.18	2.0		.0563	6.0	6.0	.288	.164	50.84	.50	.982	100	12	15	.0160
62.5	46A (Start)	.095	.2		.0335	.4		.051	2.0		.02	5.4	5.0	.129	.200	62.0	.63	.588	91	19	17	.0186
62.5	46A (End)	.095	.2		.0335	.4		.048	2.0		.02	5.4	5.0	.129	.197	61.0		.586	91	19	17	.0185
62.5	47A (Start)	.1	.2		.04	.4		.06	2.0		.01	5.4	5.0	.140	.210	65.0	.63	.632	87	25	15	.0185
62.5	47A (End)	.06	.2		.02	.4		.045	2.0		.04	5.4	5.0	.080	.165	51.2		.380	87	25	15	.0180
62.5	47B (Start)	.13	.2		.125	.4		-.072	2.0		.01	4.0	7.0	.255	.193	59.8	.59	.929	79	34	14	.0175
62.5	51A (Start)	.075	.2		.085	.4		-.018	2.0		.05	4.0		.160	.192	59.5	.64	.591				.0235

Table 10-3 Definitions

Table 10-3 is a tabulation of these parameters for the EBR-II loadings on which rod-drop experiments were done. The column headings of Table 10-3 are defined as follows:

$$H(s) = \sum_{i=1}^{I} \frac{A_i e^{-sT_i}}{1 + s\tau_i} \qquad (1)$$

Power — The power level at which the EBR-II experiments were conducted, megawatts.

A_i, i=1,4 — The weighting coefficients for Eq. (1), dollars of reactivity, defined as the product of the power level (in megawatts) and the differential power coefficient of reactivity (in dollars of reactivity per megawatt).

τ_i, i=1,4 — The estimated system time constants for Eq. (1), seconds.

T_i, i=1,4 — The deduced delay times for i-th terms in Eq. (1), units of seconds.

H(0),$ — The differential power coefficient of reactivity evaluated from Eq. (1), having same units as the weighting coefficients, A_i. H(0) is first evaluated for the first two terms of Eq. (1) only, and then for all of the terms.

H(0),Ih — The differential power coefficient of reactivity expressed in inhours using the conversion factor 310 Ih/$. This term includes all the terms of Eq. (1).

PRD/H(0) — The ratio of the PRD (power-reactivity decrement) as determined by rod-worth and the power reactivity coefficient, H(0).

$\rho^1_{FB}(0)$ — The slope of the feedback reactivity at time t=0, units of $/sec.

Fuel Row 1-7 — The number of Mark-IA driver fuel subassemblies in rows 1 through 7 for each EBR-II run.

In-Core Row 1-7 — The number of Mark-II and oxide-type experimental subassemblies in rows 1 through 7 for each EBR-II run.

Control Rods and Struct. — The number of non-fuel subassemblies in rows 1 through 7 for each EBR-II run.

Rod Worth — The worth, in dollars, of the stainless-steel drop rod used in each experiment.

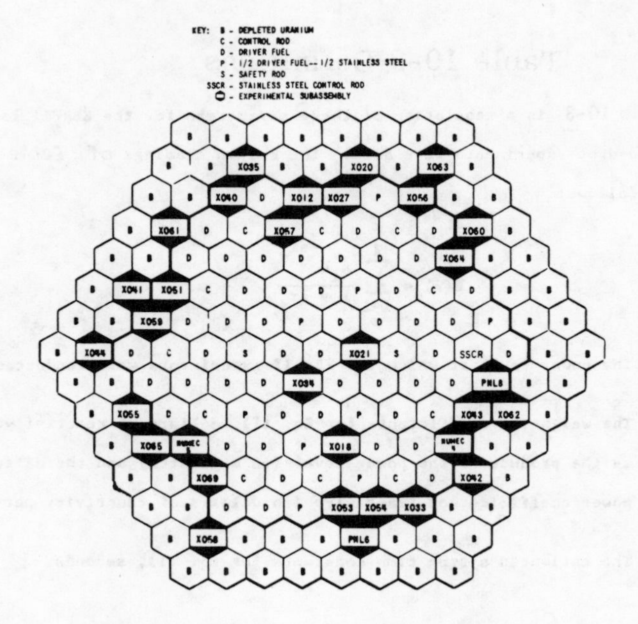

Example core loading of EBR-II as an irradiation facility.
Fig. 10-13

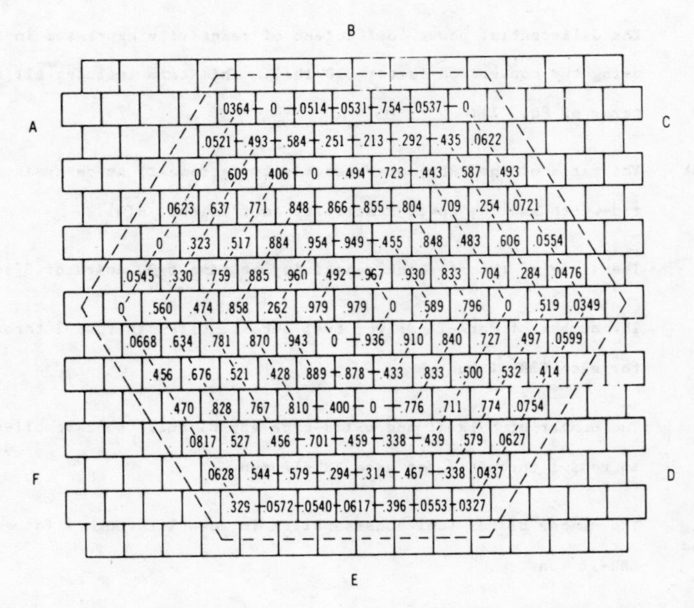

Calculated power (MWt) generated in individual subassemblies of assumed core configuration shown in Fig. 1 at 62.5 MWt operation.
Fig. 10-14

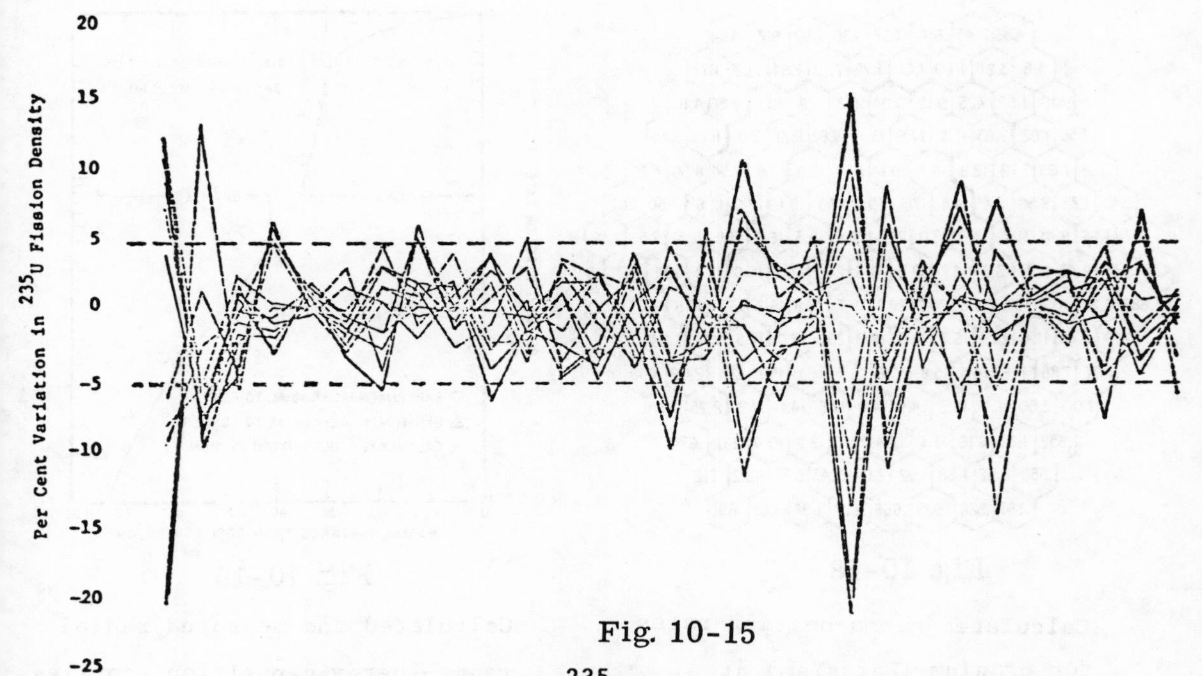

Fig. 10-15

Percent variation in ^{235}U fission density from run-to-run at each of the twelve control rod positions.

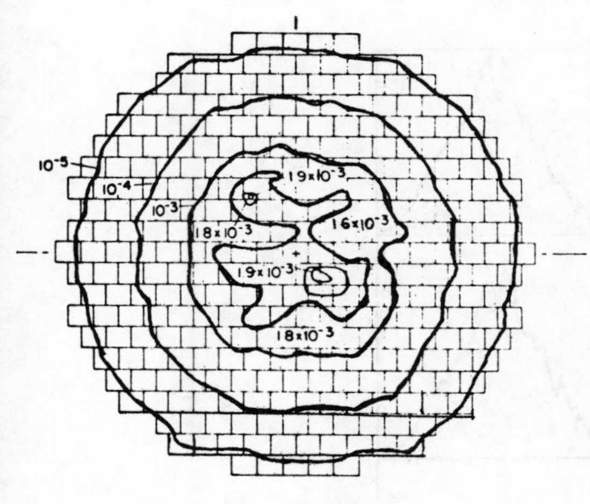

Fig. 10-16

Contour map of the relative neutron flux above 2.23 MeV for the EBR-II Run 32B. (Levels normalized to one fission neutron per centimeter height of system.)

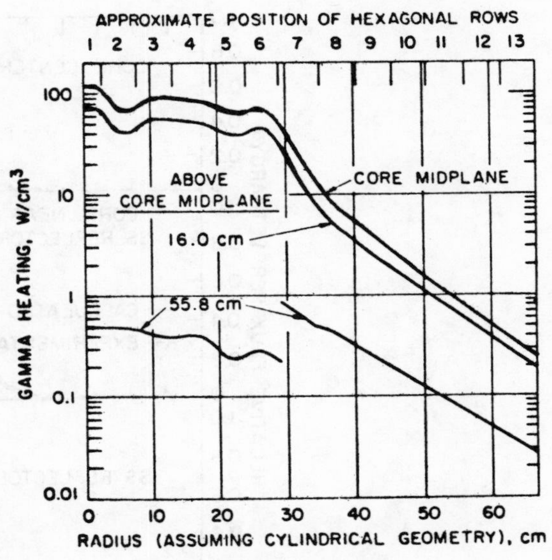

Fig. 10-17

Calculated gamma-heating rates (62.5 MWt operation) in a cylindricized representative 91-subassembly core loading.

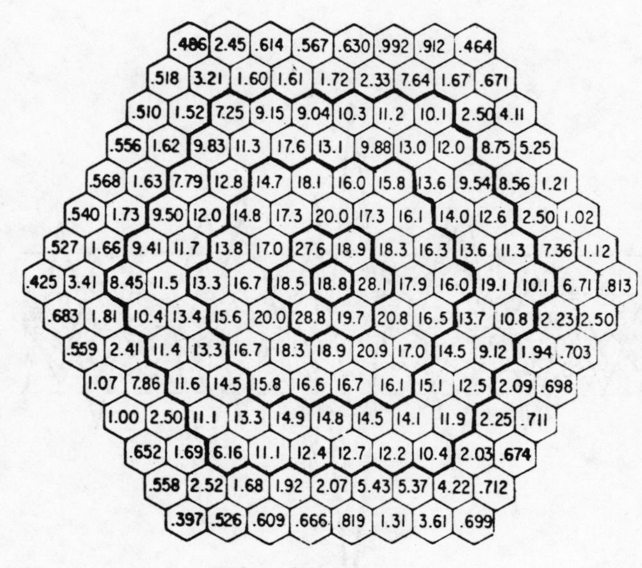

Fig. 10-18

Calculated gamma-heating rates for uranium (Watts/gm) at midplane of individual subassemblies of a representative core loading. (62.5 MWt operation)

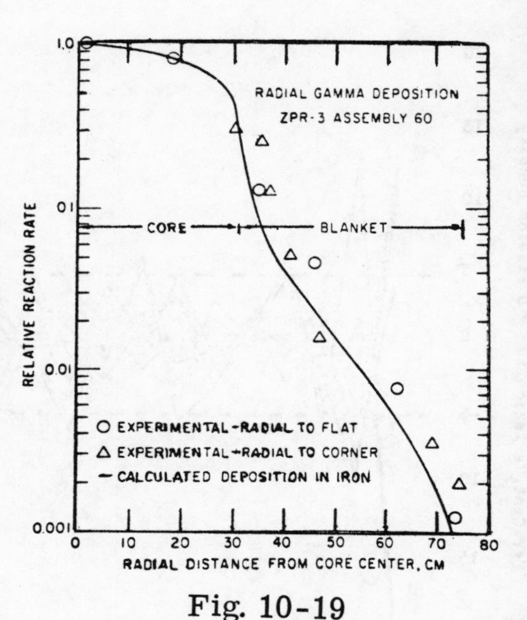

Fig. 10-19

Calculated and measured radial gamma-energy-deposition profiles.

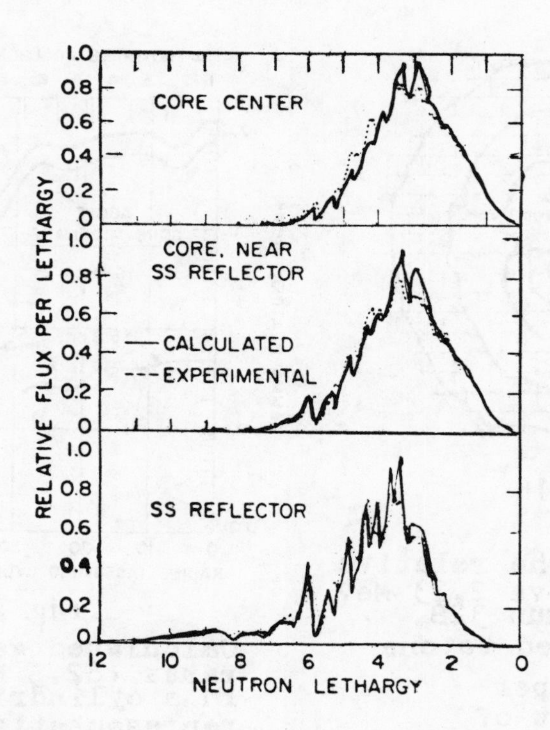

Calculated and measured neutron spectra in the steel-reflected critical assembly ZPR-3 No. 62.

Fig. 10-20

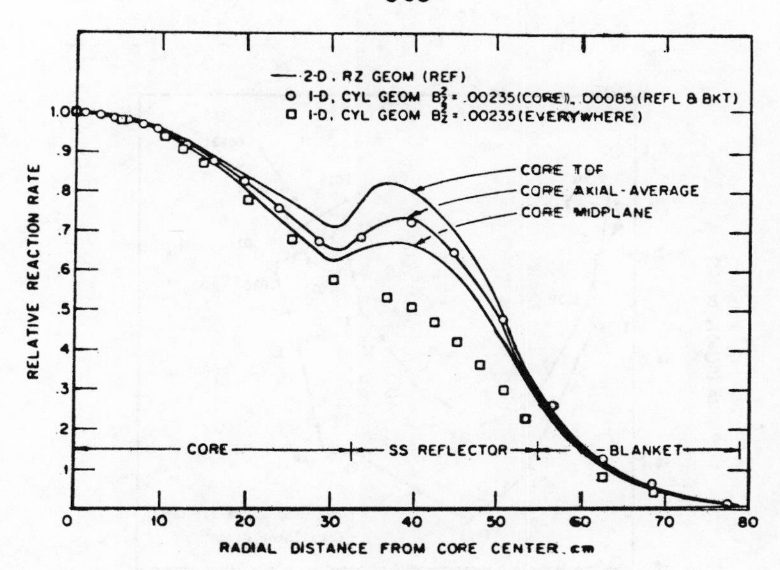

Calculated $^{235}U_f$ radial-detector traverses by 1D and 2D analyses.

Fig. 10-21

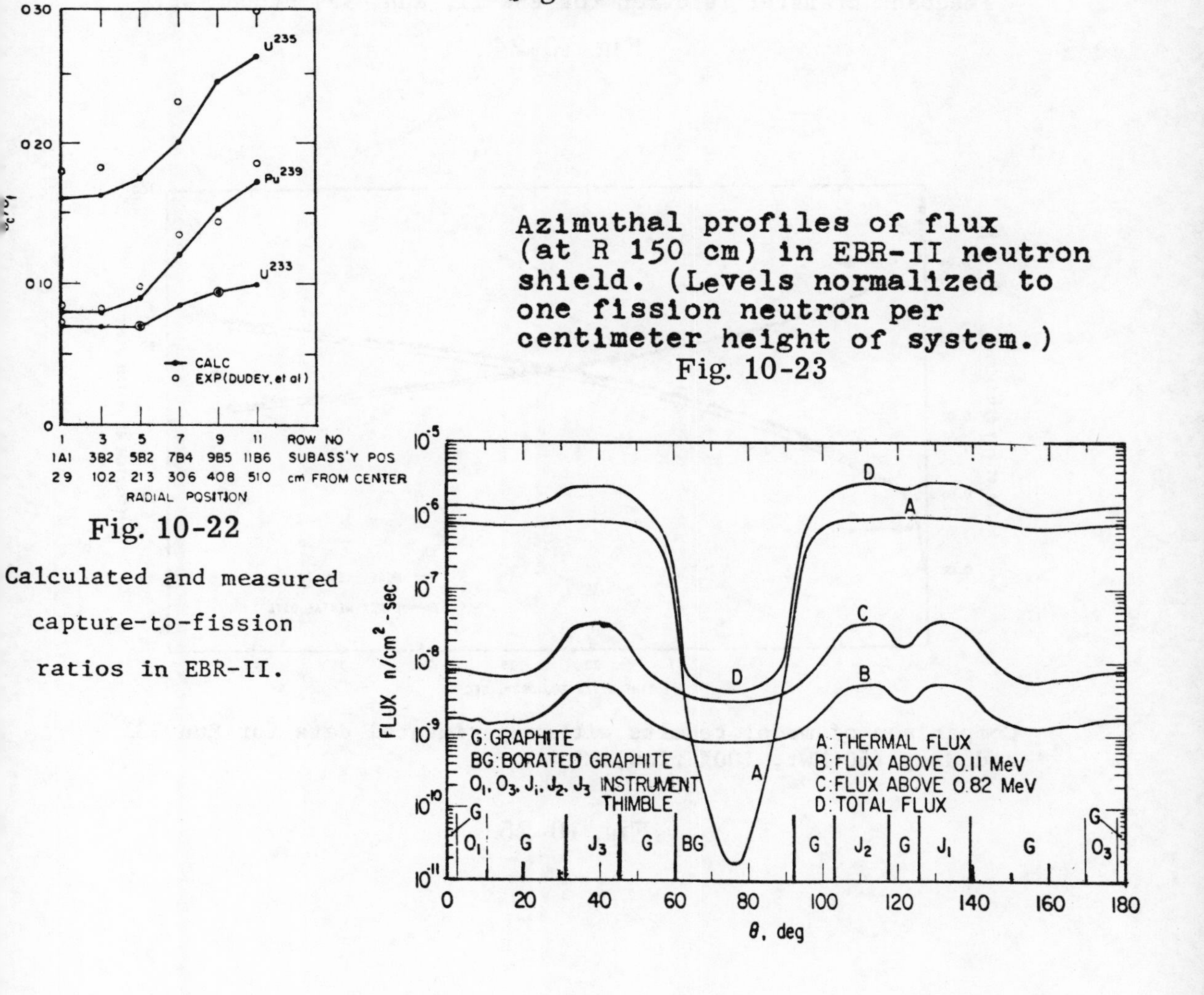

Fig. 10-22

Calculated and measured capture-to-fission ratios in EBR-II.

Azimuthal profiles of flux (at R 150 cm) in EBR-II neutron shield. (Levels normalized to one fission neutron per centimeter height of system.)

Fig. 10-23

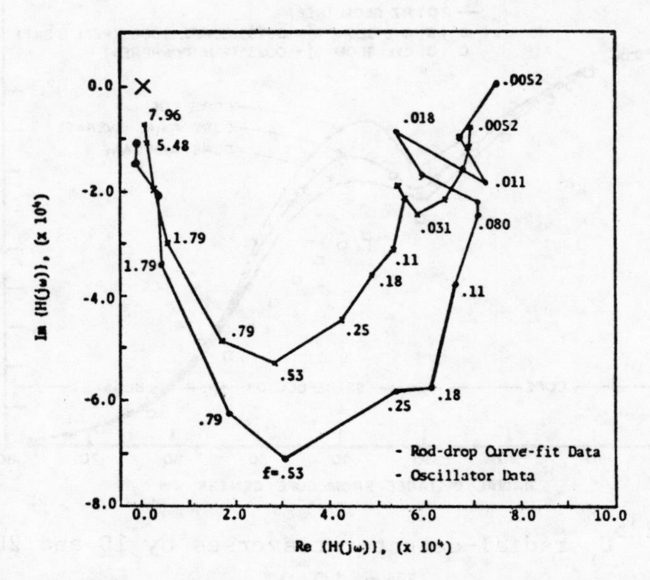

Feedback transfer function for EBR-II, Run 49F, at 58 MWt.

Fig. 10-24

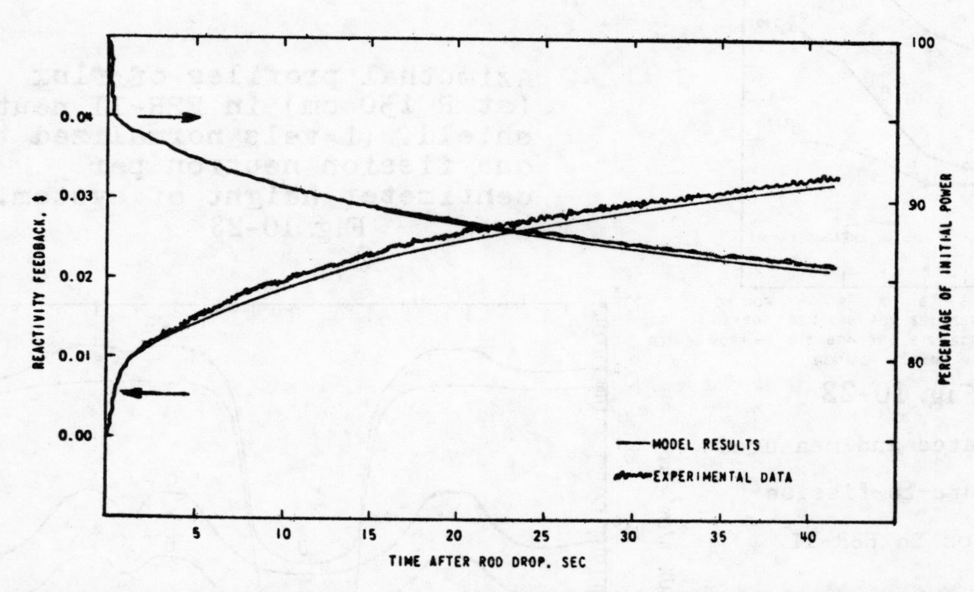

Comparison of model results with experimental data for Run 33A rod drop (50 MWt, 100% flow).

Fig. 10-25

REFERENCES

1. L. B. Miller and R. E. Jarka, "Detailed Analysis of Space-Dependent Phenomena," Trans. Am. Nucl. Soc. 13 (1), 307 (June 1970).

2. L. B. Miller and R. E. Jarka, "The Distribution of Gamma Energy Deposition in a Fast Reactor," Trans. Am. Nucl. Soc. 13 (2), 831 (November 1970).

3. L. B. Miller and R. E. Jarka, "Local Modifications of Irradiation Conditions," ANL/EBR-035 (March 1971).

4. P. L. Walker, "Reactor Development Program Progress Report," ANL-7854, pp. 1.29-1.30, Argonne National Laboratory (August 1971).

5. D. Meneghetti and K. E. Phillips, "Reactor Development Program Progress Report," ANL-7900, pp. 14.0-14.3, Argonne National Laboratory (December 1971).

6. B. R. Sehgal and R. H. Rempert, "Deep-Penetration Irradiation Environment in EBR-II Neutron Shields," Trans. Am. Nucl. Soc., 15 (1), 549 (1972).

7. D. Meneghetti, W. P. Keeney, R. O. Vosburgh, D. G. Stenstrom, K. E. Phillips, and J. M. Gasidlo, "Depleted-Uranium, Nickel, and Steel-Reflected EBR-II Critical Assemblies on ZPR-3--Experiments and Calculations," Trans. Am. Nucl. Soc., 13 (2), 733 (November 1970).

8. D. Meneghetti, W. P. Keeney, R. O. Vosburgh, G. G. Simons, D. G. Stenstrom, K. E. Phillips, and J. M. Gasidlo, "Experiments and Analyses of Heterogeneous Core and Neutron Spectra in EBR-II Critical Assemblies on ZPR-3," Trans. Am. Nucl. Soc., 14 (1), 20 (June 1971).

9. G. C. Simons, D. G. Stenstrom, D. Meneghetti, and W. P. Keeney, "Gamma-Ray Dose Evaluations for the ZPR-3/EBR-II Critical Assemblies," Trans. Am. Nucl. Soc., 13 (2), 880 (November 1970).

10. D. G. Stenstrom, "Calculations of Gamma Heating in a Fast Reactor Environment," Trans. Am. Nucl. Soc. 5 (1), 551 (June 1972).

11. D. Meneghetti and K. E. Phillips, "Analysis of Spectrally-Sensitive Foils in Depleted-Uranium, Nickel and Steel-Reflected EBR-II Criticals," Trans. Am. Nucl. Soc., 15 (1), 539 (June 1972).

12. N. D. Dudey, R. R. Heinrich, and J. Williams, "Integral Alpha Measurements for U and Pu Isotopes in EBR-II," Trans. Am. Nucl. Soc., 14 (2), 816 (October 1971).

13. D. Meneghetti and R. H. Rempert, "Sensitivities of Capture-to-Fission Ratios of U and Pu Isotopes in EBR-II Core and Blanket Regions for Various Loadings," Trans. Am. Nucl. Soc., 14 (2), 816 (October 1971).

14. D. Meneghetti and K. E. Phillips, "Resonance-Scattering Effects in Nickel- and Steel-Reflected EBR-II-Type Systems," 12 (2), 692 (November 1969).

15. D. Meneghetti and K. E. Phillips, "Structural-Subassembly Spectra Using Spatially Dependent and Spatially Independent Cross-Section Averagings," Trans. Am. Nucl. Soc., 13 (1), 256 (June 1970).

16. B. R. Sehgal and R. H. Rempert, "Reactor Development Program Progress Report," ANL-7900, 1.23-1.24, Argonne National Laboratory (December 1971).

17. B. R. Sehgal, R. K. Lo, and R. H. Rempert, "Neutronic and Thermal-Hydraulic Characteristics of a Higher-Worth Control Rod for EBR-II," ANL-7913 (to be published).

18. B. R. Sehgal and R. H. Rempert, "Tritium Production in Fast Reactors Containing B_4C," Trans. Am. Nucl. Soc., 14, 779 (October 1971).

19. W. B. Loewenstein, "Reactor Physics Division Annual Report," ANL-7010, 141-147 (January 1965).

20. R. R. Smith, et. al., "The Effects of an Over-Cooled Stainless Steel Reflector on the EBR-II Power Coefficient," ANL-7544 (May 1969).

21. D. Meneghetti and K. E. Phillips, "Reactor Development Program Progress Report," ANL-7825, 1.43-1.44, Argonne National Laboratory (April-May 1971).

22. D. A. Kucera and D. Mohr, "BOW-V: A CDC-3600 Program to Calculate the Equilibrium Configurations of a Thermally Bowed Reactor Core," ANL/EBR-014 (January 1970).

23. W. H. Sutherland and V. B. Watwood, Jr., "Creep Analysis of Statically Indeterminate Beam," BNWL-1362, UC-80 (January 1970).

24. C. C. Price and J. R. Karvinen, "EBR-II Noise-Signature Analysis During the First Half of Fiscal Year 1971," ANL/EBR-036, Argonne National Laboratory (February 1971).

25. C. C. Price and J. R. Karvinen, "Evaluation of Self-Powered Detectors in EBR-II," Trans. Am. Nucl. Soc., 15 (1), 366 (June 1972).

26. B. R. Sehgal, Personal Communication.

27. R. A. Laskiewicz, Personal Communication.

28. J. Hoover, G. K. Leaf, D. A. Meneley, and P. M. Walker, "The Fuel Cycle Analysis System, REBUS," Nucl. Sci. and Eng., 45, 1 (July 1971).

29. "DOT - Two Dimensional Discrete Ordinates Transport Code," RSIC Computer Code Collection, ORNL-K-1694.

30. W. E. Ford, III and D. H. Wallace, "POPOP4, A Code for Converting Gamma-Ray Spectra to Secondary Gamma-Ray Production Cross Sections," CTC-12, Computer Technology Center, Oak Ridge National Laboratory (1969).

31. J. R. Knight and F. R. Mynatt, "MUG, A Program for Generating Multigroup Photon Cross Sections," CTC-17, Computer Technology Center, Oak Ridge National Laboratory (1970).

32. H. A. Larson, Personal Communication.

33. H. A. Larson, I. A. Engen, and R. W. Hyndman, "Feedback Analysis by Rod-Drop and Rod-Oscillator Experiment," Trans. Am. Nucl. Soc., 15 (1), 495 (June 1972).

34. A. V. Campise, "Dynamic Simulationa and Analysis of EBR-II Rod-Drop Experiments," ANL-7664, Argonne National Laboratory (March 1970).

35. R. W. Hyndman, F. S. Kirn, and R. R. Smith, "EBR-II Self-Excited Oscillations," Trans. Am. Nucl. Soc., 8 (2), 590 (November 1965).

CHAPTER 11

FAST NEUTRON SPECTRUM MEASUREMENTS

Fast neutron spectrum measurements have played a part in the analysis of fast reactor physics since the earliest days of fast reactor technology. Indeed, some of the first such spectrum measurements were made on early GODIVA reactors using proton-recoil distributions in nuclear emulsions.[1] Since then major advancements in technique have been made, but the purpose of the studies has remained the same. The neutron spectrum is one of the most important integral characteristics of a reactor system, and its accurate prediction is essential for the prediction of other important characteristics. For example, with a fixed composition and size, the criticality of a fast system is a sensitive function of α, the capture-to-fission ratio for the fuel. This quantity, however, is rather strongly energy dependent, and therefore an accurate prediction of the spectrum is necessary in predicting criticality. Other important characteristics, such as the achievable breeding ratio, are also strongly influenced by the capture-to-fission ratio, making them spectrally dependent.

In some cases certain details of the spectrum are important, as well as the overall characterization of its shape. For example, the Doppler coefficient in a fast breeder is sensitive to the competiton for neutrons between capture in fertile materials and fission of the fuel. This competition becomes quite favorable for the fertile materials in the kilovolt energy range; therefore, this range is an important contributor to the Doppler coefficient. Thus, even when a small fraction of all neutrons occur at these energies, it is important to predict this quantity accurately when accuracy in the Doppler coefficient is required.

TECHNIQUES EMPLOYED FOR SPECTRUM MEASUREMENTS

A wide variety of techniques has been used to measure fast neutron spectra. To a large degree, the choice of technique has been dictated by the equipment or technology available at the specific laboratory. In other cases the choice has been dictated by the nature of the system being studied, and by the radiation and thermal environment in which the measurements are made.

The available techniques can quite generally be broken down into two broad classes: differential spectrometers, of which neutron time-of-flight is the best example, and integral spectrometers, of which various proton-recoil techniques are the best example.

Neutron Time-Of-Flight

The neutron time-of-flight technique can be regarded equally as either the simplest method or the most complex method for fast spectrum measurements. It is the simplest in the sense that the measurement is quite direct, and can be performed with a minimum of detection hardware. On the other hand, it is quite complex when considering the intense accelerator and substantial facility usually required.

For this measurement, the system under study is excited with an intense short pulse of neutrons (most frequently photoneutrons produced by high energy electrons). If the system is nonmultiplying, the neutron population decays in some tens of nanoseconds and a neutron beam is extracted into a long evacuated flight tube. A detector at the end of this flight tube records the time history on an energy scale, and if the energy dependence of the detector efficiency is known, then the neutron energy spectrum is obtained directly. If the system multiplies, and consequently has a slow dieaway, then the time dependence of neutron emission from the assembly becomes important. The analysis of the data is then a great deal more complicated,[2] but the principle remains the same.

Typical detectors used in this application include lithium-glass scintillators,[3] proton-recoil organic liquid scintillators,[4] ^{10}B systems viewed by gamma detectors,[5] and boron-loaded organic liquid scintillators.[6] The appropriate detector selection depends primarily on the energy range to be covered. Only the lithium-glass scintillator has been used effectively over the entire range from 1 keV to 8 MeV, which is of importance for fast reactor applications.[3]

Integral Spectrometers

A large family of integral spectrometers exists. The most important of these are the proton-recoil gas proportional chamber and the proton-recoil organic liquid scintillator. The most important characteristic of detectors in this class is the fact that they have a unique response for a given neutron energy, and the response to a continuous neutron spectrum is the integral of such responses over the spectrum. In an experiment, the spectrum is obtained from the integral data by some suitable form of unfolding.

The proton-recoil integral spectrometers are widely preferred due to the relative simplicity of their responses. However, solid-state detectors based on ^{3}He and ^{6}Li charged particle reactions[7] and activation foils[8] have been effectively used for some applications, and will also be discussed.

FAST NEUTRON SPECTRA IN BULK MATERIALS

Studies of fast spectra in bulk materials are usually made either to test the adequacy of fast cross sections, or to provide data to test radiation transport calculations for deep penetrations of shielding materials. The most active work in this area has been performed at the Rensselaer Polytechnic Institute (RPI) and Gulf Radiation Technology (Rad Tech) linear accelerator facilities. Other important work has been done at Atomics International (AI).

At RPI, the program has involved intensive studies in bulk assemblies of iron,[5] depleted uranium,[9] and aluminum.[10] At Rad Tech, studies have been made for depleted uranium,[11] iron,[12] carbon,[13] lithium hydride,[14,15] tungsten,[14] paraffin,[15] and concrete.[16]

Three different experimental designs have typically been used for studies of bulk media. The RPI assemblies have generally been cuboids, as illustrated in Fig. 11-1. The analysis of these data has essentially relied on a spherical symmetry of the flux due to the great distance to boundaries. In most of the Rad Tech work, smaller assemblies have been fabricated into spheres with re-entrant holes designed to allow the angular distribution of the spectrum to be measured at constant radius; this configuration is shown in Fig. 11-2.

A third configuration has recently been used at Rad Tech, in which the assembly under study is placed at the end of a 50 m flight path. In this case, the interactions for selected neutron energies can be studied by observing events in the assembly corresponding to a selected neutron flight time from the target. Results using this technique for concrete are described by Harris,(17) and will not be covered here.

The use of spectra in bulk material to aid in the evaluation of cross sections has been illustrated by Malaviya et al.,(5) for the case of iron. An overall comparison between measured spectra and calculations using ENDF/B-I is shown in Fig. 11-3. The agreement is seen to be quite good except for the major iron scattering resonance at 28.5 keV where ENDF/B-I data are known to be discrepant. The discrepancy in the cross section is clearly seen in the comparison to the measured spectra. This point is further illustrated by comparing the measurements to calculations using an alternative cross section, i.e., the KEDAK data(18) from Karlsruhe. This comparison is shown in Fig. 11-4 which clearly shows the superiority of the KEDAK data for this resonance.

The detail which can be achieved in measurements of this sort are illustrated in Fig. 11-5 by the data of Neill et al. for a depleted uranium sphere.(11) Here, the complete angular distribution of the spectrum is shown for a fixed radius in the sphere.

Fast spectrum studies for thick assemblies of sodium, and steel-sodium lamination, have been made at AI by Specht, in support of the LMFBR program.(19) For sodium thicknesses up to approximately 150 cm, the agreement between calculation and measurement is usually within 40% between 7 keV to 1.7 MeV. For depths of 273 cm, however, the author reports measured fluxes below 50 keV which are larger than the calculated values by factors of 4 to 8. Evidently this region should be studied carefully in assessing the leakage through thick shields of sodium

FAST NEUTRON SPECTRA IN MULTIPLYING MEDIA

Neutron spectrum studies in multiplying media have been made with a wide variety of techniques. For reactor assemblies the principal efforts have been at the Argonne National Laboratory (ANL) and Atomics International (AI), where proton-recoil proportional chambers have been used, and at Gulf Rad Tech where both time-of-flight and proton-recoil have been used. Studies of LMFBR blanket assemblies have been made at MIT for several integral spectrometers, and activation foils have been used by Hanford Engineering Development Laboratory (HEDL) in several assemblies at other laboratories. Active European programs include work at Winfrith, Aldermaston, and Karlsruhe.

Argonne National Laboratory

A systematic development of the use of proton-recoil proportional counters for in-core spectrometry was initiated at ANL over ten years ago.(21) Since that time, the program has had a broad emphasis covering the detailed instrumentation and computerization of the technique, as well as developments toward better understanding the detailed effects for which data corrections must be made to obtain accurate results. Bennett has published an extensive discussion of the measurement technique(22) used at ANL.

Much of the recent ZPR program of critical experiments at ANL has been accompanied by measurements of fast neutron spectra. Typical data from these studies are represented by results obtained for EBR-II critical assemblies in ZPR-3,(23) the FTR-3 mockup on ZPR-9,(24) and the ZPPR-2 benchmark.(25) For illustration, the spectrum measured by Simons in ZPPR-2 is shown in Fig. 11-6 along with an MC^2 calculation for the system.(25) The ANL measurements have generally been made at the center of large critical assemblies which characterize typical cores of the LMFBR program.

Over the past three years the capability for time-of-flight measurements has been built into the ZPR-6 facility. This capability uses the beam from either the FNG accelerator or a small electron linear accelerator. A flight path has been built through the cell containment system, with provisions for measurements at approximately 50 and 100 meters. When this program is initiated it will be used primarily to augment the low energy spectral studies now being done by proton-recoil.

Gulf Radiation Technology

The fast spectrum studies at Rad Tech have primarily involved the use of time-of-flight techniques, although this has recently been extended to include proton-recoil measurements in order to make careful comparisons between the two techniques. This section will concentrate on the time-of-flight work, since the proton-recoil measurements are discussed in detail by Verbinski.[26]

Much of the program of measurements on multiplying systems has been done using a split-bed subcritical assembly, the Subcritical Time-of-Flight Spectrum Facility (STSF). The general layout of this facility is shown in Fig. 7. The STSF has an aluminum matrix formerly used on ZPR-3, and is licensed for ^{235}U loadings up to 600 kg with a subcriticality of 10 dollars. It is driven by the Rad Tech Linac and is aligned with a 220 m flight path containing detector stations at 45, 110, and 220 meters.

A summary of the multiplying systems which have been studied at Rad Tech is given in Table 11-1. These results are too extensive to be discussed in detail here, and we will primarily concentrate on the most recent work involving STSF-7 and STSF-9. These two loadings were particularly clean assemblies, for which the spectrum was also measured by proton-recoil. The STSF-7 assembly is essentially identical to ZPR-III Assembly 22 which is a benchmark designated by the Cross Section Evaluation Working Group (CSEWG). Publications can be found for APFA-III,[11] STSF-1A,[27] STSF-2,[28] and TCF-1.[29]

An extensive analysis of the results for STSF-7, -9, has been performed by Neill[30] using the most up-to-date cross sections available (ENDF/B, Vers. III). The overall comparisons between calculation and measurement are shown in Figs. 11-8, 11-9, and the comparison between time-of-flight and the proton-recoil results of Verbinski[26] is shown for STSF-9 in Fig. 11-10. A significant discrepancy between calculation and measurement exists below about 20 keV and a careful analysis by Neill has shown that the difference cannot be ascribed to the measurement. His analysis of the discrepancy indicates that the most likely explanation lies in the energy distribution of secondary neutrons from inelastic scattering in ^{238}U.

Atomics International

The program of spectrum studies at AI is unique in the use of spherical proportional counters, rather than the cylindrical counters found most convenient at ANL. Measurements have been made in a large split-table critical assembly, the ECEL, and in small fast reactors for space applications.

The results obtained for the small reactor are shown in Fig. 11-11.[31] This assembly contains a mixture of $^{235}U/^{7}Li_3N/W/Hf/Ta$, and exhibits a substantial resonance structure in the spectrum. Calculations using the GAM-II group structure are compared with these measurements in Fig. 11-11. This experimental verification that there are few neutrons below 50 keV was important in confirming the

Table 11-1 . Summary of Multiplying Systems For Which Spectra Have Been Measured by Gulf Rad Tech

Assembly	ZPR Equivalent	Description
APFA-III	None	^{235}U sphere
STSF-1	III-14	$^{235}U/C$
STSF-2/2A	III-17	$^{235}U/^{238}U/C$
STSF-7, 8	III-11, 22	$^{235}U/^{238}U$
STSF-1A	III-57	$^{235}U/^{238}U/BeO$
STSF-3	VI-6	UO_2/Na LMFBR
STSF-4	VI-6	UO_2/Na Void LMFBR
STSF-5	VI-5	UC/Na LMFBR
STSF-6	VI-5	UC/Na Void LMFBR
STSF-9, 10	None	$^{235}U/^{238}U/O$
TCF-1	None	$^{235}U/CH_2$/Steel

design prediction that the temperature defect due to ^{238}U Doppler broadening would be small.

Massachusetts Institute of Technology

The MIT Blanket Test Facility has been used for a number of studies of neutron spectra, characteristic of the blankets of large LMFBR reactors; this facility is shown in Fig. 11-12.(32) It consists of a graphite/uranium converter, adjusted in composition to approximate the leakage spectrum from an LMFBR, a beam port, and a blanket assembly. The facility is driven by the thermal column of the MITR D_2O reactor.

An important feature of the MIT program is the application of three separate spectrometers to the measurements. Ortez et al.,(7) report the use of ^3He semiconductor detectors between 100 keV and 1 MeV, ^6Li semiconductor sandwich detectors between 10 keV and 3 MeV, and proton-recoil gas-proportional counters between 2 keV and 1.5 MeV. The agreement was generally good between these techniques, and is illustrated for the ^3He and ^6Li detectors in Fig. 11-13. Other work using this facility has been reported by Rogers et al.,(33) and by Leung et al.(34)

Hanford Engineering Development Laboratory

The work on neutron spectra at HEDL is a continuation of efforts by McElroy using foil activation to comprise an integral spectrometer which is unfolded in much the same way as the other integral spectrometers already described. This technique suffers somewhat from lack of an ideal set of response functions made up from the energy-dependent activation cross sections of the reactions used, but has the great advantage of being applicable in reactor facilities which can accommodate only minimal in-core components.

McElroy et al.(8) have recently published the results of spectrum measurements in the APFA-III at locations for which the spectrum had previously been determined by Neill, using the time-of-flight technique.[11] Thirty separate reactions were used in this case, and the resultant unfolded neutron spectrum agreed with the time-of-flight result within 10-20% over most of the neutron energy range.

SUMMARY

Rapid advancement has taken place during the past four years in the experimental technology for determining fast neutron spectra. In reactor systems, the most important techniques are neutron time-of-flight and proton-recoil proportional chambers. For certain shielding applications, other integral spectrometers are important.

A wealth of experimental data is now available, and most of this has been analyzed in varying degrees by the laboratories responsible for the measurements. However, the full value of the currently available data will only be realized when it has been widely distributed, and is analyzed using current cross sections by a larger number of groups, especially those involved in reactor design.

REFERENCES

1. G. M. Frye, J. H. Gammel, and L. Rosen, "Energy Spectrum fo Neutrons From Thermal Neutron Fission of U^{235} and From An Untamped Multiplying Assembly of U^{235}," LA-1670, for Los Alamos Scientific Laboratory (1954).

2. P. d'Oultremont, D. H. Houston, and J. C. Young, "Cage-Bird-Spec," Gulf-RT-10195, Gulf Radiation Technology (1970).

3. J. M. Neill, D. Huffman, C. A. Preskitt, and J. C. Young, "Calibration and Use of a 5-Inch Diameter Lithium Glass Detector," Nucl. Instr. & Methods, 82, 162-172 (1970).

4. V. V. Verbinski, et al., "Calibration of an Organic Scintillator for Neutron Spectrometry," ORNL-TM-2183, Oak Ridge National Laboratory (1968).

5. B. K. Malaviya, et al., "Experimental and Analytical Studies of Fast Neutron Transport in Iron," Nucl. Sci. Eng. 47, 329-348 (1972).

6. H. E. Jackson, G. E. Thomas, "Boron-Loaded Neutron Detector with Very Low Gamma-Ray Sensitivity," Rev. Sci. Instr. 36, 419 (1965).

7. N. R. Ortiz, et al., "Instrumental Methods for Neutron Spectroscopy in the MIT Blanket Test Facility," COO-3060-3, MITNE-129, (1972).

8. W. N. McElroy, R. J. Armani, and E. Tochilin, "The Spectral Distribution of Neutrons and Neutron Reaction Cross Sections in an Unreflected Uranium-235 Critical Assembly and the Fission Neutron Spectrum," Nucl. Sci. Eng. 48, 51-71 (1972).

9. N. N. Kaushal, B. K. Malaviya, M. Becker, E. Burns, and E. R. Gaertner, "Measurements and Analysis of Fast Neutron Spectra in Depleted Uranium," Nucl. Sci. Eng., to be published, 1972.

10. B. K. Malaviya, et al., "Time-of-Flight Measurements of Fast Neutron Spectra in Aluminum," Trans. Am. Nucl. Soc. 12, 756 (1969).

11. J. M. Neill, J. L. Russell, Jr., R. A. Moore, and C. A. Preskitt, "Fast Neutron Spectra in Multiplying and Non-Multiplying Media," Proc. of Conf. on Neutron Cross Section and Technology, Washington, D.C., March 4-7, 1968, NBS-299, p. 1183.

12. R. J. Cerbone, "Neutron Transport Measurements in Iron," Trans. Am. Nucl. Soc. 13, 260 (1970).

13. A. E. Profio, H. M. Antuney, and D. L. Huffman, "The Neutron Spectrum From a Fission Source in Graphite," Nucl. Sci. Eng. 35, 91 (1969).

14. A. E. Profio, R. J. Cerbone, and D. L. Huffman, "Experimental Verification of Neutron and Gamma-Ray Transport in Lithium Hydride and Tungsten," GA-9029, Gulf General Atomic, 1968.

15. A. E. Profio, R. J. Cerbone, and D. L. Huffman, "Fast Neutron Penetration in Paraffin and Lithium Hydride," Nucl. Sci. Eng. 44, 376-387 (1971).

16. L. Harris, et al., "Time Dependent Fast Neutron and Secondary Gamma-Ray Spectrum Measurements in Concrete," DASA 2401-1, DASA 2401-2, Gulf General Atomic (1969).

17. L. Harris, "Developments in Integral Shielding Experiments Involving Fast-Neutron Transport," Proc. this Conference.

18. I. Langer, J. J. Schmidt, and D. Wall, "Tables of Evaluated Nuclear Cross Sections for Reactor Materials," KFK-750, EUR-3715e EANDC(E)-88 "U," Kernforschungszentrum, Karlsruhe, (1968).

19. E. R. Specht, "Nonhydrogeneous Shield Test Correlations," Trans. Am. Nucl. Soc. 12, 946 (1969).

20. "Fast Reactor Spectrum Measurement and Their Interpretation," Summary of a specialists meeting sponsored by the IAEA at Argonne National Laboratory, November 10-13, 1970, Report IAEA-138.

21. W. C. Redman and J. H. Roberts, "Some Techniques of Fast Neutron Spectrum Measurements," Second Conf. Peaceful Uses Atomic Energy, P/597, 12, 72 (1958).

22. E. F. Bennett, "Fast Neutron Spectroscopy by Proton-Recoil Proportional Counting," Nucl. Sci. Eng. 27, 16-27 (1967).

23. D. Neneghetti, et al., "Experiments and Analyses of Heterogeneous Cores and Neutron Spectra in EBR-II Critical Assemblies on ZPR-3," Trans. Am. Nucl. Soc. 14, 20 (1971).

24. T. J. Yule, E. F. Bennett, "Measured Neutron Spectrum in FTR-3, A Plutonium Fueled Fast Spectrum Reactor," Trans. Am. Nucl. Soc. 13, 687 (1970).

25. G. G. Simons, R. G. Palmer, and A. P. Olson, "Applications of Neutron Spectra in the Demonstration Benchmark," Trans. Am. Nucl. Soc. 14, 834, (1971).

26. V. V. Verbinski, et al., "Proton-Recoil Proportional Counter Spectrometry for Reactor Spectrum Measurements," Proc. this Conference.

27. J. C. Young, J. M. Neill, P. d'Oultrement, E. L. Slaggie, and C. A.

Preskitt, "Measurement and Analysis of Neutron Spectra in a Fast Subcritical Assembly Containing Uranium-235, Unranium-238, and Beryllium Oxide," Nucl. Sci. Eng., 48, 45-50 (1972).

28. P. d'Oultremont, J. C. Young, J. M. Neill, and C. A. Preskitt, "Neutron Spectra and Kinetic Properties in Fast Subcritical Assemblies Containing Uranium-235, Uranium-238, and Carbon," Nucl. Sci. Eng., 45, 141-155 (1971).

29. S. C. Cohen, R. A. Moore, J. C. Young, and J. L. Russell, Jr., "Fast Neutron Spectrum for a Subcritical Section of a Homogeneous ^{235}U-Polyethylene Critical Assembly," Nucl. Sci. Eng., 37, 111-126 (1969).

30. J. M. Neill, J. C. Young, "Time-of-Flight Spectrum Measurements in the ^{235}U-Fueled Fast Subcritical Reactors," STSF-7, -9, and -10," Gulf-RT-B12154 (1972).

31. R. K. Paschall, P. G. Klann, "Neutron Spectrum Measurements in a Compact Fast Spectrum Critical Assembly," Trans. Am. Nucl. Soc., 14, 4 (1971).

32. M. J. Driscoll, Massachusetts Institute of Technology, Personal Communication.

33. V. C. Rogers, I. A. Forbes, and M. J. Driscoll, "Heterogeniety Effects in the MIT-BTF Blanket No. 2," Trans. Am. Nucl. Soc. 15, 468 (1972).

34. T. C. Leung, M. J. Diiscoll, I. Kaplan, and D. D. Lanning, "Measurements of Material Activation and Neutron Spectra in an LMFBR Blanket Mockup," Trans. Am. Nucl. Soc. 14, 383 (1971).

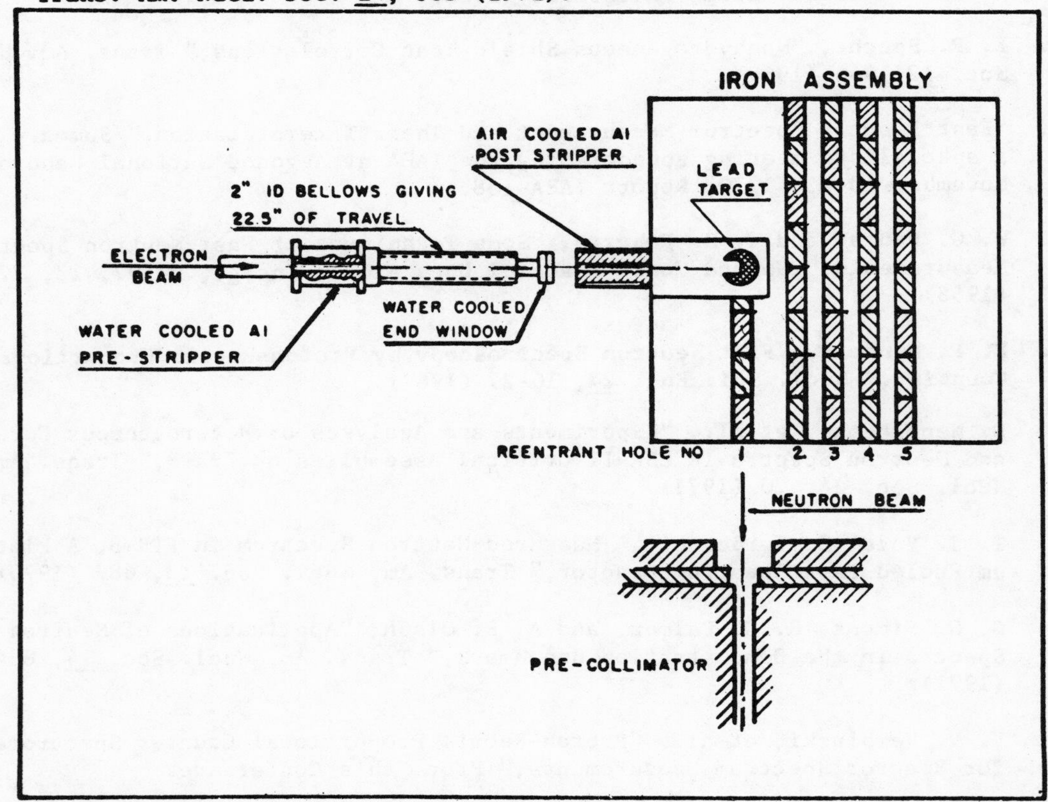

Fig. 11-1. Experimental setup used at RPI for fast spectrum measurements

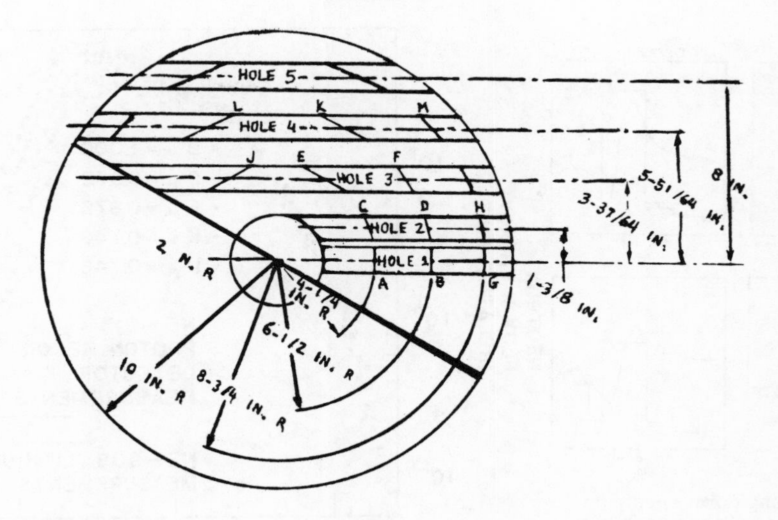

Fig. 11-2. **Geometry of spheres used by Gulf Rad Tech for fast spectrum measurements. By plugging selected holes to the appropriate depth the spectrum can be measured at various angles for fixed radius.**

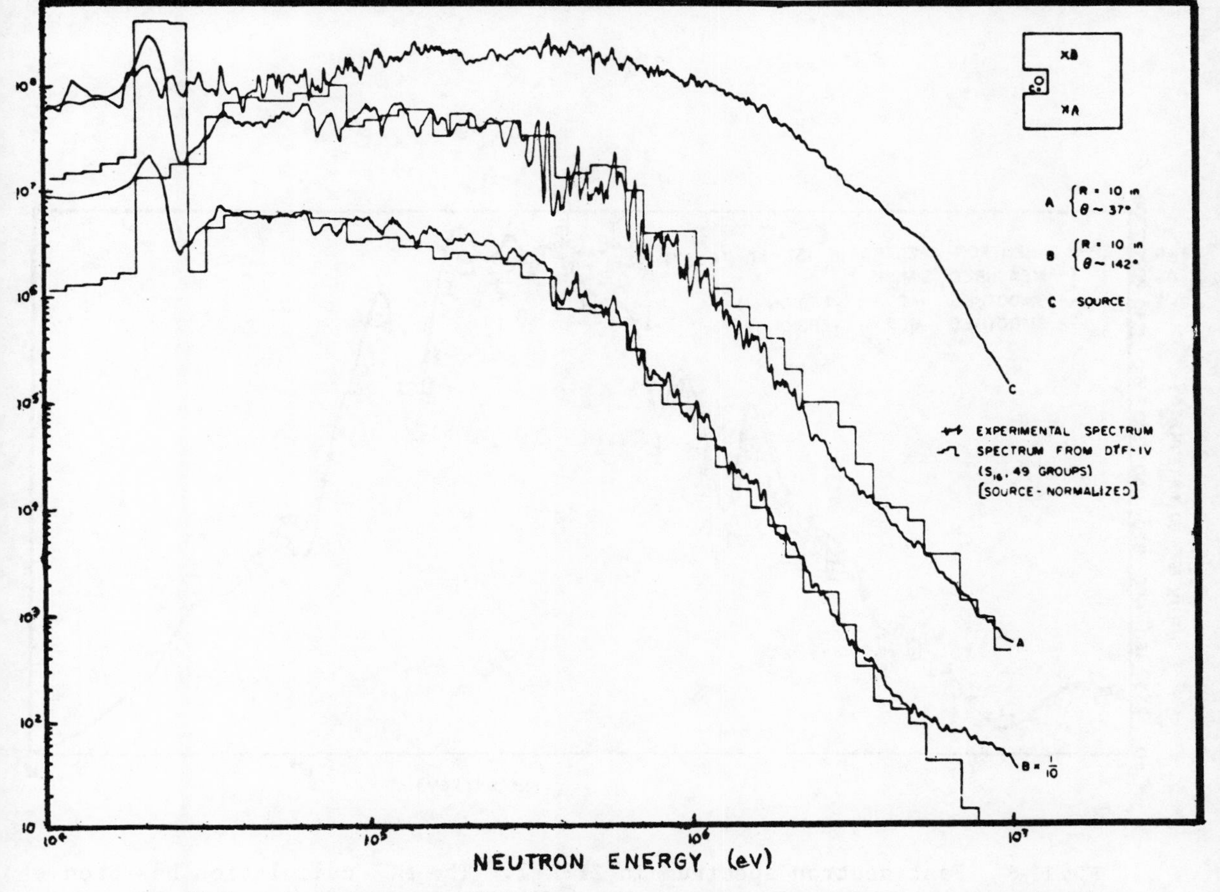

Fig. 11-3. Comparison of RPI measurements on iron to calculation using ENDF/B-I cross sections

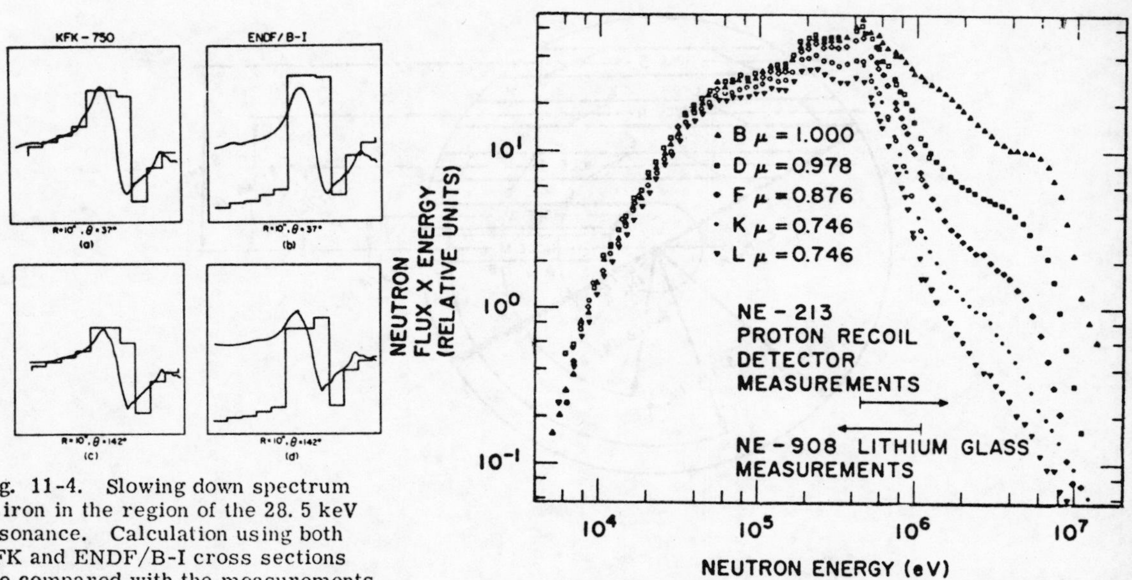

Fig. 11-4. Slowing down spectrum in iron in the region of the 28. 5 keV resonance. Calculation using both KFK and ENDF/B-I cross sections are compared with the measurements.

Fig. 11-5. Fast neutron spectrum in depleted uranium. The angular distribution of the spectrum is shown at a radius of 16. 5 c

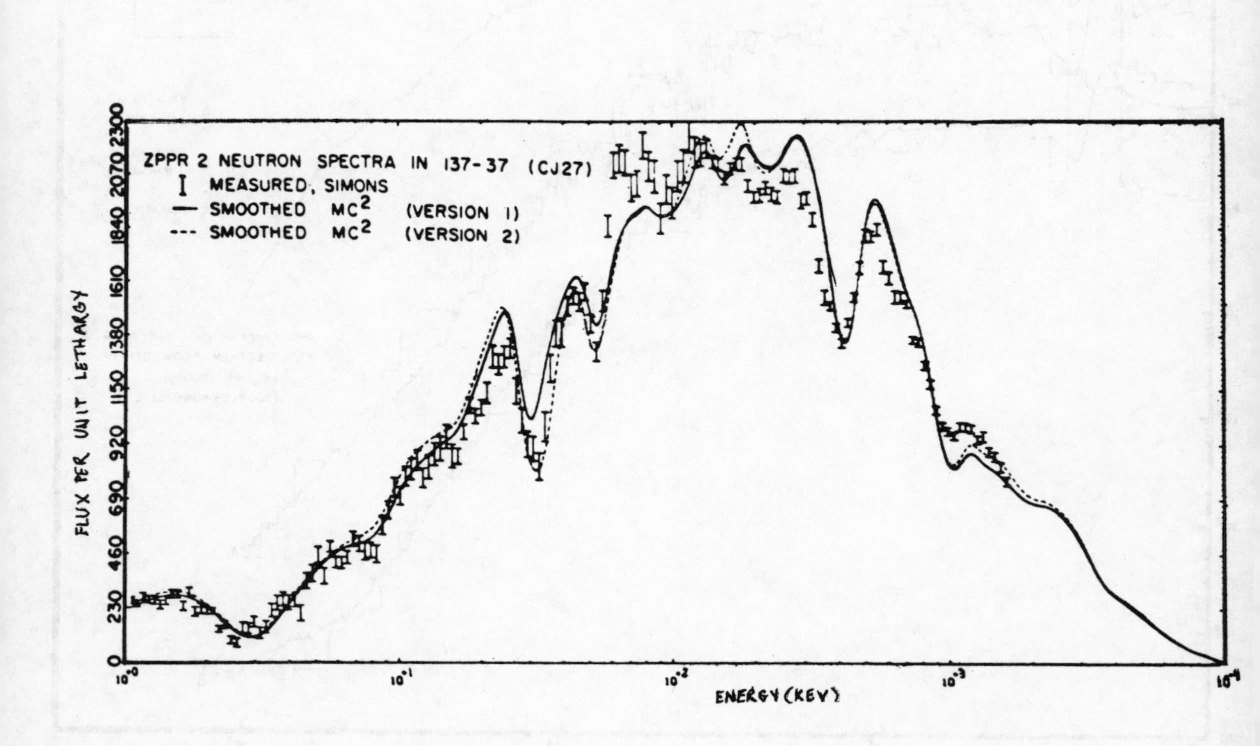

Fig. 11-6. Fast neutron spectrum in ZPPR-2. The MC^2 calculation has been smoothed with the experimental resolution function.

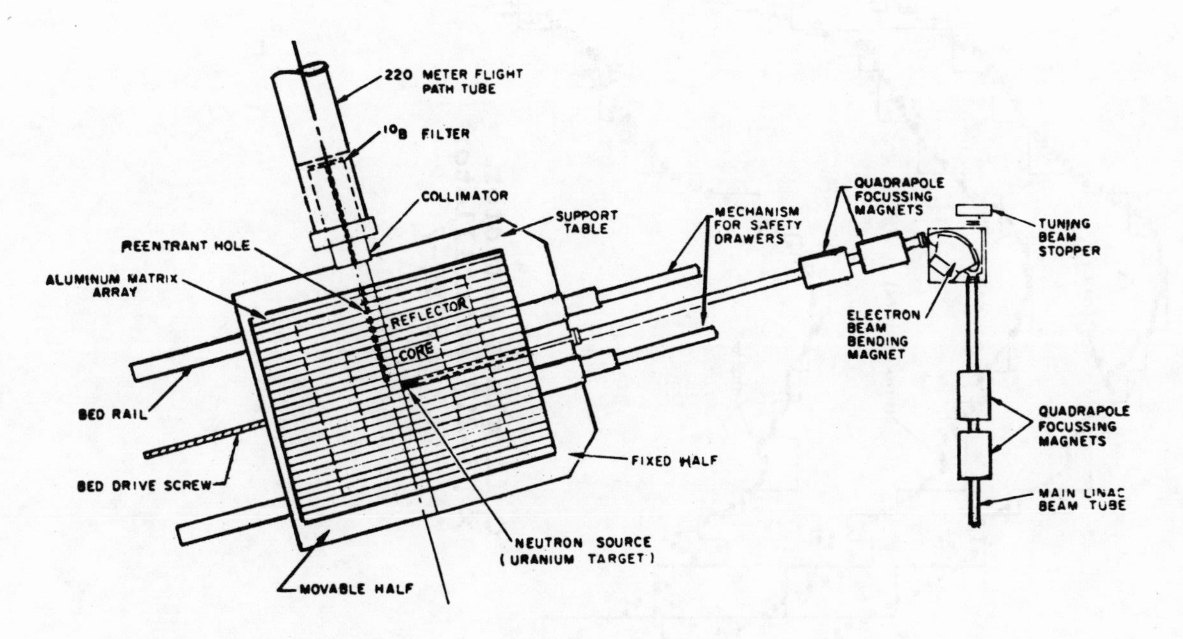

Fig. 11-7. Arrangement of the STSF at the Rad Tech Linac. The reactor is positioned at the end of a 220-meter flight path.

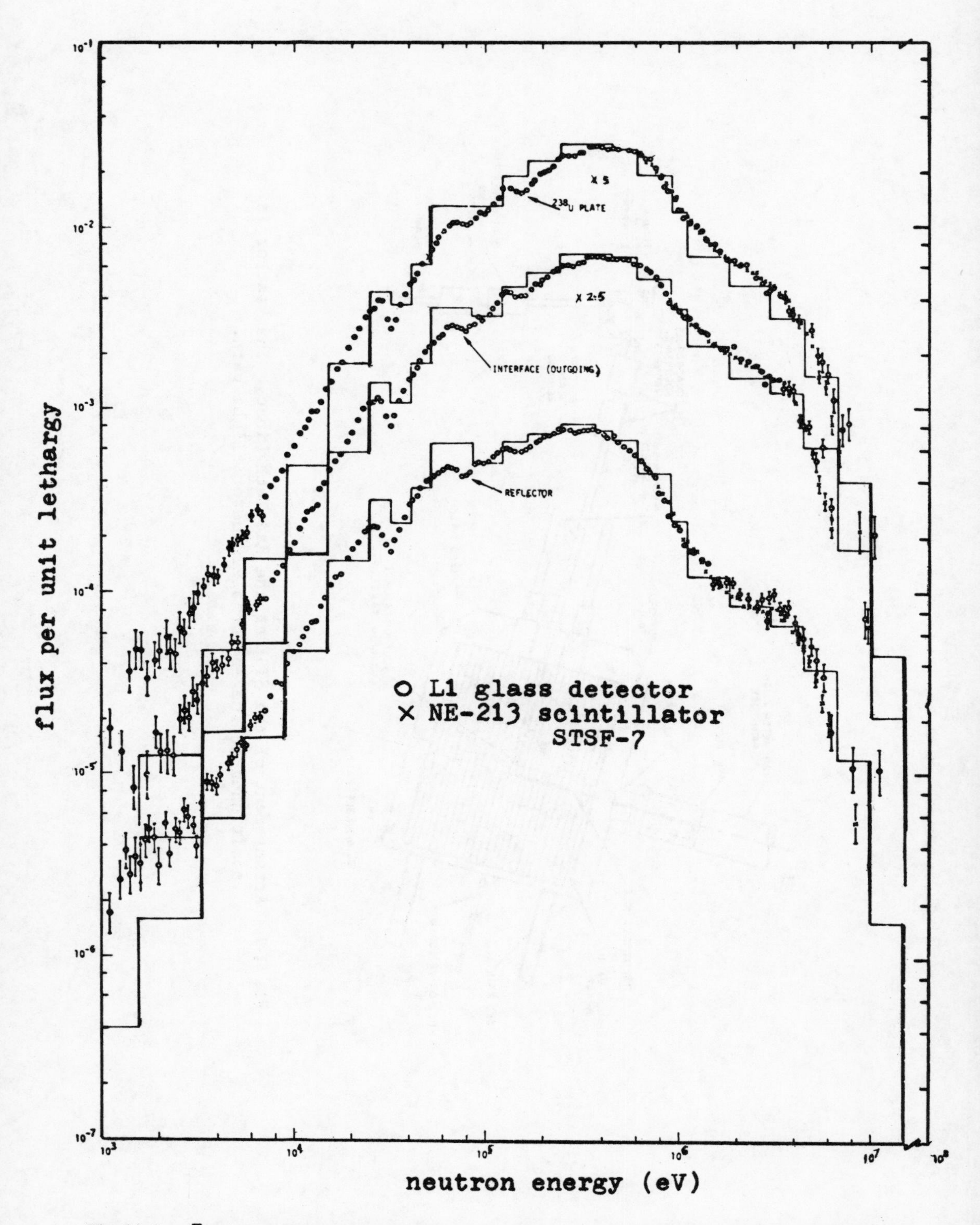

Fig. 11-8. Fast neutron spectrum in STSF-7. The calculation uses DTF-IV with ENDF/B-III cross sections.

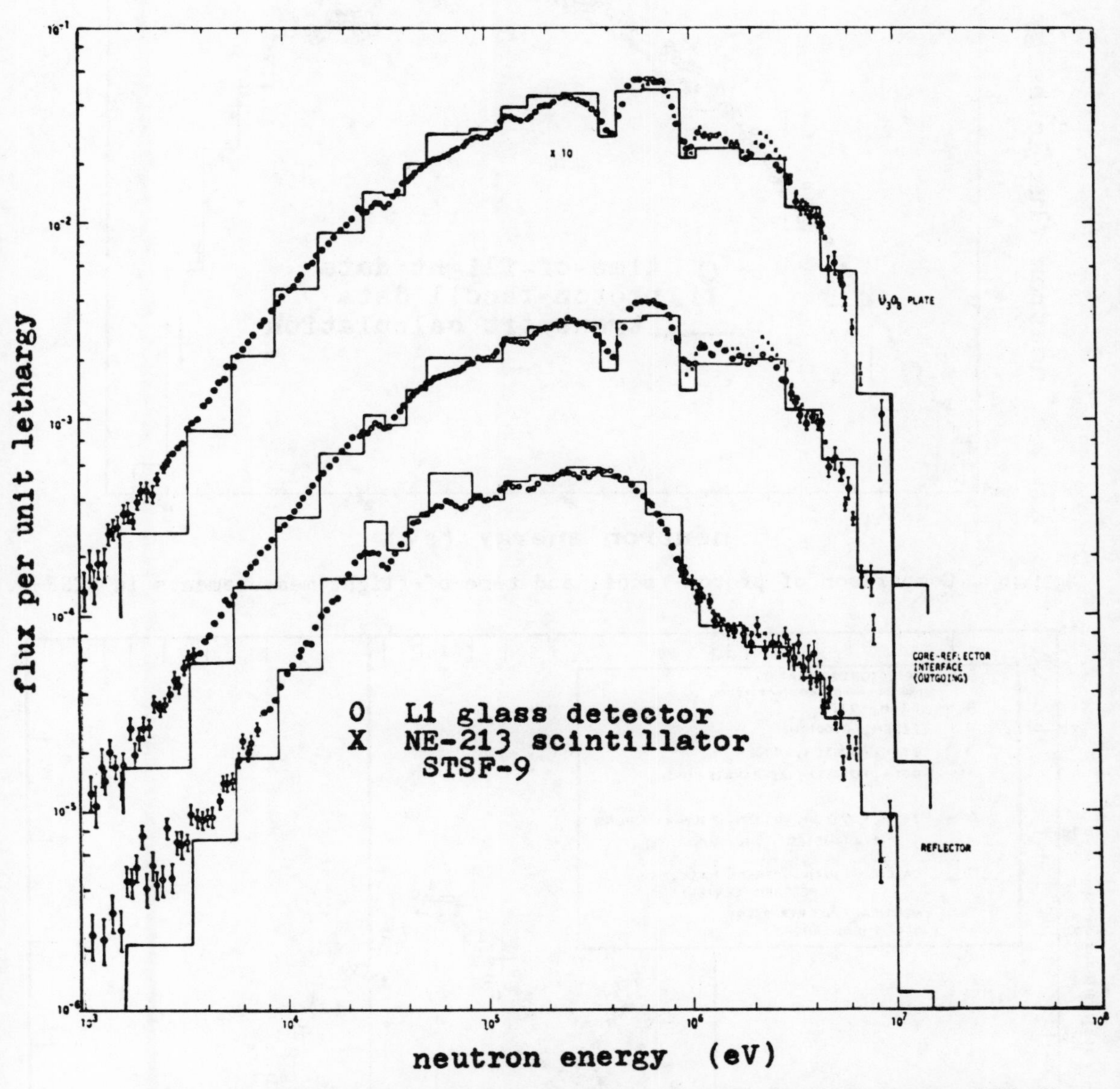

Fig. 11-9. Fast neutron spectrum in STSF-9. The calculation uses DTF-IV with ENDF/B-III cross sections.

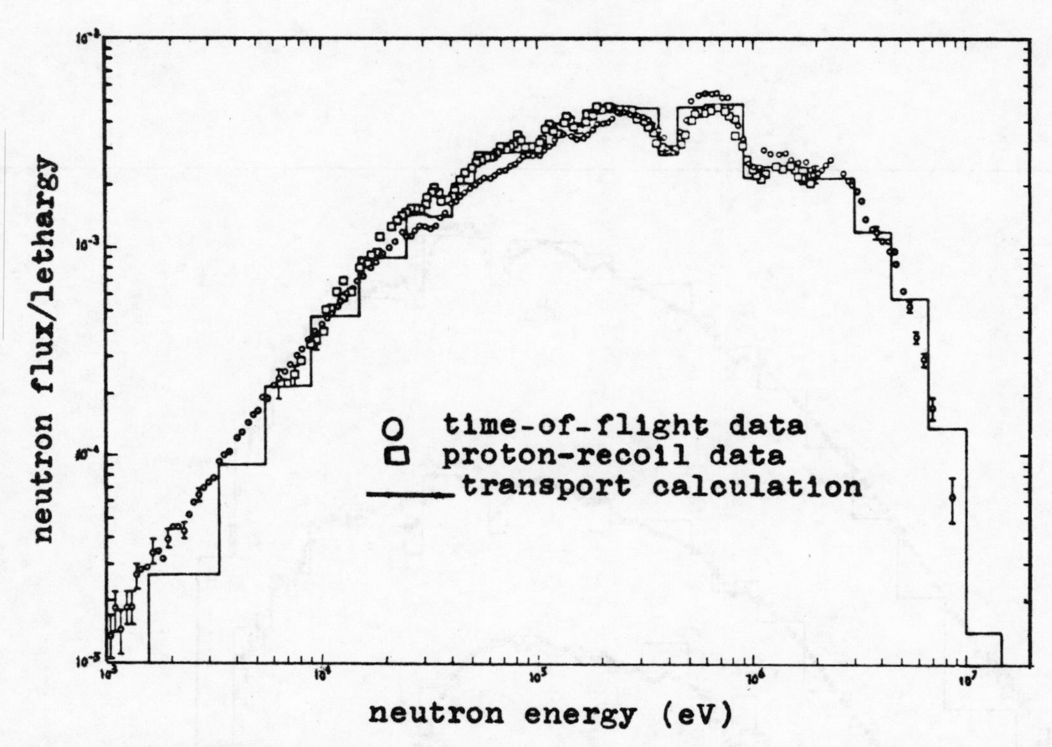

Fig11-10 . Comparison of proton-recoil and time-of-flight measurements in STSF-9.

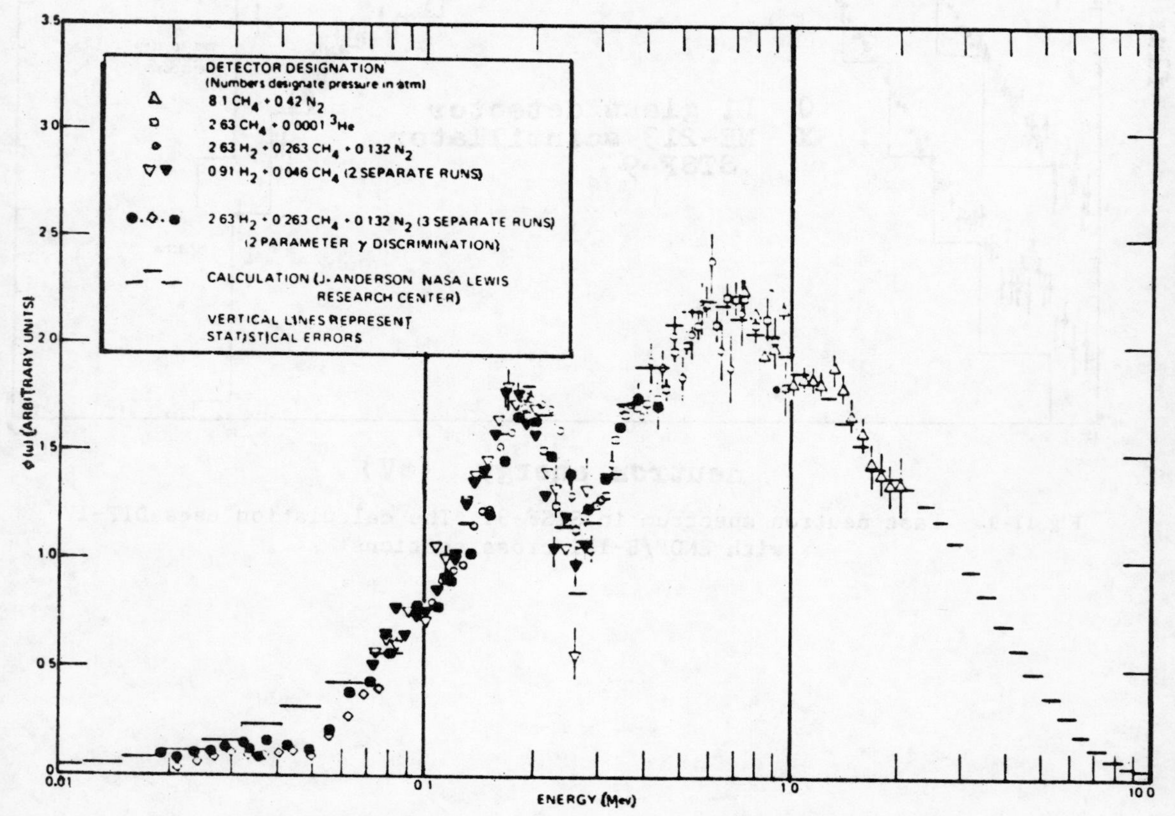

Fig. 11-11 . Fast neutron spectrum in a small fast critical assembly studied at AI.

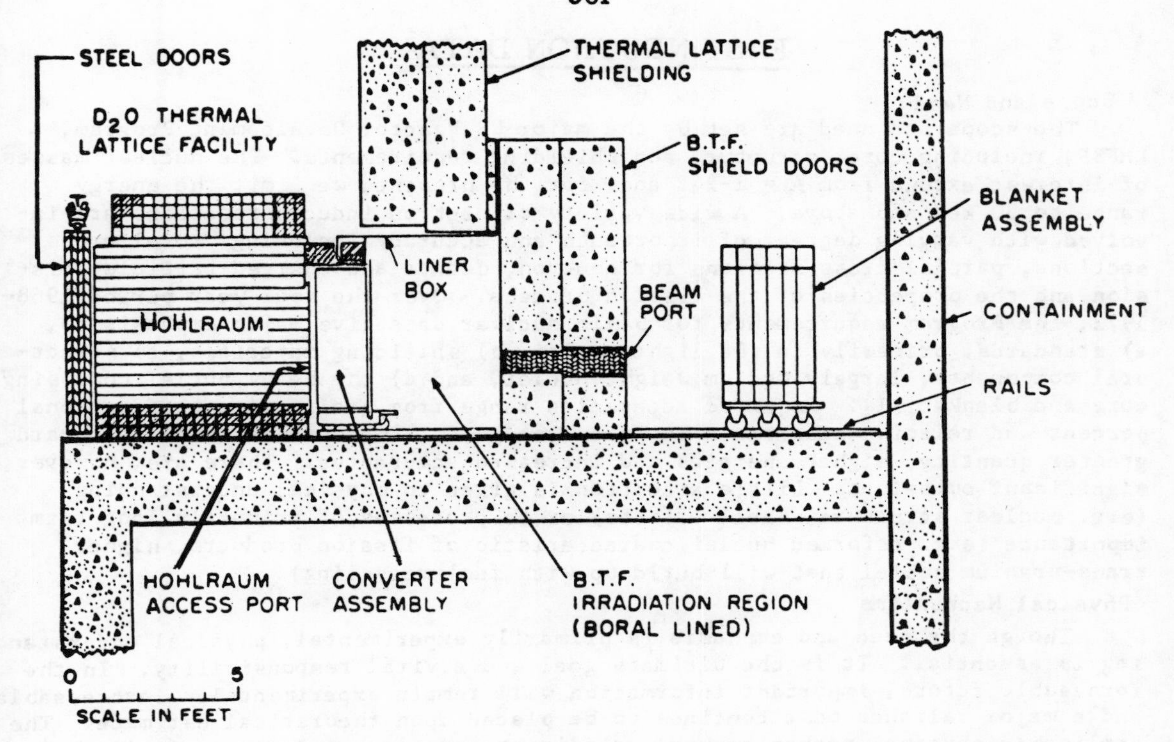

Fig. 11-12. Arrangement of the Blanket Test Facility at the MIT reactor. The converter region produces a spectrum typical of a large LMFBR.

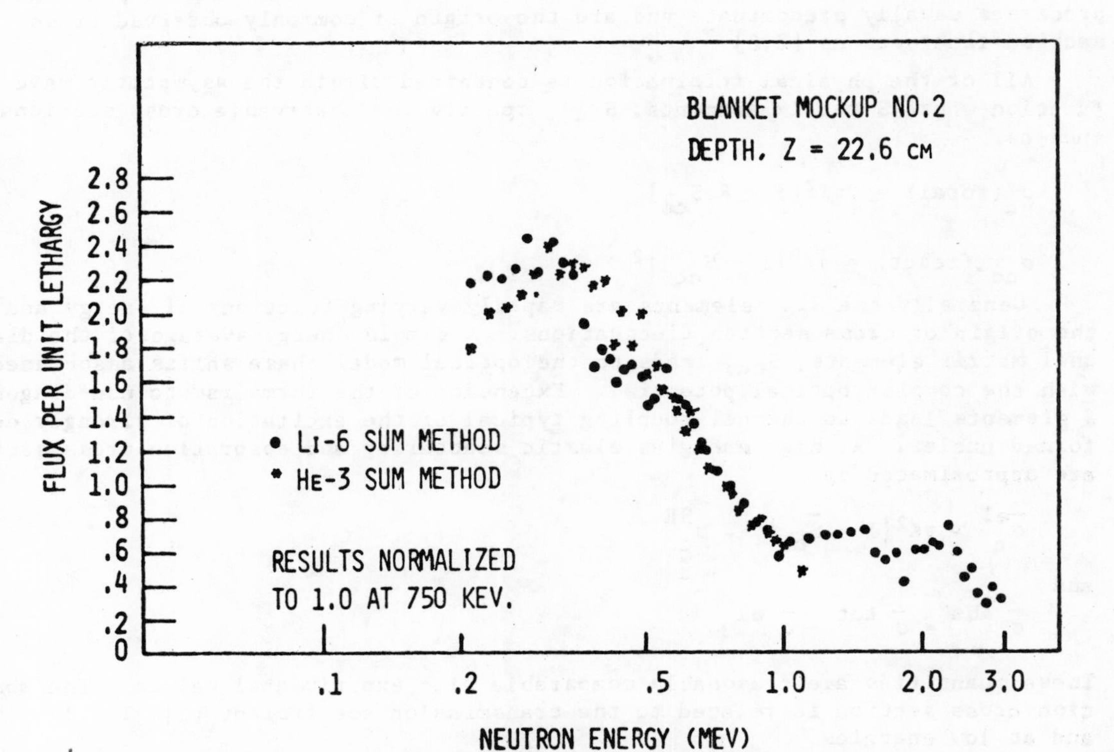

Fig. 11-13. Comparison of spectra measured with ^{3}He and ^{6}Li semiconductor spectrometers in the Blanket Test Facility.

FAST NEUTRON DATA

Scope and Need

The scope and need are set by the major US Reactor Development Program, LMFBR, including core, structure and shielding requirements. The nuclear masses of interest extend from A = 1-242 and, for the present, we limit the energy range to 30 keV and above. A wide variety of neutron induced reactions are involved with varying degrees of importance and accuracy including, total cross sections, partial cross sections for neutron, quanta and charged-particle emission and the properties of the fission process. Over the four year period 1968-1972, the Program requirements for basic nuclear data give major emphasis to, a) standards, primarily in the light nuclei, b) shielding materials, c) structural components, largely medium weight nuclei, and d) the heavy nuclei composing core and blanket [1]. Required accuracies range from qualitative to fractional percent and relative importance has been assigned. The current trend is toward greater quantity, higher energies and increased accuracies. There are, however, significant omissions. Little attention is given to physical understanding (e.g. nuclear structure, shell effects) or to processes of primarily long-term importance (e.g. deformed nuclei characteristic of fission products, higher trans-uranium nuclei that will build up with fuel recycling).

Physical Mechanisms

Though the need and emphasis is primarily experimental, physical understanding is essential. It is the ultimate goal and a vital responsibility. In the forseeable future, important information will remain experimentally inaccessable and a major reliance must continue to be placed upon theoretical estimate. The applicable physical mechanisms are, a) direct reactions, single-particle, collective, b) intermediate processes, quasi-particle configurations, and c) compound-nucleus processes including isolated- and interfering-resonances and smooth energy-averages and with a variety of exit channels. The compound-nucleus processes usually predominate and are the origin of commonly observed cross section fluctuations [2,3].

All of the physical information is contained within the asymptotic wave function whose S-matrix elements, $S_{cc'}$, specify the observable cross sections such as

$$\sigma_c(\text{total}) = 2\pi\lambda\!\!\!^{-2}[1 + \text{Re}S_{cc}]$$

$$\sigma_{cc'}(\text{react}) = \pi\lambda\!\!\!^{-2}|1 - S_{cc'}|^2 \qquad (1)$$

Generally the $S_{cc'}$ elements are rapidly varying functions of energy and are the origin of cross section fluctuations. A simple energy-average of the diagonal matrix elements, $\bar{S}_{cc}$, leads to the optical model phase shifts associated with the complex optical potential. Extension of the formalism to non-diagonal S elements leads to channel-coupling typical of the excitation of strongly deformed nuclei. At high energies elastic scattering and absorption cross section are approximated by

$$\bar{\sigma}_c^{el} \sim \pi\lambda\!\!\!^{-2}|1 - \bar{S}_{cc}|^2 \equiv \sigma_c^{SE}$$

and

$$\bar{\sigma}^{abs} = \bar{\sigma}^{tot} - \bar{\sigma}^{el} \qquad (2)$$

These quantities are reasonably comparable with experimental values. The absorption cross section is related to the transmission coefficient [5], $T = 1 - |\bar{S}|^2$, and at low energies

$$\bar{S}_{cc} = e^{-2i\phi_c}[1 - \frac{\pi\langle\Gamma\rangle}{D}] \qquad (3)$$

where $<\Gamma>/D$ is proportional to the strength function determined from low energy resonance data. Determination of the reaction cross section requires the further assumption of the reaction mechanism and a statistical model is usually adopted for this purpose [6] resulting in

$$\bar{\sigma}_{cc'} = \pi \lambda\!\!\!^{-2} \frac{T_c T_{c'}}{\sum_{c''} T_{c''}} \cdot F \text{ (Fluctuation correction)} \tag{4}$$

A widely used method for calculating the reaction cross section is the R-matrix formalism [7]. The statistical behavior of R-matrix parameters is reasonably known and it is possible to construct from them statistical synthetic cross sections which can be averaged for comparison with experimental results determined with various resolutions [4].

These theoretical concepts are not new but they continue to provide the physical framework for both experiment and evaluation. Too often they are ignored, violated or utilized in such a manner as to emphasize one facet of the process to the detriment of others. Seldom are they employed in the conceptual planning and guidance of experimental procedures.

Standards -- Relative and Absolute

The scope and need largely imply measurements relative to reference standards. Success is contingent on a precise knowledge of a few basic standard cross sections; notably H(n,n), ^{3}He(n,p), ^{6}Li(n,alpha), ^{10}B(n,alpha) and ^{235}U(n,f). For each, the objective is an accuracy of ∿1% over the useful energy range.

The H(n,n) reaction is known to ∿1% up to ∿8 MeV [8]. At higher energies the accuracy degenerates with the onset of uncertain anisotropy. In the present context this reaction is the principal and best known basic standard. In practice it is difficult to use and alternative standards are usually employed. This is the only standard that can be accurately determined without a flux measurement and, interestingly, its precise value is related to very fundamental mesonic forces.

The ^{3}He(n,p) reaction is judged known to accuracies of ∿3% to 10 keV, and to 5-10% at 100 keV [9]. These are not good accuracies and the reaction is difficult to use, thus it does not find wide application.

The ^{6}Li(n,alpha) and ^{10}B(n,alpha) cross sections are widely and easily used reference standards. Uncertainties in these values influence the experimental errors in measurements of critical reactor quantities (e.g. ^{238}U radiative capture). Both cross sections have recently been experimentally examined in detail.

Studies of ^{6}Li by Fort and Marquette [10], Coates et al. [11], Meadows [12] and others have provided a wealth of information. Some of these measurements report the (n,alpha) cross section from 10 keV to 2 MeV with uncertainties of 4-5% and in some cases excellent and internally consistent descriptions of total, (n,alpha) and scattering cross sections have been obtained using R-matrix and other resonance fitting procedures. Unfortunately, discrepancies among measurements particularly near the large resonance at 250 keV are in the range 5-10%. Uncertainties of this magnitude are not consistant with some reactor data needs.

Friesenhahn et al. [13] have recently studied the ^{10}B(n,alpha) and ^{10}B(n,alpha$_1$-gamma) cross sections with care from 10-1000 keV. The measurements are relative to the basic H(n,n) reaction. Incident neutron energy resolution was ~5% and total uncertainties were of 2-4% in the cross sections of both reactions. These values are in good agreement with a number of previous values below 100 keV [14] where the reaction is most useful and is currently a basis for some detailed fission cross section measurements. Currently, the ^{10}B(n,alpha) reaction appears to be a more accurate standard than the ^{6}Li(n,alpha) reaction particularly at energies of less than 100 keV.

The ^{235}U(n,f) reaction serves primarily as a fission standard and a flux monitor.

Beyond the basic reference cross sections there are some specialized standards. Reactor performance is sensitive to nu-bar and the standard is nu-bar of ^{252}Cf. It is not known to the desired 1% accuracy and there is no US effort in this area [15]. The half-life of ^{234}U is a key to some assays of fissile foils employed in accurate fission cross section measurements. It is probably known to only about 1% [16], no better than the desired accuracies of some fission cross sections. Certain source reactions are employed in absolute flux determinations. Few of them are sufficiently well known.

The importance of usable standards is well established and the associated problems recognized. Substantial gains are being made. However, the work is of a difficult and tedious nature and it may well be a decade before standards are generally available with an accuracy consistent with current LMFBR data requirements.

Total Neutron Cross Sections

Total neutron cross sections are seldom explicit but often implied data requirements. They are essential to consistent partial cross section measurement and evaluation and a foundation for theoretical interpretation. The total cross section is one of the few quantities that can be accurately determined without a flux measurement.

Improvements in measured total cross sections are most striking in the lighter nuclei (shielding and structural materials) where measurements to energies of $\gtrsim$ 1 MeV with resolutions of $\gtrsim$ 1 keV provide a wealth of detailed compound-nucleus resonance structure. These results provide, for the first time, an experimental basis for detailed resonance interpretation over a wide energy range usually within the framework of the R-matrix. These interpretations and measurements are well illustrated by the work of Cierjacks et al. [17], Perey et al. [18] and others with such nuclei as carbon, sodium, silicon and calcium. With increasing mass, isotopic complexity, and odd-nucleon configurations, the interpretation of the high resolution total cross section measurements becomes more complex. The observed compound-nucleus resonances are highly interfering often modulated by an intermediate resonance structure and explicit resonance interpretation becomes unmanagable. A viable alternative is a statistical interpretation. Using the known statistical properties of R-matrix parameters Moldauer has constructed synthetic total cross sections which well describe the experimental values [4]. The method has been particularly successful in the interpretation of titanium and iron total cross sections measured by Smith and Whalen [19] Interestingly, such synthesis techniques illuminate the statistical nature of intermediate resonance structure and test the validity of the optical potential in a fluctuating cross section environment.

Measurement of the total cross sections of the heavy core and blanket nuclei need not employ high resolution as the structure (when present) is too fine to be of applied significance [20]. However, precise magnitudes are essential. Care must be taken to avoid experimental perturbations leading to magnitude shifts of a few percent. It has been only recently that the total cross sections of

^{235}U, ^{238}U, and ^{239}Pu have been measured at different facilities, with different methods, and with consistencies approaching 1%. Notable of the results in this fissile and fertile area are those of Whalen et al. [21], Schwartz et al. [22], and of Langsford and Clements [23].

In addition there are certain "special" total cross section problems. An example is the definition of resonance interference minima essential to the calculation of deep penetration in shielding materials [24]. The resolution of these problems tends to be instrumental in nature.

In the next few years it is reasonable to expect a comprehensive knowledge of fast neutron total cross sections throughout the mass-energy region of interest. The accuracies should be ~1% and the energy resolutions sufficient for detailed resonance interpretation where such an approach is theoretically managable. The energy-averaged behavior of these cross sections will continue to be an essential foundation for the optical potential. The increasing availability of detailed and strongly interfering resonance cross sections should stimulate the development of statistical methods of interpretation.

Neutron Scattering Cross Sections.

Scattering is the major energy transfer mechanism in macroscopic applied systems and it constitutes a large portion of the neutron-nucleus interaction. The elastic neutron angle-differential cross sections are sensitive to theoretical premises and a key to model development. The experimental measurement of neutron scattering is not trivial since a flux determination or standard cross section is involved and the process of interest constitutes a secondary source in the environment of a primary source of far greater intensity.

Neutron scattering from light nuclei is dominated by isolated resonance structure with little or simple inelastic contributions. Recent experiments in the light region have utilized intense white-neutron sources to achieve a remarkable degree of resolution for both elastic and inelastic processes. Results such as those of Cierjacks et al. [17] and Perey et al. [18] define the resonance behavior of the differential elastic scattering distributions over wide incident energy ranges for such nuclei as carbon, sodium, silicon and calcium. Interpretation, based largely upon the R-matrix formalism, provides resonance parameters to a degree not possible from total cross section results. When present, the associated inelastic scattering is usually characterized by large and widely separated reaction Q-values. As a consequence, the observation of the gamma-ray emitted following the inelastic scattering can reasonably determine the primary inelastic neutron cross section. Measurements of this type, using the intense white-neutron sources, have provided inelastic neutron scattering cross sections from threshold to several MeV with excellent resolution. Illustrative examples are the aluminum results of Hoot et al. [25] and the sodium values of Perey et al. [26]. Interpretation is, again, on a detailed resonance basis with the requisite number of exit channels. In view of the above, it is reasonable to expect that reactor requirements for neutron scattering cross sections of light nuclei will soon be satisfied to energies of several MeV. At higher energies the problems are more difficult and the methods and results will be characterized more by those of the medium weight nuclei described below.

Neutron scattering is more complex in the middle-weight (structural) nuclei where cross sections are strongly fluctuating and interfering. Inelastic channels are often abundant and the elastic angular distributions can be highly anisotropic. As a consequence, high resolution white-source measurements are less productive in this region and most scattering measurements employ pulsed mono-energetic sources with time-of-flight techniques to achieve a good scattered neutron velocity resolution. The latter method has been utilized to satisfy most reactor needs for elastic and inelastic scattering data in the structural materials to incident energies of ∿1.5 MeV. Illustrative examples can be found in a number of papers by Smith et al. [27]. The experimental values tend to show a consistent intermediate resonance behavior and, when energy-averaged, provide a good basis for the optical potential in a region of large $\ell = 0$ strength functions. Extensive scattering measurements are now underway in the incident energy range 1.5 - 5.0 MeV [28]. It is reasonable to expect that these current efforts will provide for reactor needs in the structural region to 5.0 MeV within several years. Above 5 MeV major contributions to the knowledge of scattering have been made by Perey et al. [29], Holmquist et al. [30] and others. However, these measurements are difficult and they are in a region of transition between the inelastic excitation of discrete states and of a continuum. Consequently the experimental definitions of energy, magnitude, reaction angle and Q-value are not equivalent to that obtained at lower energies. In this high energy area, major reliance must still be placed upon theoretical interpolation and extrapolation and it is fortunately an area where the optical model readily applies.

In the heavy core nuclei, neutron scattering is less certain and the measurements are complicated by the presence of fission neutrons. Recent results of Knitter and Coppola [31] and of Smith et al. [32] reasonably describe the elastic scattering cross sections of ^{238}U to ∿5 MeV and the inelastic scattering cross sections to 1.5-2.0 MeV. At higher energies both scattering processes are far less certain. It is technically difficult, if not impossible, to clearly separate the fission neutron contribution from the inelastically scattered component. Thus there must be a continued reliance on theoretical models for determination of ^{238}U scattering over much of the higher energy region. The validity of these models has been appreciably improved by new knowledge of the excited structure of this nucleus [33]. It is reasonable to expect that the combination of theory and experiment will provide ^{238}U scattering cross sections to ∿10% accuracies over the energy range of interest in the reasonable future. For example, even now it is no longer tentable to accept 20% "across-the-board" adjustments of the ^{238}U inelastic cross section, as was suggested not too long ago.

^{240}Pu is appreciably abundant in some core configurations. This is an active nucleus with a large fission cross section and consequently it is difficult to measure its scattering properties. However, recent work by Smith et al [34] gives a reasonable knowledge of scattering from ^{240}Pu to incident energies of 1.5 MeV and provides a foundation for models suitable for extrapolation to higher energies.

Elastic scattering from the fissile nuclei ^{235}U and ^{239}Pu is reasonably known to incident energies of 5 MeV largely through the work of Knitter and Coppola [35] and Smith et al. [36]. The results are consistent to within about 10%. Inelastic scattering from this fissile nuclei remains essentially a desolate area with no clear solution in sight. Hopefully, improved knowledge of excited structures of fissile nuclei and better theoretical understanding coupled with a few very difficult and far from definitive measurements will result in reasonable estimates of fissile inelastic cross sections in the foreseeable future. However, the user should not be too optimistic and he could well give attention to alternate and more experimentally viable quantities such as neutron production and emission cross sections.

Non-neutron Exit Channels

Radiative capture decreases in importance with increasing energy, finally, becoming a small direct process. The latter is of theoretical interest, even at low energies, but of little applied reactor significance. Below 1 MeV radiative capture in ^{238}U is one of the most critical of parameters. Interpretation is largely based upon statistical mechanisms and calculation gives qualitative guidance as to both energy dependent shape and magnitude to accuracies of better than 20%. Recent summaries of the experimental status of the ^{238}U capture cross section, such as those of Fricke et al. [37] and of Konshin [38], indicate consistency to within ∿10% up to 1 MeV. Ratios of ^{238}U (n,gamma)/^{235}U (n,f) are known to ∿5% over much of the energy range of interest [39] and their evaluation is consistent with the above estimated uncertainty in the ^{238}U (n,gamma) cross section [40]. It is noted that most of the newer (n,gamma) values employ "modern" prompt capture techniques that have not generally given as accurate results as some of the older activation methods which are easily applied to the ^{238}U measurements. In the foreseeable future it is reasonable to expect 5% accuracies in important capture processes such as ^{238}U and alpha-^{239}Pu. 1% accuracies will not soon be realized.

Neutron induced charged particle reactions are of importance in the context of neutronic indexes and material damage. The former emphasize shapes and relative magnitudes and are largely activation in nature. Damage assay is less demanding of resolution but can involve small cross sections (micro-barns) and prompt processes. The qualitative systematic behavior of (n,X) reactions is well known and, in fact, is clear evidence of shell structure [41]. Quantitative magnitudes are far more uncertain and near thresholds, the shapes can be influenced by specific barrier configurations. In those instances where the activation process is favorable, the quantitative values are known to $\gtrsim$ 10% with a major contribution to this uncertainty associated with the determination of flux. Prompt processes tend to be less certain with experimental neutron intensity a continuing problem. A reasonable prognosis is 5% or 1 micro-barn accuracies for activation processes and lesser accuracies for the prompt reactions. These estimates must be qualified as they are critically dependent upon the specific characteristics of the individual processes.

The Fission Process

The difference between relative and absolute is sharp in the requirements for fission cross sections and in both contexts the accuracies are high.

The ^{235}U fission cross section is usually accepted as the standard and other fission cross sections determined relative to it. There has been considerable recent improvement in these relative values largely confined to energies of 5 MeV and below.

The ratios $^{239}Pu(n,f)/^{235}U(n,f)$ as determined by Kaeppler et al. [42], Poenitz [43], Szabo et al. [44], and Lehto [45] over the energy range 30 keV to 5 MeV are reasonably consistent and from them the uncertainty in this ratio is estimated to be ∿3-5%. These measurements confirm the step-like behavior near 1 MeV and indicate a broad minimum near 2.5 MeV. They do not support previously reported detailed structure in the range 1-5 MeV [46].

The $^{238}U(n,f)/^{235}U(n,f)$ ratios recently measured from threshold to 5 MeV by Meadows [47] and by Poenitz [48] are consistent to 1% and probably represent absolute accuracies of ∿2-3%. These newer values are generally a few percent lower than some of the results of earlier measurements [49] but do not show as low a minimum in its region 2-3 MeV as reported by Stein et al. [50].

New ratios of higher Pu-isotope fission cross sections to that of ^{235}U have been reported. Some of these do not appear appreciably better than those obtained a number of years ago with very modest facilities.

Historically, the absolute fission cross section of ^{235}U has steadily decreased and the minimum is now apparently being approached in an oscillatory manner. Below 100 KeV the values of Poenitz [51], Gwin and Silver [52], Szabo et al. [44], Diven et al. [53] and others are consistent to 5-8% though the method of normalization and/or flux determination varies widely. Generally those values in the range 100 keV-1 MeV are not quite as consistent as at lower energies with the results of White [54] an of Szabo [44] lying somewhat higher th those of Poenitz [51]. Moreover there appears to be a small but systematic difference in shape. There is reasonable agreement from the rise in cross section at 1 MeV to about 2 MeV. At higher energies, the values are less certain and the renormalization of the older Henkel et al. results [55] is not entirely consistent with 14 MeV measurements where the experimental information should be reliable.

The importance of fission cross sections is widely recognized and there has been a massive commitment of manpower to various facets of the problem. There is the promise of absolute fission cross sections of ^{235}U from 10 keV to 15 MeV with 2-5% accuracy in the near future. However, it is noted that a similar extensive effort during the past several years has resulted in relatively few substantive results and that the determination of a fission cross section to 1% accuracy is a very difficult thing indeed.

There is concern for other facets of the fission process. There remains a consistent discrepancy between macroscopic and microscopic measurements of average prompt fission neutron energy. Extensive recent microscopic studies continue to confirm the "Watt" spectral distribution [56]. There have been discrepancies and uncertainties in absolute delayed neutron yields though recent revisions in some of the later measurements apparently limit the discrepancies to ^{238}U [57]. Some fission properties, while not of direct applied importance, are vital to the physical understanding of essential applied processes such as the partially resolved resonance structure in the low energy fission cross section of ^{239}Pu. In this category are fission isomer and sub-threshold fission phenomena that lead to a basic understanding of the fission process.

Evaluation -- a Remark

The applied window to microscopic data is the evaluation -- at times it appears somewhat transluscent. Engineering motivation can lead to "adjustment" beyond physical reality. Relative, absolute and standard are now always distinguished. Experimental error assignments are taken verbatum -- obviously a dangerous course. In some instances physical ignorance is displayed with illicit shapes and reaction mechanisms. These problems are not odd as the most difficult task confronting the experimentalist is the judgment of his own experimental results. This is despite the fact that he is a specialist in the area and uniquely familiar with the particular experiment. It is difficult to understand how an evaluator not actively involved in experimental and/or theoretical research endeavors of a high order can make such sophisticated judgments. It is suggested therefore, that there is a need and a responsibility on the part of the experimental and theoretical research personnel to participate far more deeply in the evaluation process to assure that the information is properly made available to the user in an acceptable form. To do otherwise may well be a default of responsibility.

Experimental Practice and Capability

It is an axiom that "the outstanding problem of fast neutron physics is effe tive neutron intensity". The processes under study are generally secondary sources in an environment dominated by an intense primary source and the particle of primary interest is uncharged and difficult to detect. Effective intensity encompasses both source and detector and in neither aspect have the basic principles changed in a decade.

The most productive sources now in use are accelerator based, white- (electron or positive ion) or mono-energetic. The former have rapidly increased in intensity but the growth rate seems to be reaching a plateau. The white-source is a primary velocity selector with surpassing incident resolutions, multichannel productivity and good sensitivity. It is largely denied secondary velocity selection and is not compatible with inherently time-integral processes (e.g. activation). It is large and costly. The mono-energetic source is not a multichannel device nor is its spectra-integrated intensity comparable with that of the white-source. It is more economical and has the attributes of time-integral response and the capability for secondary velocity selection.

The important things to note are, 1) the importance of the effective source intensity, 2) that advanced white-sources are not grossly different in capability (ORELA, NEVIS, LAMPF) and their intensity growth-rate appears to be slowing, 3) white- and mono-energetic sources tend to be complimentary, 4) potential for order of magnitude improvement in source intensity is likely greater for the mono-energetic source than for the white-source particularly if cost is a concern, and 5) the suitability of the source is strongly prejudiced by the particular application.

A characteristic of many neutron measurements is large experimental perturbations. When the need is for high accuracy great attention must be given to detailed correction procedures. These can be both tedious and costly. The deficiencies in this area are legion and the experimentalists must give greatly increased attention to the analysis and correction of their data if they are to achieve the desired accuracy.

There is no substitute for quality and it is not necessarily synonomous with quantity or newness, nor is it always consistent with programmatic and other pressures. It is a matter for concern that only recently have massive and powerful facilities achieved a total cross section accuracy equivalent to that realized more than two decades ago with much more modest circumstances and that some of the current "flaps" tend to be due to premature or inaccurate "new" information.

References

1. LMFBR Program Plan, Vol. 9, Physics, WASH-1109 (1968); also draft revision (1972).

2. P. A. Moldauer and A. B. Smith, Proc. 3rd Conf. on Neut. Cross Sections and Technology, CONF-710301, Vol. 1 (1971). The theoretical remarks of Section II are abstracted from this reference.

3. P. A. Moldauer, Argonne National Laboratory Report, ANL-7467 (1968).

4. P. A. Moldauer, Statistical Properties of Nuclei, Plenum Pub. Co., New York, (1972).

5. W. Hauser and H. Feshbach, Phys. Rev. 78, 366 (1952.

6. N. Bohr, Nature 137, 344 (1936).

7. E. Wigner and L. Eisenbud, Phys. Rev. 72, 29 (1947).

8. J. Hopkins and G. Breit, Los Alamos Report, LA-DC-1153 (1970).

9. R. Batchelor, Conf. on Neut. Stds. and Flux Normalization, AEC Symposium Series #23 (1971).

10. E. Forte and J. Marquette, Private Communication.

11. Coates et al., Conf. on Neut. Stds. and Flux Normalization, AEC Symposium Series #23 (1971).

12. J. Meadows and J. Whalen, Nucl. Sci. and Eng. 48, 221 (1972).

13. Friesenhahn et al., Private Communication.

14. R. Lane et al., Conf. on Neut. Stds. and Flux Normalization, AEC Symposium Series #23 (1971).

15. G. Prince, Brookhaven National Laboratory Report, BNL-50168 (1969), a compilation.

16. J. Meadows, Private Communication.

17. S. Cierjacks et al., ANS 8th Annual Meeting, Las Vegas, (1972).

18. F. Perey et al., Private Communication.

19. P. A. Moldauer and A. B. Smith, Private Communication. See also Nucl. Phys A118, 321 (1968) and references 3 and 4, above.

20. D. Kopsch et al., IAEA-CN-26/12 (KFK-1199).

21. J. Whalen, Private Communication.

22. R. Schwartz, Private Communication.

23. A. Langsford et al., Private communication.

24. H. Goldstein et al., Proc. 3rd Conf. on Neut. Cross Sections and Technology CONF-710301, Vol. 1 (1971).

25. F. Perey et al., Proc. 3rd Conf. on Neut. Cross Sections and Technology, CONF-710301, Vol. 1 (1971).

26. C. Hoot et al., Proc. 3rd Conf. on Neut. Cross Sections and Technology, CONF-710301, Vol. 1 (1971).

27. A. B. Smith, Series of papers in a number of journals. See CINDA-71 for references.

28. A. B. Smith, Private Communication.

29. F. Perey et al., Series of Oak Ridge National Laboratory Reports. Illustrative is ORNL-4515 (1970).

30. B. Holmquist et al., Series of Aktiebologet Atomenerge Reports. Illustrative is AE-430 (1971).

31. H. Knitter, Private Communication.

32. A. B. Smith, Private Communication.

33. W. Poenitz, Private Communication.

34. A. B. Smith et al., Nucl. Sci. and Eng. 47, 19 (1971).

35. H. Knitter and M. Coppola, Z. Physik, 228, 286 (1969) and 232, 286 (1970).

36. A. B. Smith et al., Submitted to Journal Nucl. Energy.

37. M. Fricke et al., Proc. 3rd Conf. on Neut. Cross Sections and Technology, CONF-710301, Vol. 1 (1971).

38. V. Konshin, INDC(NDS)-18/N, (1970).

39. W. Poenitz, Nucl. Sci. and Eng. 40, 383 (1970).

40. W. Poenitz, Conf. on Neut. Stds. and Flux Normalization, AEC Symposium Series #23 (1971).

41. A. Chatterjee, Phys. Rev., 134, B374 (1964).

42. E. Pfletschinger and F. Kaeppeler, Nucl. Sci. and Eng. 40, 375 (1970).

43. W. P. Poenitz, Nucl. Sci. and Eng. 40, 383 (1970).

44. I. Szabo et al., Proc. 3rd Conf. on Neut. Cross Sections and Technology, CONF-710301, Vol. 2 (1971).

45. W. Lehto, Nucl. Sci. and Eng. 39, 361 (1970).

46. M. Savin et al., INDC(CCP)-8U (1970).

47. J. Meadows, Submitted to Nucl. Sci. and Eng.

48. W. Poenitz and R. Armani, To be published in Journal of Nucl. Energy.

49. R. Lamphere, Phys. Rev. 104, 1654 (1956).

50. W. Stein et al., Proc. 2nd Conf. on Neut. Cross Sections and Technology. NBS pub. 299 (1968).

51. W. Poenitz, Private Communication.

52. R. Givin, Private Communication.

53. B. Diven, Phys. Rev., 105, 1350 (1957) and recent private communication.

54. P. White, Journal Nucl. Energy 19, 325 (1965).

55. R. Henkel, G. Hansen, et al., Private Communication.

56. IAEA Committee Meeting on Prompt Fission Neutron Spectra, Vienna (1971). Proc. to be published.

57. G. Keepin et al., Private Communication.

Also: A. Evans and M. Thorpe. ANS 8th Annual Meeting, Las Vegas (1972).

CHAPTER 12

PRESSURIZED WATER REACTORS (PWR)

INTRODUCTION

The independent activities of at least four civil PWR manufacturers and two naval reactor laboratories makes any review of the reactor physics of such plants somewhat myopic, since the communication between different groups is not as free and frequent as it was some years ago. Although the details of recent work may be obscured by restrictions on publication, the principal changes in the physics of pressurized water reactors are recognized to have taken place largely in four general areas:

1) The refinement and extension of lattice theory to account for fuel-assembly heterogeneities and the many isotopes from fission products and depletion chains.

2) The development of various methods of describing core depletion effects for the prediction of operating characteristics, power distributions and their stability, control requirements and load-following capability during core life and in subsequent fuel cycles. This includes the use of plutonium as enriching fuel in later cycles.

3) The collection and evaluation of isotopic cross sections by many groups, coordinated by the Brookhaven Cross Section Center.

4) The exploration of the agreement between recently-acquired operating-reactor data and the results of theory as a means of improving both the theory and the techniques employed in power-reactor measurements.

Admittedly, the choice of these few categories reflects the parochial view of the reactor designer and the limited range is not intended to slight the contributions of the more academic members of the reactor-physics community whose horizons are broader and interests more ecumenical.

LATTICE SPECTRUM

The most significant change in the area of lattice theory is the realization that single fuel rod cell spectrum calculations are of questionable validity when the rod neighbors are water holes, poison shims, control rods or plutonium-loaded pins or assemblies. Various low-order approximations to the proper surroundings have been in use for some time, and several groups[1, 2, 3] have been active in developing methods, such as collision theory, to provide a systematic and more precise definition of the energy-dependent neutron leakage from each different pin cell. For speed, an approximate model of this cell coupling is desirable and the number of energy groups is usually taken to be smaller than that used in multigroup spectrum calculations.

Further work has also been necessary on the resonance shielding problem, primarily because of the growth of plutonium from zero concentration in a UO_2-fueled system to relatively high concentrations in plutonium recycle batches. This has required the derivation of equivalence principles[5] valid over a much larger range than is normally required for the more important U-238 resonance shielding. In addition, the interaction of the large U-238 resonances with those of U-235, plutonium and fission products has required some estimates of mutual shielding effects. The type of situation encountered is shown in Fig. 12-1 drawn from parts of two RABBLE calculations for fission products alone and for fuel at about one-half normal burnup. Fortunately, the importance of these interactions does not seem great enough to justify undertaking very exact solutions. Examination of the fission-product reaction rates has led us to reduce the epithermal few-group fission-product cross section by 15%, which is worth roughly 0.3 - 0.4% out of a total reactivity worth of about 5.5% at 15000 Mwd/TeU. In actual fact, this may be a smaller correction than the accuracy of the fission-product cross sections themselves warrant, but it appears, nonetheless, to be a real effect, whatever may eventually be found for the best values of the cross sections.

Similar resonance interaction effects have been identified for U-238, U-235[6] and Pu-239[7]; these arise either from true interference or from the failure of MUFT-type codes to take account of the source reduction by successive resonances in a single multigroup.

With these various resonance absorption improvements, we have been fairly successful in analyzing the ESADA plutonium lattices with eigenvalues in the range of 1.000 $\pm$ 0.005, by the use of ENDF/B-2 plutonium cross sections[8] without the need of arbitrary cross-section adjustments. Some error trend with water-to-fuel volume ratio is still apparent, however, and further improvements are needed to provide a firm design basis.

One of the complexities which has entered thermal reactor spectrum and depletion calculations, particularly for plutonium-recycle cores, is the representation of the actinide chains. These comprise a host of isotopes ranging from thorium through americium and curium, which are related by a very complicated web of transitions. Valuable work in measuring isotopic concentrations has been performed in the AEC program conducted with Yankee-Rowe fuel by Westinghouse[9]. In uranium fuel cycles, they produce typically a 2% reactivity loss at 33 Mwd/kgM, while in equilibrium plutonium recycle fuel their presence is more important and their absorption can reduce reactivity by an additional 2%, including the normal decay of Pu-241. For design calculations, a considerable truncation of the actinide chains can be made if only reactivity effects are of concern. In general, the analysis of the isotopic concentrations found in Yankee-Rowe fuel pins indicates that further work in defining the spectrum-averaged cross sections of many of these isotopes is needed before it can be said that a satisfactory explanation of the data is in hand. This is particularly true for the precursors of U-232, for which a very low concentration limit for acceptability has been set for re-enrichment in the diffusion plant.

SPATIAL POWER DISTRIBUTION CALCULATIONS

Most PWR design calculations of power distribution, control rod and boron worth, temperature coefficients and reactivity depletion are still performed in two dimensions with some batchwise iteration with axial calculations. The justification for this approximation is that PWR plants with chemical shim are largely operated with control rods out at a small bite position for vernier control of power level and that the power distribution is fairly separable into xy and z shapes, particularly during the first fuel cycle. A fine spatial mesh of one xy point per fuel rod is normally used in order to evaluate the power peaks in some detail.

Of course, as xenon appears and fuel depletion occurs, this separable representation becomes less accurate, the most noticeable error being in the eigenvalue. From three and two-plus-one dimensional depletions of several reactors, we have found that the two-plus-one dimensional reactivity error usually follows the trend shown in Fig. 12-2 . The reason for the error is easy to identify as the neglect of the correct axial weighting of the macroscopic cross sections of the various depleting isotopes. Fig. 12-3 shows a simple axial-flux weighting, inferred from a three-dimensional calculation,which provides a first-order improvement in the two-dimensional calculation. The second curve in Fig. 12-2 shows that this sort of weighting does, indeed, bring the three-dimensional and weighted two-dimensional calculation into close agreement in reactivity. Since the three-dimensional calculation places more xenon and fuel depletion in regions of high importance, the 3D calculation is less reactive than the normal unweighted 2D calculation by about 0.5% $\Delta\rho$ late in the first cycle. The assembly integrated power distribution errors between 3D and the corrected 2D calculations are apparently small.

In the analysis of experiments with either large water channels or interfaces between strongly-differing fuel assemblies, it has been found effective by Windsor and Pistella[10] to use an even finer mesh structure with four points per pin in two-dimensional diffusion calculations. Although this antidote, which can reduce the error by a factor of two, is not really feasible in full-core calculations, it can be used in generating table sets for single fuel assemblies for subsequent use in coarse-mesh full-core calculations.

The question of mesh size in full-core calculations using conventional diffusion difference equations, such as PDQ, is complicated by the variety of ways in which the non-uniform assembly and the reflector can be represented. In general, we have found that larger mesh sizes can be used if some care is taken in representing the flux peaking in water holes, the depression in poison shims relative to the fuel cell and the representation of the reflector. Fig. 12-4 shows three rather extreme cases of mesh size for the prediction of assembly power; of course. the use of four mesh points per assembly in conventional diffusion theory provides an extremely poor reflector representation and the degeneration of agreement with the fine-mesh calculation is not unexpected for this case.

Alternate forms of the spatial equations especially adapted for ultra-coarse-mesh (UCM) calculations have been developed and used for some time with varying degrees of success. Robinson [11] has recently developed a higher-order difference approximation to the diffusion equation which uses one or more mesh intervals per fuel assembly to obtain multi-dimensional flux shapes in a one-group model. A distinctive form of nearest-neighbor coupling leaves the equations simple enough to be solved for a full-core three-dimensional problem from a flat-source guess to a convergence criterion of 10^{-6} in two minutes of CDC-6600 time. Generally, the results are correct to within 5 to 6% when compared with the fine-mesh solutions examined to date, with the largest errors occurring in the low-power peripheral fuel assemblies. Two examples of the errors between fine and UCM 2D calculations are given in Fig. 12-5 for a core in a rodded and unrodded condition with one mesh point per assembly. The important core edge boundary conditions for both UCM cases were derived from the unrodded fine-mesh calculation only, and it is hoped that relatively rough estimates of the boundary conditions will suffice for many calculations.

The number of neutron-energy groups employed in full-core calculations has an appreciable influence on the predicted radial-power distribution, particularly early in cycle life. In our initial analysis of Conn-Yankee Core I, we had used two neutron groups, neglected the presence of the close-fitting core barrel and found quite good agreement with the in-core instrument signals. In refining these calculations for the second cycle, however, we found that inclusion of the core barrel raised the peripheral assembly power by about 10% and that the start-up power was more peaked at the core center than could be explained by conventional two-group calculations. The symptoms of disagreement suggested that improvement of the reflector representation was required and the usual route of using more neutron energy groups was followed. As will be shown in the subsequent discussion of plant analyses, we have found that the use of four neutron groups appears to yield a consistently improved agreement with core power distributions in all of the four reactors examined so far. In large reactors (12 ft. diameter) with mixed central zones, the differences are very pronounced, as shown in Fig. 12-6. The difference decreases quite rapidly with depletion but in shimmed cores the evolution of the differences with burnup can be appreciably more complex than the figure would suggest.

Noderer has shown by multigroup S_n calculations that the use of a typical core leakage spectrum as the source for the generation of reflector few-group constants can reduce the difference in core power shapes between two and four neutron groups by nearly a factor of two. This improved agreement results almost entirely from changes in the two-group power shape. Since the differences are all attributable to the fast-neutron spectrum, one might do even better by employing a Laplace transform in the reflector with both isotropic and linearly anisotropic source terms from the core edge included in the P-1 or B-1 MUFT equations. In lieu of such improvements, we have used the four-group calculations which appear to be much less sensitive to how the reflector constants are generated.

One last point that should be mentioned in connection with core-power distributions is the influence of thermal-hydraulic feedback. The operating reactors we have been studying are of relatively low-power density so that effects of this sort are relatively weak, although perceptible. We have used the first core of Obrigheim as an example for a 3D coarse-mesh depletion calculation with thermal feedback but without an outlet pressure-balancing iteration to redistribute flow in the core. In Fig. 12-7 the octant assembly powers are given near the top and bottom of the core at 1000 Mwd/TeU and 9400 Mwd/TeU. During this interval of operation both a moderator temperature and a core-power increase occurred, but it is doubtful that these changes alter the conclusions indicated from Fig. 12-7. Early in life the core center power rises by about 4% in scanning up the channel and an opposite shift of 2 - 3% occurs in the outer fuel assemblies. Later in life, at 9400 Mwd/TeU, the differences in radial-power pattern as a function of axial position have become smaller, particularly for those assemblies in the center of the core.

Because of the negative temperature coefficient, the axial flux and power show a slight peak toward the bottom of the core at end of cycle, the peak U-235 depletion is shifted downward from core center while the Pu-239 production is shifted slightly upward. All these effects are very mild for a core of the Obrigheim size and power density under normal operation. It is during load following that the feedbacks can be expected to show interesting effects and this type of data we have not yet collected and analyzed to any appreciable extent.

POWER REACTOR MEASUREMENTS

For those reactor physicists accustomed to the analysis of critical experiments, the study of operating reactors is at the same time frustrating and fascinating, since both the experimental conditions and the effects to be accounted for are much more complex than those encountered in laboratory experiments. Power reactor measurements involve changes in moderator and fuel temperature, uranium, plutonium, fission product and heavy element concentrations, soluble boron, shim worth and control rod insertion. Many of these changes occur simultaneously so that errors in the theoretical description of one or another are not at all easy to separate. The complexity of the puzzle makes its solution a considerable challenge, quite aside from the economic incentives of the utility or manufacturer to predict the reactor performance accurately.

On the negative side, the methods of performing operating reactor measurements are not yet developed and standardized to the point where one can always trust the results without sometimes questioning either the condition of the reactor prior to the particular measurement or the measurement operations themselves. Added to these uncertainties are those arising from obscure details of core construction, such as the shroud and core barrel geometry, the weights of spacer grids and guide tubes, the specifications of test assemblies and minor late changes in core fabrication. We have also encountered situations where the control rods were shorter than the active fuel region, where the index for control rod insertion was appreciably offset from the top of the core and where the rotations of fuel assemblies on reloading were exceedingly complex. There are no easy solutions to these many time-consuming problems but the active cooperation of the utility operating the plant can be of great help in evaluating the performance of the core.

For our initial analysis work, we chose two reactors with steel-clad fuel, Conn-Yankee(12) and San Onofre(13), and two with zirconium-clad fuel, Obrigheim (14) and Beznau(15). Since these reactors started operation early, they showed promise of providing data on reload batch fuel performance at an early date. San Onofre had four plutonium test assemblies in Cycle II and Obrigheim is expected to start full plutonium recycle in 1973. Beznau is of particular interest because of the 5% reactivity held down by Pyrex glass burnable poison shims.

The nominal core specifications for these reactors are given in Table 12-1 and some of the measured and calculated zero-power start-up data are compared in Table 12-2. The data in the second table show four persistent differences between the measurements and present predictions.

1) During the first cycle core warm-up, the reactivity error changes with temperature by $-0.1 \times 10^{-4}\ \Delta\rho/^{\circ}F$. Obrigheim is an exception in that it exhibits a much smaller error.

2) Calculations of the moderator temperature coefficient seem to be slightly too negative in the first cycle, consistent with (1) above. At the start of the second cycle the error in moderator temperature coefficient is more pronounced but is still small, -0.2 to $-0.35 \times 10^{-4}\ \Delta\rho/^{\circ}F$.

3) The calculations tend to underestimate the differential soluble boron worth by amounts ranging from 3% to 13%. Although Fig. 12-15 indicates that differential boron worth may be difficult to measure and some of the spread in the error can be attributed to this, the presence of about 5% error in our calculations for RCC cores seems fairly well established.

4) The calculated eigenvalues of the four reactors at 100°F lie in the range 1.005 ± 0.003 but show less spread at 500°F by the range 1.002 ± 0.002. The two cases examined for the second cycle at 535° F show a similar range but a 0.5% lower reactivity, 0.997 ± 0.002.

In general, the above points indicate a fairly satisfactory situation at core start-up in the first and second cycles and the error trends suggest where the remaining difficulties may occur in the theory.

As the cores mentioned above have been depleted at power, further and less systematic errors have become apparent, as can be seen in Fig. 12-9. The reactivity behavior of Obrigheim in the first cycle can be overlooked as a diagnostic tool since an appreciable amount of foreign material was deposited on the fuel, although the reactor behaved extremely well as a power producer. Conn-Yankee, on the other hand, appeared much too reactive early in its first cycle but subsequently it came back into line with the other cores. Thus, by 15 Mwd/kgU all cycles of all cores lie in the band of 0.996 ± 0.003, about 0.6% $\Delta\rho$ below the hot zero power position. The scatter in the reactivity error curves of Fig. 12-9 is appreciable but to the eye it suggests a progressive loss of reactivity at a rate of roughly 0.4% $\Delta\rho$/Mwd/kgU. It seems difficult to be more definite as to whether an error of this sort is present in the burn-up physics description because there are few, if any, reliable data points at full power early in life at 0.5 Mwd/kgU, after Sm has come into equilibrium. Consequently, the estimate of the increase of reactivity error with burnup contains the uncertainties of the power defect. These apparent depletion errors are not unique to our computational methods but appear in the predictions of other groups in much the same way.

It should be emphasized that the reactivity errors shown in Fig. 12-9 were generated by fine-mesh two-dimensional calculations and, although we have found little effect on radial fuel assembly exposure or power by performing the analysis in three dimensions, the three-dimensional depletion calculation would provide an additional 0.5% reactivity decrease, as shown in Fig. 12-2.

In describing core flux or power distributions a great number of examples could be shown; some of the most interesting examples are provided by the first two cycles of the San Onofre reactor since the second cycle contained four all-plutonium fuel rod assemblies located in the center of the outer flats of the core. The Mn aeroball detectors replace a fuel rod in thirty assemblies in a position three rods southwest of the assembly center. Thus, from symmetrical instruments, the flux tilt across the interior of some assemblies can be seen. Fig. 12-10 shows comparisons from four of these maps with calculations, two taken from each cycle, with all the instruments translated to one quadrant. In the first-cycle zero-power tests, seventeen instruments were used for measurements and the very strong flux gradient across the edge assemblies A-8 and H-1 can be seen. Some of the instruments became inoperative later in life and the data on core flux distribution is thus somewhat fragmentary. In Fig. 12-11 the frequency distribution of deviations between observed axially-integrated Mn activity and 3D calculations is summarized from fourteen maps taken at full power in Cycles I and II. The agreement is certainly quite good.

The axial flux distributions in unrodded first-cycle cores are rather uninteresting in that they do not exhibit much radial variation and they seem relatively easy to calculate either with or without moderator temperature feedback. The second cycle offers better opportunity for evaluating the accuracy of feedback models and traces, taken from three batches of San Onofre Cycle II, are shown in Fig. 12-12 as examples of the axial variations anticipated. Unfortunately, calculations for comparison have not been completed.

An extensive series of moderator temperature coefficient measurements have been made at zero and, what is more unusual, at full power in San Onofre cycles I and II and in Obrigheim during burnup of the core. The full-power values are nearly a linear function of boron concentration and, as seen in Fig. 12-13, the second-cycle values lie only slightly below those of the first cycle. The present calculations show a somewhat steeper slope than measured and the MTC calculated at end of cycle is about 0.5×10^{-4} $\Delta\rho/^{\circ}F$ below both the San Onofre measured values and those of Obrigheim, Fig. 12-14. At any given boron concentration, the Obrigheim MTC is more positive than that of San Onofre, presumably because of the larger boron worth and the wetter lattice in the former reactor.

Boron-worth measurements under similar conditions of zero and full power were also made in San Onofre cycles I and II. Comparisons with calculations at full power are shown in Fig. 12-15; here the error appears to be the reverse of that observed at start-up with the calculations overestimating the boron worth. The scatter of measured points at zero power is so large that the measurements should probably be examined.

Of the many evaluations of control-rod worth that are available, one of the most definitive is the insertion of the entire bank of 32 control rods into Obrigheim accompanied by boron dilution, shown graphically in Fig. 12-16. The calculations used a one-dimensional version of PDQ with table sets for the rodded and unrodded radially-flux-weighted core constants as functions of boron concentration. Of course, the rods could be driven in only 240 cm out of 271 cm to afford necessary remaining shut-down margin. The bank was dropped the rest of the way into the core to provide a total measured worth of 8.6% $\Delta\rho$ compared with a calculated value of 8.4% $\Delta\rho$. Considering the simplicity of the window-shade bank-insertion calculation, the accuracy attained is fairly good, although the error in the steepest part of the curve is about 0.5% $\Delta\rho$.

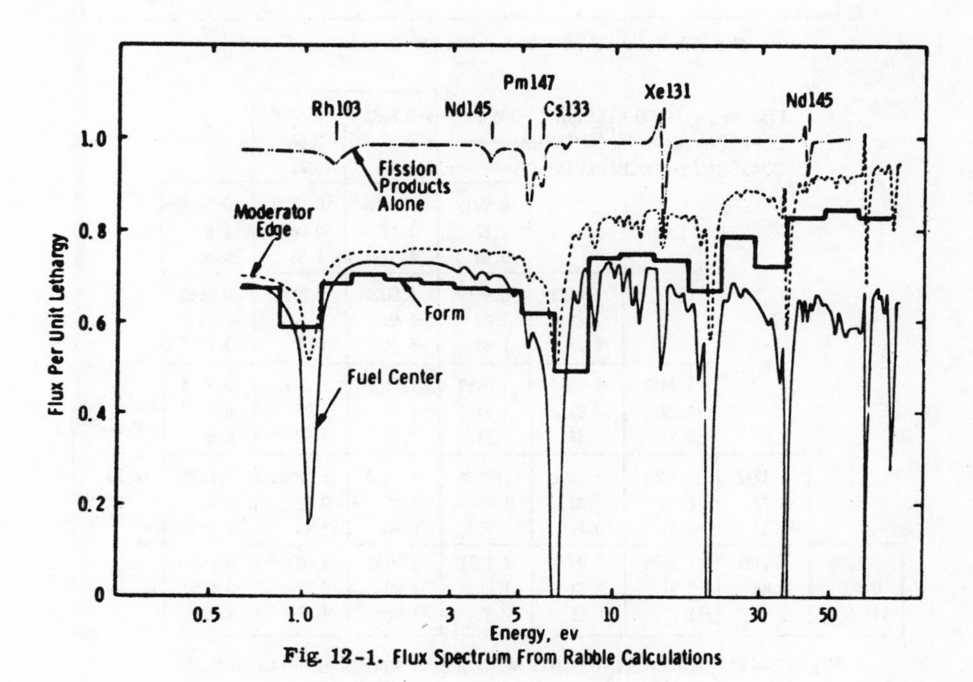

Fig. 12-1. Flux Spectrum From Rabble Calculations

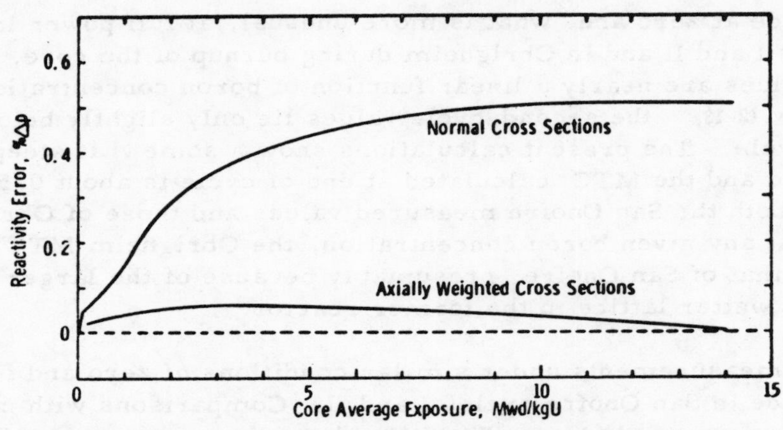

Fig. 12-2. Reactivity Error of 2D Core Calculations Relative to 3D

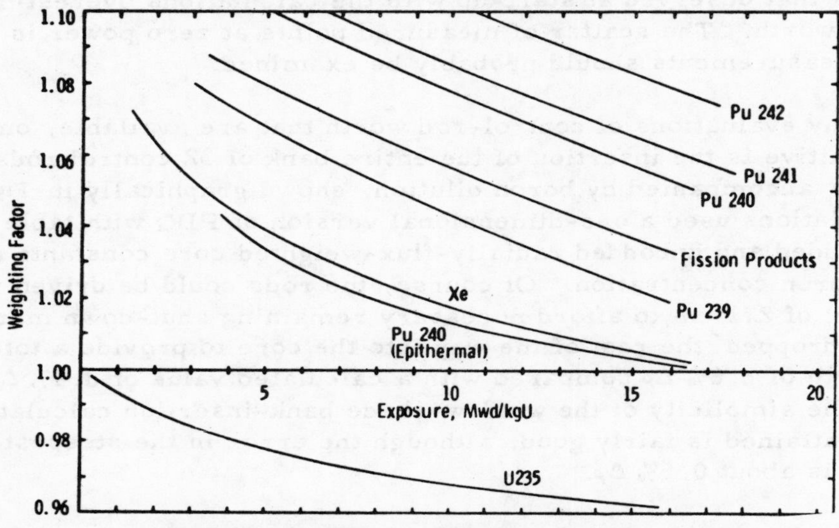

Fig. 12-3. Axial Weighting of Isotopic Cross Sections

FINE MESH (14x14) ASSEMBLY POWER →
COARSE MESH (4x4)% DEVIATION →
COARSE MESH (2x2)% DEVIATION →

					0.8638 0.37 4.91	0.6179 2.66 16.72		
				0.9915 -1.87 -5.61	0.8606 1.11 4.31	0.8579 0.062 4.88	0.5644 1.69 15.40	
			1.0932 -1.79 -8.77	0.9978 1.49 1.33	1.0005 -2.01 -5.24	0.8786 0.72 4.83	0.8166 0.19 9.09	
		1.1483 -1.25 -10.52	1.0694 1.83 -1.44	1.0969 -1.93 -8.31	1.0137 1.20 2.35	1.0455 -2.28 -3.90	0.9554 0.56 8.65	0.6714 0.87 16.10
	1.1737 -0.72 -11.38	1.1049 2.40 -2.83	1.1462 -1.43 -10.17	1.0694 1.49 -0.55	1.1054 -2.25 -6.90	1.0254 0.72 3.93	1.1123 -0.91 3.53	0.8818 -1.02 8.59
1.1806 0.40 -11.63	1.1181 2.84 -3.32	1.1679 -0.91 -11.11	1.0959 2.03 -2.13	1.1351 -1.87 -9.05	1.0598 1.20 1.50	1.0937 -2.52 -4.77	0.9867 -0.175 5.07	

Fig. 12-4. Typical Errors in Coarse-Mesh Quarter-Core Diffusion Calculations

Table 12-1. Reactor Core Descriptions

Reactor	Conn. Yankee	Obrigheim	San Onofre	Beznau
Power, Mwth.	1473/1810	907/1050	1300	1130
Core Diameter, in.	120	99	111	97
Core Height, in.	122	109	120	120
Average, kw/ft	4.53/5.57	4.62/5.35	4.61	5.22
Clad Material	SS	Zirc	SS	Zirc
No. Assemblies	157	121	157	121
No. Shims (Pyrex)	0	0	0	488
No. Control Rods (Ag, In, Cd)	45	32	45	29
W/UO_2 Volume Ratio	1.83	2.07	1.71	1.81
Enrichment W/O				
By Batch	3.02/3.23/3.67	2.50/2.80/3.11	3.15/3.40/3.85	2.44/2.78/3.48
Core Average	3.30	2.80	3.46	2.90
UO_2 Weight, kg	74750	40087	65136	45332
Kw/KgU	22.6	25.09	22.9	28.6
Avg Mod. Temp, ^oF	535/545	547/567	578	572

Table 12-2. Beginning of Cycle, Zero Power Characteristics

Reactor	Temp ^oF	Boron ppm	K_{eff}	Temp Coefficient ($10^{-4}/^oF$) Measured	Temp Coefficient ($10^{-4}/^oF$) Calculated	Boron Worth (ppm/%$\Delta\rho$) Measured	Boron Worth (ppm/%$\Delta\rho$) Calculated
Conn-Yankee	260I	2040	1.0025	0.57	0.51	--	--
	560I	2305	1.0002	1.00	0.85	125	130
	535II	1195	0.9946	-0.66	-0.88	125	129
Obrigheim	122	1727	1.0010	0.38	0.37	77.5	75.0
	536	1962	1.0005	1.13	1.14	90.0	95.6
Beznau	122	1525	1.0077	--	--	--	--
	535	1583	1.0036	-0.05	0.07	100.0	107.7
San Onofre	163I	2254	1.0072	0.34	0.38	101	114
	530I	2524	1.0039	0.74	0.69	127	144
	532II	1609	0.9986	-0.59	-0.94	137	153

Notation I, II indicates first, second fuel cycles.

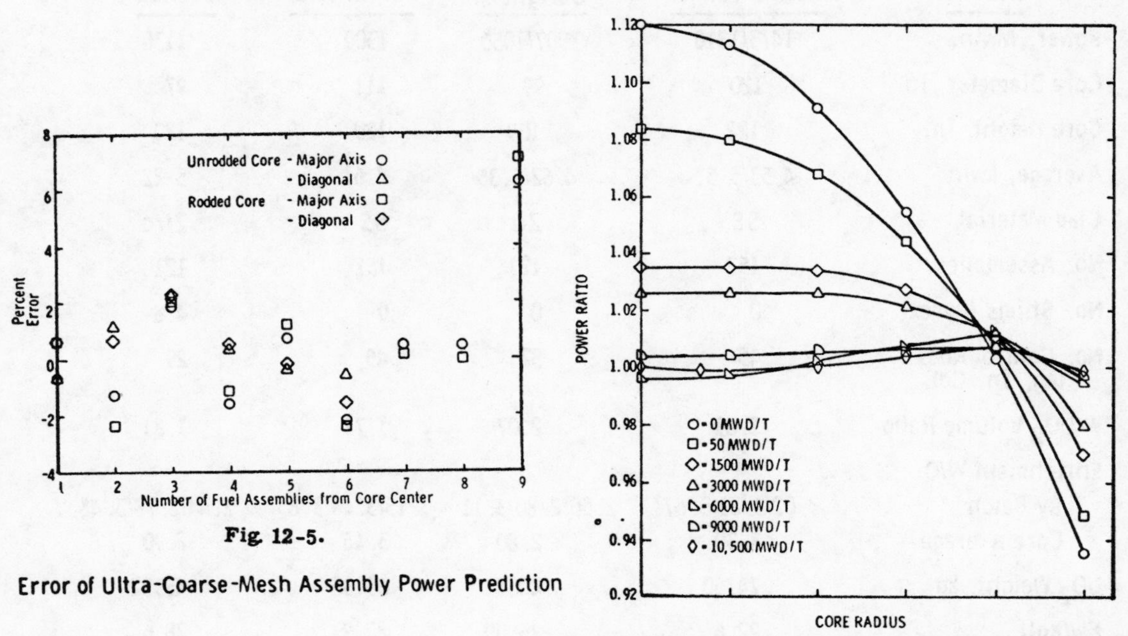

Fig. 12-5.

Error of Ultra-Coarse-Mesh Assembly Power Prediction

Fig. 12-6. Ratio of 4-Group to 2-Group Fuel Assembly Powers vs Rac 217 Assembly Equilibrium Core

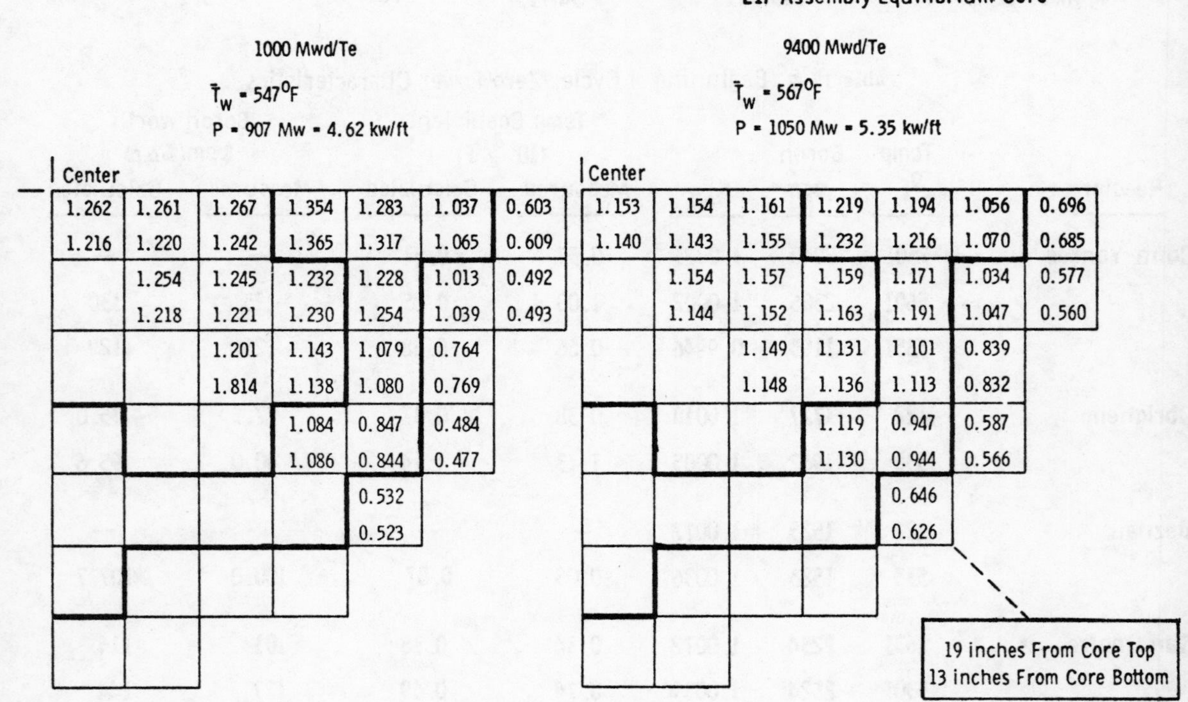

Fig. 12-7 . Effect of Moderator Density Feedback on Radial Power Patterns at Two Core Elevations and at Two Points in Life

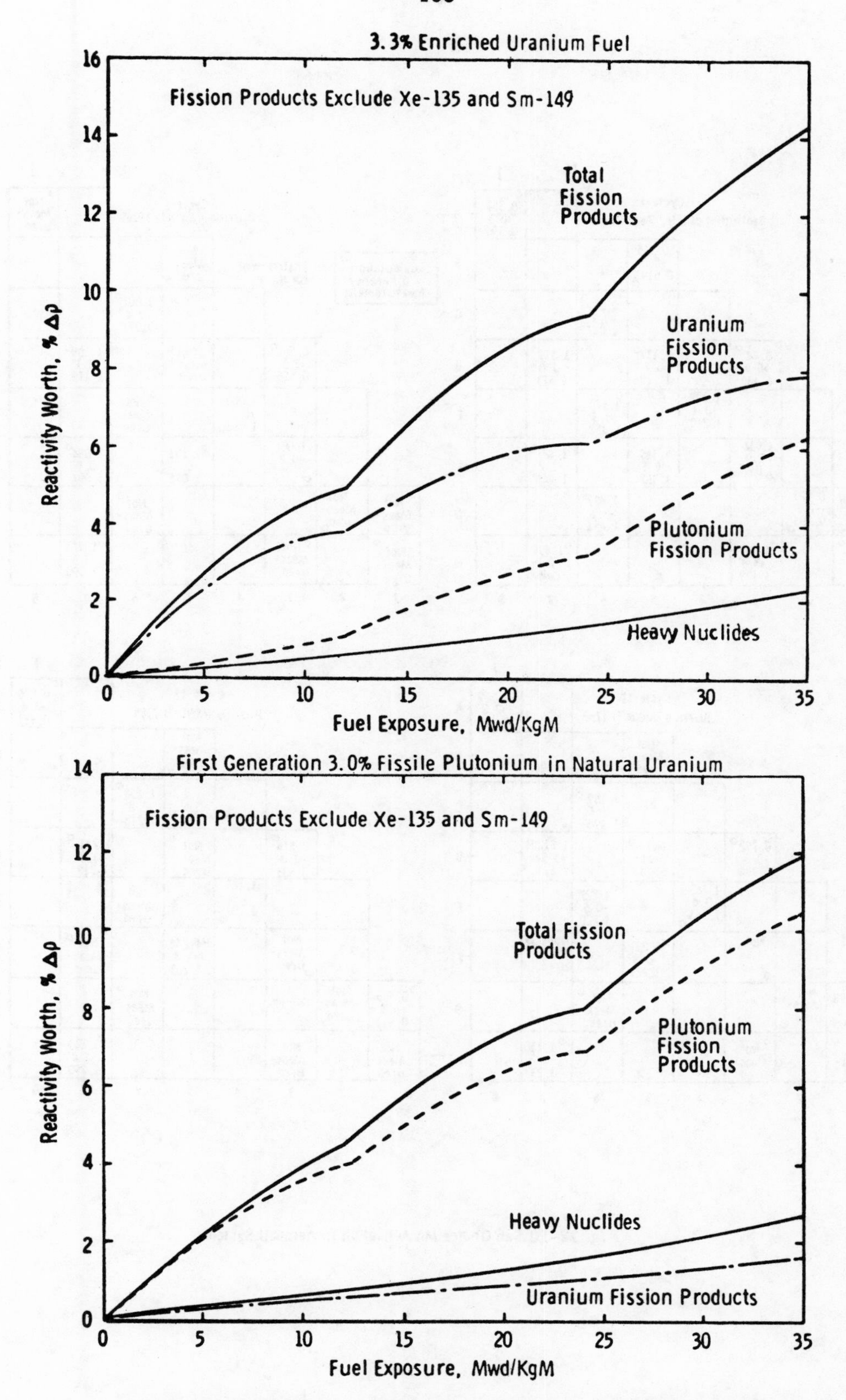

Fig. 12-8 . Buildup of Heavy Isotopes, Plutonium and Uranium Fission Products

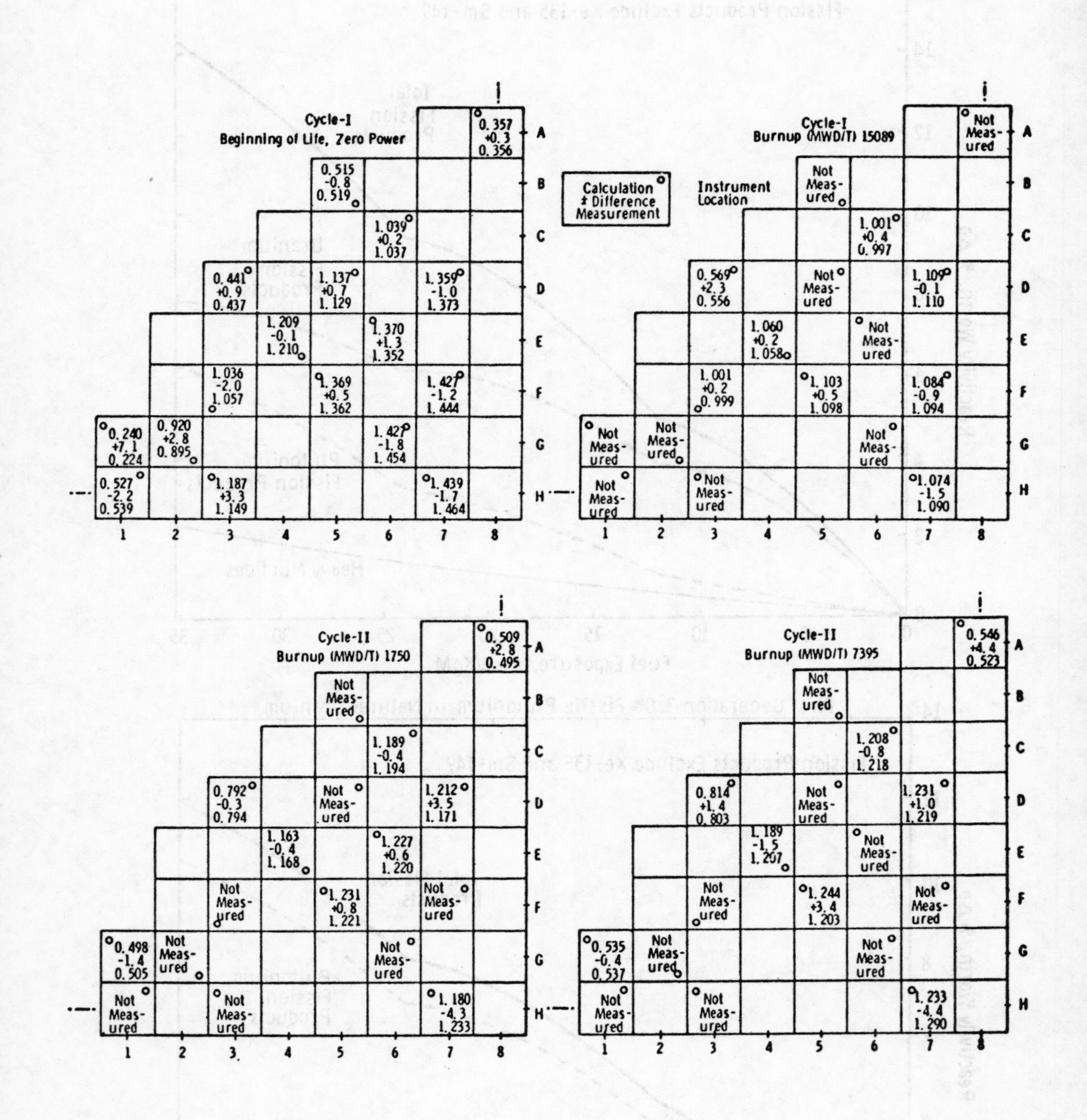

Fig. 12-10 San Onofre Mn Activation in Aeroball System

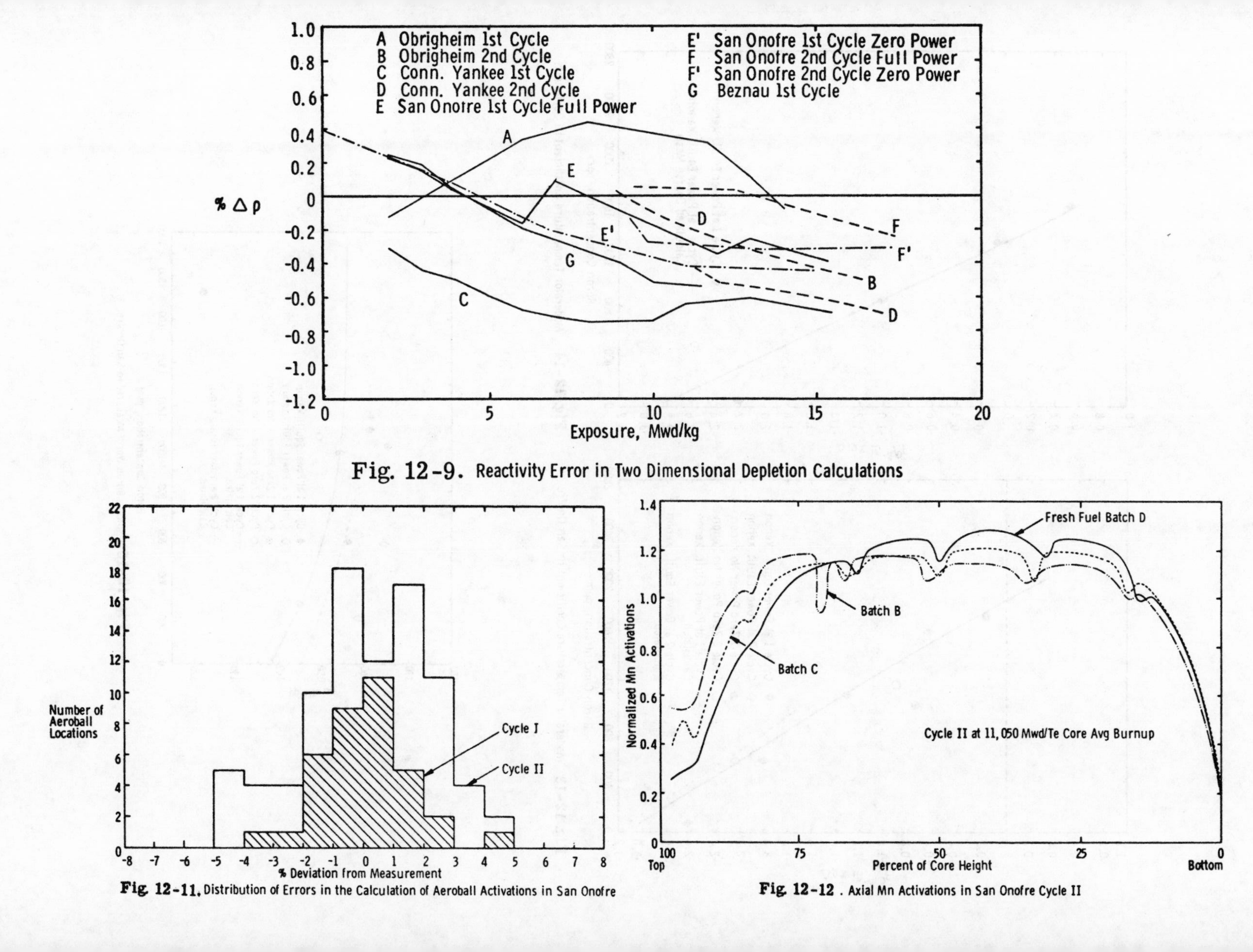

Fig. 12-9. Reactivity Error in Two Dimensional Depletion Calculations

Fig. 12-11. Distribution of Errors in the Calculation of Aeroball Activations in San Onofre

Fig. 12-12. Axial Mn Activations in San Onofre Cycle II

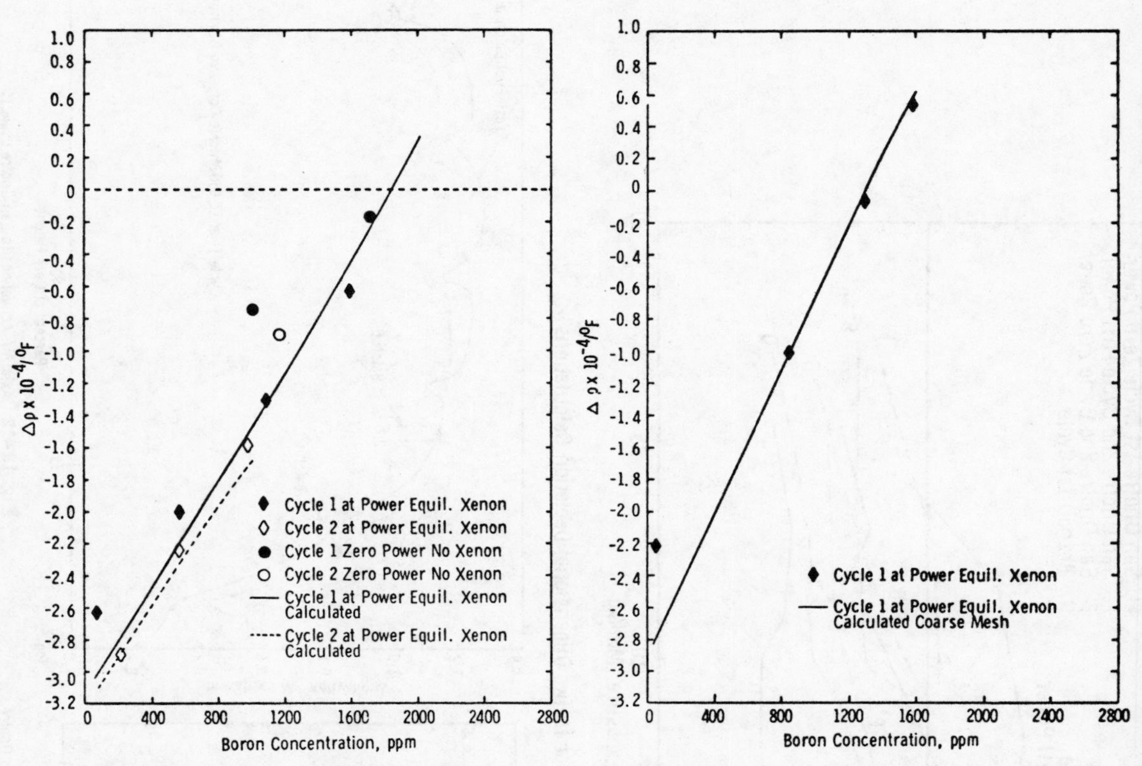

Fig. 12-13 . Moderator Temperature Coefficient in San Onofre

Fig. 12-14 . Moderator Temperature Coefficient in Obrigheim

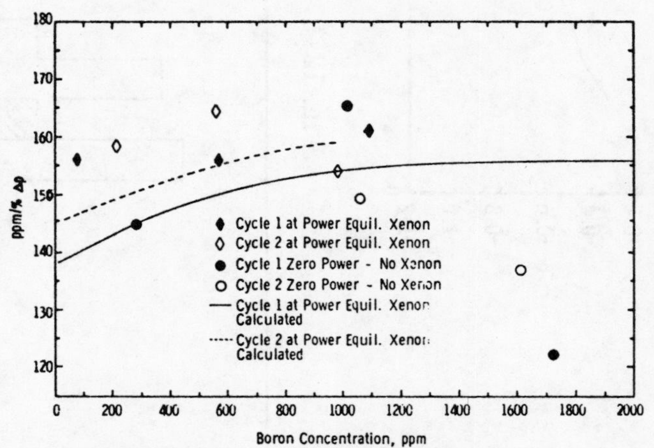

Fig. 12-15 • Soluble Boron Worth Measured in San Onofre

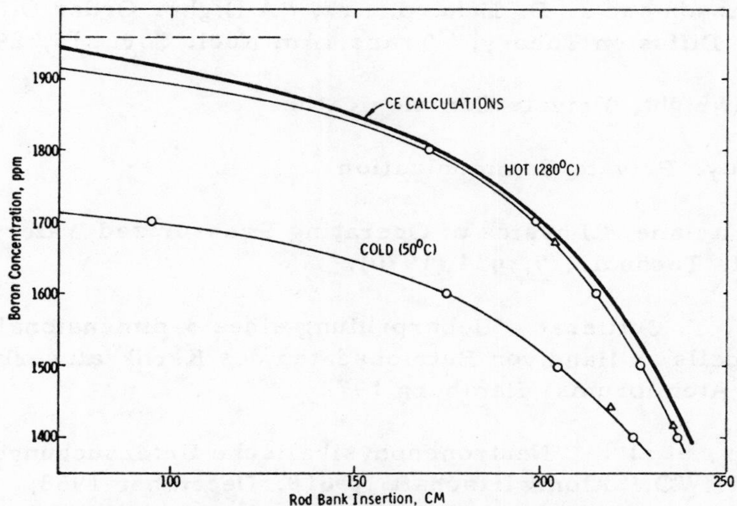

Fig. 12-16 • Obrigheim Reactor - Critical Boron Concentration at Zero Power in Fresh Fuel

REFERENCES:

1. A. Leonard et al., "A Modified Collision Probability Method for Non-Uniform Lattices," Proc. Conf. New Developments in Reactor Mathematics and Applications, Idaho Falls, Idaho, CONF - 710302 p 644 (1971)

2. H. C. Honeck, The Joshua System, DPSTM-500

3. F. J. Fayers, "UK Methods for Studying Fuel Management in Water Moderated Reactors," Jnl Br. Nucl. Energy Soc., 1972, 11 (1) 29-42.

4. K. Tsuchihashi, "CLUP77--a Fortran Program of Collision Probabilities for Square Clustered Assembly," JAERI - 1196, Japan Atomic Energy Research Institute (June 1970)

5. S. Borresen and R. Goldstein, "An Equivalence Formulation for Absorption in Plutonium," Trans. Am. Nucl. Soc., 15, 296 (1972)

6. R. L. Hellens and H. Mizuta, "The Effect of Resonance Overlap on U-238 Capture in Reactor Lattices," Trans. Am. Nucl. Soc., 11, 258 (1968)

7. R. L. Hellens, "Problems in Recycle of Plutonium in Pressurized Water Reactors," Trans. Am. Nucl. Soc., 14, 820 (1971)

8. Henry Windsor and Rubin Goldstein, "Analysis of Lattices Containing Mixed-Oxide Fuel in Particulate Form," Trans. Am. Nucl. Soc., 15, 107 (1972)

9. R. J. Nodvik, "Supplementary Report on Evaluation of Mass Spectrometric and Radiochemical Analyses of Yankee Core I Spent Fuel, Including Isotopes of Elements Thorium Through Curium," Westinghouse Atomic Power Divisions, WCAP-6086 (1969)

10. H. Windsor and F. Pistella, Private Communication

11. C. P. Robinson and J. D. Eckard, Jr., "A Higher Order Difference Method for Diffusion Theory," Trans. Am Nucl. Soc., 15, 297 (1972)

12. J. Himmelwright, Private Communication

13. W. Flournoy, Private Communication

14. A. F. McFarlane, "Physics of Operating Pressurized Water Reactors," Nucl. Appl. Technol., 9, 634 (1970);

P. Gryksa, T. Juillerat, "Ueberprüfung eines 3-dimensionalen Reaktorkern Rechenmodells an Hand von Betriebsdaten des Kernkraftwerkes Beznau I," Deutsches Atomforums, Hamburg 1972

15. G. Bronner, et al., "Neutronenphysikalische Untersuchungen bei Inbetriebnahme des KWO," Atomwirtschaft, p 618, December 1968;

C. Steinert and M. Dillig, "Parametric Study of Plutonium Recycling in Obrigheim Power Station," Proceedings of Panel on Plutonium Recycling in Thermal Power Reactors," IAEA, Vienna, Austria, 1971

CHAPTER 13

EVALUATION OF GAS COOLED FAST REACTORS

Although most of the development work on fast breeder reactors has been devoted to the use of liquid metal cooling, interest has been expressed for a number of years in alternative breeder concepts using other coolants. One of a number of concepts in which interest has been retained is the Gas-Cooled Fast Reactor (GCFR). As presently envisioned, it would operate on the uranium-plutonium mixed oxide fuel cycle, similar to that used in the Liquid Metal Fast Breeder Reactor (LMFBR), and would use helium gas as the coolant.

The long-term objective of any new reactor concept and the incentive for the government to support its development are to help provide a self-sustaining, competitive industrial capability for producing economical power in a reliable and safe manner. Successful achievement of this objective is required to permit the utilities, and others, to consider the concept as a viable option to existing power production systems and to gain public acceptance of a new form of power production. It is only after this is achieved that the utilities and industry could make the heavy, long-term commitments of resources in funds, facilities and personnel to provide the transition from the early experimental facilities and demonstration plants to full scale commercial reactor power plant systems. Consistent with the policy established for all power reactor development programs, the GCFR would require the successful accomplishment of three basic phases:

- An initial research and development phase in which the basic technical aspects of the GCFR concept are confirmed, involving exploratory development, laboratory experiment, and conceptual engineering.
- A second phase in which the engineering and manufacturing capabilities are developed. This includes the conduct of in-depth engineering and proof testing of first-of-a-kind components, equipment and systems. These would then be incorporated into experimental installations and supporting test facilities to assure adequate understanding of design and performance characteristics, as well as to gain overall experience associated with major operational, economic and environmental parameters. As these research and development efforts progress, the technological uncertainties would need to be resolved and decision points reached that would permit development to proceed with necessary confidence. When the technology is sufficiently developed and confidence in the system is attained, the next stage would be the construction of large demonstration plants.
- A third phase in which the utilities make large scale commitments to electric generating plants by developing the capability to manage the design, construction, test and operation of these power plants in a safe, reliable, economic, and environmentally acceptable manner.

Significant experience with the Light Water Reactor (LWR), the High

Temperature Gas-Cooled Reactor (HTGR) and the Liquid Metal Fast Breeder Reactor (LMFBR) has been gained over the past two decades pertaining to the efforts that are required to develop and advance nuclear reactors to the point of public and commercial acceptance. This experience has clearly demonstrated that a logical progression through each of the three phases is an extremely difficult, time consuming and costly undertaking, requiring the highest level of technical management, professional competence and organization skills. This has again been demonstrated by the recent experience in the expanding LWR design, construction and licensing activities which emphasizes clearly the need for even stronger base technology and engineering efforts than were initially provided, although these were satisfactory in many cases for the first experiments and demonstration plants. The LMFBR program, which is relatively well advanced in its development, tracks closely this LWR experience and has further reinforced this need as its applies to the technology, development and engineering application areas.

It should also be kept in mind that the large backlog of commitments and the shortage of qualified engineering and technical management personnel and prooftest facilities in the government, in industry and in the utilities make it even more necessary that all the reactor systems be thoroughly designed and tested before additional significant commitment to, and construction of, commercial power plants are initiated.

The large scale commitments to the uranium-plutonium fuel cycle through purchases of the LWRs and the substantial investment of the Nation's resources engendered by these commitments led the U. S. Atomic Energy Commission (AEC), Division of Reactor Development and Technology to initiate reviews in 1966 of the technical status and the possible benefits implicit in the development of the various reactor concepts being considered in the civilian nuclear power program for meeting future power needs. These reviews were needed to help provide guidance on making effective use of our national resources and help determine the requirements for, and allocations of, these resources. With regard to the GCFR, the AEC evaluated this concept in 1969 and issued "An Evaluation of Gas-Cooled Fast Reactors," WASH-1089, April 1969, along with a companion report "An Evaluation of Alternate Coolant Fast Breeder Reactors," WASH-1090, April 1969.

The GCFR designs evaluated in WASH-1089 were based, in large measure, on information provided by the developers of the systems, and therefore, the report generally reflected their viewpoints and enthusiasms. The information in WASH-1089 was used in large part as input to the subsequen systems analysis (WASH-1098) and cost-benefit studies (WASH-1126, USAEC, "Cost-Benefit Analysis of the U.S. Breeder Reactor Program," 1969 and WASH-1184, USAEC, "Updated (1970) Cost-Benefit Analysis of the U.S. Breeder Reactor Program," January 1972).

WASH-1098, "Potential Nuclear Growth Patterns," December 1970, was prepared by the Systems Analysis Task Force (SATF) which was concerned with the development and application of a model of the U.S. electrical power economy. The model input data was provided by the individual task forces charged with the evaluation of the reactor concepts under consideration, including GCFR. The results of this evaluation, as it applied to GCFR, indicated that largely because of the uncertainties in the cost estimates for both the LMFBR and the GCFR, it was not possible to draw a definitive conclusion concerning the soundness and desirability of conducting a parallel breeder program on the GCFR in addition to the

LMFBR. Accordingly, WASH-1098 concluded that "the current state of knowledge is not sufficient to support a definitive evaluation of the potential of the GCFR against that of the LMFBR."

The consistent conclusion reached in the cost-benefit studies (WASH-1126 and -1184), viz., sufficient information is available to indicate that the projected benefits from the LMFBR program can support a parallel breeder program, is highly sensitive to the assumptions on plant capital costs. With the recognition that even for ongoing concepts on which ample experience exists, capital costs and especially small estimated differences in costs are highly speculative for plants to be built 15 or 20 years from now, it is questionable whether analyses based upon such costs should constitute a major basis for decision making relative to the desirability of a parallel breeder effort.

In compliance with a request from the Office of Science and Technology for further review of the GCFR at this time, the AEC has undertaken this internal assessment which examines the technical developments that have taken place in the continuing research and development and design efforts on the GCFR system. This request recognized that the GCFR has been in the initial research and development phase with emphasis on the development of basic GCFR technology. The program has been carried out primarily at Gulf General Atomic (GGA) and has been supported by government-sponsored research at a level of about $1 million per year over the past several years. The utility industry is supporting research and development on this concept at a level of about $1.4 million per year. Total expenditures from 1963 to date on GCFR technology have amounted to approximately $16 million.

The GCFR is still in the early phases of an overall R&D effort notwithstanding the benefits expected to accrue to it from the HTGR and

LMFBR programs. An adequate investigation of the problems associated with the GCFR concept to move forward toward their satisfactory resolution would require a substantial research and development program along the lines outlined herein with sufficient proof-testing of major componen and systems to provide assurances commensurate with the large commitment and investment in such plants for large-scale power generation. Some prerequisites for such a program are provided by portions of major ongoi programs, e.g., some of the fuel development, physics and safety work in the Liquid Metal Fast Breeder Reactor (LMFBR) program and component development in the High Temperature Gas Reactor (HTGR) program. Also, component and plant operating and maintenance experience at the 330 MWe Fort St. Vrain reactor will provide important basic information.

However, a number of critical and unique characteristics of the GCFR are not adequately represented in research and development programs on other concepts. In addition, the flexibility for resolution of such development and engineering problems normally required by the designers, constructors, and operators of even the more proven concepts is restrict due to the compact arrangement proposed for the GCFR. Consequently, if feasibility were to be confirmed through appropriate experimental reactor and other facility operation, the additional overall technical effort needed for such a full-scale program on the GCFR could be compara in magnitude to the efforts on other major reactor development programs such as the LWR and LMFBR programs.

Experience in reactor development programs in this country and abroad ha demonstrated that different organizations in evaluating the projected costs of introducing a reactor development program and carrying it forwa to the point of large-scale commercial utilization, would arrive at

different estimates of the methods, scope of development and engineering efforts, and the costs and time required to bring that program to a stage of successful large scale application and public acceptance.
Based upon the extensive program required to bring a new concept to commercial utilization, and taking into account the benefits expected to accrue from progress in the LMFBR and HTGR programs over the next 5 years, the incremental cost to the government of a parallel breeder program of this type has been estimated by the AEC to range up to about $2 billion in undiscounted direct costs (WASH-1184). The GCFR would require magnitude of funding up to this level in order to establish the necessary technology and engineering bases; obtain the required broad industrial capability; and advance through a series of test facilities, reactor experiments, and demonstration plants to a commercial GCFR, safe and suitable to serve as a major energy option for central station power generation in the utility environment.

SUMMARY

The GCFR concept uses helium as the coolant gas, which leads to several potentially favorable attributes of the GCFR. Helium is both optically and neutronically transparent and does not become radioactive. An intermediate heat exchanger system is not required which reduces system complexity. The GCFR has a potentially high breeding ratio resulting largely from the coolant properties. This advantage may be partially offset by the higher specific fissile inventory (Kg/MWe) required due to the poorer heat transfer properties of helium as compared to sodium. Since the use of gas cooling requires a high coolant pressure to secure adequate heat transfer, the GCFR is subject to the possibility of depressurization accidents as well as loss-of-flow accidents. The potential for occurrence of these types of accidents requires resolution of significant reactor damage and safety questions which are unique to the GCFR concept. In addition, there are other outstanding engineering problems associated with safety, reliability and maintainability which differ significantly from similar considerations for the HTGR and LMFBR. The additional technical work done since the publication of WASH-1089 and WASH-1098, both by the AEC and private industry, and further definition o problem areas aided by discussions with the Regulatory staff and the Advisory Committee on Reactor Safeguards (ACRS) have served to reinforce the earlier conclusion of WASH-1089 that a substantial program would be needed to support the commercial introduction of the GCFR. This is particularly so since only limited effort on the GCFR is underway outside the United States and few benefits can accrue from significant parallel development work and related large scale commitments in other countries such as is the case for the LMFBR concept. Based upon the information currently available, as discussed in this section, changes in the GCFR development status, relative to the total effort required for its commerc application, do not appreciably alter the general conclusions of WASH-108 Although the proponents of the GCFR concept would intend to depend heavil upon the technology developed in the High Temperature Gas Reactor (HTGR) program and, to the extent possible, the technology of the LMFBR program, the GCFR can be characterized as being in the early stages of an overall development program. This should not be surprising, since the government and industrial funds expended on this concept through FY 1972 total less than $20 million. With a significantly increased commitment to an R&D program along with personnel and other resources made available on a high priority basis, the development time scale which could optimistically be expected for development of the GCFR would have to be in the context of the approach and experience with the LWR, HTGR and LMFBR programs, Fig. 13-1. With the size of the major government and industrial investment, recent experience with LWR's on problems arising in multiple commitments to full size power plants provides ample evidence of the necessity to provide a broad industrial base and to conduct extensive, in-depth development and testing efforts addressed to all critical components and systems throughout all project phases. An overall full-scale program on the GCFR would

Fig. 13-1

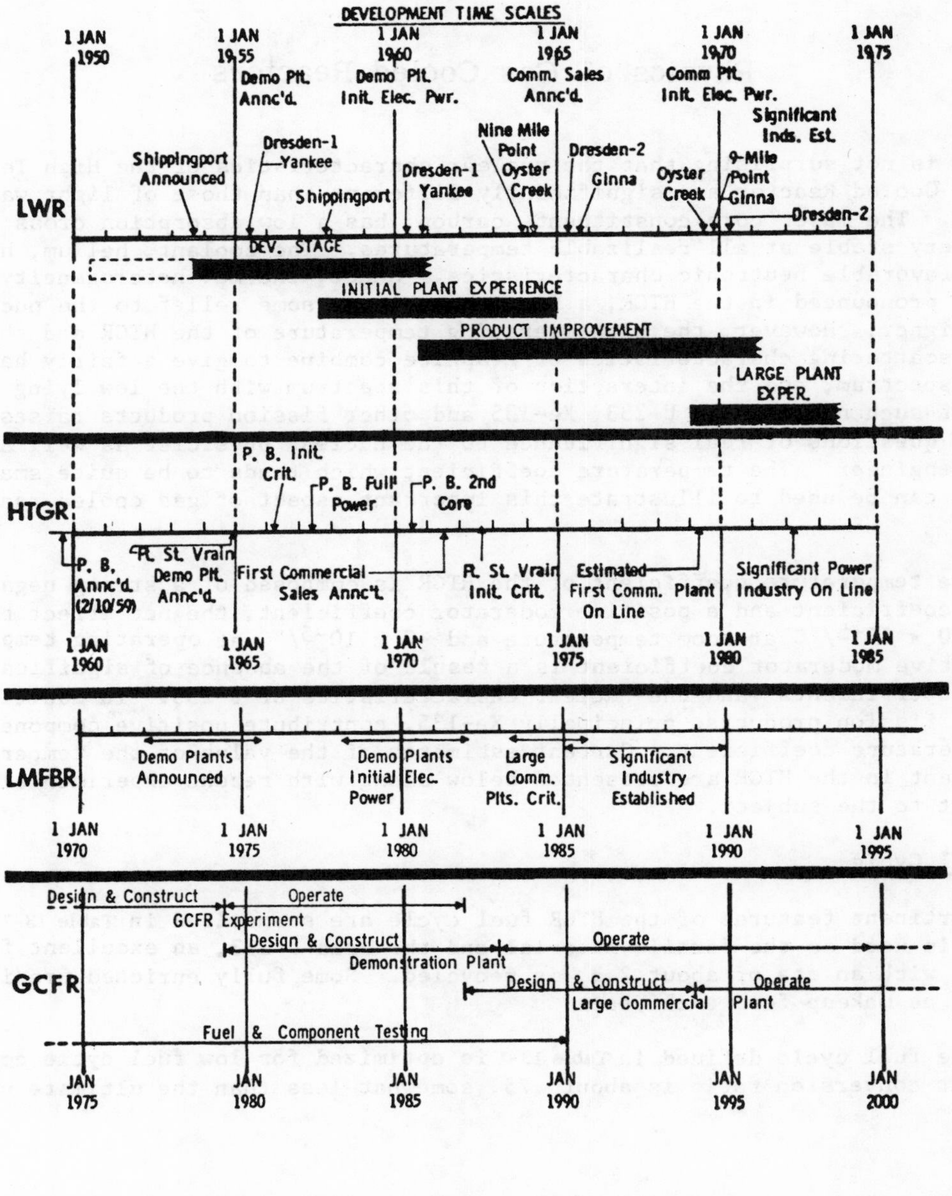

be expected to cover those phases that are currently under consideration for the LMFBR Program.

Such a program would involve costs comparable to other reactor development programs(in the range of two billion dollars from the Government) in addition to large outlays and commitments by industry and the utilities. Such a GCFR program would have to take into account any overlapping or concurrent commitments and availability of resources with the HTGR commercialization efforts. This, of course, is a problem faced by other reactor programs as well.

Physics of Gas Cooled Reactors

It is not surprising that the nuclear characteristics of the High Temperature Gas Cooled Reactor are significantly different than those of light water reactors. The major core constituent, carbon, has a low absorption cross section and is very stable at all realizable temperatures. The coolant, helium, has similar favorable neutronic characteristics. Hence, thermal heterogeneity effects are less pronounced in the HTGR, a fact that brings some relief to the nuclear core designer. However, the high operating temperature of the HTGR and the neutron scattering characteristics of graphite combine to give a fairly hard neutron spectrum, and the interaction of this spectrum with the low lying resonances of such nuclides as U-233, Xe-135 and other fission products raises problems and questions of real significance to the nuclear physicist as well as the nuclear engineer. The temperature coefficient which tends to be quite small in the HTGR can be used to illustrate this important aspect of gas cooled reactor physics.

The temperature coefficient of the HTGR is composed of a strong negative Doppler coefficient and a positive moderator coefficient, the net effect being about $-10 \times 10^{-5}/°C$ at room temperature and $-2 \times 10^{-5}/°C$ at operating temperature. The positive moderator coefficient is a result of the absence of significant density coefficients and the nuclear characteristics of U-233. In addition, selected fission products, principally Xe-135, contribute positive components to the temperature coefficient. Current estimates of the value of the temperature coefficient in the HTGR are presented below along with recent experimental data pertinent to the subject.

HTGR Fuel Cycle

Pertinent features of the HTGR fuel cycle are summarized in Table 13-1. Thorium is used as the fertile material and the bred U-233, an excellent fissile material with an eta of about 2.3, is recycled. Some fully enriched uranium is required as makeup feed each year.

The fuel cycle defined in Table 13-1 is optimized for low fuel cycle costs. The resultant conversion ratio is about .75, somewhat less than the ultimate value of .9 which could be achieved with a heavier thorium loading and on-line refueling. However, the utilization of the bred U-233 is still excellent. The fraction of the power produced in U-233 as compared to U-235 is shown below:

	Fraction of power in ...	
	U-233	U-235
Beginning of life	0	1.0
End of 1st cycle	.42	.58
Beginning of equilibrium cycle	.55	.45
End of equilibrium cycle	.66	.34

The neutron balance at the end of an equilibrium cycle is shown in Table 13-2. Most of the absorptions occur in the thorium and uranium isotopes; significant absorptions also occur in Pa-233 and the fission products; carbon and helium absorb very few of the neutrons.

Temperature Coefficient

The calculated temperature coefficient as a function of temperature at the end of an equilibrium cycle is shown on Figure 13-2. The positive moderator coefficient is due to the effect of U-233, U-234, xenon, and other fission products, as shown on Table 13-3. The principal negative contributor to the moderator coefficient is U-235.

The net thorium component is $-1.6 \times 10^{-5}/°C$, a result of a Doppler coefficient magnitude of $-3 \times 10^{-5}/°C$, and a positive thermal base sub-component of about $+1.4 \times 10^{-5}/°C$. Fortunately, the Doppler coefficient is large enough to result in a net overall negative temperature coefficient at all realizable temperatures. At the operating temperature of about 800°C, the net coefficient is about $-1.3 \times 10^{-5}/°C$.

There has been concern about the accuracy of calculations such as the ones described above, since the net coefficient is the sum of two fairly small numbers, one positive and one negative. Fortunately, kinetics calculations show that peak core temperatures during power excursions are very insensitive to the value of the temperature coefficient. One reason for this, of course, is the immense heat capacity of the core - if 100% of the heat at full power were stored within the core, none of it being removed by the coolant, then the average core temperature would rise by only about 8°F/sec. However, one of the design criteria for the HTGR is that the net temperature coefficient should be negative, and we have gone to great lengths to assure ourselves that our estimates of the temperature coefficient are accurate enough.

Three bodies of experimental data exist which are pertinent to this subject. These were generated at the

Peach Bottom HTGR
HTGR critical assembly at GGA in 1967-1969
HTLTR critical assembly at BNWL in 1970 and 1971.

A brief review of the information obtained from these sources is presented below.

Peach Bottom

Peach Bottom is a 40 MW(e) HTGR built for the Philadelphia Electric Company. It first achieved criticality in 1966. The initial and subsequent test programs have focused in a major way on temperature coefficients even though they were expected to be large and negative for this small reactor. Accurate cross section information on U-233 in the mid 1960s was not available and there was some concern over this technological gap. This gap has in the meantime disappeared However, the initial Peach Bottom temperature coefficients, and the chang in that coefficient with reactor operation, were studied in some detail, with the results shown in Tables 13-4, 13-5. In the initial core (Table 13-4) the negative temperature defect appears to be underestimated by about 10%, but the temperatu coefficient at operating temperature is less negative than estimated. Temperature coefficient measurements at specific temperatures at later times in core life have been made, but the change in the cold to hot temperature defect (Table 13-5) appear t be well estimated with conventional techniques. The impact of U-233 buildup and fission product buildup on the temperature coefficient is well shown in Table 13-5.

HTGR Critical Assembly Program

The HTGR critical assembly program developed a broad base of experimental data for use in evaluating calculational techniques (Ref. 3). It was possible to measure in a room temperature facility the Doppler coefficient of thorium in an HTGR lattice from 300°K to 600°K. The results, shown on Table 13-6, can be well predicted by theory. On the basis of these data, and similar information on the reactivity worth of thorium in the central region of a critical assembly at constant (room) temperature, we feel that the effect of thorium on the core design of an HTGR can be estimated with good accuracy. Our theoretical approach is embodied to a large extent in the calculational codes GGC (Ref. 4), GAROL (Ref. 5) and MICROX (Ref. 6) in which particle self-shielding, and spatial and energy flux variations are taken into account in the resonance calculation. ENDF/B cross section data are used nearly exclusively.

HTLTR Critical Assembly Program

A program on HTGR temperature coefficients has just recently been complet in the High Temperature Lattice Test Reactor at Battelle Northwest Laboratories. This program was funded by the Atomic Energy Commission and conducted by BNWL, with significant guidance and technical input from GGA (Ref. 7). The theoretic basis of the measurements in the HTLTR is similar but significantly more complicated than that of the PCTR (Ref. 8). The infinite multiplication constant of lattice of interest is determined by measuring the reactivity worth difference between a piece of that lattice material and a piece of copper absorber material when placed in a void region in the center of an appropriately designed critical assembly. Some correction factors, both analytical and experimental, must be applied to the measurement. The value of the HTLTR is that zero power measureme to temperatures as high as 1000°C can be made.

In the recently completed program, such measurements were made on 4 lattic of interest, defined in Table 13-7. Lattice No. 1 (Ref. 9) was the basic or benchmark lattice, containing the presumably well known materials U-235 and Th-232. Lattice No. 2 (Ref. 10) was designed to be the same as No. 1, only with the U-2 replaced by U-233. Since U-233 has a relatively lower capture cross section th U-235, fewer atoms of U-233 than U-235 were required for criticality, i.e., the C/U ratio in the second lattice was significantly greater than in the first lattice. Lattice No. 3 (Ref. 11) was the same as lattice No. 2 except for the thorium load, which was lower. Typical HTGR reactors have carbon to thorium ratios of about 225.

Lattice No. 4 utilized plutonium rather than uranium fuel, and is of long range rather than short range interest to the HTGR. The plutonium fuel was in the form of PuO_2 particles of diameter about 200 micrometers. This is in the range of interest for HTGR applications.

The results of the experiments on these lattices are summarized on Table 13-

In the U-235/Th-232 lattice (Lattice No. 1), the temperature defect was underestimated, due to an underestimate of the coefficient at low temperature. At high temperature, the coefficient is measured to be less negative than calculated, a result consistent with Peach Bottom experience. HTGR critical assembly results indicate that the Doppler coefficient can be well estimated; hence the discrepancy observed in Lattice No. 1 is probably due to the moderator coefficient.

The agreement between calculations and experiment in Lattice No. 2 is generally excellent, although the temperature coefficient at high temperature is again predicted to be more negative than measured.

In Lattice No. 3, serious disagreement between calculation and experiment was noted in the low temperature coefficient, although at higher temperatures the agreement was reasonably good. The source of the discrepancy is not known at this time; it could be due to uncertainties in the nuclear data, although this does not seem likely in view of the results obtained with Lattice No. 2. It is possible that one or two of the experimental data points at low temperature are more uncertain than currently believed, but this point has been reviewed at length by GGA and BNWL personnel with no change in the experimental data being currently indicated.

The analysis of the plutonium lattice is not yet complete, but the preliminary results shown on Table 13-8 are encouraging. Particle self shielding effects are very significant, and the agreement observed in these analyses is reassuring. Additional experiments will probably be necessary before plutonium fuels for the HTGR can be specified in detail. These experiments must focus on:

1) PuO_2 particles of different sizes;
2) particles of mixed plutonium and thorium oxides;
3) various Pu/Th ratios;
4) various lattice spacings; and
5) plutonium of various isotopic compositions.

An accurate estimate of the absorption rate in the low lying Pu resonances, particularly the Pu-240 resonance, is a physics problem of major complexity, one which requires a body of relevant experimental data for use in evaluating and normalizing analytical methods.

The three sets of experimental data discussed above, together with the supporting analyses, indicate that the temperature coefficient at operating temperature may be less negative than calculated. More information on this point will be available later this year from the initial loading and test program associated with the Fort St. Vrain HTGR. In the meantime, the temperature coefficient data used for accident analyses must be adjusted to reflect current experience.

The temperature defect between low and high temperature may be somewhat greater than calculated, although both Lattice No. 2 and No. 3 indicate this tendency,if present, could be small. Discrepancies in the temperature defect affect estimates of the shutdown margin which are, in any event, adequate in the HTGR.

The physical differences between HTGRs and LWRs result in differences in the physics characteristics of the two reactor types. One of the significant results of these physics differences is the HTGR's smaller temperature coefficient. The small coefficient is caused by the HTGR's graphite moderator and helium coolant; the neutron spectrum is fairly hard and there is essentially no density effect. The refractory nature and high heat capacity of the graphite make a large negative temperature coefficient unnecessary for safety purposes. Thus the small coefficient, which results in less cold reactivity to be covered with control rods, is a benefit. The results of on-going comparison of experiment and calculation show that the temperature coefficients, although small, can be calculated with reasonable accuracy.

Table 13-1

HTGR Fuel Cycle

Power density	8 w/cc
Fuel lifetime	4 years
Cycle duration	1 year @ 80% capacity factor
Fraction of core refueled each year	1/4
C/Th ratio:	
Initial core	225
Reload segments	250
Fuel exposure	100,000 MWD/MT
Feed uranium	U-235 (fully enriched uranium)
Bred uranium	U-233

Table 13-2

Neutron Balance - End of Equilibrium Cycle

Nuclide	Absorption Fraction
Th-232	.31
Pa-233	.01
U-233	.31
U-234	.04
U-235	.17
Xe-135	.02
Other fission products	.09
Carbon	.01
Helium	nil
Other nuclides	.04
Total	1.0

Table 13-3

Components of Temperature Coefficient at 1100°K in HTGR

Nuclide	Component, 10^{-5}/°C
Th-232	- 1.6
U-233	+ .3
U-234	+ .3
U-235	- 1.3
Np-237	- .3
Pu-239	+ .1
Pu-240	- .1
Xe-135	+ .9
Rh-103	- .1
Sm-149	+ .2
Other	+ .3
Total	- 1.3

Table 13-4

Temperature Coefficients in the Peach Bottom HTGR Initial Loading and Test Program

(References 1 and 2)

Type of Measurement	Reactivity Effect Measured	Reactivity Effect Calculated
Temperature defect		
36°C to 315°C	-.024 ±.002 Δρ	-.022
36°C to operating temperature	-.070	-.063
Temperature coefficient		
36°C	-9.0×10^{-5}/°C	-8.6×10^{-5}/°C
315°C	-8.1	-7.0
Operating temperature	-3.6	-4.9

Table 13-5

Temperature Defect in Peach Bottom - Changes With Operation in Core 1

Time in Life	Temperature Defect 100°F to Operating Temperature: Measured	Calculated
Initial test program	-.070 Δρ	-.063
168 EFPD	-.066	-.059
300 EFPD	-.060	-.054

Table 13-6

Doppler Coefficient of Thorium in HTGR Lattice

Composition of Lattice	Reactivity Change 300°K to 600°K	
	Measured	Calculated
C/Th = ∞ (i.e., no thorium)	-.029 Δρ	-.028 Δρ
C/Th = 300	-.055	-.053
C/Th = 200	-.063	-.062
C/Th = 100	-.074	-.074
No thorium, but boron equivalent to 1/v component of Th-232 in C/Th = 100 lattice	-.0124	-.0116

Table 13-7

Definition of Lattices Used in HTLTR Program

Lattice Number	Fissile Material	Carbon to Fissile Atom Ratio	C/Th Ratio	Purpose of Experiments
1	U-235	6,300	209	Doppler coefficient
2	U-233	12,200	209	U-233 coefficient
3	U-233	14,800	301	Doppler and U-233 coefficients
4	Pu	7,500	250	Pu coefficient, including particle effect

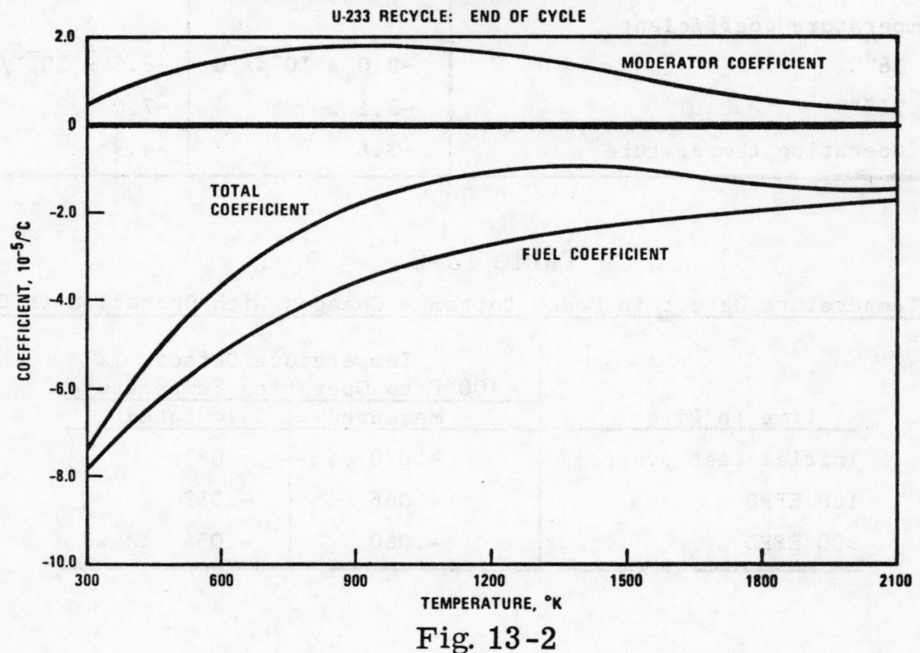

Fig. 13-2

Table 13-8

Reactivity Change: HTGR Lattices in HTLTR Experiments

	Lattice 1		Lattice 2		Lattice 3		Lattice 4	
	Calculation	Experiment	Calculation	Experiment	Calculation	Experiment	Calculation	Experiment
Δk Between 20°C and T								
Δk: 20°C to 500°C	-.040	-.058	-.029	-.029	-.0172	-.0017	-.020	-.025
Δk: 20°C to 750°C	-.051	-.075	-.033	-.034	-.0177	-.0025	-.030	-.036
Δk: 20°C to 1000°C	-.060	-.080	-.035	-.035	-.0180	-.0033	-.046	-.056
Average Temperature Coefficient, 10^{-5}/°C								
20°C to 500°C	-8.33	-12.08	-6.04	-6.04	-3.57	-0.35	-4.17	-5.21
500°C to 750°C	-4.40	- 6.80	-1.60	-2.00	-0.23	-0.32	-4.00	-4.40
750°C to 1000°C	-3.60	- 2.00	-0.80	-0.40	-0.11	-0.32	-6.40	-8.00

References

1. Brown, J. R., et al., "Isothermal Temperature Coefficient Measurement in Peach Bottom," USAEC Report GAMD-7358, Gulf General Atomic, October 1, 1966

2. Brown, J. R. and K. R. Van Howe, "Temperature and Power Coefficient Measurements During Rise to Power Program in Peach Bottom," USAEC Report GAMD-7907, Gulf General Atomic, April 11, 1968.

3. Final Safety Analysis Report for the Fort St. Vrain HTGR, Vol. 1, Public Service Company of Colorado, 1970.

4. Adir, J. and K. D. Lathrop, "Theory of the Methods Used in the GGC-3 Multigroup Cross Section Code," Gulf General Atomic Report GA-7156, July 19, 1967.

5. Stevens, C. A. and C. V. Smith, "GAROL, A Computer Program for Evaluating Resonance Absorption Including Resonance Overlap," Gulf General Atomic Report GA-6637, August 24, 1965.

6. Walti, P. and P. Koch, "MICROX, A Two-Region Flux Spectrum Code for the Efficient Calculation of Group Cross Sections," Gulf General Atomic Report GULF-GA-A10827, April 14, 1972.

7. Schultz, K. R. and D. R. Mathews, "GGA Support of the HTLTR Program During FY-71," USAEC Report GULF-GA-B10776, Gulf General Atomic (1971).

8. Lippincott, E. P., "Derivation of Corrections to K_∞ in the Two-Group Approximation," USAEC Report BNWL-1560, May 1971.

9. Lippincott, E. P., "Measurement of the Temperature Dependence of K_∞ for a $U\text{-}233O_2\text{-}ThO_2$ HTGR Lattice," BNWL-1561, May 1971.

10. Oakes, T. J., "Measurement of the Neutron Multiplication Factor as a Function of Temperature for a $U\text{-}235C_2\text{-}ThO_2$ Carbon Lattice," BNWL-SA-3621 (1971).

11. Oakes, T. J., "Measurement of K_∞ As a Function of Temperature for a $U\text{-}233O_2$-ThO_2-Carbon Lattice," USAEC Report BNWL-1601, February 1972.

CHAPTER 14

EVALUATION OF MOLTEN SALT BREEDER REACTORS

The breeding reactions of the thorium cycle are:

$$^{232}Th + n \longrightarrow {}^{233}Th \xrightarrow[22\ min.]{\beta} {}^{233}Pa \xrightarrow[27.4d]{\beta} {}^{233}U$$

Because of the number of neutrons produced per neutron absorbed and the small fast fission bonus associated with U-233 and Th-232 in the thermal spectrum, a breeding ratio only slightly greater than unity is achievable. In order to realize breeding with the thorium cycle it is necessary to remove the bred Pa-233 and the various nuclear poisons produced by the fission process from the high flux region as quickly as possible. The Molten Salt Breeder Reactor concept permits rapid removal of Pa-233 and the nuclear poisons (e.g. Xe-135 and the rare earth elements). The reactor is a fluid fueled system containing UF_4 and ThF_4 dissolved in LiF - BeF_2. The molten fuel salt flows through a graphite moderator where the nuclear reactions take place. A side stream is continuously processed to remove the Pa and rare earth elements, thereby permitting the achievement of a calculated breeding ratio of about 1.06.

The MSBR is attractive because of the following:

Use of a fluid fuel and on-site processing would eliminate the problems of solid fuel fabrication and the handling, and shipping and reprocessing of spent fuel elements which are associated with all other reactor types under active consideration.

MSBR operation on the thorium-uranium fuel cycle would help conserve uranium and thorium resources by utilizing thorium reserves with high efficiency.

The MSBR is projected to have attractive fuel cycle costs. The major uncertainty in the fuel cycle cost is associated with the continuous fuel processing plant which has not been developed.

The safety issues associated with the MSBR are generally different from those of solid fuel reactors. Thus, there might be safety advantages for the MSBR when considering major accidents. An accurate assessment of MSBR safety is not possible today because of the early state of development.

Like other advanced reactor systems such as the LMFBR and HTGR, the MSBR would employ modern steam technology for power generation with high thermal efficiencies. This would reduce the amount of waste heat to be discharged to the environment.

Selected conceptual design data for a large MSBR, based primarily on design studies performed at ORNL, are given in Table 14-1. There are, however, problem areas associated with the MSBR which must be overcome before the potential of the concept could be attained. These include development of continuous fuel processing, reactor and processing structural materials, tritium control methods, reactor equipment and systems, maintenance techniques, safety technology, and MSBR codes and standards. Each of these problem areas will now be evaluated in some detail, using as a reference point the technology which was demonstrated by the Molten Salt Reactor Experiment (MSRE) during its design, construction and operation at Oak Ridge and the conceptual design parameters presented in Table 14-1. A conceptual flowsheet for this system is shown in Fig. 14-1.

Table 14-1

Selected Conceptual Design Data for a Large MSBR

Net Electrical Power, MW(e)	1000
Reactor Thermal Power, MW(th)	2240
Steam System	3500 psia, 1000°F, 44% net efficiency
Fuel Salt	72% ^7LiF, 16% BeF_2, 12% ThF_4, 0.3% UF_4
Primary Piping and Vessel Material	Hastelloy N
Moderator	Sealed Unclad Graphite
Breeding Ratio	1.06
Specific Fissile Fuel Inventory, Kg/MW(e)	1.5
Compounded Doubling Time, Years	22
Core Temperatures, °F	1050 inlet, 1300 outlet

STATUS OF MSBR TECHNOLOGY

MSRE - The Reference Point for Current Technology

The Molten Salt Reactor Experiment (MSRE) was begun in 1960 at ORNL as part of the Civilian Nuclear Power Program. The purpose of the experiment was to demonstrate the basic feasibility of molten salt power reactors. All objectives of the experiment

Fig. 14-1

SINGLE-FLUID, TWO-REGION MOLTEN SALT BREEDER REACTOR

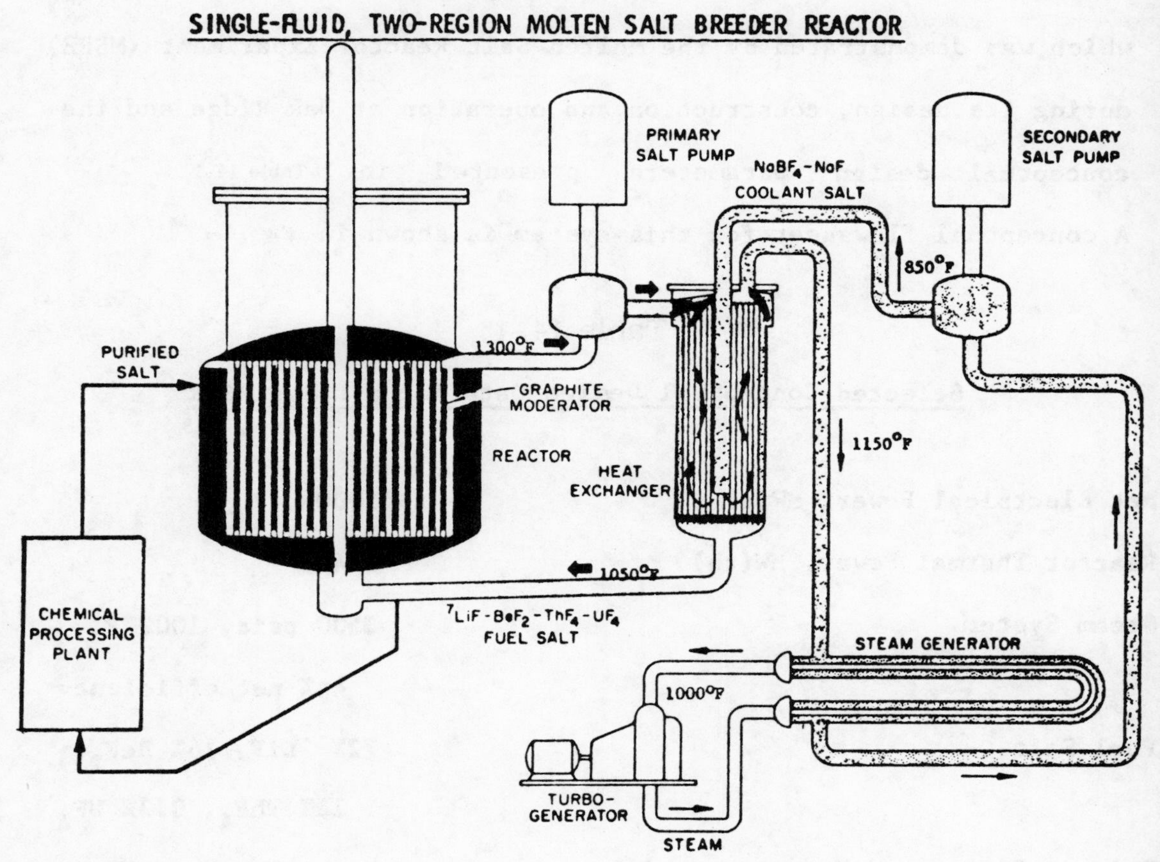

were achieved during its successful operation from June 1965 to December 1969. These included the distinction of becoming the first reactor in the world to operate solely on U-233. Some of the more significant dates and statistics pertinent to the MSRE are given in Table 14-2.

In spite of the success of the MSRE, there are many areas of molten salt technology which must be expanded and developed in order to proceed from this small non-breeding experiment to a safe, reliable, and economic 1000 MW(e) MSBR with a 30-year life. To illustrate this point, some of the most important differences in basic design and performance characteristics between the MSRE and a conceptual

Table 14-2

Important Dates and Statistics for the MSRE

Dates:

Design initiated	July 1960
Critical with ^{235}U Fuel	June 1, 1965
Operation at full power - 8 MW(th)	May 23, 1966
Complete 6-month run	March 20, 1968
End Operation with ^{235}U fuel	March 26, 1968
Critical with ^{233}U fuel	October 2, 1968
Operation at full power with ^{233}U fuel	January 28, 1969
Reactor operation terminated	December 12, 1969

Statistics:

Hours critical	17,655
Fuel loop time circulating salt (Hrs).	21,788
Equiv. full power hours with ^{235}U fuel	9,005
Equiv. full power hours with ^{233}U fuel	4,167

1000 MW(e) MSBR are given in Table 14-3. Scale-up would logically be accomplished through development of reactor plants of increasing size. Examination of Table 14-3 provides an appreciation of the scale-up requirements in going from the MSRE to a large MSBR. Some problems associated with progressing from a small experiment to a commercial, high performance power plant are not adequately represented by the comparison presented in the Table. Therefore, it is useful to examine additional facets of MSBR technology in more detail.

Table 14-3

Comparison of Selected Parameters of the MSRE and 1000 MW(e) MSBR [1/]

	MSRE	MSBR
General		
Thermal Power, MW(th)	8	2250
Electric Power, MW(e)	0	1000
Plant lifetime, years	4	30
Fuel Processing Scheme	Off-line, batch processing	On-line, continuous processing
Breeding Ratio	Less than 1.0 (No Th present)	1.06
Reactor		
Fuel Salt	$^7LiF-BeF_2-ZrF_4-UF_4$	$^7LiF-BeF_2-ThF_4-UF_4$
Moderator	Unclad, unsealed graphite	Unclad, sealed graphite
Reactor Vessel Material	Standard Hastelloy-N	Modified Hastelloy
Power Density, KW/liter	2.7	22
Exit Temperature, °F	1210	1300
Temperature Rise Across Core, °F	40	250
Reactor Vessel Height, Ft.	8	20
Reactor Vessel Diameter, Ft.	5	22
Vessel Design Pressure, psia	65	75
Peak Thermal Neutron Flux, Neutrons/cm^2-sec	6×10^{13}	8.3×10^{14}
Other Components and Systems Data		
Number of Primary Circuits	1	4
Fuel Salt Pump Flow, gpm	1200	16,000
Fuel Salt Pump Head, ft.	48.5	150
Intermediate Heat Exchanger Capacity, MW(th)	8	556
Secondary Coolant Salt	$^7LiF-BeF_2$	$NaF-NaBF_4$
Number of Secondary Circuits	1	4
Secondary Salt Pump Flow, gpm	850	20,000
Secondary Salt Pump Head, ft.	78	300
Number of Steam Generators	0	16
Steam Generator Capacity, MW(th)	0	121

Continuous Fuel Processing - The Key to Breeding

In order to achieve nuclear breeding in the single fluid MSBR it is necessary to have an on-line continuous fuel processing system. This would accomplish the following:

- Isolate protactinium-233 from the reactor environment so it can decay into the fissile fuel isotope uranium-233 before being transmuted into other isotopes by neutron irradiation.
- Remove undesirable neutron poisons from the fuel salt and thus improve the neutron economy and breeding performance of the system.
- Control the fuel chemistry and remove excess uranium-233 which is to be exported from the breeder system.

Chemical Process Development

The Oak Ridge National Laboratory has proposed a fuel processing scheme to accomplish breeding in the MSBR, and the flowsheet processes involve:

a. Fluorination of the fuel salt to remove uranium as UF_6.
b. Reductive extraction of protactinium by contacting the salt with a mixture of lithium and bismuth.
c. Metal transfer processing to preferentially remove the rare earth fission product poisons which would otherwise hinder breeding performance.

The fuel processing system shown in Fig. 14-2 is in an early stage of development at present and this type of system has not been demonstrated on an operating reactor. By comparison, the MSRE required only off-line, batch fluorination to recover uranium from fuel salt.

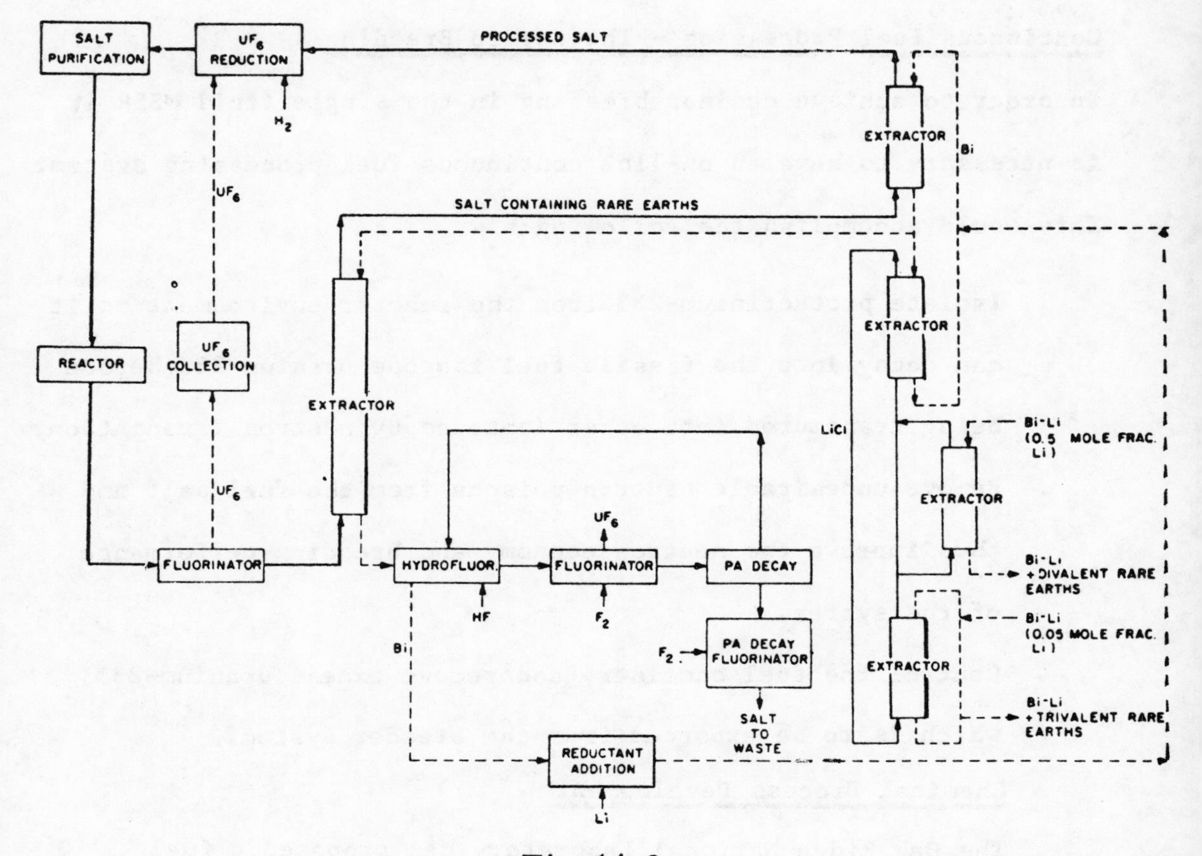

Fig. 14-2

FLOWSHEET FOR PROCESSING A SINGLE-FLUID MSBR BY FLUORINATION-REDUCTIVE EXTRACTION AND THE METAL-TRANSFER PROCESS.

At this time, the basic chemistry involved in the MSBR processing scheme has been demonstrated in laboratory scale experiments. Current efforts at Oak Ridge are being directed toward development of subsystems incorporating many of the required processing steps. Ultimately a complete breeder processing experiment would be required to demonstrate the system with all the chemical conditions and operational requirements which would be encountered with any MSBR. Not shown on the flowsheet is a separate processing system which would require injecting helium bubbles into the fuel salt, allowing them to circulate in the reactor system until they collect fission product xenon, and then removing the bubbles and xenon from the reactor system. Xenon is a highly undesirable neutron poison which will hamper breeding performance by capturing neutrons which would otherwise breed new fuel. This concept for xenon stripping was demonstrated in principle by the MSRE, although more efficient and controllable stripping systems will be desirable for the MSBR. The xenon poisoning in the MSRE was reduced by a factor of six by xenon stripping; the goal for the MSBR is a factor of ten reduction.

Fuel Processing Structural Materials

Aside from the chemical processes themselves, there are also development requirements associated with containment materials for the fuel processing systems. In particular, liquid bismuth presents difficult compatibility problems with most structural metals, and present efforts are concentrated on using molybdenum and graphite for containing bismuth. Unfortunately, both molybdenum and graphite are difficult to use for such engineering applications. Thus, it will be necessary to develop improved techniques for fabrication and joining before their use is possible in the reprocessing system.

A second materials problem of the current fuel processing system is the containment for the fluorination step in which uranium is volatilized from the fuel salt. The fluorine and fluoride salt mixture is corrosive to most structural materials, including graphite, and present ORNL flowsheets show a "frozen wall" fluorinator which operates with a protective layer of frozen fuel salt covering a Hastelloy-N vessel wall. This component would require considerable engineering development before it is truly practical for use in on-line full processing systems.

Molten Salt Reactor Design - Materials Requirements

In concept, the molten salt reactor core is a comparatively uncomplicated type of heat source. The MSRE reactor core, for example, consisted of a prismatic structure of unclad graphite moderator through which fuel salt flowed to be heated by the self-sustaining chain reaction which took place as long as the salt was in the graphite. The entire reactor internals and fuel salt were contained in vessels and piping made of Hastelloy-N, a high strength nickel base alloy which was developed under the Aircraft Nuclear Propulsion Program. Over the four-year lifetime of the MSRE, the reactor structural materials performed satisfactorily for the purposes of the experiments although operation of the MSRE revealed possible problems with long term use of Hastelloy-N in contact with fuel salts containing fission products.

The MSBR application is more demanding in many respects than the MSRE, and additional development work would be required in several areas of materials technology before suitable materials could become available.

Fuel and Coolant Salts

The MSRE fuel salt was a mixture of $^{7}LiF-BeF-ZrF_4-UF_4$ in proportions of 65.0-29.1-5.0-0.9 mole %, respectively. Zirconium fluoride was included as protection against UO_2 precipitation should inadvertent oxide contamination of the system occur. MSRE operation indicated that control of oxides was not a major problem and thus it is not considered necessary to include zirconium in future molten salt reactor fuels. It should also be noted that the MSRE fuel contained no thorium whereas the proposed MSBR fuels would include thorium as the fertile material for breeding. With the possible exception of incompatibilities with Hastelloy-N, the MSRE fuel salt performed satisfactorily throughout the life of the reactor.

The MSBR fuel salt, as currently proposed by ORNL, would be a mixture of $^7LiF-BeF_2-ThF_4-UF_4$ in proportions of 71.7-16-12-0.3 mole %, respectively. This salt has a melting point of about 930°F and a vapor pressure of less then 0.1 mm Hg at the mean operating temperature of 1150°F. It also has about 3.3 times the density and 10 times the viscosity of water. Its thermal conductivity and volumetric heat capacity are comparable to water.

The high melting temperature is an obvious limitation for a system using this salt, and the MSBR is limited to high temperature operation. In addition, the lithium component must be enriched in Li-7 in order to allow nuclear breeding, since naturally occurring lithium contains about 7.5% Li-6. Li-6 is undesirable in the MSBR because of its tendency to capture neutrons, thus penalizing breeding performance.

The chemical and physical characteristics of the proposed MSBR fuel mixture have been and are being investigated, and they are reasonably well known for unirradiated salts. The major unknowns are associated with the reactor fuel after it has been irradiated. For example, not enough is known about the behavior of fission products. The ability to predict fission product behavior is important to plant safety, operation, and maintenance. While the MSRE provided much useful information, there is still a need for more information, particularly with regard to the fate of the so-called "noble metal" fission products such as molybdenum, niobium and others which are generated in substantial quantities and whose behavior in the system is not well understood.

A more complete understanding of the physical/chemical characteristics of the irradiated fuel salt is also needed. As an illustration of this point, anomalous power pulses were observed during early operation of the MSRE with U-233 fuel which were attributed to unusual behavior of helium gas bubbles as they circulated through the reactor. This behavior is believed to have been due to some physical and/or chemical characteristics of the fuel salt which were never fully understood. Out-of-reactor work on molten fuel salt fission product chemistry is currently under way. Eventually, the behavior of the fuel salt would need to be confirmed in an operating reactor.

The coolant salt in the secondary system of the MSRE was of molar composition 66% ^{7}LiF-34%BeF_2. While this coolant performed satisfactorily (no detectable corrosion or reaction could be observed in the secondary system), the salt has a high melting temperature (850°F) and is relatively expensive. Thus, it may not be the appropriate choice for power reactors for two reasons: (1) larger volumes of coolant salt will be used to generate steam in the MSBR, and (2) salt temperatures in the steam generator should be low enough, if possible, to utilize conventional steam system technology with feedwater temperatures up to about 550°F. The operation of MSRE was less affected by the coolant salt melting temperature since it dumped the 8 MW(th) of heat via an air-cooled radiator. The high melting temperatures of potential coolant salts remain a problem. The current choice is a eutectic mixture of sodium fluoride and sodium fluoroborate with a molar composition of 8% NaF-92% $NaBF_4$; this salt melts at 725°F. It is comparatively inexpensive and has satisfactory heat transfer properties.

However, the effects of heat exchanger leaks between the coolant and fuel salts, and between the coolant salt and steam systems, must be shown to be tolerable. The fluoroborate salt is currently being studied with respect to both its chemistry and compatibility with Hastelloy-N.

Reactor Fuel Containment Materials

A prerequisite to success for the MSBR would be the ability to assure reliable and safe containment and handling of molten fuel salts at all times during the life of the reactor. It would be necessary, therefore, to develop suitable containment materials for MSBR application before plants could be constructed.

A serious question concerning compatability of Hastelloy-N with the constituents of irradiated fuel salt was raised by the post-operation examination of the MSRE in 1971. Although the MSRE materials performed satisfactorily for that system during its operation, subsequent examination of metal which was exposed to MSRE fuel salt revealed that the alloy had experienced intergranular attack to depths of about 0.007 inch. The attack was not obvious until metal specimens were tensile tested, at which time cracks opened up as the metal was strained. Further examination revealed that several fission products, including tellurium, had penetrated the metal to depths comparable to those of the cracks. At the present time, it is thought that the intergranular attack was due to the presence of tellurium. Subsequent laboratory tests have verified that tellurium can produce, under certain conditions, intergranular cracking in Hastelloy N.

Although the limited penetration of cracks presented no problems for the MSRE, concern now exists with respect to the chemical compatability of Hastelloy-N and MSBR fuel salts when subjected to the more stringent MSBR requirements of higher power density and 30-year life. If the observed intergranular attack was indeed due to fission product attack of the Hastelloy-N, then this material may not be suitable for either the piping or the vessels which would be exposed to much higher fission product concentrations for longer periods of time. Efforts are under way to understand and explain the cracking problem, and to determine whether alternate reactor containment materials should be actively considered.

In addition to the intergranular corrosion problem, the standard Hastelloy-N used in the MSRE is not suitable for use in the MSBR because its mechanical properties deteriorate to an unacceptable level when subjected to the higher neutron doses which would occur in the higher power density, longer-life MSBR. The problem is thought to be due mainly to impurities in the metal which are transmuted to helium when exposed to thermal neutrons. The helium is believed to cause a deterioration of mechanical properties by its presence at grain boundaries within the alloy. It would be necessary to develop a modified Hastelloy-N with improved irradiation resistance for the MSBR, and some progress is being made in that direction. It appears at this time that small additions of certain elements, such as titanium, improve the irradiation performance of Hastelloy-N substantially. Development work on modified alloys with improved irradiation resistance is currently under way.

Graphite

Additional developmental effort on two problems is required to produce graphites suitable for MSBR application. The first is associated with irradiation damage to graphite structures which results from fast neutrons. Under high neutron doses, of the order of 10^{22} neutrons/cm^2, most graphites tend to become dimensionally unstable and gross swelling of the material occurs. Based on tests of small graphite samples at ORNL, the best commercially available graphites at this time may be usable to about 3×10^{22} neutrons/cm^2, before the core graphite would have to be replaced. This corresponds to roughly a four-year graphite lifetime for the ORNL reference design. While this might be acceptable, there are still uncertainties about the fabrication and performance of large graphite pieces, and additional work would be required before a four-year life could be assured at the higher MSBR power densities now being considered. In any event, there would be an obvious economic incentive to develop longer lived graphites for MSBR application since a four-year life for graphite is estimated to represent a fuel cycle cost penalty of about 0.2 mills/kw-hr relative to a system with thirty year graphite life.

The second major problem associated with graphites for MSBR application is the development of a sealing technique which will keep xenon, an undesirable neutron poison, from diffusing into the core graphite where it can capture neutrons to the detriment of breeding performance. While graphite sealing may not be necessary to achieve nuclear breeding in the MSBR, the use of sealed graphite would certainly enhance breeding performance. The economic incentives or penalties of graphite sealing cannot be assessed until a suitable sealing process is developed.

Sealing methods which have been investigated to date include pyrolytic carbon coating and carbon impregnation. Thus far, however, no sealed graphite that has been tested remained sufficiently impermeable to gas at MSBR design irradiation doses, and research and development in this area is continuing.

Other Structural Materials

In addition to the structural materials requirements for the reactor and fuel processing systems proper, there are other components and systems which have special materials requirements. Such components as the primary heat exchangers and steam generators must function while in contact with two different working fluids.

At the present time, Hastelloy-N is considered to be the most promising material for use in all salt containment systems, including the secondary piping and components. Research to date indicates that sodium fluoroborate and Hastelloy-N are compatible as long as the water content of the fluoroborate is kept low; otherwise, accelerated corrosion can occur. Additional testing would be needed and is underway.

Hastelloy-N has not been adequately evaluated for service under a range of steam conditions and whether it will be a suitable material for use in steam generators is still not known.

Tritium - A Problem of Control

Because of the lithium present in fluoride fuel salts, the present MSBR concept has the inherent problem of generating tritium, a radioactive isotope of hydrogen. Tritium is produced by the following reactions:

$$(1) \quad {}^{6}Li\ (n, \alpha)\ {}^{3}H.$$

$$(2) \quad {}^{7}Li\ (n, \alpha n)\ {}^{3}H.$$

Due primarily to these interactions, tritium would be produced at a rate of about 2400 curies/day in a 1000 MWe MSBR. This compares with about 40 to 50 curies/day for light water, gas-cooled, and fast breeder reactors, in which tritium is produced primarily as a low yield fission product. Tritium production in heavy water reactors of comparable size is generally in the range 3500 to 5800 curies/day, due to neutron interactions with the deuterium present in heavy water.

To further compound the problem tritium diffuses readily through Hastelloy-N at elevated temperatures. As a result, it may be difficult to prevent tritium from diffusing through the piping and components of the MSBR system (such as heat exchangers) and eventually reaching the steam system where it might be discharged to the environment as tritiated water.

The problem of tritium control in the MSBR is being studied in detail at ORNL. The following are being considered as potential methods for tritium control:

1. Exchanging the tritium for any hydrogen present in the secondary coolant, thereby retaining the tritium in the secondary coolant.
2. Using coatings on metal surfaces in order to inhibit tritium diffusion.
3. Operating the reactor with the salt more oxidizing, thereby causing the formation of tritium fluoride which could be removed in the off-gas systems.
4. Using a different secondary coolant, e.g., sodium or helium, and processing this coolant to remove tritium.
5. Using another intermediate loop between the fluoroborate and steam to "getter" tritium.
6. Using duplex tubing in either the heat exchanger or steam generator with a purge gas between the walls.

Of these potential solutions, the use of an additional intermediate loop between the secondary and steam systems is considered the most effective method technically, but it would also be expensive due to the additional equipment required and the loss of thermal efficiency. From an economic viewpoint, the most desirable solution is one which does not significantly complicate the system, such as exchange of tritium for hydrogen present in the secondary coolant. This technique is being investigated as part of the ORNL efforts on tritium chemistry. The tritium retention problem may be eased by the low permeability of oxide coatings which occur on steam generator materials in contact with steam, and this is also being investigated at ORNL.

Reactor Equipment and Systems Development

While the MSBR would utilize some existing engineering technology from other reactor types, there are specific components and systems for which additional development work is required. Such work would have to take into account the induced activity that those components would accumulate in the MSBR system, i.e., special handling and maintenance equipment would also need to be developed. The previous discussion has already dealt with a number of these, such as fuel processing components and systems, but additional discussion is appropriate.

Components

As indicated in the Table 14-3, a number of components must be scaled up substantially from the MSRE sizes before a large MSBR is possible. The development of these larger components along with their special handling and maintenance equipment is probably one of the most difficult and costly phases of MSBR development. However, reliable, safe, and maintainable components would need to be developed in order for any reactor system to be a success.

The MSBR pumps would likely be similar in basic design to those for the MSRE, namely, vertical shaft, overhung impeller pumps. Substantial experience has been gained over the years in the design, fabrication and operation of smaller salt pumps, but the size would have to be increased substantially for MSBR application. The development and proof testing of such units along with their handling and maintainence equipment and test facilities are expected to be costly and time consuming.

The intermediate heat exchangers for the MSBR must perform with a minimum of salt inventory in order to improve the breeding performance by lowering the fuel inventory. Special surfaces to enhance heat transfer would help achieve this, and more studies would be in order. Based on previous experience with other reactor systems, it is believed that these units would require a difficult development and proof testing effort.

The steam generator for MSBR applications is probably the most difficult large component to develop since it represents an item for which there has been almost no experience to date. It is believed that a difficult development and proof testing program would be needed to provide reliable and maintainable units. As discussed previously, the high melting temperatures of candidate secondary coolants, such as sodium fluoroborate, present problems of matching with conventional steam system technology. At this time, central station power plants utilize feedwater temperatures only up to about 550°F. Therefore, coupling a conventional feedwater system to a secondary coolant which freezes at 725°F presents obvious problems in design and control. It might be necessary to provide modifications to conventional steam system designs to help resolve the problems. Because of these factors, a study related to the design of steam generators has been initiated at Foster Wheeler Corporation.

Control rods and drives for the MSBR would also need to be developed. The MSRE control rods were air cooled and operated inside Hastelloy-N thimbles which protruded down into the fuel salt. The MSBR would require more efficient cooling due to the higher power densities involved. Presumably rods and drives would be needed which permit the rods to contact and be cooled by the fuel salt.

The salt valves for large MSBR's represent another development problem, although the freeze valve concept which was employed successfully in the MSRE could likely be scaled up in size and utilized for many MSBR applications. Mechanical throttling valves would also be needed for the MSBR salt systems, even though no throttling valve was used with the MSRE. Mechanical shutoff valves for salt systems, if required, would have to be developed.

Other components which would require considerable engineering development and testing include the helium bubble generators and gas strippers which are proposed for use in removing the fission product xenon from the fuel salt. Research and development in this area is currently under way as part of the technology program at ORNL.

Systems

The integration of all required components into a complete MSBR central station power plant would involve a number of systems for which development work is still required. It should be noted that some components, such as pumps and control rod drives, would require their own individual systems for functions such as cooling and lubrication.

Given the required components and materials of construction, the basic reactor primary and secondary flow systems can be designed. However, the primary flow system would require supporting systems for continuous fuel processing, on-line fuel analysis and control of salt chemistry, reactor control and safety, handling of radioactive gases, fuel draining from every possible holdup area in components and equipment, afterheat control, and temperature control during non-nuclear operations.

The continuous fuel processing systems proposed to date are quite complicated and include a number of subsystems, all of which would have to operate satisfactorily within the constraints of economics, safety, and reliability. The effects of off-design conditions on these systems would have to be understood so that control would be possible to prevent inadvertent contamination of the primary system by undesirable materials.

The fuel drain system is important to both operation and safety since it would be used to contain the molten fuel whenever a need arises to drain the primary system or any component or instrument for maintainence or inspection. Thus, additional systems would be required, each with its own system for maintaining and controlling temperatures. The fuel salt drain tank would have to be equipped with an auxiliary cooling system capable of rejecting about 18 MW(th) of heat should the need arise to drain the salt immediately following nuclear operation. The secondary coolant system would also require subsystems for draining and controlling of salt chemistry and temperature. In addition, the secondary loop might require systems to control tritium and to handle the consequences of steam generator or heat exchanger leaks.

The steam system for the MSBR might require a departure from conventional designs due to the unique problems associated with using a coolant having a high melting temperature. Precautions would have to be taken against freezing the secondary salt as it travels through the steam generator; suitable methods for system startup and control would need to be incorporated. ORNL has proposed the use of a supercritical steam system which operates at 3500 psia and provides 700°F feedwater by mixing of supercritical steam and high pressure feedwater. This system would introduce major new development requirements because it differs from conventional steam cycles.

Maintenance - A Difficult Problem for the MSBR

Unlike solid fueled reactors in which the primary system contains activation products and only those fission products which may leak from defective fuel pins, the MSBR would have the bulk of the fission products dispersed throughout the reactor system. Because of this dispersal of radioactivity, remote techniques would be required for many maintenance functions if the reactor were to have an acceptable plant availability in the utility environment.

The MSRE was designed for remote maintenance of highly radioactive components; however, no major maintenance problems (removal or repair of large components) were encountered after nuclear operation was initiated. Thus, the degree to which the MSRE experience on maintenance is applicable to large commercial breeder reactors is open to question.

As has been evident in plant layout work on nuclear facilities to date, this requirement for remote maintenance will significantly affect the ultimate design and performance of the plant system. The MSBR would require remote techniques and tools for inspection, welding and cutting of pipes, mechanical assembly and disassembly of components and systems, and removing, transporting and handling large component items after they become highly radioactive. The removal and replacement of core internals, such as graphite, might pose difficult maintenance problems because of the high radiation levels involved and the contamination protection which would be required whenever the primary system is opened.

Another potential problem is the afterheat generation by fission products which deposit in components such as the primary heat exchangers. Auxiliary cooling might be required to prevent damage when the fuel salt drained from the primary system, and a requirement for such cooling would further complicate inspection and maintenance operations.

In some cases, the inspection and maintenance problems of the MSBR could be solved using present technology and particularly experience gained fro[m] fuel reprocessing plants. However, additional technology development would be required in other areas, such as remote cutting, alignment, cleaning and welding of metal members. Depending to some degree on the particular plant arrangement, other special tools and equipment would also have to be designed and developed to accomplish inspection and maintenance operations.

In the final analysis, the development of adequate inspection and maintenance techniques and procedures,and hardware for the MSBR hinges on the success of other facets of the program, such as materials and component development, and on the requirement that adequate care be taken during plant design to assure that all systems and components which would require maintenance over the life of the plant are indeed maintainable within the constraints of utility operation.

Safety - Different Issues for the MSBR

The MSBR concept has certain characteristics which might provide advantages relating to safety, particularly with respect to postulated major types of accidents currently considered in licensing activities. Since the fuel would be in a molten form, consideration of the core meltdown accident is not applicable to the MSBR. Also, in the event of a fuel spill, secondary criticality is not a problem since this is a thermal reactor system requiring moderator for nuclear criticality. Other safety features include the fact that the primary system would operate at low pressure with fuel salt that is more than 1000°F below its boiling point, that fission product iodine and strontium form stable compounds in the fluoride salts, and that the salts do not react rapidly with air or water. Because of the continuous fuel processing, the need for excess reactivity would be decreased and some of the fission products would be continuously removed from the primary system. A prompt negative temperature coefficient of reactivity is also a characteristic of the fuel salt.

Safety disadvantages, on the other hand, include the very high radioactive contamination which would be present throughout the primary system, fuel processing plant, and all auxiliary primary systems such as the fuel drain and off-gas systems. Thus, containment of these systems would have to be assured. Also, removal of decay heat from fuel storage systems would have to be provided by always ready and reliable cooling systems, particularly for the fuel drain tank and the Pa-233 decay tank in the reprocessing plant where megawatt quantities of decay heat must be removed. The tritium problem, already discussed, would have to be controlled to assure safety.

Based on the present state of MSBR technology, it is not possible to provide a complete assessment of MSBR safety relative to other reactors. It can be stated, however, that the safety issues for the MSBR are generally different from those for solid fuel reactors, and that more detailed design work must be done before the safety advantages and disadvantages of the MSBR could be fully evaluated.

Codes, Standards, and High Temperature Design Methods

Codes and standards for MSBR equipment and systems must be developed in conjunction with other research and development before large MSBR's can be built. In particular, the materials of construction which are currently being developed and tested would have to be certified for use in nuclear power plant applications.

The need for high temperature design technology is a problem for the MSBR as well as for other high temperature systems. The AEC currently has under way a program in support of the LMFBR which is providing materials data and structural analysis methods for design of systems employing various steel alloys at temperatures up to 1200°F. This program would need to be broadened to include MSBR structural materials such as Hastelloy-N and to include temperatures as high as 1400°F to provide the design technology applicable to high-temperature, long-term operating conditions which would be expected for MSBR vessels, components, and core structures.

CONCLUSIONS

The Molten Salt Breeder Reactor, if successfully developed and marketed, could provide a useful supplement to the currently developing uranium-plutonium reactor economy. This concept offers the potential for:

- Breeding in a thermal spectrum reactor;
- Efficient use of thorium as a fertile material;
- Elimination of fuel fabrication and spent fuel shipping;
- High thermal efficiencies.

Notwithstanding these attractive features, this assessment has reconfirmed the existence of major technological and engineering problems affecting feasibility of the concept as a reliable and economic breeder for the utility industry. The principal concerns include uncertainties with materials, with methods of controlling tritium, and with the design of components and systems along with their special handling, inspection and maintenance equipment. Many of these problems are compounded by the use of a fluid fuel in which fission products and delayed neutrons are distributed throughout the primary reactor and reprocessing systems.

The resolution of the problems of the MSBR will require the conduct of an intensive research and development program. Included among the major efforts that would have to be accomplished are:

- Proof testing of an integrated reprocessing system;
- Development of a suitable containment material;
- Development of a satisfactory method for the control and retention of tritium;

- Attainment of a thorough understanding of the behavior of fission products in a molten salt system;
- Development of long life moderator graphite, suitable for breeder application;
- Conceptual definition of the engineering features of the many components and systems;
- Development of adequate methods and equipment for remote inspection, handling, and maintenance of the plant.

The major problems associated with the MSBR are rather difficult in natur and many are unique to this concept. Continuing support of the research and development effort will be required to obtain satisfactory solutions to the problems. When significant evidence is available that demonstrate realistic solutions are practical, a further assessment could then be mad as to the advisability of advancing into the detailed design and engineering phase of the development process including that of industrial involvement. Proceeding with this next step would also be contingent upon obtaining a firm demonstration of interest and commitment to the concept by the power industry and the utilities and reasonable assurances that large scale government and industrial resources can be made availabl on a continuing basis to this program in light of other commitments to the commercial nuclear power program and higher priority energy development efforts.

REFERENCES

1. U. S. Atomic Energy Commission, "The 1967 Supplement to the 1962 Report to the President on Civilian Nuclear Power" USAEC Report, February 1967.

2. U. S. Atomic Energy Commission, "The Use of Thorium in Nuclear Power Reactors" USAEC Report WASH-1097, 1969.

3. U. S. Atomic Energy Commission, "Potential Nuclear Power Growth Patterns," USAEC Report WASH-1098, December 1970.

4. U. S. Atomic Energy Commission, "Cost-Benefit Analysis of the U. S. Breeder Reactor Program," USAEC Report WASH-1126, 1969.

5. U. S. Atomic Energy Commission, "Updated (1970) Cost-Benefit Analysis of the U. S. Breeder Reactor Program," USAEC Report WASH-1184, January 1972.

6. Edison Electric Institute, "Report on the EEI Reactor Assessment Panel," EEI Publication No. 70-30, 1970.

7. Annual Hearings on Reactor Development Program, U. S. Atomic Energy Commission FY 1972 Authorizing Legislation, Hearings before the Joint Committee on Atomic Energy, Congress of the United States p. 820-830, U. S. Government Printing Office

8. <u>Nuclear Applications & Technology</u>, Volume 8, February 1970.

9. Robertson, R. D. (ed), "Conceptual Design Study of a Single Fluid Molten Salt Breeder Reactor," ORNL-4541, June 1971.

10. Rosenthal, M. W., et al.; "Advances in the Development of Molten-Salt Breeder Reactors," A/CONF. 49/P-048, Fourth United Nations International Conference on the Peaceful Uses of Atomic Energy, Geneva, September 6-16, 1971.

11. Trinko, J. R. (ed.), "Molten-Salt Reactor Technology," Technical Report of the Molten-Salt Group, Part I, December 1971.

12. Trinko, J. R. (ed), "Evaluation of a 1000 MWe Molten-Salt Breeder Reactor," Technical Report of the Molten Salt Group, Part II, November 1971.

13. Ebasco Services Inc., "1000 MWe Molten Salt Breeder Reactor Conceptual Design Study," Final Report Task I, Prepared under ORNL subcontract 3560, February 1972.

14. "Project for Investigation of Molten Salt Breeder Reactor," Final Report, Phase I Study for Molten Salt Breeder Reactor Associates, September 1970.

15. Cardwell, D. W. and Haubenreich, P. N., "Indexed Abstracts of Selected References on Molten-Salt Reactor Technology," ORNL-TM-3595, December 1971.

16. Kasten, P. R., Bettis, E. S. and Robertson, R. C., "Design Studies of 1000 MW(e) Molten-Salt Breeder Reactors," ORNL-3996, August 1966.

17. Molten Salt Reactor Program Semiannual Reports beginning in February 1962.

CHAPTER 15

STEAM COOLED REACTORS

The coolants customarily used in atomic reactors are water under pressure or boiling water, liquid metals, and gases. It is possible, however, to use also superheated steam as a coolant for atomic reactor. Steam-cooled reactors differ from boiling-water reactors with steam superheat in that there is no boiling of the coolant in them.

Even in the combined system of a boiling reactor (BWR) connected in tandem with a reactor for separate steam superheat (SSR) [1 - 3], the second reactor is in essence steam-cooled, but does not constitute in this case an independent power installation.

As an independent power installation one can use a steam-cooled reactor with external evaporation in two variants:

with the steam coolant obtained at saturation temperature in a mixed-type boiler (Loeffler scheme);

with a surface-time steam generator.

The Loeffler scheme was previously used in steam generators using fossil fuel, and some of them are in operation to this day in a number of European countries. Such a steam generator was in operation until 1969 in the USSR at the Moscow thermal power plant TETs No 9. The Loeffler scheme for atomic electric stations is characterized by the fact that the reactor is cooled with superheated steam (Fig. 15-1) .

About 1/4 of the superheated steam from the reactor goes to the turbine, and the remaining 3/4 is diverted for intermediate superheating, after which it enters into a steam generator of the mixing type, where superheating the steam from the feedwater results in dry saturated steam that is fed with a compressor into the reactor for new superheating [4, 5]. The intermediate superheating of the steam in Loeffler's scheme is not obligatory, but the steam in this scheme must have a rather high working pressure (above 100 bar) [6], so as to obtain a higher cycle efficiency and to limit the compressor power.

The surface steam regulator makes it possible to realize a cycle with two pressures (Fig. 15-2) [5, 7].

The main advantages of the steam coolant for atomic electric stations are as follows:

steam technology is well known, this coolant is convenient when it comes to operate and repair the equipment, low cost, availability, safety, etc.;

absence of phase transitions and heat-exchange crisis in the reactor;

independence of the coolant temperature of pressure in a wide range of parameters, as is the case for any gas coolant;

economy of capital investment, owing to the use of the single-loop scheme and of well-known commercial equipment;

increased efficiency, owing to the use of a direct cycle and steam superheat.

Atomic Electric Station Plans With Steam-Cooled Thermal-Neutron Reactors

Plans for atomic electric stations with steam-cooled thermal-neutron reactors were developed in the U.S., England, West Germany, and other countrie

Table 15-1 lists data on planned steam-cooled thermal-neutron reactors for stationary and mobile installations, and also plans of reactors for separate superheat of steam in schemes with BWR.

All the considered steam-cooled thermal-neutron reactors were of the channel type.

In 1963 in Vallecitos (U.S.), an experimental steam-cooled thermal-neutron reactor EVESR [2], with water moderator, rated 17 MW (thermal), went critical. The reactor was intended for superheat of steam from an extraneous source, and prior to its shutdown in 1967 it was used for research on fuel-element irradiation, including those for fast steam-cooled reactors.

In view of the absence of phase transitions, one can classify as steam-cooled reactors also those with supercritical steam parameters. When the water temperature at the entrance is below critical, these can also be regarded as reactors with nuclear superheat [12]. Plans were developed abroad for atomic electric stations with supercritical parameters and water cooling (SPRR of the Babcock and Wilcox Company, with water moderator, and SCOTT-R of the Westinghouse Company [11] with graphite moderator); technical data on these are given in Table 15-1. Recently a group of Soviet workers published a plan of a uranium-graphite reactor of the type of the Beloyarsk Atomic Electric station, with supercritical parameters of the steam coolant of 240 kg/cm^2 pressure and superheat temperature 540°C [13]. In this design, provision is made for the use of a standard turbine and its equipment.

The relatively high temperature of the superheated steam leaving the reactor (approximately 500°C) calls for the use of stainless steel or its alloys for both the channels and the fuel element cladding. All this affects adversely the neutron balance in comparison with boiling-water reactors which have zirconium-clad fuel rods. In addition, the energy consumed in pumping the steam coolant with a compressor is relatively high. Therefore all other projects with the exception of EVESR were not realized, and main attention has been paid recently to the possibility of developing fast-neutron steam-cooled reactors.

Plans for Atomic Electric Stations with Steam-Cooled Fast-Neutron Reactors

From 1961 through 1969 there were developed in the U.S., West Germany, Belgium, and Sweden a number of plans for atomic electric stations with water-cooled fast breeder reactors (Table 15-2), mainly with power ratings 300 and

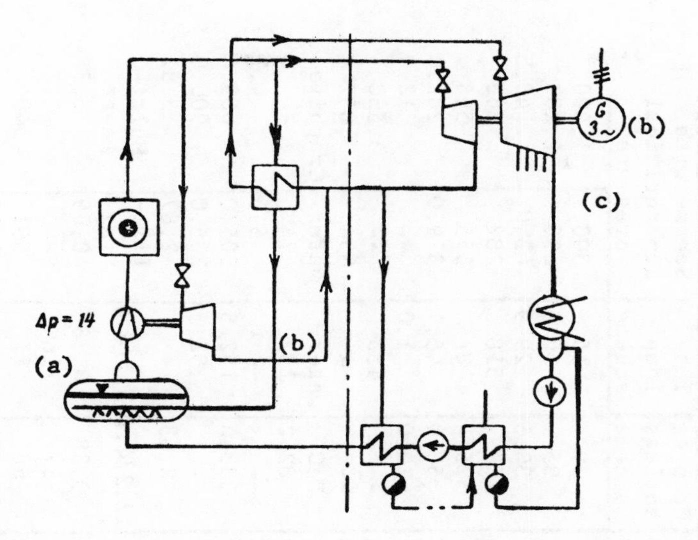

Fig. 15-1. Loeffler's scheme for a planned atomic electric station with steam-cooled fast AEG reactor (1968)

absolute pressure, kg/cm^2	temperature, °C
flow, t/hr	enthalpy, kcal/kg

Key: (a) kg/cm^2 (b) MW (c) net

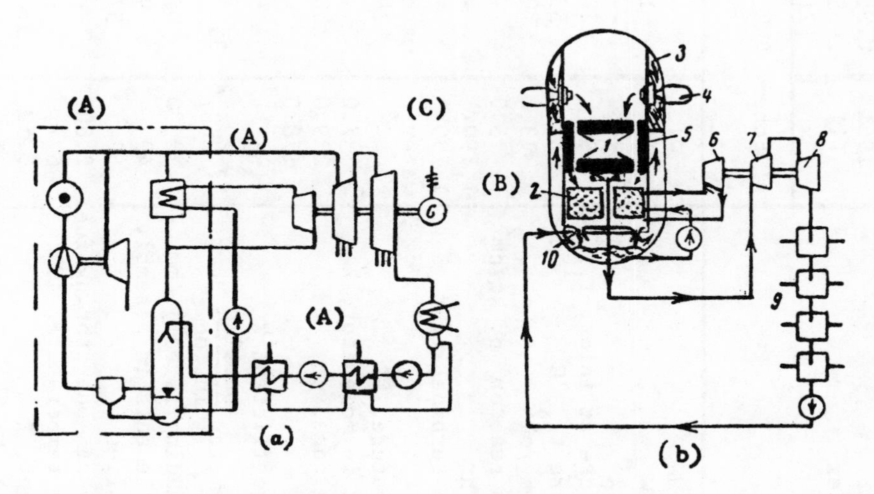

Fig. 15-2. Loeffler's scheme for atomic electric station with two-pressure cycle (second Babcock and Wilcox project -- SCBR II [5]
a - heat flow, b - arrangement of equipment. 1 - active zone, 2 - steam generator, 3 - moisture separator, 4 - compressor, 5 - conversion zone, 6 - high-pressure turbine, 7 - medium pressure turbine, 8 - low-pressure turbine, 9 - regenerative heat exchangers, 10 - injection-type evaporator.

Key: (A) kg/cm^2 (B) MW (C) net

Table 15-1

Planned atomic electric stations with steam-cooled thermal-neutron reactors

Name of index	Name of project, firm, country, and year of completion							
	SSR. GE. U.S. 1961 [1]	EVESR. GE U.S. 1963* [2]	SHR. GE. U.S. 1962 [3]	Deutsche Babcock, W. Germany 1958 [8]	England 1958 [9]	SCLMR. Mitchell England 1963 [10]	SPPR, BW U.S. 1961 [11]	SCOTT-R Westinghouse U.S. 1961 [11]
	Scheme BWR + SHR			Loeffler scheme	Compressorless scheme	Two-loop scheme	Scheme with supercritical pressure	
Electric power, MW	--	--	--	20	29	15	300	1,000
Thermal power, MW	214	17	453	78	95	50	698	2,297
Net efficiency, %	--	--	--	25.7	30.5	29.9	43.0	43.5
Coolant temperature at inlet, t_1, °C	290	285	290	250	269	318	282	263
Outlet temperature t_2, °C	482	550	510	400	525	482	565	565
Inlet pressure, kg/cm^2	73.8	70.0	74.0	40.0	50.0	45.7	319.0	262.0
Pressure loss in reactor p, kg/cm^2	5.3	--	5.5	1.0	20.0	1.0	--	12.0
Coolant flow, t/hr	1,190	--	2,260	648	--	453	--	3,530
Flow of steam to turbine, t/hr	1,190	--	2,260	110	--	--	854	3,057
Moderator	water	water	water	water	D_2O	water	water	Graphite
Moderator temperature, °C	--	285	--	249	82.2	257	74	--
Fraction of heat in moderator, %	7.0	--	7.0	5.1	--	--	5.6	4.1
Height of active zone, cm	229.0	150.0	271.0	172.0	333.0	152.5	305.0	609.6
Diameter, cm	178.0	--	263.0	175.0	--	152.5	335.0	900.0
Non-uniformity coefficient k_v	--	--	2.8	--	2.43	--	2.67	3.31
Fuel element diameter $\frac{\text{outside}}{\text{inside}}$, mm	17.6/6.6	36.0/--	12.0	12.0	12.2/4.1	13.7/8.4	Filling	Multilayer
Thickness of cladding (steel), mm	0.30	0.70	0.28	0.60	0.38	0.38	0.69	0.38
Number of fuel elements in assembly	49	9	63	16	37	116	--	--
Number of assemblies	52	32	172	404	68	21	591	540
Ratio of volumes of moderator and fuel	2.60	--	2.57	1.25	15.75	--	2.38	44.00
Maximum cladding temperature, °C	677	--	677	550	650	--	705	--
Initial enrichment, %	3.50	3.60	2.80	2.50	2.20	--	4.08	3.03
Fuel charge (UO_2), kg	10,720	--	28,200	10,400	8,760	--	40,260	121,900
Burnup, MW-days/ton	19,800	5,000	--	6,000	6,000	--	30,771	16,500

* Went critical in 1963.

Table 15-2

Planned atomic electric stations with steam-cooled fast-neutron reactors

Name of index	Name of project, firm, country, and year of completion										
	SCFBR UNC USA 1961 [14]	"Germec" GE Belgium 1964 [15]	SCBRII* BW USA 1965 [7]	D-1 KFK FRG 1966 [17,18]	ESCR GE USA 1967 [4,19]	FSB Sweden 1967 [4]	I AEG 1967 [20]	II FRG 1968 [4]	ENEA 1969 [21]	FSPPR GE USA 1965 [22]	SCBRI GE USA [16]
	Loeffler scheme									Scheme with supercritical pressure	
Electric power, MW	30	300	1,040	1,000	50	1,000	300	300	1,000	300	1,000
Thermal power, MW	876	87	2,500	2,517	139	2,500	765	847	2,700	677	2,320
Net efficiency, %	34.3	34.5	41.5	39.7	36.0	40.0	39.2	35.5	37.1	44.3	43.0
Coolant temperature at inlet, t_1, oC	320	315	297	365	320	330	350	332	340	267	400
Outlet temperature t_2, oC	510	540	565	540	510	530	490	460	500	565	538
Inlet pressure, kg/cm^2	98.0	105.0	84.4	182.0	105.0	130.0	160.0	120.0	150.0	305.0	255.0
Pressure loss in reactor Δp, kg/cm^2	-	4.5	8.4	6.6	7.0	10.0	-	10.5	-	-	12.0
Coolant flow, t/hr	-	4,285	14,850	11,200	782	11,900	-	5,976	-	-	11,350
Flow of steam to turbine, t/hr	-	1,220	-	3,000	245	-	-	1,212	-	853	3,230
Height of active zone, cm	153	100	-	151	137	75	112	-	-	153	213.5
Diameter, cm	183	-	-	263	100	145	170	-	-	88	193
Volume fraction of fuel, %	-	25.0	-	46.5	-	47.0	37.0	-	-	-	-
Volume fraction of coolant, %	25.0	25.0	-	31.6	-	33.0	36.0	-	-	-	-

[continued next page]

(Table 15-2 continued)

Conversion coefficient	1.30	1.00	-	1.15	-	1.14	1.12	-	1.19	1.14	1.08
Fuel element diameter, mm	6.35	5.80	-	7.00	5.80	6.50	7.00	-	7.00	Filling	5.85
Cladding material	Inkonel'**	Inkonel'**	-	Inkonel'**	Inkoloy** 800	Inkoloy** 800	Inkonel'** 625	-	Inkoloy** 800	Rene** 41	Inkonel'**
Classing thickness, mm	-	0.40	-	0.30	0.38	0.30	0.40	-	0.38	-	0.20
Number of fuel elements in assembly	127	151	-	469	228	-	331	-	57	-	504
Number of assemblies	326	-	-	163	64	-	82	-	-	72	108
Maximum cladding temperature, oC	732	700	705	747	-	650	650	-	700	750	732
Initial enrichment (Pu^{239})	15.4	20.0	-	10.1	16.2	-	14.0	-	-	-	10.0
Plutonium charge, kg	1,660	-	-	3,274	-	2,400	-	-	4,200	1,159	2,530
Burnup, MW-days/ton	33,000	30,000	-	55,000	-	-	-	-	65,000	90,000	80,000

* Two-pressure scheme
** Transliterated from Russian

1000 MW (electric). A characteristic feature of fast steam-cooled reactors is the possibility of the use of pressure-vessel type of reactor.

Depending on the coolant pressure and the working medium, the plans under consideration can be divided into three groups:

subcritical pressure with Loeffler scheme for the projects SCFBR [14], Hermes [15], prototypes I and II of the AEG company [4, 20], and projects ENEA [21] and D-1 [17];

subcritical pressure for the projects FSPPR [22] and SCBRI [16];

subcritical pressure of coolant and supercritical pressure of working steam of the high-pressure cylinders, for the project SCBRII [5, 7], realizing a two-pressure cycle with the aid of a surface heat exchanger and having a net efficiency that reaches 41.5 percent.

As a result of planning and experimental investigations, it was planned in 1965 to construct in West Germany a fast breeder reactor by 1980. Liquid sodium, gas, and steam were considered as equally promising coolants, and the West-German D-1 project of a steam-cooled fast-neutron reactor, at a pressure of 180 kg/cm^2 [17, 18] and 1000 MW rating appeared to be competitive.

In England, however, in 1966 they discontinued further development of a steam-cooled reactor, which differed from the analogous projects of other countries in that supercritical steam parameters were used at the reactor outlet (p_1 = 275 kg/cm^2, t_2 = 565°C [17]. This discontinuance was due to the unsolved problem of safety in the case of emergency loss of the supercritical-parameter coolant and the assumed greater promise of a gas coolant with spherical fuel elements. In addition, since steam with supercritical parameters is a significant neutron moderator, its use in fast reactors results in a low conversion coefficient.

In West Germany, where work on steam-cooled reactor continued, the question was raised of producing a fast prototype reactor and starting the construction of such a reactor in 1968 - 1969. This made it necessary to test a sufficient number of fuel elements (about 500, to a burnup of 30,000 MW-d/t). It was decided to use for this purpose the reactor in Grosswelzheim, with steam superheat.

At the October 1967 conference on fast-reactor physics [18] the British specialists reported that actually the conversion coefficient in the reactors is 4 percent lower than assumed earlier. This aggravated the problem of developing the fuel elements, particularly for the steam-cooled reactor, which has the lowest conversion coefficient. This made it necessary to have a special experimental reactor for the irradiation, planned in West Germany, of 500 fuel elements. The construction and experimental investigations of this reactor called for 200 million marks, and not 80 million as assumed earlier. The start of the construction of the prototype reactor was delayed thereby by approximately 10 years.

According to data obtained by 1967 in the U.S. on the experimental EVESR steam-cooled reactor, it turned out that the nickel steels, which are good from the point of view of anticorrosion properties, "swell up" in a stream of fast neutrons. Following this, in 1968, the General Electric Co., discontinued the development of the 50 MW steam-cooled reactor [19], which served as the prototype for the large 1000 MW reactor. The U.S. was followed by Belgium and Sweden, who shared in this project. At the same time in West Germany, in spite of the considerable disagreements concerning the feasibility of using a steam coolant, the work was continued.

This being the situation, the European Nuclear Energy Association (ENEA) organized a working group of representatives of seven countries to carry out in 1968 a technical and economic comparison of steam and gas coolants with sodium for atomic electric stations [21]. A steam-cooled fast reactor was developed (Table 15-2, Fig. 15-3) on the basis of an analysis of the Swedish, British, Belgian, and German projects. All projects were close to each other, with the exception of the British for the supercritical parameters [17, 23]. In the developed project, it was decided to use a steam pressure 150 kg/cm^2, a direct cycle, an integral assembly, a steam generator of the mixing type, compressors with water-lubricated bearings, ribbed fuel rods with helical spacing lugs, module-type fuel cassettes, and strongly alloyed nickel steels (incolloy 800, Sandwick 12x72). It was noted that it is necessary to study further the effect of "swelling" of these steels in fast-neutron fluxes. As indicated by the authors, the use of a direct cycle and a positive reactivity coefficients cause the system to become internally unstable when the pressure is reduced or when the coolant is lost, and require an effective control system.

The ENEA comparison has shown that the cost per installed kilowatt is 126 dollars for steam and 140 dollars for helium, but when account is taken of the larger fuel component for the steam, this difference is offset and the energy cost with the steam and helium coolants turns out to be approximately the same at 0.35 cents/kWh. Comparison of the physical characteristics of the three types of reactors has revealed that the conversion coefficient of the steam-cooled reactor is the lowest, the specific fuel charge is the highest, and the doubling time is considerably larger than 10 years. Account was taken also of the absence of fuel-element tests for steam- and gas-cooled fast reactors, the positive coefficient of reactivity for steam, the problem of flooding of the steam-cooled fast reactor, and a few other problems. As a result, in the final ENEA report [21], sodium was stated to be the most promising coolant for fast reactors, with gas as the standby. Steam was recognized to be less promising, and it was recommended to stop further work on the steam-cooled fast reactor.

At the order of the West-German Minister for Scientific Research, work on the steam-cooled fast reactor was stopped in February 1969, and only work on the individual fuel-element development was continued [24]. By that time, the expenditures on the steam-cooled fast reactor in West Germany amounted to about 50 million marks. Five research institute and the commercial concern AEG/GHH/MAN took part in these projects.

An open discussion on the "fast breeder" project [25] was held in West Germany in February 1971, with the Minister for Scientific Research taking part. It was noted that the start of the construction of the designed prototype reactor "Kalkar" for 300 MW, with sodium coolant, SNR, was delayed to 1972, and the need for a spare version, in the case of unforeseen difficulties in the operation of the sodium-cooled reactor, was emphasized. The spare coolant considered were helium and steam.

The adherents of the steam coolants advanced the following additional considerations in its favor:

The results of the fuel-element tests carried in West Germany from 1969 through 1971 [26] turned out to be favorable;

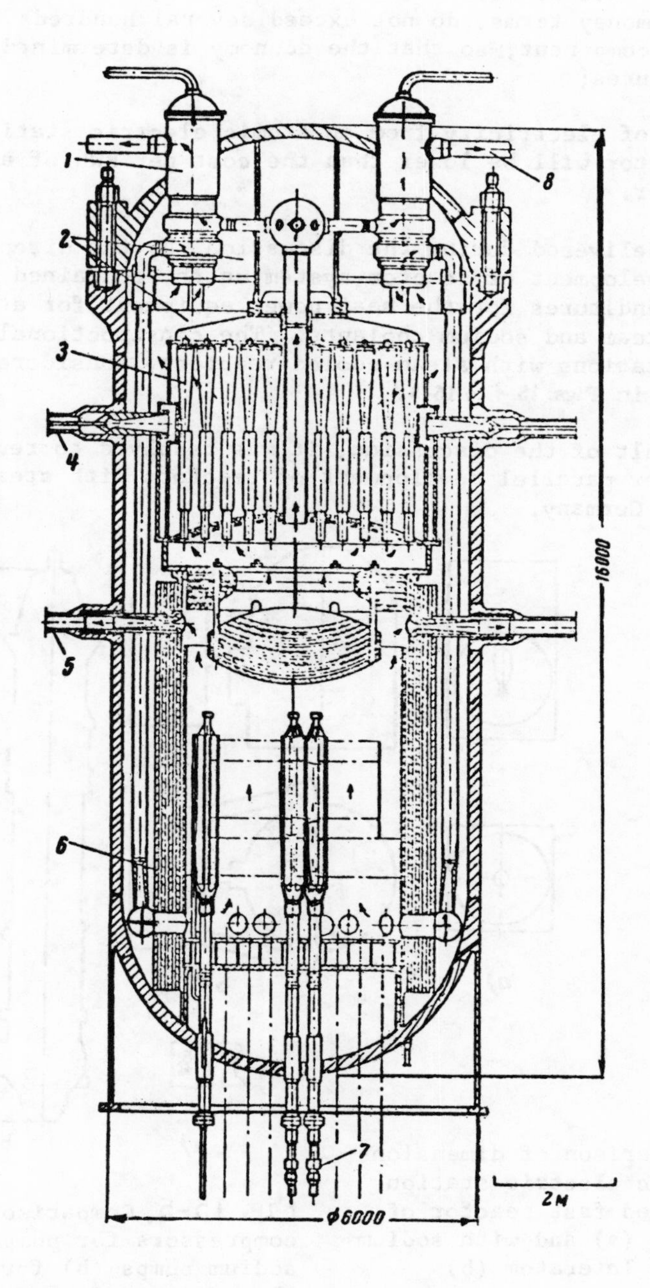

Fig. 15-3. **Longitudinal section through 1000 MW steam-cooled reactor (ENEA, 1969)**

1 - compressor, 2 - steam pipes (not shown in figure), 3 - steam generator, 4 - feedwater inlet, 5 - main steam heater, 6 - reflector, 7 - control rods, 8 - exhaust of turbocompressor.

The difference between the conversion coefficients of different types of reactors, in money terms, do not exceed several hundreds of a pfennig per kWh in the fuel component, so that the economy is determined primarily by the capital expenditures;

The cost of electricity from an atomic electric station with steam cooled fast reactor will be lower than the cost per kWh of an ordinary boiling-water reactor.

A paper delivered during the discussion, by the director of the institute for the development of reactor-system units, contained a comparison of the capital expenditures for the heat-power equipment for atomic electric stations with steam and sodium coolants. The constructional advantages of atomic energy stations with steam-cooled reactors, considered in this paper, are illustrated in Figs. 15-4 to 15-6.

As a result of the discussion, it was resolved to renew the appropriations for further parallel development of reactors with steam and helium cooling in West Germany.

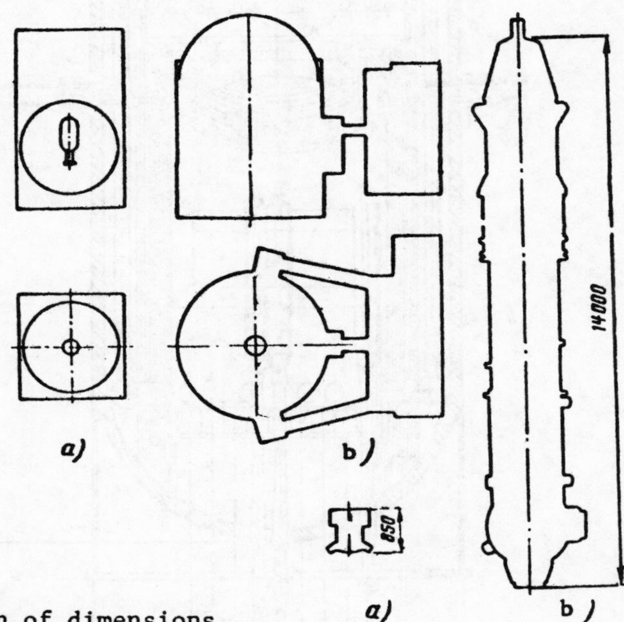

Fig. 15-4. Comparison of dimensions of 900-MW atomic electric station with steam-cooled fast reactor of the AEG company (a) and with sodium fast reactor of Interatom (b)

Fig. 15-5. Comparison of dimensions of compressors for pumping steam (a) with sodium pumps (b) for planned atomic electric stations for 300 MW.
a - 4 units, b - 6 units

It can be noted in conclusion that the failure to solve many scientific and technical problems involved in the use of superheated steam as a coolant, in spite of the advantage indicated above, has not made it possible to proceed in the construction of steam-cooled reactors, with the exception of the experimental EVESR. Nonetheless, an analysis of the unrealized plans for steam-cooled reactors, with both thermal and fast neutrons, is of interest for forecasting the possible ways of development of steam-cooled reactors and for investigating the possible use of steam as a spare coolant for fast-neutron reactors.

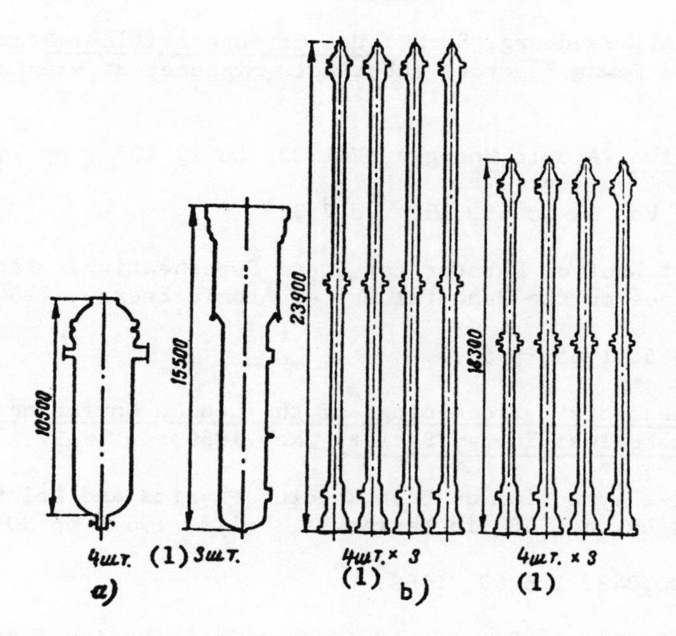

Fig. 15-6. Comparison of dimensions and of number of heat exchangers of 300-MW atomic electric station steam generator with steam coolant (a) and with sodium coolant (b).

a - 4 mixing heat exchangers; b - 3 heaters; 4 x 3 evaporators; 4 x 3 steam superheaters.

Key: (1) unit

BIBLIOGRAPHY

1. Power Reactor Technology, Vol 4, No 3, 1961, pp 71-85.

2. Atomnaya energiya za rubezhom (Atomic Energy Abroad), No 7, 1962, pp 30-32.

3. Power Reactor Technology, Vol 6, No 4, 1963, pp 97-105.

4. Fast Breeder Reactors, ANS Conference at San Francisco, April, 1967.

5. Atom und Strom, No 7, 8, 1969, pp 120-127.

6. Proceedings of the American Power Conference, Vol 29, 1967, pp 295-304.

7. Brit. Patent 1,153,075, G6c, 1969. *

8. Voprosy yadernoy energetiki (Problems of Nuclear Power), No 5, 1958, p 5 (translation).

9. U.S. Pat. 3,091,582, 176-60, 1963. *

10. Nuclear Engineering, No 2, 1963, pp 58-63.

11. Power Reactor Technology, Vol 6, No 3, 1963, pp 73-78.

12. Skvortsov, S. A., Faynberg, S. M. Use of Supercritical Steam Parameters in Water-Cooled Power Reactors (abstracts of paper at Vienna Symposium, 1961).

13. Atomnaya energiya (Atomic Energy), Vol 30, No 2, 1971, pp 149-155.

14. Nuclear Power, Vol 6, No 66, 1961, p 75.

15. "Hermes" - Fast Neutron Reactor for Steam Superheating. Paper at Third Geneva Conference on the Peaceful Use of Atomic Energy, 1964.

16. Nucleonics, No 5, 1965, p 60.

17. Fast Breeder Reactors. Proceedings of the London Conference Organized by the British Nuclear Energy Society, May 1966, pp 79-97.

18. Proceedings of a symposium on Fast Reactor Physics and Related Safety Problems, held by the IAEA in Karlsruhe, Vol 2, 1967, pp 301-398.

19. U.S. Pat. 3,400,048, 176-60, 1968. *

20. Third FORATOM Congress, Session IV (German Contribution), KFK 546, April 1967.

21. Atomwirtschaft, No 8, 1969, pp 410-413.

22. Power Reactor Technology, Vol 8, No 3, 1965, pp 186-193.

23. U.S. Patent 3,345, 66, 176-18, 1967. *

24. Atomwirtschaft, No 4, 1969, pp 190-211.

25. Atomwirtschaft, No 4, 5, 1971, pp 200-206.

26. Atomwirtschaft, No 7, 1970, pp 331-334.

---

* For patents, the country, number, class, and date of issue are indicated

CHAPTER 16

COST COMPARISON OF DRY TYPE AND CONVENTIONAL COOLING SYSTEMS

The production of electrical power requires that large amounts of waste heat from the generating process be rejected to a heat sink. The heat released by the generating process is transferred to the generating plant circulating water which then rejects the heat to the atmosphere either directly or by discharging into a body of water which, in turn, rejects the heat to the atmosphere. At plant sites where sufficient quantity of cooling water is available, such as from a river, lake, reservoir or ocean, and where thermal pollution considerations are not a factor, the water passes through the plant only once and is then returned to the source. Where once-through cooling is not feasible, the water is recirculated through the plant after having been cooled through the use of evaporative cooling towers or a cooling lake.

In the case of the conventional fossil-fueled steam-electric generating unit, a portion of the waste heat is rejected to the atmosphere in the form of products of combustion, but over 80 percent of the waste heat is rejected to the cooling water circuit. By far the greatest heat rejection is from the main steam condenser. The generator resistance losses and the mechanical losses of the turbine and the auxiliary rotating equipment contribute minor amounts of the total heat rejection. For a modern fossil-fueled plant, approximately 4,800 Btu are rejected to the circulating water for each kWh of electrical energy produced.

A pressurized-water or boiling-water nuclear generating plant rejects approximately 50 percent more heat to cooling water than a fossil-fueled plant. However, the use of more efficient nuclear plants, including HTGR (high-temperature, gas-cooled reactor) and LMFBR (liquid-metal, fast-breeder reactor) plants, will result in waste heat rejection comparable to fossil-fueled plants.

Presently Used Methods of Rejecting Heat from Generating Plants (1)

Once-through circulating water systems. Where a sufficient volume of circulating water is available from a large river such as the Ohio or the Missouri, circulating water is often taken directly from the river by means of an intake system, pumped through the condensers, and then discharged back to the river at a location selected to prevent recirculation of the heated water back to the intake.

Generally, the once-through circulating system is the least

expensive of the several types of cooling systems used. Utilities have used this method of providing circulating water whenever conditions permit.

Once-through systems are often used when cooling water is drawn from rivers, lakes, reservoirs, or the ocean.

Cooling lakes. Another method of supplying circulating water for a steam-electric generating plant is to construct a pond or lake for that purpose. Heat rejected to the circulating water is conveyed to the lake or pond and is dissipated from the water surface. Generally, the cooling lake for a nuclear plant is sized at approximately two acres of surface area per MW of generating capacity, with variations above and below this figure in specific instances.

Wet-type cooling towers. Where sufficient water is not available for a once-through circulating water system, an evaporative (wet-type) cooling tower is often used to dissipate the heat of condensation which has been transferred to the circulating water as it is pumped through the condenser. Until the present concern about thermal pollution developed, the use of evaporative cooling towers with steam-electric generating plants was generally limited to plants in locations lacking sufficient water for once-through cooling. Within the past several years, a number of large fossil-fueled and nuclear plants have been built with evaporative cooling towers either because of lack of sufficient water for once-through cooling systems or because of thermal pollution problems.

The consumptive use of water by evaporation is on the order of 0.7 gallons per kWh for a light-water-reactor nuclear plant (8). In addition to the evaporative losses from the tower, the cooling water make-up requirements include blowdown losses -- water which must be wasted and replaced with fresh make-up water in order to keep the concentration of dissolved solids in the circulating water at a level below which scaling would occur in the condensing equipment. The blowdown losses are generally between 15 and 100 percent of the evaporation losses, depending on the concentration of solids carried in the circulating water.

Based on 6:1 cooling water concentration ratio, the total make-up water requirement is 20 percent greater than the evaporation loss. The total make-up requirement for an 800-MWe nuclear plant operating at rated capacity would be approximately 16 million gallons per day, assuming evaporation losses of 0.7 gallons per kWh and a 6:1 concentration ratio.

Evaporative cooling towers are either of the cross-flow design in which air flows horizontally through the falling water, or the counter-flow design in which air flows upward through the falling water.

The driving force for moving air through the cooling tower can be either motor-driven fans or the thermal lift obtained by the use of a natural-draft tower.

Spray ponds or spray canals. Another method of rejecting heat from the condenser is to use a spray pond in conjunction with a recirculating supply of condensing water. In this method of rejecting heat to the atmosphere, the circulating water is sprayed into the air through discharge pipes and spray nozzles located above a pond which serves as a reservoir for receiving and holding the water for recirculation.

Description of Dry-Type Cooling Towers

General. There are two basic types of air-cooled condensing systems; the indirect system and the direct system. The indirect system utilizes a direct-contact condenser at the turbine to condense the exhaust steam. Water from the condenser is pumped to the dry-type tower for cooling and recirculation to the spray jets in the condenser. In the direct system, steam is condensed in the cooling coils without the use of a direct-contact condenser.

Conventional evaporative-type system. An understanding of the conventional evaporative (wet-type) tower cycle is useful in considering the two types of air-cooled condensing systems. Figure 16-1 shows the schematic arrangement of an evaporative-type cooling tower serving a condensing turbine.

Condensing water is circulated through the tubes of a surface condenser and carries away the heat of condensation of the turbine exhaust steam. The exhaust steam comes into contact with the exterior surfaces of the tubes, and condenses as it gives up heat to the water.

The warm circulating water is piped to the evaporative cooling tower where it flows over the packing or fill, which may be closely spaced strips of asbestos-cement, wood or other suitable material, to break up the circulating water into small drops through which air is pulled by the tower fan. By a combination of evaporation and convection the temperature of the circulating water is reduced and the water is again pumped through the condenser in a continuous cycle. The condensed steam (condensate) is removed from the condenser by the condensate pump and is returned to the feedwater circuit.

Indirect dry-type cooling system. Fig. 16-2 shows a diagrammatic arrangement of an indirect dry-type cooling system with a natural-draft tower.

The principal components are:

1. A direct-contact steam condenser.
2. Circulating water pumps.
3. Water recovery turbine (optional).
4. Cooling coils.

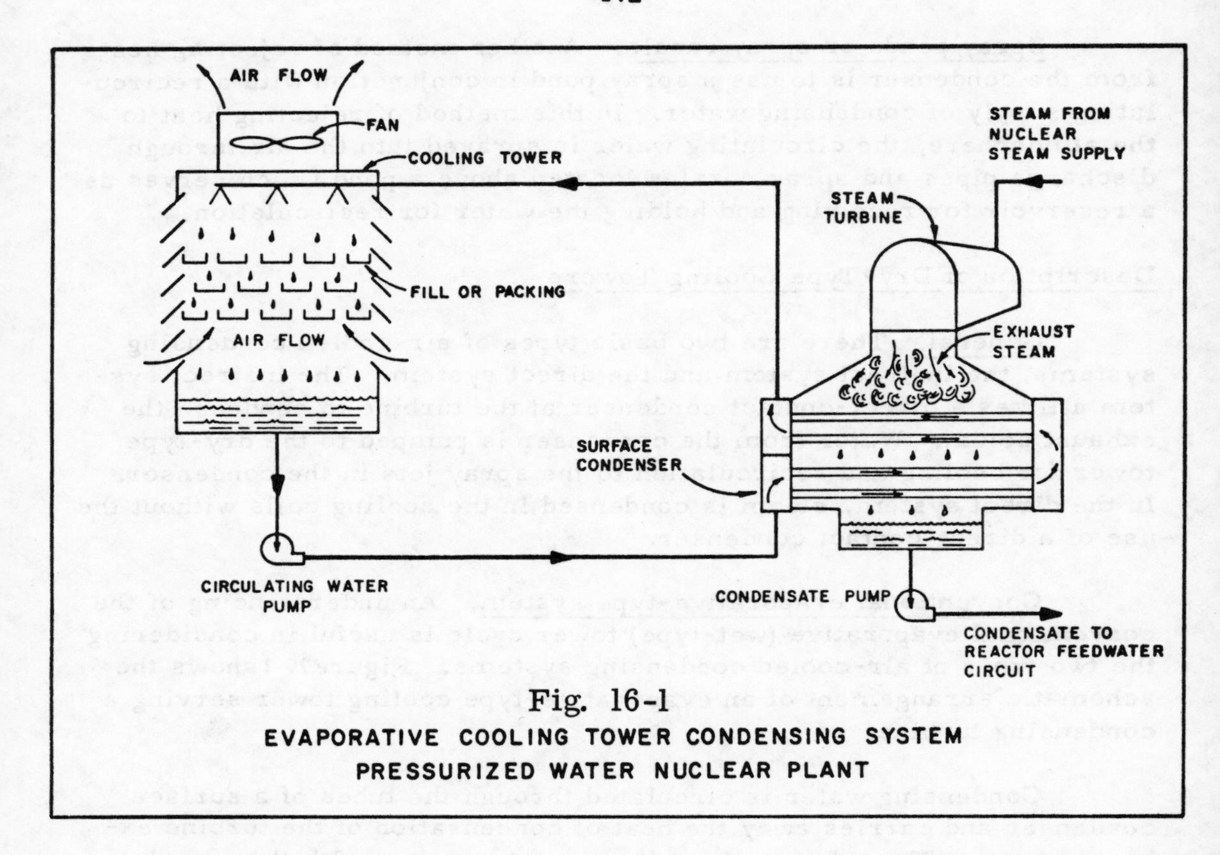

Fig. 16-1

EVAPORATIVE COOLING TOWER CONDENSING SYSTEM
PRESSURIZED WATER NUCLEAR PLANT

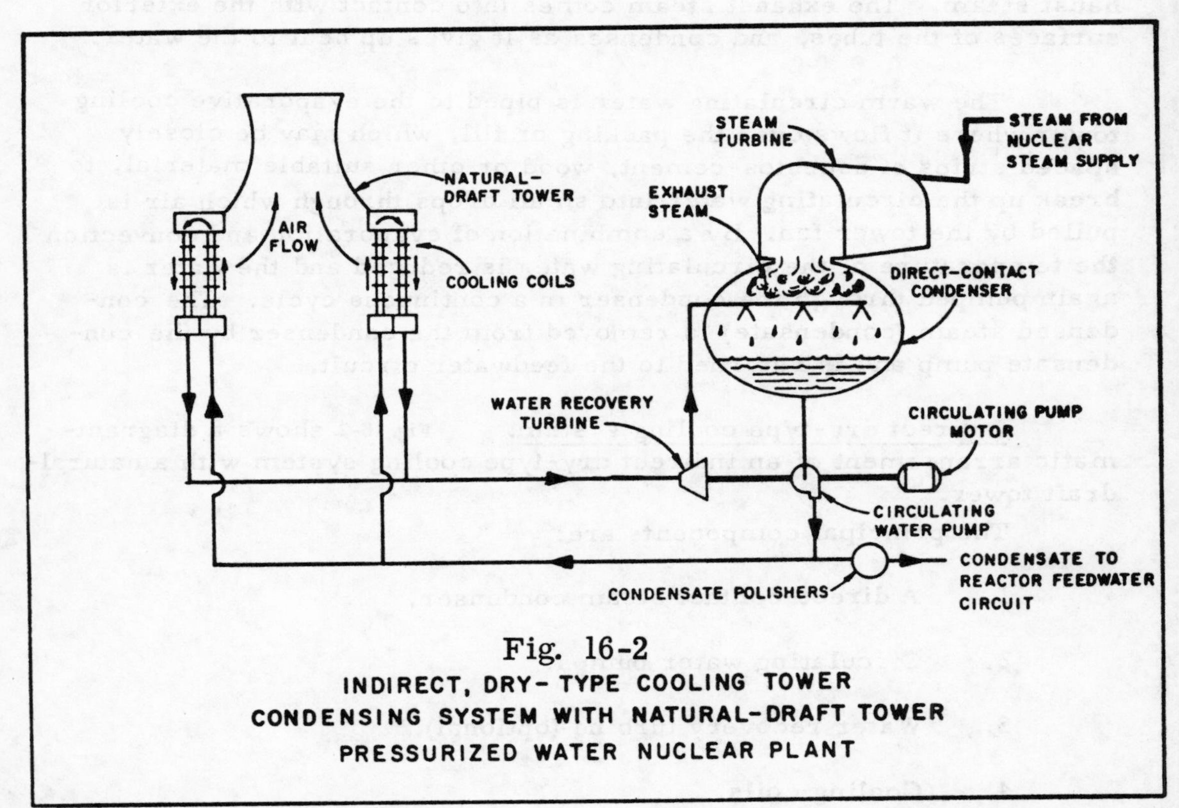

Fig. 16-2

INDIRECT, DRY-TYPE COOLING TOWER
CONDENSING SYSTEM WITH NATURAL-DRAFT TOWER
PRESSURIZED WATER NUCLEAR PLANT

5. A means for moving air across the coils; either a natural-draft tower or a mechanical-draft fan.

The choice of a mechanical-draft or a natural-draft tower is dependent upon the economics of each particular case, considering such factors as fuel cost, comparative construction costs for the two types of cooling towers and cost of money.

Cool water from the cooling tower is sprayed into the direct-contact condenser and mixes directly with the exhaust steam from the turbine. The water mixture consisting of water from the tower and condensed steam falls to the bottom of the condenser and is removed by circulating water pumps. The greater part of the water flows through pipes to the cooling coils, and an amount equal to the exhaust steam from the turbine is directed back to the feedwater circuit through condensate polishers, or demineralizers, for re-evaporation in the nuclear steam generator. Since the cooling tower circulating water and the boiler feedwater are intimately mixed, the circulating water must be of condensate purity.

The cooling coils located in the base of the natural-draft tower or beneath large mechanical-draft fans can be mounted either vertically, as shown in Figure 16-2, or horizontally.

A positive pressure head of a few pounds per square inch is imposed at the top of the cooling coils to prevent air from being drawn into the system in case of leaks in the cooling coils. This is accomplished by means of either a throttling valve in the circulating water discharge line from the tower, or, if a water-recovery turbine is used, by varying the position of the adjustable turbine vanes. In some installations, water-recovery turbines are coupled to the drive shaft of the circulating water pump in order to recover some of the pressure head between the cooling coils and the condenser.

After passing through the recovery turbine or throttling valve, the circulating water is again sprayed into the direct-contact condenser and recycled through the cooling system.

Note that the circulating water does not come into direct contact with the cooling air and, therefore, there is no evaporative loss of water as with the wet-type tower.

Direct, dry-type cooling system. Figure 16-3 shows a diagram of a typical direct air-cooled condensing system. Turbine exhaust steam is conveyed through the exhaust steam trunk, which is large in diameter to minimize the pressure drop, to the air-cooled coils where cooling air passing over the finned-coil surfaces condenses the steam. Shown here in the simplest form, steam enters the top of the coil section and condenses as it travels downward with the steam and condensate flowing in the same direction. In actual installations, provisions are made for removal of noncondensable gases and air and for prevention of freezing

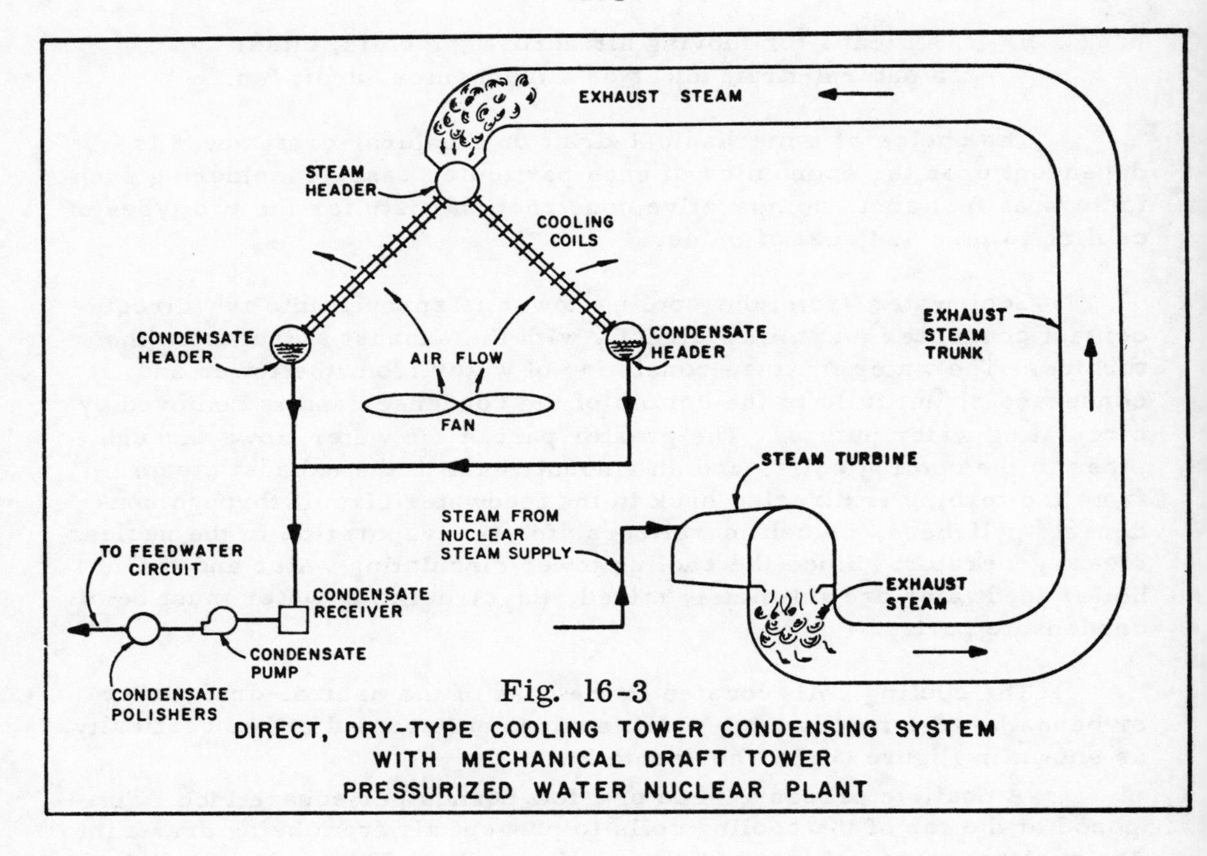

Fig. 16-3
DIRECT, DRY-TYPE COOLING TOWER CONDENSING SYSTEM
WITH MECHANICAL DRAFT TOWER
PRESSURIZED WATER NUCLEAR PLANT

during cold weather. The most common industrial direct air-cooled system in the United States uses horizontal tube bundles with 80 to 90 percent of the tubes as the main condenser and 10 to 20 percent as an after-condenser to condense the remaining steam that is not condensed in the main condenser. The steam and condensate flow in the same direction in the tubes, minimizing pressure loss and increasing the heat transfer coefficient. The purpose of the foregoing coil arrangement is to reduce to a minimum the noncondensable gases blanketing the main condenser as the residual noncondensable gases are swept out of the main condenser with the residual steam. An excessive buildup of noncondensable gases in the main condenser would be deleterious to effective condensation. Freeze protection is usually accomplished by a combination of warm air recirculation combined with fan control.

One European manufacturer uses a method of direct condensation in which a certain percentage of the cooling coils are constructed so that the remaining steam, after passing down through a condensing unit, enters the bottom headers of the after-cooling coils, and the condensate and steam flow in opposite directions in order to obtain better control of condensate temperature during cold-weather operation. Only noncondensables remain in these latter coils near the upper ends after all the steam has been condensed, thus preventing freeze-up in that region of the heat exchangers.

The condensed steam from the cooling coils flows by gravity to condensate receivers and is pumped back to the feedwater circuit by a condensate pump.

Comparison of the indirect and direct dry-type cooling systems. The principal difference between the two systems is the large volume of exhaust steam which must be handled in the direct system as compared to the smaller volume of circulating water in the indirect system.

Although discussions with users of direct-type systems did not indicate that any difficulty due to condenser air leaks has been encountered with the direct systems (1), the fact that all the cooling coils operate under a high vacuum is sometimes considered a disadvantage when compared to the indirect systems which maintain positive water pressures in the cooling coils (9).

Because of the requirements for large steam pipes from the turbine to the condenser, it has generally been considered that direct systems are limited to turbine-generator sizes of approximately 100 to 300 MW, whereas the indirect system is more economical for larger units.

Heat Rejection Characteristics of Dry-Type Towers

Designers of air-cooled heat transfer surfaces have found it more convenient to express coil performance as a function of the initial temperature difference (ITD) between the fluid entering the coil and the ambient air, rather than to use the logarithmic mean temperature difference. Ignoring the subcooling effect of the condensate, which amounts to only a few degrees Farenheit in a well-designed system, the ITD also expresses the temperature difference between the turbine exhaust steam and ambient air and, therefore, provides a direct relation between tower performance and turbine back pressure.

Figure 16-4 illustrates the temperature relationships which exist for an indirect dry-type cooling system. The left side of Fig. 16-4. represents the temperatures that exist in the direct-contact condensers where the cool circulating water from the tower mixes with the turbine exhaust steam. The upper line on the left side represents the temperature level of the condensing steam. Since condensation takes place at a constant temperature corresponding to the saturated steam temperature of the turbine back pressure, this line is horizontal at temperature T_{s1}. The lower curve on the left side of the diagram represents the temperature condition of the circulating water heated from T_{w2} to T_{w1} as the exhaust steam transfers the heat of condensation to the water.

The difference in temperature between T_{s1} and T_{w1} represents subcooling of the condensate and circulating water below the temperature of the saturated steam at the exhaust pressure and is a thermal loss to the turbine cycle. In the typical direct-contact condenser, the subcool-

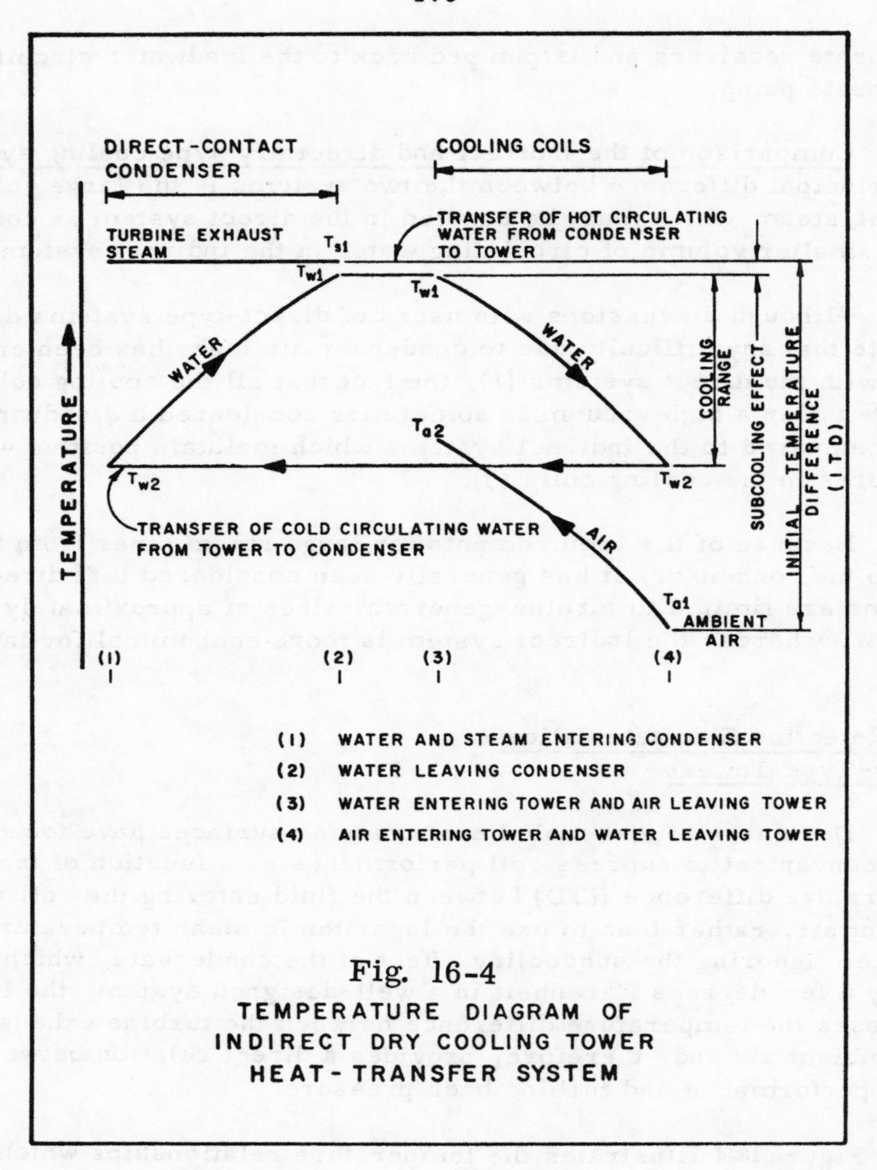

Fig. 16-4

TEMPERATURE DIAGRAM OF INDIRECT DRY COOLING TOWER HEAT-TRANSFER SYSTEM

ing is approximately 3°F, but it is possible to have a condenser in which no subcooling exists, in which case T_{s1} and T_{w1} are the same.

The right-hand side of Figure 16-4 illustrates the temperature condition of the circulating water and the cooling air as the water flows through the coils. The air at ambient temperature T_{a1} first comes into contact with the cooling coils containing cooled water at T_{w2} and is heated to T_{a2} as the water in the coils cools from T_{w1} to T_{w2}. The diagrams shown are for counterflow of air and water. In actual practice, where air flows across the cooling coils a cross-flow correction factor is used to compensate for the deviation in heat exchanger performance because of the cross-flow condition.

Figure 16-5 is a graphical representation of the heat rejection

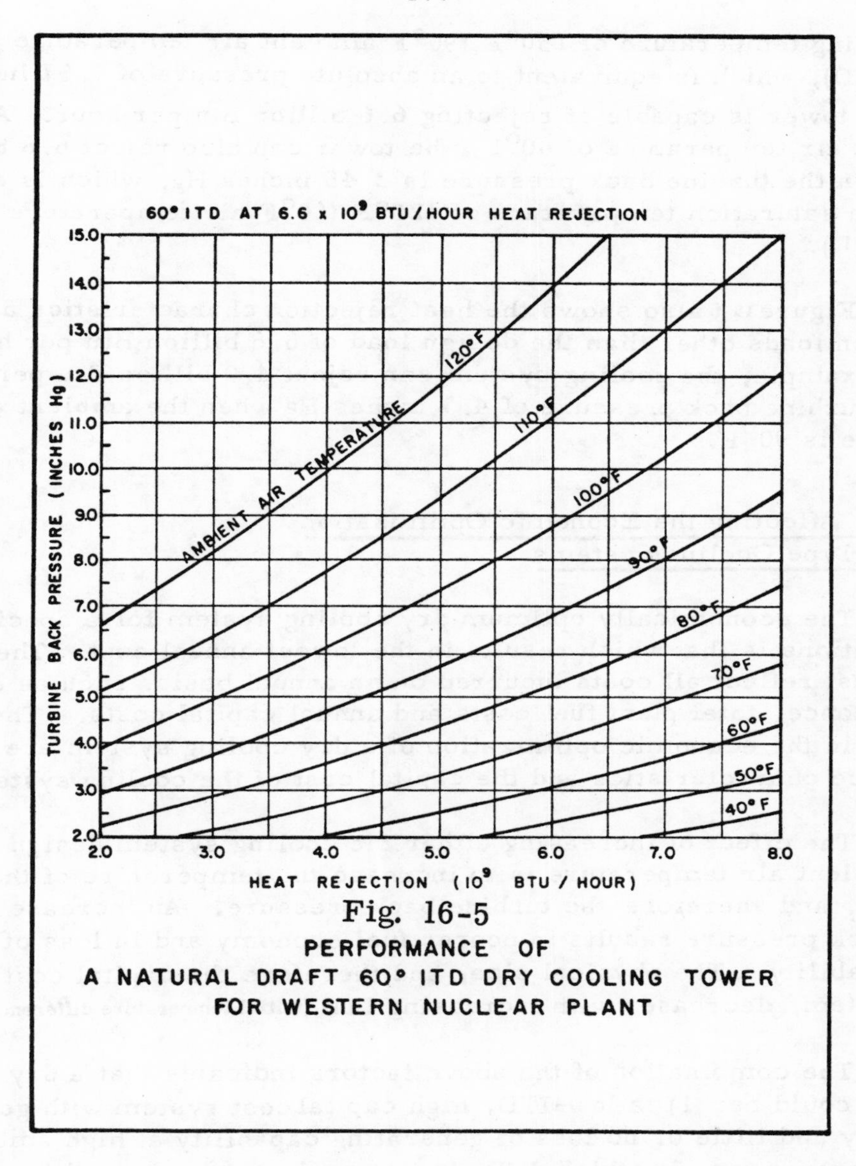

Fig. 16-5

PERFORMANCE OF
A NATURAL DRAFT 60° ITD DRY COOLING TOWER
FOR WESTERN NUCLEAR PLANT

characteristics of a natural-draft, dry-type cooling tower sized for the Western Plant.

Figure 16-5 shows the operating characteristics of a tower sized to reject 6.6 billion Btu per hour with 60°F ITD, which, by definition, means that the tower is capable of rejecting 6.6 billion Btu per hour when the condensing temperature in the turbine is 60°F higher than the ambient air temperature. This capability is indicated on the graph by the intersection of the ambient air temperature lines (curved lines) and the turbine back pressure lines (horizontal lines). The turbine back pressure plotted on the vertical scale is also representative of the turbine condensing temperature since there is a definite condensing temperature for each back pressure.

As an example, at an ambient air temperature of 90°F and a

condensing temperature of 150°F (90°F ambient air temperature plus 60°F ITD), which is equivalent to an absolute pressure of 7.57 inches Hg, the tower is capable of rejecting 6.6 billion Btu per hour. At an ambient air temperature of 60°F, the tower can also reject 6.6 billion Btu when the turbine back pressure is 3.45 inches Hg, which is equivalent to a saturation temperature of 120°F (60°F air temperature plus 60°F ITD).

Figure 16-5 also shows the heat rejection characteristics at heat rejection loads other than the design load of 6.6 billion Btu per hour. As an example, the cooling system can reject 4.0 billion Btu per hour with a turbine back pressure of 4.7 inches Hg when the ambient air temperature is 90°F.

Factors Affecting the Economic Optimization of Dry-Type Cooling Systems

The economically optimum dry cooling system for a specific set of conditions is that which results in the lowest annual cost. The annual cost must reflect all costs incurred on an annual basis, such as operation, maintenance, total plant fuel costs and annual capital costs. The key factors in the economic optimization of a dry cooling system are the performance characteristics and the capital cost of the cooling system.

The effect of increasing either the cooling system design ITD or the ambient air temperature is to increase the temperature of the turbine exhaust, and therefore the turbine back pressure. An increase in turbine back pressure results in poorer fuel economy and in loss of generating capability. The physical size, and therefore the capital cost of the dry system, decreases with increasing ITD (initial temperature difference).

The combination of the above factors indicates that a dry cooling system could be: 1) a low-ITD, high capital cost system with good fuel economy and little or no loss of generating capability at high ambient air temperatures; or, 2) a high-ITD, low capital cost system with poorer fuel economy and a significant loss of capability at high ambient air temperatures; or, 3) some intermediate cooling system.

Because of the many variables which affect the determination of the economically optimum dry cooling system, a comprehensive computer program was developed. The variables which affect the economic optimization are discussed below.

Performance related to ITD. The effect of increasing the ITD is to increase the turbine back pressure for a given air temperature. This results in poorer fuel economy and loss of generating capability.

Capital cost of the dry cooling system. The physical size and the capital cost of the dry cooling system will decrease with increasing ITD. The capital costs include all equipment and installation costs from the turbine flange outward (pumps, piping, towers, condensers, cooling coils, controls and all other components of the cooling system).

The turbines suitable for use with dry-type cooling systems have poorer heat rates than conventional turbines. The nuclear steam supply system of a dry-cooled plant must be larger than that of a conventionally cooled plant in order to achieve the same output.

Elevation. The effect of increasing ground-level elevation is to increase the capital cost of the dry cooling system, since the reduced air density makes it necessary to move a greater volume of air past the cooling elements in order to achieve the same mass flow rate.

Fixed-charge rate. The fixed-charge rate is a percentage rate applied to the capital cost which reflects the following items as defined by the Bureau of Power of the Federal Power Commission: Interest, or cost of money; depreciation, or amortization; interim replacements; insurance, or payments in lieu of insurance; taxes (federal, state and local), or payments in lieu of taxes.

The effect on the economic optimization of an increase in the fixed-charge rate is to give more weight to capital costs and less weight to annual operation, maintenance and fuel costs. Fixed-charge rates of 12, 15 and 18 percent were considered in a study.

Ambient air temperature. The effect of higher ambient air temperatures is to increase the turbine back pressure, resulting in poorer fuel economy and loss of generating capability. The full range of annual air temperatures at the site affect the fuel economy, but it is the extreme high temperature that has the most significant economic effect. The extreme high temperatures, in combination with the cooling system ITD and the turbine characteristics, sets the maximum loss of generating capability that would be experienced during the year.

Fuel costs. The effect of increasing the unit cost of fuel is to increase the weight given to fuel economy and to decrease the weight given to capital cost considerations. Increasing the fuel cost would tend to lower the optimum ITD and increase the capital cost of the cooling system.

Auxiliary power requirements. The power requirements for pumps and fans decrease with increasing ITD. The cost of the auxiliary power required for the cooling system pumps and fans was calculated assuming that the incremental nuclear plant capacity required to furnish the auxiliary power could be provided at a capital cost of $245 to $300 per kW, depending upon the site. The energy cost for the auxiliaries would be the average fuel cost in mills per kWh plus an allowance for operation and maintenance costs of 0.1 mill per kWh.

Replacement of capacity losses. Since some loss of generating capability can be expected to occur at high ambient air temperatures with a dry cooling system, the replacement of this capacity is an important consideration in the economic analysis. The relative importance of this capability loss may vary from area to area and would be of the greatest concern where the annual peak electrical demand occurs in the summer. When the greatest demand occurs in the summer, the capacity loss must be replaced from some other generating source. The loss of generating capability on a hot summer day would not affect a winter-peaking utility as much as a summer-peaking utility. The winter-peaking utility may still be economically interested in the lost capability since it may have the opportunity to sell surplus generating capacity to other interconnected systems.

Once it has been decided whether or not the replacement of lost capacity is necessary, the cost of replacing that capacity must be determined. In the optimization analysis, the economic impact of the capacity loss increases as the cost of replacing the capacity increases. The economic effect of the lost capacity is greater if the capacity is replaced at a capital cost of $150 per kW than if it is replaced at a capital cost of $100 per kW.

Economic Optimization Curve

Fig. 16-6 shows the relationship of the various costs affected by the dry-type cooling system to the tower sizes considered in the economic analysis. The economically optimum dry cooling system is indicated by the point where the sum of the individual cost curves is the lowest.

Description of Computer Printout Sheet

The method of analysis can be described by summarizing the results of one of the analyses as presented in the computer printout. Figure 16-7 shows the computer printout for a mechanical-draft dry cooling system for a Southeastern Plant. Figure 16-7 also reflects the following assumptions:

- Modified conventional turbine
- System includes hydraulic recovery turbines
- Nuclear fuel cost of 18¢ per million Btu
- Gas turbine fuel cost of 80¢ per million Btu
- Fixed-charge rate of 15% per year.

The first column of Figure 16-7 shows the initial temperature difference (ITD) which is a measure of the size or heat rejection capability of the cooling systems. The second column shows the gross energy generation for the 860.6-MWe unit with the dry-type cooling system when operating at full throttle flow for 7,500 hours per year. For some temperature conditions the dry cooling system may produce lower turbine back pressures than the conventional cooling system and at those times the generation of the plant equipped with the dry system would exceed the generation from the conventional plant. The third column shows the

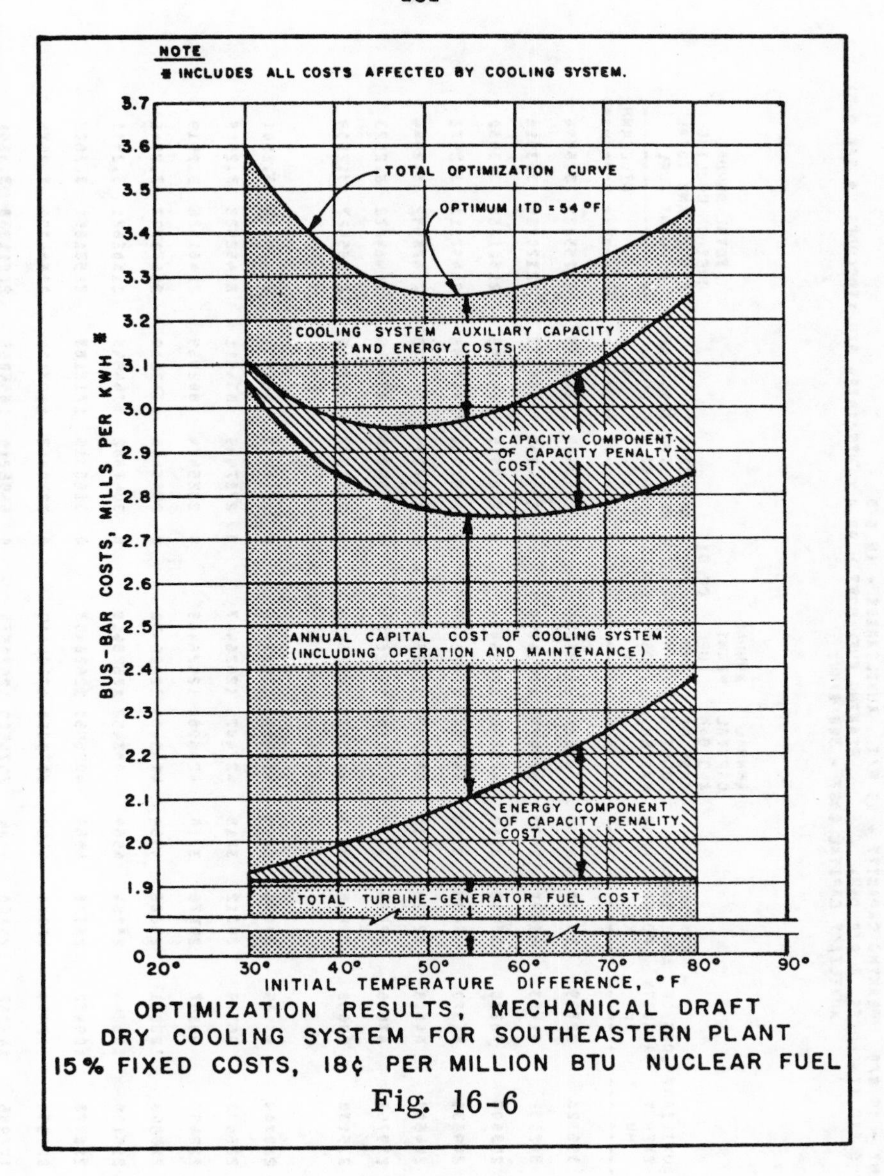

OPTIMIZATION RESULTS, MECHANICAL DRAFT DRY COOLING SYSTEM FOR SOUTHEASTERN PLANT 15% FIXED COSTS, 18¢ PER MILLION BTU NUCLEAR FUEL

Fig. 16-6

total annual amount of such excess energy, if any. The fourth column shows the energy deficiency which results when the generating capacity of the dry-cooled plant is less than that of the conventional plant. The fifth column headed "AUXILIARY ENERGY" shows the annual energy requirements of the cooling system pumps and fans. The column headed "LOSS OF CAPACITY" shows the difference in generating capability of the dry-cooled and conventional plants at temperatures equalled or exceeded 10 hours per year on the average. The column headed "MAXIMUM AUXILIARY POWER" shows the maximum generating capacity required to supply the cooling system pumps and fans. The column headed "HRS" indicates the number of hours of operation per year of the small supplemental wet cooling system which is operated to supply cooling water for the plant auxiliaries during periods when the temperature of return water from the dry tower exceeds 95°F. The next seven col-

Fig. 16-7. - COMPUTER PRINT-OUT, ECONOMIC OPTIMIZATION, SOUTHEASTERN PLANT - MECHANICAL DRAFT DRY COOLING SYSTEM

CAPITAL COST FACTORS: PLANT - 15 0/0 PEAKING CAPACITY - 15 0/0 AUXILIARIES - 15 0/0
PLANT FUEL COST -18.000 CENTS/10**6 BTU (INCLUDES 0 0/0 O+M) PEAKING FUEL COST - 80.0 CENTS/10**6 BTU (INCLUDES 0 0/0 O+M)
PEAKING CAPITAL COST - 100 $/KW AUXILIARY CAPITAL COST - 300 $/KW

INIT. TEMP. DIFF. (DEG) (F)	GROSS ENERGY OF PLANT MWH	EXCESS ENERGY DUE TO EXTRA CAPACITY MWH	CAPACITY PENALTY ENERGY MWH	AUXILIARY ENERGY MWH	LOSS OF CAPACITY KW	MAXIMUM AUXILIARY POWER KW	HRS*	ANNUAL CAPITAL AND O+M COST OF COOLING SYSTEM $	ANNUAL PLANT FUEL AND O+M COST $	CREDIT FOR EXCESS ENERGY $	CAPACITY PENALTY COST $	AUXILIARY COST $	TOTAL ANNUAL COST OF COOLING SYSTEM AND TOTAL PLANT FUEL $	TOTAL ANNUAL COST OF COOLING SYSTEM AND TOTAL PLANT FUEL MILL/KWH
30	6543663	0	56153	356323	22799	55400	1769	7432275	12626335	0	480300	3216175	23755085	3.5994
40	6501975	0	97841	302152	51463	41600	2673	5700762	12626390	0	1304062	2488973	22120186	3.3516
49	6445215	0	154601	253899	78200	33968	3718	4734568	12626454	0	2089515	2051349	21501885	3.2580
50	6436408	0	163409	249639	81370	33400	3718	4661842	12626488	0	2191094	2017688	21497111	3.2572
51	6426901	0	172916	244603	84608	32728	3718	4577257	12626523	0	2297046	1977775	21478602	3.2544
52	6416912	0	182905	239769	87666	32084	3718	4496183	12626524	0	2402016	1939549	21464272	3.2523
53	6406476	0	193341	235138	90668	31466	3718	4418619	12626496	0	2507397	1902915	21455427	3.2509
OPTIMUM:														
54	6395767	0	204049	230709	93665	30876	3718	4344565	12626455	0	2610933	1867954	21449906	3.2501
55	6384289	0	215528	226483	96749	30312	3718	4274021	12626447	0	2717449	1834611	21452528	3.2505
56	6372158	0	227659	222459	99857	29776	3718	4206988	12626445	0	2825424	1802969	21461826	3.2519
57	6359386	0	240430	218808	103021	29266	4569	4143465	12626438	0	2935605	1773289	21478797	3.2545
58	6346079	0	253738	215189	106240	28784	4569	4083452	12626418	0	3047802	1744948	21502621	3.2581
59	6332636	0	267180	211773	109472	28328	4569	4026950	12626403	0	3160315	1718183	21531851	3.2625
60	6318469	0	281348	208559	112558	27900	4569	3973958	12626449	0	3270863	1693128	21564398	3.2674
70	6151697	0	448119	177956	145262	23800	5304	3474509	12626554	0	4456219	1454056	22011338	3.3351
80	5942820	0	656997	156401	180084	20900	6279	3103473	12626430	0	5778210	1288438	22796551	3.4541

* NUMBER OF HOURS OF OPERATION OF SUPPLEMENTAL MECHANICAL DRAFT EVAPORATIVE COOLING SYSTEM. THIS SUPPLEMENTAL COOLING SYSTEM IS OPERATED ONLY TO PROVIDE COOLING WATER FOR PLANT AUXILIARIES WHEN THE TEMPERATURE OF RETURN WATER FROM DRY TOWER EXCEEDS 95F. POWER REQUIREMENT FOR THIS SYSTEM IS 200 KW.

umns show annual cooling system costs. The column "ANNUAL CAPITAL AND O&M COST OF DRY COOLING SYSTEM" is the capital cost of the dry cooling system multiplied by the fixed-charge rate plus 1 percent. The 1 percent is an allowance for operation and maintenance costs. In this case, the column is computed as the capital cost times 16 percent. The next column shows the total annual fuel cost of the 860.6-MWe unit and reflects the gross energy generation as shown in the second column. The column headed "CREDIT FOR EXCESS ENERGY" is a fuel cost credit related to the excess energy amounts shown in the third column. This credit is based on the average fuel cost per kWh of energy generated by the 860.6-MWe unit. The column headed "CAPACITY PENALTY COST" reflects the cost of replacing the capacity and energy losses which occur in the dry system as compared to the wet system under consideration. The column headed "AUXILIARY COST" reflects the cost of providing the power and energy required for the cooling system auxiliaries.

The total annual cost in dollars, the next to last column, is the sum of the preceding five columns. The last column, which is the total annual cost in mills per kWh, is the dollar sum divided by the energy target, which is the sum of Columns 2 and 4 less Column 3. This energy target is also the amount of energy generated by the conventionally cooled plant. Thus, it is seen that the computer program sets up a model of annual generation to determine the bus-bar production costs of dry cooling systems for ITD values from 30°F to 80°F in which the total energy supplied must equal the energy generated by the wet system being used as the basis for comparison.

All costs which are affected by the cooling system have been included in the analysis. These include the cost of replacing the capacity and energy losses which result from the higher turbine back pressure of the dry-cooled plant as compared to the conventionally cooled plant.

CHAPTER 17

BURNUP OF FAST REACTOR FUELS

The economical production of electrical power based on the fast breeder reactor concept requires 1) that the fuel be operated at both a high temperature and high power density to maximize the integrated power output and 2) a viable fuel reprocessing and fabrication cycle. In the development of fuels for Fast Breeder Reactors (FBRs) and in the operation of FBRs, it is imperative that there be accurate methods for the evaluation of the fuel and reactor performance. An important criterion in determining fuel performance is an accurate determination of the total fissions and the fission rate. This is accomplished by a burnup determination. Errors in a burnup measurement introduce errors in fuel design and operation, nuclear physics calculations, shielding requirements, design of transportation equipment, and fuel reprocessing equipment, all of which affect power generating costs.

Of the three chemical techniques used to determine burnup, the one based on the measurement of a fission product monitor and the residual heavy atoms in a sample of the irradiated fuel is shown to be uniquely suitable for FBR fuels. It is highly reliable for fuels of all compositions, for all irradiation levels, and requires only limited knowledge and no analysis of the unirradiated fuel. The techniques based on the measurements of heavy atom isotope ratios or the absolute heavy atom contents are limited to high burnup levels and require chemical analyses of unirradiated fuel specimens. The heavy atom isotope ratio technique also requires accurate knowledge of neutron cross section data for the heavy atom nuclides. The heavy atom content technique requires that matched specimens of the unirradiated and irradiated fuel be analyzed on an equivalent weight basis.

Emphasis in this section, is placed on the fission product

monitor-residual heavy atom technique; the factors affecting this technique as it applies to FBR fuels are treated in detail. A major need is shown to be more accurate fission yield data, both for the major and the minor fissile nuclides in the various fuels for FBR neutron spectra. A need also is shown for reliable neutron capture cross section data in FBR neutron spectra for the promising fission product burnup monitors. The selection of fission product monitors of burnup for FBR fuels is governed by the above two factors as well as by the factors of fission product migration and fission product half-life. Post irradiation studies of FBR fuel have shown that only the rare earth, and possibly zirconium, fission products do not migrate relative to uranium and plutonium and, hence, can be used as FBR fuel burnup monitors.

Three chemical techniques currently are used to determine rare earth and zirconium burnup monitors: isotope dilution mass spectrometry, X-ray spectrometry, and spectrophotometry. Isotope dilution mass spectrometry is the most accurate and sensitive of these techniques, is particularly specific, and does not require quantitative recovery of the constituent being determined in the chemical separation procedure that precedes the mass spectrometric measurements. Many laboratories have the capability of determining ^{148}Nd or the sum of the neodymium isotopes with an accuracy of 0.5%.

X-ray spectrometry has the particular advantage of enabling the entire rare earth group of fission products to be used as the burnup monitor. This technique, like isotope dilution mass spectrometry, is highly specific and does not require quantitative recovery of the rare earths in the chemical separation procedure that precedes the X-ray measurements. The X-ray technique is a factor of 10 less sensitive than mass spectrometry and the accuracy of the rare earth determination is 1.0%.

Spectrophotometry has been used to determine elemental fission product zirconium in thermal reactor thorium-uranium fuels. However, zirconium is a poorer choice than a rare earth nuclide, rare earth element, or group of rare earth elements as a burnup monitor for FBR fuels because of possible zirconium-heavy atom differential migration and variations of the fission yields with fissioning nuclides. The accuracy of the spectrophotometric zirconium method is estimated to be 3%. In spite of these detracting factors, this method, or some other one, which will require less manpower to execute than the mass and X-ray spectrometric methods and uses less expensive instrumentation would have application for routine burnup determinations of large sample loads.

The same techniques as those used to determine the fission product burnup monitor are being, or could be used, to determine the residual heavy atom content of the fuel specimen. The advantages of each method for the heavy atom determinations, including the accuracies, are about the same as those for the burnup monitor determinations. A unique feature of the mass spectrometric technique is that the uranium and plutonium isotopic compositions are also obtained in this analysis.

INTRODUCTION

The term burnup is defined several ways. In the nuclear power industry and to some extent in irradiated fuel experimental programs, burnup is defined as megawatt-days of thermal energy produced per metric ton of heavy atoms initially present in the fuel. Another definition for experimental fuels is the number of fissions per cubic centimeter of the fuel.

In this section, burnup always means:

$$\text{Burnup} = \text{Atom percent fission} = \frac{\text{number of fissions} \times 100}{\text{initial number of total heavy atoms}}$$

This definition is more fundamental in that it relates directly the number of events of interest, fissions, to the quantity of most interest the number of fissile and fertile atoms, in the fuel at the beginning of the irradiation. In other definitions it is necessary to have knowledge of either a physical constant, such as energy of fission, or a property of the fuel, such as bulk density, to calculate burnup from measurements provided by a chemist. Although the errors in these conversion factors are small, their use introduces an additional uncertainty in the burnup value obtained.

Within the megawatt-day/ton (MWd/T) and atom percent fission definitions of burnup, there are variations. These variations have, in the past, resulted in invalid comparisons being made of the performances of irradiated fuels. The megawatt-days of energy produced by a fuel is of prime concern to the power reactor operator while the megawatt-days of energy deposited in a fuel is of prime concern for an evaluation of its irradiation performance and stability. The energy produced by a fuel will, in a reactor such as EBR-II, be 5 to 10% higher than the energy deposited. In both the MWd/T and atom percent fission definitions, the quantity against which the energy or number of fissions is referenced is not always defined the same way. The "ton" in the MWd/T definition has meant, in addition to metric ton of heavy atoms, metric ton of fuel (U + Pu + O or C) and short ton of heavy atoms. In the atom percent fission definition, only the number of fissile atoms rather than the total number of heavy atoms is sometimes used as the reference quantity. In a low enrichment ^{235}U fuel, the difference between the per fissile atom and the per total heavy atom burnups is very large; for high enrichment fuels, the difference is less than 10% (relative).

From burnup measurements, both the total number of fissions and the fission rates are calculable. The total number of fissions is a measure of the thermal energy produced by a fuel. This thermal energy is required to establish fuel warranties (FBR cores will cost on the order of $100,000,000) and to evaluate the performance of experimental fuels. Fission rates are used to evaluate such fuel performance characteristics as power levels, fuel and cladding temperatures, and fuel-cladding gap conductances. Fission rates measured by foil activations also are used to establish flux distributions in critical assemblies and in test reactors.

As will be shown in this section, the only technique capable of providing highly accurate burnup values for FBR fuels is that based on chemical determinations of selected fission products and residual heavy atoms in irradiated fuel samples. The accuracy of this technique requires (1) accurate fission yields of the fission products used as the burnup monitors, and (2) accurate analytical methods to determine both the fission product monitor and the heavy atoms. Presently, the only reliable fast fission yield data are for ^{235}U and ^{239}Pu. Because these data are for a neutron spectrum much harder than expected for FBRs, the uncertainty of applying these data to the determination of burnup in FBR fuels is estimated to be 3 to 5%. There are no accurate fast fission yield data for other important fissionable nuclides such as ^{238}U, ^{240}Pu, and ^{241}Pu.

The cost factor involved in a burnup analysis merits special consideration by both the chemist doing burnup analyses and the experimente requesting analyses. All too often, because of lack of communications or inadequate knowledge on the part of the chemist or the experimenter, excessive time and money are wasted. Many experiments do not merit a ±1% burnup analysis. If a ±3% analysis is adequate because of uncertainties associated with the experiment, such information should be conveyed to the chemist. On the other hand, when a 1% reliability is merited, the chemist should be provided with complete information concerning the experiment so that he can use the most appropriate methodolo The provided information should consist of fuel and cladding composition, irradiation data including in-pile and out-of-pile time, fuel center-line temperature, estimated burnup, and any unique factors of the experiment.

TECHNIQUES FOR THE DETERMINATION OF BURNUP

Absolute burnup values are determined most accurately by destructive chemical analyses; relative burnup along the axis of fuel pins is determined by nondestructive gamma ray spectrometry. The former is discussed below:

There are three chemical techniques for determining burnup: (1) measuring a fission product monitor and the residual heavy atom contents of a dissolved fuel specimen and calculating the burnup from these values and the yield of the fission product, (2) measuring the plutonium and uranium isotopic ratios on dissolved specimens of both the irradiated and unirradiated fuel and calculating the burnup from these values and nuclear cross section data, and (3) measuring the heavy atom contents of dissolved specimens of both irradiated and unirradiated fuel and calculating the burnup from the change in these values. Each of these techniques has particular advantages and disadvantages.

The fission product monitor-residual heavy atom technique is unique. It is applicable to samples of all fuel composition regardless of the burnup level and it does not require the analysis of an unirradiated fuel specimen.

The isotopic ratio technique is applicable only when the burnup is high, when an archive sample is available for analysis, and when there are accurate neutron cross section data for the heavy atoms. It does not appear that this cross section data requirement can be met for FBR fuels until the neutron environment of these reactors has been well established. The advantage of this technique is that the measurements of the heavy atom isotopic ratios can be easily and accurately made.

The heavy atom content technique, like the heavy atom ratio technique, is applicable only when the burnup is high and when an archive sample is available for analysis. It also requires that the analyses of the unirradiated and irradiated samples be done on equivalent weight samples. This latter requirement is especially difficult to meet for most FBR fuels in which high temperature operation causes fuel slumping and reaction of cladding and fuel. The advantage of this technique is that fissions are measured directly, the decrease in the number of heavy atoms is the number of fissions.

The major emphasis in the discussion of burnup techniques that follows is devoted to the fission product-residual heavy atom technique because it has the widest range of applicability to fast reactor fuels and is the only one that can meet the accuracy needs of the Fast Breeder Program.

FISSION PRODUCT MONITOR-RESIDUAL HEAVY ATOM TECHNIQUE

The irradiated fuel specimen is dissolved and the fission product monitor and heavy atoms are determined. The computational relationship is:

$$\text{Burnup} = \text{a/o F} = 100 \frac{A/Y}{H + A/Y}$$

where

a/o F = atom percent fission

A = determined number of atoms of fission product monitor

Y = effective fractional fission yield value of A

H = determined number of residual heavy atoms.

The successful application of this technique requires accurate measurements of the fission product monitor and heavy atoms and an accurate value for the effective fission yield. Currently the uncertainty of the analytical methods used to determine the fission product and heavy atoms is as low as 0.25 relative percent. The major source of uncertainty is associated with the effective fission yield. At this time, this uncertainty is 3 to 5 relative percent for fuel specimens irradiated in FBR neutron spectra. A prime need, then, is more accurate fission yield data and a major objective of this program is to measure fission yields with an accuracy of 1 to 2% for FBR spectra. The overall accuracy of a burnup determination will then approach the 1 to 2% level.

Factors to be considered in the use of the fission product monitor-residual heavy atom technique include: a) sources of fission, b) fission yields as a function of neutron energy and fissioning nuclides, c) neutron capture cross sections of the fission products, d) fuel composition, e) fission product migration, f) fission product half-life, and g) the availability of reliable analytical methods for the determination of the fission product monitor and heavy atoms.

Sources of Fission

In reactor fuels where two or more heavy atom nuclides contribute significantly to the total number of fissions, it is desirable to select a burnup monitor whose fission yield varies as little as possible with the fissioning nuclide. If the fission yield is the same for all the fissioning nuclides, only an analysis for one fission product is required to determine the number of fissions. In low-enrichment, uranium-fueled, light-water reactors (LWRs), the fissions are primarily produced by the two nuclides ^{235}U and ^{239}Pu. Several fission products have nearly equal thermal yields for these two nuclides; the most notable among these is ^{148}Nd. The ^{235}U and ^{239}Pu values are 1.69 and 1.70%, respectively[1]. In fast reactors, all the heavy atoms are fissionable. However, the selection of the burnup monitors for FFTF and future FBRs is simplified because greater than 90% of the fissions are due to ^{239}Pu and one other nuclide. To illustrate the effect of multiple sources of fissions upon the total number of fissions, calculations of the change in the heavy element isotopic composition and the fractional source of fission have been made for a conceptual 1000 MW_e LMFBR fuel and for the FFTF fuel as a function of burnup. Core loading data and the results of these calculations for low and high burnups are given in Tables 17-1 and 17-2 .

Conceptual Fast Breeder Reactors. For the conceptual 1000 MW_e LMFBR, the major sources of fission are ^{239}Pu (~73%) and ^{241}Pu (~19%) (Table I), and the sum of the percent contribution of ^{239}Pu and ^{241}Pu fissions is relatively constant with burnup, ranging from 92.7% at 2% burnup to 92.1% at 8.0% burnup. This means that the very best burnup monitor will be one whose ^{239}Pu and ^{241}Pu fission yields are nearly equal. The contributions from the other fissioning nuclides can be calculated with small error. Based on a comparison of the thermal fission yields for ^{239}Pu and ^{241}Pu[1], it is predicted that the fission yields of one or more of the fission products or group of fission products from the fast fission of ^{239}Pu and ^{241}Pu will be nearly equal in the mass region 139 to 150.

Fast Flux Test Facility. For FFTF the major sources of fission are ^{239}Pu and ^{238}U (Table 17-2), and again the sum of the percentage fissions from these two sources is nearly constant with burnup, ranging from 93.4 to 93.0% over the range of 2 to 7% burnup. Based on fast reactor fission yield data for ^{239}Pu[1] and fission spectrum yield data for ^{238}U[2], it is predicted that one of the nuclides in the mass range 139 to 150 or the neodymium group of nuclides will be a satisfactory burnup monitor.

In the FFTF reactor the contribution of ^{238}U to the total fissions will vary significantly with the location of the fuel within the reactor because the fraction of the neutrons which have energy sufficient to fission ^{238}U will decrease in going from the center of the core to the outer rows. However, this will not affect the selection of the burnup monitor as the net effect will only be an increase in the percentage of fissions due to ^{239}Pu. The 85% figure given in Table 17-2 for the percentage due to ^{239}Pu is an expected minimum value.

Table 17-1. HEAVY ELEMENT COMPOSITION AND SOURCES OF FISSION FOR A 1000 MW_e LMFBR

Nuclide	Fuel Composition, %[3]	Source of Fission, % 2 a/o Burnup	Source of Fission, % 8 a/o Burnup
^{235}U			
^{238}U	78.1	2.6	2.7
^{238}Pu	0.26	0.9	1.0
^{239}Pu	13.0	73.1	74.8
^{240}Pu	5.2	3.3	3.5
^{241}Pu	2.6	19.6	17.3
^{242}Pu	0.88	0.5	0.6

Table 17-2. HEAVY ELEMENT COMPOSITION AND SOURCES OF FISSION FOR FFTF

Nuclide	Fuel Composition, %	Source of Fission, % 2 a/o Burnup	Source of Fission, % 7 a/o Burnup
^{235}U	0.5	2.1	2.0
^{238}U	74.4	7.3	7.6
^{239}Pu	21.7	86.1	85.4
^{240}Pu	2.9	2.1	2.4
^{241}Pu	0.4	2.3	2.5

Experimental Fuels. Most FBR experimental fuels undergoing irradiation now and in the near future in the United States and by Euratom[4] are mixed uranium-plutonium oxides with lesser numbers of mixed uranium-plutonium to uranium in these fuels ranges from 1/10 to 1/4, the uranium isotopic composition ranges from natural to 93% ^{235}U, and the usual plutonium isotopic composition is on the order of 88 to 93% ^{239}Pu, 8 to 12% ^{240}Pu, and 1 to 2.5% ^{241}Pu. The reason for using enriched ^{235}U in experimental fast reactor fuels contrasted to the expected use of natural uranium in FBR fuels is to increase the specific power per linear foot in the test irradiations. This is done because the neutron flux in EBR-II, the presently used U. S. facility, is lower than that proposed for FBRs. Another means of increasing the specific power per linear foot is to use ^{233}U. At least one laboratory has planned irradiations for mixed ^{233}U-plutonium fuels[5].

Experimental fuels with more complex compositions should be expected in the future. For example, Argonne National Laboratory experimenters have irradiated Th-U-Pu mixtures in which the major fissioning nuclides are ^{233}U, ^{235}U, and ^{239}Pu[6]. Certainly, mixed uranium-plutonium fuels in which the plutonium is recycled material will be experimentally irradiated far in advance of their use for FBRs.

Fractional Sources of Fission. The determination of the fractional sources of fission, especially in experimental fuels, is important for the verification of neutronic calculations and the design of future FBR fuels.

In fuels such as those for the FFTF and demo FBRs where there are only two fissioning nuclides of importance, the relative number of fissions from each nuclide is determined most readily by comparing the relative abundance of two fission product nuclides of the same element[7]. The determination can then be made mass spectrometrically. It is the most precise of the measurement techniques available and does not require quantitative recovery of the fission product element in the separation procedure which precedes the measurement. Maximum leverage is achieved when there is a maximum change in the ratio of the fission yields with fissioning nuclides.

A fission product ratio which appears on the basis of thermal fission yield data to provide the largest change in relative yield is $^{105}Pd/^{110}Pd$. For ^{239}Pu fission this ratio is a factor of 3 higher than it is for ^{241}Pu fission, and it is estimated that this ratio will be a factor of 1.5 higher for ^{238}U fission than it is for ^{239}Pu fission.

A pair of elements which are potentially useful for establishing the fractional source of fission are yttrium and neodymium. The yttrium to neodymium ratio is a factor of 2 higher for ^{239}Pu fission than it is for ^{241}Pu fission and it is estimated to be a factor of 1.5 higher for ^{238}U fission than it is for ^{239}Pu fission. In the X-ray spectrometric method for burnup analysis, yttrium accompanies the rare earths in the separation procedure and, therefore, is readily determinable relative to neodymium.

An important factor which will have to be considered in applying either of the above methods is the effect of neutron energy upon the fission yields of ^{110}Pd and yttrium. The fission yields of both increase significantly going from thermal to fast fission and it will have to be established whether their fission yields will significantly change over the neutron spectral range of the FBRs. It is postulated on the basis of fission yield data obtained in other fast reactor spectra that this change will be small.

Fission Yields

The most critical factor in the determination of nuclear fuel burnup by the fission product monitor method is the accuracy of the fission yield value. In the relationship

$$\text{Burnup} = \text{a/o F} = 100 \frac{A/Y}{H + A/Y}$$

A/Y is the sum of the fissions of the n fissioning nuclides, $\sum_{i}^{n} (A/Y)$, in the fuel and H is the number of post-irradiation heavy atoms. Because of the multiple sources of fission in an FBR fuel, the fission yields, Y, must be known for a variety of heavy atom nuclides; however, the accuracy requirements vary with the fractional number of fissions. Table 17-3 summarizes the required accuracies of the fission yield values to obtain an effective fission yield accurate to 1% for the FFTF, a 1000 MW_e FBR, and experimental fuels. These calculations were based on the assumption that the yield of the selected fission product nuclide was the same for ^{238}U and ^{239}Pu in the FFTF fuel, for ^{239}Pu and ^{241}Pu in the 1000 MW_e FBR fuel, and for the two major sources of fission in the experimental fuels.

Table 17-3 REQUIRED ACCURACIES OF FISSION YIELD VALUES TO OBTAIN AN OVERALL 1% ACCURACY IN THE EFFECTIVE FISSION YIELDS

Fissioning Nuclide	FFTF[a] Source of Fissions, %	FFTF[a] Fission Yield Accuracy, %	1000 MW_e FBR[b] Source of Fissions, %	1000 MW_e FBR[b] Fission Yield Accuracy, %	FBR Fuel Development Program Fission Yield Accuracy, %
^{233}U					1
^{235}U	2	10	--	--	1
^{238}U	8	5	3	10	5
^{238}Pu	--	--	1	30	30
^{239}Pu	85	1	74	1	1
^{240}Pu	2	10	4	7	5
^{241}Pu	3	10	17	2	2
^{242}Pu	--	--	1	30	30

[a] Based on a fuel composition of 0.5% ^{235}U, 74.4% ^{238}U, 21.7% ^{239}Pu, 2.9% ^{240}Pu, 0.4% ^{241}Pu, and <0.1% ^{242}Pu.

[b] Based on a fuel composition of 78.0% ^{238}U, 0.3% ^{238}Pu, 13.0% ^{239}Pu, 5.2% ^{240}Pu, 2.6% ^{241}Pu, and 0.9% ^{242}Pu.

The existent fission yield data do not meet the needs of the FBR burnup program: (1) data are available only for a limited number of fissile nuclides and the accuracy of these data is a factor of 3 to 4 less than the needs, and (2) there is, in the range of neutron energies in the FFTF and future FBRs only a limited amount of information about the effect of neutron energy on fission yield. Absolute thermal fission yields of ^{233}U, ^{235}U, ^{239}Pu, and ^{241}Pu have been measured with accuracies of 1 to 2% and absolute fast (700-keV mean fission energy) fission yields have been measured only for ^{235}U and ^{239}Pu with accuracies of 2 to 3%[1]. There are significant differences in the yields of a number of nuclides between these spectra. Although it is anticipated that fission yields in FFTF and other FBR spectra will be closer to the 700-keV values than to the thermal values, the difference between the 700-keV values and the FBR values may be several relative percent. This unknown difference combined with the 2 to 3% uncertainty in the 700-keV values would lead to an uncertainty of $\sim$4% in the burnup determination of FBR irradiated fuels if the 700-keV values were used.

The fast fission yield data for ^{238}U are quite unreliable and data for ^{240}Pu and ^{241}Pu do not exist. The ^{238}U measurements were made 15 to 25 years ago in spectra much harder than that of the FBRs. In general, the spectrum was characterized by the investigators as being a "fission neutron spectrum". In view of the fact that it is now recognized that the conditions for achieving a true fission neutron spectrum are difficult to attain, it is impossible to assess the spectral conditions in which the older irradiations were made. The reliability of the older data is also questionable because of the problems of measuring both the number of fissions and the number of fission product atoms. The latter were all radioactive nuclides. The activity assays were all made by beta counting, the standards against which the activities of these nuclides were compared were of a poorer quality than those available today, and a complete separation of the nuclide being determined from the other radioactive fission products was required. Therefore, it is highly important to measure fission yields for neutron spectra characteristic of both the FFTF and FBRs. The accuracies of presently available fission yield data, estimated FBR Program needs, and program goals are summarized in Table 17-4.

Neutron spectra, at this time, can only be approximated for the FFTF and FBRs and certainly will vary within and between reactors. Fission yield data are needed over a range of neutron energy bracketing the anticipated extremes. Recently, irradiations of ^{235}U and ^{239}Pu have been made at ANL[8] in a variety of fast neutron energy spectra and the fission yields of various short-lived nuclides were determined by gamma spectrometric analysis relative to a fission product whose yield is known to change less than 2% between a thermal and an EBR-I spectrum. In the neutron energy range of 100 to 500 keV, the changes in the yields of the major (>5%) fission products were less than 2%; however, at ∿1 keV a significant change was noted for some of the fission yields. Because the low end of the neutron energy range in the FBRs is between 100 and 1 keV, it cannot be established from these data whether there will be significant differences in fission yield values in various fast reactor spectra.

A more definitive evaluation of the variation in fission yields with neutron energy will result from the absolute fission yield measurement program in progress.

A major objective of this program, therefore, is to measure the fission yields with the accuracies shown in Table 17-4. Irradiations have been made in neutron spectra which are as high and as low as those expected in the FFTF and future FBRs and these samples will be assayed using techniques, principally mass spectrometric isotopic dilution, which will provide fission yield data having accuracies of ∿1%. Should these data show that there are significant changes in fission yields within the energy range, it may be necessary to irradiate materials at some intermediate energy value to establish the relationship between fission yield and neutron energy more definitely.

Fission Product Capture Cross Sections

In the fission product monitor-residual heavy atom technique to determine burnup, the number of atoms of the fission product monitor present in the fuel is used as the measure of the number of fissions. In the ideal case the number of fissions is directly proportional to the number of fission product monitor atoms. However, in the real case the number of fission product atoms may be altered by the fission products capturing neutrons. This is shown in the following simplified production scheme of several fission product nuclides having the same Z.

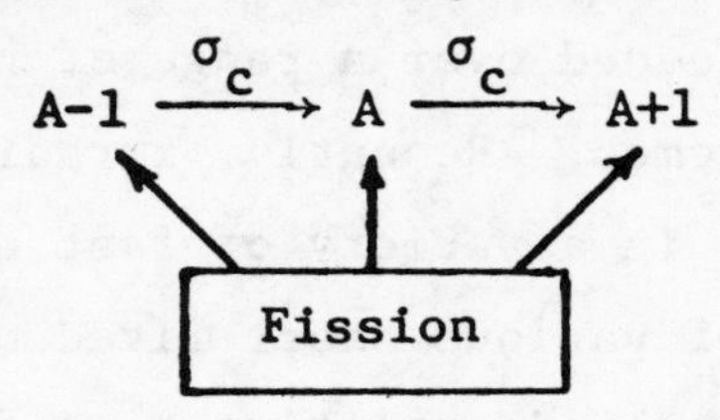

The number of A nuclide atoms is directly proportional to the number of fissions only when the capture cross sections, σ_c, of the A-1 and A nuclides are insignificant with respect to the formation cross section. The effect of the neutron capture cross section upon the abundance of

Table 17-4. ESTIMATED FAST FISSION YIELD ACCURACIES: PRESENT DATA, FBR PROGRAM NEEDS, AND EXPERIMENTAL GOALS

	Accuracies, Relative Percent			
			Goals at	
Heavy Nuclide	Present Data	Need	100 keV[c]	250 keV[c]
^{233}U	NA[a]	1	1	1
^{235}U	2-3	1	1	1
^{238}U	~15	5	ND[b]	2
^{237}Np	NA	10 - 30	ND	3
^{238}Pu	NA	10 - 30	ND	ND
^{239}Pu	2-3	1	1	1
^{240}Pu	NA	5	2	2
^{241}Pu	NA	2	1	1
^{242}Pu	NA	10 - 30	ND	3
^{241}Am	NA	10 - 30	ND	3
^{243}Am	NA	10 - 30	ND	3

[a] Reliable data not available.

[b] Data not being obtained.

[c] Estimated mean neutron energies of experimental irradiations.

a selected A nuclide is shown below. When the capture cross section of A-1 is large, the relative abundance of A increases with burnup. Conversely, when the capture cross section of A to form A+1 is large, the relative abundance of A decreases.

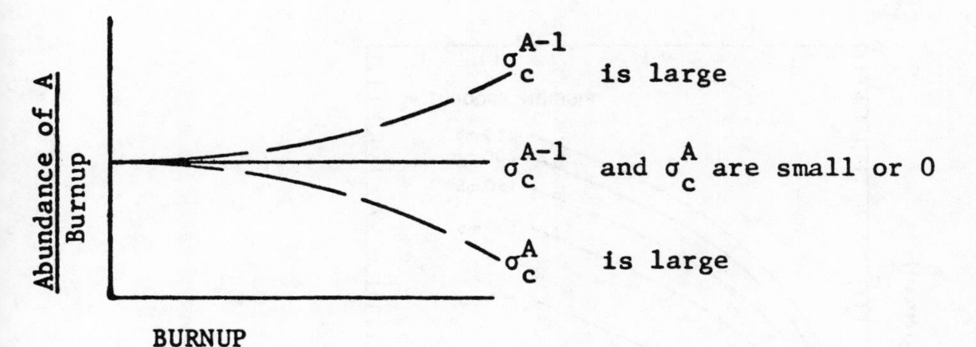

For thermal neutron irradiations, the problem of fission product burnout can be minimized by selecting a monitor whose capture cross section is small (a few barns) compared to the relatively large (500 barns or greater) fission cross section of the fissioning isotope.

For FBRs the effect is potentially more serious. The fission product cross sections may be as high as several hundred millibarns and the fission cross sections for the major fissioning isotopes are about 2000 millibarns; hence, the ratio of these cross sections can be greater than in thermal reactors. Fission product burnout as a function of burnup in fast reactor spectra is shown in Figure 17-1 for fission product capture cross sections ranging from 50 to 250 millibarns. At 10% burnup, the fission product burnout ranges from about 1 to 5%. This demonstrates the need to obtain accurate fission product capture cross sections in a FBR neutron spectrum to meet the program goal of 1% accurate burnup determinations.

Fuel Composition

It is imperative that the amount of the burnup monitor in the fuel from sources other than fission is negligible. Contaminants can be introduced in the fabrication of the fuel and in the reprocessing of the fuel. The use of zirconium fission products as monitors is precluded for zirconium clad fuels. The use of rare earth burnable poisons, characteristics of some thermal reactor fuels, or the use of rare earth neutron poisons in chemical processing plants constitutes a potential source of contamination of those rare earth isotopes which may be used as burnup monitors.

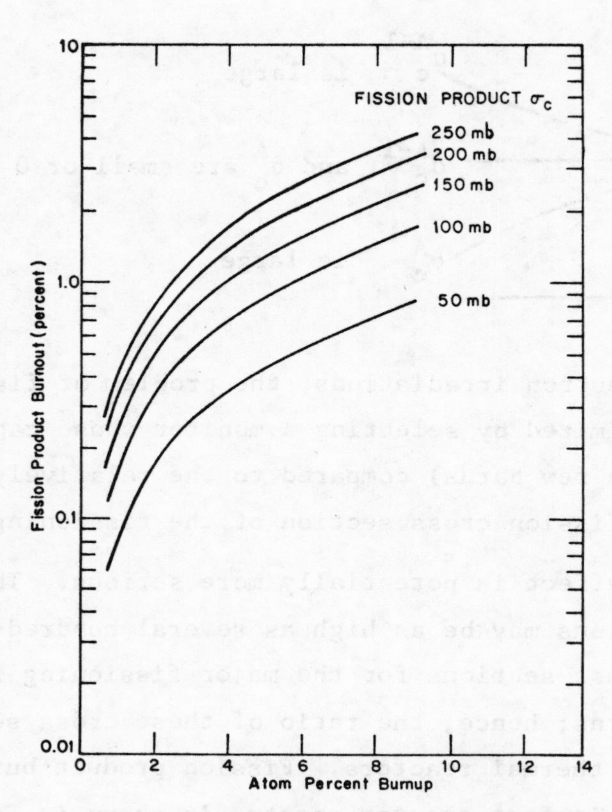

Fig. 17-1. FISSION PRODUCT BURNOUT

Fission Product and Heavy Atom Migration

Certain fission products and uranium migrate in mixed oxide fuels irradiated at high temperatures. Post irradiation examination of oxide fuels has established that cesium migrates both radially and axially with radial concentration gradients as high as 1000 having been observed. Barium also migrates axially but to a lesser degree. Technetiu and molybdenum separate from the oxide matrix and precipitate along with ruthenium, rhodium, and palladium as noble metal inclusions in the oxide matrix. These inclusions not only migrate in the fuel but are also extremely difficult to dissolve. Zirconium has been observed to separate from a UO_2-PuO_2 matrix as barium-strontium zirconates at burnups of 7% and greater. It has not been established whether or not these zirconates can be readily dissolved. Migration of zirconium has not been definitely established in FBR fuels. It has also been reported[9] that fission product zirconium in highly irradiated oxide fuels does not dissolve in such a way that a true solution of this element is obtained, i.e., the aqueous chemistry of the zirconium in a solution of the dissolver fuel is neither reproducible nor predictable. Of the fission products studied to date, only the rare earths appear stable in oxide fuels; however, rare earth migration in carbide fuels has been observed[10]. In mixed oxide fuels, the radial migration of uranium from the center of the fuel to the cooler regions has been observed to be as high as 50%[11].

The migration of fission products and heavy atoms must be considered when sampling irradiated fuel for a burnup analysis. The migration effect is most important when sampling experimental fuels for burnup analysis. Because uranium and fission products migrate, pinpoint sampling can give erroneous results.

The problem is less severe when complete fuel rods are dissolved; however, one must still consider the noble metal inclusions which are nearly impossible to dissolve. Based on the above discussion, only the rare earth and possibly zirconium nuclides can be considered as burnup monitors for FBR oxide fuels.

Fission Product Half-Life

The operation of a FBR will involve a long fuel residence time and management of fuel within the core at refueling shutdown times. Thus, stable fission product nuclides or those having a long half-life are preferred as fission monitors to avoid errors caused by corrective calculations for in-pile and out-pile decay times. Also preferred are nuclides which have short-lived precursors, again to minimize decay correction. Essentially all of the decay chains which lead to the stable rare earth fission products, except mass 144 with 284-day ^{144}Ce and 147 with 2.6-y ^{147}Pm, are characterized by short-lived precursors and stable end products.

Analytical Methodology

In the selection of the analytical method (or methods) for the determination of a fission product monitor, uranium, and plutonium, the interacting factors of accuracy and cost must be considered relative to the reliability of the burnup measurement. The factors that affect cost include man-hour effort, initial cost and amortization of the equipment, and the skill-level of the analytical and instrument maintenance personnel. Various analytical methods now in use and undergoing development are presented below. These presentations emphasize the technical aspects of each method and discuss the factors of accuracy and cost to show their relative advantages and disadvantages.

Determination of Fission Product Monitors. On the basis of the criteria presented in the preceding sections, fission product monitor for FBR fuels are limited to rare earths and zirconium. Various techniques for the measurement of these fission product nuclides are discussed below.

Neodymium by Isotope Dilution Mass Spectrometry. For thermal reactor fuels in which the major sources of fissions are ^{235}U and ^{239}Pu, the isotope dilution mass spectrometric determination of ^{148}Nd is considered to be the most accurate burnup method. The ^{235}U and ^{239}Pu fission yields of ^{148}Nd differ by less than 1% relative and the yields are accurately known. Laboratories throughout the world are proficient in its use and regard it as the reference method for interlaboratory comparisons of exchange samples and for the evaluation of new methods.

For most of the fast-reactor experimental fuels irradiated in EBR-II, the situation is the same: ^{235}U and ^{239}Pu are the principal fissioning nuclides (the uranium in these fuels is fully enriched) and the ^{148}Nd fission yields probably differ by less than 2% relative.

In the FFTF and future FBRs, mass spectrometric isotopic dilution analysis will continue to be the most accurate analytical technique for burnup determinations; however, ^{148}Nd may no longer be the burnup monitor of choice as ^{235}U will no longer be a significant source of fission in these fuels. The burnup monitor will probably be another neodymium isotope or the sum of them to better meet the criterion of constancy of fission yield with fissile nuclide.

As an analytical technique for fission product analysis, isotope dilution mass spectrometry for either a specific isotope, or the element, has the important quality of requiring only a partial recovery of the determined constituent in the separation procedure preceding the mass spectrometric analysis. It also is more accurate and sensitive than any of the other techniques. The accuracy, 0.5% or better, is a factor of two better than X-ray spectrometry and a factor of six better than spectrophotometry. The advantage afforded by the high sensitivity is that the level of radioactivity to which the analyst is exposed is less than that for the other techniques. The method is also very specific; hence, none of the large number of constituents in an irradiated fuel will interfere in the analysis. There is, therefore, a high degree of confidence in the measurement.

A disadvantage of the technique has been the use of long and complex procedures to separate the neodymium from the other fission products and the heavy atom nuclides. Present efforts to develop faster procedures are underway.

Mass spectrometric analysis is the most expensive of the burnup analytical measurement techniques with respect both to equipment cost and the caliber of personnel required to maintain and operate it. The instrument cost is about $100,000.

Rare Earths by X-ray Spectrometry. The use of the rare earth group of fission products (La, Ce, Pr, Nd, and Sm) as a burnup monitor has nuclear advantages compared to the use of a single fission product nuclide or a group of nuclides of a fission product element. Because the individual nuclides in the group are mostly sequential, any effect of fission product burnout and burnin on the linear relationship between number of fission product atoms and number of fissions is unlikely. Neutron capture on masses 139 to 145 and 148 to 151 (combined fission yield of $\sim$37%) transmute a measured rare earth nuclide to another measured rare earth nuclide. Only neutron capture on masses 146, 152, and 154 (combined fission yield of $\sim$3%) produces promethium and europium nuclides which are not measured. The only burnin reactions affecting the linearity are neutron captures on ^{139}Ba and ^{147}Pm (combined fission yield of ~7%). The accuracy of the sum of the 13 rare earth fission yields is better than the accuracy of a single fission yield value unless the error of each is totally systematic in the same direction.

In the X-ray spectrometric analysis for the rare earth group, a known amount of terbium (a rare earth not produced in fission) is added to the sample aliquot as a chemical yield monitor for the separation. The separation procedure is simple with no need for quantitative recovery and, unlike the neodymium mass spectrometric method, there is no need for a chromatographic rare earth separation. A disadvantage is that the sensitivity of the technique is about ten-fold less than the isotope dilution mass spectrometric technique. This requires the use of larger sample aliquots for the analysis. In practice, however, personnel radiation exposures are effectively controlled by strategically placed shielding. The cost of an X-ray spectrometer is about one-third the cost of a mass spectrometer and is easier to operate and maintain. The specificity is comparable to that afforded by mass spectrometry. The accuracy of the X-ray spectrometric rare earth group determination, obtained by the ANL scientists who developed the method, is ~1%.

The X-ray spectrometric technique can also be used to determine elemental neodymium. The advantage, compared to the rare earth group determination, is a reduction in the time required for X-ray measurements. Disadvantages are the use of a less accurate fission yield valu[e] and the necessity of correcting for 284-day ^{144}Ce that has not decayed to ^{144}Nd. This correction can be as high as 20% for fuel with a 1% burnup and a 30-day cooling period. For higher burnup the correction will be less. This correction can either be calculated from the irradi[-] ation history or the ^{144}Ce can be measured by a gamma spectrometric analysis. The accuracy of the neodymium measurement is ~1% including the error contribution from a measured ^{144}Ce correction.

As the monitor for total fissions, the rare earth group is considered excellent both for FFTF fuel and first generation FBR fuels. The summed rare earth group fission yields for ^{238}U, ^{239}Pu, and ^{241}Pu are expected to differ by less than 3 relative percent for FBR spectra.

Rare Earths and Zirconium Using Other Instrumental Methods.

Elemental burnup monitors such as the rare earths and zirconium, as well as uranium and plutonium, could be determined by more conventional analytical techniques; for example, spectrophotometry and microtitrimetry. If one of these techniques was used rather than mass spectrometry or X-ray spectrometry, the accuracy would be less, about 3% versus 0.5 and 1%, respectively, but there would be a large saving in equipment cost and it appears that the time required to execute a burnup determination would be less. The time reduction of a burnup determination by spectrophotometric or microtitrimetric methods would be that associated with the measurement of the fission product monitor and the uranium and plutonium after the chemical separations. Operations for spectrophotometric or titrimetric measurements are relatively fast, requiring only about 1/3 to 1/4 of the time compared to mass or X-ray spectrometry. Spectrophotometers and titrimeters cost about $5,000 compared to $100,000 for mass spectrometers and $30,000 to $40,000 for X-ray spectrometers.

When using spectrophotometric and microtitrimetric methods, two additional criteria that are not required for the mass and X-ray spectrophotometric methods are placed on the separation procedures that precede the measurements: (1) the constituent being determined must be quantitatively recovered, and (2) the separated constituent must be totally free of all other constituents that would behave or react chemically in a similar manner when the measurement is made. The first criterion need not be met in the mass and X-ray spectrometric methods because the measurements are made relative to an internal standard and the second criterion is much easier to meet because very specific properties of the constituent, isotopic abundances or the intensities of characteristic X-rays, are measured.

For the rare earths and zirconium, the criterion of quantitative recovery may be circumvented. Gamma-ray measurements can be made of the ^{144}Ce (284 days) [or ^{95}Zr (65 days)] content of the initial sample aliquot and of the separated rare earths (or zirconium) fraction under known counting geometry conditions and the recovery calculated.

Both ^{144}Ce and ^{95}Zr emit energetic gamma rays which are measurable with high precision. The time required for these relative measurements is only a few minutes.

Meeting the second criterion for spectrophotometric and microtitrimetric methods is difficult, particularly for the determination of the burnup monitors. Presently, no conventional chemical methods are available for the determination of either zirconium or rare earth burnup monitors in plutonium-containing fuels. However, an ion exchange separation-spectrophotometric measurement method has been developed for the determination of zirconium in mixed ^{232}Th-^{235}U fuels in which the fissioning nuclides are ^{233}U and ^{235}U[12]. Because the sums of the zirconium fission yields for ^{233}U and ^{235}U differ by only 10 relative percent, an accurate value for the fission yield used to calculate burnup can be made. Unfortunately, the sums of the zirconium yields for ^{238}U, ^{239}Pu, and ^{241}Pu do not agree as well. The estimated fast reactor sum yields are 25%, 19%, and 15%, respectively.

As discussed in the preceding section, total rare earths is an excellent burnup monitor for the various FBR fuels because the summed fission yields for fast-fissioned ^{238}U, ^{239}Pu, and ^{241}Pu are expected to be nearly equal. One reason for using total rare earths, rather than a single rare earth element as the burnup monitor, is that the chemical separation is faster. Separation of a single rare earth requires an additional chromatographic separation that would be time-consuming.

Because spectrophotometric and microtitrimetric methods are inherently less accurate than mass and X-ray spectrometric methods, the burnup determinations will not be as accurate. However, such methods should be very useful for the routine determination of burnup for large sample loads where less accurate results are acceptable.

Determination of Uranium and Plutonium. The number of heavy atoms originally present in the fuel (the denominator in the equation defining atom percent burnup) is obtained by determining the number of uranium and plutonium atoms in the irradiated fuel specimen and summing this with the number of fissions. (The number of heavy atoms other than uranium and plutonium in the irradiated fuel is very small and, if necessary, can be calculated.) Three techniques for the determination of the heavy atom content of the irradiated fuel are discussed below; mass spectrometry, X-ray spectrometry, and spectrophotometry. The general advantages and disadvantages of each technique were discussed in the preceding section.

Isotope Dilution Mass Spectrometry. The most accurate technique for determining uranium and plutonium in solutions of irradiated fuel is isotope dilution mass spectrometry. Known amounts of ^{233}U and ^{242}Pu are added to an aliquot of the dissolved fuel specimen, the uranium and plutonium are separated from the fission products and from each other, the isotopic ratios are measured, and the number of uranium and plutonium atoms in the aliquot are calculated from these ratios and the amounts of ^{233}U and ^{242}Pu added. Because FBRs produce a significant amount of ^{242}Pu in the fuel, it is currently necessary to perform two plutonium isotopic ratio measurements, one with and one without the ^{242}Pu added. The need for the second analysis will be eliminated when a supply of high purity ^{244}Pu becomes available. Uranium and plutonium determinations having accuracies of 0.25% are routinely obtained using this technique.

When isotope dilution mass spectrometry is used for the three determinations, the fission product burnup monitor, uranium, and plutonium, a so-called triple spike solution is used. For example, in the analysis of low ^{235}U-enriched, light-water-reactor fuel, a spike that contains a mixture of ^{150}Nd, ^{233}U, and ^{242}Pu is used rather than individual solutions of the spike isotopes. This triple spike solution is prepared in such a way that the ^{150}Nd to ^{233}U and ^{150}Nd to ^{242}Pu atom ratios are known. By using the triple spike, an important operational advantage is realized; it is no longer necessary to dispense quantitatively either the spike or the dissolved fuel solution. This contributes significantly to the accuracy of the burnup analysis as it is very difficult to operate and maintain reliable sample dispensing systems in shielded facilities. That the amount of dissolved fuel dispensed (as well as the amount of the triple spike) need not be known is the result of the fact that burnup is defined by a ratio of fission product atoms relative to the sum of the fission product atoms and the residual uranium and plutonium atoms, all measured relative to the spike isotopes.

X-Ray Spectrometry. The uranium and plutonium in the irradiated fuel specimen appear determinable by X-ray spectrometry with an accuracy of 1 to 2%. A proposed scheme is to take two aliquots of the dissolved fuel specimen, add a known amount of uranium to one of them, quantitatively separate the uranium and plutonium from the fission products, and measure the uranium to plutonium ratios by X-ray spectrometry. The amounts of uranium and plutonium in the fuel specimen are calculable using these ratios and the amount of uranium added.

If X-ray spectrometry were used to determine the fission product burnup monitor, uranium, and plutonium, a combined spike solution could be used and the same operational advantages as those discussed above for the triple isotope spike in the isotope dilution mass spectrometric technique would be realized. In the total rare earth or total neodymium burnup determination, the combined spike would contain terbium and uranium at a known ratio.

Spectrophotometry. A spectrophotometric analysis of the solution of dissolved fuel for uranium and plutonium should be less expensive than the mass spectrometric or X-ray spectrometric methods, and have an accuracy of from 2 to 3%. An even less expensive measurement, applicable to fuels with a high uranium to plutonium ratio, would be to determine only the uranium. In this case the measured quantity of uranium would be corrected for the calculated amount of uranium transmuted to ^{239}Pu by the $^{238}U(n,\gamma)^{239}Pu$ reaction and then multiplied by the pre-irradiated uranium plus plutonium to uranium ratio. At a burnup of 10%, the correction for the ^{238}U loss would be about 3±0.3 atom percent if the ^{238}U capture cross section had an uncertainty of 10%. The error in the initial heavy atom content from this source, therefore, would be only 0.5% relative.

Error Analysis

The sources of error in the determination of burnup by the fission product monitor-residual heavy atom technique are (1) the measurements of the numbers of atoms of the uranium, plutonium, and fission product monitor in the analyzed irradiated fuel specimen, and (2) the fission yield value that is used. Measurement errors are established by repeated analyses of synthesized solutions of uranium, plutonium, and the fission product elements. These solutions are prepared from NBS uranium and plutonium compounds and/or metals and highly pure and stoichiometric fission product element compounds. If biases are detected in the analyses of these solutions, modifications are made in the methods to eliminate them.

Fission yield values are reported with an associated uncertainty obtained by propagating the random measurement errors involved in their determinations and the uncertainties of the values of any factors used in the calculations. In the calculation of burnup from the fission yield and the measured fission product and heavy atom contents of a dissolved specimen, it should be recognized that the fission yield error is constant while the measurement errors are random. The uncertainty of the fission yield and the random errors of the measurements are customarily expressed as the standard deviation and the relative standard deviation, respectively. In this section both are expressed as relative standard deviation.

The overall error of a burnup determination is computed by propagating the errors of the chemical measurements and the fission yield value. Fig. 17-2 shows the magnitude of the error in the burnup determination as a function of the individual measurement errors of uranium, plutonium, and the fission product monitor for fission yield uncertainties of 1, 2, 3, and 4%. The 4% value is that of the existent fast fission yield data. The need for more accurate fission yield data is clearly demonstrated. At a 4% level of uncertainty in the yield data, the errors in the burnup values are 4.1, 4.2, and 6%, respectively, when the measurements are made mass spectrometrically, X-ray spectrometrically, and spectrophotometrically. At a 1% uncertainty in the yield data, the burnup errors would be 1.2, 1.5, and 4.3%, respectively.

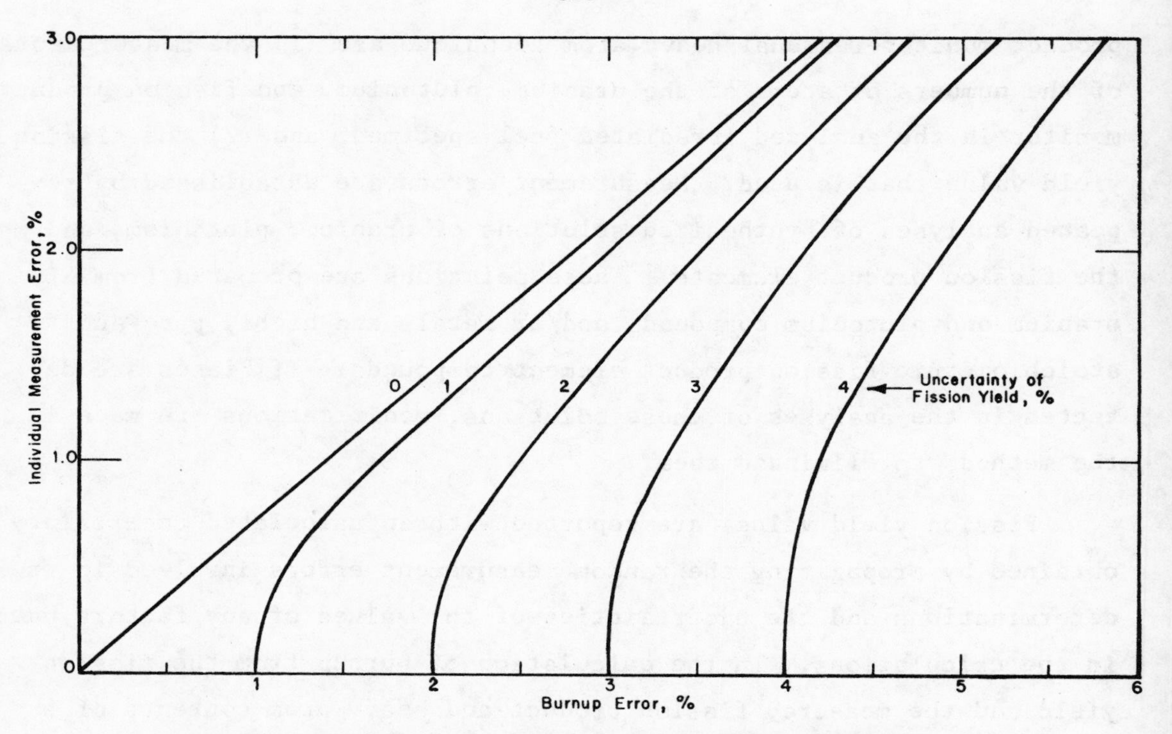

Fig. 17-2. EFFECT OF MEASUREMENT ERRORS AND FISSION YIELD UNCERTAINTY ON BURNUP ERROR

Another point to be considered is the interplay of the random errors of the chemical measurements and the constant error of the fission yield when the objective of the burnup determination is to compare a series of fuel specimens of the same composition irradiated under similar conditions. The error in the relative burnup values is due solely to the random errors of the chemical measurements of uranium, plutonium, and the fission product monitor. For this case, the line that applies in Figure 2 is the one for zero uncertainty in the fission yield value.

HEAVY ATOM RATIO TECHNIQUE

The irradiation of a fast reactor fuel to a relatively high burnup results in significant changes in certain heavy atom ratios which can serve as the basis for calculating burnup. During an irradiation, there are a multiplicity of nuclear reactions which take place, and as a result, the equations which have been derived for making the burnup calculation from the heavy atom ratios and the cross section data are quite complex. For this reason, it is not readily apparent how the accuracy of the result is affected by the errors in the measurements and the uncertainties of the nuclear constants, and as a consequence, there exists in the nuclear industry a fair degree of misunderstanding about the merits of the heavy atom ratio burnup technique. The discussion of the heavy atom ratio technique that follows presents first, the conditions on which the application is predicated; second, a summary of how the measurements are made; third, a simplified yet rigorous version of the calculational methods; and fourth, an error analysis.

Conditions of Applicability

The applicability of this technique to FBR fuels is predicated on the following:

(1) Both an irradiated sample and an unirradiated sample must be analyzed and both samples must have been equivalent prior to the irradiation. Both the uranium and plutonium isotopic compositions must have been identical and the ratio of uranium to plutonium must have been the same.

(2) The pre- and postirradiation uranium and plutonium isotopic ratios relative to one of the principal heavy atoms, the "reference" heavy atom, must be accurately measured, approximately 0.5 relative percent or better.

(3) The relative change in the amount of the reference heavy atom must be small, 10% or less.

(4) The neutron capture product of the principal fissioning atom must have a relatively long half-life, it must be an isotope of the same element, the relative increase in the amount of the product heavy atom formed during the irradiation must be large, and the amounts of this product formed or lost by other nuclear reactions must be calculable.

(5) Accurate neutron cross section data must be available for all the heavy atoms that undergo absolute changes that are 1% (relative) or more of the burnup.

For most FBR fuels, the reference heavy atom would be ^{238}U, the principal fissioning heavy atom would be ^{239}Pu, and the product heavy atom would be ^{240}Pu.

Heavy Atom Ratio Determination

The analytical method used to determine uranium and plutonium isotopic ratios relative to the reference heavy atom, ^{238}U, is isotopic dilution mass spectrometry. Specimens of both the irradiated and unirradiated fuel are dissolved, portions of a spike solution having a known $^{233}U/^{242}Pu$ ratio are added to portions of both dissolved fuels, the uranium and plutonium are separated from the fission products and each other, and the uranium and plutonium isotopic ratios are measured. Because ^{242}Pu is initially present in the fuel and is formed in the irradiation, unspiked portions of both the preirradiated and postirradiated fuel must be analyzed for their plutonium isotopic ratios. A total of six mass spectrometric measurements, two for uranium and four for plutonium, therefore, is required. From the six sets of uranium and plutonium mass spectrometric measurements and the known ^{233}U to ^{242}Pu ratio in the spike, the pre- and postirradiation atom ratios of each uranium and plutonium isotope relative to ^{238}U are calculated.

Calculations

In the derivation of the relationship for calculating burnup that follows, percentages rather than heavy atom ratios are used and the only nuclear reactions considered are those that affect ^{238}U, the reference heavy atom, ^{239}Pu, the principal fissioning heavy atom, and ^{240}Pu, the principal product heavy atom. Percentages rather than heavy atom ratios are used, because in the opinion of the authors, the nature of the calculations are more readily understood and the effect that errors in the measurements and uncertainties in the cross section data have on the accuracy of the burnup value are more apparent. The percentages are obtained by dividing each of the calculated atom ratios by the sum of the atom ratios and multiplying by 100. Examples of the pre- and postirradiation percentages are given in Columns 2 and 3, respectively, of Table 17-5. The percentages in Column 4 are the "true" postirradiation percentages on the same relative scale as the percentages in Column 2, heavy atoms per 100 heavy atoms initially present in the fuel. These are the values which the heavy atom ratio technique attempt to calculate from the values in Columns 2 and 3 and the cross section data.

Table 17-5. PRE- AND POSTIRRADIATION HEAVY ATOM PERCENTAGES AND THEIR ABSOLUTE CHANGES FOR 10% BURNUP OF FFTF FUEL

Heavy Atoms	Atom Percentages				
	Measured Preirrad.	Measured Postirrad.	"True" Postirrad. [b]	Fissions	Captures
^{235}U	0.50	0.305	0.274	0.18	0.045
^{236}U	nil	0.05	0.045	nil	nil
^{238}U	74.41	76.72	69.05	0.71	4.65
^{239}Pu	21.69	18.20	16.38	8.54	1.42
^{240}Pu	2.90	4.08	3.67	0.25	0.40
^{241}Pu	0.43	0.52	0.47	0.31	0.05
^{242}Pu	0.07	0.13	0.115	negl.	negl.
TOTALS	100.00	100.00	89.95	9.99	

[b] Values obtained when the decrease in ^{238}U caused by fission and capture is calculated with no error.

A derivation of the calculational relationships in which only the ^{238}U, ^{239}Pu, and ^{240}Pu nuclear reactions are considered has been made to provide a better understanding of the method. The relationships are simplified but the basic steps are retained. The accuracy of the heavy atom ratio burnup technique is limited by the accuracy of the measurements and calculations that relate ^{238}U, ^{239}Pu, and ^{240}Pu. The abundances and/or cross sections of the other heavy atoms are such that the atom fraction of their transmutations relative to the total is small and calculable with sufficient accuracy.

The main thrust of the burnup calculation is to normalize the postirradiation percentages to the same relative scale as the preirradiation percentages; namely, per 100 heavy atoms initially present in the fuel. This is done by an iterative calculation of the absolute percent loss of ^{238}U that occurred in the irradiation. This loss, ^{238}L, is subtracted from the preirradiation percent ^{238}U, $^{238}A_i$, and the difference is divided by the postirradiation percent ^{238}U, $^{238}A_f$, to give a factor, R. The postirradiation percentages are normalized to the preirradiation scale by multiplying them by R. The burnup is the difference between the sum of the preirradiation percentages and the sum of the normalized postirradiation percentages.

The relationships which approximate the solutions of the differential equations that are used to make the calculation of ^{238}L are:

$$^{238}L = \Phi\ ^{238}\bar{A}\ ^{238}\sigma_a \tag{1}$$

$$^{239}C = \Phi\ ^{239}\bar{A}\ ^{239}\sigma_c \tag{2}$$

$$^{240}L = \Phi\ ^{240}\bar{A}\ ^{240}\sigma_a \tag{3}$$

$$^{240}I = {}^{239}C - {}^{240}L \tag{4}$$

where

L = absolute percent loss

C = absolute percent capture

I = absolute percent increase

A_i = preirradiation percent

A_f = postirradiation percent

$\bar{A}$ = average percent

σ = cross section

Φ = integrated neutron flux

a = absorption

c = capture

An algebraic solution of these equations gives the relationship used for the iterative calculations of ^{238}L.

$$^{238}L_n = \frac{^{238}\bar{A}_{n-1}\ ^{238}\sigma_a\ ^{240}I_{n-1}}{\left(^{239}\bar{A}_{n-1}\ ^{239}\sigma_c\right) - \left(^{240}\bar{A}_{n-1}\ ^{240}\sigma_a\right)} \qquad (5)$$

where the subscript n denotes the iteration round. $\bar{A}_{n-1}$ and the I_{n-1} are the percentages calculated from the preirradiation percentages, A_is, and the product of the postirradiation percentages, A_fs, and R_{n-1}; $R_{n-1} = (^{238}A_i - {}^{238}L_{n-1})/{}^{238}A_f$. The iterative calculation of $^{238}L_n$ is carried out until the difference of $^{238}L_{n+1}$ and $^{238}L_n$ is less than or equal to 0.01.

Equation (5) can be written in the form

$$^{238}L = \frac{^{238}A\ ^{238}\sigma_a\ ^{240}I/^{239}\sigma_f}{^{239}A\ ^{239}\alpha - {}^{240}A\ ^{240}\alpha_a/^{239}\sigma_f} \qquad (6)$$

The number of rounds required to achieve this can be reduced significantly if prior to the first round, the postirradiation percentages, A_fs, are multiplied by (1-F), where F is the estimated fractional burnup. The better the estimate of the fractional burnup, the closer the A_f (1-F) values will be to those that the calculation needs to establish.

Normalized percentages obtained for several rounds of the iterative calculations as well as the pre- and postirradiation percentages are tabulated below. The burnup is 10%, and the burnup estimate is 9%, i.e., 1-F is 0.91.

Heavy Atom	Measured Percents Pre	Post	Calculated Post Percents Post(1-F)	Round 1	Round 2	Round 3
^{238}U	74.41	76.72	69.81	68.81	69.20	69.05
^{239}Pu	21.69	18.20	16.56	16.32	16.42	16.38
^{240}Pu	2.90	4.08	3.71	3.66	3.68	3.67
Others	1.00	1.00	0.92	0.90	0.91	0.91
TOTALS	100.00	100.00	91.00	89.69	90.21	90.01
PRE TOTAL-CALC. POST TOTAL			9.00	10.31	9.79	9.99

where $^{239}\sigma_f$ is the ^{239}Pu fission cross section and $^{239}\alpha$ is the ^{239}Pu capture to fission cross section ratio. Equation (6) rather than (5) should be used when the cross section ratios relative to $^{239}\sigma_f$ are more accurately known than are the individual cross section values.

Sources of Error

The accuracy of the burnup results in this technique depends on (1) the uncertainties of the neutron cross section values that are used in the computation, (2) the level of burnup, and (3) the accuracies of the isotope dilution mass spectrometric measurements.

The tabulation below shows the effects of uncertainites in the cross section values for the three cross sections involved in Equation (5) for an FFTF fuel having a burnup of 10.0%. In this computation, each cross section value was increased by 10 relative percent with the other two cross sections held constant.

Cross Section	Change in Cross Section, Relative %	Change in Burnup Result, Relative %
$^{238}\sigma_a$	+10	+4
$^{239}\sigma_c$	+10	-9
$^{240}\sigma_a$	+10	+5

The effect of the uncertainty of the mass spectrometric measurement upon the computed burnup result has been calculated for the previous example. A change in the $^{240}Pu/^{239}Pu$ postirradiated atom ratio of 0.4 relative percent, which approximates the maximum propagated error for this measurement for the pre- and postirradiated fuel in an experienced mass spectrometry laboratory, gives a relative percent change of 1% computed result at 10% burnup. This is equivalent to an absolute uncertainty in the percent burnup of 0.1. This absolute uncertainty is essentially independent of the burnup level; hence, the heavy atom ratio technique becomes less applicable as the burnup level decreases.

A prevalent source of error in the heavy atom ratio technique for the determination of burnup and one over which the chemist doing the analysis has no control is the equivalency of the pre- and postirradiated fuel specimens analyzed. It has been the observation of the authors that the equivalency requirement of the pre- and postirradiated fuel samples has not been met on a number of occasions, despite claims to the contrary by the individuals submitting the samples for analyses. For example, cases have been observed where the uranium and plutonium analyses (chemical and/or isotopic) of so-called duplicate specimens of the preirradiated fuel do not agree while replicate analyses of the same sample do. This situation is likely for fuels prepared by blending uranium and plutonium oxides, particularly when different batches of the oxides are used to manufacture a fuel lot.

The heavy atom ratio technique will be less applicable to the determination of burnup for large breeder reactor fuels than for FFTF fuel. Due to the higher atom percentages of ^{240}Pu in the LWR-produced or recycled plutonium that will be used in the manufacture of these fuels, the accuracies with which ^{238}L can be calculated will decrease. As seen from Equations (3), (4), and (5), an increase in $^{240}A_i$ leads, in turn, to a larger value of ^{240}L, a smaller value of ^{240}I, a larger error in the value of ^{240}I, and a larger error in ^{238}L.

As more accurate cross section values are experimentally determined, the accuracy of burnups calculated by this technique will improve. However, it is doubtful that accuracies comparable to those obtained by the fission product monitor-residual heavy atom technique can be achieved even at high burnup. It is important to note that the fission product monitor technique is one of the key techniques used to determine these nuclear constants and it will always be the reference method against which the heavy atom ratio technique must be evaluated.

HEAVY ATOM DIFFERENCE TECHNIQUE

In this approach, the number of fissions is established by subtracting the number of heavy atoms measured in the irradiated fuel specimen from the number of heavy atoms measured in an equivalent specimen of preirradiated fuel. In principle, this approach is the best of the three because nuclear constants (including fission yield and neutron cross section data) are not involved in the calculation. However, in practice, it is not possible to achieve the requisite accuracy for FBR fuels for two reasons. First, the number of fissions relative to the number of total heavy atoms is relatively small, not exceeding 10%. If the number of uranium and plutonium atoms is determined with a relative standard deviation of 0.25%, (this precision is routinely obtained in experienced laboratories using isotope dilution mass spectrometric techniques) the propagated error in the measured number of fissions is 18% for 2% burnup and 3.4% for 10% burnup.

The second reason is the practical difficulty of obtaining pre- and postirradiated fuel specimens that are matched on an equivalent weight basis. At higher burnups, fuel melting and swelling, heavy atom migration, and cladding-fuel reactions occur. The assumption that a weighed specimen of the irradiated fuel contains all the heavy atoms initially present less those that have fissioned and the same ratio of fuel to cladding as the preirradiated specimen is untenable.

REFERENCES

The publications referred to in chapter 17 may be found in "Burnup Determination for the Fast Reactor Fuels: A Review and Status of the Nuclear Data and Analytical Chemistry Methodology Requirements" by William J. Mack, Allied Chemical Corporation-Robert P. Larsen, Argonne National Laboratory, -James E. Rein, Los Alamos Scientific Laboratory

CHAPTER 18

FUSION POWER

Controlled fusion research is a scientific discipline which developed worldwide over the past 20 years. In the early 1950's fusion was a classified field of research, and little was known about its root science -- the physics of high temperature plasmas. In 1958, the fusion program was declassified, and by the early 1960's a number of relevant scientific problems were identified and a systematic study of them begun.

The difficulties that arose became the central problem of fusion research -- the isolation of a reacting fusion plasma from its surroundings. The principal approach to this problem, then as now, was to confine a fusion plasma through the use of specially shaped magnetic fields, which were to control the motions of its individual ions and electrons. However, in attempting to apply this technique, it was soon discovered that spontaneously arising turbulence and unstable plasma oscillations significantly weakened the confining effect of the magnetic fields. As a result of several years of intensive theoretical and experimental research, the plasma instability problem was brought under reasonable control by the late 1960's. In fact, the understanding of instabilities and means for their control was sufficient to permit experiments which exhibited confinement close to the "classical" upper limit -- the theoretical maximum possible in a completely quiescent plasma at a particular density and temperature. This achievement was obtained in several different experiments, and it provided a basis for renewed optimism with respect to ultimate success.

The scientific, technical and size limitations of present-day fusion experiments preclude any of them from simultaneously achieving all three of the plasma parameters (temperature, density and confinement time) required for a fusion reactor. This achievement, which would demonstrate scientific feasibility, will require larger, more complex facilities than presently available. In addition, it will be necessary to continue the development of relevant technologies at an expanded level in parallel with plasma experimentation.

There clearly remain significant scientific questions about plasma behavior in reactor regimes of temperature and density, but if the presently favorable trends toward better fusion plasma confinement continue, these questions should be sufficiently resolved so that preparations for scientific feasibility experiments can begin in the mid-1970's. These might be ready to begin operation near the end of the decade, and then they should be ready to test fusion scientific feasibility in the 1980-1982 period.

The laser-fusion process was recognized in the early 1960's as being of potential military interest and as a possible method of achieving practical fusion power. In the late 1960's it was possible to envision the high power lasers capable of permitting the detailed study of the important physics questions relevant to laser-fusion, and this provided part of the basis for expanding the laser-fusion program. Large lasers are now being developed in an effort to perform the key basic studies. If these studies prove favorable, still larger new laser facilities could permit the demonstration of the scientific feasibility of the laser-fusion process in the latter 1970's.

The quest for fusion power has resulted in the development of a new field of research -- high temperature plasma physics. Plasma physicists believe that in the coming decade the scientific proof that fusion reactors can indeed be built can be obtained. Given that proof, the fusion program can shift into a development phase in which the practical problems of producing economically competitive electrical power can be fully addressed.

The primary application of fusion power will be for the production of electrical and thermal energy. It is difficult to predict when this achievement will be attained because of the myriad of physical, technological, and socio-economic variables which can either accelerate or slow its future development. An analysis of what might be accomplished in an orderly aggressive program indicates that central station fusion power might become commercial about the year 2000. Assuming any of various models for its introduction into the utility market thereafter, fusion power could then have a significant impact on electrical power production in the year 2020.

INHERENT CHARACTERISTICS OF FUSION POWER SYSTEMS

Even though fusion power is not yet a reality, it is possible to assess approximately its operating, safety, economic, and environmental characteristics. Fusion systems potentially offer a number of attractive advantages and a variety of choices to the utilities and to the public. To convey the flexibility of fusion as well as to provide specific information, the approach herein is to first describe some of the inherent features of the basic system and then to present the characteristics of preliminary fusion reactor designs based on the deuterium-tritium (DT) fuel cycle.

Fusion reactors will be inherently safe against nuclear runaway. All full-scale fusion reactor concepts involve small quantities of fuel (of the order of a gram) in the core region. The working fluid of fusion is a gaseous plasma, which because of its high pressure tends to expand whenever it is not suitably confined. When a plasma expands, its fusion rate, which is proportional to the square of its density, decreases drastically. Plasma confinement is a delicate process requiring a high degree of control. These basic characteristics indicate that whenever confined fusion plasmas are perturbed in any manner other than by very special kinds of compressive forces, they will tend to expand, thereby decreasing or quenching the reaction rate.

Fusion Reactions. Fusion reactions occur when two light nuclei such as Deuterium (D), Tritium (T), or Helium (^3He) collide and rearrange themselves so as to form two other nuclei of smaller mass with a consequent release of energy. The following reactions are of primary interest in CTR research:

Energy Required	Fusion Reaction	Energy Released
~ 10 KeV	$D + T \rightarrow {}^4He + n$	17.6 MeV
~ 50 KeV	$D + D \rightarrow {}^3He + n$	3.3 MeV
	$D + D \rightarrow T + p$	4.0 MeV
~ 100 KeV	$D + {}^3He \rightarrow {}^4He + p$	18.3 MeV

There are several important features of these reactions. They take place only at very high energy, because of the strong Coulomb repulsion between approaching nuclei. To achieve particle energies of 10 KeV the mixture must be at a temperature of about 100 million degrees Kelvin. At this extreme temperature the gas atoms are fully ionized -- they have all shed their electrons. Such a fully ionized gas is called a plasma, and the fact that the particles are charged particles makes possible their containment with magnetic fields.

The reaction products of a fusion reaction carry away the energy of the fusion reactions. In the DT reaction most of the energy is carried away by a 14 MeV neutron whereas in the D^3He reaction all of the energy is released to charged particles. The form in which the energy is released determines the method of energy conversion. In the case of the DT reaction, the energy of the escaping neutrons must be degraded to heat and converted through a thermal cycle to electricity. When the energy appears in the charged particles, a unique direct conversion of their energy to electricity is possible. This has the potential for very high efficiency and low waste heat.

In the DT reaction, which is generally regarded as the easiest to achieve because of its relatively low reaction energy, tritium must also be supplied as a fuel. Breeding of tritium from lithium and thermalization of the DT neutrons would be accomplished in the blanket which surrounds the plasma of a DT fusion reactor. Energy conversion and tritium breeding in the blanket are discussed later.

Energy Balance and the Lawson Criterion. A measure of success in obtaining net power from the thermonuclear reactions is the ratio of thermonuclear power generated to the power required to create and sustain the hot reacting mixture:

$$Q = \frac{\text{thermonuclear power}}{\text{injected power}} = \frac{\left(\frac{n^2}{2}\right) < \sigma v > E_f}{\frac{(3\ n\ kT)}{\tau}}$$

where $\left(\frac{n^2}{2}\right) < \sigma v >$ is the number of fusion reactions per cubic centimeter per second and E_f is the fusion energy released per reaction. The injected power is obtained by supposing that the entire energy, 3 n kT, must be replaced every τ seconds. Thus τ is the energy replacement time, approximately the plasma confinement time.

Lawson first stated the requirement for power balance in a fusion reactor by assuming that the entire energy, both the energy from thermonuclear reactions and the energy invested in initially creating the hot plasma could be reinvested, with an efficiency of one-third in creating new plasma. This leads to the requirement $Q \geq 2$. For the DT reaction, this places a requirement on the density-containment time product:

$$n\tau > 10^{14} \text{ sec cm}^{-3} \text{ for } T \sim 10 \text{ KeV}$$

The Lawson criterion is the origin of the usually stated objective of research on magnetically confined fusion plasmas: to create and confine plasmas with temperatures of 10 KeV or more and densities of 10^{14} particles per cubic centimeter for a second. An important density-confinement time trade-off is possible: plasmas with densities of 10^{15} need to be confined for only a tenth of a second, etc. For highly compressed (1000 times solid density) laser-fusion plasmas, densities would be of the order of $10^{26} cm^{-3}$ or more so that confinement times of picoseconds (10^{-12} seconds) would be adequate. When such plasmas are achieved, an in-principle power balance is implied whether the thermonuclear energy is converted or not. Furthermore, the overall power level of experiments achieving these plasma conditions will be substantial (tenths of a megawatt per cubic meter).

Magnetic Confinement of Plasmas. Magnetic confinement of plasmas takes advantage of the fact that in strong magnetic fields individual charged particles are confined to move along field lines in tight helical trajectories. Thus individual particles are confined to one dimensional motion along magnetic field lines. The two basically different approaches to magnetic confinement shown in Figure 18-1 differ as to whether magnetic field lines, and hence the trajectories, lead directly out of the containment region as in the open systems (a), or whether magnetic lines remain largely within the containment region as in the closed toroidal system (b). In open systems, magnetic mirrors, or regions where the strength of the magnetic field increases, are used to reflect particles and reduce the loss from the ends.

All "magnetic bottles" are variants of these simple open or closed systems. Variations are desirable, indeed necessary, to provide the necessary equilibrium and stability of the plasma confinement.

Classical Plasma Losses from the Magnetic Container. With increasing plasma density, particles occasionally collide with one another, and change their orbits somewhat. A collision may, in the case of mirror systems, allow the particle to escape through the mirror directly, or in the case of closed systems, allow the particle to take a small step toward the surface of its magnetic container. In the latter case, after many such collisions the particle is lost altogether. Thus, collisions are responsible for a slow leak from the magnetic container, referred to as a classical loss. Since it is not possible to

MAGNETIC CONTAINMENT CONFIGURATIONS

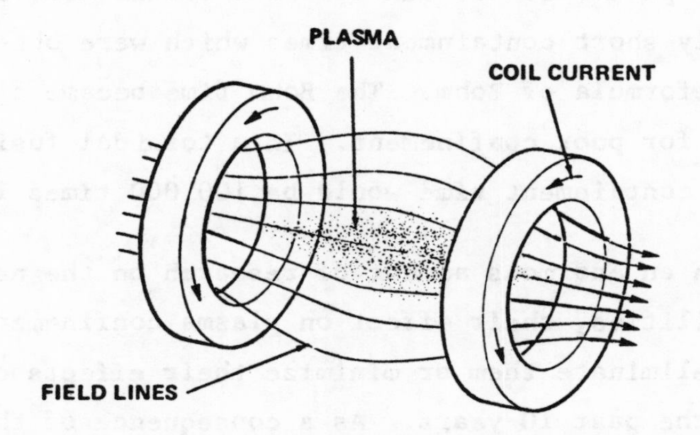

OPEN SYSTEM - SIMPLE MAGNETIC MIRROR

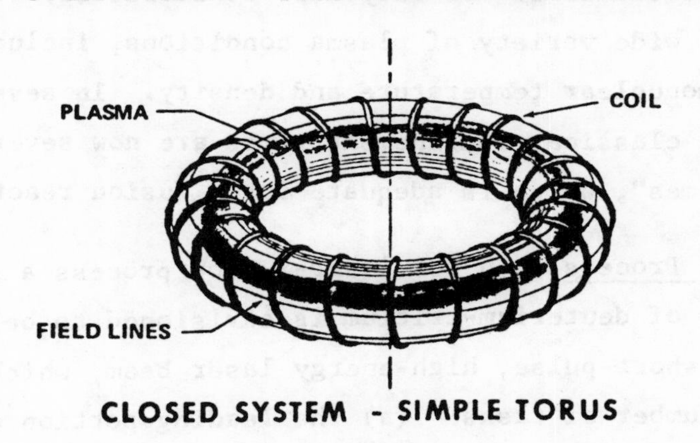

CLOSED SYSTEM - SIMPLE TORUS

Figure 18-1

eliminate these collisions, they set an upper limit on confinement time -- termed the classical confinement time -- a standard with which confinement times are generally compared. The classical confinement times are readily calculated in simple geometries and are quite long -- more than adequate for successful fusion reactors.

Anomalous Losses, Turbulence, and Plasma Instabilities. Until recently, all plasma experiments yielded loss rates considerably in excess of classical values, loss rates clearly unacceptable for a fusion reactor. Frequently these anomalous losses seemed to be associated with plasma turbulence. The existence of these fluctuating fields could often be traced to specific instabilities in the plasma which arise out of some non-equilibrium aspect of the particular plasma state. In toroidal confinement systems, the anomalously short containment times which were observed often agreed with a formula of Bohm. The Bohm time became a standard of comparison for poor confinement. In a toroidal fusion reactor the classical containment time would be 100,000 times longer.

There has been an enormous amount of research on the nature of plasma instabilities, their effect on plasma confinement, and upon ways to eliminate them or minimize their effects on confinement, during the past 10 years. As a consequence of this work, classical confinement times (or very near to classical) have been observed under a wide variety of plasma conditions, including plasmas of thermonuclear temperature and density. In several research devices classical confinement times are now several hundred "Bohm times", which is adequate for a fusion reactor.

The Laser-Fusion Process. In the laser-fusion process a solid, spherical pellet of deuterium-tritium is envisioned to be irradiated by a short-pulse, high-energy laser beam, which accomplishes a number of tasks: (a) the leading portion of the laser beam first creates a plasma surrounding the pellet such that the remainder of the beam is nearly fully absorbed; (b) the main portion of the beam then delivers energy to the pellet which not only vaporizes, ionizes, and heats the DT mixture but it also induces an implosion, which theoretically results in an ultimate compression of 1000-10,000 times the solid density; (c) the resultant highly compressed, hot DT plasma then supports fusion reactions at an extremely rapid rate in the short time (of the order of picoseconds (10^{-12})) before the pellet flies apart due to its high internal pressure. The length of time which the pellet remains in its highly compressed state is determined in large part by the inertia of the fuel mixture and the confinement is thereby called inertial.

METHOD OF ASSESSMENT

For the purposes of this section the ultimate potential of fusion power has been appraised by considering a set of reference designs for full scale fusion reactors. These reference designs were developed by personnel associated with the four major fusion concepts considered to be approaching feasibility tests. They are the tokamak, the theta pinch, the magnetic mirror, and the laser - fusion system. The reactor versions of these concepts have some common and some unique features. Three involve magnetic confinement and two of these would employ superconducting magnets to contain the hot plasma in a vacuum chamber. Laser-fusion reactions are envisioned to occur so rapidly that inertial forces provide adequate confinement. In their projected first generation reactor versions, all would utilize the DT fuel cycle and all would supply power in the range of 500 MW(e), or more, corresponding to the current size-range of central power stations. In addition, the laser and mirror systems may be suited to specialized applications requiring power outputs of as little as 50 MW(e).

All conceptual fusion reactor designs are based on the best available plasma physics information. Because the fusion reactor design activity is still in an embryonic stage, these designs differ substantially in the extent to which efforts have been made to resolve the engineering problems of both core and facility design. This is particularly true of the laser-fusion system where some thought has been given to containment vessel engineering, but little can yet be said regarding the laser. In only one case has a design gone through some iterations to factor in considerations of reactor safety. Therefore this design -- the ORNL tokamak reactor -- was chosen as the Reference Controlled Thermonuclear Reactor or Reference CTR for the purposes of this section. In many ways the conclusions drawn from the analysis of this design are believed to be representative of what might be expected for the other concepts. The choice by no means implies any favoritism towards one concept over the others.

BRIEF DESCRIPTION OF THE REFERENCE CTR

The tokamak chosen as the Reference CTR, has a torus structure divided into six sectors to facilitate construction and maintenance. Four of these are assembled and positioned around a poloidal magnet core. Figure 18-2 is a schematic of the approximately one meter thick blanket region which surrounds the toroidal plasma. It consists of a set of 60 segments, each of which consists of a 2.5 mm thick niobium shell.

These segments contain a long, slender, central "island" of graphite surrounded by a lithium-filled duct. Lithium coolant would be circulated at about 30 cm/sec around this closed loop by an electromagnetic pump at one end. Tritium is bred by neutron absorption in the lithium. A typical breeding ratio is 1.3, giving a doubling time of about a month. (Addition of neutron absorbers can easily reduce this ratio when excess tritium is no longer needed). A set of tubes installed in the lithium blanket utilizes the heat generated in the blanket to boil potassium. One set of the ring-shaped manifolds would carry the liquid potassium feed to the blanket from pipes in a duct beneath the reactor floor, and the other set carries potassium vapor to vapor pipes that extend around under the reactor and out to a potassium vapor turbine in the adjacent turbine hall (see Figures 18-3 and 18-4).

A magnet shield about 1 m thick attenuates radiation leaking from the blanket region into the liquid helium-cooled superconducting magnets so that the radiation energy deposited in them would be about 1 kW(t), and hence the power required for the liquid helium refrigeration system can be held to about 2 MW(e).

Six neutral beam injectors for plasma heating and refueling are mounted near the top of each sextant so that fuel injection takes place through the parting planes between sextants.

Figure 18-4 shows the reactor installation in a 60 m diameter evacuated shielded cell. The vacuum pumps, helium refrigeration system, and tritium recovery and handling system are located in rooms beneath the reactor.

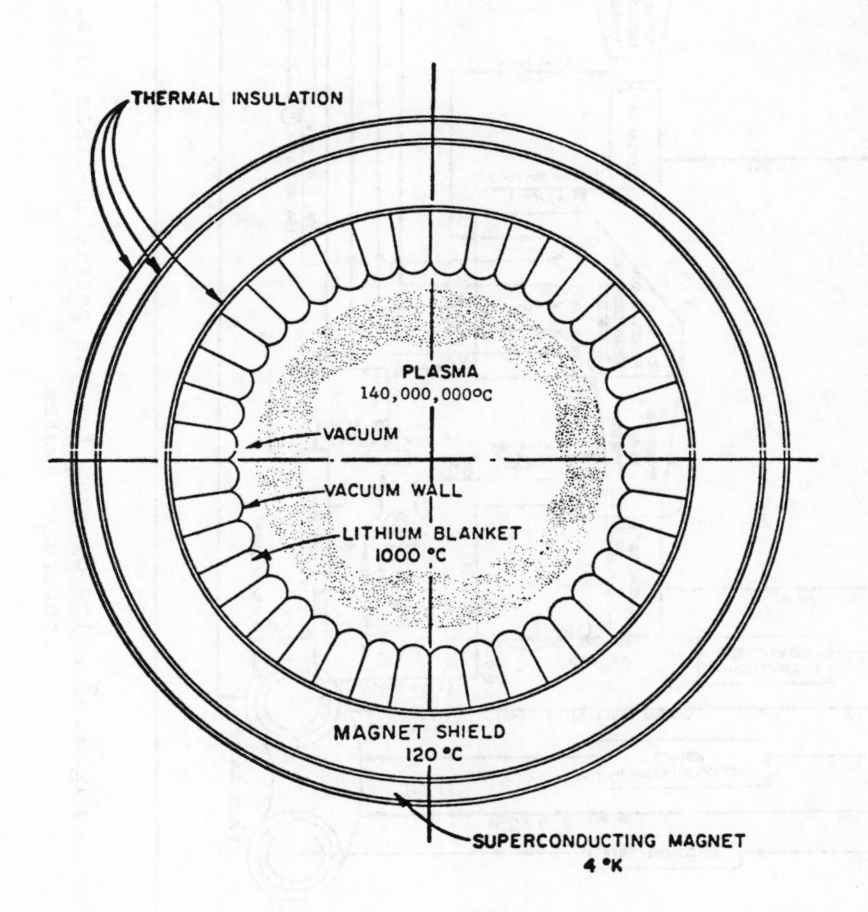

Fig. 18-2. Cross section of the toroidal core of the Reference CTR.

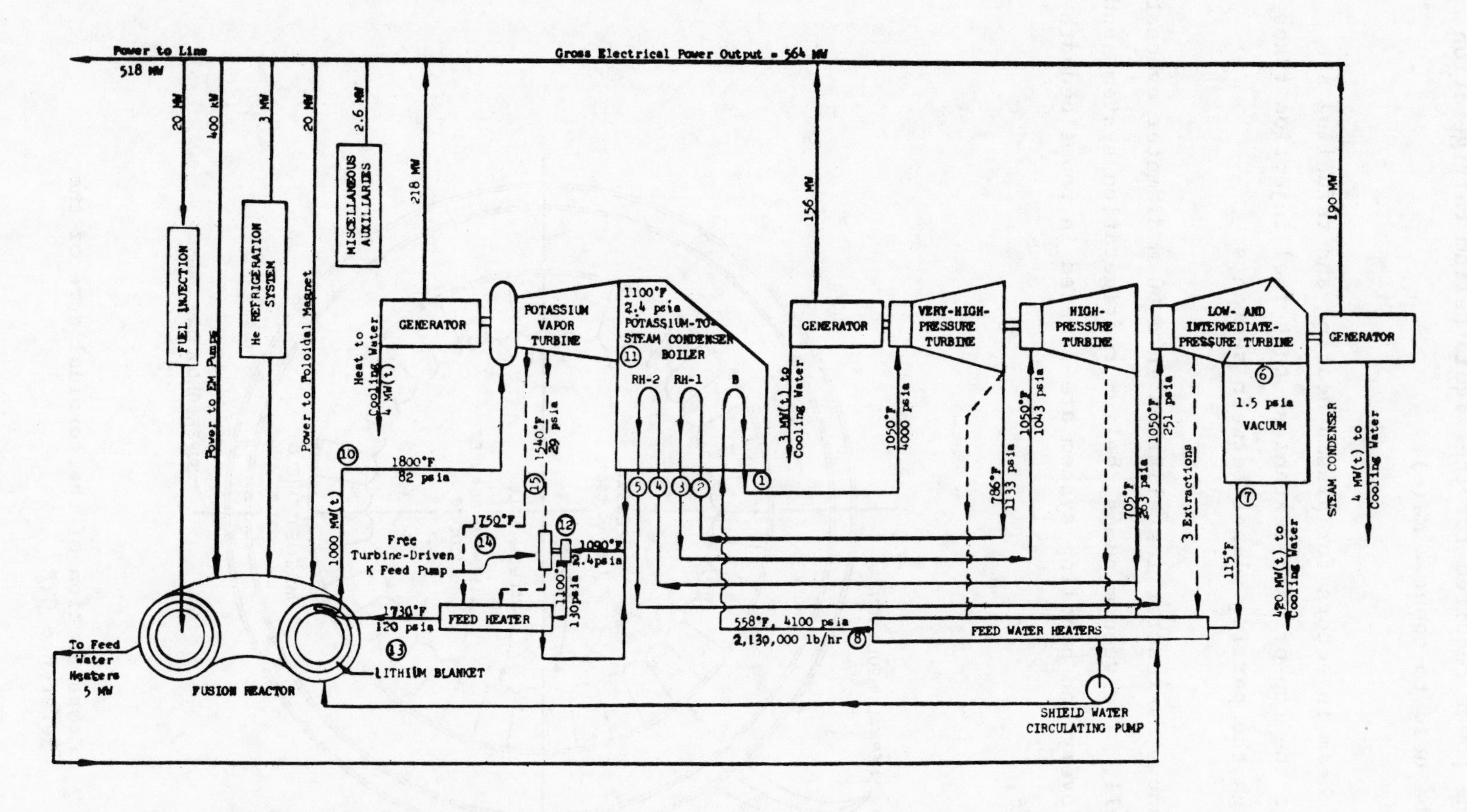

Fig. 18-3 . Flow diagram for the potassium-steam binary-vapor-cycle power conversion system.

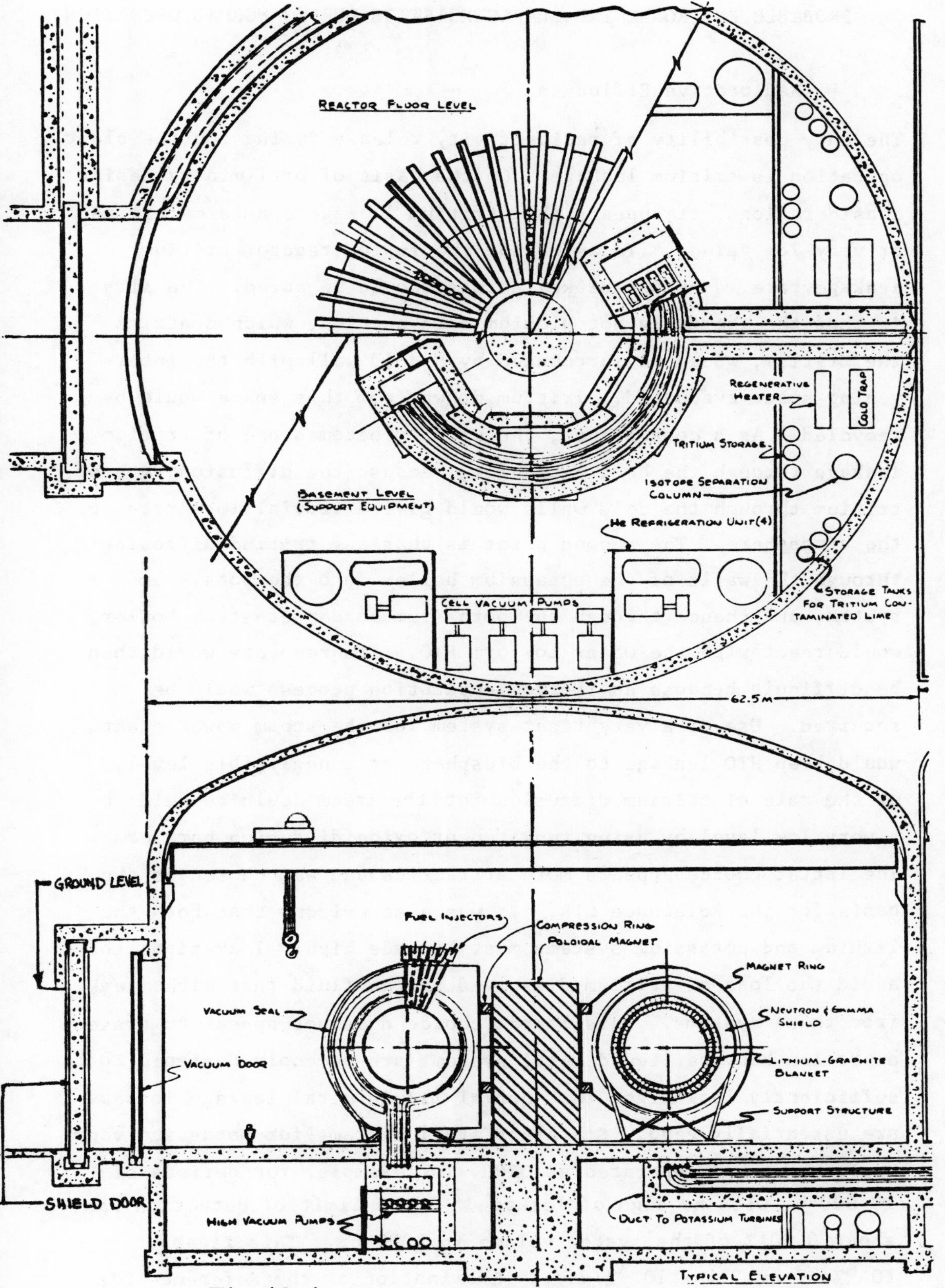

Fig. 18-4. Section through the power plant building.

PROBABLE ENVIRONMENTAL CHARACTERISTICS DURING NORMAL OPERATION

Radioactive Effluents

The only possibility of radioactivity release during routine plant operation is tritium leakage. On the basis of preliminary design considerations, it appears that tritium leakage can be maintained at very low values. In assessing the fusion reactor tritium leakage rate, a number of key points should be noted. The first is that the thermally hot niobium core section, which contains the tritium, would be surrounded by a cold wall with the intervening space evacuated. Tritium drawn from this space would be recycled. As a consequence, the problem becomes one of tritium leakage through the heat exchanger, because the diffusion rate of tritium through the cold walls would give a trivial loss rate to the atmosphere. The second point is that any tritium diffusing through the walls of the potassium boiler into the potassium system, and thence through the potassium condenser-steam boiler, would react with the water to form HTO. Its recovery would then be difficult because an isotope separation process would be required. Use of a very tight system for the steam power plant would keep HTO leakage to the biosphere at a negligible level, or the rate of tritium diffusion into the steam could be held to a very low level by using tungsten or oxide diffusion barriers. The latter choice appears more attractive and was chosen as the basis for the Reference CTR. It was also evident that both the lithium and potassium systems must be made highly leak-tight to avoid the loss of tritium dissolved in the fluid that might leak from these systems. This latter choice does not appear to present a problem because liquid metal systems are commonly designed to be sufficiently leak-tight that normal liquid metal leakage losses are essentially zero. Stainless steel systems for potassium vapor cycles have been operated at ORNL, for example, for periods of 10,000 hr with no sign of leakage.[2] (The limit of detection was about 0.001% of the system volume in 1000 hr. This is about 10^{-6}%/hr or 2.4 x 10^{-5}%/day.) Examination of the Reference CTR design and present experience with operating liquid metal systems indicates that the leakage can be kept to 0.0001%/day.

If ventilating air discharged from the reactor building is directed up through a 200 ft. stack, the maximum tritium concentration downwind at ground level would produce a dose rate of about 1 mrem/yr. This is less than 1% of the average dose to the population from natural radioactivity of 110 mrem/yr.

Long-Lived Radioactive Wastes

Fusion reactors will produce nonvolatile, long-lived radioactive wastes in modest quantities. The primary source of radioactive waste from the Reference CTR will be the activated structural material of the blanket, which will have a finite useful lifetime within the reactor owing to radiation damage. Table 18-1 shows the principal long-lived activities of the Reference CTR blanket structure (niobium or vanadium). This table gives the annual rate at which the activity is generated, normalized to one megawatt of reactor thermal power, the accumulated activity resulting from 1000 years of continuous generation, and the biological hazard potential associated with this amount of accumulated activity. Note that in Table 18-1 the Maximum Permissible Concentration (MPC) in water is used, which seems more appropriate than the MPC in air in the context of underground disposal. For niobium as the structural material the biological hazard potential associated with the accumulated Reference CTR radioactive waste is significant and would have to be treated accordingly.

The use of vanadium as the blanket structural material dramatically reduces the problems associated with radioactive waste disposal. Vanadium exhibits no known long-lived activity as a result of activation; therefore the long-lived activities result only from the activation of impurities and alloying additions within the vanadium. Niobium is typical of such an impurity and might be present in vanadium at an atomic concentration somewhere between 100 to 1000 ppm (parts per million). Assuming this concentration range, the biological hazard poten-

Table 18-1 . **Long-Lived Activities in the Blanket Structure of the Reference CTR**

Nuclide	Mean Life (yrs.)	Activity Generation Rate (curies/MW(t)-yr)	Accumulated Activity at 1000 yrs. (curies/MW(t))	Maximum Permissible Concentration* in Water (μcuries/cm^3)	Biological Hazard Potential Activity at 1000 yrs ÷ MPC (km^3 of water/MW(t))
			Reference CTR with Niobium		
^{93m}Nb	19.6	8,800	173,000	4×10^{-4}	0.4
^{94}Nb	2.9×10^4	2.9	2,900	3×10^{-6}	1.0
			Reference CTR with Vanadium		
Long-Lived Activities Due to Activation of Niobium Impurity in Vanadium					~ 0.00014 - 0.0014

* Abbreviated MPC in the text.

tial associated with the activated vanadium structure would be three to four orders of magnitude lower than that associated with the niobium structure (see Table 18-1). The same arguments would also be valid for several promising vanadium alloys, i.e., those containing titanium and chromium.

The activated structure of a fusion reactor could be reused after reprocessing if necessitated by a scarcity of niobium resources. In view of the rapidly growing use of automation in industry, the remote handling and recycling of radioactive material may prove practicable and economical, thus virtually eliminating the need for long-term radioactive waste management in a fusion power economy, e.g., recycle of the blanket structure after allowing time for radioactive decay.

Waste Heat Rejection

The DT fuel cycle requires use of a thermal power conversion system. The efficiencies of such systems are determined in large part by the maximum temperature of the heat transfer fluid, which is determined by the maximum temperature of the core structure. The Reference CTR utilizes a niobium structure which appears capable of operation at 1000°C. This may allow use of a potassium topping cycle in addition to the main steam generators, the combination of which appears to provide overall plant efficiencies greater than 50%.

The use of cooling water versus wet or dry cooling towers has not been considered in detail for fusion reactors because the choice of heat rejection mode is such a sensitive function of plant site considerations. Obviously the high operating temperatures of the Reference CTR would allow increased flexibility in system optimization using cooling towers over systems operating at lower temperatures. Because of the potential of urban siting and the high peak cycle temperature, heat can be rejected from fusion power plants at 100 - 200°C without seriously reducing plant thermal efficiency. This heat energy may then be used for building heating and cooling and/or industrial processes, and it would thereby not represent a waste.

Land Despoilment

There are three aspects to fusion power related to land despoilment. The first is the direct land use by the power plant itself, which includes buildings, switchyards, transformer yards, transmission lines, cooling equipment, etc. To a significant extent fusion reactors would be similar to fission reactors in this regard, and fusion fuel storage space requirements will be negligible.

A second aspect of land despoilment is associated with the procurement of the fuel and construction materials. DT fusion power plants would consume deuterium and lithium as fuels. Deuterium is obtained from water which is available to all countries. Its extraction results in no despoilment but rather provides useful quantities of commercial grade hydrogen and oxygen and modest quantities of purified water.

Lithium is obtainable from surface and underground brines (the least expensive extraction process) and from the oceans (a more expensive process but still relatively insignificant in cost). The land despoilment associated with the extraction of lithium and the metals incorporated in the structure of the Reference CTR are shown in Table 18-2, which shows that the residues of lead and copper are of greatest concern.

The third aspect of land despoilment is associated with the projected flexibility of fusion reactor siting. If urban siting is indeed acceptable, then the large land areas usually required for power transmission from rural to urban areas would be significantly reduced.

Transportation

To start up a fusion power plant an initial fuel charge of deuterium and tritium will be needed. Thereafter a continuous supply of deuterium and lithium will be required at the rate of about a kilogram per day. Tritium shipment will be necessary only to supply the

Table 18-2. - Yield of Required Metals from Their Ores

	Requirement for 10^7 MWE - metric megatons	Approximate average yield of metal from crude ore - percent	Ore Requirement for 10^7 MWe - metric megatons
Nb	7	2	350
Be	.6	2	30
Cr	11	5	220
Ni	5	~ 1	500
Li	5	~ 5	100
Cu	40	.9	4400
Pb	107	1.5	7100
Al	10	10	100
V	4	5	80
Mo	6	2	300
Sn	.8	10	8
Fe	170	45	380
Zr	.07	~ 5	--
		Total	13,600

initial charges to start new power plants, i.e., possibly about 10 kg quantities from each operating plant every few years on the average, depending upon the rate of growth of the fusion power industry.

The blanket structure of a fusion plant will become radioactive and will have a lifetime of the order of 10-20 years. When the blanket structure is replaced, the used activated unit will have to be shipped from the power plant to a site wherein it would be either stored or reprocessed. The structure itself will be nonvolatile and consequently its hazard potential should be relatively low. It will not require a large amount of shielding during shipment nor would it present a difficult cooling problem.

EFFECTS ON NON-RENEWABLE RESOURCES

A preliminary survey has been made of U.S. and world resources of the various materials needed for fusion reactor construction. The results are shown in Table 18-3, where the approximate quantities of materials needed to fabricate a single 1000 MWe reactor are tabulated. These figures are for current reactor designs. The development of other materials for the various components, i.e., blanket structure and superconducting magnet, is clearly possible and would alter these requirements accordingly.

To emphasize maximum resource requirements, the largest quantity of a given material required by any of the several reactor designs available (see the Appendix) has been used. For instance, a pulsed theta-pinch reactor would use more copper and less superconducting material than would a tokamak reactor. The larger needs for both materials are included in the Table. Clearly no one reactor design would use all of the material listed and this approach thereby overestimates the quantities of material needed.

In the extreme of a fully developed world fusion power economy, ten terawatts (10^7 MWe) of electric power might be generated by fusion reactors. Therefore the third column of the Table shows the mass of materials in metric megatons needed to construct and operate ten thousand 1,000 MWe fusion reactors. Plant replacement at about 5% per year would be required at a later time but is not considered here.

Also presented in the Table are estimates of the total production of the various materials projected to be required in the year 2000, along with quantities of known reserves at present prices and estimates of resources available at increased prices.

A great many evaluations of U.S. and world raw materials resources have been made but these are usually a matter of expert opinion. Consequently, values such as "known resources at current costs" vary widely from one source to another. Often the estimated quantities of a raw material available at increased costs are based on industrial projections. But when adequate reserves of a given ore are available to supply the demand for 20-30 years, exploration for additional reserves is usually curtailed with the result that total projected reserves can be underestimated to a significant degree. Most of the values quoted are from the 1970 edition of "Mineral Facts and Problems". In addition to estimating materials needs, some comments concerning environmental problems associated with a particular raw material are included in the Table.

It is apparent that the production of 10^7 MWe of fusion power would give rise to some resource use conflicts which will have to be resolved. For example the requirements for niobium could just be met by known reserves. However, additional reserves may be found or other superconducting materials developed. In addition to niobium, other possible resource conflicts exist in the projected usage of beryllium, titanium, helium, lead, vanadium, and molybdenum, and some of these problems will also be common to other power generating concepts.

FUSION POWER ECONOMICS

At the present stage of fusion development many physical and technical uncertainties clearly exist. Fusion power costs are therefore impossible to accurately predict. Nevertheless, cost estimates are of value because they indicate a general order of magnitude, and they help to identify particularly sensitive components for which further cost-reducing development could have a major impact. In this section the costs for the plant and the fuel will be considered.

The safety and environmental characteristics of fusion reactors will very likely make them acceptable for urban siting. The power costs of urban fusion power plants would be significantly reduced by savings in transmission costs as well as possible savings associated with the sale of waste heat for building heating and/or industrial processing.

Table 18-3 CTR Resource Utilization

Material	Approx. Mass In Metric Tons Per 1000 MWe Reactor	For Reactor†	Mass In Metric Megatons For 10^7 MWe	Total Estimated Production in Year 2000 In Metric Megatons		Known Resources Present Prices Metric Megatons		Resources At Increased Prices In Metric Megatons		Comments
				U.S.	WORLD	U.S.	WORLD	U.S.	WORLD	
Nb	~ 400 structural, ~ 130+180 in NbTi and Nb_3Sn	4,1,2	7	.009	.020	.07	6	.14	NA	Present mining operations are relatively nonpolluting; greatly increased demand might necessitate strip mining to obtain low grade deposits
Li	~ 900	1	9	.01	.016	5	6-8	9	250,000	100 metric megatons probable land resources; extraction from sea water possible, 1.5 lbs. of Li/100,000 gal. of sea water
Be	~ 60	2	.6	.002	.003	.026	.38	.072	1	Little information on world Be resources available, Be presents health hazards in mining and handling
Cr	~ 1100 in SS	5	11	1	4.3	0	700	1.6	NA	Resources almost entirely outside of U.S.
Ni	~ 500 in SS	5	5	.5	1.3	.2	68	5.0	NA	World estimates are based on fragmentary information and are possibly low
Ti	~ 400 structural, ~ 80 in NbTi	1,1	5	2.3	6.9	.15	6.4	.4	30	Significant quantities of mud and slimes result from dredging Ti minerals from sand deposits
He	~ 350	3	4	.012	.015	1.2*	1.2*	5*	29,000+	*In the ground +Extracted from atmosphere at up to 30 times current prices
Cu	~ 2900 coil, ~ 1100 in NbTi	3,1	40	6-12	35	77	280	180	1,100	Considerable secondary recovery possible; significant land-use conflict will result from an expanded copper industry
Graphite	~ 2200	1	22	.1	1.4	.5	> 100	NA	NA	Very rough estimates of world reserves available
Pb	~ 10,700	1	107	3	7.3	32	86	45	95	Considerable secondary recovery possible
Al	~ 570 structural, ~ 390 in Nb_3Sn	3,2	10	30	75	12	2200	275	NA	Large land areas and great amounts of energy needed to mine and process Al
V	~ 400	4	4	.03	.06	.1	9	3	NA	
Mo	~ 400 structural, ~ 200 in SS	4,5	6	.08	.24	2.9	5	NA	> 10	Substantial resources of sub-marginal-grade ore throughout the U.S. and world
K	~ 20	1	.2	11	56	120	> 10,000	770	Virtually unlimited	
Sn	~ 80 in Nb_3Sn	2	.8	.12	.41	.006	4	.042	7	Some secondary recovery possible
F	~ 500 in flibe	2	5	2.2	7.5	4.9	35	NA	NA	Increased price would stimulate expanded exploration for fluorspar
Fe	~ 12,600 steel,	1,5	170	180	800	2000	90,000	20,000	> 300,000	Potential reserves are vast

† Reactor code:

1) ORNL Tokamak, 2) PPPL Tokamak, 3) LASL Theta-Pinch, 4) LLL DT Mirror, 5) LLL D^3He Mirror

To estimate fusion power capital costs, the reactor designs developed for the various concepts were analyzed to determine the approximate amounts of the various materials used in their construction. Current prices for the required quantities of these materials in finished form were then used to estimate component costs. The prices for the superconductor material correspond to present large order levels. The unit winding costs and structure costs have been scaled somewhat less than the square of the magnetic field.

Amongst the auxiliaries for magnetic confinement systems, the greatest uncertainties are associated with injection systems and the theta-pinch reactor energy switchgear. Injector development has not yet progressed to fabrication of reactor-sized units and factors of two or so cost uncertainties are felt to exist. The switchgear estimate is based on an estimate of \$2 to 7 million for similar equipment designed for two synchrotrons requiring 1-2 second switching of 100 to 1000 megajoules. Theta-pinch reactors would require 10 msec switching of 200 gigajoules. Development and fabrication costs of \$100-200 million are considered probable.

A comprehensive projection for a fully developed superconductor industry has shown that it appears possible to obtain cost reductions for finished magnets of factors of four to five over present levels. In such a well developed situation the cost of the conductor moves from being the largest single cost to being secondary, and structure costs become dominant. Winding costs are expected to decrease from the present level of \$33 per kg of conductor to near \$10 per kg.

The results show that prototype reactor costs might be about \$500/kwe for the nuclear "island." Ultimate magnet costs would reduce mirror and tokamak reactor costs substantially. Superconductor in the theta-pinch reactor serves as an energy storage element separated from the plasma vessel, and it operates at low fields. It represents a small fraction of the system total cost and is little affected by the ultimate magnet cost patterns. Maturing of the fusion reactor industry should bring reductions associated with production quantity manufacturing and the removal of design uncertainties, further reducing costs.

Final projected fusion reactor capital costs then correspond roughly to the level projected for other types of plants in the year 2000. Because of the uncertainties, it is believed that these exercises in cost estimation serve only to suggest that fusion power capital costs could be competitive with other energy sources. To conclude any capital cost advantage at this stage of development would clearly be premature.

Fusion fuel cycle costs are determined by the costs of deuterium and lithium which are shown in Table 18-4. Fuel transportation costs will be negligible because of the small quantities of materials involved and because handling techniques for gases and liquid metals are already well developed and inexpensive.

Table 18-4

Fusion Fuel Cycle Cost Based Upon Current Prices

Element	Cost: Per Gram	Cost: Per Kilowatt-Hour
Deuterium	$0.20	6×10^{-3} mills
Lithium	0.02	10^{-3} mills
Total Cycle Cost		7×10^{-3} mills

ACCIDENT HAZARDS

Any reasonable appraisal of accident hazards requires a detailed examination of a specific design because many potential problems are in large measure dependent upon specifics of the system. As mentioned previously, only one fusion reactor design has been iterated through a number of steps in an attempt to maximize safety and minimize accident potential. That Reference CTR served as a basis for the following analysis.

A first step in appraising the possibilities and consequences of a fusion reactor accident is to determine the maximum energy stored in the system in nuclear and chemical forms and in the form of high pressure steam or gas. Table 18-5 lists the principal hazards sources in the Reference CTR and shows that the largest potential source of accidental energy release is associated with the lithium in the blanket. In the design considered here, no lithium is situated near any water, and it would require the rupture of three successive envelopes for the lithium to react with air. Further, the lithium inventory in the Reference CTR is divided into many separate segments, thus significantly limiting the energy release from a single leak. In addition, the lithium region is well protected. If, for example, an airplane were to crash into the containment shell and rupture it, the lithium region would still be well protected not only by the magnet shield but also by the massive structure of the steel reinforcing rings carrying the superconducting magnet coils.

A second concern is the possibility of the abrupt release of a substantial amount of energy via nuclear reactions. In this case the only fuel that could possibly react would be that actually inside the plasma region, i.e., about a gram. If all of the fusion energy obtainable from this charge were to be dumped into the blanket in a few seconds, the average temperature of the lithium would rise about 30°C -- a minor perturbation. A dump of such magnitude appears impossible from what is known today about plasma behavior. Further, the kinetic energy of the unburned plasma is a factor of about 1000 lower than the total available fusion energy so that a full plasma loss to the walls would have a much smaller effect. This low total available energy in the fuel charge and the low probability of liberating more than a small fraction of it in a fault situation are major factors in the inherent safety of fusion reactors.

From experience to date, a localized plasma dump onto the adjacent wall appears very unlikely. Clearly the probability of such an instability occurring must be made extremely low in a practical system. This question can be specifically studied in the larger, more energetic plasma systems to be fabricated later this decade. In any event, local wall burnout due to an inadvertent concentration of plasma would at worst cause a lithium leak into the plasma but would not cause an accident affecting the public.

Table 18-5. Energy Release Potential of Components of a Reference CTR Producing 1000 MW(e)

	Energy in Megajoules	Equivalent Gallons of Fuel Oil
Plasma, complete fusion	6.9×10^4	~ 430
Magnet	2.4×10^5	~ 1500
Lithium + water + air	6.4×10^7	~ 4×10^5
Potassium + water + air	6.4×10^5	~ 4000
Primary vacuum vessel	640	~ 4
Secondary vacuum vessel	1.6×10^4	~ 100

A third possible failure mode is associated with a magnet failure. There are two faults of concern. In a high current density coil a transition from super to normal conduction could progress over the total conductor volume in a period short compared to the overheat time in the conductor. The rate of stored energy dissipation could be handled with the insertion of an external load resistor. If, somehow, the resistor was not cut-in, the coil assembly temperature would rise to near that of the room while liquid helium evaporated and was vented with no damage to the coil. There is satisfactory experience with this type of quench fault. At low current densities in a coil the quench might not spread rapidly, and a load resistor could be inserted automatically in a time interval of the order of a minute to drive the current down, thus preventing local heating which could damage the conductor. A third possible fault mode is the breaking of a conductor in the coil. This would lead to further damage of the coil by arcing and probably a quench. In the latter two cases proper design can insure that the damage would be limited to the coil itself.

Study of the afterheat problem in connection with the Reference CTR indicates that it is possible to evolve a design that is virtually unaffected by a loss of coolant accident. A basic reason for this is given by Table 18-6 which shows the average afterheat power density at shutdown in watts per cubic centimeter for the Reference CTR and the rate of temperature rise after a cooling system failure. An analysis of the consequences of a complete loss of coolant in both the blanket and the shield region of the Reference CTR indicates that all of the afterheat could be removed by thermal radiation and conduction with a temperature rise of no more than about 100°C in the high temperature zone during the first week after the outage, assuming that no action whatsoever were taken by the plant operating personnel. This refers to a blanket structure built of niobium. If stainless steel were employed, the afterheat would be reduced by a factor of about two relative to that of niobium, or, if vanadium were employed, the afterheat immediately following shutdown would be reduced by a factor of about four. Further, in the vanadium case the afterheat would fall off much more rapidly than with the niobium.

Table 18-6 Afterheat Power Density Associated with the Niobium Structure of the Reference CTR

	Reference CTR
Average afterheat power density at shutdown	0.15 watts/cm^3
Rate of temperature rise if uncooled immediately after shutdown	0.06°C/sec

The probability of a lithium leak will be low because the lithium blanket can be designed so that the lithium pressure will differ from that in the plasma region by only about 1/10 of an atmosphere, and hence both the pressure stresses and the driving force for a leak will be small. Further, the blanket has been designed to keep all of the thermal stresses well within the elastic range both during normal operation and in the course of any of the transients that have been envisioned, and this would minimize the probability of a crack induced by thermal cycling strain.

An obvious cause for concern is a leak of lithium into the plasma region. If this occurs, even a small amount of lithium will quench the plasma because of the increased loss via bremsstrahlung radiation from the lithium atoms and/or conduction cooling.

If a lithium leak occurs in the region between the blanket and the shield, the multilayer stainless steel foil reflective insulation should prevent the lithium from reaching the titanium shield tank. If it does reach the tank, the lithium will simply solidify because the tank temperature would be below the freezing point of the lithium.

The initial design of the Reference CTR envisioned the use of a magnet shield that consisted primarily of water and lead. While the probability of lithium coming in contact with this shield water seemed exceedingly low, it was decided that with relatively little increased cost the water could be replaced with graphite or a metal oxide such as alumina or magnesia. The presence of water in the shield can be avoided completely by employing helium rather than water as the shield coolant. The energy deposition in the magnet shield as a consequence of nuclear and thermal radiation and thermal conduction will represent less than 1% of the total reactor output. Analysis indicates that the shield can be cooled easily with helium at about ten atmospheres, and thus the designer can eliminate the possibility of a substantial energy release from a lithium-water reaction.

The consequences of a lithium leak are greatly reduced by the fact that the lithium blanket is segmented into many independent elements. Any lithium leak will be quickly detected as a consequence of its effects on the plasma or the vacuum system.

The above discussion has been concerned with single point failures. It should be noted that the design is such that even a double failure would not lead to any serious difficulties. If, for example, there were a lithium leak into the region between the blanket and the shield, and a leak from this region out into the reactor cell, there would still be no serious reaction because a vacuum is maintained in the cell. Again, a leak could be readily detected. Note, too, that there is no apparent way in which a leak from the lithium system could induce a secondary leak through the walls of the high vacuum region into the reactor cell.

If a leak develops in the potassium condenser-steam generator of the Reference CTR, the steam jetting into the potassium condenser would react with the potassium to form potassium oxide and hydrogen. Inasmuch as the potassium condenser will have a large vapor volume space available, there would be adequate space to accommodate the hydrogen gas, and no explosion or even large increase in pressure would occur. (This situation differs from that in a liquid metal-heated boiler in which there is little or no free volume on the liquid metal side into which the hydrogen from the reaction can expand). As the hydrogen builds up in the condenser, it would block the flow of potassium vapor into the condenser and produce a back pressure which would provide an obvious signal to an operator or which could be used to trigger a warning signal. If a large steam leak were to develop as a consequence of a burst type of failure, the inherent nature of the inlet orificing of the reentry tube boiler is such that vapor rather than water would be injected into the potassium region, and as a consequence the rate of injection would be relatively low -- a few lb/sec per ruptured tube. In the Reference CTR this would lead to an increase in the condenser at a rate of about 1 psi/sec. Thus, if the potassium condenser were designed to take an internal pressure of 60 psi, and if the flow of either steam or potassium into the condenser could be stopped within a minute after the first evidence of the rupture, the damage would be limited to the broken tube. For the extreme case of an abrupt, complete rupture of a steam generator tube, the potassium condenser pressure would rise faster. Again this should be easily and reliably detectable and could be the basis for closing valves in the feed water supply line. If this were done in an additional 10 seconds, the inventory of superheated water in the boiler design proposed would be exhausted in another 15 seconds, and the peak pressure in the potassium condenser would be held to about 15 psig (30 psia). To protect against the contingency that no action might be taken, a rupture disc could be provided to blow off at perhaps 40 psia.

A leak from the potassium boiler into a segment of the lithium blanket will cause the liquid level in the header tank for that segment to rise and in extreme cases overflow into a dump tank. This will lead to a forced shut-down, but no other ill effects have been envisioned.

The inventory of volatile radioactive material is probably the most important factor to be considered in appraising the requirements for engineered safety features for any type of nuclear power plant. For a fusion reactor this means that the tritium inventory, particularly the active inventory in the liquid metal system, is the most vital consideration because it will be the only volatile activity in a fusion reactor. One of the systems proposed appears to be capable of holding the tritium concentration in the lithium to roughly 1-10 ppm irrespective of the type of fusion reactor, the total lithium inventory or the tritium generation rate. Thus, the tritium inventory in the lithium system would be primarily a function of the total lithium inventory.

Practically all of the tritium outside of the liquid metal systems will be contained in components in the tritium equipment room. These components will be at or close to room temperature, and the atmosphere in the room would be carefully controlled and monitored so that, if any tritium leakage occurs in that room, it would be well contained.

The only substantial inventory of radioactive material other than tritium will be that in the blanket structure, and the bulk of this activity will be in the region close to the first wall. Table 18-7 shows the estimated quantities of the principal radioactive inventories (in curies per kilowatt of reactor thermal power) for the Reference CTR, using niobium and vanadium as alternate structural materials. The biological hazard potential is provided for each item; it is defined here as the activity divided by the maximum permissible airborne concentration (MPC) as specified in the radiation protection standards for continuous exposure to individuals living in the vicinity of controlled areas.

Table 18-7. **Principal Radioactive Inventories of the Reference CTR**

Inventory	Activity (curies per kW of thermal power)	Maximum Permissible Airborne Concentration (μ curies/cm^3)	Biological Hazard Potential Activity $\div$ MPC (km^3 of air/kW thermal)
		Reference CTR 10-year Operation	
^3H (combined in H_2O)	12[a]	2×10^{-7}	0.06
		Niobium as the Blanket Structure	
^{95}Nb	155	3×10^{-9}	52
Total Niobium Structure	714	c	240
		Vanadium as the Blanket Structure	
^{48}Sc	4.20	5×10^{-9}	0.84
Total Vanadium Structure	55.1	c	0.86[b]

a The specific activity of tritium is approximately 10^4 curies per gram.

b Impurities within the vanadium might increase this number by a factor of two.

c MPC's for each individual isotope were estimated to get the composite Biological Hazard Potential.

RELIABILITY AND VULNERABILITY

As with any infant technology, when fusion reactors first become commercially available, their reliability will not be as high as that of the more mature power plant types. Areas where there will be relatively little experience include large refractory metal structures, superconducting magnets at very high fields, potassium-steam turbines, generators, and boilers, and to a lesser extent large high temperature vacuum systems. Because of this unfamiliarity, a certain amount of redundancy will be required which can be eliminated as the technologies develop. The reliability of the steam system and other standard elements of a fusion power plant should of course match the reliability of similar equipment used in other plants.

Fusion power plants, like other systems, will be vulnerable to both internal and external hazards. Clearly care in design can eliminate many potential problems. Inherently a number of potential problems will represent minimal hazards. Failure of a magnet would cause the plasma to strike the wall, extinguishing the reaction with relatively minor effects on the wall. Failure of an injector would reduce the fuel supply causing the plasma to slowly diffuse away. Failure of the on-site reprocessing system would result in an impure fuel replenishment which would markedly reduce the plasma temperature and thereby the reaction rate. At a later date when improved fusion reactor designs are developed, the matter of internal vulnerability can be considered in greater detail and the results of such analyses factored into plant design.

In terms of external influences, such catastrophies as earthquakes, tornadoes, hurricanes, lightning, aircraft crashes, etc. must be considered. Rather than attempting to consider these possibilities in any detail at this time, reference is made to the previous discussion wherein the potential sources of energy release and the radioactivity inventories were estimated. From these considerations and fission reactor experience to date, the hazards of a fusion power plant appear to be readily manageable.

SUMMARY

Deuterium for the Reference CTR is obtained directly from sea water at low cost. Tritium is bred in a blanket surrounding the plasma region by neutron absorption in lithium. Typical breeding ratios are about 1.3, giving a doubling time of about a month. With neutron absorbers this ratio can be easily reduced when excess tritium is no longer needed.

During routine power plant operation, tritium is anticipated to be the only radioactive effluent, and it appears to be readily controllable. A tritium leakage rate to the atmosphere from the Reference CTR of 0.0001%/day (based on a system inventory of 6 kG of tritium) appears reasonable from a design standpoint. Assuming that this leakage is to be discharged from the reactor building through a 200 foot stack, the maximum concentration at ground level would be reduced to the point where it would give a maximum dose rate downwind of 1 mrem/yr, i.e., less than 1% of the average dose to the population from natural radioactivity.

The primary source of radioactive waste from a fusion reactor will be the activated structural material of the blanket, which will have a finite useful lifetime within the reactor owing to radiation damage. Approximately 9000 Ci /MW yr. of long-lived radioactivity would be produced in the niobium structure of the Reference CTR. If vanadium were substituted for niobium, this activity would be reduced by a factor of 1000-10,000, depending upon the type and concentration of alloying material.

The DT fuel cycle requires use of a thermal power conversion system. The Reference CTR utilizes a niobium structure which appears capable of operation at 1000°C, which is sufficiently high to provide cycle efficiencies greater than 50%. Using stainless steel for the structure, temperatures are limited to about 500°C, which would give cycle efficiencies near 40%.

Urban siting of fusion power plants would allow rejected heat to be used for heating and cooling and industrial processing. The land despoilment associated with fusion plants appears to be similar to that for fission plants with the exception that urban siting would decrease the land requirements for power transmission.

To start up a fusion power plant, an initial fuel charge of deuterium and tritium will be needed. Thereafter, a continuous supply of deuterium and lithium will be required at the rate of about a kilogram per day. Further tritium shipment will be necessary only to supply the initial charges to start up new power plants. The blanket structure of a fusion plant will become radioactive and will have a finite lifetime of the order of 10-20 years. It will then have to be shipped for reprocessing or storage.

A projected worldwide production of 10^7 MWe from fusion and/or many other types of power will give rise to some resource use conflicts which will have to be resolved. Fusion requirements for niobium for magnets and structure could just be met by known reserves. However, additional reserves may be found or other superconducting magnet materials developed.

To estimate fusion power capital costs, reactor designs developed for the various concepts were analyzed to determine the approximate amounts of the various materials used in their construction. Current prices for the required quantities of these materials in finished form were then used to estimate component costs. These estimates yielded capital costs for the nuclear "island" of roughly the same order as projected for other types of plants in the year 2000. Because of major uncertainties, it is believed that these projections serve only to suggest that fusion power capital costs could be competitive with other energy sources.

Fusion power fuel costs are determined by the costs of deuterium and lithium, and they are essentially negligible--of the order of 0.007 mils/KWh. The safety and environmental characteristics of fusion reactors should make them potentially acceptable for urban siting, which would further reduce total fusion power costs by savings in transmission costs as well as possible savings associated with the sale of waste heat for building heating and cooling and/or industrial processing.

Fusion reactors appear very attractive when considered from the point of view of accident potential. A runaway reaction will not be possible in a fusion reactor both because of the inherent nature of plasmas and because of the low fuel inventory--about one gram--that would be resident in the core during operation.

Studies of the afterheat produced in the Reference CTR indicate that it is possible to evolve a design that is virtually unaffected by a loss-of-coolant accident. An analysis of the consequences of a complete loss of coolant in both the niobium blanket and the shield region of the Reference CTR indicates that all of the afterheat could be removed by thermal radiation and conduction with a temperature rise of no more than about 100°C in the high temperature zone during the first week after the outage, assuming no action whatsoever by automatic controls or the plant operating personnel. If stainless steel were employed for the blanket structure, the afterheat would be reduced by a factor of about two relative to that of niobium, or, if vanadium were employed, the afterheat immediately following shutdown would be reduced by a factor of about four.

The inventory of volatile radioactive material is probably the most important factor to be considered in appraising the requirements for engineered safeguards to protect against accident hazard. For a fusion reactor this means that the tritium inventory, particularly the active inventory in the liquid metal system, is the most vital consideration because it will be the only volatile activity present.

By holding the tritium concentration in the lithium to 1-10 ppm and isolating the lithium and tritium handling equipment in a single, well sealed and monitored compartment, this potential accident hazard can be kept very low.

The national security aspects of fusion power would be many-fold. The U.S. has plentiful deuterium and lithium resources and would therefore be independent of foreign sources. Fusion reactors do not utilize fissionable materials which may be subject to diversion for clandestine purposes. A mature fusion reactor industry would strengthen the country's technological base and foreign sales of fusion reactors would have a favorable effect on the balance of payments. Some reliance on foreign sources of materials such as nickel and chromium will be inherent to fusion as well as many other power sources.

SUMMARIES OF REFERENCE FUSION REACTOR DESIGNS

PPPL Tokamak Fusion Reactor

The guiding principles on which this design was based were as follows:

1. The maximum magnetic field at the superconductor of the toroidal field coils were to be limited to 160 kilogauss. This field strength is somewhat higher than the present state-of-the-art level.

2. A divertor was to be included since the reactor was expected to operate essentially on a steady state basis.

3. Inexpensive, readily available materials and common techniques were to be utilized as much as possible.

4. The "safety factor", q, was chosen to be 2.0, a reasonable expected improvement over present

experimental accomplishments.

5. The aspect ratio, A, was expected to exceed 3.0; the plasma ion density to approximate $10^{14}cm^{-3}$; the plasma temperature to be about 15 kev. The plasma composition was assumed to be equal parts of D and T. The reactor's electrical output was expected to be about 2000 MW(e) and a thermal cycle efficiency of 40% was assumed.

The resulting design (Figure 18-5) in part reflects the difficulty in placing a divertor on a tokamak reactor. The divertor windings were placed outside the neutron shield in order for them to be either superconducting or cryogenically cooled. The divertor windings also provide the vertical magnetic field that is necessary for plasma equilibrium. Furthermore, the size scale had to be sufficient to permit adequate neutron shielding between the reacting plasma and the superconducting toroidal field coils thereby limiting the heat deposition in the coils by the neutrons to acceptable levels.

In keeping with Item 3 above, stainless steel is the chief construction material. The vacuum wall is constructed of stainless steel plates welded on a steel framework. Liquid lithium is not used as a coolant to avoid associated MHD problems, but lithium in the form of flibe is used for tritium breeding. The blanket is cooled by helium gas which in turn is used to drive helium gas turbines.

The use of stainless steel limits the blanket operating temperatures to about 550°C. Thence the design foregoes the advantages of higher thermal cycle efficiencies that can be achieved with higher operating temperatures. However, the use of higher temperatures would require the use of a refractory metal, such as niobium, which is not in common use today.

LASL Theta-Pinch Fusion Reactor

A theta-pinch fusion reactor would utilize a shock-heating phase and an adiabatic compression phase. The shock-heating phase would

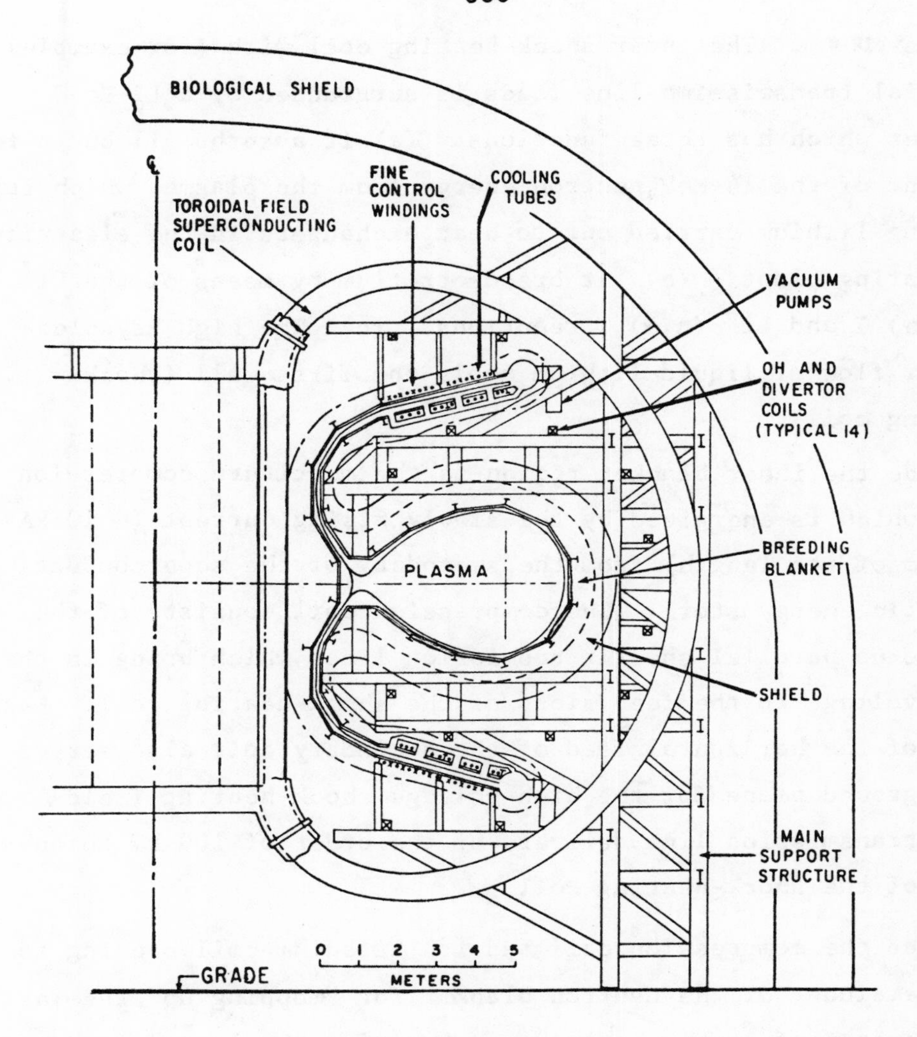

Figure 18-5

Cross Section of the Princeton Tokamak Reactor Design

have a risetime of a few hundred nsec and a magnitude of a few tens of kG to drive an implosion of a fully ionized plasma whose density is of the order of $10^{15} cm^{-3}$. After the ion energy associated with the radially directed motion of the plasma implosion has been thermalized, the plasma would assume a temperature characteristic of equilibration of ions and electrons. After a few msec the adiabatic compression field (risetime ~ 10 msec and final value B ≈ 100 to 200 kG) would be applied by energizing a compression coil.

A schematic diagram of a theta pinch reactor system is shown in

Figure 18-6 . The inner shock-heating coil with (for example) 8 radial transmission-line feeds is surrounded by a Li-Be-C blanket which has three functions: (a) it absorbs all but a few percent of the 14-MeV neutron energy from the plasma, which its flowing lithium carried out to heat exchangers in the electrical generating plant. (b) It breeds tritium by means of the Li^7 $(n,n'\alpha)$ T and Li^6 (n,α) T reactions. (c) The high Reynolds-number flow of liquid lithium cools the first wall (shock-heating coil).

Outside the inner blanket region is the multiturn compression coil which is energized by the slowly rising current (~ 10 kA per cm of its length) from the secondary of the superconducting magnetic energy store. The compression coil consists of the coiled up parallel-sheet transmission lines which bring in the high voltage to the feed slots of the shock-heating coil. Each side of the horizontal feed of the secondary coil also serves as a ground plane for the high-voltage shock-heating field. Each transmission line delivers of the order of 100 kV to one slot of the shock-heating coil.

Outside the compression coil and its titanium coil backing is the remainder of the neutron blanket for "mopping-up" the last few percent of neutron energy and breeding the last few percent of tritium. Unlike the inner blanket, which would run at ~ 800°C to provide high thermal efficiency of the generating plant, this portion of blanket could run much cooler. Surrounding the outer blanket is a neutron shield, and beyond the shield the radially emerging transmission lines are brought around to make contact with the secondary coil current feeds and the high-voltage shock-heating circuits. To the right is shown the cryogenic energy storage coil in its dewar. At the bottom of the storage coil is the variable-inductance transfer element which reversibly transfers energy from the storage coil to the compression coil and back again.

LLL Mirror Fusion Reactor

Designed to produce 500 MW(e), the LLL DT mirror reactor design

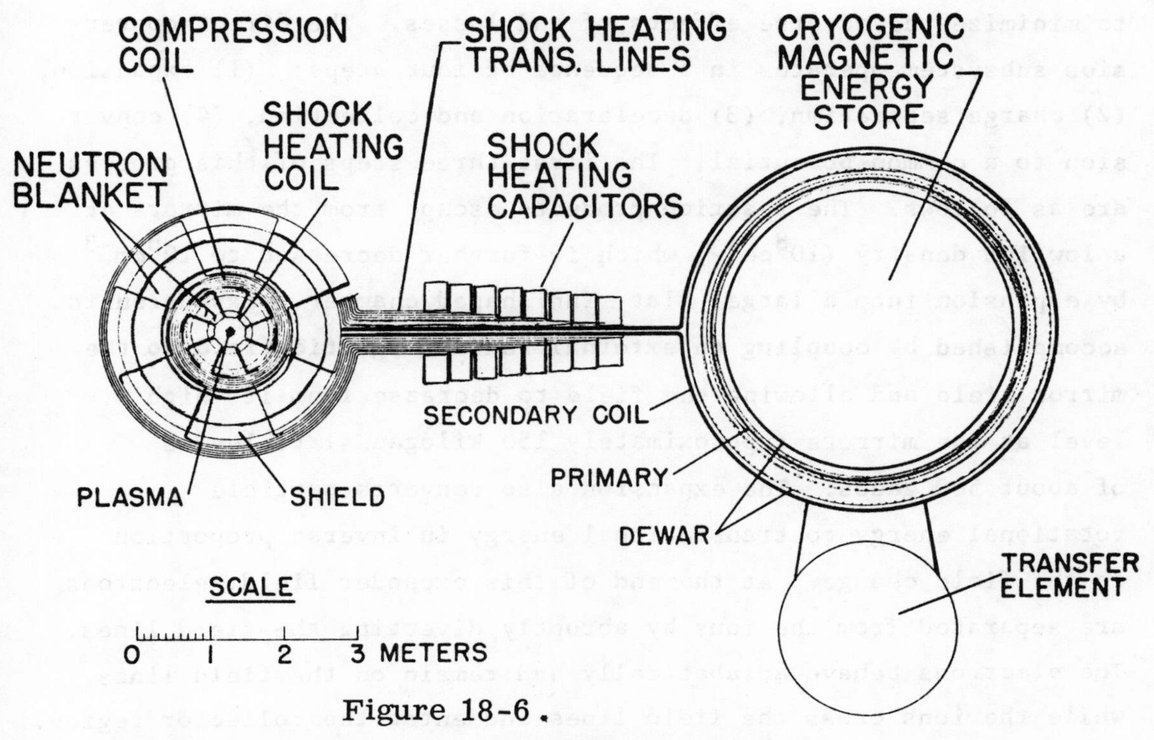

Figure 18-6.

Schematic of a Theta Pinch Fusion Reactor (Cross Section of a Torus)

may be considered as having three main parts: a magnetically contained plasma volume in which the fusion reactions take place, an ion injection and plasma heating system requiring electrical power input, and a combination thermal and direct energy converter system. The thermal portion of the converter system converts the neutron kinetic energy to thermal energy in a blanket surrounding the plasma confinement zone. The blanket breeds tritium for fuel replenishment. The second element of the energy converter system is the direct converter which accepts energetic charged particles which escape from the plasma confinement zone and it converts their energy to high voltage dc power. A fraction of this direct converter power is then fed back to the ion injection system to sustain the reaction and maintain the plasma. The reactor may be generally classified as a relatively low gain energy amplifier. This concept of combining thermal and direct conversion should be applicable to any fusion containment system; however, it is especially attractive for mirror systems because it furnishes a means to minimize the adverse effects of end losses. The direct conversion subsystem operates in a sequence of four steps: (1) expansion, (2) charge separation, (3) deceleration and collection, (4) conversion to a common potential. The first three steps of this process are as follows. The reaction products escape from the mirrors at a low ion density ($10^8 cm^{-3}$) which is further decreased to $10^6 cm^{-3}$ by expansion into a large, flat, fan shaped chamber. Expansion is accomplished by coupling an external radial magnetic field to the mirror field and allowing the field to decrease from its high level at the mirrors (approximately 150 kilogauss) to levels of about 500 gauss. The expansion also converts particle rotational energy to translational energy in inverse proportion to the field change. At the end of this expander field, electrons are separated from the ions by abruptly diverting the field lines. The electrons behave adiabatically and remain on the field lines while the ions cross the field lines and enter the collector region.

The ions emerge from the expander with a considerable spread in energy. To recover this energy at high efficiency the ions are passed through a series of electrostatically focusing collectors within which they are progressively decelerated. The ions are decelerated to a low residual energy and then diverted into a collector. Experiments at LLL have demonstrated overall collection efficiencies in excess of 80% and further improvements are expected.

The final step of direct conversion is the transformation of the electrical energy to a common potential. This is accomplished by an inverter-rectifier system using commercially available equipment.

The approximate plasma conditions are as follows: average ion energy = 400 keV, average electron energy = 40 keV, total power output = 1330 Mw, plasma beta = 0.9, plasma density = $10^{14} cm^{-3}$, and plasma radius = 4.3 meters. A schematic of the system is shown in Figure 18-7.

ORNL Laser-Fusion Concept (BLASCON)

If lasers can be economically utilized to ignite DT pellets to give small thermonuclear explosions, it may be possible to build

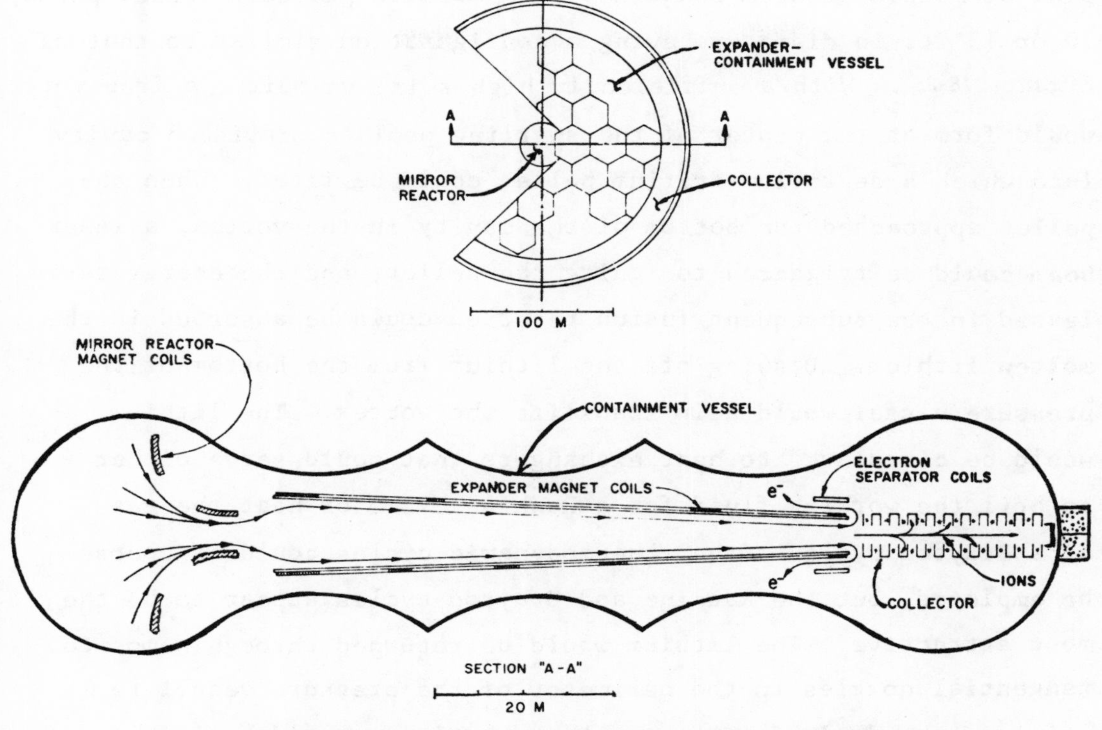

Figure 18-7. **Plan and Cross Section Views of the Livermore Mirror Reactor with Direct Converter**

reactors for central stations, ships, and spacecraft propulsion. Analyses and model tests indicate that, by igniting the pellets in the cavity of a vortex formed in a pool of liquid lithium, the explosion can be contained in conventional pressure vessels at a vessel capital cost of only about $10/kw(e). The neutron economy would be excellent -- the breeding ratio could be 1.3 to 1.5. If applied to reactors for central stations or ships, the concept would permit the construction of economic, thermonuclear reactors in sizes possibly as small as 100 MW(t). There would be no need for large cryogenic magnets, and no problem with fast neutron damage or neutron activation of structure. If applied to spacecraft propulsion the laser-exploded pellets might give a system whose propellant requirement for a typical Earth-Mars-Earth mission would be only about 10% those of a Rover-type nuclear rocket.

Frozen DT particles could be ignited at intervals of 10 to 20 sec and the energy of the explosions absorbed in a rapidly swirling pool of molten lithium contained in a massive pressure vessel perhaps 10 or 15 ft. in diameter having a configuration similar to that of Figure 18-8. With a sufficiently high swirl velocity, a free vortex would form at the center of the swirling pool to provide a cavity into which a deuterium-tritium pellet could be fired. When the pellet approached the bottom of the cavity in the vortex, a laser beam could be triggered to ignite the pellet, and the energy released in the subsequent fusion reaction could be absorbed in the molten lithium. Drawing off the lithium from the bottom of the pressure vessel would help stabilize the vortex. The lithium would be circulated to heat exchangers that could serve either to boil the working fluid for a Rankine cycle or heat the gas of a Brayton cycle. Other thermodynamic cycles could of course be employed, but the Rankine and Brayton cycles appear to be the most attractive. The lithium would be returned through pumps to tangential nozzles in the perimeter of the pressure vessel to maintain the desired vortex so that particles would be injected to a point close to the center of mass of the lithium. The operating temperature of the lithium would depend in part on the choice of containment system material, e.g., about 900°F if a chrome-moly steel were used and perhaps 1800°F if niobium were employed.

LASL Laser-Fusion Reactor

A schematic of a wetted-wall Inertial Confinement Thermonuclear Reactor (ICTR) is shown in Figure 18-9. A DT pellet is injected through a port, which penetrates the blanket, and is initiated at the center of the cavity by a laser pulse; the cavity is defined by the wetted-wall located at a radius of 1.0 m from the center. The subsequent (D+T) burn releases 200 MJ of energy. Within fractions of a microsecond, 50 MJ is deposited within the pellet and 152.5 MJ is generated within the blanket lithium and structural materials.

Within ~ 0.5 ms the pressure pulses generated by the interaction of the pellet with the lithium at the wetted-wall will subside. Within the next few milliseconds, the cavity conditions are

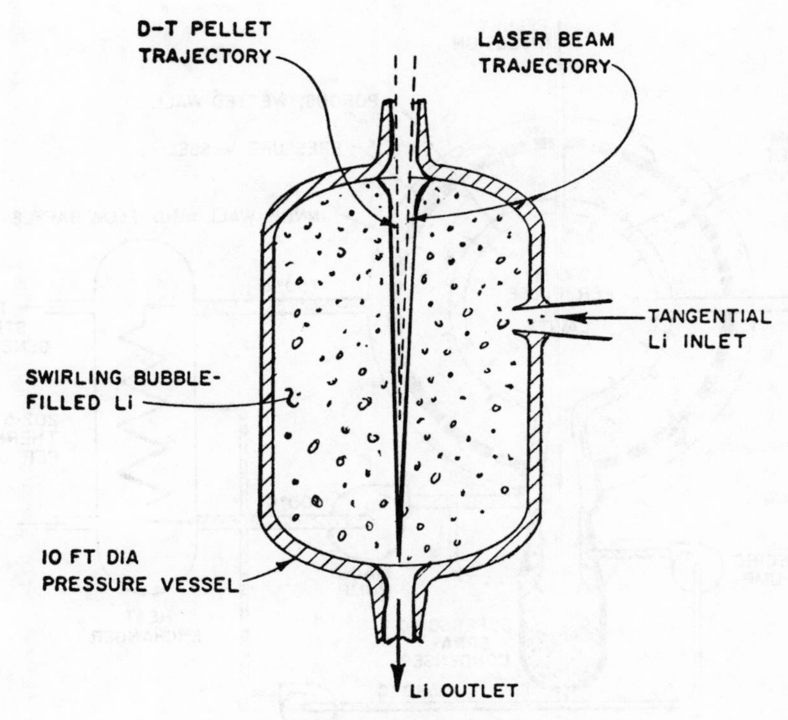

Figure 18-8

Laser-fusion reactor core employing a bubble-filled lithium vortex to absorb the energy of the explosion.

equilibrated, ~ 1.6 kg of lithium are vaporized from the protective layer at the wall, and sonic flow conditions of the cavity gases are established at the outlet port.

The flow of hot gases through the cavity outlet port is expanded in a diffuser to supersonic conditions, and the gases are then condensed in a downstream length of duct where a finely atomized spray of liquid lithium is injected. (The spray of atomized droplets is recirculated from the liquid pool at the bottom of the condenser). Downstream of the condenser duct, the mixture of gas and liquid droplets, still at supersonic velocity, is decelerated by turbulent mixing created by a spray of large lithium droplets. (The coarse-droplet spray is provided from a side-stream of the 400°C return flow from the heat exchanger.) The kinetic energy of this mixture is finally absorbed by impacting with a pool of liquid lithium at the bottom of the condenser system.

After ~ 0.2 s, the pressure within the cavity decreases to less

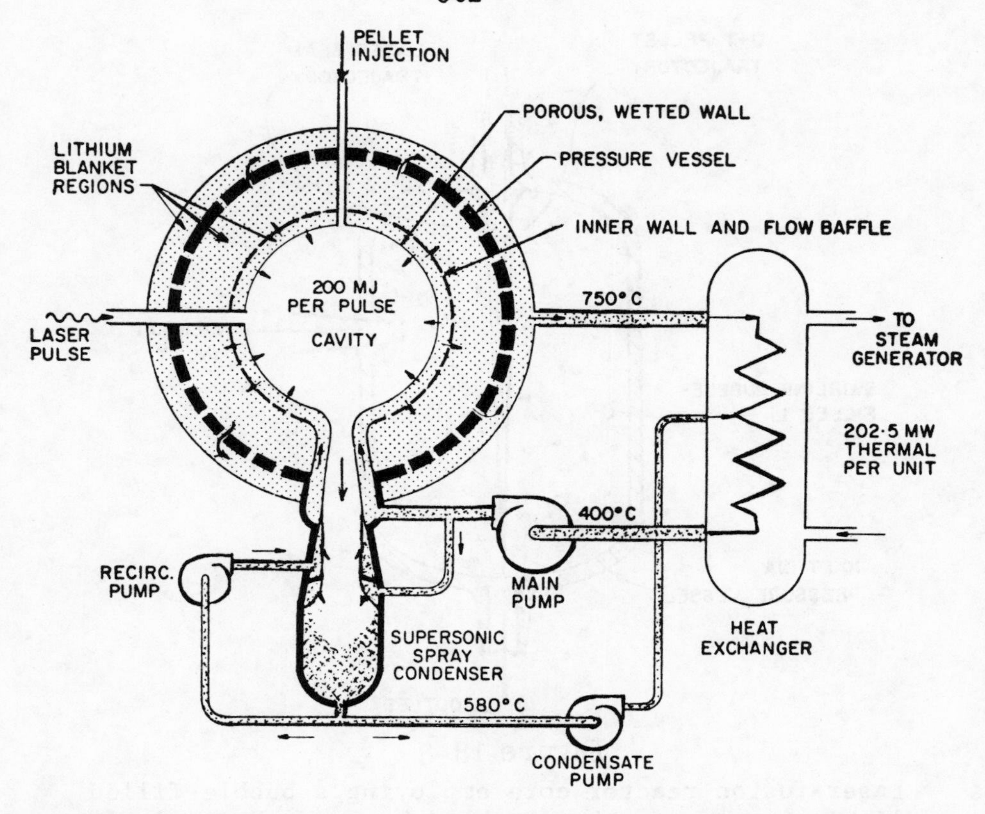

Figure 18-9. Schematic of a Laser-Fusion Reactor Design

than atmospheric, and the blow-down continues during the remaining 0.8 x of the pulse cycle, reducing the cavity pressure to less than 133 N/m^2 (1.0 mm Hg). The cycle is then repeated with the initiation of another pellet.

The energy deposited within the blanket is removed by circulating the lithium through an external heat exchanger. Lithium, flowing at 400°C from the heat exchanger, is returned to a plenum between the 1.0 cm-thick wetted-wall and the 5.0 cm-thick inner structural wall, which serves to restrain the movement of the inner blanket boundary caused by the pressure waves generated within the blanket and the cavity pressure. Located a few centimeters behind the wetted-wall, the inner structural wall also serves as a flow baffle for distributing the radial outflow. The wetted-wall moves along with the structural wall through hydrodynamic coupling, and, if needed, through mechanical attachments.

The minimum power level is based on a thermal output of ~ 200 MW, from one ICTR. Higher power levels may be obtained by combining several ICTRs in a reactor system, thereby increasing both the versatility and the overall ratio of actual operating power to full design power. The nominal thermal power level for a conceptual plant was arbitrarily chosen to be ~ 2000 MW, requiring ten modular ICTRs.

References

1. "Conceptual Design of the Blanket and Shield Region and Related Systems for a Full Scale Toroidal Fusion Reactor." A. P. Fraas. U.S. AEC Report ORNL-TM-3096, to be published.

2. "Preliminary Appraisal of the Hazards Problems of a D-T Fusion Reactor Power Plant." A. P. Fraas and H. Postma. U.S. AEC Report ORNL-TM-2822. Oak Ridge National Laboratory. December, 1970.

3. Nucleonics Week. Vol. 15, No. 44, pp. 26-31, November 3, 1969.

CHAPTER 19

FISSION RESEARCH AND DEVELOPMENT REQUIREMENTS

The central problem associated with the development of practical fusion reactors is the confinement of a plasma that has a particular minimum combination of temperature, density, and confinement time. To achieve these conditions, additional understanding of the basic scientific principles governing the confinement and heating of plasmas must be developed. At present many of the important basic principles are known as a result of systematic past efforts. This work recently resulted in a major program milestone -- the virtual elimination of premature plasma escape from experimental magnetic confinement systems. Other plasma loss mechanisms are recognized as being potential problems as fusion reactor conditions are approached, and many of these are now being investigated. The scaling of plasma phenomena with size can be predicted in a general way, but it is essential to test those scaling laws so as not to invoke unrealistically large extrapolation factors when developing a reactor design. In laser-fusion the present task is to obtain experimental conditions which approximate certain theoretical requirements and then to determine if important predictions can be physically realized.

To accomplish these tasks larger plasma volumes will be needed in magnetic systems to decrease diffusion losses or to provide appropriately shaped fields. In laser-fusion, larger, higher energy laser facilities will have to be constructed. In some magnetic confinement programs, new plasma heating techniques will have to be added to currently utilized methods.

Diagnostics with very high temporal and spatial resolution will have to be further developed for laser-fusion research. As plasma experiments are improved, it will be necessary to study plasma response and to avoid potential instability problems.

To provide these changes and improvements in future experimental systems, it will be necessary to further develop a number of existing technologies. Superconducting magnets must be extended to provide higher fields in larger volumes; neutral particle beams will have to be further upgraded in current; magnetic energy storage systems will have to be developed from their present, small, prototype stage into very large sizes, etc. Existing and new laser concepts will have to be developed to provide yet further improved capabilities. In the following, these problems are elaborated upon and paths towards their solutions are outlined. The period considered is that from the present through the demonstration of the scientific feasibility of controlled fusion in the late 1970's or early 1980's.

In the following, the present phase of the program is placed in perspective in terms of the past and the future. The major areas of research and development are described in general terms and an attempt is made to briefly describe the key problems which must be overcome in each. The steps projected to solve these problems are then outlined.

THE OVERALL TIMETABLE

Since its inception in 1952, the AEC-CTR program has concentrated on plasma physics research and supporting technological development. However, the nature of the problem has been and is such that improved plasma experimental systems require significant technological development, i.e., there is simply no method of achieving actual or scaled fusion plasma parameters short of fabricating sophisticated experimental systems. This requirement has necessitated a continuing development effort within the program.

The scientific feasibility of fission power was demonstrated in 1942 with the criticality of the "Chicago Pile". This experiment was performed by assembling and manipulating blocks of relatively readily available materials, i.e., very little technological development was required for the experiment. This will clearly not be possible in fusion development. Because of the extreme physical conditions imposed by fusion requirements, many sophisticated technological problems have had to be solved to progress to the present stage and others must be addressed prior to the fabrication of the fusion scientific feasibility experiments. This requirement imposes conditions on the current program and also means that many of the key fusion reactor technology problems will necessarily be significantly advanced when fusion scientific feasibility is demonstrated. For this reason a simple fission-fusion development comparison for the early phases of the two programs is meaningless.

Having achieved good magnetic plasma confinement, controlled fusion research now has advanced to the point of identifying potential future problems and beginning to systematically study them. In laser-fusion, critical experimental questions are just beginning to be addressed. No matter what the approach, when it becomes reasonably assured that clear paths are available, it will be necessary to fabricate new larger experiments in order to demonstrate scientific feasibility. Even after these demonstrations, a number of important physics questions will remain to be studied. Nevertheless, the principal portion of the physics problem will have been solved and the program will be in a position to begin a fusion reactor development phase.
The length of the reactor development phase will depend sensitively upon a number of important factors:

1. The extent to which relevant reactor oriented R&D can be accomplished prior to the demon-

stration of scientific feasibility. Areas of obvious importance include radiation effects on materials, superconducting magnet development, tritium handling and leakage studies, neutronics tests, etc.

2. The ease with which reactor-grade plasmas can be produced and controlled. This factor will have a profound impact on the extent and complexity of future component and subsystem development.

3. The available financial support and program priority.

Clearly a number of technologies will have to be further developed before commercial electrical power from fusion becomes a reality. But the fusion program will benefit significantly from fission reactor programs in a number of important ways:

1. A large body of knowledge and technology associated with radioactive and neutron bearing systems will be available.

2. Personnel educated and experienced in many of the technologies common to both systems will exist.

3. The management experience gleaned from the various fission development programs will be applicable to fusion development.

4. Industries exist or are being developed which will be experienced in handling many of the problems inherent to transferring fusion from the laboratory to the commercial market.

RESEARCH AND DEVELOPMENT THROUGH THE DEMONSTRATION OF SCIENTIFIC FEASIBILITY

The Major Program Elements

The present AEC fusion effort represents a multi-line attack on a very complex problem. Today four general approaches are receiving major emphasis. These are magnetic mirror systems, steady state toroidal systems (principally the tokamak), pulsed high-beta* pinch systems (principally the theta pinch), and laser-fusion systems. Each of these concepts appears to have the potential of being extrapolated to a practical power reactor, based upon preliminary systems studies.

In addition to these main lines, a broad range of supporting research which spans the range from basic plasma studies to relatively large-scale confinement systems research is conducted in national laboratories, universities, and industry. Technological development (magnets, beams, energy storage, diagnostics, etc.) is carried out in support of existing experiments and in preparation for next generation systems. Finally, a small but growing program in fusion reactor analysis and technology is supported both to provide guidance to the plasma research and to provide an understanding of the non-plasma problems which must be solved prior to the achievement of practical fusion power.

All of the major confinement concepts are to varying degrees interrelated and thereby supportive. All have uncertainties. Each would extrapolate to a somewhat different reactor type, the operating characteristics and economics of which are very difficult to estimate at this time. For these reasons it is considered very important to carry each of the major concepts to a point where a reasonable evaluation of the relative reactor merits of each can be carried out.

---

* Beta is defined as the plasma pressure divided by the magnetic pressure, i.e., $\beta \equiv \frac{nkT}{B^2/8\pi}$. High beta refers to values approaching unity.

The Major Approaches: Status and Projected Steps to Scientific Feasibility

Steady State Toroidal Systems

The tokamak concept is receiving primary emphasis in this general area of research. Tokamak plasmas in the U.S. and the U.S.S.R. have reached densities of about $3 \times 10^{13} cm^{-3}$, ion temperatures of 600 eV, and confinement times of about 20 milliseconds; near the classical value for the magnetic fields, densities, temperatures, and sizes now being utilized. A reactor-grade tokamak plasma must have a density near $10^{14} cm^{-3}$, an ion temperature above 5000 eV, and a confinement time of about one second. Theory indicates that tokamak plasma confinement time is directly proportional to the square of the minor plasma diameter, to the square of the toroidal magnetic field, and to the inverse square of the aspect ratio*. The most economical path to confinement times of one second is to increase the minor diameter from the present values of 28-47 cm to about 150 cm at near the presently applied fields of about 50 kilogauss and at practical aspect ratios of about three.

The achievement of reactor-like conditions will require a continued research effort concentrated on problems of plasma confinement and stability and on the development of auxiliary plasma heating techniques. With respect to the plasma, theory indicates that the so-called "trapped particle" instability may cause plasma loss at temperatures higher than those achieved in the present generation of tokamak experiments. The magnitude of this loss is difficult to estimate and it may be of negligible importance. Relevant data are expected from the ORMAK tokamak and the FM-1 multipole, both of which are now operational, and from the PLT tokamak, which is expected to be operational in 1975. If the trapped particle instability is troublesome, theory indicates that it can be dissipated by increasing the particle collision rate either by increasing the plasma

---

*Aspect ratio is defined as the major radius divided by the minor radius.

density or by inducing a modest level of plasma turbulence.

Another concern is tokamak plasma purity. Present tokamaks are somewhat marginal in this respect. In the future larger tokamaks, particularly the PLT, will have smaller surface to volume ratios which should accordingly be more favorable. If this is not the case, then it will be necessary to use an auxiliary component called a "divertor", a device which "skims off" plasma diffusing towards the walls before it actually contacts the walls. Divertors have been shown to be effective on both the Model C stellarator and the FM-1 multipole.

Tokamak plasmas are now heated ohmically by the same plasma current which provides a necessary part of the confining magnetic field. Ohmic heating becomes progressively less effective at higher temperatures because plasma resistivity varies inversely as the three-halves power of the electron temperature. Therefore, to reach temperatures of 5000 eV, auxiliary heating methods will be required. A number of attractive methods have been used on other concepts and will soon be tested for possible use in tokamaks. The ATC (Adiabatic Toroidal Compressor) will attempt to heat a tokamak plasma by forcing it to smaller major radius where the associated higher magnetic field should cause plasma compression*. Neutral particle beams will be added to the ORMAK during the coming year to determine whether this technique, which was originally developed for mirrors, can be of use for tokamaks. Turbulent heating allows rapid energy transfer to a plasma by momentarily increasing its resistivity to a very high level. The Texas Torus is about to test this technique.

Among the other variables of importance in tokamak design are the magnetic field strength and the shape of the plasma. The Alcator will provide a vehicle for the study of tokamak

*This was subsequently demonstrated.

plasmas at fields exceeding 100 kilogauss. Theoretically the use of a kidney-shaped plasma cross section should permit tokamak operation at magnetic fields significantly lower than needed for circular plasmas. If successful, this concept, which is being tested in the Doublet II, could result in a major savings for tokamak reactors which would be able to use lower and thereby less expensive magnetic fields

The recently authorized PLT is a significant step along the road beyond the present generation of tokamak experiments. It is estimated to cost $13 million, and, if successful, it will provide plasma conditions close to those considered necessary to establish the scientific feasibility of tokamak systems. An experiment to demonstrate tokamak feasibility is estimated to have a capital cost of about $50 million and would require 3-4 years to fabricate. Among the major supporting activities which will be necessary for future generations of tokamaks are the development of large bore, high field superconducting magnets and quite possibly the development of higher current neutral particle beam sources.

Magnetic Mirrors

A number of mirror plasma experiments have already exhibited conditions approaching those needed for a reactor, i.e., densities greater than $10^{13} cm^{-3}$ (2X-I and 2X-II) and confinement times of seconds (Baseball I and II) have been achieved. The scientific questions important to mirrors are considered to be well defined, and many of them are being actively studied at this time.
Achieving adequately stable confinement in mirror systems requires that the confined plasmas be characterized by well broadened ion energy distributions, such as those that occur when the density is high enough so that particle scattering by collisions within the plasma is well developed. These so-called "scatter-dominated" plasma conditions are now being

approached in present day mirror experiments (Baseball II and 2X-II) using different techniques of plasma formation, i.e., neutral beam plasma buildup (Baseball II), plasma injection and magnetic compression (2X-II). The stability properties of plasmas with these broadened energy distributions are being determined and compared with theoretical stability criteria. If the present scale experiments continue to show that adequate stability can be maintained as the plasma is made hotter and denser, then it will be necessary to verify that volume effects do not result in deleterious losses and this will require plasmas of still larger size, closer to those required for a reactor.

High-beta (high relative plasma pressure) will be required in mirror reactors. To date the Elmo and IMP experiments have shown that this is indeed possible when the electron pressure dominates. This result must be extended to the case wherein hot ions are the dominant component.

It has been recognized that the plasma which leaks out of the ends of mirror systems represents an undesirable energy drain. A few years ago a direct conversion technique to efficiently capture this loss was conceived. In simple terms this concept involved expanding the plasma flow to low density and then allowing the plasma to do work on an externally applied electric field, i.e., essentially an electrostatic accelerator operating in reverse. This concept has been studied on a small scale, and theory and experiment have been found to be in agreement. A larger experiment to more accurately model all of the relevant physics is now under construction to extend these results prior to testing the concept on a large plasma experiment.

Present mirror experiments are of too small a size to permit a test of mirror scientific feasibility. Looking toward an experiment that would be capable of such a test, a preliminary design for an MFX (Mirror Fusion Experiment) has been developed.

It is projected to have a capital cost of about $30 million and would represent a scale-up in magnet field intensity, plasma volume, and plasma temperature from the present Baseball II experiment. It would have a confined plasma volume of 300 liters (30 times that of Baseball II), utilize neutral beams of about 100 keV mean energy, and have a central confining field of 50 kilogauss or more (about five times that of Baseball II). Prior to the construction of MFX it will be necessary to extend the present neutral beam development from the present 10 keV, 7 ampere equivalent level to 100 keV multi-ampere sources and to upgrade present superconducting magnet technology to higher fields and larger sizes. Though these developments appear to be within the state-of-the-art, they will require a substantial effort prior to the initiation of MFX.

Theta Pinches

The theta pinch program is presently centered around the Scyllac experiment at Los Alamos. While its eventual aim is to create and confine a pulsed, high density, toroidal plasma, studies of both linear and toroidal configurations are presently underway. Scyllac's predecessor, the Scylla IV, was a linear, open-ended system whose thermonuclear plasma (5 keV ions at $5 \times 10^{16} cm^{-3}$) decayed classically in the radial direction but was dominated by end losses because of its short coil length (1 meter). One portion of the present Scyllac program involves assembly of a 5 meter straight section with magnetic mirrors to extend the Scylla IV results and to obtain a better understanding of the limits of the linear theta pinch geometry. The remainder of the Scyllac capacitor bank power supply has been assembled into an arc sector, which is actually a 5 meter section of the planned-for full torus. The purpose of this sector is to test methods of stabilizing toroidal theta pinch plasmas prior to completion of the full torus.

The sector, limited to operational times of about 15 microseconds by losses out its open ends, has nevertheless been extremely useful for studies of the effects of auxiliary stabilizing fields. Initially, without auxiliary fields the curved plasma in the sector drifted to the walls in about two microseconds as expected from theory. Two sets of auxiliary coils were then added and progressively modified over the last year until now the sector plasma can be maintained stably for near the full 15 microsecond period determined by end losses.

Beginning in the second half of FY 1973, Scyllac will begin to be converted to a full torus for operation in mid FY 1974. Its principal task will be to demonstrate control of gross plasma instabilities and drifts. This probably will involve the use of auxiliary fields now being studied on the sector and may also require the use of feedback or dynamic stabilization techniques being studied on the modified Scylla IV experiment. If Scyllac operates successfully in FY 1975, confinement will be limited only by the capacitor bank decay time of 250 microseconds, and the extension of theta pinches to the millisecond periods required to demonstrate scientific feasibility would have a high probability of success. Before starting a feasibility experiment, which is estimated to cost about $30 million, an additional concept must be verified. At present the shock heating and compression characteristics of theta pinches are accomplished in a single step. It appears advantageous to modify this to a two step sequence. This modification is necessary because it does not appear technologically or economically desirable to build fast capacitor banks (required for the single step process) larger than the one built for Scyllac. By using a two step process the compression phase could be accomplished on a slower time scale using relatively low cost magnetic energy storage systems and/or slow capacitors.

Before embarking on a theta pinch feasibility experiment, a considerable amount of development of magnetic energy storage and separated shock heating systems -- the two-step sequence -- will be necessary.

Laser-Fusion

The AEC Division of Military Application (DMA) conducts a research and development program in laser-fusion because of its potential for application to a number of military needs, such as direct application of laser radiation, weapons effect simulation, neutron and x-ray photography, and weapons. However, the basic technology has other potential applications, including the possible production of electrical power for military and civilian uses. Recently the AEC Laser-Fusion Coordinating Committee considered the long-range future of the AEC laser-fusion program. The high potential payoff was deemed sufficient to merit an energetic program even though portions of the program must be considered as having substantial uncertainty. Experimental studies coupled with a vigorous theoretical effort will be aimed specifically at resolving currently recognized questions.

Large laser facilities will pace the development of the laser-fusion program because it is clearly evident that laser energies two to four orders of magnitude higher than presently available will be necessary to first attempt to produce high target compressions and then to progress to obtain net thermonuclear energy release.

A number of physics questions basic to the laser-fusion process are presently recognized. The mechanisms of laser energy absorption by target pellets in experiments todate are not well understood. At higher laser power, theory indicates that nonlinear effects could cause pellet preheating which could limit pellet compression to values lower than deemed necessary for power production. Experiments to study this problem are clearly needed.

One prime reason for the present optimism in the laser-fusion field is the theoretical prediction that lasers may be capable of producing very high target compressions, thereby markedly reducing the laser energy requirements. To date, laser outputs have not been sufficient to produce measurable target compressions, and so related experiments to correlate with theory are a high laser-fusion program priority.

The civilian power application of laser-fusion requires efficient lasers. The present workhorse of the program is the Nd:glass laser whose practical upper energy limit is believed to be near 10^4 joules and whose overall efficiency is of the order of 0.1%, which is inadequate for power generation. Efficient lasers with outputs of 10^5 - 10^6 joules are considered to be necessary for reactor operation, and relatively few choices are presently available, in part because of the early state of the research.

CO_2 appears attractive in this regard according to theory, in part because it has already been demonstrated to be capable of efficient, high energy operation in long pulse operation (tens of microseconds) versus nanoseconds (10^{-9} seconds) for laser-fusion. Considerable development from the present tens of joules, one nanosecond levels will be required.

Understanding the important processes in a tiny, near 100 micron diameter laser irradiated pellet wherein important phenomena occur in picosecond (10^{-12} second) periods will require very sophisticated diagnostic techniques. While a number of detectors capable of the required temporal resolutions are presently available, very few provide adequate spatial sensitivity. Improved diagnostics will therefore have to be developed.

In considering future aspects of the laser-fusion program, the AEC Laser-Fusion Coordinating Committee recognized the difficulties of foretelling achievements in a field wherein there are so many uncertainties. Within this constraint it conceived that scientific breakeven (thermonuclear energy release equal to laser light input) could occur in the FY 1977 - FY 1979 time period. Net energy production (thermonuclear release greater than total energy into the laser) is expected to require larger laser facilities capable of delivering energies in the 10^6 joule regime. The Laser-Fusion Coordinating Committee considered that this milestone might be achieved in the FY 1979 - FY 1981 period.

Supporting Research and Development

Back-Up Magnetic Confinement Research

A number of relatively large confinement research programs are included in this category, which is to be differentiated from basic plasma research (described below) which involves smaller scale studies. Back-up confinement experiments are often used to study specific questions which are difficult or impossible to effectively study in main-line systems and are not considered primary approaches to the goal of developing fusion reactors. The progress and justification of back-up confinement experiments are periodically reviewed and concepts in this category are frequently terminated when new, more attractive concepts are conceived and programs are to be implemented.

At present the following experiments are included in this category. In support of steady state toroidal research are FM-1 (Floating Multipole), D. C. Octupole and Levitated Octupole; in support of high beta systems are ZT-1 (Toroidal Z Pinch) and EBT (Elmo Bumpy Torus).

The FM-1 is a toroidal system whose field is in part supplied by a floating superconducting ring around which a plasma is confined. The D. C. and Levitated Octupoles are internal ring experiments with four rings around which the plasma is confined. Both the D. C. Octupole and FM-1 have achieved near-classical plasma confinement. These systems are intimately related to and support tokamak research.

Understanding derived from theta pinch research prompted renewed interest in Z pinches in the late 1960's and considerable improvements in plasma conditions have been possible. Present efforts center around a toroidal system, the ZT-1.

As a result of mirror target plasma research, steady state mirror plasmas of the highest possible relative pressure ($\beta = 1$) were created by two frequency microwave heating. By coupling 24 of these plasmas together in a torus, the EBT will attempt to create a new type of toroidal plasma. Fabrication of EBT just began.

Research

This research category includes both theoretical and experimental studies. Research experiments are aimed at a broad range of relatively basic problems and are conducted on a smaller scale than characteristic of back-up confinement research. Research programs complement and/or extend work being done elsewhere in the program as well as helping to develop the subject of plasma physics. In recent years, budgetary pressures forced a partial curtailment of basic plasma physics research at the national laboratories, and the center of gravity of such work shifted to the "off-site" sector composed of 30-40 universities and two industrial contractors.

The scientific and technological resources of universities and industry will continue to be used to help solve many of the problems associated with the demonstration of scientific feasibility of a fusion reactor. They will do this by maintaining strong programs of research and development basic to fusion research and by serving as independent sources of new ideas on the problems associated with the major plasma confinement experiments. In addition, university fusion research programs provide engineers and physicists trained in plasma physics and fusion technology, and industrial fusion research programs help develop the industrial facilities, manpower, and technology that will be required to go from a demonstration of scientific feasibility to a practical fusion reactor.

It is often difficult to separate the contributions made to plasma physics and controlled fusion research by the AEC research program from contributions made by non-AEC sponsored programs because the solution of many of these problems has involved contributions from many quarters. However, it is possible to identify many research areas in which pioneering contributions were made by university and industrial research programs. These include the development of kinetic theory, particularly the implications of the Vlasov and Boltzmann equation to plasma phenomena; linear, quasi-linear and nonlinear theory; weak turbulence theory; MHD fluid theory and MHD energy principles; finite Larmor radius effects; drift instabilities; trapped particle and loss cone instabilities; minimum B theory, theory and experiment of linear and toroidal θ-pinches; theory and experiments related to the use of intense relativistic electron beams and theory and experiment on synchrotron radiation. In particular, basic studies of the nonlinear interaction of electromagnetic waves with non-uniform plasmas, supported under the CTR program, have provided the basis for beginning to understand the laser-solid absorption problem, a key question in laser-fusion. The present AEC-CTR research program accounts for about 20% of the total program expenditures.

Development Programs Needed for Near-Term Experiments

As mentioned throughout this section, development activities have been an inherent part of the CTR program in the past, and it will clearly have a profound impact on the speed with which the scientific feasibility experiments proceed in the future. Notable accomplishments to date include development of ultra-high vacuum systems, high current ion and neutral beam sources, large-volume, high-field copper and superconducting magnets, sophisticated plasma diagnostics, and fast, high energy capacitors.

The most important development problems which must be faced in the next few year period include superconducting magnets, magnetic energy systems, neutral beam sources, and high energy, short pulse lasers. In addition a number of other technologies will require attention including high current switches, r.f. energy sources, direct energy converters, plasma diagnostics, and feedback control systems. A brief description of the three main problems follows

The major requirements of the superconducting magnet development program are to provide both steady state and pulsed large bore, high field magnets. The need for steady state magnet systems is common to the fusion feasibility programs in magnetic mirrors and tokamaks. Pulsed magnet systems are needed for theta pinch energy storage and for ohmic heating in tokamaks.

Considerable development of superconducting magnets with bores of 1 m or larger has already occurred. NbTi units range from a 1 m bore, 41 kG solenoidal magnet at Saclay to the Baseball (minimum B design) magnet at Livermore with its 1.2 m bore and 55 kG maximum field at the windings to superconducting, levitated ring plasma experiments using Nb_3Sn ribbon with low fields at the windings (40 kG). Three large bubble chamber magnets (NAL and two at CERN) have been built but have not been tested as yet. The IMP quadrupole magnet (minimum B design) at ORNL is perhaps the most complex Nb_3Sn magnet, and it has been operated with a maximum field at the winding of about 80 kG.

Less work has been done on pulsed systems. Small magnets, mostly for synchrotron programs, have been developed which can be pulsed at 100 kG/sec up to 60 kG without undergoing a quench. The largest energy storage magnet is the 600 kJ, 76 cm bore coil built at the Laboratory de Marcoussis of CGE.

The energy storage volume needed for a pulsed θ-pinch is larger than those anticipated for tokamaks but the maximum field requirements are lower (about 20 kG). Discharge has to occur in times of the order of milliseconds. New materials with low hysteresis losses in the superconductor and low eddy current losses in the normal matrix may have to be developed for economical heavy duty systems.

The following steps in superconducting magnet development are projected to be necessary over the course of the next few year period. First develop 2 to 3 m bore magnets with B_{max} = 80kG. In parallel develop baseball-shaped magnets with plasma volumes of 500 liters and maximum fields at the windings of 125 kG for mirror systems. Also in parallel develop a magnetic energy transfer and storage system with 2 msec risetime for the compression coil in a staged θ-pinch. (Total stored energy: about 1 GJ). A magnetic energy storage system for a laser-fusion system appears necessary but its required characteristics cannot yet be specified.

The need for improvements in neutral particle beam sources has been long recognized within the CTR program. As a result, sources for specialized applications exist today which span the energy range from 1 keV to 40 keV, with currents to several amperes equivalent. Whereas there were doubts only a few years ago that single module, multi-ampere beams would be constructed, it is now generally accepted on the basis of presently available technology that units in the ten ampere equivalent range can be developed. These modules could represent the building blocks for achieving total neutral beam currents in the 100 ampere range, as would be required in a mirror feasibility experiment and may be needed for a tokamak feasibility experiment. While the extrapolation appears reasonable, the work must be completed prior to finalizing the designs of these large systems, if potentially costly delays are to be avoided at a later date.

The development of higher energy short-pulse (near one nanosecond) lasers will pace the laser-fusion program. The present emphasis is on the following activities: (1) Nd:glass laser development, in particular disk amplifiers to boost 100 psec - 1 nsec pulses from the one hundred joule level into the hundreds of joule region;

(2) to obtain nanosecond pulses from CO_2 gas lasers and to amplify these pulses into the tens of joule range. In FY 1974 these same activities will continue as follows: (1) Nd:glass lasers capable of energies to 1000 joules are expected to become operational; (2) nanosecond CO_2 gas lasers with energies approaching 1000 joules should begin operation.

A large (10^4 joule) Nd:glass laser will be needed in the near future and will require about three years to construct. A 10^5 joule CO_2 laser will then be the next planned step. These facilities are essential elements in carrying the laser-fusion program through the achievement of breakeven. Additional facilities expected to be required to demonstrate net energy return are being considered for planning purposes. However, they represent major extrapolations of both the laser and laser-fusion state-of-the-art and they cannot be clearly specified at this time. These and other uncertainties underscore the difficulty of long-range projection in this relatively new field.

Fusion Reactor Technology

The physics problem of creating, heating and confining a fusion plasma is clearly basic to the feasibility of practical fusion power but is only one of many problems which must be solved.

Recognizing the existence of the need for other developments and the fact that many will require considerable time and effort, a series of fusion reactor design studies were initiated in the late 1960's. These and subsequent efforts have served to define a number of areas which can be addressed in parallel with the physics studies.

Superconducting magnet systems will be required for magnetically confined plasma systems and will represent a significant fraction of the total capital cost. The development of large, steady state and pulsed-magnet systems for fusion feasibility experiments was described above. In addition to this near-term

development work, there is a strong incentive to develop new, ductile, high-field low-cost, easily fabricable superconducting materials. This and a back-up program on very low-resistance materials for pulsed applications also appears desirable and would be supported under fusion reactor technology.
Pulsed, cryogenic magnetic energy storage offers an essential method of energy storage for the slow compression pulse of a θ-pinch reactor, for the ohmic heating pulse of a tokamak reactor, and to energize the laser-pellet system. High field materials capable of remaining superconducting under pulsed conditions, as well as methods of low loss energy transfer must be developed. The first phases of experimental studies are underway and more effort is needed.

As in the LMFBR, fast neutron damage effects can have a profound effect on the lifetimes of certain fusion reactor components, due to embrittlement and/or swelling of structural components, owing to the accumulation of helium and displaced atoms from neutron bombardment. A small radiation damage program is now underway using existing fission reactors, (i.e., HFIR and EBR-II) and charged-particle accelerators. From LMFBR experience, an early major expansion in this work would be highly desirable, including construction of an intense source of 14 MeV neutrons.

Environmental considerations make it desirable to hold the tritium concentration in the lithium blanket to between 1 and 10 ppm, which is well below the 100 ppm limit imposed by considerations of structural metal embrittlement by hydrogen. By extrapolating the limited data presently available by about a factor of 30, it appears that this might be accomplished, but experimental data on permeation rates and concentrations at the low partial pressures involved are needed to make a good estimate of the feasibility and cost of achieving this low concentration.

Liquid lithium is attractive as a blanket coolant, but it poses

MHD problems as it flows through the magnetic fields of steady state reactors. The magnetic field has an adverse effect on the pressure drop and it suppresses turbulence, which severely reduces the heat transfer rate. These effects are roughly calculable and are heavily design-dependent, ranging from insignificant to major depending on the concept and the particular design. Molten lithium-beryllium fluoride salt and helium might be used as blanket coolants, but they have their own unique characteristics and problems which must also be better understood. Stainless steel is attractive for the blanket structure for first generation reactors but corrosion considerations limit the use of lithium in stainless steel to about 500°C. Experiments are needed to resolve uncertainties in temperature effects on corrosion and to provide a firm basis for design and system studies.

Various concepts for the blast confinement portion of a laser-fusion system have recently been proposed. When promising scientific results are available from laser-fusion research, it would be desirable to pursue the most attractive of these.

Major Planning Assumption

A major planning assumption is that scientific feasibility demonstration experiments may be necessary for each of the major lines which look promising at this time, i.e., tokamak, mirror, theta pinch, and laser-fusion. The commitment to build a large, new experiment is contingent upon the development of a sound scientific base. Should any of the major lines falter scientifically, then its next generation experiment certainly would be delayed, if not cancelled. If indeed four large experiments are constructed and if they operate successfully, then choices would be made based upon the best existing scientific and technical judgment, and thereby the program would narrow as it enters the plasma test reactor and experimental power reactor phases of the subsequent development program.

SUMMARY

The present task in fusion research is to further develop understanding of the plasma state and to prepare to move to the new experiments aimed at obtaining the full combination of temperature, density, and confinement time needed to experimentally demonstrate fusion scientific feasibility. Four general approaches are now receiving major emphasis. These are steady state toroidal systems (principally the tokamak), magnetic mirror systems, pulsed high-beta pinch systems (principally the theta pinch), and laser-fusion systems. Based upon preliminary systems studies, each of these concepts appears to have the potential of being extrapolated to a practical power reactor. In addition to these main lines, supporting research is carried out which spans a broad range from basic plasma studies to relatively large-scale confinement system research. A significant amount of technological development (magnets, beams, energy storage, lasers, etc.) is conducted in support of existing experiments and in preparation for next generation systems. Finally, a small but growing program in fusion reactor analysis and technology is supported both to provide guidance to the plasma research and to provide an understanding of related problems which must be solved prior to the achievement of practical fusion power.

To demonstrate the scientific feasibility of fusion power, large new experiments will be required. These will cost about $30-50 million, will require 3-4 years to fabricate, and are projected to establish fusion scientific feasibility in the 1980-1982 period. Presently, the program is proceeding on the assumption that four such experiments will be necessary, one for each major line. However, the initiation of each of these experiments is critically dependent upon the development of an appropriate scientific base. Should any line falter, its next generation experiment would appropriately be delayed or even cancelled. If four large experiments are built and if they operate successfully, then choices would be made based upon the best existing scientific and technical judgment, and the program would narrow as it enters the reactor development stage.

CHAPTER 20

SPECIFIC PROBLEMS OF NUCLEAR FUSION REACTORS

The Plasma of a Nuclear Fusion Reactor

Nuclear fusion reaction

There are two reactions that can be utilized in thermonuclear fusion which are given below.

$$D + T \rightarrow {}^4He\ (3.52\ MeV) + n(14.05\ MeV) \quad (1)$$

$$D + D \begin{cases} \nearrow {}^3He\ (0.82\ MeV) + n(2.45\ MeV) \\ \searrow T\ \ (1.01\ MeV) + p(3.03\ MeV) \end{cases} \quad (2)$$

The cross sections for these two reactions are shown in Fig. 20-1. A fusion reaction in a plasma that has a distribution close to the Maxwell distribution takes place in a thermonuclear fusion facility. The fraction of reaction R is obtained by averaging the cross section over the energy distribution and is given in the following manner.

$$R = n_i^2 <\sigma_V> \quad (3)$$

In the above equation, n_i is the ion density within the plasma while $<\sigma_V>$ is the averaging that is performed by multiplying the cross section with the rate of thermal movement.

Where the D-T reaction is involved, the product of the terms involving D and T, n_D and n_T, is substituted for n_i^2. If the mixture contains 50% each of D and T, the value $n_i^2/4$ is substituted.

The fraction of reaction R is frequently tied in with the parameter β that indicates the pressure of the plasma. β is given by

$$\beta = \Sigma n_i kT/(B^2/2\mu_0) \quad (4)$$

Letting the number n_i and temperature T_i of the ions in the plasma be respectively equal to the number n_e and temperature T_e of the electrons, the following relationship is obtained.

$$R = \frac{\beta^2 B^4}{64\mu_0^2} \frac{<\sigma v>}{(kT_i)^2} \quad (5)$$

The term $<\sigma v>/(kT_i)^2$ on the left side is a quantity that is determined by the temperature and is an important parameter that indicates the fraction of reaction. The relationships between this parameter and temperature for the D-T and D-D reactions are shown in Fig. 20-2. Both curves exhibit maxima close to 10 keV. On the other hand, this figure shows the value of R for the D-T reaction to be at least one order of magnitude greater than that of the D-D reaction under the same conditions. Consequently, it is apparent that the D-T reactor is more advantageous than the D-D reactor. The broken line sections in this figure show the regions in which the Bremsstrahlung loss is greater than the energy that can be obtained from the nuclear reaction.

Transfer of energy between charged particles

The energy of a thermonuclear fusion reaction that is borne by the neutrons is 14.05 MeV for the D-T reaction and 2.45 and 3.03 MeV for the D-D reaction. Since the neutrons are not held captive within the magnetic field, they escape from the confines of the plasma. It is possible to sustain the function of the fusion reactor adequately if the energy borne by the other charged particles is retained within the plasma to heat the ions that are supplemented by the fuel and replenish the energy that is dissipated by radiation. The energy of the alpha particles in the case of the D-T reactor and the alpha, tritium, and protons in the case of the D-D reactor are transmitted by collisions to D and T ions and electrons. For example, starting with an α particle of energy μ_α, the fraction of the energy that is transmitted to ions and electrons is given by the following.

$$\frac{d\mu_\alpha}{dt} = -\frac{2\times10^{-18}\mu_\alpha n_i}{T_e{}^{3/2}}\left[1+296\left(\frac{T_e}{\mu_\alpha}\right)^{3/2}\right] \tag{6}$$

The first item within the brackets represents that fraction of energy that is imparted to electrons while the second item represents the energy transmitted to ions. It is assumed in many examples of reactor design that the energy transfer from the α particles is complete, and these particles exit the plasma field after giving up sufficient energy. The Rose calculations call for the integration of this equation over a fairly long time interval and considers separately the transfer of energy to electrons and to ions. His calculations show that there is more effective transfer of energy to electrons.

Now should there be created a temperature differential between the electrons and ions, there will also result an energy transfer between these two particle types. The transfer of energy from electrons to ions is given by the following equation:

$$\frac{3}{2}nk\frac{dTe}{dt} = \frac{4.8\times10^{-18}}{T_e{}^{3/2}}T_i\left(1-\frac{Te}{T_i}\right) \tag{7}$$

Radiation loss

The radiation loss from the plasma at high temperature takes place in the guise of Bremsstrahlung and synchrotron radiation. The Bremsstrahlung per unit area is given by the following expression.

$$P_b = 4.8\times10^{-37}\, n_i^2 T_e{}^{1/2} \quad \text{watt/m}^3 \tag{8}$$

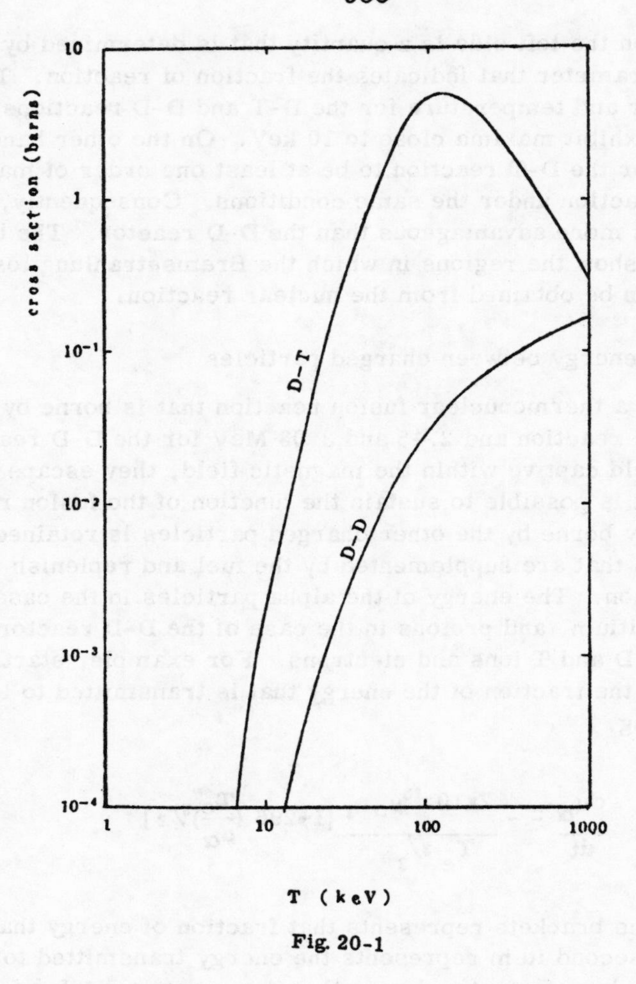

Fig. 20-1

This above relationship applies only when the plasma consists solely of hydrogen ions, and the presence of large number of charge carrying impurity particles in the plasma will result in very much larger radiation loss. The electromagnetic radiation due to Bremsstrahlung from a nuclear fusion reactor is in the region of soft x-rays which is absorbed by the vacuum wall.

The synchrotron radiation consists of electromagnetic waves of even lower frequency, and fairly effective reflection can be expected off a metal wall in this wavelength region. Furthermore, the transparency of the plasma deteriorates, and any evaluation of radiation loss becomes bothersome. When any reflection from the wall can be disregarded, the following relationship applies

$$\frac{dT_e}{dt} = 0.26B^2\, T_e\, (1+\frac{T_e}{204} \text{-----})\, K_L \tag{9}$$

Here K_L is a coefficient that represents the transparency of the plasma. K_L is a function T_e and L, and L can be described in the following manner.

$$L = L\omega p^2/e\omega_{ce} = cL\beta B/k\,(\frac{T_e + T_i}{e}) \tag{10}$$

Here L is the length of the plasma (assuming a flat plate of thickness L and infinite spread) ω_p is the plasma frequency, and ω_{ce} is the cyclotron frequency of the electrons. When 20 keV $<T_e<$50 keV, $10^3<$ $<10^5$, and the following approximation can be made.

$$L_L = 2.1 \times 10^{-3}\ T_e{}^{7}/_{4}/L^{1}/_{2} \tag{11}$$

When reflection from the walls is taken into consideration, the above equation multiplied by the coefficient C_2 gives the actual loss. If the reflective ratio of the wall is given by Γ, C_2 is given by

$$1 - \Gamma < C_2 < 1 \tag{12}$$

C_2 can be determined once the frequency components of the radiation and the absorption coefficients at the various frequencies are established. The recent evaluations by Rose and coworkers claim that $C_2 \sim 0.25$ when $\Gamma = 0.9$ which is said to be a good approximation for plasma that is of concern in nuclear fusion. As a result, the synchrotron radiation can be expressed in the following manner.

$$\frac{dT_e}{dt} = 2.41 \times 10^{-27}\ C_2\ T_e{}^{11}/_{4}\ (T_e + T_i)^{3}/_{2}(1 + \frac{T_e}{204})/D \tag{13}$$

$$D = \beta^{3}/_{2}\ (L\ B)^{1}/_{2}\ \ (Wb/m)^{1}/_{2}$$

When $C_2 = 0.2$, the synchrotron radiation for $T_e < 30$ keV is less than the Bremsstrahlung. The synchrotron radiation can be disregarded in may D-T reactors.

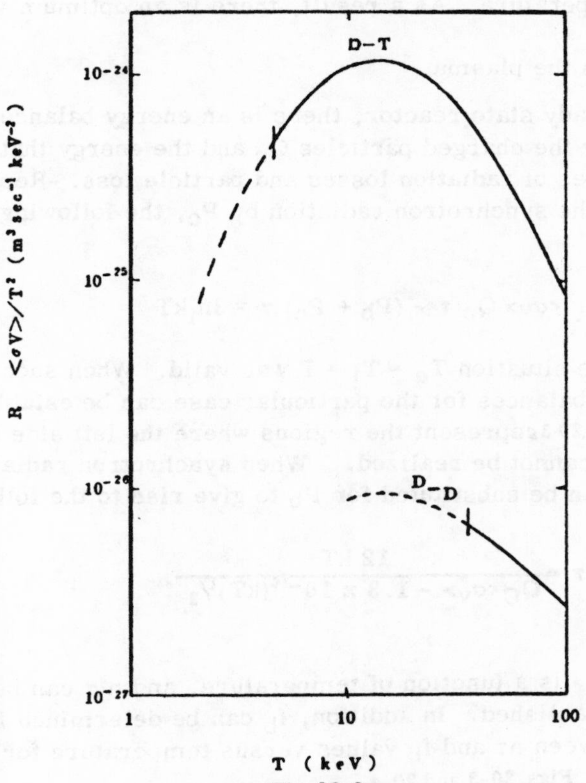

Fig. 20-2

Particle number balance

The retention of the plasma within the magnetic field is not complete, and many particles are lost by diffusion that is normal to the magnetic field that is generated by processes such as collisions between particles and other phenomena. We will designate the time for this occurrence to be τ. In addition, D and T particles can be lost to the fusion reaction. In a steady state reactor, the fraction of these two particles that is lost is normally replaced to maintain the plasma at a constant density. As a result, the particle number balance gives the following relationship.

$$\frac{dn_i}{dt} = 0 = s_i - \frac{n_i^2 \langle \sigma v \rangle}{2} - \frac{n_i}{\tau} \tag{14}$$

Here s_i is the quantity of fuel that is supplied per unit time and unit volume.

With reactors conceived in the light of present knowledge, the greater portion of the fuel that is injected is lost to the wall before it can be involved in the reaction. That portion of the s_i that is actually involved in the fusion reaction is called the fractional burnup f_b, and this value is expressed in the following manner.

$$f_b = \frac{s_i - L_i}{s_i} = \frac{1}{(1 + \frac{2}{n|\tau \langle \sigma v \rangle})} \tag{15}$$

$$L_i = n_i/\tau$$

It will be mentioned later that f_b is a parameter that determines $n_i\tau$ and which thereby determines the reactor temperature. As a result, there is an optimum value of f_b.

Energy balance in the plasma

In the case of a steady state reactor, there is an energy balance in the plasma between the energy borne by the charged particles Q_c and the energy that is lost through the above mentioned two types of radiation losses and particle loss. Representing the Bremsstrahlung by P_b and the synchrotron radiation by P_c, the following can be derived.

$$\frac{1}{4} n_i^2 \langle \sigma v \rangle Q_c \tau - (P_b + P_c) \tau = 3n_i kT \tag{16}$$

It was assumed here that the situation $T_e = T_i = T$ was valid. When such a relationship does not exist, energy balances for the particular case can be established. The broken line segments of Fig.20-3 represent the regions where the left side is negative, and steady state conditions cannot be realized. When synchrotron radiation can be disregarded, equation (8) can be substituted for P_b to give rise to the following.

$$n_i\tau = \frac{12\ kT}{Q_C \langle \sigma_V \rangle - 1.3 \times 10^{-14} (kT)^{1/2}} \tag{17}$$

As indicated in Fig. 20-1 $\langle \sigma_V \rangle$ is a function of temperature, and $n_i\tau$ can be determined once the temperature is established. In addition, f_b can be determined from $\langle \sigma_V \rangle$ and $n_i\tau$. The relationships between $n\tau$ and f_b values versus temperature for the D-T and D-D reactions are shown in Figs. 20-3 and 20-4.

It was established by this argument that it is possible to create a steady state condition at temperatures above some given temperature level. The optimum temperature, however, cannot be established before the details of the reactor construction are known. Once the temperature is established, the product $n_i\tau$ can be determined followed by the determination of f_b. The reactor construction also determines n_i and T values.

Construction and Size of a Nuclear Fusion Reactor

The researches on nuclear fusion to date have clarified to a certain degree the configuration of the magnetic field that has to be used in a nuclear fusion reactor. This magnetic facility can come under the rough classification of the torus or mirror type. The operating modes for the reactor include steady state operation and pulse operation. The primary difference between steady state and pulse operations is the comparatively low β value at which steady state operation can be maintained contrasted to the high β (~1) for the pulse mode. As a result, the pulse mode reactor can be operated with a relatively weak magnetic field. The pulse reactor is conceptually associated with a torus configuration. Either the torus or mirror configuration is possible for a steady state reactor, however, the torus mode is more advantageous as long as the torus configuration does not create unduely complex magnetic fields. On the other hand, there is no outstanding difference between these two configurations that will affect the type of reactor constructionthat is considered here. We will first consider the example of a torus configuration and then pass on to discuss problems associated with a mirror configuration and then with a pulse reactor.

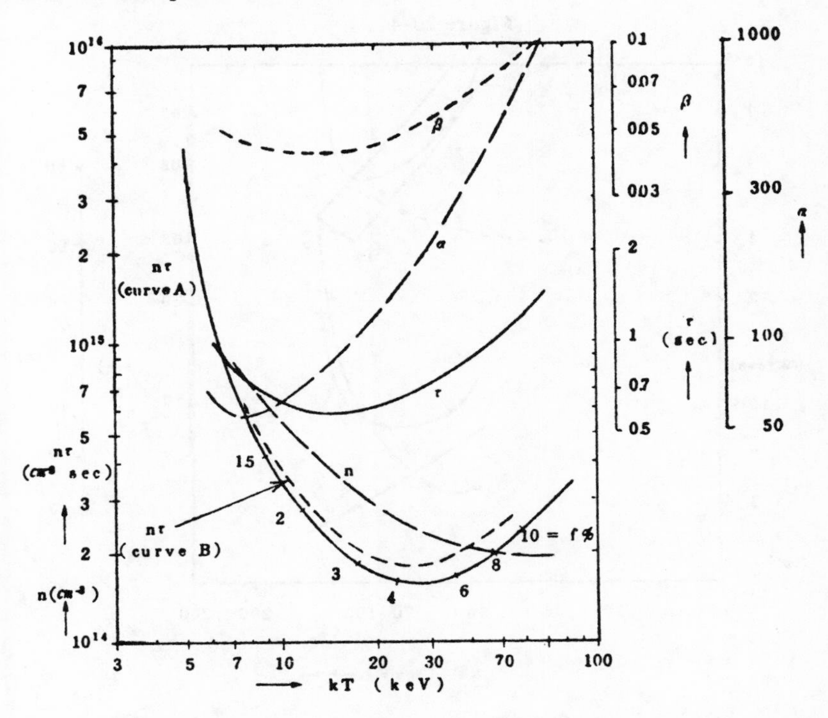

Plasma parameters for charged particle heated D-T reactor as a function of kT(keV). Q_T = 224MeV. Q_c = 352MeV B = 100 kG r_p = 125 m. r_w = 175 m. P_w = 1300 W cm⁻². β shown is not corrected for alpha particle pressure. nτ, Curve A is from eq. 4 and nτ, Curve B is from Kofoed-Hansen.

Fig. 20-3

Steady state torus reactor

We will first give a general description of the construction of the reactor. In the case of a D-T reactor, there is a plasma that is the reactor core which is enclosed at some distance by a vacuum container. Neutrons are moderated outside this vacuum container, and the (n, 2n) reaction is utilized to increase the neutrons. These neutrons are absorbed by Li to generate T in the blanket about the core. A radiation shield for neutrons and other radiation is placed outside this blanket together with a superconducting coil. The cross section of the torus is shown in Fig. 20-5. The heat generated in the blanket and vacuum wall by the neutrons and other radiation is removed by some suitable cooling system and is utilized in power generation or as heat source for other applications. The Li is circulated in the form of some suitable compound or as a liquid metal, and the T that is produced is separated.

The magnetic field for a torus facility consists of a uniform magnetic field created by a solenoid type coil wound uniformly along the axis of the torus, the socalled plasma retention magnetic field, and a stabilized magnetic field that brings about stabilization by creating a plasma equilibrium. The latter mentioned stabilizing magnetic field is expected to be established by future research For example, this field is generated by the current in the plasma in the case of the Tokamak type facility while a helical coil is used in the stellator for this purpose.

Figure 20-4

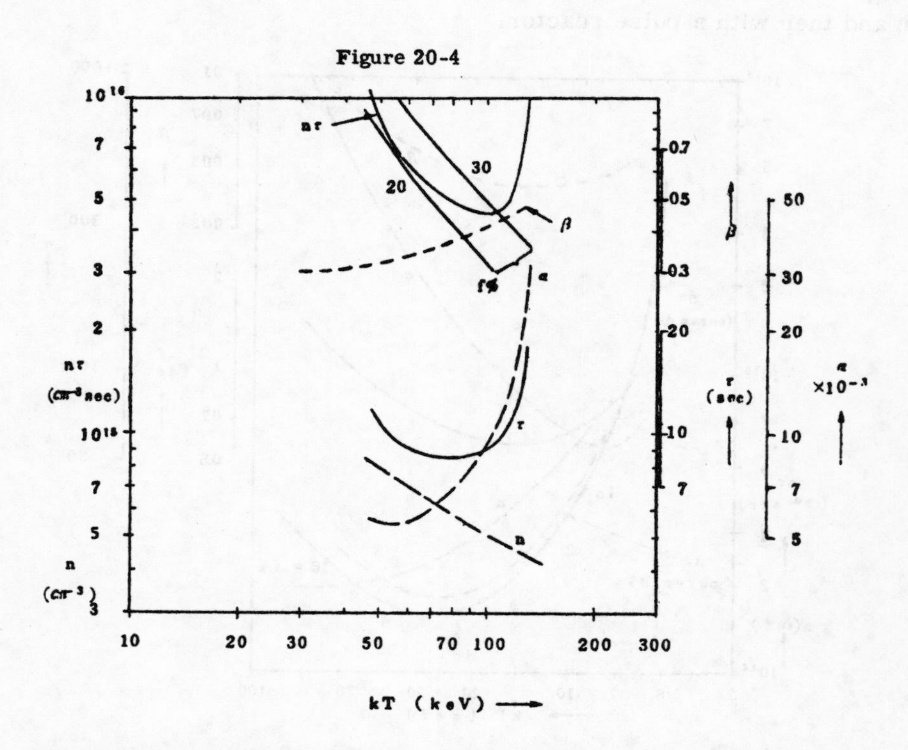

Plasma parameters for charged particle heated D-D reactor as a function of kT (keV)
The nτ-kT curve is from Kofoed-Hansen and should be compared with solid curve fo D-D

$Q_T = 684$ MeV. $B = 100$ kG. $r_p = 1.25$ m. $r_w = 1.75$ m. $P_w = 1300$ W cm^{-2}

β shown is not corrected for alpha particle pressure.

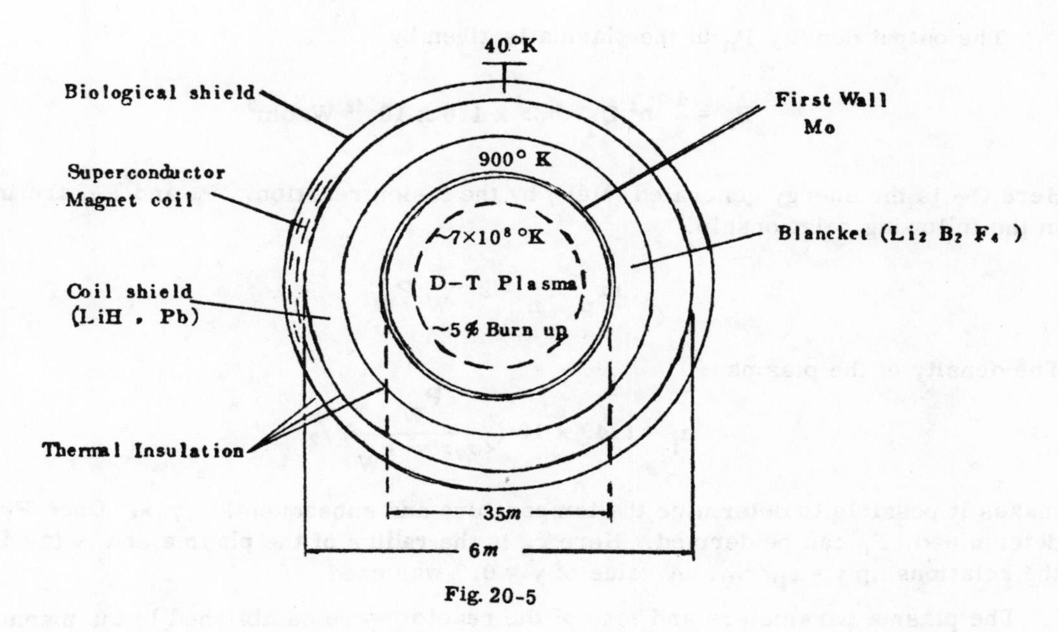

Fig. 20-5

At the present time, the primary problem that is being attacked in nuclear fusion research is the retention of the plasma at the necessary temperature for the necessary duration at the proper orientation to the magnetic field. Here we will assume that success was attained in generating a plasma of the suitable configuration and determine the size of the reactor and various plasma parameters. We cite here the design proposed by Carruthers et al. as a representative model of a steady state torus reactor.

One of the important elements that determine the size of a reactor is the problem of the heat load on the vacuum wall. We adopt here the value of 1300 W/cm² as the permissible thermal output P_W per unit area of the wall of the vacuum container. The heat absorbed by the wall for this case is approximately 250 W/cm². We assume an overall output density of P_D to calculate the radius of the vacuum wall.

$$P_D = \frac{2\pi r_W P_W}{\pi(r_W + t + s)^2} \quad \text{W/cm}^3 \tag{18}$$

Here t is the thickness of the blanket, and s is the thickness of the coil. The condition $r_W = t + s$ will make P_D maximum. Assuming t value of 125 cm and s value of 50 cm, $r_W = 175$ cm.

Substituting the above values, the value $P_D = 3.5$ W/cm³ is obtained. It can be shown that this is a reasonable value when compared to the 2-2.5 W/cm³ for a 1000 MW(e) fast reactor. Furthermore, the output density for the plasma alone is 30 W/cm³, while this value is 21 W/cm³ for the core of a fast reactor.

The magnitude of the cross section of the reactor is essentially determinable as a result of the deliberations described above. The length of the circumference of the torus is established from the output of the power plant. Here we will adopt the value R = 5.5 m as a suitable magnitude of the central radius of the torus. The output in this case is 5000 MW (thermal). The electrical output then is 2000 MW assuming 40% efficiency.

The output density P_n in the plasma is given by

$$P_n = \frac{1}{4} n^2 Q_T \langle\sigma_v\rangle \times 1.6 \times 10^{-16}\ \mathrm{W/cm^3} \tag{19}$$

Here Q_T is the energy generated (MeV) by the fusion reaction. P_n and P_W are involved in the following relationship.

$$\pi r_p^2 P_n = 2\pi r_W P_W \tag{20}$$

The density of the plasma

$$n_i = 1.49 \times 10 \left(\frac{P_W}{\langle\sigma_v\rangle y^2 r_W}\right)^{1/2} \tag{21}$$

makes it possible to determine the temperature and subsequently $\langle\sigma_v\rangle$. Once P_W is determined, P_n can be derived. Here r_p is the radius of the plasma and is involved in the relationship $y = r_p/r_W$. A value of $y = 0.7$ was used.

The plasma parameters and size of the reactor were established in the manner described above, and one remaining problem is the degree of freedom associated with the magnetic field. Although the value of the magnetic field does not appear at first glance to be an important parameter that determines the size of the reactor and related factors, it is in reality an important parameter that is associated with the retention time of the plasma. We will make the assumption here that a value of 100 kG for the magnetic field will allow the necessary retention time. Consequently, by calculating the density at some given temperature by the use of equation (21), the value of β can be calculated with the use of equation (4). Furthermore, the value of n_i can be determined from equation (17) from which the value of τ is readily obtained . The values of β, n_i, and τ are also shown in Figs. 20-3, 20-4. The term α that appears in Fig. 20-4 is the ratio between τ and loss time τ_B when the plasma is in a muddled state. The most difficult problem that confronts nuclear fusion research at the present time is how to contain the plasma for a prolonged duration, and it is expedient to select a temperature at which minimum τ is obtainable. We selected a temperature of 20 keV for this example. In any event, we derived the following values as parameters for a D-T reactor.

Table 20-1

$n\tau$	$\mathrm{cm^{-3}}$ sec	1.7×10^{14}
T	keV	20
n	$\mathrm{cm^{-3}}$	2.8×10^{14}
τ	sec	0.6
τ/τ_B		120
β		0.075
f_b		0.035

Output 5000MW(th) B = 100 kG

$r_p = 1.25$ m $r_W = 1.75$ m

The size and configuration of the reactor are depicted in Fig. 20-5.

Steady state mirror type reactor

The configuration of a mirror type facility is unlike that of a torus type and is linear, however, its cross section shows little outstanding difference from that of a torus. As a result, we will discuss below only the differences between the mirror and torus modes.

The problem item associated with the mirror type facility is the loss of plasma through both ends as a result of collisions. This mirror loss is very much greater when the plasma assumes some reasonable configuration than when there is diffusion across the magnetic field. As a result, we must incorporate this mirror loss into the entity τ that we have been discussing. The mirror loss τ is given in the following manner,

$$\tau = 1.6 \times 10^{22} \frac{T^{3/2}}{\psi n_i} \quad \text{(see)}$$

What remains here after ψ is deleted represents the 90° scattering time of the ions in the plasma. This value multiplied by a certain value ψ gives the mirror loss. A value of $\psi \sim 0.3$ is deemed normally reasonable while a value less than 0.1 makes operation very difficult. When this equation is substituted into equation (15), f_b becomes a function of temperature alone. The relationship between T and f_b is shown in Fig. 20-6. It is also possible to derive a relationship between f_b and temperature from the energy balance equations using (15) and (17). The intersection of these two curves gives the energy balance. On the other hand, the fuel particles at the time of injection have to be given a certain energy in the case of the mirror so that the conditions for energy balance can be maintained. We represent the energy of the ions by V_i at the time of injection and add the term $S_i V_i$ to the thermal equilibrium equation. When injection is made with energy imparted in the manner described, the ratio Q between the energy of the fusion reaction and the energy required for the injection assumes the following form for the D-T reaction.

$$Q = \frac{17600 \, f_b}{2\,(V_i + V_e)}$$

Electrical energy is what is usually imparted to the injected particles. Q has to assume a fairly large value to allow the desired exchange efficiency and power output. The f_b - T curves for the conditions V_i = O - 100 keV; ψ = 0.05, 0.1, 0.2; and Q = 10, 20 are illustrated in Fig. 20-6.
If, for example, we assume that Q = 10 and ψ = 0.1 are the necessary conditions, the narrow region enclosed by Q and a fixed line becomes the region in which reactor operation is possible. This situation calls for V_i = 60-100 keV, f_b = 0.07-0.2, and T_i = 55-150 keV. In this manner, the temperature of a mirror reactor tends to be high as a consequence of this mirror loss.

Pulse reactor

When we think of a pulse reactor, the first thing that comes to mind is an advanced form of high beta plasma configuration similar to the present theta pinch facility. The theta pinch applies a magnetic field in pulse form to compress the plasma and create a high temperature plasma. Consequently, the features of a pulse reactor include 1) thermal balance in the plasma is not always a necessary condition. The heat generated by the nuclear fusion reaction is discharged after each pulse, and the electrical output derived from this heat creates the magnetic field for the next pulse leaving behind a surplus. 2) The coil that generates the pulse magnetic field does not require superconducting lines and can be made of ordinary conductors. When one takes into consideration the loss and inductance, this coil needs to be placed in the inner side of the blanket. Since the coil generates a strong magnetic field, it should be able to withstand large electromagnetic forces. It should be made thin to reduce loss of neutrons. As a result, the materials problem is very important. 3) This coil requires a facility to accumulate electrical energy to be used to provide electrical current in pulse form. A condenser is presently employed for this purpose, however, there are cost problems associated with condensers of large capacity.

Engineering Problems

The basic features of nuclear fusion were discussed above. Assuming it is possible to contain the plasma by fulfilling the conditions demanded by equation (17), there still remain a number of engineering problems that have to be solved before a nuclear fusion reactor can be actually put into operation. The status of these problem items is discussed below.

Startup of a nuclear fusion reactor fuel injection, and heating

A large external source of energy has to be supplied to heat the plasma that is to serve as the reactor core to the critical temperature to initiate operation of the nuclear fusion reactor. What is considered critical here is that state at which the heat of reaction obtained from the nuclear fusion reaction heats up the fuel that is added and maintains the reaction in steady state condition without the need for supplying further energy from without. A number of approaches come to mind with respect to the method of injecting energy into the plasma, but here we will consider the following three methods; beam injection of high energy particles, resistance heating, and high frequency heating. We will also enumerate those methods that are presently being considered for fuel replenishment to a nuclear fusion reactor operating at steady state condition.

Beam injection of high energy particles

When starting up a nuclear fusion reactor for an actual run, one method of heating the plasma to the critical state is to inject a bundle of high energy particles of the order of several 10 keV to a hundred keV into the dilute plasma and utilize the ensuing collisions to heat the plasma. We will employ here a beam current (ampere) unit that is the product of the particle flux nv and the unit electrical charge e as a measure of the magnitude of the injection. A beam current of about 30 A per unit area is considered necessary from the standpoint of equilibrium between the plasma particles in actual nuclear fusion reactors. In the case of a fusion reactor of a scale of 1700 MW thermal output, the plasma volume is roughly 60 m^3 so a total beam current of 2000 A is required. If this current is to be injected through 40 ion injectors, each beam will have to deliver 50 A.

Looking at beam currents that have been reported, a beam current of 20 A Li ions (40 keV temperature) has been attained in the United States using a multislit beam source while the Soviets have reported 0.7 A and 30 keV hydrogen beam from a small type neutralizer. The extension of these facilities to adapt them to the nuclear fusion reactor can be achieved by increasing the number of slits and enlarging the entire facility if only the beam current is considered. On the other hand, there are other associated problems one of which is the diffusion that takes place before entrance to within the vacuum container takes place. Engineering problems such as the structural strength associated with a 40 unit injector facility have not even been investigated as yet.

Another method for increasing the current density is the method that has come into prominence recently which calls for accelerating cluster ions. This method is still in the stage where considerably more research is required.

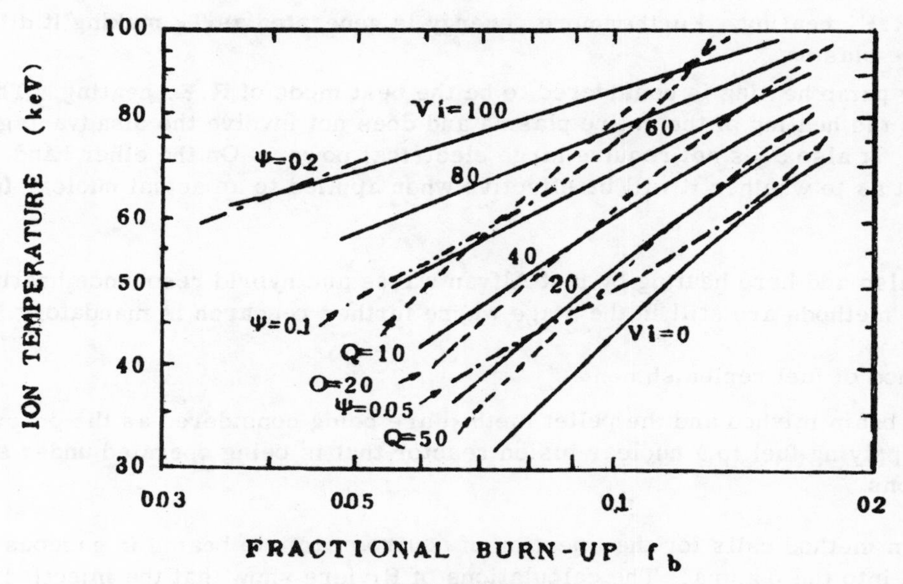

Fig. 20-6

Contours of ion temperature (solid curves) vs fractional burn-up for various ion injection energies V_i. Also shown are contours of constant Ψ (ion escape probability per effective 90° Coulomb scatter), and Q (fusion energy/injection energy).

Resistance heating

External energy is injected into the plasma through the core in the Tokamak to heat the plasma. The Joule heat of the plasma is utilized for this purpose. As the temperature of the plasma increases, its specific resistance decreases such that there is an associated decrease in the Joule heat that is generated. The specific resistance of plasma is about 1 keV which is about that of copper but decreases to about 1/40 of this value at a temperature of 20 keV of a steady state reactor. As a result, there is an upper limit to the plasma temperature that can be realized by resistance heating. Even when abnormal resistance of the plasma is considered, it is not possible to start an actual nuclear fusion reactor by resistance heat alone. Resistance heating has to be used in conjunction with some other method.

High frequency heating (RF heating)

The heating mechanism by high frequency heating can be divided into that due to collisions and that due to resonance heating. The former involves the application of a high frequency electrical field to bring about ionization heating and is used most frequently to create plasma of the order of a few eV. On the other hand, its efficiency is poor compared to the latter.

Resonance heating includes ion cyclotron resonance heating (I.C.R.H.) and magnetic pump heating. The I.C.R.H. is a method that brings about heating by applying an electrical field with the same frequency as the cyclotron frequency of the ions and has a number of associated problem areas. The presence of a magnetic field is a necessary item. If the magnetic field has curvature, the cyclotron wave is not propagated. There is also loss due to the surface skin effect as a result of high frequencies from other R.F. heating. Furthermore, energy is generated as T_1 making it difficult to contain the plasma.

Magnetic pump heating is considered to be the best mode of R.F. heating. This mode enables the heating of the entire plasma and does not involve the disadvantages of the I.C.R.H. It also does not require large electrical power. On the other hand, there is some doubt as to whether it will be effective when applied to an actual nuclear fusion reactor.

We can also add here heating by fast Alfvan waves and hybrid resonance heating, but these two methods are still in the stage where further research is mandatory.

Method of fuel replenishment

The beam method and the pellet method are being considered as the possible modes for supplying fuel to a nuclear fusion reactor that is being operated under steady state conditions.

The beam method calls for the injection of neutron particle beams in gaseous state at high speed into the plasma. The calculations of Riviere show that the injection of 100 keV neutron particles into a plasma of 2.8 m diameter will result in penetration to the center of the plasma. It is possible to adapt this method to an actual nuclear fusion reactor of the mirror type that has a small plasma diameter. On the other hand, the particles cannot penetrate to the center of a torus type reactor with large plasma diameter in which case a density peak is created close to the surface. This is not a desirable condition from the standpoint of plasma containment, and this mode is considered impractical.

The pellet method involves the high speed injection of neutron particles in solid state form (5 mmø-10 mmø) into the plasma. Rose has performed some simple cal culations with respect to this method which indicate that the injection velocity of these pellets should be of the order of 10^4 m/s. It is a very difficult technical feat to achieve such a velocity without breaking the vacuum. The pellet method still requires further research. This type of approach became possible with the advent of the T-3, and it is thought that the problem of fuel replenishment will soon be solved.

Vacuum container

The major problems associated with the vacuum container include 1) radiation damage, 2) absorption of neutrons, 3) sputter, 4) mechanical strength, and 5) heat conduction. Radiation damage, absorption of neutrons, and heat conduction will be discussed in a later section and we take up here the items of sputter and mechanical strength.

Sputter

When the plasma is contained within a vacuum container, the container is subjected to the sputter of impurities contained in the deuterium ions, α particles, neutrons, or the plasma.

According to the experiments of the A. J. Summer group, the sputter rate on niobium due to deuterium ions at temperature of 20 keV is S = 0.004 atoms/ion, and this value is essentially unaffected by the injection angle of the deuterium ions. Applying this value to the model proposed by Carruthers et al. to calculate the sputter for the case in which a particle flux of 2.8×10^{16} cm^{-2} sec^{-1} was involved, a value of 0.6 mm/yr was obtained. This is not a serious problem if one takes into consideration the diverter effect. The sputter rate of niobium due to the sputtered niobium atoms is S = 4-14 atoms/ion, and this value is greatly affected by the injection angle. This sputter due to niobium is highly dependent on the temperature of the sputtered niobium and the time that elapses before capture at the diverter, and this sputter is impossible to calculate.

On the other hand, the sputter rate tells us that the sputter of niobium becomes equal to that due to deuterium if the niobium level is 0.1% that of the deuterium. This makes a detailed study of this effect in order.

Both the sputter rate and density associated with neutron sputter are roughly one to two orders of magnitude below those of deuterium and can be disregarded.

The discussion above considered the level of sputter on the vacuum container of a nuclear fusion reactor equipped with a diverter. We next consider the sputter on the diverter itself. Assuming a diverter with an area 10% of the vacuum container wall operating at 100% efficiency, the sputter level is claimed to be 1 mm/12 days. This is an extremely large value, and there is a need to provide a construction in which the inner wall of the diverter is replaceable. The problem becomes more complex if the diverter does not operate at 100% efficiency, and it is possible that there will be considerable sputter on the wall of the vacuum container near the diverter.

Sputter is a serious problem in a nuclear fusion reactor like the Tokamak that does not have a diverter because the density of the plasma near the wall of the vacuum container is nearly equal to that at the center of the plasma.

In such a case, it becomes necessary to install a diaphragm that serves as a diverter or use some force such as magnetic force to push the plasma away from the wall of the vacuum container. The first mentioned technique has the associated problems of life and impurities while the difficulty of containing the plasma as a result of magnetic field irregularities besets the latter.

Summarizing the sputter problem, essentially no difficulties are expected of nuclear fusion reactors with the exception of the Tokamak. There is a need to introduce some technical developments aimed at separating the plasma away from the wall of the vacuum container in the Tokamak type reactor.

Detailed data on the sputter rate (1-100 keV temperature, 0°-90° angle) for D^+, T^+, He^+, Nb^+, and Mo^+ are desirable.

Mechanical strength

Thermal stress and buckling are the items that are of concern from the standpoint of the mechanical strength of the vacuum container. These two items are discussed below.

Thermal stress

There is a need to know the mechanical properties of the materials used in the construction of the vacuum container before a reasonable assessment of its thermal stress can be made. Here we assume that niobium will be used. The mechanical properties of niobium are listed in Table 20-2.

The pressure at the inside of the container is 10^{-2} Torr when the reactor is in steady state condition while the external pressure is the same as the pressure of the fluid body. This difference in pressure causes the container to buckle. The buckling pattern is different between the mirror type reactor with its cylindrically shaped vacuum container and the torus shaped vacuum container for the Tokamak. In the case of the cylindrical vacuum container with its both ends closed, external forces operate in the form of axial and surface forces. Here we will disregard the axial force and consider that only the surface force operates in which case the buckling load is given by the following equation.

$$P_k = 0.807 \frac{Et^2}{l_r} \sqrt{\frac{-t}{1 - \nu^2) r}}$$

In the above E is the Young modulus, r is the radius, t is the plate thickness, 1 in the length, ν is the Poisson ratio, and P_k is the pressure differential between the outside and inside.

If we now fit the mirror type model of Fraas to this equation, $P_k = 7.3 \times 10^{-6}$ kg/mm^2 for 1 = 20 m, r = 10 m, E = 1.065 x 10^4 Kg/mm^2, $\nu = 0.38$, and t = 3 mm. Consequently, reinforcing rings are absolutely necessary. We will omit the calculations for the determination of the buckling load when reinforcing rings are inserted in a close pattern because very complicated calculations are involved. When the spacings between these reinforcing rings are made very small, buckling does not occur but the vacuum container gives to bending forces and is destroyed. Representing the width of the vacuum container bracketed by the reinforcing rings by 1 assuming a flat plate of infinite length, the relationship between the stress σ that is created and the surface pressure q is given in the following manner.

$$q = \frac{t^2 \sigma}{4 \times 0.833\, l^2}$$

Here we envision a bending moment that is a completely plastic moment. When σ = 15 Kg/mm^2, 1 = 100 mm, and t = 3 mm, q = 4.0 Kg/cm^2. The pump pressure for the coolant pump is determined on the basis of the thermal conductivity, but it is difficult to remove heat at the rate of 250 W/cm^2 with the above values.

When the vacuum container is a torus shaped affair, there is also a uniform force operating in the direction of the central axis of the torus because of the large curvature in the toroidal direction which arises in addition to the buckling phenomenon that is seen with the cylindrical type vacuum container. This force is given as follows.

$$F = 2\pi^2 r^2 q$$

Here r is the small torus radius and q is the pressure differential between the inside and outside. We will consider here a vacuum container that is supported at four sites in the toroidal direction. Assuming a straight beam between the support points, the bending moment that is generated at the support points is given by the following.

$$M = \frac{\pi^3 r^2 kq}{48}$$

q must assume a value close to the following

$$q \leq 1.14\, E\, \tau\, t^2 \times 48/(1 - \nu^2)\, \pi^3\, r^2\, R \sim 10^{-2}\ \mathrm{Kg/mm^2}$$

to prevent the cylinder from buckling to this bending moment assuming the Carruthers model and plate thickness of 3 mm. The actual construction of a vacuum container for a nuclear fusion reactor is fraught with severe technical difficulties if one takes into account thermal buckling of the vacuum container in addition to the above.

The need for Nuclear Fusion Reactors

We will discuss in this section the need for nuclear fusion as well as the energy demand as viewed from the standpoint of environmental contamination.

Energy demand and cost

According to the report of the Demand Section of the Total Energy Investigating Group, the expected future demand for power in Japan is 439 x 10^6 kl in 1975 and 10^9 kl in 1985 when converted to petroleum equivalent.

It is expected that firepowered power generation in 1975 will deliver 456 x 10^9 kWh and 0.9 x 10^{12} kWh in 1985 (197 x 10^9 kWh in 1968). The petroleum required to fulfill such a power demand will be 0.7 x 10^9 kl in 1985. This is equivalent to ten 200,000 ton tankers delivering their load each day. The problem here not only involves the transport of the petroleum but also takes in the environmental contamination that will result from the exhaust gas from the firepowered power plants.

Let us now see what would happen if we used nuclear fusion reactors to generate the power. Nuclear fusion reactors do not release sulfur dioxide, carbon monoxide, and nitric oxide that are the usual products of the combustion of petroleum. The fuel for these reactors is deuterium that is extracted from sea water which alleviates the need for 3000 tankers of 200,000 ton capacity each to ply the sea lanes between Japan and the middle east every year. There will also be no need to provide large storage tanks for this fuel.

What was described above will also apply if the fusion reactors were replaced by fission type nuclear reactors. On the other hand, these two types of nuclear plants differ in the quantity of radioactive products they evolve and the fuel. The output of radioactive material from the fusion reactors is roughly 10^6 orders of magnitude smaller than that from fission reactors. There can be siting restrictions to the placement of fission reactors, and there are some technological problems associated with the disposal of its waste products.

There also are limitations to the available fuel if the present energy demand were to be supplied by the presently available nuclear reactors even when breeder reactors are employed. In contrast, the deuterium that is the fuel for fusion reactors is present in 0.02% level in sea water which offers a semi-permanent source.

Table 20-2. Mechanical Properties of Niobium

Temperature	Yield Stress tons/m²	Tensile Stress tons/in²	Elongation %	
20	18.6	22.1	19.2	
200	15.1	23.9	14.2	Material annealed 30 minutes at 1100°C
300	13.1	20.0	13.2	
400	14.3	21.9	13.3	
500	12.6	22.0	9.6	
600	8.0	20.8	17.5	
660	7.1	20.8	22.4	
800	5.8	20.1	20.7	
970	5.2	12.3	37.5	
1050	4.4	7.2	42.5	
20	35.0	37.5	4.7	
200	27.2	31.2		Cold worked material
300	32.6	37.8		
400	33.0	34.6		
500	31.9	35.4	2.4	
600	33.6	34.6	3.2	

1 ton/in^2 1.54 Kg/mm^2

The thermal flux q that passes through the vacuum container in the Carruthers model is 250 W/cm^2 sec. Since the coefficient of thermal conductivity λ of niobium is 0.14 cal/cm sec °C, the temperature gradient ΔT that is created at the wall of the vacuum container is $T = q/\lambda$ from which the value 426°C/cm is obtained. A thickness of 1 cm is considered necessary in an actual nuclear fusion reactor from the standpoint of structural strength. Assuming here that the thermal stress on the wall of the vacuum container is created only from the temperature differential, the maximum thermal stress σ for plate thickness of 1 cm becomes 18.2 kg/mm^2. This applies only to an idealized model of a vacuum container, and large thermal stresses occur in the actual case in the neighborhood of reinforcing rings used to prevent buckling and support points. Should the temperature of the coolant exceed 500°C, the temperature at the inner surface of the vacuum container comes close to 1000°C, and the tensile strength of the wall material becomes extremely small. This is the reason why thermal stress is such an important problem with respect to the vacuum container.

Application of Superconducting Magnets to Nuclear Fusion

General Description of Problem Items

Superconductor technology had its inception with the development of magnets for MHG and bubble chambers in high energy physics. Its present applications extend to various areas including superconducting accelerators, superconduction transmission, superconducting rotary units, magnetic floating craft, and weak electric technology. There have been great advances in superconductor technology accompanying these varied applications. When one considers the application of superconductor technology to a nuclear fusion reactor, the following items of difference become apparent compared to the other applications.

i) The superconducting magnets will be even larger than any of the other large superconducting units.

For example, the design of a D-T reactor of 5000 MW output calls for a magnetic field of 100 KG and a magnetic field capacity of 1000 M^3[1]. As will be discussed later, the largest superconducting magnet presently on the design board has a volume of about 200 m^3 (bubble chamber of the National Accelerator Company).

ii) A strong magnetic field of 100 KG is called for. As will be discussed later, a magnetic field of 100 KG poses little problem with respect to critical magnetic field for linear material, however, the only magnet over 100 KG to date is a medium sized magnet (inner diameter 1 5 cm, outer diameter 48.5 cm) that was test produced for NASA, and manufacturing experience is extremely limited.

iii) The superconducting magnet of a fusion reactor has to operate within a radiation field.

iv) The coil windings are relatively complex. The coils used in present day magnets are of either the saddle or solenoid type.

A part of these areas of differences is compensated by advances in superconductor technology in other fields, but the major part will probably have to await developments in superconductor technology specifically in the nuclear fusion reactor field. As a result, the greater part of the diverging items will remain as problems that have to be solved exclusively in the realm of fusion reactor development. These items include the following.

It is difficult to conceive that the very large and strong magnetic field magnet as called for in i) and ii) will become necessary in other fields in the near future. Technological problems that are associated with the development of superlarge and high intensity magnetic fields will be discussed later.

The problems associated with the use of superconducting magnets in a radiation field that is item iii) above will most likely be somewhat clarified by developments in designs of superconducting accelerators. It should be noted that a nuclear fusion reactor entails much longer exposure to radiation than in an accelerator.

Item iv) above is tied with coil winding, forming, and cooling technology, and it is probable that this problem will be lessened by advances in other fields.

Stabilization of Superconducting Wire

Rigid superconductor wires give rise to sharp superconductor, normal transmission transition (S-N transition) so that the unit wire for superconductor use has to be stabilized. A sudden S-N transition causes the magnetic field energy to be released in very short time to result in a helium explosion and subsequent damage to the coils. There is also the possibility of current deterioration phenomenon in the unit wire from instabilities like flux jump making it impossible for the specified current to flow and thereby make the specified magnetic field unattainable.

As illustrated in Fig. 20-7 the method of stabilization calls for the embedding of superconducting wire elements within a large quantity of ordinary conductor (copper, aluminum, etc.). The principle of this stabilization is illustrated in Fig. 20-7 (a).

i) When there is localized quenching of the wire element as a result of flux jump, a part of the current is shunted to the ordinary conductor.

ii) The propagation of the quenching effect is prevented by making the cooling surface large to minimize localized temperature rise. These are the basic elements of this principle[7]. The ordinary conductor used for this purpose should have low intrinsic resistance and a small magnetic resistance effect. Some examples of such conductors include oxygen free copper (OFHC), aluminum, and indium. The electrical and thermal contact resistances between the ordinary conductor and superconductor should be small.

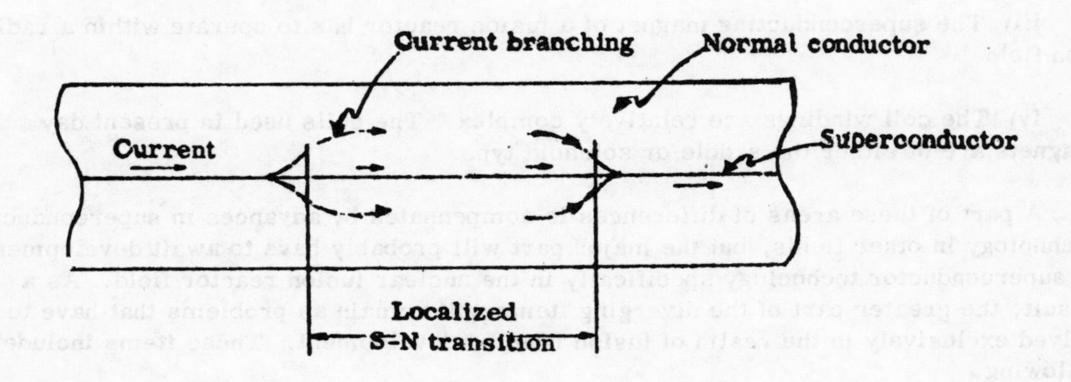

(a) Principle of stabilization

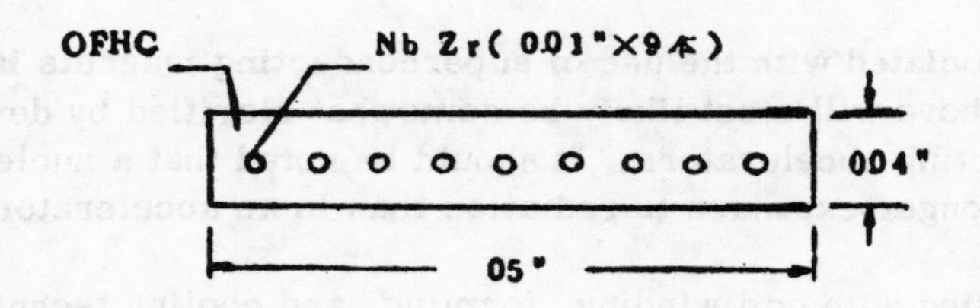

(b) (AVCO stabilized conductor)

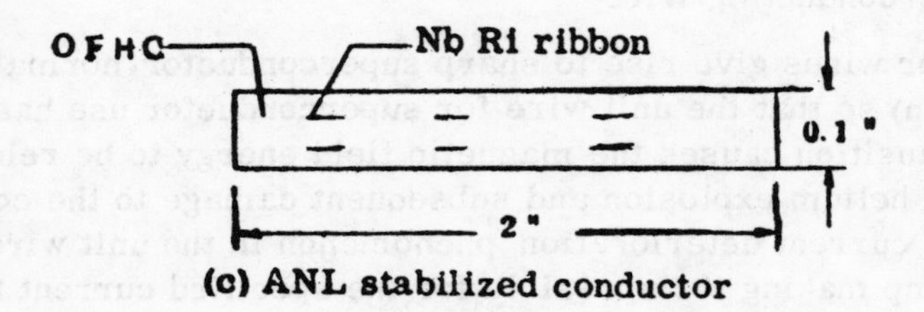

(c) ANL stabilized conductor

Fig. 20-7 . Stabilized Conductor

The current Ie (H) of the stabilized conductor according to the principle of stabilization described above is represented in the following manner.

$$\frac{q_M}{A} \geq \sqrt{\frac{I_e^2(H)\ \rho_N(H)}{S}}$$

Here q_{M}/A -: maximum thermal flux from conductor to helium (W/cm^2) (less than the critical flux for the transition from nuclear boiling to skin boiling) This value varies with the cooling conditions of the coil

A : Area of cooling surface

$I_e(H)$: Conductor current, function of magnetic field

N(H) : Intrinsic resistance of ordinary conductor magnetic resistance effect and thereby a function of the magnetic field

S : Cross sectional area of ordinary conductor

As a result, the stabilized current will vary with the maximum magnetic field, cooling conditions, and intrinsic resistance of the ordinary conductor of this magnet. This is the reason why the large magnets of today are equipped with stabilized conductors which are made to order to fulfill the particular design. For example, Fig. 20-7 (b) shows the stabilized conductor used in the saddle type (MHD magnet made by the AVCO Company (United States) while Fig. 20-7 (c) shows the conductor used for the bubble chamber at ANL in which the electromagnetic force that operates on the coil is borne by the mechanical strength of a copper piece.

Examples of basic data used in the design of stabilized conductors are listed in Figs. 20-8 to 20-12. Fig. 20-8 shows the values of the thermal conductivity coefficient K of superconductors made from different conductors at low temperature in which the K of the superconductor is seen to vary with the magnetic field. In making plans for a large magnet such as 100 KG (= 10T), K data up to this level become necessary. Fig. 20-9 shows the residual electrical resistance of a superconductor that has been quenched and the changes with temperature in the electrical resistance of copper that is used as stabilizing material. The magnetic resistance effect of stabilized material is shown in Fig. 20-10 and there is need for data above the 100 KG level. The changes in specific heat of superconductors and ordinary conductors at low temperature are shown in Fig. 20-11 and the stress-strain curves for ordinary conductors are given in Fig. 20-12. These are examples of the basic data that are required. When designing and manufacturing stabilized conductors for the large magnets to be used in nuclear fusion reactors, these basic data have to be supplemented by measurements made under conditions of large load and irradiation at low temperature up to large magnetic field intensities of about 150 KG. Data under intense magnetic field and irradiation are practically non-existent at present.

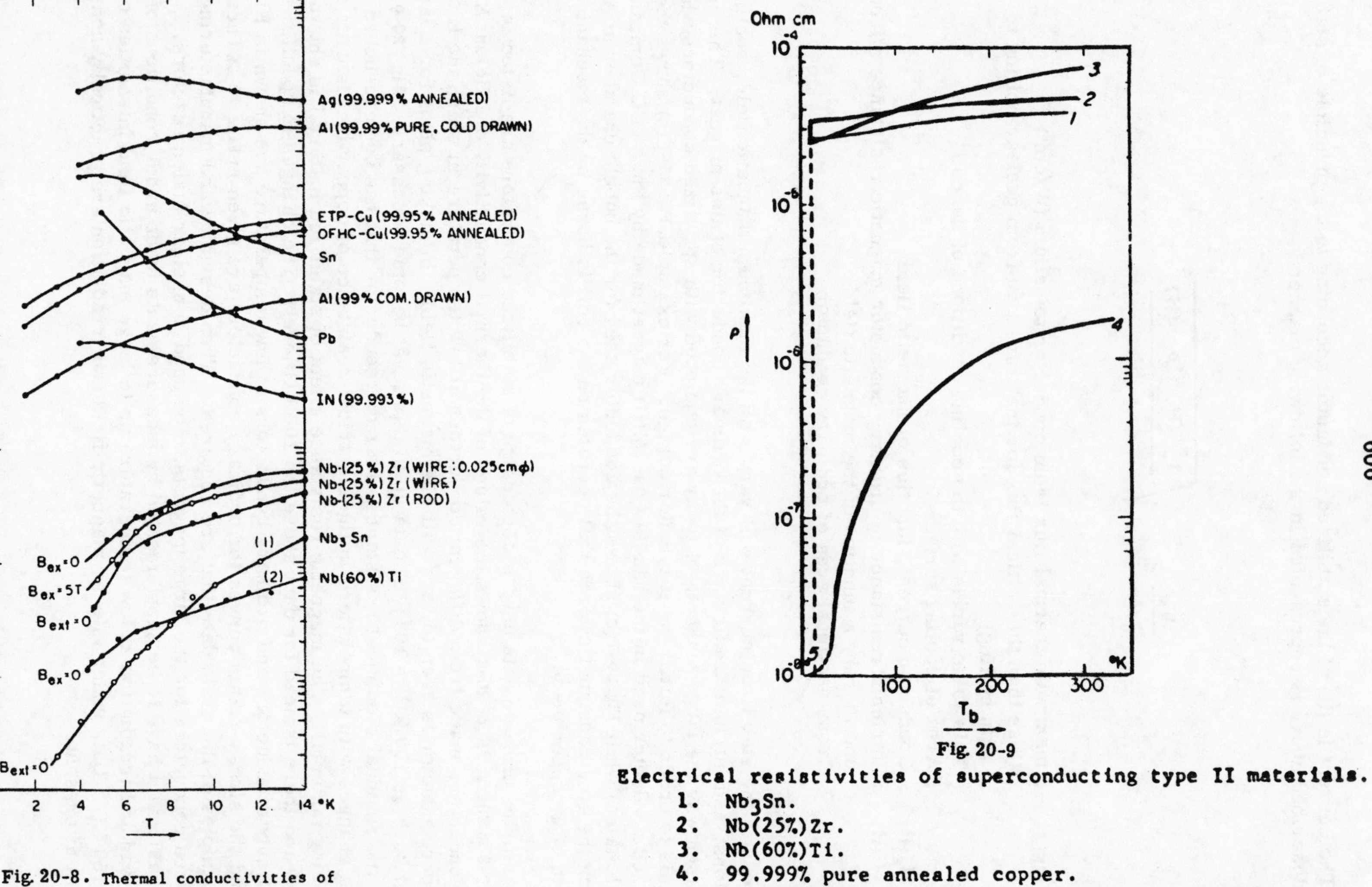

Fig. 20-8. Thermal conductivities of superconductors type II and normal metals.

Fig. 20-9

Electrical resistivities of superconducting type II materials.

1. Nb_3Sn.
2. Nb(25%)Zr.
3. Nb(60%)Ti.
4. 99.999% pure annealed copper.
5. 99.995% OFHC copper as received, not annealed, $\rho_{4.2^oK,B=0} = 1.2 \times 10^{-8}\ \Omega\cdot\text{cm}$.

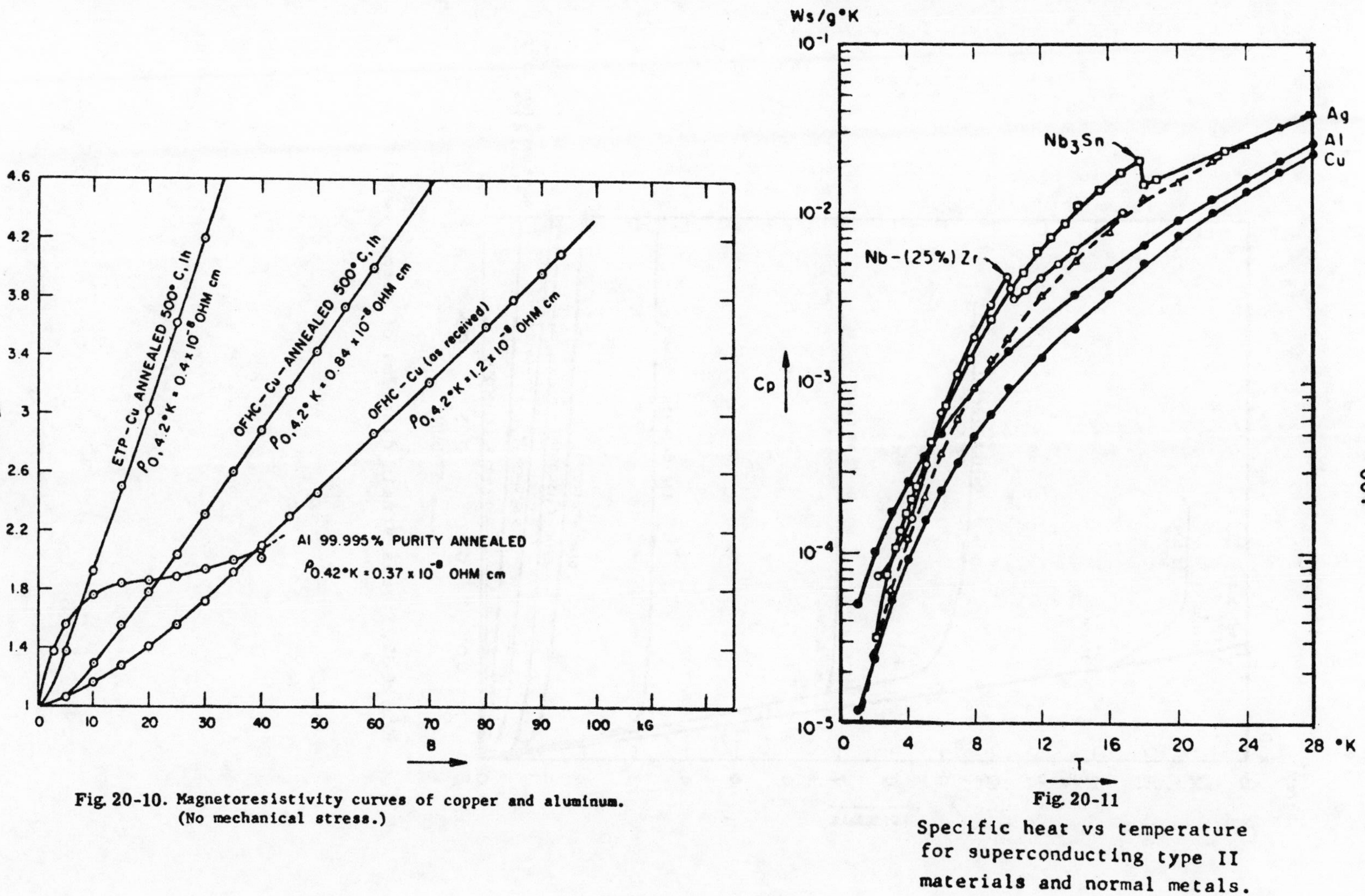

Fig. 20-10. Magnetoresistivity curves of copper and aluminum. (No mechanical stress.)

Fig. 20-11
Specific heat vs temperature for superconducting type II materials and normal metals.

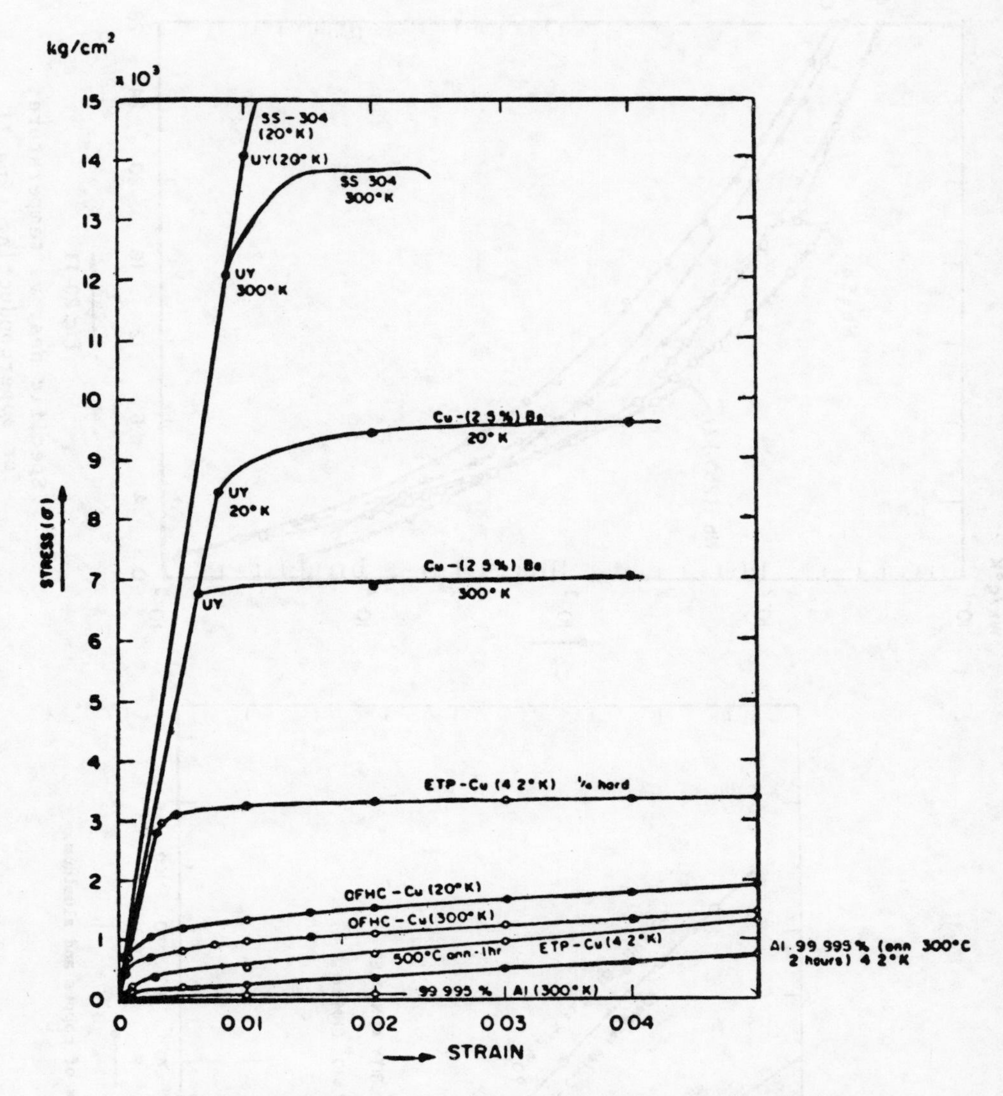

Fig. 20-12. Stress-strain diagrams of normal metals.

CHAPTER 21

HIGH ENERGY PHYSICS

The primary goal of High Energy Physics is to achieve an understanding of the fundamental nature of matter and energy and their relationship to space and time through the study of interactions of elementary particles. The discovery of new ways in which matter behaves is a central activity of the experimental program. The reason for the use of elementary particles as tools determining the structure and properties of matter is that these are the simplest probes presently available to man in his attempt to gain a clear picture of the basic forces in the universe. At our present level of understanding, only four such basic forces—the electromagnetic, strong, weak, and gravitational—seem necessary to describe qualitatively the myriad of physical processes known to occur in the world. High energy physics has placed major emphasis on the first three of these forces since the masses of the elementary particles are so small that the relatively very weak gravitational force appears to play no significant role in these processes.

The principal method currently used for achieving the goals outlined above consists of a program of experimentation directed towards determining the properties of the known particles and their mutual interactions, and towards the discovery of new phenomena. This experimentation deals with phenomena which man has never seen before because the required conditions do not occur ordinarily on earth. Physical research has steadily progressed beyond those domains of nature which can be directly sensed by humans and, as a result, complex means are required both to produce the necessary physical conditions and to make the experimental observations. Therefore, the design and construction of the necessary facilities and detection systems often comprise a major portion of the effort required. Consequently, these technical undertakings should be viewed as an integral part of the experimental process. In parallel with the experimentation, a continual development of theoretical ideas is pursued in order to incorporate new information into the framework of existing theories and to construct entirely new theories founded on the experimental results. This process of systemization and correlation of facts provides unifying simplifications, thereby improving comprehension and enabling the prediction of further new phenomena. Consequently, through the continual confrontation of theory with data, the high energy physics community seeks to focus the research activities at existing facilities, and the planning of new facilities, toward the solution of the most significant problems pertaining to the nature of matter.

ATOMIC, NUCLEAR, AND SUBNUCLEAR MATTER

The world of elementary particles is a novel one, filled with new forms of matter and energy and new transformations between these forms, having an unexpected richness and variety. Only some dozen or so years ago, very few of the currently known particle states and processes had been found. Today, it is uncertain how many new particles will be found as we continue into the higher energy domain and what new discoveries will affect our view of matter. Experimentation is difficult because we are dealing with matter under extreme conditions which are realized in nature probably only under cataclysmic conditions such as exploding or collapsing stars, or which may have been present at the very beginning of an expanding universe. A few of these phenomena are produced on earth by cosmic rays which, themselves, are believed to be products of stellar cataclysms.

The present situation concerning the large number of species of particles which are considered as elementary may have historical precedent. By the middle of the nineteenth century, chemists had discovered over sixty elements or substances which could not be chemically decomposed. An atom, the smallest bit of an element, was held at that time to be indivisible and, in that sense, was regarded as an elementary particle. The molecules of all chemical substances were correctly thought to be built up from these "elementary particles."

An important clue that the atom was not in fact "elementary" was provided by the periodic table of the elements. It was observed that certain chemical properties recur with a definite periodicity if the elements are listed in order of increasing atomic mass. This regularity in the properties of the elements had been recognized for a half century before experiments conducted at higher energies established the underlying atomic structure.

Near the beginning of the twentieth century, the atom had thus been revealed as a complex entity consisting of a massive central nucleus and an orbiting cloud of electrons. The observed chemical periodicity resulted from the recurring properties of the electron structure. The nuclei themselves, however, exhibited a somewhat more subtle periodicity in mass and charge, suggesting a still deeper substructure. Experiments at still higher energies eventually revealed the more basic building blocks, protons and neutrons, of which atomic nuclei are composed.

Over the last fifteen years, attempts to study the basic properties of protons and neutrons ("nucleons") have revealed a bewildering array of new particles. Some of them are relatively stable, while others are extremely short-lived, but none appear to be significantly less "fundamental" than the nucleons themselves. In addition, an intriguing set of symmetries and periodicities have been discovered in the properties of these particles. As before, the existence of such periodicities is very difficult to understand except as the manifestation of a more basic substructure.

A very satisfactory description of many of these properties has, indeed, been realized in the quark model of elementary particles. This theory hypothesizes three basic entities called quarks, of which all strongly interacting particles are composed; protons, neutrons, and other heavy particles (baryons) are composed of three quarks, while the lighter mesons contain one quark and one antiquark.

As the accompanying illustration indicates, we may thus be again in the familiar situation of looking for the next level of substructure in matter. To catch the elusive quark has been the goal of intensive research. No particle having its properties has yet been observed; nevertheless, the success of the model in explaining many basic regularities of the observed elementary particles, and in elucidating their interactions, suggests that a quark-like structure will be a component of any future successful theory.

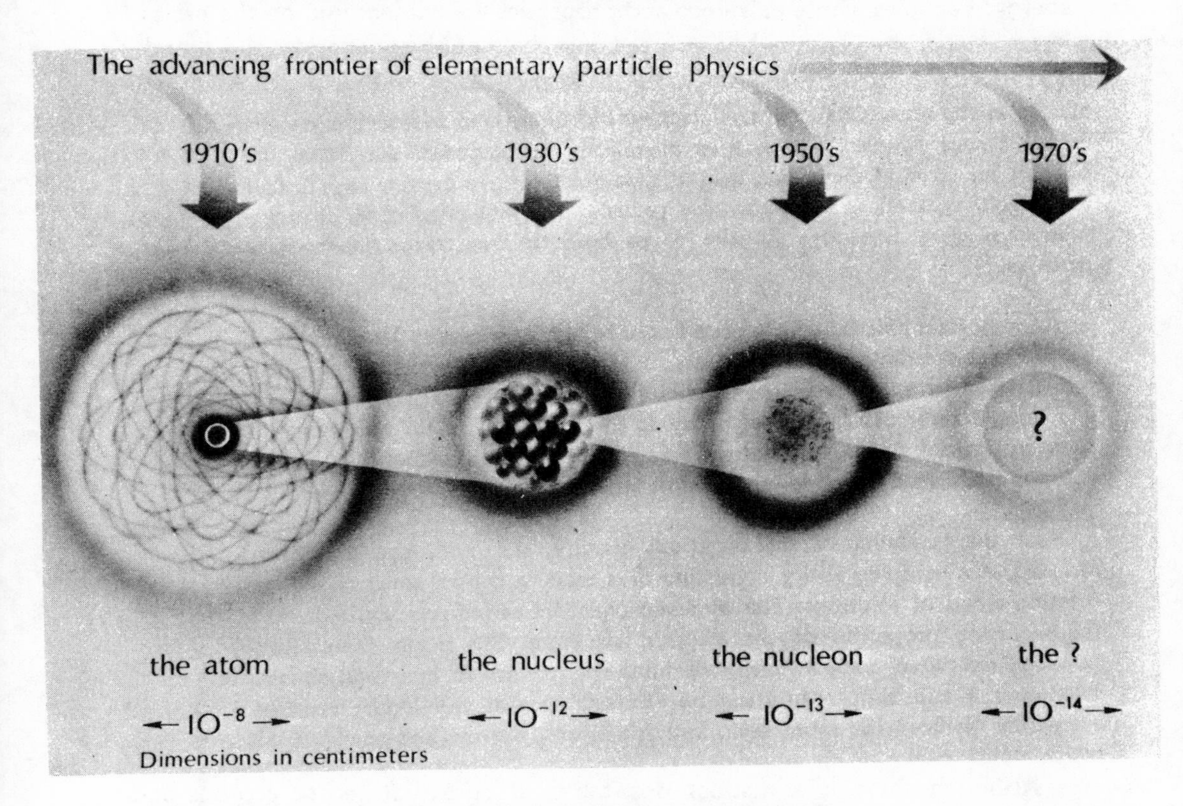

SOME TRENDS IN HIGH ENERGY PHYSICS

What is the nature of the forces that rule the behavior of matter? It is, of course, possible that new forces will be uncovered as advances are made in the particle physics program. Recent experiments involving the decay of K-mesons have, in fact, hinted at the presence of a possible new fifth "superweak" force. The major historical trend, however, has been toward a unification of the types of forces. Maxwell's electromagnetic theory, for example, unified electricity, magnetism and optics. This trend toward bridging and unifying continues today; for example, the weakly interacting particles (leptons) and the strongly interacting particles (hadrons) are each known to carry both weak and electromagnetic properties. Recently important progress has been made in the direction of unifying the weak and electromagnetic interactions within a single theoretical framework. Present plans include a series of neutrino and muon experiments at ANL, BNL, NAL, and SLAC, covering a broad range of energies so as to obtain further information on the weak interaction to clarify its relationship to the electromagnetic interaction. In addition, there are theoretical attempts in progress to tie the gravitational interaction to the strong, and the gravitational to the electromagnetic. Thus, it may be that all four of the forces will be joined onto a common framework, and the unification of the forces, a dream pursued by Einstein thirty years ago, appears to be a more realistic goal today.

One of the major thrusts in advancing this field of research is toward higher energy. Particles whose existence and properties are conjectured on the basis of lower energy results may be produced copiously at higher energies and thereby permit studies otherwise not possible. Processes observed at lower energies may be better understood by studying their behavior over a larger range of energies since higher energy permits exploration at much smaller distances. It is also conjectured that phenomena will become relativly simple at very high energies. In addition to examining the trends with energy, entirely new and unexpected phenomena will undoubtedly emerge. For these and many other reasons the prospect of experimentation in a higher energy range not previously explored has always had an immensely stimulating effect on research activity. History has taught us that new steps in energy bring new breakthroughs in our understanding of the laws of nature.

The need for higher energy to probe the smaller structures in matter arises from one of the most fascinating facts of modern physics. The French theorist, deBroglie, in 1924, was the first to recognize that in certain aspects the behavior of matter might be similar to that already known for light waves. His conjecture, now overwhelmingly verified, associates a wavelength with the energy of every particle, smaller wavelengths corresponding to higher energies. The wavelength sets the scale for the size of a physical system. To explore the more minute substructure of matter requires, at each stage, smaller wavelengths and consequently higher energies.

Although there is a distinct possibility of substructure for protons and neutrons, some of the successful descriptions of particle collisions at the Bevatron and the other accelerators are derived from a principle of "particle democracy." This notion is based upon the hypothesis proposed by G. Chew that all particles are equally fundamental and that there are no basic building blocks such as the quark. The spectrum of elementary particles can be calculated through a "boot-strap" procedure by starting out with any one of the multitude of choices for basic states (all particles being equivalent). The results of these "boot-strap" calculations also predict successfully the type of particles which are found to exist experimentally. The fundamental quark and the "democratic" view of the family of elementary particles, although appearing to represent contradictory points of view, yield similar predictions in several areas. It is, consequently, both correct and essential to permit the competition of ideas and experiments so as to probe the details of each point of view and decide on the basis of experimental fact which unifying principles are the most useful. It is perfectly possible, although not obvious, that both points of view are correct.

DEEP INELASTIC ELECTRON-PROTON SCATTERING

About sixty years ago, Rutherford performed a series of experiments in which a beam of positively charged alpha particles, obtained from radioactive material, was directed against a thin metal-foil target. The object was to observe the manner in which the energetic alpha particles were deflected, or "scattered," by the electrical forces present in the atoms of the metal foil. It was expected on the basis of the current models of the atom that many of the particles would be deflected through small angles, but that large-angle scattering events would be very rare. In the actual experiment, however, large-angle scattering was found to be much more common than expected, and from this observation, Rutherford was eventually able to draw a most important conclusion about the structure of the atom: that its positive electrical charge and nearly all of its mass was concentrated in a very small central core, or nucleus, with the remaining volume diffusely occupied by swirls of planetary electrons.

Rutherford's nuclear atom was a great advance over previous conceptions. About 20 years later, in the early 1930's, another major step was taken with the observation that a new kind of particle, the neutron, was occasionally ejected as "debris" from the nucleus when the energy of the bombarding particles was high enough. The discovery of the neutron was direct evidence that the nucleus had a composite structure, and these early neutron experiments are a prime example of the fact that information about the composition of an unknown structure can be obtained not only by observing the pattern of scattered beam particles, as Rutherford had done, but also by observing the other particles that often appear during the scattering process. Indeed, treasures are often found in the debris; and many high energy physics experiments are basically studies of the nature and behavior of the particles that make up this debris.

Electron scattering experiments are particularly useful for the study of structure of nucleons since no strong interaction is introduced by the electron probe. The first series of inelastic electron scattering experiments at the Stanford Linear Accelerator Center, carried out by scientists from SLAC and MIT beginning in 1968, were similar to Rutherford's experiments in both the technique used and the results obtained; only the scattered beam particles were measured, and large-angle scattering events were found to occur with surprising frequency. Analysis of the inelastic experiments is more complex than that of Rutherford's alpha-particle experiments, but their results may indicate that the

large-angle events are caused by electrons which scatter not from the proton as a whole, but rather from individual sub-units of concentrated electric charge within the larger volume of the proton. The evidence thus points to a picture of the proton as a composite structure. For convenience, these hypothetical constituents of the proton have been given the name "partons," and much effort in high energy physics is aimed at trying to find these elusive particles and to understand their properties. Theoretical effort is underway to determine how these "partons" might be related to the "quarks".
The SLAC-MIT group has continued their work by extending the study of inelastic electron scattering to electron-neutron collisions in the hope that a detailed comparison of the results from the neutron experiments with those from the initial proton work will help to illuminate the physical processes involved.

Two most recent SLAC experiments have been designed to extend the earlier work by observing the "debris" of particles that emerges from the collision in addition to the scattered beam particles. In this sense, they are comparable to the neutron experiments of 40 years ago. These experiments have provided additional indirect evidence that the proton consists of partons; however, the partons themselves were not found, for the debris did not contain any particles that had not been observed before. Thus, while the direct observation of any new constituent particle seems to require higher energies, the continuing efforts at lower energies are expected to shed further light on the structure problem.

One of the significant experimental advances in the last decade has been the exploration of the mechanisms of particle interactions through their dependence on the magnetic properties of the particles involved. Just as a magnet orients itself in the direction of the magnetic field, a particle may have one of several possible orientations in a magnetic field depending on its intrinsic property known as "spin." A collision between particles exhibits different characteristics which are dependent upon the spin orientation, called polarization, of the particles involved. Many of the significant advances in the verification or elimination of models has been accomplished through polarization measurements. The development of a polarized electron beam at SLAC and a polarized proton beam at ANL, in conjunction with the use of polarized proton targets, offer new research opportunities to understand the spin dependence of the forces involved and offers new testing grounds for theories.

ELECTRON-POSITRON COLLIDING BEAMS

In a colliding-beam machine, two particles with opposite velocities are made to collide head on. In such a head-on collision much more energy is available for reaction than in a collision using a conventional accelerator of comparable energy, wherein one particle is moving (the accelerated projectile) and the other is at rest (target). In the latter case a fraction of the energy is required to keep the system in motion so that less energy is available for the interaction. In the U.S. the colliding beam experimental program at the Cambridge Electron Accelerator is nearing completion and the Stanford Positron-Electron Assymetric Ring (SPEAR) colliding beam experimental program at SLAC is in its preliminary stage.

Interactions between colliding beams of electrons and positrons have recently been pursued with particular vigor because of the relative simplicity of interpretation of the process. The positron is the antiparticle of the electron, and, as a consequence, annihilation, a process in which the electron and positron both disappear and new particles are created, can be studied. In electron-positron reactions annihilation is expected to proceed through the electromagnetic interaction, which is the best understood of the four fundamental forces. It provides a pure electrodynamic initial state (no hadrons) and, consequently, can be used to probe for any high energy deviations from electromagnetic theory. It is not surprising, therefore, that electron-positron annihilation is an important area of colliding-beams research.

PROTON-PROTON COLLISIONS

If the interpretation of the deep inelastic electron scattering experiments in terms of internal nucleon structure is correct, experiments using other suitable nuclear probes, under proper conditions, should give results which require a similar interpretation. Scientists from Rockefeller University, Columbia University, and the Center of European Nuclear Research (CERN) in Geneva, Switzerland, have undertaken an experiment at the Intersecting Storage Rings (ISR) at CERN to examine this possibility. The ISR is a proton-proton colliding beams facility which permits two beams of protons, each accelerated to 28 GeV by the CERN Proton Synchrotron, to undergo near head-on collisions. The energy available for reaction in these interactions is equivalent to that obtained from a conventional accelerator (firing particles into a stationary target) having an energy of about 2000 GeV. Consequently, the ISR provides opportunities for experimentation not available elsewhere in the world. One of the objectives of this work is to study the production of neutral pi-mesons (π°) with large transverse velocities. Qualitatively, one might expect that the probability with which such pi-mesons are produced would drop off extremely rapidly with increasing transverse velocity if the nucleon is a soft object without an internal core or structure. Conversely, one may expect that the presence of "point-like" scattering centers within the nucleon would greatly enhance the probability of producing pi-mesons with large transverse velocity. Preliminary results from this experiment indicate that the number of π° mesons produced with large transverse velocity is much larger than expected on the basis of theory.

These results are difficult to understand without invoking structure within the nucleon. Indeed, there may well be a close relationship to the SLAC results. These new and dramatic features of the data indicate that this experimental approach will give new insights into the basic nucleon structure problem adding impetus to further applications of the colliding beams experimental technique.

UNIVERSAL BEHAVIOR AT HIGH ENERGIES

A common feature of high energy particle collisions is the transformation of some of the incident energy of motion into matter. This newly created matter appears in the form of particles, primarily pions and kaons, that emerge from the collisions with high velocities. Some years ago, R. Feynman and C. N. Yang suggested that a great deal of information about this transformation process could be gained from systematic studies of the behavior of any one of the particles produced. The basis for this suggestion is that at extremely high energies particle collisions seem to take on a universal character that is independent of the detailed nature of the target or projectile. Recent studies at the University of Rochester, based on a summary of available data pertinent to this question, indicate that this conjecture is approximately correct.

Although these experimental results appear to be consistent with the somewhat intuitive notions of Feynman and Yang, this apparent universal behavior is also fascinating from the point of view that it implies a basic similarity for all matter. One possible simple explanation for the emerging fact that particle production has a rather universal character, independent of the properties of the colliding particles, is that these familiar "elementary" particles are themselves made up out of as yet undiscovered building blocks, and it is these new basic constituents which are the mediators of interactions at the highest energies.

HEAVY ION RESEARCH- BEAM FRAGMENTATION

The Bevatron at the Lawrence Berkeley Laboratory has recently achieved a capability for accelerating heavy ions (nuclei of elements helium through neon) to high energies. The results from the ensuing heavy ion research program are promising to be of major importance to the study of cosmic rays and cosmology. It is found that in the collision of a heavy ion with a stationary target the beam particle breaks apart and fragments emerge from the collision with very nearly the same velocity as that of the initial beam. In addition, detailed studies of these fragments suggest that the kinds of fragments produced depend only on the type of beam particle and are independent of the identity of the target. This particularly simple behavior in a potentially very complicated situation shows a striking similarity to the behavior of very high energy nucleon-nucleon collisions.

Advances in our knowledge of the elements and isotopes present in the cosmic rays near the earth have been obtained from recent satellite experiments, and theoretical techniques are available to interpret these observations and obtain an understanding of the primordial composition of cosmic rays. However, the lack of comprehensive laboratory data on the fragementation cross sections for the complex nuclei involved has become a principal impediment to progress in this area of cosmic ray research. With the heavy ion capability of the Bevatron, this fragmentation of nuclei can be studied in the laboratory. The heavy nuclei in the primary cosmic rays comprise somewhat less than one percent of the total cosmic ray flux, but they contain much information on the inter-stellar medium, the regions of cosmic ray confinement, and the motions of magnetic field lines in the galaxy. Clues to the sources of the cosmic rays and the nature of the radiation at these sources are being obtained from studies of the identity and energy of arriving heavy nuclei. The interpretation of these observations in terms of theories of propagation, given the relevant fragmentation cross sections of the heavy ions in the hydrogen and helium inter-stellar gases, should have a significant impact on astrophysics.

NEUTRINOS

Among the many forms which matter and energy can assume, perhaps none is more fascinating than the family of particles known as the neutrinos. These particles are the most penetrating form of radiation which has yet been discovered. If a beam of neutrinos (of say 1 Gev energy) were formed and directed at the sun, over 99 percent of the beam would pass directly through the sun and would emerge undeflected from the opposite side. As a comparison, the time required for ordinary light to "travel" from the center of the sun to its surface is estimated to be greater than 10 million years because of the vast numbers of collisions, absorptions and re-emissions which it would undergo. Most neutrinos, on the other hand, would transit the sun in less than five seconds without collisions or deflections.

How could particles with such a small probability of interaction ever be discovered? Historically, the neutrino was postulated in 1930 because of experiments on radioactive disintegrations in which electrons were emitted from the disrupting nucleus. It was found that some of the nuclear energy disappeared in the decay and rather than abandon the concept of the conservation of energy, the physicists of that period accepted the hypothesis of W. Pauli, one of the leading theoretical physicists of the 20th century, that an undetected, electrically neutral particle—later called the neutrino—was also emitted in the decay process and was carrying off the missing energy.

Direct experimental verification of the neutrino was not accomplished until 1953 when two American physicists, F. Reines and C. Cowan, using a high power nuclear reactor as an ultra-intense neutrino source, were able to detect the minute fraction of neutrinos which did interact in their very large detector.

The modern period of neutrino physics began in 1960 when M. Schwartz, then at Columbia University, and B. Pontecorvo of the USSR, pointed out that direct detection of nuetrinos produced using high energy accelerators was also possible. Theory predicted that the probability of neutrino interaction would increase dramatically as the neutrino energy was increased. Neutrinos in the billion electron volt range could be produced at the high energy accelerators as contrasted with a few million volt energies of the neutrinos from nuclear beta-decay. In addition, physicists had learned how to produce beams of neutrinos by manipulating the particles which, in decaying, produce them. The advantage of a neutrino beam for these experiments is substantial since it requires the construction of a very thick detector only in the beam direction. Two years after the proposal for high energy neutrino experiments was published, these ideas were verified when the first successful experiment to detect high energy neutrinos was carried out.

High energy physicists use the term "leptons" to describe the entire collection of muons, muon neutrinos, electrons, and electron neutrinos. The muon is an electrically charged particle which appears to be very much like the electron in its properties and interactions except that it has a mass about 200 times that of an electron. The electron and muon are so similar that the large mass difference between them has been a long-standing enigma to theoretical physicists. Additionally, there are two kinds of neutrinos, electron neutrinos and muon neutrinos. The mass of both types of neutrinos is known to be zero or very small. Current theoretical ideas hold that the neutrino masses are identically zero and that neutrinos always travel at the speed of light.

The character of the force governing the interactions of leptons, the so-called "weak" force, continues to be one of the central concerns of contemporary physics. The great penetrating power of the neutrinos illustrates the effective weakness of the force. Perhaps the most significant unknown associated with the weak interaction is the distance over which one particle can affect another. Is the effective weakness of the interaction due to a very short range rather than an intrinsic weakness of the force? What is the mechanism of the weak force? Is there another as yet undetected agent (the "intermediate boson") which transmits the weak force between particles? Despite the apparent differences, is there a deeper connection between the electromagnetic and weak interactions, as suggested by theorists?

UNIFIED THEORIES OF WEAK AND ELECTROMAGNETIC INTERACTIONS

A most significant question is whether there are basic relationships between the fundamental forces in nature. As mentioned earlier, only four different fundamental forces are known by which all physical bodies interact—gravitational, electromagnetic, and the strong and weak nuclear forces. The latter is responsible for radioactive processes. It is known to have a short range because of the way it is able to affect nuclear processes; however, the precise range of the force is uncertain. At large distances (1 cm) there are only electromagnetic and gravitational forces between particles, and the electromagnetic force dominates the gravitational force, between protons for example, by a factor of 10^{36} At nuclear dimensions (about 10^{-13} cm) the dominant force between particles is the strong force. The strong force in this range is about 100 times the electromagnetic force between protons. And the electromagnetic force is, perhaps, about 10^{10} times the weak force. At smaller dimensions, perhaps in the $10^{-14}-10^{-17}$ cm range, the weak force must become increasingly significant since we know that the neutrino-proton probability of interaction keeps increasing linearly with energy in the experiments performed to date. This fact also explains some of the rationale for experimentation at higher energies since the latter permits probing of smaller dimensions.

Current theoretical ideas involve a particle as the carrier or mediator of each force. There is a connection between the range of a force and the mass of the particle which carries the force. The shorter the range, the higher the mass of the carrier. Thus, the electromagnetic force has a very large range and its carrier, the photon, has zero mass. The strong force, having a range of about 10^{-13} cm has, as a carrier, the pion, which has a mass roughly one-seventh that of the proton. The graviton, a hypothetical particle presumed to mediate the force of gravity, has not been detected. The very short range of the weak interaction implies that a high mass should be associated with its carrier, commonly called the W, or intermediate boson.

Not too long ago, S. Weinberg proposed a theory in which the electromagnetic field and the weak interaction field are, in fact, two aspects of the same interaction. The excitement of a plausible, unified theory, which also surmounted many of the calculational difficulties which have plagued earlier field theories, has led to a number of theoretical models which predict, among other things, several new particles. Verification of these predictions awaits the results of experimental searches. More recently, this line of theoretical research has found an additional application in attempts to explain the unsolved problem of determining the origin of the proton-neutron and electron-muon mass difference. Additionally, the mathematical structure of the theory bears a strong formal resemblance to some existing strong interaction theories, raising the hope that a still deeper unification may be possible.

HIGH ENERGY PHYSICS AND ASTRONOMY

An understanding, or theory, of astronomical phenomena generally requires the knowledge of several fields of physics. Relativity, fluid mechanics, plasma physics, and nuclear physics are all heavily involved. Recently, high energy physics and astronomy have been developing close connections in areas of great concern to both fields. The newly discovered pulsing stars where the intensity of radiation varies periodically, in addition to a number of other phenomena, have been interpreted, in part, in terms of properties of "elementary" particles at high energy. Conversely, since many elementary particle properties are still uncertain, it is possible that astronomical discoveries will shed light on the particles.

Pulsing stars, "pulsars," have been convincingly interpreted as giant globs of "nuclear" material. The largest stable atomic nucleus found is that of the element lead with 82 protons and 126 neutrons. It is about 10^{-12} cm across. The tendency of larger nuclei to break up via radioactive decay is so strong that they are not found naturally beyond Uranium 238 although some heavier nuclei have been produced artifically. The possible existence in space of gigantic nuclei miles across composed mainly of neutrons–neutron stars–was suggested some time ago.

A sphere about a mile across is, of course, very small for a star. Its density, however, would be very, very great. Such nuclear matter has about a thousand billion times the density of ordinary matter. The neutron star does not break up as does a tiny radioactive nucleus. Because of its enormous concentration of matter, the force of gravity holds the object together. In summary, we have a picture in which there are stable nuclei from hydrogen to lead, then there is a great zone of nuclear size with no stable nuclei, and then nuclear objects a mile or so in diameter are, again, stable.

Larger neutron stars are believed to collapse under their enormous gravitational pressure into even denser objects. How compressible is nuclear matter? It is like rubber or steel? Here high energy physics enters. The compressibility of matter at these extremely high densities depends on the fundamental structure of nuclear particles, which is the realm of study of high energy physics. We can look at very small particles only through their interactions with other very small particles. Only if particles undergo hard, i.e., high energy, collisions will they penetrate each other and reveal information about their internal structure. The degree of compressibility at high density depends on whether or not nuclear particles are made of more fundamental constituents or are more or less basic in their own right. High energy physics can help to find answers to questions like: What is the maximum size of a neutron star, and is the neutron star interior solid or liquid? (The answer is probably vital for understanding "starquakes.")

It is believed that neutron stars may originate as remnants of supernova which are exploding stars emitting tremendous amounts of energy. For example, there is a pulsar in the "Crab Nebula." The Crab is the remnants of a supernova which illumined the sky–about as brightly as the moon–for some months in 1054 A.D.

Neutrino production is an aspect of high energy physics which enters the description of supernova events as well as other phases of stellar evolution. Since neutrinos are particles which interact very infrequently, once produced, they are almost certain to escape without being reabsorbed. Thus, "neutrino cooling" is an important process for energy transfer out of stellar interiors. The process resembles cooling by evaporation of perspiration, except that the neutrinos come right out in the interior rather than from the surface.

Stellar neutrinos come from the interior, while all our other information about a star comes from its surface. Thus the sun's neutrinos are being observed in order to gain information about the sun's interior. Considerably fewer neutrinos were detected than had been predicted. One possible explanation is that the sun is not as hot inside as has been thought. Another explanation may have something to do with the intrinsic properties of neutrinos—a high energy physics problem.

Other profound aspects of astronomy are related to high energy physics. The state of matter in the superdense "beginning" of the universe, the possibility of astronomical systems made of antimatter, a possible connection between high energy processes sensitive to the direction of the flow of time and cosmology, and fundamental relations between particle structure and general relativity all are important topics for both high energy physics and astronomy.

The relationship between astronomy and high energy physics has been growing rapidly and changing in unexpected ways. The history of science teaches us that the establishment of connections between very different kinds of observations is likely to prove most fruitful. Thus, we may look forward to even greater continued benefits of this cross-fertilization of two areas of modern science.

CHAPTER 22

CONSTRUCTION AND OPERATION OF NUCLEAR POWER PLANTS

INTRODUCTION

Under the Atomic Energy Act of 1954, as amended, AEC licenses the construction and operation of nuclear power plants. AEC's licensing activities are carried out under the Director of Regulation who is responsible for ensuring that the construction and operation of nuclear facilities and the licensed use of radioactive materials will not result in undue risk to the health and safety of the public.

Within the regulatory organization of AEC (see organization chart on Fig. 22-1),the primary responsibility for reviewing, processing, and evaluating applications for permits to construct nuclear power plants and licenses to operate them has been placed in the Division of Reactor Licensing (DRL). The review and evaluation performed by DRL is directed toward the health and safety aspects of the design, location, and operation of the nuclear plants.

The safety aspects of a proposed power reactor are reviewed by the Advisory Committee on Reactor Safeguards (ACRS), in addition to DRL, prior to issuance of a construction permit or an operating license. ACRS, consisting of a maximum of 15 members, is a committee established by the Congress and is statutorily required to conduct a safety review of reactor applications.

The decision to issue a construction permit is made only after a public hearing is held under the direction of a three-member atomic safety and licensing board composed of two technical experts and one lawyer who acts as chairman of the board for the hearing. Members of the board are appointed by AEC from private life or from AEC or other Federal agencies. With respect to the issuance of an operating license, a hearing is required to be held only if the issuance of such a license is contested or if AEC so directs.

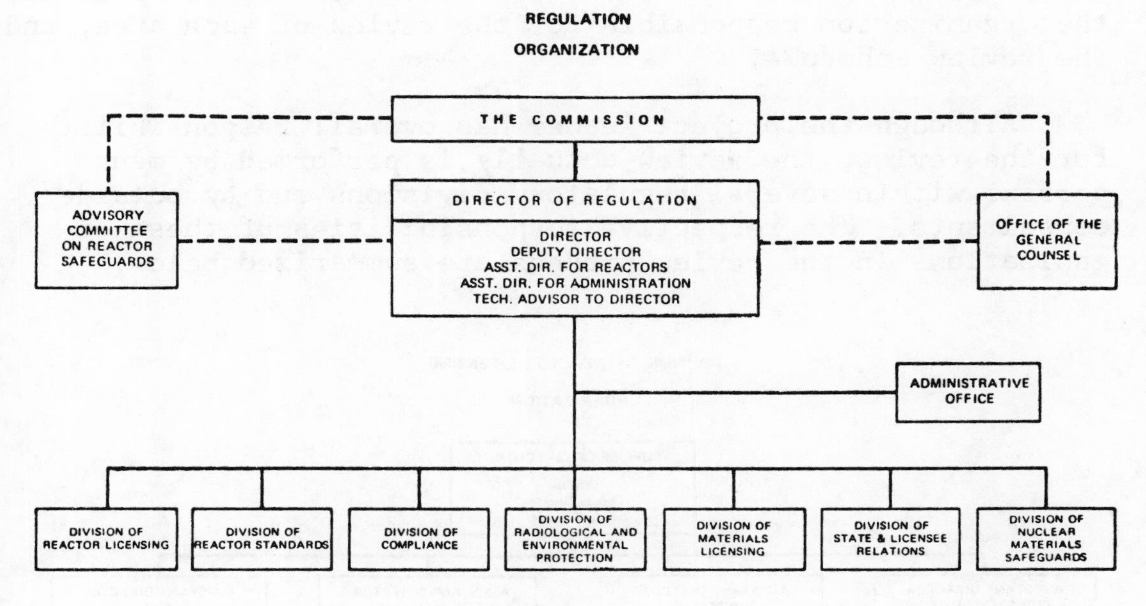

Fig. 22-1

DESCRIPTION OF INTERNAL APPLICATION REVIEW PROCESS

The licensing process begins when an application for a construction permit or an operating license is filed with AEC. The application must cover, among other things, the financial qualifications of the applicant, the design of the facility, and a safety analysis report. The safety analysis report discusses various accident situations and the engineered safety features which will be provided to prevent accidents or, if they should occur, to mitigate the consequences of such accidents.

Under the National Environmental Policy Act of 1969, an environmental impact report also must be submitted with each application.

AEC advises ACRS when it receives an application, and ACRS assigns a subcommittee representing various technical disciplines to study the application. This subcommittee studies the application concurrently with the review by the AEC regulatory staff.

The organizational structure of DRL is shown in Fig. 22-2. Within DRL the overall responsibility for conducting and coordinating the review of each application is assigned to a project leader. The project leader is required to prepare a review plan which identifies the areas to be reviewed, the organization responsible for the review of each area, and the review schedule.

Although the project leader has overall responsibility for the review, the review actually is performed by many persons within several regulatory divisions and by outside consultants. The respective responsibilities of these organizations in the review process are summarized below.

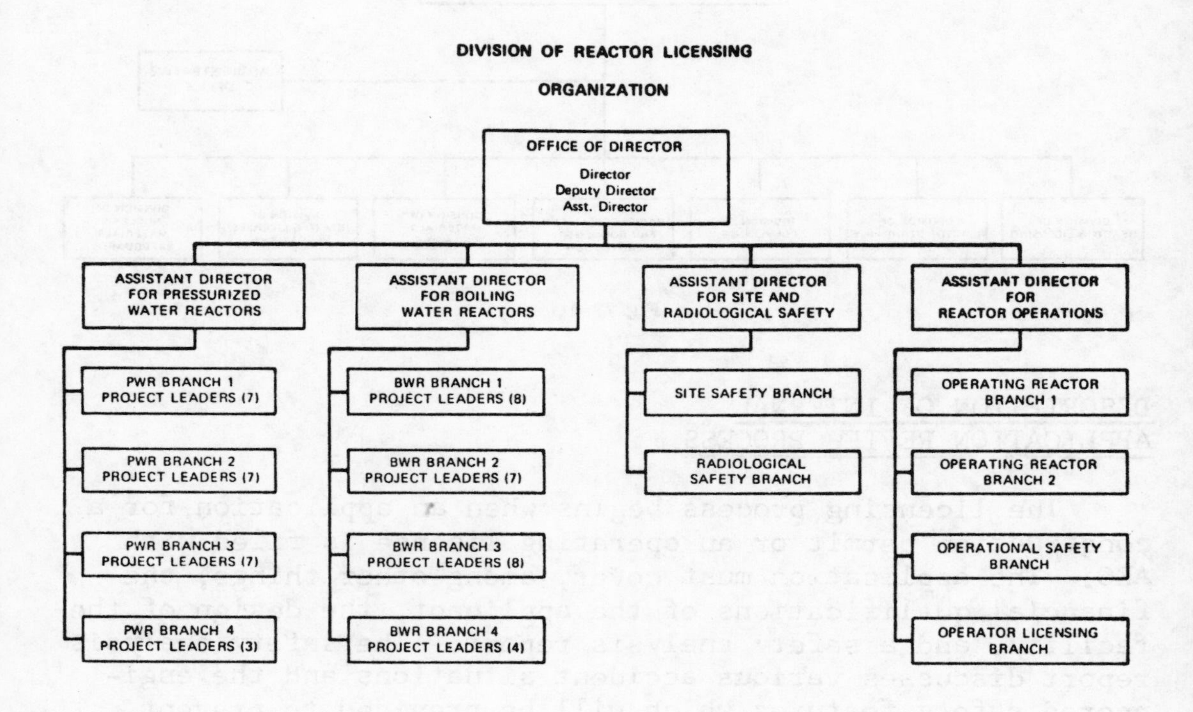

Fig. 22-2

Division of Reactor Licensing--Certain aspects of the review are performed directly by the project leader. In addition, the Assistant Director for Site and Radiological Safety is responsible for the safety review and evaluation of certain aspects of proposed sites for nuclear facilities and their radiological systems and components as well as proposed programs and limits for facility operation and control. His group (the DRL site group) reviews such items as population distribution, site meteorology, effluent monitoring, radioactive waste controls, and radiological consequences of potential accidents.

Division of Reactor Standards (DRS)--This division provides technical assistance to DRL by analyzing and evaluating the electrical, mechanical, structural, and material components and systems of the proposed nuclear power plant as well as the geological and hydrological aspects.

Division of Radiological and Environmental Protection--This division is responsible for administering the regulations governing the implementation of the National Environmental Policy Act of 1969 (42 U.S.C. 4321) and the Water Quality Improvement Act of 1970 (33 U.S.C. 1151) for all AEC-licensed activities. This responsibility includes the review, evaluation, and processing of the environmental impact reports submitted with applications for licenses to construct and operate nuclear facilities and the preparation of environmental impact statements.

Consultants--During each review the regulatory staff utilizes the capabilities of private firms and other Government agencies as consultants to review portions of applications for which the staff does not have the in-house expertise. These reviews relate generally to site geology, meteorology, and seismology and to the seismic design of the reactor. The regulatory staff also has arranged to use, as needed, specific employees of AEC's national laboratories or of other Government agencies to review unique aspects of individual applications.

The initial efforts by these various organizations are directed toward reviewing and evaluating those sections of an application for which they are responsible and identifying the additional technical information needed to permit them to complete their evaluations. AEC regulations describe the broad technical information required in an application. In practice, however, AEC has found that additional technical information is needed and must be requested from the applicant.

This additional information is requested from the applicant through a series of formal questions. The replies received from the applicant become amendments to its original application. Generally several sets of questions must be sent to the applicant before all the technical information needed to complete the evaluation process is received. The evaluation of the application continues during this question-

and-answer process; however, AEC has stated that the missing information, when supplied, may necessitate reevaluation of much of the previously submitted material.

When answers to the final set of questions have been received and evaluated, the various organizations involved in the review and evaluation process are in a position to develop their final reports. These reports are consolidated by the project leader into a final report to ACRS that presents DRL's evaluation of the safety aspects of the proposed nuclear power plant.

ACRS considers the applicant's safety analysis report, together with the evaluation prepared by DRL. Representatives of the applicant; members of the technical staffs of DRL and DRS; and, when necessary, AEC consultants meet with ACRS to deal with questions that arise during ACRS' review. When ACRS reaches a conclusion as to the safety aspects of the proposed reactor, it reports its views to the AEC Commissioners. After the ACRS report has been received, DRL prepares an evaluation of the safety aspects of the proposed reactor that is made available to the public. This evaluation takes into account the recommendations and advice of ACRS.

The above discussion relates to the formal steps in the licensing process; however, during the course of the entire review process, there are many meetings with an applicant as well as with ACRS for the purpose of seeking additional information and clarification on the many technical matters involved in approving a license application.

INCREASE IN LICENSING WORK LOAD

The role of nuclear reactors in the production of electricity is growing rapidly. In the last several years, there has been substantial growth in the size and number of nuclear power plants being constructed and operated for the production of electrical energy. Correspondingly there has been a significant increase in the number of license applications under review by AEC and in the manpower resources used to perform the review.

Under the provisions of the National Environmental Policy Act of 1969 and the Water Quality Improvement Act of 1970, AEC is required to review license applications from an environmental as well as a safety standpoint. Until recently AEC sought advice on environmental questions from the appropriate Federal and State agencies; however, a recent court ruling stated that AEC had failed to appropriately implement the National Environmental Policy Act and that AEC must conduct its own investigation of all environmental aspects of commercial nuclear facilities and must make its own judgments on all environmental questions, even when a plant is in compliance with other Federal, State, or local environmental standards. This decision has had a significant impact on the work load of AEC's regulatory staff.

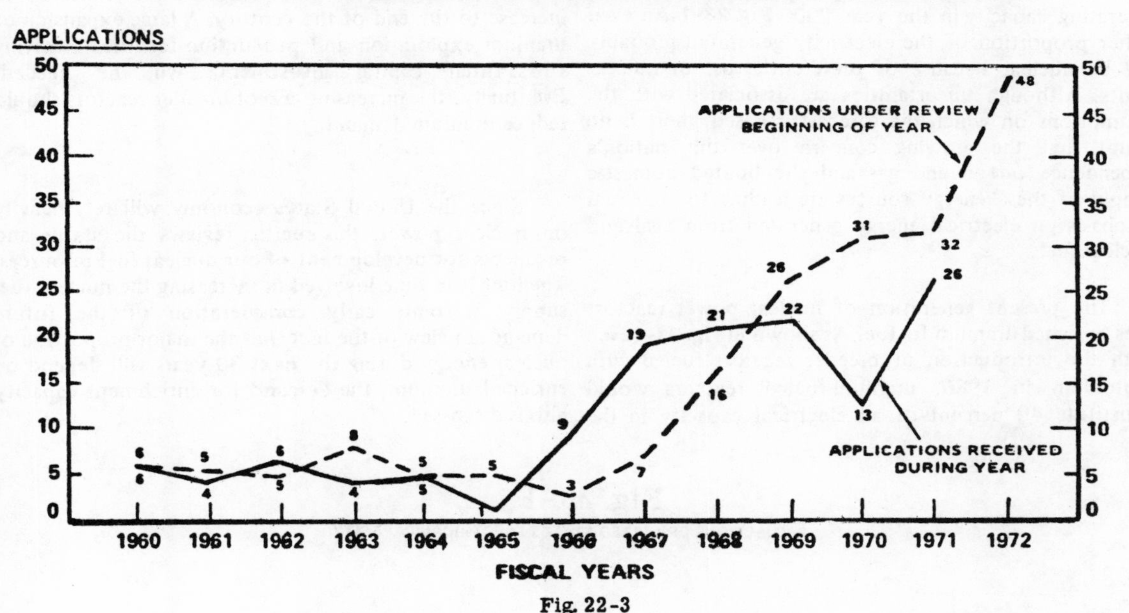

Fig. 22-3

CHAPTER 23

NUCLEAR FUEL SUPPLY

Energy consumption in the United States is projected to double between 1970 and 1985; it will continue to rise to three times that in 1970 by the year 2000. Projections of electrical energy use indicate an even more rapid growth. AEC forecasts that nuclear energy will become the dominant source of electricity in the 1990's and represent 60 percent of the electrical generating capacity in the year 2000 **Fig. 23-1.** An even higher proportion of the electricity generated probably will be nuclear because of preferential use of nuclear plants. Although uncertainties are associated with the assumptions on which the forecast is based, there is no doubt that the growing concern over this nation's dependence on oil and gas and the limited domestic supply of these energy sources are leading to increased emphasis on electrical energy generated from coal and nuclear fuel.

The present generation of nuclear power reactors uses enriched uranium for fuel. As shown in **Fig. 23-1** even with the introduction of breeder reactors fueled with plutonium in 1986, uranium-fueled reactors would constitute 40 percent of all electrical capacity in the year 2000. Presently known domestic resources and existing production capacity are inadequate to meet the corresponding uranium demand. For example, the projected annual uranium production needed in 1990 would be about six times the peak rate previously achieved (in 1961) after the major expansion of uranium for defense purposes. Requirements will continue to increase to the end of the century. A large expansion of uranium exploration and production facilities involving substantial capital investments will be needed. Eventually, the increasing use of breeder reactors should reduce uranium demand.

Since the United States economy will rely heavily on nuclear power, this section reviews the status and prospects for development of our nuclear fuel resources. The long lead time involved in increasing the nuclear fuel supply warrants early consideration of the future demand. In view of the fact that the major proportion of nuclear energy during the next 30 years will depend on enriched uranium, the demand for enrichment capacity also is discussed.

Fig. 23-1

PROJECTIONS OF UNITED STATES ELECTRICAL CAPACITY

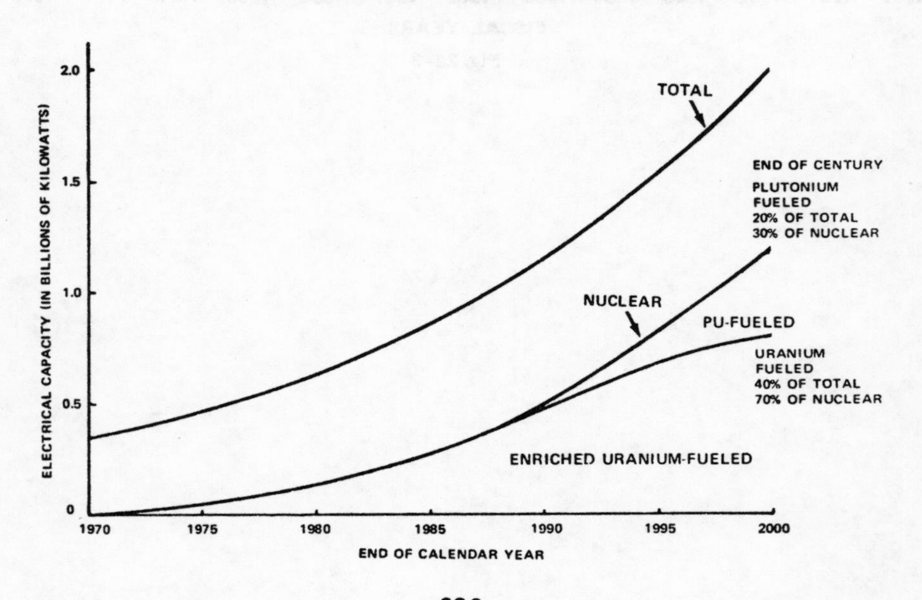

NUCLEAR POWER DEVELOPMENT

The electrical generating capacity of nuclear plants in operation, under construction and on order in the United States already exceeds 140,000 megawatts. During 1972 alone, 35 reactors with a capacity of 38,000 megawatts were ordered. These nuclear power plants are very large, commonly over 1,000 megawatts electric capacity each, and take up to 8 or 10 years to complete and place in operation.

Beyond the year 2000 the dominant type of nuclear power plant will be the light-water-cooled reactor that derives its heat primarily from the fissionable uranium isotope U^{235}. The U^{235} content of natural uranium is 0.7 percent. The remainder of natural uranium (99.3 percent) is the isotope U^{238}. The U^{235} content of uranium must be "enriched" to fuel light-water reactors as well as the high-temperature gas-cooled reactors which will also be used.

In the last decade of the century, breeder reactors should come into general use. They will initially utilize plutonium generated from the uranium isotope U^{238} in light-water-cooled reactors. The breeders will subsequently generate their own plutonium from U^{238}. Uranium depleted in U^{235} accumulated during uranium enrichment will be one source of U^{238}. The plutonium surplus to operating breeders will fuel new breeders. After the end of the century commercial use of controlled thermonuclear (fusion) reactors, burning deuterium and tritium (derived from lithium), may be achieved.

URANIUM REQUIREMENTS

A typical light-water reactor of 1,000 megawatts electric capacity needs enriched uranium produced from 550 to 625 tons of uranium oxide (U_3O_8) for the initial loading and about 200 tons of U_3O_8 per year for refueling. Recycling of plutonium recovered from the spent reactor fuel elements could reduce annual makeup requirements to about 170 tons of U_3O_8. Power plants normally have an operating life of 30 years or more. The lifetime requirement of the 140,000 megawatts of capacity on order is about 700,000 tons of U_3O_8.

Breeder reactors can greatly expand the energy derived from natural uranium and hence will reduce uranium demand. One ton of uranium can produce 50 million kilowatt hours of electricity in light-water reactors but could produce 3 billion kilowatt hours in breeders. For comparison, if coal were used to generate the same amounts of electricity, 18,000 and 1,100,000 tons, respectively, would be needed.

The recent AEC report "Nuclear Power 1973-2000" (WASH-1139) forecasts increasing annual uranium demand through the year 2000 based on certain assumptions on U.S. population growth, electrical share of energy consumption, and the rate of penetration of nuclear power into the energy market. Case A (Fig. 23-2), considered the most likely case, assumes 1986 for initial use of commercial breeder reactors. No other reactor types are available which could reduce significantly uranium demand. Annual uranium requirements would

Fig. 23-2

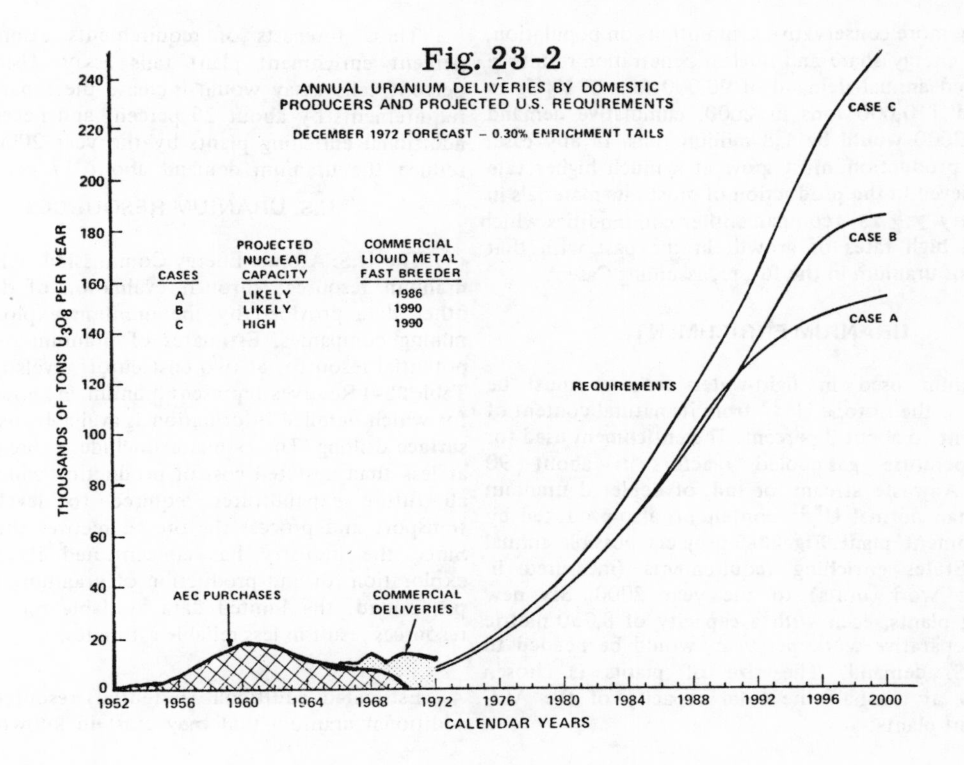

Fig. 23-3

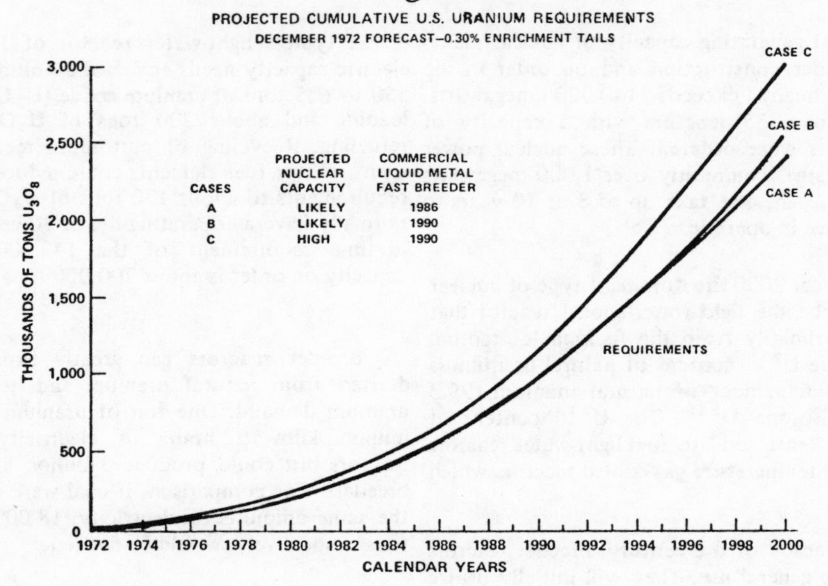

increase from 18,000 tons of U_3O_8 in 1975 to 120,000 tons in 1990 and 150,000 tons in 2000. Case B is based on the same assumptions but with delay of the breeder to 1990. Case C shows a still higher possible demand assuming a greater population growth and a greater electrical share of total energy consumption and nuclear penetration. Cumulative requirements by the year 2000 for the three cases (**Fig. 23-3**) range from 2.4 to 3.1 million tons of U_3O_8.

Using more conservative assumptions on population, electrical energy share and nuclear penetration results in a projected annual demand of 90,000 tons of U_3O_8 in 1990 and 110,000 tons in 2000; cumulative demand through 2000 would be 1.8 million tons. In any case, uranium production must grow at a much higher rate than achieved in the production of most raw materials in this century **Fig. 23-4** compares other commodities which have had high rates of growth in the past with that required of uranium in the future, assuming Case A.

URANIUM ENRICHMENT

Uranium used in light-water reactors must be enriched in the isotope U^{235} from its natural content of 0.7 percent to about 3 percent. The enrichment used for high-temperature gas-cooled reactors is about 90 percent. A waste stream, or tail, of depleted uranium (lower than normal U^{235} content) is also produced by the enrichment plant. **Fig. 23-5** projects possible annual United States enriching requirements (measured in separative work units) to the year 2000. Six new enriching plants, each with a capacity of 8,750 metric tons of separative work per year, would be needed to meet U.S. demand. The size of plants is chosen arbitrarily at one-half the total capacity of the AEC enrichment plants.

The projected foreign demand for both enriching services and U_3O_8 is somewhat higher than the U.S. demand. In **Fig. 23-5** the proportion of the foreign enriched uranium market furnished by the U.S. has been estimated only to 1983, the first year that production is expected from new plants. It is too early to predict accurately for the entire period the portion of foreign requirements for enriched uranium that may be supplied by U.S. plants.

These forecasts of requirements assume a 0.30 percent enrichment plant tails assay. Using a 0.20 percent tails assay would increase the separative work requirements by about 25 percent and necessitate two additional enriching plants by the year 2000; it would reduce the uranium demand about 17 percent.

U.S. URANIUM RESOURCES

The U.S. Atomic Energy Commission estimates U.S. uranium resources through evaluation of drilling and other data provided by the uranium exploration and mining companies. Estimates of uranium reserves and potential resources at two cost cutoff levels are given in Table **23-1** Reserves represent uranium in known deposits for which detailed information is available, usually from surface drilling. The estimates include all ore producible at less than a stated cost of production which includes all future expenditures required to develop, mine, transport and process the ore to recover the uranium. Since the industry has concentrated its efforts on exploration for and production of uranium at $8 or less per pound, the limited data available on higher cost resources result in less reliable estimates.

Estimated additional (potential) resources refer to additional uranium that may exist in known favorable

Fig. 23-4

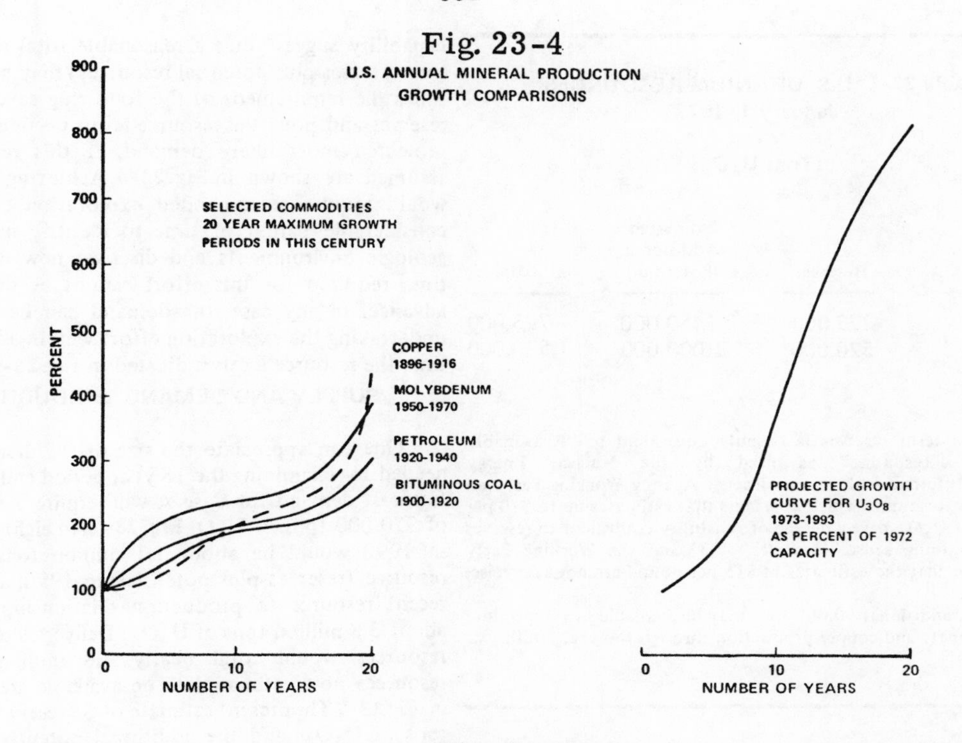

Fig. 23-5

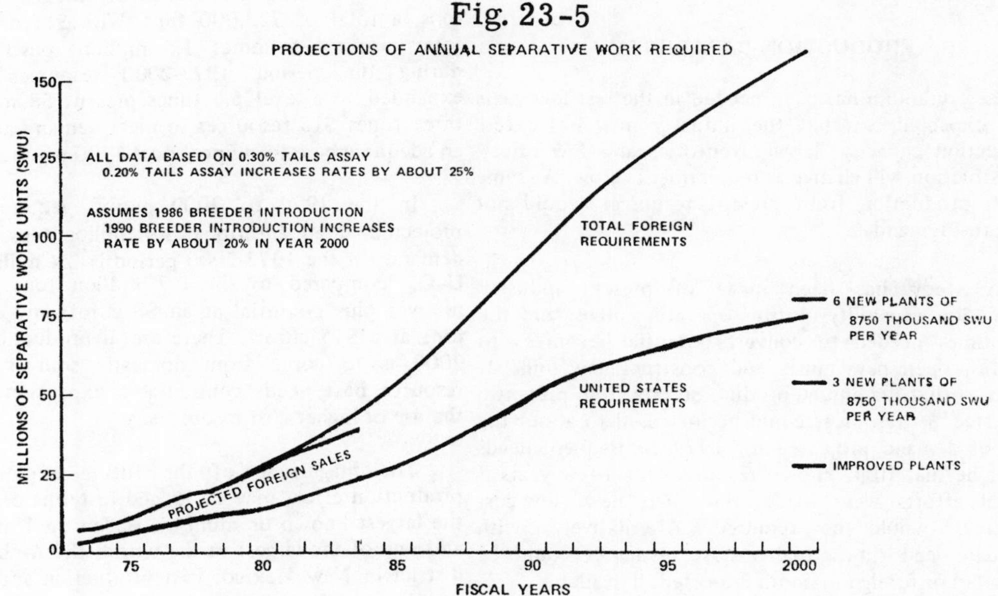

geologic environments, primarily in or around the areas containing ore reserves. Estimates of potential are inherently less accurate than those of reserves. Moreover, because geologic knowledge of the United States is incomplete and the estimates have been largely confined to producing areas, they do not represent a complete appraisal of U.S. resources. Judging by past experience with uranium and other mineral commodities, further exploration should both extend known uranium districts and disclose new areas.

Domestic reserves at a cutoff cost of $8 per pound of U_3O_8 are mostly in New Mexico and Wyoming, but there also are important reserves in Texas, Colorado, Utah and Washington. Most uranium deposits are small, containing less than 100 tons of U_3O_8. The relatively few large deposits have the bulk of the reserves; 10 percent, or about 70 deposits, contain 85 percent of the $8 reserves. The average uranium content of ore mined in 1972 was about 0.21 percent U_3O_8.

Table 23-1 U.S. URANIUM RESOURCES
January 1, 1973

(Tons U_3O_8)

Cost Cutoff $/lb. U_3O_8	Reserves[1]	Estimated Additional (Potential)	Total
$8	273,000	450,000	723,000
$15	520,000[2]	1,000,000	1,520,000

[1] The term reserves is roughly equivalent to "Reasonably Assured Resources" as used by the Nuclear Energy Agency/International Atomic Energy Agency Working Party on Uranium Resources which considers that estimates up to $10 per pound of U_3O_8 have a degree of reliability equivalent to reserves in the mining sense. However, AEC and the Working Party recognize that the estimates at $15 per pound are not as precise.

[2] An additional 90,000 tons may be available as a byproduct of phosphate and copper production through the year 2000.

PRODUCTION CAPABILITY

Less uranium has been needed in the last few years than expected, so that the industry now has excess production capacity, large inventories and low prices. The situation will change as requirements grow. At some point, production from present resources would not meet the demand.

A study has been made of present industry production capability, future operating plans, and the lead times needed to convert potential resources to reserves, open new mines and construct new mills. It indicates that the annual production rate from presently estimated $8 resources could be increased to about the level of demand projected for 1979. Subsequent needs might be met from known resources for a few years if special efforts were made, but then new, low-cost resources would be required. Alternatively, with adequate lead times, higher cost resources could be exploited or foreign uranium imported, if available.

DESIRABLE RESOURCE LEVELS

Some level of uranium reserves and potential resources must be maintained to assure adequate future production rates. Determination of adequate levels is difficult. However, the time required to find new ore, construct new mines and mills, amortize investments and permit contracting for deliveries to consumers over a reasonable period of time in the future indicates the need for an ore reserve equal to at least the following eight years' requirements. In addition, analyses of production capability suggest that a reasonable total resource level (ore reserves plus potential resources) may be about four times the requirement of the following seven years. The reserves and potential resource levels needed to meet the projected most likely demand, if this relationship is assumed, are shown in **Fig. 23-6.** Achieving these levels would require an expanded exploration effort over a considerable period of time to identify new favorable geologic environments and discover new deposits. The time required for this effort cannot be determined in advance. In any case, the demand can be met only by undertaking the exploration effort well in advance of the time the resource levels indicated in Fig. **23-6** are needed.

SUPPLY AND DEMAND THROUGH 1990

One can appreciate the size of the uranium supply needed by examining the 18-year period ending in 1990. Forecast demand for Case A will require the production of 970,000 tons of U_3O_8 **Fig. 23-7.** An eight-year reserve in 1990 would be about 1.1 million tons. The total resource (reserves plus potential) in 1990, assuming the recent resource to production relationship, should be about 3.8 million tons of U_3O_8. Deliveries plus desirable resources would total nearly 4.8 million tons. The resources now estimated to be available are also shown in Fig **23-7.** The present estimate of $8 reserves is 273,000 tons of U_3O_8, and the additional potential is 450,000 tons, a total of 723,000 tons. With $15 uranium the total resource becomes 1.5 million tons. Therefore, during the period 1973-2000 resources must be expanded to a level 6.6 times present $8 resources or three times $15 resources to meet demand and provide an adequate base to support post-1990 requirements.

In the 1990 to 2000 period, requirements are projected at an additional 1.4 million tons. Thus, the demand for the 1973-2000 period is 2.4 million tons of U_3O_8 compared to the 0.7 million tons of known reserves plus potential at an $8 cutoff and 1.5 million tons at a $15 cutoff. Therefore, if production through 2000 is to come from domestic sources, the U.S. resource base needs considerable expansion even with the use of higher cost resources.

The magnitude of the future exploration and production effort may be judged in terms of the size of the largest known uranium district in the United States, and one of the largest in the world, the Ambrosia Lake district in New Mexico. Past production and estimated remaining $8 resources in this district total about 150,000 tons of U_3O_8 **(Table 23-2).** The 1990 demand is equivalent to producing in one year 80 percent of the entire resources of Ambrosia Lake. The cumulative demand through 1990 (18 years) will require mining the equivalent of 6.5 Ambrosia Lake districts or four times the total United States production of 244,000 tons during the past 23 years.

HIGHER COST RESOURCES

Low-grade material corresponding to a production cutoff cost of $15 or more per pound of U_3O_8 may well

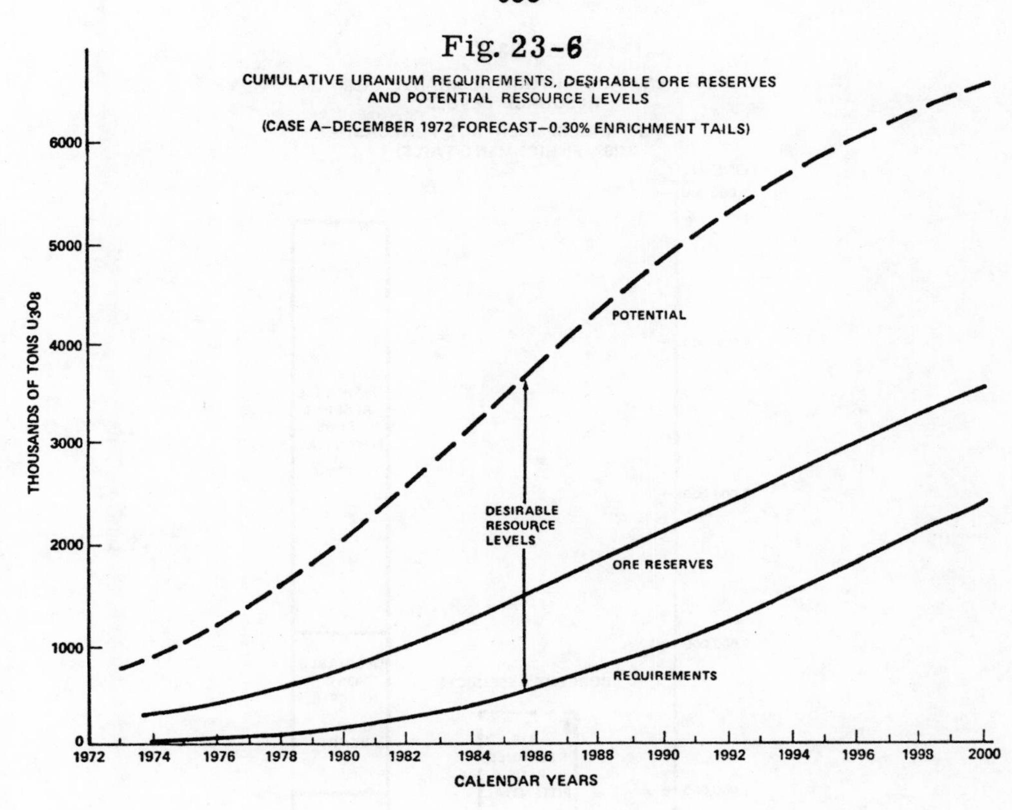

Table 23-2. URANIUM PRODUCTION TO DATE, RESERVES AND POTENTIAL COMPARED TO REQUIREMENTS

(Tons U_3O_8 in Ore)

	Production 1948-1972	Resources, January 1973 Cutoff Cost	Reserves	Potential	Production Plus Resources Total
Ambrosia Lake,	67,000	$ 8	54,000	30,000	151,000
New Mexico		$15	97,000	59,000	223,000
Total U.S.A.	244,000	$ 8	273,000	450,000	967,000
		$15	520,000	1,000,000	1,764,000

Requirements (December 1972 Forecast–0.30% Enrichment Tails)

1973-1990–970,000 tons U_3O_8 OR 6.5 Ambrosia Lakes – 4 Times Total Production to Date

1973-2000–2,400,000 tons U_3O_8 OR 16 Ambrosia Lakes – 10 Times Total Production to Date

have to be mined eventually to meet requirements. However, the effort to develop a production capability for low-grade ores will not begin until there is positive evidence of a market for higher cost uranium. A cutoff cost of $15 would reduce the average grade of ore mined to about half the grade mined at $8. Therefore, almost twice as much of the lower grade material would have to be mined and processed just to maintain the existing production level.

A major proportion of these higher cost resources is in the same deposits as the lower cost ore. So long as mining continues at the cutoff appropriate for lower costs, and the higher cost material is bypassed, much of it will become economically unavailable. Fig. 23-8 shows the erosion of $15 reserves as a result of mining $8 reserves. On the other hand, estimates of $15 uranium are probably conservative for lack of data because industry's effort has been directed to development of higher grade ore.

The additional cost to the consumer of using lower grade ore also would be considerable. For example, by 1990 each one-dollar increase in the price of U_3O_8

Fig. 23-7

CURRENT AND FUTURE DESIRABLE URANIUM RESOURCE LEVELS

(CASE A—DECEMBER 1972 FORECAST—0.30% ENRICHMENT TAILS)

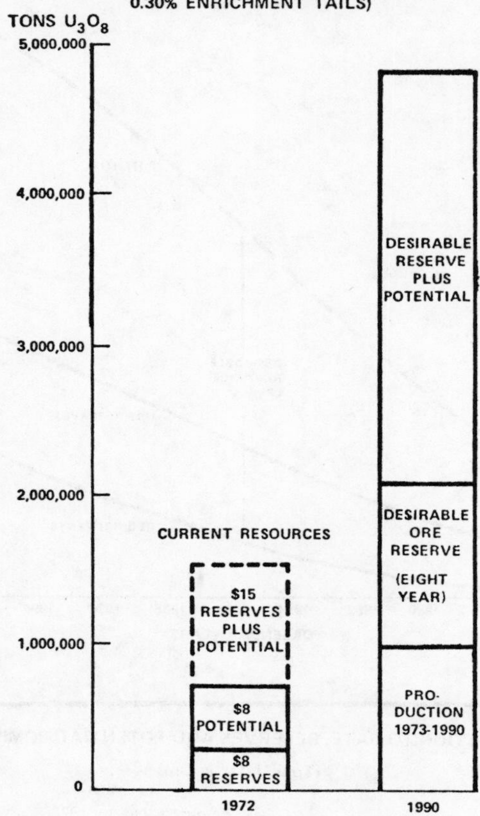

Fig. 23-8

DEPLETION OF $15 RESERVES AS $8 RESERVES ARE MINED

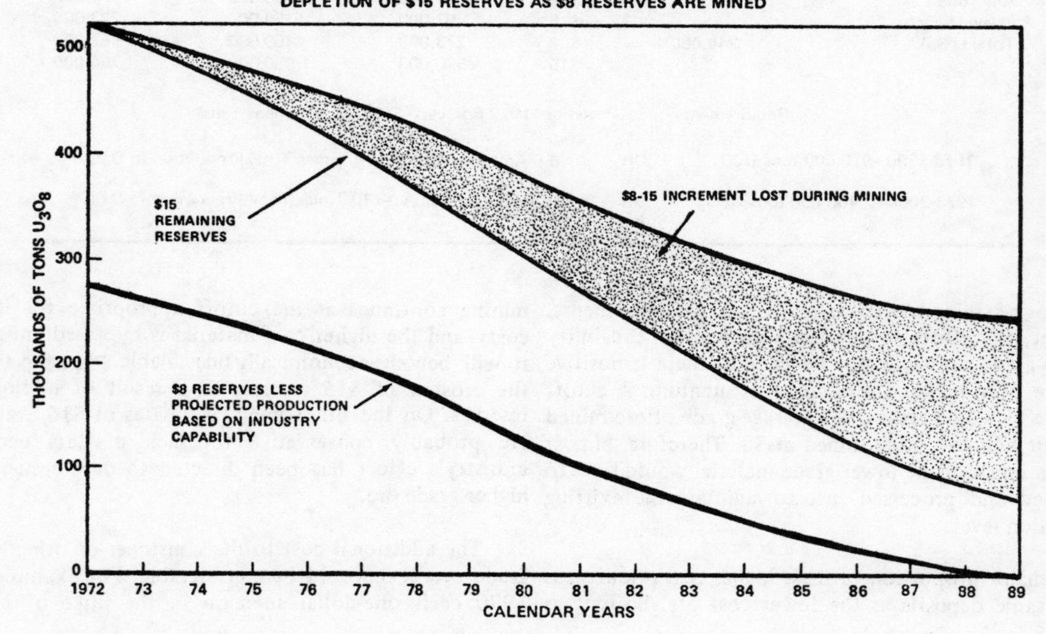

Fig. 23-9

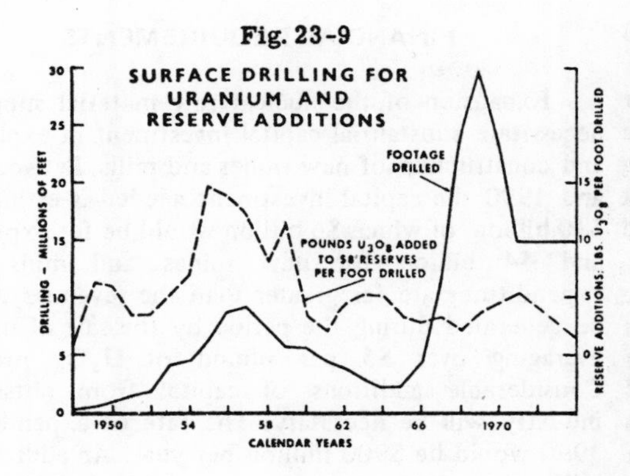

would cost utilities about an additional 230 million dollars per year.

EXPLORATION

As noted, AEC evaluation of U.S. uranium resources has been mostly concentrated in and around the established mining areas. A question remains as to the magnitude of resources which may exist in addition to those included in AEC estimates. Probably the most common view among those familiar with the uranium industry is that reserves in known districts can be expanded and some additional districts will be discovered. Uranium deposits are relatively small compared to those of many other metals and the fossil fuels, and most near-surface deposits presumably have been found. Thus, discovery of new deposits will become increasingly difficult and expensive. The resources that can be produced at a given cost are limited and cannot be expanded indefinitely. In any case, over the next 30 years many new uranium areas must be found and many new mines opened to achieve production at the projected demand level.

United States exploration history is illustrated in Fig. 23-9 Since the late 1940's more than 750,000 holes, totaling about 195 million feet, have been drilled in the search for uranium. In the 1950's much was learned about the geology of uranium. New districts were found, and discovery rates of $8 reserves greater than ten pounds of U_3O_8 per foot of drilling were obtained. As delineation of the known uranium districts progressed, discovery rates fell to less than five pounds per foot.

The projected increasing demand for uranium greatly stimulated exploration activity in the United States in the late 1960's, reaching a peak of about 30 million feet in 1969. Since that time, exploration has declined to about 15.5 million feet in 1971 and 1972, largely because of an oversupply and a soft market. Discovery rates continue to decrease in the late 1960's and early 1970's, reaching 1.8 pounds per foot in 1972. These figures relate to $8 reserves for which, as indicated, the data are best. There is inadequate information to judge the effect on the per-foot discovery rate if exploration were directed toward large, low-grade deposits such as those which might be exploited at $15 per pound.

Fig. 23-10

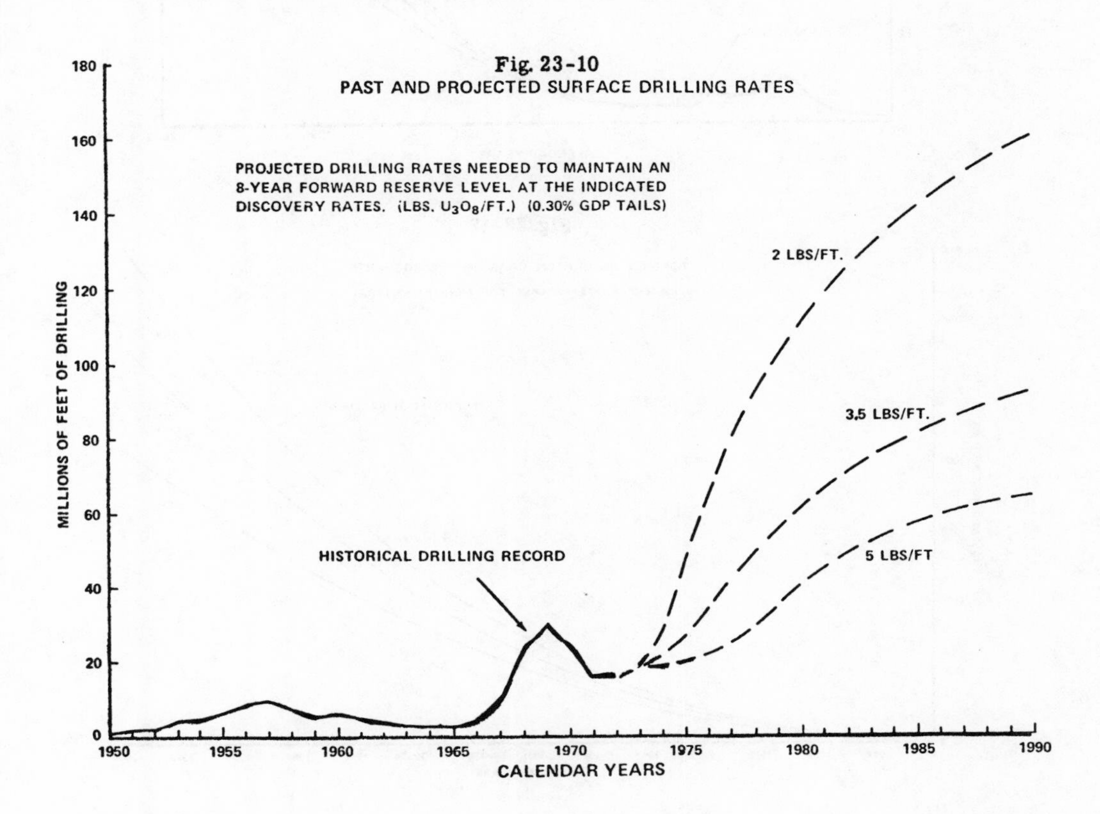

Attaining the desirable resource levels through 1990 will require additions to $8 reserves totaling 1.8 million tons of U_3O_8. If discovery rates through the 18-year period averaged 3.5 pounds per foot of drilling, the same as during the last five years, 1 billion feet of drilling would be required, an average of about 60 million feet annually. This compares to the 15.4 million feet drilled in 1972. While higher rates of exploration success are possible, recent experience has not been encouraging. Considering the increasing difficulty of finding deposits in new areas or at greater depth, discovery rates may in fact decrease in the future, requiring an even higher level of exploration effort to fill the need. Past drilling levels and needed drilling rates at recent and possible higher or lower discovery rates are shown in **Fig. 23-10** .

FINANCIAL REQUIREMENTS

Expansion of the nuclear raw material supply will necessitate substantial capital investment in exploration and construction of new mines and mills. Between 1973 and 1990 the capital investment needed is estimated at $10 billion, of which $6 billion would be for exploration and $4 billion for new mines and mills. These expenditures are far greater than the revenues that will be generated during the period by the sale of uranium, averaging over $5 per pound of U_3O_8 produced. Considerable additions of capital from outside the industry will be necessary. The rate of expenditure in 1990 would be $900 million per year. An additional $8 billion investment is required in the period 1991-2000.

Fig. 23-11

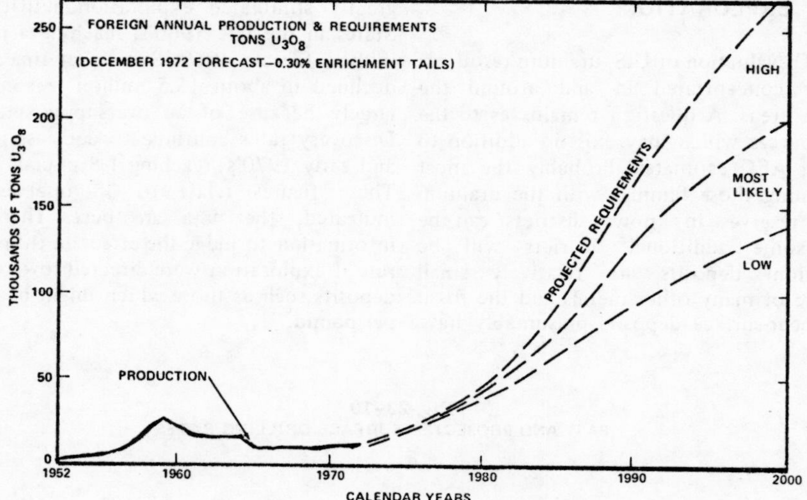

Fig. 23-12

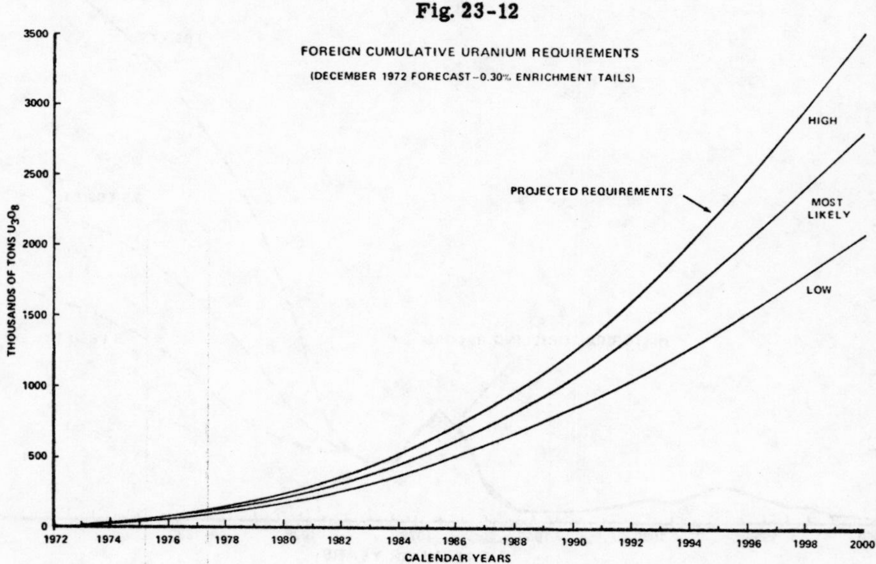

FOREIGN URANIUM SUPPLY

Foreign uranium represents a possible additional source of supply for the United States. Reserves at $10 per pound of U_3O_8 are estimated at about 800,000 tons of U_3O_8 and potential resources at 500,000 tons. Australia, Canada, South Africa and South West Africa have about 75 percent of the reserves; the remainder is primarily in central Africa and Europe. AEC forecasts that foreign requirements will grow more rapidly than in the United States (Fig. 23-11, 23-12). Foreign resources are larger than those in the U.S. but, despite an apparently better supply and demand position, there are limitations on the production rates attainable from foreign resources. South African uranium is a byproduct of gold mining, while Canadian and South West African resources are contained in a few deposits for which there are physical and economic limitations on production levels. Therefore, the foreign supply-demand situation is expected to be much like that projected for the United States with no excess of supply unless major new discoveries are made.

CHAPTER 24

REACTOR FUEL CYCLE COSTS

Today's fuel cycle industry is based largely on the use of slightly enriched uranium in light water reactors. A light water reactor fuel cycle is shown diagrammatically in Fig. 24-1. The main steps consist of (1) mining of uranium ore; (2) milling and refining to U_3O_8 "yellow cake"; (3) conversion to UF_6; (4) enrichment of ^{235}U content to about 2.5% isotopic concentration; (5) preparation of UO_2 fuel pellets; (6) fabrication into fuel assemblies; (7) production of useful energy in a nuclear reactor; (8) removal, cooling, and transportation to a reprocessing plant; (9) chemical reprocessing to recover uranium and plutonium as nitrate solutions free of waste and fission products; and (10) reconversion of uranyl nitrate to UF_6 for reenrichment and reuse. The recovered plutonium may be sold directly or used in plutonium-bearing fuel assemblies for insertion in light water reactors (plutonium recycle).

Typical fuel cycles that may be employed in the future are shown in Figs. 24-2 to 24-4. Fig. 24-2 shows a fuel cycle for a light water reactor operating with plutonium recycle. In the cycle shown, plutonium from the reactor is recycled in combination with plutonium makeup so that natural uranium oxide can be used in the fuel pellets. Another possibility is to recycle only the plutonium from the reactor, omit the makeup plutonium, and use slightly enriched uranium for a portion of the fuel.

Fast breeder reactors, including the liquid metal fast breeder reactor, may operate on a cycle similar to that shown in Fig. 24-3. In the case of the heterogeneous breeder reactors, the reactor contains both core–axial-blanket fuel elements and radial-blanket fuel elements. The blanket material is fabricated from

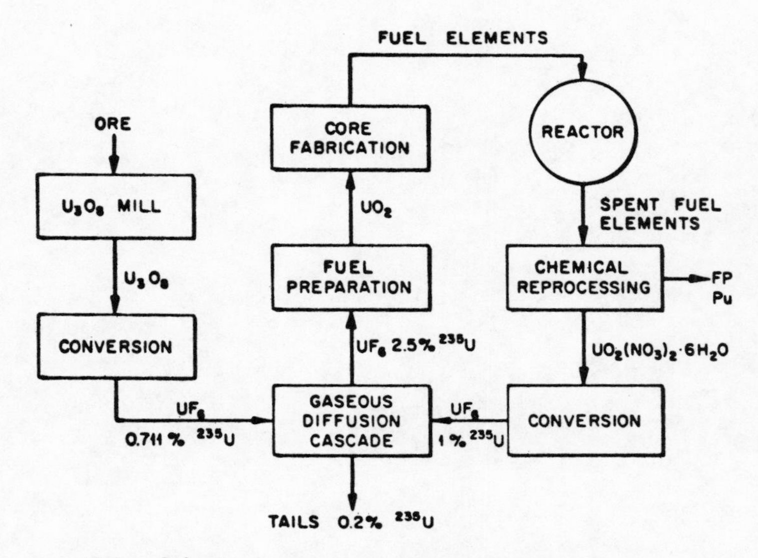

Fig. 24-1. Fuel Cycle for LWR on Uranium Cycle.

depleted UF_6, and the core materials are fabricated from recycled plutonium and depleted UF_6. Some excess plutonium is produced. This plutonium can be used to start up new fast breeder reactors, or it can be recycled in existing light-water reactors. In the chemical reprocessing operation, both radial-blanket elements and core–axial-blanket elements are reprocessed for recovery of plutonium.

A typical fuel cycle for the high-temperature gas-cooled reactor, operating at equilibrium on the thorium-^{233}U cycle, is shown in Fig. 24-4. This fuel cycle uses both ^{233}U, which is bred from the fertile thorium in the reactor, and ^{235}U makeup. Thorium may also be recovered in the chemical reprocessing operation for recycle to the reactor. Another distinguishing characteristic of the fuel cycle shown is the use of two different types of particles, a fissile particle and fertile particle, which by their geometry and properties enable the chemical reprocessor to keep the ^{235}U separate from the ^{233}U and thorium. This allows the spent ^{235}U, which contains relatively large amounts of ^{236}U, to be removed from the cycle if desired.

The Fuel Cycle Industry and Its Future Development

The present status of the fuel cycle industry has been described in several AEC reports[1-3] and in a study of the nuclear industry by Arthur D. Little, Inc.[4] For the purpose of price projection the information of most value would be the capacities of existing privately owned plants, their capital and operating costs, and the prices charged to customers. Generally, however, true capacity, cost, and price data are treated as proprietary information and are not released to the public.

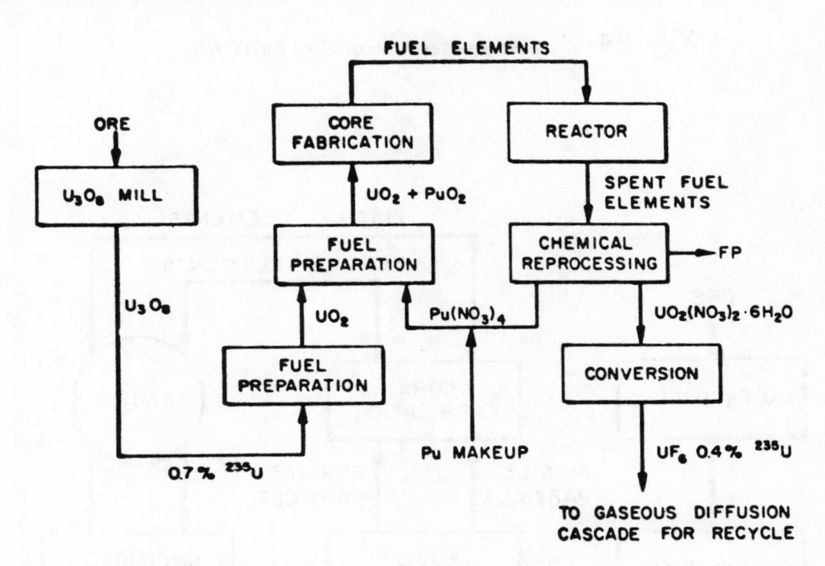

Fig. 24-2. Fuel Cycle for LWR with Pu Recycle.

In regard to the future development of the fuel cycle industry, the most striking single factor is the tremendous growth expected over the next 50 years. The exact pattern of growth is, of course, difficult to predict. It will be dictated partly by the impact of various new reactor technologies and their economics and partly by engineering and management policy decisions made in the industry itself. Because of the highly capital-intensive nature of the industry and the economies of scale inherent in many fuel cycle processes, there will be a considerable economic incentive for building larger plants. The risk of technological obsolescence must be taken into account when planning new facilities, however, and this will have a moderating influence on plant size. The possibility of excessive overcapacity in the industry must also be considered, as has already become apparent in connection with the expansion of fuel reprocessing

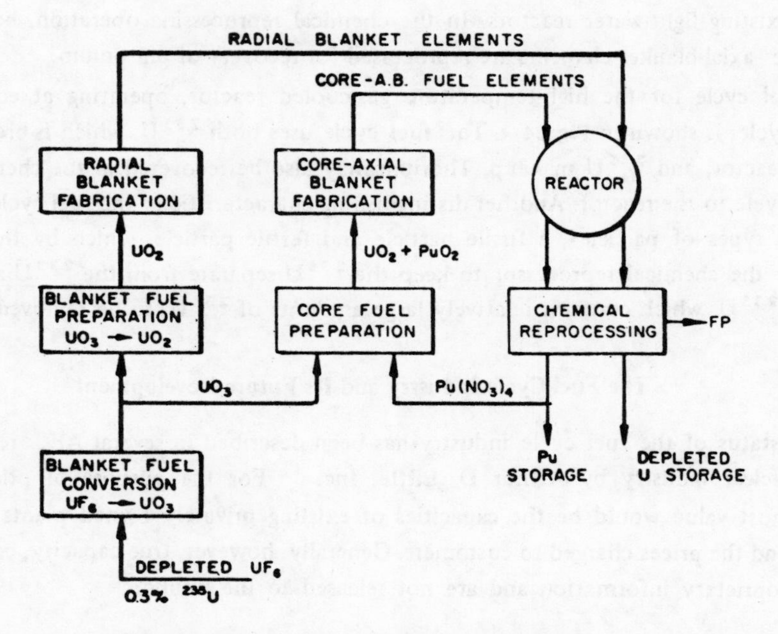

Fig. 24-3. **Fast Breeder Reactor U-Pu Fuel Cycle.**

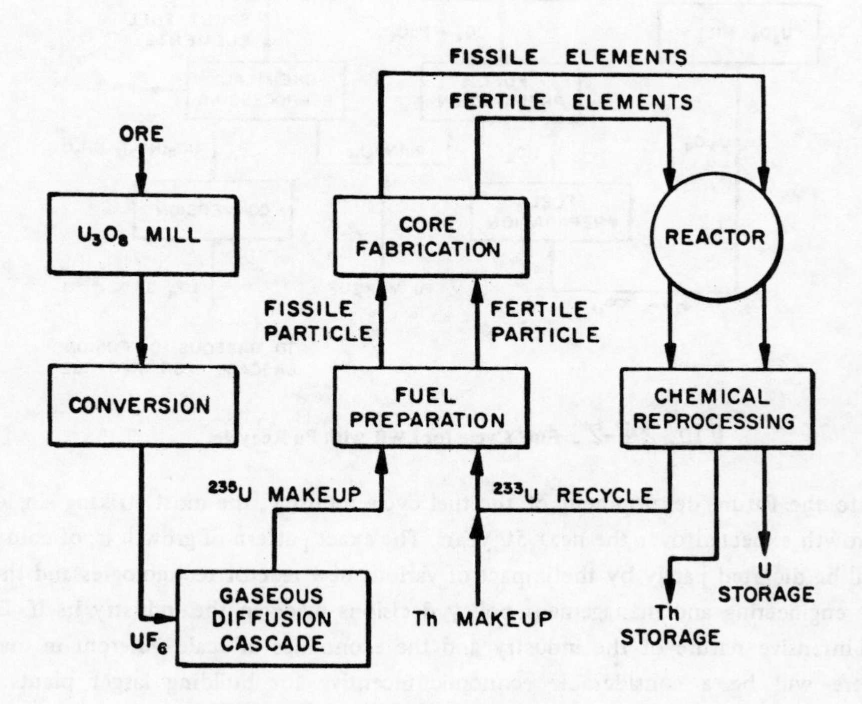

Fig. 24-4. **HTGR Th-^{233}U Fuel Cycle.**

facilities. Such factors will be important considerations in making new investment decisions.

Regardless of the exact path of expansion, however, it is clear that over the next 50 years the total demand for fuel cycle services will expand enormously. Table 24-1 shows the total projected annual demand for LWR and LMFBR fuel fabrication based on SATF Case No. 40, which included light-water reactors (both U and Pu recycle), the reference oxide LMFBR, and the reference carbide LMFBR. A list of the SATF case descriptions is given on p. 123. For comparison, the Arthur D. Little forecast[4] made in 1968 for LWR fuel is also shown. Total fuel throughput projected in Case No. 40 grows from 1020 metric tons of U per year in 1970 to about 90,000 metric tons/year in 2020.

Table 24-1. Estimated Fuel Fabrication Demand
Metric tons of U per year

Year	FRTF Results, Case No. 40		Arthur D. Little 1968 Estimate (Ref. 4)
	LWR	LMFBR	LWR
1970	1020		1100
1975	3150		3600
1980	6300	200	7100
1990	10100	6200	
2000	12600	18200	
2020	17300	62000	

Table 24-2. Estimated Fuel Reprocessing Demand
Metric tons of U per year

Year	FRTF Results, Case No. 40, 1967	Arthur D. Little (Ref. 4)	AEC 1967[a]	NFS 1968 (Ref. 1)	AEC 1969 (Ref. 1)
1970	155	80	110	109	55
1975	1330	943	1100	1288	1125
1980	3950	3088	3600	3551	2950
1990	17600[b]				
2000	38700[b]				
2020	121100[b]				

[a]AEC report, *Forecast of Growth of Nuclear Power, December 1967.*
[b]These are in terms of equivalent metric tons per year, 1 ton of LMFBR fuel being counted as 2 equivalent tons of LWR fuel. See Table E.7.

Table 24-2 shows the estimated total fuel reprocessing demands obtained in SATF Case No. 40 and compares these with the published forecasts available from various other sources. The total load shown for Case No. 40 includes both LWR and LMFBR fuel and is stated in equivalent metric tons of LWR fuel per year, 1 ton of LMFBR fuel being considered as 2 equivalent tons of LWR fuel.

The growth of fuel reprocessing demand is even more striking than that of fuel fabrication, increasing from 155 metric tons/year in 1970 to about 121,000 equivalent metric tons/year in 2020.

It is expected that this tremendous growth of industry throughput will result in appreciable reductions in the prices of fuel cycle services. There is, of course, considerable disagreement as to the exact amount of reduction that will take place. Fig. 24-5 shows the projections of price vs time made by the FRTF working group for typical processes. These examples are also for SATF Case No. 40; the LMFBR prices shown are for the reference carbide design. All prices are in terms of 1967 dollars.

Table 24-3 compares the projected prices of chemical reprocessing estimated by the FRTF working group with those estimated by various other forecasters. For further comparison, the prices estimated by Rodger and Reese[5] are of interest. Their estimates indicate that a 5-ton/day plant coming on stream in 1974 could charge a levelized price of $27.50 per kilogram of U for the period 1974–1988, with price expressed in terms of 1972 dollars. Assuming 5% per year escalation from 1967 to 1972, this corresponds to a levelized price of $21.60 per kilogram of U for the period 1974–1988, in terms of 1967 dollars. The FRTF prices for the same time period give a levelized average (weighted by annual industry throughput) of $20 per kilogram of U, in 1967 dollars.

In evaluating the price projections shown here, one must consider the potential effects of possible future regulations regarding siting, effluent control, waste disposal, and safeguards against the diversion of special nuclear materials. It is quite possible that such regulations may become more stringent and may therefore tend to increase fuel cycle costs. Such increases could result either directly, from the additional expense of meeting new requirements, or indirectly, through the need to limit plant sizes.

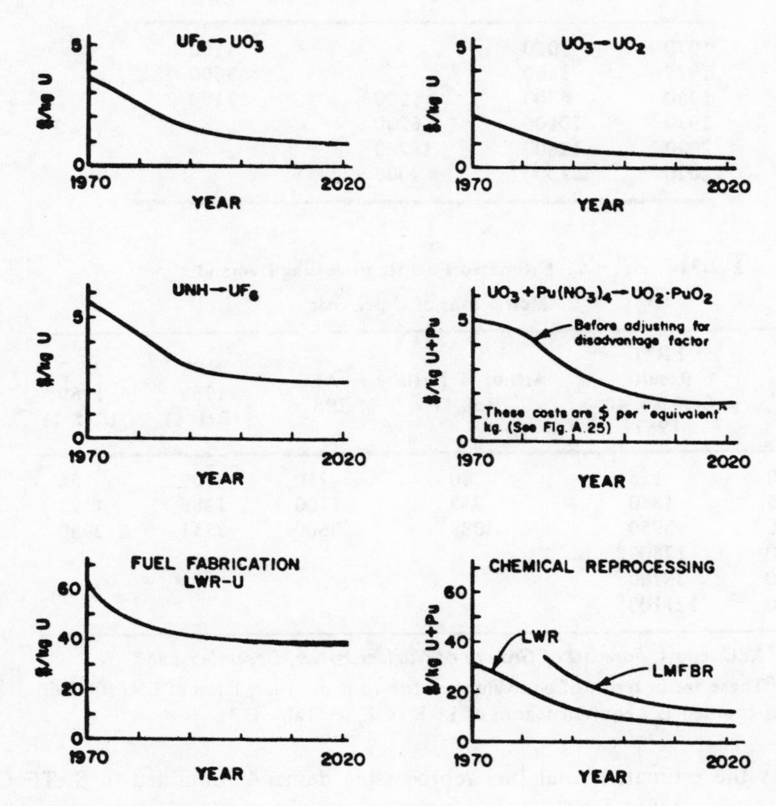

Fig. 24-5. **Estimated Prices of Fuel Cycle Services in Terms of 1967 Dollars, SATF Case No. 40. Chemical reprocessing price includes waste disposal.**

The effect of possible plant size limits on the price of chemical reprocessing was investigated. Fig. 24-6 shows the prices that result from various assumptions as to the maximum allowable plant size. Curves *A*, *B*, and *C* had size limits of 2600, 6500, and 10,600 metric tons per year respectively. The resulting prices of reprocessing in year 2000 were $14, $12, and $11, respectively, including waste disposal. The results shown are based on SATF Case No. 55, which included advanced converters and gas-cooled fast breeders, in addition to light water reactors and LMFBR's. Curve *B* represents the prices used as input to the SATF calculations.

The SATF study covered many cases and many types of reactors, and fuel throughputs varied appreciably from case to case. A full discussion of potential nuclear power growth patterns and associated fuel cycle costs is presented in WASH-1098. The choice of Case Nos. 40 and 55 for illustrating typical results here should not be interpreted as an implication that these were the best cases studied.

In the sections that follow, the methods used by the FRTF working group to make projections such as those summarized here are explained.

Table 24-3. **Estimated Chemical Reprocessing Prices for Light Water Reactors[a]**
Dollars per kilogram of U

Year	FRTF Results, Case No. 40, 1967 (Table E.7)	Arthur D. Little, Dec. 1968 (Ref. 4)	LWR Task Force, March 1968 (Ref. 3)		Bonanni *et al.*, 1968[b]
			Case No. 1 (LWR Only)	Case No. 2 (LWR-LMFBR)	
1970	30.87	31.30	31.50	31.50	31.50
1975	26.80		27.44	27.44	30.30
1980	22.01	14–24[c]	18.79	18.79	23.30
1985	16.66		14.76	14.76	
1990	13.09		12.31	13.16	
1995	11.55		9.95	13.16	
2000	9.65		9.83	13.16	
2010	7.65				
2020	6.96				

[a]Cost of waste disposal is included.

[b]M. Bonanni *et al.*, "Nuclear Fuel Cycle Cost Trends Up to 1980," in *Economics of Nuclear Fuels*, IAEA, Vienna, 1968.

[c]Reference 4 shows estimated prices of $14/kg for three 5-ton/day plants with a 22% fixed charge rate and $24/kg for five 3-ton/day plants with a 30% fixed charge rate.

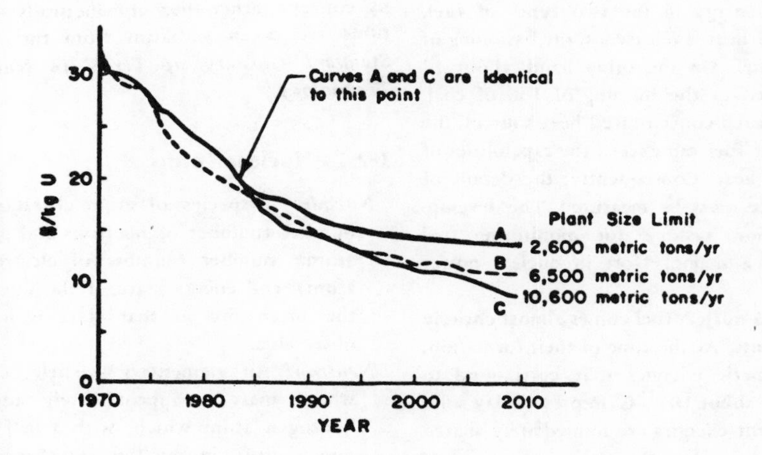

Fig. 24-6. Estimated Price of Chemical Reprocessing of LWR Fuel Using Various Plant Size Limits. Prices are in terms of 1967 dollars and include waste disposal charges. SATF case No. 55.

CHAPTER 25

NUCLEAR REACTOR INSTRUMENTATION

SECTION 1
FUNDAMENTAL CONSIDERATIONS

1-1 NUCLEAR AND FOSSIL FUELS

Many instrumentation systems are required to monitor, control, and regulate the nuclear and chemical processes of an operating nuclear power plant. Although some of the systems are identical to those used in fossil-fueled power plants, many are entirely different.

Perhaps the most important reason for differences between the instrumentation systems of conventional (fossil-fueled) and nuclear power plants is that conventional plants operate with continuous fuel feed and nuclear plants operate with stored feed. In a conventional plant fuel is fed continuously to the combustion chamber; in present-day nuclear plants all the nuclear fuel necessary for many months of operation is in the reactor all that time. Because of the large fuel inventory, an increasing nuclear reaction rate caused by equipment failure or malfunction will not be stopped by exhaustion of fuel. Perhaps in future nuclear plants (e.g., those using fluid fuels), the inventory of fuel in the reacting region can be reduced and the associated hazard thereby diminished. In the meantime nuclear power plant instrumentation must be dependable to prevent damage.

Another reason for instrumentation differences is the concentration of heat energy in the two types of fuel. About 40 billion Btu of heat is released in the fissioning of 1 lb of pure nuclear fuel. On the other hand, about 14 thousand Btu is released in the burning of 1 lb of coal. Since nuclear fuels are such concentrated heat sources, the heat flux from a nuclear fuel can exceed the capabilities of the coolant to remove heat. Consequently, the details of nuclear-fuel performance must be measured. The development of instrumentation systems for monitoring fuel performance has been a major effort in nuclear power technology.

The heat energy of a nuclear fuel comes almost entirely from the fission fragments. At the time of their formation, the fragments have kinetic energies that correspond to particle temperatures of about 10^{12}°C. In present-day solid fuels, the fission-fragment energies are immediately shared with the surrounding approximately 10^{22} atoms/cm^3 of the fuel (usually uranium metal or oxide or carbide), and the average temperature of the nuclear fuel is many orders of magnitude less than the initial fission-fragment temperature. In fact, the rate of coolant flow is regulated to keep the interior temperature of the nuclear fuel in the same general range (a few thousand degrees centigrade) as typical fossil-fuel temperatures. Should coolant flow be impeded, however, the nuclear fuel temperature could become much higher. One of the primary purposes of nuclear power reactor instrumentation is to prevent this.

Another difference between nuclear and conventional power plants that affects instrumentation is the presence of strong nuclear radiation fields in certain regions of a nuclear reactor. The interaction of these fields with sensors and with electrical components can cause a deterioration in signal and system performance. Because of this, the materials and techniques used in nuclear power plant instrumentation often differ from those used in fossil-fueled power plants.

1-2 DEFINITIONS

Nuclear power plants are based on concepts that require careful definition. In subsequent chapters many definitions of specialized terms are presented whenever relevant to the discussion. In this section definitions of a number of fundamental terms basic to an understanding of nuclear power plant operation are given. The definitions are listed by concept rather than alphabetically. Most of the definitions are taken verbatim from the *American National Standard Glossary of Terms in Nuclear Science and Technology.*[1]

1-2.1 Nuclear Terms

Nuclide. A species of atom characterized by its mass number (number of neutrons and protons in nucleus), atomic number (number of electrons in the neutral atom), and energy state of the nucleus, provided that the mean life in that state is long enough to be observable.

Neutron. An elementary particle, electrically neutral, whose mass is approximately equal to that of a hydrogen atom which, with a half-life of about 11.7 min, decays, in the free state, into a proton and an electron.

Thermal neutrons. Neutrons essentially in thermal equilibrium with the medium in which they exist.

Fast neutrons. Neutrons of kinetic energy greater than some specified value. In reactor physics the value is frequently chosen to be 0.1 MeV.

Beta particle. An electron, of either positive charge (β^+) or negative charge (β^-), which has been emitted by an atomic nucleus or neutron in the process of a transformation.

Radioactive decay. A spontaneous nuclear transformation in which the nucleus emits particles or gamma radiation, or undergoes spontaneous fission, or in which the atom emits x-radiation or Auger electrons following orbital electron capture or internal conversion.

Decay constant (or disintegration constant). For a radioactive nuclide (radionuclide), the probability per unit time for the spontaneous radioactive decay of a nucleus. It is given by

$$\lambda = -\frac{1}{N}\frac{dN}{dt}$$

in which N is the number of nuclei of concern existing at time t.

Curie. The special unit of activity (nuclear disintegration rate). One Curie equals 3.7×10^{10} disintegrations per second, exactly. "Curie" is abbreviated as Ci.

Half-life (radioactive half-life). For a single radioactive decay process, the time required for the activity (dN/dt or λN) to decrease to half its value by that process. The half-life is related to the decay constant: $T_{1/2} = (\log_e 2)/\lambda = 0.69315/\lambda$.

Cross section. A measure of the probability of a specified interaction between an incident radiation and a target particle or system of particles. It is the reaction rate per target particle for a specified process divided by the particle-flux density of the incident radiation *(microscopic cross section).* In reactor physics the term is sometimes applied to a specified group of target particles, e.g., those per unit volume *(macroscopic cross section),* or per unit mass, or those in a specified body. [*Note:* Unless otherwise qualified the term "cross section" means "microscopic cross section."]

Macroscopic cross section. The cross section per unit volume of a given material for a specified process. It has the dimension of reciprocal length. For a pure nuclide, it is the product of the microscopic cross section and the number of target nuclei per unit volume; for a mixture of nuclides, it is the sum of such products.

Microscopic cross section. The cross section per target nucleus, atom, or molecule. It has the dimension of area and may be visualized as the area normal to the direction of an incident particle which has to be attributed to the target particle to account geometrically for the interaction with the incident particle. Microscopic cross sections are often expressed in *barns,* where 1 barn = 10^{-24} cm^2.

Particle-flux density. At a given point in space, the number of particles or photons incident per unit time on a small sphere centered at that point divided by the cross-sectional area of that sphere. It is identical with the product of the particle density and the average particle speed. The term is commonly called *flux.*

Neutron-flux density. Particle-flux density for neutrons. Also commonly called *neutron flux.* Often denoted by nv or ϕ.

Particle fluence. At a given point in space, the number of particles or photons incident during a given time interval on a small sphere centered at that point divided by the cross-sectional area of that sphere. It is identical with the time integral of the particle-flux density. Often denoted by nvt.

Particle density. At a given point in space, the number of particles or photons per unit volume in a small sphere centered at that point.

Ionizing radiation. Any electromagnetic or particulate radiation capable of producing ions, directly or indirectly, by interaction with matter.

Indirectly ionizing particles. Uncharged particles or photons which can liberate directly ionizing particles or can initiate a nuclear transformation.

Directly ionizing particles. Charged particles having sufficient kinetic energy to produce ionization by collision.

Exposure. A measure of the ionization produced in air by x or gamma radiation. It is the sum of the electrical charges on all of the ions of one sign produced in air when all electrons liberated by photons in a volume element of air are completely stopped in the air, divided by the mass of the air in the volume element. The special unit of exposure is the roentgen.

Roentgen. The special unit of exposure. One roentgen = 1 R = 2.58×10^{-4} coulomb per kilogram of air.

Dose. A general term denoting the quantity of radiation or energy absorbed in a specified mass. For special purposes, its meaning should be appropriately stated, e.g., absorbed dose.

Absorbed dose. The energy imparted to matter in a volume element by ionizing radiation divided by the mass of irradiated material in that volume element. The special unit of absorbed dose is the rad. (Absorbed dose is often called dose.)

Rad. The special unit of absorbed dose. One rad equals 100 ergs/gram.

Dose equivalent (radiation protection). The product of absorbed dose, quality factor, dose distribution factor, and other modifying factors necessary to express on a common scale, for all ionizing radiations, the irradiation incurred by exposed persons. The special unit of dose equivalent is the rem.

Rem. The dose equivalent in rems is numerically equal to the absorbed dose in rads multiplied by the quality factor, the distribution factor, and any other necessary modifying factors.

Quality factor (radiation protection). A linear-energy-transfer-dependent factor by which absorbed doses are to be multiplied to obtain the dose equivalent. (*Note:* The term "RBE" should be used only in the field of radiobiology.)

Linear energy transfer (LET). The average energy locally

imparted to a medium by a charged particle of specified energy per unit distance traversed. [*Notes*: (1) The term "locally imparted" may refer either to a maximum distance from the track or to a maximum value of discrete energy loss by the particle beyond which losses are no longer considered as local. In either case, the limits chosen should be specified. (2) The concept of LET is different from that of stopping power. The former refers to energy imparted within a limited volume, the latter to loss of energy from the particle regardless of where this energy is absorbed.]

Dose distribution factor (radiation protection). A factor used in computing dose equivalent to account for the nonuniform distribution of internally deposited radionuclides.

Maximum permissible dose equivalent (MPD) (radiation protection). The largest dose equivalent received within a specified period which is permitted by a regulatory agency or other authoritative group on the assumption that receipt of such dose equivalent creates no appreciable somatic or genetic injury. Different levels of MPD may be set for different groups within a population. (By popular usage, "maximum permissible dose" is an accepted synonym.)

Kerma (kinetic energy released in material). The ratio of the sum of the initial kinetic energies of all the charged particles liberated by indirectly ionizing particles in a volume element to the mass of the matter in the volume element.

1-2.2 Fission-Process Terms

Nuclear fission. The division of a heavy nucleus into two (or, rarely, more) parts with masses of equal order of magnitude; usually accompanied by the emission of neutrons, gamma rays, and, rarely, small charged nuclear fragments.

Fast fission. Fission caused by fast neutrons.

Thermal fission. Fission caused by thermal neutrons.

Spontaneous fission. Fission which occurs without the addition of particles or energy to the nucleus.

Fissionable. Of a nuclide, capable of undergoing fission by any process.

Fissile. Of a nuclide, capable of undergoing fission by interaction with slow neutrons. (In reactor physics, slow neutrons are frequently defined as those of kinetic energy less than 1 eV.)

Fertile. Of a nuclide, capable of being transformed, directly or indirectly, into a fissile nuclide by neutron capture. Of a material, containing one or more fertile nuclides.

Fission fragments. The nuclei resulting from fission and possessing kinetic energy acquired from that fission.

Fission products. The nuclides produced either by fission or by the subsequent radioactive decay of the nuclides thus formed.

Fission yield. The fraction of fissions leading to fission products of a given type.

Prompt gamma radiation. Gamma radiation accompanying the fission process without measurable delay.

Prompt neutrons. Neutrons accompanying the fission process without measurable delay.

Delayed neutrons. Neutrons emitted by nuclei in excited states which have been formed in the process of beta decay. (The neutron emission itself is prompt, so that the observed delay is that of the preceding beta emission or emissions.)

1-2.3 Nuclear-Reactor Terms

Nuclear chain reaction. A series of nuclear reactions in which one of the agents necessary to the series is itself produced by the reactions so as to cause similar reactions. Depending on whether the number of reactions so caused directly by one reaction is on the average less than, equal to, or greater than unity, the reaction is convergent *(subcritical)*, self-sustained *(critical)*, or divergent *(supercritical)*.

Nuclear reactor. A device in which a self-sustaining nuclear fission chain reaction can be maintained and controlled (fission reactor). The term is commonly called "reactor" or "pile."

Fast reactor. A reactor in which fission is induced predominantly by fast neutrons. (Also called fast-neutron reactor.)

Thermal reactor. A reactor in which fission is induced predominantly by thermal neutrons.

Multiplication factor. The ratio of the total number of neutrons produced during a time interval (excluding neutrons produced by sources whose strengths are not a function of fission rate) to the total number of neutrons lost by absorption and leakage during the same interval. When the quantity is evaluated for an infinite medium or for an infinitely repeating lattice, it is referred to as the *infinite multiplication factor* (k_∞). When the quantity is evaluated for a finite medium, it is referred to as the *effective multiplication factor* (k_{eff}) (The term is also called *multiplication constant*.).

Critical. Fulfilling the condition that a medium capable of sustaining a nuclear chain reaction has an effective multiplication factor equal to unity. (A nuclear reactor is critical when the rate of neutron production, excluding neutron sources whose strengths are not a function of fission rate, is equal to the rate of neutron loss.)

Delayed critical. Identical with critical; the term is used to emphasize that the delayed neutrons are necessary to achieve the critical state.

Prompt critical. Fulfilling the condition that a nuclear chain-reacting medium is critical utilizing prompt neutrons only.

Prompt-neutron fraction. The ratio of the mean number of prompt neutrons per fission to the mean total number of neutrons (prompt plus delayed) per fission.

Delayed-neutron fraction. The ratio of the mean number of delayed neutrons per fission to the mean total number of neutrons (prompt plus delayed) per fission.

Effective delayed-neutron fraction. The ratio of the mean number of fissions caused by delayed neutrons to the mean total number of fissions caused by delayed plus prompt neutrons. (*Note:* The effective delayed-neutron fraction is generally larger than the actual delayed-neutron fraction.)

Reactivity. A parameter, ρ, giving the deviation from criticality of a nuclear chain-reacting medium such that positive values correspond to a supercritical state and negative values to a subcritical state. Quantitatively, $\rho = 1 - (1/k_{eff})$, where k_{eff} is the effective multiplication factor.

Excess reactivity. The maximum reactivity attainable at any time by adjustment of the control members.

Built-in reactivity. The reactivity of a system as a function of design excluding the experimental and control inserts of the system.

Reactor control. The intentional variation of the reaction rate in a reactor or adjustment of reactivity to maintain a desired state of operation.

Reactivity coefficient. The change in reactivity caused by inserting a small amount of a substance in a reactor. The reactivity coefficient of a substance may depend upon the amount and distribution of the substance inserted, but is usually quoted as the reactivity change per unit mass of the substance at specific positions in the reactor or as a uniform distribution.

Void coefficient. The partial derivative of reactivity with respect to a void (i.e., the removal of the material) at a specified location within a reactor. It is equal to the reactivity coefficient of the material removed.

Isothermal temperature coefficient of reactivity. The change of reactivity caused by a one-degree increase in the uniform temperature of a reactor at zero power.

Power coefficient of reactivity. The change of reactivity per unit change of reactor thermal power when other variables are not independently changed.

1-3 NUCLEAR-REACTOR KINETICS

The design of instrumentation systems for a nuclear power plant must take into account the specific properties of the reactor for that plant. Of particular importance is the kinetic behavior of the reactor. Many textbooks and monographs have been written on nuclear reactor kinetics; the reader is referred, for example, to Refs. 2 through 6 for details. The following paragraphs summarize basic material particularly relevant to instrumentation systems in nuclear power reactors.

1-3.1 Point Kinetics Without Delayed Neutrons

The symbol n (neutrons/cm^3) is used to designate the neutron density at a given position in a nuclear fission chain reactor. If the reactor is just critical, the effective multiplication factor, k, is exactly 1 and the neutron density, n, is constant. If the effective multiplication factor is increased by $\delta k = k - 1$ (with $\delta k > 0$), then n increases with time.

The rate of increase of n, dn/dt, is the number of extra neutrons in the next generation, $n\,\delta k$, divided by the time between generations l:

$$\frac{dn}{dt} = \frac{n\,\delta k}{l} = \frac{n(k-1)}{l} \tag{1.1}$$

Integrated, this is

$$n = n_0 e^{(\delta k/l)t} \tag{1.2}$$

where n_0 is the neutron density at $t = 0$. The reciprocal of the first factor, $\delta k/l$, in the exponential has the dimensions of time and is known as the reactor period:

$$n = n_0 e^{t/T} \qquad T = l/\delta k \tag{1.3}$$

These equations have been developed on the assumption that there is a single characteristic time between generations in a nuclear fission chain reaction. This is the same as assuming that only prompt neutrons participate in the chain reaction.

1-3.2 Point Kinetics with Delayed Neutrons

Some of the neutrons participating in the chain reaction are emitted at various times after the fission event. When certain fission-product nuclides decay by emitting beta particles, the resultant nuclides are unstable, and each of the nuclides emits a neutron immediately after the beta decay. The rate of neutron emission therefore is the same as the rate of beta decay of these "precursor" nuclides, i.e., a rate that decreases exponentially with time.

Tables 1.1 and 1.2 list the half-lives and decay constants [$\lambda = (\ln 2)/T_{1/2} = 0.693/T_{1/2}$] of the delayed-neutron emitters resulting from the fissioning of ^{233}U, ^{235}U, and ^{239}Pu by thermal neutrons and by fast neutrons, respectively. The tables also list the absolute yields of delayed neutrons (number of delayed neutrons per fission emitted by each precursor type) and the relative abundances of the delayed neutrons (number of delayed neutrons emitted by each precursor type divided by the total number of delayed neutrons emitted in a fission process).

To introduce the effect of delayed neutrons on the nuclear chain reaction, we consider the effective multiplication factor to be the sum of two terms:

$$\begin{aligned} k &= (\text{multiplication factor for prompt neutrons}) \\ &\quad + (\text{multiplication factor for delayed neutrons}) \\ &= k(1-\beta) + k\beta \end{aligned} \tag{1.4}$$

where β is the delayed-neutron fraction, or the number of delayed neutrons per fission divided by the total prompt and delayed neutrons per fission. The delayed-neutron

Table 1.1 — Delayed-Neutron Half-Lives and Yields in Thermal-Neutron Fission[7]

Isotope	Delayed neutrons/fission	Group index (i)	Half-life ($T_{1/2}$), sec	Decay constant* (λ), sec^{-1}	Relative abundance (a)	Absolute group yield, %
^{233}U	0.0066 ± 0.0003	1	55.00 ± 0.54	0.0126 ± 0.0002	0.086 ± 0.003	0.057 ± 0.003
		2	20.57 ± 0.38	0.0337 ± 0.0006	0.299 ± 0.004	0.197 ± 0.009
		3	5.00 ± 0.21	0.139 ± 0.006	0.252 ± 0.040	0.166 ± 0.027
		4	2.13 ± 0.20	0.325 ± 0.030	0.278 ± 0.020	0.184 ± 0.016
		5	0.615 ± 0.242	1.13 ± 0.40	0.051 ± 0.024	0.034 ± 0.016
		6	0.277 ± 0.047	2.50 ± 0.42	0.034 ± 0.014	0.022 ± 0.009
^{235}U	0.0158 ± 0.0005	1	55.72 ± 1.28	0.0124 ± 0.0003	0.033 ± 0.003	0.052 ± 0.005
		2	22.72 ± 0.71	0.0305 ± 0.0010	0.219 ± 0.009	0.346 ± 0.018
		3	6.22 ± 0.23	0.111 ± 0.004	0.196 ± 0.022	0.310 ± 0.036
		4	2.30 ± 0.09	0.301 ± 0.012	0.395 ± 0.011	0.624 ± 0.026
		5	0.61 ± 0.083	1.13 ± 0.15	0.115 ± 0.009	0.182 ± 0.015
		6	0.23 ± 0.025	3.00 ± 0.33	0.042 ± 0.008	0.066 ± 0.008
^{239}Pu	0.0061 ± 0.0003	1	54.28 ± 2.34	0.0128 ± 0.0005	0.035 ± 0.009	0.021 ± 0.006
		2	23.04 ± 1.67	0.0301 ± 0.0022	0.298 ± 0.035	0.182 ± 0.023
		3	5.60 ± 0.40	0.124 ± 0.009	0.211 ± 0.048	0.129 ± 0.030
		4	2.13 ± 0.24	0.325 ± 0.036	0.326 ± 0.033	0.199 ± 0.022
		5	0.618 ± 0.213	1.12 ± 0.39	0.086 ± 0.029	0.052 ± 0.018
		6	0.257 ± 0.045	2.69 ± 0.47	0.044 ± 0.016	0.027 ± 0.010

*The decay constants are related to the half-lives by the equation $\lambda = (\ln 2)/T_{1/2} = 0.693/T_{1/2}$.

fraction can be expressed as the sum of the delayed-neutron fractions for each group of the delayed-neutron emitters:

$$\beta = \sum_{i=1}^{m} \beta_i \tag{1.5}$$

where m is the number of delayed-neutron groups. Table 1.3 lists the values of β_i for the fission of ^{233}U, ^{235}U, and ^{239}Pu by thermal and fast neutrons. The table also lists ν, the number of prompt neutrons per fission. Values of β_i are obtained by multiplying the relative abundance values in Tables 1.1 and 1.2 by the values of β in Table 1.3.

The average energy of the delayed neutrons is not the same as the average energy of the prompt neutrons. Thus, in any chain-reacting system, the effectiveness of the delayed neutrons in propagating the nuclear fission chain reaction differs from that of the prompt neutrons. The factor β used in Eq. 1.4 does not take this into account since β is a simple ratio of numbers of neutrons. To take the neutron energies into account, replace β with

$\gamma\beta$ = effective delayed neutron fraction

= (number of fissions caused by delayed neutrons)/(number of fissions caused by delayed plus prompt neutrons) (1.6)

where γ is the delayed-neutron effectiveness. The value of γ depends on the chain-reacting system and is generally slightly greater than 1. For the power reactors discussed in this book, it is a good approximation to assume $\gamma = 1$. Likewise, the delayed-neutron effectiveness for the individual delayed-neutron groups can be assumed to be 1, i.e., $\gamma = \gamma_i = 1$.

The basic kinetic equations for a nuclear fission chain reaction in which delayed neutrons are taken into account are obtained by writing the rate of change of the neutron density (n = neutrons/cm^3) as a sum of two terms:

$$\frac{dn}{dt} = \text{(rate of change of prompt-neutron density)} + \text{(rate of change of delayed neutron density)}$$

$$= \frac{n}{l}[k(1-\beta) - 1] + \sum_{i=1}^{m} \lambda_i C_i \tag{1.7}$$

where C_i is the density (number/cm^3) of delayed-neutron emitters of the *ith* group and λ_i is the decay constant (fraction decaying/sec) of the *ith* delayed-neutron emitter group. The number of groups, m, is 6 (see Tables 1.1 and 1.2). The first term of Eq. 1.7 is obtained by substituting the multiplication factor for prompt neutrons (Eq. 1.4) for k in Eq. 1.1.

The density of each of the delayed neutron-emitting groups, C_i, is obtained from the equation

$$\frac{dC_i}{dt} = \text{(rate of production of } ith \text{ group of delayed-neutron emitters)} - \text{(rate of decay of } ith \text{ group of delayed-neutron emitters)}$$

$$= \frac{k\beta_i n}{l} - \lambda_i C_i \tag{1.8}$$

The first term is the multiplication factor for delayed neutrons (Eq. 1.4) divided by the time interval between generations, or the prompt-neutron lifetime, *l*. If there is a source of neutrons present other than the fissionable isotopes and the fission products that emit neutrons, then a source term must be added to the right-hand side of Eq. 1.8.

Equations 1.7 and 1.8 are the basic neutron kinetics equations. They are important in the design of the instrumentation and control systems for nuclear power reactors. In Sec. 6 the equations are used to show how transfer-function measurements can yield useful information on power-reactor behavior. In Sec. 7 the equations are shown to be basic to the design of reactor control systems.

1-3.3 Reactivity

When a nuclear power-reactor plant is generating electrical energy at a steady or constant rate, the reactor is in a steady state in which the neutron density is fixed, the temperatures at various positions in the reactor are constant, etc. Equations 1.7 and 1.8 show that, since $dn/dt = dC_i/dt = 0$ in this steady state, the effective multiplication factor is just 1. If the effective multiplication factor increases above 1, n will increase with time. Similarly, when $k < 1$, n decreases with time.

As k is increased above 1, it reaches a value where the first term of Eq. 1.7 becomes zero; then, for higher values of k, the term becomes positive. When $k(1 - \beta) - 1$ is positive, the neutron density increases with time at a rate depending on the ratio of the prompt-neutron lifetime, l, to $k(1 - \beta) - 1$. Under these conditions the reactor is prompt critical, and, because l is so small ($l \leqslant 10^{-4}$ sec for a thermal reactor, $l \leqslant 10^{-7}$ sec for a fast reactor), n increases very rapidly with time for any appreciable positive value of $k(1 - \beta) - 1$. The value of the effective multiplication factor when the reactor is just prompt critical is $1/(1 - \beta)$. Since power reactors are always kept below prompt criticality, the practical range of the effective multiplication factor is between 1 and $1/(1 - \beta)$ when the reactor is operating and between 0 and 1 when the reactor is being started up or shut down.

In place of the effective multiplication factor it is more convenient to refer to the reactivity, or the fractional deviation of the effective multiplication factor from unity, $(k - 1)/k$,

$$\text{Reactivity} = \rho = \frac{k-1}{k} = \frac{\delta k}{k} \tag{1.9}$$

When the effective multiplication factor varies from 1 to $1/(1 - \beta)$, the reactivity varies from 0 to β. In other words, the reactivity increases by β as the reactor goes from delayed critical to prompt critical. It is convenient to designate this change of reactivity as "one dollar" = \$1 = unit of reactivity equal to the reactivity difference between the prompt critical ($k = 1/1 - \beta$) and the delayed critical ($k = 1$) conditions of a reactor. The dollar is further subdivided into 100 cents; a change in reactivity of 1¢, for example, is $\Delta\rho = 0.01\beta$. From the values of β in Table 1.3, it follows that a 1¢ reactivity increment is 0.000064 for a ^{235}U-fueled reactor.

Frequently the terms "excess k," "δk/k," and "reactivity" are used interchangeably. As Eq. 1.9 shows, the second and third terms are exactly equivalent. The use of "excess k" or "δk" as equivalent to reactivity is approximately correct, since $\rho = \delta k/k = \delta k/(1 + \delta k) = \delta k - (\delta k)^2$, etc., and $\delta k \ll 1$ in all practical cases.

1-3.4 The Inhour Equation

The basic kinetic equations, Eqs. 1.7 and 1.8, can be solved for constant k (e.g., following a step change in reactivity). The neutron density as a function of time is:

$$n = \sum_{j=1}^{m+1} A_j e^{\omega_j t} \tag{1.10}$$

where the values of A_j are determined by the initial values (at $t = 0$) n_0 and C_{i0}, and where the values of ω_j are the $m + 1$ roots of the equation:

$$\rho = \frac{l\omega}{k} + \sum_{i=1}^{m} \frac{\omega\beta_i}{\omega + \lambda_i} \tag{1.11}$$

The β_i and λ_i are the delayed neutron fractions and decay constants for the m groups of delayed-neutron emitters.

The roots of ω in Eq. 1.11 have the following properties: For ρ = constant > 0, m roots are negative and 1 is positive. The m negative roots are approximately $-\lambda_1$, $-\lambda_2, \ldots, -\lambda_m$, the decay constants of the delayed-neutron emitters. For ρ = constant < 0, all $m + 1$ roots are negative.

Thus, for constant positive values of the reactivity, the neutron density is the sum of one positive exponential and m negative exponentials. After an interval of time large compared to the delayed-neutron periods, the positive exponential remains

$$n = n_0 e^{\omega_0 t} = n_0 e^{t/T} \tag{1.12}$$

The quantity T ($= 1/\omega_0$) is the stable reactor period or asymptotic period, and $1/\omega_1$, $1/\omega_2$, ..., $1/\omega_m$ are the transient periods. Figure 1.1 shows the stable and transient periods vs. reactivity for ^{235}U with the prompt-neutron lifetime as a parameter. Note that for δk small and positive the stable period is independent of l (for $l \leqslant 10^{-3}$ sec); in fact, T is approximately $\bar{l}/\delta k$ where $\bar{l} = l + \tau_{av}$ and τ_{av} is the average decay period of the delayed-neutron emitters, $\tau_{av} = (1/\beta)\Sigma(\beta_i/\lambda_i)$. The quantity $\bar{l}$ is the effective neutron lifetime. Figure 1.1 also shows that for large δk the stable period is approximately $l/\delta k$.

The relation between the reactivity and the stable reactor period is obtained by substituting 1/T for ω_0 in Eq. 1.11,

$$\rho = \frac{l}{Tk} + \sum_{i=1}^{m} \frac{\beta_i}{1 + T\lambda_i} \tag{1.13}$$

This is the inhour equation. Reactivity can be expressed in "inverse hours" or "inhours," where 1 inhour is defined as the amount of reactivity that makes the stable reactor period equal to 1 hr. Substituting T = 3600 sec and the values of β_i and λ_i from Table 1.1 and noting that $l/T \leqslant 3 \times 10^{-6}$, we find the following:

Table 1.2—Delayed-Neutron Half-Lives and Yields in Fast Fission*[7]

Isotope	Delayed neutrons/ fission	Group index (i)	Half-life ($T_{1/2}$), sec	Decay constant (λ), sec⁻¹	Relative abundance (a)	Absolute group yield, %
^{233}U	0.0070 ± 0.0004	1	55.11 ± 1.86	0.0126 ± 0.0004	0.086 ± 0.003	0.06 ± 0.003
		2	20.74 ± 0.86	0.0334 ± 0.0014	0.274 ± 0.005	0.192 ± 0.009
		3	5.30 ± 0.19	0.131 ± 0.005	0.227 ± 0.035	0.159 ± 0.025
		4	2.29 ± 0.18	0.302 ± 0.024	0.317 ± 0.011	0.222 ± 0.012
		5	0.546 ± 0.108	1.27 ± 0.266	0.073 ± 0.014	0.051 ± 0.010
		6	0.221 ± 0.042	3.13 ± 0.675	0.023 ± 0.007	0.016 ± 0.005
^{235}U	0.0165 ± 0.0005	1	54.51 ± 0.94	0.0127 ± 0.0002	0.038 ± 0.003	0.063 ± 0.005
		2	21.84 ± 0.54	0.0317 ± 0.0008	0.213 ± 0.005	0.351 ± 0.011
		3	6.00 ± 0.17	0.115 ± 0.003	0.188 ± 0.016	0.310 ± 0.028
		4	2.23 ± 0.06	0.311 ± 0.008	0.407 ± 0.007	0.672 ± 0.023
		5	0.496 ± 0.029	1.40 ± 0.081	0.128 ± 0.008	0.211 ± 0.015
		6	0.179 ± 0.017	3.87 ± 0.369	0.026 ± 0.003	0.043 ± 0.005
^{238}U	0.0412 ± 0.0017	1	52.38 ± 1.29	0.0132 ± 0.0003	0.013 ± 0.001	0.054 ± 0.005
		2	21.58 ± 0.39	0.0321 ± 0.0006	0.137 ± 0.002	0.564 ± 0.025
		3	5.00 ± 0.19	0.139 ± 0.005	0.162 ± 0.020	0.667 ± 0.087
		4	1.93 ± 0.07	0.358 ± 0.014	0.388 ± 0.012	1.599 ± 0.081
		5	0.49 ± 0.023	1.41 ± 0.067	0.225 ± 0.013	0.927 ± 0.060
		6	0.172 ± 0.009	4.02 ± 0.214	0.075 ± 0.005	0.309 ± 0.024
^{239}Pu	0.0063 ± 0.0003	1	53.75 ± 0.95	0.0129 ± 0.0002	0.038 ± 0.003	0.024 ± 0.002
		2	22.29 ± 0.36	0.0311 ± 0.0005	0.280 ± 0.004	0.176 ± 0.009
		3	5.19 ± 0.12	0.134 ± 0.003	0.216 ± 0.018	0.136 ± 0.013
		4	2.09 ± 0.08	0.331 ± 0.012	0.328 ± 0.010	0.207 ± 0.012
		5	0.549 ± 0.049	1.26 ± 0.115	0.103 ± 0.009	0.065 ± 0.007
		6	0.216 ± 0.017	3.21 ± 0.255	0.035 ± 0.005	0.022 ± 0.003
^{240}Pu	0.0088 ± 0.0006	1	53.56 ± 1.21	0.0129 ± 0.0004	0.028 ± 0.003	0.022 ± 0.003
		2	22.14 ± 0.38	0.0313 ± 0.0005	0.273 ± 0.004	0.238 ± 0.016
		3	5.14 ± 0.42	0.135 ± 0.011	0.192 ± 0.053	0.162 ± 0.044
		4	2.08 ± 0.19	0.333 ± 0.031	0.350 ± 0.020	0.315 ± 0.027
		5	0.511 ± 0.077	1.36 ± 0.205	0.128 ± 0.018	0.119 ± 0.018
		6	0.172 ± 0.033	4.04 ± 0.782	0.029 ± 0.006	0.024 ± 0.005
^{232}Th	0.0496 ± 0.0020	1	56.03 ± 0.95	0.0124 ± 0.0002	0.034 ± 0.002	0.169 ± 0.012
		2	20.75 ± 0.66	0.0334 ± 0.0011	0.150 ± 0.005	0.744 ± 0.037
		3	5.74 ± 0.24	0.121 ± 0.005	0.155 ± 0.021	0.769 ± 0.108
		4	2.16 ± 0.08	0.321 ± 0.011	0.446 ± 0.015	2.212 ± 0.110
		5	0.571 ± 0.042	1.21 ± 0.090	0.172 ± 0.013	0.853 ± 0.073
		6	0.211 ± 0.019	3.29 ± 0.297	0.043 ± 0.006	0.213 ± 0.031

*"Fast fission" is defined as fission induced by a continuous neutron spectrum similar to a prompt fission neutron spectrum.

Table 1.3—Delayed-Neutron Fractions and Yields*†

Fission nuclide	Fast fission (E_{eff} ~ fission spectrum)			Thermal-neutron-induced fission		
	n/F	ν	β	n/F	ν	β
^{239}Pu	0.0063 ± 0.0003	3.08 ± 0.04	$0.0020_4 \pm 0.0001_1$	0.0061 ± 0.0003	$2.82_6 \pm 0.02_1$	$0.0021_6 \pm 0.0001_1$
^{233}U	0.0070 ± 0.0004	2.61 ± 0.03	$0.0026_8 \pm 0.0001_6$	0.0066 ± 0.0003	$2.46_9 \pm 0.02_0$	$0.0026_7 \pm 0.0001_2$
^{240}Pu	0.0088 ± 0.0006	3.3 ± 0.2	$0.0026_6 \pm 0.0002_4$			
^{241}Pu				0.0154 ± 0.0015	3.14 ± 0.06	0.0049 ± 0.0005
^{235}U	0.0165 ± 0.0005	2.59 ± 0.03	$0.0063_7 \pm 0.0002_2$	0.0158 ± 0.0005	$2.43_0 \pm 0.001$	$0.0065_0 \pm 0.0002_1$
^{238}U	0.0412 ± 0.001	2.80 ± 0.13	0.0147 ± 0.0009			
^{232}Th	0.0496 ± 0.0020	2.42 ± 0.20	0.0205 ± 0.0019			

*From H. C. Paxton and G. R. Keepin, *The Technology of Nuclear Reactor Safety,* Vol. 1, p. 267, The M.I.T. Press, Cambridge, Mass., 1964.

†Symbols: n/F = delayed neutrons per fission; ν = average total neutrons per fission; β = n/Fν = fraction of total neutrons that are delayed.

1 inhour ≅ 2.4 × 10^{-5} for a ^{235}U-fueled thermal reactor
≅ 1 × 10^{-5} for a ^{239}Pu-fueled thermal reactor
≅ 1.4 × 10^{-5} for a ^{233}U-fueled thermal reactor.

The inhour equation is shown graphically in Figs. 1.2, 1.3 and 1.4, where the reactivity is plotted against the stable period for various values of l and for various isotopes of uranium and plutonium.

1-3.5 Effects of Reactivity Insertions*

When a power reactor is operating in a steady state (constant coolant flow, constant temperatures, etc.), the effective multiplication factor is 1 and the reactivity is zero. If any of the basic parameters, such as coolant flow or temperature, are changed (e.g., to increase or decrease the power level or to compensate for changes in fuel reactivity), then reactivity must be added or subtracted. The most common situations are those in which the reactivity is inserted at a steady rate or as a step function.

Equations 1.7 and 1.8 can be solved for the case where the reactor is taken from delayed critical ($\rho = \delta k/k = 0$) to prompt critical ($\rho = \delta k/k = \beta$) by inserting reactivity at a constant rate (ramp insertion). Figure 1.5 shows how the relative neutron density, $n/n_0 = n(t)/n(0)$, increases with time for several reactivity insertion rates and for several values of the neutron lifetime. Table 1.4 presents similar data in tabular form.

In Fig. 1.6 the effect of inserting a step change in k is shown. The reactor is at delayed critical at t = 0.

1-3.6 Reactivity Changes

For curves, tables, and equations presented in the preceding sections, we assumed that the reactivity was only being altered by some control mechanism that "inserts reactivity." There are other ways that reactivity is altered in an operating power reactor. The most important are: (1) variation of fission-product concentrations, (2) burnup or depletion of fuel, and (3) variations in reactor temperatures, pressures, and densities.

An increase in the concentration of fission products reduces reactivity because the fission products absorb some of the neutrons that carry on the chain reaction. The products of fission comprise a large variety of radioactive and stable nuclei whose relative concentrations in a reactor vary with time, power level, and prior operating history. Two thermal-neutron-absorbing fission products have strong effects on the reactivity of thermal reactors (the pressurized-water reactors, the boiling-water reactors, and the gas-cooled reactors of Sec. 15, 16, and 18, respectively), namely, ^{135}Xe (a 9.2-hr beta emitter) and ^{149}Sm (a stable nuclide). Because both these nuclides are strong absorbers of thermal neutrons, they are referred to as fission-product poisons or simply poisons. The absorption of thermal neutrons by ^{135}Xe is about 5000 times more probable, on an atom-for-atom basis, than the absorption of thermal neutrons by ^{235}U. Similarly, ^{149}Sm absorbs thermal neutrons nearly 90 times as easily as ^{235}U. In almost all present-day power reactors, the chain reaction is propagated almost entirely by thermal-neutron fission processes. Consequently, the presence of thermal-neutron absorbers in the fuel reduces the reactivity of these reactors by absorbing neutrons that otherwise would be available to carry on the chain reaction.

In the following paragraphs, the basic effects of these two fission products are briefly summarized. For details see Refs. 3, 5, and 7.

(a) Xenon-135. In 6.1% of the fissions of ^{235}U (or 5.1% of ^{233}U fissions, or 5.5% of ^{239}Pu fissions), one of the fission fragments has a mass number of 135. It decays to stable ^{135}Ba in the following chain:

$$\text{Fission} \rightarrow {}^{135}\text{Te}(0.5\ \text{min}) \xrightarrow{\beta} {}^{135}\text{I}(6.7\ \text{hr})$$
$$\xrightarrow[0.3]{\beta} {}^{135}\text{Xe}(15.3\ \text{min})$$
$$\xrightarrow[\beta]{0.7} {}^{135}\text{Xe}(9.2\ \text{hr}) \xrightarrow{\beta} {}^{135}\text{Cs}(2.6 \times 10^6\ \text{yr})$$
$$\xrightarrow{\beta} {}^{135}\text{Ba(stable)}$$

Because of the short half-life of ^{135}Te, the above decay scheme can be simplified for most purposes to:

$$\text{Fission} \rightarrow {}^{135}\text{I}(6.7\ \text{hr}) \xrightarrow{\beta} {}^{135}\text{Xe}(9.2\ \text{hr}) \xrightarrow{\beta} {}^{135}\text{Cs}$$

In addition to being produced via the above chain, ^{135}Xe is produced directly in 0.3% of the fissions of ^{235}U.

The rate of change in the concentration of ^{135}Xe is the difference between its production rate (per cm^3) and its loss (per cm^3). It is produced from the decay of ^{135}I and directly from the fission process. It is lost by decay to ^{135}Cs and by neutron absorption to ^{136}Xe(stable). The equation is thus:

$$\frac{dX}{dt} = (\lambda_I I + Y_X \Sigma_f \phi) - \lambda_X X - \delta_X X\phi \qquad (1.14)$$

where X = ^{135}Xe concentration (nuclei/cm^3)
I = ^{135}I concentration (nuclei/cm^3)
λ_X = ^{135}Xe decay constant (fractional change in concentration attributable to beta decay) = 0.693/9.2 hr = 2.1 × 10^{-5}/sec

---

*The terminology "reactivity insertion," adding or subtracting reactivity, is used here because it is commonly considered proper language of the trade. More exactly, reactivity can be negative, positive, or zero at any operating instant, and adding reactivity could mean, for example, decreasing the negative reactivity toward zero as in startup or in going critical. If, during reactor operation, power is falling and we do not want it to, we say that we add reactivity. If the power is steady but low and we want to increase it, we again say that we add reactivity. If the power is high and we want to lower it, we say that we decrease reactivity, which more exactly means inserting negative reactivity to cause the power to fall. To level the power at a lower point, however, we again say that we add reactivity to compensate or return to a condition of zero reactivity.

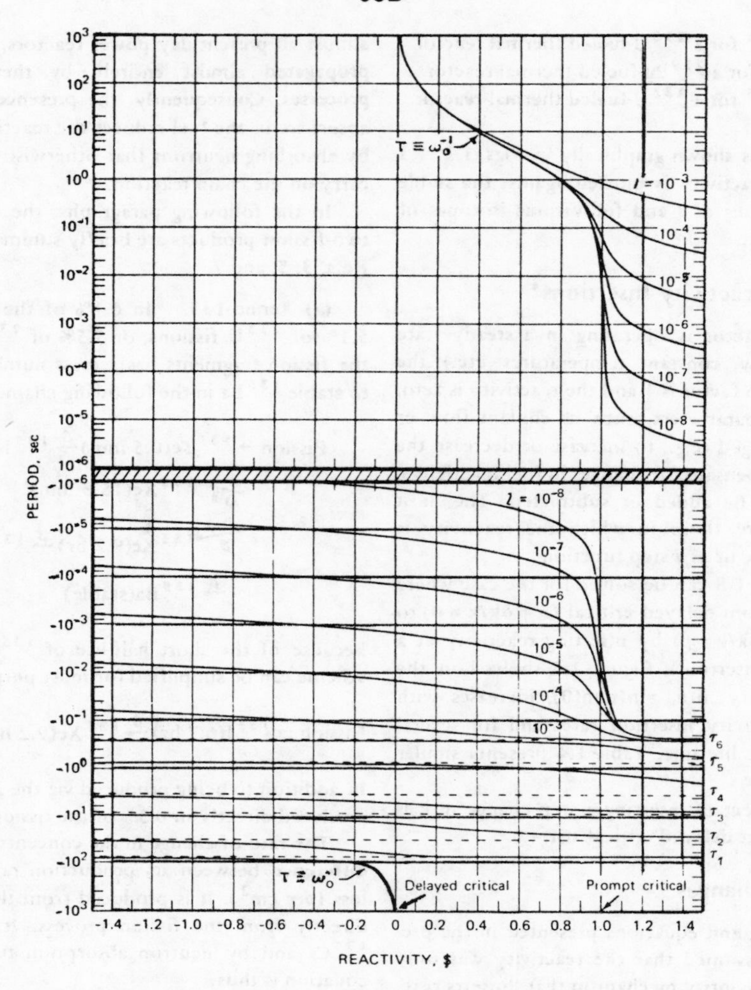

Fig. 1.1—Reactor stable and transient periods vs. reactivity for ^{235}U. The broken lines are drawn at the delayed-neutron mean lives ($\tau_1 = 1/\lambda_1, \tau_2, \ldots, \tau_6$) for ^{235}U. The parameter l is the prompt-neutron lifetime in seconds. (From H. C. Paxton and G. R. Keepin, *The Technology of Nuclear Reactor Safety*, Vol. 1, p. 262, The M.I.T. Press, Cambridge, Mass., 1964.)

λ_I = ^{135}I decay constant (fractional change in concentration attributable to beta decay) = 0.693/6.7 hr = 2.9×10^{-5}/sec

ϕ = thermal-neutron flux (neutrons cm^{-2} sec^{-1})

σ_X = microscopic thermal-neutron-capture cross section of $^{135}Xe = 3.5 \times 10^{-18}$ cm^2

Σ_f = macroscopic thermal-neutron-fission cross section of fuel = concentration of fuel (nuclei/cm^3) times the microscopic thermal-neutron-fission cross section

Y_X = fractional yield of ^{135}Xe directly from fission

The quantity $\lambda_I I$ can be determined by considering the rate of change in the ^{135}I concentration. The ^{135}I is produced from the decay of ^{135}Te, which, in turn, is produced directly from the fission process. Since the ^{135}Te is so short-lived, it is valid to consider the ^{135}I as produced directly from fission. In this case the equation for the ^{135}I concentration is

$$\frac{dI}{dt} = Y_{Te}\Sigma_f\phi - \lambda_I I - \sigma_I I \phi \qquad (1.15)$$

where Y_{Te} is the yield of ^{135}Te (6.1% for ^{235}U fission, etc.). The final term is the loss of ^{135}I because of neutron capture ($\sigma_I \ll \sigma_X$).

Equations 1.14 and 1.15 can be solved for various initial (t = 0) conditions and for various values of the thermal-neutron flux. One important solution is the equilibrium concentration of ^{135}Xe. The last term of Eq. 1.15 can be neglected; so $\lambda_I I \cong Y_{Te}\Sigma_f\phi$ at equilibrium conditions, and

$$\frac{dX}{dt} = 0 = Y_{Te}\Sigma_f\phi + Y_X\Sigma_f\phi - \lambda_X X - \sigma_X X\phi$$

Solving for X yields

$$X_{eq} = \frac{Y\Sigma_f\phi}{\lambda_X + \sigma_X\phi} \qquad (1.16)$$

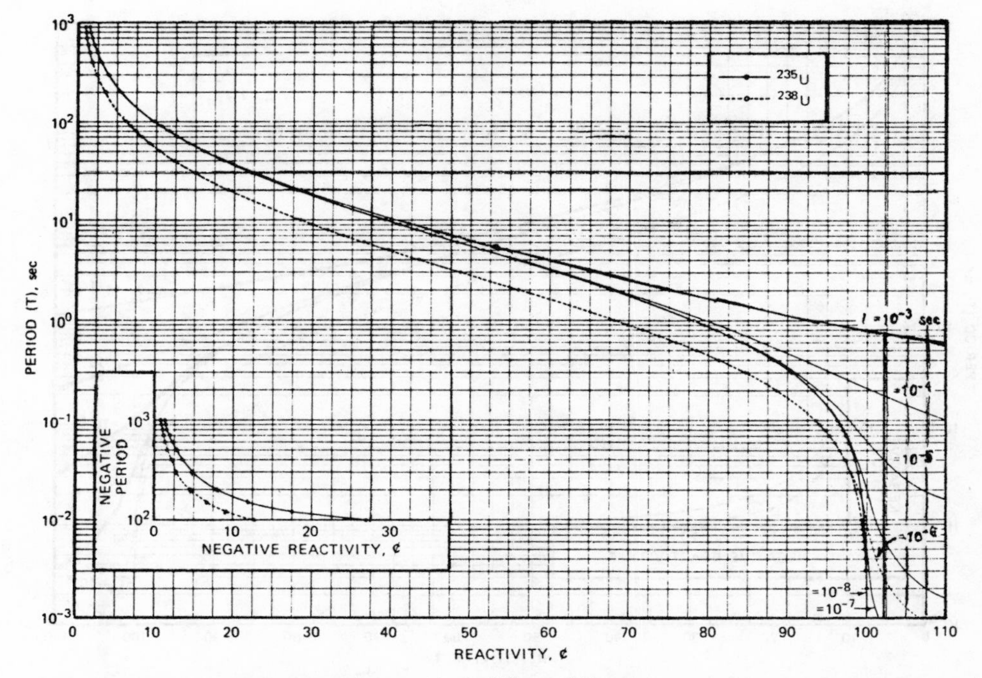

Fig. 1.2—Stable (asymptotic) period vs. reactivity for ^{235}U and ^{238}U. The parameter l is prompt-neutron lifetime. Heavy curves are calculated from Laplace-transformed prompt burst decay data; corresponding points are calculated from delayed-neutron periods and abundances. (From H. C. Paxton and G. R. Keepin, *The Technology of Nuclear Reactor Safety*, Vol. 1, p. 263, The M.I.T. Press, Cambridge, Mass., 1964.)

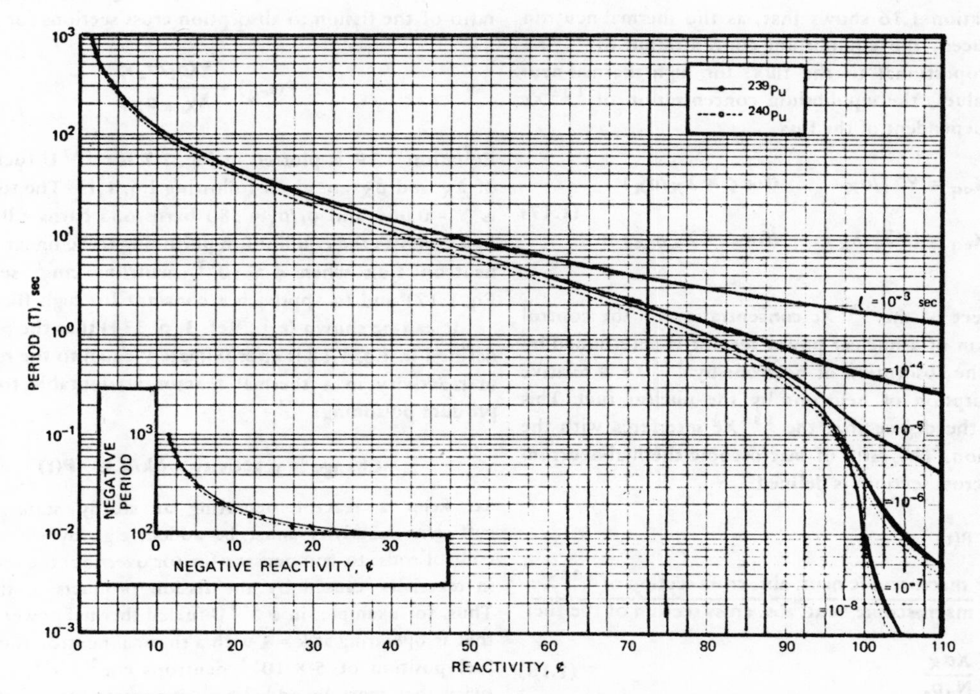

Fig. 1.3—Stable (asymptotic) period vs. reactivity for ^{239}Pu and ^{240}Pu. See caption for Fig. 1.2. (From H. C. Paxton and G. R. Keepin, *The Technology of Nuclear Reactor Safety*, Vol. 1, p. 264, The M.I.T. Press, Cambridge, Mass., 1964.)

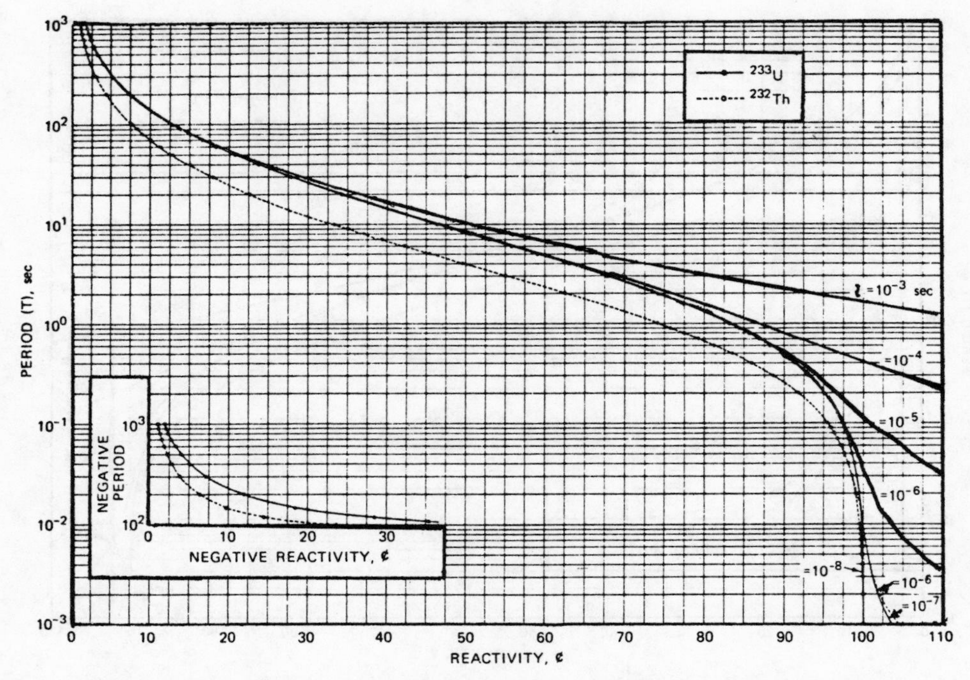

Fig. 1.4—Stable (asymptotic) period vs. reactivity for ^{233}U and ^{232}Th. See caption for Fig. 1.2. (From H. C. Paxton and G. R. Keepin, *The Technology of Nuclear Reactor Safety*, Vol. 1, p. 265, The M.I.T. Press, Cambridge, Mass., 1964.)

where Y is $Y_X + Y_{Te}$, the total fractional yield of ^{135}Xe per fission (i.e., the yield via the ^{135}Te chain and the direct yield). Equation 1.16 shows that, as the thermal-neutron flux is reduced, the equilibrium concentration of ^{135}Xe becomes proportional to the flux; for high thermal-neutron-flux values, the equilibrium concentration of ^{135}Xe becomes independent of the flux:

$$X_{eq} \cong Y\Sigma_f/\sigma_X \qquad (\text{for } \phi \gg \lambda_X/\sigma_X)$$
$$X_{eq} \cong (Y\Sigma_f/\lambda_X)\phi \qquad (\text{for } \phi \ll \lambda_X/\sigma_X) \tag{1.17}$$

The effect of the ^{135}Xe concentration on the control and operation of a nuclear power reactor is determined by how large the absorption of neutrons by ^{135}Xe is relative to the absorption of neutrons by the nuclear fuel. This determines the degree that the ^{135}Xe interferes with the chain reaction. The ratio of macroscopic thermal-neutron-absorption cross sections is defined:

$$\text{Poisoning} = P(t) = \frac{\text{macroscopic neut. abs. cross section of } ^{135}\text{Xe}}{\text{macroscopic neut. abs. cross section of the fuel}} = \frac{X\sigma_X}{N_u\sigma_a} \tag{1.18}$$

where N_u is the number of fuel nuclei/cm^3 and σ_a is the thermal-neutron-absorption cross section for the fuel. The poisoning, when the ^{135}Xe concentration is the equilibrium concentration, is obtained by substituting Eq. 1.16 into Eq. 1.18 and noting that $\Sigma_f/N_u\sigma_a = N_u\sigma_f/N_u\sigma_a$ = the ratio of the fission to absorption cross sections for the fuel:

$$P(t_{eq}) = \frac{Y(\sigma_f/\sigma_a)\phi}{\lambda_X + \sigma_X\phi} \tag{1.19}$$

Equation 1.19 is plotted in Fig. 1.7 for ^{235}U fuel. Values of λ_X and σ_X are given following Eq. 1.14. The total yield is Y = 0.064 and σ_f/σ_a = 580 barns/685 barns = 0.85. The figure shows the equilibrium poisoning to be linear with the neutron flux when $\phi \lesssim 10^{12}$ neutrons cm^{-2} sec^{-1} (see Eq. 1.17) and to approach a constant for high flux values.

It can be shown (e.g., Ref. 3, p. 334) that the poisoning defined in Eq. 1.18 is approximately equal to the reduction in reactivity in a thermal reactor attributable to fission-product poisoning:

$$\text{Change in reactivity} = \delta k/k \cong -P(t) \tag{1.20}$$

To keep a reactor operating at steady state (k = 1), sufficient reactivity must be added, e.g., by withdrawing control rods, to compensate for (or override) the reduction in reactivity caused by the fission products in the fuel. Thus, for example, in a ^{235}U-fueled thermal power reactor that is operating at k = 1 with a thermal-neutron flux at the fuel position of 5×10^{13} neutrons cm^{-2} sec^{-1}, the reactivity that must be added to compensate for the effect of the equilibrium concentration of ^{135}Xe is about $\sigma k/k$ = 0.049 (see Fig. 1.7).

The effect of ^{135}Xe poisoning is most pronounced when a reactor is shut down after it has been operating at

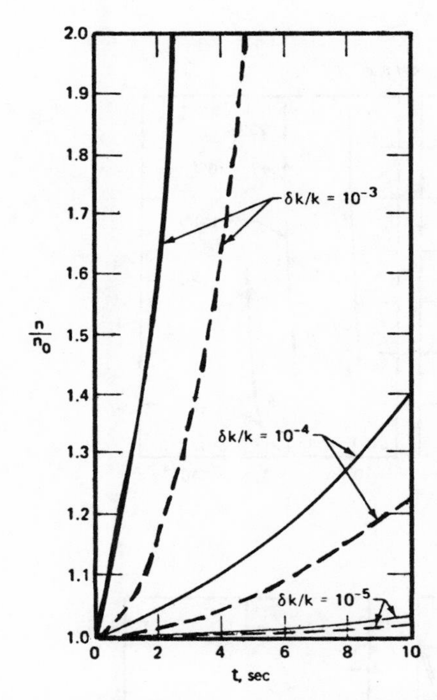

Fig. 1.5—Relative neutron density, n/n_0, vs. time for reactivity-insertion rates of 10^{-3}, 10^{-4}, and 10^{-5} $\delta k/k$ per second for ^{235}U. Solid curve is for neutron lifetime of 10^{-6} sec, broken curve is for 10^{-2} sec. (From J. M. Harrer, *Nuclear Reactor Control Engineering*, p. 92, D. Van Nostrand Company, Inc., Princeton, N. J., 1963.)

full power for a time sufficiently long that the equilibrium concentration of ^{135}Xe (Eq. 1.16) is present. In this case the xenon concentration increases considerably above its equilibrium value since it is no longer being removed by thermal-neutron capture. The ^{135}Xe is being produced by the decay of the equilibrium concentration of ^{135}I (6.3 hr) and being lost by its own 9.2-hr beta decay. The net result is shown in Fig. 1.8, where the poisoning is plotted as a function of time after shutdown from equilibrium for several values of the thermal-neutron flux. The $t = 0$ values of Fig. 1.8 are obtained from the equilibrium curve shown in Fig. 1.7. The ^{135}Xe poisoning builds up to a maximum after shutdown. For low values of the flux, the time to reach maximum poisoning is only a few hours. For the higher flux values normally encountered in power-reactor operation, the poisoning reaches a maximum about 10 hr after shutdown. The value of the poisoning does not return to its preshutdown value until 30 or 40 hr after shutdown. As Fig. 1.8 shows, the value of the poisoning at maximum can be many times the equilibrium (before shutdown) value. The excess reactivity required to overcome the maximum poisoning may be more than is available in power reactor, particularly if the fuel has been depleted by prior operation. If this is the case, then the reactor shutdown time has to be limited to less than a few hours or more than 30 or 40 hr must be allowed.

The full shutdown from equilibrium shown in Fig. 1.8 is not the only situation of practical interest. Often the power level is cut back or increased by some fraction of full power. Initially the ^{135}Xe concentration has a value corresponding to the initial power level; after the change in power level, the ^{135}Xe concentration changes until it reaches a new equilibrium value corresponding to the final power level. Figures 1.9 and 1.10 show the time to reach the maximum poisoning following a step decrease or a step increase of the thermal-neutron flux (which is directly proportional to the reactor power level). Figure 1.9 shows, for example, that a 50% cutback from 4×10^{13} neutrons cm^{-2} sec^{-1} creates a maximum ^{135}Xe poisoning about 23,000 sec (6.4 hr) after the cutback. In the reverse process, Fig. 1.10 shows that when the flux level is doubled from 2×10^{13} neutrons cm^{-2} sec^{-1}, the maximum ^{135}Xe poisoning effect occurs about 11,600 sec (3.2 hr) after the increase. From the initial values of the neutron flux, the initial equilibrium concentration of ^{135}Xe and ^{135}I, and the value at the time the maximum effect occurs, the maximum poisoning or maximum reduction in $\delta k/k$ can be calculated.

(b) Samarium-149. In 1.13% of the fissions of ^{235}U (or 0.66% of ^{233}U fissions, or 1.9% of ^{239}Pu fissions), one of the fission fragments has a mass number 149. Some of the fissions form ^{149}Pm and others form ^{149}Nd:

$$\text{Fission} \rightarrow {}^{149}\text{Pm(54 hr)} \xrightarrow{\beta} {}^{149}\text{Sm(stable)}$$

$$\text{Fission} \rightarrow {}^{149}\text{Nd(2 hr)} \xrightarrow{\beta} {}^{149}\text{Pm(54 hr)}\ {}^{149}\text{Sm(stable)}$$

Because the ^{149}Nd half-life is small compared to the ^{149}Pm half-life, the first chain above is a good approximation for both chains. As noted earlier, ^{149}Sm strongly absorbs thermal neutrons. The other nuclides in the chain are not anomalous in this respect.

The rate of change of the ^{149}Sm concentration is just equal to its production rate from ^{149}Pm decay minus its rate of loss from thermal-neutron capture (which converts it to stable ^{150}Sm):

$$\frac{d(\text{Sm})}{dt} = \lambda_{Pm}\,(\text{Pm}) - (\text{Sm})\sigma_{Sm}\phi \qquad (1.21)$$

where (Pm) and (Sm) are the concentrations of ^{149}Pm and ^{149}Sm, respectively, λ_{Pm} is the decay constant of $^{149}Pm \doteq 3.56 \times 10^{-6}$/sec, σ_{Sm} is the thermal-neutron-capture cross section of $^{149}Sm = 50{,}000$ barns $= 5 \times 10^{-20}$ cm^2, and ϕ is the thermal-neutron flux (neutrons cm^{-2} sec^{-1}). Note that, unlike ^{135}Xe, the ^{149}Sm is removed only when it captures thermal neutrons. The rate of change of ^{149}Pm is its production rate from fission (neglecting the intermediate ^{149}Nd) minus its loss by beta decay (loss by neutron capture is negligible):

$$\frac{d(\text{Pm})}{dt} = Y_{Pm}\phi\Sigma_f - \lambda_{Pm}\,(\text{Pm}) \qquad (1.22)$$

where (Pm) is the concentration (atoms/cm^3) of ^{149}Pm, Y_{Pm} is the yield of ^{149}Pm in fission, and Σ_f is the macroscopic thermal-neutron-fission cross section of the nuclear fuel.

When a power reactor has been operating at a steady state ($k = 1$) for many hours, the equilibrium concentrations of ^{149}Pm and ^{149}Sm (from Eqs. 1.21 and 1.22) are

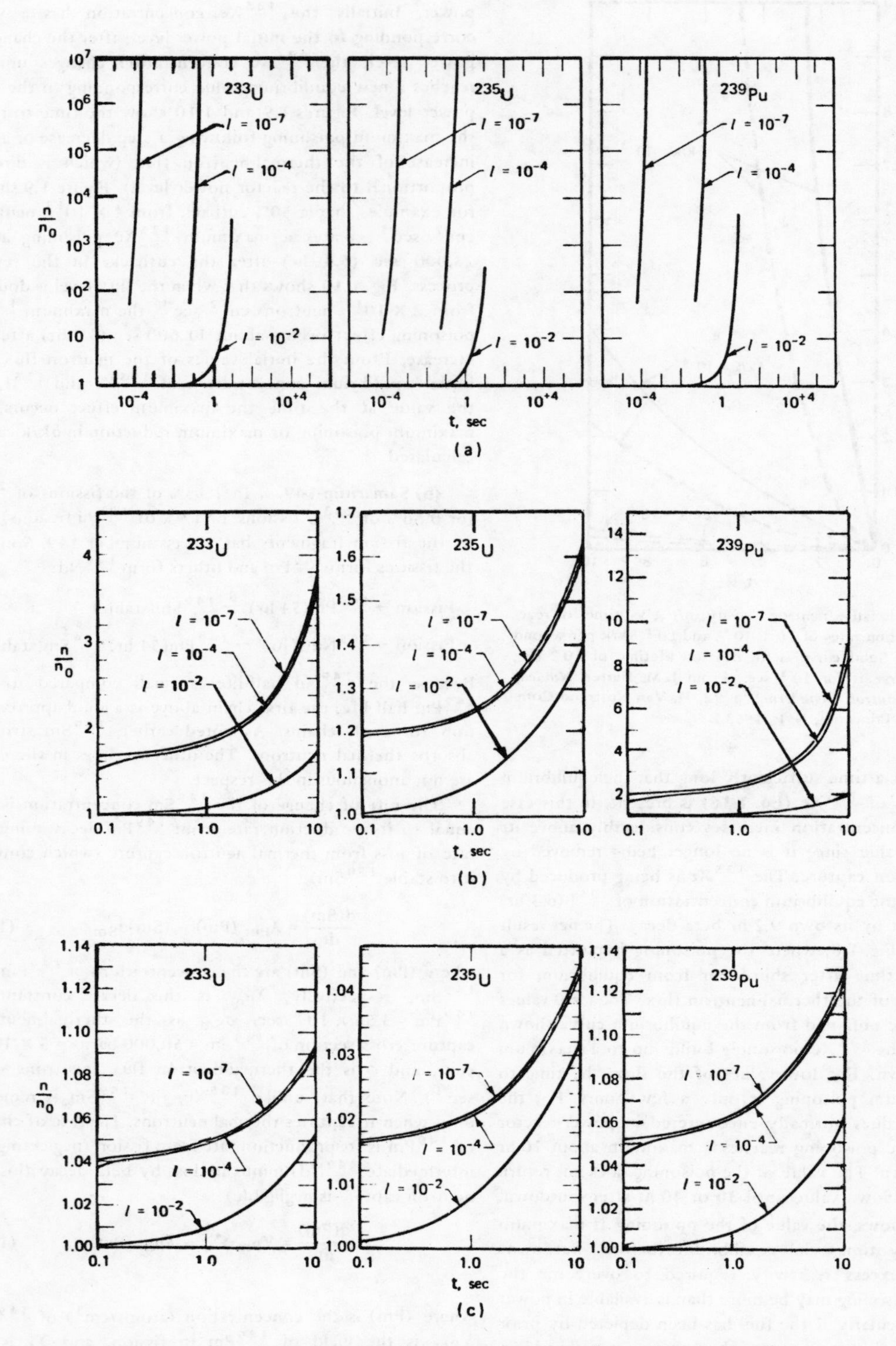

Fig. 1.6—Relative neutron density, n/n_0, vs. time for step insertions of (a) 10^{-2} $\delta k/k$; (b) 10^{-3} $\delta k/k$; and (c) 10^{-4} $\delta k/k$ starting at delayed critical in ^{233}U, ^{235}U, and ^{239}Pu. (From J. M. Harrer, *Nuclear Reactor Control Engineering*, pp. 88 and 89, D. Van Nostrand Company, Inc., Princeton, N. J., 1963.)

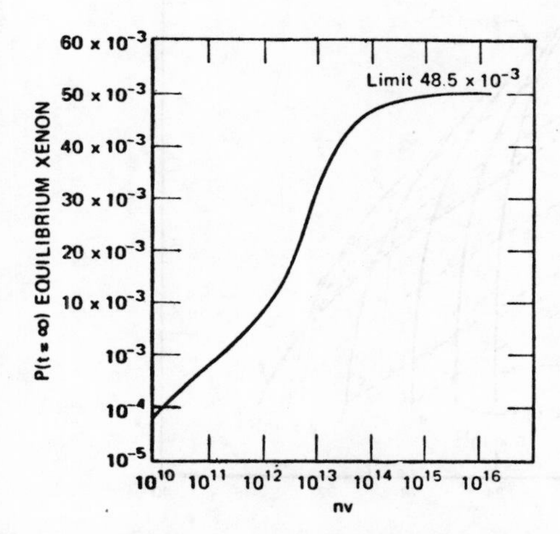

Fig. 1.7—Poisoning, P(t), vs. thermal-neutron flux for equilibrium ^{135}Xe concentration in ^{235}U. (From J. M. Harrer, *Nuclear Reactor Control Engineering,* p. 393, D. Van Nostrand Company, Inc., Princeton, N. J., 1963.)

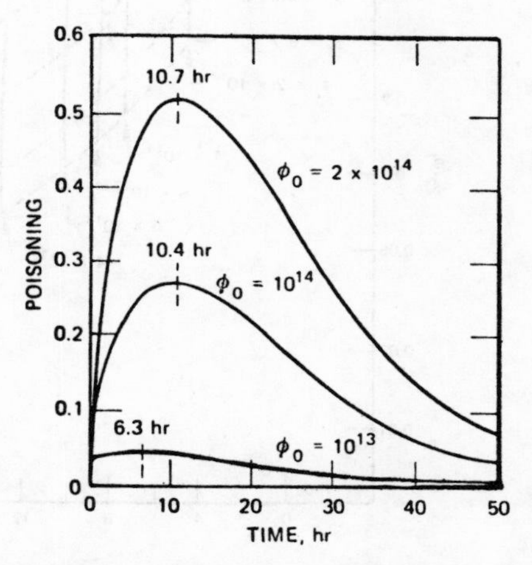

Fig. 1.8—Poisoning vs. time for various thermal-neutron-flux values, assuming equilibrium concentration of ^{135}Xe at t = 0. (From J. M. Harrer, *Nuclear Reactor Control Engineering,* p. 394, D. Van Nostrand Company, Inc., Princeton, N. J., 1963.)

Table 1.4—Relative Neutron Density (n/n_0) as a Function of Time During Ramp Insertions of 10^{-3}, 10^{-4}, and 10^{-5} $\delta k/k$ Per Second*

Time, sec	($\delta k/k$)/sec ramp	^{233}U		^{235}U		^{239}Pu	
		$l = 10^{-2}$	$l = 10^{-6}$	$l = 10^{-2}$	$l = 10^{-6}$	$l = 10^{-2}$	$l = 10^{-6}$
0.5	A = 10^{-3}	1.012	1.254	1.011	1.092	1.012	~1.35
	10^{-4}	1.001	1.021	1.001	1.009	1.001	1.026
	10^{-5}	1.0001	1.002	1.0001	1.0009	1.0001	1.003
1.0	A = 10^{-3}	1.047	1.723	1.042	1.218	1.048	2.112
	10^{-4}	1.005	1.045	1.004	1.018	1.005	1.057
	10^{-5}	1.0005	1.004	1.0004	1.002	1.0005	1.005
1.5	A = 10^{-3}	1.106	2.768	1.090	1.388	1.108	(1.4 = 3.84) (1.6 = 6.16)
	10^{-4}	1.010	1.071	1.009	1.029	1.010	1.093
	10^{-5}	1.001	1.007	1.0009	1.003	1.001	1.009
2.0	A = 10^{-3}	1.189	6.350	1.156	1.626	1.195	62.5
	10^{-4}	1.017	1.101	1.014	1.041	1.018	1.133
	10^{-5}	1.002	1.009	1.001	1.004	1.002	1.012
5.0	A = 10^{-3}	2.644	3.0 sec = 3.7×10^{37}	2.069	15.9	2.792	2.5 sec = 7.7×10^{37}
	10^{-4}	1.096	1.358	1.069	1.137	1.103	1.517
	10^{-5}	1.009	1.028	1.007	1.012	1.010	1.037
10.0	A = 10^{-3}	46.5	∞	15.7	6.8 sec = 1.4×10^{38}	59.0	∞
	10^{-4}	1.366	2.401	1.229	1.406	1.411	3.675
	10^{-5}	1.031	1.069	1.020	1.031	1.034	1.094

*From J. M. Harrer, *Nuclear Reactor Control Engineering,* p. 91, D. Van Nostrand Company, Inc., Princeton, N. J., 1963.

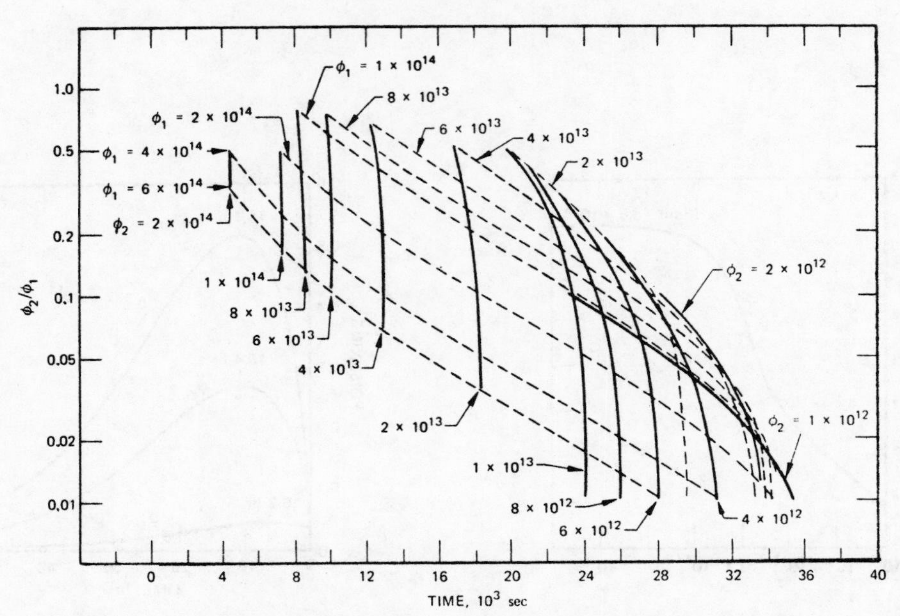

Fig. 1.9—Time to reach maximum xenon poisoning after a decrease of thermal-neutron flux from ϕ_1 to ϕ_2.[7]

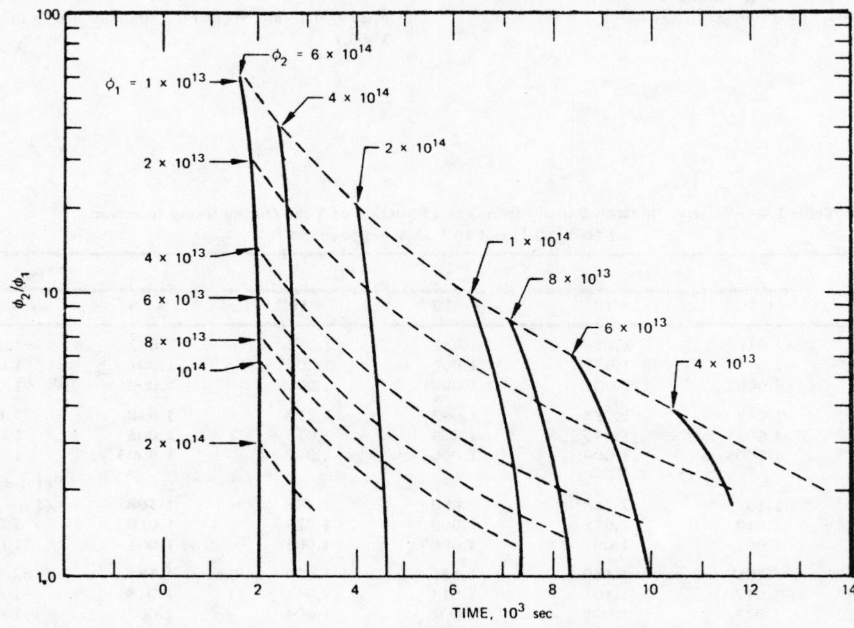

Fig. 1.10—Time to reach maximum xenon poisoning after an increase of thermal-neutron flux from ϕ_1 to ϕ_2.[7]

$$(\mathrm{Pm})_{eq} = \frac{Y_{Pm}\phi\Sigma_f}{\lambda_{Pm}}$$

$$(\mathrm{Sm})_{eq} = \frac{\lambda_{Pm}(\mathrm{Pm})_{eq}}{\sigma_{Sm}\phi} = \frac{Y_{Pm}\Sigma_f}{\sigma_{Sm}} \tag{1.23}$$

Note that the equilibrium concentration of ^{149}Pm is proportional to the thermal-neutron flux, ϕ, while the equilibrium concentration of ^{149}Sm is independent of the flux.

The equilibrium ^{149}Sm poisoning (Eq. 1.18) is

$P(t_{eq})$ = poisoning of ^{149}Sm at equilibrium concentration

$$= \frac{(\mathrm{Sm})_{eq}\sigma_{Sm}}{N_u\sigma_a} = \frac{Y_{Pm}\Sigma_f\sigma_{Sm}}{N_u\sigma_a\sigma_{Sm}} = Y_{Pm}\left(\frac{\sigma_f}{\sigma_a}\right) \tag{1.24}$$

which is also independent of the thermal-neutron flux. Substituting into Eq. 1.24 the values of the yields and the fission/absorption cross section ratios gives:

	^{235}U	^{233}U	^{239}Pu
Y_{Pm}	0.0113	0.0066	0.019
σ_f/σ_a	580/685	524/593	860/1220
$P(t_{eq})$ for ^{149}Sm	0.0096	0.0058	0.077

These correspond approximately to $\delta k/k$ values of −0.96% for ^{235}U, −0.58% for ^{233}U, and −7.7% for ^{239}Pu. Comparison of the equilibrium value of ^{149}Sm poisoning in ^{235}U-fueled reactors with the equilibrium values of ^{135}Xe poisoning (Fig. 1.7) shows the former to be only about one-fourth of the latter.

When a power reactor that has been operating at steady state (k = 1) for some hours is shut down, the concentration of ^{149}Sm increases from its initial (t = 0) value by the creation of ^{149}Sm from ^{149}Pm decay:

$$^{149}\text{Sm conc. after shutdown} = {}^{149}\text{Sm conc. at shutdown} + (1 - e^{-\lambda_{Pm}t}) \times {}^{149}\text{Pm conc. at shutdown} \qquad (1.25)$$

Since the half-life of ^{149}Pm is 54 hr, the ^{149}Sm concentration is increased by one-half the ^{149}Pm shutdown concentration during the first 54 hr after shutdown. After a few hundred hours the ^{149}Sm concentration is equal to the sum of the ^{149}Sm concentration at shutdown and the ^{149}Pm concentration at shutdown.

When a power reactor has been operating at a steady state (k = 1) and at constant flux for a few hundred hours, both the ^{149}Pm and ^{149}Sm concentrations have their equilibrium values (Eq. 1.23). If the reactor is then (at t = 0) shut down, the ^{149}Sm concentration builds up according to Eq. 1.25. Substitution of the equilibrium concentrations into Eq. 1.25 gives

$$^{149}\text{Sm conc. after shutdown from equilibrium} = (\text{Sm})_{eq}\,[1 + (\phi\sigma_{Sm}/\lambda_{Pm})(1 - e^{-\lambda_{Pm}t})] = (\text{Sm})_{eq}\,[1 + 1.40 \times 10^{-14}\phi \times (1 - e^{-\lambda_{Pm}t})] \qquad (1.26)$$

where ϕ is the thermal-neutron flux in neutrons $cm^{-2}\ sec^{-1}$ and λ_{Pm} is the disintegration constant of ^{149}Pm (= 3.56×10^{-6}/sec = 0.693/54 hr). It is apparent that the poisoning effect of ^{149}Sm becomes quite important in high flux operation. A shutdown from operation at a flux of 10^{14} can increase the poisoning effect by a factor of 1 + 1.40, or 2.40 times the equilibrium poisoning (Eq. 1.24), if the shutdown continues for a few hundred hours. Sufficient excess $\delta k/k$ must be available to compensate for this poisoning when the reactor is started up.

(c) Fuel Burnup. During the operation of a nuclear power reactor, fuel (^{235}U, ^{233}U, or ^{239}Pu) is continually being burned up (i.e., fissioned) so the remaining fuel becomes depleted in the fissionable nuclide. The effect of this burnup, or depletion, is to reduce the reactivity available to compensate for fission-product poisoning or for other reactivity-reducing effects. Eventually the depletion becomes intolerable, and the reactor has to be refueled.

At a point in the reactor fuel where the thermal-neutron flux is ϕ, the absorption of neutrons decreases the concentration of fissile material (^{235}U, ^{233}U, or ^{239}Pu) exponentially with time:

$$\text{Fuel conc. at time t} = (\text{Fuel conc. at t} = 0)\exp\left(-\sigma_a \int_0^t \phi\, dt\right) \qquad (1.27)$$

where σ_a is the neutron-absorption cross section of the fissile material. The neutron flux is assumed to vary with time. (For convenience the integral can be written as $\phi_{av}t$, where ϕ_{av} is the average flux during the time from t = 0 to t = t.)

The fractional burnup is defined as the change in fuel concentration divided by the initial concentration. From Eq. 1.27 it follows that

$$\text{Fractional burnup} = F = 1 - e^{-\sigma_a\phi_{av}t} \qquad (1.28)$$

As an example, the fractional burnup of ^{235}U in three months (7.8×10^6 sec) in a power reactor with an average thermal-neutron flux at the fuel of 5×10^{13} neutrons $cm^{-2}\ sec^{-1}$ is $1 - \exp\,[(-685\ \text{barns})(5 \times 10^{13})(7.8 \times 10^6\ cm^{-2})] = 1 - \exp(-0.267)$, or 0.234, i.e., a fractional burnup of 23.4%.

Figure 1.11 shows the relation between the fractional burnup of fuel and the resulting loss in reactivity $\delta k/k$. For a 23.4% burnup, the loss in reactivity is about 2.7%. This is to be compared with the reactivity loss of 4.9% (see discussion after Eq. 1.20) attributable to equilibrium xenon poisoning at the same average neutron flux and the 1% (see discussion after Eq. 1.24) loss in reactivity from samarium poisoning.

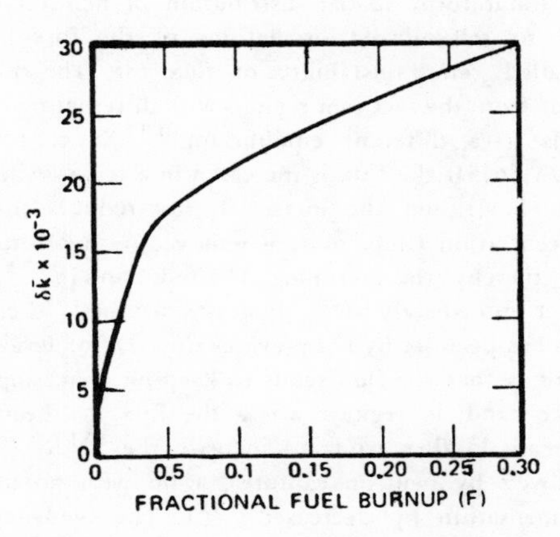

Fig. 1.11—Poisoning as function of fractional fuel burnup. (From J. M. Harrer, *Nuclear Reactor Control Engineering*, p. 399, D. Van Nostrand Company, Inc., Princeton, N. J., 1963.)

In any actual power reactor, calculations of burnup must take into account many factors not considered in the foregoing. These include the presence of fertile material (^{238}U or ^{232}Th) in the fuel, the energy and spatial distributions of neutron flux, the geometry and composition of the fuel elements, and the operating history of the reactor. The data presented here are intended to provide a semi-quantitative indication of the effect of fuel depletion on the reactivity that is needed for instrumentation system design.

1-3.7 Three-Dimensional Kinetics

The reactor kinetics equations derived in 1-3.2 are labeled "point" kinetics since the neutron density, n, was considered only as a function of time. Actually, n should be written as n(x,y,z,t) to emphasize its spatial dependence. Similarly, the density of delayed-neutron emitters should be written C_i(x,y,z,t), and each of the fission-product concentrations should be written as functions of x,y,z,t. The spatial dependence of the neutron flux depends on many factors, such as fuel loading, primary-coolant-system structure, and reactor-vessel penetrations. It also depends on the past history of the individual fuel elements.

Since the neutron flux in the fuel determines the heat generated, it follows that knowledge of the spatial distribution of the flux is necessary if the reactor is to be used safely and efficiently as a heat source. Instrumentation systems must be included to provide this knowledge to the operator.

If the neutron flux is not constant throughout the reactor, then the kinetics of the chain reaction is not the same in all parts of the reactor. The preceding section shows that many reactivity effects are flux-dependent. For example, if a reactor is operating with the neutron flux less in one region of the reactor than in another region, then the equilibrium concentrations of the fission-product poisons in the two regions will not be the same, nor (if the reactor has run this way for any length of time) will the fuel burnup be the same in the two regions.

Nonuniform spatial distribution of neutron flux can lead to self-induced oscillations of the flux level, the so-called xenon instability or flux tilt. The oscillations result from the fact that regions with different neutron-flux levels have different equilibrium ^{135}Xe concentrations (Eq. 1.16). If the flux is increased in a region where it has been low, then the increased flux reduces the ^{135}Xe concentration (since more is removed by neutron capture) and, thereby, the poisoning. The reduction in ^{135}Xe is not offset immediately by ^{135}I decay since the ^{135}I concentration has been set by the previous (lower) flux level. The net result is that the flux tends to keep on increasing. On the other hand, in regions where the flux has been high, a decrease in flux tends to increase the ^{135}Xe (fewer are removed by neutron capture), again with no immediate compensation by decreased ^{135}I. The tendency of the increasing flux to keep on increasing and the decreasing flux to keep on decreasing is eventually reversed by the ^{135}I decay—the peaking of the xenon poisoning shown in Fig. 1.8. The net result is an oscillation in ^{135}Xe poisoning between the regions of the reactor with a period given by

$$\text{Period of xenon oscillation} = \frac{2\pi}{[\lambda_I(\lambda_X + \sigma_X\phi)]^{1/2}} \quad (1.29)$$

where λ_I = disintegration constant of ^{135}I (2.9×10^{-5}/sec)
λ_X = disintegration constant of ^{135}Xe (2.1×10^{-5}/sec)
σ_X = microscopic thermal-neutron cross section of ^{135}Xe
ϕ = thermal-neutron flux

If the flux is 5×10^{13} neutrons cm^{-2} sec^{-1}, the period of the oscillation is 23 hr, somewhat longer than either the ^{135}I half-life (6.7 hr) or the ^{135}Xe half-life (9.2 hr). As the flux is lowered, the period of the xenon oscillation approaches 70 hr; at 10^{14} neutrons cm^{-2} sec^{-1}, it is about 17 hr.

The possibility of such flux oscillations must be taken into account in designing the reactor control system. The possibility of high and low regions of heat generation can introduce potential hazards if there is a natural mechanism that makes the high higher and the low lower. Instrumentation must be provided to sense the onset of any such oscillations.

1-4 NUCLEAR POWER PLANTS

1-4.1 Types of Plants

Nuclear power plants are categorized according to the type of nuclear reactor that is the primary heat source. In this book, reactor types are identified by the coolant used to extract heat from the nuclear fuel:*

Pressurized-water reactors: reactors cooled by water in the liquid state
Boiling-water reactors: reactors cooled by water in the liquid and gaseous states
Sodium-cooled reactors: reactors cooled by liquid sodium
Gas-cooled reactors: reactors cooled by gas (helium in the United States)

Every nuclear power plant presently in operation in the United States derives its heat from a reactor in one of these four categories.

Classification of nuclear power plants according to the primary reactor coolant is particularly appropriate to a consideration of power-reactor instrumentation systems since the coolant properties determine many aspects of instrumentation design. This is not surprising since the basic function of the nuclear reactor in a power plant is to generate the heat and to transfer it to a coolant that can then transfer heat to the steam that drives a turbogenerator.

*Reactors can also be classified in other ways: according to the energy spectrum of the neutron population (thermal, intermediate, and fast), according to use (research, development, test, plutonium production, and power), according to fuel arrangement (homogeneous or heterogeneous), or according to whether the fuel fissioned is less than or greater than the fuel generated (breeder, nonbreeder, and converter).

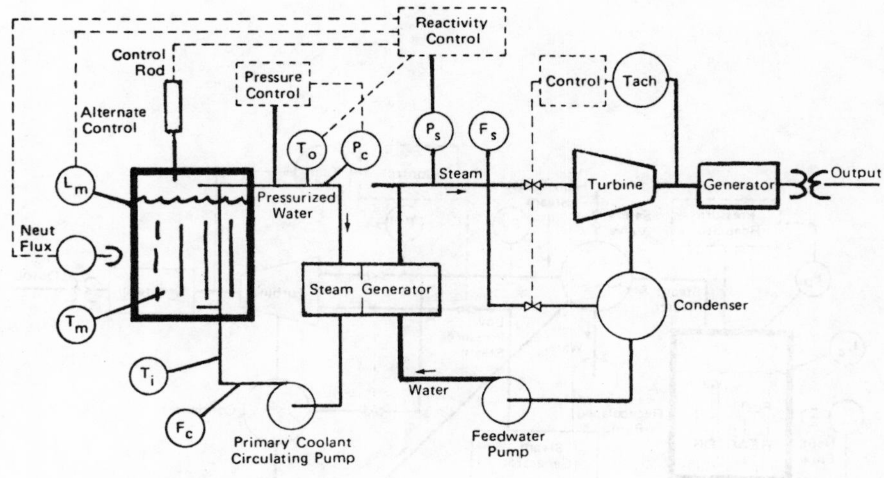

Fig. 1.12—Pressurized-water reactor.

The coolant that extracts the heat from the nuclear fuel is the key link in the sequence of operations that converts nuclear energy to electrical energy. Moreover, because material constraints are critically important in any heat engine, the properties of the coolant have a strong influence on the plant design.

Figures 1.12 through 1.15 illustrate the basic configurations of the four categories of nuclear power plants. It must be emphasized that the figures do not purport to show any actual plant configuration (see Sec. 15 through 18) but rather show those features of each reactor type which are relevant to the design of the principal instrumentation systems.

1-4.2 Sensed Variables

The nuclear chain reaction produces heat (primarily from the dissipation of the kinetic energy of the fission fragments) and nuclear radiations. Consequently, nuclear-power-reactor instrumentation depends primarily on thermal sensors and nuclear-radiation sensors. The former (thermocouples and resistance thermometers) are discussed in Sec. 4 and the latter in Sec. 2 and 3. Although a number of nuclear radiations are associated with the fission process, only neutrons can be unambiguously related to the occurrence of fissions; because of this, neutron sensors are the most important of the nuclear-radiation sensors. The neutron sensors are used to determine the rate of fissions, the time derivative of the fission rate, and the fission rate as a function of position in the reactor. (The circuits required to convert the signals from neutron sensors into outputs that are directly related to nuclear reactor performance are described in Sec. 5.)

The primary coolant that transfers heat from the nuclear fuel to the turbogenerator (or to a heat exchanger coupled to the turbogenerator) must be examined by suitable sensors to determine such important parameters as (1) the temperature of the coolant entering the reactor (T_i in Figs. 1.12 through 1.15), (2) the temperature of the coolant leaving the reactor (T_o in the figures), (3) the temperature of the coolant at other positions in the reactor, (4) the rate of flow of coolant into and out of the reactor (F_c in the figures), (5) the rate of flow of coolant in various coolant channels in the reactor, (6) the radioactivity of the coolant after leaving reactor, (7) the purity of the coolant, and (8) the presence of water vapor in the coolant when the coolant is a gas. To sense these parameters, there must be temperature sensors, flowmeters, humidity detectors, nuclear-radiation (gamma in this case) sensors, etc.

Reactor operation itself involves a number of parameters, including (1) the position of the control rods (L_a in Fig. 1.15), (2) the water level of the moderator (L_m in Fig. 1.12), (3) the water level in the reactor (L_c in Fig. 1.13), (4) the pressure in the primary system (P_p in Fig. 1.13), (5) the pressure at the coolant outlet (P_c in Figs. 1.12 and 1.15), and (6) the temperature of the moderator (T_m in Fig. 1.15). Temperature sensors, position indicators, pressure transducers, etc., are required to take data on these parameters.

The steam system is characterized by such parameters as steam flow rate (F_s in the figures), steam pressures (P_s in the figures), steam quality, and feedwater flow (F_f in Fig. 1.13). Thermal sensors, pressure and differential-pressure transducers, flowmeters, water-level indicators, and other sensors (Sec. 4, Pt. 4-6), must be used.

In addition, there will be sensors associated with the important components of the plant. Thus, for example, tachometers to sense turbine rotation, meters to sense electrical generator output, thermal and mechanical devices to sense the performance of primary-coolant-pump drive motors, etc., must be installed.

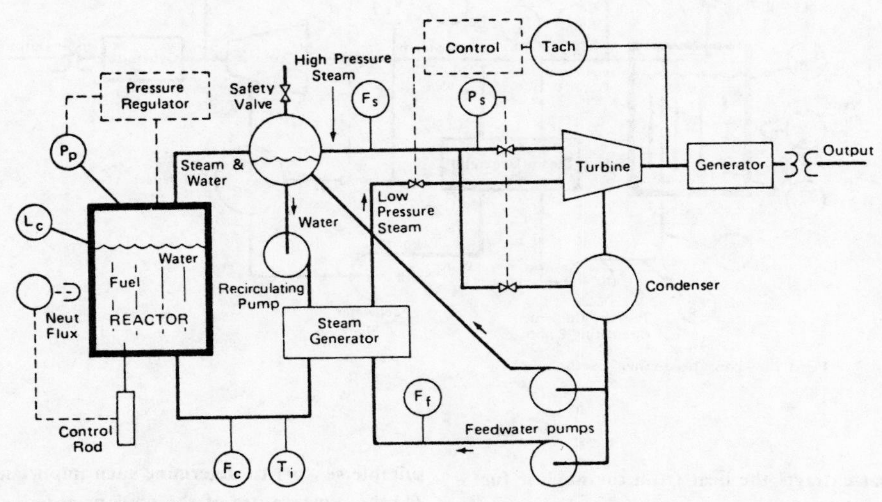

Fig. 1.13 —Boiling-water reactor.

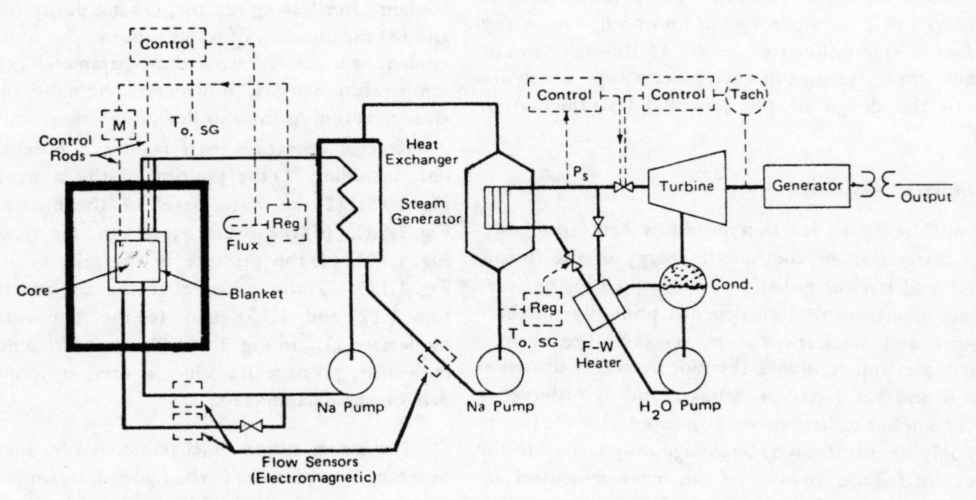

Fig. 1.14—Sodium-cooled reactor.

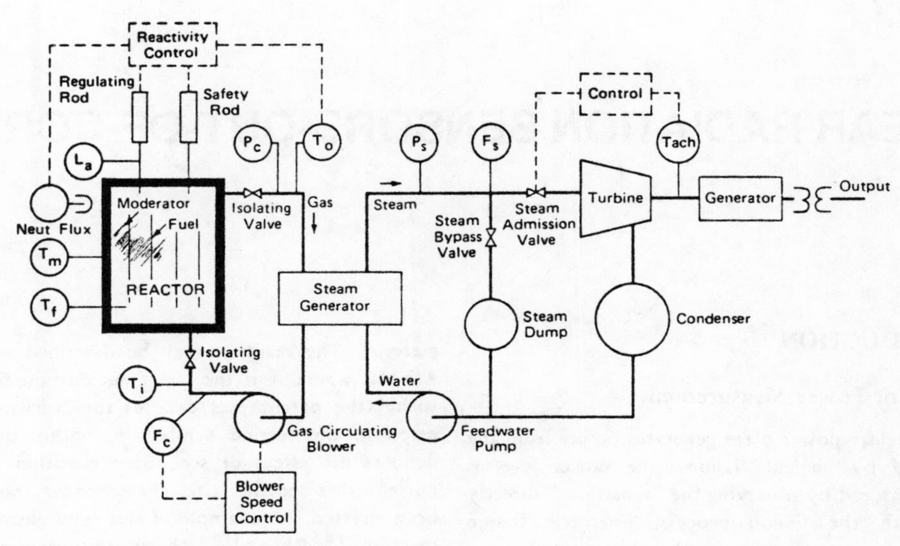

Fig. 1.15—Gas-cooled reactor.

REFERENCES

1. American National Standards Institute, *American National Standard Glossary of Terms in Nuclear Science and Technology,* N1.1-1967, American National Standards Institute, New York, New York.
2. A. M. Weinberg and E. P. Wigner, *The Physical Theory of Neutron Chain Reactors,* University of Chicago Press, Chicago, 1958.
3. S. Glasstone and M. C. Edlund, *The Elements of Nuclear Reactor Theory,* D. Van Nostrand Company, Inc., Princeton, N. J., 1952.
4. R. V. Meghreblian and D. K. Holmes, *Reactor Analysis,* McGraw-Hill Book Company, Inc., New York, 1960.
5. J. M. Harrer, *Nuclear Reactor Control Engineering,* D. Van Nostrand Company, Inc., Princeton, N. J., 1963.
6. S. Glasstone and A. Sesonske, *Nuclear Reactor Engineering,* D. Van Nostrand Company, Inc., Princeton, N. J., 1963.
7. Reactor Physics Constants, USAEC Report ANL-5800(2nd Ed.), Argonne National Laboratory, Superintendent of Documents, U. S. Government Printing Office, 1963.

SECTION 2

NUCLEAR RADIATION SENSORS-OUT-OF-CORE

2-1 INTRODUCTION

2-1.1 Reactor-Power Measurement

Since a nuclear power plant generates power from the heat produced by nuclear fissions, the power level is commonly measured by observing the "radiations" directly associated with the fission process. Energetic fission fragments, neutrons, photons, and other particles are produced at the time each fission occurs.[1,2] The number of these radiations, or components of these radiations, is proportional to the number of fissions. The rate of appearance of these radiations is proportional to the fission rate and, thus, to the reactor power level.

Most fission fragments are radioactive and continue to emit betas and gammas long after the fission events in which they were created. In addition, the fission neutrons and some of the more energetic photons can induce radioactivity. This induced radioactivity also persists long after the creating process. The radiations from the total residual radioactivity, variously called afterglow, decay heat, or fission-product activity, contribute up to 5% of the reactor heat. However, the decay heat is not an indicator of reactor power but is related to the history of operation of the reactor.

The most desirable way to make any measurement is to use the most direct method. Reactor power therefore should be measured by detecting the prompt fission radiations. The fission fragments and beta radiations are short range and are stopped in the reactor fuel. However, the neutrons and gamma rays accompanying fission are sufficiently penetrating to be detected at some distance. The technology for reactor power measurement is based on the detection of neutrons or gammas or both.

2-1.2 Interactions with Matter

Fission gammas and neutrons interact with surrounding material in many ways. Three interaction processes are of special interest in reactor power measurement; nuclear reactions, recoils or collisions, and ionization. The interaction of any single gamma photon or neutron may involve more than one of these three basic processes. (For details on these interactions, see Refs. 3 and 4.)

A nuclear reaction results from sufficiently energetic collisions of a specific radiation with the nuclei of a specific material. The consequence of a nuclear reaction is a nuclear excitation or transmutation or the formation of a new material. The reaction may be described symbolically as A(a,b)C where A is the *target* nucleus, the first symbol, a, within the parentheses denotes the radiation causing the reaction, the second symbol, b, within the parentheses denotes the effect or secondary radiation, and C is the nucleus that remains after the secondary radiation, b, has been emitted. An example of this symbolism is the nuclear reaction $^{103}Rh(n,\gamma)^{104}Rh$; in this reaction ^{103}Rh has been converted to ^{104}Rh by the capture of a neutron, n, and the emission of a capture gamma, γ.

If the principal interest lies not in ^{104}Rh but in its decay product, the expression may be expanded to $^{103}Rh(n,\gamma)^{104}Rh \xrightarrow{\beta} {}^{104}Pd$. Here the arrow with β superposed indicates radioactive decay by beta-particle emission to ^{104}Pd.

Usually there are other radiations associated with the nuclear reaction, but the symbolism is restricted to the principal reaction or the reaction of interest. Reactions of interest in nuclear instrumentation are neutron induced and result in the emission of fission fragments and of alpha particles, namely, (n,f) and (n,α) reactions.

Most fission-neutron interactions with the nuclei of material in and around a nuclear reactor core are capture reactions [(n,γ) reactions] and elastic collisions [(n,n) reactions]. Some interactions are transmutations of the (n,p) type, and some are inelastic collisions [(n,n') reactions]. The products of these interactions, i.e., the neutrons, gammas, and protons, also interact with the nuclei of material in and around the reactor core.

Fission gammas and the gammas produced in neutron-capture or -scattering reactions interact with the electrons in surrounding material, usually creating energetic electrons (Compton or photo-effect).

Energetic particulate radiations (protons or nuclei recoiling after a nuclear reaction), if not stopped in nuclear reactions or nuclear collisions, ultimately lose their energy by ionizing atoms. Ionization results in a "cloud" of free electrons and positive ions. The cloud or track of ionization is sharply defined by the trajectory of the initiating particle and usually has a definite length, or range, that depends on the particle energy and the density of the medium. A heavy, highly charged particle, such as a fission fragment, has a short, densely ionized range. An electron, or beta, has a longer range with less-dense ionization.

The neutral radiations (neutrons and gamma rays) associated with fission travel much farther than the charged particles and have poorly defined ranges. They can penetrate thick layers of matter. Neutrons usually terminate in some nuclear reaction; gamma rays produce secondary electrons, which, in turn, are stopped by ionizing other atoms. The difference in the penetrating power of neutron and gamma rays makes possible a partially selective detection process in monitoring nuclear reactors.

All the interactions of the radiations accompanying fission produce heat. Thus the heat generated is a direct measure of the power of a nuclear reactor. (It is also a nuisance in instruments used to detect radiation.) Unfortunately, heat is a very slow indicator (because of thermal inertia). In the steady state, the reactor heat or thermal power can be accurately measured and used in calibrating the nuclear instrumentation.

2-1.3 Accepted Detection Principles

In principle, any fission-neutron or gamma interaction with matter which produces measurable effects can be used for reactor-power measurements. Practical considerations, however, limit the choice to a few. The commonly used detectors have evolved through a selection process based on considerations of signal strength, response time, and tolerance of interfering radiations.

The radiation detector produces an electrical signal that overrides typical electronic-circuit noise levels. If the detector is to be used in reactor protection systems, its sig[illegible] must be reliable and its response time should be sh[illegible] event of a reactor accident or major component failure, the detector may have to respond rapidly to initiate a shutdown before damage can occur. The time constants of signal-conditioning circuits are longer than detector response and normally are limiting for protection system considerations. In addition, a strong gamma background can obscure the neutron signal. Fission-product gamma radiation is always a large contributor to the background gammas in power reactors.

Despite considerable research and development on other types most operating detectors depend on gas ionization. Generally, gas ionization detectors[5] can be made to have sufficient sensitivity without excessive size and to have a wide operating range, a fast response time, and adequate radiation selectivity.

Most neutron sensors utilize gas ionization caused by charged particles emitted in neutron-induced fission reactions in ^{235}U or in (n,α) reactions in boron. Gamma sensors detect the ionization caused by Compton electrons.

2-1.4 "Out-of-Core" Defined

Figure 2.1 shows a typical location for an out-of-core neutron sensor. In this example, the sensor is also outside the reactor vessel. The figure also shows the magnitude of the neutron flux, the gamma exposure rate, and the temperature in the out-of-core location typical of a boiling-water or pressurized-water reactor during operation at rated power.

In today's power reactors the neutron flux inside the core boundary is always greater than 10^{11} neutrons cm^{-2} sec^{-1}. Consequently, it is current practice to define an out-of-core sensor as one that is not exposed to a neutron flux greater than 10^{11} neutrons cm^{-2} sec^{-1}. The temperature and gamma exposure rate are not involved in this definition. Both the temperature and the gamma exposure rate in Fig. 2.1 are at least an order of magnitude less in out-of-core locations than they would be within the core

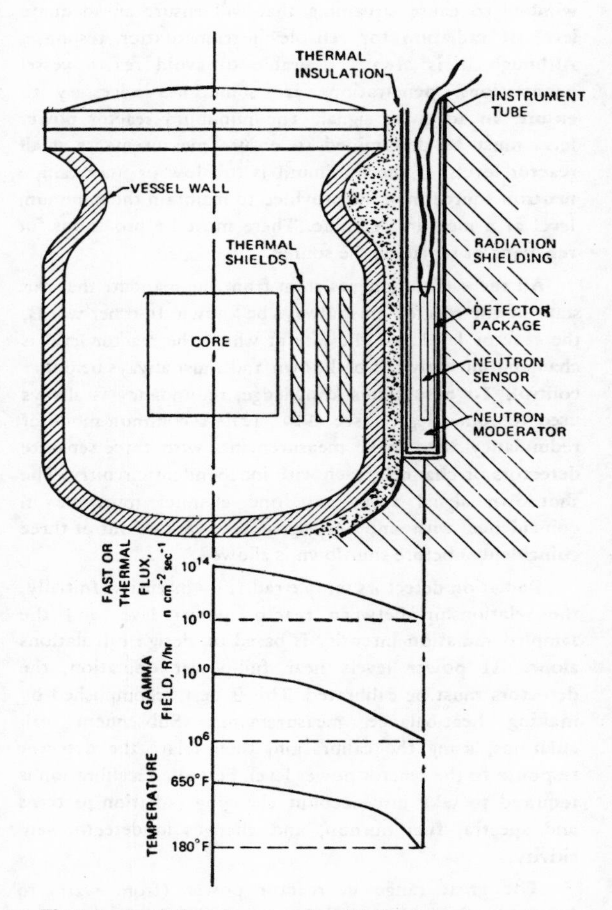

Fig. 2.1—Typical environmental profile for neutron sensors in a pressurized-water reactor at full power.

boundary. An out-of-core sensor, however, can be located inside the reactor pressure vessel, e.g., in the region of the thermal shield, provided the neutron flux does not exceed 10^{11} neutrons cm^{-2} sec^{-1}

2-1.5 Use of Out-of-Core Sensors in Reactors

The application of nuclear instrumentation demands an understanding of the behavior of a reactor. Because power density varies with position in the reactor, an average power measurement is needed. Out-of-core detectors are considered to be spatially averaging and are discussed here from this viewpoint. Detectors for measuring spatial variations in nuclear fluxes are discussed in Sec. 3. Out-of-core detectors are reasonably good averaging devices for most reactors if their installation is properly designed, especially in regard to shadowing by movable objects in or around the reactor.

To a limited extent, nuclear instrumentation influences reactor design. For example, it may be necessary to adjust the location of control elements to avoid shadowing effects on radiation sensors. It may be desirable to introduce a window to cause streaming that will ensure an adequate level of radiation for reliable instrumentation response. Although it is always desirable to avoid reactor-vessel penetrations, penetrations are sometimes necessary to ensure an adequate signal. The minimum reactor power level must be determined to ensure measurements at all reactor levels. If the minimum is too low or uncertain, a neutron source must be provided to maintain the minimum level at a measurable value. There must be provisions for renewing or replacing the source.

All these requirements stem from the mandate that the state of the reactor must always be known. In other words, the reactor level and the rate at which the reactor level is changing must always be known and must always be under control. To ensure this knowledge, redundancy is always used to some degree (see Sec. 12). A common mode of redundancy is to make measurements with three separate detectors or channels, each with independent circuitry. The shutdown signal from any one channel must be in coincidence with another signal (i.e., a two-out-of-three coincidence) before shutdown is allowed.

Radiation detectors sample radiation intensity. Initially, the relationship between reactor power level and the sampled radiation intensity is based on design calculations alone. At power levels near full-power operation, the detectors must be calibrated. This is best accomplished by making heat-balance measurements. Subsequent calculations, using the calibration, then relate the detector response to the reactor power level. Periodic recalibration is required to take into account changing radiation patterns and spectra, fuel burnup, and changes in detector sensitivity.

The great range in reactor power (from watts to hundreds of megawatts) makes it impossible to use one set of detectors and circuits, despite the wide range of the detectors. Research has produced detector and circuit arrangements capable of measuring over a range of 10 decades. Signal-conditioning circuits are the key to success (see Sec. 5).

A single set of detectors can be used to measure only a part of the reactor range and must be complemented by additional sets of detectors. For safety and reliability, a part of the range of the detectors is sacrificed by having them duplicate the measurements of a part of the range of other detectors. This duplication, or overlapping, is needed for a smooth transfer of control and safety functions from one detector set to another. The amount of overlap is typically one to two decades.

The most common way of dividing the power-level range is to use three ranges: source, intermediate, and power ranges.[6] This nomenclature is used in commercial practice. Figure 2.2 shows a typical selection of neutron detectors to cover these ranges.

Each range has peculiarities that depend on the radiation levels corresponding to that range and on whether or

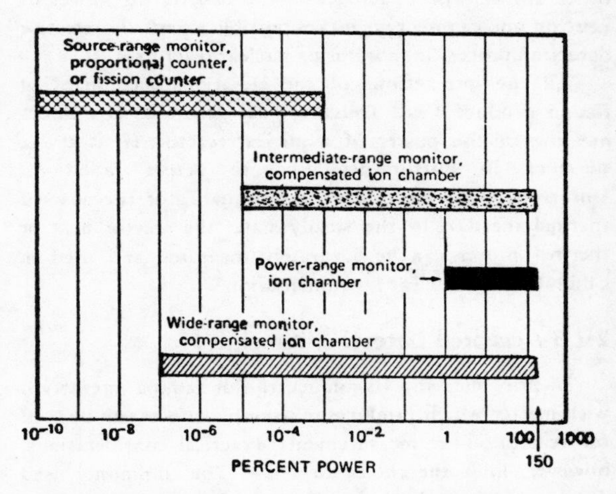

Fig. 2.2—Typical detectors used in out-of-core systems to cover the source, intermediate, and power ranges.

not a reactor has been operated. Special features, sometimes temporary, must be incorporated into the instrumentation design to ensure reliable performance during the initial period of a reactor when it is "clean and cold." The same instrumentation must operate when the reactor has accumulated its full burden of radioactivity and at every condition in between.

In the source and intermediate ranges, the reactivity of the reactor is limited by controlling or limiting the rate (period) at which the power can be increased. In the power range, instrumentation must prevent the reactor from exceeding its rated or licensed operating limit.

In the fully shutdown condition, the neutron density to which the detectors are exposed is frequently quite low, in fact, so low that individual neutrons are counted in order to gain information about the reactor status. Counting is also the only way to detect neutrons in the relatively high gamma fields that may be present.

The limits of the source range (or counting range) are determined by permissible counting rates, expressed in counts per second. The low end of the source range is determined by the counting rate needed to achieve a safe condition, as specified in the safety review (see Chap. 12). This minimum counting rate is normally from 1 to 10 counts/sec. The counting rate is established during the preliminary design and is fixed by consideration of the statistical nature of the neutron population and the time interval needed to achieve a measurement of prescribed accuracy. The counter must be located where the flux density is sufficiently large to ensure that this counting rate is achieved. The magnitude of the neutron source is selected to attain (at the detector location) a neutron flux that results in at least the minimum counting rate at all times. The maximum or high end of the source range is determined by the ability of the counter and the associated electronic circuits to resolve the individual counts. If the counting rate is too high, the resolution loss produces a serious error in the signal. Typical maximum counting rates are 0.5 to 1 × 10^6 counts/sec, and the allowable resolution loss is less than 10%.

The source range presents an adverse situation for the detection of neutrons. The detector used must be carefully selected for its sensitivity to neutrons in the presence of a large gamma background. The condition of few neutrons and many gammas exists immediately following a scram from full power.

The intermediate range overlaps the source range, and its gamma background is not as severe. However, since the neutron flux is high, individual neutrons are no longer resolved, and the signal takes on a direct-current aspect, becoming indistinguishable from the gamma background (also a d-c signal). Here again, it is essential to know the gamma level at the low end of the intermediate range. Through sensor design the gamma contribution at the low end is normally kept below 10% of the neutron signal. Again, the worst condition exists during a start-up immediately following full-power scram. The intermediate range usually extends into and completely overlaps the power range.

The power range covers from 1 to 150% of full power to provide some allowance for small power excursions. In the power range there is normally no great difficulty with interfering radiation. Neutron detectors that are not gamma compensated are satisfactory, but gamma-compensated sensors may be used for uniformity. These are similar to those used in the intermediate range.

2-2 PRINCIPLES OF GAS IONIZATION SENSORS

Most detectors in common use in nuclear power reactors are ionization chambers.[5,7] Since the principles of the ionization chamber apply broadly to gas-filled detectors, only ionization chambers are considered in this section. Specific characteristics of other types of gas-filled detectors are discussed in pt. 2-4.

2-2.1 Ionization Chambers

Ionization chambers are used to collect and measure the electric charge of ions and electrons that result from the interaction of incident radiation and secondary radiation from the chamber structure with the fixed, known volume of gas in the chamber. The quantity of collected charge is a measure of the incident radiation.

The sensitivity of ionization chambers can be increased by incorporating materials[8] into their structure which interact with the incident reactor radiation to create energetic ions or electrons. These materials can be coated as a thin film on the electrodes of the chamber or they can be included in the chamber as a gas. For the detection of thermal neutrons, ^{235}U and ^{10}B in various degrees of isotopic enrichment are used. Other materials can be used if it is desired to increase the chamber sensitivity to fast neutrons. Table 2.1 is a partial listing of materials that can be used in fast-neutron detectors. Of the isotopes listed, ^{238}U is available commercially in a limited number of chamber configurations; ^{239}Pu and ^{237}Np can be obtained by special order.

Table 2.1—Threshold Energies of Materials for Fast-Neutron Detection

Thermal	<1 MeV	>1 MeV
^{233}U	^{234}U	^{232}Th
^{235}U	^{237}Np	^{238}U
^{239}Pu		

Many isotopes of the heavy elements are of potential value in fast-neutron detection. Of these, the most likely candidates are ^{236}U, ^{241}Am, ^{240}Pu, and ^{241}Pu. The list is limited by the availability of accurate fission cross-section data, by interfering alpha radiation, and, in some instances, by spontaneous fission.

Fast-neutron measurements must be interpreted with caution. Some of the above isotopes are available only by separation from thermally fissionable isotopes. For example, ^{236}U and ^{240}Pu may have traces of ^{235}U and ^{239}Pu, respectively. These traces may cause errors. A second reason for caution is that these isotopes are sensitive to the entire neutron spectrum above the fission threshold. Thus, information about the entire neutron spectrum is contained in the signal.

Basically, an ionization chamber consists of at least two insulated electrodes sealed within a metallic case or enclosure. Connections are made through electrical seals designed for minimum charge leakage. Typical resistance between electrodes and case is high, of the order of 10^{12} ohms or greater. A guarded structure can be used for the seals if very low signal levels are anticipated. Such a structure consists of insulated conducting guard rings or cylinders surrounding each lead. The guard rings are maintained at the potential applied to the corresponding electrode. Figure 2.3 shows the structure of a fission chamber.

The electrodes must be designed and placed so there is a uniform electric field within the sensitive volume of the chamber. Auxiliary electrodes can be used to help attain field uniformity, which is particularly important when the sensitive volume of the chamber must be well defined and when uniform ion collection rates are needed. In practical detector designs for reactor service, electrode spacing is small compared to the electrode dimensions. As a consequence, the fringing electric field at the edges of the electrodes does not seriously compromise performance. Possible errors are usually small compared to other uncertainties in reactor flux measurements. Because of this, auxiliary electrodes for field shaping are not used in normal commercial practice.

High voltage, the magnitude depending on the chamber design and application, is applied to one of the electrodes, and the other electrode is operated close to ground potential. This high voltage may be quite low for a chamber of small physical dimensions and low gas pressure (<100 volts). For good performance, the minimum voltage on a large chamber with high gas pressure may be many hundreds of volts. Since the gas pressure is fixed to

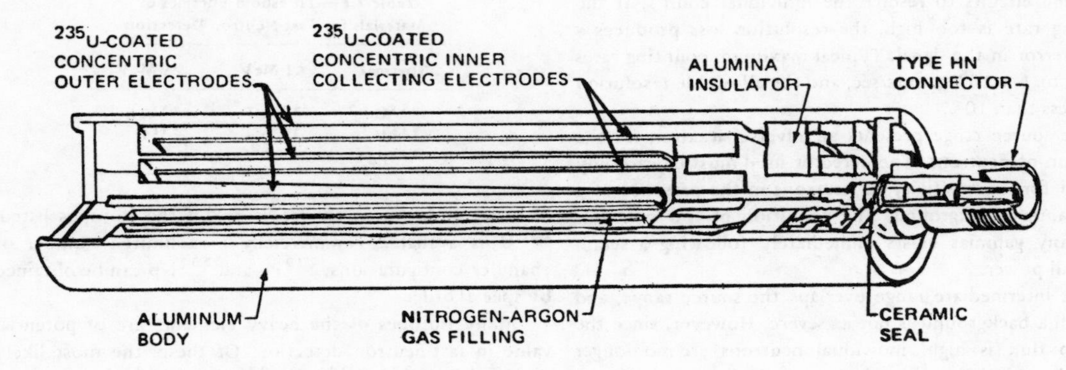

Fig. 2.3—Fission ionization chamber.

optimize performance (Fig. 2.4), the high-voltage limits are also fixed. The maximum voltage is limited by the danger of breakdown. The applied voltage is usually negative for counters since the collection of electrons is desired at the low-voltage electrode. For other types, the voltage may be positive or negative. The value of the applied voltage depends on the plateau of the chamber.

Free electrons and positive ions are produced in the gas fill of the chamber. Because ions are much heavier than electrons and not as mobile, electrons are normally collected faster.[9] However, in some gases there is a tendency for the free electrons to associate with the molecules of the gas, producing negative ions. These negative ions are also relatively immobile. The gas fill of the chamber is therefore commonly selected from one of the electron-free gases, such as the inert gases helium and argon. Nitrogen also has good properties and is frequently used. A number of gas mixtures promote electron mobility to an extent much greater than possible with any pure gas and have found great favor in ionization chambers.[10]

In general, gases whose chemical activity is enhanced by radiation are avoided since they may be gettered by the metals of the electrodes and the chamber enclosure. Gases that are chemical compounds may be dissociated in a high radiation field; recombination is not spontaneous in most of these gases, and their use is avoided. A BF_3 counter or chamber, for example, may have a short life.[11]

When a voltage is applied to an ionization chamber and the resulting current is measured in a constant radiation field, the current is found to vary with the voltage. A graph of pulse amplitude vs. voltage (Fig. 2.5) shows the pulse amplitude increasing with voltage in a very characteristic fashion. At low voltages competing gas processes hold the current down. As voltage is increased, the current increases rapidly, and, then, as the voltage is increased further, the current remains nearly constant, increasing only slightly. The flat region of nearly constant current is known as the plateau, and the region where the current begins to flatten is called the knee of the plateau. The voltage just above the knee is the minimum operating voltage. Normally, the operating voltage is selected, somewhat arbitrarily, to be considerably above the minimum.

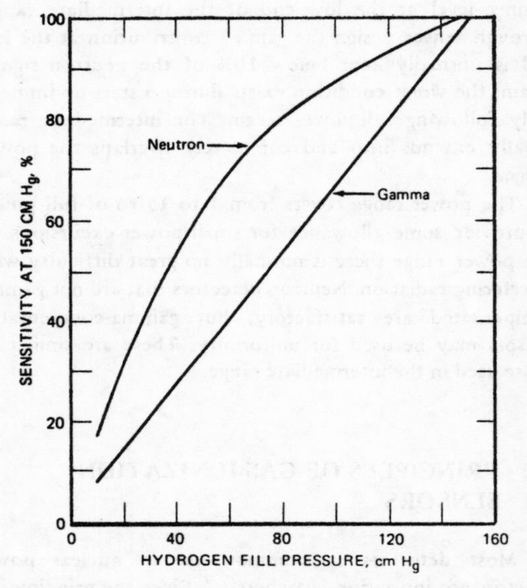

Fig. 2.4—Variation of neutron and gamma sensitivity with pressure for boron-coated electrodes spaced 0.25 in. apart. [From W. Abson and F. Wade, Nuclear-Reactor-Control Ionization Chambers, *Proc. Inst. Elec. Eng. (London)*, 103B(22): 590 (1956).]

As the chamber voltage is increased along the plateau, assuming that voltage breakdown does not occur, a point is reached where the current again begins to increase. Voltages above this point cause gas multiplication. This voltage or the breakdown voltage, whichever is smaller, determines the maximum voltage. In turn, the maximum rated voltage must normally be considerably below the breakdown voltage for reliable operation, again a somewhat arbitrary choice. Since the reactor radiation field tends to promote voltage breakdown, operation at minimum voltage is preferred.

If the voltage plateau for a given chamber[12] is determined for a number of radiation-field values, it is found that the voltage at the knee increases with radiation intensity, as shown in Fig. 2.6. This is caused by recombination and space-charge effects, which increase rapidly with radiation

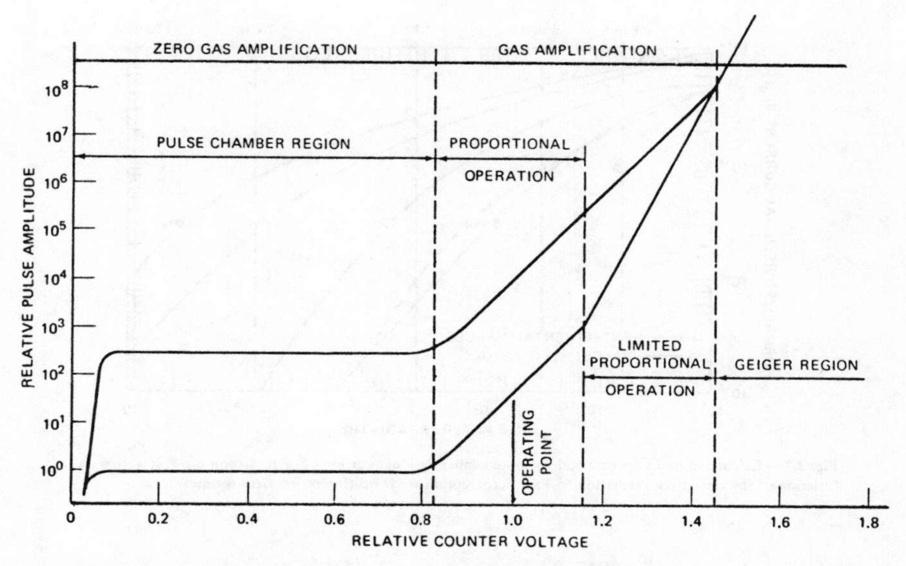

Fig. 2.5—Counter pulse height vs. supply voltage. [From H. Etherington (Ed.), *Nuclear Engineering Handbook*, p. 8-54, McGraw-Hill Book Company, Inc., New York, 1958.]

intensity. It is important to recognize this fact and to select an operating voltage that is on a plateau at the maximum radiation intensity experienced during operation. At this voltage, again somewhat arbitrary, the magnitude of the chamber current is a faithful indicator of radiation intensity.

Radiation emitted by neutron-induced radioactivity within the chamber may, under some circumstances, increase the background and limit the range of usefulness of the chamber.[13] Figure 2.7 shows the residual current due to the induced activity in the fission chamber.

2-2.2 Compensated Ionization Chambers

One of the problems with ionization chambers is that the detector is indiscriminate and will detect any ionizing radiation. If, for example, neutrons are to be detected in the presence of a strong gamma field and the neutron flux is to be related to the average current, then it is necessary to take into account the component of the current that is due to the gamma field.

Part of the gamma-induced current is due to prompt gammas and is proportional to the neutron-induced current. This part is indicative of reactor power and is not detrimental. However, the remainder of the gamma-induced current is relatively unchanging and creates a spurious signal, i.e., a signal not indicative of the reactor power level. This, in turn, is not a problem at high power levels when the neutron field is much more intense than the background gammas. At low power levels, the gamma contribution to the chamber current may be a large fraction of the chamber current and might exceed the neutron-induced current. Thus the range of reactor power in which an ionization chamber can be used to measure the power level may be severely reduced.

Two ionization chambers can be used to decrease the

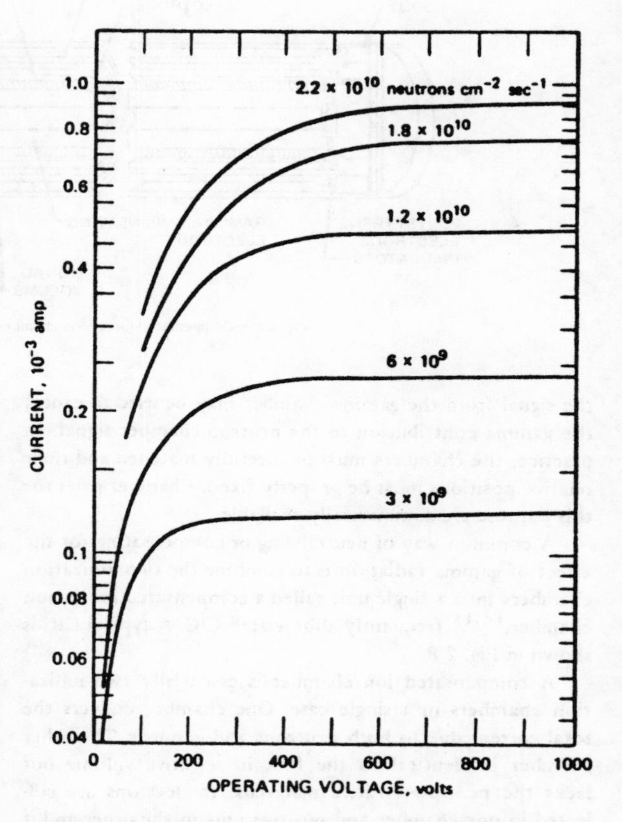

Fig. 2.6—Typical neutron saturation curves, illustrating the change in plateau with increasing current. (*Courtesy Westinghouse Electric Corp.*)

effect of the gamma background. If an ionization chamber sensitive to gamma radiation only is installed near an ionization chamber that is primarily sensitive to neutrons,

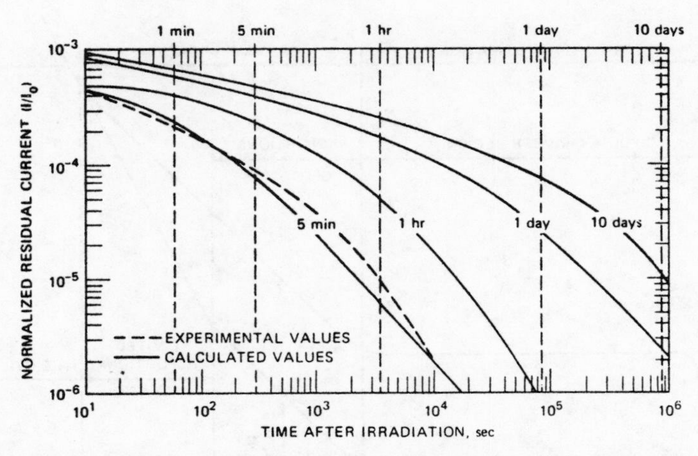

Fig. 2.7—Calculated and experimental fission-chamber residual currents due to fission-product activity as a function of the time after irradiation.[13] Parametric curves are for different irradiation times.

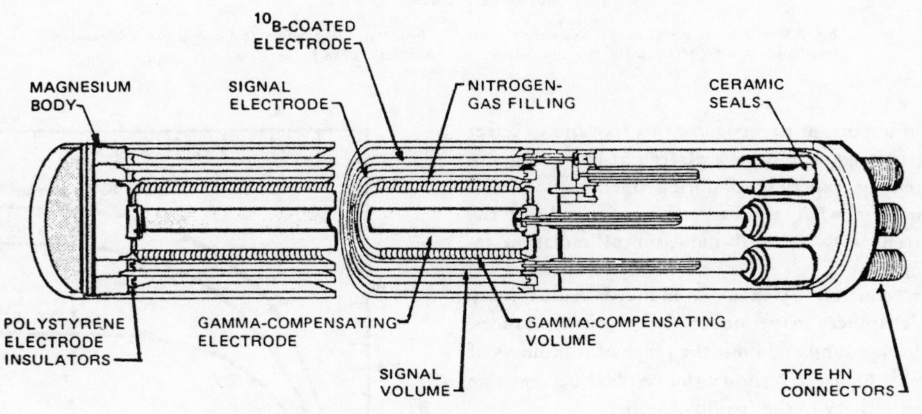

Fig. 2.8—Compensated ionization chamber.

the signal from the gamma chamber may be used to cancel the gamma contribution to the neutron-chamber signal. In practice, the chambers must be carefully matched and their relative positions must be properly fixed. Chamber pairs for this purpose are commercially available.

A common way of neutralizing or compensating for the effect of gamma radiation is to combine the two ionization chambers into a single unit called a compensated ionization chamber,[14,15] frequently abbreviated CIC. A typical CIC is shown in Fig. 2.8.

A compensated ion chamber is essentially two ionization chambers in a single case. One chamber collects the total current due to both neutrons and gammas. The other chamber is identical to the first in sensitive volume but lacks the neutron-sensitive materials. If electrons are collected in one chamber and positive ions in the other and if the resulting currents are summed, the gamma-induced currents are cancelled. In practice, it is normally immaterial whether the signal is obtained by collecting electrons or positive ions.

Theoretically, the cancellation could be complete. In practice, cancellation can be made complete at a given reactor power level. However, over the range of reactor power levels, most of the gamma effects can be nulled but a residual should be expected. The residual may be less than 1% of the signal or as great as 10% and depends on the specific detector and the effort made to achieve good compensation. A reasonable state-of-the-art number is 2 to 3%, i.e., 97 to 98% compensation.

It is also possible to overcompensate. Figure 2.9 illustrates the manner in which compensation varies as a function of operating voltage. Because of this voltage-sensitive feature, it is possible to use variable compensation.[16] The variable-compensation feature has been commercially used.

Figure 2.10 shows one way in which compensation might vary over a portion of the reactor range. With fixed voltages, compensation is exact at only one point. Figure 2.11 is an illustration of the improvement gained from using a compensated ionization chamber. The figure also provides insight for measuring and testing compensation in an operating reactor.

The use of a gamma-compensated detector extends the range, compared to that of an uncompensated detector, by

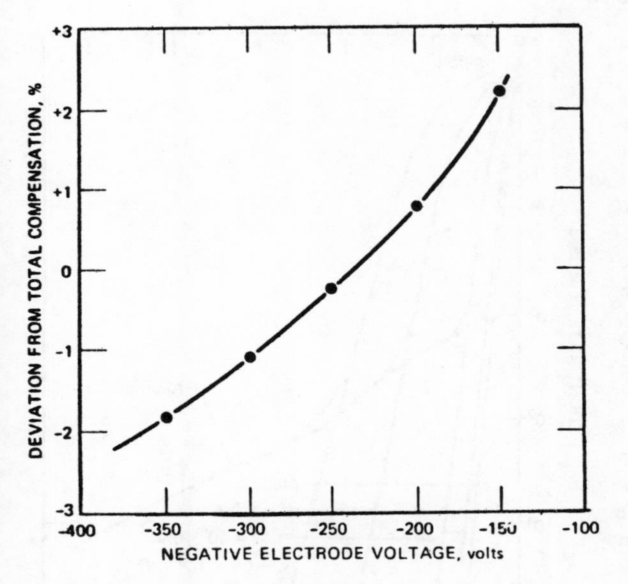

Fig. 2.9—Gamma compensation as a function of electrode voltages. Measurements were made in a hot cell with a ^{60}Co source.[13] Fixed positive electrode voltage, 300 volts. Basic compensation, −1.2%.

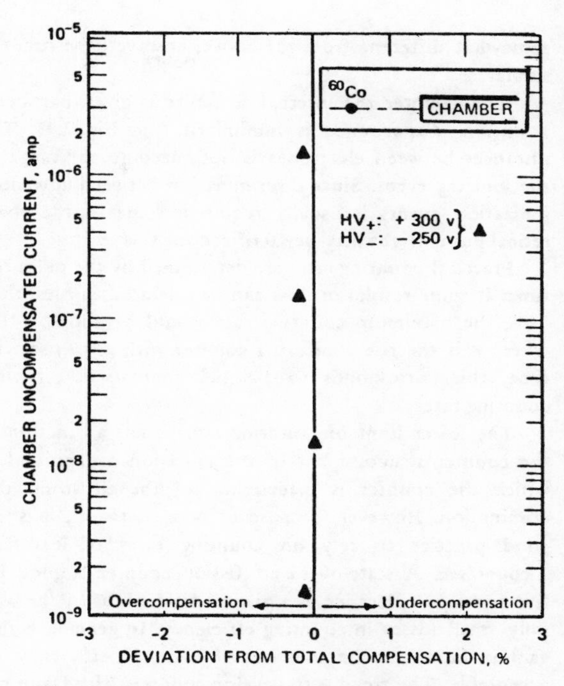

Fig. 2.10—Gamma compensation as a function of chamber current.[13]

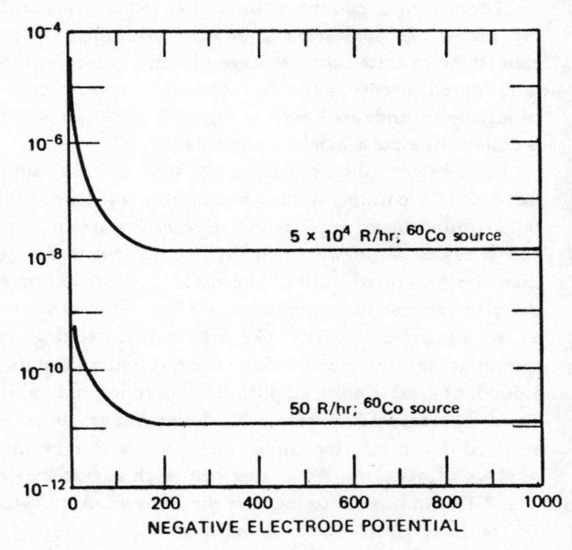

Fig. 2.11—Compensation characteristics in a compensated ionization chamber. (*Courtesy General Electric Co.*)

about two decades. Depending on the reactor, an uncompensated detector may be expected to cover three or more decades; a properly located compensated chamber can reasonably be expected to measure the reactor power over at least five decades. Recent practice has generally been to operate with fixed voltage and to design safety systems that avoid dependence on as much as two decades of compensation.

2-2.3 Counters

A counter is an ionization chamber designed to deliver a current pulse for each ionizing event.[17,18] A number of features are common with those of the more general ionization chamber. In fact, superficially, chambers designed for the current and counting modes are indistinguishable. There are commercial chambers designed to perform in both the current and counting modes, which, if the application is not demanding, can serve as well as a detector limited to a single mode.

The difference stems from basic differences in the signal, i.e., a pulse vs. direct current. For large pulses that are easily distinguishable from one another, the ionization of each ionizing event must be collected quickly. This is accomplished by careful design. In addition, the gas fill of the chamber must be selected to maximize electron mobility. There are a number of gas mixtures that are better in this respect than pure inert gases. A high performance counter would use one of these gas mixtures.

Because the signal is a pulse, the insulation quality of the chamber is not as critical. The pulse height depends primarily on the capacitance, C, and resistance, R, of the system in which the counter is used. Thus, if the time constant, RC, of the output circuit is 100 or more times the pulse rise time, negligible attenuation occurs. Since a typical installation might have a capacitance of 2000 pF, a counter with a pulse rise of 0.2 μsec requires an output resistance of only 10,000 ohms to meet this test. The insulation resistance then needs to be 10^6 ohms or more for satisfactory performance, a criterion that is easily met. The faster the pulse rise, the less a high resistance is required. If an electronic amplifier is introduced, the analysis is somewhat different from the above; however, the result is similar.

In a counter the internal structure is given particular attention. Capacitance is minimized (see Fig. 2.4). The distances between electrodes is optimized to the range of the ionizing event. Since discrimination between unwanted ionization events is usually required, some of the theoretical pulse height may be sacrificed for this end.

Practical counting rates are determined by the pulse rise time. If some resolution loss can be tolerated, say less than 10%, the maximum counting rate would be about $1/10\tau$, where τ is the rise time. In a counter with a 0.2-μsec rise time, this corresponds to 5×10^5 counts/sec, a typical counting rate.

The lower limit on counting rates is not a function of the counter structure but of the radiation background in which the counter is placed and of the electronic discrimination. However, because of poor statistics, it is not good practice to rely on counting rates of less than 1 count/sec. A state-of-the-art fission counter is good for five and a half decades in a gamma field of 10^5 R/hr with only small losses in counting efficiency. In general, higher gamma fields can be tolerated if losses in efficiency are acceptable. The trend is to develop counters with faster rise times and with corresponding improvements in maximum counting rate and gamma tolerance.

The use of a counter is somewhat more involved than the use of an ionization chamber. Not only must the counter be operated on a voltage plateau, but counts from the desired events must be separated from counts attributable to undesired events. Figure 2.12 shows how this is done with a pulse-height discriminator.

Pulse-height discrimination is one of the simpler methods of separating wanted pulses from unwanted pulses. It has found many applications in reactor start-up systems and is easiest to apply when the wanted pulses are larger than the unwanted pulses. The usual way of determining the effectiveness of a counter and of the counting system is by an integral bias curve. The discriminator setting corresponds to the number of counts, expressed in counts per second, of greater pulse amplitude. In practice, only a single curve is measured. Figure 2.12, however, has been constructed by measuring alpha and gamma curves in the absence of neutrons. An arrangement such as that shown in Fig. 2.13 can be used to measure these curves. A knowledge of the integral bias curves is indispensable in selecting and using a fission counter.

With a knowledge of the gamma background, possibly by measuring the integral bias curve, one can select the operating point to eliminate any desired function of the unwanted counts. There is, of course, a corresponding loss of efficiency since the neutron integral bias curve has a slope. This loss of efficiency is so well known that it is commonly equated with gamma discrimination. Ideally, and in some radiation detecting instruments, this is nearly the case, the wanted counts form a horizontal line (zero slope). In this ideal situation there is no loss of efficiency. One way to approach this ideal is to use a thin coating of

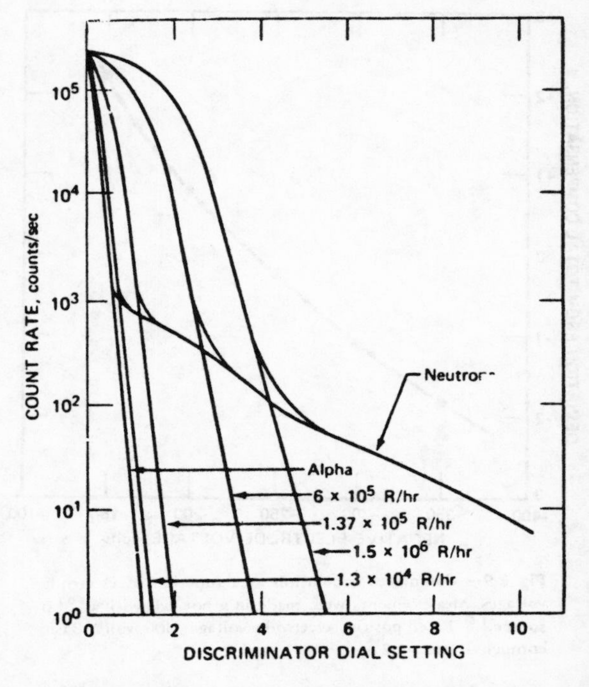

Fig. 2.12—Integral bias curve for a fission chamber modified to show typical channel count rate limiting. *(Courtesy Reuter-Stokes Electronic Components, Inc.)*

sensitive material. Figure 2.14 shows the effect of sensitive material thickness on discriminator response. It is seen that very thin films do approach the ideal.

Unless the chamber size were increased (with a consequent increase in gamma sensitivity) a great loss in sensitivity would occur if very thin films were used. Thus, most practical designs favor a thicker film of sensitive material, typically 1 to 2 mg/cm^2, as a workable compromise. The very thin films, however, are advantageous if absolute measurements must be made.

2-3 MECHANICAL FEATURES OF GAS IONIZATION SENSORS

2-3.1 Structural Design

In common with other reactor instruments, nuclear radiation detectors must be rugged. This entails construction features that should be recognized by the user.

The mechanical design is determined by the requirements of the manufacturing process, by the realities of handling and installation, and by the rigors of the environment in which it is to function. During construction, a gas ionization detector must withstand the mechanical stresses when it is evacuated before being filled with gas. It must also withstand the high temperature and vacuum used in the outgassing process. After it is filled with gas, the detector must be able to withstand the pressure and temperature changes in the reactor environment. These requirements in themselves are sufficient to ensure a high

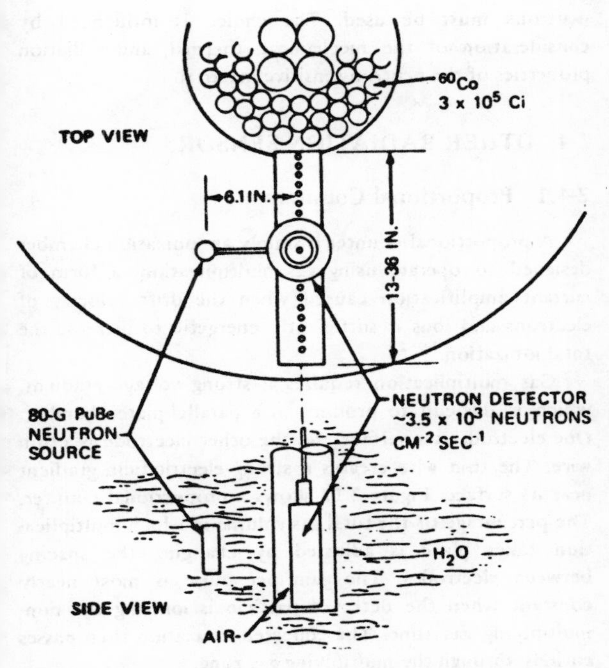

Fig. 2.13—Arrangement for measuring counter characteristics.

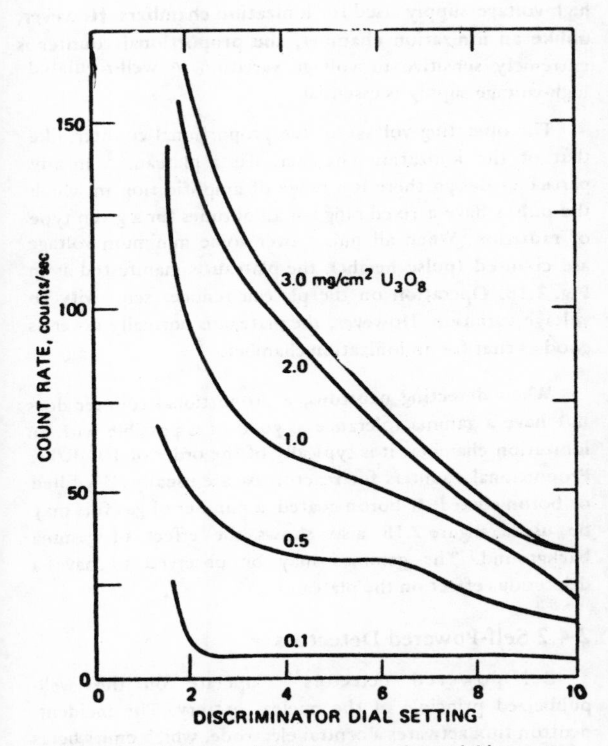

Fig. 2.14—Variation of fission-counter integral bias curve with coating thickness. [Adapted from William Baer and O. F. Swift, Some Aspects of Fission Counter Design, *Rev. Sci. Instrum.*, 23(1): 55 (1952).]

degree of ruggedness.

The detector must also withstand mishandling and possible abuse during shipping and installation. Some applications may require that the detector be moved during operation or between operating periods. Unless the accelerations associated with movements are very small, the detector must have electrodes with exceptional mechanical stability to prevent vibration. This must be ensured by proper design, e.g., including adequate electrode supports.

The detector environment is frequently subject to changes in temperature and pressure and to vibration or mechanical noise. Differential thermal expansion within the detector is minimized by using materials with similar thermal expansion coefficients, by providing adequate constraints, or by permitting relative motion. Any of these methods may make the detector microphonic. The high voltage and considerable interelectrode capacitance of the detector greatly enhance the generation of microphonics. This tendency can be minimized by careful design of the electrode assembly.

2-3.2 Materials of Construction

The choice of materials of construction for a gas ionization detector is a compromise between radiological, mechanical, and thermal requirements. For example, in a light structure made of materials of low atomic number, both the self-absorption and the buildup of detector radioactivity are reduced. Reduction of both of these effects is desirable in all nuclear radiation detectors. However, such a structure is not optimum for a gamma detector; gamma detection is based on generation of Compton-scattered (recoil) electrons, a process that is most efficient in heavier structures made of materials with high atomic number. Moreover, a light structure may not meet the requirements for detector ruggedness.

The primary criterion in materials selecting is to maximize the generation of the signal. Selecting materials to maximize sensitivity to incident gammas involves different considerations from those involved in selecting materials for a neutron detector.

In ionization chambers for gamma detection, construction is very important. Sensitivity to incident gammas is maximized by increasing the thickness of chamber walls and electrodes, by using structural materials of high atomic number, and by using high-density gases of high atomic number. There are optimum values for most design features. For example, chamber-wall thickness must not be so great that self-shielding causes excessive gamma attenuation. In fact, wall thickness need not greatly exceed the range of the most energetic Compton electrons. Gas pressure should not be so great as to cause excessive recombination of the ion pairs that are generated by the passage of Compton electrons through the gas. The materials selected for a gamma chamber should have low cross sections for reaction with any neutrons that may accompany the incident gammas. Moreover, it is preferable that any neutron reactions that do occur should generate a minimum of energetic ionizing radiations. Figures 2.15 and 2.16 show the effect of material selection and construction features.

On the other hand, if the ionization chamber is designed for maximum neutron sensitivity (with minimum gamma sensitivity), neutron-sensitive materials, i.e., materials that create ionizing radiations when exposed to

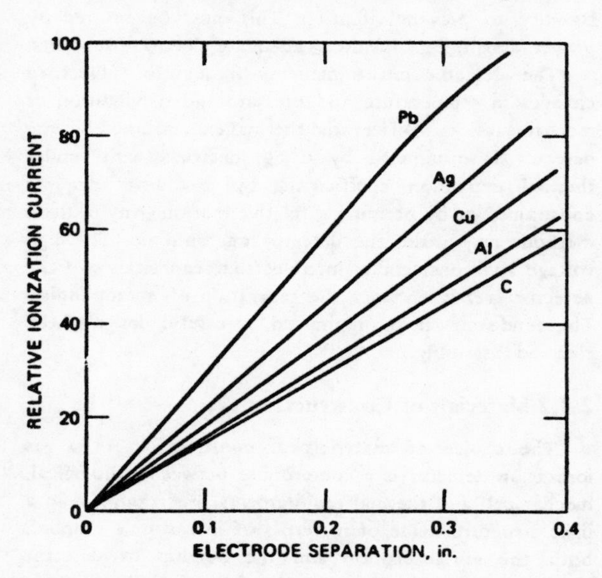

Fig. 2.15—Variation of ionization current with size of air volume and chamber material using ^{60}Co radiation. [From D. V. Cormack and H. E. Johns, The Measurement of High-Energy Radiation Intensity, *Radiat. Res.*, 1(2): 151 (1954).]

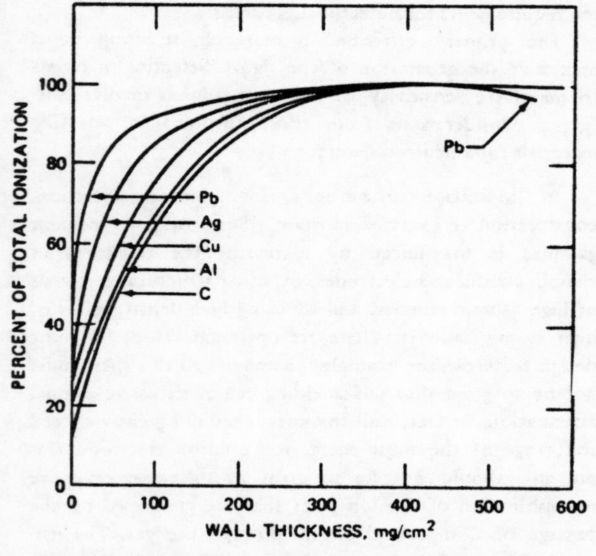

Fig. 2.16—Variation of ionization current with wall thickness and chamber material using ^{60}Co radiation. [From D. V. Cormack and H. E. Johns, The Measurement of High-Energy Radiation Intensity, *Radiat. Res.*, 1(2): 146 (1954).]

neutrons must be used. The choice is influenced by consideration of the mechanical, thermal, and radiation properties of the neutron-sensitive material.

2-4 OTHER RADIATION SENSORS

2-4.1 Proportional Counters

A proportional counter is simply an ionization chamber designed to operate using gas multiplication, a form of current amplification caused when the drift velocity of electrons and ions is sufficiently energetic to increase the total ionization.[19-22]

Gas multiplication requires a strong voltage gradient, which is difficult to produce in a parallel-plate chamber. One electrode is a surface and the other electrode is a thin wire. The thin wire creates a strong electric-field gradient near its surface. Figure 2.17 shows a proportional counter. The percentage of the total gas volume in which multiplication takes place is adjusted by changing the spacing between electrodes. The gain per pulse is most nearly constant when the detected particle is ionizing the nonmultiplying gas zone. The collected ionization then passes entirely through the multiplying gas zone.

Gas amplification is electrically adjusted by changing the applied high voltage. The sharp gradient about the wire electrode permits operation at moderate voltage. Thus, the high-voltage supply may be interchangeable with the high-voltage supply used for ionization chambers. However, unlike an ionization chamber, the proportional counter is extremely sensitive to voltage variation. A well-regulated high-voltage supply is essential.

The operating voltage of the proportional counter, like that of the ionization chamber, has a plateau.[23] In any particular design there is a range of amplification in which the pulses have a fixed range of amplitudes for a given type of radiation. When all pulses over some minimum voltage are counted (pulse height), the plateau is manifested as in Fig. 2.18. Operation on the plateau reduces sensitivity to voltage variation. However, the plateau is normally never as good as that for an ionization chamber.

When detecting neutrons, a proportional counter does not have a gamma tolerance as good as is possible with an ionization chamber; it is typically of the order of 10^3 R/hr. Proportional counters for reactor use are usually BF_3-filled or boron-coated. If boron-coated, a number of gas fills may be used. Figure 2.18 also shows the effect of gamma background. The gammas may be observed to have a deleterious effect on the plateau.

2-4.2 Self-Powered Detectors

Self-powered detectors[24] operate on the well-publicized principle of the nuclear battery. The incident-neutron flux activates a central electrode, which emits betas that are collected by a surrounding electrode. This type detector is usually designed for in-core neutron-flux sensing. It is discussed in Sec. 3, pt. 3-3.3.

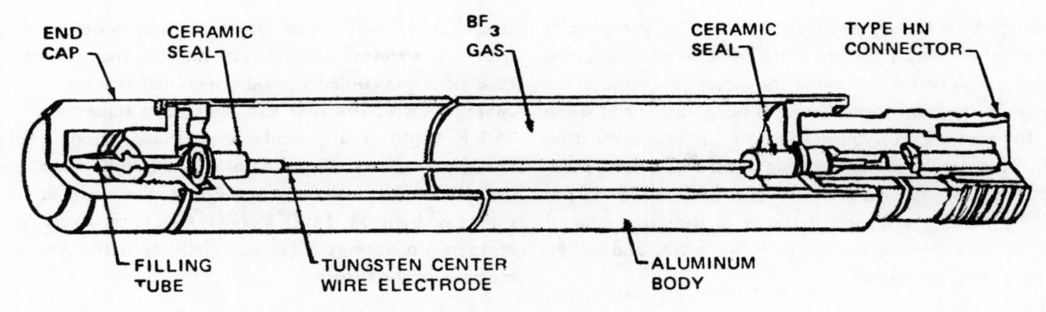

Fig. 2.17—A proportional counter.

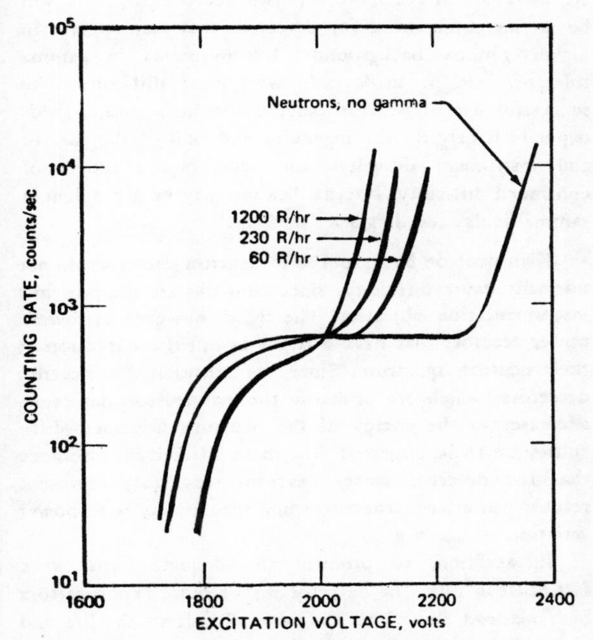

Fig. 2.18—Plateau characteristics of a BF_3 proportional counter under ^{60}Co gamma irradiation. [From O. F. Swift and R. T. Bayard, A Rugged BF_3 Proportional Counter, *Nucleonics*, 17(5): 126 (1959).]

2-4.3 Activation Detectors

Neutron flux at a given position in a reactor can be measured by exposing a material object[25] to the flux, removing it from the flux, and determining the activity that has been induced by exposure to the flux. From the exposure time and the known properties of the exposed material, the incident-neutron flux can be determined. This method can be used in in-core neutron-flux mapping. The exposed material can be in the form of wire, foil, ribbon, etc. Even liquids and gases[26] can be used.

2-4.4 Solid-State and Scintillation Detectors

Solid-state and scintillation detectors[27-30] can only be used where the neutron- and gamma-flux levels are low, for example, in radiation-monitoring systems. Solid-state detectors convert directly to an electrical signal, while scintillation detectors require an intermediate photoelectric stage. There is a wide variety of types. Most, however, are not applicable at neutron fluences above 10^{15} neutrons/cm^2 and integrated gamma exposures of 10^{12} rad.

2-5 INSTALLATION

Once a nuclear radiation sensor has been selected, installation arrangements must be made.

2-5.1 Cables

A mild environment (low temperature and noncorrosive media) poses few problems. Sensors are available with cable fittings, and the cable choice can be made easily. Over extended periods of time, the nuclear radiation field can interact with the gaseous medium in the vicinity of a sensor and cause corrosion. If the medium is air, elimination of moisture during installation and prevention of moisture accumulation after installation is advisable.

In a severe environment, a detector with an integral cable should be used. In this case, the electrical parts of the sensor are effectively separated from the medium.

2-5.2 Hardware

Radiological and mechanical requirements pose different installation problems. Frequently, some radiation conditioning must be provided, i.e., gamma or neutron shielding at the detector location. If the exact conditions are known, permanent shielding or neutron thermalizers can be built. Usually, some permanent shielding is installed, and flexibility is provided by adjusting the detector position. In this case, all possible locations of the detectors must be considered and shielded as needed.

It is often possible to provide some gamma shielding or neutron thermalization in the detector package. A detector assembly is then needed. The detector assembly must be positioned as a unit, which requires some provision for storing excess cable in or around the reactor shield. This is particularly difficult for a detector with integral cable. For counters a preamplifier must also be mounted as close to the detector as possible. Frequently, the excess cable and preamplifier space can be combined.

For a detector that is positioned vertically, gravity can usually be relied on. A load-bearing cable is then required to support the detector, relieving the electrical connections of any possible stress. If positioning is horizontal or at some angle to the vertical, constraints in both directions must be provided. This can be done by using a fixed or a temporary rigid member. If the position might be disturbed, for example, by moving a second detector in the same general location, a rigid connection or possibly a line-and-pulley arrangement may be needed.

2-5.3 Circuits

The more-specialized circuits are covered in other sections, particularly Sec. 5, and the effects of reactor-safety requirements are covered in Sec. 12. Some special points are considered here.

Interlocking circuits are customarily provided with each detector. These interlocks provide a scram signal if the high voltage or signal connections to a detector are interrupted regardless of the cause of the interruption.

Also it is generally desirable to prevent the establishment of a ground loop, i.e., to prevent ground currents from flowing in the return or ground lead. This is done by preventing the exposure of any bare metallic members of the return path. Thus, cable connectors and the chamber itself are sometimes insulated. The signal circuits are sensitive and of high impedance; so ground loops are a frequent source of trouble (see Sec. 10).

2-5.4 Immersion in Coolant

Dry thimbles provide a common and convenient method of installing detectors in coolants. The cost of installing dry thimbles is quite high, however, and, in addition, some installations may require cooling or circulating air to ensure dryness. Then too, there is always some concern over possible leakage, particularly if the thimble is a penetration of a reactor.

These difficulties can be reduced by using detectors suitable for immersion in the particular medium involved. Detectors with integral cable have been developed for immersion in water and hot gas, and a detector for immersion in liquid sodium is being developed.

Direct immersion, while generally advantageous, has some disadvantages. There must still be some sort of channel to restrict the path of motion of the detector. Also, the difficulty of making adjustments in position is increased. Nevertheless, detectors for direct immersion are finding increased favor.

2-6 ENVIRONMENT

Detector environment is largely determined by the type of reactor being instrumented. In general, out-of-core detectors can be placed in a milder environment than in-core detectors, especially if out-of-vessel locations are suitable. The various environments are described in Sec. 15 to 18.

The most important single environmental condition is temperature, which can vary from near room level remote from the reactor core to very high in the vicinity of the core of a gas-cooled or sodium-cooled reactor. Ionization-chamber detectors that can operate in temperatures up to 750°F (400°C) are available and are suitable for most existing reactors. Experimental ionization chambers have been operated with varying degrees of success at temperatures as high as 1400°F (760°C). It is likely that the temperature ratings of commercially available detectors will increase in the future.

Gamma background is also an important environmental condition for neutron detectors. As reactor power densities are increased in the more sophisticated designs, there will be an increasing need for detectors that can operate in higher gamma backgrounds. Improvements in gamma tolerance can be made only with great difficulty. The successful use of neutron detectors in high gamma fields depends largely on the ingenuity and skill of the reactor and instrument designers and represents a source of continued difficulty. Internal heating may be a problem in gamma fields exceeding 5×10^7 R/hr.

High neutron backgrounds in neutron detectors do not normally cause difficulty, since neutrons are the principal instrumentation objective. The trend, however, is toward power reactors that have a larger fast-neutron fraction in their neutron spectrum. Since the sensitivity of neutron detectors, which are primarily thermal-neutron detectors, decreases as the energy of the neutrons increases, difficulties are to be expected. The main difficulty is assurance that the detector senses neutrons that truly represent reactor power, i.e., that vary in a regular way with power variation.

In addition, to produce an adequate signal in a fast-neutron flux, the detector may require exposure to a high neutron flux. This can adversely affect the life and physical characteristics of the detector.

Adverse environmental conditions can be avoided by moving the detector. For example, start-up detectors that do not have to operate when the reactor is at power can be moved to regions of lower radiation intensity once they are out of range. Similarly, other sensors may be moved to a more favorable environment as reactor power is increased. The use of detector withdrawal might enable reliable instrumentation where the background is otherwise too severe. Safety instruments should not be moved unless they are no longer required. Satisfactory operation in the new position must be assured.

Excessive neutron flux and gamma gradients may contribute to instrumentation faults. In general, it is desirable to operate with a large signal to extend the sensor range. In a high flux gradient, a part of a sensor may be operating in excess of its rating. This implies operation with a lower input signal than proper use would provide. In addition, the most effective gamma compensation is obtained in a low-gradient gamma field. Thus, to the extent that the gamma gradient is related to the neutron gradient, it is likely that compensation may be less effective, and there will be a loss of range.

2-7 LIFE AND RELIABILITY

Valid statistical data on the performance of out-of-core neutron and gamma sensors are scarce. However, it may be inferred from examination of sensor designs and from general knowledge of the field that reliability has been good. The most frequently reported difficulties have been from spurious signals attributable to microphonic and electrical effects. (These are discussed in Sec. 10.) Since these effects can usually be observed in the debugging period, they can be rectified by modifying the system design.

Ionization detectors generally have long lives. The life of a neutron detector is, however, limited by consumption of sensitive materials. Since the consumption is directly related to the neutron fluence, the loss may be calculated and compensated for by recalibration. In early designs, flaking of sensitive material with subsequent deposition in insensitive parts of the chamber volume was occasionally experienced. Generally, flaking is not a problem at present. As might be expected, gamma detectors have an indefinitely long life since they contain no sensitive materials.

In a number of ionization-chamber detectors using special gas mixtures, gradual degradation results from radiologically induced changes in gas composition. Since this type detector is normally used only for very special purposes, it is expected that the user would be alert to any possible difficulties. Certain proportional counters fall in this category and are normally used only at moderate or low radiation levels.

Scintillation and solid-state detectors are useful at low radiation levels only and can withstand only a limited total radiation exposure.

2-8 TYPICAL SPECIFICATIONS OF COMMERCIAL GAS IONIZATION SENSORS

Typical specifications are summarized in Table 2.2. Since many different types are available, the values in the table have been entered as ranges.

Tables 2.3 through 2.7 indicate the variety of neutron and gamma sensors available from one manufacturer. The sensors listed are limited to those discussed in this chapter. Commercial in-core neutron sensors are described in Sec. 3.

REFERENCES

1. S. Glasstone and M. C. Edlund, *The Elements of Nuclear Reactor Theory*, D. Van Nostrand Company, Inc., Princeton, N. J., 1955.
2. Reactor Physics Constants, USAEC Report ANL-5800 (2nd Rev.), Argonne National Laboratory, Superintendent of Documents, U. S. Government Printing Office, Washington, D. C., 1963.
3. J. A. Crowther, *Ions, Electrons and Ionizing Radiations*, 8th ed., Edward Arnold, Ltd., London, 1949.
4. D. R. Bates (Ed.), *Atomic and Molecular Processes*, Academic Press, Inc., New York, 1962.
5. B. B. Rossi and H. H. Staub, *Ionization Chambers and Counters, Experimental Techniques*, McGraw-Hill Book Company, Inc., New York, 1949.
6. *USA Standard Glossary of Terms in Nuclear Science and Technology*, USAS N 1.1–1967, United States of America Standards Institute, New York, 1967.
7. W. Abson and F. Wade, Nuclear-Reactor-Control Ionization Chambers, *Proc. Inst. Elec. Eng. (London)*, 103B(22): 590 (1956).
8. M. L. Awcock, U^{235}-Coated Ionization Chamber, Type 1Z-400, Canadian Report AECL-805, pp. 44-45, August 1959.
9. L. Colli and V. Facchini, Drift Velocity of Electrons in Argon, *Rev. Sci. Instrum.*, 23: 39 (1952).
10. V. Facchini and A. Malvicini, Argon–Nitrogen Fillings Make Ion Chambers Insensitive to O_2 Contamination, *Nucleonics*, 13(4): 36 (1955).
11. W. M. Trenholme, Effects of Reactor Exposure on Boron-Lined and BF_3 Proportional Counters, *IRE Trans. Nucl. Sci.*, NS-6(4): 1 (1959).
12. J. L. Kaufman, High Current Saturation Characteristics of the ORNL Compensated Ionization Chamber (Q-1045), USAEC Report CF-60-5-104, Oak Ridge National Laboratory, May 25, 1960.
13. D. P. Roux, Parallel-Plate Multisection Ionization Chambers for High-Performance Reactors, USAEC Report ORNL-3929, Oak Ridge National Laboratory, April 1966.
14. E. B. Hubbard, Compensated Ion Chamber, in *Proceedings of the 1959 Biannual National Nuclear Instrumentation Symposium, Idaho Falls, Idaho, June 24–26, 1959*, ISA Vol. 2, pp. 99-106, Instrument Society of America.
15. W. H. Todt, A Gamma Compensated Neutron Ionization Chamber Detector for the NERVA Reactor, *IEEE (Inst. Elec. Electron. Eng.) Trans. Nucl. Sci.*, NS-15(1): 9-1 (1968).
16. H. S. McCreary, Jr., and R. T. Bayard, A Neutron Sensitive Ionization Chamber with Electrically Adjusted Gamma Compensation, *Rev. Sci. Instrum.*, 25: 161 (1954).
17. W. Baer and R. T. Bayard, A High Sensitivity Fission Counter, *Rev. Sci. Instrum.*, 24: 138 (1953).
18. W. Abson, P. G. Salmon, and S. Pyrah, The Design, Performance and Use of Fission Counters, *Proc. Inst. Elec. Eng. (London)*, 105B(22): 349 (1958).
9. S. A. Korff, Proportional Counters, *Nucleonics*, 6(6): 5 (1950).
0. S. A. Korff, Proportional Counters, II, *Nucleonics*, 7(5): 46 (1950).
21. W. Abson, P. G. Salmon, and S. Pyrah, Boron-Trifluoride Proportional Counters, *Proc. Inst. Elec. Eng. (London)*, 105B(22): 357 (1958).
22. N. M. Gralenski and J. E. Schroeder, Description and Analysis of a Sensitive BF_3 Filled Uncompensated Ionization Chamber, USAEC Report DC-60-11-73, General Electric Company, Nov. 10, 1960.
23. R. B. Mendell and S. A. Korff, Plateau Slopes and Pulse Characteristics of Large, High-Pressure BF_3 Counters, *Rev. Sci. Instrum.*, 30: 442 (1959).
24. J. W. Hilborn, Self-Powered Neutron Detectors for Reactor Flux Monitoring, *Nucleonics*, 22(2): 64 (1964).
25. J. Moteff, Neutron Flux and Spectrum Measurement with Radioactivants, *Nucleonics*, 20(12): 56 (1962).
26. W. C. Judd, Continuous Flux Monitoring of a High-Flux Facility with A^{41}, in Reactor Technology Report No. 14, USAEC Report KAPL-2000-11, p. III-1, Knolls Atomic Power Laboratory, 1960.
27. V. Adjacic, M. Kurepa, and B. Lalovic, Semiconductor Measures Fluxes in Operating Core, *Nucleonics*, 20(2): 47 (1962).
28. R. Babcock, Radiation Damage in SiC, *IEEE Trans. Nucl. Sci.*, NS-12(6): 43-47 (1965).
29. R. V. Babcock and H. C. Chang, SiC Neutron Detectors for High Temperature Operation, in *Neutron Dosimetry*, Symposium Proceedings, Harwell, Eng., December 1962, pp. 613-622, International Atomic Energy Agency, Vienna, 1963 (STI/PUB/69).
30. R. R. Ferber and G. N. Hamilton, Silicon Carbide High Temperature Neutron Detectors for Reactor Instrumentation, *Nucl. Appl.*, 2: June 1966.

Table 2.2—Typical Specifications for Commercial Out-of-Core Gas Ionization Detectors

	Gamma chambers	Ionization chambers	Compensated ionization chambers	Fission counters	Proportional counters
Sensitivity:					
amp/(R/hr)	10^{-13} to 10^{-9}	10^{-12} to 10^{-10}	10^{-12} to 10^{-11}		
amp/nv*		10^{-14} to 5×10^{-13}	10^{-15} to 10^{-14}		
(counts/sec)/nv*				10^{-4} to 2	3 to 40
Operating voltage	100 to 1500	200 to 1200	200 to 1500	200 to 1200	800 to 5000
Max. temp., °F	175 to 600	175 to 850	175 to 750	250 to 850	175 to 500
Diameter, in.	1 to 3	1 to 3.5	3 to 4	0.1 to 3	1 to 6
Length, in.	12 to 16	10 to 16	8 to 25	5 to 300	10 to 40

*Where nv is measured in neutrons cm^{-2} sec^{-1}.

Table 2.3—Uncompensated Ionization Chambers*

Neutron sensitive material	Thermal-neutron sensitivity, amp/nv†	Gamma sensitivity, amp/(R/hr)	Max. oper. thermal-neutron flux, nv†	Typical oper. voltage, volts (d-c)	Min. signal resistance, ohms	Signal capacitance, pF	Max. oper. temp., °F	Detector insulator ‡	Nominal dimensions: Length: Sensitive, in.	Nominal dimensions: Length: Overall, in.	Nominal dimensions: Detector O.D., in.
^{235}U	1.4×10^{-13}	4.2×10^{-11}	1.4×10^{10}	300–1000	10^{9}	150	300	Al_2O_3	6	11½	2
^{10}B	4.4×10^{-14}	4.5×10^{-11}	5.0×10^{10}	200–1000	10^{11}	170	300	Al_2O_3	7	13⅜	3
^{235}U	2.6×10^{-14}	3.0×10^{-11}	8.5×10^{10}	300–1000	10^{9}	140	300	Al_2O_3	6	11½	2
^{235}U	3.0×10^{-14}	4.2×10^{-11}	6.0×10^{10}	300–1000	10^{9}	150	300	Al_2O_3	6	11½	2
^{235}U	1.4×10^{-14}	4.2×10^{-11}	1.4×10^{11}	300–1000	10^{9}	150	300	Al_2O_3	6	11½	2
^{10}B	4.4×10^{-14}	4.5×10^{-11}	5.0×10^{10}	200–1000	10^{10}	170	575	Al_2O_3	7	13⅜	3
^{235}U	4.0×10^{-14}	4.0×10^{-11}	2.7×10^{10}	200–1000	10^{9}	160	575	Al_2O_3	7	13⅜	3
^{235}U	1.4×10^{-13}	4.2×10^{-11}	1.4×10^{10}	300–1000	10^{9}	150	300	Al_2O_3	6	11½	2
^{10}B	4.4×10^{-14}	4.5×10^{-11}	5.0×10^{10}	200–1000	10^{11}	170	300	Al_2O_3	7	13⅜	3
^{10}B	1.5×10^{-14}	3.5×10^{-12}	5.0×10^{10}	300–800	10^{13}	110	175	Rex	5½	10½	3½
^{235}U	2.8×10^{-14}	4.0×10^{-11}	5.0×10^{10}	300–1000	10^{9}	170	500	Al_2O_3	7	13⅜	3
^{235}U	5.1×10^{-14}	5.0×10^{-11}	2.7×10^{10}	300–1000	10^{7}	283	850	Al_2O_3	10	15⅞	1⅞
^{235}U	4.0×10^{-14}	4.0×10^{-11}	2.7×10^{10}	300–1000	10^{7}	150	700	Al_2O_3	7	13⅜	3
^{235}U	1.4×10^{-13}	4.2×10^{-11}	1.4×10^{10}	300–1000	10^{8}	1000	390	Al_2O_3	6	276	3
^{10}B	1.2×10^{-14}	3.0×10^{-12}	1.0×10^{10}	200–1000	10^{12}	1880	175	Al_2O_3	10	15⅛	1
^{10}B	3.0×10^{-13}	1.8×10^{-10}	2.5×10^{10}	300–1100	10^{13}	1850	175	Rex	108	113⅜	3½

**Courtesy Westinghouse Electric Corp.*

†Nv is expressed in neutrons cm^{-2} sec^{-1}.

‡Al_2O_3 is a high-alumina-content ceramic; Rex is a cross-linked styrene.

Table 2.4—Compensated Ionization Chambers*

Thermal-neutron sensitivity, amp/nv†	Uncomp. gamma sensitivity, amp/(R/hr)	Max. oper. thermal-neutron flux, nv†	Typical oper. voltage, volts (d-c)	Min. signal resistance, ohms	Signal capacitance, pF	Max. oper. temp., °F	Insulation type‡		Nominal dimensions		
									Length		
							Detector	Conn.	Sensitive, in.	Overall, in.	Detector O.D., in.
4.4×10^{-14}	2.3×10^{-11}	2.5×10^{10}	300–1000	10^{14}	275	175	Rex	Rex	14	$23\frac{5}{8}$	$3\frac{1}{8}$
4.4×10^{-14}	2.3×10^{-11}	2.5×10^{10}	300–1000	10^{14}	275	175	Rex	Rex	14	$24\frac{1}{8}$	$3\frac{1}{8}$
4.4×10^{-14}	2.5×10^{-11}	2.5×10^{10}	300–1000	10^{12}	315	575	Al_2O_3	Al_2O_3	14	$23\frac{3}{4}$	$3\frac{1}{8}$
4.4×10^{-14}	2.3×10^{-11}	2.5×10^{10}	300–1000	10^{13}	275	175	Rex	Rex	14	$23\frac{5}{8}$	$3\frac{1}{8}$
1.5×10^{-14}	3.5×10^{-12}	2.5×10^{10}	300–800	10^{13}	130	175	Rex	Rex	$5\frac{1}{2}$	$10\frac{1}{2}$	$3\frac{1}{2}$
1.5×10^{-14}	3.5×10^{-12}	2.5×10^{10}	300–800	10^{13}	135	175	Rex	Rex	$5\frac{1}{2}$	$10\frac{1}{2}$	$3\frac{1}{2}$
4.4×10^{-14}	2.3×10^{-11}	2.5×10^{10}	300–1000	10^{13}	290	400	Al_2O_3	Al_2O_3	14	$19\frac{1}{8}$	$3\frac{1}{8}$
1.0×10^{-15}	1.5×10^{-13}	1.5×10^{12}	25–250	10^{11}	155	660	Al_2O_3	Al_2O_3	$2\frac{5}{16}$	$7\frac{5}{16}$	3
4.4×10^{-14}	2.3×10^{-11}	2.5×10^{11}	300–1500	10^{13}	290	300	Al_2O_3	Al_2O_3	14	$19\frac{1}{8}$	$3\frac{1}{8}$

**Courtesy Westinghouse Electric Corp.*

†Nv is expressed in neutrons cm^{-2} sec^{-1}

‡Al_2O_3 is a high-alumina-content ceramic; Rex is a cross-linked styrene.

Table 2.5—Fission Counters*

Thermal-neutron sensitivity, (counts/sec)nv†	Max. oper. thermal-neutron flux, nv†	Typical oper. voltage, volts (d-c)	Min. signal resistance, ohms	Signal capacitance, pF	Max. oper. temp., °F	Insulator type‡		Nominal dimensions		
								Length		
						Detector	Conn.	Sensitive, in.	Overall, in.	Detector O.D., in.
0.7	1.4×10^5	200–800	10^9	150	300	Al_2O_3	Al_2O_3	6	$11\frac{1}{2}$	2
0.2	5.0×10^5	200–800	10^9	140	300	Al_2O_3	Al_2O_3	6	$11\frac{1}{2}$	2
0.14	7.0×10^5	200–800	10^9	150	300	Al_2O_3	Al_2O_3	6	$11\frac{1}{2}$	2
1.25×10^{-3}	1.0×10^8	250–500	10^9	55	575	Al_2O_3	Rex	$\frac{7}{8}$	$53\frac{5}{16}$	0.210
0.52	2.0×10^5	200–800	10^9	150	300	Al_2O_3	Al_2O_3	6	$11\frac{1}{2}$	2
0.07	1.4×10^6	200–800	10^9	150	300	Al_2O_3	Al_2O_3	6	$11\frac{1}{2}$	2
0.7	1.4×10^5	200–800	10^9	160	575	Al_2O_3	Al_2O_3	7	$13\frac{3}{4}$	3
0.7	1.4×10^5	200–800	10^9	150	300	Al_2O_3	Al_2O_3	6	$11\frac{1}{2}$	2
0.7	1.4×10^5	200–800	10^7	150	700	Al_2O_3	Al_2O_3	6	$11\frac{1}{2}$	2
0.1	1.0×10^6	300–800	10^9	30	575	Al_2O_3	Al_2O_3	$4\frac{3}{4}$	$7\frac{7}{8}$	1
0.14	7.0×10^5	200–800	10^9	170	500	Al_2O_3	Al_2O_3	7	$13\frac{5}{8}$	3
0.5	2.0×10^5	200–800	10^7	283	850	Al_2O_3	Al_2O_3	10	$15\frac{7}{8}$	$1\frac{7}{8}$
0.25	4.0×10^5	200–800	10^7	150	700	Al_2O_3	Al_2O_3	7	$13\frac{5}{8}$	3
5×10^{-3}	2.0×10^8	350–650	10^1	40	750	Al_2O_3		1	5	2
0.7	1.4×10^5	75	10^9	160	500	Al_2O_3	–	7	14	3
0.35	2.8×10^8	200–800	10^9	150	300	Al_2O_3	Al_2O_3	6	$11\frac{1}{2}$	2
§	1.0×10^8	200–800	10^9	150	300	Al_2O_3	Al_2O_3	6	$11\frac{1}{2}$	2
0.7	1.4×10^5	200–800	10^8	1000	390	Al_2O_3	Al_2O_3	6	276	3
2.2×10^{-4}	5.0×10^8	250–800	5×10^8	2	500	Al_2O_3		$\frac{5}{16}$	$1\frac{1}{16}$	0.220
1.5×10^{-3}	1.0×10^8	300–500	10^8	260	500	Al_2O_3	Al_2O_3	$\frac{3}{8}$	243	0.210
1×10^{-5}	1.0×10^{11}	100–200	10^9	125	250	TE	Rex	$\frac{11}{32}$	$63\frac{5}{16}$	0.090
0.18	6.0×10^5	200–800	10^9	45	575	Al_2O_3	Al_2O_3	$8\frac{7}{8}$	12	1
0.5	2.0×10^5	200–800	10^8	150	390	Al_2O_3	Al_2O_3	6	$11\frac{1}{2}$	2
0.35	3.0×10^5	200–800	10^8	90	390	Al_2O_3	Al_2O_3	3	$8\frac{1}{2}$	2

**Courtesy Westinghouse Electric Corp.*

†Nv is expressed in neutrons cm^{-2} sec^{-1}

‡Al_2O_3 is a high-alumina-content ceramic: Rex is a cross-linked styrene.

§Sensitive material is ^{238}U. Sensitivity to $\gtrsim 1.5$ MeV neutrons = 10^{-3}; to thermal neutrons = 1.4×10^{-4}.

Table 2.6—Proportional Counters*

Thermal-neutron sensitivity, (counts/sec)/nv†	Max. oper. thermal-neutron flux, nv†	Gas fill press., cm Hg	Typical oper. voltage, volts (d-c)	Min. signal resistance, ohms	Signal capacitance, pF	Max. oper. temp., °F	Insulator type‡		Nominal dimensions		
									Length		
							Detector	Conn.	Sensitive, in.	Overall, in.	Detector O.D., in.
					BF_3 Counters						
4.5	1.0×10^5	55	2000	10^{11}	10	250	Al_2O_3	Rex	8 5/8	12	1
13	4.0×10^4	55	2000	10^{11}	20	250	Al_2O_3	Rex	26	30 7/8	1
40	1.25×10^4	55	2000	10^{11}	610	175	Al_2O_3	Rex	26	39 5/8	5 7/8
13	4.0×10^4	55	2000	10^{10}	30	175	Al_2O_3	Rex	8 5/8	15 7/16	2 7/8
13	4.0×10^4	55	2000	10^{11}	15	250	Al_2O_3		26	29 7/8	1
40	1.25×10^4	55	2000	10^{11}	610	175	Al_2O_3	Rex	26	39 5/8	4 13/16
35	1.25×10^4	55	2000	10^{11}	70	310	Al_2O_3	Rex	26	33	3
45	1.0×10^4	55	2000	10^{10}	780	175	Al_2O_3		26	33 1/2	§
19	2.6×10^4	160	2200	10^{11}	10	250	Al_2O_3	Rex	12	15 7/16	1
13	4.0×10^4	160	4100	10^{11}	10	250	Al_2O_3	Rex	8 5/8	12	1
35	1.4×10^4	70	2100	10^{12}	10	300	Al_2O_3	Rex	12	15 7/16	2
6.5	7.5×10^4	55	2000	10^{11}	10	250	Al_2O_3	Rex	12	15 7/16	1
20	2.5×10^4	167	4700	10^{11}	10	250	Al_2O_3	Rex	12	15 7/16	1
					^{10}B Counters						
10	5.0×10^4	20	800	10^{12}	20	400	Al_2O_3	Rex	26	30 7/8	1
3	1.7×10^5	20	800	10^{12}	12	400	Al_2O_3	Rex	8 5/8	12 13/16	1
4	1.25×10^5	20	800	10^{12}	14	400	Al_2O_3	Rex	10 1/2	14 15/16	1
3	1.7×10^5	20	800	10^{12}	12	500	Al_2O_3	Al_2O_3	8 5/8	12	1
31	1.7×10^4	20	800	10^{11}	70	400	Al_2O_3	Rex	26	33	3
					^3He Counters						
22	2.5×10^4	760	4500	10^{11}	10	300	Al_2O_3	Rex	6	10 7/8	1
16	3.0×10^4	152	900	10^{11}	10	300	Al_2O_3	Rex	8 5/8	12	1

Courtesy Westinghouse Electric Corp.

†Nv is expressed in neutrons cm^{-2} sec^{-1}.

‡Al_2O_3 is a high-alumina-content ceramic; Rex is a cross-linked styrene.

§Oval case: 7 15/16 in. and 3 7/16 in.

Table 2.7—Gamma Chambers*

Gamma sensitivity, amp/(R/hr)	Max. oper. gamma flux, R/hr	Typical oper. voltage, volts (d-c)	Min. signal resistance, ohms	Signal capacitance, pF	Max. oper. temp., °F	Detector insulator†	Nominal dimensions		
							Length		
							Sensitive, in.	Overall, in.	Detector O.D., in.
3.0×10^{-12}	4×10^5	200–1000	10^{12}	1880	175	Al_2O_3	10	15 1/8	1
1.0×10^{-11}	5×10^7	100–1200	10^{13}	125	575	Al_2O_3	8	12 7/8	2
1.0×10^{-10}	3×10^7	100–1200	10^{13}	125	575	Al_2O_3	8	12 7/8	2
2.5×10^{-9}	2×10^3	200–1500	10^{11}	170	300	Al_2O_3	7	13 7/8	3

†Al_2O_3 is a high-alumina-content ceramic.

Courtesy Westinghouse Electric Corp.

SECTION 3

NEUTRON SENSORS-IN-CORE

3-1 INTRODUCTION

In-core neutron sensors accomplish one or more of the following: (1) confirm calculated core performance, (2) confirm core operating safety margins, (3) provide input data for fuel management, and (4) detect the existence of xenon-induced power asymmetries or oscillations (see sec. 1, pt. 1–3.6(a) for a discussion of xenon).

The first-of-a-kind power reactor core is usually more thoroughly instrumented than cores of duplicate or similar reactors because of the need to confirm calculated nuclear and thermal performance. This practice has become prevalent as core designers have become more dependent on computer codes for nuclear and thermal data and less dependent on critical-experiments data. The first-of-a-kind plant therefore serves as a field laboratory to confirm design calculations. It is not intended to generate new experimental data.

When reactor complexity, size, power output, power density, or neutron-flux level increases beyond certain limits, safe operation of the reactor over its lifetime cannot depend on data from out-of-core instruments. The reactor operator must have available to him the outputs from in-core neutron sensors so he can determine if the core is operating within prescribed safety limits.

In large power reactors in-core neutron sensors provide the data needed for carrying out an effective fuel-management program and monitor for the existence of xenon-induced power asymmetries or instabilities. It is normally possible to detect the presence of such asymmetries or instabilities with out-of-core instruments, but in-core sensors must be used to ascertain their exact nature and to provide the data from which an operator can carry out effective control actions (i.e., equalizing asymmetries or damping out oscillations). In-core neutron sensors also provide the data for correlating core performance with the response of the out-of-core sensors.

To be effective, the in-core sensors should provide data continuously or, at a minimum, at periodic intervals while the reactor is operating in its normal mode. Reactor shutdown should not be required to collect the data, such as would be the case if activation foils were used for flux-distribution measurements or if gamma scanning of fuel elements were used for determining irradiation history.

3-2 IN-CORE ENVIRONMENT

The environment in which an in-core sensor operates is hostile to both the sensor materials and the means of transmitting signals to the readout instruments. In most cases the environment includes high neutron flux ($>10^{12}$ neutrons cm^{-2} sec^{-1}), intense gamma fields ($>10^8$ R/hr), elevated temperature ($>500^\circ$F or 210°C), and high pressure (>1000 psi), along with other undesirable effects, such as vibration induced by coolant flow or boiling. Although it is not reasonable to expect in-core sensors to last through the entire life of the reactor in such a hostile environment, nevertheless the sensors should be designed so they do not require removal or replacement more frequently than once every time the reactor is refueled. Most in-core neutron sensors can last through several refueling cycles.

Another factor in the design of in-core sensors is the space limitation. Power reactors have closely spaced lattices of fuel rods and fuel assemblies; seldom is more than ½ to ¾ in. available for installing an in-core sensor. As a result in-core sensors must be rugged enough to withstand the rigors of the nuclear radiations and the thermal and mechanical environment and small enough to fit in the available space.

At full power the thermal-neutron flux in the core of a power reactor is more than 10^3 times the out-of-core neutron flux. Typical values are 3 to 5×10^{13} neutrons cm^{-2} sec^{-1}, with peak values of over 10^{14} Materials used in the construction of in-core neutron sensors must be resistant to neutron damage during the expected lifetime of the sensors. Similarly, the gamma field is more than 10^3 times that in out-of-core positions, where the exposure rate is usually 3 to 5×10^4 R/hr. Heating of the sensor materials by absorption of gamma energy must be considered in both the design and location of the sensor, and adequate cooling must be provided. Damage to the material from gamma exposure must also be taken into account.

The temperature in a power reactor at the location of an in-core neutron sensor is generally determined by the temperature of the reactor coolant at that location. Boiling-water reactors operate at saturated steam conditions that average 550°F (288°C). In some instances where higher pressure is used to extend unit capacity, the temperatures range up to 595°F (313°C). In pressurized-water reactors the core operating temperatures vary with load and location in the core but seldom fall below 520°F

(271°C) or exceed 630°F (332°C). Gas- and sodium-cooled reactors operate at temperatures considerably higher than those in water reactors. Gas-cooled reactor temperatures range from 650°F (343°C) to 1450°F (788°C), depending on load and location in the core, and sodium-cooled reactor temperatures vary from 700°F (370°C) to 1000°F (540°C).

Water reactors characteristically operate at a higher pressure than gas- and sodium-cooled reactors because of the higher vapor pressure of water. Nominal operating pressure for boiling-water reactors is 1000 psi. Some of the early boiling-water reactors were capable of operation up to 1500 psi to obtain increased steam-flow capability at the turbine generator.

Pressurized-water reactors operate at pressures that maintain subcooled conditions in the reactor coolant system. Most pressurized-water reactors operate at 2250 psi. The external sheath or enclosure of an in-core neutron sensor must withstand these pressures without collapsing and must be watertight so moisture cannot degrade the insulation resistance of the sensor and cables. Likewise, the high operating pressure calls for careful design and installation of penetrations through the reactor vessel to preclude coolant leakage to the atmosphere.

In gas-cooled reactors operating pressures vary from 300 to 700 psi, and in sodium-cooled reactors the operating pressures are 200 psi. Pressure does not present as significant a problem in designing in-core neutron sensors and cables and seals for gas- and sodium-cooled reactors as does the high operating temperature.

The velocity of the reactor coolant through the core of pressurized-water reactors averages 15 ft/sec (4.6 m/sec). Bulk boiling takes place in the core of a boiling-water reactor. The dynamic forces resulting from coolant flow and bulk boiling must be factored into the design of in-core sensors.

3-3 IN-CORE NEUTRON-FLUX SENSING

In-core neutron sensors are most important because of the direct relation between the neutron-flux distribution and the thermal-power distribution in the reactor core.

Systems for determining neutron-flux distribution fall into two broad categories: systems using *fixed sensors* at a large number of fixed locations to provide data for generating one-, two-, or three-dimensional power-distribution information, and systems using *traveling (mobile) neutron-sensing devices* to provide a large number of neutron-flux scans of the core from which the desired power-distribution information can be derived. There are advantages and disadvantages to each system.

Fixed sensors can provide the operator with neutron-flux data at all times during reactor operation. They can also be adapted to sound an alarm or to control or protect against any power-distribution anomaly that develops during the time interval between successive scans of a traveling sensor. Because the sensors are fixed in position, they must be made so they require no maintenance; in fact, generally, no maintenance or replacement can be performed on a fixed in-core sensor without shutting down the reactor. However, because fixed sensors are continuously exposed to the in-core environment during plant operation, they suffer radiation degradation or damage and must be replaced at planned intervals during refueling periods. Fixed sensors distributed throughout the reactor volume provide data at discrete points; data at all other points must be obtained by interpolation through curve fitting, usually with a computer (either on-line or off-line). The errors in the interpolated data depend on the sensor spacing and the precision of the computer curve-fitting routines.

Traveling or mapping flux-sensor systems, although unable to provide flux-distribution information at all times for alarm, control, or protection, can provide a spatially continuous flux plot along the entire path over which they travel. Traveling sensors thus can detect flux perturbations not picked up by fixed sensors, such as the flux disturbances at fuel spacer grids and at the ends of control rods. Although these perturbations are seldom of great significance in reactor operation, since there is not much one can do about them, the ability to sense them can lend confidence that the entire neutron-flux distribution is being observed, i.e., an accurate one-dimensional picture of the real flux distribution is being observed. The other two dimensions must still be filled in by interpolation through computer calculations unless, of course, there are traveling flux sensors in the other dimensions as well (which is not the case in present-day power reactors).

Although traveling or mapping neutron-flux-sensing systems incorporate motors and gear boxes that may require periodic maintenance, they are located where maintenance can be performed with minimum difficulty. The flux sensors themselves may last the life of the reactor since flux maps are run only at relatively infrequent intervals and since the sensors are withdrawn from the core when not in use.

All neutron-flux-sensing systems measure the properties of the products of interactions between neutrons and the sensor materials (see Sec. 2). When a neutron sensor is exposed for a long time to a neutron flux, its neutron sensitivity (output signal per unit neutron flux) usually decreases and its gamma sensitivity (output signal per unit gamma flux) remains unchanged. This results in a steady decrease in the signal-to-noise ratio with neutron exposure. When the sig. al-to-noise ratio decreases below a specified value, the lifetime of the neutron sensor is ended (by definition). For a given neutron sensor exposed to a mixed neutron and gamma flux, it follows that any design action that increases the initial value of the neutron to gamma signal ratio also increases the lifetime of the sensor.

3-3.1 In-Core Fission Chambers

Fission chambers are most commonly used as in-core neutron-flux sensors. They are the backbone of the in-core neutron-detection systems in the majority of boiling-water reactors. The fission chamber is used as the neutron sensor in most traveling in-core probe systems for both pressurized-water reactors and boiling-water reactors.

Fission chambers feature relatively slow burnup of the uranium liner. They provide satisfactory operation in all three basic modes: (1) the pulse-counting mode, (2) the mean-square-voltage mode, and (3) the mean-current (d-c) mode (see Sec. 5, pt. 5-5, and Sec. 2, pt. 2-2). In-core fission chambers are thus suitable for use in source-range channels (where pulse counting is required because of the low value of the neutron flux), in the intermediate-range channels (where mean-square-voltage techniques extend the operating range), and in the power-range channels (where mean-current techniques provide accurate power measurements for both fixed sensors and traveling probes). For each of these modes of operation, the optimum design is different with respect to size, materials, fill-gas pressure, emitter–collector gap dimensions, neutron sensitivity, etc.

Just as in out-of-core neutron detectors, both the neutron and the gamma fluxes contribute to the total output of an in-core fission chamber. Many of the design parameters which may be changed to achieve a specific neutron-sensitivity characteristic also affect the gamma sensitivity. Since the output signal attributable to incident gamma radiation is not unambiguously related to the reactor power level, the design of an in-core fission chamber is optimum if at the end of detector life the ratio of the output current due to neutron flux to the output current due to gamma radiation is still acceptable.

The main design parameters that can be varied to meet the specific requirements for an in-core fission chamber are (1) the physical form of the uranium used, (2) the enrichment of the uranium, (3) the surface area and thickness of the uranium, (4) the type of fill gas used, (5) the fill-gas pressure, (6) the dimensions of the gap between the emitter and the collector, and (7) the dimensional tolerances. Each of these is discussed below.

(a) Uranium Form. Two basic types of in-core fission chambers have been developed and manufactured. One type incorporates an enriched uranium oxide layer plated on the inside of the detector housing that forms the outside wall of the sensitive volume (see Fig. 3.1). The second type

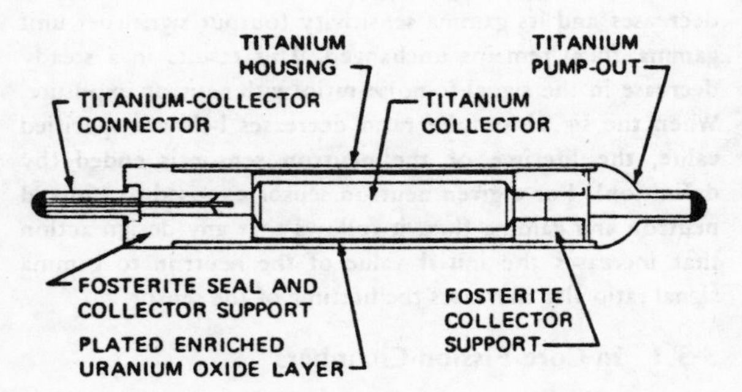

Fig. 3.1—Uranium oxide-coated in-core fission chamber.

contains a machined sleeve of enriched uranium–aluminum alloy at the outer surface of the sensitive volume (see Fig. 3.2). The more carefully the weight and thickness of the uranium coating or uranium–aluminum sleeve are controlled, the more accurately the neutron sensitivity of the detector can be controlled. The majority of commercial detectors are manufactured with a ±20% tolerance on initial neutron sensitivity. Under very special circumstances a ±5% tolerance on initial neutron sensitivity can be achieved by carefully controlling the uranium plating process or the uranium–aluminum alloy machining.

(b) Uranium Enrichment. Because the enrichment of the uranium in a neutron sensor has no effect on the gamma sensitivity (total mass of uranium is constant), the best way to increase the signal-to-noise ratio is to increase the enrichment of the uranium used in the sensor. Fully enriched uranium provides the maximum neutron sensitivity while maintaining the same gamma sensitivity.

(c) Uranium Surface Area. For a given total mass of the enriched uranium layer, the neutron sensitivity of an in-core fission chamber depends on the surface area of the uranium. Surface area is varied by changing the chamber diameter and length. The gamma sensitivity also varies when the chamber geometry is changed. Consequently, there is a combination of sensor diameter and length which yields the highest signal-to-noise ratio.

(d) Type of Fill Gas. Argon is the most commonly used fill gas. It has all the desired properties (chemically inert, good thermal conductivity, low thermal-neutron cross section, and suitable ionization properties). Other commonly used fill gases are helium and nitrogen or mixtures of argon and nitrogen.

Chemical inertness is desirable since a gas that is not inert may combine with chamber materials (particularly in presence of an intense nuclear radiation field) and thus reduce the gas available for ionization. High thermal conductivity is desirable to remove heat developed in the chamber by the signal-generating processes. If the thermal-neutron cross section is high, the fill gas will be depleted by nuclear transformation; in addition, neutrons absorbed by the fill gas are not available for reaction with the uranium.

(e) Fill-Gas Pressure. Neutron and gamma sensitivities of an in-core fission chamber are directly proportional to the fill-gas pressure as long as the range of fission fragments and gamma photons is greater than the gap between the emitter and collector. Most in-core fission chambers operate at several atmospheres pressure to achieve higher neutron and gamma sensitivities. Because both the gamma and neutron sensitivities are similarly increased, detector life is not appreciably affected by varying the fill-gas pressure.

(f) Emitter–Collector Gap. Of all the factors involved in in-core fission-chamber design, the most critical is the sizing of the gap between the neutron-sensitive emitter and the positively charged collector. Since the ionization current is a function of the number of fill-gas atoms, a large gap produces a large detector current. This characteristic is especially important at low flux levels, such as those existing in the source range. At higher flux levels the gap must be reduced to ensure detector saturation [pt. 3-3.1(h)] up to and exceeding the highest neutron

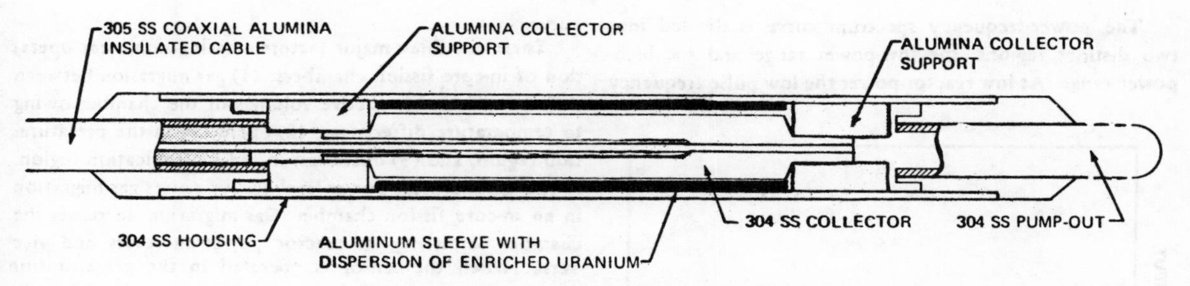

Fig. 3.2—Uranium–aluminum alloy sleeve in-core fission chamber.

flux the detector is designed for.

(g) Dimensional Tolerances. As noted earlier, the accuracy of initial detector sensitivity is directly related to the dimensional tolerances applied to the neutron-sensitive material and to the emitter–collector gap. Because so many of the characteristics of in-core fission chambers are related directly to dimensions, the effects of tolerance accumulation are extremely important and must be carefully considered.

(h) Operating Characteristics. In-core fission chambers exhibit most of the operating characteristics of out-of-core neutron sensors. As pointed out in sec. 2, pt. 2-2.1, variation of chamber voltage provides three regions of detector performance: the low-voltage (presaturation) region, the plateau (saturation) region, and the multiplication region. The exact shape of the current–voltage curve depends on chamber construction parameters.

In view of this characteristic behavior of fission chambers, it follows that the voltage applied to the chamber should be high enough to keep it on the plateau region at or above the highest radiation flux in which it is expected to operate. If there is any question about the chamber voltage required to achieve saturated operation, it is preferable to err on the high side, since the chamber current is proportional to the incident radiation flux in both the saturation and multiplication regions but not in the presaturation region. For most in-core fission chambers being used in power reactors today, the neutron flux never exceeds 2×10^{14} neutrons cm^{-2} sec^{-1}, so an operating voltage of 125 volts d-c is sufficient to guarantee saturated operation. Figure 3.3 shows saturation curves for a typical in-core fission chamber.

The ideal detector plateau would be flat, but this is never achieved. Below 10^{13} neutrons cm^{-2} sec^{-1}, the plateau has a slope that is generally attributable to detector-cable leakage current. As the neutron flux increases, the plateau starting voltage (the low end of the plateau) increases and the multiplication starting voltage (the low end of the multiplication region) decreases. When the neutron flux increases to the point where the plateau starting voltage equals the multiplication starting voltage, the plateau disappears. This point is generally defined as the *upper flux limit* of the detector.

It is more important for an in-core fission chamber operating in the mean-square-voltage mode to stay in the plateau region than for one operating in the pulse-counting mode or the mean-current (d-c) mode. This results from the fact that the signal is a function of chamber current squared rather than chamber current alone. Figure 3.4 shows that, as a result, the plateau starting voltage of a mean-square-voltage chamber is somewhat higher and the multiplication starting voltage somewhat lower than a d-c chamber.

In the mean-square-voltage mode of operation, adjustment of the chamber voltage must be related to the band pass of the signal amplifier into which the chamber operates. Pulses from the chamber are distributed in energy in accordance with the power-frequency spectrum curves in Fig. 3.5. The breaks in the curves occur at the frequencies corresponding to the ion collection time, T_i, and the electron collection time, T_e. Reducing the chamber voltage shifts the entire power-frequency spectrum curve to lower frequencies because the time required to collect the ions and electrons in the chamber increases.

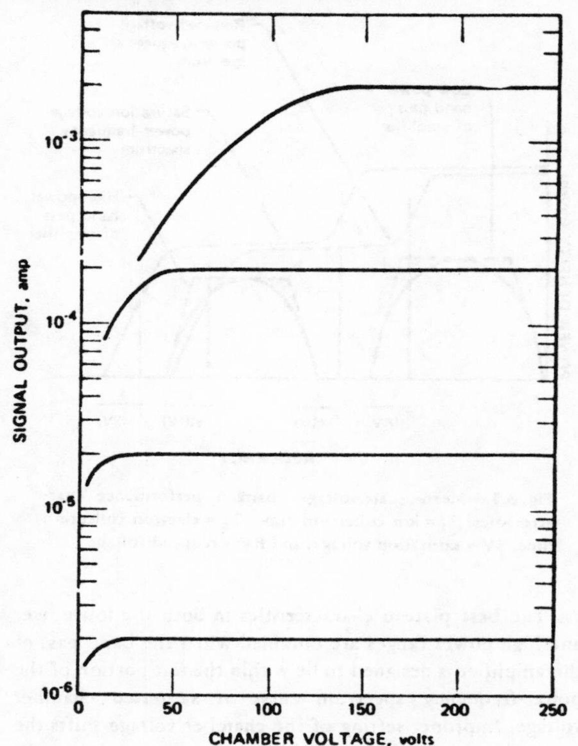

Fig. 3.3—Typical in-core fission chamber saturation characteristics.

The power-frequency spectrum curve is divided into two distinct regions, the low-power range and the high-power range. At low reactor power the low pulse frequency

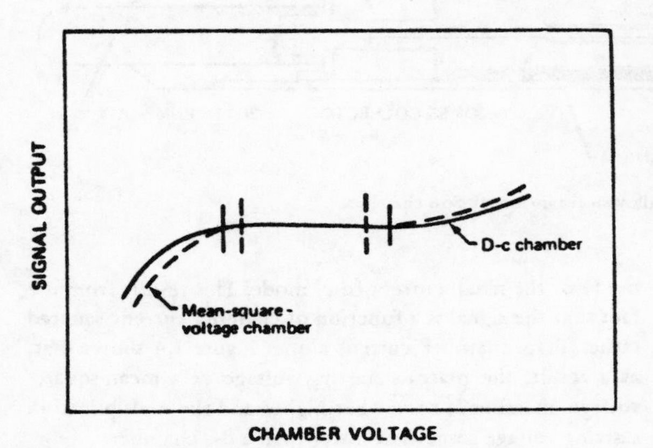

Fig. 3.4—Saturation curves for d-c chambers vs. mean-square-voltage chambers.

allows ample time for both the ions and electrons to be collected in the chamber. At high reactor power the pulse frequency is so high that the ions are not collected; only electrons are collected. The break in the curve occurs at the point where the transition from the collection of ions plus electrons to the collection of electrons alone is made.

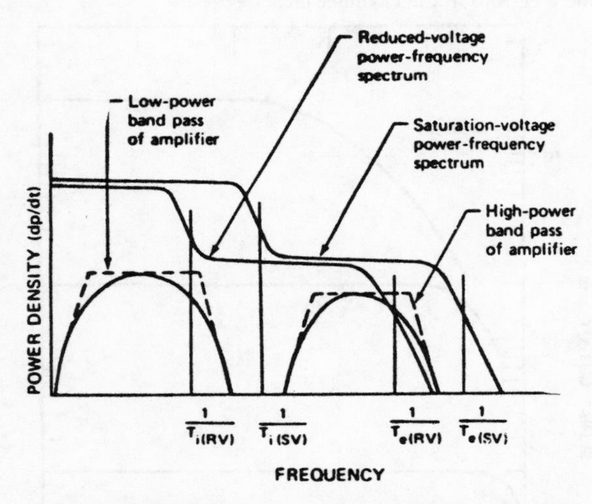

Fig. 3.5—Mean-square-voltage chamber performance characteristics. T_i = ion collection time; T_e = electron collection time; SV = saturation voltage; and RV = reduced voltage.

The best plateau characteristics in both the low-power and high-power ranges are obtained when the band pass of the amplifier is designed to be within the flat portion of the power-frequency spectrum curve at saturated chamber voltage. Improper setting of the chamber voltage shifts the frequency of the power-frequency spectrum curve and drives the amplifier response off the flat portion of the curve.

There are three major factors creating nonlinear operation of in-core fission chambers: (1) gas migration between the active and the inactive volumes of the chamber owing to temperature differences, (2) operation in the presaturation region, and (3) operation in the multiplication region. A change in reactor power level always causes gas migration in an in-core fission chamber. Gas migration decreases the chamber sensitivity as reactor power increases and vice versa. When the sensor is operated in the presaturation region, a further decrease in sensitivity occurs because the chamber current is not linear with neutron flux.

In the saturation region the chamber current is linear with neutron flux, so the only source of nonlinearity is gas migration. In the multiplication region the decrease in current due to gas migration is counteracted by the increase in multiplication due to the gas migration. It may be desirable to set the chamber voltage somewhere in the multiplication region rather than in the saturation region to obtain maximum linearity on chambers with large gas migration.

(i) Operating Ranges. Table 3.1 summarizes the operating characteristics of typical in-core fission chambers used in the pulse-counting mode, the mean-square-voltage mode, and the direct-current mode. Figure 3.6 shows the upper and lower boundaries of the operating ranges for in-core fission chambers used to monitor reactor power from below source level to above the overpower trip level.

Operating Range of Pulse-Counting Fission Chambers. The normal operating range of an in-core pulse-counting fission chamber is from 10^3 to 10^9 neutrons cm^{-2} sec^{-1}. The lower limit is determined by the statistics of pulse counting. The variations in the neutron counting sensitivity of a pulse-counting fission chamber noted in Table 3.1 result from variations in gamma flux at the detector. The integral bias curves, Fig. 3.7, show the reduction in neutron counting sensitivity as the gamma exposure rate increases from 10^5 to 2.5×10^7 R/hr while the neutron flux remains constant. Figure 3.7 also demonstrates the importance of the proper discriminator setting. Any discriminator setting less than 3 would result in operation to the left of the plateau at higher gamma levels with attendant errors in count-rate information. At a neutron flux of 5×10^2 neutrons cm^{-2} sec^{-1}, the chamber indicates approximately 1 count/sec, somewhat less than is desirable for statistical confidence. Accordingly, the neutron source in the reactor should be sized to provide from 2 to 10 counts/sec.

The upper limit of the pulse-counting chamber is determined by the 1-MHz bandwidth of the signal amplifier and the 300-nsec rise time and collection time of the chamber pulses. At a true random input of 10^6 counts/sec, the counting loss is 23% of the true count rate. Since the pulse-counting chamber has a neutron counting sensitivity of greater than 10^3 counts/sec per unit flux (1 neutron cm^{-2} sec^{-1}) in a 10^5 R/hr gamma field, the upper limit of

Table 3.1—Operating Characteristics of Typical In-Core Fission Chambers

	Pulse-counting fission chamber	Mean-square-voltage fission chamber	Mean-current (d-c) fission chamber
Electrode coating	U_3O_8 enriched to >90% ^{235}U	U_3O_8 enriched to >90% ^{235}U	U_3O_8 enriched to >90% ^{235}U
Neutron sensitivity:			
Pulse-counting, counts/sec/nv	0.5 to 2.5 x 10^{-3}		
Mean current (d-c), amp/neutrons cm^{-2} sec^{-1}		7 x 10^{-18}	2.15 x 10^{-17} ± 20%
Mean-square-voltage, amp^2/neutrons cm^{-2} sec^{-1}		1.5 to 4.5 x 10^{-31}	
Gamma sensitivity:			
Mean current (d-c), amp/(R/hr)		2.5 x 10^{-14} ± 20%	2.0 x 10^{-14} ± 30%
Mean-square-voltage, amp^2/(R/hr)		1.5 x 10^{-30}	
Neutron flux (max.), neutrons cm^{-2} sec^{-1}	1 x 10^{10}	1.5 x 10^{13}	1.8 x 10^{14}
Gamma flux (max.), R/hr	2.5 x 10^7		1.68 x 10^8
Operating voltage, volts (d-c)	200 to 700	100 to 200	100 to 200
Temperature (max.), °C	540	540	315
Fill gas	Argon	Argon	Argon
Dimensions:			
Sensitive length, in.	1.00	1.00	1.00
Case diameter, in.	0.250	0.250	0.230
Case material	Titanium	Titanium	Stainless steel
Collector material	Titanium	Titanium	Stainless steel
Collector-to-emitter insulator	Alumina	Alumina	Alumina
End seal	Titanium and Fosterite	Titanium and Fosterite	Titanium and Fosterite
Lifetime, neutrons/cm^2	10^{19}		3.8 x 10^{21}

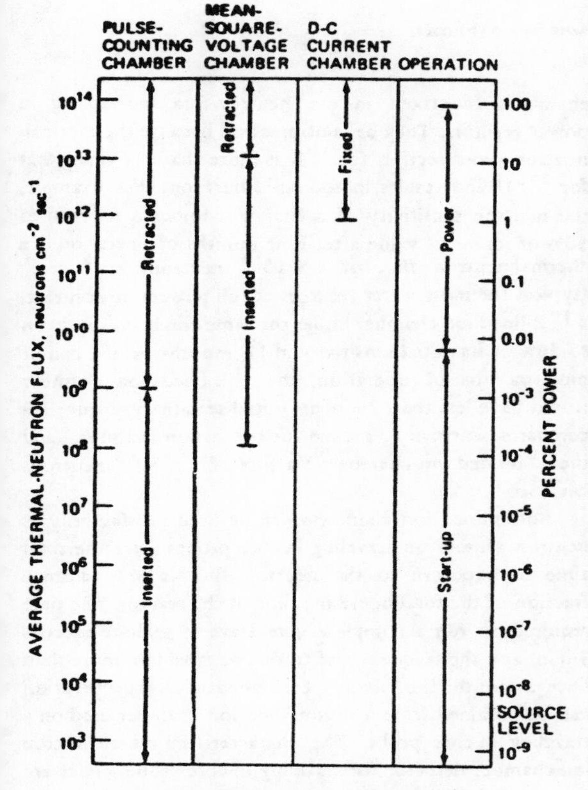

Fig. 3.6—Ranges of in-core fission chamber.

the neutron flux is 10^9 neutrons cm^{-2} sec^{-1} at 10^6 counts/sec.

When the in-core neutron flux exceeds 10^{10} neutrons cm^{-2} sec^{-1}, the pulse-counting chamber should be removed from the core to prevent depletion of the fissionable material.

Operating Range of Mean-Square-Voltage Fission Chambers. The operating range of the mean-square-voltage fission chamber is 10^8 to 10^{13} neutrons cm^{-2} sec^{-1}. The lower limit is set by the detector noise resulting from alpha emission from the uranium coating, prompt gamma radiation, and delayed neutrons at low reactor start-up levels. When the neutron flux is 10^8 neutrons cm^{-2} sec^{-1} or greater, the neutron-flux signal exceeds the noise signal.

The upper limit of the range of a mean-square-voltage chamber is reached when the plateau starting voltage becomes equal to the multiplication starting voltage, as described in pt. 3-3.1(h). Because the plateau is shorter for chambers operating in the mean-square-voltage mode, the upper limit of their range is lower than the same chamber operating in the mean-current (d-c) mode.

When the in-core flux exceeds 10^{11} neutrons cm^{-2} sec^{-1}, the mean-square-voltage chamber should be removed from the core to prevent depletion of the uranium.

Operating Range of Mean-Current (d-c) Fission Chambers. The operating range of in-core fission chambers used in the mean-current mode is from less than 10^{12} neutrons cm^{-2} sec^{-1} to greater than 10^{14} neutrons cm^{-2} sec^{-1}. The

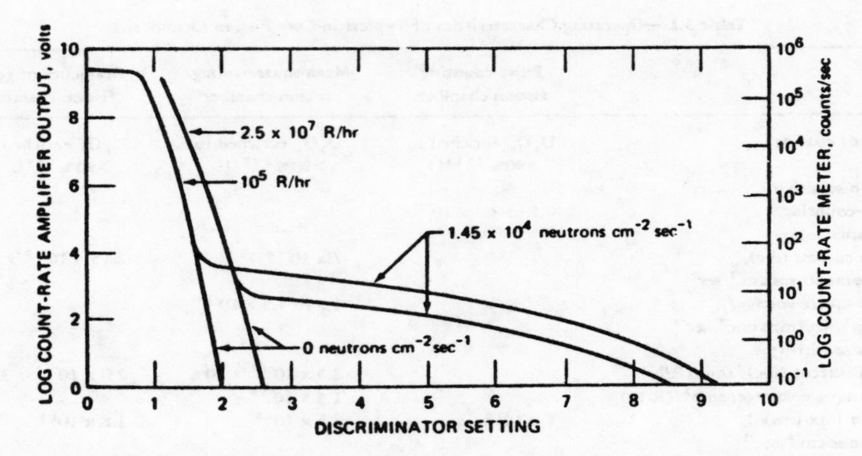

Fig. 3.7—Integral bias curve for pulse-counting chamber. Chamber voltage, 400 volts. Chamber sensitivity, 2×10^{-3} (counts/sec)/(neutrons cm^{-2} sec^{-1}) at 10^5 R/hr and 5×10^{-4} (counts/sec)/(neutrons cm^{-2} sec^{-1}) at 2.5×10^7 R/hr.

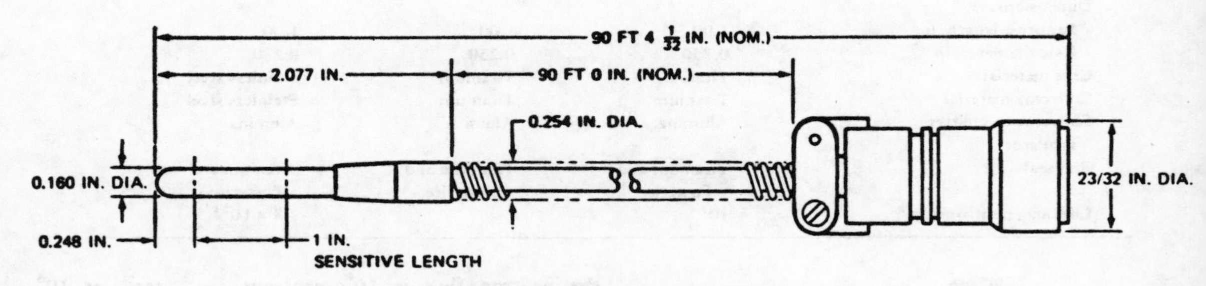

Fig. 3.8—Traveling in-core fission chamber.

lower limit is set primarily by the leakage current in the ceramic-insulated cable. The upper limit is reached when the plateau starting voltage is equal to the multiplication starting voltage. Because mean-current (d-c) chambers normally provide the signals that initiate reactor overpower trip and are calibrated against a reactor heat balance, they must have good linearity over at least one decade of their operating range.

(j) Traveling In-Core Fission Chambers. The design requirements for traveling in-core fission chambers are the same as those for fixed in-core mean-current chambers except that the traveling chambers must withstand the rigors of periodic insertion into and withdrawal from the core. The traveling in-core fission chambers have the same requirements for operating range and linearity as the fixed in-core mean-current chambers. Table 3.2 summarizes the characteristics of traveling in-core fission chambers commonly used in power reactors. Figure 3.8 shows a typical traveling in-core probe. The helically wound outer sheath of the drive cable engages the drive gears to move and position the sensor.

3-3.2 Boron-Lined Chambers

It is possible, but not practical, to use ^{10}B-lined chambers for fixed in-core neutron-flux monitoring in power reactors. They are not practical because the thermal-neutron cross section for ^{10}B is more than six times that for ^{235}U and results in too-rapid burn out. For example, the neutron sensitivity of a fission chamber is reduced to 50% of its initial value after nine months of operation in a thermal-neutron flux of 4×10^{13} neutrons cm^{-2} sec^{-1} (typical for most water reactors at full power). In contrast, a ^{10}B-lined ion chamber under the same conditions is down to 50% of its initial sensitivity in 1½ months. At the end of nine months of operation, the ^{10}B-lined ion chamber would have less than 2% of its initial sensitivity. Figure 3.9 compares sensitivity vs. time for the fission chamber and the ^{10}B-lined ion chamber in a flux of 4×10^{13} neutrons cm^{-2} sec^{-1}.

Boron-lined ion chambers can be used satisfactorily as neutron sensors on traveling in-core probes since the total time of exposure to the neutron flux is only a small fraction of the total operating time of the reactor. The time required to run a complete core traverse seldom exceeds 3 min, and the frequency of traversing is seldom more than once a month; thus many years of satisfactory operation can be obtained from a boron-lined ion chamber used on a traveling in-core probe. The characteristics of an in-core ion-chamber detector for traveling in-core probe service are described in Table 3.2.

Table 3.2—Operating Characteristics of Traveling In-Core Fission Chambers*

Neutron-sensitive material	Thermal-neutron sensitivity, amp/neutrons cm^{-2} sec^{-1}	Gamma sensitivity, amp/(R/hr)	Maximum thermal-neutron flux, neutrons cm^{-2} sec^{-1}	Typical operating voltage, volts (d-c)	Maximum operating temperature, °F	Detector insulator	Nominal Dimensions: Sensitive length, in.	Nominal Dimensions: Detector O. D., in.
^{235}U	2.0×10^{-17}	1.5×10^{-14}	1×10^{14}	25 to 200	750	Al_2O_3	1	1/4
^{10}B	1.0×10^{-17}	1.0×10^{-14}	1×10^{14}	25 to 200	750	Al_2O_3	1/2	7/16
†	3.0×10^{-18}	2.0×10^{-14}	5×10^{14}	25 to 200	650	Al_2O_3	1	1/4
^{235}U	6.8×10^{-18}	1.0×10^{-14}	3×10^{13}	25 to 300	650	Al_2O_3	1	0.090

*From Westinghouse brochure *Radiation Detectors. Quick Reference Guide*, November 1967.
†The neutron-sensitive material is a breeder mixture of 90% ^{234}U and 10% ^{235}U.

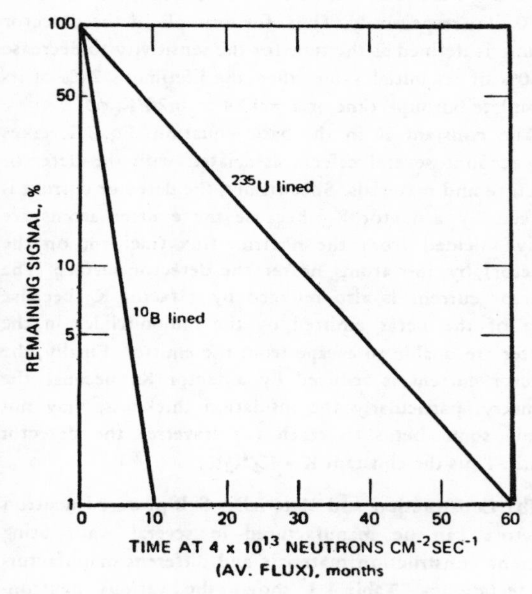

Fig. 3.9 —In-core chamber burnup curves.

3-3.3 Self-Powered Neutron Detectors

The problems of manufacturing in-core fission chambers and ion chambers small enough to fit the allowable space, rugged enough to withstand the in-core environment, and inexpensive enough to permit extensive coverage of a power-reactor core have led to efforts to develop other types of devices to measure neutron flux at fixed locations in a power reactor. One device developed to solve these problems is the self-powered neutron detector.

The self-powered neutron detector uses the basic radioactive decay process of its neutron-activated material to produce an output signal. As the name implies, no external source of ionizing or collecting voltage is required to produce the signal current. The construction of the detector is quite different from that of an in-core fission chamber. There is no gas-filled region where ionization takes place; instead, the detector is of solid construction with a neutron-sensitive material connected to a lead wire, both being separated from the detector outer sheath by hard-packed ceramic insulation. The resulting detector is like a mineral-insulated coaxial cable, small in size and rugged.

The simplicity of construction and operating principle lead to a number of advantages, including low cost, simplicity of readout equipment, low burnup rate, long lifetime, and ease of reproducing sensitivity.

(a) Operating Principles. As shown in Fig. 3.10, a typical self-powered neutron detector consists of four parts: emitter, insulation, lead wire, and sheath (or collector). The emitter is a material of reasonably high thermal-

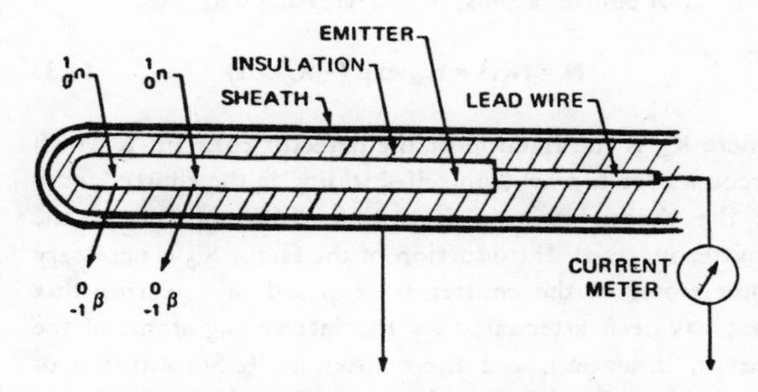

Fig. 3.10—Operating principle of a self-powered neutron detector.

neutron-activation cross section which, after activation, decays by emission of high-energy betas with a reasonably short half-life. The insulation is a solid that must maintain high electrical resistance in the in-core temperature and nuclear-radiation environment; it should ideally emit no betas or electrons (e.g., from neutron activation). Both the lead wire and sheath, or collector, must emit few betas or electrons compared to the emitter so that undesirable background signals are minimized.

When the self-powered neutron detector is placed in a neutron flux and connected to ground through an electrical current meter, the current measured is proportional to the net beta escape from the emitter. If the detector remains in the neutron flux until equilibrium beta emission is reached, the current is directly proportional to the incident-neutron flux.

When the activity induced by thermal-neutron absorption in the emitter is from a single radionuclide with a single half-life, the detector output after exposure to the neutron flux for a time t is

$$I(t) = K\sigma_{act}QN\left[1 - \exp\left(-\frac{0.693t}{T_{½}}\right)\right]\phi \qquad (3.1)$$

where $I(t)$ = detector current (amp) at time t (sec)

K = a dimensionless constant determined by detector geometry and materials

σ_{act} = thermal-neutron-activation cross section of emitter material (cm^2)

Q = charge emitted by emitter per neutron absorbed (coulombs)

N = number of emitter atoms

$T_{½}$ = half-life of radionuclide generated in emitter (sec)

ϕ = thermal-neutron flux (neutrons cm^{-2} sec^{-1})

Steady state is reached when the exposure time is several times the half-life, $T_{½}$, of the radionuclide:

$$I_0(t) = K\sigma_{act}\, QN\, \phi \qquad (t \gg T_{½}) \qquad (3.2)$$

where $I_0(t)$ is the steady-state detector current at time t. The inclusion of t is made necessary by the fact that the number of emitter atoms, N, is decreasing with time:

$$N = N(t) = N_0 \exp(-K_\phi \phi\sigma t) \qquad (3.3)$$

where K_ϕ is the fraction of the detector constant K which accounts for the neutron self-shielding in the emitter and σ is the thermal-neutron-absorption cross section of the emitter material. Introduction of the factor K_ϕ is necessary since atoms of the emitter are exposed to a neutron flux that has been attenuated by the intervening atoms of the sheath, insulation, and the emitter itself. Substitution of Eq. 3.3 into Eq. 3.2 gives the steady-state detector current:

$$I_0(t) = K\sigma_{act}QN_0 \exp(-K_\phi\phi\sigma t)\,\phi \qquad (t \gg T_{½}) \qquad (3.4)$$

The sensitivity of a self-powered detector is defined as the change in the steady-state detector current per unit change in the thermal-neutron flux:

$$S(t) = \frac{\Delta I_0}{\Delta\phi} = [K\sigma_{act}QN_0 \exp(-K_\phi\phi\sigma t)] \times (1 - K_\phi\phi\sigma t) \qquad (3.5)$$

The detector sensitivity is seen to decrease (because of burnup of the emitter material) approximately exponentially with time. The rate of sensitivity decrease is determined by the thermal-neutron flux that the detector is exposed to, the thermal-neutron-absorption cross section of the emitter material, and the neutron self-shielding in the emitter.

If the detector life is defined as the time during which the sensitivity decreases to a fraction f of its initial value when the detector is in a constant thermal-neutron flux, then Eq. 3.5 shows that

$$f = e^{-T/\tau}\left(1 - \frac{T}{\tau}\right) \qquad (3.6)$$

where T = detector life in constant thermal-neutron flux (ϕ)

$f = S(t = T)/S(t = 0)$ = relative sensitivity at $t = T$

$\tau = 1/K_\phi\phi\sigma$ = time to complete burnup of emitter material assuming K_ϕ and ϕ are constant

The lifetime T can be calculated as a function of f from Eq. 3.6. The values of T corresponding to f = 0.9, 0.8, 0.7, 0.6, and 0.5 are found to be $T = 0.05\tau, 0.11\tau, 0.17\tau, 0.24\tau$, and 0.31τ, respectively. Thus, for example, if the detector lifetime is defined as the time for the sensitivity to decrease to 60% of its initial value, then the lifetime is 24% of its "complete burnup" time or $T = 0.24\tau = 0.24/K_\phi\phi\sigma$.

The constant K in the basic equation, Eq. 3.1, takes into account several effects associated with the detector structure and materials. Specifically, the detector current is reduced by a factor K_ϕ because the emitter atoms are partly shielded from the neutron flux (incident on the detector) by the atoms nearer the detector surface. The detector current is also reduced by a factor K_β because some of the betas emitted by the radionuclides in the emitter are unable to escape from the emitter. Finally, the detector current is reduced by a factor K_g because the geometry, particularly the insulation thickness, may not permit some betas to reach (or traverse) the detector sheath. Thus the constant $K = K_\phi K_\beta K_g$.

(b) Construction and Materials. Self-powered neutron detectors can be manufactured in several ways using different construction materials and different manufacturing techniques. Table 3.3 shows the various neutron-activated beta emitters that have been considered as potential candidates for the emitter material. Of the materials listed, only ^{103}Rh and ^{51}V have been used in commercial applications. Each of the others has been rejected for one or more undesirable characteristics. Because each has a low thermal-neutron-activation cross section, ^{7}Li, ^{11}B, and ^{27}Al yield unacceptably low signal-to-noise ratios. With a 2.58-hr half-life, ^{55}Mn results in too long a time constant if used in a detector. Since ^{99}Tc does not occur naturally, it is not readily available; ^{115}In is unsatisfactory because 76% of its beta decay has a 54-min half-life. Silver has both an acceptable cross section value and acceptable half-lives, but it would be difficult to compensate for burnup with the 24-sec ^{109}Ag burning up three times as fast as the 2.4-min ^{107}Ag.

Neither ^{103}Rh nor ^{51}V has any of these undesirable characteristics. Because of its larger thermal-neutron-activation cross section, ^{103}Rh is used where short detectors are needed for measuring local flux. Many ^{103}Rh detectors distributed throughout the reactor core can provide three-dimensional power distribution information. The relatively lower signal level of ^{51}V is more suited to

Table 3.3—Emitter Materials for Self-Powered Neutron Detectors

Material	Abundance, %	Activation cross section, barns	Half-life	Maximum beta energy, MeV	Burnup at 10^{13} neutrons cm^{-2} sec^{-1}
^{7}Li	92.58	0.036	0.855 sec	13.0	Negligible
^{11}B	80.4	0.005	0.025 sec	13.4	Negligible
^{27}Al	100	0.23	2.30 min	2.87	Negligible
^{51}V	99.76	4.9	3.76 min	2.47	0.013%/month
^{55}Mn	100	13.3	2.58 hr	2.85	0.035%/month
^{99}Tc	Artificial	22	16 sec	3.37	0.058%/month
^{103}Rh	100	150 (11 + 139)	4.4 min; 42 sec	2.44	0.40%/month
^{115}In	95.72	203 (4*,154,45	54 min; 14 sec	1.0 3.25	0.53%/month
^{109}Ag	48.18	92	24 sec	2.87	0.24%/month
^{107}Ag	51.82	35	2.4 min	1.64	0.092%/month

*Isomeric transition.

long detectors designed to average the neutron flux over the full core height. Such detectors cannot be used to determine power distribution along the axis of the reactor core, although several ^{51}V detectors dispersed throughout the core can provide data on the radial power distribution.

Figures 3.11 and 3.12 show the two types of construction commonly used for self-powered neutron detectors in commercial use today.

The earliest, and in many respects the simplest, type of detector construction is shown in Fig. 3.11. The 0.020-in.-diameter rhodium wire emitter is fastened to the Inconel lead wire of a standard 0.040-in. magnesium oxide insulated, Inconel sheathed, coaxial cable. All parts are baked out before assembly and are maintained scrupulously clean during assembly. The section of the detector sensitive to neutrons has a larger diameter than the coaxial cable. The construction of a neutron-detector assembly containing several of these detectors can be complicated by the change in diameter.

Figure 3.12 shows an alternate construction. The neutron-sensitive emitter is fastened to the Inconel lead wire before the insulation is installed. Magnesium oxide insulators are threaded over the emitter and lead wire. The Inconel sheath is slid over the insulators, and the entire assembly is swaged down to a finished diameter of 0.062 in. Bakeout before assembly and rigorous cleanliness during

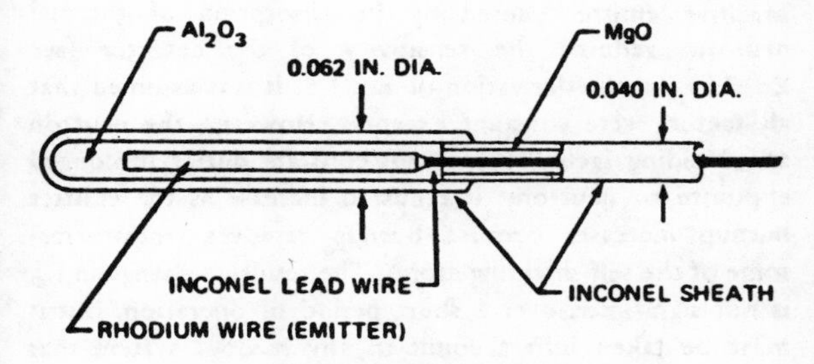

Fig. 3.11—Self-powered neutron detector.

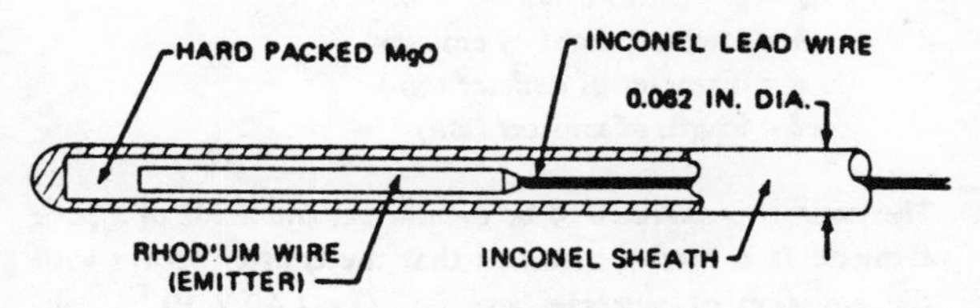

Fig. 3.12—Self-powered neutron detector.

assembly are important procedures. The resulting detector has a constant diameter over its length, and the magnesium oxide insulation is compacted tightly around the emitter.

Only three insulation materials have been considered for self-powered neutron detectors: aluminum oxide, beryllium oxide, and magnesium oxide. Aluminum oxide has been most frequently used when the detector is assembled as shown in Fig. 3.11. Aluminum oxide is not suitable for detectors assembled as shown in Fig. 3.12 because of the danger of damaging the emitter or lead wire during the swaging operation. In addition, aluminum is activated by thermal neutrons and emits high-energy betas (see Table 3.3) which contribute to background and reduce the signal-to-noise ratio. The characteristic 2.3-min beta decay of aluminum has been observed during irradiation of cables insulated with aluminum oxide. Whether or not this is tolerable depends on the accuracy desired in the detector output signal. Beryllium oxide has no significant advantage other than its extremely low thermal-neutron cross section. This advantage, however, is more than offset by its high cost and toxicity. Magnesium oxide is the most satisfactory of the three insulation materials because of its low cost, high resistivity, workability, and low noise potential.

The two most commonly used sheath materials in detectors for pressurized-water and boiling-water reactors are type-304 stainless steel and Inconel because both are compatible with the reactor coolant. Inconel is now standard in all commercial detectors. The use of 30, stainless steel in the sheath of detectors intended for high-accuracy applications is problematical since manganese, which constitutes about 0.5 wt. % of most stainless steels, is activated by thermal neutrons and emits betas (see Table 3.3.)

Inconel is used universally as the lead-wire material, although Nichrome was used successfully in earlier detectors.

(c) Sensitivity. The initial sensitivity of a self-powered detector is given by Eq. 3.5, with t = 0:

$$S(0) = K\sigma_{act}QN_0\phi \tag{3.7}$$

The number of emitter atoms at t = 0 is

$$N(t = 0) = N_0 = \rho\left(\frac{A_0}{A}\right)\left(\frac{\pi}{4}\right)d^2\, l \tag{3.8}$$

where ρ = density of emitter material (g/cm³)
A_0 = Avogadro's number = 6.02×10^{23}
A = atomic weight of emitter
d = diameter of emitter (cm)
l = length of emitter (cm)

The emitter is assumed to be cylindrical and made of a pure element. It is further assumed that the emitter decays with the emission of a single beta, i.e., $Q = 1.60 \times 10^{-19}$ coulomb, the electron charge.

As noted at the end of pt. 3-3.3(a), the constant K is the product of K_ϕ, K_β, and K_g, the constants that take into account the neutron self-shielding, beta self-shielding, and geometric effects. Figure 3.13 shows the values of these constants as a function of emitter diameter for rhodium with 10 mils (0.25-mm) of MgO insulation. Figure 3.14 shows the same constants for vanadium emitters.

The initial sensitivity of a rhodium detector with a 20-mil (0.51-mm) emitter and 10 mils of MgO insulation can be calculated from Eqs. 3.7 and 3.8 using K values from Fig. 3.13 and the rhodium cross section from Table 3.3. (The density of rhodium is 12.4 g/cm³.) The result is:

$$\frac{S}{l} = 1.3 \times 10^{-21} \text{ amp/(neutrons cm}^{-2}\text{ sec}^{-1})$$

A similar calculation for a vanadium detector with a 20-mil emitter and 10 mils of MgO insulation yields (vanadium density is 6.0 g/cm³):

$$\frac{S}{l} = 7.1 \times 10^{-23} \text{ amp/(neutrons cm}^{-2}\text{ sec}^{-1})$$

This value, as well as the one for the rhodium detector, is in good agreement with experimental values.

Because of the rapid decrease in K with increasing emitter diameter, detector sensitivity depends more on emitter surface area than on emitter volume. This is evident from Fig. 3.15, where the detector sensitivity is shown as a function of emitter diameter. The relation is nearly a straight line.

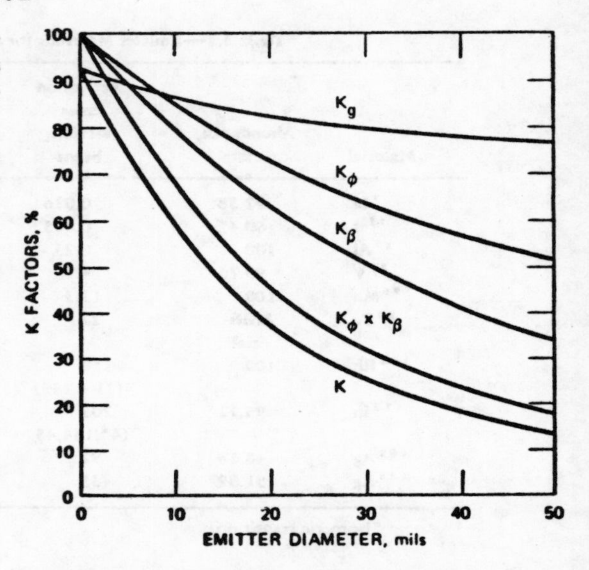

Fig. 3.13—K factors for rhodium detectors with 10 mils of MgO.

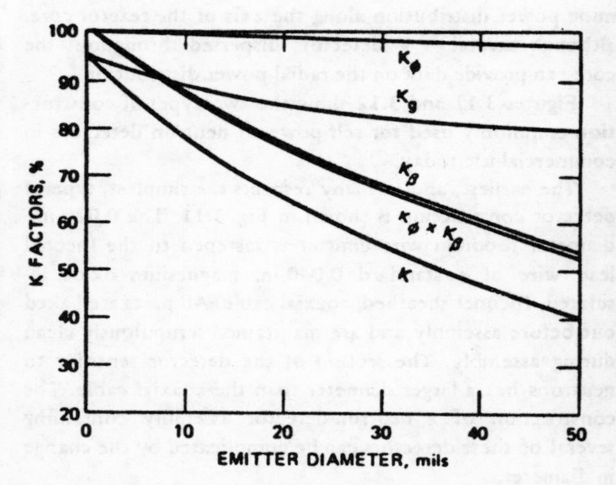

Fig. 3.14—K factors for vanadium detectors with 10 mils of MgO.

Figure 3.16 shows the change in detector sensitivity when the detector length and diameter are varied but the emitter mass is kept constant.

(d) Emitter Burnup. The depletion of the neutron-sensitive emitter caused by its absorption of thermal neutrons reduces the sensitivity of the detector (see Eq. 3.5). In the derivation of Eq. 3.5, it was assumed that all factors were constant except t. However, the neutron self-shielding factor, K_ϕ, is not constant during prolonged exposure to neutrons; it tends to increase as the emitter burnup increases because burnup removes (transforms) some of the self-shielding atoms. The resulting change in K_ϕ is not significant over a short period of operation, but it must be taken into account in any readout system that automatically compensates for detector burnup.

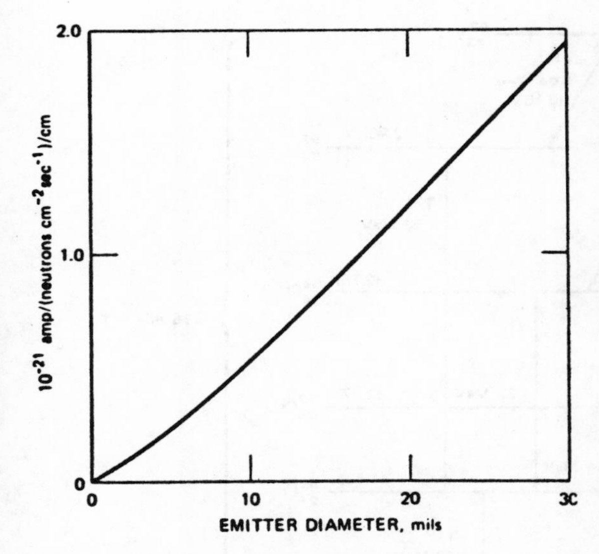

Fig. 3.15—Rhodium-detector sensitivity vs. emitter diameter.

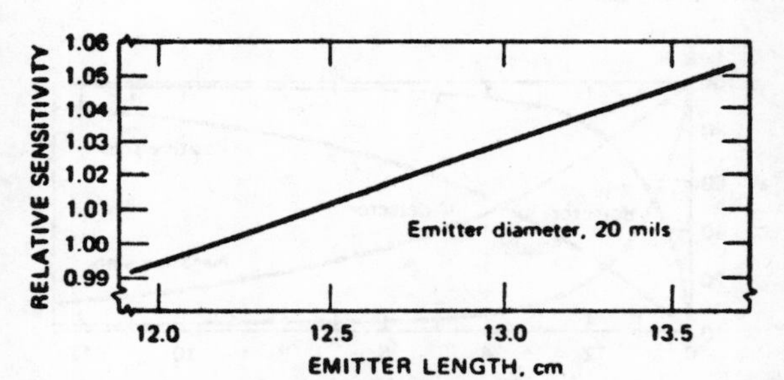

Fig. 3.16—Rhodium-detector sensitivity vs. length at constant emitter mass.

From Eq. 3.5 the relative detector sensitivity is

$$\frac{S(t)}{S(0)} = \exp(-K_\phi \phi \sigma t)(1 - K_\phi \phi \sigma t)$$

$$= e^{-t/\tau}\left(1 - \frac{t}{\tau}\right) \qquad (3.9)$$

where $\tau = 1/K_\phi \phi \sigma$.

For a rhodium detector with a 20-mil emitter and 10 mils of MgO insulation, the characteristic burnup time, τ, is $0.913 \times 10^{22}/\phi$. In an average thermal-neutron flux, $\phi = 4 \times 10^{13}$ neutrons cm^{-2} sec^{-1}, the value of τ is 23×10^7 sec, or 7.18 years. When the detector is exposed to this flux for one year, its sensitivity relative to its initial value decreases to $\exp(-1/7.18) \times [1 - (1/7.18)] = 0.75$. In other words, its sensitivity is now 75% of its initial value. In two years, the sensitivity decreases to 55% of its initial value. From this, it follows that the emitter burnup time is longer than the normal reactor refuelling cycle. The emitter burnup rates given in Table 3.3 are based on the assumption that $K_\phi = 1$. As noted earlier, K_ϕ is less than 1; so the actual burnup rates will be lower than those given in the table.

The product NQ in Eqs. 3.1 through 3.4 is equal to the total charge (betas) generated in the emitter before it is used up (all atoms activated). Since K_β and K_g are the factors that identify how many betas escape to become useful detector output current, the total useful charge generated by the detector is

$$q = K_\beta K_g Q N \qquad (3.10)$$

If we use the K values from Fig. 3.13 and calculate N from Eq. 3.8, we find the total useful charge generated by a 20-mil rhodium detector with 10 mils of MgO insulation to be

$$q = 12.1 \text{ coulombs/cm of emitter length}$$

which is in good agreement with experimental values.

(e) Response Characteristics. The response characteristics of a self-powered neutron detector are directly related to the radioactive scheme of the radionuclides formed in the emitter.

Figure 3.17 shows the decay scheme for self-powered detectors that use vanadium as the emitter material. All neutron absorptions in the emitter material, ^{51}V, which has a thermal-neutron-activation cross section of 4.9 barns, result in the creation of ^{52}V. The latter decays by beta emission to four excited states of ^{52}Cr, which then immediately go to the ground state by gamma emission. The half-life of this decay scheme is 3.76 min (226 sec). The 2.4-MeV beta accounts for almost 99% of the beta emission and provides energetic betas for a good signal-to-noise ratio.

Because the emitter radionuclide has a single beta decay period, the time response of the vanadium detector to a *step change in thermal-neutron flux from ϕ to zero* is

$$I(t) = I_0 e^{-0.693t/226} = I_0 e^{-t/326} \qquad (3.11)$$

and the time response to a *step change in thermal-neutron flux from zero to ϕ* is

$$I(t) = I_0(1 - e^{-t/326}) \qquad (3.12)$$

where t is in seconds, I(t) is the detector signal current, and I_0 is the steady-state signal given in Eq. 3.2. The mean life, 326 sec, is the half-life divided by 0.693; it is the time for I(t) to decrease by 1/e. Figure 3.18 shows the response of a vanadium self-powered detector to a step decrease in neutron flux (Eq. 3.11) on a semilogarithmic scale. T'e vanadium detector response follows an exponential with a characteristic 3.76-min half-life (time constant = 3.76/0.693 = 5.4 min) and reaches 90% of the step change in 12.5 min. Figure 3.19 shows the response of a vanadium self-powered detector to a step decrease and a step increase in neutron flux; the curves are plotted on a linear scale.

Figure 3.20 shows the decay scheme for rhodium emitters. Neutron absorption by rhodium creates two radioisotopes of ^{104}Rh. The total neutron activation cross section of rhodium is 150 barns. The cross section for the creation of the ground state ^{104}Rh is 139 barns (92.7% of

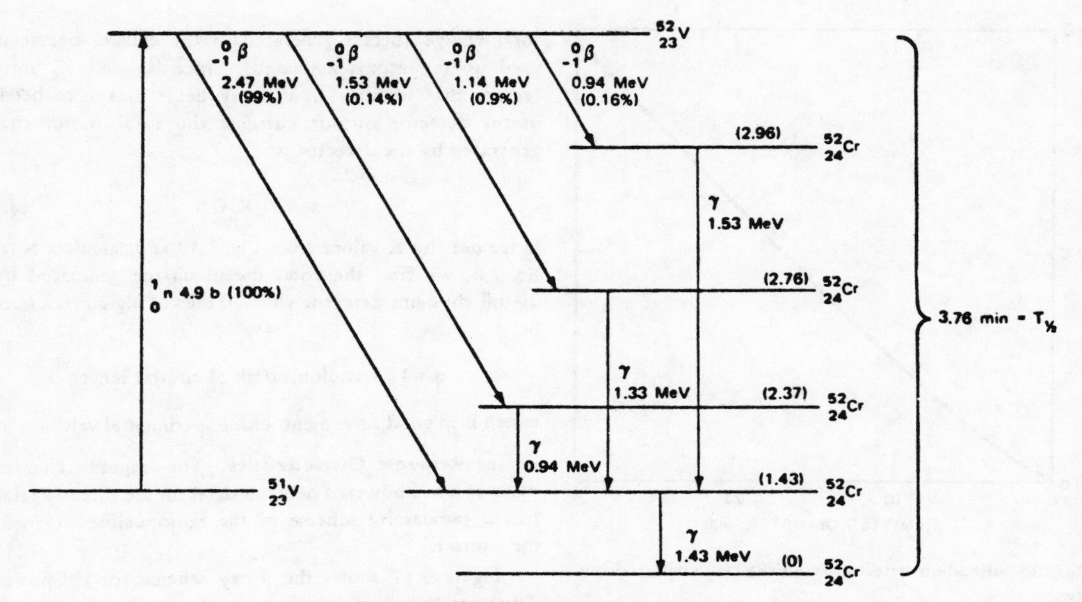

Fig. 3.17—Vanadium-emitter decay scheme.

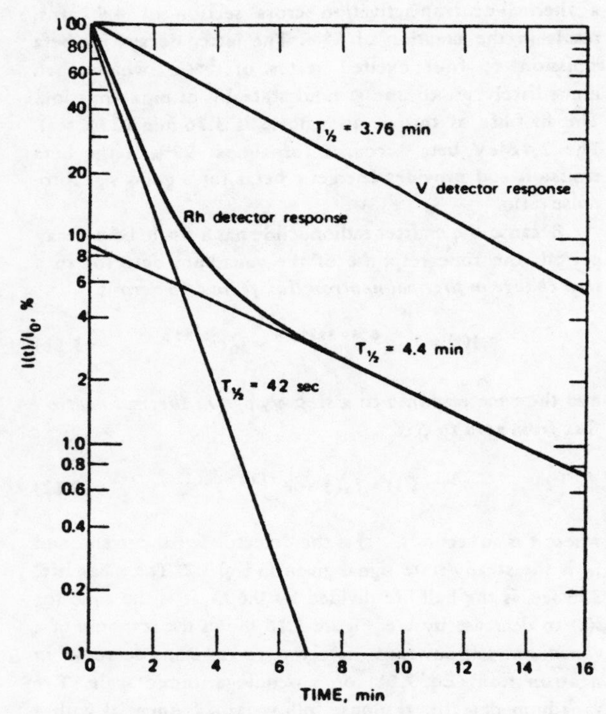

Fig. 3.18—Self-powered neutron-detector response to step change in neutron flux from $\phi = \phi$ to $\phi = 0$.

the 150 barns), and, for the creation of the metastable ^{104m}Rh, the cross section is 11 barns (7.3% of 150 barns). The metastable state decays by gamma emission to the ground state with a characteristic half-life of 4.4 min (264 sec). The ground state ^{104}Rh decays by beta emission to the ground state ^{104}Pd with a half-life of 42 sec. In this final beta-decay process, 98% of the emitted betas have an

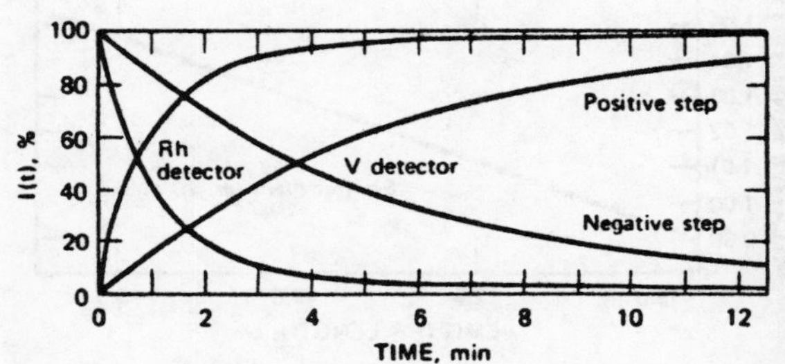

Fig. 3.19—Self-powered neutron-detector response to step change in neutron flux.

energy of 2.44 MeV, sufficiently high to provide adequate signal-to-noise ratio.

Because it has an emitter with two beta-decay constants, the time response of a rhodium self-powered detector to step changes in neutron flux involves two exponentials. For a *step change in neutron flux from ϕ to zero*,

$$I(t) = I_0\left[\left(1 - 0.073\,\frac{381}{381-61}\right)e^{-t/61} + 0.073\,\frac{381}{381-61}\,e^{-t/381}\right]$$

$$= I_0(0.913\,e^{-t/61} + 0.087\,e^{-t/381}) \qquad (3.13)$$

where t is in seconds and 61 (= 42/0.693) and 381 (= 264/0.693) are the mean lives in seconds of ^{104}Rh and ^{104m}Rh, respectively. Similarly, for a *step change in the thermal-neutron flux from zero to ϕ*, the time response is

$$I(t) = I_0[1 - (0.913\,e^{-t/61} + 0.087\,e^{-t/381})] \quad (3.14)$$

Figure 3.18 shows $I(t)/I_0$ vs. time on a semilogarithmic plot for a step decrease in thermal-neutron flux on a rhodium

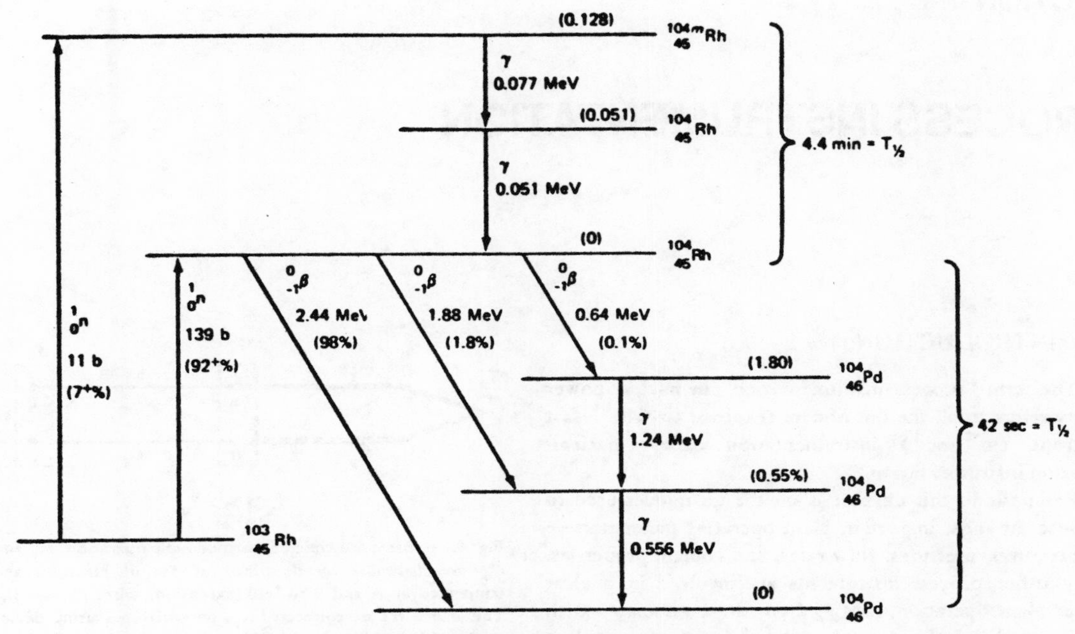

Fig. 3.20—Rhodium-emitter decay scheme.

detector. For the first 3 min after the step change, the detector signal is dominated by the 42-sec half-life (61-sec time constant); beyond 6 min after the step change, the rhodium detector signal follows the 4.4-min decay of ^{104m}Rh. The signal reaches 90% of the step change in less than 3 min. The y-intercepts of the two half-life curves in Fig. 3.18 correspond to the coefficients of each exponential in Eq. 3.13. Figure 3.19 shows the time responses of a rhodium detector to step decreases (Eq. 3.13) and to step increases (Eq. 3.14) in flux, plotted on a linear scale.

(f) Connecting Cables. The connecting cables required to bring the self-powered detector signals out of the core should produce as little background signal as possible. As shown in Figs. 3.11 and 3.12, the cables commonly used are 40 or 62 mils in outside diameter with magnesia insulation and Inconel lead wire and sheath. The neutron cross section for Inconel is negligible, so neutron-induced signals are negligible. The major sources of noise signals are Compton electrons and photoelectrons generated when gamma rays are absorbed or scattered by the lead wire and sheath. The noise signals are prompt in responding to changes in incident gamma flux.

Compton electrons and photoelectrons originating in the wire and absorbed in the sheath produce positive background signals. Electrons originating in the sheath and absorbed in the lead wire produce negative background signals. Equal absorption of electrons in the wire and sheath would result in a zero background signal, ideal for maximum neutron signal. The larger mass and surface area of the sheath tend to make a negative background signal for most connecting cables. Appropriate selection of emitter and sheath materials and dimensions should make it possible to build cables with essentially zero background signal.

SECTION 4

PROCESS INSTRUMENTATION

4-1 INTRODUCTION

The term "process instrumentation" in nuclear power plants refers to all the out-of-core (exposed to $<10^{11}$ fast neutrons cm^{-2} sec^{-1}) instrumentation except nuclear-radiation instrumentation.

Emphasis in this chapter is on the techniques used to measure the most important plant operating parameters—temperatures, pressures, flow rates, and coolant properties. Many other process instruments are involved in nuclear power plant operation, ranging from those associated with auxiliary chemical processes (waste treatment, coolant purification, etc.) to those involved in maintaining the plant atmosphere (air filtration and conditioning, ventilation, etc.).

Certain process instrumentation is so closely associated with specific plant components (e.g., control-rod position sensing) or with a particular reactor type (e.g., moisture detection in gas-cooled reactors) that it is more properly discussed in other chapters. Cross references are provided in these cases.

No single book (and certainly no single chapter in a book) can cover all significant aspects of even one category of process instrumentation in nuclear power plants.

4-2 TEMPERATURE SENSING

4-2.1 Thermocouples

(a) Basic Considerations. A thermocouple (Fig. 4.1) is a device that generates a small electromotive force (millivolts) almost directly proportional to the temperature difference between its "cold" and "hot" junctions (Seebeck effect). The degree of departure of the voltage–temperature relation from linearity depends on the materials used in the thermocouple (Fig. 4.2).

In a circuit made up of wires of dissimilar metals, the existence of emf's other than those at the junctions was discovered by Thomson (Lord Kelvin). An examination of this effect led to the conclusion that in general an emf exists between any two regions at different temperatures in a single conductor. In other words, a *thermal gradient* in a homogeneous material creates an emf; the direction of the emf is reversed when the thermal gradient is reversed. The effect is usually negligible in thermocouple applications since there are equal and opposite thermal gradients around the circuit. Moreover, the practice of keeping both legs of

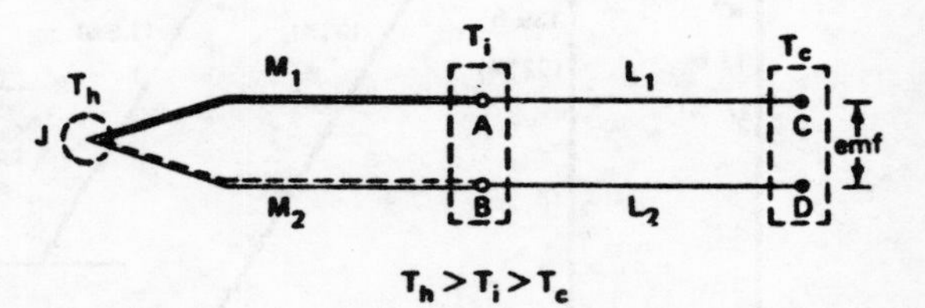

Fig. 4.1—Basic principle of thermocouple operation. M_1 and M_2 are dissimilar metals joined at J (hot junction) and connected at A and B to lead (extension) wires L_1 and L_2. The lead wires are connected to a potential-measuring device at C and D (cold junction). The electromotive force (Seebeck emf) across C and D depends on M_1, M_2, and on $T_h - T_c$ provided: (1) C and D are at the same temperature, (2) A and B are at the same temperature, and (3) the materials L_1 and L_2 are such that emf's across L_1 and L_2 due to the temperature difference $T_i - T_c$ are the same as the emf's that would exist if L_1 were replaced by M_1 and L_2 by M_2. (Note: Terminals A and B do not, in general, have to be at exactly the same temperature. However, it is good practice to have them at the same temperature.)

thermocouple circuits in the same thermal gradients ensures that the Thomson effect can be neglected.

Of greater importance are changes in the properties of thermocouple materials after prolonged exposure to elevated temperatures and steep temperature gradients such as exist near the hot junction. The thermocouple wires become inhomogeneous along their length, the degree of inhomogeneity depending on temperature, time, and the environment (air, steam, water, etc.). If there is a thermal gradient along the region where the homogeneity varies, then an emf can be generated (as if there were a thermocouple composed of fresh and altered material). This effect can be taken into account by thermocouple calibration or replacement. However, the calibration must be made in the same thermal gradient as that which will exist when the thermocouple is used.

(b) Thermocouple Lead (or Extension) Wire. The cold junction of a thermocouple circuit is almost always located at the indicating or controlling instrument itself. The instrument usually has means for automatically compensating for variations of the cold junction caused by variations in the ambient temperature. Indicating instrument and sensor are sometimes separated by substantial distances (as much as a few hundred feet). The thermocouples should be made long enough to extend from the

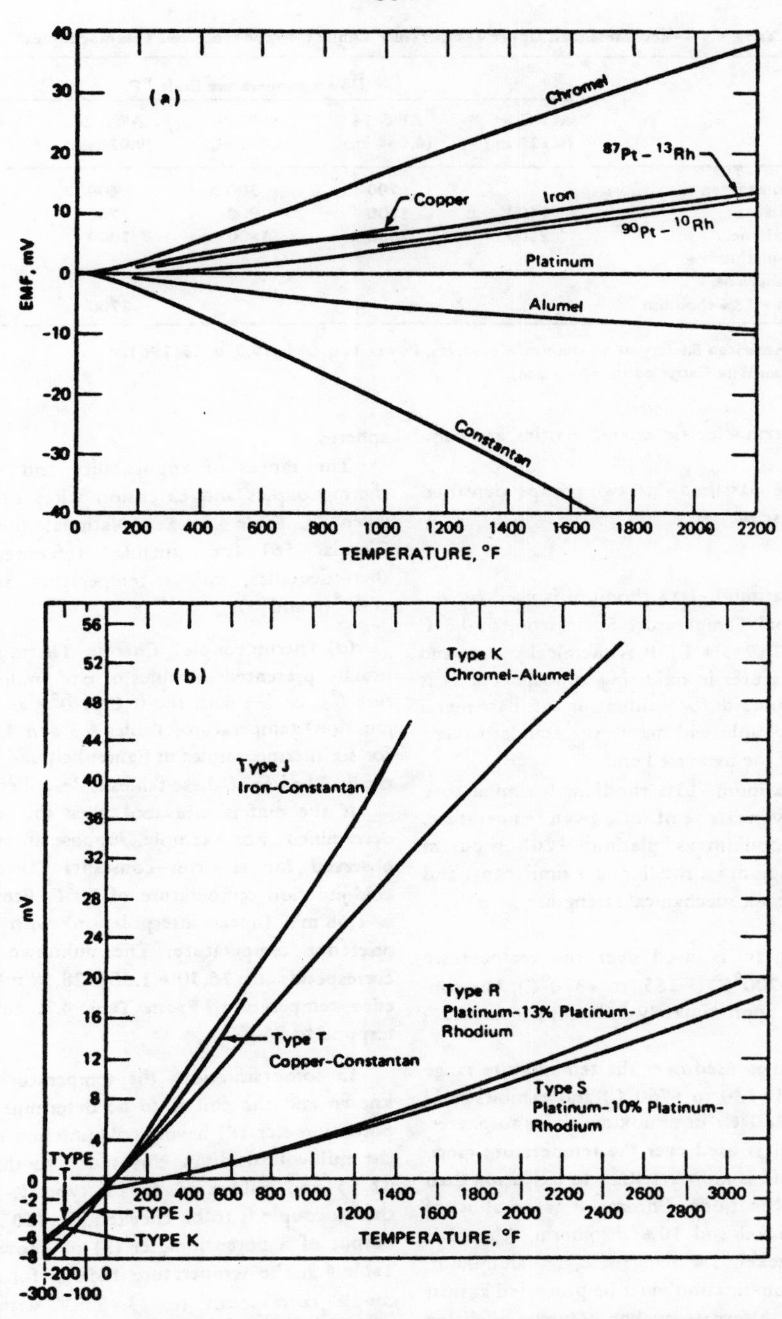

Fig. 4.2—Electromotive force vs. temperature difference for various materials (a) and thermocouples (b). In (b) the reference junction is at 32°F. [From: (a) Instrument Society of America, R.P. 1.6, Aug. 1, 1954; (b) American Society of Mechanical Engineers, Power Test Code 19-3, 1961.]

point of heat measurement to the indicators. If the thermocouple were made of platinum–rhodium, it would be prohibitively expensive. Therefore, lead (or extension) wires of cheaper metals that have thermoelectric characteristics matching the thermocouple over a limited temperature range must be used. This range is based on the ambient temperature expected at the point where the thermocouple extension wires connect to the thermocouple and where the greatest range of ambient temperature is expected.

(c) Thermocouple Materials. Thermocouple materials are available for use within the approximate limits of −300 to +3200°F (−185 to +1760°C). No single thermocouple meets all application requirements, but each possesses characteristics desirable for selected applications. Platinum is generally accepted as the standard reference material

Table 4.1—Recommended Upper Temperature Limits (°F) for Protected Thermocouples*

	Upper temperature limit, °F				
	AWG 8† (0.128 in.)	AWG 14 (0.064 in.)	AWG 20 (0.032 in.)	AWG 24 (0.020 in.)	AWG 28 (0.013 in.)
Copper–constantan		700	500	400	400
Iron–constantan	1400	1100	900	700	700
Chromel–alumel	2300	2000	1800	1600	1600
Platinum–platinum + 10% rhodium or platinum + 13% rhodium				2700	

*From American Society of Mechanical Engineers, Power Test Code 19.3, p. 23, 1961.
†American Wire Gauge number and size.

against which the thermoelectric characteristics are compared.

Thermocouples are classified into two groups identified as *noble metals* and *base metals.*

Noble metals

1. Platinum vs. platinum–10% rhodium is used for defining the International Temperature Scale from 630.5°C (1166.9°F) to 1063°C (945.4°F). It is chemically inert and stable at high temperatures in oxidizing atmospheres. It is widely used as a standard for calibration of base-metal thermocouples. This couple will match the standard reference table to ±0.5% of the measured emf.

2. Platinum vs. platinum–13% rhodium is similar to 1 and produces a slightly greater emf for a given temperature.

3. Platinum–5% rhodium vs. platinum–20% rhodium and platinum–6% rhodium vs. rhodium are similar to 1 and 2 but show slightly greater mechanical strength.

Base metals

1. Copper–constantan is used over the temperature range of −300 to +700°F (−185 to +370°C). The constantan is an alloy of approximately 55% copper and 45% nickel.

2. Iron–constantan is used over the temperature range of −200 to +1400°F (−130 to +760°C) and exhibits good stability at 1400°F (760°C) in nonoxidizing atmospheres.

3. Chromel–alumel is used over the temperature range of −200 to 2300°F and is more resistant to oxidation than other base-metal combinations. Chromel-P is an alloy of approximately 90% nickel and 10% chromium. Alumel is approximately 94% nickel, 3% manganese, 2% aluminum, and 1% silicon. This combination must be protected against reducing atmospheres. Alternate cycling between oxidizing and reducing atmospheres is particularly destructive.

4. Chromel–constantan produces the highest thermoelectric output of any conventional thermocouples. It is used up to 1400°F (760°C) and exhibits a high degree of calibration stability at temperatures not exceeding 1000°F (538°C).

The upper temperature limits for the various thermocouple materials depend on the wire sizes and the environment in which the thermocouples are used. Table 4.1 lists the recommended upper temperature limits for thermocouples protected from corrosive or contaminating atmospheres.

The ranges of applicability and limits of error for thermocouples and extension wires of standard sizes are given in Table 4.2. See National Bureau of Standards Circular 561 for expanded reference tables of these thermocouples, emf vs. temperature, and temperature vs. emf, °F, and °C.

(d) Thermocouple Charts. Thermocouple data are usually presented as tables of emf (millivolts) vs. temperature (°C or °F) with the 0°C or 0°F as the reference (cold junction) temperature. Tables 4.3 and 4.4 present the data for six thermocouples in Fahrenheit and Centigrade, respectively. The use of these tables is described below.

If the emf is measured, then the temperature can be determined. For example, suppose an emf of 26.50 mV is observed for an iron–constantan thermocouple with a cold-junction temperature of 70°F. From Table 4.3, 70°F is 1.68 mV (linear interpolation) with 0°F cold-junction reference temperature. The unknown temperature thus corresponds to 26.50 + 1.68 = 28.18 mV with 0°F reference temperature. From Table 4.3, this is seen to correspond to 960°F.

In some situations the temperature is assumed to be known and the emf is to be determined. For example, a potentiometer (P) having cold-junction compensation (i.e., the millivolt readings correspond to the cold junction at 32°F) and calibrated for a type K (chromel–alumel) thermocouple is to be checked at 1450°F by means of the output of a potentiometer (S) of known accuracy. From Table 4.3, the temperature 1450°F for a type K thermocouple corresponds to 33.48 mV with a cold-junction temperature of 0°F. The emf between 0°F and 32°F is also seen to be 0.69 mV (linear interpolation in the table); so the temperature 1450°F with a cold-junction temperature of 32°F corresponds to 33.48 − 0.69 = 32.79 mV. This is the potential that should be observed on potentiometer "S" when potentiometer P is checked at 1450°F.

In the charts usually supplied with thermocouple devices, the temperature interval is smaller than the 50°F (or 50°C) interval of Tables 4.3 and 4.4. This reduces the labor of interpolation.

(e) Series-Connected Thermocouples. In one method

Table 4.2—Range of Applicability and Limits of Error* for Commercially Available Thermocouples and Extension Wires†

	Thermocouples			Extension wire		
		Limits of error			Limits of error, °F	
Type	Temperature range, °F	Standard	Special	Temperature range, °F	Standard	Special
Copper–constantan	−300 to −75		±1%			
	−150 to −75	±2%	±1%			
	−75 to +200	±1 1/2 °F	±3/4 °F	−75 to +200	±1 1/2	±3/4
	200 to 700	±3/4 %	±3/8 %			
Iron–constantan	0 to 530	±4°F	±2°F	0 to 400	±4	±2
	530 to 1400	±3/4 %	±3/8 %			
Chromel–alumel	0 to 530	±4°F		0 to 400	±6	
	530 to 2300	±3/4 %				
Platinum–platinum + 10% rhodium or platinum + 13% rhodium	0 to 1000	±5°F		75 to 400	±12	
	1000 to 2700	±0.5%				

*Does not include use or installation errors.
†From American Society of Mechanical Engineers, Power Test Code 19.3, p. 23, 1961.

Table 4.3—Electromotive Force vs. Temperature (°F) for Thermocouples*

°F	Chromel–alumel (type K)	Iron–constantan (type J)	Copper–constantan (type T)	Chromel–constantan (type E)	Platinum–platinum + 10% rhodium (type S)	Platinum–platinum + 13% rhodium (type R)
0	0.00	0.00	0.000	0.00		
50	1.08	1.39	1.059	1.61	0.156	0.154
100	2.20	2.83	2.187	3.29	0.321	0.319
150	3.34	4.30	3.381	5.06	0.501	0.499
200	4.50	5.80	4.637	6.89	0.695	0.695
250	5.65	7.31	5.950	8.78	0.900	0.906
300	6.77	8.83	7.317	10.73	1.117	1.129
350	7.88	0.37	8.734	12.73	1.342	1.360
400	8.99	11.92	10.195	14.77	1.574	1.603
450	10.11	13.46	11.700	16.86	1.812	1.853
500	11.25	15.01	13.245	18.97	2.056	2.111
550	12.39	16.54	14.827	21.11	2.305	2.376
600	13.54	18.07	16.443	23.27	2.558	2.646
650	14.70	19.61	18.091	25.46	2.816	2.922
700	15.86	21.15	19.770	27.67	3.077	3.202
750	17.03	22.68	21.475	29.88	3.340	3.486
800	18.21	24.21		32.11	3.606	3.776
850	19.38	25.74		34.34	3.875	4.069
900	20.57	27.29		36.59	4.146	4.363
950	21.75	28.84		38.84	4.419	4.662
1000	22.94	30.41		41.08	4.696	4.967
1050	24.12	32.01		43.33	4.974	5.275
1100	25.31	33.61		45.58	5.256	5.587
1150	26.49	35.25		47.83	5.540	5.904
1200	27.66	36.90		50.06	5.826	6.224
1250	28.83	38.60		52.29	6.115	6.545
1300	30.00	40.32		54.52	6.407	6.872
1350	31.17	42.08		56.73	6.701	7.202
1400	32.33	43.85		58.94	6.997	7.535
1450	33.48	45.64		61.13	7.269	7.873
1500	34.61	47.42		63.32	7.598	8.215
1550	35.75	49.20		65.49	7.903	8.559

*From Instrument Society of America, Recommended Practices 1.7, Aug. 1, 1954.

Table 4.4—Electromotive Force vs. Temperature (°C) for Thermocouples*

°C	Chromel–alumel (type K)	Iron–constantan (type J)	Copper–constantan (type T)	Chromel–constantan (type E)	Platinum–platinum + 10% rhodium (type S)	Platinum–platinum + 13% rhodium (type R)
0	0.00	0.00	0.000	0.00	0.000	0.000
50	2.02	2.58	2.035	3.04	0.299	0.298
100	4.10	5.27	4.277	6.32	0.643	0.645
150	6.13	8.00	6.703	9.79	1.025	1.039
200	8.13	10.78	9.288	13.42	1.436	1.465
250	10.16	13.56	12.015	17.18	1.868	1.918
300	12.21	16.33	14.864	21.04	2.316	2.395
350	14.29	19.09	17.821	24.97	2.778	2.890
400	16.40	21.85	20.874	28.95	3.251	3.399
450	18.51	24.61		32.96	3.732	3.923
500	20.65	27.39		37.01	4.221	4.455
550	22.78	30.22		41.05	4.718	5.004
600	24.91	33.11		45.10	5.224	5.563
650	27.03	36.08		49.13	5.738	6.137
700	29.14	39.15		53.14	6.260	6.720
750	31.23	42.32		57.12	6.790	7.315
800	33.30	45.53		61.08	7.329	7.924
850	35.34	48.73		64.99	7.876	8.544
900	37.36			68.85	8.432	9.175
950	39.35			72.68	8.997	9.816
1000	41.31			76.45	9.570	10.471
1050	43.25				10.152	11.138
1100	45.16				10.741	11.817
1150	47.04				11.336	12.503
1200	48.89				11.935	13.193
1250	50.69				12.536	13.888
1300	52.46				13.138	14.582
1350	54.20				13.738	15.276
1400					14.337	15.969
1450					14.935	16.663
1500					15.530	17.355
1550					16.124	18.043

*From Instrument Society of America, Recommended Practices 1.7, Aug. 1, 1954.

for averaging temperatures, each thermocouple is connected in series with the others (Fig. 4.3) using extension wires of the correct materials. Note that the extension wires are connected at the instrument from each couple in the series. This permits proper cold-junction compensation. The emf developed at the terminal G is the sum of the emf's developed by all the thermocouples. Consequently, instruments used with this type of circuit must be calibrated for the total emf of all the thermocouples, and the cold-junction compensation must be adjusted to compensate for the greater millivoltage change due to ambient temperature changes.

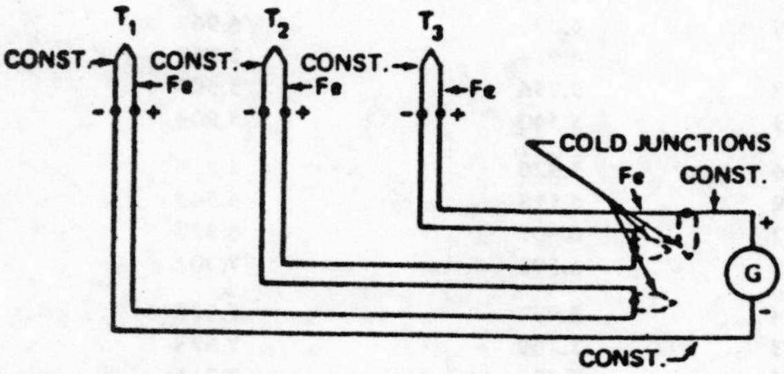

Fig. 4.3—Series-connected thermocouples (iron = Fe; constantan = const.).

The advantages of this method are (1) a large emf is developed for a given temperature and (2) burnout of any single thermocouple is immediately apparent.

The disadvantages are (1) a special calibration is required; (2) a short circuit, which might materially reduce the emf of one couple, might not be detected by observation of total emf; (3) on a multipoint installation the same number of thermocouples must be used in series at each point; and (4) grounded thermocouples cannot be used.

(f) Parallel-Connected Thermocouples. Temperatures can be averaged by connecting thermocouples in parallel (Fig. 4.4). Here the net emf developed at G is the average of the potential drops across each individual branch of the

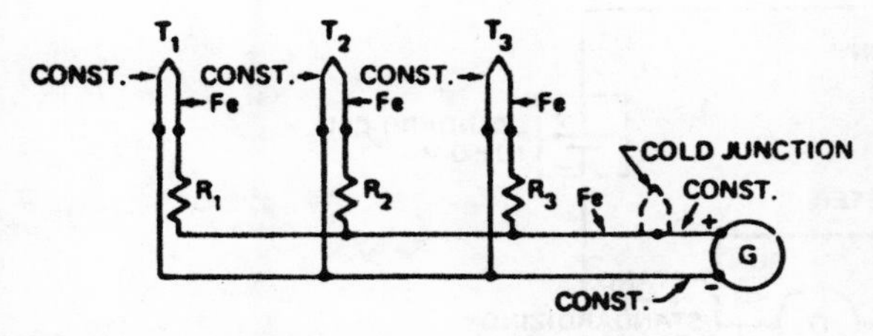

Fig. 4.4—Parallel-connected thermocouples (iron = Fe; constantan = const.).

circuit rather than an average of the emf's. Since current circulates among the thermocouples when the temperatures T_1, T_2, and T_3 are different, the resistance of each individual thermocouple circuit must be equalized by resistors R_1, R_2, and R_3 (swamping resistors).

The resistance of the actual thermocouple will also vary with its temperature. The effect of this variation can be minimized by making the values of R_1, R_2, and R_3 high (500 ohms typical) in comparison with the resistance changes encountered. R_1, R_2, and R_3 are of equal value. The resistances of the swamping resistors are limited by the sensitivity of the indicator (amount of current required to actuate it) and the number of thermocouples in the arrangement. Maximum sensitivity is achieved when the total impedance (resistance in this case) of the thermocouple circuit (all thermocouples and swamping resistors) is equal to the internal impedance (resistance) of the indicator. The resistance of a thermocouple circuit can be measured with a millivoltmeter and rheostat (Fig. 4.5).

The advantages of parallel connection are (1) standard instrument calibrations and cold-junction compensation for a single couple can be used and (2) if one couple fails, operation can be continued.

The disadvantages are (1) failure of a single couple is not readily apparent and (2) grounded thermocouples cannot be used.

These thermocouples are used for measuring the efficiency of heat exchangers.

(g) Differentially Connected Thermocouples. Two typical arrangements for differentially connected thermocouples are shown in Fig. 4.6. Figure 4.6(a) illustrates the basic circuit in which one couple senses one temperature and a second couple senses another temperature. Note that the similar metals are interconnected and that *it is not necessary to refer the cold junction back to the galvanometer.* This is because the one couple constitutes a hot junction and the second the cold junction. Also, both leads to the galvanometer are of the same material, and each of these leads joins a third metal at a point where there is no temperature difference. As a result, even though two junctions of dissimilar metals are formed, they have no effect because there is no temperature difference. There are no unbalanced thermocouple emf's created when the copper leads of the galvanometer are both connected to the constantan leads at a point where no temperature difference exists.

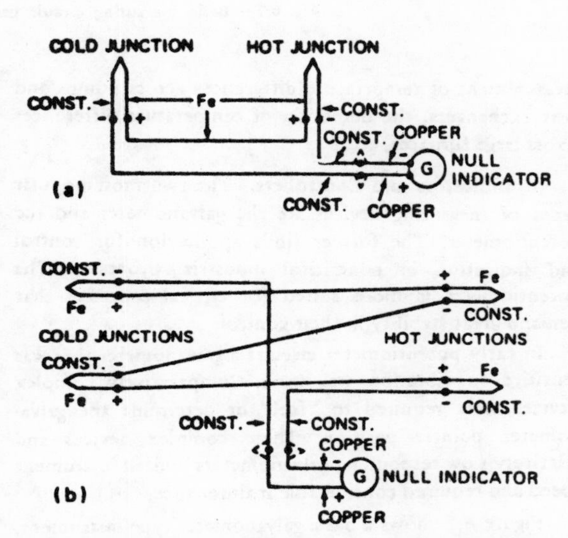

Fig. 4.6—Two typical arrangements of differentially connected thermocouples (iron = Fe; constantan = const.).

Figure 4.6(b) shows a circuit using differentially connected couples in series to provide larger emf's with small temperature difference. Note that the basic differential circuit in Fig. 4.6(a) is duplicated in Fig. 4.6(b). Any number of couples may be connected similarly to produce a desired emf.

When differentially connected couples are used, cold-junction compensation is not required in the indicator or controller since, from the basic nature of the circuit, there is no cold junction located at the indicator as in all the other circuits discussed.

Applications of differentially connected couples include

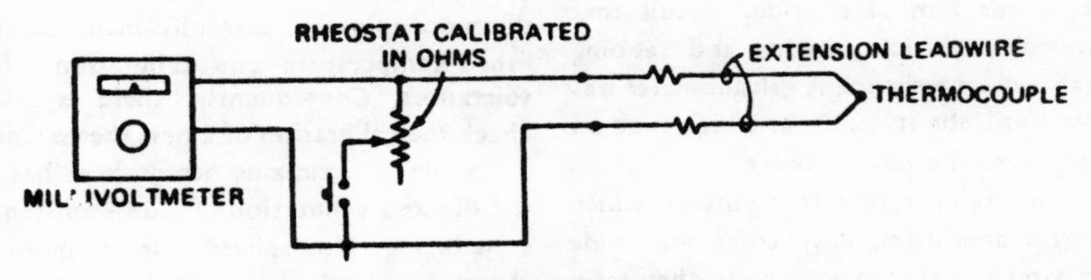

Fig. 4.5—Method for measuring thermocouple-circuit resistance.

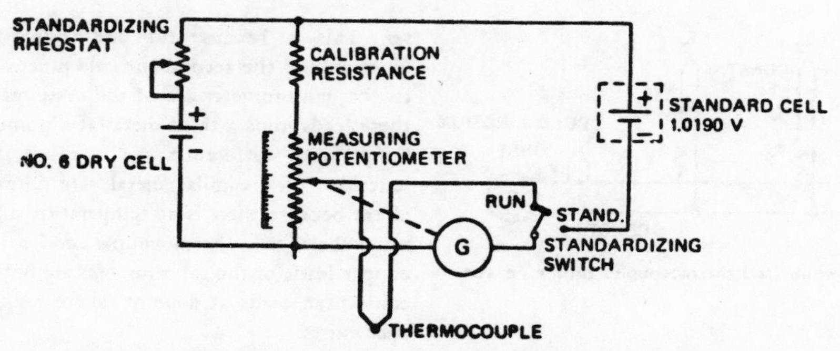

Fig. 4.7—Basic measuring circuit using potentiometer and galvanometer.

measurement of temperature differences across pumps and heat exchangers, the detection of temperature differences across large furnaces, etc.

(h) Indicators and Controllers. The two most popular types of measuring devices are the galvanometer and the potentiometer. The former finds application for control and indication in noncritical industrial processes. The potentiometer is more suited for critical processes that demand great stability in their control.

In early potentiometer circuits a galvanometer that was sensitive to vibration was used. Comparatively complex devices were required to "feel" or determine the galvanometer pointer position. These complex devices and relatively slow responding galvanometers limited instrument speed and required considerable maintenance.

Figure 4.7 shows a basic galvanometer type instrument. The No. 6 dry cell provides a constant current through the slide-wire. This current is held constant by periodically shifting the standardizing switch to the "standard" position and comparing the voltage of the No. 6 dry cell with that of the standard cell. If they are not equal (the galvanometer being deflected from its null position), the standardizing rheostat is adjusted until the galvanometer does not deflect in switching from the No. 6 dry cell to the standard cell. The "standard" switch is then set at the "run" position.

The thermocouple is now in the galvanometer circuit, and, by moving the slide-wire contactor along the slide-wire, a point will be reached at which the thermocouple voltage is matched by the voltage drop across the slide-wire. The scale located above the slide-wire is calibrated in terms of temperature in place of voltage; thus a conversion from voltage to temperature is avoided.

The circuit of Fig. 4.7 shows the essentials of the potentiometer type measuring instrument. In practice the measuring slide-wire is one arm of a bridge circuit that includes temperature-compensation circuits and ratioing resistors as well as calibration resistors. A galvanometer was used in these circuits until about 1940, at which time an electronic amplifier replaced the galvanometer.

At the present time, temperature transmitters, which are solid-state high-gain amplifiers, have come into wide use. Having no slide-wires or other moving parts, they may, for all practical purposes, be considered maintenance-free devices. They are very useful in a modern coordinated plant control system because their output (typically 4 to 20 mA), which is proportional to the input, can be used in many ways, e.g., in conversion to pneumatic signals and in driving recorders, indicators, and controllers. Temperature transmitters may also serve as one of the inputs to Btu-calculating devices and as inputs for temperature compensation in flow-measuring and -recording meters. Figure 4.8 is typical of a 4- to 20-mA temperature transmitter.

(i) Thermocouple Response Time. The rate at which a thermocouple reaches the temperature of the medium in which it is immersed (or which it contacts) depends primarily on the rate of heat transfer (conduction) through its protecting sheath.

In Fig. 4.9 the time lag between the thermocouple temperature and the medium temperature is shown for three thermocouple arrangements. The medium is stirred water heated at a constant rate. The time lag is shown for a bare 20-gage thermocouple, a thermocouple embedded in a silver plug that is in contact with the bottom section of the thermocouple well, and a thermocouple (14-gage) butt-welded and forced against the bottom of the well. With respect to the bare couple, the average time lags are about 20 sec for the well-and-plug arrangement and over 90 sec for the couple forced against the bottom of the well.

The silver-plug arrangement (so-called "high-speed couple") reduces the response time considerably. Further reduction could be achieved by reducing the wall thickness of the well; however, operating conditions may not permit such reduction.

In pt. 4-2.3(c) the structure of thermometer wells is discussed in detail.

(j) Testing Thermocouples. To realize the degree of accuracy obtainable with a modern industrial pyrometer, thermocouples are carefully manufactured to match published temperature–emf calibration tables within specific tolerances. Consequently, there is seldom any need to check the calibration of a new thermocouple.

While an oxidizing atmosphere has a greater effect on the life and calibration of iron–constantan thermocouples, a reducing atmosphere has a more severe effect on chromel–alumel couples.

It is *not* recommended that a used thermocouple be

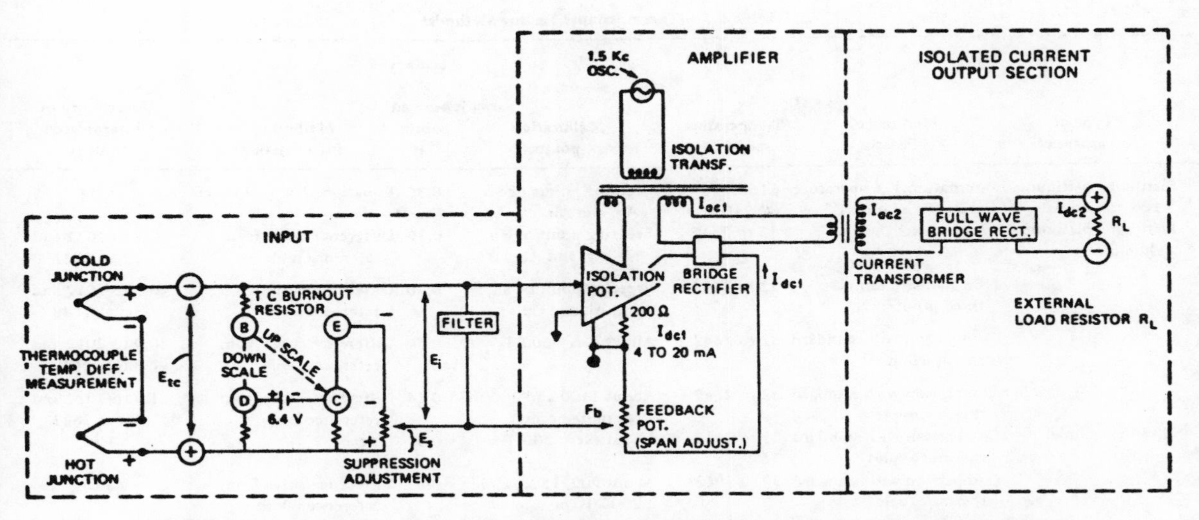

Fig. 4.8—Typical temperature transmitter circuit. This is a simplified schematic of a solid-state transmitter.

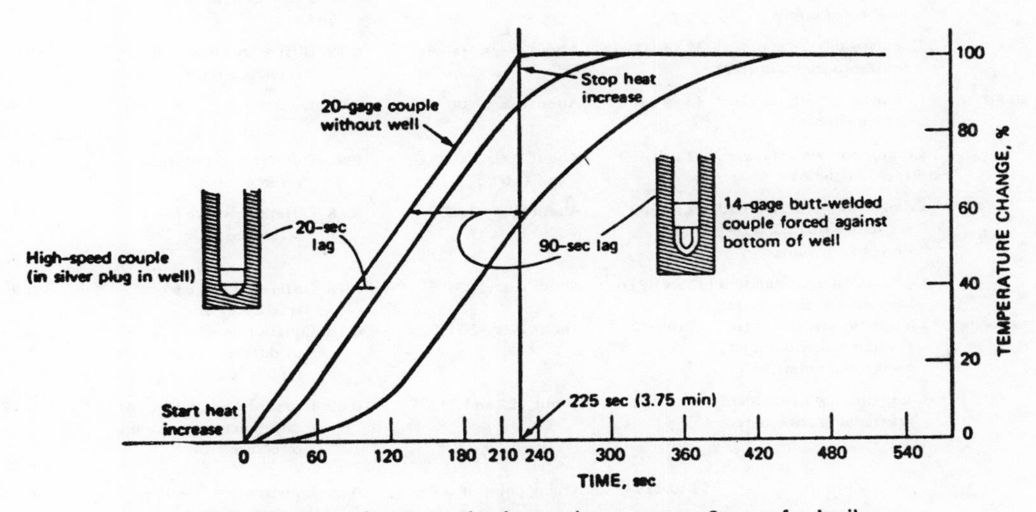

Fig. 4.9—Thermocouple response time in several arrangements. See text for details.

removed from the installation for testing in a laboratory furnace since it is practically impossible to duplicate in the laboratory furnace the temperature gradients of the actual installation. It is advisable to test a used thermocouple under the same conditions and in the same installation where it is normally used.

Two basic methods are used for checking the accuracy of thermocouples: (1) the fixed-points method in which the emf is measured when the couple is immersed in standard liquid metals at their freezing points or in water at the boiling point or in subliming CO_2 or boiling liquid oxygen and (2) the comparison method in which the emf of the couple is compared with the emf of a standard couple. Table 4.5 summarizes the methods used for testing the major types of thermocouples. Note that in all except high-precision laboratory work the comparison method is used to test and calibrate thermocouples.

To test a used thermocouple, you must have a reference standard that is known to be accurate. A new thermocouple whose calibration has been determined by comparison with a primary standard or directly against platinum may be used as a reference standard. It should then be labeled accordingly and reserved for this purpose. Since the original characteristics of the reference couple (like that of any used thermocouple) will change as it is used in testing, such a reference standard should be tested at intervals determined by the frequency of use and replaced when it is beyond the desired limits of accuracy.

The method selected for testing used thermocouples depends on the type of installation. The success of each method depends primarily on the stability of the temperature during tests. The following methods are recommended in the order in which they are listed.

Table 4.5—Thermocouple Testing Methods*

Type of thermocouple	Methods of calibration	Temperature range, °F	Calibration points	Accuracy at observed points, °F	Method of interpolating	Uncertainty of interpolated values
Platinum–platinum + 10% rhodium	International temperature scale (fixed points)	1166.9 to 1945.4	Freezing points of Sb, Ag, and Au	0.36	Equation: $E = a + bt + ct^2$	0.54
Platinum–platinum + rhodium†	Fixed points	32 to 2642	Freezing point of Zn, Sb, Ag, and Au	0.36	Difference curve from reference table	0.9 to 2012 and 3.6 at 2642
	NBS standard samples, fixed points	32 to 2642	Freezing point of Sn, Zn, Al, and Cu	0.36	Difference curve from reference table	0.9 to 2012 and 3.6 at 2642
	Comparison with standard thermocouple	32 to 2642	About every 200° F	0.54	Difference curve from reference table	0.9 to 2012 and 3.6 at 2642
	Comparison with standard thermocouple	32 to 2642	About 1100 and 2000° F (or more points)	0.54	Difference curve from reference table	1.8 to 2012 and 5.4 at 2642
Chromel–alumel	Comparison with standard thermocouple†	32 to 2012	About every 200° F	0.9	Linear	1.8
	Comparison with standard thermocouple†	32 to 2012	About 900, 1500, and 2000° F (or more points)	0.9	Difference curve from reference table	3.6
	Comparison with standard resistance thermometer‡ or at fixed points	32 to 662	About every 200° F	0.18	Difference curve from reference table	0.9
	Comparison with standard resistance thermometer‡	32 to −310	About every 100° F	0.18	Difference curve from reference table	0.9
Iron–constantan	Comparison with standard thermocouple‡	32 to 1400	About every 200° F	0.9	Linear	1.8
	Comparison with standard thermocouple‡	32 to 1400	About 200, 600, 900, and 1400° F	0.9	Difference curve from reference table	1.8
	Comparison with standard resistance thermometer‡ or at fixed points	32 to 662	About every 200° F	0.18	Difference curve from reference table	0.9
	Comparison with standard resistance thermometer‡	32 to −310	About every 100° F	0.18	Difference curve from reference table	0.9
Copper–constantan	Comparison with standard resistance thermometer‡ or at fixed points	32 to 572	About every 200° F	0.18	Equation: $e = at + bt^2 + ct^3$ or difference curve from reference table	0.36
	Comparison with standard resistance thermometer‡	32 to 212	About 122 and 212° F	0.09	Equation: $e = at + bt^2$ or difference curve from reference table	0.18
	Fixed points	32 to 212	Boiling point of water	0.09	Equation: $e = \frac{a(t-32)}{1.8} + \frac{0.04(t-32)^2}{3.24}$	0.36
	Comparison with standard resistance thermometer‡	32 to −310	About every 100° F	0.18	Equation: $e = at + bt^2 + ct^3$ or difference curve from reference table	0.36
	Fixed points	0 to −310	Sublimation point of CO_2 and boiling point of O_2	0.18	Difference curve from reference table	0.54

*From American Society of Mechanical Engineers, Power Test Code 19.3, Chap. 9, 1961.
†Either 10 or 13% rhodium.
‡In stirred liquid bath.

1. Insert a reference standard into the same protecting tube the used thermocouple is in if the size of the tube permits. Connect each thermocouple to a portable potentiometer through a selector switch, and compare the emf's developed.

2. Install a reference standard adjacent to the fixed thermocouple. Drill a hole as close to the fixed installation as practicable, and install the standard in such a manner that the ends of the two protecting tubes are as close together as possible. To ensure a fair comparison, use thermocouples of the same size and protecting tubes of the same size and type. Connect the couples to the instrument as in method 1, and compare readings.

3. Compare readings of successive installations of used and reference thermocouples. Test the used thermocouple first by reading the emf developed at a selected tempera-

ture. Remove it from its protecting tube, and replace it with the reference standard. Note that the standard should always be inserted in the protecting tube to the same depth as the used thermocouple. Insert the assembly to the same depth as the one tested, wait for the reference thermocouple to come to equilibrium, and then read the emf developed and compare it with that of the used thermocouple to come to equilibrium, and then read the emf others because it requires that the temperature remain constant for a longer period.

4-2.2 Resistance Thermometers

(a) Basic Considerations. The resistance thermometer is based upon the inherent characteristic of metals to change electrical resistance when they undergo a change in temperature. The electrical resistance of very pure metals varies with temperature from about 0.3 to 0.6% resistance change per degree at room temperature (or about 0.17 to 0.33% per degree Fahrenheit). Industrial resistance-thermometer bulbs are usually made of platinum, copper, or nickel.

An impurity or alloying constituent in a metal decreases the temperature dependence markedly except for a few unusual alloys. Pure platinum in a fully annealed and strain-free state has a resistance–temperature relationship that is especially stable and reproducible. For this reason, pure platinum has been chosen as the international standard of temperature measurement in the temperature range from the liquid oxygen boiling point to the antimony melting point. For the resistance element, platinum is drawn into wire with utmost care to maintain high purity, and the wire is formed into a coil that is carefully supported so that it will not be subjected to mechanical strain caused by differential thermal expansion. Rugged designs are required in military and other applications so that vibration and mechanical shocks will not give momentary or permanent detrimental strain to the platinum coil.

Pure nickel has also been widely used for industrial and many military applications where moderate temperature ranges are involved. Tungsten, copper, and some other metals are also used.

The fractional change in electrical resistance of a material per unit change in temperature is the temperature coefficient of resistance for the material. The coefficient is expressed as the fractional change in resistance (ohms per ohm) per degree of temperature change at a specific temperature. For most metals, the temperature coefficient is positive.

For pure metals the change in resistance with temperature is practically linear, at least over a substantial range of temperature. The relationship can be expressed as

$$R_t = R_0 \, (1 + \alpha t) \qquad (4.1)$$

where R_t equals the resistance in ohms at temperature t, R_0 equals the resistance in ohms at 0°C (or some other reference temperature), and the coefficient α is the temperature coefficient of resistance. In differential form the relationship is

$$\alpha = \frac{1}{R_0}\frac{dR}{dt} \qquad (4.2)$$

When the resistance does not vary linearly with the temperature, it is customary to include quadratic and cubic terms:

$$R_t = R_0 \, (1 + \alpha t + bt^2 + ct^3) \qquad (4.3)$$

where the coefficients α, b, and c are determined from measurements of the resistance at three or more temperatures uniformly spaced over the working range of temperature.

The resistance–temperature relation for platinum is given by the Callendar–VanDusen equation:

$$\frac{R_T}{R_0} = 1 + \alpha\left[T - b\left(\frac{T}{100} - 1\right)\left(\frac{T}{100}\right) - \beta\left(\frac{T}{100} - 1\right)\left(\frac{T}{100}\right)^3\right] \qquad (4.4)$$

where T is the temperature in degrees Centigrade and β is taken as zero for T above 0°C.

(b) Comparison of Resistance Materials. In Fig. 4.10 the resistance R and dR/dT vs. temperature T for a typical platinum resistance sensing element are normalized to 1.00 ohm at 0°C.

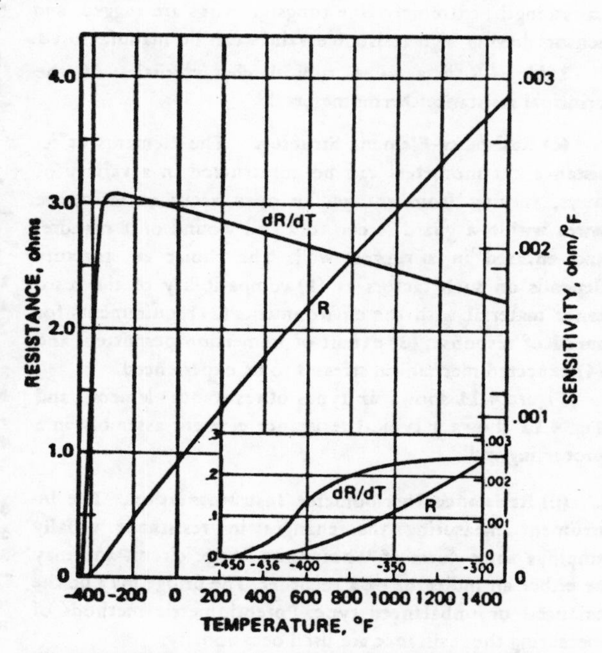

Fig. 4.10—Resistance and sensitivity versus temperature for various materials. Figure is for platinum with a resistance of 1.00 ohm at 32°F.

Tables 4.6 and 4.7 give the values of resistance vs. temperature for platinum, nickel, and copper.

Platinum. As noted earlier, platinum is the standard reference material for resistance thermometers. Recently, sensors made of very thin platinum films deposited on a substrate (usually a ceramic) have come into use. This method of constructing resistance thermometers leads to small sensing elements with high impedance (resistance) values.

Copper. Copper is inexpensive and has the most nearly linear relation of known metals over a rather wide temperature range. Copper has low resistance to oxidation above moderate temperatures and has much poorer stability and reproducibility than platinum in most applications. The low resistance of copper is a disadvantage when a high-resistance element is desired.

Nickel. Nickel has been widely used as a temperature sensing element over the range from about −100 to +300°C (−150 to 570°F), principally because of its low cost and the high value of its temperature coefficient. Above 300°C (570°F), the resistance−temperature relation for nickel changes character. Nickel is very susceptible to contamination by certain materials, and the relation of resistance to temperature is not as well known nor as reproducible as that of platinum.

Tungsten. The resistance vs. temperature relation of tungsten is not as well known as that of platinum. Full annealing of tungsten is impractical, and therefore tungsten sensors have been found to be less stable than well-made platinum sensors. Tungsten has been shown to have good resistance to very high nuclear-radiation levels and compares with platinum in this respect. Because of its mechanical strength, extremely fine tungsten wires are rugged, and sensors having high resistance values can be manufactured.

Table 4.8 lists some typical characteristics of the principal resistance thermometers.

(c) Resistance-Element Structure. The elements of resistance thermometers can be constructed in a variety of ways, varying from a cage-like open array of resistance wires within a guard screen to a coil wound on a mandrel and encased in a rugged well. The choice of structure depends on such factors as (1) compatibility of the resistance material with the environment, (2) requirements for speed of response, (3) extent of immersion permitted, and (4) expected mechanical stresses to be experienced.

Figure 4.11 shows six types of resistance elements, and Fig. 4.12 shows a typical resistance element assembly in a protecting well.

(d) Resistance-Thermometer Instrumentation. The instrument measuring the changes in resistance usually employs some form of Wheatstone bridge circuit and may be either an indicator or a recorder. The bridge may be the balanced or unbalanced type. Potentiometric methods of measuring the resistance are used occasionally.

Figure 4.13 is a diagram of a typical Wheatstone bridge used for resistance-thermometer measurement: a and b are ratio arms of equal resistance; and r is a variable resistance,

Table 4.6—Resistance vs. Temperature (°F) for Platinum, Nickel, and Copper Resistance Elements*

°F	Platinum 10 ohms	Platinum 100 ohms	Nickel 100 ohms (type I)†	Nickel 200 ohms (type II)‡	Copper 10 ohms at 25°C
0	9.290	91.165	89.94	227.190	8.358
32	10.000	98.129	100.00	235.116	9.042
50	10.398	102.030	105.84	239.696	9.428
100	11.496	112.807	122.79	252.890	10.498
150	12.585	123.495	140.92	266.811	11.568
200	13.665	134.095	160.34	281.498	12.638
250	14.736	144.605	181.16	296.993	13.708
300	15.798	155.027	203.51	313.341	14.778
350	16.851	165.361	227.51	330.589	
400	17.895	175.606	253.26	340.787	
450	18.930	185.762		367.986	
500	19.956	195.829		388.242	
550	20.973	205.808		409.614	
600	21.981	215.699		432.162	
650	22.980	225.500			
700	23.970	235.213			
750	24.951	244.838			
800	25.922	254.374			
850	26.885	263.821			
900	27.839	273.179			
950	28.783	282.449			
1000	29.719	291.630			
1050	30.646	300.723			
1100	31.563	309.727			

*From Scientific Apparatus Makers Association Standard RC 21-4-1966.

†Type I nickel resistance thermometers include a series padding resistor to match a specific curve with nickel of varying purity.

‡Type II nickel resistance thermometers include a series and shunt padding resistor to facilitate linear temperature readout.

Table 4.7—Resistance vs. Temperature (°C) for Platinum, Nickel, and Copper Resistance Elements*

°C	Platinum 10 ohms	Platinum 100 ohms	Nickel 100 ohms (type I)†	Nickel 200 ohms (type II)‡	Copper 10 ohms at 25°C
0	10.00	98.129	100.00	235.116	9.042
50	11.976	117.521	130.62	258.923	10.968
100	13.923	136.625	165.20	285.141	12.894
150	15.841	155.442	204.44	314.013	14.820
200	17.729	173.972	249.02	345.809	
250	19.588	192.215		380.825	
300	21.418	210.171		419.386	
350	23.218	227.840			
400	24.990	245.221			
450	26.732	262.315			
500	28.444	279.122			
550	30.128	295.642			
600	31.782	311.875			

*From Scientific Apparatus Makers Association Standard RC 21-4-1966.

†Type I nickel resistance thermometers include a series padding resistor to match a specified curve with nickel of varying purity.

‡Type II nickel resistance thermometers include a series and shunt padding resistor to facilitate linear temperature readout.

Table 4.8—Characteristics Typical of Resistance Thermometers*

Element	Temperature range, °F	Tolerance	
		Standard	Special
10- and 100-ohm Pt	−330 to +300	±1½°F	±¾°F
	Above +300°F	±½% of temp. rdg.	±¼% of temp. rdg.
Ni (type I)	−40 to 400	±1°F or ±½ of temp. rdg., whichever is greater	
Ni (type II)	−150 to −40	±2.0°F	
	−40 to 400	±0.5°F	
	400 to 600	±¼% of temp. rdg.	
Cu	−100 to 300	±½°F	±⅛°F

*From Scientific Apparatus Makers Association Standard RC 21-4-1966.

the value of which can be adjusted to balance the bridge so that, except for lead resistance, r = x, x being the resistance of the thermometer resistor.

Copper lead wires have a temperature coefficient of the same order of magnitude as that of a thermometer resistor, and, if their resistance is appreciable in comparison with that of the thermometer resistor, the lead wires may introduce large and uncertain errors into the measurement of temperature. Since the thermometer resistor usually must be placed at a considerable distance from the bridge, the resistance of the lead wires must be compensated. Figure 4.13 illustrates one method of accomplishing this result. Three wires (A, B, and C) connect the measuring instrument and the thermometer resistor (x). Of these, A and C should be identical in size, length, and material and should be placed side by side throughout their length so as to be at the same temperature. The B wire, which is one of the battery wires, need not be similar to the others; however, it is common practice to form the three wires into a cable and make them all alike. A and C are in the thermometer resistor arm (x) and the variable resistance arm (r), respectively. Their resistance remains equal although their temperature conditions may change, and, hence, with a one-to-one bridge ratio, such changes have no effect on the bridge reading.

No variable contact resistances should be included in the bridge arms, because the variations in bridge balance introduced at the contacts may be sufficient to affect the reliability of the measurements. The effect of these variations, as well as those resulting from unequal lead resistances, may be reduced by using a resistor of several hundred ohms resistance in the thermometer.

(e) Comparison with Other Sensor Types. *Thermocouples.* A comparison of thermocouples with platinum resistance temperature sensors or any other resistance sensor would indicate that thermocouples have certain advantages. For thermocouples the temperature-sensitive zone can be extremely small, and the measurement can be made with an extremely sensitive potentiometric device. Thermocouples are also well suited for relatively high temperatures and are relatively easy to install. However, at low temperatures, higher output and higher accuracy are much in favor of resistance sensors.

Some of the principal advantages of resistance sensors over thermocouples are

1. A much higher output voltage can be obtained.
2. Related recording, controlling, or signal-conditioning equipment can be simpler, more accurate, and much less expensive because of the higher possible bridge output signal.
3. The output voltage per degree for resistance sensors can be chosen to be exactly as desired over wide limits by adjusting the excitation current and/or the bridge design.
4. A reference-junction temperature or compensating device is unnecessary.
5. The shape of the curve of output vs. temperature can be controlled, within limits, for certain resistance sensor bridge designs.
6. The output of a resistance sensor bridge can be made to vary with temperature and another variable by causing the excitation to vary with the second variable.
7. Because of the higher output voltage, more electrical noise can be tolerated with resistance sensors; therefore, longer lead wires can be used.
8. Sensitivity to small temperature changes can be much greater.
9. In moderate temperature ranges, absolute accuracy and calibration and stability of calibration for resistance elements can be better by a factor of 10 to 100.

Thermistors. Thermistors are relatively inexpensive and are very sensitive to temperature. The change in resistance per unit change in temperature is large. They are available in small sizes and are available with unusually high resistance values when desired. Thermistors have a particularly nonlinear resistance–temperature relation. Because of the nonlinear relation, relatively numerous calibration points are necessary, and the expense of calibration at many points is frequently a major part of the cost of a thermistor temperature sensor.

Semiconductors. The resistance–temperature relation for the semiconductors consisting of alloy combinations is very complex and therefore requires many more calibration points than platinum sensors. At very low temperatures semiconductor thermometers consisting of doped germanium sensors have been looked upon with much favor, at least for applied thermometry, as compared to all

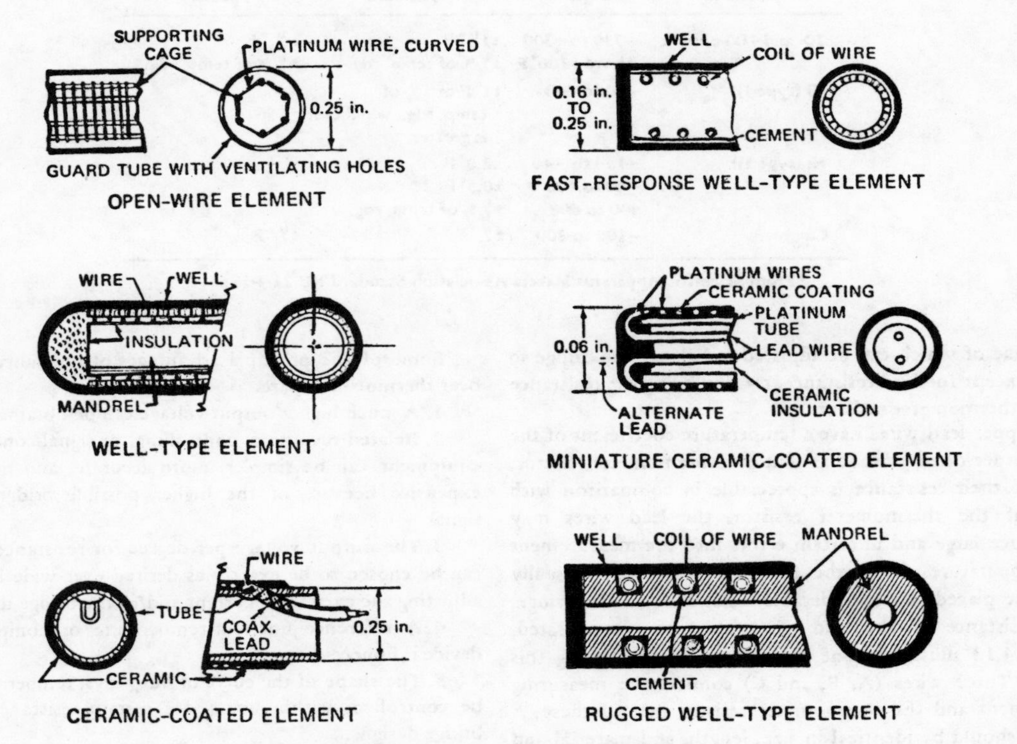

Fig. 4.11—Typical structures used in resistance thermometers.

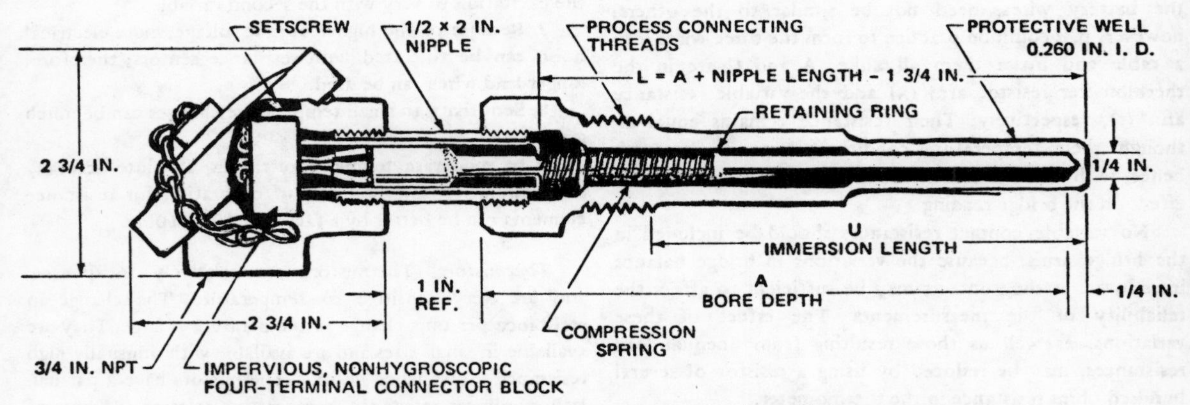

Fig. 4.12—Resistance element assembly in a protecting well. This spring-loaded design keeps the element and well in constant firm contact, assuring vibration as a single unit and improving response time. Installation and removal is simplified since it is not necessary to disassemble anything. The installation instructions for this typical resistance element assembly are (a) thread well, nipple, and head firmly into process equipment; (b) unscrew head over and insert element assembly down into well (connector block will be approximately ½ in. from seating position when element tip bottoms in well); and (c) push connector block against spring, force to seating position, and screw setscrew into mating hole in connector block.

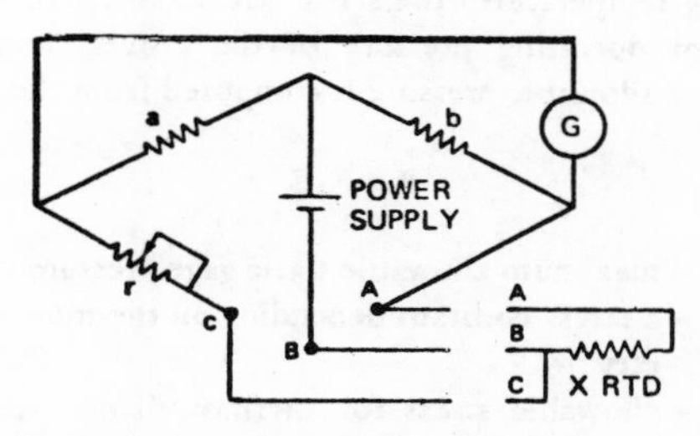

Fig. 4.13—Wheatstone bridge circuit.

other methods of measuring temperature. When it is necessary to make continuous measurements over the range from approximately 1 to 40°K, they can be used to good advantage.

Carbon Resistors. At extremely low temperatures carbon resistors are very sensitive to temperature. They have been widely used, mainly for research purposes, for temperature measurements from about 0.1 to 15 or 20°K with good results. Their stability is less than might be desired.

4-2.3 Thermowells

Thermowells are protective devices for the sensors of temperature indicating, recording, and controlling instruments. As used in out-of-core locations in a nuclear power plant, temperature sensors may be exposed to a wide range of pressures and temperatures and to a variety of potentially corrosive materials.

This section includes a description of the basic types of thermowells and their materials of construction, a summary of methods for ensuring that the thermowell design will survive the mechanical stresses met in service, and a guide to the selection of thermowell materials.

(a) Connection to Process Vessel. A thermowell is usually secured to a process vessel by threads, flanges, or welding (Fig. 4.14).

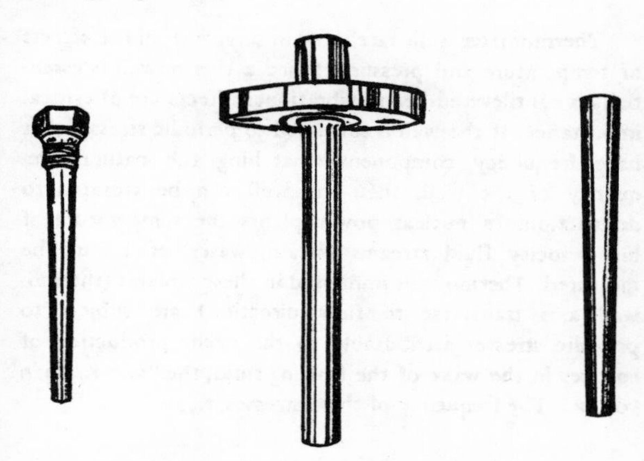

Fig. 4.14—Commonly used connections for securing thermowells to process equipment.

The threaded connection, normally using standard-taper pipe threads, is most popular owing in large measure to its simplicity and low cost. Standard threaded well connections range in size from ½ in. to 1 in. NPT, with specials ⅛ in. to 2 in. NPT meeting most requirements.

Flanged assemblies of any size and/or pressure rating are available. Normal means of well mounting are provided by ASME-approved welding techniques, with follow-up machining to provide any standard sealing-face configuration. Flanges are commonly used to seal long thermowells or those wells which are inserted into large vessels. An alternate flange type well is the nonwelded Van Stone well with integral flange, using a lap-joint flange to hold it in place. Also available is the ground-joint type with a machined ball that mounts in a socket between a pair of mating flanges. These latter two designs have an advantage in that as thermowell replacement becomes necessary, flanges may be reused with the new assembly.

Welded connections are normally used where process pressures are too great for flanged or threaded assemblies or where long-term inexpensive connections are desirable. The welded-in type is commonly used in conjunction with high-pressure, high-velocity steam lines. This type well is frequently furnished with close tolerance limits on outside diameters in the area to be welded. These are tapered-stem wells with greater wall thickness in the weld area but with relatively low mass at the end to improve response with tip-sensitive temperature-measuring devices.

(b) Length, Bore, and Wall Thickness. Overall well length is determined not only by desired insertion length but also by external extension of the connection end. Most threaded connection wells require an additional 2 in. of nonimmersed length to provide threads and wrenching surface. Welded or flanged wells normally require at least 1.25 in. of extra length for instrument-connection threading and welding surface. Where there are layers of thermal insulation, a lagging extension should be added between the process connection and the instrument connection.

Bore size (both length and diameter) depends on the thermal sensing element to be used. The fit between the sensor and the inner wall of the thermowell must be good if accuracy and rapid response are to be achieved [¶. 4-2.1(i)]. Care should be taken to prevent heat loss to surroundings and to avoid variations caused by stratification of process fluids. Where clearances between measuring element and bore are minimal and welding must be performed in the field, a counter bore of 10 to 20 mils greater diameter than the bore should be made. This counter bore should be carried sufficiently far past the welded area to avoid distortion in the bore due to heat of welding.

To withstand mechanical stresses, the thermowell wall should be thick. However, to provide rapid response to process-temperature changes, the wall should be thin (and the immersed well mass should be minimum). These conflicting requirements have been met by using tapered thermowells, in which the tip has a thin wall for optimum heat transfer and a thick mounting for improved strength. The design of these wells is discussed in the next section.

(c) Design of Power Test Code Thermowells. The American Society of Mechanical Engineers recommends a standardized Power Test Code thermometer well, as shown in Fig. 4.15. Wells of this design, with $\frac{3}{16}$ in. minimum wall thickness, are expected to satisfy 95% of the present needs.

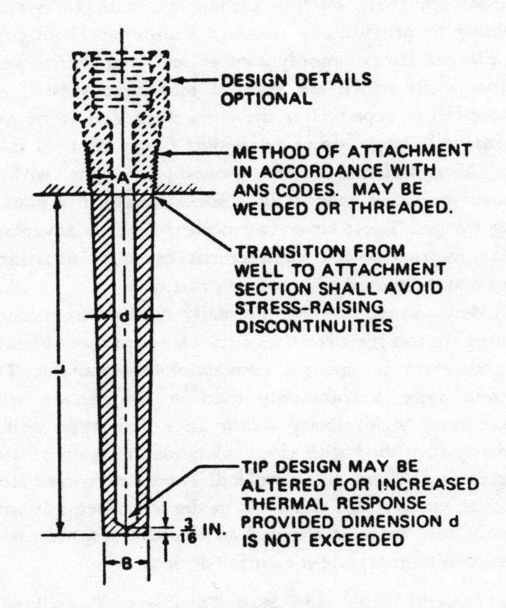
DESIGN DETAILS OPTIONAL
METHOD OF ATTACHMENT IN ACCORDANCE WITH ANS CODES. MAY BE WELDED OR THREADED.
TRANSITION FROM WELL TO ATTACHMENT SECTION SHALL AVOID STRESS-RAISING DISCONTINUITIES
TIP DESIGN MAY BE ALTERED FOR INCREASED THERMAL RESPONSE PROVIDED DIMENSION d IS NOT EXCEEDED
3/16 IN.
A
d
L
B

Fig. 4.15—Power Test Code thermometer well.

The following design procedure enables a user to determine if a well selected for thermometry considerations is strong enough to withstand specific application conditions of temperature, pressure, velocity, and vibration. This design procedure does not allow for effects due to corrosion or erosion. If corrosion or erosion is anticipated, additional wall thickness must be allowed in all exposed sections to prevent premature well failure.

The nominal size of the sensing element is considered here to vary between $\frac{1}{4}$ in. (6.35 mm) and $\frac{7}{8}$ in. (22.225 mm). For this range the dimensions of the thermowell are assumed to be those given in Table 4.9.

Table 4.9—Thermowell Dimensions (in.)*

Dimension	Nominal size of sensing element				
	$\frac{1}{4}$	$\frac{3}{8}$	$\frac{9}{16}$	$\frac{11}{16}$	$\frac{7}{8}$
A (min.)	$\frac{13}{16}$	$\frac{15}{16}$	$1\frac{1}{8}$	$1\frac{1}{4}$	$1\frac{7}{16}$
B (min.)	$\frac{5}{8}$	$\frac{3}{4}$	$\frac{15}{16}$	$1\frac{1}{16}$	$1\frac{1}{4}$
d (min.)	0.254	0.379	0.566	0.691	0.879
d (max.)	0.262	0.387	0.575	0.700	0.888

*From Scientific Apparatus Makers Association Standard RC 21-4-1966.

A thermometer well must be able to withstand (at the operating temperature) the static stress associated with the maximum operating pressure of the process vessel. The maximum allowable pressure is computed from the formula

$$P = K_1 S \tag{4.5}$$

where P = maximum allowable static gage pressure (psig)
K_1 = a stress constant depending on thermowell geometry
S = allowable stress for thermowell material at the operating temperature as given in the ASME Boiler and Pressure Vessel or Piping Codes (psi)

For wells constructed as shown in Fig. 4.15 with dimensions as given in Table 4.9, the stress constant has the values listed in Table 4.10. For wells of other dimensions, the stress constant is given by

$$K_1 = \left(\frac{B - d}{2B}\right)^2 \frac{F_B}{2} \tag{4.6}$$

where (see Fig. 4.15) B is the minimum outer diameter (inches) at the well tip and F_B is a factor varying between 2.0 and 1.0 as shown in Table 4.11.

Table 4.10—Values of the Stress Constants K_1, K_2, and K_3*

Stress constant	Nominal size of sensing element				
	$\frac{1}{4}$	$\frac{3}{8}$	$\frac{9}{16}$	$\frac{11}{16}$	$\frac{7}{8}$
K_1	0.412	0.334	0.223	0.202	0.155
K_2	37.5	42.3	46.8	48.7	50.1
K_3	0.116	0.205	0.389	0.548	0.864

*From Scientific Apparatus Makers Association Standard RC 21-4-1966.

Thermometer wells rarely fail in service from the effects of temperature and pressure. Since a thermowell is essentially a cantilevered beam, vibrational effects are of critical importance. If the well is subjected to periodic stresses that have frequency components matching the natural frequency of the well, then the well can be vibrated to destruction. In nuclear power plants the temperature of high-velocity fluid streams (steam, water, etc.) must be measured. Thermowells immersed in these streams (thermowell axis transverse to flow direction) are subject to periodic stresses attributable to the cyclic production of vortices in the wake of the flowing fluid, the "von Kármán vortex." The frequency of these stresses, f_w, is

$$f_w = 2.64 \frac{V}{B} \text{ (in Hz)} \tag{4.7}$$

where V = fluid velocity (ft/sec) and B = well diameter at tip (in.), see Fig. 4.15. The natural frequency of the thermowell (cantilever structure) is

Table 4.11—Values of F_B*

(Note: t = B − d, D = 2B)

t/D From	t/D To	F_B	t/D From	t/D To	F_B
0.084	0.091	2.0	0.150	0.169	1.5
0.092	0.099	1.9	0.170	0.199	1.4
0.100	0.114	1.8	0.200	0.219	1.3
0.115	0.129	1.7	0.220	0.239	1.2
0.130	0.149	1.6	0.240	0.249	1.1
			0.250	Up	1.0

*From Scientific Apparatus Makers Association Standard RC 21-4-1966.

$$f_n = K_f \left(\frac{E}{\gamma L}\right)^{1/2} \text{ (in Hz)} \tag{4.8}$$

where E = elastic modulus of well material at the operating temperature (psi)
γ = specific weight of well material (lb/in.^{3})
L = length of well (in.) (see Fig. 4.15)
K_f = a factor depending on well dimensions (Table 4.12)

The wake frequency f_w should not go above 80% of the natural well frequency, f_n,

$$r = \frac{f_w}{f_n} \leq 0.8 \tag{4.9}$$

If the ratio r is over 0.8, the well will tend to vibrate to failure.

Table 4.12—Values of K_f*

Well length (L), in.	Nominal size of sensing element: 1/4	3/8	9/16	11/16	7/8
2½	2.06	2.42	2.97	3.32	3.84
4½	2.07	2.45	3.01	3.39	3.96
7½	2.08	2.46	3.05	3.44	4.03
10½	2.09	2.47	3.06	3.46	4.06
16	2.09	2.47	3.07	3.47	4.08
24	2.09	2.47	3.07	3.48	4.09

*From Scientific Apparatus Makers Association Standard RC 21-4-1966.

In any practical situation, the fluid velocity, V, is fixed, and the parameters under the instrumentation engineer's control are the well dimensions. Once the size of the sensing element is decided on (e.g., on the basis of speed of response, ruggedness, etc.), the thermometer-well outer diameter B is fixed (Table 4.9), and the wake frequency (Eq. 4.7) is determined. The only well parameter remaining (except materials of construction, see next section) is the well length, L. Since f_n decreases with increasing length (Eq. 4.8), the requirement for f_w/f_n to be less than 0.8 imposes a limitation on the length, L.

The maximum length of a thermometer well for a given service depends not only on the vibratory stresses imposed by the flowing limit but also on the steady-state stresses (drag) of the flowing fluid. These stresses limit the well length according to the following formula:

$$L_{max} = \frac{K_2}{V}\left[\frac{v(S - K_3 P_0)}{1 + F_m}\right]^{1/2} \text{ (in in.)} \tag{4.10}$$

where V = fluid velocity (ft/sec)
v = specific volume of fluid (ft^3/lb)
S = allowable stress for well material at operating temperature per codes (psi)
P_0 = static operating pressure (psig)
K_2, K_3 = stress constants (Table 4.10)

The factor F_m is a "magnification factor" dependent on the ratio r of wake frequency to the natural frequency of the well:

$$F_m = \frac{r^2}{1 - r^2} \qquad r = \frac{f_w}{f_n} \tag{4.11}$$

Example. To clarify the use of the above formulas, consider the following example. It has been determined that a 4½-in. well is required to accommodate a 9/16-in. sensing element that will measure the temperature of superheated steam at 2000 psia, 1050°F, and flowing at a velocity of 350 ft/sec. If the thermometer well is to be made of type 316 stainless steel dimensioned according to Table 4.9, will the well be safe?

Step 1: Obtain the necessary data as follows:

v	Specific volume of steam	0.4134 ft³/lb	ASME Steam Tables, 1967
E	Modulus of elasticity at 1050°F	22.35 × 10⁶ psi	B31.1.0-1967, App. C
γ	Specific weight of metal at 70°F	0.290 lb/in.³	
S	Allowable stress at 1050°F	9725 psi	ASME Code, Sec. VIII

Step 2: Maximum static pressure (Eq. 4.5)

$$P = 0.223 \times 9725 = 2170 \text{ psig} > 2000 \text{ psia (satisfactory)}$$

Step 3: Frequency calculations
(a) Natural frequency (Eq. 4.8)

$$f_n = \frac{3.01}{4.52}\left(\frac{22.35 \times 10^6}{0.290}\right)^{1/2} = 1305 \text{ Hz}$$

(Use of dimensions and specific weight values for 70°F instead of for 1050°F is partially compensatory and causes no significant error.)
(b) Wake frequency (Eq. 4.7)

$$f_w = 2.64 \times \frac{350}{15/16} = 986 \text{ Hz}$$

(c) Frequency ratio (Eq. 4.9)

$$r = 986/1305 = 0.755 < 0.8 \text{ (satisfactory)}$$

Step 4: Maximum length calculation:
(a) Magnification factor (Eq. 4.11)

$$F_m = \frac{0.755}{(1 - 0.755^2)} = 1.325$$

(b) Maximum length (Eq. 4.10)

$$L_{max} = \frac{46.8}{350}\left[\frac{0.4134(9725 - 0.389 \times 2000)}{(1 + 1.325)}\right]^{1/2}$$

$$= 5.33 \text{ in.} > 4\tfrac{1}{2} \text{ in. (satisfactory)}$$

Conclusion: The well selected is satisfactory for the application.

An example of the installation of a typical thermocouple or resistance thermometer is shown in Fig. 4.16. The thermowell shown is a heavy-duty weld-in well.

(d) Materials Used in Thermowells. Because of their ability to withstand chemical attack from process fluids, stainless steels are most frequently used in thermowells. Customarily, stainless steels are put in three groups: martensitic, ferritic, and austenitic.

The martensitic steels contain 11.5 to 18% Cr, <2.57% Ni, and 0.06 to 1.20% C. They have a ferritic structure when annealed but take on the properties of a martensite when cooled. They can be heat-treated, hardened, and tempered to provide a wide range of mechanical properties for use in abrasive environments or where particularly high strength is required. Martensitic steels are in the 400-series stainless steels, excepting the ferritic grades, and are strongly magnetic. Examples are the AISI types: 403, 410, 414, 416, 420, 431, and the 440 letter series.

The ferritic steels contain 11.5 to 28% Cr, no Ni, and 0.06 to 0.35% C. They are always magnetic and do not respond to heat treatment. They are strong and ductile when properly annealed and are generally more corrosion-resistant than the martensitics. Examples are the AISI types: 405, 430, 430F, 442, and 446.

The austenitic grades are chrome–nickel alloy steels with a maximum carbon content of 0.25% and with 7 to 30% Cr and 6 to 36% Ni. They are nonmagnetic in a fully annealed condition but become slightly magnetic with cold working. They are generally tougher and more ductile than martensitic and ferritic steels and have a much higher corrosion resistance. The austenitic steels belong to the 300-series.

Selection of a stainless steel requires consideration of which properties are most desirable for the application: corrosion resistance, strength at operating temperature, oxidation resistance, particularly at elevated temperatures, availability in a form suitable for fabrication, or ease of fabrication (machinability, weldability).

The principal properties of commonly used grades of stainless steels are summarized in Table 4.13.

Other materials may be used in thermowells. Tables 4.14 and 4.15 give the recommended allowable stress values and maximum operating temperatures for a number of thermowell materials.

4-2.4 Temperature Sensors in Gas-Cooled Reactors

The measurement of temperatures in gas-cooled reactors requires certain specialized sensors, e.g., sensors based on the transmission of acoustic energy through gases.

4-3 PRESSURE SENSING AND TRANSMITTING

4-3.1 Sensors

This section deals with elastic sensing elements that respond to a system pressure change and, in so doing, generate a measurable physical quantity, such as position or mechanical or electrical force. Each sensor is a differential element, and atmospheric pressure is constantly applied in opposition to the system pressure. To sense the absolute pressure, you must apply a second element (e.g., a calibrated spring) in opposition or place that part of the

Table 4.15—Recommended Maximum Operating Temperatures of Common Thermowell Materials*

Material	Maximum operating temp., °F	Melting point, °F
Copper	600	1980
Aluminum	700	1200
Monel	1000†	2450
Carbon steel	1200	2760
304 s.s.	1650	2600
309–310 s.s.	2000	2550
316 s.s.	1650	2525
321–347 s.s.	1600	2575
430 s.s.	1550	2725
446 s.s.	2000†	2725
Inconel 600‡	2100†	2575
Hastelloy X§	2300†	2350
Nickel	2300†	2625
Inconel X750	2400†	2570
Tantalum	4500†	5425

*From Pall Trinity Micro Corporation, *Thermocouple Guidebook*, TT-335, Courtland, N. Y.

†At high temperature, the effect of process atmosphere on the thermowell may cause severe limitations in service life. The values listed constitute mill recommended maximums under average circumstances.

‡Huntington Alloys Division, International Nickel Company.

§Material Systems Division, Union Carbide Corporation.

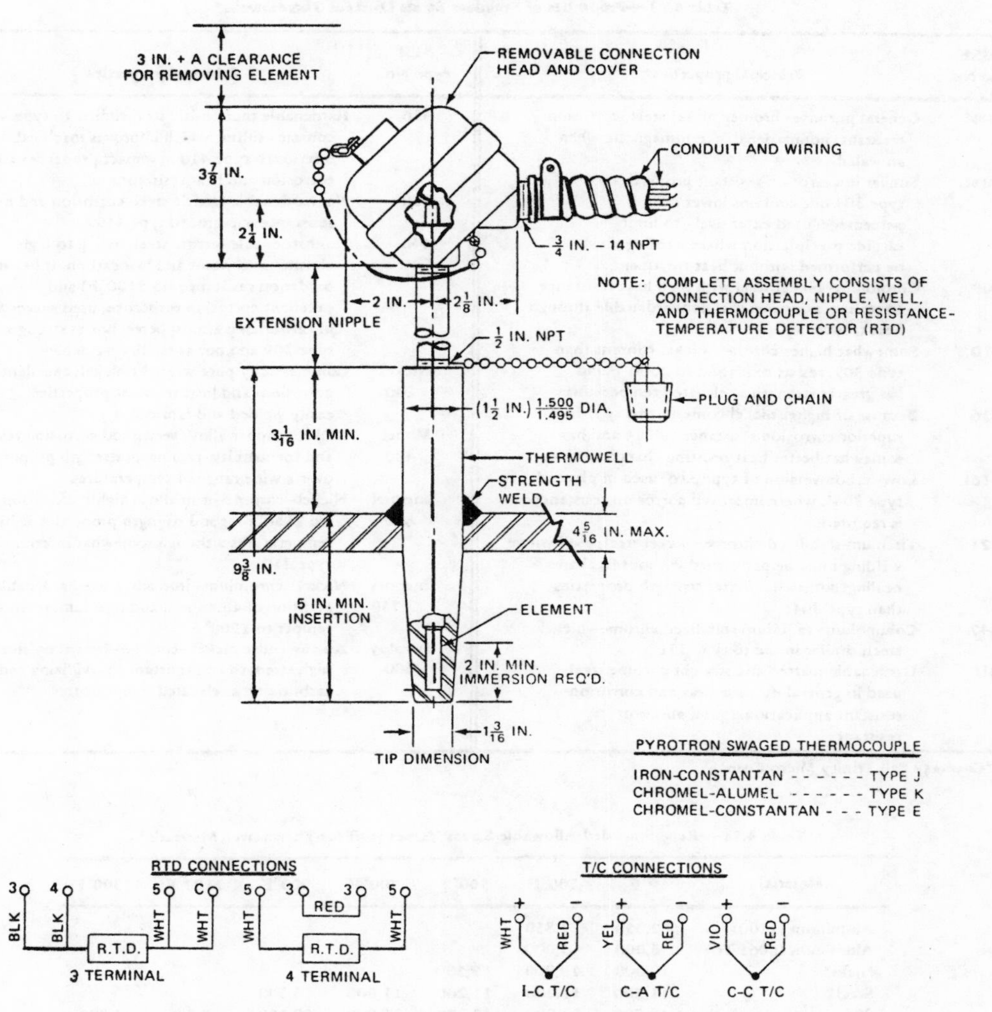

Fig. 4.16—Typical thermocouple and resistance element installed in heavy-duty thermowell.

sensor that is normally at ambient (atmospheric) pressure within an evacuated containment.

(a) Materials. In out-of-core pressure sensors, materials coming in contact with the measured fluid must be noncorrosive, must not otherwise deteriorate, and must not contain elements that may become dangerously radioactive by accidental exposure to neutrons. The objective is a device capable of continuous, dependable pressure sensing over an extended period of time. Sensor materials contacting the measured fluid should be compatible with the fluid. This is the same problem that is involved in choosing thermowell materials. Stainless steels, type 304 or better, are frequently used. Sometimes Inconel is used. Teflon materials for seals and O-rings are avoided as are any components containing cobalt. If the most highly desirable materials are not available at the sensor, diaphragm seals described in ¶. 4-3.4(b) are used.

(b) Basic Types. *Elastic metal sensors*, available in a variety of forms, consist of slack and rigid diaphragms, multiple or stacked diaphragms, corrugated bellows, and the Bourdon tube in a variety of forms, from single-turn and torsion-bar to helical and spiral multiple-turn designs. Each manufacturer has his own series of ranges for the various designs based upon the sizing of components and the required performance of a complete linkage system or other device that depends on this initiating element for its successful operation. Table 4.16 gives some typical ranges, and ¶. 4-3.6 gives a sample set of performance specifications.

***Strain gages* consist of a fine wire or an array of fine wires usually bonded into an assembly for mechanical strength. Under an applied stress the array of fine wires is stretched; this results in an increase in its electrical resistance. If this array is incorporated in a suitable**

Table 4.13—Properties of Stainless Steels Used for Thermowells*

AISI type No.	Principal properties
304	General purpose chrome–nickel steel, corrosion resistant; nonhardenable, nonmagnetic when annealed
304L	Similar in corrosion-resistant properties to type 304 but contains lower carbon percentage; used extensively to limit carbide precipitation where welding must be performed without heat treatment
309	Chromium–nickel steel with high heat resistance to scaling; nonmagnetic, nonhardenable through heat treatment
310	Somewhat higher chrome–nickel content than type 309; resists oxidation to 2000° F and has greater strength at elevated temperatures
316	Because of higher nickel content, this type has superior corrosion resistance to 304 and has somewhat better heat-resisting characteristics
316L	Low-carbon version of type 316; used in place of type 304L where improved corrosion resistance is required
321	Titanium-stabilized chrome–nickel steel; used where welding must be performed without final annealing; somewhat better strength properties than type 304L
347	Columbium–tantalum stabilized chrome–nickel steel; similar in use to type 321
410	Hardenable martensitic straight chrome steel; used in general-purpose heat and corrosion-resistant applications; good abrasion resistance
416	Hardenable martensitic steel similar to type 410; contains sulfur, which improves machinability; inferior to type 410 in impact properties and corrosion and heat resistance
430	Nonhardenable ferritic steel; corrosion and heat resistance superior to type 410
446	Nonhardenable ferritic steel; owing to high chromium content and low carbon, it has superior oxidation resistance (to 2100° F) and excellent corrosion resistance; used successfully in carburizing atmospheres; not as strong as type 309 and not as readily weldable
Nickel 200	Commercially pure wrought nickel; excellent corrosion- and heat-resistant properties; easily welded and fabricated
Monel 400	Nickel–copper alloy; very good corrosion resistance and formability; retains its strength properties over a wide range of temperatures
Inconel 600	Nickel–chrome–iron alloy; highly oxidation resistant (to 2150° F); good strength properties at high temperatures although somewhat inferior to AISI type 310
Inconel X750	Nickel–chromium–iron alloy; age hardenable by addition of aluminum and titanium; retains spring temper to 1200° F
Incoloy 800	An austenitic nickel–chrome–iron alloy steel; high strength and resistant to oxidizing and carburizing at elevated temperatures

*Courtesy Pall Trinity Micro Corp.

Table 4.14—Recommended Allowable Stress Values (psi) for Thermowell Materials*

Material	0° F	300° F	500° F	700° F	900° F	1100° F	1300° F
Aluminum (1100)	2,350	1,850					
Aluminum (6061-T6)	6,000	5,000					
Nickel	10,000	10,000	9,500				
Steel†	11,200	11,200	11,200	11,000	6,500		
304 s.s.	18,700	14,000	12,100	11,000	10,100	8,800	3,700
310 s.s.	18,700	15,800	14,100	12,700	11,600	5,000	700
316 s.s.	18,700	14,600	12,400	11,300	10,800	10,300	4,100
347 s.s.	18,700	16,000	14,000	12,900	12,600	9,100	2,200
410 s.s.	16,200	14,900	13,900	13,100	10,400		
446 s.s.	17,500	16,100	15,000				
A182-F11 (Chrome-Moly)	17,500	17,500	17,500	16,100	13,100	4,000	
A182-F22 (Chrome-Moly)	17,500	17,500	17,500	17,500	14,000	4,200	
Copper	6,000	5,000					
Admiralty brass	10,000	10,000					
Monel 400	16,600	13,600	13,100	13,100	8,000		
Inconel 600	20,000	18,800	18,500	18,500	16,000	3,000	
Incoloy 800‡	15,600	12,100	10,400	9,600	9,100	8,800	4,150
Hastelloy B§	25,000	24,750	21,450				
Hastelloy X¶	23,350	18,850	16,000	15,500	15,500	15,500	9,500

*Courtesy Pall Trinity Micro Corp. Values from ASME Boiler and Pressure Vessel Code, Sec. VIII—Pressure Vessels, 1971.

†ASME Spec. Min. Tensile = 45,000 psi.

‡ASME Code, case 1325 (special ruling).

§ASME Code, case 1323 (special ruling).

¶ASME Code, case 1321 (special ruling).

arrangement, the resistance change can be made directly proportional to the imposed pressure. Close temperature control must be maintained by comparing the strain wire with unstressed wire (or compensation); electrical shielding of the sensor wire is also important. In some designs the strain wire may be mounted (bonded) on a Bourdon tube, bellows, or mechanical structure, such as a beam or ring. In Fig. 4.17 the strain gage is in the form of a short tube sealed by a diaphragm. Note the variable resistance is applied as a leg of a conventional Wheatstone measuring bridge. Because the fractional change in strain-gage resistance is very small, electrical amplification and signal conditioning are usually required before use in readout and action modules.

Piezoelectric sensors are similar to strain-gage sensors with a crystal used for stress sensing instead of a wire. The crystal responds to a pressure change (usually expressed by a force in a predetermined direction with respect to the crystal axes) by generating a small electrical potential difference. The latter depends on the magnitude of the imposed stress and on the crystal properties. Again, temperature control or compensation, as well as amplification and signal conditioning of the output, are essential.

Silicon wafer piezoelectric sensors now available are capable of sensing range spans from 0 to 6 psig to 0 to 1500 psig. Output is 10 to 50 mA d-c. Features include a range span adjustment of 4 : 1 for a given diaphragm and the capability of elevating the range span (zero suppression) to the maximum pressure range for the unit. Thus a 0- to 1500-psi range device would be expected to calibrate for 1100- to 1500-psig range input for a 10- to 50-mA d-c output.

4-3.2 Range Selection

A common practice is to specify the pressure interval in which the sensor is to be used so that the sensor is normally operating at a pressure between ⅓ and ⅔ of the range span. Data for the elastic-metal sensors are given in Table 4.16. The basic ranges shown in the table may be modified. Modifications include ranges having an elevated zero (compound range) and suppressed ranges where the minimum pressure is greater than atmospheric zero.

Motion balance refers to the mechanical system whereby the sensor motion is transferred by linkage or other means to a pointer, recording pen, or transducer mechanism, such as an armature or core (see Fig. 4.24). Typical motion-balance mechanisms are shown in Figs. 4.18, 4.19, and 4.20. Where restraining members are used, they are limited to providing means for adjusting zero or permitting calibration within the initial range capability of the sensing element; in no way should the restraining members restrict the sensing action within the intended range span of the device.

Force balance refers to the system whereby the free motion of the sensor is limited and actively opposed by some mechanical or electrical means. In effect this reduces the actual mechanical motion of the sensor to a very few thousandths of an inch throughout its range. An example of such a device is shown in Fig. 4.21. The mechanism is capable of highly elevated ranges. One design features four interchangeable diaphragms or capsules (Table 4.17).

4-3.3 Installation

An in-line sensor or a tap to an adjacent mounted sensor must be located in a position where errors due to local disturbances, such as turbulence and vibration created by the process or adjacent machinery, are avoided. For accuracy in lower pressure ranges, the sensor should include provisions for compensating for the weight of liquid in connecting lines so that the transmitted or observed pressure is that in the main piping or containment.

An ANSI Piping Code recommends that the pressure take-off size not be less than ⅜-in. pipe for operating up to 900 psi and 800°F and not less than ¾-in. pipe above these values. An acceptable method for installing a take-off is to weld a Weld-O-let or similar adapter to the main pipe or vessel and then drill through the adapter and pipe or vessel wall a ¼-in.-diameter hole (or larger if desired) to produce a sharp clean edge at the inner wall. Actual size of the hole

Table 4.16—Typical Ranges of Elastic-Metal Pressure Sensors

Design	Material available	Range span		Maximum over-range pressure, psig
		Min.	Max.	
Thin diaphragm	316 s.s.	0.5 in. H_2O	10 in. H_2O	50
Thick diaphragm	316 s.s.	25 psig	2000 psig	500 to 4000*
Bellows	Inconel or AM350 steel†	10 in. H_2O	1000 in. H_2O	6000‡
Bourdon tube	Ni-Span C,§ 316 s.s., or beryllium–copper	30 psig	8000 psig	Max. range × 1.25

*See discussion of force balance in Sec. 4-3.2.
†Modified type-304 stainless steel adapted for welding and stress relieving.
‡And 100 psig on gases.
§Constant-modulus alloy.

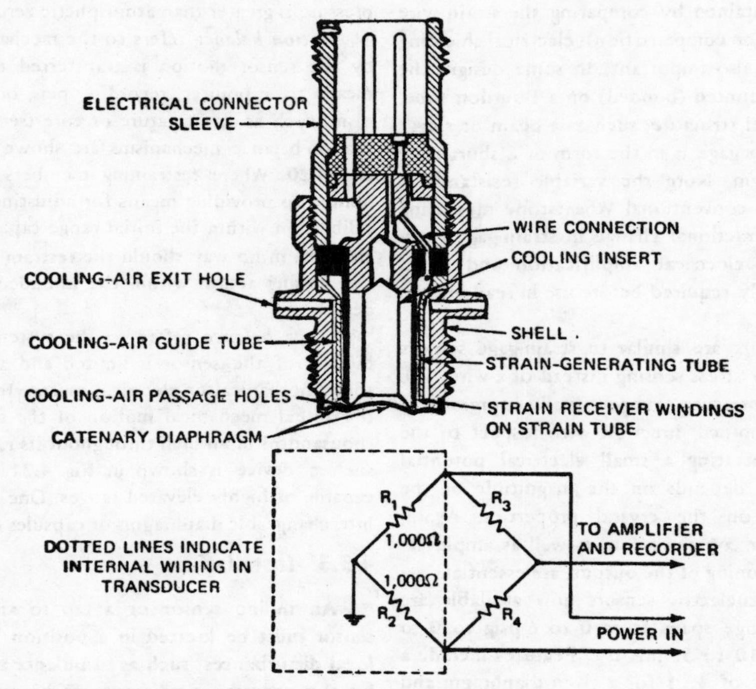

Fig. 4.17—Catenary-diaphragm sensing element and typical bridge circuit.

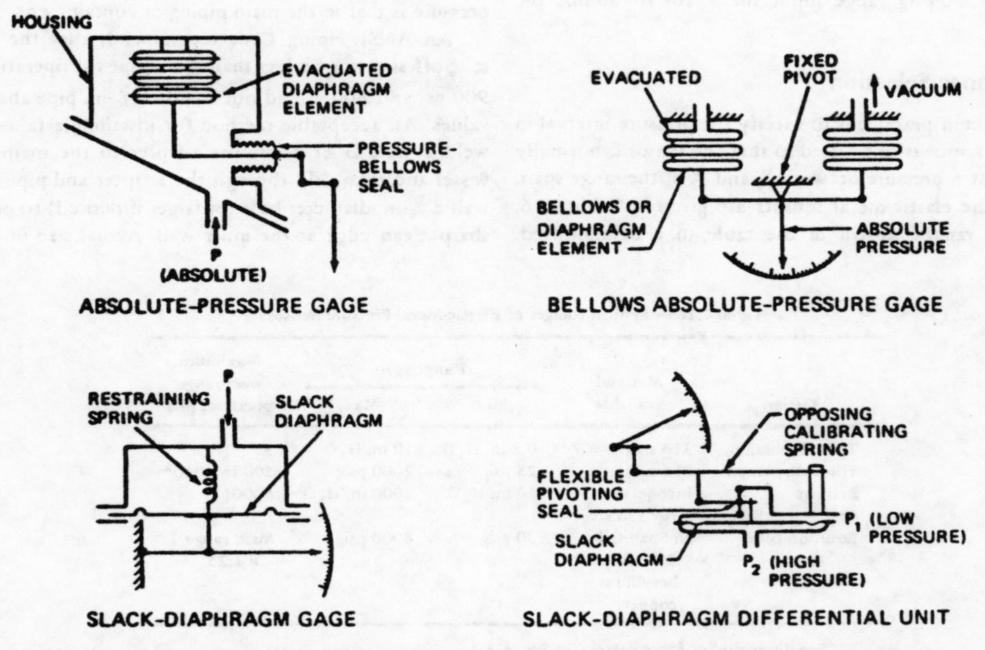

Fig. 4.18—Typical bellows and diaphragm pressure gages. Sensor motion is transmitted to indicator by mechanical linkage (motion-balance mechanisms).

should be large enough to avoid plugging. Alignment of the axis of the opening should be perpendicular to the direction of flow to avoid false pressure readings due to impact velocity effects. Material specification and controls should comply with ASME Nuclear Piping Systems of proper class 1, 2, or 3.

The pressure sensor is mounted adjacent to the take-off in such a way as to reduce transmission of piping- or vessel-expansion strains, process heat, or system vibrations to the sensor mechanism. Figure 4.22 illustrates a common installation practice featuring ½-in.-O.D. tubing or equivalent pipe pitched to facilitate draining and maintenance. Full support of connecting tubing is recommended; unsupported tubing must take a lower pressure rating. Root valves at take-offs must be the same size as the take-off. Above 900 psi and 800°F, the take-off may be swaged or reduced to allow a ½-in. root valve, but the size at the main piping or vessel may not be reduced. Blow-down valves for drain must be at least ⅜-in. pipe size. Instrument shut-off valves may be ¼-in. pipe size and threaded to match standard instrument casing connections; this latter practice facilitates disassembly and routine maintenance and calibration.

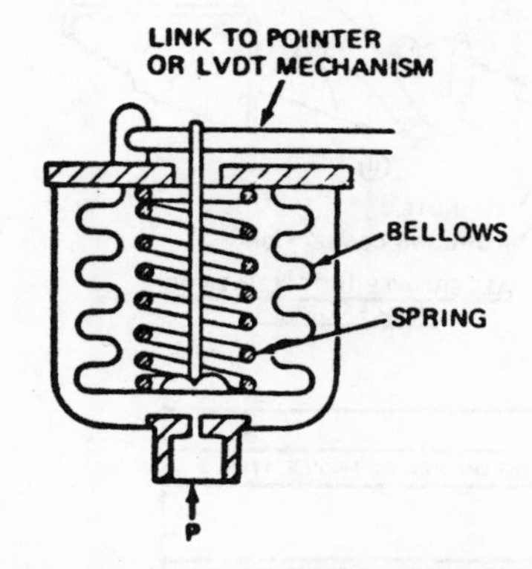

Fig. 4.19—Sectional view of a motion-balance pressure gage.

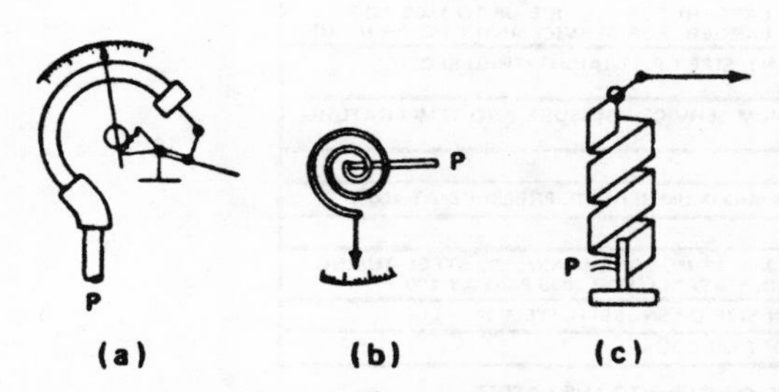

Fig. 4.20—Types of Bourdon gages. (a) C type; (b) spiral; (c) helical.

4-3.4 Accessories

(a) Pulsation Dampeners. Dampeners may be included in sensing take-off lines and are available in stainless-steel construction in ¼-in. and ½-in. pipe sizes. (In Fig. 4.22 item 14 is a typical pressure dampener.) One design consists of a sintered stainless-steel disk or cylinder held in a stainless-steel body; another a captive stainless-steel pin in an orifice opening. Plugging may present a problem; so periodic cleaning may be required. Some sensors have electronic dampening of the output signal to avoid a mechanical dampener.

Table 4.17—Force Balance with Four Interchangeable Capsules*

Capsule	Range limit,† psi	Range-span limits,† psi	Max. over-range pressure, psi
A	−15 to 350	25 to 250	500
B	−15 to 750	50 to 500	1000
C	−15 to 1500	100 to 1000	2000
D	−15 to 3000	200 to 2000	4000

*With this design, for example, an A capsule may be adjusted for an operating range span of 210 to 250 psi with an expected accuracy to ±2 psi (±0.5% range span). A range span of −15 to +10 psig involves the lowest range and narrowest range span possible using an A capsule.

†Basic industry terminology is given in Bailey Meter Company Instruction pt. E41-6.

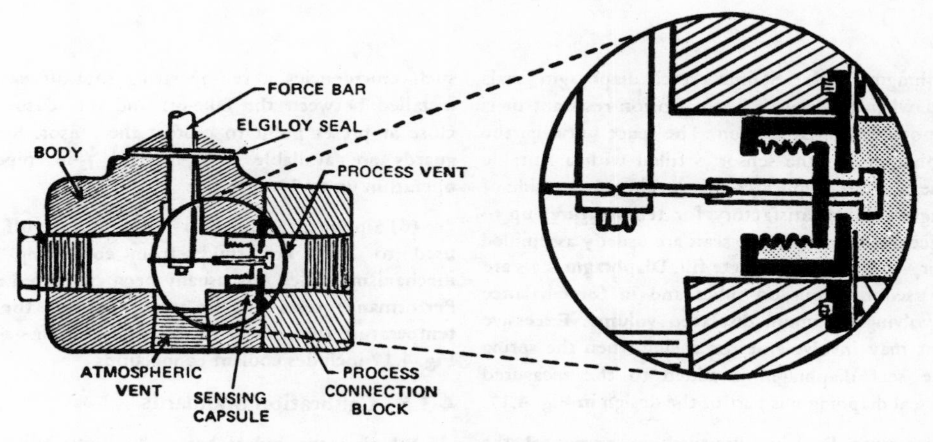

Fig. 4.21—Sectional view of a force-balance pressure transmitter.

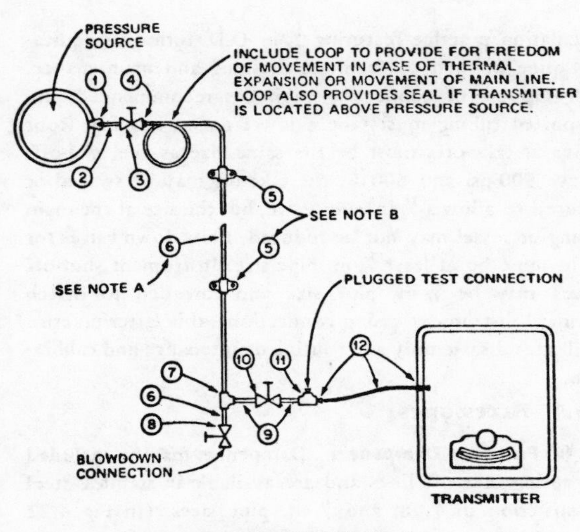

CONNECTING TUBING NOTES

A. CONNECTING LINE SHOULD BE AS SHORT AS IS PRACTICAL. EQUIVALENT VERTICAL HEAD (VERTICAL DISTANCE BETWEEN PRESSURE SOURCE AND TRANSMITTER) IN PSIG MUST BE LESS THAN 20% OF RANGE OF TRANSMITTER.

B. DO NOT ANCHOR TUBE SO TIGHT THAT IT CANNOT EXPAND DURING BLOWDOWN.

C. USE PRESSURE SNUBBER FOR ALL WATER FLOW, GAS FLOW, HIGH PRESSURE (1500 PSIG OR ABOVE) HIGH VELOCITY STEAM FLOW, ANY FLOW WHERE A PUMP IS USED, OR WHEREVER RAPID PRESSURE OSCILLATIONS ARE ANTICIPATED.

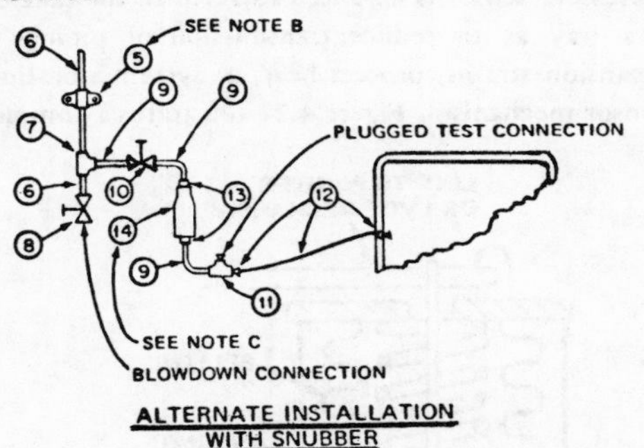

ITEM	MATERIAL
1	1/2-IN., 3/4-IN., OR 1-IN. WELDING ADAPTER, SIZE DEPENDING ON SIZE OF NIPPLE, ITEM 2
2	1/2-IN. NIPPLE, FOR SERVICE UP TO 900 PSIG OR 800°F 3/4-IN. NIPPLE } FOR SERVICE 901 PSIG OR 801°F OR HIGHER 1-IN. NIPPLE }
3	1/2-IN., 3/4-IN., OR 1-IN. GLOBE VALVE, SUITABLE FOR MAXIMUM SERVICE PRESSURE AND TEMPERATURE, SIZE DEPENDING ON SIZE OF NIPPLE, ITEM 2
4	FITTING OR BUSHING, IF REQUIRED, SIZE DEPENDING ON SIZES OF VALVE, ITEM 3, TUBING, ITEM 6
5	ANCHORING CLIP OR OTHER DEVICE
6	TUBING, WITH NECESSARY FITTINGS: 1/2-IN. O.D. TUBING OR 3/8-IN. PIPE (OR LARGER) FOR SERVICE UP TO 1500 PSIG 5/8-IN. O.D. TUBING OR 1/2-IN. PIPE (OR LARGER) FOR SERVICE 1501 PSIG OR HIGHER
7	REDUCING TEE, 1/4-IN. OUTLET TO INSTRUMENT, SIZE OF STRAIGHT-THRU SECTION DEPENDING ON SIZE OF TUBING, ITEM 6
8	GLOBE NEEDLE VALVE, SUITABLE FOR MAXIMUM SERVICE PRESSURE AND TEMPERATURE, SIZE DEPENDING ON SIZE OF TUBING, ITEM 6
9	1/4-IN. STEEL NIPPLE, LENGTH AS REQUIRED
10	1/4-IN. GLOBE NEEDLE VALVE, SUITABLE FOR MAXIMUM SERVICE PRESSURE AT 100°F
11	1/4-IN. STEEL TEE AND PLUG
12	FLEXIBLE TUBING CONNECTOR, PT. NO. 681853A1, 18-IN.-LONG STAINLESS STEEL TUBING WITH 1/4-IN. MALE CONNECTORS AT EACH END, SUITABLE FOR 5000 PSIG AT 100°F
13	STEEL BUSHING, IF REQUIRED, DEPENDING ON SIZE OF SNUBBER, ITEM 14
14	1/4-IN. OR 1/2-IN. PRESSURE SNUBBER, ASCROFT OR EQUAL

ALL SIZES AND MATERIALS LISTED CONFORM TO THE LATEST REVISION OF THE CODE FOR PRESSURE PIPING, ASA B31.1. WHERE MATERIALS ARE NOT NOTED OR WHERE DIFFERENT FITTINGS ARE TO BE USED, ALWAYS SELECT MATERIALS THAT CONFORM TO SAID CODE. DO NOT CHANGE ANY SIZE TO ONE WHICH WILL NOT MEET THE CODE.

Fig. 4.22—Recommended connecting tubing or piping for pressure transmitters.

(b) Diaphragm Seals. Stainless-steel diaphragm seals can be used when a sensor is not corrosion resistant or is subject to possible contamination. The space between the sealing diaphragm and the sensor is filled with a suitable liquid whose pressure duplicates that on the process side of the diaphragm. Fluids satisfactory for temperatures up to 350°F service are common. The seals are usually assembled in the factory to ensure a complete fill. Diaphragm seals are commonly used on Bourdon tubes and in force-balance capsules involving minimum displaced volume. Excessive displacement may involve an error arising when the spring rate of the seal diaphragm is added to the measured pressure. A seal diaphragm is part of the design in Fig. 4.17.

(c) Overpressure Devices. Pressures in excess of the normal design rating of the sensor may be encountered. For such emergencies a self-operating shut-off valve may be installed between the take-off and the sensor and set to close at preset point to protect the sensor. Stainless-steel guards are available in ¼-in. and ½-in. pipe sizes for operation up to 9000 psi.

(d) Siphons. Siphons or loops in take-off piping are used to keep hot fluids from contacting the sensor mechanism, which has usually been calibrated at ambient. Performance tests on the sensor indicate the maximum temperature that can be tolerated. The strain-gage sensor in Fig. 4.17 includes coolant connections.

4-3.5 Calibration Standards

Whether the sensor has readout capability or not, its proper calibration involves subjecting it to precise pressures and reading out on accurate gages. Correct readout voltmeters, ammeters for use with component signal conditioners, and other necessary accessories are recommended by the various manufacturers. Equipment for developing pressure varies according to the magnitude of the desired pressure.

For very low pressures, water, oil, and mercury manometers can be used. The bore should be large enough to provide an accurate column reading of the deflection scale. Air pressure from a compressor or a vacuum-pump source is also needed for a complete setup.

For medium pressures, transfer gages having calibration traceable to the U. S. Bureau of Standards can be used. Where water is used in the gage, the readings must be corrected for the weight of water unbalance between the sensor and the master gage.

For high pressures, a deadweight tester must be used. These are available for pressures from 15 to 10,000 psi. Constructed of stainless steel, using distilled water, and including a self-contained hand pump, the deadweight tester is an important calibrating device and may be used to calibrate master transfer gages for medium-pressure work.

4-3.6 Dynamic Testing and Performance Standards

Sensor manufacturers subject their designs to a series of tests simulating actual operating conditions to determine the on-line operating characteristics. Standard definitions are given in the Scientific Apparatus Makers Association (SAMA) publication PMC-20, *Measurement and Control Terminology.*

A sample performance report on a motion-balance sensor is given below:

Description
- Pressure-sensing mechanism: Bourdon tube (316 stainless steel)
- Electric transmission
- Output-signal ranges:
  - ±10 volts d-c; ±50 mV d-c; 0 to 100 mV d-c

Operating Conditions
- Ambient temp.: nominal, 75°F; reference, calibration ±5°F; normal, 40 to 140°F, operative limits, −10 to 200°F
- Supply voltage: nominal, 118 volts a-c; normal, 107 to 127 volts; operative limits, 100 to 135 volts
- Frequency: nominal, 50 or 60 Hz; normal, 48 to 62 Hz; operative limits, 45 to 75 Hz
- Ambient temp. effect: Zero-shift error/100°F temp. change, −1% range span; Range-shift error/100°F temp. change, +1% range span

Reference Performance Characteristics (% range span):
- Accuracy: 0.5%
- Dead band: 0.2%
- Hysteresis: 0.5%
- Linearity: 0.25%
- Repeatability: 0.25%

Design Data:
- Source impedance: a-c signal coil, 200 ohms; d-c signal demodulator, 180 ohms
- Minimum external load: a-c transmitted signal, 2000 ohms; d-c transmitted signal, 30,000 ohms
- Maximum ripple: 0.15% a-c ripple
- Case classification: NEMA (National Electrical Manufacturers Association) type 2 or NEMA type 7D
- Over-range protection: $1\frac{1}{4}$ times max. scale measured pressure

Performance data on a force balance sensor is given as:

Description:
- Pressure-sensing capsule: 316 stainless steel
- Electric transmission: 2 wire d-c
- Output signal range: 10 to 50 mA d-c

Operating Conditions:
- Power supply: 63 to 85 volts d-c
- Supply voltage effect: 0.25% per 10-volt variation

Performance characteristics (% range span):
- Accuracy: 0.5%
- Dead band: 0.005%
- Repeatability: 0.15%

Design Data:
- Output load limits: 600 ohms (+10%, −20%)
- Case classification: NEMA type 4; hazardous area Class I Group D, Div. 1

4-3.7 Transmitting Devices

(a) Pressure Switches. These are widely used to actuate alarms or initiate sequential operations. A Bourdon tube or similar sensor is linked to a snap-acting mechanical switch. (In some cases an enclosed mercury switch is used.) The switch may be indicating or nonindicating, have range-setting capability, and provide necessary logic at predetermined pressures.

(b) Electric Modulating Transmitters. These produce an electrical output proportional to input pressure (or force) applied to the sensor. Either the motion-balance or the force-balance principle may be involved. The output may be a voltage or a current of suitable value and range for input to readout devices, such as recorders, indicators, computers, and control loops to action equipment. A sample circuit for the motion-balance example of pt. 4-3.6 is shown in Fig. 4.23. Forms of the linear voltage differential transformer (LVDT) mechanism and a sample output curve are shown in Fig. 4.24.

(c) Pneumatic Modulating Transmitters. Differential-pressure sensors installed with one side open to the atmosphere and the other side connected to a pressure source can be used. The device shown in Fig. 4.35 can be used and the pneumatic force-balance principle applied to obtain a pneumatic output proportional to sensor gage pressures at connection H (or L, as desired).

4-4 FLOW SENSING

4-4.1 Differential-Pressure Flowmeters

(a) Basic Considerations. All flowmeters are considered to consist of two parts: a primary element, which contacts the flowing fluid, and a secondary element, which indicates or otherwise displays the desired information.

In the common differential-pressure or head-type flowmeter, the primary element is an obstruction placed in the pipe to create a pressure drop, and the secondary element is a device to measure this pressure drop and convert it to rate of flow. The secondary element itself may consist of two parts, a transmitter and a receiver, in case the information is to be displayed at some distance from the point of

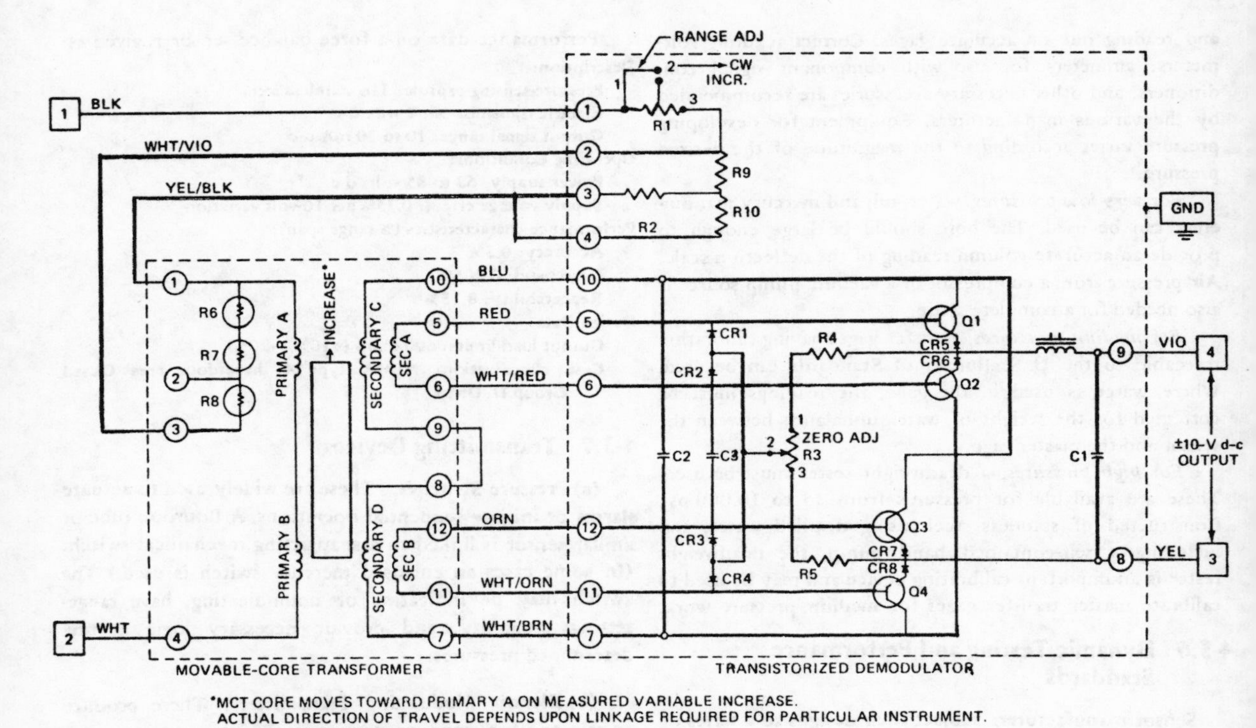

Fig. 4.23 — Schematic of a linear voltage differential transformer (LVDT) and demodulator circuit.

measurement. The primary element is usually one of three types: orifice plate, flow nozzle, or Venturi tube.

(b) Orifice Plates. The orifice plate is a thin disk clamped between gaskets in a flanged joint, with a usually concentric circular hole smaller than the internal pipe diameter. This is the simplest type of primary element and the most easily reproducible. It can be used without individual calibration with the greatest assurance of accuracy.

Flow through a sharp-edged orifice plate is characterized by a change in velocity, which reaches a maximum at a point slightly downstream from the orifice. At this point, called the vena contracta, the flowing stream has its greatest convergence. Beyond this point the flowing stream diverges until it again fills the entire pipe area; the velocity is reduced back to its original value (assuming fluid density and pipe cross section are the same upstream and downstream of the orifice), and the pressure increases to a value less than its original value.

Several locations for metering connections are shown in Fig. 4.25.

Vena Contracta Taps. The high-pressure connection is one pipe diameter upstream from the orifice, and the low-pressure connection is at the vena contracta. This tap arrangement is commonly used, particularly in the power industry. It has a slight advantage in that the connections are in regions of nearly constant velocity, thereby offering some tolerance in tap location without a noticeable change in differential pressure.

1D and ½D Taps. These are an approximation of vena contracta taps. The high-pressure connection is one pipe diameter upstream from the orifice, and the low-pressure connection is one-half pipe diameter downstream from the orifice inlet.

Flange Taps. These have connections drilled through the edges of the flanges 1 in. from the adjacent orifice surface. Flange taps are widely used in the gas and chemical industries because they are convenient to use and install. "Orifice flanges" are readily available with pressure connections and jackscrews to spread the flanges to facilitate orifice replacement.

Corner Taps. These have effective connections immediately adjacent to the orifice plate. Corner taps are commonly used in Europe but are seldom used in this country because they require special flange machining. They are used in some small orifice pipe assemblies furnished by several instrument manufacturers.

(c) Flow Nozzles. The flow nozzle, usually of ASME long-radius high-ratio design (Fig. 4.26), has an elliptically flared inlet and a cylindrical throat section. There is no vena contracta effect, because maximum velocity takes place in the throat. The flow nozzle passes approximately 60% more flow than an orifice with the same differential pressure and the same ratio of throat diameter to internal pipe diameter. (This latter ratio is the β of the flow nozzle, see Sec. (e) below.) A flow nozzle can be installed in welded piping but cannot be used to meter flow in either direction. The flow nozzle can be used successfully in some

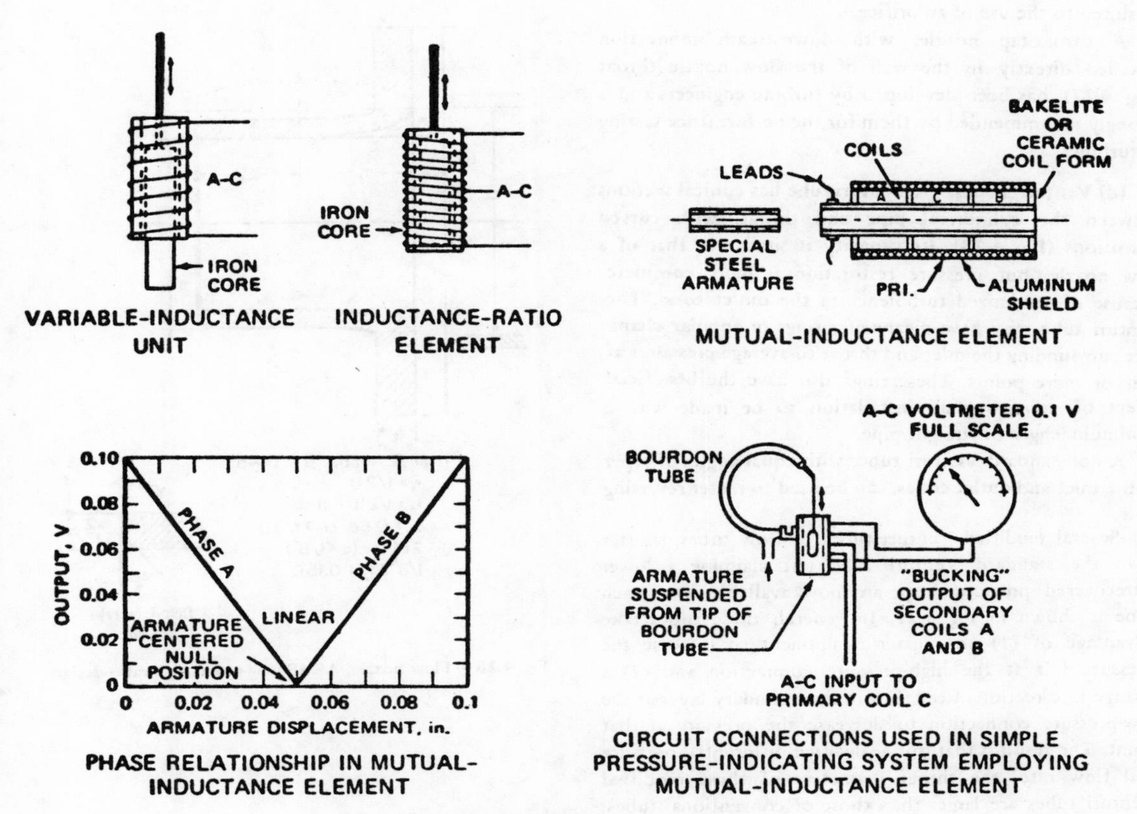

Fig. 4.24—Examples of inductance elements for LVDT's.

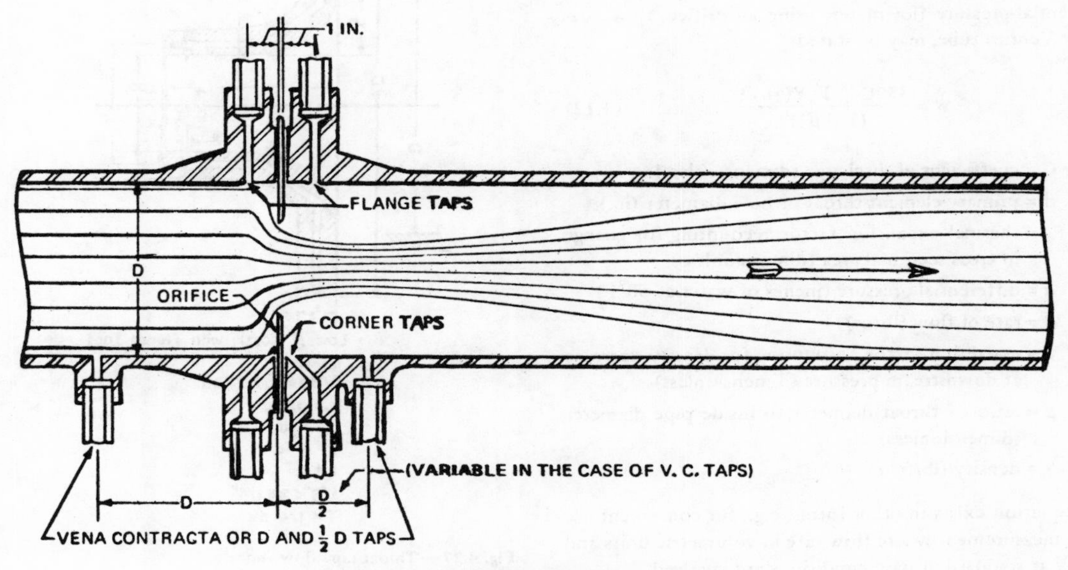

Fig. 4.25—Tap locations for orifice meters.

installations where limited length of straight pipe would not be suited to the use of an orifice.

A throat-tap nozzle, with downstream connection installed directly in the wall of the flow nozzle throat (Fig. 4.27), has been developed by turbine engineers and is strongly recommended by them for the performance testing of turbines.

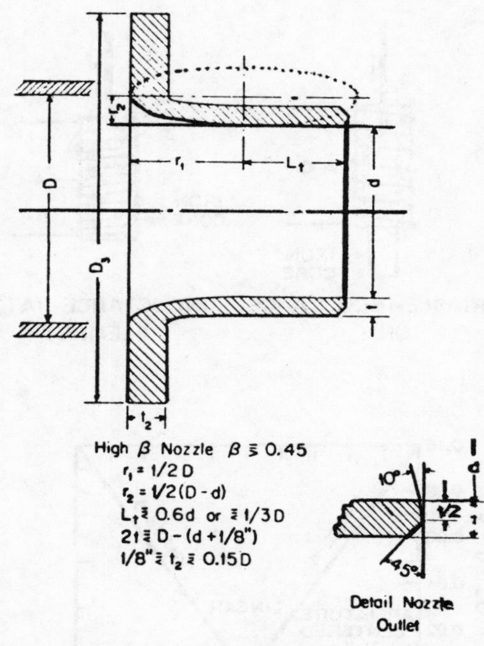

Fig. 4.26—Flow nozzle, ASME long-radius high-ratio design.

(d) Venturi Tubes. A Venturi tube has conical sections between the cylindrical pipe and throat, with curved transitions (Fig. 4.28). Its capacity is similar to that of a flow nozzle, but pressure restoration is more complete because of minimized turbulence in the outlet cone. The Venturi tube may have piezometer rings or annular chambers surrounding the inlet and throat to average pressures at four or more points. These rings also have the beneficial effect of permitting an installation to be made with a minimum length of straight pipe.

A nonstandard Venturi tube, with equal angles of taper in the inlet and outlet cones, can be used to meter reversing flow.

Several modified Venturi tubes or flow tubes, shorter than the standard Venturi tube but claiming a lower unrecovered pressure drop, are now available. One such tube is shown in Fig. 4.29. In general, these tubes take advantage of (1) an impact component to increase the pressure felt at the high-pressure connection and (2) a change in direction of the flow in the boundary layer at the low-pressure connection to decrease the pressure at that point. The result is that, for a specified differential pressure and flow rate, the throat diameters of these modified Venturi tubes are larger than those of conventional tubes; thus the overall pressure loss for a given flow rate may be less.

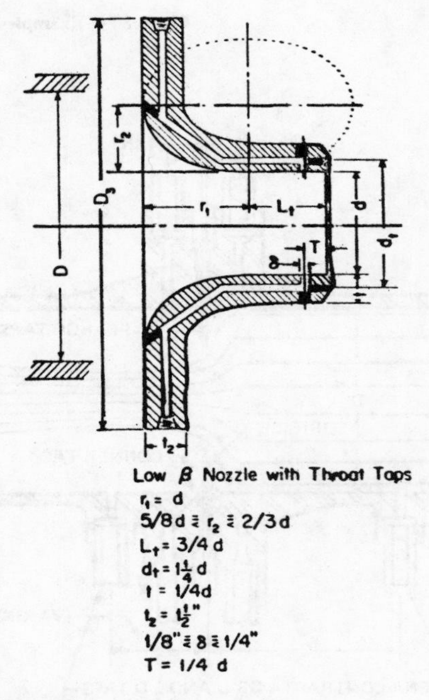

Fig. 4.27—Throat-tap flow nozzle.

(e) Sizing Primary Elements. The basic equation for a differential-pressure flowmeter, using an orifice, flow nozzle, or Venturi tube, may be stated:

$$W = \frac{359Cd^2\ F_a Y(\gamma h)^{1/2}}{(1-\beta^4)^{1/2}} \qquad (4.12)$$

where C = coefficient of discharge (dimensionless)

d = primary-element throat or hole diameter (in.)

F_a = thermal expansion factor accounting for change in cross sectional area (dimensionless)

h = differential pressure (inches of water at 68°F)

W = rate of flow (lb/hr)

Y = expansion factor accounting for density decrease at downstream pressure (dimensionless)

β = ratio of throat diameter to inside pipe diameter (dimensionless)

γ = density (lb/ft^3)

This equation exists in other forms, e.g., for convenient use in gas measurement where flow rate in volumetric units and density at standard or base conditions are involved.

The coefficient of discharge, C, may be presented by curves or in tables as a function of pipe Reynolds number, R_D, or of throat Reynolds number, R_d. Sometimes C is

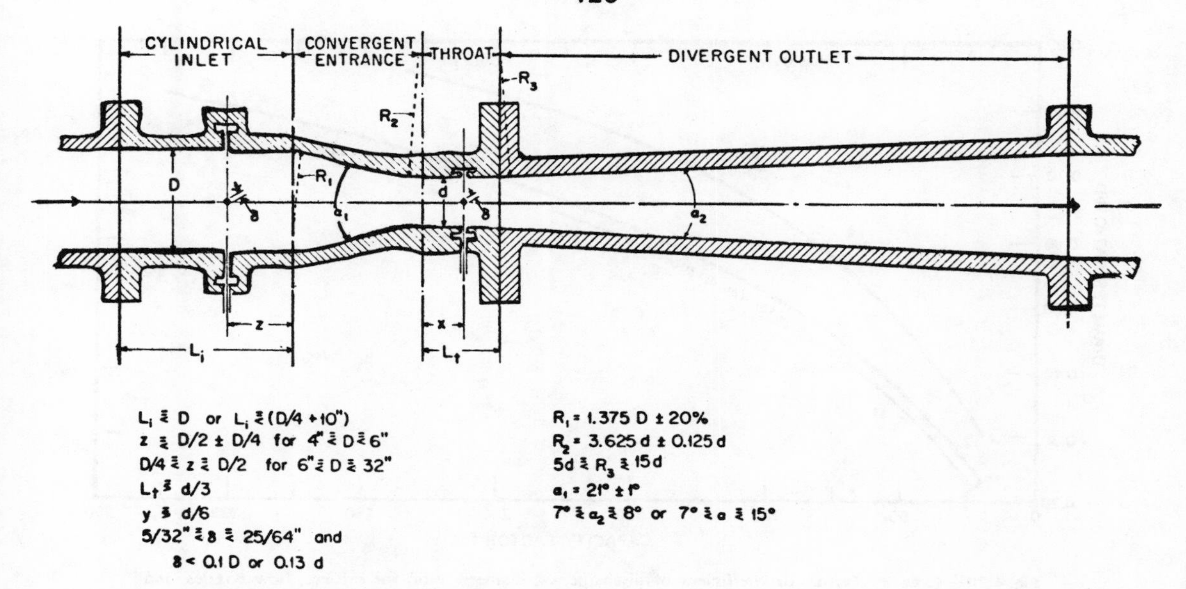

Fig. 4.28—Herschel or classical Venturi tube.

replaced by a combined coefficient, already divided by the denominator of Eq. 4.12; then it is labelled the "coefficient K with velocity-of-approach factor included." Rather than including here sufficient information to permit complete and accurate flow determinations to be made by measuring differential pressure across a primary element of any type and dimensions, this text is limited to presenting a method for determining approximate dimensions. These are suitable for estimating purposes or for assessing the acceptability of a proposed piping installation for flow-measurement purposes.

The diameter ratio, β, is determined from Fig. 4.30 with a calculation of the capacity factor, I, from the applicable equation:

For liquids:

$$I = \frac{W}{D^2 (h\gamma_f)^{1/2}}$$

or

$$I = \frac{Q_L \gamma_s}{7.48 D^2 (h\gamma_f)^{1/2}} \qquad (4.13a)$$

For steam:

$$I = \frac{W}{D^2 (h/v)^{1/2}} \qquad (4.13b)$$

For gases:

$$I = \frac{Q_G \gamma_s}{D^2 (h\gamma_f)^{1/2}}$$

or

$$I = \frac{Q_G [G(T + 460)]^{1/2}}{21.5 D^2 (hP)^{1/2}} \qquad (4.13c)$$

where D = internal pipe diameter (in.)
G = specific gravity (dimensionless)
I = capacity factor (dimensions consistent with above equations)
P = pressure at primary-element inlet (psia)
Q_G = gas flow rate [ft^3/hr at standard or base conditions, often abbreviated "scfh" (standard conditions are usually 30 in. Hg and 60°F)]
Q_L = liquid flow rate [gal/hr at standard or base temperature (usually 60°F for petroleum products, but flowing temperature for water)]
T = temperature (°F)
v = specific volume (ft^3/lb)
γ_s = density (lb/ft^3 at standard or base conditions)
γ_f = density (lb/ft^3 at flowing conditions)

In general, a primary-element β ratio of 0.50 to 0.70 may be considered normal. Higher ratios may be expected in high-pressure or high-temperature applications where pipeline economics suggests using the smallest acceptable

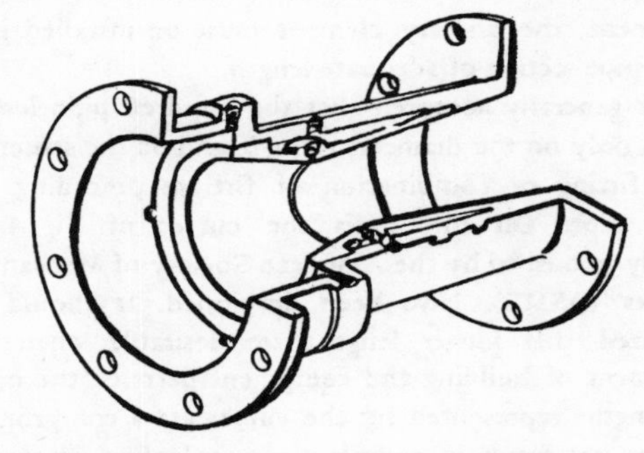

Fig. 4.29—Typical short modified Venturi tube.

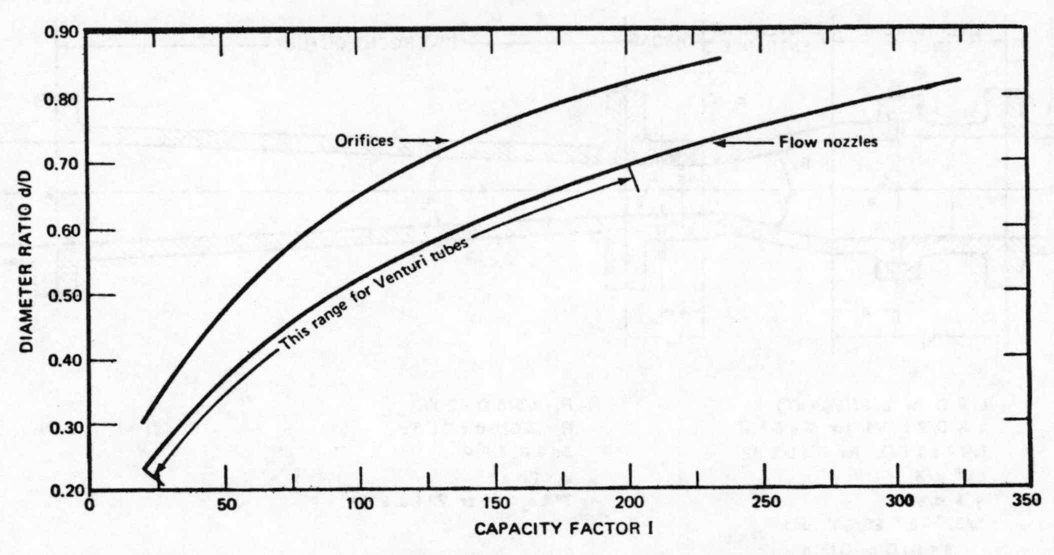

Fig. 4.30—Capacity factor, or coefficient of discharge, vs. diameter ratio for orifices, flow nozzles, and Venturi tubes.

size with consequent high velocities. An extremely low β value might indicate that the pipe size is larger than required or that a meter of lower differential-pressure rating would be preferable. The low β ratios of throat-tap nozzles are desirable in turbine testing to provide high differential pressure which may be read with accuracy on a manometer. The high differential pressure can be tolerated during the short duration of a test (after which the nozzle is removed).

The minimum and maximum limits of the diameter ratio, β, for various primary elements are shown in Table 4.18. These β values are important in the selection of maximum meter differential pressure. For compressible fluids (steam, air, and gases), the value of the maximum differential in inches of water should not exceed the operating pressure in psia in order to avoid inaccuracies due to variations in the expansion factor, Y, that are not taken into account.

Table 4.18—Maximum and Minimum Diameter Ratios for Orifices, Flow Nozzles, and Venturi Tubes

Type of element	Connection location	Pipe size, in.	Beta Min.	Beta Max.
Concentric orifice	Flange	$1\frac{1}{2}$	0.17	0.70
		2	0.125	0.70
		3	0.10	0.70
		>4	0.10	0.75
Concentric orifice	Vena contracta	$1\frac{1}{2}$	0.17	0.70
		2	0.125	0.75
		>3	0.10	0.80
Eccentric orifice		>4	0.30	0.80
Segmental orifice		>4	0.35	0.85
Flow nozzle	Pipe wall		0.20	0.80
	Throat		0.25	0.50
Venturi tube		>2	0.25	0.75

(f) Installation. Accurate flow measurements require that the primary element be in a "normally turbulent" flow pattern, with fully developed velocity profile and without spiral motion or velocity stratification. Because of this requirement, the primary element must be installed in a straight pipe section of adequate length.

It is generally accepted that the required pipe length depends only on the diameter ratio, β, and on the structure of the fitting or combination of fittings preceding the straight pipe. On this basis the curves of Fig. 4.31, originally published by the American Society of Mechanical Engineers (ASME), have been developed. It should be emphasized that longer lengths are desirable when the arrangement of building and equipment permits; the minimum lengths represented by the curves are a compromise and may not result in complete removal of an abnormal inlet turbulence.

(g) Accuracy of Primary Elements. Any meter manufacturer can be expected to furnish conventional primary elements (orifices, flow nozzles, or Venturi tubes) sized in accordance with the standard equations and discharge coefficients published by ASME. These coefficients are average values that may be considered to be correct within tolerances that vary with β, Reynolds number, and pipe size. The approximate accuracy of these coefficients is as follows:

±1.0 for concentric orifices with flange or vena contracta taps; $D \geqslant 2.0$ in., $0.20 \leqslant \beta \leqslant 0.70$
±2.25% for same, $\beta = 0.75$
±0.75% for Venturi tubes
±2.0% for flow nozzles
±1.4% for eccentric orifices, $D > 4$ in.
±2.0% for segmental orifices

When greater accuracy is required, the primary element in its pipe section may be calibrated at a hydraulic laboratory to establish, as nearly as possible, the true value of its individual coefficient of discharge. Ideally the calibration should be performed at Reynolds numbers

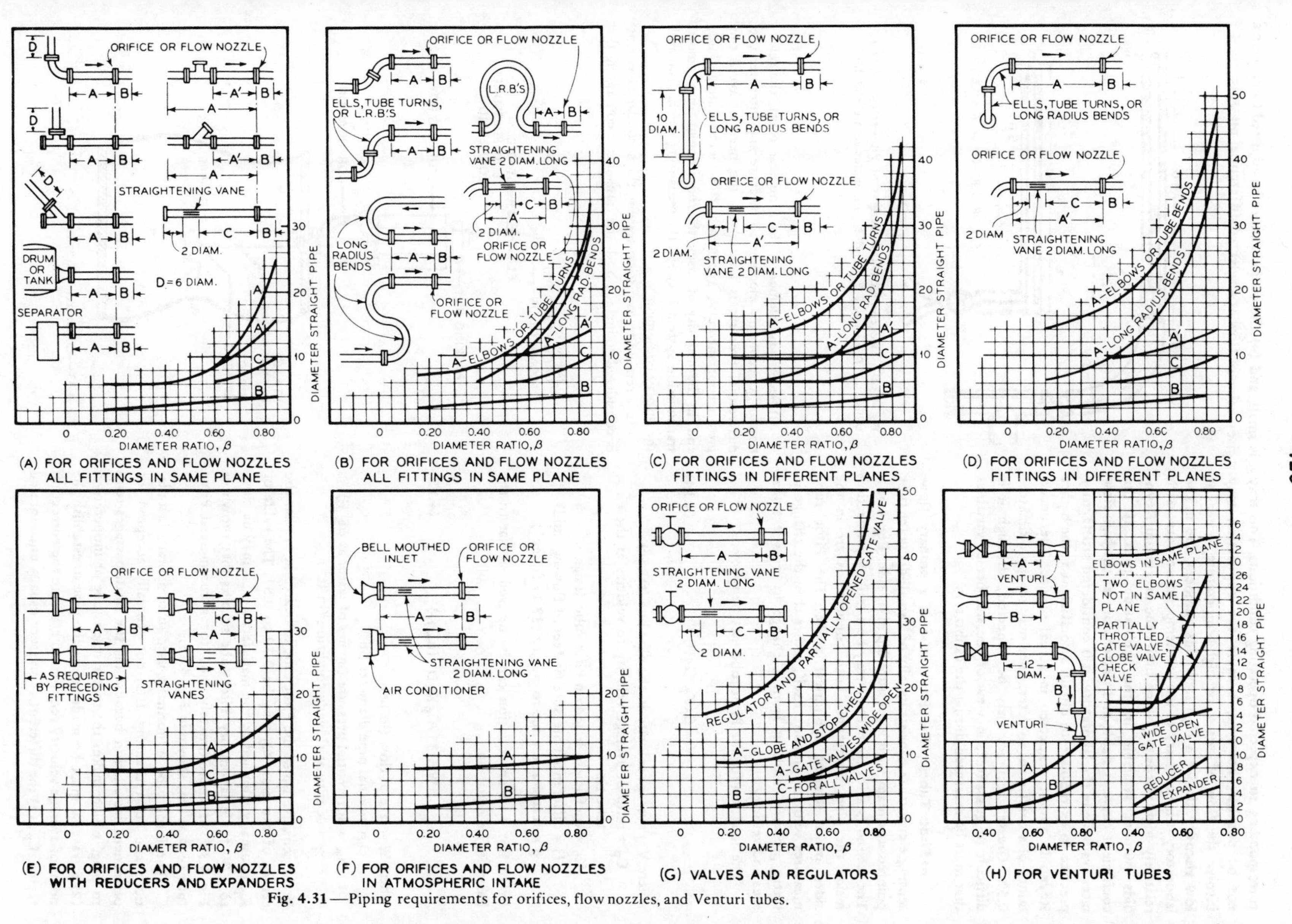

Fig. 4.31—Piping requirements for orifices, flow nozzles, and Venturi tubes.

corresponding to expected operating conditions. This may not be possible, however, because the application may involve the flow of steam, or high-temperature water, at low viscosity (hence a high Reynolds number), and in the laboratory it is necessary to use low-temperature water with relatively high viscosity, which limits the maximum Reynolds number available in the calibration runs. If the coefficient curve is stable and flattens out at the high flow rates available in the laboratory, it is considered satisfactory practice to extrapolate to obtain a coefficient at the higher Reynolds number expected in actual service. The uncertainty of a coefficient so determined may be approximately 0.5%. Greater accuracy may be expected, although it is difficult to guarantee, in view of the number of variables that must be measured during the calibration.

(h) Pitot Tubes. The Pitot tube is a primary flow-sensing element. It consists usually of a small-diameter tube pointed upstream with a pressure tap in its side (Fig. 4.32). The second (static) connection may be a trailing connection inside the pipe. Sometimes both connections are in the sides of a straight cylindrical probe. The Pitot tube measures velocity pressure itself, rather than the change in static pressure resulting from a change in cross-sectional area and velocity.

The equation for the Pitot tube may be written:

$$V = C_p K_p \, (2gH)^{1/2} \tag{4.14}$$

where V = average velocity in the pipe line (ft/sec)
C_p = ratio of average velocity to velocity at the Pitot tube tip
K_p = coefficient based on Pitot tube design
H = differential pressure (in feet of flowing fluid)
g = acceleration due to gravity (32.17 ft/sec^2)

From Eq. 4.14 an equation can be derived similar to the basic orifice equation:

$$W = 359 C_p K_p D^2 \, (h_w \gamma_f)^{1/2} \tag{4.15}$$

where W = rate of flow (lb/hr)
D = internal pipe diameter (in.)
h_w = differential pressure (inches of water at 68°F)
γ_f = density at flowing condition (lb/ft^3)

If the sensitive tip of the Pitot tube is located at or near the center of the pipe, C_p may be 0.70 to 0.81. The value of K_p is usually supplied by the manufacturer; it may be close to 1.00 for a laboratory Pitot tube such as that shown in Fig. 4.32, or approximately 0.82 for a commercial Pitot tube such as that shown in Fig. 4.33.

The Pitot tube is normally used in temporary installations and for estimating. It is not generally accepted for permanent installation because (1) the small passages tend to plug, which makes the meter insensitive or inoperative; (2) a traverse must be made across the pipe to establish the point of average velocity required to evaluate the correction factor, C_p; (3) the differential pressure usually encountered is small, and (4) the device cannot be adjusted to provide a given value of flow at a selected differential pressure.

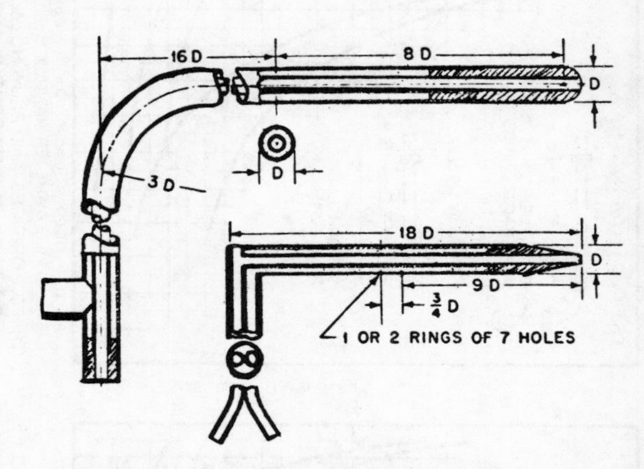

Fig. 4.32—Two commendable designs of Pitot-static tubes. Values of D between $^3/_{16}$ and $^5/_{16}$ in. inclusive are suitable.

(i) Secondary Elements. In a nuclear plant, the selection of a secondary element to measure and interpret the differential pressure is limited by the need to avoid the presence of mercury, a common flowmeter sealing fluid, and, in the majority of cases, the requirement for a remote transmission system between the measuring mechanism and the display equipment.

In a remote transmission system, the secondary element is considered to include both the transmitter containing the

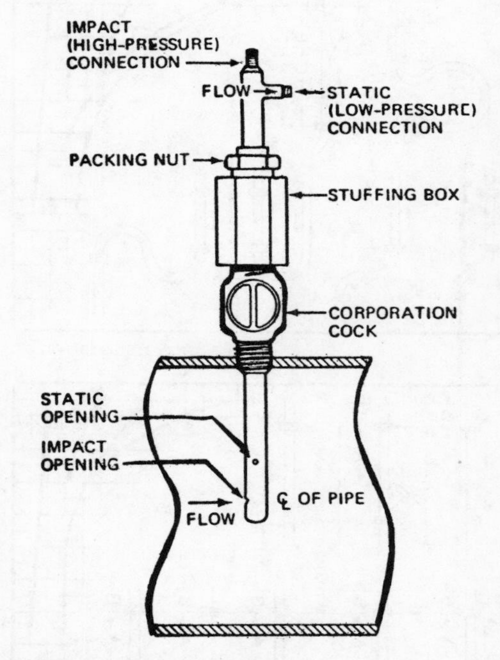

Fig. 4.33—Commercial Pitot tube.

measuring mechanism and the receiver (i.e., the components for recording, indicating, integrating, etc.). In effect, the direct mechanical connection of a self-contained secondary element is replaced by a pneumatic or electric position-transmitting mechanism.

As noted in pt. 4-3.2, measuring mechanisms of differential-pressure transmitters are of two general classes: motion balance and force balance. In a motion-balance system, the differential pressure exerted on a bellows or diaphragm displaces it, and the displacement is opposed by a spring, the force exerted by the spring being directly proportional to the applied differential pressure. A typical motion-balance-type electric transmitter is shown in Fig. 4.34.

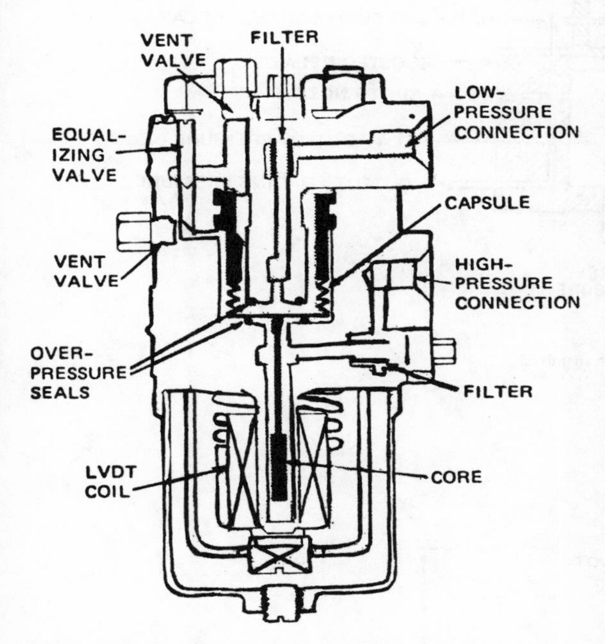

Fig. 4.34—Electric motion-balance transmitter.

In a force-balance system, the output signal is supplied to a pneumatic bellows or electromagnet that opposes the force exerted by the measuring mechanism. In a typical pneumatic force-balance transmitter (Fig. 4.35), a change in differential pressure changes the air gap at a nozzle tip, thereby changing the nozzle pressure, which is then amplified to provide the output signal. The changed output pressure changes the force opposing the measuring force, restoring equilibrium at a new value of the opposing forces with only an imperceptible change in nozzle air gap.

The receiver of a flowmeter may contain two features peculiar to flow measurement, a square-root extractor and an integrator. The square-root extractor is required if the pointer motion is to be directly proportional to flow rate instead of to differential pressure. The integrator is required to convert flow rate to total flow for accounting purposes.

Because the rate of flow is proportional to the square root of differential pressure, a flow scale on a differential-pressure meter is not uniform; the divisions at the low end of this scale are compressed close together and difficult to read. The square-root extractor interposes a variable gain between the measuring mechanism and the display, magnifying the motion significantly at lower flow rates to make the scale divisions uniform. The extractor may be in the transmitter of a transmitting system, in the receiver, or in a separate unit installed between transmitter and receiver. A simple mechanism that performs this function pneumatically is shown in Fig. 4.36. In the figure, the rise d is proportional to the differential pressure, and the pointer motion indicating flow is proportional to the angle α. A cam follower operates through a vane and nozzle to maintain a light contact with the cam. For low values of α (i.e., d is very small compared to the beam length R) the value of α is almost exactly proportional to the square root of d.

The integrator is a counter that is usually driven by a constant-speed motor through a friction clutch and cam-operated escape mechanism in such a way that the counter rotates during a portion of cam rotation, the duration being proportional to rate of flow or point of position. The amount registered during any time period therefore is the total flow in that period. A diagram of a typical integrator is shown in Fig. 4.37.

(j) Accuracy of Secondary Elements. Secondary-element accuracy may be represented as a single tolerance or as an individual tolerance for each component, such as transmitter, square-root extractor, receiver, and integrator. For several tolerances to be combined, they must be expressed on the same basis, i.e., in terms of differential pressure or of flow, and as percent of maximum or of actual reading. The square root of percent maximum differential is percent maximum flow; and a tolerance in percent of maximum flow is divided by percent of scale to determine the percent of actual flow.

For instance, at mid-scale on a differential-pressure transmitter, a tolerance of ±1% of maximum would represent an actual flow tolerance of 1% since the square root of $(0.50 - 0.01)/0.50 = 0.99$; but, at 25% scale, the flow tolerance is 2% since the square root of $(0.25 - 0.01)/0.25 = 0.98$. The addition of individual tolerances may be made on a root-mean-square basis and stated as such.

Because the high gain action of the square-root extractor at low flow rates magnifies uncertainties as well as the flow signal itself, the rangeability, or turn-down ratio, of a differential-pressure meter is practically limited to about 25% of maximum flow (6.25% of maximum differential). When operation at wider range is required, multiple primary elements or multiple transmitters can be used with automatic switching, which simultaneously changes meter capacity and integrator speed. Alternately, a meter with linear output might be specified.

4-4.2 Linear Flowmeters

(a) Area Meters. In an area flowmeter the fluid flows upward, displacing an obstructing float or piston. The float or piston is arranged so that the unobstructed area increases

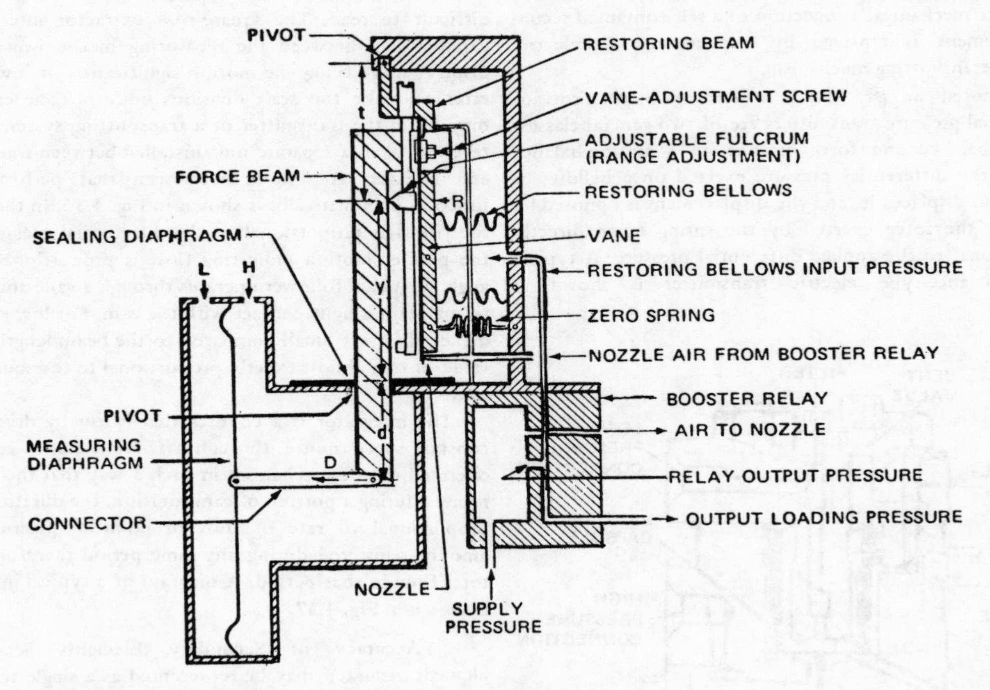

Fig. 4.35—Schematic of pneumatic force-balance transmitter.

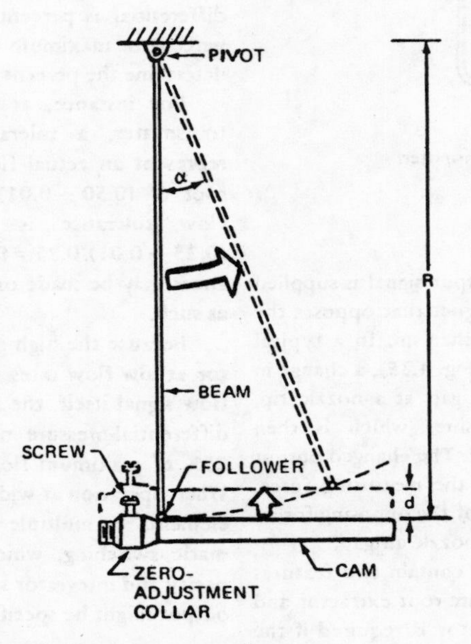

Fig. 4.36—Square-root extraction.

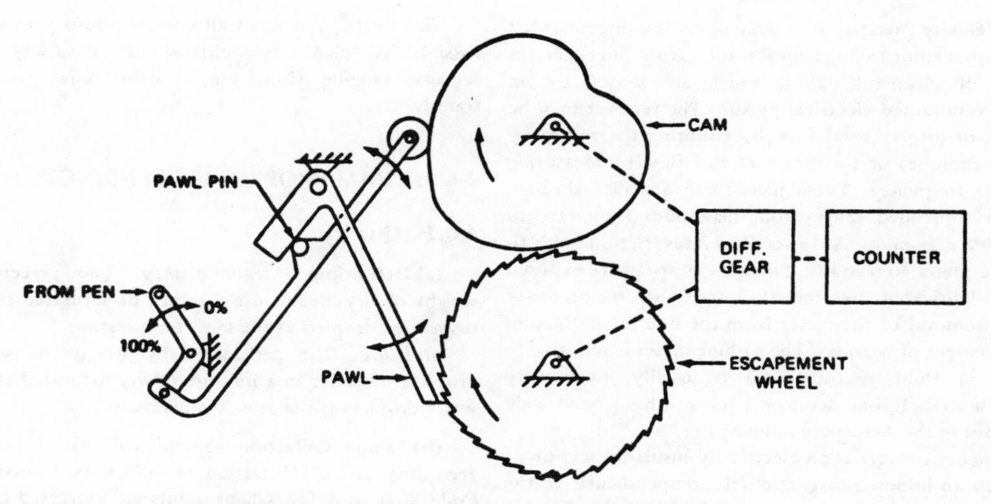

Fig. 4.37—Schematic diagram of integrator with cam and escapement mechanism.

with upward displacement; the float or piston then moves until the area is open enough to permit the flow to pass. The basic theory of area flowmeters is the same as that of a differential-pressure meter, but, since differential pressure is held constant, or reasonably so, square-root extraction is not required. A measurement of float or piston position is a measurement of unobstructed area, and thus of flow. Area meters are of two general types: rotameters and piston-type meters.

A *rotameter* (Fig. 4.38) consists of a float inside a tapered tube, the small end of the tube being at the bottom. The force exerted in the tube by the flow moves the float upward until the area of the annular space between float and tube is sufficient to permit a flow-created differential pressure to balance the weight (less the buoyancy force) of the float. Differential pressure is determined by the weight of the float and its cross-sectional area. A scale is marked on the outside of the transparent, tapered tube so that float position can be read directly. Since the unobstructed area permitting flow is almost exactly proportional to float rise (for a slightly tapered tube), the scale markings indicating flow can be uniform. For high-pressure applications or for transmitters, the tube can be metal, and the float position can be sensed by a magnetic pickup.

In a *piston-type area meter,* upward motion of the piston or plug uncovers ports in the sleeve or cage, increasing the area of the opening in direct proportion to plug movement and to flow rate. A spring pulls downward on the plug to increase the differential pressure beyond that which might be obtained by the weight of the parts. The vertical position of the plug establishes the rotational position of a spindle, which extends through a packing gland to operate an indicating pointer and transmitting mechanism (either pneumatic or electric).

(b) Positive-Displacement Meters. The flowing fluid is divided into separate discrete volumetric portions that are counted by a mechanical register built into the meter. Alternately, the rotation of the meter mechanism may be made to generate an electrical signal with frequency proportional to the rotational speed; the signal can then be transmitted to a remote register or recorder.

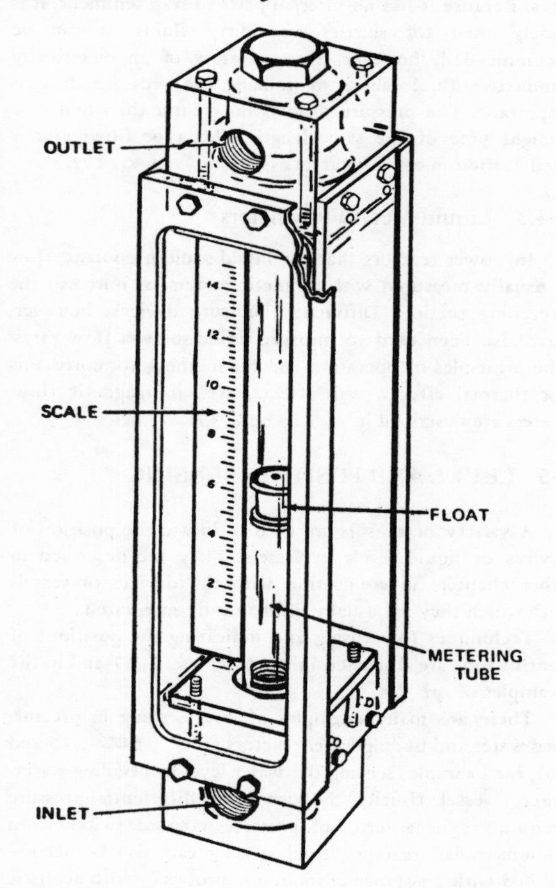

Fig. 4.38—Rotameter.

(c) Velocity Meters. A *turbine meter* is a line-mounted meter with a rotor having helical blades. Rotation generates a series of electrical pulses, which are sensed by an externally mounted electrical pickup. The receiver may be arranged to display total flow by counting the pulses by digital techniques or to display rate of flow by measuring the pulse frequency. These meters are accurate through their recommended range from maximum flow rate to about 10% maximum. At lower flow rates friction tends to cause the meter to read low. Since the bearings are exposed to and lubricated by the flowing fluid, maintenance involves removal of the meter from the line for inspection or replacement of bearings. The turbine meter is sensitive to changes in fluid viscosity and is usually individually flow-calibrated. It was developed for and has gained wide acceptance in the Aerospace industry.

A *magnetic meter* is an electrically insulated section of pipe with an imposed magnetic field, perpendicular to the pipe axis, through which a conductive fluid develops an electrical potential (perpendicular to magnetic field and pipe axis) directly proportional to its average velocity through the pipe section. Electrodes flush with the pipe wall are connected to a circuit for measuring the generated voltage. The magnetic meter is accurate and linear through a wide range of flow rates and is available in a wide range of sizes. Because it has no internal parts to trap sediment, it is widely used for slurries and dirty fluids; it can be recommended, however, for any flow of an electrically conductive fluid where minimizing the pressure drop is important. The pressure drop is no greater than that of a straight pipe of the same length. This type flowmeter is used in sodium-cooled reactors (see sec. 17).

4-4.3 Liquid-Metal Flowmeters

In power reactors that use liquid-sodium coolant, flow is usually measured with magnetic meters, as noted in the preceding section. Differential-pressure devices, however, have also been used to measure liquid-sodium flow rates. The principles of operation and the methods for correcting for thermal effects, wall effects, etc., in magnetic flowmeters are described in sec. 17.

4-5 LEVEL AND POSITION SENSING

A variety of sensors are used to locate the position of devices or liquid levels in vessels. They are described in other chapters in connection with the devices or vessels with which they are usually mechanically integrated.

Techniques for sensing and indicating the positions of control rods are discussed in sec. 7, pt. 7-3.7 and in the examples of pt 7-4.

There are many examples of level sensing in pressurized-water and boiling-water reactors (sec. 15 and 16), for example, sensing the water level in a boiling-water reactor vessel. Usually, the sensors are differential-pressure transducers or a series of pressure-actuated switches. In sodium-cooled reactors, level sensing can also be accomplished with resistance or induction probes or with acoustic devices. These sensors are described in sec. 17, pt. 17-4.3.

The steam systems of all nuclear power plants (and of fossil-fueled plants as well) include a variety of level sensors, ranging from simple sight tubes to pressure transducers.

4-6 STEAM PROPERTIES SENSING

4-6.1 Quality

(a) Definitions. *Steam quality:* The percentage by weight of dry steam in a mixture of saturated steam and suspended droplets at the same temperature.

Moisture: The percentage by weight of suspended droplets of water in a mixture of dry saturated steam and water droplets at the same temperature.

(b) Sample Collection. Sample collection is carried out according to ASTM D1066 or ASME Performance Test Code, Part II. A few salient points are extracted here. It is important to note that the ASME Performance Test Code does not recommend using the electrical conductivity method for determining the moisture content of steam. A recommended form of sampling nozzle is shown in Fig. 4.39.

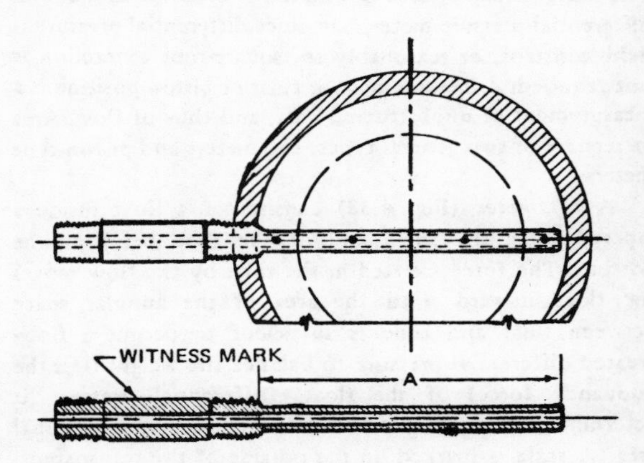

Fig. 4.39—Recommended sampling nozzle.

The pipe or tube in the sampling nozzle extends across the pipe on a diameter to within 0.25 in. of the opposite wall. The drilled holes face upstream in the pipe and are spaced so that each port represents an equal area of pipe section. For a representative steam sample, the hole size in the sample tube must be chosen so the rate of sample flow is equal to the rate of steam flow. The shortest possible connection should be used between the sampling nozzle and the calorimeter or cooling coil.

(c) Moisture Determination. *Throttling Calorimeter.* This is a simple device (Fig. 4.40). Its essential details are a throttling orifice admitting steam to an expansion chamber and a thermometer well entirely surrounded by the low-pressure steam from the throttling orifice. The principle of operation is the equality of initial and final enthalpies when steam passes through an orifice from higher to lower pressure, provided there is no heat loss and the difference between initial and final kinetic energies is negligible. Two conditions are necessary in the use of a throttling calorimeter: (1) there must be a significant pressure difference between steam in the sample and steam in the expansion chamber and (2) the quality of the sample must be high enough to produce a measurable degree of superheat (8°F) in the calorimeter.

For example, if the pressure in the expansion chamber is atmospheric, the temperature is 280°F, and the sample pressure is 135 psia, the constant enthalpy line on a Mollier chart shows the initial moisture to be 1%, or 99% quality.

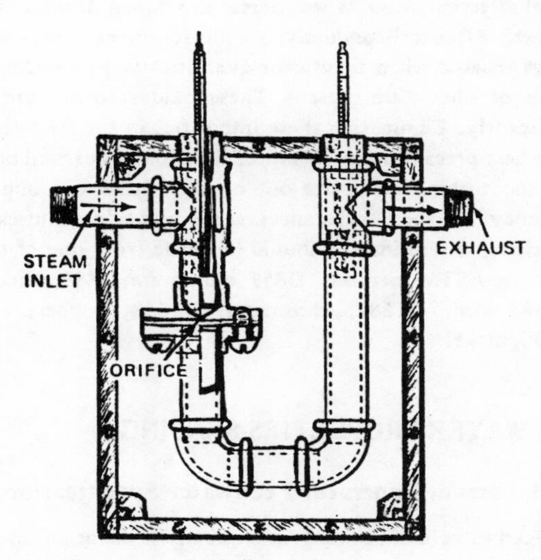

Fig. 4.40—Throttling calorimeter.

Quality can be calculated from the following formula:

$$X = \frac{h_2 - h_f}{h_{fg}} \times 100 \qquad (4.16)$$

where X = the quality (%)

h_2 = the enthalpy of superheated steam at the calorimeter pressure and temperature

h_f = the enthalpy of saturated liquid in the mixture prior to throttling

h_{fg} = the enthalpy of vaporization of steam entering the calorimeter

Separating Calorimeter. The throttling calorimeter cannot be used when the enthalpy in the calorimeter chamber is equal to or less than the enthalpy of saturated steam. Either a separating or universal instrument must be used. In the separating calorimeter (Fig. 4.41), water is separated out from the steam and read in a graduated gage glass. The quality (%) is

$$X = \frac{M + R}{W + M} \times 100 \qquad (4.17)$$

where M = the weight of dry steam condensed after passing through the calorimeter

W = the weight of water as read from the gage-glass scale

R = the weight of water corresponding to heat loss by radiation

With an insulated calorimeter, the radiation loss can be neglected.

The accuracy of the separating calorimeter is somewhat less than that of the throttling calorimeter.

Throttling Separating Calorimeter. The throttling separating calorimeter is made up of two calorimeters, a

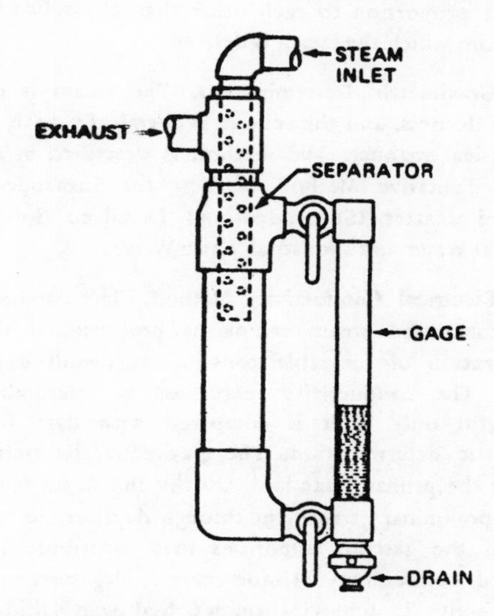

Fig. 4.41—Separating calorimeter.

throttling calorimeter and a separating calorimeter of low and high range, respectively, in series. The steam first passes through a throttling orifice and, if the moisture is not excessive, the quality is determined as in a throttling calorimeter. If the moisture is outside the throttling-calorimeter range, the separating calorimeter is at once available and no delay is caused by its use.

Separating Throttling Superheating Calorimeter. A throttling calorimeter can be connected to the exhaust of a separating calorimeter if the separation of moisture in the separating calorimeters is not complete. The quality of the original sample can be found by multiplying the qualities for each process.

Radioactive Tracers. Radioactive tracers can be used to determine steam quality from a boiling-water reactor. This method is generally limited to steam pressures under 1000 psi because of the occurrence of volatilized salts in the steam. Concentrations of specific radionuclides in condensed steam and boiler-water samples are determined with a multichannel analyzer. Steam quality, in percent, is calculated from the tracer activities as follows:

$$X = 100\left(1 - \frac{A_s}{A_w}\right)$$

where A_s is the activity in steam and A_W is the activity in boiler water.

4-6.2 Purity

(a) Definition. *Steam purity* refers to the solid matter in steam. Solid matter is defined as materials in steam which are solids at room temperature and which are capable of deposition as solids in superheaters, steam lines, turbines, or other steam-utilizing apparatus so as to reduce capacity, efficiency, or utility. Solids in the steam may be in different proportion to each other than the solids in the water from which the steam is derived.

(b) Gravimetric Determination. The steam is evaporated to dryness, and the residue is chemically analyzed in the classical manner. The method is described in ASTM D1069, Tentative Method of Test for Suspended and Dissolved Matter (Suspended and Dissolved Solids) in Industrial Water and Industrial Waste Water.

(c) Electrical Conductivity Method. The conductivity of a condensed steam sample is proportional to the concentration of ionizable constituents dissolved in the sample. The conductivity, expressed in micromhos, is meaningful only if it is compared with data from a gravimetric determination. The gravimetric-determination data are the primary standard. Usually the steam sample is given a preliminary treatment through degassers to remove most of the gaseous impurities that contribute to the measured conductivity. Unfortunately, degassers are not very effective in removing amines, hydrogen sulfide, and sulfur dioxide.

There are four sources of interference with respect to the calibration of electrical conductivity measurements:

1. Dissolved volatile substances, such as ammonia, amines, H_2S, H_2, CO_2, and SO_2, increase conductivity. Ammonia and hydrazine are commonly used in once-through boilers for pH adjustment and dissolved-oxygen scavenging.
2. Dissolved solids, such as the oxides of silicon, copper, and iron, ionize very little and therefore have little influence on conductivity.
3. The conductivity of pure water, as small as it is, must be subtracted from the combined conductivity.
4. The current-carrying capability of each ion species is different at any one temperature, and temperature coefficients are different; therefore, calibration depends on an assumed composition that may change in any one system and is almost certain to be different in different systems.

(d) Sodium Tracer Method for High-Purity Steam. The method is described in ASTM D2186. Generally, this method assumes that the ratio of sodium concentration to impurity concentration in the steam is equal to the ratio of sodium concentration to impurity concentration in boiler water:

$$S_t = S_s \frac{W_t}{W_s} \qquad (4.18)$$

where S_t = concentration of impurities in steam
S_s = concentration of sodium in steam
W_t = concentration of total solids in boiler water
W_s = concentration of sodium in boiler water

The values on the right-hand side of this formula are determined by ASTM methods.

The principal advantages of this method are its freedom from interferences, its ability to measure extremely small concentrations of impurities, and its rapid response to transient conditions. Sample temperature control is not required in the method.

(e) Determination When Silica and Metal Oxides Are Present. Electrical-conductivity measurements are not always reliable when significant quantities of the oxides of metals or silicon are present. These oxides do not ionize significantly. Eliminating these impurities in the feedwater is the best precautionary measure. Metal oxides carried over into the turbine can plate out on the blades and impair efficiency. If these substances are present in significant quantities, determination should be made from one of the following ASTM methods: D857, aluminum; D859, silicon; D1068, iron; D1687, chromium; D1688, copper; and D1886, nickel.

4-7 WATER PROPERTIES SENSING

4-7.1 Steam-Generator Feedwater Specifications

Feedwater conditioning is required to maintain operational capability of the steam generator. The steam-generating surfaces remain clean and heat-transfer capabilities are favorable if good water quality is maintained. The minimum standards given in Table 4.19 should be maintained for satisfactory feedwater quality.

4-7.2 Usual Impurities in Water Supply

Table 4.20 lists the impurities usually found in water supplies and indicates their properties, effects, and methods for treatment and removal.

4-7.3 Effect of Impurities in Steam-Generator Feedwater

(a) Total Solids (Dissolved and Undissolved). The total solids in the feedwater is a general indicator of how much material is collecting in the steam generator. Insoluble materials are deposited on the steam-generator surfaces. The soluble material (e.g., NaCl, NaOH, and Na_2SO_4) is carried over in the steam with the remaining material and tends to collect on the steam-generator tubes. The types of constituents in the feedwater depend on the preboiler characteristics. The quantity of soluble material should be larger than the quantity of insoluble material. Turbidity is a measure of the undissolved constituents, and electrical conductivity is a measure of the dissolved constituents.

(b) Dissolved Oxygen. Dissolved oxygen promotes corrosion in the steam generator and therefore should be kept as low as possible. Deaeration removes much of the dissolved oxygen. Hydrazine, a good oxygen scavenger, can eliminate the remaining oxygen. A high level of dissolved oxygen could indicate a malfunctioning deaerator or an air leak in the area of the condenser.

Table 4.19—Feedwater Quality Standards*

Maximum total solids (dissolved and suspended), ppb	50
Maximum dissolved oxygen, ppb	7
Maximum total silica (as SiO_2), ppb	20
Maximum total iron (as Fe), ppb	10
Maximum total copper (as Cu), ppb	2
pH at 77°F (adjusted with ammonia)	9.3 to 9.5
Total hardness†	No specification listed
Organics‡	0
Lead§	0

*From Babcock and Wilcox Nuclear Power Generation Division, Water Chemistry Manual, Part 8, p. 8-1.

†Hardness constituents should be eliminated because of deposition on steam generator surfaces.

‡Organic contamination can lead to resin fouling.

§Lead contamination should be kept below the lowest value detectable by acceptable methods to avoid problems with Inconel-600 in oxygenated water.

(c) Total Silica. Silica should be maintained at a low level for two reasons: (1) silica can concentrate in the steam generator and subsequently plate out on heat-exchange surfaces thereby reducing steam generation and (2) silica may carry over and plate out in the turbine causing turbine inefficiency. A higher than allowable silica concentration implies a demineralizer breakthrough. Switching to the spare demineralizer while the exhausted resin is regenerated or changed can probably eliminate the high silica level.

(d) Total Iron. Iron tends to build up in the steam generator and reduces its efficiency by degrading the heat-transfer characteristics. The level of iron in the feedwater affords some measure of the degree and rate of corrosion in the system.

(e) Total Copper. Copper should be avoided where possible in the feedwater system. Equipment in contact with the feedwater should be ferritic or austenitic stainless steel. Copper in the feedwater system can be carried into the steam generator in solution and plate out there. The copper plate can then make it necessary to clean the steam generator in a two-stage process: one for copper and another for iron. Copper carryover in the steam can plate out on the turbine and lower its efficiency. Copper alloy tubes in the condenser should be satisfactory because the temperatures and pressures are reduced and the dissolution of the copper is less likely.

(f) Total Lead. Lead in the feedwater concentrates in the steam generator. This can result in problems with Inconel in oxygenated water containing lead. Satisfactory instrumentation for monitoring traces of lead is not available at the present time.

(g) Conductivity. Cation (positive-ion) conductivity cells can be used to monitor the feedwater. Measurements should be made after removal of the ammonia that is used to regulate the pH.

(h) Corrosion. The principal accelerators of corrosion are dissolved oxygen; acids; surface deposits, especially those electronegative to steel; dissimilar metals in contact; and electrolytes.

Common methods to prevent corrosion are removal of dissolved gases, especially oxygen; neutralization of acids and maintenance of desirable alkalinity and pH; periodic mechanical cleaning; and avoiding excessive salt concentrations.

4-7.4 Acidity (pH)

(a) Definition. The pH is defined as the logarithm (to the base 10) of the reciprocal of the hydrogen-ion concentration in moles per liter:

$$pH = \log \frac{1}{H^+ \text{ concentration in moles per liter}}$$

Figure 4.42 shows pH vs. hydrogen-ion concentration. Points indicate the pH values of various common acids and bases.

All water solutions owe their chemical activity to their relative H^+ and OH^- concentrations. In water, the equilibrium product of the H^+ and OH^- concentrations is a constant 10^{-14} at 25°C. When concentrations of H^+ and OH^- in pure water at 25°C are equal, the H^+ concentration is 10^{-7} and, from the definition, the pH is 7.0. Note that the scale of pH values is not linear with concentration. A change of one unit in pH represents a 10-fold change in the effective strength of the acid or base.

The pH value depends only on the concentration of hydrogen ions actually dissociated in a solution and not on the total acidity or alkalinity. Therefore, because dissociation of water increases with temperature and pH is a measure of H^+ concentration only (and not the ratio of H^+ to OH^-), the pH of pure water increases above 7.0 if the temperature is increased above 25°C. There is no simple way to predict the pH of a solution at a desired temperature from a known pH reading at some other temperature.

(b) Measurement Techniques. *Chemical Indicators.* The pH of a sample may be determined by adding a small quantity of an indicator solution to the sample and comparing the color with that of a color standard. When good color standards are available in steps of 0.2 pH unit and observations are made in a comparator, the limit of accuracy is considered to be 0.1 pH unit. Turbid and colored solutions cannot be observed with accuracy, and indicators are not stable in many strongly oxidizing or reducing solutions. Table 4.21 lists some common pH indicators and their range of use.

Potentiometric pH Measurement. A potentiometric pH-measuring system consists of (1) a pH-responsive elec-

Table 4.20—Impurities in Water Supplies*

Impurity	Formula	Molecular weight	Equivalent weight	Solubility	Probable effect in boiler	Methods of treatment and removal
Calcium bicarbonate	$Ca(HCO_3)_2$	162.10	81.05	Moderate	Scale and sludge; liberates CO_2	In external treatment of calcium and magnesium compounds, lime and soda softeners plus coagulation and filtration give partial removal; Zeolite softeners and evaporators give more complete removal, the former replacing calcium and magnesium with sodium; corrosive compounds require alkali treatment
Calcium carbonate	$CaCO_3$	100.08	50.04	Slight	Scale and sludge; liberates CO_2	
Calcium hydroxide	$Ca(OH)_2$	74.10	37.05	Slight	Scale and sludge	
Calcium sulfate	$CaSO_4$	136.14	68.07	Moderate	Hard scale	
Calcium silicate	Variable			Slight	Hard scale	
Calcium chloride	$CaCl_2$	110.99	55.50	Very soluble	Corrosive; scale and sludge	
Calcium nitrate	$Ca(NO_3)_2$	164.10	82.05	Very soluble	Corrosive; scale and sludge	
Magnesium bicarbonate	$Mg(HCO_3)_2$	146.34	73.17	Moderate	Deposits; liberates CO_2	
Magnesium carbonate	$MgCO_3$	84.32	42.16	Slight	Deposits; liberates CO_2	
Magnesium hydroxide	$Mg(OH)_2$	58.34	29.17	Very slight	Deposits	
Magnesium sulfate	$MgSO_4$	120.38	60.17	Very soluble	Corrosive; deposits	
Magnesium silicate	Variable			Slight	Hard scale	
Magnesium chloride	$MgCl_2$	95.23	47.62	Very soluble	Corrosive; deposits	
Magnesium nitrate	$Mg(NO_3)_2$	148.34	74.17	Very soluble	Corrosive; deposits	
Sodium bicarbonate	$NaHCO_3$	84.00	42.00	Very soluble	Increases alkalinity and soluble solids; liberates CO_2	Excess sodium alkalinity may be reduced by boiler blowdown; it sometimes is neutralized with sulfuric acid externally; phosphoric acid and acid phosphates also are used; evaporation is best practical means of removing sodium compounds from feedwater, boiler blowdown is used for internal reduction of soluble solids
Sodium carbonate	Na_2CO_3	106.00	53.00	Very soluble	Increases alkalinity and soluble solids; liberates CO_2	
Sodium hydroxide	NaOH	40.00	40.00	Very soluble	Increases alkalinity and soluble solids	
Sodium sulfate	Na_2SO_4	142.05	71.03	Very soluble	Inhibitor for caustic embrittlement; increases soluble solids	
Sodium silicate	Variable			Very soluble	Increases alkalinity; may form silica scale	
Sodium chloride	NaCl	58.45	58.45	Very soluble	Increases soluble solids; encourages corrosion	
Sodium nitrate	$NaNO_3$	85.01	85.01	Very soluble	Increases soluble solids	
Iron oxide	Fe_2O_3	159.68	26.61	Slight	Deposits; encourages corrosion	Coagulation and filtration, evaporation, blowdown
Alumina	Al_2O_3	101.94	16.99	Slight	May add to deposits	Coagulation and filtration, evaporation, blowdown
Silica	SiO_2	60.06	30.03	Slight	Hard scale; acts as binder for deposits	Precipitation with aluminates, coagulation and filtration, evaporation, blowdown
Dissolved oxygen	O_2	32.00	16.00	Slight	Corrosive	Deaeration preferred
Carbonic acid or dissolved CO_2	H_2CO_3	62.02	31.01	Very soluble	Retards hydrolysis of carbonates; reduces alkalinity	Deaeration and alkali treatment
Hydrogen sulfide	H_2S	34.08	17.04	Very soluble	Corrosive	Deaeration and alkali treatment
Acids, organic and mineral				Very soluble	Corrosive	Neutralization by alkali treatment
Oil and grease				Slight	Corrosive; deposits; foaming and priming	Coagulation and filtration, skimming
Organic matter				Very soluble	Corrosive; deposits; foaming and priming	Coagulation and filtration, evaporation

*From R. T. Kent, *Mechanical Engineers' Handbook*, Power, 12th ed., p. 7-51, John Wiley & Sons, Inc., New York, 1950.

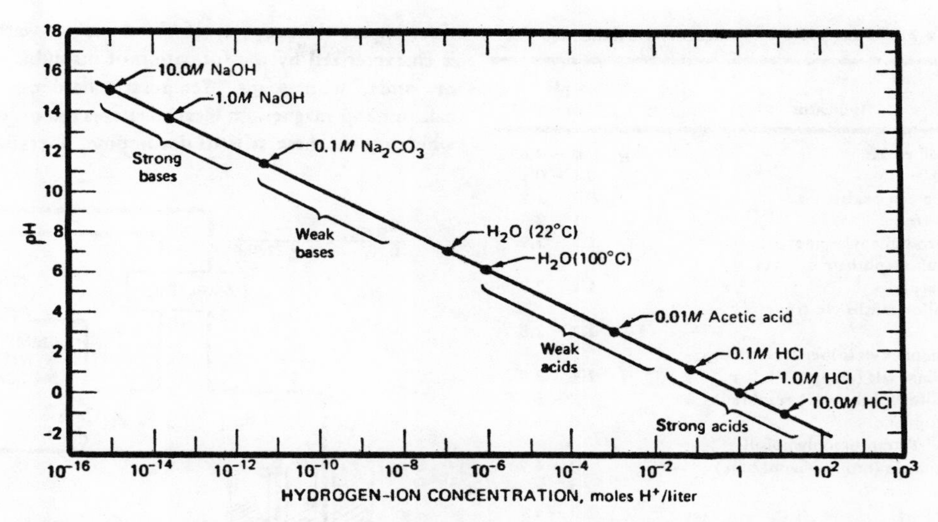

Fig. 4.42 —pH vs. hydrogen-ion concentration. (From D. M. Considine, *Process Instruments and Controls Handbook*, p. 6-96, McGraw-Hill Book Company, Inc., New York, 1957.)

trode, such as glass, antimony, quinhydrone, or hydrogen; (2) a reference electrode, usually calomel or silver–silver chloride; and (3) a potential-measuring device, such as a pH meter, usually some form of vacuum-tube voltmeter. Figure 4.43 shows a typical potentiometric system.

Table 4.22 lists the characteristics of six pH-measuring electrodes. Glass electrodes are electrically sensitive to hydrogen-ion concentration. The voltage response to hydrogen-ion concentration is:

$$E = E^0 - 0.0591 \log H^+ \text{ (at 25°C)}$$

where E^0 is the voltage of the particular glass electrode at pH zero.

In Figs. 4.44 to 4.47, some pH-measuring meters are illustrated. The feedback type pH meter (Fig. 4.49) has a circuit capable of good performance if matched tubes are employed to minimize drift. The electrodes must be checked periodically against standards for asymmetry.

Figure 4.48 shows the theoretical curve for the pH at 25°C vs. the concentration of ammonia. Figure 4.49 gives the temperature correction for the ammonia curve.

Figure 4.50 shows the theoretical curve for the pH at 25°C vs. the conductivity of ammonia. Conductivity measurements can be used to monitor the pH of the feedwater or the ammonia concentration in the feedwater (Fig. 4.51).

Limitations and Practical Considerations.

1. Glass electrodes can develop cracks, which allow some diffusion between the inner filling solution and the sample. When diffusion occurs responses are erratic and nonreproducible.

2. Glass is soluble in strongly alkaline solutions and thus has a shorter service life. Special alkali-resistant electrodes should be used for these applications.

3. If the glass becomes coated, the response is sluggish.

4. High sodium-ion concentration for extended periods of time results in loss of sensitivity.

5. Avoid temperature transients.

6. New electrodes should be soaked several hours before use to improve stability.

7. Avoid electrical leakage in the high-impedance input circuit by preventing moisture buildup on the glass electrode body and lead. Electrical leakage is sometimes caused by the buildup of humidity and dust inside the instrument case.

8. Grounding problems: Many pH meters provide for separate grounding of the amplifier chassis and case. The ground of the amplifier is maintained at the glass-electrode potential by connections with feedback circuits.

9. Shorting of the electrodes causes polarization. The pH reading drifts under these conditions.

10. Colloids are sensitive to salt and may precipitate at the liquid junction as the result of the diffusion of the salt-bridge electrolyte or may form a film on the glass-electrode bulb. Slurries cause similar trouble.

11. Glass is attacked by soluble silicates and by acid fluorides. Special alkali-resistant electrodes are available.

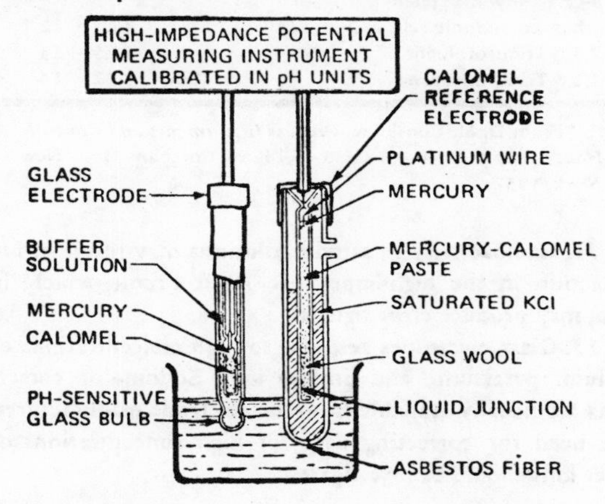

Fig. 4.43—A potentiometric pH-measuring system.

Table 4.21—Range of Use of Chemical pH Indicators*

Indicator	pH range
p-Naphtholbenzein	0 – 0.8
Picric acid	0.1 – 0.8
Malachite green oxalate	0.2 – 1.8
Quinaldine red	1 – 2
4-Phenylazodiphenylamine	1.2 – 2.9
m-Cresolsulfonephthalein (meta cresol purple)	1.2 – 2.8
Thymolsulfonephthalein (thymol blue)	1.2 – 2.8
p(*p*-Anilinophenylazo)benzenesulfonic acid sodium salt (orange IV)	1.4 – 2.8
o-Cresolsulfonephthalein (cresol red)	2 – 3
2,4-Dinitrophenol	2.6 – 4.4
3′,3″,5′,5″-Tetrabromophenolsulfonephthalein (bromophenol blue)	3 – 4.7
Congo red	3 – 5
Methyl orange	3.2 – 4.4
3-Alizarinsulfonic acid sodium salt	3.8 – 5
Propyl red	4.6 – 6.6
3′,3″-Dichlorophenolsulfonephthalein (chlorophenol red)	4.8 – 6.8
p-Nitrophenol	5 – 7
5′,5″-Dibromo-*o*-cresolsulfonephthalein (bromocresol purple)	5.2 – 6.8
3′,3″-Dibromothymolsulfonephthalein (bromothymol blue)	6 – 7.6
Brilliant yellow	6.6 – 7.9
Neutral red	6.8 – 8
Phenolsulfonephthalein (phenol red)	6.8 – 8.4
o-Cresolsulfonephthalein (cresol red)	7.2 – 8.8
m-Cresolsulfonephthalein (meta cresol purple)	7.4 – 9
Ethyl bis(2,4-dinitrophenyl)-acetate	7.5 – 9.1
Thymolsulfonephthalein (thymol blue)	8 – 9.6
o-Cresolphthalein	8.2 – 9.8
Phenolphthalein	8.3 – 10
Thymolphthalein	9.4 – 10.6
5-(*p*-Nitrophenylazo)salicylic acid sodium salt (alizarin yellow R)	10 – 12
p-(2-Hydroxyl-l-naphthylazo)benzenesulfonic acid sodium salt (orange II)	10.2 – 11.8
p-(2,4-Dihydroxyphenylazo)-benzenesulfonic acid sodium salt	11.2 – 12.7
2,4,6-Trinitrotoluene	11.5 – 13
1,3,5-Trinitrobenzene	12 – 14

*From D. M. Considine, *Process Instruments and Controls Handbook*, p. 6-104, McGraw-Hill Book Company, Inc., New York, 1957.

12. Radioactivity in sample solutions may result in ion collection in the high-impedance input circuit, which, in turn, may produce error signals.

13. Glass electrodes respond to high concentrations of sodium, potassium, and lithium ions. Sodium-ion corrections are usually available from the electrode manufacturer. The need for correcting data for high concentrations of other ions should be investigated.

4-7.5 Hardness

(a) Definition. Dissolved salts of calcium and magnesium impart the property of "hardness" to water. Hardness is characterized by the formation of insoluble precipitates, or curds, with soaps. Temporary hardness, caused by calcium and magnesium bicarbonate, is removed by boiling, which causes these salts to decompose, liberating CO_2 and precipitating carbonates. Permanent hardness is due to the sulfates, chlorides, and all the soluble calcium and magnesium salts other than the bicarbonates. Permanent hardness is not removed by boiling.

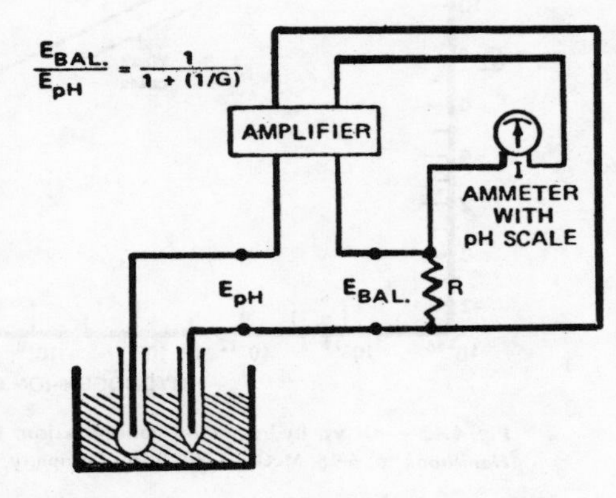

Fig. 4.44—Feedback-type pH meter.

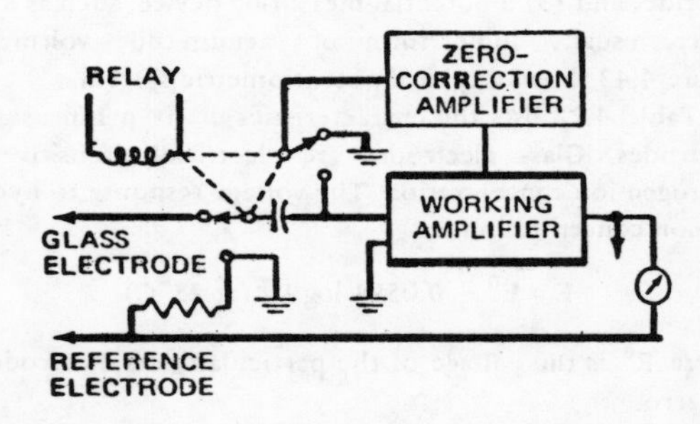

Fig. 4.45—Direct-current feedback pH amplifier with automatic zero adjustment.

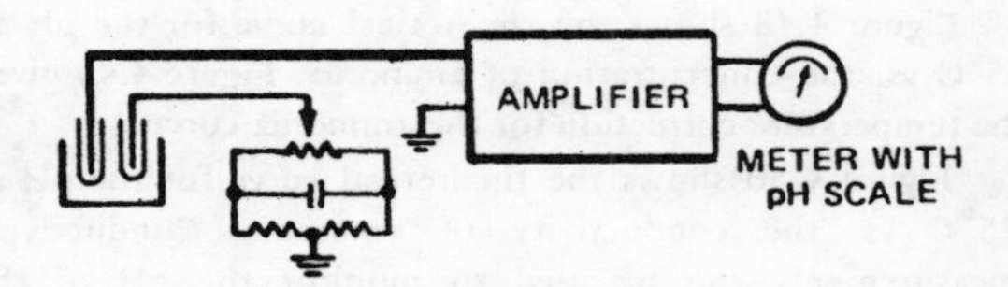

Fig. 4.46—Direct-current amplifier direct-reading pH meter.

(b) Measurement Techniques. *Total hardness* is determined by adding a standard soap solution to a measured amount of sample and shaking the mixture vigorously between additions of the soap solution until an unbroken

Table 4.22—Characteristics of pH-Measuring Electrodes*

Electrode type	Operating range			Limitations	Advantages
	pH	Temp., °C	Pressure, psi		
Glass (pH-sensitive)	0–13	0–100	0–100	Has high internal resistance; requires shielding, excellent insulation, and electrometer-type voltmeter; error occurs in high conc. of alkali; attacked by fluoride solutions	Wide pH and temperature range; not affected by oxidizing or reducing solutions, dissolved gases, or suspended solids; not affected by moving liquids except at high velocity
Antimony (pH-sensitive)	4–11.5	0–60	Not limited	Electrode poisoned by Bi, As, Cu, Ag, Hg, and Pb; affected by some oxidizing and reducing solutions; tartrates and citrates cause errors; dissolved O_2 must be present to maintain pH-sensitive oxide coating; active surface must be periodically scraped and reformed	Very rugged and durable for use in abrasive slurries; has low cell resistance; shielding and special voltmeter not required
Quinhydrone (pH-sensitive)	0–8.5	0–37	Not limited	Limited pH range, "salt error," cannot be used in presence of oxidizing and reducing agents, "protein errors," quinhydrone may change pH of unbuffered solution	Simple electrode, low resistance
Hydrogen (pH-sensitive)	Not limited	Not limited	Atmospheric pressure	Cannot be used in presence of oxidizing or reducing agents; cannot be used in presence of elements below hydrogen; slow to reach equilibrium; large samples required	Standard of reference, no alkaline error
Calomel (reference)	Not limited	Life shortened at high temperatures	Atmospheric pressure or below except for special designs	No interferences except from contamination from high-pressure test solutions	Can be used with any pH-sensitive electrodes
Silver–silver chloride (reference)	Not limited		Atmospheric pressure or below except for special designs	Interference by contamination from high-pressure solutions	Mercury-free, may be used with any pH-sensitive electrodes

*From D. M. Considine, *Process Instruments and Controls Handbook*, p. 6-105, McGraw-Hill Book Company, Inc., New York, 1957.

lather is maintained for 5 min on the water surface. The volume of soap used is referred to a chart or multiplied by a factor. The result is expressed in parts per million.

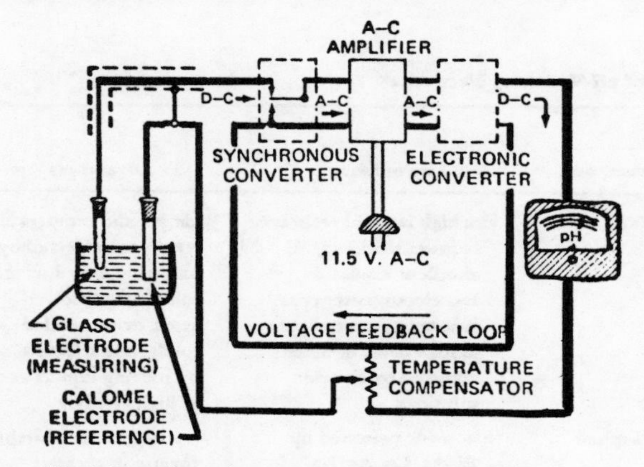

Fig. 4.47—Alternating-current amplifier pH meter.

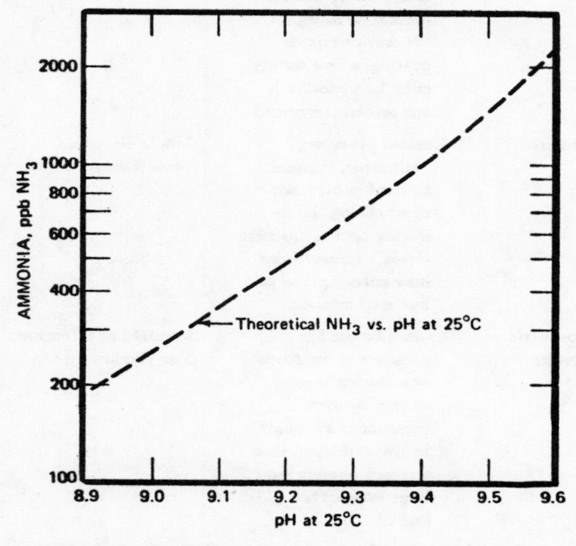

Fig. 4.48 —Theoretical relationship for ammonia concentration vs. pH at 25°C.

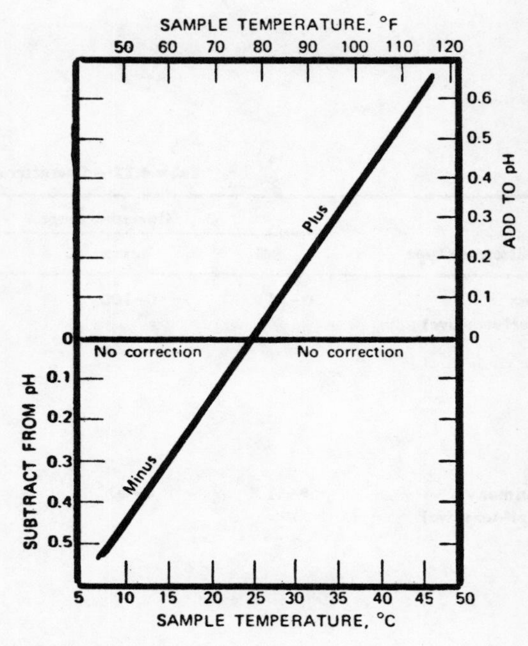

Fig. 4.49—Temperature correction to 25°C for pH (ammonia solutions).

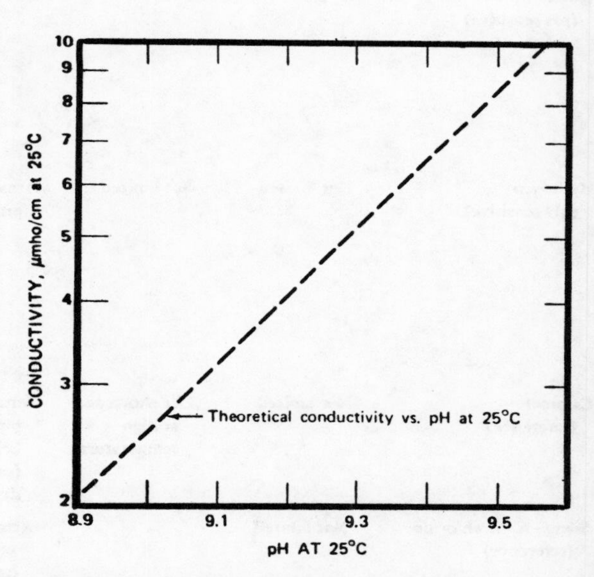

Fig. 4.50 —Theoretical relationship between pH and conductivity for ammonia.

Chloride concentration is determined by titrating a measured volume of sample with standard $AgNO_3$ solution, using potassium chromate as an indicator. The end point is red coloration.

Equivalent sodium sulfate determination is made by titrating with sodium hydroxide solution, with phenolphthalein as indicator, after adding an excess of benzidine sulfate to a measured sample. The benzidine sulfate precipitates the sulfate in the sample. After standing, filtering, and washing, the precipitate is titrated with NaOH. Turbidity-measuring instruments, described in pt. 4-7.6, can be used for sulfate determination. Barium chloride and hydrochloric acid added to the measured sample cause a white precipitate of barium sulfate to form. The resulting turbidity is measured and is an indication of sulfate concentration.

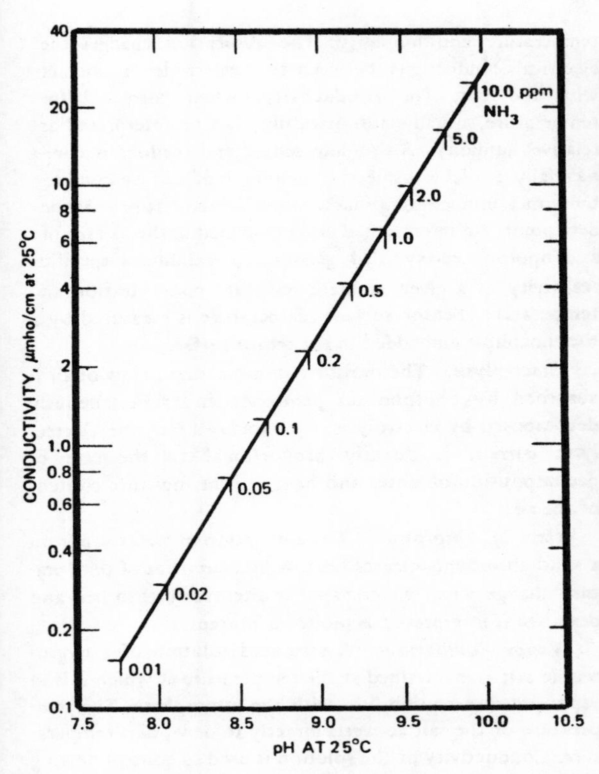

Fig. 4.51—Relationship between conductivity and pH for ammonia solutions.

4-7.6 Turbidity

A turbidity transmitter depends upon the principle that suspended solids in a liquid absorb and scatter part of any light passing through the liquid. The transmitter is an integral assembly of a light source, a flow tube, and a light-intensity detector (Fig. 4.52). The change in radiant energy reaching the detector varies the resistance in one arm of a Wheatstone bridge. A compensating filament in the second arm of the bridge corrects for ambient temperature.

4-7.7 Electrical Conductivity

(a) Discussion. Electrical conductivity (see also ¶. 4-9.1) is related to the concentration of total dissolved solids. The purest water is a very poor conductor. Pure water has a conductivity approaching very closely the theoretical minimum of approximately 0.05 μmho/cm at 25°C, which is due to the dissociation products of water itself.

(b) Measurement Techniques. Alternating current is generally used in a measuring system because direct current produces progressive changes in concentration near the electrodes. Also, products of the electrode reactions (with direct current) may set up a voltaic cell and an appreciable back emf. Figure 4.53 shows an a-c Wheatstone bridge circuit for measuring electrical conductivity. In the figure

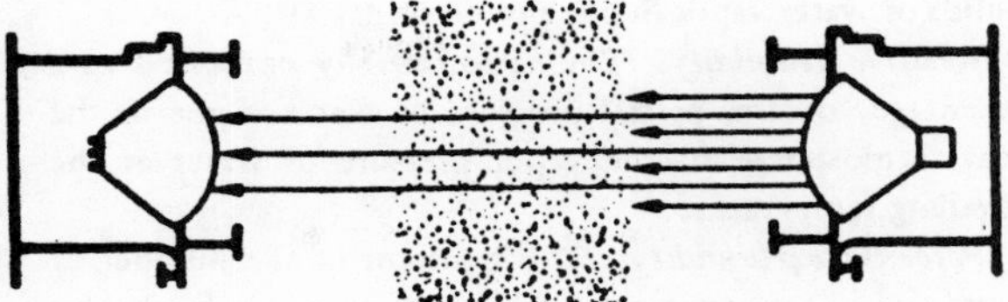

Fig. 4.52—Turbidity measurement. The liquid is passed between the light source (right) and the bolometer (left). The amount of radiant energy detected by the bolometer has a definite relation to the concentration of suspended solids.

R_x is the resistance of the electrolyte measured between two electrodes of a conductivity cell, R_3 and R_4 are end resistors whose function is to establish the limits of bridge calibration, and R_5 is a calibrated slidewire which does not enter into the arms of the bridge, and therefore variable values cause no error in bridge readings. The condition for balance of the Wheatstone bridge is $A/B = R_5/R_x$, and this

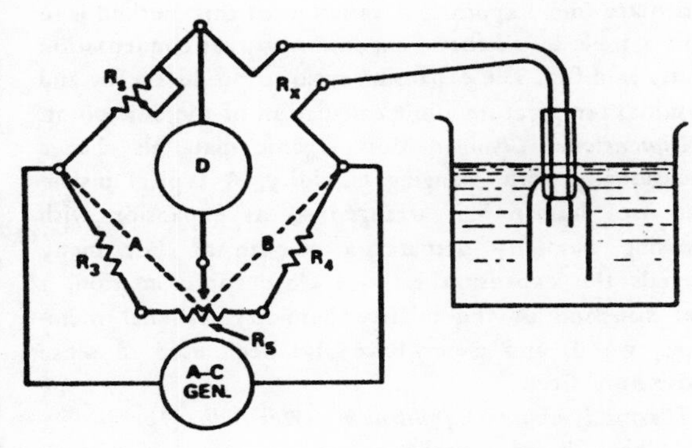

Fig. 4.53—Alternating-current Wheatstone bridge.

condition is indicated by no current flow through detector D (a galvanometer or microammeter).

4-7.8 Specific Weight of Compressed Water

The specific weight of water varies markedly with temperature. The effect of pressure on specific weight is less dramatic but still significant. The curves of Fig. 4.54 illustrate the effects of temperature and pressure on the specific weight (lb/ft³) of water.

4-8 GAS PROPERTIES SENSING

4-8.1 Humidity and Dew Point

(a) Definitions. *Absolute Humidity.* The number of

pounds of water vapor in one pound of dry air.

Relative Humidity. The ratio, usually expressed as a percentage, of the partial pressure of water vapor in the actual atmosphere to the vapor pressure of water at the prevailing temperature.

Percentage Humidity. The quotient of the number of pounds of water vapor carried by 1 lb of dry air divided by the number of pounds of water vapor which 1 lb of dry air would carry if it were completely saturated at the same temperature, multiplied by 100.

Dew Point. The temperature at which a given mixture of air and water vapor is saturated with water vapor.

Dry-Bulb Temperature. The temperature of an atmosphere. The qualification "dry-bulb" is used to distinguish the normal temperature measurement from the temperature measured by the wet bulb.

Wet-Bulb Temperature. The dynamic equilibrium temperature attained when the wetted surface of an object of small mass (bulb of a thermometer) is exposed to an air stream. Evaporation of water causes cooling, which is counterbalanced by heat absorbed from the air.

(b) Measurement Methods. *Condensation.* The surface in contact with the atmosphere is cooled until condensate (dew) appears. A variation of this method is to cool a sample by adiabatic expansion so that condensation appears as a fog. The expansion ratio to produce a fog and the initial temperature allow calculation of the dew point.

Dimensional Change. Most organic materials change dimensionally with changing humidity. A typical instrument uses human hair arranged so its expansion with increasing humidity actuates a mechanism. For many materials the expansion is, to a close approximation, a linear function of the relative humidity. Animal membranes, wood, and paper have also been used to sense relative humidity.

Thermodynamic Equilibrium (Wet-Bulb Thermometer). The bulb of a thermometer is wrapped in a cloth wick that is kept wet with water. The wrapped bulb is exposed to an air stream, and the temperature observed is the wet-bulb temperature. This reading, in combination with a reading of the air temperature (dry-bulb temperature) is a measure of the moisture content of the air. Variations of this basic design use ceramic sleeves instead of cloth wicks. In one design, the wick is eliminated, and the temperature of the air stream is measured after cooling by saturation from a water spray. The basic design is known as a wet- and dry-bulk psychrometer. A sling psychrometer has glass thermometers in a frame designed to be swung through the air rapidly to secure sufficient air velocity.

Absorption (Gravimetric). A measured volume of air is passed through a water-absorbing material, such as phosphorous pentoxide. The gain in weight of the absorbent is the moisture content of the known volume of air. In a variant of this technique, the change in pressure when the absorbent is brought into contact with air in a sealed vessel is measured.

Absorption (Conductivity). The amount of water absorbed by a quantity of a hygroscopic salt varies with temperature and humidity. The absorption changes the electrical conductivity between two electrodes in contact with the salt. The conductivity, when corrected for temperature, usually automatically, can be interpreted as relative humidity. An on-line sensor and readout is commercially available using the technique of surface conductivity measurement of an inert non-conductor at or near the dew point. An intermeshed grid embedded in the surface of a nonporous epoxy-filled glass cloth exhibits a specific resistivity at a given ambient moisture concentration and temperature. Sensor surface temperature is measured by a thermocouple embedded in the sensor surface.

Electrolysis. The moisture in a measured flow of air is absorbed by phosphorous pentoxide and simultaneously decomposed by electrolysis. By Faraday's law, the electrolytic current is directly proportional to the rate of decomposition of water and hence to the moisture content of the air.

Heat of Absorption. The absorption of water vapor on a solid absorbent releases heat. A measurement of temperature change when water vapor is alternately absorbed and desorbed is interpreted as moisture content.

Vapor Equilibrium. A saturated solution of a hygroscopic salt is maintained at the temperature at which it is in vapor-pressure equilibrium with the atmosphere. The temperature of the salt converts directly to dew-point temperature. Conductivity of the solution is used to control heating and to maintain the equilibrium temperature.

Absorption (Infrared). Water vapor and most other compounds absorb radiation in certain portions of the infrared region. A measurement of the infrared absorption can be interpreted in terms of moisture content.

4-8.2 Chemical Composition

(a) Physical Methods. *Condensation and Fractional Vaporization.* Separation of condensable vapors, generally in groups. Identification and quantitative evaluation is performed by other methods. Curves of vapor pressure vs. temperature of possible components must be known.

Fractional Distillation. Separation, identification, and quantitative evaluation of condensable hydrocarbons even in complex mixtures.

Adsorption or Absorption and Desorption (Chromatography). Separation, identification, and quantitative evaluation of many gases and vapors even in complex mixtures.

Diffusion. Separation of hydrogen and some isotopes.

Thermal Diffusion. Separation of some isotopes.

Electric Discharge. Separation of nonionizable gases.

(b) Chemical Methods. *Selective Absorption.* Separation and quantitative evaluation of gases and vapors already known. There is a need for selective reagents. Quantitative analysis can be accomplished by (1) volumetric or barometric methods, (2) gravimetric methods, (3) titrimetric methods, (4) electrical conductivity, (5) colorimetry, or (6) calorimetry.

(c) Combustion Analysis. *Fractional Combustion.* Separation, identification, and quantitative evaluation, mainly

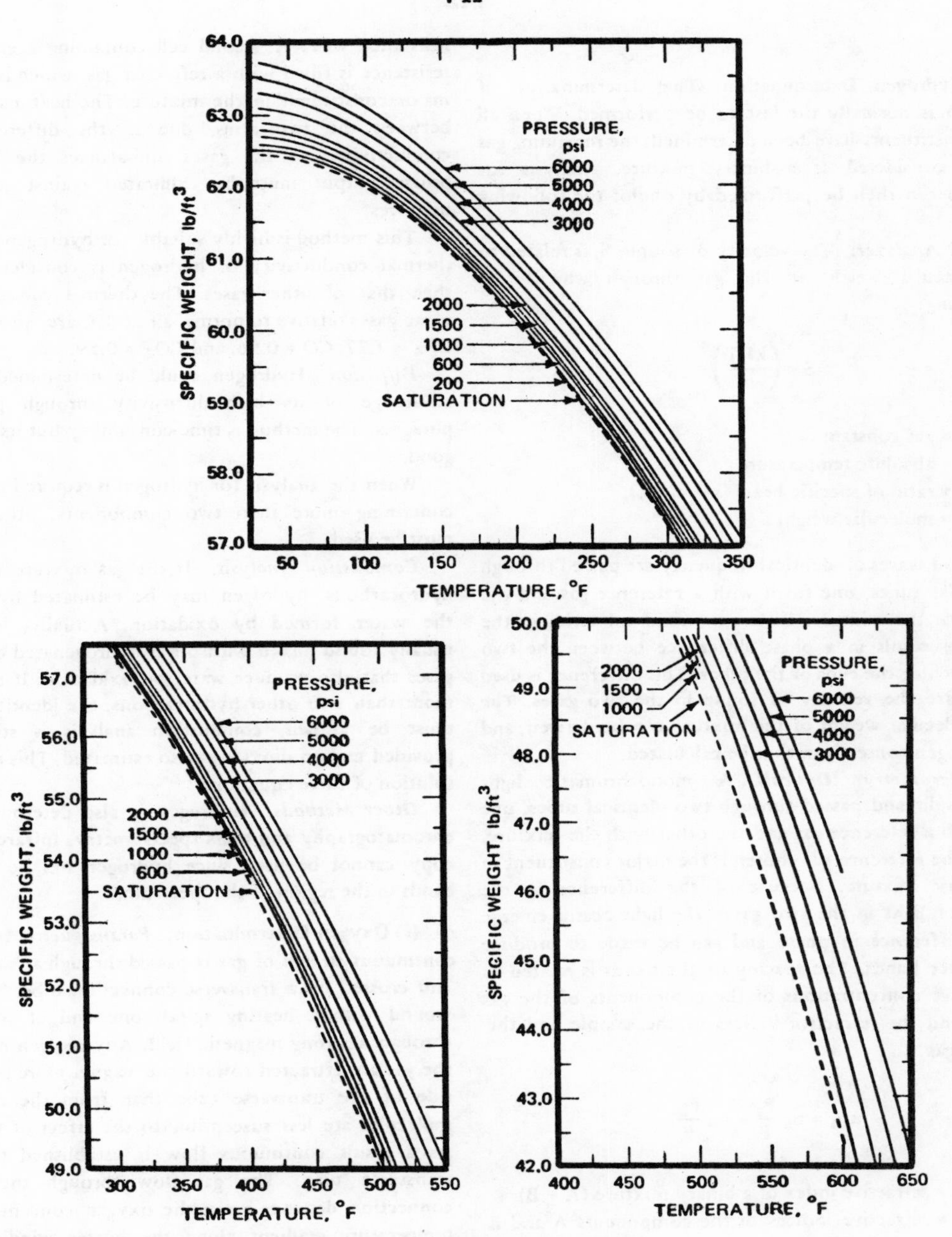

Fig. 4.54—Effects of temperature and pressure on the specific weight of water.

of H_2, CO, and hydrocarbons.

Complete Combustion. Quantitative evaluation of H_2, CO, and hydrocarbons.

(d) Absorption of Electromagnetic Radiation. *Magnetic Susceptibility.* Quantitative evaluation of O_2, NO, ClO_2, and NO_2, not mixed with each other. Mainly used for O_2.

Visible. Quantitative evaluation of colored gases, not mixed with each other (NO_2, Cl, etc.)

Ultraviolet. Quantitative evaluation of O_3, NO, C_6H_6, $C_6H_5(CH_3)$, etc.

Infrared. Identification and selective quantitative evaluation of CO, CO_2, hydrocarbons, NH_3, SO_2, SO_3, etc.

Visible Spectroscopy. Identification of various substances; quantitative evaluation doubtful.

Ultraviolet Spectroscopy. Identification of various substances; quantitative evaluation doubtful.

Infrared Spectroscopy. Identification and quantitative determination of H_2O, CO, CO_2, hydrocarbons, organic compounds, etc. Can detect composition of highly diluted mixtures.

Mass Spectrometry. Identification and quantitative evaluation of a large number of substances. Precise and capable of detecting composition of highly diluted mixtures.

(e) Hydrogen Determination. The determination of hydrogen is normally the last to be performed. When all other constituents have been determined, the remaining gas can be considered as a binary mixture. Analysis for hydrogen can then be performed by one of the following methods:

Sonic Analyzer. The velocity of sound, S, is related to the molecular weight of the gas through which it is propagating:

$$S = \left(\frac{kRT}{m}\right)^{1/2}$$

where R = gas constant
T = absolute temperature
k = ratio of specific heats ($k = c_p/c_v$)
m = molecular weight

Two sound waves of identical frequency are passed through two similar tubes, one filled with a reference gas and the other with the mixture. Different sound velocities in the two tubes result in a phase difference between the two waves reaching the ends of the tubes. This difference is used to compare the velocity of sound in the two gases. The mean molecular weight of the mixture can be derived, and the hydrogen content can thus be calculated.

Interferometry (Optical). A monochromatic light beam is split and passed through two identical tubes, one filled with a reference gas and the other with the mixture. Usually the reference gas chosen is the major constituent in the binary mixture. Because of the difference in the velocity of light in the two gases, the light beams emerge with a difference in phase and can be made to produce interference bands. The spacing of the bands is related to the relative concentrations of the components of the gas mixture and the refractive indices of the sample and the reference gas:

$$n_{ab} = n_a \frac{P_a}{P} + n_b \frac{P_b}{P}$$

where n_{ab} = refractive index of a binary mixture (A + B)
n_a and n_b = refractive indices of the components A and B
P = pressure of the gas mixture
P_a and P_b = partial pressures of the components

When the refractive index of the mixture is known, the partial pressure of one component can be deduced and hence the concentration can be estimated. Interferometers are capable of great accuracy provided they are used with skill and provided pressure and temperature corrections are applied.

Thermal Conductivity. In binary mixtures thermal conductivity can vary linearly with the concentration of one component. Absolute evaluation of thermal conductivity is very difficult, and normally only relative values are determined. For this purpose a hot-wire Wheatstone bridge is used. Sample gas passes through a cell that contains a resistance wire. A second cell containing a compensating resistance is filled with a reference gas, which is usually the major constituent in the mixture. The heat-loss difference between the two arms, due to the different thermal conductivities of the gases, unbalances the bridge. The bridge output must be calibrated against standard gas mixtures.

This method is highly suitable for hydrogen because the thermal conductivity of hydrogen is considerably higher than that of other gases. The thermal conductivities of some gases relative to normal air at 0°C are: air = 1, H_2 = 7, CH_4 = 1.27, CO = 0.96, and CO_2 = 0.59.

Diffusion. Hydrogen could be determined by taking advantage of its high diffusivity through porous diaphragms. The method is time-consuming, but its accuracy is good.

When the analysis for hydrogen is required in a mixture containing more than two components, other methods must be used.

Combustion Analysis. If the gas mixture contains no hydrocarbons, hydrogen may be estimated by measuring the water formed by oxidation. Actually, hydrogen is usually found mixed with other hydrogenated combustible gases that also produce water on oxidation. If there are no more than two other hydrocarbons, the identity of which must be known, combustion analysis is still possible provided carbon dioxide is also estimated. This requires the solution of three equations.

Other Methods. Hydrogen is also determined by gas chromatography and mass spectrometry. Infrared spectroscopy cannot be used since hydrogen has no absorption bands in the region of the spectrum.

(f) Oxygen Determination. *Paramagnetic Analyzer.* A continuous stream of gas is passed through an annular tube and crossed by a transverse connection tube. The latter is wound with a heating spiral, one end of which passes through a strong magnetic field. Any oxygen molecules in the gas are attracted toward the magnet more from the left side of the transverse tube than from the right. Warm molecules are less susceptible to the effect of the magnet. As a result continuous flow is established through the transverse tube. The gas flow through the transverse connection depends upon the oxygen concentration. The temperature gradient along the heater winding depends upon gas flow. Therefore, oxygen concentration is measured by temperature gradient.

Electrochemical Gas Analyzer. A heated zirconium oxide tube sets up a current when there are different concentrations of oxygen in two gases that flow inside and outside the tube. The unknown gas is passed inside the tube, and the reference gas (air) is passed outside the tube. The electrical output of the tube is proportional to the logarithm of the ratio of the oxygen concentration of the two gases. The advantages of the method are that it is accurate, requires no fuel, is unaffected by high SO_2 or SO_3 concentrations, is unaffected by high CO_2 concentrations, reads net O_2, and has a fast response.

(g) Summary of Methods Used for Gas Analysis. Table 4.23 summarizes the various methods that may be used for gas analysis. Instrumental procedures for gas analysis tend to supersede classical laboratory methods. Laboratory methods are often used as standard references. Instruments are used as the ordinary tools of the investigation.

For simple analysis (CO_2, CO, O_2, H_2, and CH_4), the nondispersive infrared analyzer, the magnetic oxygen analyzer, and the sonic analyzer for hydrogen are suitable. For mixtures of increased complexity, chromatography is recommended. It is economical and quick. It can fractionate mixtures into single or groups of components, which can be useful before applying infrared or mass spectrometry.

(h) Mass Spectrometry. The material to be studied is subjected to an ionizing process, separated according to mass by electromagnetic means, and the resulting mass spectrum is analyzed, quantitatively and qualitatively, by comparing it with the spectra of known calibrating materials.

Ions are produced by four methods: (1) electron bombardment, in which the unknown, if it is gaseous, is bombarded in an evacuated chamber by electrons; (2) direct emission of ions from the surface of some solid materials by heating a filament that is covered with a thin layer of the material to be analyzed; (3) the crucible method, in which materials (e.g., halides) are evaporated from a small furnace, and subsequently the vapor is ionized by electron bombardment; and (4) the spark method, in which a high-voltage spark between electrodes of the material to be analyzed yields ions of that material. Positive ions are accelerated by electric fields between a system of electrodes. Ions are focused in their passage through slits or apertures in the electrodes. The ion source in a mass spectrometer is a combination of the region where the ions are generated (this region usually has an electron gun to provide the electron bombardment) and the ion-accelerating region.

The ions are separated by one of these four basic methods: a magnetic analyzer (masses separate according to their momenta in a magnetic field); a time-of-flight analyzer (ions with same kinetic energy but different masses have velocities inversely proportional to the square root of the mass and become separated if injected into a field-free "drift" region); a linear-accelerator analyzer (accelerated ions are segregated by electrostatic deflection); and an ion-resonance analyzer (ions move in a region where a radiofrequency electric field is set up at right angles to a magnetic field and, if the frequency is in resonance with the spiralling ions, the ions spiral out of the field and are collected).

4-9 OTHER SENSORS

4-9.1 Electrical Conductivity

(a) Discussion of Applications. Electrical conductivity of a solution is a measure of all ions present. Pure water is a very poor conductor. The conductivity of a water solution is, in practice, almost exclusively due to ions other than the hydrogen (H^+) and hydroxyl (OH^-) ions. Figure 4.55 shows the conductivity of certain electrolytes as a function of their concentration in water.

Most practical applications of electrical-conductivity measurements fall into one of the following categories:

1. Concentration in simple water solutions. Common examples are sodium chloride, sodium hydroxide, and sulfuric acid. In such cases the concentration–conductivity curve must be known in advance, or the system must be experimentally calibrated.

2. Boiler steam quality detection. The exact nature of the electrolyte is usually less important than its magnitude. Nuclear installations require extremely pure feedwater.

3. Measuring the extent of a reaction. Reactions such as precipitation, neutralization, and washing soluble electrolyte from insoluble materials can be monitored by conductivity measurements. These procedures require calibration or a comparison between conductivities of streams before and after the reaction.

4. Detecting contaminations. Leaks in heat exchanger with the resultant contamination. Any sudden change in conductivity of the heat-exchange medium is taken as leakage. Salt-water contamination of freshwater can be detected as well as breaks in condenser tubes.

(b) Measurement Methods. *Measuring Circuits.* The a-c Wheatstone bridge circuit (Fig. 4.53) is the most widely used technique. It is sensitive, stable, and accurate. An ohmmeter circuit (Fig. 4.56) can also be used. The current is a function of cell resistance; the system is sensitive to

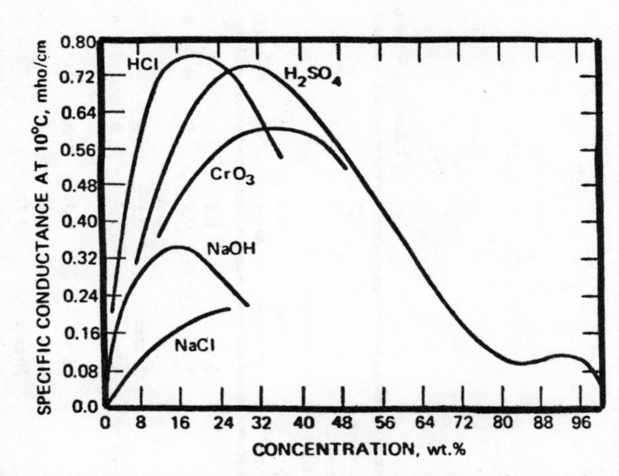

Fig. 4.55—Conductivity–concentration curves for certain electrolytes. (From D. M. Considine, *Process Instruments and Controls Handbook*, p. 6-159. McGraw-Hill Book Company, Inc., New York, 1957.)

voltage variations. In addition to these circuits, an a-c crossed-coil electrodynamometer can be used in a conductivity-measuring circuit. One of the two crossed moving coils responds to the current flow in the conductivity cell circuit; the other responds to the source voltage. It is a relatively simple technique, but it is not as accurate or sensitive as the Wheatstone bridge.

Conductivity Cells. The first criterion in selecting a conductivity cell is that the *cell constant* must be such that

Table 4.23—Summary of Gas-Analysis Methods*†

Method	Average sample size† (s.t.p.), cm^3	Average time required‡	Average accuracy†	H_2	O_2	CO_2	SO_2	SO_3	CO	Alcohols	Ethers	Aldehydes	Organic acids	Organic peroxides	CH_4	Other hydro-carbons
Fractional distillation	10^4	6 hr	0.5													A
Gas chromatography	10	5 min–1 hr	0.1 T.A.	B	B	B			B	A	A	A	A		A	A
Diffusion	200	20 min	0.1	A												
Combustion and gravimetric analysis	1 to 4×10^3	6 hr	T.A.	B		A			A						A	A_1
Cambridge	10	16 hr	0.01	A	B	A			A			A_2			A	A_1
Modified Cambridge	10	3 hr	0.01			A			A			A_2			A	A_1
Simplified Cambridge	10	20 min	0.01			A										
Interferometry§	4000	5 min	0.01	A	B	A			A						A	
Sound velocity§	500	5 min	1	A	B	B									B	
Thermal conductivity bridge§	f.s.	i.d.	0.001	A		B			B							
Paramagnetic detector	f.s.	i.d.	0.05		A											
Nondispersive infrared	f.s.	i.d.	0.001			A	A		A	A		B	B		A	B
Infrared spectrometry	20	20 min–1 hr	0.1			B	B		B	A		B	B		A	A
Mass spectrometry	0.01	15 min–1 hr	T.A.	B	B	B	B		B	A	A	A	A	B	B	A
Special laboratory methods							A	A		A	A	A	A	A		

*From G. Tine, *Gas Sampling and Chemical Analysis in Combustion Processes*, p. 86, Pergamon Press, Inc., New York, 1961.

†Abbreviations: i.d. = instrumental delay (sec)

A = particularly suitable method

B = possible method

A_1 = undifferentiated estimation

A_2 = estimation of formaldehyde only

T.A. = trace analysis

f.s. = analysis can be performed also on flowing streams

‡The actual sample sizes, time consumptions, and accuracies depend upon the particular apparatus that is used and, in many instances, upon the gas to be detected.

§Only suitable for binary mixtures of known components.

the resistance of the solution under test falls within the limits of the cell range. When high electrolytic resistance is being measured, as in the determination of steam purity, a capacitative impedance in series with the cell has a negligible effect on bridge readings; on the other hand, capacitative impedance in parallel impairs the sharpness of bridge balance. Impedance varies inversely with frequency. Therefore, low bridge frequencies are desirable when measuring high resistance. A relatively low cell constant, such as K = 0.1, has large electrodes close together and is suitable for measuring high resistance systems. Spreading the plates apart and constricting the electrolyte cross section increases the cell constant.

The mechanical features of conductivity cells are illustrated in Fig. 4.57. There are four basic types: *dip cells*, designed for dipping or immersing in open vessels; *screw-in cells*, designed for permanent installation in pipelines and tanks; *insertion cells* with removal devices, designed to permit removal of the element without closing down the line in which they are installed; and *flow cells*, glass or plastic with internal electrodes close to the wall to offer little resistance to the flowing medium. (In small sizes, the flow-cell tubes are connected to the system with rubber or plastic tubing; in large sizes standard pipe flanges are used.)

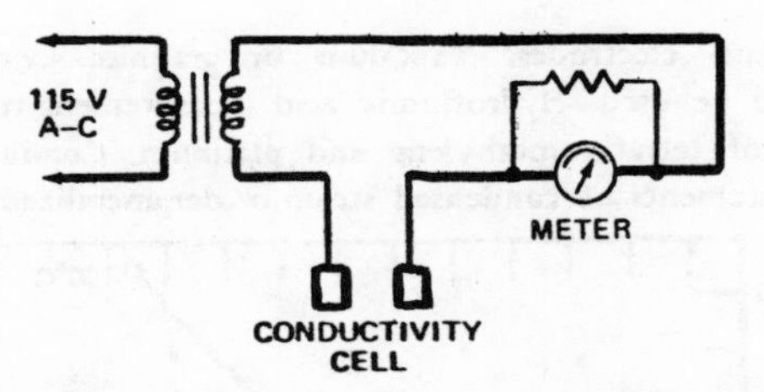

Fig. 4.56 — Conductivity-measuring system using a simple ohmmeter circuit.

Temperature, flow velocity, and presence of solids have significant effects on conductivity-cell performance. Temperature should be held as nearly constant as possible. The conductivity of most solutions increases about 2.5% for each 1°C rise in temperature. The flow velocity should be sufficient to ensure circulation of liquid between the electrodes. Entrained solids and high velocity increase the scouring effect. Low velocity can result in the accumulation of solids and the plugging of the cell chamber.

Chemical considerations are important. Strong electrolytes, such as hydrochloric acid, can slowly dissolve

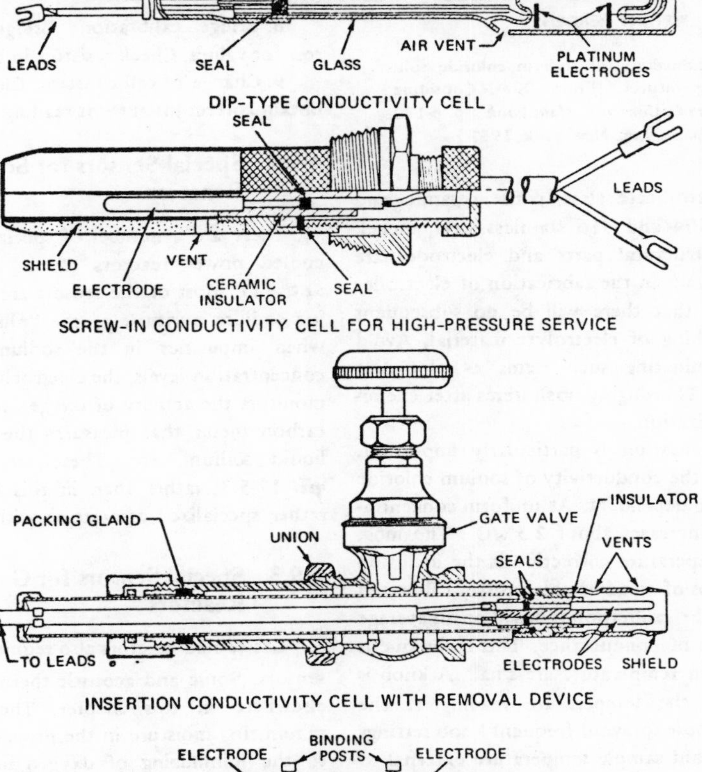

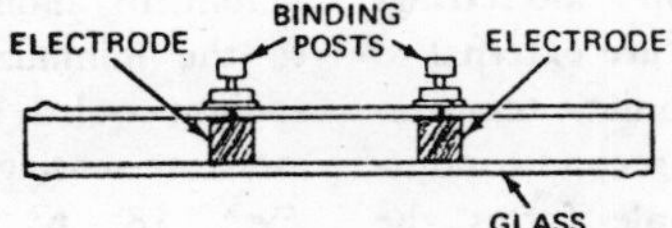

Fig. 4.57 — Basic types of conductivity cells.

platinum electrodes. Tantalum or graphite electrodes should be used. Hydrofluoric acid measurement requires cells of tetrafluoroethylene and platinum. Conductivity measurements of condensed steam or demineralized water

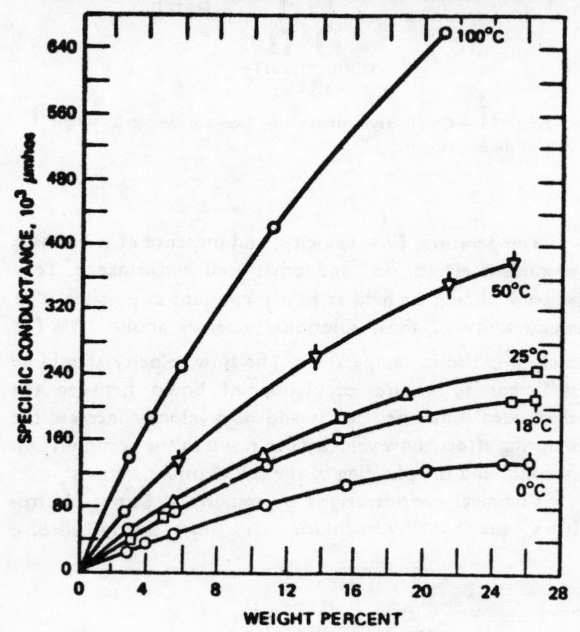

Fig. 4.58—Specific conductance of sodium chloride solutions for various temperatures. (From D. M. Considine, ***Process Instruments and Controls Handbook***, p. 6-170, McGraw-Hill Book Company, Inc., New York, 1957.)

can be made with borosilicate glass, dense ceramic, and most plastics. Types 304 and 316 stainless steel, nickel, gold, and platinum structural parts and electrodes are suitable. The use of fluxes in the fabrication of electrodes should be avoided so that there will be no subsequent contamination by leaching of electrolyte material. Avoid contamination by eliminating such items as pipe-joint compounds and dopes. Thoroughly wash items after chemical cleansing or replatinization.

Temperature compensation is particularly important. As shown in Fig. 4.58, the conductivity of sodium chloride solutions is temperature dependent. At uniform concentration the conductivity increases about 2.5%/C°. The most common means of temperature correction is the inclusion in one of the bridge arms of an adjustable resistor calibrated in temperature units. The calibration is based on an average temperature coefficient of conductance. This technique is used where variations in temperature are small. A knob is set to correspond to the temperature reading at the conductivity cell. It is best to avoid frequent knob settings by maintaining a constant sample temperature external to the cell by the use of throttling valves. Automatic temperature-compensation methods can be used. A second conductivity cell dipping into an isolated sample forms the variable-resistance arm of the bridge. Changes in temperature will affect both the thermal cell and the measuring cell to the same extent, canceling out the temperature effect. Other automatic temperature compensators include: bimetallic strip electrodes, expanding or contracting metallic bellows coupled to the variable resistor, a rising mercury column in a special thermometer to shunt the standard arm of the bridge, a resistance thermometer that automatically adjusts the standard arm resistance, and thermistors of high negative temperature coefficient.

(c) Sources of Error. Errors in conductivity measurements may be attributable to:

1. Insufficient circulation. Sluggish response is a symptom.
2. Contaminated cell. Sluggish response to great concentration changes.
3. Need of electrode revitalization. Characterized by broad null point or stepwise change in recorder.
4. Electrical leakage in conductivity cell, characterized by erratic results.
5. Leaching of electrolytes. Characterized by drift toward higher conductance.
6. Temperature errors. Characterized by drifting when concentration is known to be constant.
7. Reference temperature. Characterized by inability to obtain check reading from two different instrument systems even though each bridge and cell checks out against data.
8. Bridge calibration. Bridge will not check fixed resistor values. Check resistors in bridge circuit.
9. Change of cell constant. Characterized by inability to obtain correct instrument reading in known solution.

4-9.2 Special Sensors for Sodium-Cooled Reactors

There are a number of special sensors used in sodium-cooled power reactors. These are discussed in Sec. 17. Most of the sensors are associated with monitoring sodium purity, e.g., the "plugging meter" that senses when impurities in the sodium have reached specific concentration levels, the electrochemical oxygen meter that monitors the activity of oxygen in the sodium coolant, the carbon meter that measures the carburizing potential in liquid sodium, etc. These are discussed in Sec. 17, pt. 17-5.3, rather than in this chapter because of their rather specialized association with sodium-cooled reactors.

4-9.3 Special Sensors for Gas-Cooled Reactors

Gas-cooled reactors also require a number of specialized sensors. Sonic and acoustic thermometry have been noted elsewhere in this chapter. The critical importance of minimizing moisture in the primary gas coolant (analogous to the minimizing of oxygen in the sodium coolant in sodium-cooled reactors) has led to the development of moisture-sensing systems. These are discussed in Sec. 18, pt. 18-3.3. In addition to moisture, coolant purity is also required. This has led to the development of gas-analyzing systems in the high-temperature gas-cooled reactors. These are also discussed in pt. 18-3.3.

SECTION 5

NEUTRON-FLUX SIGNAL CONDITIONING

5-1 INTRODUCTION

Neutron-flux measuring channels in nuclear power plants are generally classified according to the neutron-flux range involved:

Start-up channel	10^0 to 10^5 neutrons cm^{-2} sec^{-1}
Intermediate channel	10^4 to 10^{10} neutrons cm^{-2} sec^{-1}
Power channel	10^6 to 10^{10} neutrons cm^{-2} sec^{-1}

The channels successively overlap one another (Fig. 5.1) to provide continuous measurements of the neutron-flux level at all power levels (see also sec. 2, pt. 2-2.3).

Control circuits are provided in conjunction with one or more of the channels. Reactor power is varied by the action of circuits associated with the neutron-flux-measuring equipment. High-power operation is controlled with the

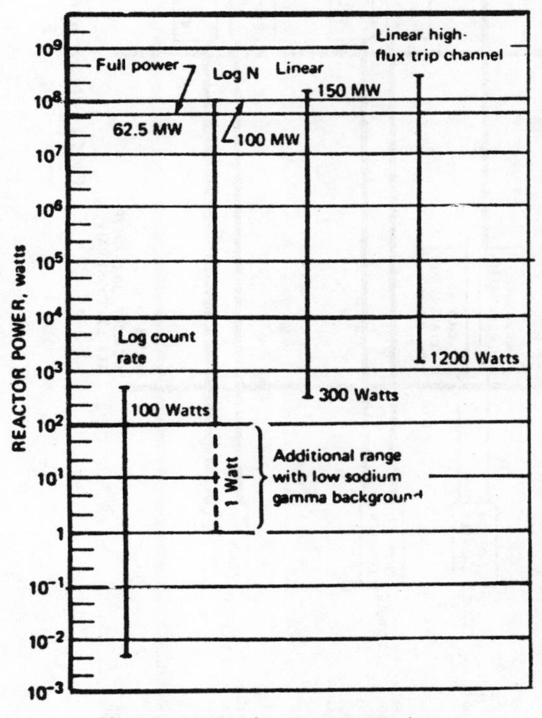

Fig. 5.1—Nuclear-instrument-range chart.

power channel. The neutron-measuring systems must be capable of providing alarm set points and annunciator units to warn the reactor operator.

In conjunction with the alarm features are the reactor-protection mechanisms. When an alarm set point is exceeded, appropriate automatic action must be taken to reduce the reactor operating power to a safe level. The type of protective circuits and action (see sec. 12) varies from reactor to reactor, from a simple reduction in reactivity (cutback) to a scram where essentially all the negative reactivity available is inserted.

The neutron-measuring equipment and circuits must be designed to minimize reactor downtime caused by unit maintenance, component failures, and spurious noise alarms. Redundancy and independent systems are used for each power-level channel.

5-2 START-UP CHANNEL

5-2.1 Introduction

A typical start-up channel, shown in block form in Fig. 5.2, consists of the following major components: (1) sensor, (2) pulse amplifier, (3) high-voltage power supply, (4) amplifier–discriminator unit, (5) count-rate meter with control functions, and (6) readout equipment.

5-2.2 Sensor

Because the neutron flux in a shutdown reactor is low, the output of the neutron detector (see sec. 2) is a series of pulses proportional to the neutron flux resulting from a neutron source in the reactor. The detector must have a high neutron sensitivity with a very low sensitivity to gammas.

The fission chamber is widely used because of its inherent ability to discriminate against gamma-generated signals. A fission chamber 2 in. in diameter and 12 in. long typically will have a neutron sensitivity of 0.7 (count/sec)/(neutron cm^{-2} sec^{-1}) with a gamma sensitivity of 4×10^{14} amp/(R/hr). The neutron signal generated in the fission-chamber circuit is of the order of 100 μV with a pulse width of approximately 0.5 μsec. The gamma pulses produced in the chamber are smaller in amplitude. A typical ratio of gamma to fission pulse height is of the order of 10^{-2} to 1. This ratio makes successful discrimination possible.

Gamma pulses generated in the chamber can become a problem when the detector is located in a gamma field of

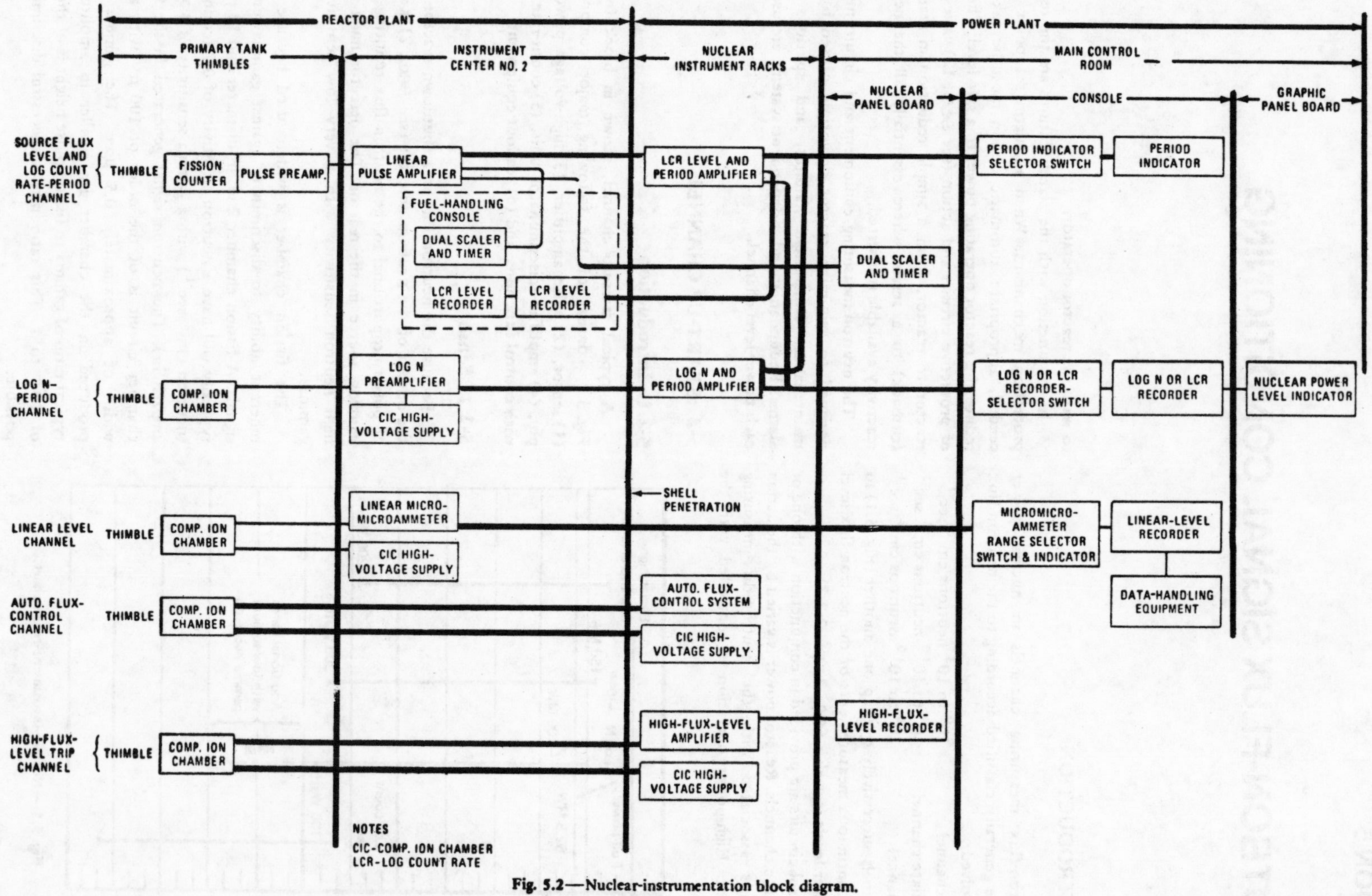

Fig. 5.2—Nuclear-instrumentation block diagram.

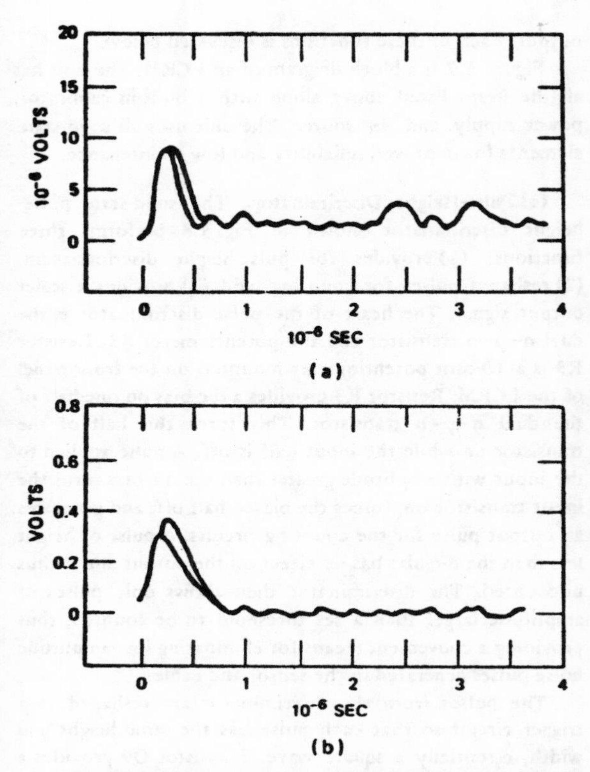

Fig. 5.3—Typical pulse shapes. (a) Background noise and low energy particles. (b) Preamplifier output pulse.

5×10^4 R/hr or more. A pulse pileup in the discriminator circuit causes gamma pulses to be counted as neutrons. Special techniques, such as pulse clipping or reshaping of the pulses, are used to minimize and protect against this. Typical pulse shapes are shown in Fig. 5.3.

5-2.3 Pulse Preamplifier

(a) Introduction. The pulse generated in the sensor must be amplified for transmission and for driving standard counting electronics. The ideal pulse preamplifier must have high gain, wide bandwidth, stability, and low noise characteristics.

Until recently fission sensor preamplifiers used vacuum tubes and were located as close to the sensor as possible. In the last few years, good-quality solid-state preamplifiers have replaced the vacuum-tube units. There are advantages and disadvantages with both types.

A vacuum-tube preamplifier can be placed near the fission detector, i.e., in the same radiation field as the sensor. However, radiation damage to the preamplifier components decreases their life and increases spurious noise so that maintenance is required. Vacuum-tube preamplifiers produce a large output pulse, ideal for counting equipment, but the vacuum tubes deteriorate with steady use and must be replaced regularly or suffer loss of gain. Generally, substantial maintenance is required per hour of successful operation.

Solid-state preamplifiers, usually charge-sensitive, can be located at considerable distance (up to 90 ft) from the sensor but, in any event, must be out of the radiation field. This makes maintenance more convenient and increases channel availability. However, this arrangement is susceptible to noise pickup in the cable between the sensor and the preamplifier. The output pulse from solid-state preamplifiers, typically 10V, is generally smaller than that from a vacuum-tube preamplifier. Because of the lower signal output, the counting equipment must be capable of accepting a low-level signal and processing it for use in the readout and control equipment that follows it in the channel.

(b) Vacuum-Tube Preamplifier. A vacuum-tube preamplifier is shown in Fig. 5.4. The amplification is provided by the four RCA 7586 (nuvistor) vacuum tubes. The preamplifier has a gain of 30, a rise time of 5×10^{-8} sec, and a fall time of 2×10^{-7} sec. Because vacuum tubes are used in this preamplifier, it can operate in a radiation field. The tube-filament heaters are d-c powered to minimize a-c noise pickup. The capacitors are ceramic, both for improved temperature stability and for reduced susceptibility to radiation damage. Capacitor C1 is used to isolate the detector high voltage from the preamplifier and to couple the sensor pulse to the preamplifier input circuit. The life expectancy of this preamplifier is approximately 600 hr in a 10^6 R/hr gamma flux (0.5- to 1.5-MeV gammas).

(c) Solid-State Preamplifier (Current Input). The solid-state preamplifier shown in Fig. 5.5 is a voltage amplifier, as opposed to the charge-sensitive amplifier to be discussed in the next subsection. The active elements are transistors. The input stage is a grounded-base transistor, which has low input impedance, high output impedance, wide bandwidth (high frequency), high bias stability, and good voltage gain.

The interconnecting coaxial cable between the sensor and the preamplifier is terminated into its characteristic impedance. A resistor in series with the emitter is selected to do this matching. This circuit eliminates cable ringing or reflections and provides a low impedance path for the current pulse generated in the sensor. The pulse is capacitance-coupled to the input stage, and the capacitor also blocks the sensor d-c voltage. The input current pulse from the detector is converted in transistor Q1 to a voltage pulse. The remainder of the preamplifier is a high-gain standard operational amplifier with feedback. The preamplifier can amplify pulses at a repetition rate of 10^6 Hz or 10^6 neutrons/sec. Typical preamplifier characteristics are:

Gain	0.5 volt/μA
Input impedance	50 to 120 ohms
Output	±3 volts into 50 ohms
Rise time	<5% for 200-nsec pulse width
Power	±15 volts at 40 mA

As noted above preamplifiers mounted at a distance (20 to 40 ft) have noise problems associated with cable pickup. Every effort must be made to shield against stray noise in the form of electrostatic or electromagnetic voltages between the sensor and the preamplifier.

In summary, the principal advantages of locating the preamplifier at a distance from the sensor are: (1) sensor cooling is not so critical, (2) maintenance is simplified, and (3) system availability is increased.

(d) Solid-State Preamplifier (Charge-Sensitive). A third type of preamplifier is a charge-sensitive unit (Fig. 5.6). This preamplifier is a fast-rise-time charge-sensitive preamplifier with dual-polarity output. The input signal is coupled to the pulse-shaping and amplifier input module A1 by capacitor C4, which blocks the high voltage. A Shockley diode connected to ground at the input to amplifier A1 protects the input from momentary breakdown of the detector or cable shorts. The output of A1 feeds a cable driver with dual-polarity output.

Amplifier module A1 is a special fast-pulse amplifier connected in a charge-sensitive configuration. This is accomplished by connecting a stable small-value capacitor from the amplifier output back to its inverting input. This negative feedback will attempt to keep the amplifier input very near zero volts, thus making it a virtual ground.

Incoming charge is collected on the input plate side of the feedback capacitor, C_f. This will cause some voltage shift at the input of A1 and, as a result of its inverting gain, a much larger inverted shift at the output. Negative feedback action through C_f will restore the input to its normal zero-volt level. The magnitude of the output voltage shift is directly proportional to the amount of charge received, $V = Q/C_f$.

It is very important that the amplifier have a very short rise time with respect to the incoming pulses, otherwise the virtual ground cannot be maintained, and charge may be diverted to ground through shunt protective devices or other circuit elements.

When only the feedback capacitor, C_f, is used, the circuit has become a charge-to-voltage converter. However, this configuration is limited by the ultimate saturation of its output as more and more charge is accumulated. For this reason the negative feedback resistor, R_f, has been added to discharge C_f between incoming input pulses. The parallel combination of these two components, C_f and R_f, forms the clipping-time constant of the preamplifier, $T = R_f \times C_f$. This clipping time also serves to provide the best pulse shape required by subsequent circuits.

The output of A1 is fed through C5 to Q1, which provides two outputs of opposite polarity by means of the output driver amplifiers. These driver amplifiers consist of parallel-connected Q2–Q3 and Q4–Q5 transistors that function as emitter followers to provide low output impedance suitable for driving any reasonable length of terminal coaxial cable.

5-2.4 Log Count-Rate Meter

The log count-rate meter (LCRM) is a pulse-counting component used to convert input pulses from the detector and preamplifier analog signal for use in control components. The LCRM has five basic functions: (1) pulse-height discrimination, (2) count-rate indication, (3) period indication, (4) scaler output signal, and (5) adjustable alarm output. Each of these functions is discussed below.

Figure 5.7 is a block diagram of an LCRM. The unit has all the items listed above along with a built-in calibrator, power supply, and test source. The unit uses all solid-state elements for improved reliability and low maintenance.

(a) Pulse-Height Discriminator. The solid-state pulse-height discriminator shown in Fig. 5.8 performs three functions: (1) provides for pulse-height discrimination, (2) reshapes pulses for counting, and (3) provides a scaler output signal. The heart of the pulse discriminator is the dual n–p–n transistor and the potentiometer R3. Resistor R3 is a 10-turn potentiometer mounted on the front panel of the LCRM. Resistor R3 provides a d-c bias on one-half of the dual n–p–n transistor. This turns this half of the transistor on while the input half is off. A pulse applied to the input with amplitude greater than the d-c bias turns the input transistor on, forces the biased half off, and generates an output pulse for the counting circuits. A pulse of height less than the d-c bias has no effect on the output and is thus uncounted. The discriminator then allows only pulses of amplitude larger than a set threshold to be counted, thus providing a convenient means for eliminating low-amplitude noise pulses generated in the sensor and cable.

The pulses from the discriminator are reshaped in a trigger circuit so that each pulse has the same height and width, essentially a square wave. Transistor Q9 provides a pulse output for a scaler, and transistor Q8 provides a pulse output for the LCRM counting circuitry.

(b) Count-Rate Indication. Figure 5.9 shows a Cook–Yarborough log circuit. The function of this circuit is to convert the constant-width and constant-height pulses into an analog signal. Since five or six decades of counts are to be covered by the LCRM, the circuit provides a logarithmic signal.

The diode pump is composed of CR5 to CR25, C7 to C27, and resistor R10 to R20. The purpose of the diode pump is to convert from a count rate to a d-c voltage. This is accomplished in the following manner. Consider components R10, C7, CR4, CR5, and CR6. Component CR6 is a pulse-coupling capacitor. Component CR4 is a positive-voltage clipper which prevents a positive voltage from appearing at the cathode of diode CR5. Diode CR5 and capacitor C7 form a low-impedance charge circuit for negative pulses. After the negative pulse has passed through CR5, the diode prevents C7 from discharging back through CR1 or the input circuit. Hence the capacitor C5 must discharge through resistor R10. The time constant is then R10 times C5. Should a second pulse follow the first pulse rapidly, the capacitor will not have time to discharge, and, as a result, the input to the d-c amplifier through R10 is essentially a d-c or rectified a-c voltage.

There are 11 such circuits with various time constants, varying from 40 sec to 8 msec. Resistors R10 to R20, C7 to C27, and CR5 to CR25 serve the same purpose.

The pulses are thus converted to a d-c voltage output proportional to different count rates. The d-c outputs for high count rates are provided for by the shorter time-

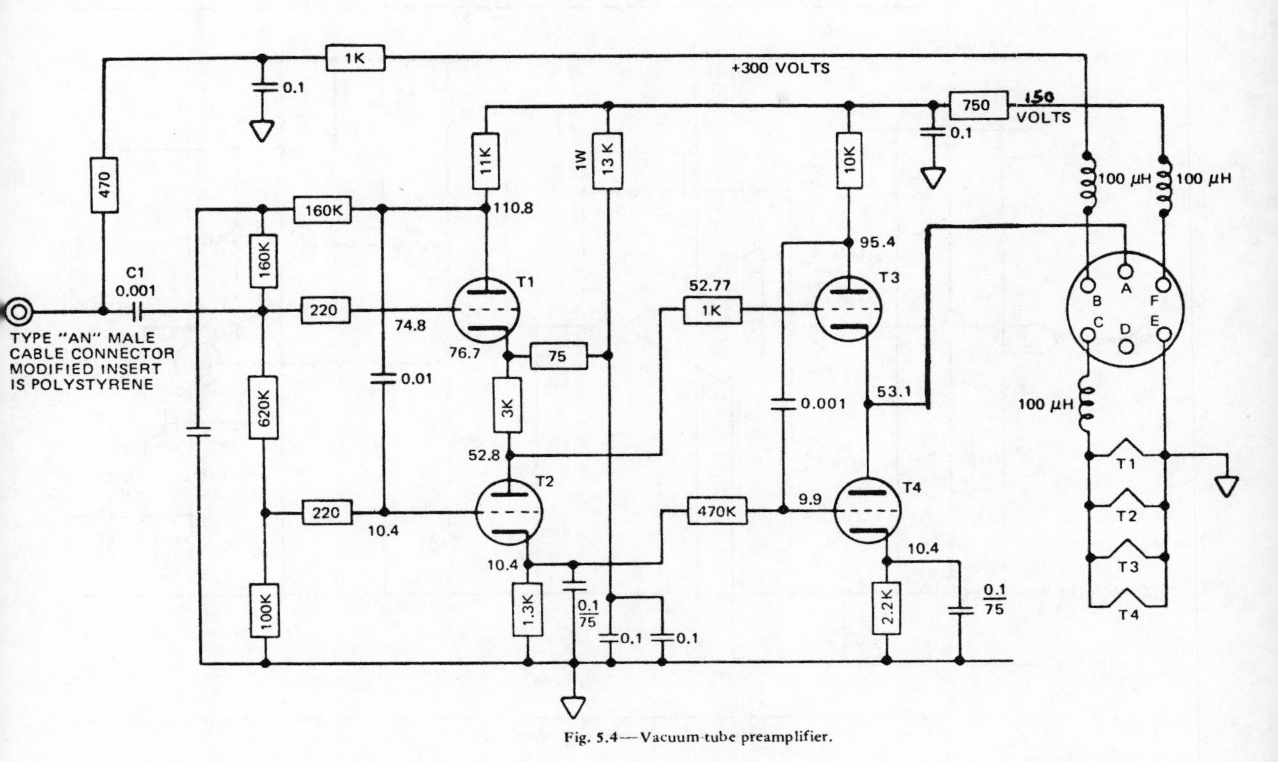

Fig. 5.4—Vacuum-tube preamplifier.

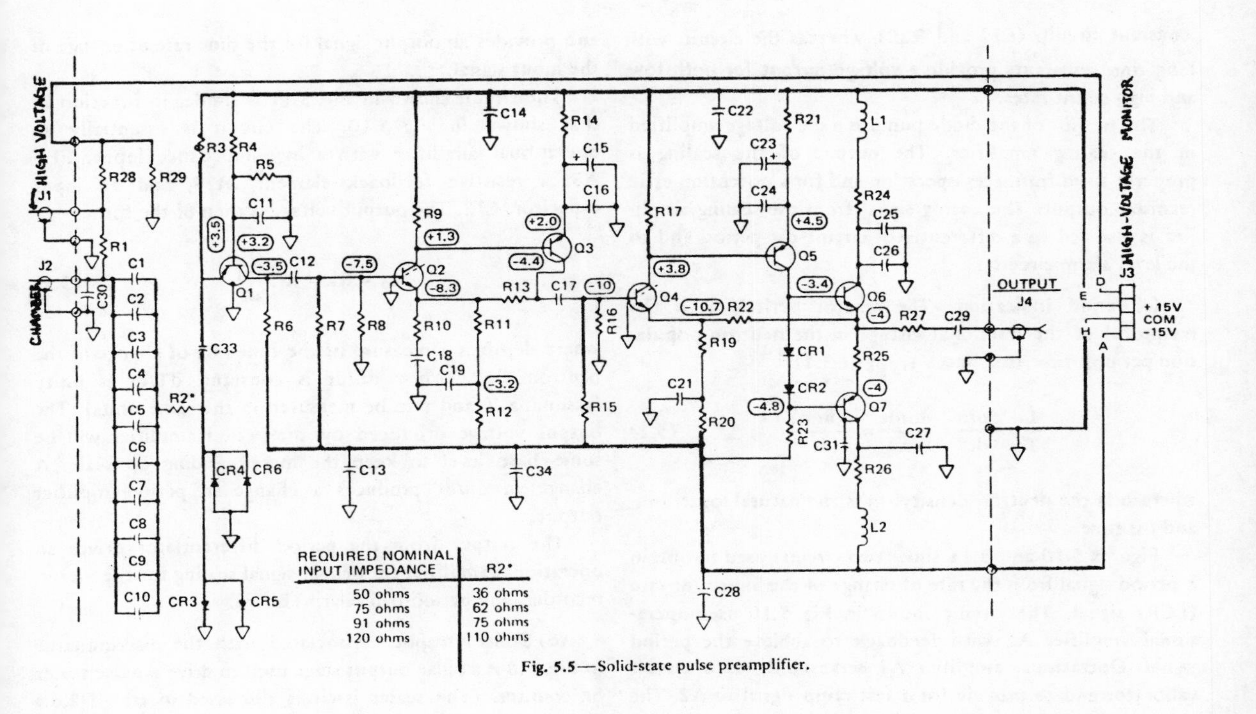

REQUIRED NOMINAL INPUT IMPEDANCE	R2*
50 ohms	36 ohms
75 ohms	62 ohms
91 ohms	75 ohms
120 ohms	110 ohms

Fig. 5.5—Solid-state pulse preamplifier.

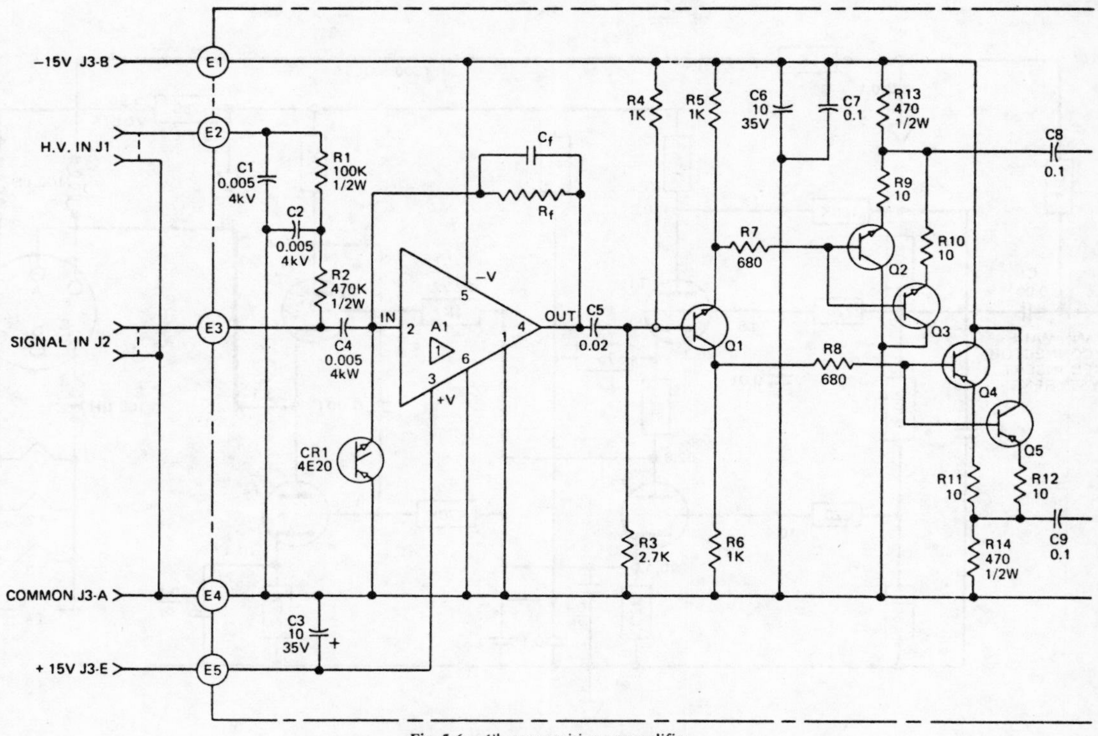

Fig. 5.6—Charge-sensitive preamplifier.

constant circuits (C27 and R20), whereas the circuits with long time constants provide a voltage output for both low and high count rates.

The output of the diode pump is a d-c voltage amplified in the scaling amplifier. The output of the scaling is properly sized for meter operation and for a potentiometric recorder output. The analog signal from the scaling amplifier is also fed to a differentiator circuit for period and to the level alarm circuits.

(c) Period Indication. The reactor period, T, is the reciprocal of the fractional change in the neutron population per unit time (see Sec. 1, pt. 1-3.1):

$$\frac{1}{T} = \frac{dn/n}{dt} = \frac{dn/dt}{n} = \frac{d(\ln n)}{dt} \tag{5.1}$$

where n is the neutron density, ln is the natural logarithm, and t is time.

Figures 5.10 and 5.11 show two circuits used to obtain a period signal from the rate of change of the log-count-rate (LCR) signal. The circuit shown in Fig. 5.10 uses operational amplifier A2 with feedback to achieve the period signal. Operational amplifier A1 serves only as a circuit calibrator and to provide for a test ramp signal to A2. The output signal from the diode pump is supplied to A2. Amplifier A2 differentiates any input-signal changes in level and provides an output signal for the time rate of change of the input signal.

The circuit shown in Fig. 5.11 is similar in function to that shown in Fig. 5.10. The circuit is essentially an operational amplifier with a high-impedance input, FET A8, a resistive feedback element, R29, and an input capacitor, C12. The output voltage is then of the form

$$e_0 = Rk = RC\frac{de_i}{dt} \tag{5.2}$$

where de_i/dt is a measure of the time rate of change of the neutron flux. When dn/dt is constant, dT/dt is unity (assuming T and t to be measured in the same units). The output voltage produced by the period amplifier will be some base level to keep the meter reading up scale. A change in dn/dt produces a change in period-amplifier output.

The output from the period differentiator drives an operational amplifier for proper signal scaling for the meter, recorder, and period-trip (alarm) board.

(d) Scaler Output. Associated with the discriminator in Fig. 5.8 is a pulse output stage used to drive a pulse scaler or counter. (The scaler itself is discussed in pt. 5-2.6.) The pulse-generating equipment consists of the circuit described in Sec. (a) above and also transistor Q9. Transis-

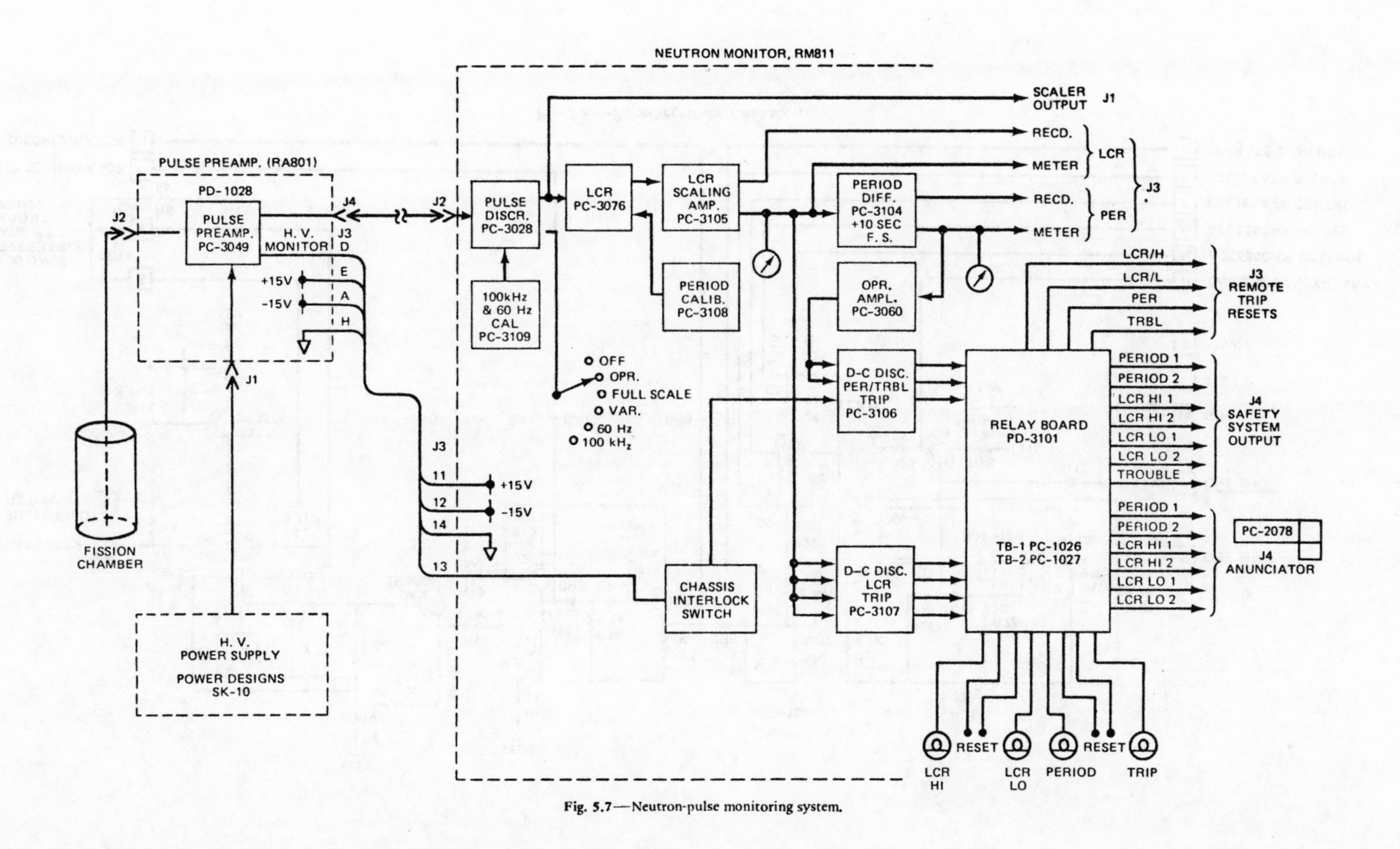

Fig. 5.7—Neutron-pulse monitoring system.

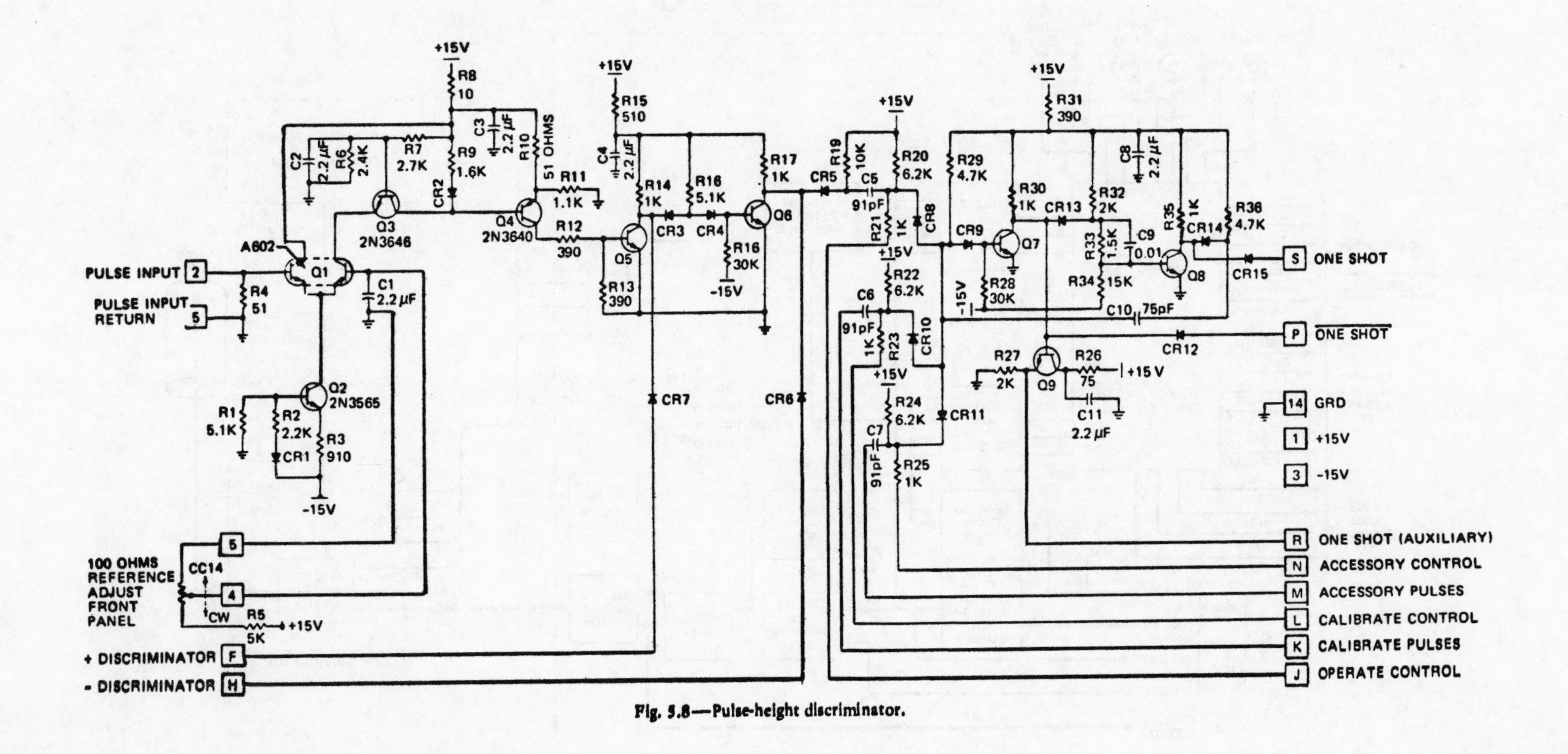

Fig. 5.8—Pulse-height discriminator.

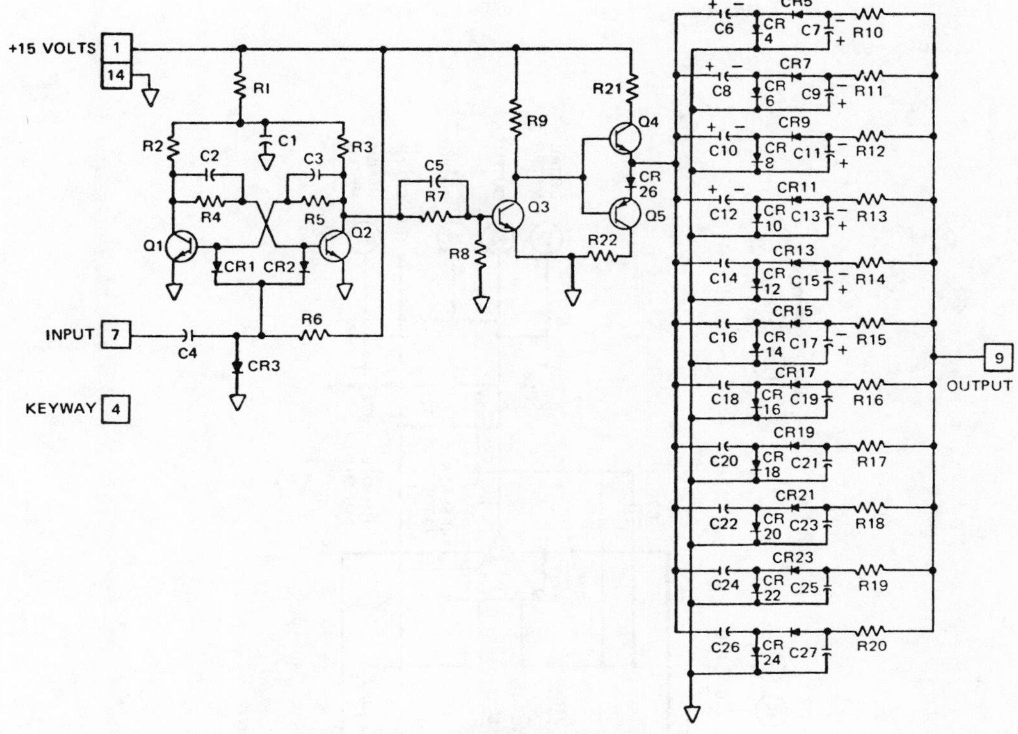

Fig. 5.9—Printed log diode circuit board.

tor Q9 is used to isolate the scaler output from the LCR circuit.

(e) Alarm Unit. The alarm unit provides a signal at selected and adjustable values, such as low count rate, high count rate, period, or loss of chamber high voltage. A solid-state alarm (trip) unit is shown in Fig. 5.12. Relays R1 and R2 are deenergized when the set point is exceeded. Transistors Q3 and Q7 must change from the saturated to the unsaturated (off) condition. The output of Q1 or Q2 will change any time the input-signal voltage levels exceed the set-point voltage as determined by potentiometers R28 and R29.

Each alarm function has two identical trip amplifiers similar to the one shown. The redundancy is necessary to decrease the probability of a failure in the operation of the alarm unit. This is very important for reactor protection since the output transistor Q3 or Q7 could short between collector and emitter, in which case the relay would not deenergize (fail to trip) when the set point was exceeded. The contacts of the trip relays are connected in series so that either relay deenergizing will cause an alarm.

5-2.5 Interconnecting Cables and Grounding

Proper cables must be used for interconnecting nuclear instrumentation to reduce noise and to transmit the best possible signal to the readout equipment. See Sec 10 for additional information on grounding, shielding, and selection of cables.

Noise-free cables must be used in the start-up instrumentation. Noise pulses will be amplified and counted as neutron pulses. Besides the inaccuracies in counting, noise bursts can cause period and level scrams at low reactor power levels.

When a vacuum-tube preamplifier is used, the sensor and preamplifier are usually connected together as a unit or with a few feet of coaxial cable. The cable used must have an impedance that matches the preamplifier input. The signal cable used between the vacuum-tube amplifier and the LCRM should be coaxial and match the impedances of the two units. If the cable is routed through high-noise areas, a triaxial cable should be used with the outer braid connected to ground.

Power-supply cables between the power supply and vacuum-tube preamplifier should be shielded conductors to minimize the noise level. The power-supply ripple voltage should not exceed 10 mV for satisfactory operation of the sensor and attached preamplifier.

The greater the distance between the detector and the preamplifier, the more important is the cable quality. For solid-state preamplifiers the importance of the cable cannot be overemphasized. There is one positive procedure for noise reduction: complete the installation, and, using one or more of the methods outlined in Sec. 10, experiment until the noise has been reduced to a minimum. Neutron sensors for use at high temperatures are under development. Cables and connectors for use with these sensors must operate at high temperature without producing noise

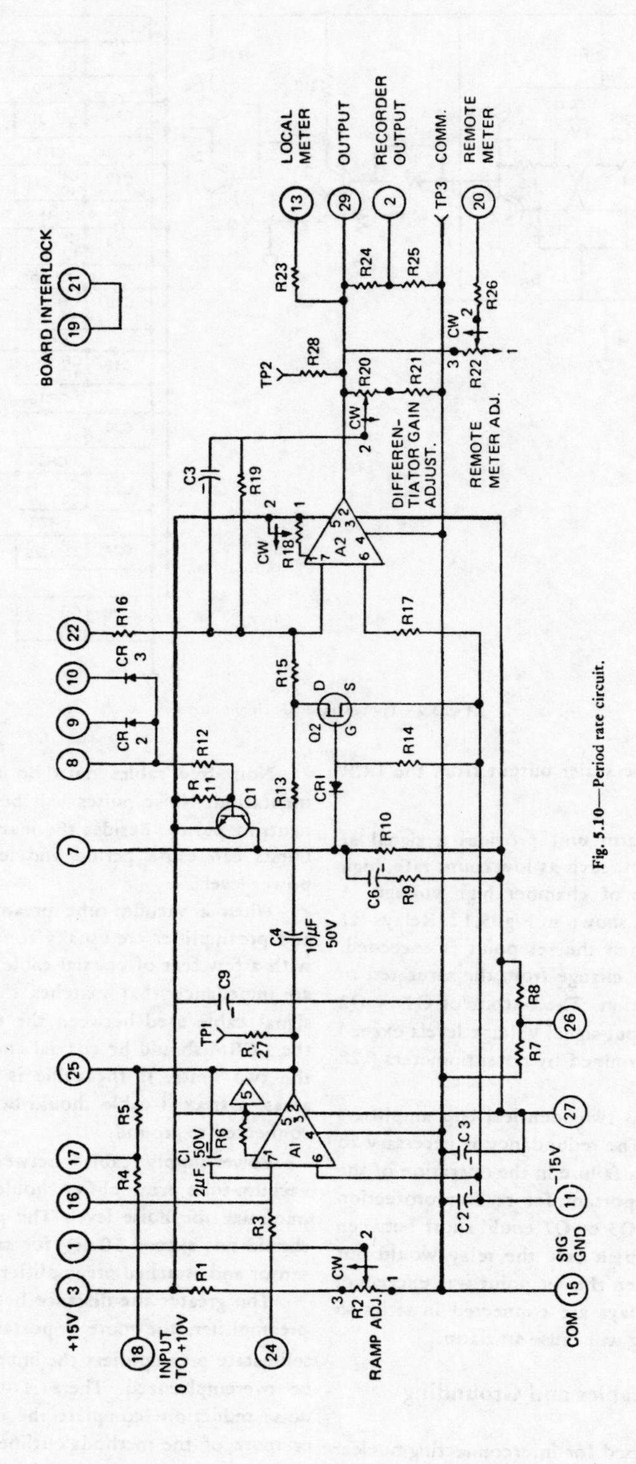

Fig. 5.10—Period rate circuit.

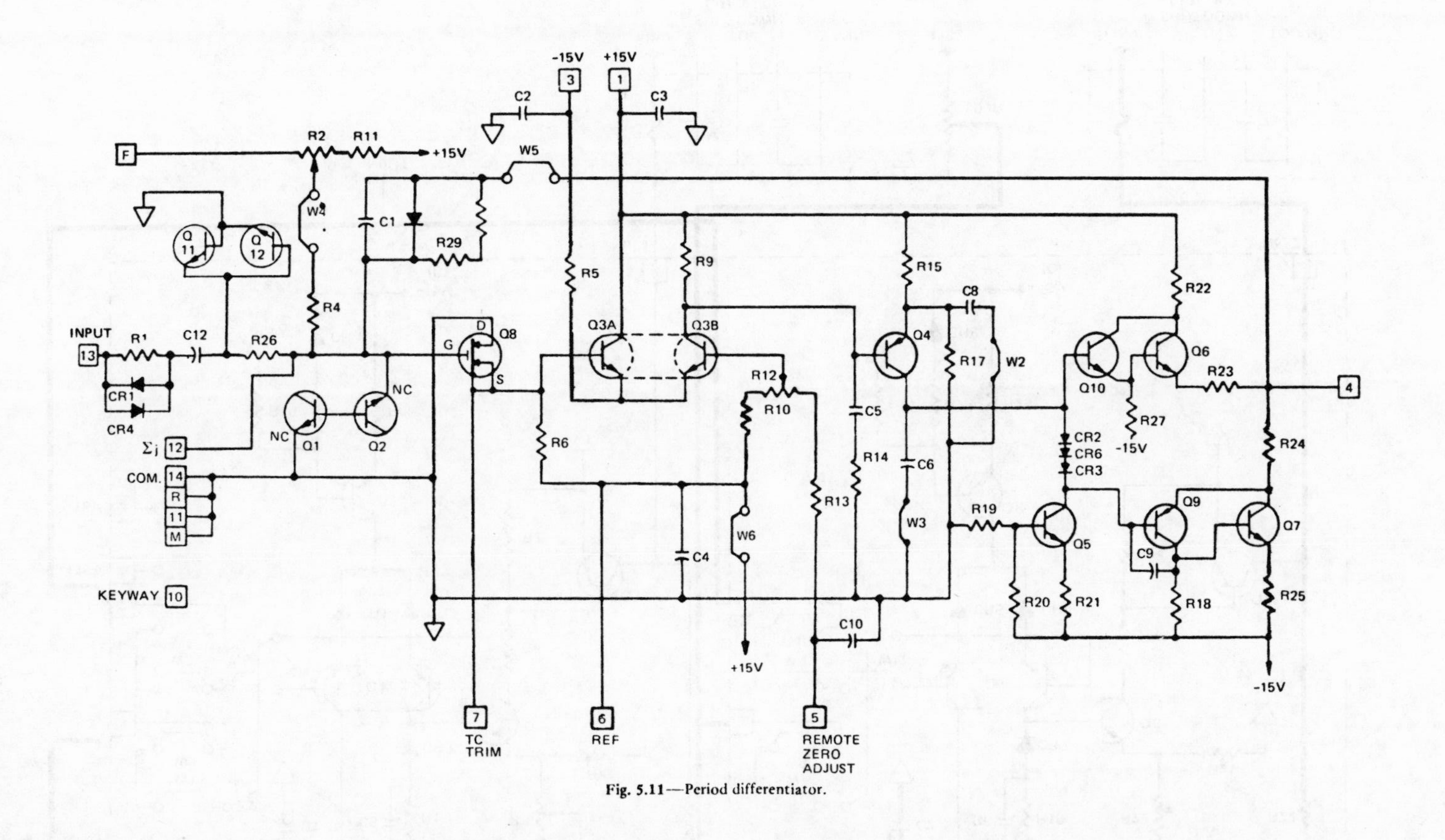

Fig. 5.11—Period differentiator.

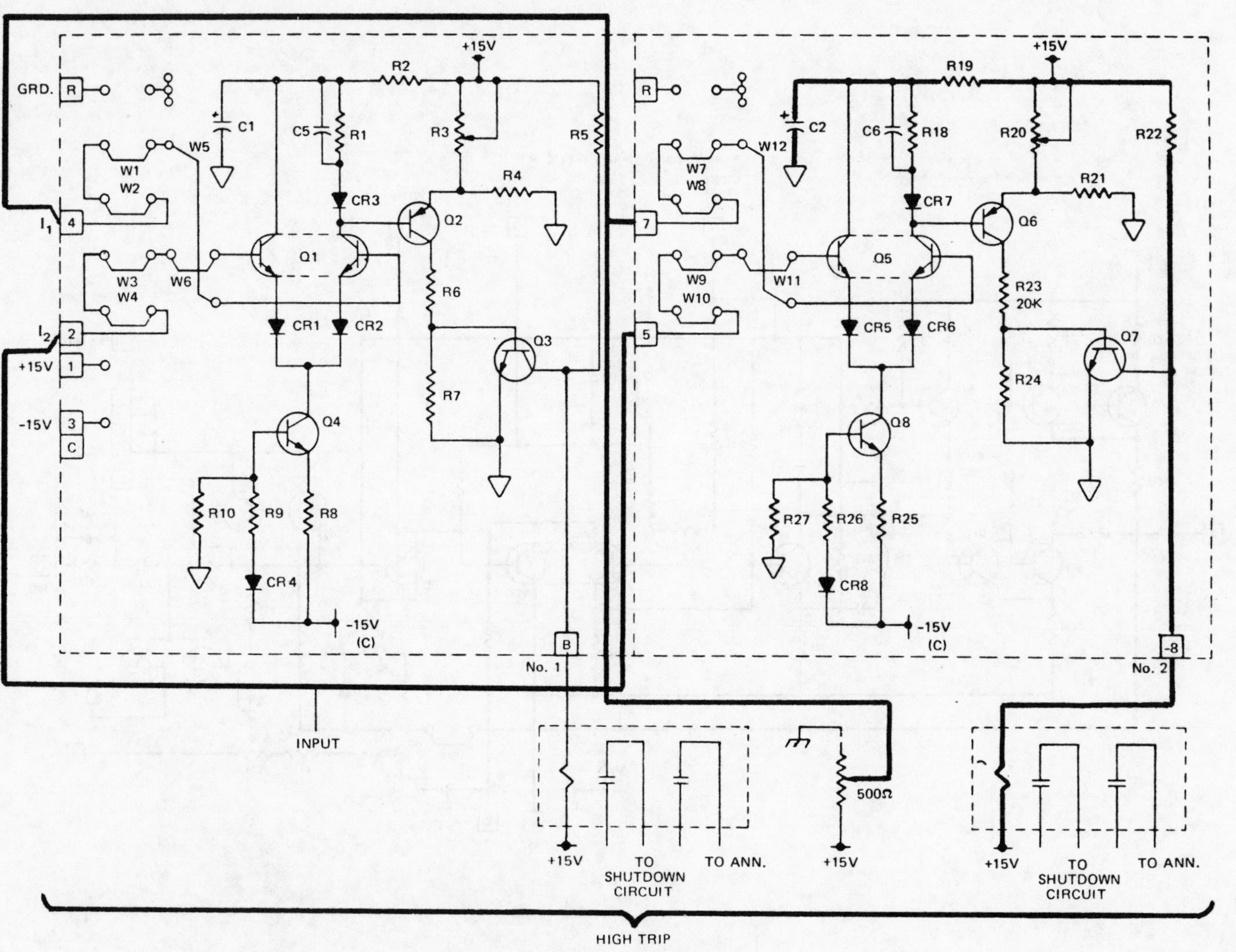

Fig. 5.12—Solid-state alarm (trip) circuit.

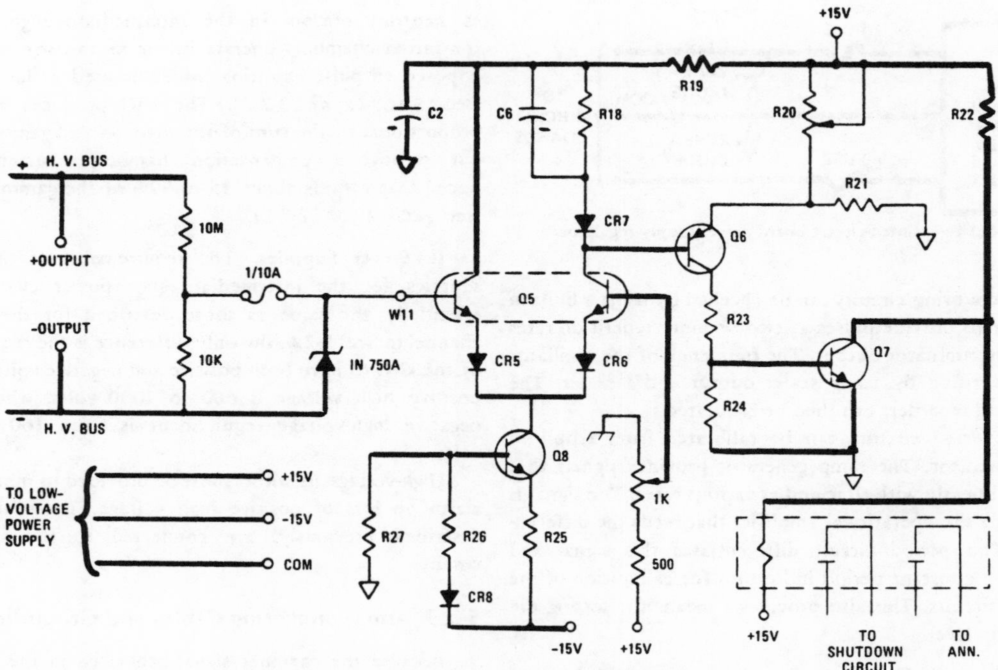

Fig. 5.13—High-voltage monitor with variable set-point alarm.

pulses. Radiation-resistant cables should be used to increase the time between replacement, to reduce maintenance costs, and to improve availability. Cables and connectors are available for high-temperature use, but extreme care must be exercised in their use. Avoid use of high-temperature components if at all possible; they are subject to changes in resistance which adversely affect both signals and high-voltage cables.

A triaxial cable should be used to minimize the noise introduced between the sensor and solid-state preamplifier. The outer braid of the triaxial cable should be tied at both ends to the inner braid. The impedance of the cable must match the input impedance of the preamplifier to reduce pulse reflections on the cable.

5-2.6 Power Supplies for Detectors and Electronic Equipment

(a) High-Voltage Power Supplies. Solid-state power supplies with excellent characteristics are available for powering the nuclear detectors. Solid-state power supplies reduce the package size, the heat generated, and the maintenance time. Many power supplies are capable of supplying power to two or more detectors connected in parallel. When power supplies are used to provide for two or more detectors, the detectors must be isolated from ground. Also, note that on a common-failure basis the units are not independent and cannot be used as redundant units in a shutdown circuit. It is desirable that the loss of one power supply will not cause a scram.

The high-voltage power supplies must have a voltage monitoring circuit or an output voltage proportional to the high-voltage output which can be used to monitor the high voltage. The high-voltage monitor is used to give an alarm when the high voltage drops below a predetermined value. Figure 5.13 shows a high-voltage monitor unit with a variable set-point alarm. The 10-turn potentiometer dial reading corresponds to the trip voltage.

The high-voltage power supplies should have short-circuit protection. Protection against voltage surges at the input is also desirable.

(b) Low-Voltage Power Supplies. Power supplies used in the counting equipment must have excellent regulation and stability. The low-voltage supplies should have internal short-circuit protection to prevent supply damage due to overcurrent demands from the unit. Because the counting equipment is of solid-state design, it is important that the voltage supplied to the printed circuit boards be limited below a value that would cause component damage.

Power supplies with built-in protection are commercially available and should be specified when ordering the nuclear equipment. An inexpensive method of providing overvoltage protection is shown in Fig. 5.14. The Zener diodes prevent the voltage from exceeding a given value. Excessive current drawn from the supply through the Zener diodes will cause the fuse to open before circuit damage has occurred.

5-2.7 Calibration and Checkout Control

(a) Log Count-Rate Meter. The LCRM has a count-rate circuit for control and readout; it will also have period and level circuits that must be calibrated periodically to ensure proper alignment of the unit.

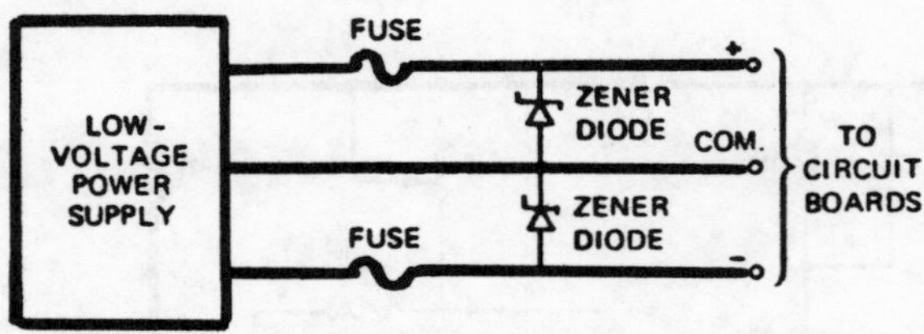

Fig. 5.14—Printed-circuit board power-supply regulator.

The counting circuits can be checked by using a built-in oscillator to provide pulses at two or more repetition rates to the discriminator output. The frequency of the oscillator can be verified by using scaler output and a scaler. The meters and recorders can then be calibrated.

The period circuits can be calibrated from a built-in ramp generator. The ramp generator provides signals that increase linearly with time and at various rates. The signal is applied to the operational amplifier that feeds the differentiator. The period circuit differentiates this signal and provides a constant period indication for calibration of the readout circuits. This also provides a means for setting the period trip level.

(b) Channel Checkout. Some means should be provided at the preamplifier input for checking and calibrating the complete channel. Ideally, a neutron source inserted near the neutron sensor would provide complete system checkout. Since this is generally not possible, the system is checked from the preamplifier. A calibrated pulser provides the input signal to the preamplifier, and meter, recorder, and scaler readings are verified. Level and period trips are also verified.

5-2.8 Control and Safety Circuits

A start-up channel system is shown in Fig. 5.15. The system consists of independent channels for control and local indication but with a common recorder for switch selection of a desired readout channel. The three channels provide the redundancy necessary to satisfy safe operation of the reactor and also provide for sufficient channels so that loss of a single instrument need not result in lost reactor operating time. The shutdown circuits associated with the three channels are arranged in a two-out-of-three shutdown logic; i.e., two instruments must trip before scram is initiated. The initiating shutdown circuit is shown in Fig. 5.15.

5-3 INTERMEDIATE POWER CHANNEL

5-3.1 Introduction

An intermediate power channel, shown in Fig. 5.16, consists of a compensated ionization chamber (CIC), a dual-voltage power supply, a log N unit, and readout equipment. Each of these units is discussed below.

5-3.2 Ionization Chamber and Power Supplies

(a) Sensor. Compensated ionization chambers are used as neutron sensors in the intermediate-range channel. Ionization chambers operate in the mean-current mode as opposed to pulse counting, which is used at lower fluxes (see Sec. 2, pt. 2-2.3). The CIC produces a current proportional to the sum of the neutron and gamma fluxes; but, through a compensation chamber, a current is produced that cancels about 95 to 99% of the gamma current (see Sec. 2, pt. 2-2.2).

(b) Power Supplies. The requirements of the power supplies for the intermediate-range power channels are essentially the same as those described for the start-up channel in Sec. 5-2.6; the only difference is the requirement of the CIC to have both positive and negative voltages. The positive high voltage is 600 to 1000 volts, whereas the negative high-voltage requirement is from 100 to 1600 volts.

High-voltage monitors must be provided to monitor and alarm on loss of positive high voltage. These alarms, as described previously, are connected to the shutdown circuits.

5-3.3 Interconnecting Cables and Grounding

Because the chamber signal generated in the CIC is a current and not a pulse, noise generated in the cable between the chamber and log N amplifier is not nearly so critical as that with a pulse channel. Noise pulses are suppressed in the log N amplifier–integrator circuits. No electronics are installed near the CIC in the high-radiation area. If cables with high radiation resistance are used, the operating time between cable changes can be increased.

The output signal produced in the CIC varies from about 10^{-15} to 10^{-2} amp. Ground loops between the detector and the log N amplifier should be avoided in order to reduce stray signal pickup. The detector is insulated from the detector thimble by ceramic standoff insulators, which are impervious to radiation damage. The shields of the coaxial cables are connected to the chassis, which should be grounded to nuclear instrumentation ground at one point, effectively reducing or eliminating the ground loops (see Sec. 10).

5-3.4 Logarithmic Amplifier

A block diagram of a logarithmic amplifier (log N) is shown in Fig. 5.17. The log N amplifier has two essential circuits for signal conditioning. The first is the log section, which converts the detector signal to a logarithmic output. The second circuit is the period differentiator circuit.

The input signal is biased to range over nine decades. So that this signal can be continuously monitored, it is converted to a log signal in the log amplifier circuit. The circuit consists of an operational amplifier with an active feedback element, namely, the grounded-base transistor. The circuit can be described by the following equation:

$$e = -E_0 \log \frac{i}{I_0} \tag{5.3}$$

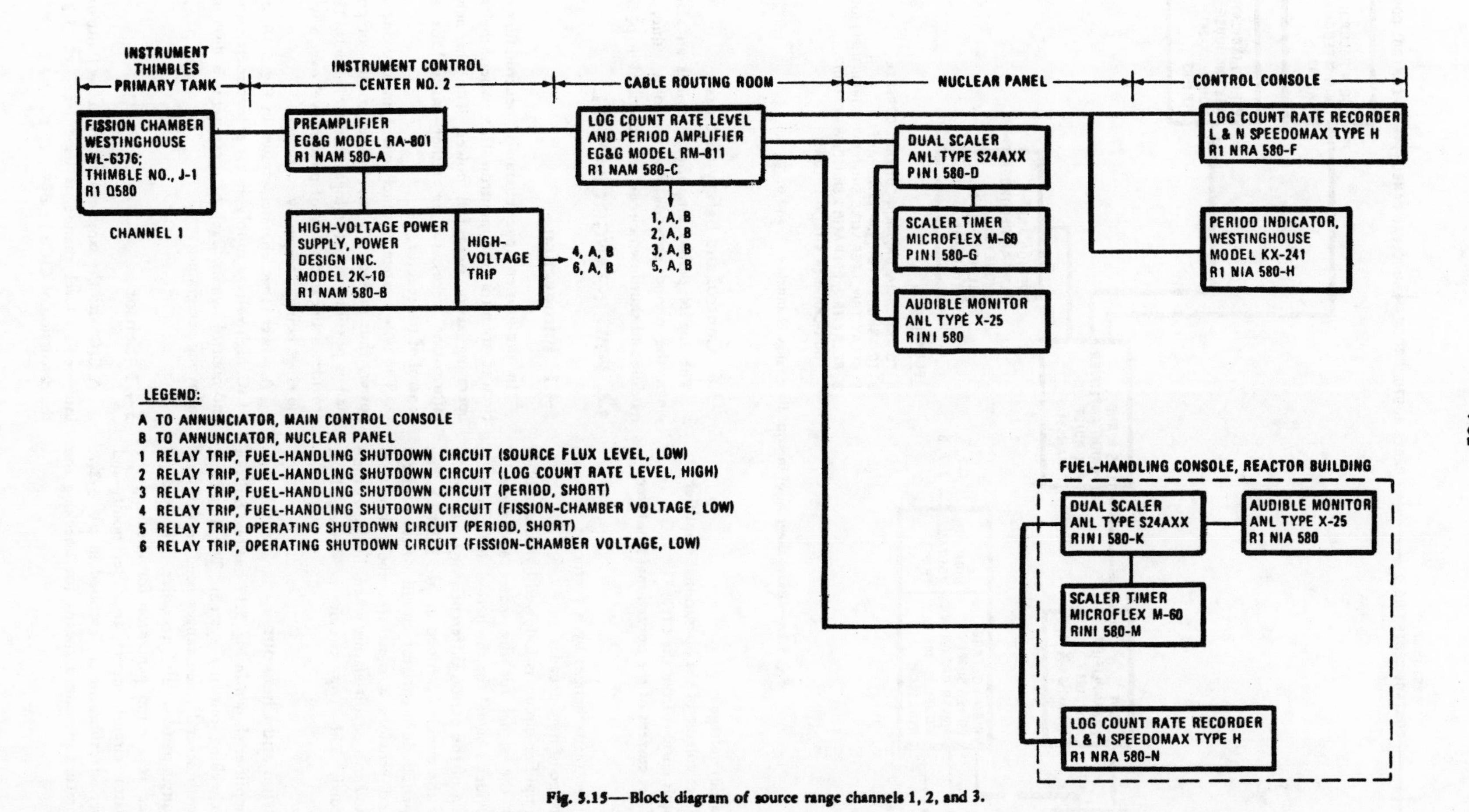

Fig. 5.15—Block diagram of source range channels 1, 2, and 3.

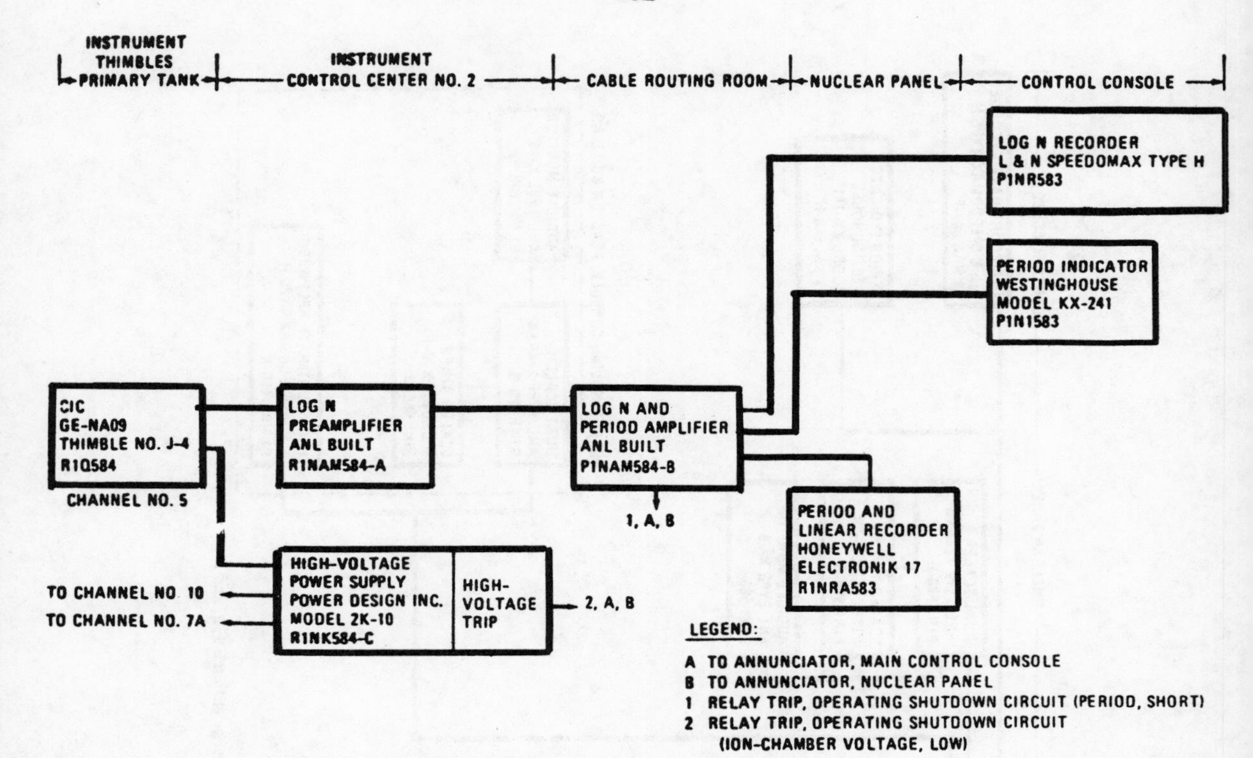

Fig. 5.16— Block diagram of intermediate-range channels 4, 5, and 6.

where e = output voltage

E_0 = offset voltage of the operational amplifier

i = input current from the CIC

I_0 = offset current of the operational amplifier

The output voltage then changes by a factor of 9 for an input current change of nine decades.

The log N amplifier circuit is followed by an amplifier that conditions the signal for the meter and recorder outputs and provides a signal for the period differentiator.

The operation of the period differentiator circuit is the same as that for the circuit described in pt. 5-2.4(c). The output of the period differentiator circuit drives a meter and recorder and provides a signal for the period trip circuit. The period trip circuit alarms when the preset time value is exceeded. The trip circuit is described in pt. 5-2.4(e).

5-3.5 Calibration and Checkout

The log N amplifier shown in Fig. 5.17 has two built-in calibrators for checking out the system. The first is a current source with several fixed ranges used to calibrate the log N circuits, meters, and recorder. The second calibration device is a ramp generator for checking the period-differentiator circuit, meter trip, set point, and recorder. The period calibrator is described in pt. 5-2.7. These two calibrators provide a means for checking the entire system.

5-3.6 Control and Safety Circuits

The log N period amplifier provides an alarm (trip) when the period exceeds the set-point value. The trip module and shutdown circuit are described in pt. 5-2.8.

5-4 POWER-RANGE CHANNEL

5-4.1 Introduction

In the power-range channels, neutron-flux-measuring circuits monitor the neutron flux when the reactor is operating at or near full power. The channels provide information necessary for either manual or automatic control of the reactor.

The power-range channels cover linearly one decade of power, thus providing a much finer control of power level than is possible with the logarithmic channels. The system provides a linear display of power level over a wide range, the range being changed by switching.

A power-range channel, shown in Fig. 5.18, consists of a CIC, high-voltage supply, linear picoammeter, and readout and control signals. Each of these items is discussed in the following paragraphs.

5-4.2 Sensor

A CIC with the same characteristics as those used in the intermediate-level channel is used [see pt 5-3.2(a) and the discussion of CIC's in sec. 2, pt. 2-2.2].

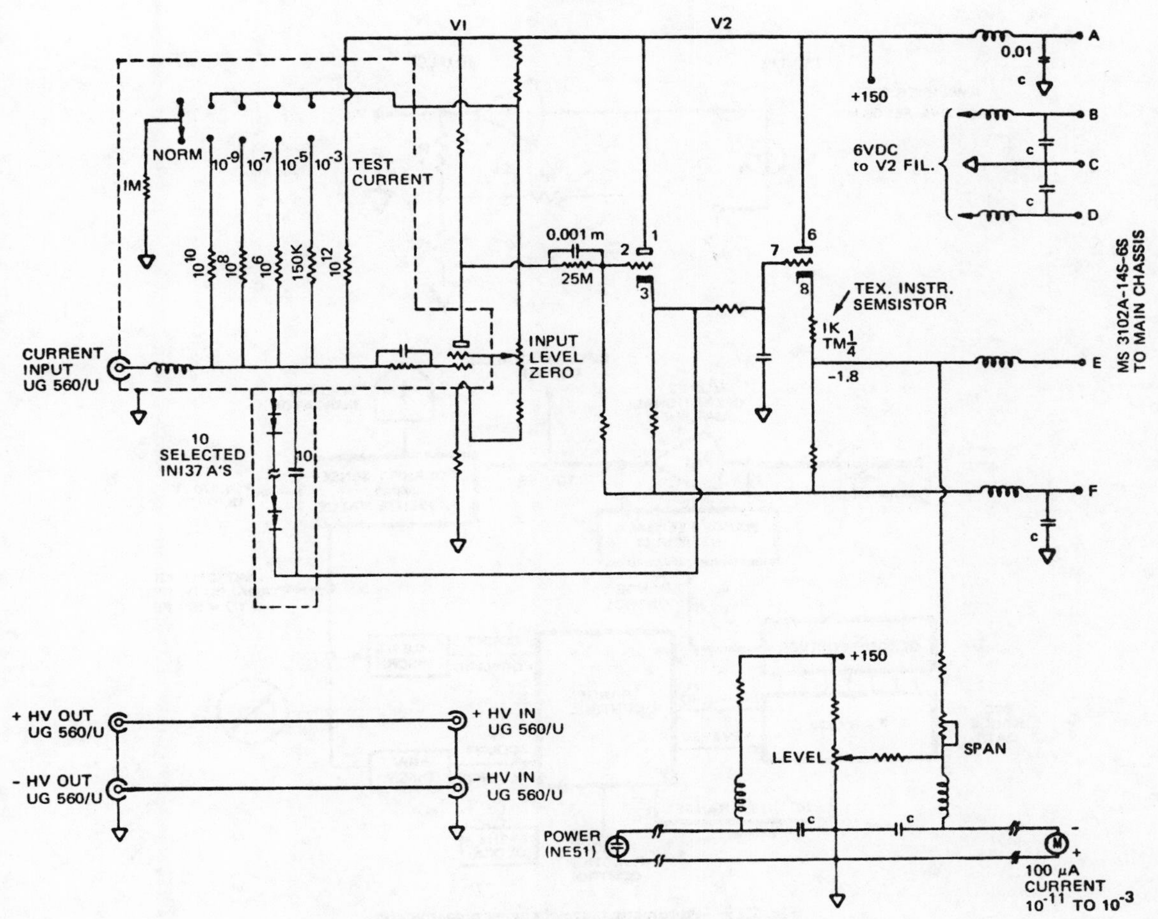

Fig. 5.17—Circuit diagram of preamplifier for log N channel.

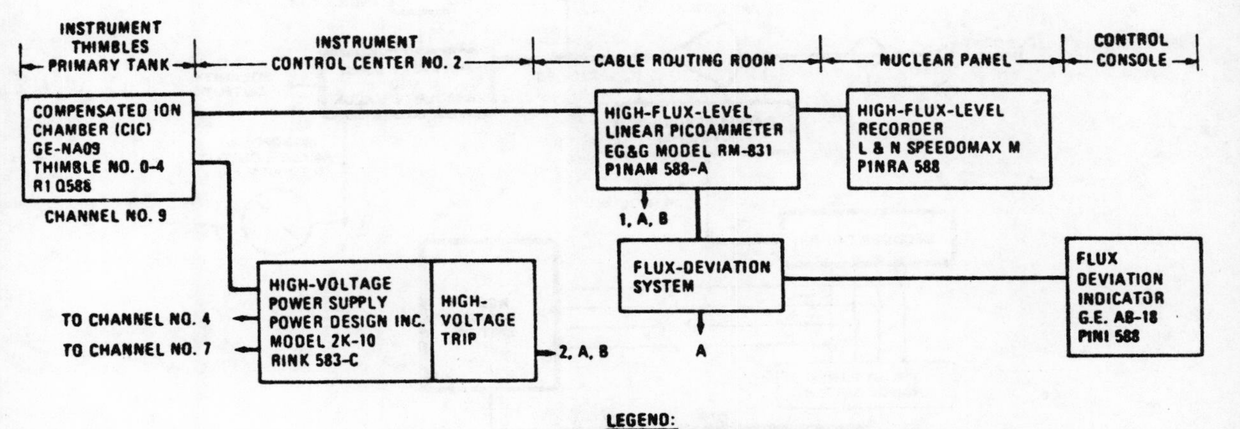

Fig. 5.18—Block diagram of high-flux channels 9, 10, and 11.

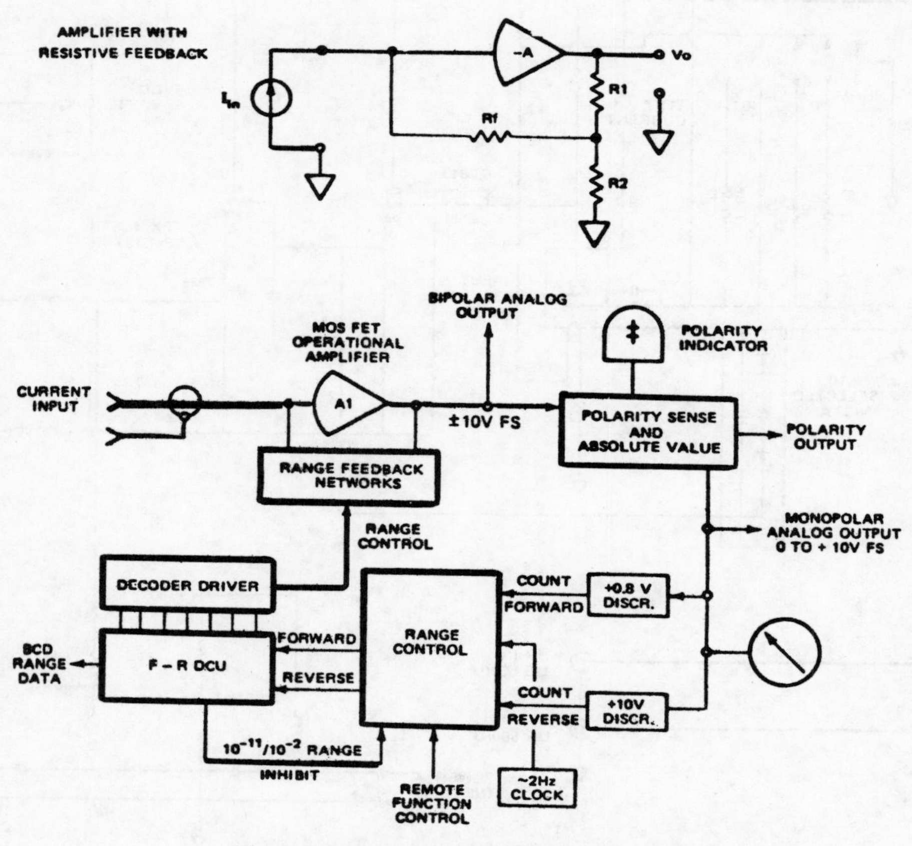

Fig. 5.19—Automatic range-changing picoammeter.

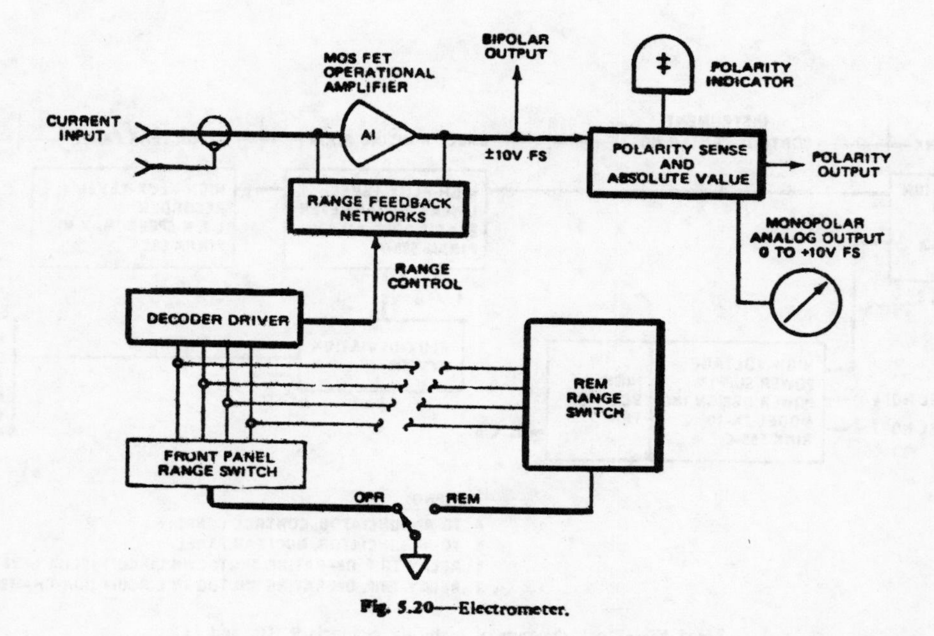

Fig. 5.20—Electrometer.

5-4.3 High-Voltage Supplies

See pt. 5-3.2(b).

5-4.4 Linear Picoammeter

The signal from the CIC is monitored by a linear picoammeter with either manual or automatic range control. In a manually operated unit, the current range is predetermined and normally provides an accurate indication of power level only for the highest two decades of power. Figure 5.19 shows an automatic range-changing picoammeter that gives continuous coverage from a few watts to full power. The unit automatically changes range on an increasing and decreasing signal at the range set points.

The basic current-measuring device of the picoammeter is an electrometer with appropriate feedback resistors to determine the range of operation. The electrometer, shown schematically in Fig. 5.20, consists of a very high input impedance operational amplifier.

5-4.5 Readout Indication

Signals generated from the picoammeter are presented to the operator in several ways. The range is indicated by readout lamps; the level is displayed on meters and recorders marked 0 to 100%. The level can also be displayed on a digital readout unit.

5-4.6 Control and Safety Circuits

Alarm units are provided in the picoammeter for initiating a reactor shutdown if the power level exceeds a predetermined maximum value. The alarm unit is described in pt. 5-2.4(c). The relays are connected in a two-out-of-three sequence to increase system reliability against single failures causing reactor shutdown (see sec. 11).

5-4.7 Calibration and Checkout

As with the other channels described in this chapter, a built-in test source is provided to check the system response from the picoammeter. The test source also provides a means for setting and checking the alarm units.

5-5 MEAN-SQUARE-VOLTAGE (MSV) METHODS

5-5.1 Basis of Method

The MSV method depends on the fact that, if the time distribution of pulses from a nuclear radiation sensor is a Poisson distribution, the variance (mean of the squares of the deviations from the mean) is a direct measure of the mean. (See Table 5.1 for relevant definitions and formulas.) For all practical purposes this condition is met by boron and fission-counting chambers.

There are at least three advantages to be gained from using MSV methods: increased gamma discrimination (compared to compensation), improved operation when chambers and cables are exposed to elevated temperatures, and more efficient use of chambers (sensors).

So that the advantages of the method can be realized, a measure of a quantity proportional to the square of the charge (or current) is made. One way to do this is to subtract the mean signal with a differentiator and measure the temperature rise in a resistor, as is done in a true root-mean-square meter commonly used in the shop or laboratory. Another way is to pass the variable (a-c) signal through a half- or full-wave rectifier measuring, as a result, the average magnitude or average magnitude squared. This latter technique is accurate at only one frequency for which a correction factor can be applied. More generally, the pulses can be passed through an electronic squaring amplifier and the output read without correction as a linear measure of the mean square. If a mean-square signal is sent to a log converter circuit, the output is again proportional to the log of the mean.

5-5.2 Gamma Discrimination

Gamma discrimination must be compared with "compensation" for gammas as accomplished in the CIC discussed in pt. 2-2.2 of sec. 2. In a CIC a compensating signal generated in a volume not sensitive to neutrons is subtracted electrically from the gamma-plus-neutron signal. In practice the compensating volume and the mass of material forming the neutron and compensating volumes cannot be matched exactly; through engineering compromise the compensation is usually between 95 and 99% of the gamma signal. Although in theory the compensation could be much better, it just cannot be achieved. Commercial units use two concentric volumes that are adjusted so that overcompensation in some gamma range is avoided. Overcompensation would result in negative readings and confusion to control-system functions. Manufacturers usually guarantee a gamma/neutron signal ratio of 1/20 and a maximum of 1/100, the latter to avoid overcompensation.

With MSV methods the compensation or discrimination is not dependent on the mechanical construction of the chamber. Only one ionization volume is involved, and advantage is taken of the charge ratio of a fission fragment to a gamma-scattered beta particle. If the number of events is N per unit time and the charge collected per event is Q, then an average voltage $E_{d\text{-}c}$ is developed:

$$E_{d\text{-}c} = \overline{N}\,\overline{Q} \int_0^\infty h(t)\,dt \qquad (5.4)$$

where $\overline{N}$ is the mean number of events, $\overline{Q}$ is the mean charge per event, and h(t) is the circuit response to a single pulse of unit charge. By definition the mean-square voltage is

$$E_{ms} = (E_{d\text{-}c})^2 + \overline{N}\,\overline{Q^2} \int_0^\infty [h(t)]^2\,dt \qquad (5.5)$$

The voltage $E_{d\text{-}c}$ is made zero if h(t), the circuit response, is

Table 5.1—Poisson Distribution: Definitions and Formulas

Definitions

n = number of events observed

$\bar{n}$ = average (mean) number of events observed

$n - \bar{n}$ = deviation from mean
= deviation of the observed number of events from the average

$\overline{(n - \bar{n})^2} = \sigma^2$
= variance
= mean of the squares of the deviations

Formulas

Poisson distribution

$$P(n) = \frac{e^{-\bar{n}}\bar{n}^n}{n!} \quad (1)$$

$$\sum_{n=0}^{\infty} P(n) = 1 \quad (2)$$

For Poisson distribution

$$\bar{n} = \sum_{n=0}^{\infty} nP(n) = \bar{n} \quad (3)$$

$$\overline{n^2} = \sum_{n=0}^{\infty} n^2 P(n) = \bar{n}^2 + \bar{n} \quad (4)$$

$$\sigma^2 = \sum_{n=0}^{\infty} (n - \bar{n})^2 P(n) = \overline{n^2} - \bar{n}^2 \quad (5)$$

Substitute (4) into (5) and obtain

$$\sigma^2 = \bar{n} \quad (6)$$

Note: If n is the number of counts indicated by a sensing device in a given time interval, then n = count rate × time interval.

limited to acceptance of a-c signals alone. In this case

$$E_{ms} = \bar{N}\,\overline{Q^2} \int_0^\infty [h(t)]^2\, dt \quad (5.6)$$

It is not difficult, as indicated earlier, to make the circuit such that E_{ms} is the lone acceptable result. A differentiation circuit at the input of the squaring circuit will do the job.

The relation between the resulting signals can be established if a chamber operating in the d-c mode is compared to a similar chamber operating in the MSV mode. If a subscript n is used for neutron events and a subscript β for gamma effects (since gammas scatter β particles or electrons into the chamber), the discrimination in the d-c mode for gammas is (from Eq. 5.4)

$$D_{d\text{-}c} = \frac{\bar{N}_n\bar{Q}_n}{\bar{N}_\beta\bar{Q}_\beta} \quad (5.7)$$

and the discrimination for the MSV mode is

$$D_{ms} = \frac{\bar{N}_n\overline{Q_n^2}}{\bar{N}_\beta\overline{Q_\beta^2}} \quad (5.8)$$

The ratio of the discriminations is thus

$$\frac{D_{ms}}{D_{d\text{-}c}} = \frac{\overline{Q_n^2}\,\bar{Q}_\beta}{\bar{Q}_n\,\overline{Q_\beta^2}} \simeq \frac{\bar{Q}_n}{\bar{Q}_\beta} \quad (5.9)$$

The ratio $\overline{Q_n^2}/\overline{Q_\beta^2}$ has been set equal to $(\bar{Q}_n)^2/(\bar{Q}_\beta)^2$ in deriving Eq. 5.9 since it can be shown that in practical cases this is a good approximation.

The ratio $\bar{Q}_n/\bar{Q}_\beta$ is about 10^3 for a fission chamber. Compared to a compensated chamber, this means an assured 10^3 discrimination against gammas instead of the 20 to 100. In practice, experimental results show a nearly hundredfold improvement in gamma discrimination, mainly because the 1/20 ratio of gamma to neutron signal is more realistic for a CIC than the 1/100 ratio.

5-5.3 Temperature Effects

In practice, temperature effects are not entirely separable from the gamma effect. If it is assumed that there is no need for gamma discrimination, at elevated temperature the cable insulation resistance is seen to decrease as the temperature increases whereas the breakdown voltage, which creates noise or a mean deviation, is relatively constant. Thus the effect of noise is minimal whereas the mean current attributable to neutrons can be completely obscured by direct-current leakage.

Quantitative results have not been reported. The first reported use of MSV methods was for in-core units of about 1 cm^3 volume operating at about 600°F (316°C). The mean signal was obscured by leakage, but the MSV signal was still present. The gross effect has been discussed by DuBridge et al.(see the Bibliography at the end of this chapter).

From these observations it can be concluded that, if neutron-sensitive chambers and their associated cables are to be operated at elevated temperatures, the proper method is to use MSV measurements instead of mean-current measurements.

5-5.4 Wide-Range Neutron Monitoring Channels

As noted in Sec. 5-1, the large range of neutron-flux values to be monitored in a power reactor requires the use of several separate channels with two decades of overlap between the channels. With the MSV technique it has been possible to develop monitoring systems that cover 10 decades of reactor power. The signal from a single fixed-position fission chamber is used with what has come to be called counting-MSV or counting-Campbelling circuits. The combination of count rate and MSV can be read out as a linear indication of power, much as is done with a range-switched picoammeter (see Sec. 5-4.4), or a log output can be taken from which period can be derived.

The wide-range instrumentation permits the use of fewer sensors, recorders, meters, cables, and sensor thimbles. This represents a significant saving in power-reactor instrumentation costs.

5-5.5 System Descriptions and Components

Figure 5.21 is a block diagram of an average-magnitude-squared channel that covers 10 decades of reactor power. The MSV portion of the channel uses the combination pulse-charge or current preamplifier, band-pass amplifier, rectifier, log amplifier, and d-c amplifier. The noise level for the Campbell part of the channel is reduced by special shielding of the 40-ft cable between the chamber and the double-shielded preamplifier. The impedance and signal levels out of the preamplifier are such that essentially any reasonable length of conventional coaxial cable to the rest of the circuit may be used with good results. As shown in Fig. 5.21, the wide-range log power channel consists of an LCR circuit and a log-Campbell circuit working out of the same fission chamber and preamplifier. Their output signals are combined to give a single output indication of log power and, at the same time, to eliminate normal limiting errors of each one.

The high-frequency components of the signal pass through a conventional LCR circuit using a discriminator, flip-flop, and multiple-diode log pump circuit. This circuit gives an output voltage, E_1, that is proportional to the log of the count rate from about 0.3 to 300,000 counts/sec. A biased diode, D_1, limiter is used to cut off that portion of the response which is adversely affected by resolution counting loss (usually about 2×10^5 neutrons cm^{-2} sec^{-1}). As is customary with LCR circuits, the log diode pump uses fixed low-temperature-coefficient components to obtain the log relation. The pump circuit output passes through a d-c operational amplifier with adjustable gain and bias; so output volts per decade and volt level for a particular flux level can be established.

The low-frequency components governed by the Campbell band-pass amplifier pass through the linear rectifier, filter (T = 50 msec), and through a stable d-c log amplifier to give a log indication of power from 2×10^5 to 2×10^{10} neutrons cm^{-2} sec^{-1}. The diode D_2 allows only signals above zero to pass; therefore, if the output of the log amplifier is biased, the lower end of the response curve below 2×10^5 neutrons cm^{-2} sec^{-1} can be eliminated. This portion of the curve may be adversely affected by gamma and alpha background, noise, imperfect rectification, and lack of pulse overlap. A gain adjust at the output of the log amplifier allows the slope of E_2 to be properly adjusted to match that of E_1. Since the bias cutoff points can be adjusted to coincide, a smooth, fast, all-electronic transition is made from counting to Campbelling or vice versa.

Since misalignment or drift might produce an offset at the crossover point (2×10^5 neutrons cm^{-2} sec^{-1}) and cause an exaggerated false period indication in that region, great emphasis has been placed on stability of the circuits and on a built-in calibration and test circuit that will allow proper alignment without the use of a reactor. High-gain solid-state operational amplifiers (some integrated circuits) with feedback through stable circuit elements were used throughout. The principal temperature problem, as usual, was associated with the log amplifier in the Campbell circuit, and this was solved by using two operational amplifiers and two sets of logging diodes thermally coupled to obtain a temperature-compensated log response. Temperature tests were performed between 50°F and 150°F (10°C and 66°C). Temperature drift results in an output error of less than 1.5% from 50°F to 120°F and only a slight further degradation up to 150°F.

Another average-magnitude-squared channel is shown in Fig. 5.22. The circuit is divided into two portions, as shown in the figure. The signal from the radiation sensor enters the instrument and is routed both to the log count rate circuit (LCRM) and the statistical level amplifier (MSV). Pulses within the range of 1 per second to 10^6 per second pass through the LCRM discriminator into a flip-flop, a driver network, and then to a Cooke–Yarborough diode pump log converter, where the pulse rate is converted to a d-c output. At that point the signal is amplified to the desired output level and routed to the front panel meter and other instrument inputs. When the input signal enters the MSV portion of the instrument, it is first applied to a filter that passes only the portion of the signal between 4 and 8 kHz. This signal is then applied to a wide-range rectifier, thus providing a half-wave rectified signal to a smoothing filter, which then routes the d-c output to a log amplifier and hence to the output meter and other external devices. The output of the MSV circuit is scaled to read two decades of output for each decade of input current to provide the square in the mean-square measuring technique.

A third meter on the instrument measures the rate of change of both the LCRM and MSV channels alternately, switching from the LCRM channel to the MSV channel automatically as the LCRM reaches near full scale and as the MSV channel begins to operate. Since this portion of the circuit is somewhat novel, it is detailed in Fig. 5.23.

In operation the output levels from the LCRM and MSV circuits are each applied to a separate differentiator circuit that operates continuously. The two outputs are routed through a solid-state switching network that is controlled by the magnitude of the MSV input signal such that below a preset level of the MSV circuit the signal from the LCRM differentiator enters the output scaling amplifier and above this level the MSV differentiator controls the output scaling amplifier. The output of the scaling amplifier is fed to the rate-of-change meter, which reads from −1 to +9 decades per minute.

Another feature of the rate-of-change circuit is the "LCRM suppress" feature. This involves a circuit to suppress rate-of-change movements on the lower portion of the LCRM channel and thus has the effect of preventing rate-of-change meter movements where counting statistics cause an erratic signal.

In addition to the rate-of-change circuits, the instrument generates pulses for calibrating the LCRM, ramp functions for checking the rate-of-change circuits, and a variable-level 5-kHz signal for calibrating the MSV circuit. Furthermore, all meters have electronic level trip circuits that read the output level of a particular circuit and produce both an automatic resetting alarm and one that latches and must be reset manually after an alarm is produced by an excessively high or low level reading on any one of the three front panel meters.

5-5.6 MSV Vs. AMS Systems

The circuits described in the previous section operate

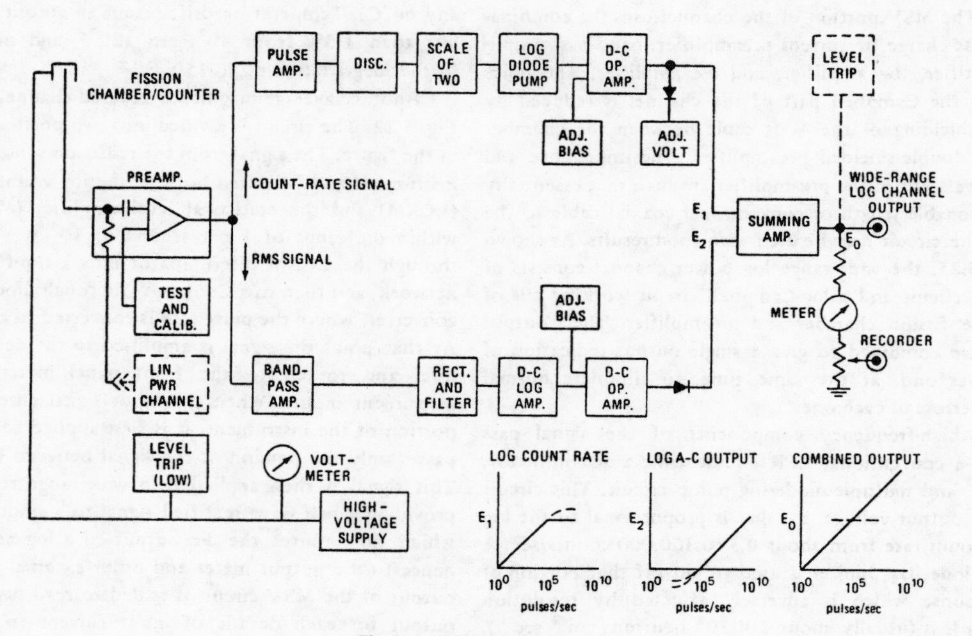

Fig. 5.21—Wide-range log channel.

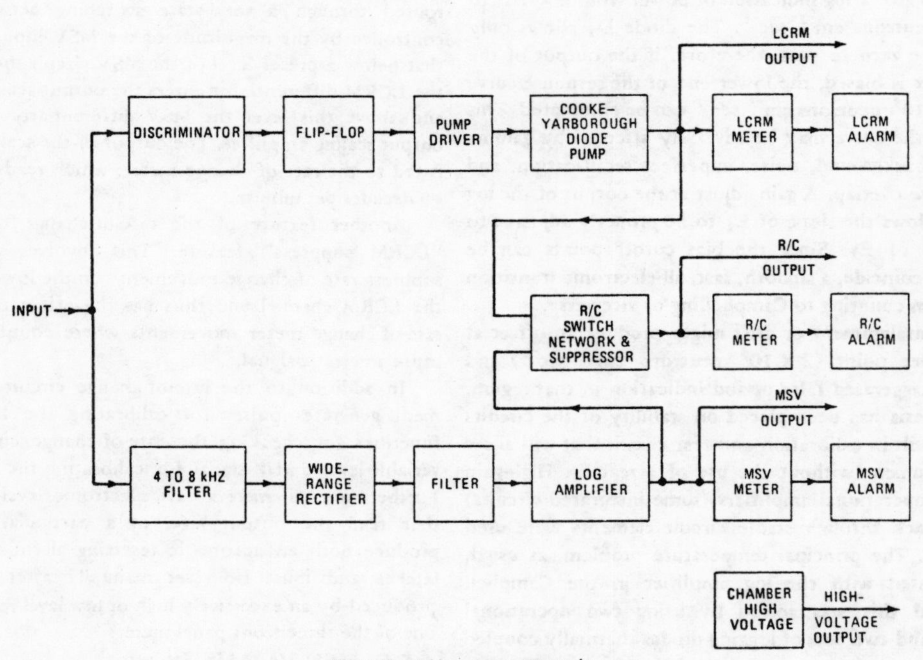

Fig. 5.22—Wide-range power monitor.

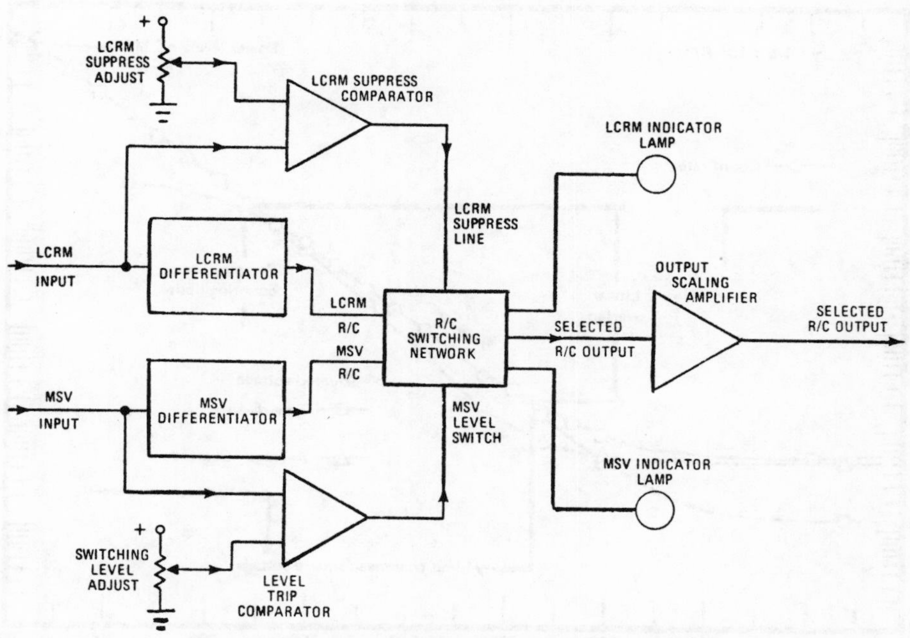

Fig. 5.23—Rate-of-change switching circuit for wide-range power monitor.

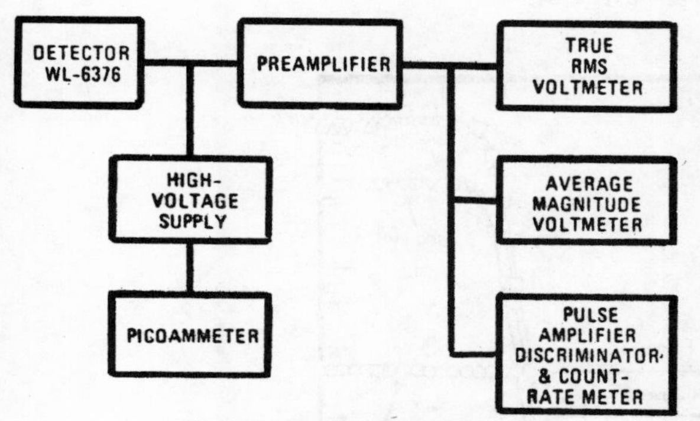

Fig. 5.24—System block diagram.

on the average-magnitude-squared (AMS) principle and are therefore frequency-sensitive. Adjustments must be made in the band-pass filter to make the system fit a particular reactor. The lower cutoff at 4 kHz, indicated above, corresponds to about 25,000 radians/sec; this is below the β/l cutoff of a fast reactor (where the roll-off in the transfer function is at about 70,000 radians/sec.).

The principle involved is to use the lower frequency components of the band to assist in pulse overlap. Thus there is an incentive to use a lower frequency cutoff. (The high-frequency cutoff is selected for noise rejection.) As the low-frequency cutoff is raised in the AMS system, the overlap with the pulse-counting range decreases.

In a test run in 1967, the AMS and true MSV systems were compared using the instrument setup shown in Fig. 5.24. The data shown in Fig. 5.25 were obtained with a band pass of about 250 to 500 kHz. The EBR-2 was put on a 59-sec period, and, because of its wide flux range before reaching power feedback conditions, this is constant over four decades or more. As shown by the data, the MSV system had a correct signal for nearly two decades before the AMS signal became correct. Had a lower band pass been used, this could have been corrected, but the low pass of the band selected is not important to the true MSV system and bands above 250 kHz are used for other reasons.

If on Fig. 5.21 the blocks representing the band-pass filter and rectifier were replaced with a Hewlitt–Packard true RMS convertor, the signal would be as shown in Fig. 5.25. At present, the use of this converter adds materially to the number of components in the system, and commercial suppliers are hesitant to supply such systems.

5-5.7 MSV Vs. Mean-Current CIC

Data taken on a commercial CIC were compared with commercial log-average-magnitude-squared (LAMS) units operating off fission counters in the EBR-2 shutdown gamma field of about 2×10^5 R/hr. The resulting curves,

Table 5.2—Typical Discrimination Ratios*

	$\frac{R/hr}{nv\dagger}$
Fission chamber	
D-c mode	0.0062
True-mean-squared mode	5.8
Average-magnitude-squared mode	5.8
Compensated ionization chamber	
95% compensation	0.027
99% compensation	0.13

*Calculated by G. F. Popper, Argonne National Laboratory.
†Neutrons cm^{-2} sec^{-1}.

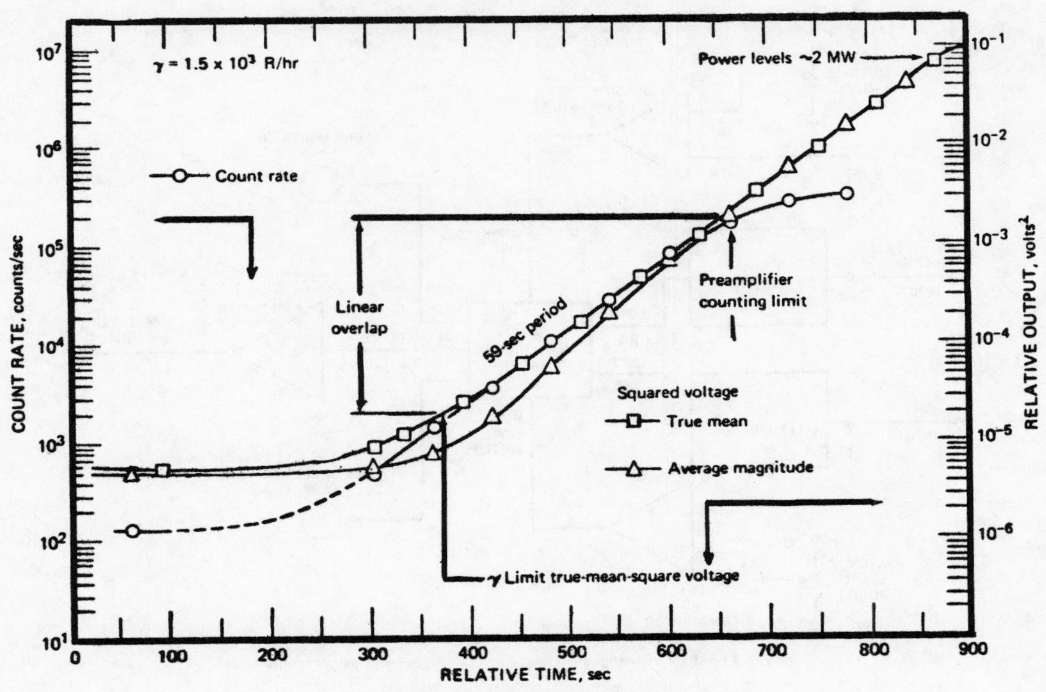

Fig. 5.25—EBR-2 period response.

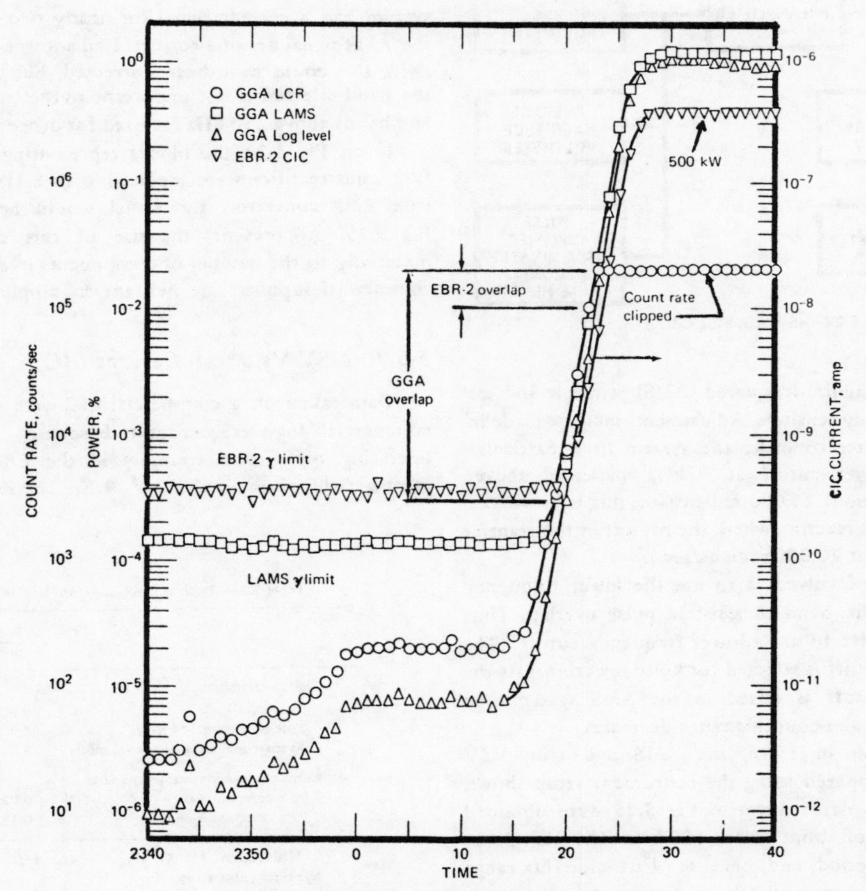

Fig. 5.26—Comparison of CIC and LAMS units (Run 1).

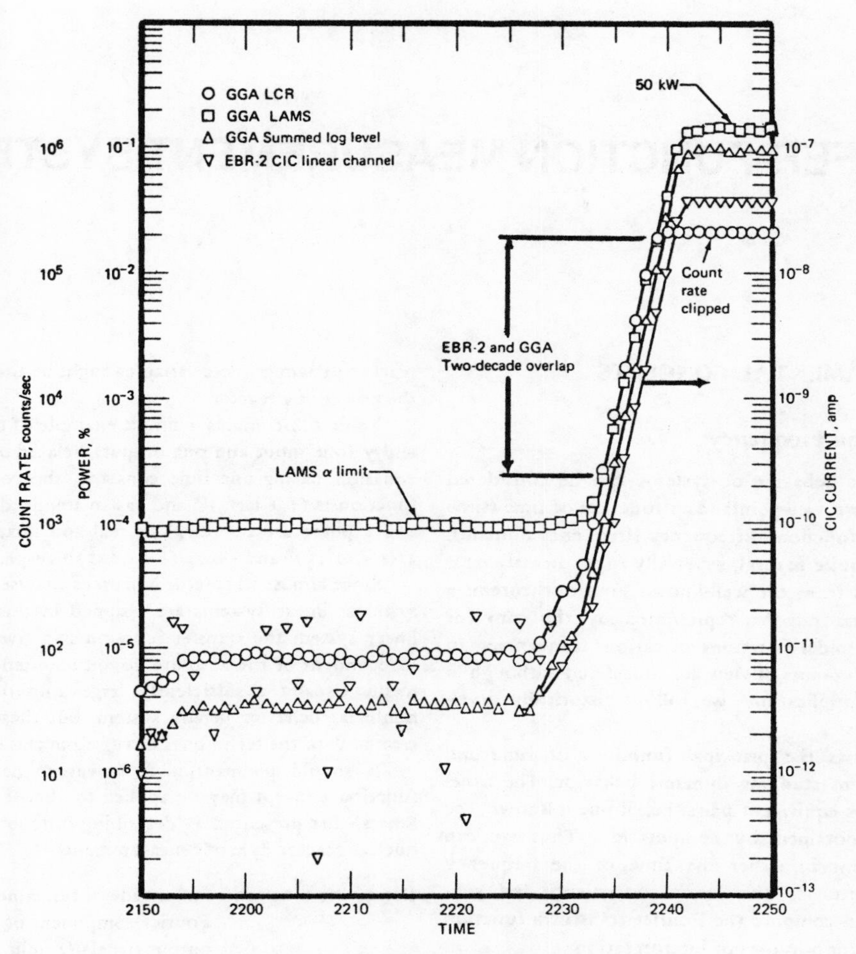

Fig. 5.27—Comparison of CIC and LAMS units (Run 2).

shown in Figs. 5.26 and 5.27, are multivalued on the ordinate to emphasize the overlap. As shown, the overlap for the CIC is only a factor of 2 compared to the overlap of the Gulf General Atomic counters of 100. If a true LRMS system had been used, the overlap would be the same or greater. The improvement over a CIC is evident.

Had the gamma field been more intense, the CIC overlap would be nil but the LAMS or log mean square voltage would still be 150. Table 5.2 shows comparative values of the discrimination ratio for the various systems.

BIBLIOGRAPHY

Campbell, N. R., The Study of Discontinuous Phenomena, *Proc. Cambridge Phil. Soc.*, 15: 117, 310, 513 (1909-1910).

Chase, R. L., *Nuclear Pulse Spectrometry*, McGraw-Hill Book Company, Inc., New York, 1961.

Harrer, J. M., *Nuclear Reactor Control Engineering*, D. Van Nostrand Company, Inc., Princeton, N. J., 1963.

Murphy, G., *Elements of Nuclear Engineering*, John Wiley & Sons, Inc., New York, 1961.

Popper, G. F., Counting and Campbelling: A New Approach to Neutron-Detection Systems, *Reactor Fuel-Process. Technol.*, 10(3): 199 (Summer 1967).

DuBridge, R. A., J. P. Neissel, L. R. Boyd, W. K. Green, and H. W. Pielage, Reactor Control Systems Based on Counting and Campbelling Techniques, Final Progress Report, USAEC Report GEAP-4900, General Electric Company, July 1965.

Elmore, W. C., and M. Sands, *Electronics Experimental Techniques*, McGraw-Hill Book Company, Inc., New York, 1949.

Fowler, E. P., and R. W. Levell, The Use of Current Fluctuations from a Neutron Detector with Logarithmic Amplifiers as a Measure of Neutron Flux Levels, British Report AEEW-R-375, June 1964.

Green, W. K., Ten-Decade Wide Range Neutron Monitoring System, Final Test Report, USAEC Report GEAP-11094, General Electric Company, October 1970.

Gwinn, D. A., and W. M. Trenholme, A Log-N and Period Amplifier Utilizing Statistical Fluctuation Signals from a Neutron Detector, *IEEE (Inst. Elec. Electron. Eng.), Trans. Nucl. Sci.*, NS-10(2): 1-9 (April 1963).

——, W. C. Lipinski, and J. M. Harrer, Ten Decades of Continuous Neutron Flux Monitoring from a Single Fixed Position Detector, *Trans. Amer. Nucl. Soc.*, 9(1): 316-317 (June 1966).

Thomas, H. A., and A. C. McBride, Gamma Discriminations and Sensitivities of Averaging and RMS Type Detector Circuit for Campbelling Channels, Report GA-8035, Gulf General Atomic.

Trenholme, W. M., and D. J. Keefe, A Neutron Flux Measuring Channel Covering Ten Decades of Reactor Power with a Single Fixed Position Detector, *IEEE (Inst. Elec. Electron. Eng.), Trans. Nucl. Sci.*, NS-14(1): 253-260 (February 1967).

SECTION 6

TRANSFER-FUNCTION MEASUREMENT SYSTEMS

6-1 FUNDAMENTAL CONCEPTS

6-1.1 Time and Frequency

The dynamic behavior of systems can be considered from either of two viewpoints: as a function of time (time domain) or as a function of frequency (frequency domain). This dualism is quite natural, especially to mathematicians, because it stems from the well-known Fourier theorem: a function of time can be represented by the sum (or integral) of sinusoidal functions of various frequencies. In this chapter both points of view are considered, although in certain specific applications we follow historically developed conventions.

Table 6.1 lists the principal functions of time and frequency used in studying dynamic behavior. The functions are given as equivalent pairs; i.e., if one is known, the other can be obtained by computation. Thus we can measure functions in either the time or the frequency domain, whichever is the more convenient, and subsequently we can compute the Fourier transform function if it is preferred for purposes of interpretation.

6-1.2 Transfer Functions

The concept of transfer functions was introduced in reactor plant analysis because of its proven utility in electrical engineering. As defined in Table 6.1, the transfer function is the ratio of output complex amplitude to input complex amplitude; this ratio is a complex number that depends on the amplitude ratio and the phase difference between two sinusoidal signals in a system of two or more dynamically related variables, all of which are oscillating at a given frequency. We may speak of a transfer function between any two variables. However, if the input driving function is oscillatory, it is conventionally used as one of the variables, in which case the transfer function is the output amplitude per unit input amplitude of sine-wave excitation. Several zero-power transfer functions are given in Fig. 6.1.

Complete specification of a transfer function involves both an amplitude value and a phase difference given as a function of frequency. A complete specification of the dynamics of a system would involve all the transfer functions between all the pairs of variables given for the entire band of frequencies of physical interest. Often, however, the two most meaningful variables are related. In reactor dynamics these variables might be the reactivity and the power of a reactor.

Table 6.2 contains a simple example of two variables, x and y (one input and one output), related by a differential equation having one time constant: the complex transfer function is $(1 + i\omega\tau_c)^{-1}$ and has an amplitude $(1 + \omega^2\tau_c^2)^{-1/2}$ and a phase $\arctan(-\omega\tau_c)$ or real and imaginary parts of $1/(1 + \omega^2\tau_c^2)^{1/2}$ and $-\omega\tau_c/(1 + \omega^2\tau_c^2)^{1/2}$, respectively.

Since almost all reactor dynamics analyses involve linear systems, linear systems are assumed in this chapter. In a linear system the transfer function at a given frequency is independent of the absolute magnitude used in its measurement. Usually a sufficiently large amplitude will cause nonlinear behavior in any system, but these cases are not treated with the techniques discussed in this chapter.

It should be mentioned, however, that the transfer-function concept may be applied to almost-linear systems. Smets[1] has presented a "describing function" approach to nuclear-reactor dynamic measurements:

Describing function = (amplitude of fundamental Fourier component of output signal)/(amplitude of sinusoidal input signal) (6.1)

where the input signal is $x(t) = a \sin \omega t$ and the output signal is

$$y(t) = A_1 \sin(\omega_1 t + \phi_1) + A_2 \sin(\omega_2 t + \phi_2) + \ldots \quad (6.2)$$

The describing function is thus A_1/a. If A_1 is not linear in a, then the describing function depends on the magnitude of the input amplitude a. If the system is linear, the describing function is synonymous with the transfer function.

6-1.3 Impulse Response

As implied by its name and defined in Table 6.1, the impulse response, h(t), is the system output, y(t), when its input, x(t), is a very narrow pulse (i.e., a unit pulse of time duration much less than the smallest important time constant of the system). The impulse response is also the Green's function or weighting function. It is appropriate to discuss the impulse-response function in connection with transfer functions since, as shown in Table 6.1, it is the Fourier transform of the transfer function. To date the impulse-response function has not enjoyed the popularity of the transfer function as an analytical tool. Recently,

Table 6.1—Principal Frequency-Domain Functions and Their Corresponding Time-Domain Functions

Symbol	Name	Definition	Relation to corresponding time or frequency function
$G(f)$	Transfer function	(Output complex amplitude)/ (input complex amplitude)	$= \int_{-\infty}^{\infty} h(t)\, e^{-i\omega\tau}\, dt$
$P(f)$	Spectral density	$\frac{1}{T}\left\lvert \int_{-T/2}^{T/2} x(t)\, e^{-i\omega t}\, dt \right\rvert^2$	$= \int_{-\infty}^{\infty} C(\tau)\, e^{-i\omega\tau}\, d\tau$
$P_{xy}(f)$	Cross spectral density	$\frac{1}{T}\int_{-T/2}^{T/2} x(t)\, e^{i\omega t}\, dt \int_{-T/2}^{T/2} y(t')\, e^{-i\omega t'}\, dt'$	$= \int_{-\infty}^{\infty} C_{xy}(\tau)\, e^{-i\omega\tau}\, d\tau$
$Co_{xy}(f)$	Cospectrum	Fourier cosine transform of $C_{xy}(\tau)$	$= \int_{-\infty}^{\infty} C_{xy}(\tau) \cos \omega\tau\, d\tau$
$Qu_{xy}(f)$	Quad-spectrum	Fourier sine transform of $C_{xy}(\tau)$	$= \int_{-\infty}^{\infty} C_{xy}(\tau) \sin \omega\tau\, d\tau$
$h(t)$	Impulse response	Time response to a narrow pulse	$= \int_{-\infty}^{\infty} G(f)\, e^{i\omega t}\, df$
$C(\tau)$	Autocorrelatic function	$\frac{1}{T}\int_{-T/2}^{T/2} x(t)\, x(t+\tau)\, dt$	$= \int_{-\infty}^{\infty} P(f)\, e^{i\omega\tau}\, df$
$C_{xy}(\tau$	Cross-correlation function	$\frac{1}{T}\int_{-T/2}^{T/2} x(t)\, y(t+\tau)\, dt$	$= \int_{-\infty}^{\infty} P_{xy}(f)\, e^{i\omega\tau}\, df$

Table 6.2—Illustration of Dynamic Functions for a System Having a Single Time Constant, τ_c, One Input, and One Output

Description	Frequency-domain expression	Time-domain expression
Differential equation of the system and its Laplace transform	$x(s) = (1 + i\tau_c\omega)\, y(s)$	$x(t) = y(t) + \tau_c \frac{dy(t)}{dt}$
Transfer function	$G(\omega) = \frac{y(\omega)}{x(\omega)} = \frac{1}{1 + i\tau_c\omega}$	
Impulse response		$h(t) = 0 \quad (\tau < 0)$ $h(t) = \frac{e^{-t/\tau_c}}{\tau_c} \quad (\tau > 0)$
Spectral density of output for an arbitrary input	$P_y(\omega) = \frac{P_x(\omega)}{1 + \tau_c^2\omega^2}$	
Cross spectral density of input and output (for an arbitrary input)	$P_{xy}(\omega) = \frac{P_x(\omega)}{1 + i\tau_c\omega}$	
Autocorrelation function of output for a constant spectral-density (uncorrelated) input		$C(\tau) = e^{-\lvert\tau\rvert/\tau_c}$
Cross-correlation function of output with a constant spectral-density input		$C_{xy}(\tau) = 0 \quad (\tau < 0)$ $C_{xy}(\tau) = e^{-\tau/\tau_c} \quad (\tau > 0)$

however, Dorf[2] pointed out that since digital computers greatly facilitate time-domain analyses of systems the impulse-response function should become more popular.

Table 6.2 gives an example of the impulse response of a system with a single time constant. Evidently excitation by an input pulse, x(t), having a width much less than the system time constant stretches this pulse to a width or duration of the order of the system time constant. A physical interpretation of this is that the output, y(t), "remembers" an input pulse and shows its effect (up to about the time constant) after the input pulse.

Finally, another insight into the nature of the impulse response comes from using it as a weighting function in relating arbitrary input and output time functions:

$$y(t) = \int_0^\infty h(t')\, x(t - t')\, dt' \qquad (6.3)$$

Here the integral may be viewed as a sum of sequential pulses, $x(t - t')$, weighted according to how long ago they occurred. The Fourier transform of this equation allows one the corresponding viewpoint in the frequency domain:

$$y(\omega) = G(\omega)\, x(\omega) \qquad 6.4)$$

i.e., the transfer function, $G(\omega)$, is a weighting factor that, when applied to the input amplitude at each frequency, gives the output amplitude at that frequency.

6-1.4 Spectral Density

Whereas the transfer function treats the dynamic relation between two variables in the frequency domain, the spectral density characterizes a single variable, also in the frequency domain. Table 6.1 gives its definition in terms of the Fourier amplitude of a signal as well as its relation to the autocorrelation function (discussed below). Figure 6.2 shows the physical interpretation of the spectral density: it is the power, P(f) df, in the frequency band df present in the signal x(t). In this conceptual measurement it is the time average of the current and voltage across a unit resistor.

A signal considered as a function of time may consist of three contributions:

1. A steady-state, or d-c, value, which is the time average of the signal.

2. Discrete frequency components, $a_1 \sin(\omega_1 t + \phi_1)$, $a_2 \sin(\omega_2 t + \phi_2)$, etc.

3. A continuum of frequency components make up the randomly fluctuating or nonperiodic part of the signal.

The spectral-density concept applies primarily to the last. Spectral densities associated with the first two contributions are additive with the total of the third:

$$P_{total} = (\bar{x})^2 + \left(\frac{a_1^2}{2} + \frac{a_2^2}{2} + \ldots\right) + \int_{-\infty}^{+\infty} P(f)\, df \qquad (6.5)$$

If the signal is a current through a resistor, the three terms are, respectively, the d-c power, the a-c power of discrete

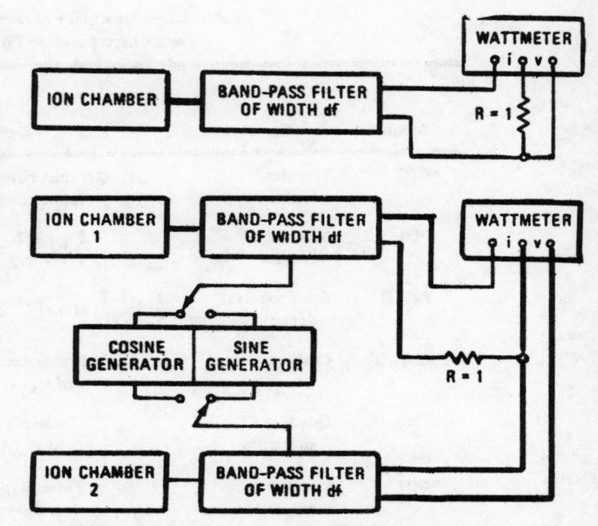

Fig. 6.2—Conceptual experimental arrangements to measure the spectral density of an ion chamber (upper) or the cross spectral density of two chambers (lower) with the current, i, and voltage, v, inputs of a time-averaging wattmeter.

frequencies, and the a-c power of random noise. The usefulness of the spectral-density concept is in characterizing the last term, and hence P(f) is sometimes called a random-noise spectrum.

Fundamental relations associated with spectral density are given in Table 6.3. As also indicated in Table 6.1, the

Table 6.3—Formulas Associated with Spectral-Density Analysis of a Random Signal, x(t), Having a Zero Mean Value

Description	Formula
Fourier integral relations between x(t) and its transform X(f)	$x(t) = \int_{-\infty}^{\infty} X(f) \exp(i\omega t)\, df$
	$X(f) = \lim_{T \to \infty} \int_{-T/2}^{+T/2} x(t) \exp(-i\omega t)\, dt$
Spectral density from Fourier amplitude	$P(f) = \lim_{T \to \infty} \frac{\lvert X(f) \rvert^2}{T}$
Total spectral power = variance = square of standard deviation = autocorrelation function at zero lag	$P_t = \int_{-\infty}^{\infty} P(f)\, df$
	$= \lim_{T\to\infty} \frac{1}{T} \int_{-T/2}^{T/2} [x(t)]^2\, dt = \overline{x^2}$
	$= \sigma^2 = C(0)$

spectral density may be obtained from Fourier amplitudes or, alternatively, by integration of an autocorrelation function. Ideally the signal duration, T of Table 6.3, would be infinite. In practice, the finite duration of the signal available for spectral analysis is an important experimental limitation (see pt. 6-7).

6-1.5 Cross Spectral Density

Just as the spectral-density function, P(f), is used to

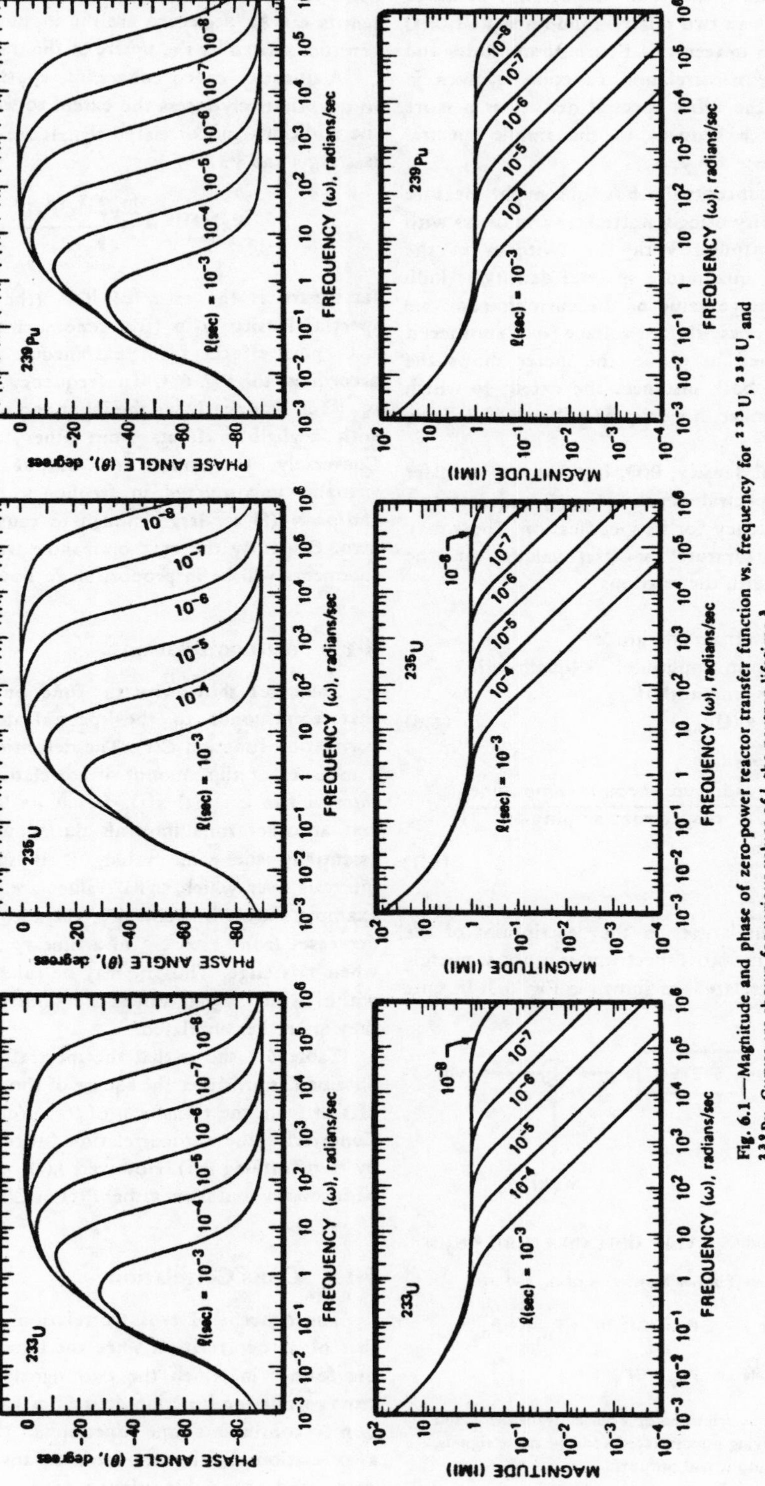

Fig. 6.1—Magnitude and phase of zero-power reactor transfer function vs. frequency for ^{233}U, ^{235}U, and ^{239}Pu. Curves are shown for various values of l, the neutron lifetime.[3]

display the relative importance of various frequency components in a single random signal, the cross spectral density, $P_{xy}(f)$, is used to show the joint importance of these frequency components in two related random signals, x(t) and y(t). Its definition in terms of Fourier amplitudes and in relation to the cross-correlation function is given in Table 6.1. Evidently the cross spectral density is a more general concept, which reduces to the simple spectral density, P(f), for the case x = y.

Figure 6.1 shows conceptually how one might measure the cross spectral density using a wattmeter and filters with switchable phased outputs. With the switches in the positions indicated, a quadrature spectral density is indicated by the time-average value of the current from one chamber and the 90° phase-shifted voltage from another; if the filters are switched in phase, the meter shows the cospectral density. In both instances the extent to which the two signals are similar in a frequency band df is being measured.

Unlike the spectral density, P(f), but like the transfer function, the cross spectral density, $P_{xy}(f)$, requires two numbers at each frequency for its specification. These may be the "co" and "quadrature" spectral values or the amplitude and phase, with the relations

$$\begin{aligned} |P_{xy}|^2 &= (\text{cross-spectrum amplitude})^2 \\ &= (\text{cospectrum amplitude})^2 + (\text{quadrature spectrum amplitude})^2 \\ &= (Co_{xy})^2 + (Qu_{xy})^2 \end{aligned} \quad (6.6)$$

$$\begin{aligned} \theta &= \text{phase angle} \\ &= \arctan \left| \frac{\text{quadrature spectrum amplitude}}{\text{cospectrum amplitude}} \right| \\ &= \arctan \frac{Qu_{xy}}{Co_{xy}} \end{aligned} \quad (6.7)$$

Because of the similarities in the descriptions of the transfer function and the cross spectrum, it is not surprising to find that these are related, as shown in Fig. 6.3. In spite

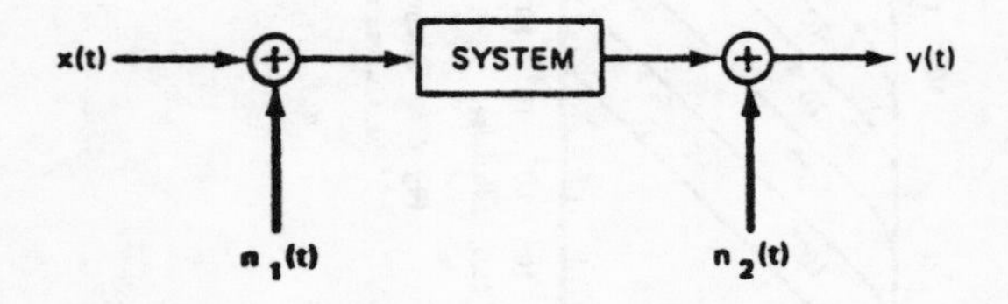

Transfer-function relation: $Y(s) = G(s)\,[X(s) + N_1(s)] + N_2(s)$

where $N_1(s)$ and $N_2(s)$ are Fourier transforms of n_1 and n_2

Spectral-density relation: $P_y = |G|^2\,[P_x + P_{n_1}] + P_{n_2}$

Cross-spectral-density relation: $P_{xy} = GP_x$

Fig. 6.3—Input–output relations of Fourier transforms and spectra in a system having uncorrelated additive noise signals, n_1 (t) and n_2 (t), at its input and output.

of additional signals (such as unwanted noise) at the input and output, a simple relation exists: the transfer function, G, times the input spectral density, P_x, is the cross spectral density, P_{xy}. On the other hand, only when the unwanted signals can be neglected are the input and output spectral densities related by the square of the transfer function.

A quantity called coherence, $c_{xy}(f)$, has been defined to quantitatively assess the extent to which the presence of the additional uncorrelated signals are not negligible when relating x and y:

$$|c_{xy}(f)|^2 = \frac{|P_{xy}|^2/|P_x|^2}{P_y/P_x} = \frac{|P_{xy}|^2}{P_x P_y} \quad (6.8)$$

Its square is the ratio of $|G|^2$ (the numerator) to the spectral-density ratio (the denominator), the latter being $|G|^2$ plus effects from extraneous signals n_1 and n_2, according to Fig. 6.3. In frequency ranges over which $c_{xy}(f)$ is 1 or nearly so, the input and output can be related with negligible effects from other uncorrelated signals. Conversely, the input and output can be considered virtually uncorrelated in frequency ranges where $N_1(f)$ and/or $N_2(f)$ are large enough to cause $c_{xy}(f)$ to be near zero. Evidently the ease of making transfer-function measurements will be in proportion to how near $c_{xy}(f)$ is to 1.

6-1.6 Autocorrelation

Table 6.1 shows that the function in the time domain that corresponds to the spectral density is the autocorrelation function $C(\tau)$. The definition indicates that it is a measure of the amount of correlation existing at a time interval τ in a signal x(t). It has its largest values at $\tau = 0$ and at other time intervals during which the signal has essentially the same value; it is smallest during time intervals over which signal values are uncorrelated. In the example shown in Table 6.2, the autocorrelation function decreases from 1 to e^{-1} in a time τ_c and approaches zero when τ is large. Thus τ_c may be called a correlation time within which signal values are similar and beyond which they are rather unrelated.

Table 6.1 shows that the spectral density, P(f), can be obtained from either the square of the Fourier transform of x(t) or from the transform of its autocorrelation function. Conversely, the autocorrelation function can be obtained by transforming P(f). However, x(t), when random, cannot be reconstructed from either P(f) or $C(\tau)$.

6-1.7 Cross Correlation

The concept of cross correlation is more general than that of autocorrelation since the latter is a special case of the former in which the two signals are the same. The cross-correlation function defined in Table 6.1 is an application to continuous time functions of the digital concept of a correlation coefficient of statisticians: if x_i and y_i are two time series of variable values spaced in time (x_i being at the same time as $y_{i+(\tau/\Delta t)}$) in which the degree of correlation is sought, then

$$C_{xy} = \sum_{i=1}^{N} \frac{x_i y_i}{N} \qquad (6.9)$$

is a measure of this. However, it is customary to define a normalized correlation coefficient in terms of fluctuations from means:

$$\phi_{xy} = \frac{1}{N\sigma_x\sigma_y} \sum_{i=1}^{N} (x_i - \bar{x})(y_i - \bar{y})$$

$$= \frac{C_{xy} - \overline{xy}}{\sigma_x\sigma_y} \qquad (6.10)$$

where σ_x and σ_y are the standard deviations $\phi_{xx}^{1/2}$ and $\phi_{yy}^{1/2}$. This is +1 or −1 for perfect correlation or anticorrelation, respectively, and is 0 if there is no correlation. The integral expression for cross correlation in Table 6.1 is evidently digitally evaluated in Eq. 6.9.

In the example in Table 6.2, C_{xy} is zero for $\tau < 0$ because the output cannot "know" ahead of time what the perfectly random input, x(t), will be. A significant input–output correlation, however, does exist for values of τ up to the order of τ_c, the correlation time of the system. In other more complex systems, the maximum value of C_{xy} might occur at some time other than zero, in which case a time-lag effect between x and y will have been identified.

Table 6.1 shows the frequency-domain function corresponding to $C_{xy}(\tau)$ to be the cross spectral density, $P_{xy}(f)$. These are Fourier-transform pairs, and, if one is known, the other can be found from the relations shown.

6-2 REACTOR APPLICATIONS

6-2.1 Neutron Kinetics

For a study of the time behavior of reactors, the equations giving the time dependence of the neutron density, N + n (mean value plus deviations therefrom), and the *jth* group of delayed-neutron precursors, $C_j + c_j$, are

$$l\frac{dn}{dt} = [k(1-\beta) - 1](N + n) + \sum_j l\lambda_j(C_j + c_j) + lS \qquad (6.11)$$

$$\frac{dc_j}{dt} = -\lambda_j(C_j + c_j) + \frac{\beta_j k(N + n)}{l} \qquad (6.12)$$

where the sum of the delayed-neutron fractions, β_j, is the total fraction, β; λ_j is the decay constant of a precursor; S is a source; l is the prompt-neutron lifetime; and k is the effective multiplication constant that, when not unity, represents the departure of the reactor from exact criticality,

$$\rho = 1 - \frac{1}{k} \qquad (6.13)$$

being the excess reactivity.

The solution to these equations, under conditions of all variables undergoing small oscillations about their mean values, is the zero-power transfer function, G_0, defined as

$|G_0|$ = [(amplitude of power oscillation)/(average power)] /(amplitude of reactivity oscillation) (6.14)

and having the phase

Phase angle = 360° × (fraction of a cycle that the power lags behind the reactivity oscillation) (6.15)

The zero-power transfer function can also be regarded as the quotient of the Fourier transforms of the power and reactivity divided by the average power.

Table 6.4 gives explicit formulas for this transfer function in terms of reactor constants and the frequency.

Table 6.4—Forms of the Complex Amplitude of the Zero-Power-Reactor Transfer Function $G_0(\omega)$

Conditions	Formula for complex amplitude of $G_0(\omega)$
No approximations	$\dfrac{1 - i\omega \sum_{j=1}^{6} \beta_j/(\lambda_j + i\omega)}{1 - k + i\omega\left[l + k \sum_{j=1}^{6} \beta_j/(\lambda_j + i\omega)\right]}$
$\omega > 2\bar{\lambda}$	$[1 - k(1-\beta) + i\omega l]^{-1}$
One delay group	$\left[1 - k + i\omega\left(l + \dfrac{k\beta}{\bar{\lambda} + i\omega}\right)\right]^{-1}$
One delay group and $\omega < \dfrac{1}{2l}[1 - k(1-\beta)]$	$\left[1 - k + i\omega\dfrac{k\beta}{\bar{\lambda} + i\omega}\right]^{-1}$

Also, G has been tabulated in detail in Ref. 3. In essentially all but subcritical reactor applications, k may be set equal to 1 to further simplify the approximations there. At mid-frequencies, where $2\bar{\lambda} < \omega < 0.5\beta/l$, a very simple result, $G \cong 1/\beta$, exists. At these frequencies the physical interpretation is

Percent power oscillation about its mean

$$= \frac{1}{\beta} \times \text{(percentage reactivity amplitude)}$$

$$= \text{reactivity amplitude in cents} \qquad (6.16)$$

where β is typically 0.007.

6-2.2 Zero-Power Measurements

Historically the rod-oscillator measurement of the zero-power transfer function is one of the oldest reactor-dynamics measurements,[4] dating back to CP-2. Since then it has been repeated many times on many reactors with the techniques detailed below.

Basically, a sinusoidal reactivity excitation provided by

a rotating or reciprocating control rod causes a power oscillation that is detected by ion chambers. The purpose of these measurements may be any of the following:

1. To determine β/l for a particular reactor by fitting the measurement to a formula in Table 6.4.

2. To verify experimental techniques on a known transfer function.

3. To compare reactivity effects of small samples by the amplitude of the resulting power oscillation.

A considerable number of methods in addition to the rod oscillator have proved useful in obtaining dynamics information, i.e., in determining quantities closely related to the reactor transfer function. These methods are listed and classified in Table 6.5. Methods having no reactor excitation by external equipment depend on the random fluctuations or noise in the neutron population, as detected by counters or ion chambers, to provide information about reactor characteristics. The following incentives for applying such methods differ slightly from those for the rod oscillator:

1. To determine specific reactor parameters, such as β/l at critical, the subcritical reactivity, or the absolute reactor power.

2. To verify experimental techniques on a known system.

3. To investigate spatial neutron effects.

Table 6.5—Experimental Methods of Obtaining Dynamic Information from Zero-Power Reactors

Method	External excitation	Detection equipment
Rod oscillation or pseudorandom motion	Control rod	Ion chamber or gamma detector
Source oscillation or pseudorandom changing	Neutron-source generator	Ion chamber
Rossi alpha	None	Coincidence counting
Correlation and spectral analysis	None	Ion chamber or gamma detector (or pair for correlation)
Variance to mean	None	Gate scaler counting, or ion-chamber current integrating
Probability of neutron events	None	Coincidence counting

Experiments on the dynamic behavior of zero-power reactors have been numerous and also somewhat repetitious owing to similarities among the reactors and kinds of equipment used. The particular reactors studied along with the classes of information obtained are given in Table 6.6. It will suffice here to point out that the quantities being measured are the constant parameters appearing in the neutron kinetics equations and the transfer function. (The numerous cases in which samples have been oscillated to make reactivity measurements are omitted since the emphasis here is on transfer functions.) Part 6-3, Methods of Measurement, summarizes the features of the various methods and how the results in Table 6.6 are obtained.

Although neutron detectors have been used in virtually all the zero-power dynamics studies to date, there has been some theoretical[73a] and experimental[23,69a] work involving gamma detectors. Čerenkov detectors and liquid scintillators have been found suitable for monitoring the fluctuations of prompt gammas from fission. The emission of these gammas, like the neutrons, exhibits statistical fluctuations that depend on the reactor's zero-power transfer function. Since gammas travel farther than neutrons in a reactor, the prompt gammas offer a means of using peripheral detectors to monitor deep into the core of a large reactor. In one novel application[69a] two widely separated gamma detectors were pointed in collimated fashion at various fuel elements, and relative local power values were obtained by cross correlation.

6-2.3 Power-Reactor Feedback

The neutron kinetics equations (Eqs. 6.11 and 6.12) that were applied to zero-power reactors also hold for power reactors. However, additional relations exist here between reactor power and reactivity because the former is high enough to cause effects on the latter (see Fig. 6.4). This feedback of reactor power to reactivity is the important characteristic of power-reactor dynamics. As might be expected, transfer-function measurements are used to obtain information about this feedback.

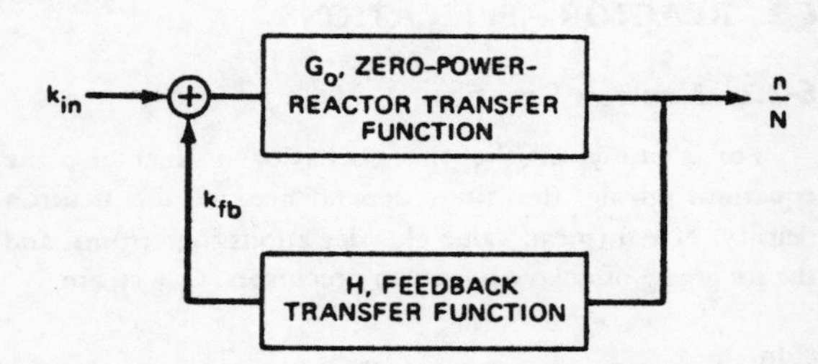

Fig. 6.4—Relations between the feedback and the zero-power transfer functions which make up the transfer function of a power reactor.

Equations that relate power to reactivity can be numerous and complex, depending on the system. Specific forms for boiling- and pressurized-water reactors have been summarized.[74] Where commonly a set of linear differential equations adequately represent the system, the feedback transfer function from power to reactivity may be written

$$\frac{k_{fb}(s)}{n(s)/N} = H(s) = \sum_i \gamma_i \frac{\Pi_j (1 + \tau_{ij} s)}{\Pi_k (1 + \tau_{ik} s)} \qquad (6.17)$$

Table 6.6—Results Achieved in Zero-Power Dynamic Studies

Method and excitation	Reactors used	Results
Random pulsing of a neutron source	GE Critical Assembly,[5] Rubeole and Ulysse,[6] UFTR,[7,8] University of Florida subcriticals,[9,10]	β/l, subcritical reactivity
Oscillating control rod	CP-2,[4] DFR,[11] EBR-1 and EBR-2,[12,13] EBWR,[14] Fermi,[15] GTRR,[16,17] KEWB,[18] Kiwi-A,[19] LPTR,[20] LAMPRE-1,[19] NORA,[21] Pathfinder,[22] Penn. State,[23] PTR,[24] SHE,[25] SPERT,[26] Zephyr and Zeus[11,27]	β/l, verification of equipment performance
Variance to mean (no excitation)	AGN-201,[28] Cornell,[29] Ford,[30] Homogeneous D_2O Facility,[31] NORA,[21,32] Russian subcritical,[33] SPERT-2 and SPERT-4,[34,35] Tokyo Inst. of Technology,[36] ZPR-4,[37,38] ZPR-5[37]	β/l, absolute power level
Pseudorandom control rod	ATSR,[39] Brookhaven,[40] Godiva-II,[41] JRR-3 and SHE,[42-44] MSRE,[44a] Saxton,[45] UFTR[8]	β/l, verification of equipment performance
Spectral analysis and/or autocorrelation (no excitations)	Atomics Int.,[46] Babcock & Wilcox Test Reactor,[47] BSR,[48-49] Battelle Plutonium Assembly,[50] Brookhaven High Flux,[50a] EBWR,[51] Daphne and Jason,[52,53] GLEEP,[54], HFIR,[55] HTR, JRR-1, SHE, and TRR-1,[42,43,55a,55b] Iowa State Univ., UTR-10,[56] KEWB and SRE,[57] LFR,[58] MSRE,[59] NORA,[21] ORNL Pool,[60,61] Penn State,[23] Savannah,[62] Saxton,[63] SPERT-1,[34] SPERT-3, ATRC, ARMF-1, and PBF,[64] SNAP-10A and ETR,[65,66] UFTR,[7,8] Westinghouse CES,[67] ZPR-3, ZPR-4, ZPR-5, ZPR-7, and Argonaut,[68] ZPR-6[68a]	β/l, subcritical reactivity, absolute power level
Two-detector cross correlation (no excitation)	GTRR,[16] JPDR,[55a] Karlsruhe Argonaut,[68b] ORNL Pool,[69] Penn State,[69a] STARK,[70] Savannah SR-305 Pile,[71] TRR-1,[72] ZPR-9[73]	β/l, subcritical reactivity, local power distribution

Table 6.7—Results Achieved in At-Power Dynamics Studies

Method and excitation	Reactors used	Results
Oscillating control rod	BORAX-4 and BORAX-5,[75,76] DFR,[11] EBR-1. and EBR-2,[12,13,77] EBWR,[14,78] Fermi,[15] GE BWR,[79] HTR,[80] JPDR,[80a] SPERT-1,[26] SRE,[81] VBWR[82]	Stability study; feedback determination
Pseudorandom control rod	Kiwi-A3,[41] MSRE,[44a,83] Saxton,[84,85] JRR-3[85a]	Stability study; control-system study
Spectral analysis (no excitation)	BORAX-1, -2, -4, and -5,[75,76,86] Chapelcross,[87] Dresden,[88] EBR-2,[89] EBWR,[90-92] Elk River,[93] GTR,[94] HBWR,[95] HFIR and ORR,[59,96,97] HFBR,[98] HRE-2,[99] HTR and JRR-1 and JRR-2,[42,43,80,100,100a] Indian Point, Savannah, and Yankee,[62] JPDR,[55a,100a] Kyoto Univ.,[101] ML-1,[102] Pluto,[53] Saxton,[63] SPERT-4,[103] SRE,[81] THOR,[104] Trino,[85] VBWR[105,106]	Stability study; noise-source determinations
Pseudorandom plant control	Nerva,[107] Phoebus 1A,[108] SNAP[109]	Intervariable transfer functions
Cross correlation and spectral analysis	DFR,[110] DMTR,[110a] SNAP,[46] Pathfinder[111]	Stability study, intervariable transfer functions; noise-source determinations

where $s = i\omega$ and the values of τ are time constants associated with a system variable that has a feedback reactivity change of γ_i percent for a 1% power change (in the limit of very slow transients).

The transfer function between an external excitation reactivity, k_{in}, such as an oscillator rod, and the reactor power is the solution of Eqs. 6.14 and 6.17 if $k = 1 + k_{in} + k_{fb}$ is used:

$$G(s) = \frac{G_0(s)}{1 - G_0(s)\,H(s)} \qquad (6.18)$$

Since G_0 is well known, this equation is generally used to obtain H(s) or G(s) from the other.

6-2.4 Power-Reactor Measurements

Measuring the dynamic characteristics of power reactors is commonly accepted to be an integral part of testing during reactor commissioning. Transient response to externally induced system changes is perhaps the most popular category of such tests. However, transfer-function and noise-spectrum measurements are also common. It is not unusual for the latter to be required by AEC licensing in the interest of safety.[74a] The incentive for transfer-function and noise testing usually stems from the desire for a more detailed knowledge than is possible from the transient tests.

The possibility that the denominator of Eq. 6.18 will approach zero and result in an unstable oscillatory system is one important reason for measuring and understanding G(s). On the other hand, for reactors known to be quite stable, measurements of G(s) can give H(s) if Eq. 6.18 is used. The term H(s) gives information about reactor and plant parameters, such as the constants in Eq. 6.17.

Transfer-function measurements in power reactors are not restricted to rod-oscillator tests. The next section shows that a considerable variety of methods involving variables other than just reactivity and power are used. Table 6.7, listing the many power-reactor transfer-function and related spectral-analysis experiments, gives an idea of the wide applicability of the techniques given in Table 6.8.

In power-reactor dynamics it is often desirable to know transfer functions among a variety of variables, not necessarily just between reactivity and power. Thus it is not uncommon to simultaneously measure a number of transfer functions, or spectral-density functions, by simultaneously measuring pairs of system variables over a period of time. In the experiments listed in Table 6.8, the primary interest is usually in the neutron-flux fluctuations, but other fluctuation variables have also been analyzed, namely, pressure, flow, acoustical noise, temperature, gamma flux, valve position, and pump speed. These, along with the reactivity and power, represent principal variables of interest in dynamics analyses.

6-3 METHODS OF MEASUREMENT

6-3.1 Reactor Excitation

To determine the dynamic characteristics of a system, one must measure variables that are changing with time. Furthermore, the variations must have the following properties:

1. Amplitudes sufficient to override unwanted effects that could reduce accuracy.
2. A sufficiently long duration or a sufficient number of repetitions to provide the desired accuracy.
3. Frequencies in the ranges to be investigated.

If the intrinsic variations, or random noise, of a system are used, the system is said to be self-excited. On the other hand, a system is said to be externally excited if a perturbation is introduced by a signal-generating device. In both instances transducers responsive to the variations of interest provide the experimental data.

Table 6.9 lists the kinds of excitation that have been used to date. The relative popularity of the various forms of excitation (Tables 6.6 and 6.8) is indicated somewhat arbitrarily by the number of dynamics experiments that have used each form. Transfer functions and related functions have been emphasized in Tables 6.6 and 6.8. Many other* dynamics tests (such as valve-position changes in power plants, rod drops, positive-period tests, and Rossi-alpha coincidence counting) are not represented even though they may be somewhat related to the tests discussed here. With this understanding the predominance of self-excitation experiments indicated in Table 6.9 can be attributed, at least in part, to their experimental simplicity

Table 6.8—Experimental Methods of Obtaining Dynamic Information from Power Reactors

Method	External excitation	Detection equipment
Rod oscillation or pseudorandom motion	Control rod	Ion chamber and other transducers
System excitation	Valve, pump, etc.	Ion chamber and other transducers
Correlation and spectral analysis	None	One (or more for correlation) ion chamber and other transducers
Event analysis	Any cause	All detectors that respond to the transient event

(from the standpoint of reactor hardware). Also, although this table shows more periodic devices than random devices used as externally induced excitation, a trend in recent years to increased use of random excitation must be noted.

For externally excited experiments, a variety of methods is used (see Table 6.10). Except for the occasional use of on-line electronic analyzer methods, most experiments do not give transfer functions until the recorded data are processed off-line. Both electronic analyzers and digital computers are used for this processing.

As noted in Table 6.8, the excitation may be sinusoidal or pseudorandom and either a control rod or some other plant control device may be used to excite the system. In

Table 6.9—Relative Use in Reactor Dynamics of Excitation Devices for Measurement of Transfer Functions and Allied Functions

Type of fluctuations	Percentage of zero-power-reactor (ZPR) or power-reactor (PR) experiments using a particular excitation device							
	Neutron source		Control rod		Other		Self-excited	
	ZPR	PR	ZPR	PR	ZPR	PR	ZPR	PR
Periodic–sinusoidal	0	0	22	23	0	0	0	0
Nonperiodic–random	6	0	8	6	0	6	64	65
Total, %	6	0	30	29	0	6	64	65

sinusoidal excitation the transfer function between the excitation variable (such as the reactivity of a control rod) and the system output variable (such as the reactor power) may be obtained at the excitation frequency by any one of the following approaches:

1. Using the separately measured fundamental frequency amplitudes and phases of input and output, applying Eq. 6.1 or Table 6.1.

2. Applying the appropriate electronic gain and phase to the output and using it to "null out" the input signal (see pt. 6-5.2).

3. Cross-correlating the input and output signals, using Eq. 6.9 in digital processing or Table 6.1 in continuous processing (see pt. 6-5.4).

The last approach is commonly used at present.

Table 6.10—Data-Acquisition and Data-Processing Techniques Used in Externally Excited Reactor Dynamics Experiments

On-line acquisition device	Off-line processing device	References to typical applications
Chart or film recorder	Digitizer, digital computer	18, 76
Electronic analyzer	None	11, 14
F-m tape recorder	Electronic analyzer	44, 85
F-m tape recorder	Digitizer, digital computer	44, 107
Digitizer, tape recorder	Digital computer	53, 78

Figure 6.5 shows typical rod-oscillator test results for a boiling-water reactor. In such tests one is interested in the difference between the at-power transfer function and the zero-power transfer function for this can determine the feedback, H, in Eq. 6.17. In addition, the height and width of the resonance (in this example at $\omega = 7$ radians/sec) are of interest because they indicate the extent to which an instability of self-sustained oscillations is being approached.

In pseudorandom excitation you attempt to introduce all frequencies in the band of interest into the system at once rather than sequentially as in sinusoidal testing. In the commonly used binary excitation, an input control signal has two values, such as +1 and −1; however, ternary signals (having values +1, 0, and −1) have been suggested.[112] Rather than letting the duration of +1 and −1 values be determined by an ideal random process, it is more advantageous to use a repetitive almost-random signal,[41] such as that shown in Fig. 6.6.

In analyzing data in pseudorandom excitation experiments, you obtain the cross-correlation function by either on-line or off-line integration

$$C_{xy}(\tau) = \frac{1}{T} \int_{-T/2}^{T/2} x(t)\, y(t + \tau)\, dt \qquad (6.19)$$

using the time shifted product of the input variable, x(t), and the output variable, y(t). Figure 6.7 shows a typical experimental result. Using the relations in Table 6.1, you obtain the transfer function from $P_{xy}(f)$, the Fourier transform of $C_{xy}(\tau)$:

$$G(f) = \frac{P_{xy}(f)}{P_x(f)} \qquad (6.20)$$

With x(t) perfectly random, it can be shown that $C_{xy}(\tau)$ in Eq. 6.19 is the system impulse-response function[41] and $P_x(f)$ in Eq. 6.20 is a constant. The transfer function in Fig. 6.7 was obtained in this manner.

It is not usual to measure transfer functions by pulse or step excitation, although this is possible.[59] In the method the quotient of Fourier amplitudes is taken from an analysis of the input and the output signals to obtain the transfer function. However, it is preferable to have a series of pulses or steps, such as pseudorandom excitation, because the added signal energy helps overcome unwanted noise.

6-3.2 Noise Methods

Noise techniques, the class of reactor dynamics experiments in which no external excitation signal is used, are among the categories listed in Table 6.6, namely:

1. Variance-to-mean method (sometimes called the Feynman method).

2. Spectral analysis or the time-domain equivalent, autocorrelation.

3. Cross spectral analysis or the time-domain equivalent, cross correlation.

Other noise-analysis techniques (such as various kinds of probability analysis of individual pulses) are not sufficiently related to transfer functions to warrant discussion here, but they have been treated in Refs. 113, 113a, and 113b along with the three techniques cited above. In power reactors only the last two methods are used, whereas in zero-power

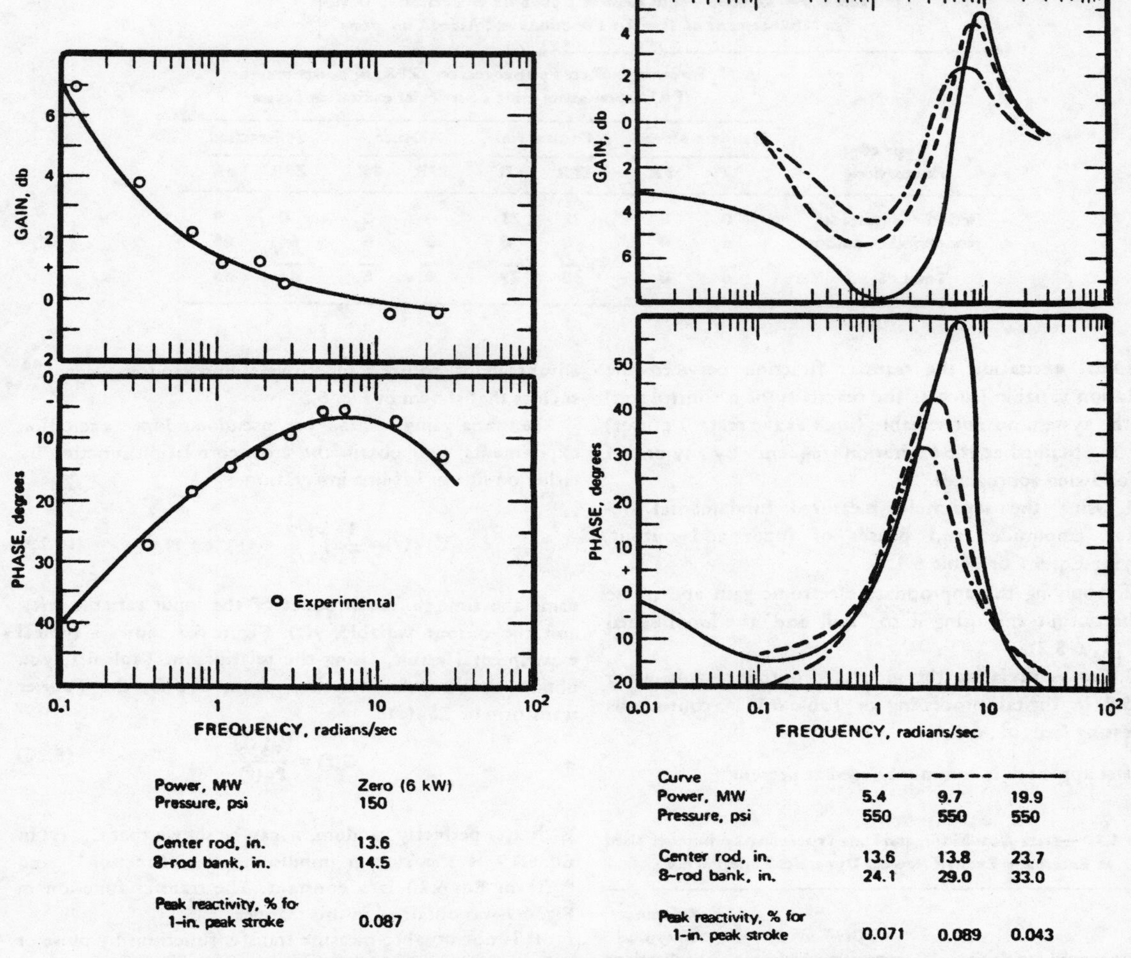

GAIN, db
PHASE, degrees
O Experimental
FREQUENCY, radians/sec
Power, MW
Zero (6 kW)
Pressure, psi
150
Center rod, in.
13.6
8-rod bank, in.
14.5
Peak reactivity, % for 1-in. peak stroke
0.087
Curve
Power, MW
5.4
9.7
19.9
Pressure, psi
550
550
550
Center rod, in.
13.6
13.8
23.7
8-rod bank, in.
24.1
29.0
33.0
Peak reactivity, % for 1-in. peak stroke
0.071
0.089
0.043

Fig. 6.5—Transfer functions of the Experimental Boiling Water Reactor at various power levels as obtained by the rod-oscillator null-balance method.[14] Curves at left show frequency response at zero power. Curves at right are for 5.4, 9.7, and 19.9 MW(th).

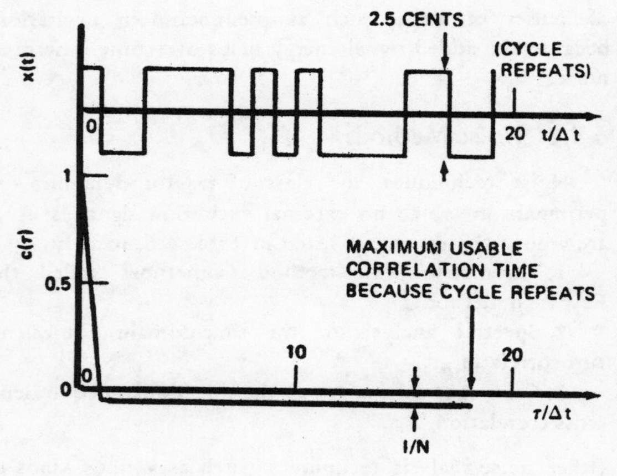

Fig. 6.6—A simple pseudorandom signal (above) that has been tailored to give the almost ideal autocorrelation function (below).[113]

reactors all three methods can be used.

Whether reactor dynamics are studied by introducing external excitation or by relying on the reactor's intrinsic self-induced noise, the data-acquisition and data-processing hardware are almost the same. Noise methods, of course, process no signals from excitation equipment. Table 6.11, having much in common with Table 6.10, shows the types of equipment used in the various noise-analysis experiments described briefly here. Most of the equipment is used for spectral and cross spectral analysis, and only that involving the gate scaler or ion-chamber current integrator is used for the variance-to-mean method.

In the variance-to-mean method, the dynamic constants in the neutron kinetic equations can be determined by using a digital computer to give

1. The variance of neutron-detector counts $[\overline{c^2} - (\bar{c})^2]$ taken many times over a time interval or "gate," τ.

2. The average count, $\bar{c}$, during τ.

Results for various gate times[30] can be shown to conform to the following equation:

$$\frac{\overline{c^2} - (\overline{c})^2}{\overline{c}} = 1 + 1.59\epsilon \sum_{j=1}^{7} \frac{A_j}{\gamma_j} G_0(\gamma_j) \left(1 - \frac{1 - e^{-\gamma_j \tau}}{\gamma_j \tau}\right) \quad (6.21)$$

Figure 6.8 shows an example of this relation.

Table 6.11—Data-Acquisition and Data-Processing Techniques Used in Reactor-Noise-Analysis Experiments

On-line acquisition device	Off-line processing device	References to typical applications
Chart recorder	Digitizer, digital computer	90, 111
Electronic analyzer	None	67, 72, 113c
F-m tape recorder	Electronic analyzer	46, 63, 110
F-m tape recorder	Digitizer, digital computer	109
Digitizer, tape recorder	Digital computer	92, 95
Digital computer	None	68a, 73
Pulse tape recorder	Gate scaler, digital computer	30
Gate scaler	Digital computer	34, 35
Ion-chamber current integrator	Digital computer	38

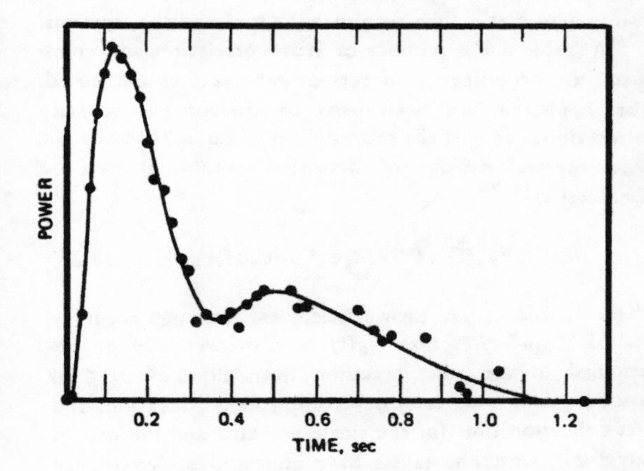

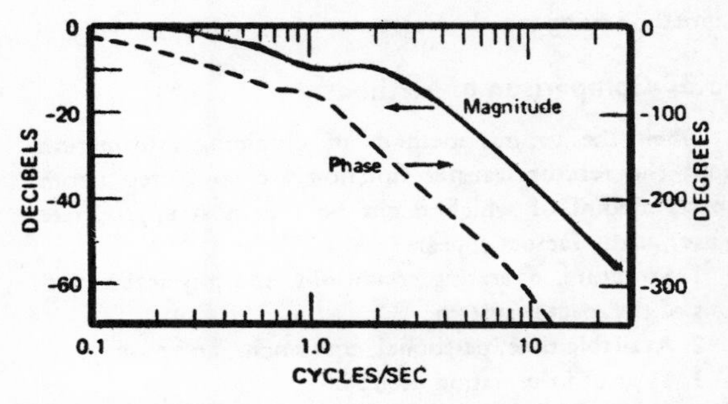

Fig. 6.7—Impulse response obtained by cross-correlating reactor power and a pseudorandom control rod in Kiwi-A3 (upper curve) and the transfer function of this reactor (lower curves).[113]

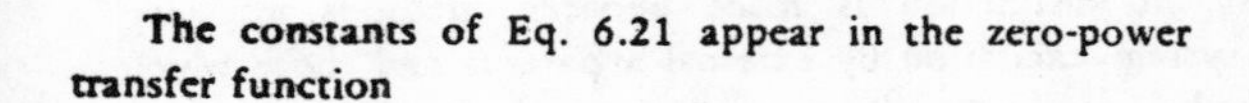

The constants of Eq. 6.21 appear in the zero-power transfer function

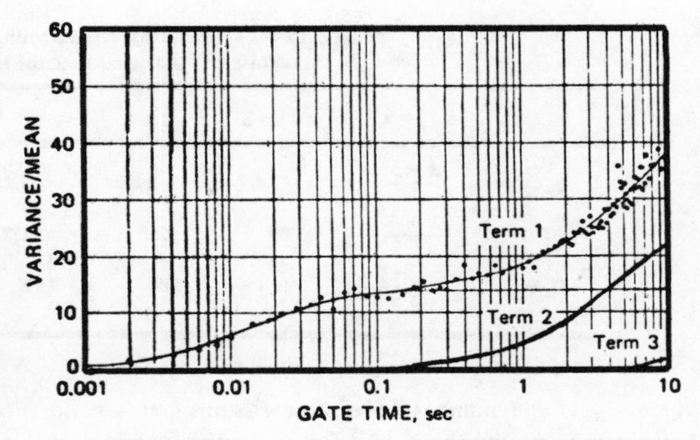

Fig. 6.8—Results of determinations of the variance-to-mean ratio for many counter gate times on the Ford reactor.[113]

$$G_0(\omega) = \sum_{j=1}^{7} \frac{A_j}{\gamma_j + i\omega} \quad (6.22)$$

and are given in Table 6.12. By fitting Eq. 6.21 to data, as in Fig. 6.8, you can evaluate the constants, especially $\gamma_1 = (\beta - \rho)/l$.

Related to this method are others, such as the Mogil'ner method,[33] based on the probability of no counts in a time interval, and a count interval-distribution method of Babala.[32] These and many similar techniques of time-domain analysis of neutron pulses in zero-power reactors have been extensively reviewed in Refs. 113a and 113b.

The constant ϵ in Eq. 6.21 is important in noise measurements of zero-power reactors. It is the detector efficiency and is defined as

$$\epsilon = \frac{\text{number of counts/sec}}{\text{number of reactor fissions/sec}} \quad (6.23)$$

Evidently ϵ is the probability of detecting an individual fission. Counters are located in or quite near the reactor core to obtain the values above about 10^{-5} which are needed for a successful experiment. For very large reactors only a zone near the detector contributes neutrons and determines an effective efficiency.

Spectral analyses of ion chambers or fluctuations of other variables are usually accomplished experimentally in one of two ways: (1) by passing the signal through a narrow band-pass filter tuned sequentially to the various desired frequencies or (2) by obtaining the autocorrelation function of the signal and then performing a Fourier analysis at the various desired frequencies. (Direct Fourier analysis of the signal is rarely done.) Table 6.11 indicates the various combinations of equipment that can be used to accomplish one or the other of these approaches.

In spectral analysis of the signal from an ion chamber in a zero-power reactor, the shape of the transfer function, G_0, is obtained directly from the measured P(f) of the detection of particles by using the relation[113]

$$\frac{P(f)}{\epsilon F_0} = 1 + 0.795\, \epsilon |G_0(f)|^2 \quad (6.24)$$

Table 6.12—Constants Associated with the Zero-Power Transfer Function for a ^{235}U-Fueled Reactor Near Delayed Criticality

	j = 1	j = 2	j = 3	j = 4	j = 5	j = 6	j = 7
γ_j	$\frac{\beta - \rho}{l}$	2.89	1.02	0.195	0.068	0.0143	-11.6ρ
A_j	$\frac{1 - \beta}{l}$	29	20	11.2	6.1	1.2	11.6
$G_0(\gamma_j)$	$\frac{1 - \beta}{\lambda(\beta - \rho)}$	164	186	237	284	343	$415 - (0.5/\rho)$

where F_0 is the number of reactor fissions per second. Again ϵ must exceed about 10^{-5} for successful experiments.

For the ion-chamber noise in a power reactor, the spectrum $P_n(f)$ of $n(t)$ in Fig. 6.4 is measured, its fluctuations being induced by an internal noise source, $k_{in}(t)$. Evidently,

$$P_n(f) = |k_{in}(f)|^2 \; |G(f)|^2 \tag{6.25}$$

where G(f) is given by Eq. 6.20. Thus ion-chamber noise analysis in a power reactor gives information about both the transfer function and the input reactivity noise. Figure 6.9 shows typical results for power operation and how the spectrum differs from the zero-power spectrum. Figure 6.10 indicates that large pressurized-water reactors of similar structure have similar noise spectrums.

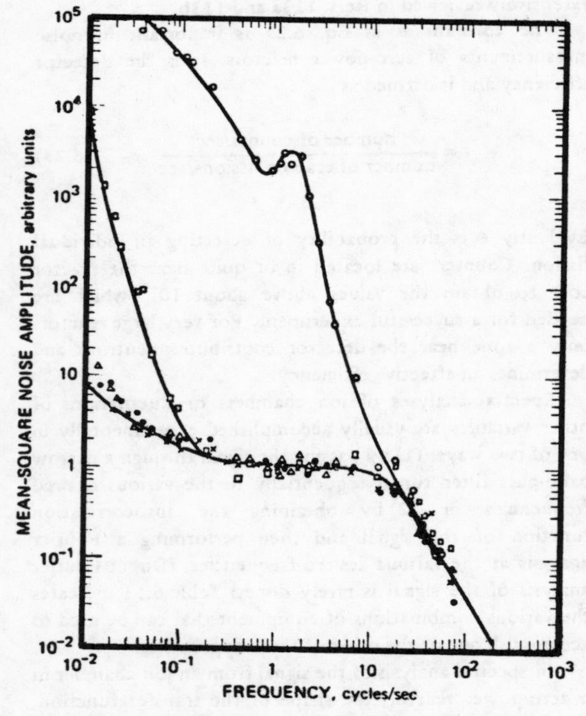

Fig. 6.9— Spectral-density measurements of ion-chamber noise in the Hanford Test Reactor at powers of 1 watt (x), 5 watts (•); 500 watts (△); 5 kW (□); and 100 kW (○).[80]

In the more informative power-reactor experiments, two signals are observed simultaneously and correlated to obtain a transfer function between them. One way to do this follows from Eqs. 6.19 and 6.20. The terms x(t) and y(t) are any two system variables whose fluctuations are related. After their cross-correlation function has been measured, their cross spectrum can then be determined. An accurate transfer function can be obtained from Eq. 6.20 if x(t) and y(t) depend primarily on the same noise-source excitation, i.e., if they have a high coherence, Eq. 6.8.

As indicated in Table 6.11, a cross-spectrum analyzer can be used directly; it is not necessary to determine the cross-correlation function first. Equation 6.20 is still used to obtain a transfer function. As in cross correlation, only two variables at a time are treated in multivariable systems.

In Table 6.6 a number of cross-correlation and cross-spectrum experiments in zero-power reactors are noted. This approach has been used to measure a quantity proportional to just the second term of Eq. 6.24 since the cross spectral density of detection events in two ion chambers is[70,72]

$$P_{xy}(f) = 0.795\epsilon_1 \epsilon_2 F_0 \; |G_0(f)|^2 \tag{6.26}$$

where ϵ_1 and ϵ_2 are their efficiencies. Although accuracy analysis[70] indicates that $G_0(f)$ may theoretically be determined to the same precision from Eq. 6.24 or 6.26 (assuming the same total detection volume, location, and data-collection time for the single detector and the pair of detectors), experimentalists have indicated preference for the method using two detectors.

6-3.3 Comparison of Methods

When the various methods of obtaining information about the reactor transfer function are compared (from the standpoint of which might be the most appropriate to use), many factors appear:

1. Structure, operating conditions, and physical limitations of the reactor system.
2. Available time, personnel, equipment, and money.
3. Type of information desired.
4. Accuracy desired.
5. Established operating and experimental policies of the plant.

The various methods described in preceding sections are treated here with this set of determining factors in mind.

A distinction is made between methods involving system excitation by external apparatus and those which rely on internally generated noise. In Table 6.13 an

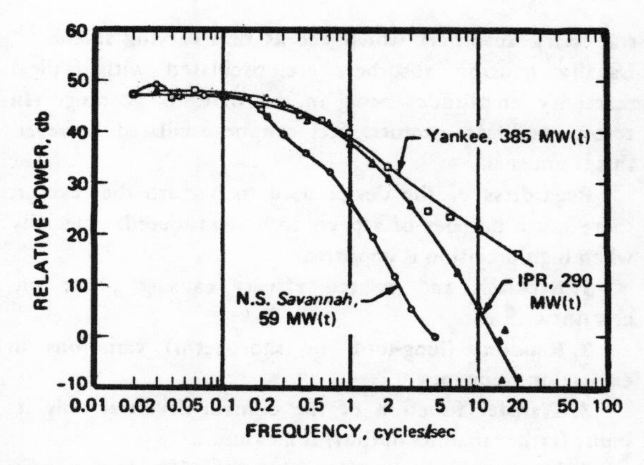

Fig. 6.10—Spectral-density measurements in three large pressurized-water reactors.[62] □, Yankee, 385 Mw(th). △, Indian Point, 290 MW(th). ○ N.S. *Savannah*, 59 MW(th). All show the usual reduction in frequency content at higher frequencies caused by intrinsic values of system time constants.

attempt is made to evaluate the methods in a general fashion; however, it should be noted that variations in the methods or special "ground rules" for comparison may lead to exceptions.

Whether an input excitation signal is applied to a neutron absorber or to a plant control device will depend primarily on the information desired and secondarily on convenience. When this selection has been made, the signal may then be chosen to be either sinusoidal or pseudorandom. In recent years there has been some preference for pseudorandom signals which simultaneously measure all frequencies in the band of interest. Pseudorandom excitation has been preferred[107] because smaller perturbing amplitudes can be used and less reactor time is required for obtaining a given frequency resolution (see also Sect. 6-7).

Table 6.13—A General Qualitative Comparison of Excitation Experiments with Intrinsic Noise-Analysis Experiments

	Excitation	Noise
Experimental complexity and cost	More	Less
Interpretation of data	Easy	Difficult
Disturbance to reactor system	Some	None
Measures transfer function	Always	Sometimes
Measures spectra	No	Yes
Typical precision	High	Medium

For the noise methods there is no input signal injection. However, there must be sufficient internally generated noise in the frequency band of interest to excite the one or more variables being investigated. If this is the case, then selection of the appropriate data-acquisition and data-processing devices is the major consideration, as is also true for excitation experiments. In the following sections the types of equipment are described and their relative merits are assessed.

Whether one or more than one signal is used in noise analysis depends on the information desired. Transfer functions in power reactors normally require two signals. Furthermore, these must have sufficient coherence, Eq. 6.8, in the frequency band covered to achieve the accuracy desired, i.e., the effects of the same noise source—to a larger extent than separate independent noise sources—must be seen in both signals. Besides giving the amplitude and phase of transfer functions, these multisignal experiments also may provide insight into the cause of the intrinsic noise.

6-4 REACTOR EXCITATION EQUIPMENT

6-4.1 Excitation Signal

In experiments using external excitation, a plant control device, often a reactor control rod, is moved in accordance with a prescribed signal while the prescribed signal and one or more output variables are recorded. The signal is usually either sinusoidal or pseudorandom (such as Fig. 6.6). With the former the transfer function expressed in complex form is simply

$$G(\omega) = \frac{B}{A} e^{i\phi} \qquad (6.27)$$

for an output variable $B \sin(\omega t + \phi)$. With a pseudorandom signal the input and output must be cross correlated and a Fourier transform carried out on the result (see Sect. 6-3).

As a signal generator in these experiments, one of the following may be used: a function generator (of sine waves usually); a stored function on tape (usually pseudorandom); or a random or pseudorandom electronic noise generator. It is important that the function-generator output not deviate significantly from a constant frequency and that it contain no significant harmonics. The function generator might take any of the following forms: a constant-speed motor driving a control actuator, an electronic oscillator driving a control-positioning circuit, or a square wave ("up" or "down") or ternary signal ("up," "hold," or "down") driving a control switch. Regarding pseudorandom noise generation, a stored tape has been used[41] as well as an on-line generator.[107] Some experiments use commercial random-noise generators.[84]

An important characteristic of the excitation variable is its amplitude. The amplitude must be large enough to overcome unwanted noise without requiring excessively long experimental times. It will be shown in Sect. 6-7 that the amplitude, unwanted noise, and test duration all combine to determine the accuracy of the result.

The amplitude cannot be made arbitrarily large, however, since nonlinear aspects of the system can complicate data analysis. It can be shown[91] that, because of nonlinearities in the neutron kinetics equations, Eqs. 6.11 and 6.12, one type of nonlinearity resulting from a sinusoidal reactivity excitation is a power function

$$N(t) = N_0 + N_1 \sin(2\pi f t) + N_2 \sin(4\pi f t + \theta) + \ldots \qquad (6.28)$$

where the harmonic-to-fundamental ratio is

$$\frac{N_2}{N_1} = \frac{1}{2}\frac{N_1}{N_0}\frac{|G_0(2f)|}{|G_0(f)|} \qquad (6.29)$$

This usually means that the harmonic content is of the same order of magnitude as the modulation fraction, N_1/N_0.

Another important characteristic of the excitation signal is its frequency band. The frequency content desired depends on the information to be obtained. It is sufficient that frequencies cover the band

$$\frac{0.5}{\tau_{max}} < 2\pi f < \frac{2}{\tau_{min}} \qquad (6.30)$$

if τ_{max} and τ_{min} are the largest and smallest system time constants of interest. In practice the amplitudes and frequencies attainable may not be those desired because of equipment limitations; so compromises may be required. Although there is little difficulty in attaining the lowest frequency desired, either the available power or the limitations of materials are likely to place a limit on the highest frequency attainable. Also, the inverse relation between amplitude and frequency, if constant-speed motors are used (e.g., to drive a control rod in and out in a periodic triangular wave shape), places limits on the maximum frequency

$$\text{Peak-to-peak amplitude} = \frac{\text{linear rod speed}}{2 \times \text{frequency}} \qquad (6.31)$$

6-4.2 Control Device

In reactor dynamics experiments the variable directly excited affects reactivity either directly, as in the motion of a control rod, or indirectly, as in the alteration of some system parameter (flow, pressure, etc.). Some ways in which reactivity can be varied are:

1. By a specially installed rotary or reciprocating control rod.

2. By moving the normal control rod of the reactor through special switching or signal injection into the automatic control system.

3. By changing valve position, pump output, etc., usually by signal injection into its control system.

It is not surprising that 2 and 3 are more commonly encountered than 1, except perhaps in special-purpose experimental reactors, since they require relatively little modification of the existing system.

Among the specially installed rod oscillators, there is some preference for the rotary type over the reciprocating type, usually because of the higher frequencies attainable. Rotary types use the following methods to vary reactivity (see Fig. 6.11): eccentric neutron absorbers rotating in a flux gradient,[19,26] eccentric fuel rotating in a flux gradient,[11] or neutron absorbers rotating past similar stationary absorbers which act as time-varying shields.[18] Usually neutron absorbers are oscillated with typical reactivity amplitudes being in the 0.5¢ to 5¢ range. In zero-power fast reactors, fuel can be oscillated; however, this is unusual.

Regardless of the device used to perturb the reactor, there are a number of aspects to be considered, especially when high precision is important:

1. Backlash and related effects causing phase uncertainty.

2. Random (long-term or short-term) variations in excitation frequency.

3. Transfer function of the control device if only its input (rather than its output) is measured.

4. Unwanted harmonics when striving for a pure sinusoidal input.

5. Reactor conditions affecting the excitation device.

In the last of these considerations, it should be remarked that more than just the integrity of materials is desired. For example, the reactivity worth of an oscillator rod can depend on reactor power and flux distribution; for this reason it is often desirable to have a sufficiently high frequency available to calibrate the oscillator rod against the zero-power transfer function in a frequency range where feedback is expected to be negligible.

6-5 TRANSFER-FUNCTION ANALYZERS

6-5.1 Usage

Experimental data from transfer-function tests consist of records of signals from which transfer functions or related quantities are to be extracted. Tables 6.10 and 6.11 indicate that there are a variety of approaches to data acquisition and analysis. This section treats those which operate on the signals in the time domain to obtain a transfer function. Section 6-6, Frequency Analyzers, is devoted to analysis in the frequency domain.

The signals to be analyzed can consist of any or all of the following components: a single frequency, a continuum of frequencies having meaning to the test, and unwanted frequencies, either discrete or continuous. In general, the purpose of an analyzer is to separate these three with the primary purpose of picking out the first two in the presence of the third.

Results from time-domain analyzers can be expressed ultimately in the frequency domain. For tests where a single frequency is excited, the combined results from a number of sequential tests at various frequencies constitute a transfer function evaluated at those frequencies. If a continuous frequency is used in external excitation or self-excitation, the correlation function obtained in the time domain may be subsequently Fourier analyzed, as indicated in Table 6.1, to obtain the desired frequency function.

6-5.2 Null-Balance Analyzer

The principle behind the null-balance method of analyz-

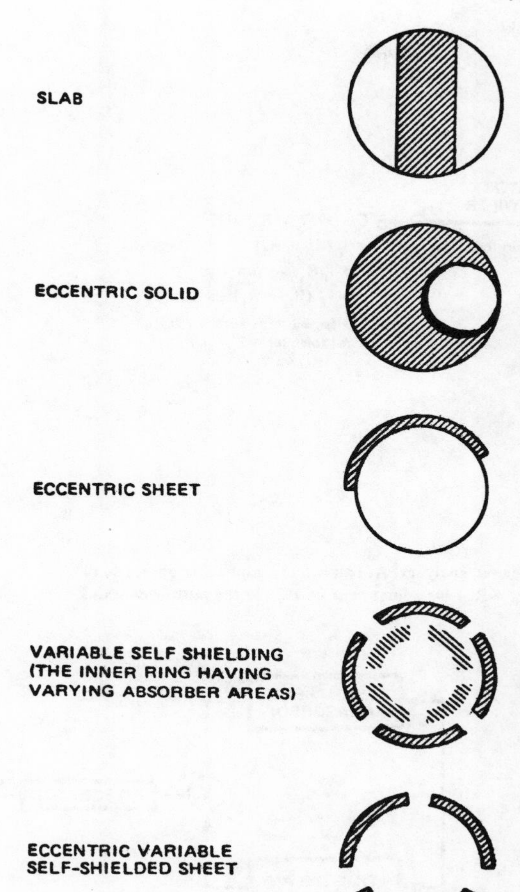

Fig. 6.11—Top view of some typical cylindrical rotary oscillator rods.

ing sinusoidal excitation experiments is one of nulling or bucking out the output signal with the input signal. Here the transfer function is simply the gain and phase adjustment used on one signal or the other to achieve this cancellation. A number of rod-oscillator tests have used this method successfully.[14,15,26] However, the method is not applicable to analysis of a continuum of frequencies, as in pseudorandom excitation.

Figure 6.12 shows schematically how the oscillating component of the ion-chamber current, $I_1 \sin(\omega t + \theta)$, is nulled against a mechanical oscillating signal to a sine potentiometer from the excitation device. By having a sinusoidal resistance variation in the potentiometer through which the ion-chamber current flows, you obtain a mixed signal whose sin ωt components are nulled by resistance and mechanical phase adjustments. Usually the output (Fig. 6.12) is sent through a band-pass filter (for frequencies near that of the oscillator and observed by an operator making the nulling adjustments. Some skill is required for high precision.

6-5.3 Synchronous Transfer-Function Analyzer

A specially constructed analyzer has been used in rod-oscillator experiments[11,15] and has been found to give high-precision results. The basic principle, as indicated in Fig. 6.13, is to multiply the ion-chamber signal (with its steady-state component bucked out) by sin ωt or cos ωt using a synchro-resolver whose mechanical input signal is precisely in phase with the rod-oscillator device. Since

$$I_1 \sin(\omega t + \theta) = I_1 \sin \omega t \cos \theta + I_1 \cos \omega t \sin \theta \qquad (6.32)$$

integration of sin ωt or cos ωt times the right-hand side of Eq. 6.32 over an integral number of cycles gives a result proportional to $I_1 \cos \theta$ or $I_1 \sin \theta$, respectively. The amplitude and phase of the ion-chamber current may then be obtained from these two integrated outputs:

$$\text{Amplitude} = [(I_1 \cos \theta)^2 + (I_1 \sin \theta)^2]^{1/2} \qquad (6.33)$$

$$\text{Tangent of phase} = \frac{I_1 \sin \theta}{I_1 \cos \theta} \qquad (6.34)$$

As indicated in Fig. 6.13, the signal at the point of the modulator modulates a carrier (typically several hundred Hertz), which is later demodulated at the demodulator. Regarding the noise accompanying the signal $I_1 \sin(\omega t + \theta)$, the analyzer acts as a sharply tuned filter at ω. The noise near ω will cause randomness in successive determinations of I_1 and θ, the randomness being proportional to the noise and inversely proportional to the integration time.

6-5.4 Cross Correlators

Devices involving the principles of Figs. 6.12 and 6.13 are usually restricted to sinusoidal excitation experiments. A technique that is more generally applicable in transfer-function determinations involves the use of cross correlation. In addition to being used with sinusoidal excitation, cross correlation is commonly used with self-induced noise or types of excitation other than sinusoidal.

The method consists in determining the cross-correlation function, $C_{xy}(\tau)$, over a range of the time-lag τ. Either the digital definition, Eq. 6.9, or the continuous integral (Table 6.1) definition may be used, depending on the experimental approach. It is convenient to handle only the fluctuating parts of variables, i.e., $x(t) - \overline{x}$ and $y(t) - \overline{y}$, such as in Eq. 6.10. After $C_{xy}(\tau)$ is known, either from an on-line or off-line device, it must be Fourier analyzed to obtain the cross spectrum, $P_{xy}(f)$ (see Table 6.1), and the transfer function,

$$G(f) = \frac{P_{xy}(f)}{P_x(f)} \qquad (6.35)$$

where $P_x(f)$ is the spectrum of the input variable, $x(t)$.

The three operations of time delay, multiplication, and integration required to obtain $C_{xy}(\tau)$ are indicated in Fig. 6.14. Here $x(t)$ may be either a fluctuating signal in the system or an excitation signal. The operations shown have been done on-line[41] for pseudorandom excitation by using "0" and "1" signals (read at the appropriate values of τ from a tape containing the input sequence) in a simple switching multiplier.

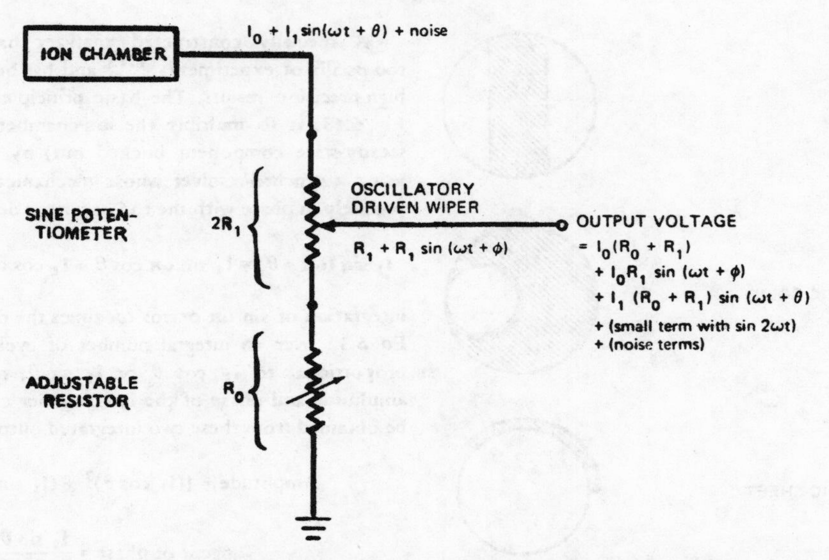

Fig. 6.12—Simplified schematic operation of a null-balance analyzer. At balance the adjustable phase, ϕ, of the potentiometer wiper equals $\theta + \pi$ and $I_0R_1 = I_1(R_0 + R_1)$ by adjustment of R_0; so the output contains no sin ωt component.

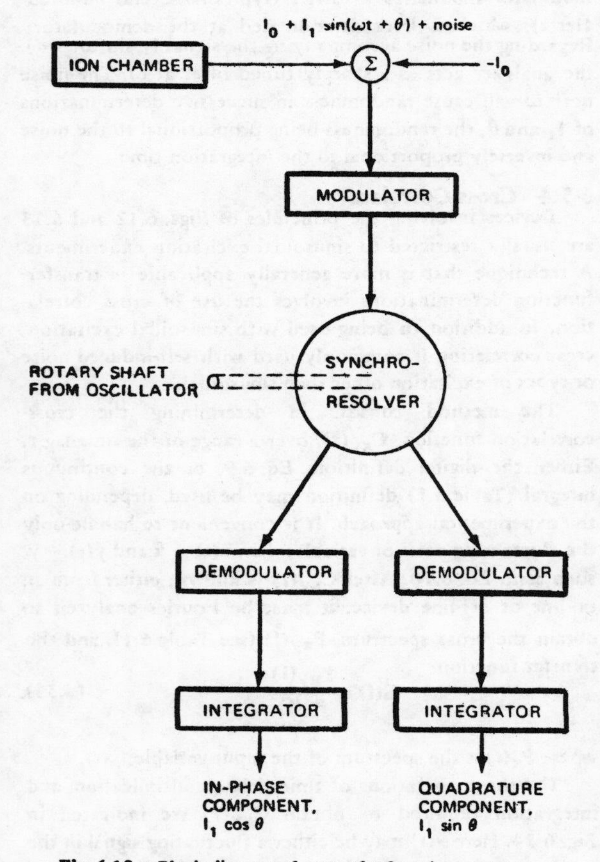

Fig. 6.13—Block diagram of a transfer-function analyzer.

Frequently, however, the two signals x(t) and y(t) are recorded on a frequency-modulation (f-m) magnetic-tape system.[40,95,100] Off-line playback is carried out using

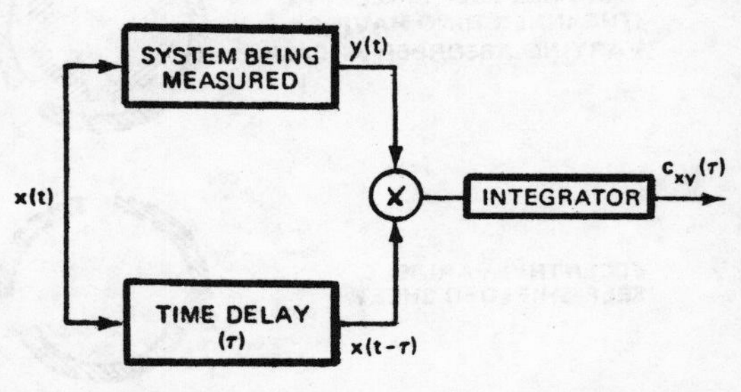

Fig. 6.14—Block diagram of a cross correlator of two signals, x(t) and y(t).

tape heads that are displaced to give the y(t) and $x(t - \tau)$ input to the multiplier–integrator combination of Fig. 6.14. Analog-computer components are typically used to perform the operations required to give $C_{xy}(\tau)$. In the special case where x(t) = y(t), the autocorrelation function may be obtained in this manner.

6-5.5 Digital Techniques

Although on-line digital-computer analysis of reactor dynamics is possible[73] and perhaps will be prevalent in the future, it has been the practice until now to perform digital analysis off-line. As shown in Table 6.11, the digitizing process may be on-line (creating the proper magnetic-tape format for a computer) or off-line. The off-line digitizing may be automatic from an f-m tape or semiautomatic, as in the case of manually operated strip-chart readers whose electrical output punches cards. In any event the result is that one or more sequences (x_i, y_i, etc.) of variables at time spacings Δt are generated in a form suitable for input to a digital computer.

The selection of a digitizing interval, Δt, and of a total duration of the data collection, T, is discussed in Sec. 6-7.4.

It will suffice here to note that the digitizing interval determines the upper frequency limit, $f_{max} = 1/(2\Delta t)$, of the analysis and the total duration is associated with the frequency resolution (i.e., minimum frequency interval between independently determined spectral values) and accuracy of results. The quotient, $T/\Delta t$, is the number of digital values per signal and may be 10^3 to 10^5 in typical experiments.

A number of versatile programs are available to users of the various commercial computers for statistical analysis of large quantities of data. Typical of these are the Biomedical Computer Programs,[115] a series of 42 programs that are useful not only in biomedical research but also in any field requiring analysis of data for frequency counts, variances, correlations, and related functions. Table 6.14 lists the available computer outputs from just one of these 42 programs (BMD-02T) if one has, for example, three related system variables. All possible time-correlation functions and their Fourier transforms are computed with x(t) regarded as an input signal. Evidently there is sufficient versatility and generality to permit adaption to almost any type of transfer-function experiment.

Even more versatile than the Biomedical Computer Programs series is the BOMM system of programs.[116] Here the user describes in few-word control statements the step-by-step data-handling operations to be performed on a time series, such as finding the mean, doing a cross correlation, or plotting an answer. These control statements call in standard subprograms to the computer that perform all the detailed calculations. Thus individualized data-processing needs can be satisfied with BOMM, although more effort is required to list the control statements than in the Biomedical Computer Programs.

Although a computer is a virtual necessity for performing the required analysis on random fluctuations, it is not necessarily required in the special case of obtaining transfer functions from strip-chart recordings of sinusoidal oscillations where the signal-to-noise ratio is good.[15,76] For example, the transfer function of the BORAX-4 reactor[113] could be obtained to an accuracy of ±5% by chart reading and simple hand calculations, even though the root-mean-square oscillatory amplitude, $|G|Nk_0/(2)^{1/2}$, excited by $k_0 \sin \omega t$ was only about twice the root-mean-square boiling noise. In this technique the digitally determined (Eq. 6.9) normalized cross-correlation function, C_{kn}, of the reactivity and the power [which is $|G|Nk_0 \sin(\omega t + \theta)$ + noise] is equated to its theoretical expectation

$$C_{kn}(\tau) = \frac{1}{2} |G|Nk_0^2 \cos(\omega\tau - \theta) \qquad (6.36)$$

$|G|$ and θ may be determined from as few as two values, $C_{kn}(0)$ and $C_{kn}(\pi/2\omega)$.

Whether the digital approach discussed here or the continuous-signal approach discussed in previous sections should be used depends on a variety of factors, some of which are mentioned in Table 6.15. The digital approach has been more common in recent years as digitizing costs and computer rental costs per data point decrease and as demands for computer versatility (see Table 6.14) increase. However, for applications requiring many repetitive determinations of a single function, the special-purpose continuous analyzer is strongly entrenched. The considerations of Table 6.15 also apply to frequency-domain analysis, as discussed in the following sections.

Table 6.15—General Comparison of Digital and Continuous Analysis Methods

	Digital	Continuous
Usual use of equipment	Rental	Own
Relative amount of use to date	Little	Much
On-line results	Rarely	Often
Versatility of analysis	High	Medium to low

As the use of on-line digital computers becomes more prevalent and accepted in reactor operation, on-line digital analysis of noise may be expected to be used competitively with other methods. Cohn[68a,73] has demonstrated the ability to sample noise as often as every 0.5 μsec and to do on-line correlations with a digital computer. Polarity correlating (i.e., replacing the noise amplitude value by +1 or −1 for its fluctuation about an average of zero in computing correlation functions) was found useful in this application.

6-6 FREQUENCY ANALYZERS

6-6.1 Usage

In the preceding section analyzers suitable for handling data in the time domain were considered. These analyzers

Table 6.14—Functions Generated in Computer Analysis[115] of Three Variables

Function	Variables used in computing the functions				
	x(t)	y(t)	z(t)	x(t) and y(t)	x(t) and z(t)
Autocorrelation function	x	x	x		
Power spectrum	x	x	x		
Cross-correlation function				x	x
Cross-spectrum amplitude and phase				x	x
Transfer-function amplitude and phase				x	x
Coherence				x	x

are used mostly for correlations or for obtaining transfer functions from single-frequency excitation. However, there is an alternative approach, used primarily for data having a continuum of frequencies, in which the various frequencies in the signal are separated and analyzed in the frequency domain. Correlators seek the amount of correlation at various time intervals, whether within one signal or between two signals. Frequency analyzers, on the other hand, seek the relative amounts of various frequencies in one signal or existing in a related fashion in two signals. In both approaches Table 6.1 is used to compute the corresponding functions in the time or frequency domain from those functions which are obtained by experiment.

The analyzers discussed in this section are most often used for directly obtaining the spectral power density, $P_x(f)$, and the cross spectral power density, $P_{xy}(f)$. This implies that, in the system being analyzed, a continuum of excitation frequencies exists and has been generated by either internal or external excitation sources. In many instances the spectral power density contains the required information about the dynamics of the reactor system. In other instances both $P_{xy}(f)$ and $P_x(f)$ are determined, and then, by using Eq. 6.20, the transfer function between x and y is obtained.

6-6.2 Spectrum Analyzers

In contrast to a cross-spectrum analyzer, a spectrum analyzer of a single variable has just one input, x(t), and a single-number output, P(f), at each frequency. A multichannel analyzer displays $P(f_1)$, $P(f_2)$, . . ., simultaneously,[46] but the more commonly encountered single-channel analyzer obtains $P(f_1)$, $P(f_2)$, . . ., sequentially, requiring an analysis time T for each. This means that the same recorded signal is reanalyzed off-line at various frequency values, whereas on-line the same reactor conditions are preserved during each analysis time T to maintain the same spectral characteristics.

The three basic parts of an analyzer are its filter, detector, and averager. In the bottom half of Fig. 6.1 the detection–averaging function is shown as a wattmeter receiving the output of the filter. In Fig. 6.15 these functions are separated since they correspond to distinct components in most analyzers. The slowly varying amplitude, E(t), of the filter output is detected first, usually by squaring or rectification. After detector fluctuations have been averaged out, $P(f_1)$ or $[P(f_1)]^{1/2}$ is obtained.

It is worth noting that the output of an analyzer is, in fact,

$$P(f_1) = \frac{\int_{f_1-(1/2)\Delta f}^{f_1+(1/2)\Delta f} P(f)\, B(f)\, df}{\int_{f_1-(1/2)\Delta f}^{f_1+(1/2)\Delta f} B(f)\, df} \qquad (6.37)$$

where B(f) is the spectral-window function of the analyzer. If P(f) is a function varying substantially within the resolution band, Δf, then it is desirable that B(f) resemble the ideal filter of Fig. 6.16 as closely as possible. (This is discussed further in pt. 6-6.4.)

Spectrum analyzers that have been used in reactor experiments can be classified as follows: specially designed and constructed,[62] based on commercially available analog-computer components,[48] and a commercially available spectrum analyzer.[117] Some spectrum analyzers use electrical filters resonant at the desired frequency, f_1; others use a heterodyne technique in which the signal modulates a

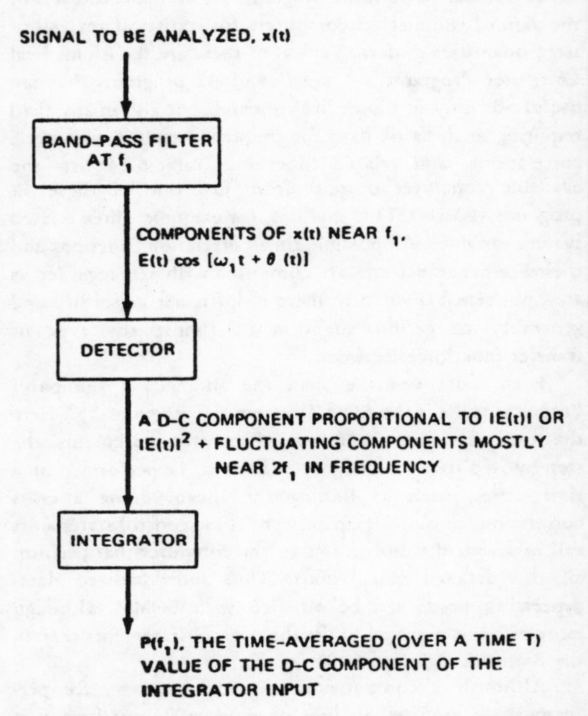

Fig. 6.15—Operation of the three essential elements (filter, detector, and averager) in a spectrum analyzer.

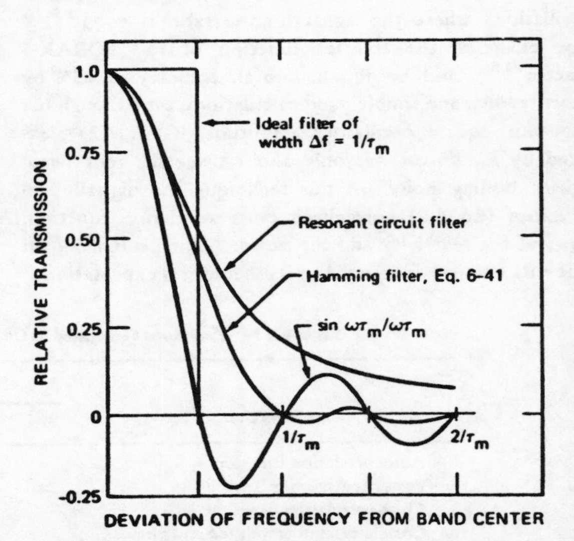

Fig. 6.16—Shapes of some spectral windows used in spectral analysis. Here τ_m is the maximum lag used in the correlation function associated with the spectra.

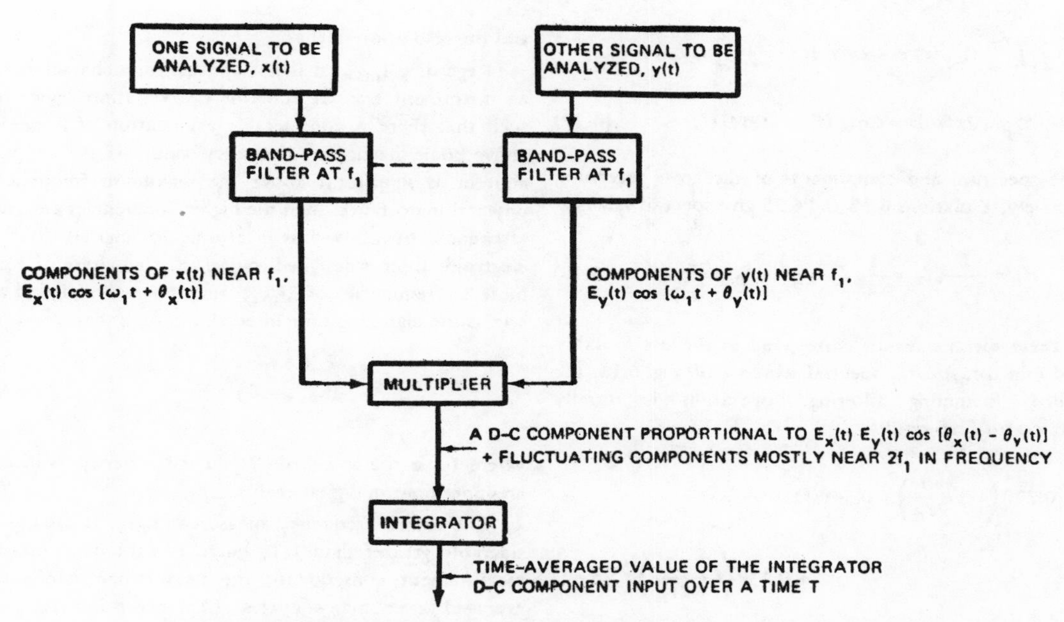

Fig. 6.17—Operation of the four essential elements in a cross-spectrum analyzer.

high-frequency carrier, f_c, that is subsequently filtered in a narrow band. Since all these approaches perform the same P(f) determination, such factors as cost, convenience, and availability are usually the deciding factors for experimenters.

6-6.3 Cross-Spectrum Analyzers

The purpose of a cross-spectrum analyzer is to measure the cross spectral density, P(f), of two correlated signals, x(t) and y(t). This is done, one frequency at a time, by integrated multiplication of band-pass-filter outputs, as indicated in Fig. 6.17. A comparison of Figs. 6.15 and 6.17 shows that this cross-spectrum analyzer is a slight generalization of an ordinary spectrum analyzer using a multiplier as a detector.

In some analyzers[70,72] tuned-circuit band-pass filters are used. In others[84,109] the input to the multiplier is the result of passing a modulated pair of signals, x(t) cos $(\omega_1 t + \phi_x)$ and y(t) cos $(\omega_1 t + \phi_y)$ (constructed by multiplying the signals by oscillator outputs), through a low-pass filter. The broken line in Fig. 6.17 indicates the ability to control the relative phase in the two modulations at ω_1: in-phase operation gives the cospectral density, $Co_{xy}(f)$; with 90° between the two signal channels, the output is the quad-spectral density, $Qu_{xy}(f)$. Then Eqs. 6.6 and 6.7 can be used to determine the cross spectrum, $P_{xy}(f) = Co_{xy}(f) - i\,Qu_{xy}(f)$.

As in the discussion of the spectrum analyzer (pt. 6-6.2), the major parameters selected by the experimentalist are the frequencies at which $P_{xy}(f)$ is determined, the frequency resolution, Δf (defined in Fig. 6.16), and the analysis time, T. Quantitative criteria for making these selections are given in pt. 6-7.

6-6.4 Digital Spectrum Analysis

In pt. 6-5.5 it was noted that digitizing plus subsequent computer analysis can be used as an alternative to continuous-signal analysis. The computer programs in use give not only time-domain functions (usually computed first in the program) but also their Fourier transforms, as indicated in Table 6.14. Thus the programs discussed may be regarded as frequency analyzers too.

Although not indicated in Figs. 6.15 and 6.17, the incoming signal may be "conditioned" before analysis. In the continuous analysis this could often consist in filtering out frequencies above and/or below those of interest in the analysis. Similarly, in digital analysis it is not only customary to remove any nonzero mean values (i.e., d-c components) from the signals but also to do some of the following: detrending, i.e., removing a linear trend in time; filtering, such as prewhitening; and normalizing signal magnitudes by dividing deviations from the mean by the signal's standard deviation.

Digital filters,[113] computer arithmetic operations on the sequential data points of x(t) that have the effect of changing its spectrum, can be used to advantage. Thus, in prewhitening, the filter characteristic is modified so that the spectrum is one more nearly like white noise [i.e., P(f) is made to be approximately constant] and more amenable to analysis.

Computer programs usually calculate autocorrelation and cross-correlation functions from the data and for values of τ only up to τ_m; this value is generally some small fraction (typically 10% or less) of the data duration T in order to secure good accuracy. Then the computer Fourier analyzes the correlation functions:

$$P_x(f) = 2\int_0^{\tau_m} C_x(\tau)\cos 2\pi f\tau\, d\tau \qquad (6.38)$$

$$P_{xy}(f) = \int_{-\tau_m}^{\tau_m} C_{xy}(\tau) \cos 2\pi f\tau \, d\tau - i \int_{-\tau_m}^{\tau_m} C_{xy}(\tau)$$

$$\times \sin 2\pi f\tau \, d\tau = Co_{xy}(f) - i\, Qu_{xy}(f) \qquad (6.39)$$

for the spectrum and components of the cross spectrum, respectively. Equations 6.38 and 6.39 give spectral values at

$$f_1 = \frac{1}{2\tau_m},\ f_2 = \frac{1}{\tau_m},\ f_3 = \frac{3}{2\tau_m},\ f_4 = \frac{2}{\tau_m} \ldots \qquad (6.40)$$

Since these spectral results correspond to the less desirably shaped $(\sin \omega\tau_m)/\omega\tau_m$ spectral window of Fig. 6.16, the so-called "hamming filtering" operation is usually performed for final results:

$$P(f) = 0.23P\left(f - \frac{1}{2\tau_m}\right) + 0.54P(f)$$

$$+ 0.23P\left(f + \frac{1}{2\tau_m}\right) \qquad (6.41)$$

to obtain the more desirable window shown.

6-7 EXPERIMENTAL CONSIDERATIONS

6-7.1 Error Sources

When the transfer function, $G(f)$, between $x(t)$ and $y(t)$ is measured, the transducers of the signals to the analyzer may have a frequency dependence. In addition, the analyzer may have a transfer function of its own, e.g., if its amplifier gains are frequency dependent. As a consequence significant corrections may have to be applied to transfer-function measurements to obtain the ideal function desired. Calibration, using known sinusoidal amplitudes or white-noise generators, is frequently necessary.

Unwanted signals, such as random noise (as from instrumentation or a digitizing process) or periodic signals (as in 60-Hz hum), are usually introduced by the devices measuring a system. Moreover, the system itself may mix unwanted signals with the signals observed at the transducer inputs. Since all these effects influence accuracy, they must be eliminated or lessened. Some techniques for coping with the problem are:

1. Reduce or eliminate the unwanted signals at or near their source.
2. Increase the desired signal levels.
3. Filter the signals, emphasizing frequencies of interest over others.
4. Correlate pairs of related signals, as in cross-correlation and cross-spectral analysis.
5. Use long durations of data or many repetitions.

6-7.2 Frequency Limits

There are three important frequencies in a transfer-function or spectrum measurement: the highest frequency of interest (f_{max}), the lowest frequency of interest (f_{min}), and the resolution (Δf).

Regarding f_{max}, it is not unusual for a characteristic of an instrument transfer function or excitation device to be such that there is considerable attenuation of frequencies above some particular frequency value. If the frequency content is significant above the maximum frequency of physical importance, then the higher frequencies are usually attenuated by a low-pass electronic (or digital) filter. The electronic filter is designed to give an attenuation of 3 db or more at frequencies of $2f_{max}$ and above. The digital filter selects the digitizing time interval, Δt, such that

$$2f_{max} = f_N = \frac{1}{2\Delta t} \qquad (6.42)$$

where f_N is the so-called "Nyquist frequency" which sets an upper limit in digital analysis.

The lowest frequency measured, f_{min}, is usually considerably greater than $1/T$, where T is the duration of the measurement; consequently the measurement is effectively averaged over many cycles. (An exception to this is sinusoidal excitation at f_{min} with an amplitude well in excess of system noise levels; the duration of the measurement may be kept as small as $1/f_{min}$ in this case.) Usually a high-pass filter is used to prevent frequencies below f_{min} from entering the analyzer.

Reference has already been made to the resolution Δf in Eq. 6.37 and Fig. 6.16. Continuous spectra are averaged over Δf. In excitation with discrete frequencies, the interval between frequencies is effectively the resolution. The value of Δf is usually selected to be just small enough to obtain the detail required in the spectrum. Too small a value is disadvantageous from the standpoint of accuracy, as is seen in Table 6.16. Two common situations affecting a choice of Δf are:

1. The gross variation of a nonresonant spectrum over a few decades of frequencies is desired, in which case Δf can be as large as a half to one octave, thus giving about 3 to 5 points per decade of frequency.
2. The details of a resonant peak in the spectrum are desired, in which case Δf must be somewhat less than the width of the peak to obtain several points across the peak.

6-7.3 Statistical Accuracies

Formulas for computing the expected variation due to the statistical nature of an experiment are vital in optimum planning. Table 6.16 contains formulas useful in ascertaining the precision of the various functions involved in noise analysis. The meaning of the fractional-error formulas is that, if many values of a function were determined (at a particular f or τ), 68.3% would lie within the average ± this fractional error.

In all cases the error varies inversely as the square root of the measuring time and inversely as the square root of either the bandwidth, B, or the resolution, Δf:

B = upper frequency limit of P(f), which is

Table 6.16—Statistical Errors of Correlation and Spectral Measurements[113c] Expressed in Terms of Signal Bandwidth (B), Duration of Data (T), and Resolution of Analyzer (Δf)

Function (computed from x and y having zero means)	Standard deviation of function / Function	Condition of applicability
Autocorrelation, $C(\tau)$	$\frac{1}{(2BT)^{1/2}}\left[1+\frac{C(0)^2}{C(\tau)^2}\right]^{1/2}$	$BT > 5$; $T > 10\lvert\tau\rvert$
Cross correlation, $C_{xy}(\tau)$	$\frac{1}{(2BT)^{1/2}}\left[1+\frac{C_x(0)\,C_y(0)}{C_{xy}(\tau)^2}\right]^{1/2}$	$BT > 5$; $T > 10\lvert\tau\rvert$
Spectrum, P(f)	$\frac{1}{(T\,\Delta f)^{1/2}}$	$T\,\Delta f > 5$
Cospectrum or quad-spectrum	$\frac{<1^*}{(T\,\Delta f)^{1/2}}$	$T\,\Delta f > 5$

*An upper bound to the error. The true error is well below this for highly coherent signals in which statistical effects are minor.

approximately constant from 0 to B and thereafter near zero (6.43)

or

$$B = \frac{1}{2\tau_c} \text{ if } C(\tau) \text{ is approximately } C(0)e^{-\tau/\tau_c} \quad (6.44)$$

$$\Delta f = \frac{1}{\tau_m} \text{ for the ideal and hamming windows of Fig. 6.16} \quad (6.45)$$

or

$\Delta f = \pi$ times the half-power bandwidth of a sharply tuned circuit (6.46)

The half-power bandwidth is also defined as the resonant frequency divided by the so-called resonance Q.

The above pertains primarily to continuous frequency analysis; however, there is also statistical error associated with sinusoidal excitation experiments because random noise is also present. For determination of a transfer function by the cross-correlation-function method of Eq. 6.36, the presence of noise having a variance of σ_y^2 along with the output causes the following fractional error[118]

$$\frac{\text{Standard deviation of } |G|}{|G|} = \frac{\sigma_y/[|G|Nk_o/(2)^{1/2}]}{(BT)^{1/2}[1 + (\omega^2/4B^2)]^{-1/2}} \quad (6.47)$$

The numerator is evidently the ratio of noise of bandwidth, B, to signal; the denominator is the square root of the number of effectively "independent" measurements.

6-7.4 Spectral-Analysis Data Planning

In a spectrum or cross-spectrum measurement, the desired accuracy and three frequency parameters (highest and lowest frequency of interest and the resolution) must be chosen. For digital analysis:

$$\text{Fractional error in } P(f) = \frac{1}{(T\Delta f)^{1/2}} = \frac{1}{(N/M)^{1/2}} \quad (6.48)$$

$$f_{max} = \frac{1}{2\Delta t} = \frac{N}{2T} \quad (6.49)$$

$$f_{min} = \Delta f = \frac{1}{\tau_m} = \frac{1}{M\Delta t} \quad (6.50)$$

where N is the number of data points, $T/\Delta t$, and M is the maximum number of lag intervals, $\tau_m/\Delta t$. If, when these equations are applied, it is found that some of the parameters so determined are not readily attainable (if, for example, too many digits are required), then obviously suitable compromises must be made between the limitations of the analysis and the desired results. In continuous analysis one evidently optimizes just Δf and T of Eqs. 6.48 and 6.50 somewhat independently of the f_{max} selected.

An illustration of selections made in an ion-chamber noise analysis[90] of the Experimental Boiling Water Reactor to measure a resonance at 1.7 Hz is:

N = 3331	Fractional error in P(f) = 0.3
M = 300	f_{max} = 10 Hz
Δt = 0.05 sec	f_{min} = Δf = 0.067 Hz

REFERENCES

1. H. B. Smets, The Describing Function of a Nuclear Reactor, *IRE Trans. Nucl. Sci.*, NS-6: 32 (1952).

2. R. C. Dorf, *Modern Control Systems*, Addison-Wesley Publishing Company, Inc., Reading, Mass., 1967.

3. Argonne National Laboratory, Reactor Physics Constants, USAEC Report ANL-5800 (2nd ed.), 1963.

4. J. M. Harrer, R. E. Boyer, and D. Krucoff, Transfer Function of Argonne CP-2 Reactor, *Nucleonics*, 10(8): 32 (1952).

5. P. Meyer and E. Garelis, Use of a Pseudorandom Source Input in the Measurement of Impulse-Response Functions, in *Neutron Noise, Waves, and Pulse Propagation*, Gainesville,

Fla., Feb. 14–16, 1966, Robert E. Uhrig (Coordinator), AEC Symposium Series, No. 9 (CONF-660206), p. 333, 1967.

6. T. E. Stern and J. Valat, Highly Negative Reactivity Measurement Using Pseudorandom Source Excitation and Cross Correlation, in *Reactor Kinetics and Control*, Tucson, Ariz., Mar. 25–27, 1963, Lynn E. Weaver (Coordinator), AEC Symposium Series, No. 2 (TID-7662), pp. 27-45, 1964.
7. R. E. Uhrig, Measurement of Reactor-Shutdown Margin by Noise Analysis, in *Reactor Kinetics and Control*, Tucson, Ariz., Mar. 25–27, 1963, Lynn E. Weaver (Coordinator), AEC Symposium Series, No. 2 (TID-7662), pp. 1-26, 1964.
8. R. E. Uhrig, Nuclear-Noise Research Program at the University of Florida, in *Noise Analysis in Nuclear Systems*, Gainesville, Fla., Nov. 4–6, 1963, Robert E. Uhrig (Coordinator), AEC Symposium Series, No. 4 (TID-7679), pp. 251-284, 1964.
9. C. D. Kylstra and R. E. Uhrig, Measurement of the Spatially Dependent Transfer Function, in *Noise Analysis in Nuclear Systems*, Gainesville, Fla., Nov. 4–6, 1963, Robert E. Uhrig (Coordinator), AEC Symposium Series, No. 4 (TID-7679), pp. 285-300, 1964.
10. R. E. Uhrig and M. J. Ohanian, Pseudorandom Pulsing of Subcritical Systems, in *Neutron Noise, Waves, and Pulse Propagation*, Gainesville, Fla., Feb. 14–16, 1966, Robert E. Uhrig (Coordinator), AEC Symposium Series, No. 9 (CONF-660206), p. 315, 1967.
11. A. R. Baker, Oscillator Tests in British Fast Reactors, in Proceedings of the Conference on Transfer Function Measurements and Reactor Stability Analysis held at Argonne National Laboratory, Argonne, Ill., May 2–3, 1960, USAEC Report ANL-6205, p. 201, Argonne National Laboratory, 1960.
12. R. R. Smith, Recent EBR-I Stability Studies, in Proceedings of the Conference on Transfer Function Measurements and Reactor Stability Analysis held at Argonne National Laboratory, Argonne, Ill., May 2–3, 1960, USAEC Report ANL-6205, p. 232, Argonne National Laboratory, 1960.
13. R. R. Smith et al., EBR-II Dynamics, *Trans. Amer. Nucl. Soc.*, 8(2): 478 (1965).
14. J. A. DeShong, Jr., Power Transfer Functions of the EBWR Obtained Using a Sinusoidal Reactivity Driving Function, USAEC Report ANL-5798, Argonne National Laboratory, 1958.
15. A. Klickman, R. Horne, and H. Wilber, Oscillator Tests in the Enrico Fermi Reactor, USAEC Report APDA-NTS-11, Atomic Power Development Associates, Inc., 1967.
16. W. Graham et al., Kinetics Parameters of a Highly Enriched Heavy-Water Reactor, USAEC Report TID-23037, Georgia Institute of Technology, 1966.
17. R. J. Johnson and R. N. Macdonald, Calculation of Space-Dependent Effects in Pile-Oscillator and Reactor-Noise Measurements, in *Neutron Noise, Waves, and Pulse Propagation*, Gainesville, Fla., Feb. 14–16, 1966, Robert E. Uhrig (Coordinator), AEC Symposium Series, No. 9 (CONF-660206), p. 649, 1967.
18. R. N. Cordy, Measurement and Analysis of the KEWB Transfer Function by Reactor Modulation Techniques, in Proceedings of the Conference on Transfer Function Measurements and Reactor Stability Analysis held at Argonne National Laboratory, Argonne, Ill., May 2–3, 1960, USAEC Report ANL-6205, p. 115, Argonne National Laboratory, 1960.
19. R. M. Kiehn, Progress in Frequency Response Measurements at Los Alamos, in Proceedings of the Conference on Transfer Function Measurements and Reactor Stability Analysis held at Argonne National Laboratory, Argonne, Ill., May 2–3, 1960, USAEC Report ANL-6205, p. 214, Argonne National Laboratory, 1960.
20. C. Cowan, Measurement of the Livermore Pool-Type Reactor (LPTR) Transfer Function Using Reactor Oscillation Techniques, USAEC Report UCRL-12476, Lawrence Radiation Laboratory, 1965.
21. H. Christensen et al., A Review of NORA Project Noise Experiments, in *Neutron Noise, Waves, and Pulse Propagation*, Gainesville, Fla., Feb. 14–16, 1966, Robert E. Uhrig (Coordinator), AEC Symposium Series, No. 9 (CONF-660206), p. 503, 1967.
22. G. C. Rudy and J. T. Stone, Pathfinder Atomic Power Plant: Pile Oscillator Rod Assembly and Measurements of Reactor Transfer Functions, USAEC Report ACNP-67522, Allis-Chalmers Manufacturing Company, 1967.
23. E. S. Kenney, Noise Analysis of Nuclear Reactors with the Use of Gamma Radiation, in *Neutron Noise, Waves, and Pulse Propagation*, Gainesville, Fla., Feb. 14–16, 1966, Robert E. Uhrig (Coordinator), AEC Symposium Series, No. 9 (CONF-660206), p. 399, 1967.
24. R. T. Frost and R. J. Schemel, PTR Zero-Power Reactor Transfer Function, USAEC Report KAPL-M-RTF-3, Knolls Atomic Power Laboratory, August 1955.
25. T. Hoshino, H. Yoshikawa, and J. Wakabayashi, Pile-Oscillator Measurement and Theoretical Study of Space-Dependent Transfer Function of a Reflected Assembly, in Japan–United States Seminar on Nuclear Reactor Noise Analysis, Tokyo and Kyoto, Sept. 2–7, 1968, pp. 321-336, Japan Society for the Promotion of Science and National Science Foundation, Washington, D. C.
26. A. A. Wasserman, High and Low Power SPERT-I Transfer Function Measurements, in Proceedings of the Conference on Transfer Function Measurements and Reactor Stability Analysis held at Argonne National Laboratory, Argonne, Ill., May 2–3, 1960, USAEC Report ANL-6205, p. 156, Argonne National Laboratory, 1960.
27. G. Ingram, D. B. McCulloch, and J. E. Sanders, Neutron Lifetime Measurements in the Fast Reactors Zeus and Zephyr, *J. Nucl. Energy*, 10: 22 (1959).
28. A. Lindeman and L. Ruby, Subcritical Reactivity from Neutron Statistics, *Nucl. Sci. Eng.*, 29: 308 (1967).
29. D. T. Austin, B. Z. Zolotar, and K. B. Cody, Comparison of the Waiting-Time Alpha with the Rossi-Alpha, *Trans. Amer. Nucl. Soc.*, 10(2): 591 (1967).
30. R. W. Albrecht, The Measurement of Dynamic Nuclear Reactor Parameters by Methods of Stochastic Processes, Ph. D. Thesis, University of Washington, 1961.
31. Y. Gotoh, Measurement of Neutron Life in a D_2O-System by Neutron Fluctuation, *J. Nucl. Sci. Technol. (Tokyo)*, 1(6): 193 (1964).
32. D. Babala and R. Ogrin, Measurement of the Prompt Neutron Decay Constant of the NORA Reactor by the Interval Distribution Technique, *Nucl. Sci. Eng.*, 29(3): 367 (1967).
33. A. I. Mogil'ner and V. G. Zolotukhin, Measuring the Characteristics of Kinetics of a Reactor by the Statistical p-Method, *At. Energ. (USSR)*, 10(4): 377 (1961).
34. F. Schroeder (Ed.), SPERT Project Quarterly Technical Report, April, May, June, 1960, USAEC Report IDO-16640, Phillips Petroleum Company, Apr. 7, 1961.
35. F. Schroeder (Ed.), SPERT Project Quarterly Technical Report, October, November, December, 1960, USAEC Report IDO-16687, Phillips Petroleum Company, June 1, 1961.
36. A. Furuhashi and S. Izumi, A Proposal on Data Treatment in the Feynman-2 Experiment, *J. Nucl. Sci. Tech.*, 4(2): 55 (1967).
37. J. Bengston, Determination of Prompt Neutron Decay Constants of Multiplying Systems, in *Proceedings of the Second United Nations International Conference on the Peaceful Uses of Atomic Energy, Geneva, 1958*, Vol. 12, pp. 63-71, United Nations, New York, 1958.
38. E. F. Bennett, The Rice Formulation of Pile Noise, *Nucl. Sci. Eng.*, 8: 53 (1960).

39. E. F. Bennett and R. L. Long, Noise Measurements on a Reactor Servo-Control System, *Trans. Amer. Nucl. Soc.*, 5(1): 184 (1962).
40. V. Rajagopal, Determination of Reactor Transfer Functions by Statistical Correlation Methods, USAEC Report BNL-5456, Brookhaven National Laboratory, March 1961.
41. J. D. Balcomb, H. B. Demuth, and E. P. Gyftopoulos, A Cross-Correlation Method for Measuring the Impulse Response of Reactor Systems, *Nucl. Sci. Eng.*, 11: 159 (1961).
42. J. Miida, K. Sumita, and Y. Kuroda, Noise Analysis of Nuclear Reactor Systems in Japan, in *Noise Analysis in Nuclear Systems*, Gainesville, Fla., Nov. 4–6, 1963, Robert E. Uhrig (Coordinator), AEC Symposium Series, No. 4 (TID-7679), pp. 155-170, 1964.
43. J. Miida (Ed.), Some Engineering Studies on the Analysis of Kinetics and Control of Tokai Atomic Power Station, in *Proceedings of the Third United Nations International Conference on the Peaceful Uses of Atomic Energy, Geneva, 1964*, Vol. 4, p. 128, United Nations, New York, 1965.
44. M. Hara and N. Suda, Some Investigations into the Application of Pseudorandom Binary Signals to Reactor-Dynamics Measurements, in *Neutron Noise, Waves, and Pulse Propagation*, Gainesville, Fla., Feb. 14–16, 1966, Robert E. Uhrig (Coordinator), AEC Symposium Series, No. 9 (CONF-660206), p. 247, 1967.
44a. T. W. Kerlin, Deterministic Signals for Frequency Response Measurements in Reactor Systems, in Japan–United States Seminar on Nuclear Reactor Noise Analysis, Tokyo and Kyoto, Sept. 2–7, 1968, pp. 249-266, Japan Society for the Promotion of Science and National Science Foundation, Washington, D. C.
45. V. Rajagopal and G. E. Swen, Dynamic Measurements on the Saxton Reactor Using Random Perturbations, *Trans. Amer. Nucl. Soc.*, 7(2): 281 (1964).
46. R. L. Randall and C. W. Griffin, Application of Power Spectra to Reactor-System Analysis, in *Noise Analysis in Nuclear Systems*, Gainesville, Fla., Nov. 4–6, 1963, Robert E. Uhrig (Coordinator), AEC Symposium Series, No. 4 (TID-7679), pp. 107-134, 1964.
47. M. A. Schultz, Measurement of Shutdown Reactivity in Large Gamma Fields, in *Neutron Noise, Waves, and Pulse Propagation*, Gainesville, Fla., Feb. 14–16, 1966, Robert E. Uhrig (Coordinator), AEC Symposium Series, No. 9 (CONF-660206), p. 413, 1967.
48. C. W. Ricker, D. N. Fry, E. R. Mann, and S. H. Hanauer, Investigation of Negative Reactivity Measurement by Neutron-Fluctuation Analysis, in *Noise Analysis in Nuclear Systems*, Gainesville, Fla., Nov. 4–6, 1963, Robert E. Uhrig (Coordinator), AEC Symposium Series, No. 4 (TID-7679), pp. 171-182, 1964.
49. C. W. Ricker, S. H. Hanauer, and E. R. Mann, Measurement of Reactor Fluctuation Spectra and Subcritical Reactivity, USAEC Report ORNL-TM-1066, Oak Ridge National Laboratory, April 1965.
50. R. W. Albrecht and G. M. Hess, Reactor Noise Experiments in Reflected Plutonium Assemblies, in *Neutron Noise, Waves, and Pulse Propagation*, Gainesville, Fla., Feb. 14–16, 1966, Robert E. Uhrig (Coordinator), AEC Symposium Series, No. 9 (CONF-660206), p. 381, 1967.
50a. C. Sastre, Fission-Rate Measurements Using Reactor Noise, *Trans. Amer. Nucl. Soc.*, 9(2): 507 (1966).
51. L. C. Schmid, J. H. Lauby, W. P. Stinson, and V. O. Votinen, Summary of Results of EBWR Critical Experiments, in USAEC Report HW-84608, Hanford Atomic Products Operation, 1965.
52. L. G. Kemeny, Random Fluctuations in a Nuclear Fission Reactor, *Nature*, 189: 130 (1961).
53. L. G. Kemeny and W. Murgatroyd, Stochastic Models for Fission Reactors, in *Noise Analysis in Nuclear Systems*, Gainesville, Fla., Nov. 4–6, 1963, Robert E. Uhrig (Coordinator), AEC Symposium Series, No. 4 (TID-7679), pp. 29-60, 1964.
54. O. R. Frisch and D. J. Littler, Pile Modulation and Statistical Fluctuations in Piles, *Phil. Mag.*, 45: 126 (1954).
55. D. N. Fry, D. P. Roux, C. W. Ricker, S. E. Stephenson, S. H. Hanauer, and J. R. Trinko, Neutron-Fluctuation Measurements at Oak Ridge National Laboratory, in *Neutron Noise, Waves, and Pulse Propagation*, Gainesville, Fla., Feb. 14–16, 1966, Robert E. Uhrig (Coordinator), AEC Symposium Series, No. 9 (CONF-660206), p. 463, 1967.
55a. E. Suzuki, T. Nomura, and T. Tsunoda, Application of Noise Analysis to Power Reactor Control, in Japan–United States Seminar on Nuclear Reactor Noise Analysis, Tokyo and Kyoto, Sept. 2–7, 1968, pp. 87-103, Japan Society for the Promotion of Science and National Science Foundation, Washington, D. C.
55b. N. Suda and S. Shirai, Estimate of Subcriticality by Optimal Filtering Method, in Japan–United States Seminar on Nuclear Reactor Noise Analysis, Tokyo and Kyoto, Sept. 2–7, 1968, pp. 123-140, Japan Society for the Promotion of Science and National Science Foundation, Washington, D. C.
56. R. A. Danofsky, Cross Power Spectral Measurements in the Two-Core University Training Reactor-10, in *Noise Analysis in Nuclear Systems*, Gainesville, Fla., Nov. 4–6, 1963, Robert E. Uhrig (Coordinator), AEC Symposium Series, No. 4 (TID-7679), pp. 229-250, 1964.
57. C. W. Griffin and J. G. Lundholm, Jr., Measurement of the SRE and KEWB Prompt Neutron Lifetime Using Random Noise and Reactor Oscillation Techniques, USAEC Report NAA-SR-3765, Atomics International Division, North American Aviation, Inc., Oct. 15, 1959.
58. J. B. Dragt, Analysis of Reactor Noise Measured in a Zero-Power Reactor and Calculations on Its Accuracy, in *Neutron Noise, Waves, and Pulse Propagation*, Gainesville, Fla., Feb. 14–16, 1966, Robert E. Uhrig (Coordinator), AEC Symposium Series, No. 9 (CONF-660206), p. 591, 1967.
59. D. N. Fry, D. P. Roux, C. W. Ricker, S. E. Stephenson, S. H. Hanauer, and J. R. Trinko, Neutron-Fluctuation Measurements at Oak Ridge National Laboratory, in *Neutron Noise, Waves, and Pulse Propagation*, Gainesville, Fla., Feb. 14–16, 1966, Robert E. Uhrig (Coordinator), AEC Symposium Series, No. 9 (CONF-660206), p. 463, 1967.
60. C. W. Ricker, D. N. Fry, and B. R. Lawrence, Determination of Reactor Parameters by Spectral Density Measurements, *Trans. Amer. Nucl. Soc.*, 7(2): 280 (1964).
61. J. R. Trinko et al., Reactor Noise Analysis Using a Pulse-Type Detector, *Trans. Amer. Nucl. Soc.*, 10(1): 286 (1967).
62. R. M. Ball and M. L. Batch, Measurement of Noise in Three Pressurized-Water Reactors, in *Noise Analysis in Nuclear Systems*, Gainesville, Fla., Nov. 4–6, 1963, Robert E. Uhrig (Coordinator), AEC Symposium Series, No. 4 (TID-7679), pp. 387-404, 1964.
63. V. Rajagopal, Reactor-Noise Measurements on Saxton Reactor, in *Noise Analysis in Nuclear Systems*, Gainesville, Fla., Nov. 4–6, 1963, Robert E. Uhrig (Coordinator), AEC Symposium Series, No. 4 (TID-7679), pp. 427-448, 1964.
64. R. Morrell and J. Hamilton, Experimental Measurement of β/l for SPERT III, ARMF-1, and PBF Reactors, in Nuclear Technology Branch Quarterly Report, April–June, 1964, USAEC Report IDO-17042, Phillips Petroleum Company, 1964.
65. R. L. Randall and G. Grayban, Measurement of SNAP-10A SNAPTRAN Kinetics Parameters by Reactor Noise Techniques, USAEC Report NAA-SR-Memo-9160, Atomics International Division, North American Aviation, Inc., 1963.
66. N. Wilde and J. Hamilton, Experimental Measurement of β/l for the ETR Critical Facility; Experimental Measurement of the Kinetic Parameters for the SNAP 10A SNAPTRAN Reactor, in Materials Testing Reactor–Engineering Test Reactor Technical Branches Quarterly Report, January 1–

Mar. 31, 1964, USAEC Report IDO-16994, Phillips Petroleum Company, 1964.

67. M. A. Schultz, Shutdown Reactivity Measurements Using Noise Techniques, in *Noise Analysis in Nuclear Systems*, Gainesville, Fla., Nov. 4–6, 1963, Robert E. Uhrig (Coordinator), AEC Symposium Series, No. 4 (TID-7679), pp. 135-154, 1964.

68. C. E. Cohn, Determination of Reactor Kinetic Parameters by Pile Noise Analysis, *Nucl. Sci. Eng.*, 5: 331 (1959).

68a. C. E. Cohn, Reactor-Noise Studies with an On-Line Digital Computer, in Japan–United States Seminar on Nuclear Reactor Noise Analysis, Tokyo and Kyoto, Sept. 2–7, 1968, pp. 141-160, Japan Society for the Promotion of Science and National Science Foundation, Washington, D. C.

68b. R. W. Albrecht and W. Seifritz, The Coherence Function: A Measure of Spatially Dependent Nuclear Reactor Properties, in Japan–United States Seminar on Nuclear Reactor Noise Analysis, Tokyo and Kyoto, Sept. 2–7, 1968, pp. 285-306, Japan Society for the Promotion of Science and National Science Foundation, Washington, D. C.

69. R. C. Kryter, D. N. Fry, and D. P. Roux, Two-Detector Cross Correlation for Shutdown Margin Measurements in Gamma Fluxes, *Trans. Amer. Nucl. Soc.*, 10(1): 283 (1967).

69a. E. S. Kenney and M. A. Schultz, Local In-Core Power Measurements with Out-of-Core Gamma Detectors, in Japan–United States Seminar on Nuclear Reactor Noise Analysis, Tokyo and Kyoto, Sept. 2–7, 1968, pp. 103-122, Japan Society for the Promotion of Science and National Science Foundation, Washington, D. C.

70. W. Seifritz, D. Stegemann, and W. Vath, Two-Detector Cross-Correlation Experiments in the Fast–Thermal Argonaut Reactor (STARK), in *Neutron Noise, Waves, and Pulse Propagation*, Gainesville, Fla., Feb. 14–16, 1966, Robert E. Uhrig (Coordinator), AEC Symposium Series, No. 9 (CONF-660206), p. 195, 1967.

71. A. R. Buhl, S. H. Hanauer, and N. P. Baumann, Experimental Investigation of Spatial Effects on the Power Spectra of a Large Reactor, *Trans. Amer. Nucl. Soc.*, 10(1): 288 (1967).

72. T. Nomura, S. Gotoh, and K. Yamaki, Reactivity Measurements by the Two-Detector Cross-Correlation Method and Supercritical-Reactor Noise Analysis, in *Neutron Noise, Waves, and Pulse Propagation*, Gainesville, Fla., Feb. 14–16, 1966, Robert E. Uhrig (Coordinator), AEC Symposium Series, No. 9 (CONF-660206), p. 217, 1967.

73. C. E. Cohn, Fast-Reactor Noise Analysis with an On-Line Digital Computer, *Trans. Amer. Nucl. Soc.*, 10(1): 285 (1967).

73a. R. K. Osborn, Gamma-Ray Fluctuation Measurements Versus Neutron Fluctuation Measurements, in Japan–United States Seminar on Nuclear Reactor Noise Analysis, Tokyo and Kyoto, Sept. 2–7, 1968, pp. 25-34, Japan Society for the Promotion of Science and National Science Foundation, Washington, D. C.

74. J. A. Thie, Water Reactor Kinetics, in *The Technology of Nuclear Reactor Safety*, p. 446, The M.I.T. Press, Cambridge, Mass., 1966.

74a. Northern States Power Company, Pathfinder Requests Substitution of Noise Analysis for Reactivity Oscillator for Transfer Function Measurements, Docket-50130, 1966.

75. R. A. Cushman, D. Mohr, R. N. Curran, and D. H. Brown, Stability Measurements on BORAX-V Boiling Core B-2, in *Reactor Kinetics and Control*, Tucson, Ariz., Mar. 25–27, 1963, Lynn E. Weaver (Coordinator), AEC Symposium Series, No. 2 (TID-7662), pp. 150-168, 1964.

76. B. S. Maxon, O. A. Schulze, and J. A. Thie, Reactivity Transients and Steady-State Operation of a Thoria–Urania-Fueled Direct-Cycle Light-Water Boiling Reactor (BORAX-IV), USAEC Report ANL-5733, Argonne National Laboratory, 1959.

77. R. R. Smith, C. B. Doe, R. O. Haroldsen, F. D. McGinnis, and M. Novick, Terminal Report for the Mark IV (Plutonium) Loading in EBR-I, USAEC Report ANL-6865, Argonne National Laboratory, 1965.

78. W. G. Lipinski, T. P. Mulcahey, and C. Michels, EBWR–Pu Transfer Functions, *Trans. Amer. Nucl. Soc.*, 10(1): 218 (1967).

79. L. K. Holland, R. O. Niemi, and L. H. Youngborg, Nuclear Boiler Dynamic Analysis and Testing, *Trans. Amer. Nucl. Soc.*, 10(1): 216 (1967).

80. S. Yamada and H. Kage, Reactor Noise Caused by Coolant-Flow Fluctuation, in *Neutron Noise, Waves, and Pulse Propagation*, Gainesville, Fla., Feb. 14–16, 1966, Robert E. Uhrig (Coordinator), AEC Symposium Series, No. 9 (CONF-660206), p. 455, 1967.

80a. T. Hoshi, Transfer Function Measurements by the Rod Oscillator Tests and Dynamic Analysis at JPDR, in Japan–United States Seminar on Nuclear Reactor Noise Analysis, Tokyo and Kyoto, Sept. 2–7, 1968, pp. 337-350, Japan Society for the Promotion of Science and National Science Foundation, Washington, D. C.

81. C. W. Griffin and R. L. Randall, At-Power, Low-Frequency, Reactor-Power-Spectrum Measurements and Comparison with Oscillation Measurements, *Nucl. Sci. Eng.*, 15(2): 131 (1963).

82. General Electric Company, VBWR Stability Test Report, USAEC Report GEAP-3971, June 1963.

83. T. W. Kerlin and S. J. Ball, Power-Dependent Frequency Response Measurements on the Molten-Salt Reactor Experiment, *Trans. Amer. Nucl. Soc.*, 10(1): 219 (1967).

84. V. Rajagopal and J. M. Gallagher, Jr., Some Applications of Dynamic (Noise) Measurements in Pressurized-Water-Reactor Nuclear Power Plants, in *Neutron Noise, Waves, and Pulse Propagation*, Gainesville, Fla., Feb. 14–16, 1966, Robert E. Uhrig (Coordinator), AEC Symposium Series, No. 9 (CONF-660206), p. 487, 1967.

85. V. Rajagopal, Dynamic Measurements in PWR Power Plants, *Trans. Amer. Nucl. Soc.*, 10(1): 217 (1967).

85a. M. Hara et al., Application of the Pseudo Random Binary Signals to JRR-3 High Power Dynamics Measurements, in Japan–United States Seminar on Nuclear Reactor Noise Analysis, Tokyo and Kyoto, Sept. 2–7, 1968, pp. 267-274, Japan Society for the Promotion of Science and National Science Foundation, Washington, D. C.

86. J. A. Thie, Elementary Methods of Reactor Noise Analysis, *Nucl. Sci. Eng.*, 15: 109 (1963).

87. P. Bentley and E. Burton, The Measurement and Analysis of Core Temperature Fluctuations in Chapelcross Reactors, *Reactor Sci. Technol.*, 19: 313 (1965).

88. E. S. Beckjord, Dresden Reactor Stability Tests, *Trans. Amer. Nucl. Soc.*, 3(2): 433 (1960).

89. R. Hyndman, F. Kern, and R. Smith, EBR-II Self Excited Oscillations, *Trans. Amer. Nucl. Soc.*, 8(2): 590 (1965).

90. J. A. Thie, Statistical Analysis of Power-Reactor Noise, *Nucleonics*, 17(10): 102 (1959).

91. J. A. Thie, Dynamic Behavior of Boiling Reactors, USAEC Report ANL-5849, Argonne National Laboratory, May 1959.

92. E. Wimunc et al., Performance Characteristics of the EBWR from 0 to 100 Mwt, in *Operating Experience with Power Reactors*, Vol. 1, Symposium Proceedings, Vienna, 1963, International Atomic Energy Agency, Vienna, 1963 (STI/PUB/76).

93. J. A. Thie, Noise Sources in Power Reactors, in *Noise Analysis in Nuclear Systems*, Gainesville, Fla., Nov. 4–6, 1963, Robert E. Uhrig (Coordinator), AEC Symposium Series, No. 4 (TID-7679), pp. 357-368, 1964.

94. E. Jordan, Detection of In-Core Void Formation by Noise Analysis, *Trans. Amer. Nucl. Soc*, 9(1): 317 (1966).

95. T. Eurola, Reactor-Noise Experiments on Halden Boiling Water Reactor, in *Noise Analysis in Nuclear Systems*, Gainesville, Fla., Nov. 4–6, 1963, Robert E. Uhrig (Coordinator), AEC Symposium Series, No. 4 (TID-7679), pp. 449-468, 1964.

96. J. C. Robinson and D. N. Fry, Application of a Theoretical Model for Interpreting Observed Spectral Density of Power Fluctuations in Water Reactors, *Trans. Amer. Nucl. Soc.*, 10(2): 590 (1967).

97. R. F. Saxe, Acoustic Characteristics of the Oak Ridge National Laboratory Reactor, in *Neutron Noise, Waves, and Pulse Propagation*, Gainesville, Fla., Feb. 14–16, 1966, Robert E. Uhrig (Coordinator), AEC Symposium Series, No. 9 (CONF-660206), p. 475, 1967.

98. C. Sastre, Noise Measurements in a Reactor Start-Up, *Trans. Amer. Nucl. Soc.*, 10(1): 216 (1967).

99. J. Hirota, Statistical Analysis of Small Power Oscillations in the HTR, USAEC Report CF-60-1-107, Oak Ridge National Laboratory, Jan. 29, 1960.

100. H. Kataoka and M. Kubo, Identification of Transfer Function of Reactor Control System by Noise Analysis, in *Neutron Noise, Waves, and Pulse Propagation*, Gainesville, Fla., Feb. 14–16, 1966, Robert E. Uhrig (Coordinator), AEC Symposium Series, No. 9 (CONF-660206), p. 439, 1967.

100a. S. Yamada, H. Kage, and T. Hoshi, Reactor Noise Behavior at Power Operation, in Japan–United States Seminar on Nuclear Reactor Noise Analysis, Tokyo and Kyoto, Sept. 2–7, 1968, pp. 227-232, Japan Society for the Promotion of Science and National Science Foundation, Washington, D. C.

101. M. Utsuro and T. Shibata, Power Noise Spectra of a Water Reactor in Low Frequency Region, *J. Nucl. Sci. Tech.*, 4(6): 267 (1967).

102. D. H. Crimmins and H. J. Snyder, ML-1 Reactor Stability Tests, USAEC Report AGN-TM-403, Aerojet-General Nucleonics, June 1963.

103. J. G. Crocker, Z. R. Martinson, R. M. Potenza, and L. A. Stephan, Reactor Stability Tests in the SPERT IV Facility, USAEC Report IDO-17088, Phillips Petroleum Company, 1965.

104. Y. Ko, P. Ma, and J. Chien, The Measurement of Reactor Noise by Means of a Multiscaler Analyzer, *Trans. Amer. Nucl. Soc.*, 9(2): 517 (1966).

105. P. R. Pluta, VBWR Noise Analysis, USAEC Report APED-4285, General Electric Company, Mar. 30, 1964.

106. P. R. Pluta, Preliminary Results of Vallecitos Boiling Water Reactor Noise Analysis, in *Noise Analysis in Nuclear Systems*, Gainesville, Fla., Robert E. Uhrig (Coordinator), Nov. 4–6, 1963, AEC Symposium Series, No. 4 (TID-7679), pp. 405-426, 1964.

107. A. A. Wasserman, G. H. Steiner, C. A. Bodenschatz, and E. K. Honka, NERVA Reactor Transfer-Function Measurements with Cross-Correlation Techniques, in *Neutron Noise, Waves, and Pulse Propagation*, Gainesville, Fla., Feb. 14–16, 1966, Robert E. Uhrig (Coordinator), AEC Symposium Series No. 9 (CONF-660206), p. 285, 1967.

108. J. A. Johnson, Measurement of Pressure-to-Pressure Transfer Functions Through the Phoebus-1A Nuclear Reactor, in *Neutron Noise, Waves, and Pulse Propagation*, Gainesville, Fla., Feb. 14–16, 1966, Robert E. Uhrig (Coordinator), AEC Symposium Series, No. 9 (CONF-660206), p. 271, 1967.

109. R. L. Randall and P. J. Pekrul, Applications of Analog Time and Frequency Correlation Computers to Reactor-System Analysis, in *Neutron Noise, Waves, and Pulse Propagation*, Gainesville, Fla., Feb. 14–16, 1966, Robert E. Uhrig (Coordinator), AEC Symposium Series, No. 9 (CONF-660206), p. 357, 1967.

110. F. D. Boardman, Noise Measurements in the Dounreay Fast Reactor, in *Noise Analysis in Nuclear Systems*, Gainesville, Fla., Nov. 4–6, 1963, Robert E. Uhrig (Coordinator), AEC Symposium Series, No. 4 (TID-7679), pp. 469-500, 1964.

110a. F. D. Boardman, A Theory of Noise in DMTR, British Report TRG-Memo-1128, June 1962.

111. Northern States Power Company, Northern States Power Company Pathfinder Atomic Power Plant Six-Month Report No. 2, 1967.

112. R. J. Hooper and E. P. Gyftopoulos, On the Measurement of Characteristic Kernels of a Class of Nonlinear Systems, in *Neutron Noise, Waves, and Pulse Propagation*, Gainesville, Fla., Feb. 14–16, 1966, Robert E. Uhrig (Coordinator), AEC Symposium Series, No. 9 (CONF-660206), p. 343, 1967.

113. J. A. Thie, *Reactor Noise*, American Nuclear Society/U.S. Atomic Energy Commission Monograph Series, Rowman and Littlefield, Inc., New York, 1963.

113a. N. Pacilio, *Reactor-Noise Analysis in the Time Domain*, AEC Critical Review Series, USAEC Report TID-24512, 1969.

113b. R. E. Uhrig, *Random Noise Techniques in Nuclear Reactor Systems*, Ronald Press Company, New York, 1970.

113c. J. S. Bendat and A. G. Piersol, *Measurement and Analysis of Random Data*, John Wiley & Sons, Inc., New York, 1966.

114. R. L. Randall, On-Line Reactor Safety Monitor for Rapid Detection and Diagnosis of Reactor-System Instabilities, *Trans. Amer. Nucl. Soc.*, 6: 74 (1963).

115. W. Dixon (Ed.), *BMD Biomedical Computer Programs*, University of California at Los Angeles, 1964.

116. E. Bullard et al., *Users Guide to BOMM*, University of California, 1966.

117. Milletron Transfer Function Computer, Milletron, Inc., Pittsburgh, Pa., 1962.

118. C. Hsu and W. Pipinski, Statistical Error Estimation for the Transfer Function Measurement of a Noisy Reactor System, *Nucl. Sci. Eng.*, 21(3): 407 (1964).

SECTION 7

CONTROL-ROD DRIVES AND INDICATING SYSTEMS

7-1 INTRODUCTION

7-1.1 Reactor Kinetics

A nuclear reactor that is generating heat at a constant rate is a chain-reacting system in which the number of neutrons being produced in nuclear fission processes exactly balances the number of neutrons being absorbed in or escaping from the system. If it is desired to change the rate of heat generation (number of fissions per second), means must be provided to increase or decrease the absorption and escape of neutrons. Once the heat-generation rate has reached the desired new level, means must be provided to restore the neutron balance so the system will once again generate heat at a constant rate. The specific means used in present-day power reactors to alter the heat-generation rate upward or downward or to keep it constant are discussed in this chapter.

In the steady state (constant rate of generating heat), the reactor is critical when the effective multiplication constant k (sometimes written as k_{eff}) is just equal to 1. To increase or decrease the power level of the reactor requires that k be increased or decreased above or below unity during the interval when the power level is changing. Once the desired power level has been reached, k must be restored to unity so the reactor can again operate in a steady state, albeit at a new power level. The fractional deviation of the effective multiplication constant from unity is defined as the reactivity[1]:

$$\text{Reactivity} = \rho = \frac{k-1}{k} = 1 - \frac{1}{k}$$

or

$$\text{Reactivity} = \frac{\delta k}{k} \quad (\text{with } \delta k = k - 1)$$

Because k is very close to 1 at all times in an operating power reactor, the reactivity ρ or $\delta k/k$ is often abbreviated to δk or "excess k" if $\delta k > 0$. In terms of reactivity, the basic types of reactor performance are

$\rho = \delta k/k = 0$ constant power level
$\rho = \delta k/k > 0$ power level increases
$\rho = \delta k/k < 0$ power level decreases

When the reactivity is not zero, the reactor power level increases or decreases with a characteristic time constant (reactor period) that is primarily dependent on the value of the reactivity, the prior operating history of the reactor, and the reactor configuration (arrangement and composition of fuel, moderator, coolant, etc.). Period is the time required for the neutron level to increase (or decrease) by a factor of "e" (2.718) (see sec. 1). The reactor period becomes too short for practical control if the reactivity is increased above zero by an amount equal to the delayed-neutron fraction, β. For most presently operating power reactors, $\beta \simeq 0.007$. This means that a positive ρ or $\delta k/k$ is always between zero and about 0.06%. During operation at power, reactor control systems normally make adjustments at rates in the general range from 10^{-3}/sec to 10^{-5}/sec in $\delta k/k$ per second. Although the control adjustments during operation involve relatively small changes in reactivity, this is not necessarily true during reactor start-up and shutdown. The control system must be capable of "adding negative reactivity" to balance out the reactivity excess built into the reactor, and, if an emergency exists, it must do so rapidly. Under certain conditions reactivity changes of ~ -0.1/sec in $\delta k/k$ per second may be required. The excess reactivity built into a power reactor depends on many factors; it can be more than 10% in $\delta k/k$. For this reason control (and safety) systems must be capable of accomplishing large changes in reactivity during reactor start-up and shutdown. In addition, they must be capable of compensating for the effects of changing concentrations of the fission products ^{135}Xe and ^{149}Sm. These can involve reactivity changes of several percent (see pt. 1-3.6 of sec. 1).

7-1.2 Reactivity Variations During Operation

A number of inherent reactivity variations occur during the operation of a power reactor and must be considered in designing reactor control systems. Some of these can be used to assist in reactor control; others require attention in the control-system design.

A key effect to be considered is the temperature coefficient of reactivity. Temperature variations change neutron cross sections and dimensions of reactor materials and thus change the reactivity. A desirable condition is for the reactivity to decrease as the reactor temperature increases. Such a negative temperature coefficient has a stabilizing effect since, as the power increases and raises the reactor temperature, the negative temperature coefficient reduces the reactivity and tends to limit or level out the rise in power. Most of today's reactors have a negative temperature coefficient of reactivity, usually $\sim -10^{-4}/°F$ in $\delta k/k$ at operating temperatures.

In boiling-water reactors (section 16), the reactivity changes as the void volume in the core changes. Since the water (steam) acts both as neutron moderator and absorber, the void coefficient of reactivity can be either positive or negative.

A generalization often used in preliminary calculations is to lump all the reactivity effects into a single power coefficient of reactivity, the change in reactivity resulting from a unit change in reactor power. This coefficient is typically $\lesssim 10^{-4}$ decrease in $\delta k/k$ per megawatt (thermal) change in power level.

A power reactor is inherently stable. The degree of stability is temperature dependent since the reactivity coefficients are temperature dependent. Thus a reactor may be less stable when cold than at operating temperatures. When inherently stable, the negative temperature effects dominate the positive, and the requirements placed on the control system are less stringent. In fact, with a net negative coefficient of reactivity, some reactors may be controlled over a limited range without resorting to control-rod movement.

Long-time inherent changes in reactivity are those attributable to the increase of fission-product poisoning and fuel depletion. These can be controlled by chemical shimming or by incorporating burnable poisons in the fuel.

7-1.3 Methods of Reactivity Control

Most power reactors are controlled by rods that are inserted into or withdrawn from the reactor core. The rods contain neutron-absorbing or fissionable material or a combination of the two. Some power reactors have been controlled by the rotation of control drums on the core periphery, the control drums being made of combinations of neutron-reflecting and neutron-absorbing materials.

For power reactors where changes in neutron level may be accomplished over relatively long time periods, but where constant power levels are wanted once full power is achieved, burnable poisons and chemicals dissolved in the coolant (so-called "chemical shimming") have proved effective. Burnable poisons can be added to the fuel elements to decrease fuel-element absorption of neutrons in proportion to the decrease in the fissionable material content of the fuel elements. Either type of control, chemical shimming or burnable poisons in the fuel, reduces the total reactivity that must be offset by the control-rod system.

7-2 REACTOR CONTROL SYSTEM

A basic reactor control system is shown schematically in Fig. 7.1. During operation at power, the demand signal is an output of the plant control system. This demand signal is compared with the measured neutron level, and the reactivity is correspondingly adjusted by programming the control-rod actuators to increase or decrease reactor power.

In reactors that are inherently stable, it is possible, though not necessarily preferable, for an operator to keep the power at the demand level by manually adjusting the control-rod position. When a reactor is not inherently stable, a continuous feedback control system, usually a servo-controlled rod, is essential.

There are four distinct phases of reactor operation: the approach to criticality, power increase or decrease, power operation, and shutdown. Each phase imposes different requirements on the reactor control system.

7-2.1 Approach to Criticality

The reactor is manually controlled by an operator during the approach to criticality. The control rods are moved intermittently to add reactivity until the reactor is critical. At this juncture the reactivity is zero and the rate of neutron production from fission is just equal to the rate of neutron loss. The magnitude and rate of rod motion is governed by the need for maintaining the reactor period longer or slower than some predetermined value.

The following requirements are imposed on the control system in this phase of reactor operation:

1. The control rods must be capable of motion in increments small enough to insert very low values of δk. Rod drives subject to uneven motion, e.g., because of friction, must be avoided.

2. The rod actuation system must be capable of measuring necessary reactor performance data during the initial start-up. Typical measurements required are total and incremental rod worths. (Ganging several control rods which are then driven by one actuator sometimes makes this measurement difficult.)

3. The control system must be able to insert reactivity at a maximum rate consistent with the start-up time requirements.

7-2.2 Power-Increase Phase

The power-increase phase is normally started by establishing coolant flow, adjusting power and flow to equal the demanded values, and closing the automatic control loops. As the coolant flow is then increased, the power level is also increased, and the reactor temperature rises to the operating value. The reactor control system responds to the plant control demand by causing appropriate motion of the control rods. The coolant flow and power level may be increased simultaneously or separately, with the flow reaching full value before power.

The primary requirements on the control system during this phase are:

1. The control rods must smoothly adjust the reactivity in accordance with the plant controller demands. The amount of rod motion depends on the reactivity change associated with the incremental rod worth and the reactor temperature and pressure conditions. Particular attention must be given to the transient conditions during changes of reactor power and flow, since excessive overshoots in temperature or pressure might cause intolerable damage to the reactor core.

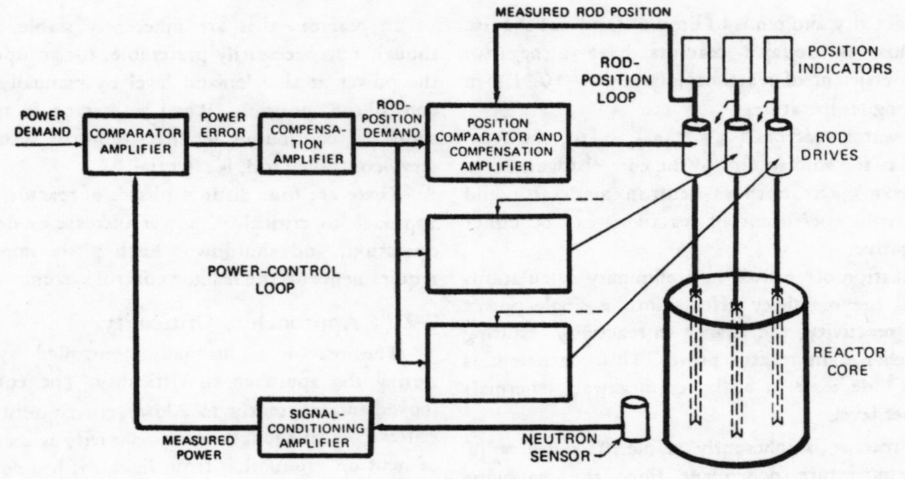

Fig. 7.1—Basic reactor control system.

2. Since full-power conditions are being approached, the relative control-rod position becomes important to ensure uniform power distribution throughout the core.

7-2.3 Power Operation Phase

During this phase the reactor must respond to the plant control-system demands to deliver the desired power, maintain the reactor operating conditions, and remain within predetermined reactor parameter limits. The required dynamic characteristics of the control system are different for each reactor type (PWR, BWR, and gas-cooled or liquid-metal-cooled reactors).

Some of the important factors that must be considered in the design of the control system for this phase of reactor operation are:

1. Since the reactor in this phase is at or near full power, the control system must respond to the plant power-system demands rapidly enough to meet plant requirements and yet maintain the core temperatures within prescribed limits (such as the hot-spot temperature limit). To derive the maximum power, the designer faces a trade-off between the desirability of operating close to the maximum temperature limits of the reactor and the associated requirements of more accurate temperature measurement and dynamic response of the control system.

2. In addition to the requirement that the reactor rod-actuation system be accurately positioned in response to a demand, the banks of rods must also be positioned accurately relative to each other. Inaccuracies in relative positioning of control rods causes local power increases in the region adjacent to those rods that are farthest withdrawn. Positioning inaccuracies can result in a smaller margin between the limiting temperatures and the normal operating conditions of the reactor core.

3. Increments of control-rod motion must be of such magnitude that thermal transients do not increase the temperatures or temperature gradients to undesirable values. For example, the reactivity insertion steps resulting from unit rod motion in stepping-motor systems must be kept within allowable limits. Excessive friction and control deadbands must also be considered in designing the overall control system.

4. Some of the fission products produced during power operation absorb neutrons and necessitate the addition of reactivity by control-rod movement. The most important of these are ^{135}Xe and ^{149}Sm (see pt. 1-3, section 1). The reactivity, $\delta k/k$, to overcome the fission-product poisoning under equilibrium operating conditions is a function of reactor design but normally varies from 0.3 to 3.0%. The control rods must be capable of compensating for the buildup of these neutron absorbers. About 10 hr after shutdown, the ^{135}Xe poisoning increases to a value higher than the equilibrium value at full power. The poisoning peaks at about 10 hr after shutdown and then diminishes. This leads to the requirement that the control rods be capable of introducing sufficient neutron absorber at a rate that will maintain the reactor in a shutdown condition as the ^{135}Xe poisoning is reduced.

5. The reactivity must be compensated for the reduction of fissionable material attributable to burnup during power operation. This compensation is normally provided by the control system, which automatically positions the control rods to maintain full power. The control-rod worth designed into a reactor is a function of the amount of fuel depletion anticipated during the interval between reloadings. In some reactors a burnable poison, such as ^{10}B, is introduced into the core. As neutrons are absorbed by the boron, the number of remaining boron atoms available for neutron absorption decreases. The amount of burnable poison introduced into a reactor is governed by the desire to have this effect compensate for fuel depletion. In this manner the total reactivity control requirement of the rod system is reduced.

7-2.4 Shutdown Phase

The shutdown phase of operation is usually accomplished by a controlled insertion of the control rods in response to a reduction in power demand. A second type of shutdown is a scram in which the reactivity and reactor power must be reduced in a short time to prevent exceeding the reactor or plant limits. The control system must be designed such that sufficient negative reactivity is available to shut down the reactor under all conditions.

To simplify the control-rod drive system for normal power operation, the rate of control-rod motion is normally the same in either direction. However, the rate of change of reactivity can still be varied significantly by using only one or a few rods at a time when increasing the reactivity and using all rods simultaneously when decreasing the reactivity.

7-3 SELECTION OF REACTIVITY-CONTROL METHOD

7-3.1 System Requirements

The requirements that must be met by the reactor control system, including sensors and control rods, may be divided into the following categories:

1. Amount of reactivity controlled and rod-positioning accuracy.
2. Rate of change of reactivity for normal operation, including initial start-up, planned shutdown, and restart.
3. Emergency shutdown.
4. Reliability.

The excess or maximum reactivity requirements of a reactor are dependent on the planned rate of fuel depletion, fission-product buildup, inherent reactivity effects (e.g., temperature), and control range desired. Table 7.1 shows that the range of total rod reactivity for a number of typical power reactor plants varies from 6 to 25% in $\delta k/k$.

The control-rod drive must be capable of making small changes in reactivity to maintain a flat neutron-flux distribution (uniform generation of heat throughout the core) and to be able to adjust the power level of the reactor with sufficient precision throughout full-power operation. Typically, changes in $\delta k/k \lesssim 10^{-5}$ are required. In terms of positioning accuracy, this means that the control-rod drive system must be able to position a rod to an accuracy of about 0.02 in. This value can vary widely, however, depending on the particular reactor design and the individual rod worth; values from 0.01 in. to several inches are possible.[2]

The reactor designer determines the required rate of change of reactivity by examining how rapidly power must be changed to maintain proper operation during both normal and emergency operation. For conventional start-up and power phases, periods of 30 sec or longer are normal. This requires reactivity adjustments in the range of 10^{-3} to 10^{-5} δk/sec, with an average value of 2×10^{-4} δk/sec.* This same rate is usually satisfactory for steady-state control. For linear control rods, this corresponds to about 10 in./min as an average, although it can vary[2] (with reactor design) from 3 to 300 in./min. For a directly coupled rotational system, where the control drum may rotate 180°, this rate corresponds to about 0.2°/sec. During shutdown the reactivity (and thus the reactor power) is usually decreased more rapidly than its rate of increase during start-up. As noted earlier, this requirement is often satisfied by moving the control rods at the same rate used for start-up, but moving all rods simultaneously rather than a few at a time.

For scram or emergency shutdown, the required rate of reactivity reduction normally exceeds the insertion rate for power control by a factor of 10 to 100. Higher rates of reactivity reduction do not yield significant benefits since, after the reactivity becomes about 1% below critical ($\rho = \delta k/k = -0.01$), the reactor power decreases as the delayed-neutrons decrease. Of more importance in shutdown is the release time or turnaround time following receipt of a shutdown command. Small delays in beginning the reduction of reactivity can result in significant power excursions. The usual practice is to design for release times of about 10 to 50 msec. When this rapid initiation is coupled with a reactivity insertion rate of about 5×10^{-2} δk/sec, the reactor can be shut down on a nominal negative period of 5 sec or less.

It is common practice to satisfy both normal performance and safety requirements with one reactivity adjustment mechanism. It is also common practice to design the scram mechanism to operate in a fail-safe mode, i.e., to operate in the event of loss of primary power. (The primary-power loss may be inadvertent or it may be initiated by another part of the scram system.) There are many ways of designing the rod-drive mechanism to fail safe. One general practice is to use gravitational force to store energy for scram. A simple example of this practice is to place a coupling device between the control rod and its drive mechanism that has the same primary-power source as that which supplies the control-rod actuator. If the primary power is lost, the control rod is released and gravity forces the rod into the reactor. Springs and hydraulic devices can also be used to store energy and to release it on power failure. If higher scram velocities are desired, a spring may be incorporated to increase the acceleration of the control rod into the reactor. If springs are used, the actuator may be designed such that the rod is held against force, and, if the primary power is lost, the spring returns the control rod rapidly to the shutdown position. The spring can also serve to eliminate backlash and thus reduce the deadband in the control loop.

The reliability requirements for the control-rod drive system are influenced by considerations of safety and maintainability. For safety reasons there must be a high level of confidence that the scram system will operate correctly, i.e., it will be reliable (see Sec. 12). The required confidence that this system will work is significantly higher than that required of the control system during normal operation. In addition, the availability of the control-rod drive system is essential. This means the system must be available for operation during all scheduled operational periods, except during normal preventive-maintenance periods when the reactor is shut down. A failure in the control-rod drive system normally requires

Table 7.1—Typical Drive-System Characteristics for Nuclear Power Reactors

Reactor	Thermal power, MW	Temp. compensation, % δk	Poisons, % δk	Fuel depletion, MWd/ton % δk	Rod total, % δk	Rod-control velocity, δk/sec	Rod-scram velocity, δk/sec	Rod-position accuracy, δk	Total number rods	Rod length, ft
Boiling-water										
Dresden	686	7.7	3.6	12,000/6.0	14	1.3×10^{-4}	0.136	Continuous	80	8.5
EBWR	100	3.13	3.36	11,000/6.0	15	1.4×10^{-4}	0.38	Continuous	9[a]	5.0
Humboldt Bay	165[b]	3.6	3.1	10,000/12.3	17.3	4×10^{-4}	0.042	0.0004	32[c]	6.6
Pressurized-water										
Shippingport 1	231	2.6	3.8	4500/11.0	25.6	1.07 to 1.38×10^{-4}		Continuous[b]	32[d]	6.0
Shippingport 2	505	4.3	2.8		16.0				20	8.0
Yankee	485[b]	4.1	3.0	7830/7.1	15.1	0.8×10^{-4}		0.0003	24	7.54
Gas-cooled										
Bradwell	538	2.0	1.7		7.5[e]	2×10^{-4}		Continuous	108[f]	28.00
Calder Hall 2	285	2.19	1.7		6.6	2×10^{-4}	0.01	Continuous	48[g]	33.3
Hunterston	535	1.9	1.7	3600/10.1	8.5	3×10^{-4}			156	
Berkeley	565	1.64	~1.76		4.5	3×10^{-4}		Continuous	132	25.5
Peach Bottom	115	2.6	3.0		2.3 (12.0)	7.7×10^{-5}	0.23 (0.0048)	Continuous	36[h]	7.0
Sodium-cooled										
EBR-2	62.5	1.5	negl.	–/2.4	6.1	3.8×10^{-5}	0.055	Continuous	14[i]	1.16

Reactor	Actuator control velocity, in./sec	Actuator scram velocity, in./sec	Actuator type	Actuator position interval, in.	Actuator position indicator	Actuator position readout	Actuator type scram
Boiling-water							
Dresden	6.0	102.0	Hydraulic		Magnet in piston	Mag.-actuated switches	Hydraulic
EBWR	0.1–0.467	41.0 (av.)	Rack and pinion	±0.025	Syn. trans. and speed reducer	Syn. receiver	Gravity
Humboldt Bay	3 (0.7)	28.4 (av.)	Locking piston	3 steps @±1.0	Magnet in piston	Mag.-actuated switches	Hydraulic/gravity
Pressurized-water							
Shippingport 1	0.046–0.133	56.4 (av.)	Lead screw/col. Rotor/v.f.ind. motor	±1.5 or ±0.125	Magnetic coil or inverter reluctance		Gravity
Shippingport 2							
Yankee			Magnetic jack	±3.0	30 transformers	Lights in secondaries	Gravity
Gas-cooled							
Bradwell	0.00085	8.5 (av.)	Var. freq. motor				Mag. clutch/gravity
Calder Hall 2	0.00834 out 0.00167 in	48.0	Var. freq. motor	±1.0	2-in. mag. slip transmitter	Mag. slip receiver	Gravity
Hunterston							
Berkeley	0.00721		Var. freq. motor				Gravity
Peach Bottom	0.72	120.0					Hydraulic (battery and motors)
Sodium-cooled							
EBR-2	2.85		Rack and pinion	±0.01	Syn. trans. and gear reduction	Syn. receiver	Mag. clutch/gravity

[a]Plus 1 oscillator. [b]Initially. [c]Plus 8 peripheral. [d]Four groups. [e]From 80 rods. [f]Plus 11 shutoff. [g]46 coarse. [h]Plus 19 emergency. [i]12 shim.

(Table 7.1 continues on next page.)

Table 7.1—(Continued)

Reactor	Controller type	Controller feed-back signal	Controller rods controlled	Coupling type	Scram time, sec
Boiling-water					
Dresden	Manual	N.a.	All	Mech., 90° rotation	3.0
EBWR	Manual	N.a (signals cause loss of power)	All in; one out	Mag. clutch	1.35
Humboldt Bay	Manual	N.a.	All	Mech.	3.0
Pressurized-water					
Shippingport 1	Manual or temp.	Coolant resistance thermometers	All for scram, all by sequence	Roller out dir. coupling	1.5
Shippingport 2	Temp.				1.35
Yankee	Temp.		All		2.0
Gas-cooled					
Bradwell	Zone outlet temp. reactor power	Gas temp. turbine speed	28 any of 80	Chain and sprocket	5.0
Calder Hall 2	Manual	N.a.	All or one	Steel cable/drum	5.0
Berkeley	Zone outlet temp. reactor power	Gas temp. turbine speed	9	Chain and sprocket	6.0
Peach Bottom			All (one at a time above 10% power)		1.0
Sodium-cooled					
EBR-2	Manual	N.a.			0.6

that the reactor be shut down in view of the possibility of unsafe operation (loss of control of output power or loss of scram capability). The reactor is then unavailable for the total time required to shut down, correct the failure, and restart. Hence, unscheduled maintenance must be avoided. Nuclear power plants are designed for a life of about 30 years, with scheduled shutdowns for maintenance at intervals of 6 to 18 months. Since the control-rod drive systems can be serviced during the scheduled maintenance periods, their reliability requirements are correspondingly reduced. In essence, control-rod drive systems must be highly reliable, but only for relatively short periods of time. For an increase in overall reactor reliability, some systems are designed so a failure of one control-rod drive does not require shutdown. In these systems failures may occur during operation, but corrective maintenance is not required until the next scheduled maintenance period. If unscheduled maintenance is required because a rod-drive mechanism fails and forces a shutdown, it is very desirable to minimize the time required for corrective maintenance. This time can be reduced if the designer has considered this requirement during the initial design phase.

The requirements discussed above result from nuclear design operational considerations and are applicable to reactors controlled through a primary control loop on nuclear power. In establishing the requirements for a control-drive mechanism, the designer must first consider the reactivity span and the rate of change needed; these are used to calculate the desired period and steady-state operating conditions. However, since an automatic control loop is normally used, the designer must also ensure that the control system can operate in a stable closed-loop manner. The requirements on the control-drive mechanism that must be met to provide stable closed-loop operation during all feasible transients and perturbations may be more severe than those for satisfactory period and steady-state operation. These requirements are identified by dynamic control analysis of the complete reactor system.

As shown in Fig. 7.1, the control-rod position loop is usually an inner loop of the automatic power-control loop. Although the speed of response of the power-control loop is selected to provide the desired reactor performance, a dynamic control-systems analysis would indicate that the speed of response of the rod-position loop must be from 2 to 10 times more rapid to result in stable (nonoscillatory and nondiverging) operation when all control loops are closed.

The basic power-control loop, as shown in Fig. 7.1, is sometimes supplemented by a trim loop that controls the temperature of the primary-coolant flow. This latter would be an outside loop on power control which compares a measured temperature with a demanded value and produces a supplemental power-demand signal. More stringent requirements are placed on the drive mechanism when the reactor controllers are complicated by introducing coolant temperature, or variables from the steam side of the power plant loop, because of the interaction of these parameters and the normal dynamic requirement to make all inner loops respond about six times faster than outer loops.

The requirements of the amount of reactivity controlled, the rod-positioning accuracy, reactivity rate of change under various situations, and the reliability and availability have been met in many power reactors by using low-maintenance, high-reliability a-c motors for the control-rod drive. The coupling to the control rod is usually by a rack-and-pinion mechanism or a lead screw and nut. However, other techniques have been used, including d-c motors, hydraulic cylinders and motors, linear induction motors, and magnetic jacks. The designer must establish the requirements for a given reactor design and then review all available systems and components to select the most appropriate. The advantages and disadvantages of a number of systems and typical applications are discussed in the following sections.

7-3.2 Means of Control

The control rods selected for water reactors are usually linear structures of a neutron-absorbing material designed to be moved vertically into and out of the core. The amount of neutron absorber in the core is determined by the position of the rods with respect to the core. At shutdown, the rods are positioned with the maximum amount of neutron-absorbing material within the core. As the rod is withdrawn from the core, the reactivity and neutron population increase by an amount generally proportional to the amount of neutron-absorbing material removed.

Although control rods of neutron-absorbing material are used in most reactor installations, in some instances neutron-reflecting and neutron-moderating or fuel-bearing rods are used. Another type of control, a cylindrical device called a drum, has its surface comprised partly of neutron-absorbing material and partly of neutron-reflecting material, or there can be combinations of absorber–fuel or absorber–moderator. Several of these drums are located in a vertical position around the core periphery and have rotary control motion, the extent of drum rotation determining the core reactivity.

Reactivity control by a neutron-absorbing liquid may be heterogeneous, with the liquid flow in a sealed pipe or pipe system adjacent to the reactor core, or homogeneous, with the liquid mixed with the reactor coolant water and extracted by ion-exchange equipment. For example, mercury can be used in a heterogeneous system and boric acid can be introduced into the reactor coolant-water system in a homogeneous system.

7-3.3 Materials

The conventional control rod for water-cooled reactors is made of stainless steel with neutron-absorbing material either supported and clad by the rod structure or alloyed with the rod material. The neutron-absorbing material absorbs thermal neutrons, which reduces the effective multiplication constant, k, below unity. The most commonly used neutron-absorbing or poison materials in control rods are cadmium, boron, and hafnium. Other less commonly used materials are silver, europium, and indium. Factors determining the usefulness of a control-rod material include not only thermal-neutron absorption properties but also availability, cost, and structural and machinability properties. The material must be fabricated into various shapes and must not be affected appreciably by the temperature or pressure of the particular environment in which it is to be used. Its susceptibility to corrosion and nuclear-radiation damage must be low. In some instances

Table 7.2—Control Worth for Various Materials*

Material	Relative effectiveness in a water-cooled reactor
3.0 wt.% ^{10}B in stainless steel (dispersion of minus-100 mesh particles of 90% enriched ^{10}B)	1.12
Dispersion† containing 10 vol.% B_4C (90% enriched ^{10}B)	1.06
Hafnium	1.00
0.97 wt.% ^{10}B in stainless steel (alloy)	0.98
Ag–22 wt.% In alloy	0.96
15 wt.% Eu_2O_3 in stainless steel (dispersion)	0.96
Indium	0.93
Silver	0.88
Cadmium	0.80
8.7 wt.% gadolinium–titanium	0.77
Tantalum	0.71
2.7 wt.% Sm_2O_3 in stainless steel (dispersion)	0.70
Haynes Stellite 25(Co–20 wt.% Cr–15 wt.% W–10 wt.% Ni)	0.68
Titanium	0.24
Zircaloy-3	0.05
2S aluminum	0.02

*Based on data from C. R. Tipton, Jr. (Ed.), *Reactor Handbook*, Vol. 1, Materials, 2nd ed., p. 779, Interscience Publishers, Inc., New York, 1960, and W. K. Anderson and J. S. Theilacker (Eds.), *Neutron Absorber Materials for Reactor Control*, p. 117, Superintendent of Documents, U. S. Government Printing Office, Washington, D. C., 1962.

†Dispersion assumed to have a nonabsorbing matrix and to be clad with 0.02 in. of nonabsorbing materials.

alloys of the poison materials with other materials improve their suitability for control-rod application.

Table 7.2 indicates the relative worth of commonly used control-rod absorber materials.

7-3.4 Rod Shape

The shape, dimensions, and number of reactor control rods are dependent on the core mechanical design and the amount of negative reactivity needed for shutdown. To function efficiently as a neutron poison, a control-rod material must have sufficient thickness to absorb most of the flux at the rod surface. In light-water reactors the slowing-down length, i.e., the distance required for a fission neutron to be reduced to thermal energy, is short. Therefore, to be effective in absorbing thermal neutrons, the poison material must be physically close to the fuel surface and have a high surface-to-volume ratio. A cruciform shape with thin wide blades of poison material fulfills these requirements. Figure 7.2 shows typical control-rod configurations for power reactors. Since a light-water reactor has a neutron spectrum with an appreciable fraction of epithermal neutrons, materials with large absorption cross sections for these energies, such as hafnium and indium, are used in addition to the usual thermal-neutron absorbers, such as cadmium or boron.

7-3.5 Rod Configuration

Control rods are arranged in symmetrical patterns within the core structure and around the core periphery. The total absorption capability in all the rods largely determines the shutdown reactivity of the core. Shutdown reactivity is defined as the increase in reactivity required to bring the reactor to critical from a fully shutdown condition.

Safety rods are specifically designed to effect rapid shutdown or scram should a hazardous reactor or plant condition occur. These rods normally contain poison material and are withdrawn to a maximum degree before start-up and are kept in that position during power operation. They are designed for rapid release and acceleration into the reactor. Separate safety rods are seldom used in the present generation of power reactors, but their scram function is combined into the shim rods (Sec. 12).

Shim rods comprise the greatest number of rods and control the greatest amount of reactivity. Shim rods are used to remove shutdown reactivity during start-up and to offset the effects of temperature, xenon, samarium, and fuel depletion during power operation. Controlled shim-rod motion is very slow during power operation and may be only a few inches or tenths of inches per day. Shim rods are usually arranged in groups or banks, only one of which can be moved at a time. In any group, one or several rods can be moved at the same time, at the option of the operator.

The regulating rod, similar in design to a shim rod, is used for fine control of reactor power level. Small changes of reactivity are needed. Regulating-rod motion may be

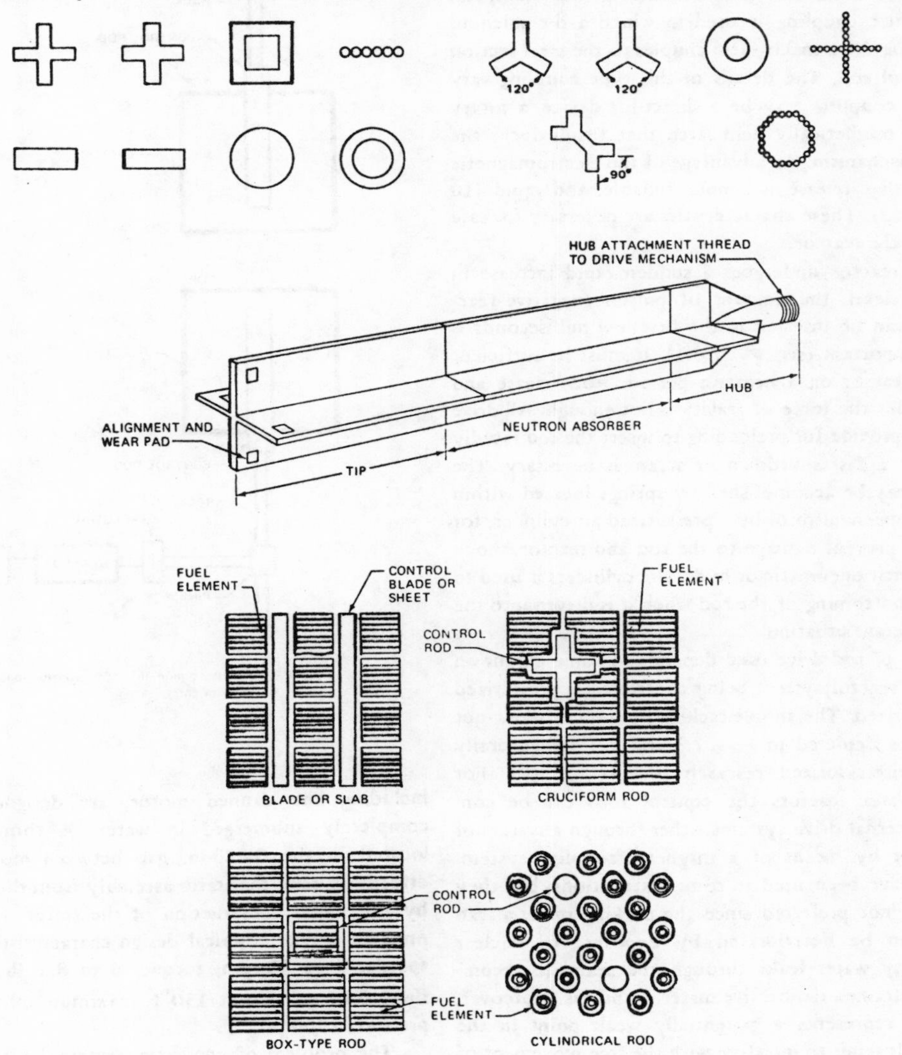

Fig. 7.2—Representative cross sections (not to scale) of control rods. Heavily outlined cross sections indicate clad absorber sections. The center figure shows the typical structure of cruciform rods used in some water-cooled power reactors. (

manually controlled, but usually the regulating rod is driven by a servomechanism in a feedback control loop that finely regulates power or flux. To satisfy the performance requirements of automatic control, the drive mechanism for the regulating rod has more stringent performance requirements than the drives for shim or safety rods.

7-3.6 Types of Drives

In the usual water reactor configuration, the control rods may be raised or lowered out of the core. Each rod may have a follower section of some inert material to fill the void left when the rod poison section is moved out of the core. Without the follower section, the void would fill with water, a good moderator material, and a neutron-flux peak would develop in that region.

The control rod can be coupled to a drive mechanism by any one of a number of different devices. Commonly, an electromagnetic coupling is used in which a d-c solenoid section on the drive mechanism couples to the steel section of the control rod. The details of this type coupling vary widely. The coupling may be a direct-lift device, a rotary clutch, or a magnetically held latch that firmly locks the rod to the mechanism. An advantage of the electromagnetic coupling is that release is simple, reliable, and rapid (10 msec is typical). These characteristics are necessary for safe operation of the reactor.

When a reactor undergoes a sudden rapid increase in neutron-flux level, the amount of poison (negative reactivity) that can be inserted in the first few milliseconds is extremely important (see Pt. 7-3.1). It must be sufficient to put the reactor on a negative period. Rod release and insertion under the force of gravity is not enough. All drive mechanisms provide for preloading to insert the rod rapidly in the event a fast shutdown or scram is necessary. The preloading may be accomplished by springs located within the latching mechanism or by a pressurized-air cylinder, for example. To prevent damage to the rod and reactor, shock absorption with pneumatic or hydraulic cylinders is used to cushion the bottoming of the rod when it is driven into the reactor in a scram situation.

The type of rod drive used depends to some extent on whether the reactor system being controlled is pressurized or nonpressurized. The simple rack-and-pinion or screw-nut control drives depicted in Figs. 7.3 and 7.4 are generally used for nonpressurized research or test reactors. For pressurized-water reactors the control rods can be connected to external drive systems either through a system of shaft seals or by means of a magnetic coupling system. Shaft seals have been used in some installations, but they are generally not preferred since the packing material can leak and can be deteriorated by exposure to nuclear radiation. Any water leaks through the seals are accompanied by entrained radioactive material and gas. Moreover, a shaft seal represents a potentially weak point in the system since it tends to interfere with the free movement of the control rod. Shaft seals have been usually applied in installations where rotary shaft motion is translated through the seals to a rack-and-pinion or screw-nut driven assembly located in a pressurized environment. Magnetic coupling through a pressure-containment material in the path of the magnetic flux can be accomplished in a number of ways. The containment material must, of course, be nonmagnetic.

Another method for driving control rods in pressurized reactors is by means of the "canned" motor. Figure 7.5 shows the construction of this type drive. It was originally developed for very low speed operation, enabling the gearing between the rotor shaft and the ball-screw mechanism for the control rod to be minimized. All moving parts,

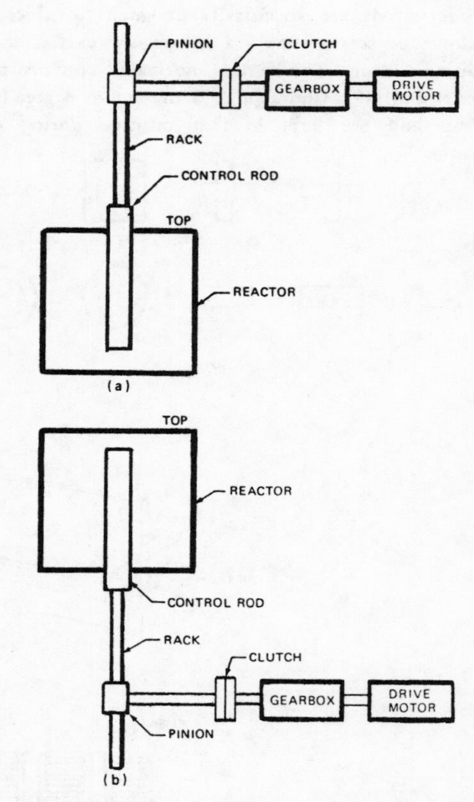

Fig. 7.3—Rack-and-pinion drive for control rod. (a) Top entry. (b) Bottom entry.

including the canned motor, are designed to operate completely submerged in water. A thin Inconel shell located in the 0.020-in. gap between motor and stator effectively seals the stator assembly from the water. Backed by the metal construction of the stator, this shell is the pressure barrier. Typical design characteristics are: normal speeds, 0 to 50 rpm; torque, 6 to 8 ft-lb; cooling water flow, 0.15 gal/min at 130°F maximum inlet temperature; pressure, 2500 psi.

The problem of converting external rotary motion to linear motion inside a reactor pressure vessel, as required with the canned motor, is eliminated by the linear inductance motor. This motor has the unusual capability of

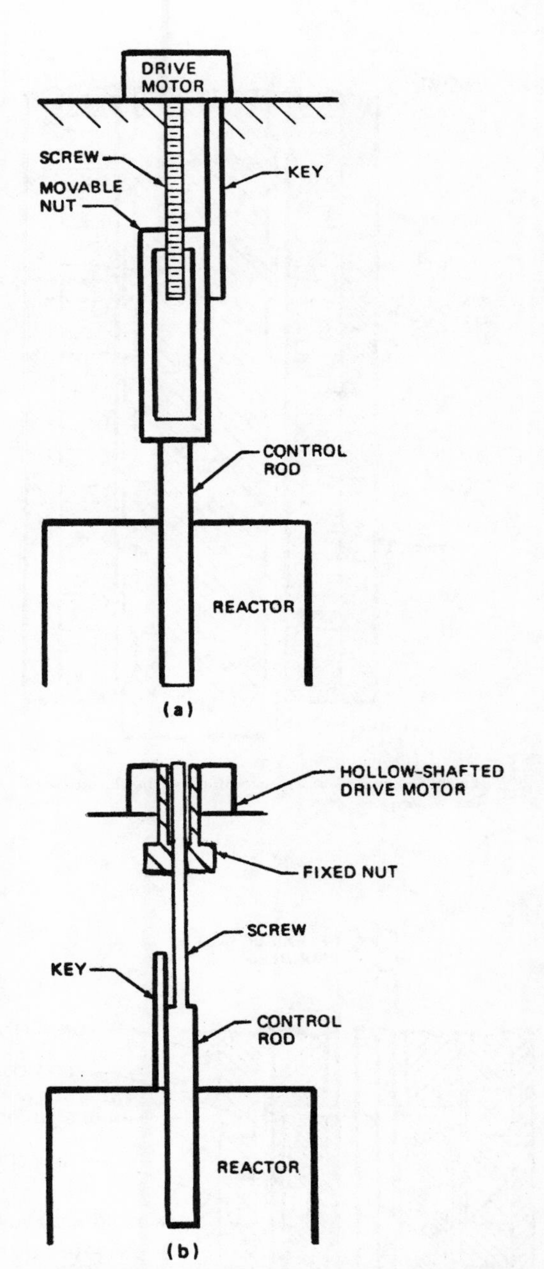

Fig. 7.4—Screw-and-nut control-rod drives. (a) Fixed screw. (b) Fixed nut.

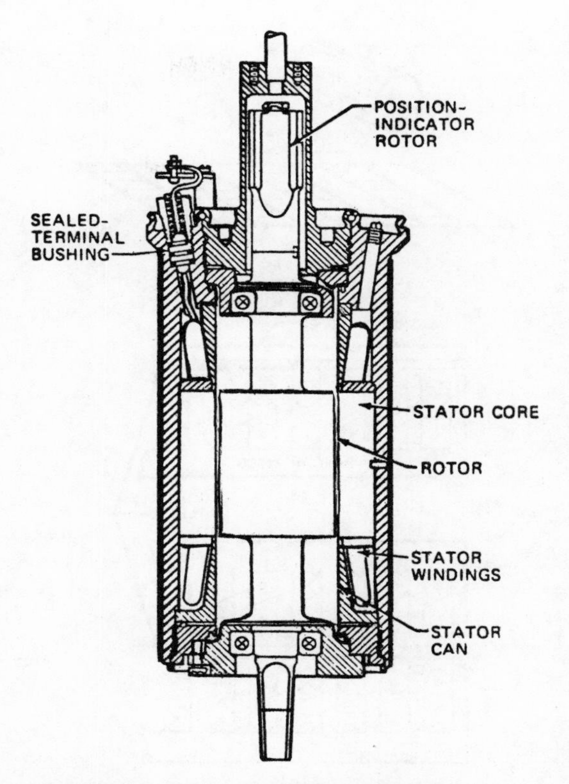

Fig. 7.5—Pressurized synchronous reluctance motor.

producing linear motion without an intermediate rotary motion.

The design of this motor can be visualized if the synchronous reluctance motor (used in the canned motor)[3] is considered. As described in *Control of Nuclear Reactors and Power Plants* by Schultz,[2] "if this cylindrical motor were figuratively to be sliced open at one axial place, flattened out, and then rolled up again at right angles to the previous cylinder, a structure similar to the cross section of Fig. 7.6 would be obtained. Here the field coils are nothing but circular doughnuts around a long tube. The armature consists of a series of ringlike poles on a long bar. The winding slots are cut into a thick magnetic cylinder, and the inner walls of this cylinder are backed up by the windings to take the pressure from the inside of the motor. To complete the magnetic path and also to assist in restraining internal pressures, a magnetic sleeve is placed outside the field-coil structure. The windings may be connected either in two phase or three phase, and the operation and drive systems are identical with those of the rotating synchronous reluctance motor." A complete linear motor drive system for a pressurized reactor is indicated in Fig. 7.7.

Another type of rod drive where magnetic coupling is used between the rod and the drive unit is the magnetic jack. The most successful version, shown in Fig. 7.8, has magnetic moving and holding devices. The pressure shell contains the control rod and a sliding armature section. Power applied to the lift, grip, hold-down and hold coils in certain sequences (see Pt. 7-5.3) holds the rod stationary and lifts or lowers it in definite increments; when all power is released, the rod can be suddenly dropped or scrammed. The power unit controlling rod movement consists of a reversible sequential switching device and a d-c power supply.

7-3.7 Rod-Position Indicators

Two aspects of position indication are essential for the

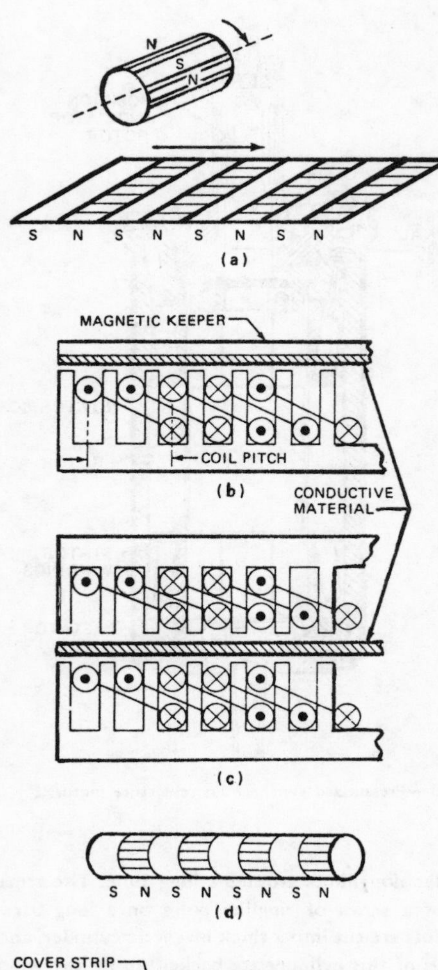

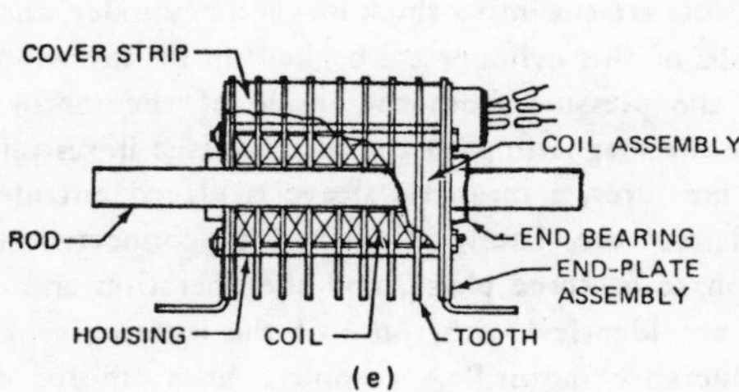

Fig. 7.6—Flat and round-rod linear induction motors. Development from cylindrical structure to a flat structure shown in (a) and to a rod structure shown in (d). Cross section of single-sided type of flat linear induction motor shown in (b); cross section of double-sized type shown in (c). Typical construction of round-rod linear induction motor shown in (e).

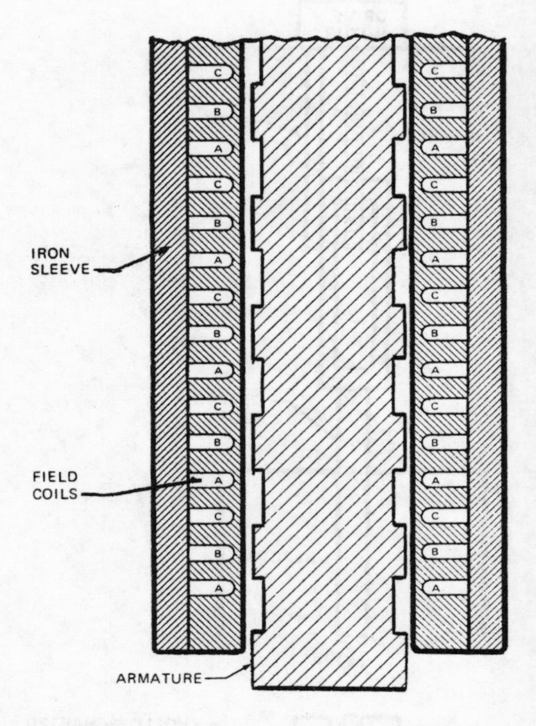

Fig. 7.7 —Linear induction-motor drive system for a pressurized-water reactor.

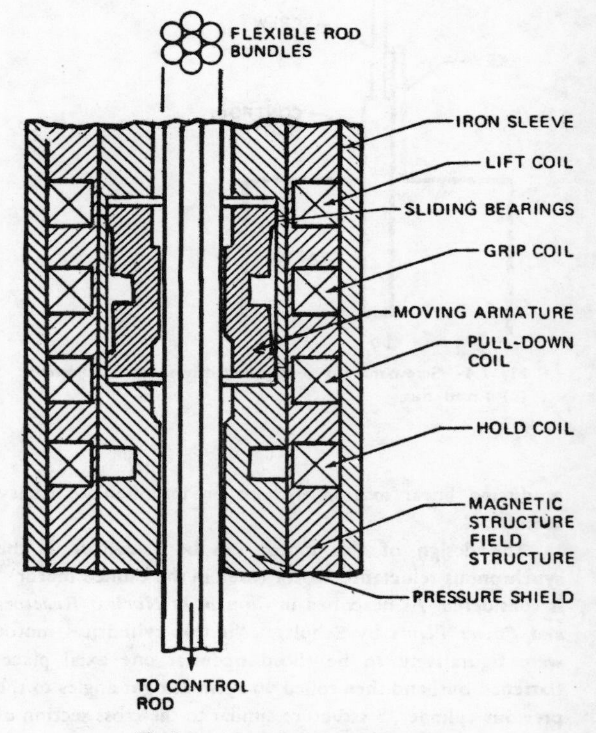

Fig. 7.8—Magnetic-jack configuration.

operation of control rods in a reactor: the measurement of control-rod position with respect to the core and the determination of rod position with respect to the drive mechanism. The first is far more important. The second is relevant to systems where the rod can be separated from the drive, as is often the case in drive systems that satisfy both the shim or control and safety requirements.

Pickup or release of a control rod and full insertion or full withdrawal are usually indicated by panel lights controlled by relays or sensing switches actuated by magnetic or mechanical coupling to the control rod. These switches or relays generally have auxiliary contacts used in the control logic circuits. An example is the logic that initiates movement of a bank of shimming control rods when the automatic regulating rod has exceeded its normal control range. Another is a permissive logic circuit that often takes the form of withholding control power at start-up until all rods are completely down or fully inserted in the reactor and all drives are down and engaged to their respective rods.

Whenever more than one group of shim rods is used on reactor start-up, limit-sensing switches can be used to permit or initiate movement of a second bank of rods when the first has reached a programmed "full-out" position.

Once the condition of major rod position and attachment to drives has been satisfied, knowledge of the position of rods with respect to the core is essential to reliable control of a reactor. Both absolute position and relative movement must be measured with an accuracy sufficient to establish safe conditions during start-up and power adjustment and to establish predetermined power distribution patterns in the core. The position of rods relative to each other can also be used to diagnose reactivity anomalies within the core. A study of the history of the reactor operating characteristics makes it possible to predict rod position under various operating conditions. If an indication differs markedly from its expected value, it is usually a sign of anomalous behavior in the core, control system, instrumentation, or rod-drive system. Analysis can often pinpoint the malfunctioning item.

Position indication may be direct, when the control rod actuates the sensing device, or indirect, when the position sensor is coupled to the drive mechanism. The control system is arranged so that after rod release the drives are automatically returned to zero position.

In all reactor operation, indication of rod position is displayed at the control console. Panel-mounted dial indicators driven by synchro-receiver units with the synchro-transmitters directly geared or coupled to the driven mechanisms, synchro-driven bar indicators, tapes, or digital counters are conventional. The bar or tape indicators give an immediate and graphic nonambiguous indication of absolute rod position, thereby minimizing possible operator error.

In conjunction with the previously mentioned "all-in" and "all-out" position indication and indication of relative position, position switch sensing of intermediate position is valuable for interlocking with rod-programming circuits and for giving the operator an independent indication of where the rods are under all circumstances. Intermediate position measurements can be used to signal definite rod positions on a bar or tape indicator, thereby providing a backup measurement that also prevents operator misinterpretation of absolute rod position.

Rod-position sensing in sealed systems, in which the drive as well as the rod is in a pressurized high-temperature coolant environment, is usually accomplished magnetically. Here extreme in/out positions and also intermediate points may be indicated by having magnetic coupling to coils located at the various points along the pressure-seal housing around a control-rod extension. Accurate and continuous sensing of rod position along the entire length of travel, however, is more difficult.

An example of rod-position sensing and indication in a sealed system is the magnetic-jack control rod in the San Onofre Atomic Power Plant (see Sec. 7-4.2). Lights on the control panel indicate rod position. Thirty transformers are mounted around the control-rod-extension pressure housing. Each of the 30 secondary windings is connected to its own individual light on the control panel. As the magnetic portion of the control-rod extension passes each transformer, the coupling between primary and secondary windings is increased to the point where each lamp in the row is successively lighted and stays lit as rod movement progresses. Although these lights give only approximate position, there is no ambiguity. A secondary scheme for securing indirect position indication uses synchro-repeater indication of the rotation of the cam shaft that actuates the jacking mechanism from the power supply.

Another magnetic readout device for a sealed system uses a winding distributed around the thimble (pressure housing) around the rod extension. The winding is part of an induction bridge with readout on a panel meter. As the rod is withdrawn, the magnetic portion enters the solenoid and changes the inductance of the winding as it progresses. Through suitable design of the coil, its spacing and number of turns, the indication can be made approximately linear with rod displacement.

7-4 EXAMPLES OF REACTIVITY-CONTROL SYSTEMS

7-4.1 PWR Power Plant at Shippingport

The Shippingport reactor is a pressurized-water reactor with 32 vertical hafnium control rods, both manually and automatically controlled, to adjust the power level (Figs. 7.9 and 7.10). Since the reactor coolant system is completely sealed, the control rod and the driving element must be within the pressure barrier of the coolant system. For this application the "canned" motor shown in Fig. 7.6 is used. The rod-drive mechanism is a roller nut attached directly to the motor rotor; it operates on the lead-screw portion of a control-rod extension as shown in Fig. 7.11. Since the coupling between the rotor and roller nut is direct, i.e., no reduction gearing is used, a special very slow (22 rpm) motor is required.

The motor torque is applied by magnetic coupling through the pressure barrier (the thin "can" between the rotor and stator). For the control-rod lead screw to disengage quickly from the roller-nut mechanism, the nut is split into halves that are held together by a magnetic flux generated in the stator winding. Thus cutting the power to the motor also scrams the rod. The split roller nut must be kept from reengaging until the rod has fully inserted to prevent damage to the mechanism.

The motor winding is designed for a three-phase power

supply to produce a rotating field similar to that of the ordinary 60-Hz induction motor, except that it operates on alternating current with a frequency variable from zero (direct current) to a few hertz. The d-c power (zero frequency) is required, as pointed out earlier, to keep the nut engaged to the lead screw and to maintain the rod at the selected position. Even though the frequency can be reduced to zero to stop rod motion, the voltage must be kept applied to maintain the latch of the roller nut to the rod. This implies the requirement that the alternating current be changed to direct current without any change in value. Another requirement is for phase sequence to reverse the direction of rod motion.

A low-frequency d-c to a-c converter is used to meet these unusual power supply requirements (see Fig. 7.12). Basically, this converter is a circuit of series-connected resistors formed into a closed ring and mounted on an insulated commutator disk. Each junction point of the resistors is connected to one of the commutator segments arranged in a ring. Two diametrically opposite points on the ring of resistors are permanently connected to a d-c power supply. A rotating brush structure, with three insulated brush segments 120° apart, picks off a three-phase voltage from the commutator segments. Shorting-type brushes are used to prevent circuit interruption as the brushes move from one segment to the next. The a-c voltage is taken from the brushes through slip rings on the rotating brush structure. The frequency of the rod-motor voltage is determined by the speed of a small d-c motor that drives the brush structure. The peak a-c voltage is determined by the value of the d-c voltage applied to the series-resistor rings. Stopping the brush rotation automatically results in applying a d-c voltage to the motor stator fields so that the motor is held stationary.

In the Shippingport reactor two sets of brushes are incorporated into each resistor-commutator assembly. Each set of brushes supplies power to two rod-drive motors in parallel. With this arrangement four control rods are moved simultaneously whenever a commutator assembly is rotated. If four rods from different portions of the core are selected to operate off each commutator assembly, the core power and the desired symmetrical distribution of neutron flux can be more easily maintained at the various power levels. For the 32 rods 8 inverters are used; 2 spare inverters are available.

In addition to four rods being controlled by one inverter, the rod programming plan for this reactor divides the rods into two or more groups, each having subgroups of four rods. Group 1 is necessary to bring the reactor from shutdown to initial criticality. Sixteen rods were selected for this group. They are moved individually in multiples of four in sequence such that any subgroup is never more than 3 in. beyond the position of the remainder of the group. This procedure, together with a maximum limit on rod speed, meets the criteria for the maximum rate of insertion of reactivity. The remaining 16 rods are programmed in groups of 8, 4, and 4. The rod-programming equipment is designed to place 16, 20, 24, 28, or 32 rods in the first group, leaving any remaining rods grouped in multiples of 4.

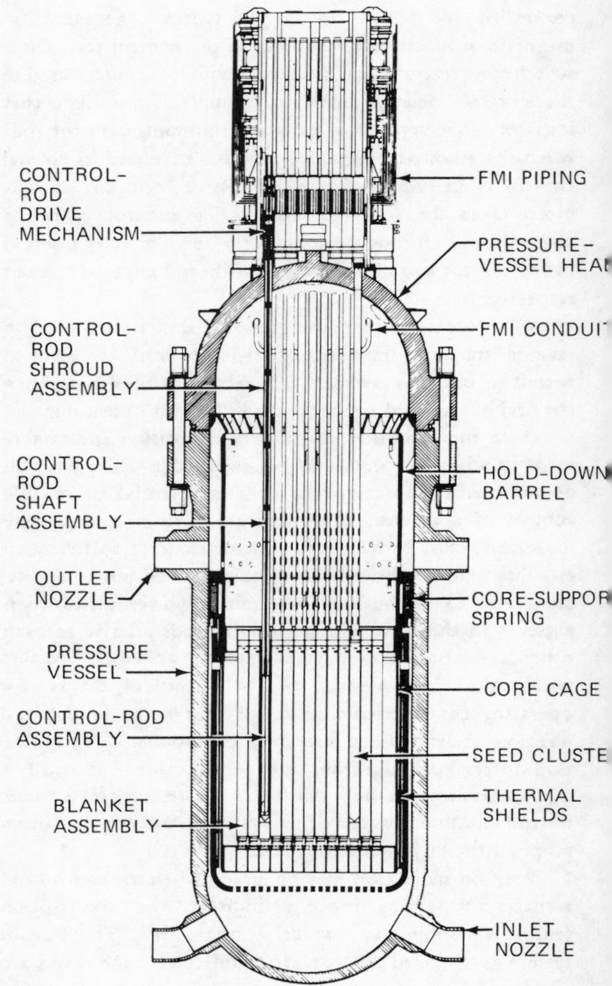

Fig. 7.9—Shippingport Pressurized Water Reactor, center-lines section.

Versatile combinations of rod speed and grouping are necessary to provide for a rapid and uniform burnout of ^{135}Xe after a scram or a power reduction. Although a scram causes all rods to drop, there are some situations demanding a fast or "safety" insertion in which power must be rapidly reduced a relatively small amount without a full-scale scram. (The latter, incidentally, also demands a correlative reduction or shutdown of the power-plant output.)

Each rod position is displayed on the control console by an individual column of indicator lights. Each light is connected to a small transformer on the control-rod-extension housing. As the magnetic material of the control-rod extension passes through each transformer, the coupling between the windings increases, and the lamp is lit.

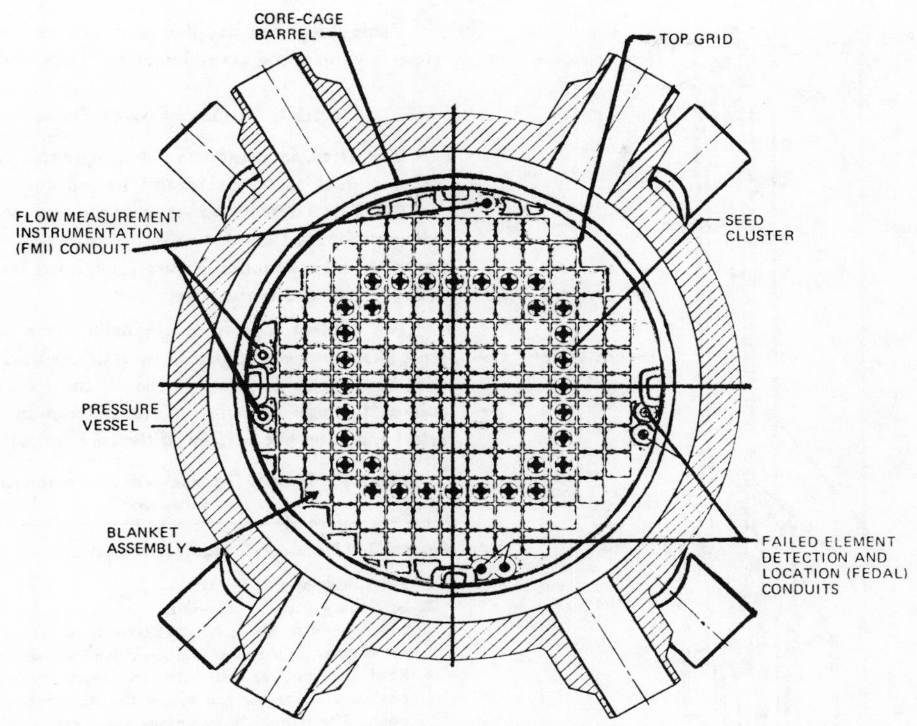

Fig. 7.10—Shippingport Pressurized Water Reactor, cross section through nozzle.

Automatic control of reactor power is provided by a power and temperature control system. The reactor is inherently self-regulating through a negative temperature coefficient of reactivity that compensates for reactor coolant-temperature variations caused by variations in steam load at the turbine. The accumulation of fission-product poisons makes gradual adjustment necessary. As in most reactors, rod movements must be fairly frequent immediately after start-up or a load change.

7-4.2 San Onofre Atomic Power Plant

The San Onofre Atomic Power Plant in San Clemente, Calif., has a pressurized-water reactor similar to that at Shippingport except for the method of rod drive and control. Each of the 45 control rods for this reactor (Fig. 7.13) consists of a 5-in.-square spider-like cluster of 16 absorbing rods, 126-in. long, each fabricated from an alloy of silver, indium, and cadmium and hermetically sealed within a stainless-steel sheath. The cluster of absorber rods fits into guide thimbles in a fuel-rod assembly (Figs. 7.14 and 7.15) and is attached to a driven vertical shaft extending through the top of the reactor to the rod-drive mechanism. Boric acid is added to the coolant water to control reactivity during both operating and shutdown periods to reduce the required number of control rods and to achieve uniform neutron absorption throughout the core.

Whereas each rod drive of the Shippingport reactor uses a canned motor to rotate the nut mechanism attached to a lead-screw extension of the control rod, the San Onofre reactor rod-drive mechanism uses a form of the magnetic jack. There is a similarity between the two in that motion is produced magnetically through the pressure barrier. The motion is rotary in the Shippingport reactor; it is linear with the magnetic jack.

The rod-drive mechanism (Fig. 7.16) consists of a latch assembly and a rod-drive assembly that operate within a thimble, all at reactor pressure. A stack of operating coils surrounds the rod-drive portion of the thimble. A 120-in.-long coil stack over the upper part of the thimble, into which the control-rod extension travels as the control is raised, is used for rod-position indication. The reactor coolant water fills the pressure-containing parts of the mechanism and cools and lubricates the moving parts of the drive.

The drive shaft has circular grooves, spaced ⅜ in. apart, machined into its surface along its entire length. Magnetically operated gripper latches lock into the grooves to hold the drive shaft stationary with the drive. The operating coil stack consists of the lift coil, movable gripper coil, and stationary gripper coil. These are energized in a fixed sequence by cam switches actuated by a rotating cam shaft. The coils induce magnetic flux through the pressure housing and operate the latch components. Within the pressure housing, two sets of latches lift or lower the grooved drive shaft. Turning the cam shaft one revolution causes the lifting mechanism to cycle once and quickly moves the rod out in one ⅜-in. step. Reversing the rotation of the cam shaft reverses the direction of rod motion. Latch actuation is produced by pole piece motion for the three

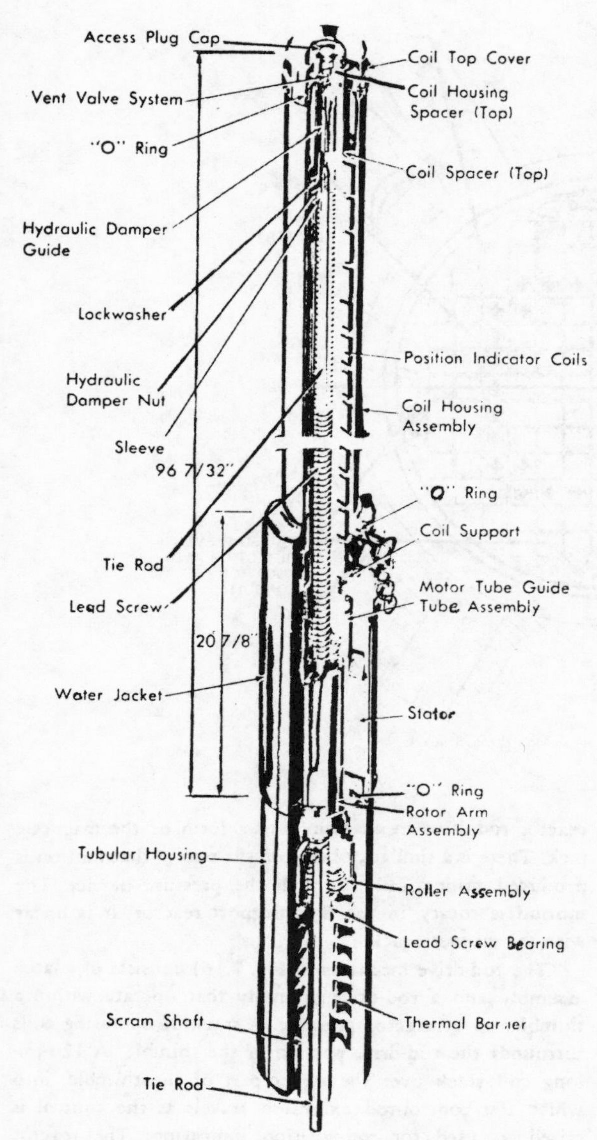

Fig. 7.11—Shippingport Pressurized Water Reactor, control-rod drive mechanism.

magnets.

Since the magnetic jack develops a lifting force of 400 lb and the total drive shaft weight is 144 lb, there is ample capacity to overcome friction in the system. For rod insertion, gravity furnishes the driving force and overcomes the friction load.

The upper set of latches operates to raise or lower the drive shaft in ⅜-in. steps. After one lift step the lower latches are engaged in a shaft groove and raise the rod 1/32 in. to unload the upper latches so they may be reset to lift the rod another increment. The latches are actually linked to sliding armature or pole pieces that move when their respective magnet coils are energized or deenergized.

Table 7.3 gives a detailed description of the sequence of steps in control-rod actuation at the San Onofre plant.

7-4.3 Dresden Nuclear Power Plant

The Dresden plant has a boiling-water reactor system using a dual steam cycle and forced circulation. Figures 7.17 and 7.18 depict two views of the reactor vessel. Eighty cruciform-shaped control rods are interspersed through the array of fuel assemblies. Figure 7.19 shows how the control rods are arranged in the core.

The control rods were originally made of 2% boron–steel alloy but were replaced by stainless-steel sheet packed with boron carbide powder. The 80 control rods enter tne reactor through thimbles in the bottom of the pressure vessel and enter the bottom of the core through holes in the

Table 7.3—Detailed Description of San Onofre Control-Rod Actuation*

Control-Rod Withdrawal: Sequence of Operations

1. Movable gripper coil—On.
2. Stationary gripper coil—Off.
3. Lift coil—On. The ⅜-in. gap between the lift armature and the lift magnet pole closes, and the drive rod rises one step length.
4. Stationary gripper coil—On. The stationary gripper armature rises and closes the gap below the stationary gripper magnet pole. The three links, pinned to the stationary gripper armature, swing the stationary gripper latches into a drive-shaft groove. The latches contact the shaft and lift it 1/32 in. In this manner, the load is transferred from the movable to the stationary gripper latches.
5. Movable gripper coil—Off. The movable gripper armature separates from the lift armature under the force of three springs and gravity. Three links, pinned to the movable gripper armature, swing the three movable gripper latches out of the groove.
6. Lift coil—Off. The gap between the lift armature and lift magnet pole opens. The movable gripper latches drop ⅜ in. to a position adjacent to the next groove.
7. Movable gripper coil—On. The movable gripper armature rises and swings the movable gripper latches into the drive-shaft groove.
8. Stationary gripper coil—Off. The stationary gripper latches, and the armature moves downward by gravity until the load of the drive shaft is transferred to the movable gripper latches. It then swings out of the shaft groove.

The sequence described above, where the control rod moves ⅜ in. for each cycle, is termed "one step" or "one cycle." The sequence is repeated at a rate of up to 40 steps per minute. The control rod is thus withdrawn at a rate of up to 15 in. per minute.

Control-Rod Insertion: Sequence of Operations

1. Stationary gripper coil—On.
2. Movable gripper coil—Off.
3. Lift coil—On. The movable gripper latches are raised to a position adjacent to a shaft groove.
4. Movable gripper coil—On. The movable gripper armature rises and swings the movable gripper latches into a groove.
5. Stationary gripper coil—Off. The stationary gripper armature moves downward and swings the stationary gripper latches out of the groove.
6. Lift coil—Off. Gravity separates the lift armature from the lift magnet pole, and the control rod drops down ⅜ in.

*From the operating manual for the San Onofre plant.

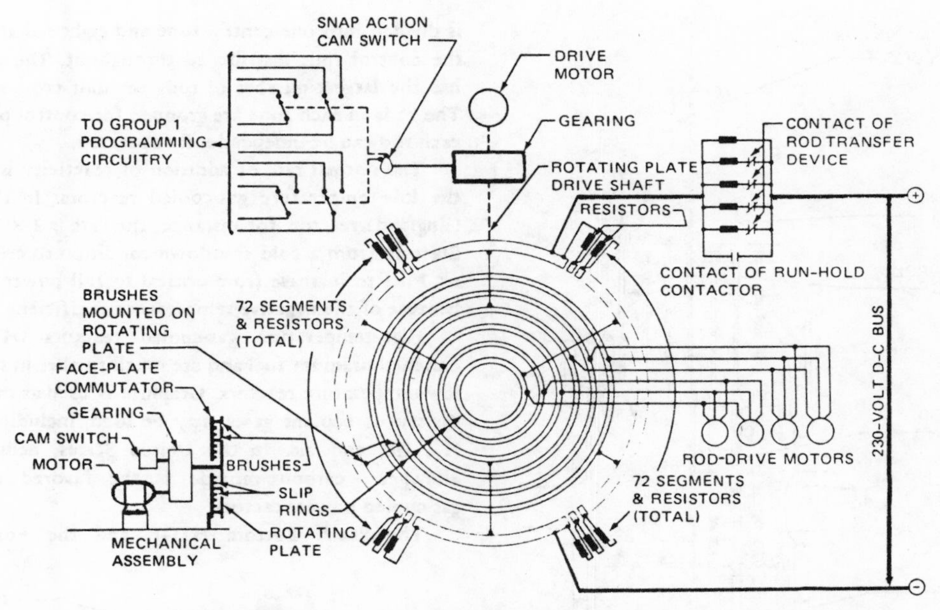

Fig. 7.12—Shippingport Pressurized Water Reactor, schematic diagram of rod-control d-c to a-c power inverter.

core support plate. A tube for each control rod extends upward from the bottom of the vessel to the bottom core support plate to guide the rods in this region. Each rod is equipped with its own drive mechanism located within the pressure-vessel thimble and operating in the reactor pressure environment.

In most other reactors using vertical control-rod motion, upward rod movement increases the core reactivity by moving the poison section out of and above the core. The Dresden reactor, however, has its control rods fully in the reactor in the "up" position. Reactivity is increased by lowering the control rod below the core. The control-rod drive mechanism is basically a hydraulic cylinder. Normal controlled movement of the rod is attained by applying reactor feedwater regulated at 200 psia above reactor pressure to either the top or bottom surface of the piston rod and simultaneously connecting the opposite side to a vent tank held at 30 psia above reactor pressure. This movement, however, cannot be effected until the locking ball located in a slot between the control-rod piston and the cylinder wall is released by the unlocking piston (Fig. 7.20). For rod "up" motion (direction toward decreasing reactivity) pressure is applied on the bottom of the piston rod. The control-rod piston, secured by the ball to the spring-loaded locking piston, moves upward a small but sufficient distance against the spring to release the ball into the annulus in the unlocking piston and free the rod. The rod continues to move upward as long as pressure is kept applied to the "up" inlet port. When the inlet pressure is shut off, reactor pressure is applied through the shuttle valve, and upward movement continues until the next notch in the piston reaches the ball. At this point, the spring loading of the locking piston is enough to lock both pistons together and prevent further movement.

For "down" motion, the high pressure is applied to the "down" inlet port and the "up" inlet port is switched to the vent tank. In this condition, the high pressure is applied to the top surfaces of both the rod piston and the unlocking piston. The unlocking piston moves down against a spring load and exposes an annulus that frees the locking ball, allowing the main rod piston to be moved. If the down pressure is maintained, rod down movement continues. If the pressure is applied only momentarily, the ball is freed long enough to allow the piston to move to the next slot where it again engages the locking piston and stops.

There are 12 slots along the 8.5-ft travel of the control rod, and rod movement in 8-in. steps can be obtained. The maximum rate of reactivity insertion from this control rod is 1.3×10^{-4} δk/sec. The interlocking circuits allow movement of only one rod at a time.

For scramming, a 1400-psi accumulator tank is supplied for every three drives. For each drive two solenoid valves provide pressure to the "up" inlet port and open the "down" inlet port to a dump tank. The rod movement for scram is completed in 2.5 sec.

Rod position is indicated by a series of magnetically operated switches located inside an inner cylinder along the drive stroke. A magnet built into the rod piston actuates the switches as it passes them, and corresponding indicating lights on the control panel show the rod position.

The effect of scram-system failure is minimized by the use of independent systems for every three rods. If a large-scale failure occurs, however, a chemical poison-injection system containing sodium pentaborate at a pressure much higher than that of the reactor is actuated.

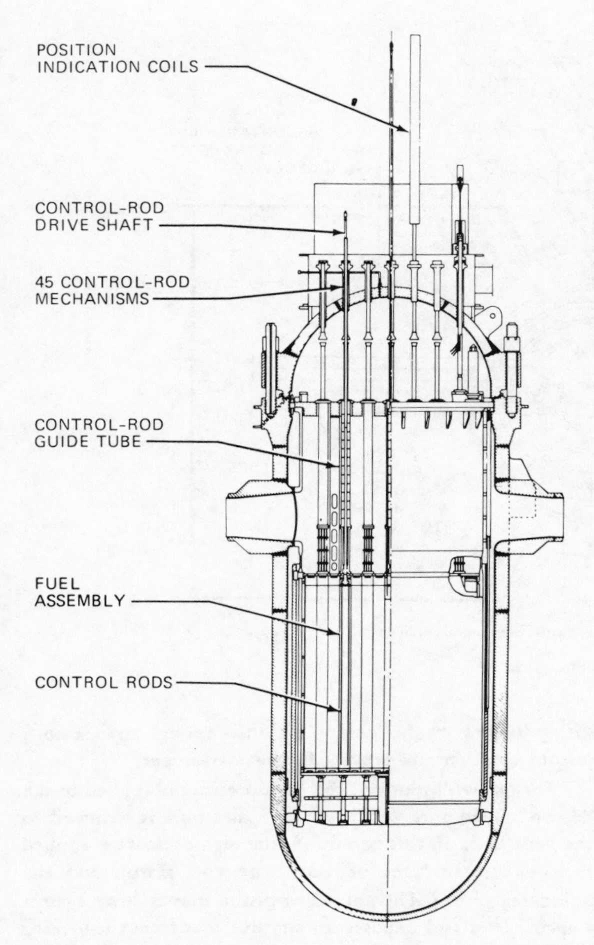

Fig. 7.13—Vertical section of the San Onofre reactor. The reactor vessel has 8-in. walls clad on the inside with 0.109 in. of type 304 stainless steel.

7-4.4 Gas-Cooled Reactors

Gas-cooled reactors can be considered as being either low-temperature or high-temperature gas-cooled reactors.

The low-temperature gas-cooled reactors are typified by the natural-uranium graphite-moderated CO_2-cooled units in the United Kingdom and similar installations in Italy, France, and Japan. These reactors are very large structures owing to the low excess reactivity available from natural-uranium fuel. For example, 40- to 70-ft diameter and 40-ft height are typical. The number of control rods is correspondingly large, 100 or more. The control rods (usually boron steel) are generally mounted and operated vertically. They are suspended by steel cables wound around a drum which is driven by a low-speed motor equipped with induction braking. A typical control-rod drive mechanism is shown in Fig. 7.21.

The control problems in low-temperature gas-cooled reactors arise chiefly from spatial variations in fission-product (especially ^{135}Xe) poisoning. To cope with these problems, for example, the Hunterston reactor in Scotland is divided into one central zone and eight radial zones, with the control rods distributed throughout. The central zone has the largest number of rods per unit cross section area. The rods in each zone are grouped for control purposes and each rod can be independently operated.

The normal rate of addition of reactivity is very low in the low-temperature gas-cooled reactors. In the Berkeley (England) reactor, for instance, the rate is 2×10^{-6} δk/sec. Start-up from a cold shutdown condition to critical takes 6 or 7 hr; to increase from critical to full power takes 11 hr because of the negative temperature coefficient.

High-temperature gas-cooled reactors (HTGR's) use enriched-uranium fuel and are much smaller in size than the low-temperature reactors. Graphite is used as moderator. A variety of coolant gases may be used, including N_2, CO_2, H_2, He, and air. In the United States, helium is used; elsewhere, carbon dioxide is the favored coolant for gas-cooled power reactors.

The Peach Bottom HTGR and the Fort St. Vrain

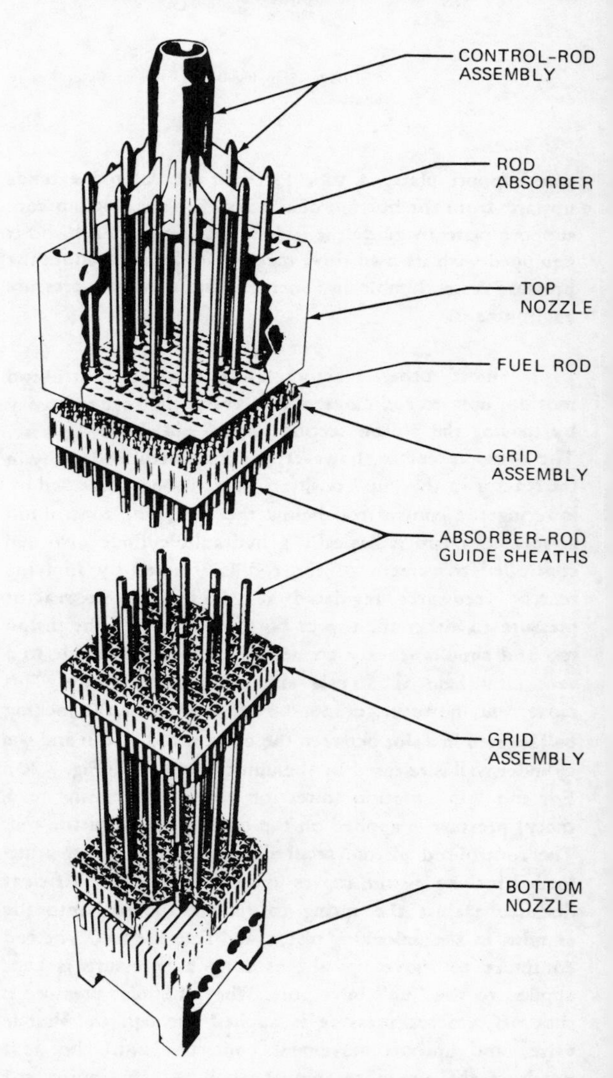

Fig. 7.14—Rod-cluster assembly, San Onofre Atomic Power Pla

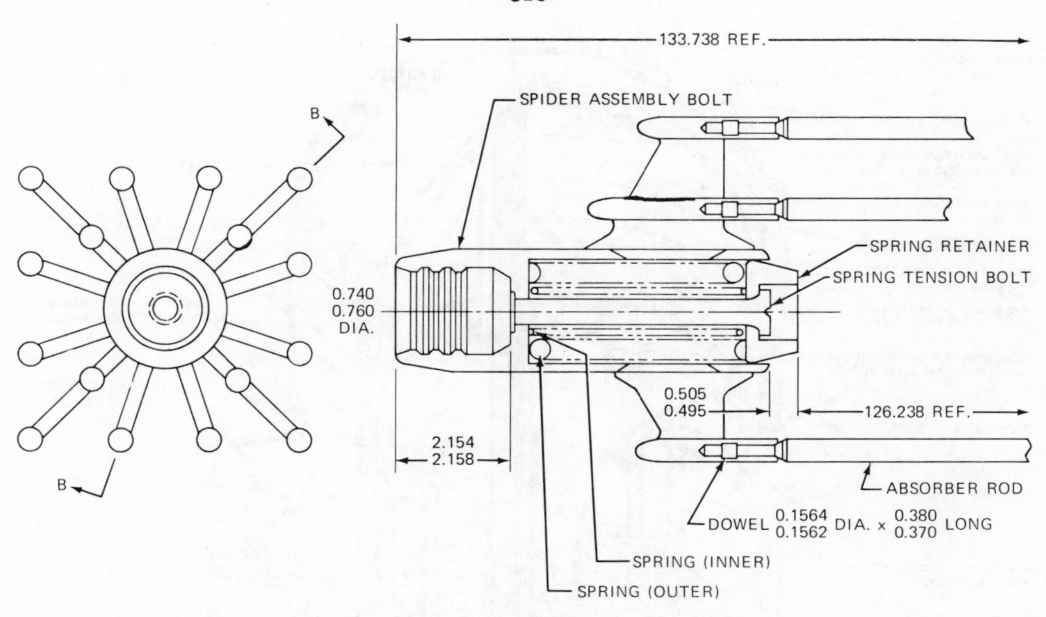

Fig. 7.15—Control rods for San Onofre reactor.

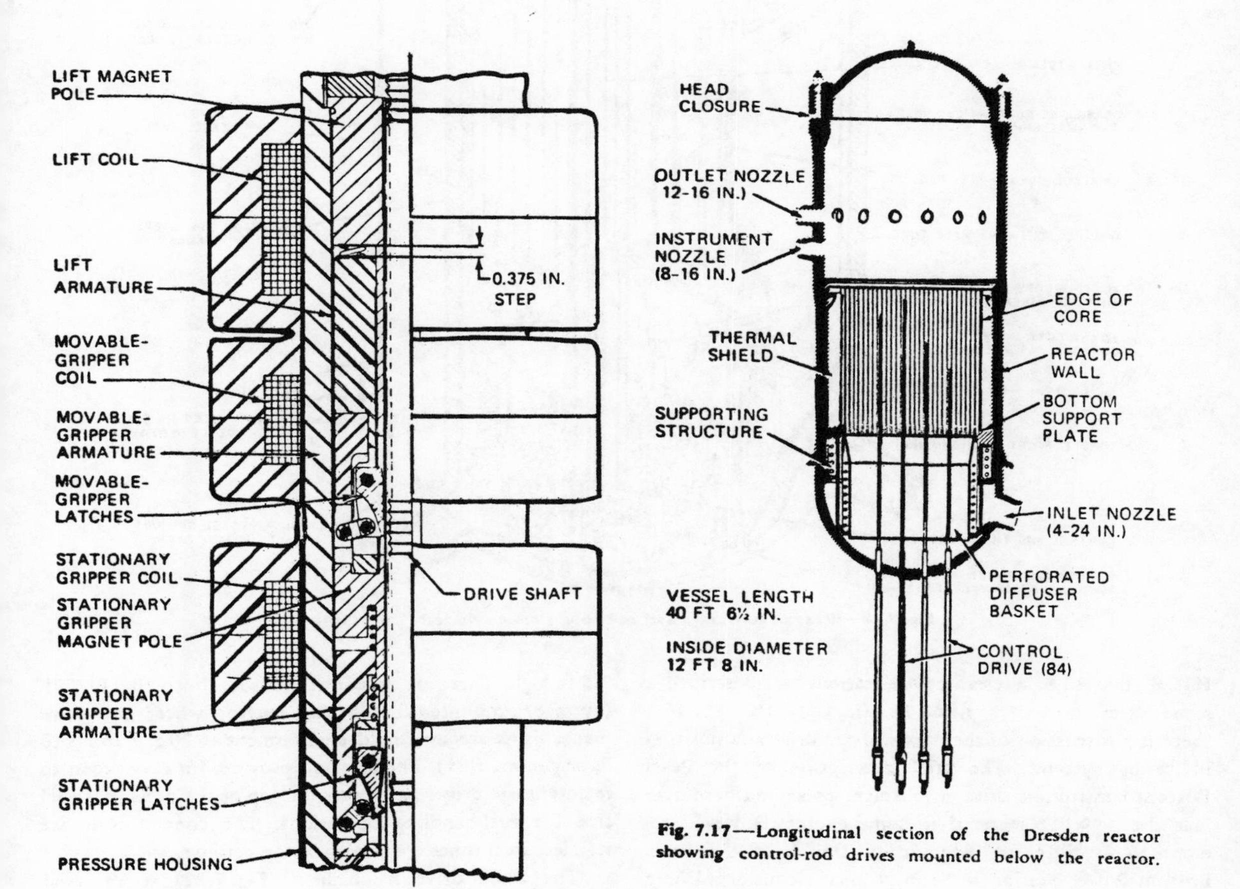

Fig. 7.16—San Onofre reactor rod-drive mechanism.

Fig. 7.17—Longitudinal section of the Dresden reactor showing control-rod drives mounted below the reactor.

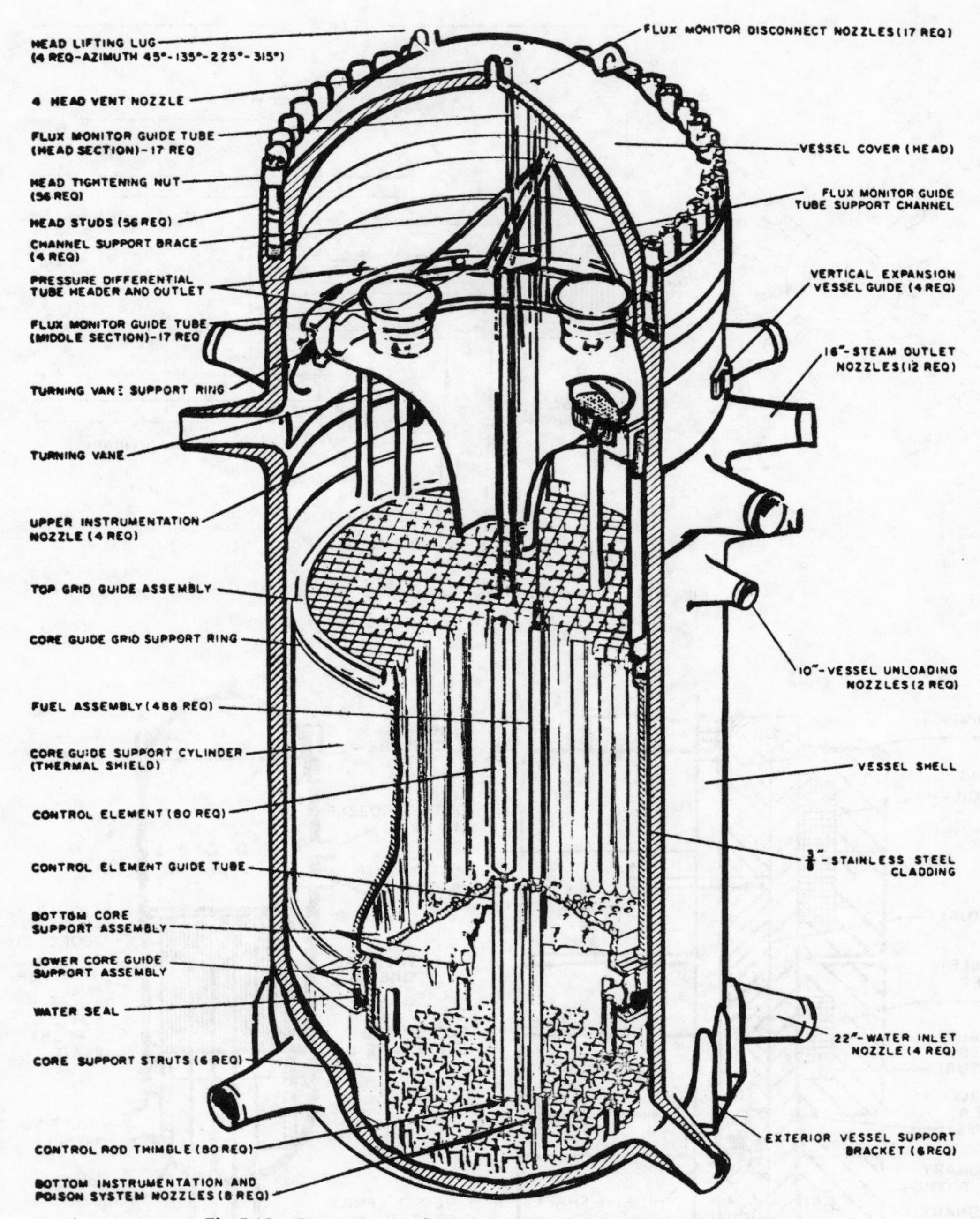

Fig. 7.18—Reactor core and vessel assembly, Dresden Nuclear Power Plant.

HTGR, two U. S. nuclear power stations, are described in some detail in sec. 18. In sec. 18, pt. 18-6, there is a discussion of their control-rod drive and position-indicating systems. The brief description of the Peach Bottom control-rod drive mechanism presented here overlaps the sec. 18 material to some extent. However, the emphasis is different. Specifically Unit 1 of the Peach Bottom Power Station in Pennsylvania is considered here. The core, located near the bottom of a 25-ft-high cylindrical pressure vessel, is approximately 9 ft in diameter and 7.5 ft high. There are 36 control rods, each worth 0.007 δk (average), mounted below the reactor where the drive mechanisms are in a mild environment ($\leqslant 200^\circ$F and ~10 R/hr gamma flux). The location provides for easy access to maintain the drives and leaves the top of the pressure vessel free for fuel-handling operations. The control rods are stainless-steel tubes containing boron carbide.

The basic drive mechanism (Fig. 7.22) is an axial piston-type hydraulic motor that turns a ball screw and produces linear motion of a ball-nut assembly. The ball-nut

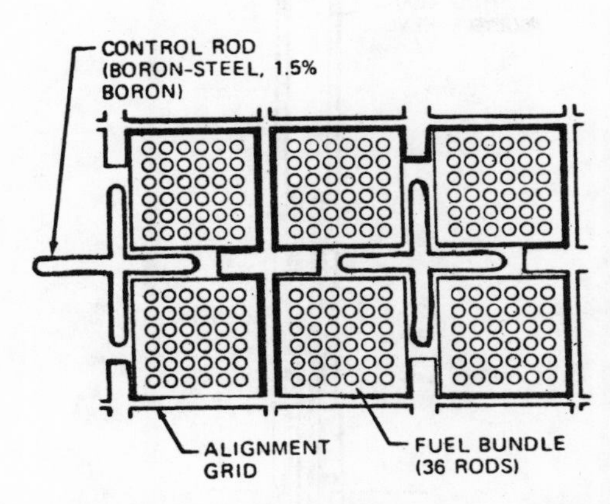

Fig. 7.19—Dresden reactor fuel-element and control-rod array. Fuel elements are 36-rod assemblies of UO_2 contained in Zircaloy-2 tubes. Cruciform control rods fit in spaces between the elements.

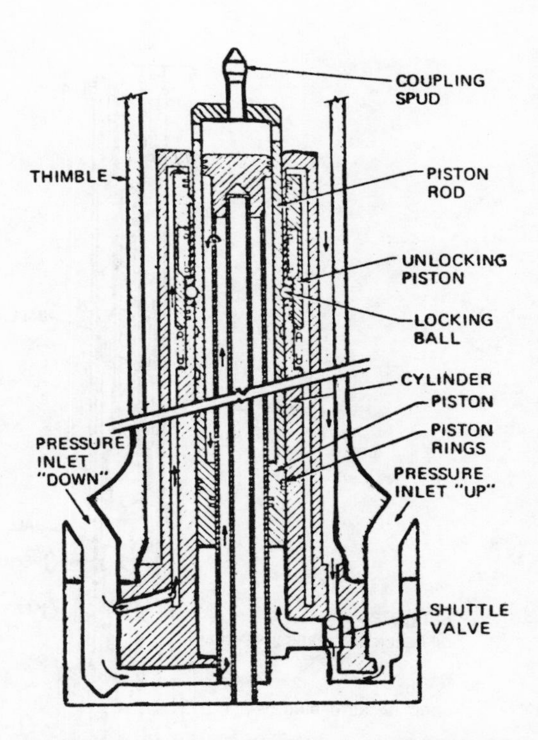

Fig. 7.20—Control-rod drive mechanism, Dresden Nuclear Power Plant.

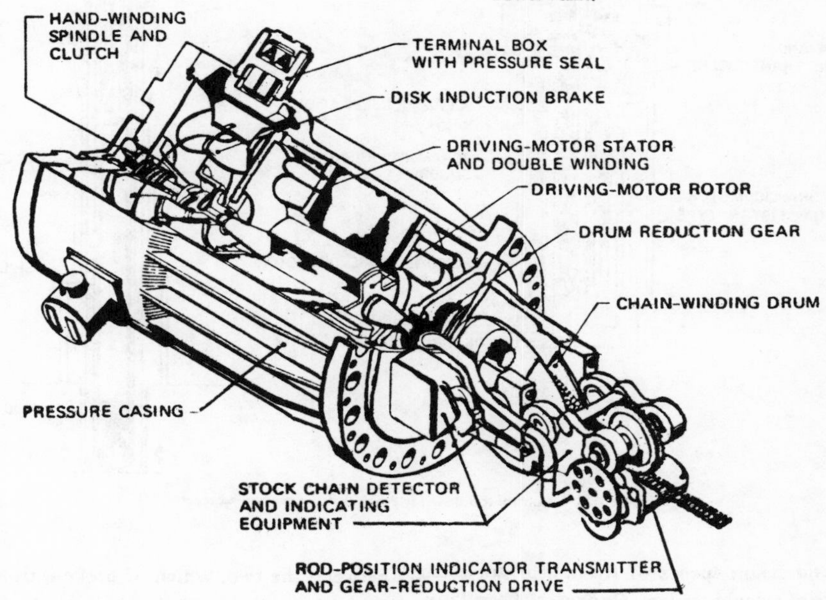

Fig. 7.21—Typical control-rod drive mechanism for a low-temperature gas-cooled reactor.

moves a push rod, which, in turn, raises or lowers the control rod. Each control-rod port (extension of the pressure vessel) contains the entire drive mechanism including the hydraulic motor, regulating and scram valves, position transmitters, rotary-to-linear motion ball-nut screw device, scram-energy accumulator, and scram-action snubber.

Hydraulic turbine oil is used for the motor and is supplied to the drive from two header connections. One is a low-pressure supply that produces the normal operating speed of the control rod. The high-pressure line maintains the fluid level in a scram accumulator, pressurized by helium, which drives the rod at the scram speed of 10 ft/sec. A return header is supplied for the effluent from the motor.

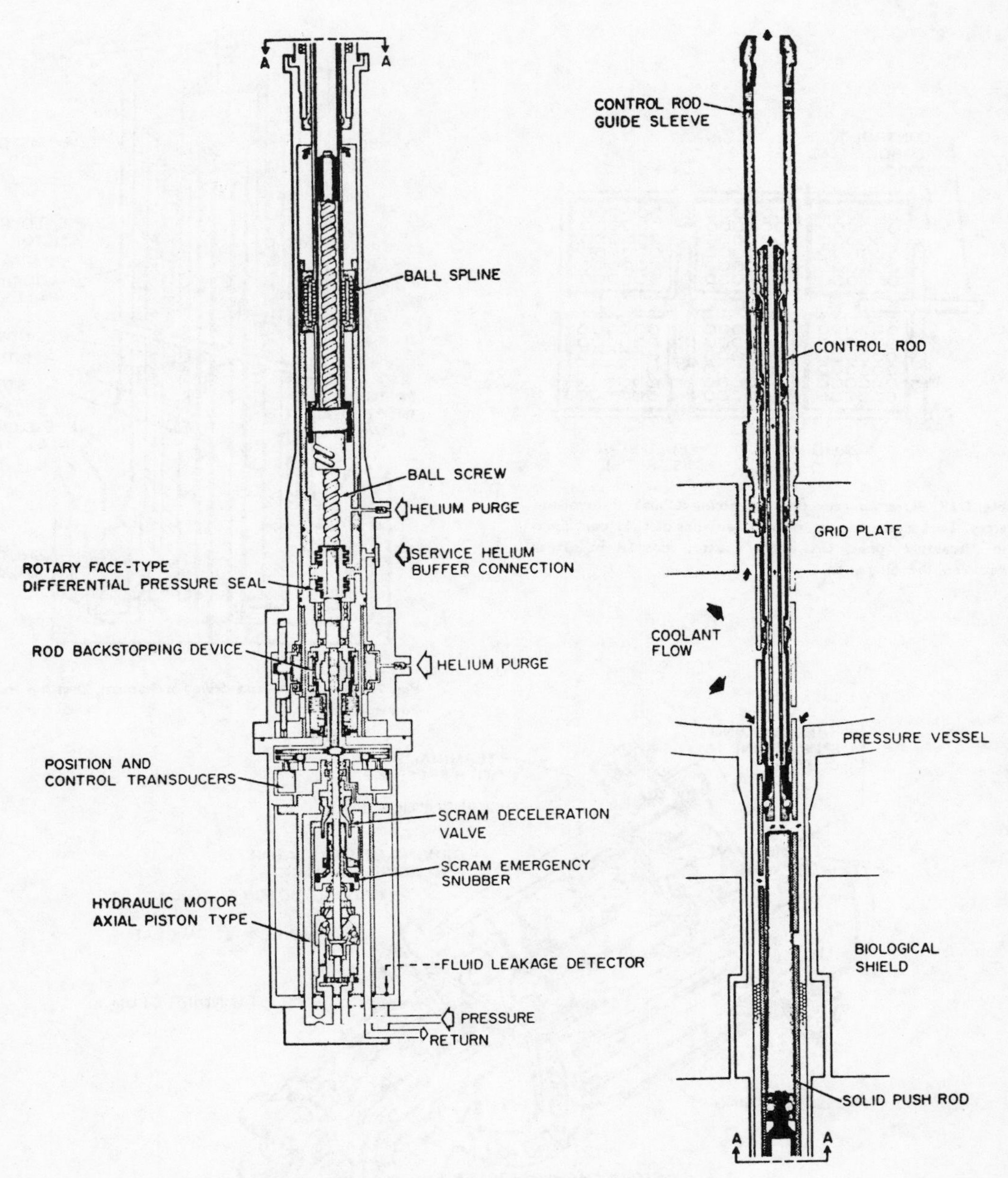

Fig. 7.22—Control-rod drive for Peach Bottom gas-cooled reactor.

The regulating and scram speeds of the motor are established by two sets of solenoid valves. One set, the regulating valves, admit low-pressure hydraulic oil for rotating the motor in either direction. The other set, scram valves, apply high-pressure oil to the motor for one direction of rotation only, namely, that for driving the rod upward into the core. Total rod movement is 7 ft. The regulating or control speed is 0.06 ft/sec or 0.72 in./sec, corresponding to a rate of change of reactivity of 1.1×10^{-4} δk/sec (maximum). Except for rod-removal operations, the drive is never detached from the rod. Monitoring of the coupling between the drive and the rod depends on an electrical circuit between the two, which, if broken, trips an annunciator at the operator's console. Downward motion of the rod increases core reactivity, and scram motion is upward. Control of rod direction is by manual activation of solenoid valves through remote switches or from an automatic "on-off" control circuit.

A clutch-brake on the hydraulic-motor shaft prevents downward drift of the rod when no hydraulic pressure is being applied to the motor. This is accomplished by means of a friction brake and an over-running clutch that allows completely free rotation of the shaft to produce rod motion upward (decreasing reactivity) but applies a reverse torque at 1.75 to 2.5 times that produced by the deadweight of the rod and attached drive parts acting vertically. The operation of the hydraulic motor, however, easily overcomes this friction when rod-down motion is desired. A mechanical latch holds the rod in the "up" (full-in) position after insertion. The latch has to be actuated when the rod is withdrawn down past the latch. Mechanically, the latch cannot be withdrawn if the rod weight is on the latch. Safety requires that no more than three rods at one time be in a position intermediate between latched in and full out.

There are two independent emergency shutdown systems. One is a group of 19 electrically driven emergency shutdown rods. These rods are operated by conventional Acme thread screw-nut mechanisms driven by simple and rugged d-c motors. Batteries, located in an extension of the rod port housing, supply power to drive the motors in the scram direction. External d-c supplies are used to provide withdrawal power to these drives. Operation of these rods is by operator manual control only. They are used only in the event that normal scram operation fails. Complete rod insertion is attained in 24 sec, and the design provides very large torques that can overcome a resistive force as high as 10,000 lb, in which case insertion takes 1 min. Such operation was provided for the remote possibility that reactor damage (warping or debris) could restrict the normal movement of the rods. The other emergency shutdown system, located in the top section of the guide tubes for each of the control and emergency shutdown rods, is a set of 55 gravity-drop neutron absorbers that are thermally released by abnormally high core temperature. This system is intended to operate in the event the other rod systems fail to act.

Rod-position indication is provided by four selsyn receivers and by limit lights. Each control-rod drive mechanism contains a selsyn transmitter geared to 170° rotation for full rod travel. Each position indicator has a graphical display (column type, ~5½-in. scale corresponding to 0 to 84-in. rod travel, position accuracy ±2%) and a digital display (three-digit counter calibrated in inches, indication of rod position ±0.12 full scale, full-scale slewing time <6 sec).

When operating at power, the reactor is controlled by a group of three rods in the center region of the core. The particular rods in the group are continuously monitored by the synchro position indicators at the reactor operator's console. The fourth indicator is switched to the individual rod being moved and thus serves as a check on the synchro-receiver normally on that particular rod. The remainder of the rods not being used for control are either full in or full out, as indicated by a green or red background light in the particular control-rod window on the operating console; the lights are operated by upper and lower limit switches. In addition, a precision potentiometer is driven from each rod-drive gear train to provide for testing rod response and operation and for checking rod-position indication if a synchro-transmitter malfunctions.

7-4.5 Fast Reactors

The control systems for fast reactors (EBR-2 and Fermi) are described in Section 17, pt. 17-2.

REFERENCES

1. United States of America Standards Institute, *USA Standard Glossary of Terms in Nuclear Science and Technology*, USAS N1.1-1967, USA Standards Institute, New York, 1967.
2. M. A. Schultz, *Control of Nuclear Reactors and Power Plants*, 2nd ed., McGraw-Hill Book Company, Inc., New York, 1961.
3. W. H. Esselman and W. H. Hamilton, Position Control in Sealed Systems, in *Proceedings of the 1953 Conference on Nuclear Energy*, University of California Press 1953

BIBLIOGRAPHY

Argonne National Laboratory, The EBWR Experimental Boiling Water Reactor, in *AEC Nuclear Technology Series*, USAEC Report ANL-5607, May 1957.

Freund, George A., *Materials for Control Rod Drive Mechanisms*, Rowman and Littlefield, Inc., New York, 1963.

Harrer, Joseph M., *Nuclear Reactor Control Engineering*, D. Van Nostrand Company, Inc., Princeton, N. J., 1963.

Hutter, E., Fast Reactor Control Mechanisms, AEC, Washington, D. C., 1964.

International Atomic Energy Agency, *Directory of Nuclear Reactors*, Vol. I-VI, International Atomic Energy Agency, Vienna, Austria.

Lipinski, W. C., J. M. Harrer, and R. L. Ramp, in *Reactor Handbook*, Vol. IV, Engineering, Chap. 8, Interscience Publishers, a division of John Wiley & Sons, Inc., New York, 1964.

Loftness, Robert L., *Nuclear Power Plants: Design, Operating Experience, and Economics*, D. Van Nostrand Company, Inc., Princeton, N. J., 1964.

Nuclear Power, pp. 282-286, November 1956.

Nuclear Power, p. 53 ff, May 1962.

Nuclear Power, pp. 90-97, April 1962.

Weaver, Lynn E., *Systems Analysis of Nuclear Reactor Dynamics*, Rowman & Littlefield, Inc., New York, 1963.

SECTION 8

COMPUTER APPLICATIONS AND DATA HANDLING

8-1 INTRODUCTION

The meaning of "automatic control" has changed over the years along with the advance of control technology, the change being mainly one of scope. At one time describing a servomechanism, the term was broadened to include feedback control of a single process parameter; now it usually denotes the untended operation of complex processes or entire plants. The word "automation" was coined to indicate the elimination of the actions and decisions of human operators; it will be applied in this chapter to the internal control of nuclear-reactor facilities. However, just as advances in automatic control were expansions of scope based on previous technology, we may expect that automation will soon be taken to imply optimum operation under changing external influences, such as product markets and raw materials costs.

The instrument engineer has recently been given the chance to begin applying several decades of theory to the control of real industrial processes.[1] The means were provided by those digital-computer manufacturers who designed or modified their equipment for on-line data handling and control and who developed the beginnings of process program systems. Because of the important part computers play in automation, this chapter will deal mainly with the special problems that arise in the design, procurement, and application of digital-computer control systems.

The past growth of the field of computer control has shown an exponential trend that is typical of a newly introduced technology.[2] The increase may eventually be expected to slow down to more of a linear rise as the demand stabilizes; however, it is likely that the limiting influence will not be the usual market saturation but rather the lack of a sufficient number of knowledgeable applications engineers.

The history of computers in control shows an early and continuing preponderance in the petroleum and other chemical industries with applications in electric-power and metals production running not far behind.[3] On the other hand, computer control of nuclear facilities is among the least developed of the industrial areas, a state of affairs due largely to the plant builders' and the operators' uneasiness over acceptance by licensing authorities.[4] Once this barrier has been surmounted, we expect nuclear-plant applications to catch up quickly with the rest of industry. Moreover, the combination of an expanding competitive nuclear power business, the unusually thorough mathematical representations of nuclear reactors. the continuing improvements in computer speed and reliability, and the ability of the computer to solve complex problems affords an unprecedented opportunity for the engineer to apply advanced control methods.[5]

Control systems can be described in terms of the extent to which they provide automatic plant operation and the kind of equipment used to do it. First, a system type may be classed by whether its primary implementation method is analog, digital, or a combination of both. This classification provides an indication of automation since it implies certain automation capabilities. A second classification concerns the degree of automatic control provided, the criterion being how completely the equipment replaces the human operator's decisions and actions. Third, the system can be described according to its scope, i.e., the proportion of the total plant that is under automatic control. These three categories will be discussed further in the following section.

An underlying objective of automation is improved economics of plant operation, and the anticipated amount of improvement is the measure of the justification for control-system cost. This objective will be achieved in different ways depending on the mission of the facility. A power-producing plant with proven components needs a redundant system with moderate data handling to provide maximum operating continuity, whereas a prototype requires less emphasis on high plant factor and more on data acquisition and analysis. The effect of facility mission on control-system design will be treated later in some detail because of its heavy influence on system size and cost.

8-2 SYSTEM COMPARISONS

The shift from analog to digital hardware is providing greater capability for automation.[6] The use of more digitally oriented process-control equipment tends to overcome the disadvantages in cost or reliability of an analog predecessor.

8-2.1 Analog Control Systems

Before the advent of the transistor, analog controllers, independently serving individual process loops, became the design standard. As shown in Fig. 8.1, this system provides process control under one set of conditions. Changes in the

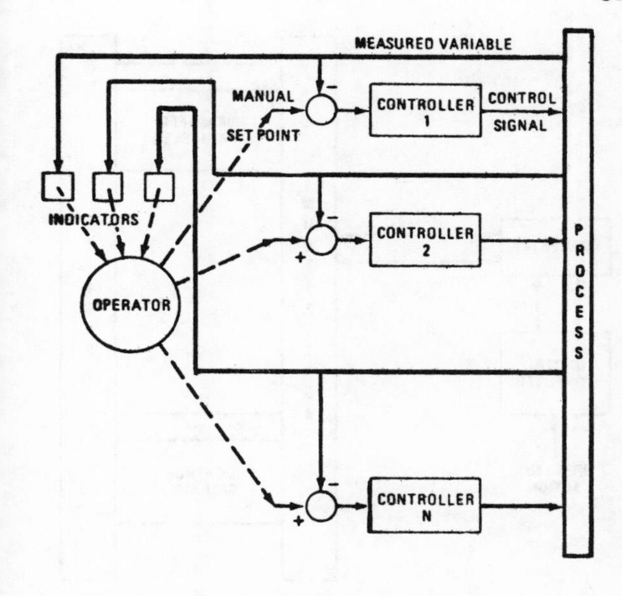

Fig. 8.1—Analog feedback control system.

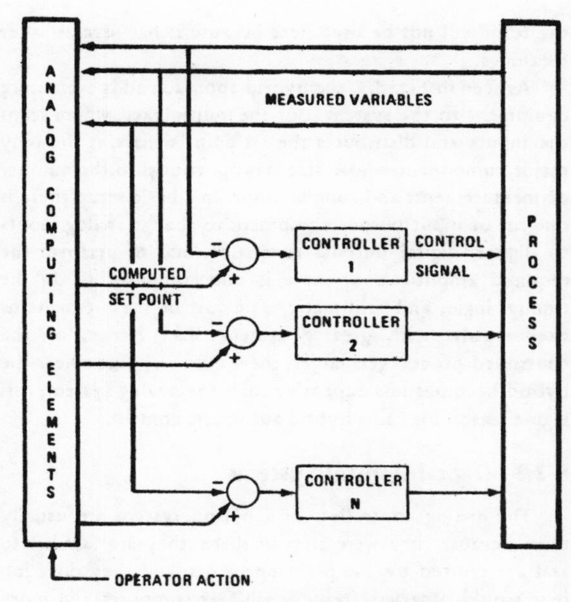

Fig. 8.2—Analog automatic control system.

operating status of the process are accommodated by manually adjusting the set point and the analog controller characteristics. These devices incorporate proportional, integral, and derivative control action as appropriate to the process.

When the transistor introduced solid-state control electronics, more-elaborate analog systems became feasible. Figure 8.2 shows schematically the components of such a system. Automation is achieved through relieving the operators of having to adjust set points for changes in process conditions. A distinguishing feature of this system is that the analog computer has available to it all significant process-variable signals. This permits feed-forward, cascade, and multivariable control modes.

In contrast to the controller bank of Fig. 8.1, the number of operators needed does not necessarily grow with increasing size and complexity of the process under control. The control system grows, and the cost of installing and maintaining an extensive configuration limits large-scale analog automation even with solid-state hardware.

8-2.2 Hybrid Control Systems

Although the transistor gave needed improvement to analog devices, its effect in the digital field was outstanding. Vacuum-tube digital computers required a great deal of maintenance and almost continuous adjusting. As transistors improved in time response and reliability, larger and faster machines were built. The continued advance into integrated circuits lowered the cost of smaller models into the realm of process control.

The utility of the digital computer lies in its inherent time-shared nature and its memory. One set of arithmetic elements serves the calculating needs of all control loops one at a time, and the computer remembers the inputs,

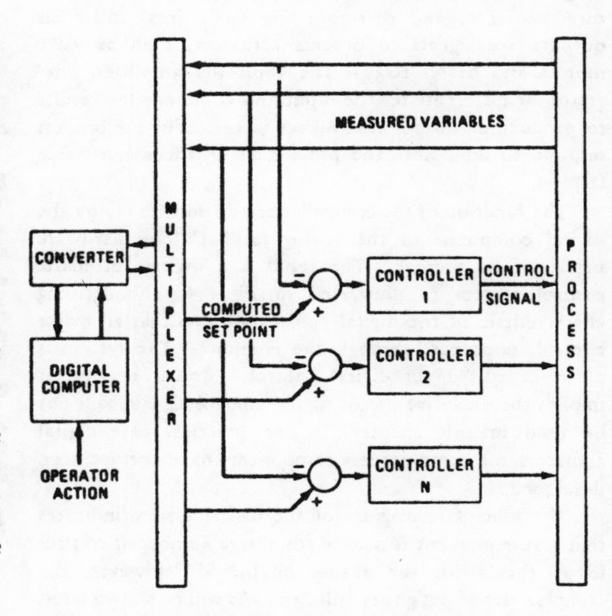

Fig. 8.3—Hybrid automatic control system.

outputs, limit values, and decisional requirements of each loop between turns. These features make the digital computer superior to analog computing elements; the configuration shown in Fig. 8.3 was that seen in most of the early digital applications. Since the digital computer was added to feedback control systems already in operation, it was visualized as "supervising" the process by monitoring and alarming out-of-limits conditions, adjusting analog controller set points, and generally performing the simpler tasks of a human operator. Thus the hybrid arrangement is often called "supervisory control," although the term will not be used here because it has several other meanings.

As seen in Fig. 8.3, the hybrid approach adds even more equipment to the system. But the multiplexer, which scans the inputs and distributes the set-point signals, is the only major component whose size is proportional to the number of measurements and control loops. In a basic setup there is one set of input/output equipment to change analog inputs to digital, digital outputs to analog, and to perform the required amplification; there is one computer to do the timing, logic, and arithmetic. The cost of these two items rises slowly with greater system size. Hence, as the controlled process gets larger, there exists a point where the hybrid becomes less expensive than the analog system; this is one reason for using hybrid automatic control.

8-2.3 Digital Control Systems

The analog controllers in a hybrid system are usually there because they were already there, they are familiar to and are trusted by the plant operators, and they do a job that would otherwise require a bigger computer and more programming. Their inputs are differences between measured variables and corresponding set points, and their outputs are signals to process actuators, such as valve motors and heater relays. The controller amplifies, integrates, or differentiates the input and combines the results to produce an output that makes a portion of the process respond to set points and process disturbances in a stable fashion.

The function of the controller can be done easily by the digital computer; so the analog feedback-loop hardware need not be present.[7] The result is a digital automatic control system as illustrated in Fig. 8.4. The defining characteristic of the digital configuration is that all major control loops pass through the computer. The system is often called "direct digital control," which, because it implies the exclusive use of digital control signals, will not be used in this chapter. In the practical case digital actuators for some process components have not yet been developed.

The schematic diagram of the digital system indicates that less equipment is needed for a large number of control loops than with the analog or hybrid. However, the complexities of design are still there. As will be shown later, they have been largely transferred from the hardware to the computer programs.

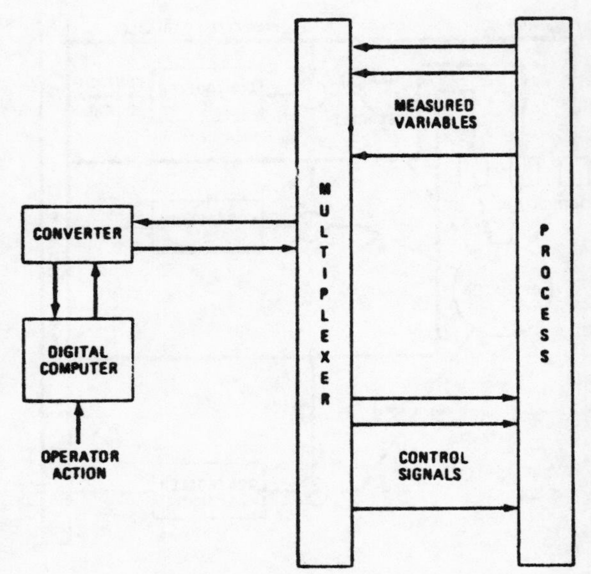

Fig. 8.4—Digital automatic control system.

8-3 DEGREE OF AUTOMATION

The preceding section treated control systems from the standpoint of the kind of hardware used, analog or digital; the amount of automatic operation provided was a secondary consideration. In this regard it should be noted that the presence of a digital computer implies no special level of automation. For example, many digital systems have been installed which perform only the control functions of a set of analog feedback controllers with the added benefit of automatic data logging.

To avoid hardware implications, we will take the functional approach and use the degree of automation to define the extent to which the operator is removed from the immediate closed loop.

8-3.1 Operator Control

The defining property of this system is that no significant logical decisions are made by the hardware, either on the display side or in the control channels. In a system like that shown in Fig. 8.1, the feedback controllers provide only process control at a given static operating level. All control actions are taken by the operator and are based on the information he obtains by observing the process variables displayed at the consoles and his interpretation, through experience and training, of this information in terms of needed process changes.

8-3.2 Monitored Operator Control

Hardware logic may be placed in the path from the operator to the process to prevent a control action at the wrong time or under adverse conditions. These may range from simple electrical interlocks to a complicated set of prerequisites in a plant start-up sequence. Such systems are now seldom seen in the absence of other automatic control features. They are mentioned here only to illustrate one aspect of automation: easing the operator's decision-making burden.

8-3.3 Operator Guidance

In a large plant the amount of data generated cannot be assimilated by one operator or even a staff of operators. In many installations digital computers are used only to get process information to the operator quickly and in meaningful form. This involves data acquisition, on-line analysis, and information display, and it often results in a highly complex instrumentation system made practical only through the use of a digital computer. The essence of the operator guidance system is that it comprises nearly all the hardware elements needed for fully automatic control, the only missing item being the part of the computer–process interface which sends control signals back to the process.

The operator guidance system appears to produce a low degree of automation; nevertheless, it provides all the aspects of automatic control except direct feedback action. No data analysis or decision making is required of the operator. In practice, however, the operator is seldom reduced to a robot. On the contrary his effectiveness is enhanced by his possession of a continuous and up-to-date knowledge of the process. This includes energy balances, anticipatory alarms based on predicted plant behavior, or instructions for optimizing the process, all requiring that data be processed faster and more accurately than is possible by humans.

8-3.4 Automatic Control

When under a given set of conditions a process is run efficiently without human direction, its operation represents the highest degree of automation. Input data are analyzed, decisions are made, and control actions are taken entirely under the guidance of the control equipment.

Automating a facility to a high degree may allow a significant reduction in the size of the control-room staff, although resulting operating cost savings may be partly offset by the need for computer programmers and more highly skilled maintenance specialists. More importantly the operators are relieved of trivial and routine tasks requiring continual alertness and are allowed to perform more complex functions, such as general surveillance and emergency intervention, which are beyond the capability of a control system of reasonable cost.[8]

8-4 SCOPE OF AUTOMATION

To complete a description of the extent of plant automation, one must discuss the proportion of facility operation that is under automatic control. This is here called "scope" and comprises two distinct aspects: (1) the physical portion of the facility involved and (2) the number of different plant operating conditions included.

Reactor safety circuits (see Sec. 12) are not treated in this chapter since they are not considered a part of the control system.

8-4.1 Conventional Processes

A reactor facility can usually be divided into two parts according to whether or not the operation of the component processes is markedly affected by using a nuclear reactor for a source of heat. Thus the generators, turbines, and steam loops in a pressurized-water-reactor power-generating unit could be considered conventional equipment. Their operation is not substantially different from that of a fossil-fuel plant; so automatic control systems that have proven successful for this part of a nonnuclear station can be applied to a nuclear plant. Nevertheless, extensive computer control of only the conventional part of a nuclear plant is uncommon. There are several interrelated reasons for this: (1) computers are installed primarily for data acquisition and display; (2) basic feedback control by computer to date has shown little or no cost advantage over analog controllers; and (3) the more serious control problems involve complex modes (such as multivariable, feed-forward, or cascade) which include controlling a part of the nuclear plant as well, again introducing concern by the designer over acceptance by licensing authorities.

8-4.2 Nuclear Processes

When computer control extends into the nuclear part of the facility, the engineer becomes involved with design procedures not common to a conventional plant. Radiation and radioactive materials reduce access to many components of the control system and constitute potential hazards that demand extra care in system design and strict conformance to safety criteria. The division between conventional and nuclear processes is made to emphasize those control functions requiring the most thorough reliability analysis.[9]

8-4.3 Auxiliary Processes

During the early planning of a process computer system, the following question arises: "How much of the facility is going to be under computer control?" One must decide whether or not to include equipment not directly related to the main plant process, such as coolant makeup units, coolant-purification loops, and standby power plants. Added functions of this kind are not being included in current plant designs. The hardware cost savings are outweighed by the expense of greater system size and complexity.

An exception to the above is the control of a certain class of on-line instrumentation comprising coolant samplers, chromatographs, mass spectrometers, neutron-flux scanners, etc. These are characterized by their need for precisely timed control signals with a resulting data input to the computer. The control signals are normally of a fixed-sequence kind, independent of the reading of the instrument, and do not effect closed-loop control actions in the usual sense.

A second exception is monitoring and procedural control of reactor refueling operations. If so programmed, the control system keeps track of the reactivity status of the reactor at all times. Given the incremental reactivity of a replacement fuel element, the computer can predict the new reactivity status at each stage in the refueling sequence. This, coupled with a step-by-step comparison between fuel-handling machine control actions by an operator and a prescribed checklist in the computer, can substantially improve the charge–discharge process.

8-4.4 Start-Up and Shutdown

The preceding sections were concerned with the processes in a facility that were placed under computer direction. There remains the question of how extensively should the control system cover the full range of operating conditions.

If our only worry were to keep a reactor plant going in stable fashion at full power, it would be difficult to justify computer control. The task of regulating processes under minor perturbations usually can be done adequately by analog controllers. Even operation at different predetermined power levels can be provided, although in a large plant this begins to reach the practical limit of analog capability. An example is automatic power reduction, sometimes called "power setback," where several combinations of out-of-limits measurements can automatically cause the plant to go to a selected lower power, thus avoiding the stresses of a full scram and the consequent restart troubles.

The greatest need for computer control arises during operation at reduced or changing power levels. Changes in power level may be planned, as in start-up or shutdown, or unforeseen, as in recovery from transients and response to equipment failure. The superiority of a computer-based system under such conditions is due to one or more of the following:

1. There are a large number of interdependent procedural steps to be taken that a computer can execute in considerably less time and with a lower probability of erro than with human operators using conventional controls. Reactor start-up and steam-turbine run-up are in this class.[10] Both are being included in current reactor plant designs.

2. The controlled variables are changed at different operating levels. One way to start up a reactor is to maintain a fixed and safe period up to about 1% of power, then to raise the level while keeping below a predetermined maximum power rate of rise to avoid damaging thermal stresses, and finally to control at full power with power level as the input to the control program. The decision-making ability of the computer can be used to make these changes in control organization at the optimum points in the ascent-to-power routine.

3. Control elements, processes, and sensors are nonlinear. Fixed controller settings can provide good regulation over a limited part of the plant operating range but produce inefficient or unstable behavior at others. Again the ability of the computer to make decisions allows it to adjust its own transfer function automatically to fit changing plant characteristics and to provide nearly optimum control under all conditions.

4. Corrective actions often require a faster speed of response than human operators can provide. Automatic power reduction is a good example. The combination of fast logical analysis of a large amount of data and the ability to take quick remedial action permits the computer system to reduce reactor power or to shut down the plant in a controlled manner and to avoid situations that would cause reactor scrams. This results in two very real benefits: (1) reducing reactor scrams lessens the chance of damage by thermal and mechanical stresses, thereby reducing maintenance and increasing the useful plant lifetime, and (2) lowering power only as far as is necessary allows rapid recovery to full power and raises the plant factor.

8-4.5 Nonprocess Functions

The time-shared nature of a process computer system makes it capable of doing nonessential off-line tasks at the same time that it is controlling the plant. The central processor idle time could be used to compile and run FORTRAN calculating routines, assemble new subroutines, or modify and expand the process control programs.[11] However, outweighing the benefit of greater machine utilization are the following drawbacks:

1. The operating speed of a nuclear-plant control system does not permit interchanging whole programs between core and drum (or disk); so the core memory must be larger.

2. Greater keyboard–printer capacity is needed.

3. A much more complex monitor is required.

4. There is a finite probability that the time-shared systems software will derail the control program, although this probability can be made small through a computer memory-protect feature.

Taking all these into consideration, with nuclear safety to emphasize the last item, power-reactor control engineers have not included on-line program preparation as a system function. An exception, discussed in pt. 8-5.4, is a configuration of redundant central processors where one is on standby and may be used by a programmer until interrupted to take over the control task.[12]

8-4.6 Typical Applications

Examples of computer applications to specific power-reactor functions are given in the chapters on instrumentation systems in boiling-water reactors (see Sec. 16, pt. 16-9), sodium-cooled reactors (see Sec. 17, pt. 17-5.5), and gas-cooled reactors (see Sec. 18, pt. 18-6.3).

8-5 SYSTEM DESIGN

The relative importance of the steps in control-system design and the order in which they should be taken depend on the kind and size of the facility, the project components to which the design tasks are assigned, and the time and money constraints imposed. No matter what the design procedure, experience has shown the importance of formulating and documenting certain design procedures. A schedule and a set of rules must be established early in the project, even if they will be changed many times later.

The design of a computer-based data-acquisition system or a computer-based control system differs in many ways from that of their analog counterparts. The main reason for these differences is that a large part of the system design resides in the computer programs. This fact goes a long way toward explaining why the schedule, vendor responsibility, costing, and specification are so unlike those for analog equipment.[13]

8-5.1 Schedule

One aspect of the difference between the two systems is apparent from examination of the control-system part of a

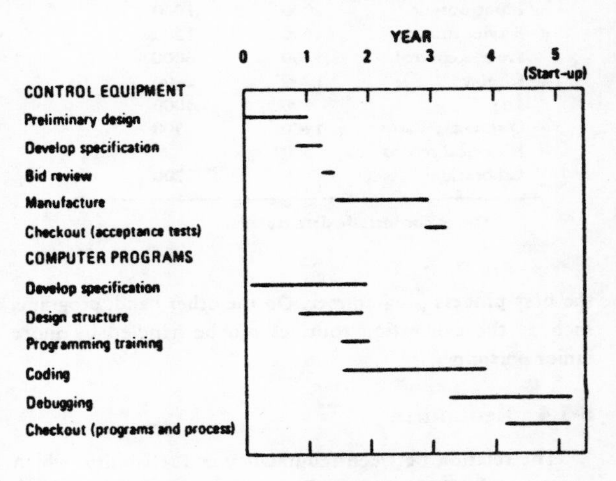

Fig. 8.5—Sample project schedule (see text).

project schedule. A typical schedule (Fig. 8.5) is developed by taking the target date for reactor start-up as the completion time for the control system. The previous 12 to 18 months are then reserved for the major programming effort, which is on-site program development and debugging after equipment delivery. Coding starts as soon as the computer model is chosen and finishes when the programs have been tried out on the computer after it has been installed in the plant. Still working backward in time, a year or two is allowed for equipment manufacture and checkout. The length of this period depends on the complexity of the system and how much of nonstandard hardware is to be put together by the supplier. Finally, a few months are added for vendor selection and contract negotiation and a few more for writing the specification. At this point the control engineer finds that he is already behind schedule or, at best, that he will reach the specification deadline having dangerously little information about the processes that his system is going to control.

Then begins the task of shortening the various segments of the chart to produce a properly timed, but realistic, set of goals. To be done right, this job requires familiarity and experience with both computer systems and reactor projects. Factors that must be considered are:

1. The minimum amount of plant design data needed to develop a sound specification.

2. The extra control-system capacity that should be specified as a contingency against further plant design changes.

3. The anticipated delays because of prevailing procurement policies.

4. The probable system suppliers and their reputations for making accurate delivery-schedule estimates.

5. The available programming personnel.

6. The likelihood of a postponement of the plant start-up date and the estimated number of months extension.

These and many other factors influence the control-system schedule, which becomes a firm guideline for acquiring, programming, and installing the equipment. It will, of course, be modified from time to time as the project progresses toward completion.

8-5.2 Preliminary Design

It is unlikely that complete plant automation will come about in big jumps. It is interesting to contemplate and discuss an imagined power station with everything from control rods to switchgear under computer control; but that stage will be reached in small steps over a long period of time, just as analog control technology has developed.

Lacking a charter to automate the whole plant, the engineer must select the type of control (analog, hybrid, or digital) to be applied to the constituent processes and the degree and scope of automation appropriate to those under the command of the computer. Every facility has one or more process elements for which conventional control devices are inadequate; these are the prime targets for computer control.[14] Next come those elements for which computer control promises a substantial cost savings either in capital investment or in greater operating efficiency. Finally, there are those functions where the advantages of either digital or analog control are comparable. The scope of computer applications will depend heavily on the strength and completeness of the control engineer's system analysis. Examples of the results of this stage of planning can be found in recent and current reactor plant designs, several of which are described in Sect. 8-7.

Up to now this section has dealt with computer control, but from here on our discussion of control equipment and software must include their other functions: data acquisition and display. All new reactor plants will have a computer-based data-handling and data-display system; the control functions can be considered as an extended use of equipment already planned. A small percentage in hardware cost will buy computer-control capability. However, adding control programs raises the level of complexity of the system software in proportion to the number of functions provided.

8-5.3 Program Description

The programs discussed in this section are those necessary to perform data handling, control, and routine self-checking. They are called process programs and are recognized by their being always in the computer system and routinely or potentially operative when the computer is on line as a part of the operating plant. A second kind, called systems programs, are those used to write and debug the process programs. These include the programmers console routines, loaders, assemblers, compilers, and editors. They are discussed in the section on equipment specifications because they are always part of the computer procurement.

For an analog control system, the engineer can develop a set of functional specifications, most of which can be translated into equipment requirements based on a one-to-one correspondence among the set of input-data points, control-system channels, and control-signal outputs. This simple relation between function and hardware is shown in Fig. 8.2. However, there is no such correspondence in a computer-based system. As seen in Fig. 8.4, only the process interfaces can be sized according to the system functions; the rest of the hardware must be specified on the basis of the computer programs needed to perform the data acquisition and display and control tasks.

So the computer-system designer starts by compiling the usual functional specification, setting forth what the system must do, and the process and operator inputs that will be needed to implement them. The next step is to develop a detailed description of the computer programs required to make the hardware function as specified. This task is the most demanding part of the design procedure. If it is not done well, an adequate system procurement specification cannot be written.

The program description can be developed in three stages for convenience in planning and carrying out the separate activities.

1. The process software is divided into major programs. The level of the division is dictated by the need for each program to have a distinct and identifiable function. The number of programs on the list should be such as to make it easy to assign them to different engineers on the programming staff. Too few program elements will overload the programmers; too many will make it difficult to combine them into a working whole.

2. The memory required for each program is estimated. An accurate estimate requires a detailed knowledge of process requirements in terms of precision and response times, the programs that provide the control functions, and how they are to be accomplished by computer-program instructions. The engineer may have to write trial routines in a typical process assembly language.

3. Each program is labeled as to whether it will normally reside in core, in auxiliary memory, or partly in each. The results of this step determine the proportion of computer memory which must be provided by high- and low-speed devices.

The results of the preceding three activities are summarized in a deceptively brief compilation similar to that shown in Table 8.1. These data, along with roughly estimated execution times, form the basis for specifying memory sizes, computation speed, and capabilities of peripherals, such as line printers and high-speed operator displays.

It is important that one recognize the disparity in level of the different programs on the list. The one titled "executive-monitor," for example, is of the highest level. It is the master program that ties all the others into a coordinated system. Its development requires the talents of the best process programmer. On the other hand, programs such as the calibration routines can be handled by more junior personnel.

Table 8.1—Major Control Programs

Program	Number of words and residence	
	Core	Drum or Disk
Executive-monitor	1500	
Input/output	2000	1000
Service routines	1500	1500
Process control	3000	6000
Display	1200	2400
Log	500	5000
Diagnostic/alarm	1400	500
Historical record	300*	
Calibration		2200

*Does not include data storage.

8-5.4 Redundancy

The relation between redundancy and reliability, which is treated in sec. 11, is discussed here only on a large scale. The minimum addition in providing a dual or backup computer control system is adding another central processor; further duplication of equipment then depends on plant-design concepts, such as whether or not the reactor will be operable if the computer fails. Also to be considered is duplication of critical functions by independent hardware. Analog and manual backup are examples.[15]

Table 8.2 shows some of the reasons for having or not having redundant computers. The use of a nonredundant system is usually justified by the requirement that the reactor and processes be capable of at least steady-state operation if the computer fails. This implies either extensive analog backup control or a plant small enough to be handled manually by the control-room staff. A second reason is that a high plant-availability factor is not essential; therefore the reactor safety instrumentation can take care of computer failure. These are the cautious approaches to computer control which place little reliance on the computer part of the system. They are becoming less evident since experience and familiarity with digital systems have

Table 8.2—Some Reasons for Single and Redundant Systems

Plant	Computer	Justification
Single	Single	Analog or manual backup Low plant factor
Single	Dual	High plant factor Plant data essential Off-line service
Dual	Dual	High plant factor
Dual	Triple	High plant factor Off-line service

influenced plant designers to allow shutdown in case of total control-system failure if redundancy is used.[16]

Restating the above: if a high plant factor is important, then frequent shutdown cannot be tolerated and a redundant control system is required. This is the predominant reason for dual computers. A second reason, especially applicable to prototype facilities, is the importance of maintaining continuity of plant operating data, a criterion that applies to pure data-acquisition equipment as well as control systems. A third justification is that a standby computer can be used for off-line program preparation and data processing but can be automatically interrupted when the unit is called into service to replace the operating computer.

It has become common to design nuclear power stations with two complete reactor–generator units operating independently. At first sight it seems that reliable computer control could be effected by placing a control system on each reactor unit using the computers as backup for each other, at least for the essential control functions. Although this configuration has been given much consideration, few, if any, such systems have been built. Both computers would have to be larger, peripheral switching to one or the other would be difficult, and developing the complex switching programs would be very costly. The result is that dual plants usually have three computers, one as standby for an operating system on each reactor unit.[17]

The basic justifications for the triple control system are the same as for redundancy in general: the importance of reliability, as it affects plant factor, and the convenience of having a standby for off-line programming.

The final objective of redundancy is to improve plant availability or, in project terms, to ensure that the system will permit attaining the target plant factor. The immediate objective is to make the control system more reliable. However, a precise estimate of the reliability of a redundant computer system cannot be made because the reliabilities of the constituent parts—processor, peripherals, interfaces, and displays—are not precisely known. So it is not surprising that past and present justifications for replicating components are commonly based on judgment and inference from statistically inadequate data on past operating experience. This state of affairs will prevail until reactor power plants with computer control are commonplace, at which time the need for stringent justification will be far less.

8-5.5 Cost Estimates

After the program description and the equipment configuration have been developed, the control engineer has enough information to make a preliminary estimate of system cost. He compares the conceptual computer system with a conventional analog layout designed to perform the same basic tasks, which shows whether or not the computer control system is economically acceptable. The analog equipment will include panels, panel meters, data loggers, strip-chart recorders, alarm annunciators, and controllers. Some controllers must be capable of three-mode, feed-

Table 8.3—Cost-Comparison Chart

Item	Number required	Cost, $
Analog System		
Panels		20,000
Meters	500	10,000
Recorders	12 (multipoint)	50,000
Annunciators	4000 (points)	240,000
Control hardware*		260,000
Data logger	4000 (points)	420,000
Spare parts		100,000
	Total	$1,100,000
Computer System		
Central processors	2	150,000
Process input/output	4000 (points)	450,000
Mass storage	Drums, disk, and magnetic tapes	125,000
Operator input/output	Printers, typers, and consoles	75,000
Displays	Cathode-ray tubes and interfaces	100,000
Spare parts		25,000
Programming		350,000
	Total	$1,275,000

*Only for those functions which the digital system would perform.

forward, and cascade operation. The systems should have equivalent orders of reliability and redundancy.

In all likelihood the computer control hardware required for a reactor power plant will have a lower cost estimate than the analog configuration but will have a higher estimate when the cost of computer programming is included (see Table 8.3). If so, the decision to use a computer system has to be justified on the basis of services beyond those which analog systems can provide. The values of such advantages are difficult to assess; they depend on such things as future savings in plant operating cost, benefits in case of a plant accident of low probability, and capability of readily accommodating later plant modification. Examples of some advantages are:

1. The computer can lessen the time and effort applied to system installation and checkout. For instance, several hundred sensors and cables can be connected and tested all at once by using a short computer program to detect and print out faulty channels. Such procedures substantially lower installation time and cost compared with the standard method of "ringing out" each signal path by hand.

2. The control-room staff can possibly be reduced because of greater efficiency in data acquisition and display. Considering the cost of an operator over the lifetime of the plant (diminished by the initial cost of training personnel in the capability, programming language, and structure of the computer-based system), this factor can tip the scales in favor of a computer system. Large offsetting increases in other activities usually are not expected; for example, the number of maintenance person-

nel should be about the same with either kind of hardware, although they will generally be more highly paid for maintaining computer hardware.

3. At some time during its life, the plant will undergo changes in components, operating mode, or power level. The cost of altering the instrument and control system to provide for such changes will vary greatly but will be less with a computer system. The cost of adapting to plant modification is hard to forecast except when such action is planned from the beginning. The latter is illustrated by a full-scale prototype power plant that is currently under construction.[18] At start-up and during initial testing, the plant will run with saturated steam, later to be changed to superheated steam to produce full-design electric power. It is estimated that the changeover will be done in a few hours by loading new computer programs, in contrast to a many-month-long job of altering conventional analog equipment.

4. The ability to format and output both transitory and permanent data in report form is an asset. These data include postincident data, fuel exposure and inventory, and management reports on plant operation.

5. Reactivity balance, control-rod calibration, reactivity coefficient, and other complex on-line calculations can materially aid in efforts to achieve optimum reactor operation.

After all factors such as the above are taken into account and given conservative cost values, the total is incorporated into the initial system cost for comparison against an analog configuration.

The reader is cautioned that the figures in Table 8.3 are composite and illustrative. The cost ratios of the separate items will differ greatly for the various types of reactors and different plant operating modes.

8-6 SYSTEM SPECIFICATION

The basic criteria for judging a system specification are the same for a computer-based control system as for analog equipment. There are, however, additional items in a computer system which even an experienced designer may sometimes overlook. It is therefore advisable for the specification writer to have a checklist of items available or to consult the procurement documents of a successful control system.

A complete outline of the sample specification is presented in Table 8.4. In this section a sample specification is presented and discussed section by section. The sample shows what should be included in a specification that requires the manufacturer to provide computer systems programs but not the control software. Not all the items shown in the sample will appear in every specification. Many systems will not require double precision and floating-point hardware, memory protect, automatic restart, etc. On the other hand, the sample specification does not include special-purpose electronics that might be supplied by the computer manufacturer.

We stress the point that the following specimen is for illustration only. It would not be used for an actual procurement because:

1. There will usually be some items that are undergoing improvement and field testing of which the specification writer is unaware. A "functional" specification for such items may allow a supplier to offer equipment that would not meet strict performance specifications but is actually best suited to the application.

2. The detailed requirements in the specimen represent performance at or near the state of the art. In the interest of a lower price and better competition, these should be relaxed wherever possible to correspond to the needs of the plant.

3. System suppliers are continually developing executive programs and adding applications programs to their software packages. By using as much of the supplier's proven control software as appropriate to the application, the designer can realize a considerable project cost saving.

4. Design engineers often prefer to evaluate computer speed by finding the running time of a bench-mark program that includes arithmetic, logic, transfer, and input/output instructions in about the proportions that they will occur in the actual application. Besides providing a good functional criterion for acceptance, this method can reveal deficiencies in computer architecrure that might not show up if only the operating speeds of the individual computer functions were analyzed.

8-6.1 Introductory Sections

The specification should begin with a short description of the reactor the computer is to be associated with and what the computer is expected to do. The bidder must be given a thorough and unambiguous concept of the whole system. Following this summarizing statement, the broad specifications of the computer itself should be described. A list of the major equipment should be presented in the order in which the equipment is described in the body of the specification.

The sample section below shows how the introductory sections of the system specification might be written.

SPECIFICATION FOR XPR-1 COMPUTER SYSTEM

SUMMARY

This specification details the requirements for the digital-computer system to be installed as an integral part of the XPR-1 instrumentation and display system. The computer system will operate on-line to the reactor and provide the necessary equipment for reactor systems support, analysis, control, and reporting.

This application supports the operation of XPR-1 (Experimental Power Reactor Number One), a 1000 MW(e) nuclear facility designed for the generation of electric power. The computer system will support this facility by providing data acquisition, analysis, control, display, and reporting. The reactor is characterized by both on–off and continuous data-acquisition and -control processes. On–off operations are typified by block valves, start–stop pump, supply utilities, etc. Continuous monitoring and control is applied to neutron flux, temperature, flow, etc. The success of this installation will depend on the reliability of the system; thus, meeting the reliability and quality assurance requirements is important.

1.0 GENERAL DESCRIPTION

This specification describes a general-purpose digital computer of the binary, core-memory, parallel, single-address type with indirect and indexed addressing. The computer system is to be used as the basis of a real-time control system performing control functions in the operation of a nuclear power reactor. The computer system is required to respond to both analog and digital inputs, provide analog control signals, and store data in on-line bulk storage. Operator communication is by keyboard and interactive cathode-ray-tube display.

2.0 MAJOR EQUIPMENT LIST

2.1 24-bit general-purpose computer with 16K of core memory.
2.2 250K word disk storage unit.
2.3 100 line-per-minute line printer.
2.4 300/100 character-per-second paper-tape reader/punch.
2.5 Two 21-in. color cathode-ray-tube display consoles.
2.6 An analog input facility for 2000 points.
2.7 A digital input facility for 2000 points.
2.8 An analog output facility for 20 points.

8-6.2 Central Processor

The requirements of the central processor are then listed. An acceptable range of word lengths, the result of a detailed analysis of data-handling and computation needs, is stated. The instruction set, memory requirements, and the other items in this section must reflect the type of application: whether it emphasizes data acquisition, data analysis, or process control. Taken together, these items should also force the bidders to confine their offerings to heavy-duty industrial-grade equipment to the exclusion of light laboratory computers, desk computers, and machines designed for scientific or business data processing.

The central-processor section of the sample specification is as follows·

3.0 CENTRAL PROCESSOR

3.1 Core Memory

3.1.1 Word Length
The computer shall utilize a basic word length of 24 bits excluding memory parity and memory protect.

3.1.2 Word Capacity
A minimum of 16,384 twenty-four-bit words of core memory shall be provided. The computer shall be capable of field expansion to at least 32,768 words.

3.1.3 Speed
The maximum read/restore memory cycle time shall not exceed 2.0 μsec.

3.1.4 Parity
Memory parity shall be provided for each word in memory such that each transfer to, or from, memory is checked for correct parity. An error shall cause an interrupt signal which identifies the location of the word in error.

3.1.5 Protect
Memory protect shall be provided for each word in memory. This bit shall be selectable under program control for each individual word. When an attempted violation is detected by the computer, an interrupt signal shall be generated which identifies the location of the attempted violation.

3.2 Arithmetic Unit

3.2.1 Hardware Arithmetic
Hardware arithmetic shall be provided to perform (1) single precision; (2) double precision; and (3) floating-point add, subtract, multiply, and divide.

3.2.2 Execution Times
Execution times shall not exceed those listed below:

	Single precision, μsec	Double precision, μsec	Floating point, μsec
Add	4.0	4.0	20
Subtract	4.0	4.0	20
Multiply	20	20	100
Divide	30	30	100

3.3 Addressing

3.3.1 Direct
The computer shall be capable of directly addressing a minimum of 2048 memory locations.

3.3.2 Indirect
Multilevel indirect addressing shall be provided with a capability of reading a minimum of 32,768 memory locations. Each level of indirect address shall add no more than one (1) memory cycle to an instructions execution time.

3.3.3 Indexing
At least three (3) dedicated index registers shall be provided.

3.4 Priority Interrupts
At least 32 channels of multilevel hardware interrupt shall be provided such that any higher priority channel can interrupt the processing of a lower priority channel. Each interrupt shall have a separate dedicated memory location (32 total) that contains space for the necessary instructions to initiate a device service routine. All interrupts except those assigned to the stall alarm and the power fail safe shall be capable of being individually turned on or off under program control.

3.5 Direct Memory Access
A minimum of two (2) direct-memory-access (DMA) ports shall be provided. Each port's transfer rate shall be at least 500,000 twenty-four-bit words per second. Multiplex capability for two channels at 250,000 twenty-four-bit words per port shall be provided.

3.6 Clocks
Two (2) basic clocks shall be provided, a real-time clock and an interval timer. The real-time clock shall have a basic frequency of 60 Hz. The interval timer shall have a crystal-controlled rate of 100 kHz with an accuracy of ±10 Hz per day. Additional registers shall be provided with the interval timer such that they can be loaded from memory under program control and incremented or decremented by the clock. The timer shall provide an output signal (for use as an interrupt) when the register reaches zero.

3.7 Stall Alarm
A stall alarm shall be provided that detects machine looping or stalls. The method used shall be discussed in the response to bid.

3.8 Operator's Console
An operator's console shall be provided and shall include a tabletop working surface of at least 200 sq in.

3.8.1 Console Switches
A data entry switch corresponding to each bit in a

Table 8.4—Specification Outline (Summary Description)

1. General description

2. Major equipment list

3. Central processor
- **3.1 Core memory**
  - **3.1.1 Word length**
  - **3.1.2 Word capacity**
  - **3.1.3 Speed**
  - **3.1.4 Parity**
  - **3.1.5 Protect**
- **3.2 Arithmetic unit**
  - **3.2.1 Hardware arithmetic**
  - **3.2.2 Execution times**
- **3.3 Addressing**
  - **3.3.1 Direct**
  - **3.3.2 Indirect**
  - **3.3.3 Indexing**
- **3.4 Priority interrupts**
- **3.5 Direct memory access**
- **3.6 Clocks**
- **3.7 Stall alarm**
- **3.8 Operator's console**
  - **3.8.1 Console switches**
  - **3.8.2 Display**
  - **3.8.3 Console teletype**

4. Process input/output
- **4.1 Input/output channels**
- **4.2 Speed**
- **4.3 Input/output parity**
- **4.4 Input/output addressing**
- **4.5 Digital inputs**
  - **4.5.1 Capacity**
  - **4.5.2 Logic definition**
  - **4.5.3 Speed**
- **4.6 Analog outputs**
  - **4.6.1 Number of channels**
  - **4.6.2 Output range**
  - **4.6.3 Data input**
  - **4.6.4 Accuracy and linearity**
  - **4.6.5 Monotonicity**
  - **4.6.6 Sag**
  - **4.6.7 Slew rate**
  - **4.6.8 Settling time**
  - **4.6.9 Short-circuit capability**
- **4.7 Analog inputs**
  - **4.7.1 Multiplexer**
    (Subheadings: Input switches, Number of channels, Input configuration, Sampling rate, Input impedance, Full-scale input voltage, Crosstalk, Scatter, Common-mode rejection, Maximum common-mode voltage, Full-scale output voltage, Address modes)
  - **4.7.2 Analog-to-digital converter**
    (Subheadings: Number of bits, Conversion speed, Aperture time—sample and hold, Acquisition time—sample and hold, Sag—sample and hold, Accuracy, Monotonicity, Linearity, Overscale indicator, Display)

5. Standard peripherals
- **5.1 Disk memory**
  - **5.1.1 Capacity**
  - **5.1.2 Access time**
  - **5.1.3 Error control**
  - **5.1.4 Write lock**
- **5.2 Line printer**
  - **5.2.1 Print speed**
  - **5.2.2 Number of columns**
  - **5.2.3 Character spacing**
  - **5.2.4 Character size**
  - **5.2.5 Character registration**
  - **5.2.6 Character set**
  - **5.2.7 Character replacement**
  - **5.2.8 Line spacing**
  - **5.2.9 Line registration**
  - **5.2.10 Paper handling**
  - **5.2.11 Ribbon**
  - **5.2.12 Printer-cabinet soundproofing**
  - **5.2.13 Paper storage**
- **5.3 Paper-tape reader/punch**
  - **5.3.1 Speed**
  - **5.3.2 Code**
  - **5.3.3 Tape take-up and supply**
- **5.4 Display—Alphanumeric and graphic**
  - **5.4.1 Display area**
  - **5.4.2 Phosphor**
  - **5.4.3 Spot size**
  - **5.4.4 Random-positioning time**
  - **5.4.5 Positioning repeatability**
  - **5.4.6 Jitter**
  - **5.4.7 Resolution**
  - **5.4.8 Contrast ratio**
  - **5.4.9 Brightness**
  - **5.4.10 Vector data format**
  - **5.4.11 Vector end-point registration**
  - **5.4.12 Vector writing rate**
  - **5.4.13 Character plot**
  - **5.4.14 Character sizes**
  - **5.4.15 Number of characters displayed**
  - **5.4.16 Aspect ratio**
  - **5.4.17 Intensity**
  - **5.4.18 Light pen**
  - **5.4.19 Keyboard**
  - **5.4.20 Memory**
  - **5.4.21 Enclosure**

6. Software
- **6.1 Executive**
- **6.2 Compiler**
- **6.3 Assembler**
- **6.4 Correction program**
- **6.5 Diagnostic and utility programs**
- **6.6 Input/output programs**
- **6.7 Maintenance programs**
- **6.8 Delivery form**
- **6.9 Documentation**

7. Environmental and miscellaneous characteristics
- **7.1 Temperature**
- **7.2 Humidity**
- **7.3 Power**
- **7.4 Enclosure**
- **7.5 Spare parts**
- **7.6 Documentation**
- **7.7 Reliability**
- **7.8 Quality-assurance program**

word shall be provided. It shall be possible to enter data "manually" to "memory" and to all registers that are software accessible.

3.8.2 Display

The console shall provide for the display of the status of the following registers or their equivalent:

A-register (A-accumulator)
B-register (B-accumulator)
P-register (program counter)
I-register (instruction register)
M-register (memory address register)
X-register (index register)

In addition, the console display shall provide a run–halt indicator, an input/output hold indicator, and a protect–violation indicator.

3.8.3 Console Teletype

A KSR-35 teletype, or equivalent heavy-duty machine, shall be interfaced to the computer for use as an operator's console.

8-6.3 Process Input/Output

The process input/output are the communicating links between the plant and the computer. The multiplexer and converters are first specified individually as to number of points, speed, and addressing. Then, because of the complex interrelations involved, a tendency toward functional specification is introduced; the requirements are placed on the entire input or output channel rather than on individual components. Since the input/output list (see sample below) is a summary of many of the process interfaces, it is important to make sure that the two sets of requirements agree.

The input/output sample specification is as follows:

4.0 PROCESS INPUT/OUTPUT

4.1 Input/Output Channels

In addition to the direct-memory-access channel specified in 3 (Central Processor), the system shall provide a shared input/output (I/O) bus such that all peripheral devices can communicate directly with the computer.

4.2 Speed

The I/O bus shall support I/O transfers at rates up to 30 kHz.

4.3 Input/Output Parity

I/O parity shall be provided. The system shall provide a hardware parity test for each I/O transfer and indicate all I/O parity errors by program interrupts.

4.4 Input/Output Addressing

Capability for addressing a minimum of sixty-four (64) peripheral devices shall be provided. Each bidder shall indicate the standard I/O assignment by logical device number and hardware address number for all standard peripherals available for the computer.

4.5 Digital Inputs

4.5.1 Capacity

The system shall provide for the input of at least two thousand (2000) binary signals. These signals take the following form:

Type a: 500 twelve (12)-bit voltage words.
Type b: 1000 one-bit binary voltages.
Type c: 500 one-bit contact closures.

4.5.2 Logic Definition

Voltage inputs shall be positive true (1 = positive voltage). The following voltage range is required:

$$\text{Logic } 0 = 0.0 \text{ volt} {}^{+0.5 \text{ volt}}_{-0.0 \text{ volt}}$$

$$\text{Logic } 1 = 2.3 \text{ volts} {}^{+2.7 \text{ volts}}_{-0.0 \text{ volt}}$$

(The values specified are for conventional diode-transistor logic. It should be noted that high-level logic with improved noise immunity has recently become available from several sources. This high-level logic should be used whenever possible.) Contact signals shall be input as closure true.

4.5.3 Speed

The minimum transfer rates are as follows

50 type a inputs: 1 kHz/channel	(50,000 channels/sec)
450 type a inputs: 0.033 Hz/channel	(3 channels/sec)
48 type b inputs: 30 kHz/channel (2 computer words)	(60,000 words/sec)
952 type b inputs: 1 Hz/channel (40 computer words)	(40 words/sec)
500 type c inputs: 0.5 Hz/channel (21 computer words)	(11 words/sec)

4.6 Analog Outputs

4.6.1 Number of Channels

Twenty (20) channels of digital-to-analog output shall be provided.

4.6.2 Output Range

The output shall be ±10 volts full scale.

4.6.3 Data Input

The data input shall be twelve (12) bits per channel fully buffered.

4.6.4 Accuracy and Linearity

Accuracy and linearity shall be at least ±0.05% of full scale 5 mV.

4.6.5 Monotonicity

The converter output shall be monotonic for each input bit change from negative (−) to positive (+) full scale.

4.6.6 Sag

The output sag shall be less than 1 mV/μsec.

4.6.7 Slew Rate

The analog output rise time (10 to 90%) shall be 3 μsec or less for a full-scale step change (digital) at the input.

4.6.8 Settling Time

The time required to settle to within 0.1% of the final value shall be less than 15 μsec for a full-scale step change (digital) at the input with 1000 pF capacitive load.

4.6.9 Short-Circuit Capability

The output amplifier(s) shall be capable of sustaining a continuous short circuit to ground without damage.

4.7 Analog Inputs

The analog input system shall consist of a multiplexer(s), sample and hold amplifier(s), and an analog-to-digital converter. It shall include all interface hardware required to make the analog system a functional part of the computer system.

4.7.1 Multiplexer

4.7.1.1 Input Switches

The input switches shall be field-effect transistors, either junction type (J-FET) or metal-oxide-semiconductor (MOSFET). If MOSFET devices are supplied, each gate shall be protected from oxide rupture due to overvoltage.

4.7.1.2 Number of Channels
A minimum of 1000 input channels shall be provided. At least 500 channels shall be low level; the balance shall be high level (as defined in Sec. 4.7.1.6).

4.7.1.3 Input Configuration
Each input shall be differential-guarded (three-wire). All three inputs shall be commutated. A minimum of two (2) levels of subcommutation shall be provided to isolate the input.

4.7.1.4 Sampling Rate
The following minimum sampling rates shall be provided:
100 low-level channels, 5000 channels/sec
100 high-level channels, 2500 channels/sec
400 low-level channels, 10 channels/sec
400 high-level channels, 10 channels/sec

4.7.1.5 Input Impedance
The input impedance of an "off" channel shall be greater than ten (10) megohms when measured differentially or from either input to ground.

4.7.1.6 Full-Scale Input Voltage
The full-scale input range of the multiplexer shall be as follows:
Low-level inputs, ±10 mV
High-level inputs, ±10 volts

4.7.1.7 Crosstalk
Crosstalk shall be less than ±0.01% of full scale on any channel when a 100% overload is present on an adjacent channel.

4.7.1.8 Scatter
Channel-to-channel scatter shall be less than ±0.1% of full scale for the same input on all channels.

4.7.1.9 Common-Mode Rejection
Common-mode rejection shall be at least 120 db from direct current to 60 Hz with a balanced source impedance. It shall be at least 85 db from direct current to 60 Hz for a 500-ohm unbalanced source impedance.

4.7.1.10 Maximum Common-Mode Voltage
The multiplexer(s) shall be capable of sustaining ±20 volts direct current or peak alternating current on any input without damage to the input switches and without turning on deselected channels.

4.7.1.11 Full-Scale Output Voltage
The full-scale output voltage shall be ±5 volts or greater for both low- and high-level inputs.

4.7.1.12 Address Modes
Three separate address modes shall be provided: random, sequential, and dwell. Each mode shall be program initiated. The random access mode shall permit an external binary word to select any address at random. The sequential mode shall provide a fixed sampling pattern and be capable of operating from an internal or external clock. The dwell mode preselects one channel for continuous duty.

4.7.2 Analog-to-Digital Converter

4.7.2.1 Number of Bits
The converter shall provide 12 bits of information with the most significant bit representing the sign of the input data.

4.7.2.2 Conversion Speed
The total conversion time shall not exceed 10 μsec including sample time and hold time.

4.7.2.3 Aperture Time–Sample and Hold
The converter shall incorporate a sample and hold amplifier. The aperture time shall be 100 nsec or less.

4.7.2.4 Acquisition Time–Sample and Hold
The acquisition time shall not exceed 6 μsec for a full-scale step input.

4.7.2.5 Sag–Sample and Hold
The permissible decay during hold shall be less than ½ of the least significant bit (lsb).

4.7.2.6 Accuracy
The transfer accuracy of the converter shall be at least ±0.05% ±½ lsb.

4.7.2.7 Monotonicity
The converter output shall be monotonic increasing (or decreasing) for a change from plus (+) to minus (−) full scale and from minus (−) to plus (+) full scale.

4.7.2.8 Linearity
The deviation from a straight line through plus (+) and minus (−) full scale shall not exceed ±0.1% ±½ lsb.

4.7.2.9 Overscale indicator
One bit shall be provided to indicate an overscale input.

4.7.2.10 Display
A front-panel display that indicates the output word status shall be provided for troubleshooting purposes.

8-6.4 Standard Peripherals

Whereas the input/output includes the peripherals that are peculiar to process control systems, the standard peripherals include those items and their interfaces which are common to most computer systems. The characteristics of each item depend on the application and often are compromises between programming efficiency and hardware cost. This is particularly true of punched-card units and line printers. Only by experience can one estimate accurately the tradeoff between the programmers' man-hours and the machine cost involved.

Although cathode-ray-tube (CRT) displays are becoming more common, they have been used in computers in so many different ways that no standard set of design criteria has yet been developed. It is therefore necessary, once the display content and formats have been decided on, to make a detailed analysis of the required data storage and transfer rates. These are then related to the known capabilities of currently marketed hardware. Equipment to display both alphanumeric and graphic data should be studied carefully; a raster method may be required for one and beam steering for the other, with the result that two separate spares are needed if standby redundancy is a system requisite.

The standard-peripherals section of the sample specification is given below. The intermediate-speed bulk storage device may be either disk or drum, the specifications being quite similar. The disk was arbitrarily chosen for the example.

5.0 STANDARD PERIPHERALS

5.1 Disk Memory

A disk memory shall be interfaced to the computer. It shall utilize a fixed head per track design. The read/record heads shall not contact the disk surface.

5.1.1 Capacity

Disk capacity shall be a minimum of 12,000,000 bits.

5.1.2 Access Time

Worst case access time shall be less than or equal to 16 msec.

5.1.3 Error Control

Parity generation and error detection shall be provided for all transfers to and from the disk. The nominal error rate shall be no greater than 1 bit lost in 10^{10} data transfers.

5.1.4 Write Lock

Write lock shall be provided for at least 50% of the total storage capacity.

5.2 Line Printer

A line printer shall be interfaced to the computer. The printer will be used for on-line printing of alphabetic, numeric, and symbolic data.

5.2.1 Print Speed

Print speed shall be not less than 300 lines per minute.

5.2.2 Number of Columns

A minimum of 132 print positions (columns) across shall be provided.

5.2.3 Character Spacing

Character spacing shall be ten (10) characters to the inch (horizontally). The maximum cumulative error shall not exceed ±0.02 in. for the 132-column line.

5.2.4 Character Size

Nominal character size shall be 0.1 in. (vertically) by 0.066 in. (horizontally).

5.2.5 Character Registration

Vertical and horizontal registration shall be within a ±0.005-in. tolerance.

5.2.6 Character Set

The following 64 characters shall be supplied: A, B, C, D, E, F, G, H, I, J, K, L, M, N, O, P, Q, R, S, T, U, V, W, X, Y, Z, 1, 2, 3, 4, 5, 6, 7, 8, 9, ϕ, (,), ′, ;, :, ., ., /, -, +, &, ?, $, <, >, ↑, !, ″, =, #, @, , %, *, [,], \, ←

5.2.7 Character Replacement

It shall be possible to individually remove and replace characters.

5.2.8 Line Spacing

Line-to-line spacing shall be 6 lines per inch.

5.2.9 Line Registration

Line-to-line registration shall be within ±0.005 in. nonaccumulative.

5.2.10 Paper Handling

The printer shall accept paper widths ranging from 2.5 to 18.5 in. The paper used shall be standard Z-fold sprocket feed. It shall be possible to use 6-part paper (multiple copy).

5.2.11 Ribbon

The printer ribbon shall be equipped with an automatic reversal feature.

5.2.12 Printer-Cabinet Soundproofing

The printer cabinet shall contain sound-deadening material with a minimum thickness of ½ in.

5.2.13 Paper Storage

Both input and output paper storage shall be provided.

5.3 Paper-Tape Reader/Punch

A paper-tape reader/punch shall be interfaced to the computer.

5.3.1 Speed

The system shall read/punch at a minimum speed of 300/100 characters per second.

5.3.2 Code

The system shall read and punch the standard 7-level ASCII code as described in ASA Standard X3.18-1967.

5.3.3 Tape Take-Up and Supply

The system shall utilize fanfold paper tape. Both supply and take-up bins for fanfold tape shall be supplied.

5.4 Display—Alphanumeric and Graphic

Two independent displays shall be interfaced to the computer. Each display system consists of a cathode-ray-tube assembly, vector generator, character generator, light pen, keyboard, and interface electronics.

5.4.1 Display Area

The display screen shall be not less than 19 in. across the color cathode-ray tube (diagonal or diameter measurement). The usable viewing area shall be not less than 12 sq in. All specifications shall be met everywhere in this area.

5.4.2 Phosphor

The CRT shall utilize a P22 phosphor (or equivalent). The faceplate shall have a boned safety shield installed.

5.4.3 Spot Size

The display-spot size shall not exceed 0.020 in. at the half-light points.

5.4.4 Random-Positioning Time

It shall be possible to display points at random. The maximum time required to position the beam and settle to within one spot width of the final position shall not exceed 16 μsec.

5.4.5 Positioning Repeatability

The display-spot position repeatability shall not exceed ±0.3% of full scale independent of the previous position.

5.4.6 Jitter

The display-spot jitter shall not exceed ±0.2% of full scale over a time interval of 1 hr.

5.4.7 Resolution

The resolution in the x and y directions shall be ten (10) binary bits. The matrix defined is 1024 by 1024 points.

5.4.8 Contrast Ratio

The contrast ratio for black and white shall be no less than 4 : 1 in an ambient light level of 20 ft-c.

5.4.9 Brightness

The brightness shall exceed 20 ft-lamberts in an ambient light level of 8 ft-c.

5.4.10 Vector Data Format

The display shall provide at least two (2) vector formats, one long and one short. Short vectors are defined as less than 1 in. The time required to display short vectors shall not exceed 5 μsec. It shall be possible with the long-vector mode to specify either (inclusive) relative vector or absolute vector. In relative mode the x and y coordinates of the end point are given and the vector is drawn from the previous point. In absolute mode the x and y coordinates are specified and the vector is drawn relative to the origin of the position matrix.

5.4.11 Vector End-Point Registration

End-point registration shall be within ±0.035 in.

5.4.12 Vector Writing Rate

The time required to write a vector shall be less than 50 μsec regardless of length. The display intensity shall be constant irrespective of the length of the vector.

5.4.13 Character Plot

The character generator shall contain at least sixty-four (64) alphanumeric and symbolic characters. The character set shall include those characters listed for the line printer in pt. 5.2.6. The character generation time shall not exceed 10 μsec.

5.4.14 Character Sizes

At least two (2) programmable character sizes shall be provided.

5.4.15 Number of Characters Displayed

The display system shall be capable of displaying and refreshing not less than 1500 characters (50 Hz refresh rate).

5.4.16 Aspect Ratio

The character aspect ratio shall be not less than 4 : 3 (height to width) or greater than 3 : 2.

5.4.17 Intensity

Three programmable intensity levels shall be supplied for each primary color (red, blue, and green).

5.4.18 Light Pen

A photoelectric light pen shall be supplied for each display. It shall transmit an interrupt signal to the computer whenever a point on the display screen, within view of the pen, is intensified.

5.4.19 Keyboard

An alphanumeric and symbolic keyboard shall be supplied with each display such that all defined characters can be entered from the keyboard. In addition, a function keyboard with at least 16 keys shall be supplied. A keyboard overlay that permits the selection of at least eight (8) different function groups shall be provided for all function keys. A separate overlay shall be supplied for each of the eight groups. Overlay code names shall not be supplied.

5.4.20 Memory

A local refresh memory capable of storing one frame of data shall be supplied with each display.

5.4.21 Enclosure

The display system shall be enclosed in a console cabinet. It shall have a table extending in the front below the cathode-ray tube. The cathode-ray tube shall be tilted slightly toward the back to facilitate operator viewing.

8-6.5 Software

The software part of the specification is the most difficult to state in precise terms. This is partly attributable to the prevalence of an imprecise vocabulary in which several words can mean the same thing (monitor, executive, organizer), a practice that compels the specification writer to describe programs by function. Some of the difficulty may also be credited to the existence of the two distinct kinds of software: *systems programs* (sometimes called utility programs or operator routines), which implement programming, and *process programs*, which implement data handling and control. Avoid confusion by specifying programs in terms of the tasks that each is to perform.

The designer must be very firm about receiving the systems programs in working order. Although many of the programs are fixed for all systems, those which require the operation of peripherals may never have been tried on the exact configuration being specified.

Several of the programs listed in the sample specification (below) are optional. Many users will not make the compiler a firm requisite, although most U. S. manufacturers include some version of FORTRAN in the software package. The editor program is very valuable when paper tape is the programmer's only high-speed communication with the computer, but the editor program loses value if punched-card equipment is available. The process programs are useful only to the extent to which they can be adapted to the application at hand. The current difficulty in this regard is simply that a compact and efficient software package cannot support a variety of applications, nor can a computer, properly sized for the application, support an oversized software package that has been designed to suit all occasions. It is hoped that in the future a truly modular approach to process software will allow the designer to specify a control system, equipment, and programs as easily as he has specified analog equipment in the past.[19] The software section of the sample specification follows:

6.0 SOFTWARE

Specific software requirements to be furnished with this equipment are listed in the following sections. In addition, each bidder shall include a description of all software available (including extra price where applicable) for the specific configuration (core memory, peripherals, etc.) described in this specification.

A complete listing of the software documentation available for the specified configuration shall be supplied.

6.1 Executive

An executive monitor oriented to on-line real-time processing shall be provided. The monitor shall reside on the disk, along with the processor and library routines, and shall utilize the disk for scratch storage if necessary. Foreground—background processing shall be required. Foreground operations shall include provision for both resident (in-core) and nonresident (core-disk swapping) real-time programs. Both the monitor and real-time programs shall be protected against inadvertent destruction by a background program.

The monitor shall provide automatic interrupt, context switching and storing, programmable priority-interrupt structure, nested-interrupt inquiries, program-priority queries, and linkages and facilities for handling all devices included in this specification. Device-independent I/O programming shall be provided. The monitor shall provide for reentrant subroutine execution. The maximum time that the monitor disables interrupts shall be stated.

6.2 Compiler

A FORTRAN compiler shall be provided to allow programming in English and mathematical-like statements. The compiler shall be capable of operating in a real-time environment and, as a minimum, of meeting the following requirements:

Compliance with ASA Standard X3.9-1966 FORTRAN.

Intermixing of FORTRAN statements and assembly-language statements by macros or other suitable means.
Provision for reentrant subroutines.
Provision for queuing and utilization of priority interrupts in real time.

6.3 Assembler
A symbolic assembler program shall be provided which processes a machine-oriented language. The assembler shall provide pseudo instructions for the purpose of defining symbols, reserving memory, linking subroutines, and controlling input/output options. The assembler shall provide macroinstruction capability.

6.4 Correction Program
A correction program shall be provided which shall enable corrections and additions to be made on source programs by inputting the source program into memory (limited by the capacity of the memory) and making corrections through the keyboard. The output shall be a new program taken from memory which includes the corrected statements.

6.5 Diagnostic and Utility Programs
Programs shall be provided to assist programmers during the debugging phase of program development. These programs shall include, but not be limited to, the following features:
Clearing all or part of memory.
Modifying memory from the keyboard.
Printing all or part of memory under specified conditions.
Inserting and deleting breakpoints.
Initiating a jump on condition to any part of memory.
Additional features shall be listed in the bid response.

6.6 Input/Output Programs
Input/output drivers shall be provided for all peripheral devices required in this specification. The drivers shall provide for testing device status and executing data transfers. The tests used for each device, the device address (logical and hardware), and the number of interrupts, by level, shall be indicated in the bid response.

6.7 Maintenance Programs
Maintenance programs shall be provided which enable testing of the entire central processor and all peripheral equipment. Testing shall include the memory, instruction set, central control, arithmetic section, priority interrupts, and each peripheral.

6.8 Delivery Form
All software shall be delivered as individual paper tapes. Where more than one program (e.g., loader and monitor) resides on a tape, the individual tapes will also be supplied. This requirement does not apply for the library decks for FORTRAN and assembly language.

6.9 Documentation
Three sets of all software manuals shall be provided. All paper tapes shall be accompanied by a source program listing.

8-6.6 Environmental and Miscellaneous Characteristics

The environmental and miscellaneous characteristics comprise the temperature and humidity ranges under which the system is to operate and such various items as power requirements, documentation, cabinets, and quality assurance.

The hardships that have been caused by inadequate or erroneous circuit and schematic diagrams are well-known. The same pains can be experienced if the software is incorrect. Complete program descriptions and updated listings of all supplied software, as well as accurate as-built drawings of the hardware, must be required.

A subsection on the manufacturer's quality-assurance practices should be included in the specification. It is usually a requirement of any large system. The specification should require that all proposals describe the seller's current quality-control programs to an extent commensurate with the size of the system. Incoming-materials inspection, manufacturing procedures, quality control, documentation control, and use of existing codes and standards should be covered. It should be made clear that response to this section will influence the selection of the supplier.

The environmental and miscellaneous characteristics section of the sample specification is as follows:

7.0 ENVIRONMENTAL AND MISCELLANEOUS

7.1 Temperature
The computer and all peripheral devices shall meet the requirements of this specification over a temperature range of 50 to 90° F.

7.2 Humidity
The computer and all peripheral devices shall meet the requirements of this specification over a humidity range of 50 ± 30%.

7.3 Power
The computer shall operate from 120/208-volt, 60-Hz power, single or multiple phase. Each manufacturer shall provide literature that describes the following items:
1. Power required—by voltage, current, and phase.
2. Power dissipation in British thermal units per hour and kilovolt-amperes separately stated for the central processor and each peripheral device plus the total power consumed by the system.
3. Connector wiring diagrams for every input power connector, cross labeled by manufacturer type.

7.4 Enclosure
Each item specified is to be supplied in a fully enclosed cabinet with access doors for ease of maintenance. Peripheral devices may be grouped to share a single cabinet. Each manufacturer shall provide literature that describes the following items:
1. Cabinet configuration and mechanical position assignment for each peripheral device and for each portion of the central processor. Unused rack space shall be identified and dimensioned.
2. Service clearances required for all racks and cabinets.
3. Installation dimensions for assessing doorway, elevator, and loading-dock clearances.
4. Total system weight and weight by cabinet (console or rack). Shipping weight.

7.5 Spare Parts
Standard spare parts package and terms for later exchange of faulty circuit boards shall be provided. Extender boards shall be provided for each type of printed-circuit connector supplied (where applicable).

7.6 Documentation
Four copies, unless otherwise noted, of the following

documentation shall be supplied:
1. Final as-built drawings.
2. System block diagram.
3. Instruction and maintenance manuals.
4. Parts list.

7.7 Reliability
Reliability data shall be provided, both calculated and tested. The method of calculation (MIL-METHOD etc.) shall be specified. Where data are not available, so indicate.

7.8 Quality-Assurance Program
The vendors shall supply in the proposals evidence that a Quality-Assurance Program is maintained that will assure the purchaser that all articles procured from the vendors will satisfy the purchaser's requirements. This evidence shall consist of either the vendor's quality-control procedures manual or references to applicable public documents which constitute the procedures used in the vendor's quality program.

The vendor's quality program shall show evidence of an organized and documented approach to the attainment of quality both in inspection and manufacturing procedures and in document control as it pertains to the purchased system.

8-6.7 Other Sections of the Specification

There are several other items to consider when procuring a computer system for the automation of a nuclear power plant. The specification should provide for appropriate vendor action on each item.

(a) Acceptance Tests. The complex nature of a computer control system demands a close and continuing check on its operability. A thorough set of checkout programs should be run at the factory and again after delivery to the site to disclose any faults that might have occurred in transit. These tests must encompass all the requirements of the specification and will be time consuming and tedious.

(b) Installation. It is impractical in a full-size power plant to require the control-system manufacturer to install his equipment. He will, however, be expected to supervise the placement of cabinets, interconnecting cables, and the running of final acceptance tests.

(c) Training. There has been, to date, no surplus of experienced process programmers. The quality of programming and maintenance training that the system manufacturer offers will depend on the availability of his personnel. It is advisable to leave the training schedule open to negotiation, within specified limits, to make best use of the teacher's time. However, the specification should include a request for outlines of the training courses.

(d) Appendixes. A list of control-system inputs and outputs can be appended to the specification to aid the bidder in sizing the equipment required. A rough layout of the control console will also help him visualize the display configuration and perhaps offer suggestions for improvement. A floor plan will aid in placing components and in determining the lengths of interconnecting cables.

8-7 CURRENT PLANT DESIGNS

A few of the world's nuclear power plants are described briefly in this section to give the reader an idea of present computer applications. These plants were selected because they are completed or soon will be, have some degree of computer control, and have been adequately discussed in meetings or publications. Pertinent data are summarized in Table 8.5. The dates listed are years during which the plant goes on-line as a power-generating station; in some cases these have been inferred from other planned milestones, such as plant completion and date critical.

Table 8.5—Computer-Controlled Power Plants

Plant	Country	Year on-line	Reactors	Computers	Computer control functions
Douglas Point	Canada	1967	1	1	Flux tilt control rods; power-level demand; and safety logic modification
Marviken	Sweden	1969	1	1	Start-up and shutdown sequence; refueling; and superheater throttle valves
Wylfa	United Kingdom	1969	1	1	Turbine run-up
Dungeness "B"	United Kingdom	1970	2	3	Reactor start-up; reactor-coolant outlet temperature; and turbine run-up
Hinkley "B"	United Kingdom	1971	2	3	Reactor start-up; reactor-coolant outlet temperature; and turbine run-up
Prototype Fast Reactor	United Kingdom	1971	1	2	Start-up sequence; fueling sequence; and power and temperature regulation
Pickering	Canada	1972	1*	2	Zone reactivity; reactor power; boiler pressure; and refueling
Gentilly	Canada	1972	1	2	Zone reactivity; reactor power; and coolant flow

*Initially.

8-7.1 Douglas Point

The CANDU reactor[20] at Douglas Point, Ontario, is probably the earliest planned use of digital-computer control in a nuclear power plant. At present the computer directly drives neutron absorber rods to control the reactor flux profile and indirectly adjusts power by providing the set point to the analog moderator level controller.

As an assist to higher plant factor, the computer also modifies safety-circuit operation. The safety circuit alone will initiate scram if low coolant flow in any fuel channel is signaled; however, when the computer is operating, it inhibits the trip unless low flow is accompanied by high coolant outlet temperature.

8-7.2 Marviken

The Marviken Nuclear Power Plant,[18] a 200-MW(e) station, will have a comprehensive set of sequence programs for start-up, running, shutdown, and refueling. These are largely automatic with interspersed stopping points that require a manual command to proceed. The control system will also automatically adjust the throttle valves in 32 superheat channels in the reactor core, a difficult procedure because a high-channel temperature requires resetting not only the affected channel valve but also those in several surrounding channels. This is a good example of applying computer control where other means are inadequate.

8-7.3 Wylfa

Wylfa,[21] a MAGNOX reactor station, is the first in a series of three plants in which computer-based data and display systems have been applied progressively more toward direct control as United Kingdom experience grows. Wylfa has automatic turbine run-up.

8-7.4 Dungeness "B"

In addition to automatic turbine run-up, the Dungeness "B,"[22] an advanced gas-cooled reactor station, will have complete start-up, from subcritical to power, under computer control with manual intervention required if the system encounters abnormal conditions. The computer will also control, by rod movement, the ratio of outlet gas temperatures among five reactor zones. This is a difficult control problem under all conditions of coolant flow and reactor power level.

8-7.5 Hinkley "B"

Except for a different array of inputs and outputs, the Hinkley "B"[13] is controlled similarly to the Dungeness "B." Both stations are examples of the "dual-plant three-computer" configuration discussed in pt. 8-5.

8-7.6 Prototype Fast Reactor

Automatic computer control of reactor flux, coolant outlet temperatures, and steam-generator outlet temperatures is being considered for the prototype fast reactor,[16] a 250-MW(e) sodium-cooled fast reactor facility. This is the plant, cited before, in which a detailed economic and technical study resulted in a redundant computer system and a revision of principles to allow plant shutdown on complete control-system failure.

8-7.7 Pickering

Nearly all the major reactor variables in the Pickering Nuclear Power Station,[23] as in other Canadian plants, will be computer controlled. This includes flux profile, overall reactivity, boiler pressure, and the fueling process.

8-7.8 Gentilly

As in the Pickering, most control of the Gentilly Nuclear Power Station[24] will be by computer. Of interest is the reactor's large positive void coefficient, which will be compensated by moving a set of booster rods according to changes in plant power level. Primary-coolant-flow valves will also be automatically controlled as a function of power.

REFERENCES

1. T. J. Williams, The Application of University Research to Industrial Process Control, in 22nd Annual ISA Conference, Chicago, 1967, Part 3, Paper No. 5-1-ACOS-67, p. 3, Instrument Society of America, Pittsburgh, 1967.
2. Control Engineering Company, Compilations of Process Control Applications in Control Engineering, May 1962, September 1963, August 1965, September 1966, March 1967, July, 1968.
3. M. A. Schultz and F. C. Legler, Application of Digital Computer Techniques to Reactor Operation, in *Proceedings of the Third International Conference on the Peaceful Uses of Atomic Energy, Geneva, 1964,* Vol. 4, pp. 321-330, United Nations, New York, 1965.
4. M. A. Schultz, Automatic Digital Computer Control, Sec. 4.2, in Small Nuclear Power Plants, USAEC Report COO-284(Vol. 2), pp. 126-130, Chicago Operations Office, March 1967.
5. W. C. Lipinski, Optimal Digital Computer Control of Nuclear Reactors, USAEC Report ANL-7530, Argonne National Laboratory, January 1969.
6. J. T. Tou, *Digital and Sampled-Data Control Systems*, McGraw-Hill Book Company, Inc., New York, 1959.
7. W. D. T. Davies, Control Algorithms for DDC, *Instrum. Pract.*, 21: 70-77 (January 1967).
8. K. L. Gimmy, On-Line Computers at the Savannah River Plant, in *Application of On-Line Computers to Nuclear Reactors*, Seminar held at Sandefjord, Norway, September 1968, pp. 727-737, Organization for Economic Cooperation and Development, Paris, 1968.
9. A. Pearson and C. G. Lennox, Sensing and Control Instrumentation, in *The Technology of Nuclear Reactor Safety*, Vol. 1, Reactor Physics and Control, T. J. Thompson and J. G. Beckerley (Eds.), pp. 285-416, M.I.T. Press, Cambridge, Mass., 1964.
10. J. R. Howard, Experience in DDC Turbine Start-Up, *ISA (Instrum. Soc. Amer.) J.*, 13: 61-65 (July 1966).
11. T. J. Glass, Current Trends in Process Computer Software, paper presented at Annual ISA Conference, Chicago, 1967, Paper No. D2-3-DAHCOD-67.
12. R. G. Basten, Impact of Nuclear Reactor Control on the Structure of Computer Systems, in *Application of On-Line Computers to Nuclear Reactors*, Seminar held at Sandefjord, Norway, September 1968, pp. 517-533, Organization for Economic Cooperation and Development, Paris, 1968.

13. M. W. Jervis, On-Line Computers in Central Electricity Generating Board Nuclear Power Stations, in *Application of On-Line Computers to Nuclear Reactors*, Seminar held at Sandefjord, Norway, September 1968, pp. 51-78, Organization for Economic Cooperation and Development, Paris, 1968.
14. Computers in Control, *Nucl. Eng.*, 11: 618-620 (August 1966).
15. J. C. Kite, How to Assure Maximum Performance in Redundant Computer Control Systems, paper presented at Annual ISA Conference, Chicago, 1967, Paper No. D4-4-DAHCOD-67.
16. N. T. C. McAffer, The Computer Instrumentation of the Prototype Fast Reactor, in *Application of On-Line Computers to Nuclear Reactors*, Seminar held at Sandefjord, Norway, September 1968, pp. 351-379, Organization for Economic Cooperation and Development, Paris, 1968.
17. Hinkley Point B, *Nucl. Eng.*, 12: 26-28 (January 1967).
18. J. Akesson, Techniques of Computer Application For the Marviken Nuclear Power Plant, in *Application of On-Line Computers to Nuclear Reactors*, Seminar held at Sandefjord, Norway, September 1968, pp. 301-318, Organization for Economic Cooperation and Development, Paris, 1968.
19. J. C. Spooner, Real Time Operating System for Process Control, paper presented at Annual ISA Conference, Chicago, 1967, Paper No. D1-1-DAHCOD-67.
20. E. Siddall and J. E. Smith, Computer Control in the Douglas Point Nuclear Power Station, in *Heavy Water Power Reactors*, Symposium Proceedings, Vienna, 1967, International Atomic Energy Agency, Vienna, 1968 (STI/PUB/163).
21. D. Wellbourne, Data Processing Control by a Computer at Wylfa Nuclear Power Station, in *Advances in Automatic Control*, Convention held at Nottingham, England, Apr. 5-9, 1965, Paper 16, Institute of Mechanical Engineers, London.
22. A. R. Cameron, The On-Line Digital Computer System For The Dungeness "B" Nuclear Power Station, in *Application of On-Line Computers to Nuclear Reactors*, Seminar held at Sandefjord, Norway, September 1968, pp. 273-300, Organization for Economic Cooperation and Development, Paris, 1968.
23. J. E. Smith, Digital Computer Control System Planned For Pickering Nuclear Station, *Elec. News Eng.*, 39-41 (March 1967).
24. W. R. Whittal and K. G. Bosomworth, *Dual Digital Computer Control System For the Gentilly Nuclear Power Station*, 4th International Federation for Information Processing Congress, Edinburgh, Scotland, Aug. 5-10, 1968, North-Holland Publishing Company, Amsterdam, 1969.

BIBLIOGRAPHY

A comprehensive bibliography covering the general field of computer control is presented in T. J. Williams, Computers and Process Control, *Ind. Eng. Chem.*, 62(2): 28 (February 1970).

Argonne National Laboratory, Liquid Metal Fast Breeder Reactor (LMFBR) Program Plan, Vol. 4, Instrumentation and Control, USAEC Report WASH-1104, August 1968.

Bullock, J. B., and H. P. Danforth, The Application of an On-Line Digital Computer to the Control System of the High Flux Isotope Reactor (HFIR), in *Application of On-Line Computers to Nuclear Reactors*, Seminar held at Sandefjord, Norway, September 1968, pp. 459-478, Organization for Economic Cooperation and Development, Paris, 1968.

Demuth, H. B., J. Bergstein, K. H. Duerre, and F. P. Schilling, Digital Control System for the UHTREX Reactor, in *Application of On-Line Computers to Nuclear Reactors*, Seminar held at Sandefjord, Norway, September 1968, pp. 621-642, Organization for Economic Cooperation and Development, Paris, 1968.

Etherington, H. (Ed.), *Nuclear Engineering Handbook*, McGraw-Hill Book Company, Inc., New York, 1958.

Freymeyer, P., and H. Stein, Automation in the Control of Nuclear Power Stations, *Kerntechnik*, 11(9/10): 514 (September/October 1969).

Harrer, J. M., *Nuclear Reactor Control Engineering*, D. Van Nostrand Company, Inc., Princeton, N. J., 1963.

Holland, L. K., On-Line Computer Experience with Boiling-Water Reactors, *Trans. Amer. Nucl. Soc.*, 9(1): 264 (June 1966).

Moen, H., et al., Computer Control of the Halden Boiling Heavy Water Reactor, in *Application of On-Line Computers to Nuclear Reactors*, Seminar held at Sandefjord, Norway, September 1968, pp. 647-667, Organization for Economic Cooperation and Development, Paris, 1968.

Morin, R., Utilization of Digital Computers for Starting and Running the EDP-3 Atomic Power Plant, Report AEC-TR-691, December 1967.

Nuclear Reactors Built, Being Built, or Planned in the United States as of December 31, 1968, USAEC Report TID-8200 (19th Rev.), December 1968.

Pearson, A., Computer Control on Canadian Nuclear Reactors, in *Application of On-Line Computers to Nuclear Reactors*, Seminar held at Sandefjord, Norway, September 1968, pp. 123-144, Organization for Economic Cooperation and Development, Paris, 1968.

Schultz, M. A., *Control of Nuclear Reactors and Power Plants*, 2nd ed., McGraw-Hill Book Company, Inc., New York, 1961.

Williams, T. J., *A Manual for Digital Computer Application to Process Control*, Purdue University Press, Lafayette, Indiana, 1966.

SECTION 9

POWER SUPPLIES

9-1 INTRODUCTION

The safe operation of nuclear power reactors requires that many of the instrumentation and control systems have a high degree of reliability. One factor in reliability is the integrity of the power source for the instrumentation and control power buses. This section provides the designer of instrumentation and control systems with sufficient information to choose the power-source system best suited to his specific application and giving maximum support to the overall system reliability.

9-1.1 System Requirements

The power sources most commonly used are electrical. Accordingly, this section deals mainly with them and only briefly with nonelectrical systems. The common combinations of static, rotating, and stored-energy system components are discussed. Battery-supported static inverter systems, which are among the most frequently used electrical source systems, are emphasized.

The systems discussed differ with respect to their degree of noninterruptibility, i.e., their capability to continue operating under emergency conditions after normal source failure, the quality of their output, and their cost. The designer of instrumentation and control systems must first determine his requirements and establish the relative importance and criticality of each part of the system before choosing the most economical power-source systems to meet his needs. The various systems described offer phase changing, direct-current transformation, line-frequency and line-voltage transformation, isolation, and stabilization. Capability for short-time operation with stored-energy sources and long-time operation with engine-driven energy sources can also be included.

9-1.2 Design Objectives

To design the power-source system for nuclear-reactor instrumentation and control and to achieve maximum reliability requires, at the outset, a thorough evaluation of the load characteristics. A designer of electronic systems is painfully aware of the risks in having instrumentation and control systems depend on plant auxiliary-power sources. Complete independence from outside power sources would provide ideal integrity. The best power system for a particular application cannot be designed by just selecting available components to obtain complete power-source independence. The designer must take into account, in his effort to satisfy the overall design objectives economically, such aspects as the allowable outage time of the power source, allowable transfer time between normal and standby power sources, initial cost, maintenance expense, and operating cost. The system that provides the desired reliability with the minimum cost is usually based on a compromise between many design considerations.

9-2 TYPES OF POWER SUPPLY

9-2.1 System Similarities

Each available power-source system provides a different degree of power continuity and independence, or isolation, from the plant auxiliary-power system. In general, each system has a form of stored energy for providing power to the critical bus during an unacceptable frequency or voltage excursion and during the time interval when it becomes necessary to disconnect the normal source of power and transfer to the standby power source. Each system has a means of isolating the unacceptable normal power source and initiating the alternate source (if one is provided) before the stored energy of the system is depleted. Automatic reconnection to the normal source, after it has returned to a stable condition for a given period, is also a common feature.

9-2.2 Energy Storage Methods

The five major ways to provide stored energy for a standby power system are:

1. Pneumatic (stored air or gas) systems.
2. Hydraulic accumulation systems.
3. In-house or on-site steam-driven turbogenerator systems.
4. Inertia, flywheel with and without eddy-current coupling systems.
5. Storage-battery-supported systems.

The first four systems are commonly used with rotating machinery, and the last is associated with static systems.

The last two methods are electrical in nature and have had the greatest acceptance. Each of the last two methods is used to improve the quality of the power normally provided by the plant auxiliary-power source by acting as a filter or buffer. When they are combined with a diesel-driven generator backup, they also provide reliable protection from prolonged power outages. Storage-battery systems are normally combined with static battery-charging rectifiers and either static inverters or d-c to a-c motor-generator sets to provide short-time power continuity. Static inverters, with and without output transfer switching, have also gained wide acceptance as reliable sources of a-c power for critical loads.

9-3 REQUIREMENTS FOR POWER SUPPLY

9-3.1 General System Categories

Instrumentation and control power systems may be categorized as either interruptible or noninterruptible.

(a) Interruptible Systems. Interruptible systems are those in which power is obtained directly from the plant auxiliary-power system through suitable filters and regulating equipment, if they are required to improve the power quality. An interruption of plant auxiliary power results in total loss of instrumentation power. This type of system is applicable only where the instrumentation and control loads are noncritical, i.e., absolute continuity of operation is not essential. Usually the total interruption time is a maximum of approximately 15 sec, i.e., the time in which the plant standby auxiliary-power source, several diesel-engine-driven generators, reestablishes power to the essential auxiliary buses.

(b) Noninterruptible Systems. Noninterruptible systems are those in which any interruption of power results in loss in continuity of operation or in erroneous operation of critical instrumentation and control loads. Since noninterruptible power systems are the type most commonly required for nuclear-reactor instrumentation and control applications, the remaining sections of this chapter concentrate on power-source systems that are noninterruptible for a discrete length of time after a plant auxiliary-power failure.

There are two categories of noninterruptible power systems, *nonisolated* and *isolated.*

Nonisolated systems are those in which plant auxiliary power is imposed directly on the essential auxiliary-power buses. Normally the noninterruptible power supply is also fed from these essential power buses and is storing up potential energy. If a plant auxiliary system fails or if there is a severe transient, the noninterruptible power supply releases the stored energy to the critical bus, thus affording a short-time carry-over of stable power.

The more commonly used type of noninterruptible power supply is the *isolated* system. This differs from the nonisolated system in that the instrumentation power buses do not receive power directly from the plant auxiliary-power system but rather through the noninterruptible power supply. The noninterruptible isolated power supply thus acts as a filter or buffer separating the instrumentation power buses from the plant auxiliary-power system and any transients that appear on that system.

9-3.2 Load Characteristics and Causes of Trouble

If, for economic reasons, the reactor instrumentation and control system is designed for use on a single d-c source, then the power-source system is simple. If a single d-c power source is not satisfactory, the choices of power-source configurations increase greatly. Factors to be considered include: alternating source voltage and frequency stability, harmonic distortion, plant auxiliary-power outage time and allowable transfer time, maximum allowable rate of change of voltage and frequency, and load power factor. A specification that is overly restrictive, just to be safe, results in an unnecessarily high cost for the power supply and should be avoided.

Momentary loss and surges and dips of voltage are relatively frequent in plant auxiliary-power systems. Voltage dips and surges are caused by switching, failure of equipment remote from the critical bus, and starting large motors on the same power system. The duration and intensity of the undesirable transient depend on its proximity and the clearing time of the protective equipment ahead of the faulted section. A single cycle can be severely distorted without a power interruption. (Such distortion might occur when heavy loads are placed or pulsed on a radial feeder remote from the critical bus.) Wave-shape distortion can also result from momentary faults that, in fact, become equivalent phase shifts. Large banks of capacitors and intermittent reactive loads can give rise to wave-shape distortion.

The most widely accepted method of avoiding or minimizing these effects is to insert a noninterruptible power source to act as a filter between the plant auxiliary-power supply and the critical instrumentation load. In addition, the system cable routing and installing must be designed to segregate the redundant systems and methods so that the clean output from the noninterruptible power source will not be contaminated by induced noise from cables in other noncritical systems. The imposed static and dynamic seismic loading postulated for the specific area within the nuclear plant in which the noninterruptible power supply is to be located must be taken into account. The seismic criteria should be incorporated within the equipment specification. In addition, proper seismic design of equipment foundations and anchors is essential; electric cables and conduit between all critical items of equipment must be flexible.

The reliability of a noninterruptible power system is only as good as the weakest part of the total installed system. This point should be emphasized not only during the design and installation phases but also throughout the life of the system. A continuous and conscientious maintenance and testing program is essential.

9-4 COMPONENTS OF POWER-SUPPLY SYSTEMS

All the power-supply systems discussed in this chapter are comprised of individual components, or building blocks.

The degree of power continuity and overall system reliability and cost achieved in any system depends on how the basic building blocks are combined. The following sections describe the building-block components of the system. Some necessary aspects of properly specifying the components as part of the overall system are also covered.

9-4.1 Static Inverters

Static inverters become an essential part of a high-reliability power system if short transfer time and an a-c output are required. These inverters are usually used in conjunction with batteries and a static rectifier to provide, on failure of the plant auxiliary-power source, instantaneous transfer of power from the plant auxiliary-power source to the battery system.

Static inverters consist of three basic parts: a low-power oscillator, a power-switching section, and (usually) an output ferroresonant transformer. The low-power oscillator determines the operating frequency of the inverter and can be independent of or synchronized with the plant auxiliary-power system. Once the inverter has been running for several hours, an output-frequency stability within ±0.25% of the desired frequency is obtained. The major factors tending to change the frequency are long-term drifts in components and variations in the ambient temperature.

The power-switching section is probably the most critical portion of the overall inverter and usually consists of bridge-connected silicon-controlled rectifiers (SCR). The four main SCR's are alternately switched in pairs, which converts the input d-c source into a square-wave alternating voltage that is applied across the primary of the output transformer. The peak amplitude of the alternating voltage is essentially equal to the direct input voltage. At the end of each half cycle, the two conducting SCR's are shut off by momentarily providing a reverse voltage bias. This process, called commutation, is an extremely critical function in the proper operation of the inverter. If during any half cycle the commutation should fail, the system would be left with more than two SCR's in the conducting state. This would result in an effective short circuit across the battery and would shut down the unit. Ensuring proper commutation even under the most adverse conditions is essential in designing reliable inverters. In addition to proper commutation, the associated circuit must limit the SCR rate of rise of current (di/dt) or voltage (dv/dt) and the peak forward and reverse voltages that appear across the SCR's.

In most applications the desired output is a regulated low-distortion sine wave. Usually both regulation and filtering are provided by the ferroresonant output transformer, a passive magnetic system similar to the commonly used constant-voltage transformer. An important feature of the ferroresonant transformer is that, as the load current is increased, the output voltage remains essentially constant up to a point in excess of rated load. Above this point the characteristic becomes a very nearly constant current mode. As a consequence of this, the inverter can be operated continuously into any overload, up to and including a short circuit, without affecting the square-wave switching portion of the inverter. A sine-wave inverter of this type can therefore satisfactorily handle load transients that might otherwise cause misoperation or lack of commutation in the square-wave section.

The normal regulation that can be expected is ±3% for all conditions of input voltage and for loads between zero and rated maximum at unity power factor. Loads at other than unity power factor have an additional effect on the output. Generally, inductive loads reduce the output voltage whereas capacitive loads increase the output voltage. Loads with a power factor below 0.8 lagging increase the harmonic distortion in the output. For these reasons it is preferable to operate with a load having a power factor as near unity as practical; unity power factor also corresponds to minimum d-c input drain. Where the load has an inherent low power factor, the designer must provide suitable correction either at the load or the inverter.

In applications where other critical loads are also fed from the battery system, it is desirable to use a filter on the input to the inverter. With a sine-wave output from the inverter, the input current resembles a half-wave rectified sine wave. Superimposed on this direct current is a large a-c component that may modulate the battery voltage sufficiently to cause an undesirable hum in the input of other equipment fed from the battery. The filter eliminates this problem. A second and probably more important function of an input filter is the elimination of spikes that are generated across the battery by other equipment, such as d-c motors and solenoids. Such equipment, commonly used in nuclear power plants, is notorious in generating large voltage transients during operation. Inverter input protection should be provided for short-term transients, in the order of 100 μsec up to 4000 volts, when fed from large-station battery systems.

Table 9.1—Input and Output Ratings of Typical Static Inverters

	Single-phase output	Three-phase output
Output voltage, volts (a-c)	120	208Y/120 230/399
Frequency, Hz	50 or 60	50 or 60
Output, kVA	2.5 to 28	9 to 150
Input voltage, volts (d-c)	48 or 125	125 or 250
Output voltage and frequency regulation, %	±1	±1 to ±2
Output harmonic distortion, %	<5	<5

Inverters are readily available in a variety of standard output and input ratings, the most common of which are listed in Table 9.1. The typical standard ratings in Table 9.1 do not, of course, represent the limits of the manufacturers' capabilities. Nonstandard inverters of different output and input voltage, kilovolt-ampere rating, and output quality are available for a premium price on request. The following is an outline of the major areas of importance which should receive attention when specifying a static inverter.

The range of input voltage over which the required output must be maintained for a stated time during normal and emergency operation is most important. In most cases the inverter supplier does not have control over the input source, which is usually the nuclear-power-plant station battery. It is not sufficient just to determine the normal long-term variations of input voltage. Transient input voltages are also important. A common design error is to focus on the large-magnitude short-time transients and neglect the higher energy transients of low frequency (0.5 to several Hz). Input voltage transients due to starting motors may not show up on either a long-time source voltage recorder or an oscilloscope set up to detect switching transients. Comprehensive knowledge of the characteristics of the input source is prerequisite to proper inverter application and protection.

The precise definition of source impedance is not always necessary. However, it is relevant to note the length and size of conductors between the inverter and the d-c source and between the inverter and any switching or protective devices in the incoming lines. Because of input-current pulsing during the SCR switching, an input filter may be desirable to remove unwanted modulation of the d-c source. This adds to the inverter cost, and, the actual need for it should be determined before specifying an input filter.

The load characteristic should be carefully defined. Of specific importance are the load power factor, the variation of load, and the maximum load to be switched at one time. Static inverters have definite limits to their momentary overload capacity, and the limits cannot be exceeded. This characteristic is different from the characteristic of a rotating inverter that has inertia and becomes very important where motor loads or other high inrush loads are present. The possibility of load short circuits should also be considered. It may be concluded that a current-limited output is desirable, and, if so, this should be specified.

The efficiency of static inverters may be defined in two ways. For the amount of heat to be removed due to losses in the inverter, efficiency may be expressed as the ratio (in percent) of output power losses to rated output power losses. However, if the purpose is to determine the source current, efficiency is expressed as the ratio of rated output a-c power to rated input d-c power (in percent). It is important to indicate which definition is to be submitted by the manufacturer in his bid proposal. This is particularly important for load power factors, which differ significantly from unity.

The output of the static inverter may be synchronized with another source or with a frequency standard. It is important to indicate the impedance, potential variation, and transient noise capability of the synchronizing signal to be used.

In most cases an extremely low harmonic output distortion level is not necessary. If the a-c output is to be rectified and filtered, a square-wave output would even be desirable. Should a square-wave inverter output be used in conjunction with external transformer loads, the transformers must be capable of handling the additional 11% swing in flux without overheating. In general, a wider harmonic distortion tolerance in the specification results in reduced size, cost, and weight of the inverter.

9-4.2 Storage Batteries

Two principal types of lead–acid batteries are in use today: Plante and Fauré. Plante invented the lead–acid battery about 100 years ago using positive and negative plates of pure-lead sheets. Today batteries that have pure lead in their positive plates are called Plante types. The original Plante battery was expensive because the lead peroxide, which is the active material of the positive plate, was extremely difficult to form on the surface of the positive plate. Some time later, Fauré invented an economical method of pasting lead peroxide on the positive plates and lead oxide on the negative plates. This is the type of construction used today.

Fauré batteries can be either lead–antimony or lead–calcium alloy plate grid types. The names are derived from the hardening alloy material used in the manufacture of the plate grids. Of the many types of batteries available, the lead–antimony and lead–calcium Fauré types are the most common in noninterruptible power supplies. A lead–calcium battery costs up to 15% more than a lead–antimony battery; however, the lead–calcium battery offers several advantages over the lead–antimony battery. The lead–calcium battery requires much less water replacement and therefore generates a proportionately smaller amount of hydrogen gas during the recharge cycle. The floating voltage is less critical, 2.17 to 2.25 volts per cell (compared to 2.15 to 2.17 volts per cell for the lead–antimony battery). If the lead–calcium batteries are float-charged between 2.20 and 2.25 volts per cell, they are reported to never require an equalizing charge. These advantages must be balanced against the higher initial cost of the lead–calcium battery since satisfactory operation for equal lifetimes can be expected from both types of batteries when properly maintained.

The current produced by a lead–acid battery results from the chemical reaction of dilute sulfuric acid on the active materials in the plates. Lead dioxide reacts with the sulfuric acid at the anode to produce a positive charge. At the cathode, metallic lead reacts with the acid to produce a negative charge. During the discharge process, sulfuric acid is consumed and replaced by a corresponding amount of water. During charging the process is reversed, acid being formed at the plates with a corresponding consumption of water and generation of oxygen and hydrogen gas.

As the discharge of a lead–acid battery progresses, the water formed is absorbed into the electrolyte, resulting in a reduction of the specific gravity of the acid. The battery open-circuit voltage depends on the concentration of the acid in contact with the active plate materials. The voltage available for useful work is the voltage across the battery terminals during discharge. This latter voltage is equal to the sum of the internal cell voltages minus the drop due to the internal resistance of cells. The reduction of acid concentration as the cell discharges is accompanied by an increase in internal resistance; the increase is gradual at first and then rapid as the cell approaches full discharge. The internal resistance may increase by as much as a factor of 2 to 3 on approaching full discharge. The internal resistance of a fully charged battery is so small that it has little effect on the terminal voltage except when high discharge rates are encountered. During discharge lead sulfate accumulates on the plates. Lead sulfate is a nonconductor and has a tendency to block the pores of the plates, thereby impeding the chemical reaction. Sulfate is a contributing factor in the reduction of the terminal voltage during discharge.

When the battery is discharged at low rates, the formation of water and lead sulfate proceeds slowly, allowing the acid in the electrolyte to be readily absorbed into the pores of the plates and resulting in a gradual decrease of the terminal voltage. When the battery is discharged at a very high rate, the depletion of acid at the plates takes place so rapidly that the rate of acid replacement cannot keep pace, which causes a greater reduction in terminal voltage. The increased depletion of acid at the plates on high discharge is the primary reason that batteries have a lower ampere-hour capacity at high discharge current rates.

The capacity of the lead–acid battery is reduced as the ambient temperature decreases. The amount of reduction also depends on the rate of discharge and cell design. The major reasons for the reduction in capacity with decreasing temperature are the increased electrolyte viscosity, which impedes the diffusion of the acid at the plates, and the higher internal cell resistance caused by increased resistance of the electrolyte. The common practice is to refer to the capacity of a battery at reduced temperatures as a percentage of its capacity at 77°F.

Batteries are rated on the basis of ampere-hours, which means the amperes that the battery will deliver for a given time at a specified temperature to reach a specified "final" voltage. The standard battery terminal voltages commonly used are 24, 48, 120, and 240 volts. These correspond to 12, 24, 60, and 120 cells, respectively, at a nominal 2 volts per cell. Fauré batteries are float-charged at a cell voltage ranging from 2.15 to 2.25 volts, depending on the type of cell. Determining the size or ampere-hour capacity of a battery involves a detailed procedure that is beyond the scope of this test. (Details of the methods are given in the bibliography at the end of this chapter.) Sizing of the battery can be done by the design engineer, who is aware of the battery limitations, or by the manufacturer or supplier of the noninterruptible supply package. In general, the battery must have sufficient ampere-hour capacity to carry momentary loads plus continuous or basic loads for a specified length of time before reaching its final voltage, commonly referenced at 1.75 volts per cell.

A lead–acid battery must receive the correct charge to give optimum performance and life. It is difficult and impractical to obtain this precisely on every charge or under floating operation. Lead–acid batteries do not need a full charge on every recharge to obtain satisfactory operation. For long life, however, they must be brought to the fully charged condition periodically, the period depending on the degree and frequency of discharge. On daily or frequent recharges, it is common practice to charge slightly short of a fully charged condition; a complete charge must be made every 1 to 3 months. This complete charge is commonly referred to as an "equalizing charge." Most modern battery-charging equipment is designed to effect a periodic equalizing charge.

The equalizing charge is intended to be sufficient to equalize any minor differences among the cells and should be continued until each cell of the battery reaches maximum voltage and specific gravity. To achieve this state requires manual attention and is, therefore, impractical. However, the same effect is obtained by giving the entire battery an additional amount of charge for a limited period, which is not harmful to the battery. Most charging is controlled by an automatic timing switch incorporated in the charging equipment. The most practical means of giving an equalizing charge is to set the time switch for an additional period of time. For batteries in normal full-floating service, which is the most common service in standby power systems, a common practice is to raise the floating voltage above its normal value by about 5 to 10% for a period of 8 to 24 hr, depending on the type of battery and application. It is important to ensure that any noninterruptible loads on the battery during the equalizing charge are rated for operation at the higher voltage levels.

The efficiency of a battery may be expressed as ampere-hour efficiency, voltage efficiency, or watt-hour efficiency. Ampere-hour efficiency is the ratio of the number of ampere-hours a battery yields on discharge to the number of ampere-hours required to fully recharge the battery. A typical lead–antimony battery requires a recharge in ampere-hours about 10% greater than the previous discharge, and the ampere-hour efficiency is 91%. Voltage efficiency is similarly defined as the ratio of discharge voltage to charge voltage. For the same typical lead–antimony battery, the average voltage on charge is approximately 17% higher than on discharge; the voltage efficiency is 85%. The product of ampere-hour and voltage efficiencies gives a watt-hour, or total, efficiency. For the lead–antimony battery, the watt-hour efficiency is 0.91×0.85, or 77%. This is a representative value for such batteries. (The watt-hour efficiency is slightly higher for a typical lead–calcium battery.)

The Fauré battery is the most reliable single component used in noninterruptible power systems today. Properly designed, battery-supported systems afford the optimum in overall reliability.

9-4.3 Stored-Energy Eddy-Current Coupling

The systems described in Pts. 9-5.6 and 9-5.7 use an eddy-current magnetic coupling in combination with a stored-energy flywheel to obtain limited sustained operation (i.e., 10 sec to 2 min) after failure of the normal source of power. An eddy-current coupling consists of rotor and stator assemblies, quite similar in many aspects to the common squirrel-cage induction motor. The rotor assembly is an input shaft on which is mounted a field coil and pole pieces, the coil being energized from a d-c source through slip rings. The stator is simply a hollow soft-iron cylinder with the output shaft attached to one of the cylinder bases.

The air gap between the rotor pole pieces and the inner surface of the stator is small. As a result, the magnetic field established by energizing the rotor coil is concentrated in the soft-iron cylinder. As the energized rotor rotates within the iron cylinder, the magnetic flux of the rotor sweeps through the stator cylinder and induces eddy currents. The eddy currents in the stator set up magnetic fields that interact with the rotor fields, and, as a result, torque is developed which tends to drag the stator along with the rotor. There must be relative motion between the rotor and stator to develop any torque. If there is no relative motion, no eddy currents are produced and no torque is created. The amount of torque produced by the coupling is a function of the rotor field strength and the speed difference between rotor and stator. The output torque (stator) increases with increased rotor excitation and also with increased slip.

The use of a rotating field coil and slip rings with brushes creates maintenance and reliability problems that can be eliminated by using a brushless stationary field coil. In this type of eddy-current coupling, the excitation coil is rigidly mounted in a frame. The input shaft carries a smooth cylindrical drum designed so that there is a small air gap between the stationary field assembly and the drum. The output shaft carries the rotor, which is fitted into the input shaft cylindrical drum and separated by a small air gap. The flux path is from the stationary field poles to the cylindrical drum on the input shaft to the rotor and then axially in the rotor back to the cylindrical drum and to the field poles. The flux actually traverses two air gaps, as opposed to only one when slip rings and brushes are used. The output torque is developed by the interaction of eddy-current-induced magnetic fields on the inner surface of the cylindrical drum with the main field concentrated in the rotor. The main field flux is prevented from being short-circuited in the cylindrical drum by a nonmagnetic strip that separates the two halves of the drum. The stationary field design has increased reliability and reduced maintenance. The efficiency is less than that of a brush and slip-ring coupling because the double air gap requires more excitation for equal output torque.

As noted earlier, the eddy-current coupling is similar to a squirrel-cage induction motor. In an induction motor a rotating magnetic field is established in the air gap by means of a polyphase winding on the stator. In the eddy-current coupling of the rotating field type, a rotating field is established by mechanical rotation of the energized rotor assembly by a prime mover. The soft-iron cylindrical rotor of the coupling is analogous to the squirrel-cage rotor bars of an induction motor. In addition to these similarities, the slip-torque characteristic of an eddy-current coupling is similar to that of a squirrel-cage induction motor. The slip-torque characteristic of an eddy-current coupling can be modified in the same manner as that of an induction motor. Use of high-resistance material for the soft-iron cylinder affects the slip-torque curve of an eddy-current coupling in the same manner as the use of high-resistance rotor bars affects the slip-torque curve of a squirrel-cage motor.

Eddy-current couplings are noted for their low efficiencies, especially for large differences between input and output shaft speeds. Whenever the output speed is different from the input speed, heat is generated. This loss, called slip loss, is essentially equal to the difference between input shaft power and output shaft power. Slip loss is the major source of heat in an eddy-current coupling, and the heat must be dissipated by cooling fluid or air. At rated torque and output speed, the slip loss of a typical eddy-current coupling will be about 2 to 4%. Considering other losses, such as friction, windage, magnetic drag, and excitation, the peak efficiency is about 92%. At reduced speeds the slip loss increases, and the efficiency becomes essentially equal to the ratio of output speed to input speed.

The application of an eddy-current coupling in a noninterruptible power supply with a drive motor, stored-energy flywheel, and output generator requires a large difference between the input and output eddy-current coupling shaft speeds. This results in a large coupling slip loss and hence poor efficiency; this can be relieved by operating normally without energizing the eddy-current coupling, see pt. 9-5.7.

If the poor efficiency experienced when operating with the coupling normally energized is discounted, extremely close speed control and hence output frequency control can be attained with an eddy-current-coupled system. Upon coast-down, after losing the prime mover input power, the eddy-current coupling, used in conjunction with a flywheel, is able to dissipate the flywheel stored energy at a finely controlled rate just sufficient to maintain a constant output shaft speed (thereby providing an acceptable generator output) for intervals as long as several minutes.

9-4.4 Engine-Driven Alternators

The systems discussed in pt. 9-5.4 to 9-5.7 use an a-c generator driven by an internal-combustion engine in combination with other components to provide a sustained power supply. Once operation is established with the engine as the prime mover, the time during which the alternator will operate is limited only by the available fuel supply.

The simple system shown in Fig. 9.8 is the basic configuration of a typical large standby auxiliary-power supply used in nuclear power plants. For a two-unit nuclear power station, a number of such large standby power sources sufficient to deliver as much as 20 MW when the normal auxiliary power is lost are required to maintain safe shutdown conditions or to provide power for the engineered safeguard auxiliaries during a loss-of-coolant accident.

The principles involved in the design and application of such power supplies are well known and are defined for nuclear power generating stations by standards. These principles or criteria are also applicable to the much smaller engine–generator units discussed elsewhere in this section.

9-4.5 A-C and D-C Drive Motors

The normal drive motor for a noninterruptible rotating power supply can be either induction or synchronous. The choice is usually determined by the frequency requirement of the output generator. The standard a-c squirrel-cage induction motor is considered more reliable and can operate for a longer period with minimum maintenance. The slip rings and secondary excitation required by a synchronous motor require maintenance. However, brushless synchronous motors are available and approach the reliability of induction motors, although at a greater cost.

The a-c induction motor can accelerate a greater load from rest than a synchronous motor can accelerate. The induction motor can be designed to have a very low slip value and thus can provide an output speed as little as 1% below that of a synchronous motor.

Synchronous motors provide a constant output frequency with constant input frequency. If a brushless synchronous motor is used, it is necessary to furnish an external exciter with voltage control and an external out-of-step protective relay; these are not required by an induction motor. Usually a large synchronous motor, when used with a stored-energy flywheel, is required to accelerate the flywheel into step. Thus, when large flywheels are used for stored energy, a synchronous motor that has several times the horsepower needed to drive the load may be required. In addition to the higher initial cost of the larger motor, the motor would be running normally at light load with decreased efficiency.

The emergency drive motor for a rotating noninterruptible power supply can be a standard d-c motor with the possible addition of auxiliary fields for speed control. Usually the motor can have an intermittent rating since its duration of operation is relatively short, being limited by the d-c battery source used. Because the characteristics of common loads and supply voltages vary, automatic speed-regulating equipment should be furnished as part of the motor control.

9-5 DESIGN OF POWER-SUPPLY SYSTEM

The following various combinations of the basic power-supply building blocks are representative of the power-supply systems in use. In any system different possibilities exist for improving one characteristic at the expense of others. However, the vast majority of practical systems incorporate the principles and characteristics represented by the power supplies described in this section.

9-5.1 Simple A-C/D-C System

The basic a-c/d-c system (Fig. 9.1) consists of a single voltage-regulating and harmonic-filtered stepdown transformer and a static rectifier. This system is usually fed from the facility essential auxiliary-power buses and, of course, is subject to interruptions in the order of 15 sec before the large standby diesel generators can reestablish power to the essential buses. At best the simple a-c/d-c system provides filtering and voltage regulation for noncritical instrumentation loads. When the system is required to supply only d-c power, the regulating transformer is usually incorporated within the static rectifier.

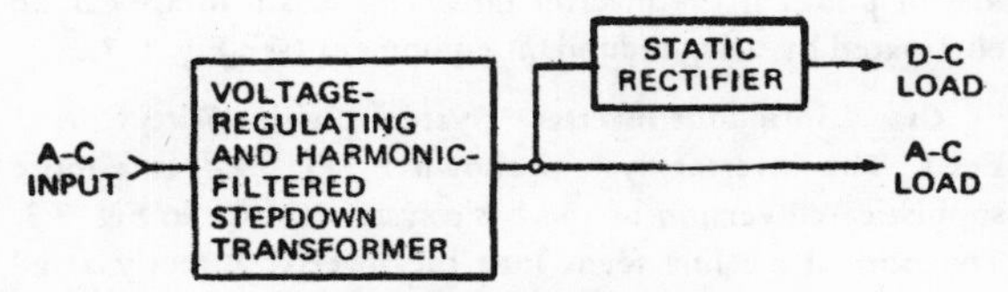

Fig. 9.1—Simple a-c/d-c system.

9-5.2 Rectifier–Battery System

The basic system (Fig. 9.2) consists of a static rectifier, a battery charger, and a battery. The normal d-c load is supplied from the rectifier. If the a-c line fails, the load is automatically transferred to the battery without interruption. Although the system is highly reliable, the capacity of the batteries limits the length of possible emergency operation. Of course, the system can only serve d-c loads.

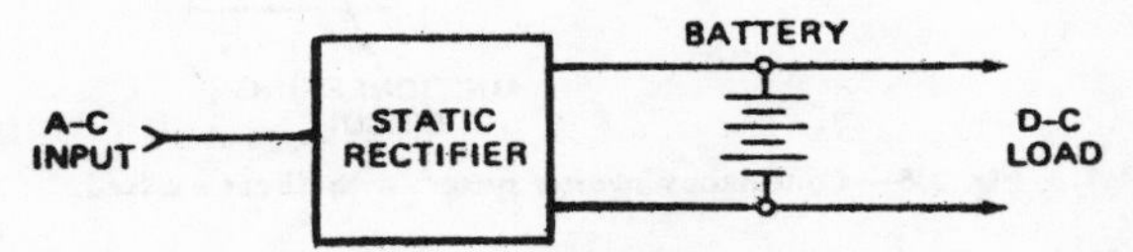

Fig. 9.2—System using static rectifier and battery.

9-5.3 Rectifier–Battery–Static Inverter Systems

(a) Basic Continuous-Inverter System. The basic continuous-inverter system (Fig. 9.3) consists of a static rectifier, battery charger, battery, and static inverter. The inverter carries the a-c load at all times. Under normal

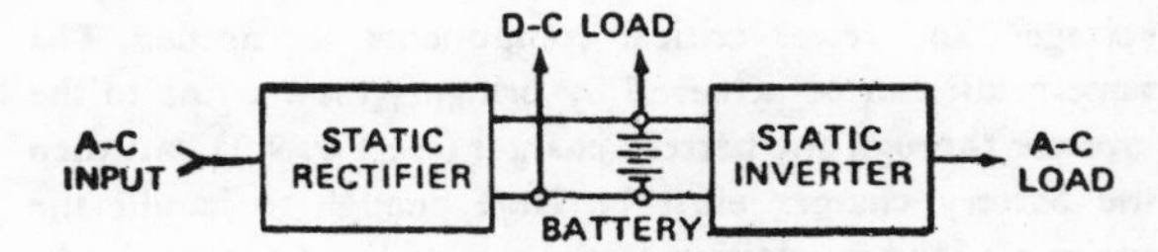

Fig. 9.3—Basic continuous-inverter system.

circumstances the charger provides the d-c load plus the input to the inverter and floats the battery or recharges it as required. The charger in turn is fed from the plant auxiliary source of a-c power. If the plant auxiliary power fails, the inverter continues to run on the battery for a period of time dependent upon the reserve battery capacity. In this mode of operation, as long as the charger is functioning properly, there is no drain on the battery. Input current for the inverter is derived from the charger and does not pass through the battery. If the a-c line fails, the inverter drain is transferred to the battery without any interruption or disturbance in the inverter output. Thus this simple system functions as a complete no-break backup power source. Some capacitance is generally required in the input circuit of the inverter, and this provides a small amount of energy that, because there is a fast voltage regulator in the inverter, enables the system to handle the input transient from float voltage on the battery to discharge voltage without a significant transient in the output. Loss of voltage on the input of the rectifier has no effect on the inverter output. The advantages of this system are offset by the complete loss of power if the inverter fails. This disadvantage can be eliminated by using redundant equipment (see Fig. 9.7).

(b) Continuous-Inverter System with Direct A-C Feed. The inverter system shown in Fig. 9.4 is a more sophisticated version of the basic system shown in Fig. 9.3. The normal a-c line feeds into the inverter directly at all times and is rectified. The resulting value of the d-c voltage is compared to the d-c voltage from the battery. Solid-state circuitry within the inverter determines the highest d-c source, battery or rectified a-c line, and then inverts it to supply the a-c load. This is commonly referred to as

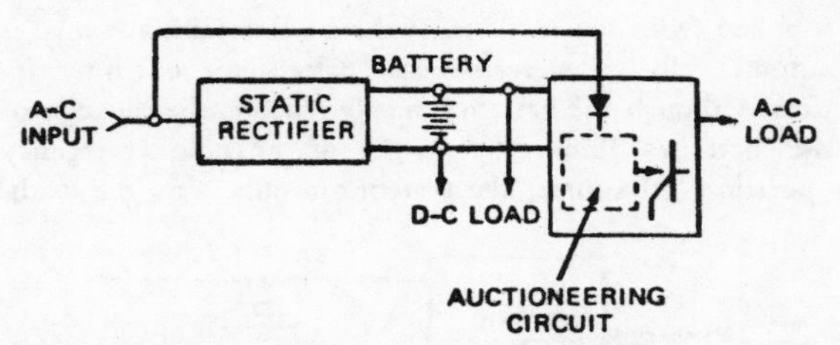

Fig. 9.4—Continuous-inverter system with direct a-c feed.

auctioneering. The internal transfer between the two sources of d-c power occurs instantaneously and is virtually undetectable. In this way the a-c load never sees the incoming a-c line, only the a-c output from the static inverter. Line synchronization problems, which are often a cause of difficulty (when external switching methods are used to effect the transfer between the inverter output and plant auxiliary a-c line), are avoided. Reliability is enhanced because there are no switching operations to cause transient voltages, and fewer critical components are needed. The same result can be achieved by bringing the a-c line to the inverter through the battery charger (see Fig. 9.3), but then the battery charger must be large enough to handle the entire a-c load in addition to the d-c load and battery loads. This would necessarily increase system cost.

Another inherent advantage of the system where the a-c line is fed into the inverter directly is built-in stabilization of line frequency and voltage. Since the incoming a-c supply is rectified at once, input frequency is of little concern. The output oscillator of the inverter can have frequency stability to almost any accuracy desired, being only a function of inverter design. The typical 60-Hz system maintains frequency to a tolerance of ±1.0%. The addition of an oscillator standard in the inverter can reduce the variation in output frequency to ±0.01%. The input frequency does not necessarily have to define the inverter output frequency, and therefore the same power supply can be used for frequency conversion. Similarly, phase changing can also be accomplished since the number of phases in does not determine the number of phases out.

(c) Continuous-Inverter System with Electromechanical Transfer Switch. In certain applications it is desirable to feed the a-c load from a source other than the inverter. This can be accomplished by switching from inverter to line (see Fig. 9.5) or switching from line to inverter. The two methods differ in the length of interruption of a-c power to the load and should be used where short-term interruptions can be tolerated. Conventional transfer switches are used and are generally supplied as electrically held, mechanically interlocked contactors. In the commoner mode of operation, the inverter is considered as the normal source. The inverter carries the a-c load until it is manually transferred by an operator or until an inverter failure occurs. The chief advantage of this arrangement is that the inverter failure rate is substantially lower than that of the plant auxiliary-power source. Thus transfers occur substantially less often than they would if the a-c line were considered the normal source. Since the inverter is operating continuously, there is assurance that both sources are available as long as the a-c line is present.

A disadvantage of the inverter-to-line switching mode of operation is that the charger must have sufficient capacity not only to feed any d-c loads and recharge the battery but also to supply the input current to the inverter. In the alternate arrangement, where the plant auxiliary a-c line is considered the normal source and transfers are made to the inverter on-line failure, the charger capacity need only be sufficiently greater to recharge the battery and feed any d-c loads.

In most transfer arrangements of this type, a delay is provided to prevent the emergency from transferring to the normal source after an outage until the normal source has been present for some period of time. This time interval ranges from a few seconds to as much as several minutes, depending on the application.

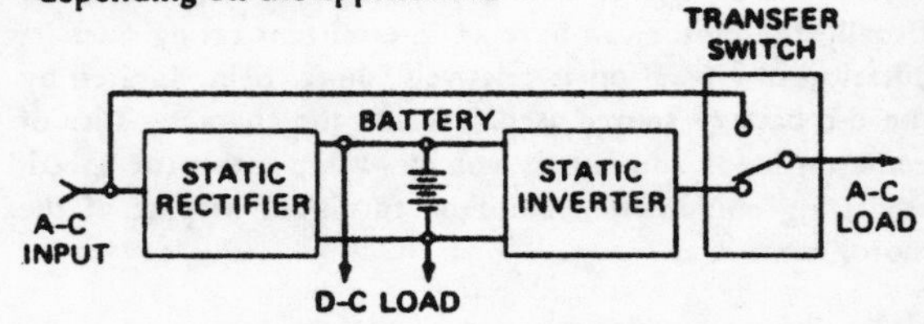

Fig. 9.5—Continuous-inverter system with electromechanical transfer switch.

With standard contactors operated in a conventional manner, outage times on transfer are of the order of 0.1 sec. The outage time depends primarily on the time for failure sensing and transfer of the contactor. In switching from line to inverter, however, an additional period of reduced output voltage is caused by the response of the inverter to a step-load change. For most commercially available inverters, the time to respond to a full-load step change is less than 0.04 sec. Because of the relatively long transfer time, it is generally not necessary for the inverter to be synchronized in phase with the a-c power line. Frequency synchronization may be desirable where the inverter carries the load continuously to keep certain classes of loads, such as clocks and chart drives, in step with local time.

In certain applications the transfer from normal to emergency a-c power source need not be automatic. By substituting a make-before-break manual transfer switch for the automatic transfer switch indicated in Fig. 9.5, the load can be successfully removed from the inverter without loss of potential. The manual make-before-break transfer is an inexpensive concept of switching to the plant auxiliary source of power while still maintaining the continuity of power flow to the load. This system suffers the same problems as the system shown in Fig. 9.5. The load experiences complete loss of power should the inverter fail. (One could, however, manually transfer to the plant auxiliary a-c source after detection of the inverter failure.)

(d) Continuous-Inverter System with High-Speed Transfer Switch. In certain cases extremely sensitive a-c instrumentation and control systems cannot tolerate the finite outage time given by the transfer arrangement of Fig. 9.5. The system shown in Fig. 9.6 uses special transfer-switch drive circuitry that allows transfer to be effected in less than 1 cycle, or 0.016 sec, in a standard 60-Hz system. Depending on the sensitivity of the sensing circuitry and

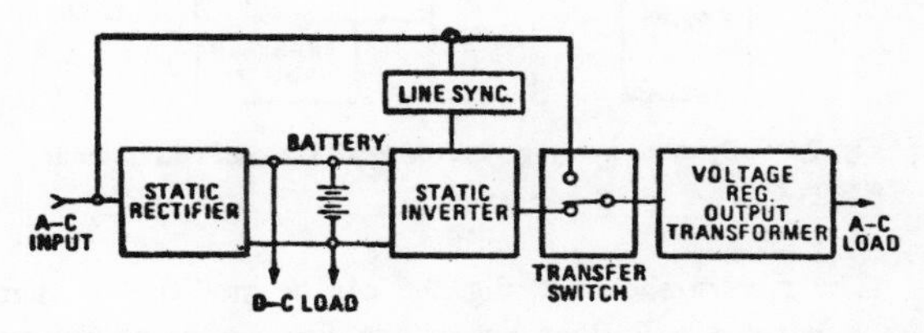

Fig. 9.6—Continuous-inverter system with high-speed transfer switch.

the size of the transfer switch, transfer times below 0.008 sec can be obtained. In all short-time transfer schemes, special attention must be given to line-phase synchronization and transformer saturations.

Because of the short transfer time, the inverter must be operated in phase synchronism with the commercial power line. It must also be recognized that the transfer time from line to inverter is increased by the load response time of the inverter. For these reasons it is generally preferable to use the inverter as the normal source of power and transfer to the line only if an inverter fails. Retransfer, after correction of the inverter failure, can be made either by allowing for the increased transfer time or by providing an auxiliary manual make-before-break switch which momentarily parallels the inverter output with the plant auxiliary-power system on retransfer.

To achieve a high-speed transfer, the sensing circuit must detect departure of the source from its standard value in periods as short as a millisecond. Most plant auxiliary-power systems experience short-term transients. It is extremely difficult to design transfer-sensing circuits that avoid transferring on a momentary line transient that is not necessarily followed by a complete line failure. The inverter output is relatively free of such spikes unless they are generated by the load, and therefore the likelihood of false transfer is avoided when operating with the inverter as the normal source. However, most inverters have a current-limited output characteristic; so any overload exceeding the output capability of the inverter is regarded as a failure and causes transfer to the plant auxiliary-power system.

In several available commercial systems of the type shown in Fig. 9.6, the high-speed electromechanical transfer switch is replaced by a solid-state silicon-controlled rectifier (SCR) a-c switch. The advantages gained include a decrease in transfer time down to 0.002 sec or less and, since the switch has no moving parts, a virtual elimination of maintenance requirements. A disadvantage is that the static transfer switch does not provide the same degree of positive isolation from the plant auxiliary system as does the mechanical transfer switch. Therefore, in a situation where a plant auxiliary-power-system failure is accompanied by a high-voltage transient, the static switch could fail and cause a complete system breakdown involving both a portion of the plant auxiliary power and the inverter. An important consideration when assessing the relative merits of a static vs. electromechanical transfer switch is that the inverter response time must be added to the transfer time when transferring from line to inverter.

The system shown in Fig. 9.6 uses the stored energy of the output transformer to overcome the relatively slow response of the inverter and provides improved transfer-time characteristics. By transferring on the primary of the ferroresonant output transformer with a high-speed transfer switch, one can achieve transfer times in the range of 0.008 to 0.016 sec. Because of the stored energy in the output transformer, there is no interruption of supply to the load during the transfer period. As previously mentioned, the inverter must be operated in phase synchronism with the plant auxiliary-power system to effect the desired uninterrupted transfer.

The system indicated in Fig. 9.6 requires the inverter to be the normal source of power. Several advantages result from this mode of operation. The frequency of transfer during operation is substantially reduced since the inverter output is comparatively clean. During plant-auxiliary-power-system fault conditions, when transfers from the line normally take place, the plant auxiliary power is characterized by erratic phase shifts and voltage excursions, and, if the transfer from source to inverter is to be successful at these times, the inverter must be maintained in phase with the faulted line. It is questionable if the inverter could respond properly to this mode of operation. Either the transfer would be made out of phase or the inverter would malfunction. In addition, the voltage transients characterized by the failing a-c line may make clearing of the transfer-switch contacts difficult. For example, a sufficiently large voltage transient on the line prior to switching could cause an arc to persist in the opening contact of the transfer switch for more than 0.008 sec. Since this would, in essence, short the inverter output to the failing line, an inverter malfunction would be likely. Using the inverter as the normal source avoids these undesirable possibilities.

(e) Continuous-Inverter System with Redundant Inverter and Transfer Switch. In the systems shown in Figs. 9.5 and 9.6, the switching is from the inverter back to the commercial a-c line to provide a backup source of power. The system shown in Fig. 9.7 represents a significant improvement with respect to isolation from the plant auxiliary a-c line. During normal operation both of the inverters are operated in parallel and are sized so that either could carry the entire a-c load. The two inverters are connected through normally closed transfer switches to the common a-c load. The logic and synchronizing circuits ensure that under all circumstances the inverters are operating in phase synchronism with each other and with the commercial a-c line if required. As long as both units are operating in phase with each other, the load is shared. Should either inverter fail, by either a reduction in output voltage or a shift in phase, the logic circuit disconnects the faulty inverter from the system, thereby transferring the entire load to the remaining inverter. The transfer switch need not operate with extreme rapidity since the surviving inverter can drive the output transformer of the failing inverter without adverse effects. The failure of either inverter would cause some load disturbance attributable to

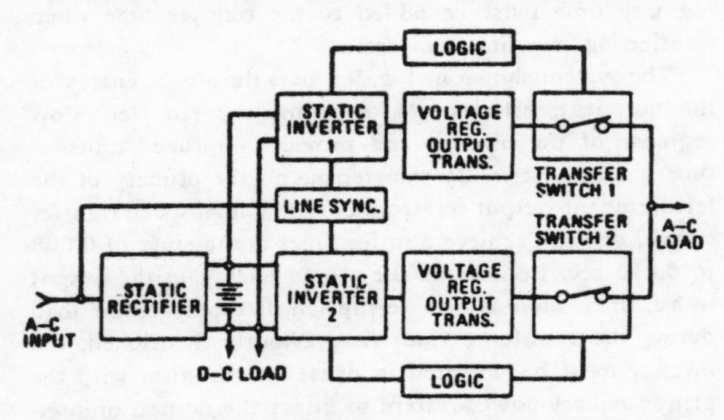

Fig. 9.7—Continuous-inverter system with redundant inverter and transfer switch.

the response of the surviving inverter to the 50% step-load increase (assuming each inverter to be normally 50% loaded). The overall output performance of this system is, therefore, essentially identical to that of the system represented by Fig. 9.6. The principal advantage is that both sources of power can be considered extremely reliable. Since there is no connection to the commercial power line except that provided by the charger, the possibility of introducing large voltage transients from external sources is minimal.

It is important to realize there is a substantial increase in cost as the complexity of the transfer circuits is increased. The transfer-circuit complexity increases as the permissible load-interruption time decreases. Usually the power source for the instrumentation and control system in a nuclear reactor facility is only one part of a larger system. In fact, because redundancy is necessary in the instrumentation and control system, several separate isolated power systems may be required. Therefore, in overall operation an inverter failure may be indistinguishable from a failure in a portion of the instrumentation and control system that the inverter feeds. In such circumstances justification of an elaborate (and expensive) power supply must take into account that the additional cost might better be used to improve some part of the system other than the power source.

9-5.4 Generator and Internal-Combustion-Engine System

In the system using a generator and an internal-combustion engine (Fig. 9.8), the normal flow of power is from the plant auxiliary a-c system. If the commercial power fails, the internal-combustion engine is started. As soon as the proper voltage and frequency are established at the generator terminals, the transfer switch connects the load to the generator. This system is widely used in nuclear power plants as the emergency auxiliary-power source. The main disadvantage of this system is that the load is interrupted for the time interval required to start the engine and transfer the load to the generator; typically this is 10 to 15 sec.

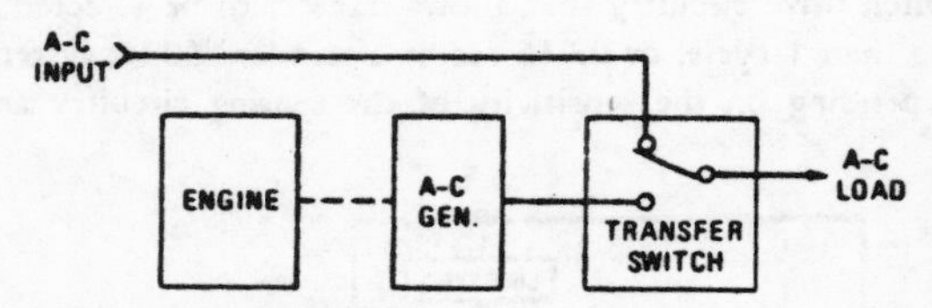

Fig. 9.8—System using generator and internal-combustion engine.

The system shown in Fig. 9.8 can be modified so that the generator will float across the line by being driven continuously by the engine. Whenever the normal a-c source failed, the generator would deliver power immediately without interruption. With this scheme the engine starting period is eliminated. However, the engine runs constantly, and transients are present on the line when switching. Power directional relays and synchronizing equipment are required for proper operation of the transfer switch. Since the engine is running continuously, increased maintenance and operating costs are involved. This is a distinct disadvantage.

9-5.5 Synchronous Motor-Generator–Flywheel–Clutch–Internal-Combustion-Engine Systems

(a) Nonisolated System. In the nonisolated system shown in Fig. 9.9, the critical load is normally fed directly from the plant auxiliary a-c power system in parallel with a synchronous machine operating as a motor driving a flywheel. Whenever normal plant auxiliary power is interrupted or a frequency or voltage anomaly in excess of preset tolerances is experienced, the synchronous motor and critical bus are disconnected from the system. The synchronous motor instantaneously converts to generator operation to supply the critical bus with interim power, with the stored energy of the flywheel being transferred to drive the generator. The engine, when furnished, would

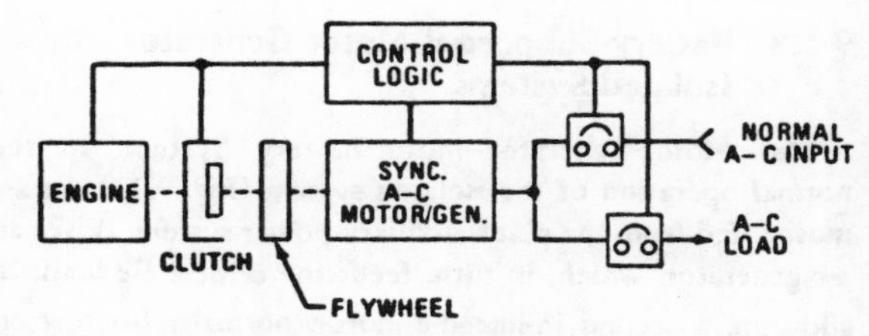

Fig. 9.9—Nonisolated system using synchronous a-c motor and generator, stored-energy flywheel, clutch, and internal-combustion engine.

simultaneously be started and then connected to the load when it is up to speed. The engine is recommended only for those systems requiring operating time, after plant auxiliary-power failure, in excess of the stored-energy capability of the flywheel.

Under fault conditions this type of system is subject to a power dip during the time required to isolate the critical a-c load from the plant auxiliary source. In addition, since the critical load is normally fed from the plant auxiliary source, it is subjected to any transients occurring on that system. At best this system (with the engine) is justified only where the plant auxiliary source is very unreliable.

(b) Isolated System. The isolated system shown in Fig. 9.10 offers a significant improvement over the nonisolated system (Fig. 9.9) in that complete isolation from the plant auxiliary source is obtained. The system consists of a synchronous motor with a stored-energy flywheel unit, which is fed from the plant auxiliary source and drives a synchronous generator that feeds the critical bus. A standby engine is used whenever the duration of the outage exceeds the capability of the inertial unit. During normal operation voltage- and frequency-sensing devices monitor the incoming power line for variations beyond acceptable limits.

Should an unacceptably large voltage or frequency excursion occur, the synchronous motor is disconnected from the plant auxiliary-power source, and the stored

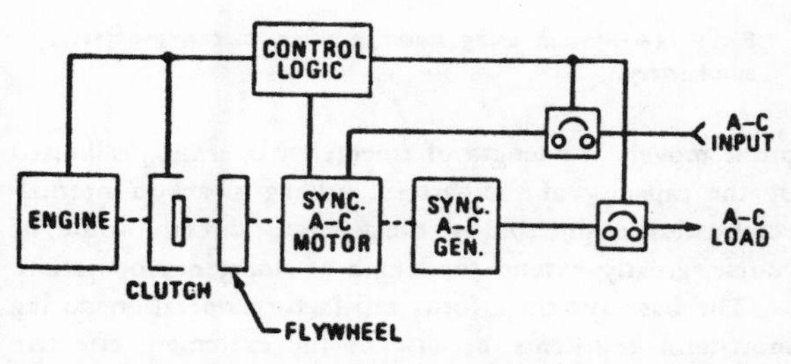

Fig. 9.10—Isolated system using synchronous a-c motor and generator, stored-energy flywheel, clutch, and internal-combustion engine.

energy in the flywheel is used to drive the synchronous generator. The standby engine, if furnished, is simultaneously started and brought up to synchronous speed in about 10 to 60 sec, depending on the stored-energy content of the flywheel, at which time the engine is connected to the generator shaft by the magnetic clutch. The frequency stability under engine operation is maintained by a highly sensitive load- and frequency-sensing governor that closely controls the speed of the engine. Should the voltage and frequency of the plant auxiliary power return to acceptable values and remain for a preset time period, the synchronous motor is synchronized to the source, the clutch is deenergized, and the engine is returned to standby condition.

On loss of plant auxiliary power, the system begins to draw energy from the flywheel and causes it to lose speed. Since the frequency is directly related to the speed of the flywheel-driven motor–generator unit, the frequency is soon reduced to a value below an acceptable limit to the critical bus.

The overall frequency regulation of this system is equal to that of the plant auxiliary-power system.

9-5.6 Induction Motor-Generator—Stored-Energy Eddy-Current-Coupling—Internal-Combustion-Engine System

In normal operation of the system (Fig. 9.11), an induction motor drives a flywheel at a higher speed than the speed of the generator shaft. The induction motor–flywheel system is coupled to the generator shaft through an eddy-current coupling. The variable slip provided by the coupling allows the generator to be maintained at synchronous speed under all normal load conditions (see Sec. 9-4.3). When the frequency or voltage of the plant auxiliary-power system deviates from preset limits, it is interrupted. Through the use of speed sensing and slip control of the eddy-current coupling, the generator is maintained at synchronous speed as the flywheel speed drops to that approaching the generator synchronous speed. Simultaneously, the incoming plant auxiliary-power system voltage- and frequency-sensing relays determine whether a momentary transient or a complete loss of power has occurred. At a predetermined flywheel speed, the standby engine is started and brought up to speed, but it is not connected to the system at this time. If the speed of the flywheel continues to drop and reaches a preset minimum value prior to the return of the plant auxiliary-power system, the engine is connected to the system by energizing the magnetic clutch. The energy storage capability of this system varies between 10 sec and a maximum of approximately 2 min. If the allowable outage time for which protection is desired is less than 2 min, the engine standby unit can be eliminated.

This system gives excellent results in speed control, output voltage, and frequency regulation, and, when combined with the engine backup, it is able to operate for extended periods of time. However, eddy-current-coupling systems are inefficient and generate considerable heat that has to be dissipated. Other disadvantages include high maintenance cost and, when combined with an engine backup, the requirement for fuel storage, exhaust ventilation, and scheduled exercising.

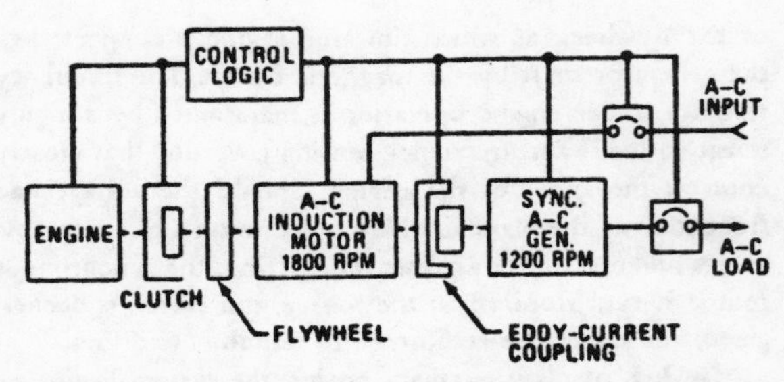

Fig. 9.11—System using induction motor, synchronous generator, stored-energy flywheel, eddy-current coupling, and internal-combustion engine.

9-5.7 Synchronous Motor-Generator—Stored-Energy Eddy-Current-Coupling—Internal-Combustion-Engine System

This system (Fig. 9.12) consists of a synchronous motor, fed from the utility power system, which drives a synchronous generator, which, in turn, feeds the critical a-c load. Simultaneously, a small induction motor is driving a flywheel at a speed considerably in excess of the synchronous generator speed. The flywheel is not coupled to the generator under normal operation since the eddy-current coupling between the generator and flywheel is not energized. When an unacceptable excursion in voltage or frequency occurs on the plant auxiliary-power system, the power-source feed is disconnected and the eddy-current coupling is energized, thereby coupling the flywheel to the generator. The stored energy of the flywheel then serves to drive the generator at synchronous speed in the same manner as described for the system shown in Fig. 9.11. The present system (Fig. 9.12), by excluding the eddy-current coupling from being energized during normal operation, operates at a much greater efficiency than the system shown in Fig. 9.11.

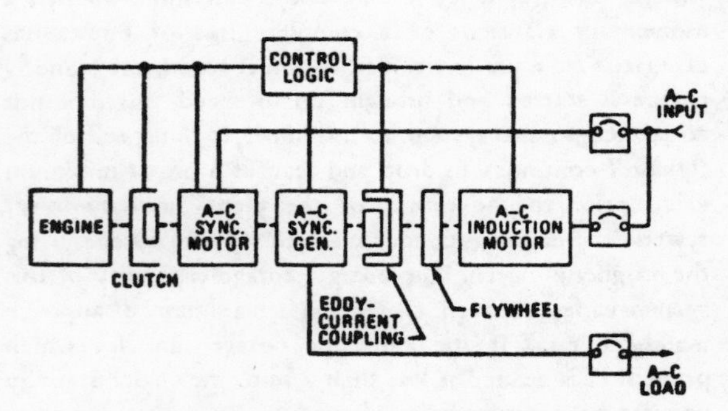

Fig. 9.12—System using synchronous motor-generator, induction motor, stored-energy flywheel, eddy-current coupling, and internal-combustion engine.

With the exception of improved efficiency, this system has the same characteristics as the system described in pt. 9-5.6, and, in addition, the frequency excursion of the generator output during transfer from normal utility supply to flywheel operation may exceed acceptable limits.

9-5.8 Battery-Supported Motor-Generator Isolated Systems

(a) Motor-Generator—Motor-Battery System. In the normal operation of the isolated system (Fig. 9.13), an a-c motor, fed from the plant auxiliary-power system, drives an a-c generator, which, in turn, feeds the critical a-c load. In addition, a second in-line d-c motor, normally floating on the battery system, is instantaneously available to drive the system if the plant auxiliary-power system fails. This system suffers from the ills common to all systems with rotating equipment, including increased maintenance and wear, when compared to static systems. In addition, the duration of operation after the failure of the utility source is limited by the battery capacity.

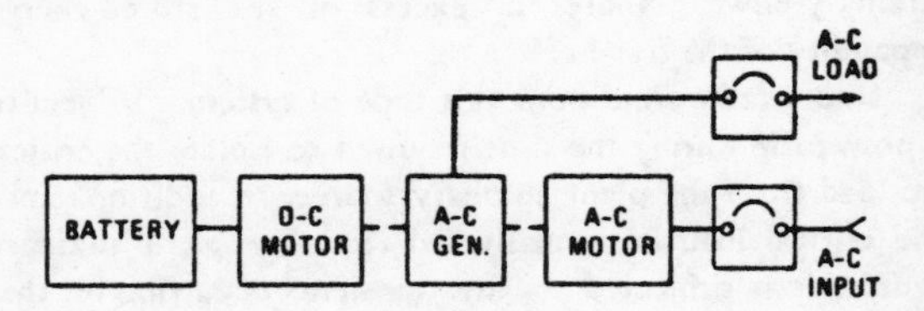

Fig. 9.13—System using motor-generator and motor and battery.

(b) Static Rectifier—Motor-Generator—Battery System. During normal operation of the system (Fig. 9.14), utility-line power is rectified and applied to the d-c motor that drives the a-c generator supplying power to the critical a-c load. The rectifier is sized to accommodate any normal d-c load in addition to the power required by the d-c motor and the power needed to maintain the battery at full charge. On failure of the plant auxiliary-power system, the d-c motor is supplied with power from the batteries, thereby maintaining the continuity of the a-c generator

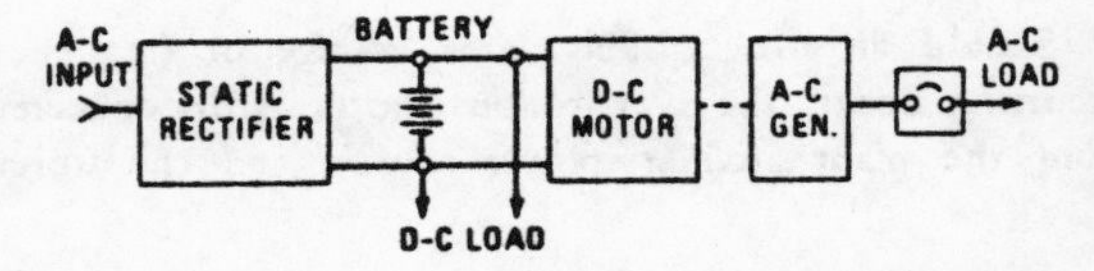

Fig. 9.14—System using static rectifier, motor, generator, and battery.

prime mover. The length of emergency operation is limited by the capacity of the battery. Adding a backup internal-combustion engine to drive the generator directly would, of course, greatly extend the length of emergency operation.

The base system affords satisfactory operation during short-term transients because of the extremely effective filtering action of the battery and motor—generator combination.

9-6 CONCLUSIONS

The material of this chapter aids in selecting and specifying high-reliability power sources. Only the major electrical systems have been discussed because they are the most commonly used in nuclear power reactor plants.

In determining which type of system is best, an important aspect is responsibility. A given system may be designed and specified and the component parts purchased and assembled by the purchaser. The purchaser thereby assumes the responsibility for satisfactory system operation.

An alternative procedure, and one that usually guarantees satisfactory results, is to specify the required parameters of the power source. The system supplier then submits a quotation and assumes the responsibility for system operation to meet the specifications. This latter procedure is recommended.

BIBLIOGRAPHY

Bleikamp, R. P., Load Factors in Selection of Eddy-Current Drives, *Elec. Mfg.*, 63(4): 92–98 (April 1959).

C & D Batteries, Division of Eltra Corporation, Longer Life For Lead Acid Stationary Batteries, Technical Bulletin RS-15810M, November 1, 1964.

Dunsmore, C. L., Integrated Emergency Power Supply For Nuclear Plants, *Power Eng.*, 72(8): 39–41, 89 (August 1968).

Everson, H. K., Uninterrupted Electric Power Systems Utilizing A DC motor As Emergency Drive, *IEEE Transactions on Aerospace Support*, pp. 1371–1384, Institute of Electrical and Electronics Engineers.

Farber, J. D., and D. C. Griffith, Static Inverter Standby AC Power for Generating Station Controls, IEEE Paper 31PP67-15, Institute of Electrical and Electronics Engineers.

General Electric Company, Static Inverters and SCR Regulated Battery Chargers, *Technical Bulletin GEA-7522b.*

General Electric Company, Static Uninterruptible Power For Critical Loads, *Technical Bulletin GEA-8631.*

Gould-National Batteries, Inc., *Gould Invert-A-Stat, DC to AC Static Inverter*, Technical Bulletin.

Grooms, F. H., and P. D. Wagner, Which Standby Power System Is The Best For You?, *Hosp. Management* (June 1963).

Hoxie, E. A., Some Discharge Characteristics Of Lead–Acid Batteries, AIEE Paper 54-177, presented at AIEE Winter General Meeting, New York, Jan. 18–22, 1954, American Institute of Electrical Engineers.

Ideal Electric and Manufacturing Company, Eddy Current Coupling Variable Speed Drives, Bulletin 100.

Jackson, S. P., Application of Static Inverters In Control And Instrumentation Systems, *IEEE Transactions on Industrial Electronics and Control Instrumentation*, Vol. IECI-13, No. 1, Institute of Electrical and Electronics Engineers, April 1966.

Jackson, S. P., Standby Power, *Instrum. Contr. Syst.*, 39(6): 135 (June 1966).

Jackson, S. P., The Use of Static Inverters in the Gas Industry, *Gas Mag.*, 41(12): 48–53 (December 1965).

Mueller, George V., *Alternating-Current Machines*, McGraw-Hill Book Company, Inc., New York, 1952.

Rubenstein, L., Precise Continuous Power, *Actual Specif. Eng.*, pp. 60–66 (August 1967).

Taylor, W. H., Reliable Power Packages For Switchgear Tripping Control And Emergency Diesel Engine Starting, AIEE Paper CP-62-484, presented at AIEE Winter General Meeting, New York, January 1962, American Institute of Electrical Engineers.

SECTION 10

INSTALLATION OF INSTRUMENTATION SYSTEMS

10-1 INTRODUCTION

The designers and constructors of the first nuclear power plants generally followed installation practices already established for conventional process and chemical industries. Adapting process instrumentation systems to nuclear power plants was reasonably successful. However, nuclear radiation systems had no conventional counterparts. During installation and preoperational tests in nuclear power plants, noise problems were found to be commonplace, and extensive modifications to the signal, control, power, and ground cables were required before nuclear radiation instrumentation systems could satisfy the established safety criteria.

Since the nuclear power plants being built today involve many different geometric configurations and a number of different basic materials, it is impractical to recommend a standard installation for all plants. However, the material presented here should provide engineers in the design and construction fields with a set of installation practices that will enable them to avoid many problems and pitfalls.

10-2 REACTOR INSTRUMENTATION SYSTEMS

10-2.1 Control Room

The control room in a nuclear power plant has many features in common with fossil-fueled generating stations. Many practices developed in fossil-fueled plants are directly applicable to nuclear plants.

The control room should be designed and installed so that it can be safely operated and occupied under any external-hazard condition, such as fire, smoke, contaminated atmosphere, flood, seismic disturbance, or major electrical fault.

An acceptable nuclear-power-plant control-room installation requires not only that equipment and components be integrated into a compatible system capable of the necessary overall performance but also that the human operator and his relation to this equipment be considered. If the man–machine relation is to provide maximum efficiency in operation, human factors must be considered as part of the initial engineering criteria. The control console and panel must be arranged so the reactor can be operated in a reliable manner with a minimum number of personnel.

Good design and installation practices dictate that all important variables in the plant operation be available for display and control in the station control room. Variables associated with reactor and heat-transfer control are generally installed on the main control console. Variables associated with the auxiliary equipment, such as the turbine and generator control, electric switchgear control, and several process-instrumentation control systems, are usually installed on the control panel.

A representative control-room installation is shown in Fig. 10.1. A plan view of this control room is shown in Fig. 10.2.

10-2.2 Control Console

The wraparound control console is widely accepted by the nuclear industry. The console shown in Fig. 10.1 is one example of good installation practices. The accessibility to instruments, switches, controls, and terminal boards at the rear of the console is excellent. The console is installed on a base structure that provides a step-down passageway behind the terminal boards and termination points at the rear of the control console. Access to this passageway is through the rear doors of the console. The passageway is wide enough to accommodate important test equipment, such as oscilloscopes.

In some nuclear power plant installations the main control console is incorporated into and made an integral part of the vertical control board (see pt. 10-2.3). Accessibility to instruments, switches, and controls is made at the rear of the vertical control board.

The following practices, based on observations of control consoles in several nuclear plants and on experience in nuclear-plant maintenance, are recommended:

1. Accessibility to the rear of the console must be provided for maintenance and test.

2. All field cabling coming into the control console and cabinets must be brought through a suitable dust seal, such as a penetration sealed by a compound. This provision will help to maintain the control room at a slightly higher pressure than ambient to prevent such hazards as fire, smoke, and noxious fumes from spreading into the control center.

3. Access must be provided to all components and electrical connections to facilitate maintenance (see also pt. 10-4).

10-2.3 Control and Monitoring Panels and Cabinets

The control panel is referred to as the "control board" or "vertical board" in the electric utility industry. A variety of control instruments, recorders, indicators, meters, dials, and knobs are installed on the control panels. The more familiar instrument systems include (1) switchgear and substation control, (2) turbine and generator control, (3) critical-temperature-measuring recorders and scanners, (4) process instruments for controlling and monitoring critical pumps and valves in the auxiliary heat-transfer loops, (5) emergency shutdown and monitoring for water reactors, (6) water treatment, and (7) annunciator alarm windows for all control panels involved. The neutron-monitoring instrumentation cabinets and drawers and the radiation-monitoring system may be installed as part of the control panel or in separate cabinets.

The depth of a control panel should be no more than required for easy access to all terminals and components at the rear. Avoid stacking instruments and conduit boxes behind panel-board instruments at the rear of the control panels. It limits access to terminals and components at the rear of the panels and can adversely affect the safety of plant operation.

Field cables entering the control panels either from floor conduits or from overhead trays should be bundled and installed in a manner that will not inhibit access to any instruments, components, controls, or termination points at the rear of the control panel. The installation of racks and cabinets is discussed in pt. 10-3.

Graphic panel installations already generally accepted for control in fossil-fueled power plants are being used in nuclear power plants, particularly for generator output bus, switch gear, substations, and heat-transfer systems. Opinions differ as to the effectiveness of graphic panels since the plant operator gets accustomed to the control knobs, lights, indicators, etc. (see pt. 10-6). However, a survey of operating personnel in several generating plants revealed a decisive preference for graphic panels in the central control room.

10-2.4 Nuclear-Instrument Systems

Slide-out drawers containing electronic circuits which are housed in some modular arrangement have been accepted as a standard for nuclear industry. Either a Nuclear Instrument Module System (NIMS) bin configuration is used or modules are removed from the top of the chassis. The installation and interfacing of nuclear instruments with the control and signal cabling coming into the cabinet must be done properly.

Very serious operating problems will affect the performance of the nuclear-instrument channels if there is

Fig. 10.1—Control room of the San Onofre nuclear generating station.

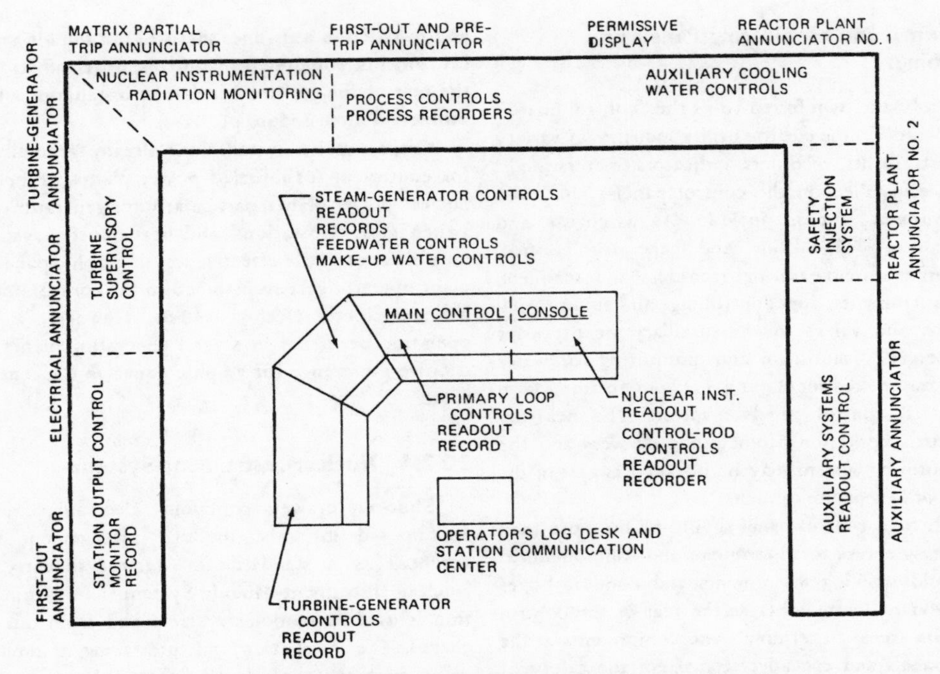

Fig. 10.2—Plan view of the control room at the San Onofre nuclear generating station.

improper installation and bundling of critical coaxial, triaxial, and multiconductor cables. Cables that are improperly installed frequently break off after they have been flexed a few times, causing open circuits. Adherence to the following practices will reduce this problem:

1. Avoid supporting a cable, wire, or bundle of wires by the terminal point. Good engineering practice provides a solid support fastened to the cable, wire, or bundle such that there is no stress on the terminal. Support as much of the cable length as possible by such mechanical means as cable retractors and springs.

2. Avoid using a single-point support. Distribute the support points over as wide an area as possible.

3. Mount the cable, wire, or bundle so that kinks do not develop. Use mechanical stiffening, such as nylon spiral wrap, wherever possible to prevent sharp bends.

10-2.5 Plant-Protection-System Cabinets

Cabinets for the plant protection system contain relays, solid-state devices, and other components that make up the logic circuits of the plant protection system. The plant protection system should be totally enclosed, either the cabinets themselves or the area in which open cabinets are located.

The equipment should be designed so that any component can be replaced or repaired without disturbing any other component. Relays and other remotely operated equipment should be accessible for authorized maintenance and troubleshooting and protected against unauthorized access. Each component should be clearly marked to prevent a mistake in identification.

The cabinet terminal blocks, used to interface the field wiring to the cabinet wiring, should be accessible to the incoming cables as well as to the internal wiring. The wiring on the terminal block should be arranged so that the internal wiring is terminated on one side of the terminal block and the field wiring on the other side.

Terminal blocks that do not contain field wires may utilize both sides of the terminal block for internal wiring, particularly where it is convenient to install a series of shorting bars. In installations where more than one row of terminal blocks is used, the internal wiring should be terminated on the terminals facing a common space between the terminal blocks, thereby leaving a space common to two rows of terminal blocks for the incoming cable terminations.

All terminal blocks installed in the plant-protection-system cabinets should be clearly identified by both block and terminal point.

10-3 INSTRUMENT RACKS AND CABINETS

10-3.1 Instrument-Rack Structure

The instrument-rack structure is defined, for the purposes of this chapter, as a structure in which the components of an instrument system are mounted. This structure may be a completely enclosed cabinet with the components mounted inside the enclosure or a panel with the components mounted in a cutout in the face of the panel. It may be a rack used to mount and support the equipment but which does not function as an enclosure for the components.

(a) Structural Materials. The structural materials used

in an instrument-rack structure must have sufficient strength to support all equipment mounted in the structure. Steel and aluminum, the materials most used for panel structures, are easily worked and readily available. The thickness of the structural material used for the front panel depends on the type and weight of the instruments being installed, the structural material used, and the panel design. Stiffening members may be required; however, they must not limit access to instruments and terminal points at the rear of the panels.

Since instrument systems are generally assembled at a vendor's plant rather than at the location where the system is to be used, the instrument-rack structure must have sufficient strength to allow the handling of the structure with all the equipment mounted. Lifting eyes should be provided to permit moving completed rack assemblies.

(b) Standard Modular Enclosures. Modular enclosures are made by several manufacturers to meet the requirements of both stationary and mobile instrumentation systems. They are designed to conform to an industry standard, such as the Electronic Industries Association (EIA) Mounting Standards, thus eliminating by the standardization of parts the need for many special mounting devices. Because of the modular design, a series arrangement of almost any configuration can be developed. Manufacturers of modular enclosures also make many special features and accessories for the enclosures, such as radio-frequency interference (RFI) shielding, equipment-cooling blowers, and special enclosure trims.

Modular enclosures allow flexibility in panel configuration and equipment layout (see Fig. 10.3).

(c) Mounting Practices. Instrument cabinets are generally designed to be free-standing structures which may or may not be fully enclosed. Instrument cabinets that are stationary are mounted on bases or curbs. The bases or curbs are either steel or concrete and are designed so that the instrument structure with all the equipment installed can be placed on them and attached and held in place by bolts or clamps.

A general practice is to provide each rack group with a

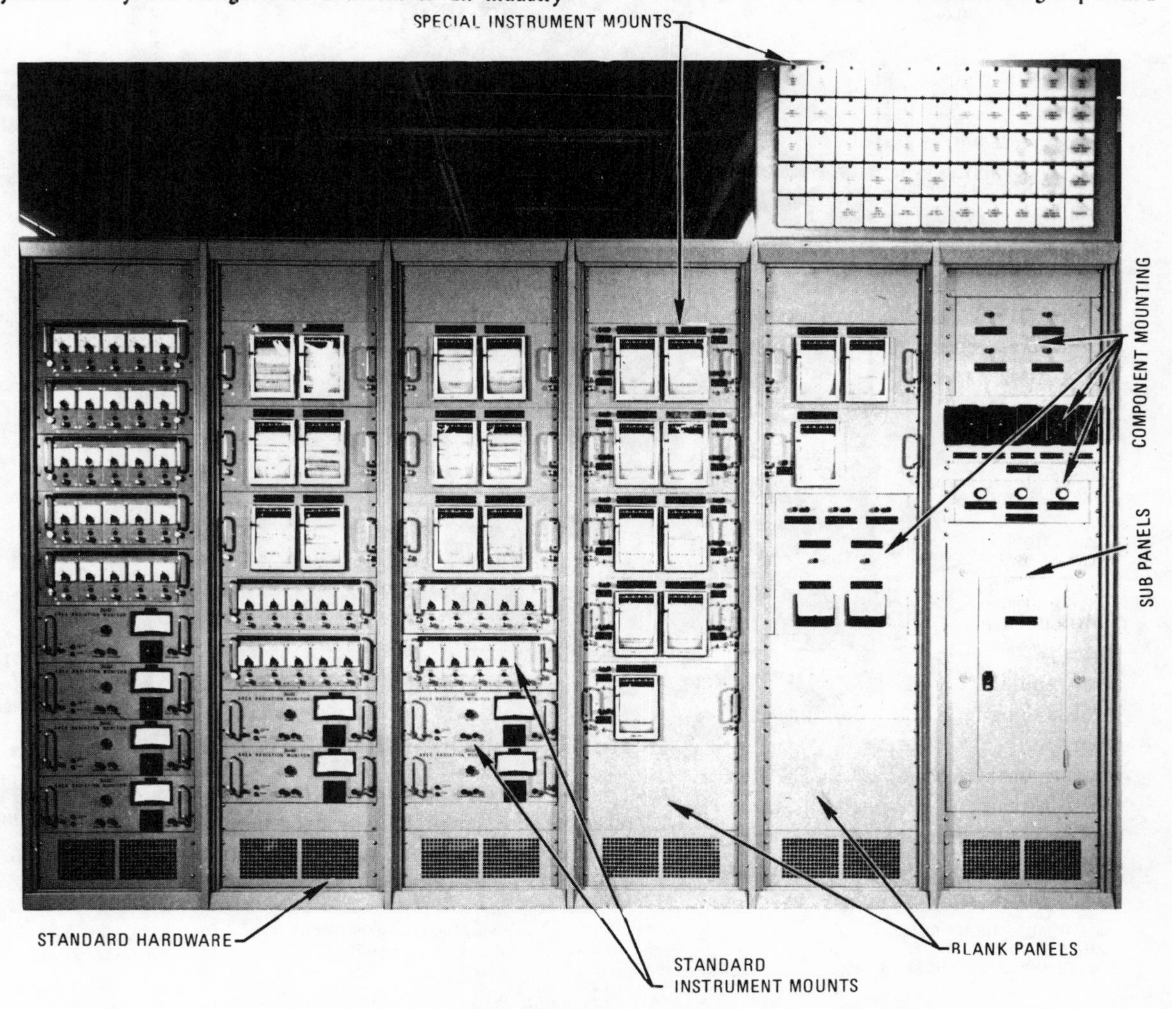

Fig. 10.3—Installation practices for standard modular enclosures. Note the flexibility of the panel configuration.

fabricated, 3- to 5-in. channel iron base. This allows for easy mounting to curbs with bolts and clamps sufficiently strong to withstand nominal seismic forces.

10-3.2 Fabrication and Assembly

(a) Mounting Major System Components. Each instrument and piece of equipment in the instrument-rack structure should be mounted, wired, or piped, where possible, so it can be removed without interruption of service to adjacent instruments and equipment. The instruments and equipment should be located and mounted so all wiring terminals and piping connections are readily accessible (see Fig. 10.4).

(b) Mounting Electrical Equipment and Hardware. The electrical equipment and hardware, as well as the installation of these items, should conform to some standard, such as the National Electrical Code. The ambient conditions at the location where the system is to operate and the type of system will determine which sections of the electrical code are applicable.

Terminal Blocks. Terminal blocks are arranged so that one side of each block is reserved for field connections.

Spare terminals (10% minimum) should be provided for future use and should be distributed throughout the terminal blocks. Where a field cable may fill all the spaces on a terminal block, the spares requirement should

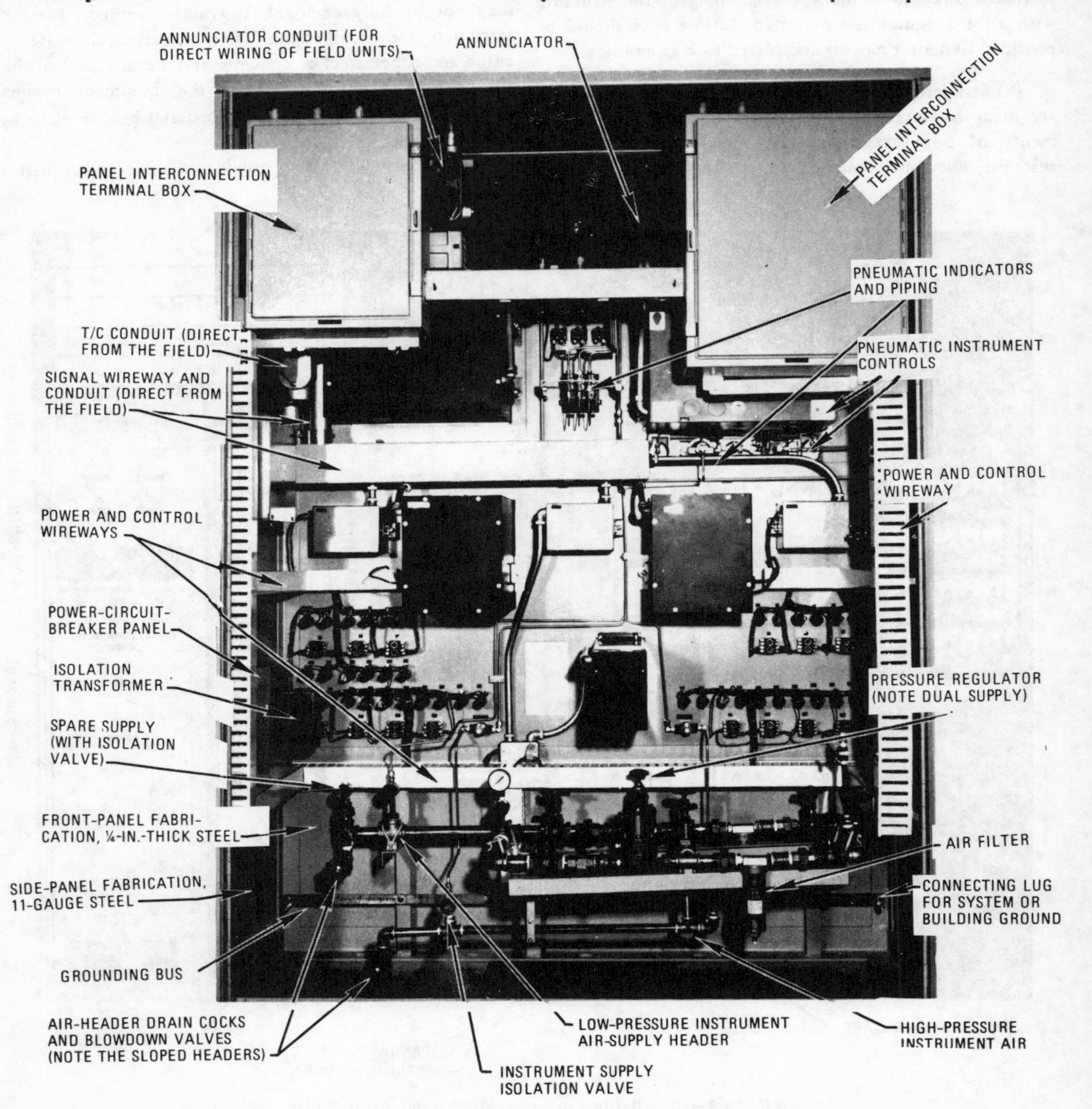

Fig. 10.4—Installation practices for major system components.

be overruled to prevent the need for splitting the cable between two terminal blocks. Terminal blocks should be located in the panels and cabinets to facilitate maintenance and testing without impairing access to other equipment mounted in the structure. Terminal blocks and terminal points should be identified and labeled.

Wiring Installation in Panels and Cabinets. The wire used for power distribution in the instrument cabinets should be adequate to carry the current used by the circuit. Design and installation engineers must adhere to the National Electrical Code in sizing wire for power-distribution circuits. No wire smaller than No. 14 American Wire Gauge (AWG) should be used in power-distribution circuits. Control circuits with less than 5 amp of maximum operating current may use No. 16 AWG copper wire. Wire sizes smaller than No. 16 AWG could handle the current requirement of most control circuits; however, mechanical strength becomes an overriding consideration, and these wires are not recommended for installation in panels and cabinets.

Architect–engineers and reactor designers have specified both stranded and solid wires for instrumentation and control applications in nuclear plants. For power and control circuits, stranded wire is preferred and is much more widely used than solid wire. The flexibility of stranded wire facilitates installation and maintenance. Either stranded or solid wire can be used without affecting the electrical characteristics or performance of the circuit.

All wiring should have a minimum of 600-volt insulation and should be resistant to heat, oil, moisture, flame, and corrosive vapors. Insulation materials have been developed which meet the requirements for switchboard and control-panel wiring without requiring braid or fibrous coverings. This results in a smaller overall diameter with fewer stripping and terminating problems.

All wiring connections in the instrument panels and cabinets should be made with preinsulated compression-type terminals unless a solder connection is required. For solder connections, insulated sleeves should be used to snugly cover the finished solder joint.

Wires entering or leaving the instrument cabinet should be terminated in terminal boxes to facilitate maintenance. However, some wires, such as coaxial, triaxial, and thermocouple lead wires, should be terminated through appropriate connectors directly to the instruments or thermocouple junction boxes.

Multiconductor or twisted-pair shielded cable should be used for analog signals (low-level, millivolt or milliampere) in instrumentation circuits. Wire no smaller than No. 18 AWG is recommended to minimize wire breakage during installation. Each conductor and the outer jacket or sheath of the shielded cable should have a flame-resistant insulation. The shield is carried as a separate conductor at all cable junction points.

The signal wires are run in wireways separate from the control and power wiring to minimize noise pickup in signal wiring. Separate terminal boxes are recommended for the signal and the power wiring. Lacing of low-level signal cables into bundles with power or control wiring should be avoided. Wiring in the instrument racks should not be spliced; each wire should run unbroken from terminal to terminal.

Wiring between panel-mounted instruments and terminal boxes should be grouped in a neat and orderly manner and run in enclosed metal wireways. Exposed wiring should be laced or bundled together with lacing, tie straps, or similar means.

Each wire should be properly identified. There are several coding or identification methods, such as nonconductive markers, color-coding the wires, and providing label identifications on the terminal blocks. Proper identification facilitates testing and maintenance. Wiring identification should correspond to that shown on the elementary and connection diagrams.

Terminations. The termination of conductors, whether they carry low-current signals or high-current power, is an important part of installation. Whether the termination is made by a simple "crimp-on" lug or a complex triaxial connector, good workmanship is of the utmost importance. Careful adherence to the manufacturer's mounting instructions, including the use of proper tools, can save many hours of troubleshooting and wire tracing.

Instructions and procedures for installing lugs and connectors on wire and signal cables are shown in Figs. 10.6 to 10.9. The most widely used hardware for terminating

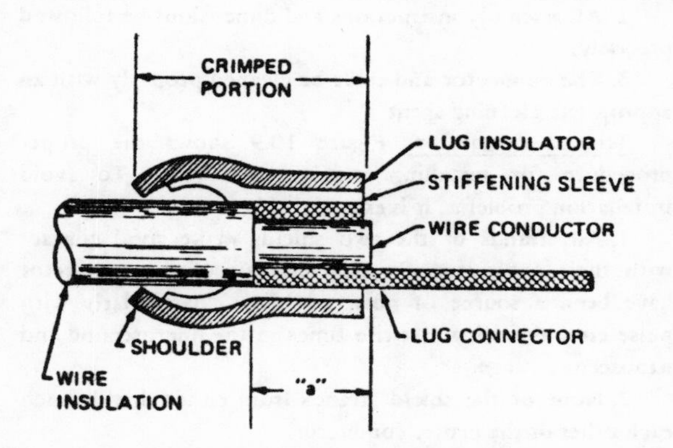

Fig. 10.6—Assembly of crimp-on lugs.

wiring and cabling is shown in these drawings. Recommendations are included on how to avoid common problem areas in the installation of connectors.

Crimp-on lugs. Figure 10.6 shows the proper procedure for installing crimp-on lugs. To avoid installation problems, it is essential that:

1. The proper type of lug (insulated or noninsulated, ring or spade, etc.) be used.

2. The proper size lug for the wire and terminal be used.

3. The insulation be stripped to the proper length (refer to "a" in Fig. 10.6). The conductor should be inserted completely through the lug with the insulation butted up against the shoulder and the conductor cut so that it protrudes just past the crimped portion of the lug.

4. The lug be properly crimped, using the proper crimping tool, with the wire conductor and insulator, where applicable, completely compressed to the lug. It is recommended that a fixed-release crimping tool be used. This tool assures proper crimping of the lug every time by not allowing release of the lug until the full amount of crimping pressure has been applied.

Standard Coaxial (BNC) Connector. Figure 10.7 shows the proper procedure for installing standard coaxial (BNC) connectors. To avoid installation problems, it is essential that:

1. All strands of the shield be free of the center conductor.

2. All strands of the shield make a good contact with the connector shell.

3. All dimensions on the assembly drawing be followed precisely so that the connector will fit together properly.

4. A good solder connection be made between the contact tip and the center conductor.

5. The connector and cable be cleaned properly with an appropriate cleaning agent.

Crimp-On Coaxial (BNC) Connector. Figure 10.8 shows the proper procedure for installing crimp-on coaxial (BNC) connectors. To avoid installation problems, it is essential that:

1. All strands of the shield make good contact with the connector shell.

2. All assembly instructions and dimensions be followed precisely.

3. The connector and cable be cleaned properly with an appropriate cleaning agent.

Triaxial Connector. Figure 10.9 shows the proper procedure for installing triaxial connectors. To avoid installation problems, it is essential that:

1. All strands of the two shields make good contact with their conductor. Strands left out of the conductor have been a source of noise problems, particularly with pulse circuits having fast rise times in the microsecond and nanosecond range.

2. None of the shield strands from either shield touch each other or the center conductor.

3. All assembly instructions and dimensions be followed precisely.

4. The connector and cable be cleaned properly with an appropriate cleaning agent.

(c) Mounting Pneumatic Equipment and Hardware. The pneumatic instrumentation system uses compressed air for the operation of the measuring devices, indicators and controllers, and final control elements.

Relatively trouble-free operation can be realized. In systems requiring 100% availability, a backup or dual system as shown in Figs. 10.4 and 10.12 should be used. The installation of an instrument air system should conform to a standard of the industry, such as the *American Standard Association Code for Pressure Piping*, A.S.A. B31.1.

Instrument Air Supply. In a pneumatic system the air is supplied by a compressor to a storage tank, and the system is supplied from the tank. The compressor and storage tank are sized so that the air usage of the system does not require continuous operation of the compressor. These items are generally located in a service equipment area. The air is then piped to the instrumentation-rack structure.

Condensation in a compressed-air piping system must be limited because moisture can damage instruments and make the system inoperative. Several air-drying techniques are available to remove the moisture from compressed air.

Desiccant dryer. A desiccant dryer is located in the piping between the compressor storage tank and the filter–regulator station. The dryer consists of two identical units; each unit has a desiccant chamber, check valve with a reduced-area bypass, and a solenoid valve, connected as shown in Fig. 10.10. Part of the dried air from the chamber in service is used to regenerate the other chamber; this amounts to about one-third of the dried air produced. Because additional air is required for regeneration of the system, the compressors must be sized to supply the regeneration air in addition to the air required by the instrument system. As the chamber drying the air becomes saturated, the chamber on the regeneration cycle is dried out. An electric timer controls the solenoid valves and periodically switches them, reversing the operating cycle of the system.

After-condensers. After-condenser air dryers are also located in the air line between the compressor-storage tank and the filter–regulator station. The after-condenser consists of a heat exchanger that uses water as a cooling agent. Figure 10.11 is a simplified diagram of a water-cooled moisture condenser. The use of chilled water for cooling increases the capacity of the condenser.

Filter–Regulator Station. A filter is located upstream of the pressure regulator and is used to remove foreign matter or contaminants from the air stream. The pressure regulator reduces the air pressure to the level required by the instrument system.

Where instrument air must be available to the system 100% of the time, a dual filter–regulator station is used. A typical dual station is shown in Fig. 10.12 in schematic form, and an actual installation, in Fig. 10.4. Each of the parallel filter–regulator stations is sized to handle the total requirements of the system. Isolation valves in each of the parallel piping arrangements allow either of the filter–regulator stations to be isolated from the system for repair and maintenance without shutting down the entire system. Each instrument using air is connected to the instrument air header through an isolation valve. Each instrument air header should have spare air takeoff points (10% minimum). The spare takeoff points should be equipped with isolation valves to allow the addition of new instruments to the system without requiring system shutdown. The header is sloped to the output, and so any condensation collects at the drain cock (see Fig. 10.4).

Pneumatic Signal Lines. The air supply and signal lines downstream of the instrument air header are plastic or

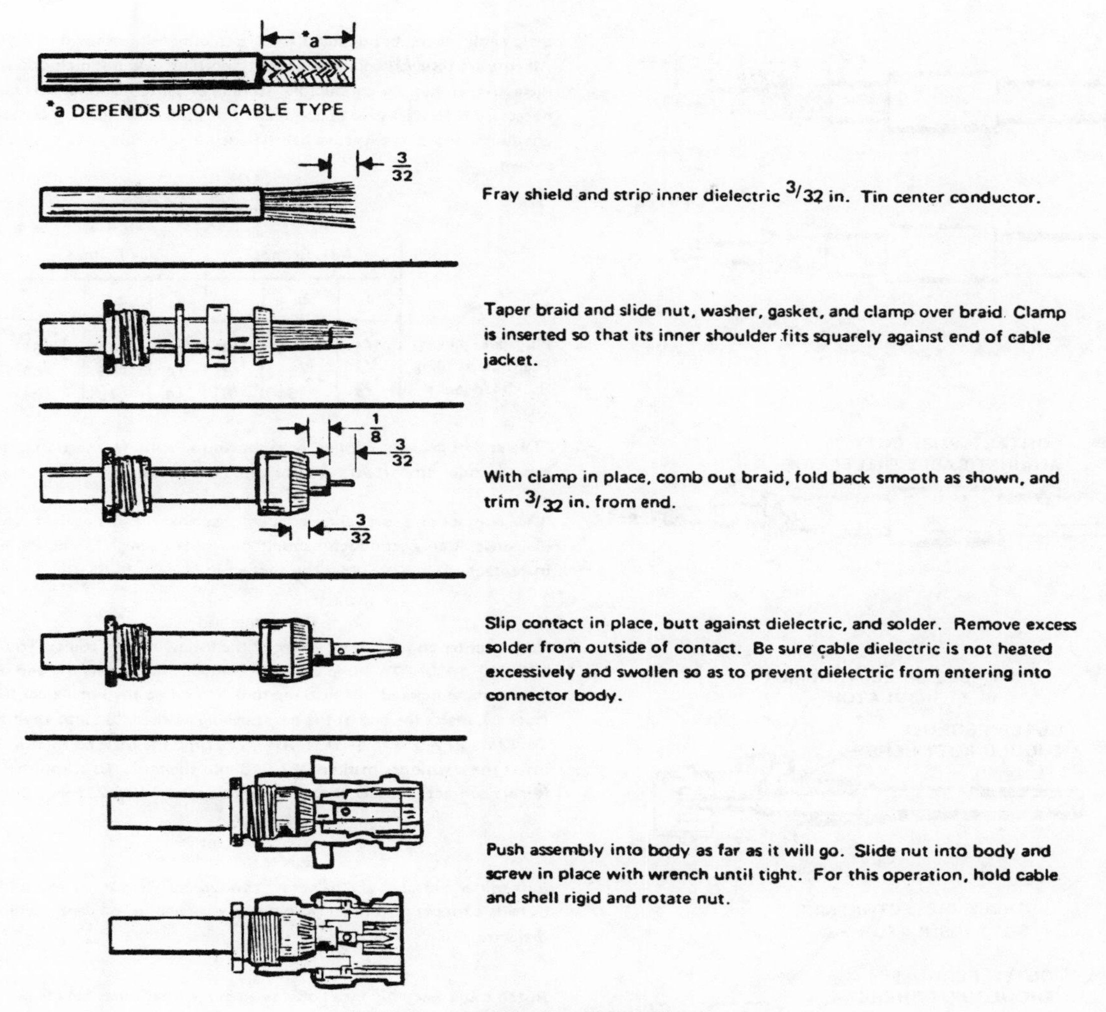

*a
*a DEPENDS UPON CABLE TYPE
3/32
Fray shield and strip inner dielectric 3/32 in. Tin center conductor.
Taper braid and slide nut, washer, gasket, and clamp over braid. Clamp is inserted so that its inner shoulder fits squarely against end of cable jacket.
1/8
3/32
3/32
With clamp in place, comb out braid, fold back smooth as shown, and trim 3/32 in. from end.
Slip contact in place, butt against dielectric, and solder. Remove excess solder from outside of contact. Be sure cable dielectric is not heated excessively and swollen so as to prevent dielectric from entering into connector body.
Push assembly into body as far as it will go. Slide nut into body and screw in place with wrench until tight. For this operation, hold cable and shell rigid and rotate nut.

Fig. 10.7—Assembly of standard coaxial connector.

copper tubing. Runs of this tubing should be straight, parallel, accessible, and logical with vertical runs plumb and horizontal runs dropping away slightly from the instruments. Tubing runs must be rigidly supported and fastened to the instrument structure or supporting braces. These installation requirements ensure that the tubing installation will not only have a pleasing appearance but also will be easily maintained.

Each instrument signal-input line should be equipped with an isolation valve and a test tee with a shutoff valve. This arrangement allows maintenance or testing without shutting down the whole system.

Pneumatic Input–Output Terminal Panel. The fact that instrumentation systems are generally assembled and tested at the vendor's plant and not at the location where the system will be used requires that provisions be made for terminating the pneumatic input–output lines. One method of providing a terminal for both instrument structure lines and field-installed lines is to use a bulkhead tubing connector (see Fig. 10.13). The bulkhead connectors are mounted on the enclosure surface or on a mounting plate in a panel. The connectors are located in an area accessible to the field lines.

After the pneumatic systems have been installed, each system must be pressure-tested to be certain that leakage in the system does not affect the operation of the system. The Instrument Society of America (ISA) Pneumatic Control Circuit Pressure Test, ISA-RP7.1, is one test procedure for verifying the leakage in pneumatic systems and establishing the criteria for acceptance of the work.

10-3.3 Installation for Environmental Control

(a) Temperature Control. Temperature control of instrumentation systems is usually not given sufficient attention. For example, although the overall average temperature of all components may not be excessive, many "hot spots" can develop through improper attention to cooling requirements. Even though such hot spots may not cause immediate failure, they eventually show up in system failures and poor mean time before failure (MTBF), with resulting high maintenance costs. Although temperature

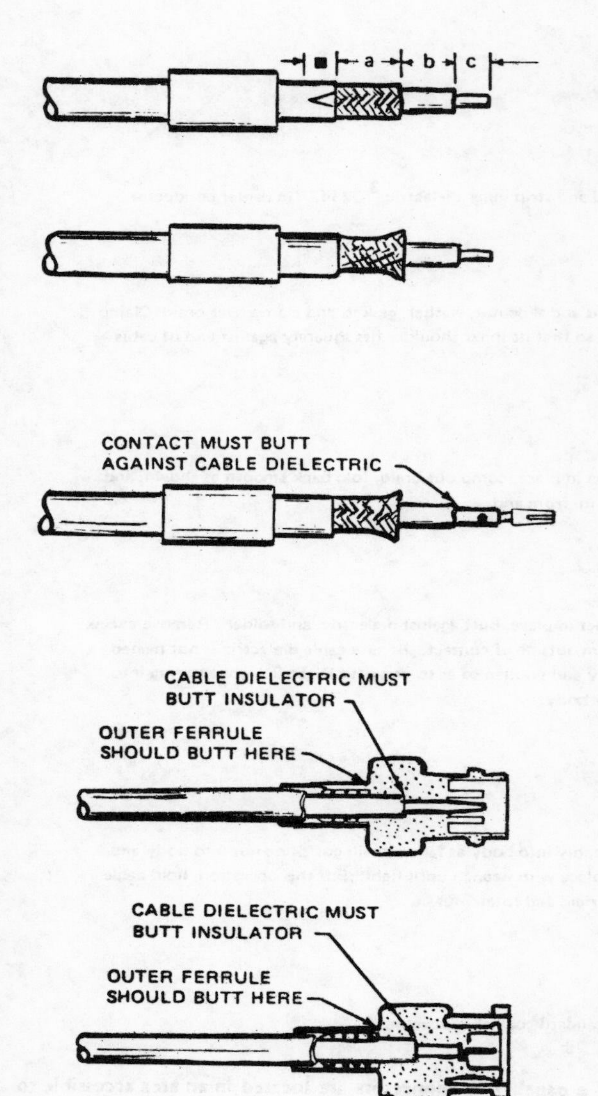

Strip cable jacket, braid, and dielectirc to dimensions shown in table. All cuts are to be sharp and square. Important: Do not nick braid, dielectric, and center conductor. Tinning of center conductor is not necessary if contact is to be crimped. For solder method, tin center conductor avoiding excessive heat. Slide outer ferrule onto cable as shown.

Stripping dimensions (± 1/64)

	MIL-Crimps			Quick-Crimps		
	a	b	c	a	b	c
Plugs and jacks	1/4	13/64	1/8	1/4	7/32	11/64
Right-angle plugs	1/4	13/64	1/8	1/4	3/16	11/64
Bulkhead jacks	1/4	13/64	1/8	1/4	1/4	11/64

Flare end of cable braid slightly as shown to facilitate insertion onto inner ferrule. Important: Do not comb out braid.

Place contact on cable center conductor so that it butts against cable dielectric. Center conductor should be visible through inspection hole in contact. Crimp or solder the contact in place as follows:

Crimp method

Crimp center contact using either of the following two tools: Tool No. 227-912-1000—To crimp the male contact (pin), insert the end of the nest bushing marked "P" into the tool. To crimp the female contact (socket), insert the end of the nest bushing marked "S" into the tool. Tool No. 227-917 (MS-3191-A)—To crimp the male contact (pin), insert the positioner marked 227-918 into the tool. To crimp the female contact (socket), insert the positioner marked 227-919 into the tool.

Solder method

Soft solder contact to cable center conductor. Do not get any solder on outside surfaces of contact. Avoid excessive heat to prevent swelling of dielectric.

Install cable assembly into body assembly so that inner ferrule portion slides under braid. Push cable assembly forward until contact snaps into place in insulator. Slide outer ferrule over braid and up against connector body. Crimp outer ferrule with tool specified in table.

Fig. 10.8—Assembly of crimp-on coaxial connector.

considerations are basically a design function, no designer's product can function effectively if operated in an environment in which it was never intended to operate. For this reason those in charge of the installation must make certain that all equipment is operated within the designed environmental limits, whether thermal, vibrational, radiation, or any other.

Since most instrumentation equipment is located in enclosed cabinets, proper ventilation must be provided to avoid convection cooling that may allow heat from a lower chassis in a cabinet to pile up in the top chassis, thereby effectively "baking" every component in the equipment. Under these conditions some units that functioned well in a 55°C test oven have been known to fail when operating in a "room-temperature" relay rack. For this reason each chassis placed in a cabinet must be checked for power consumption before installation and provisions made for the total cooling load demanded per cabinet. The type of equipment in the cabinet must also be considered; for example, a log amplifier requires closer temperature control than a relay.

If temperature might be a problem (either high temperature for indoor installation or low temperature in winter at outdoor locations), provisions for correcting the problem should be made before the equipment is installed, not after. Since the instruments in control rooms are usually a large source of heat, air may be drawn from the room through vents placed in the bottom of each cabinet and then drawn out of the top of the cabinet into the building's central air-conditioning return. Additional cooling capacity may be required in the air-conditioning system at the time of plant construction to handle the load of the control room. If proper size ducts are used from the cabinets to the air return, the airflow through the equipment and in the room will be silent, low speed, and unobtrusive. If the air input to

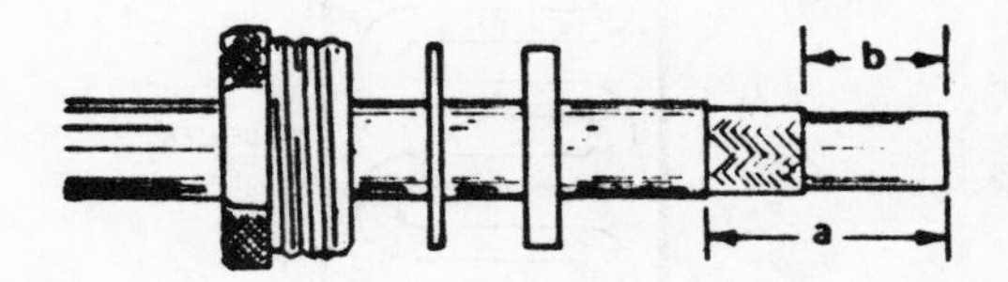

Slide nut, washer, and gasket over cable. Cut off outside jacket (using razor blade or wire strippers) to dimension a. Make a clean cut, being very careful not to nick braid. Cut first braid to dimension b.

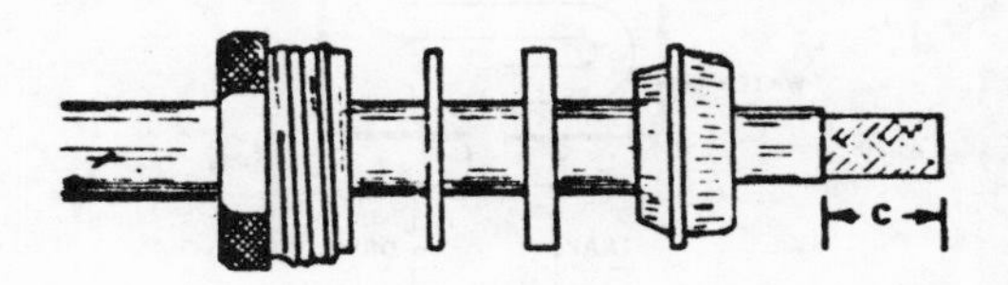

Slide first braid clamp over braid up to jacket of cable. Fold first braid back over clamp, making sure braid is evenly distributed over the surface of the clamp. Trim second jacket to dimension c, again being very careful not to nick braid.

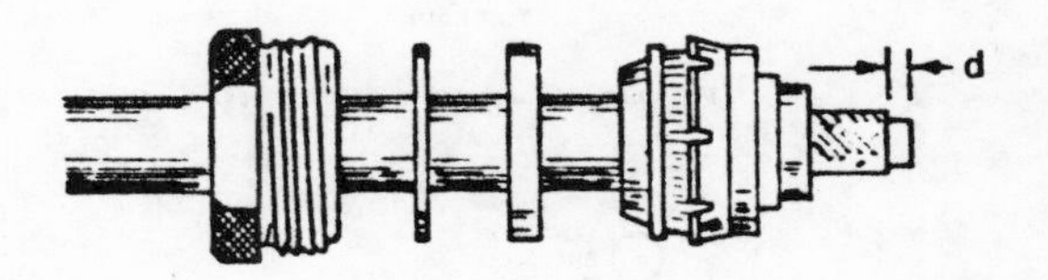

Trim second braid to dimension d. Slide on outer ground washer, Insulator, and second braid clamp. Fold second braid back over braid clamp, again making sure that braid is evenly distributed over surface of clamp.

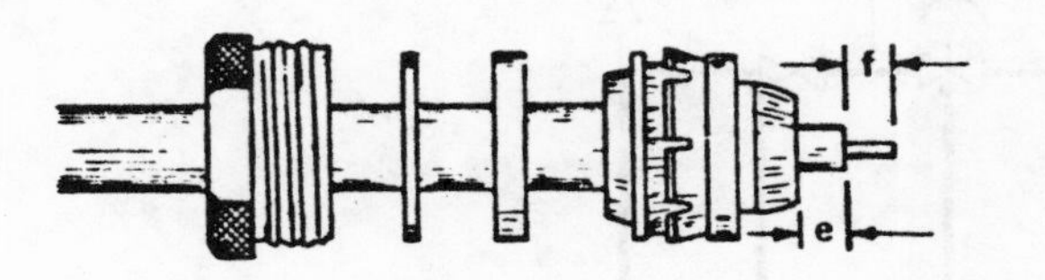

Plug only: Place front insulator and outer contact assembly into back of connector body and push into proper place. Insert cable contact assembly into body. Screw nut into body with wrench until moderately tight.

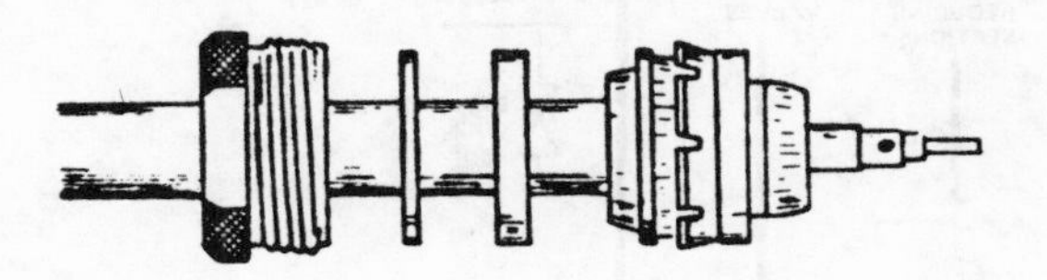

Trim cable dielectric to dimension e.

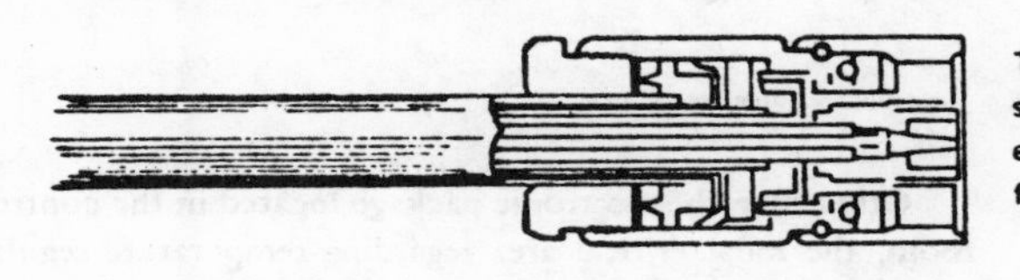

Tin the inside hole of the contact. Tin wire and insert into contact and solder. Remove any excess solder. Be sure cable dielectric is not heated excessively and swollen so as to prevent dielectric from entering body of fitting.

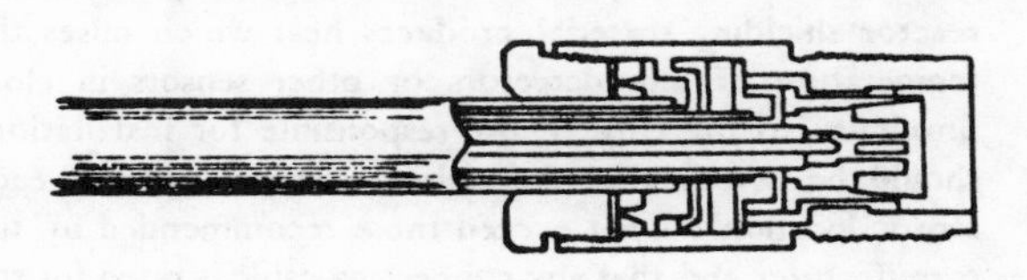

NOTE: "a" thru "f" dimension depends on cable and connector type

Fig. 10.9—Assembly of triaxial connector.

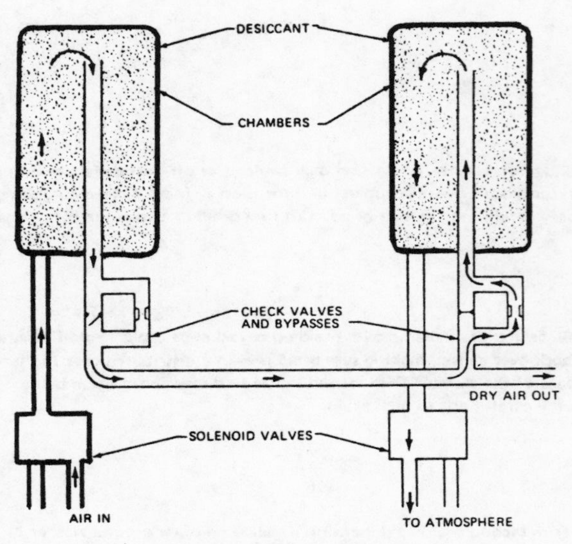

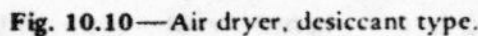

Fig. 10.10—Air dryer, desiccant type.

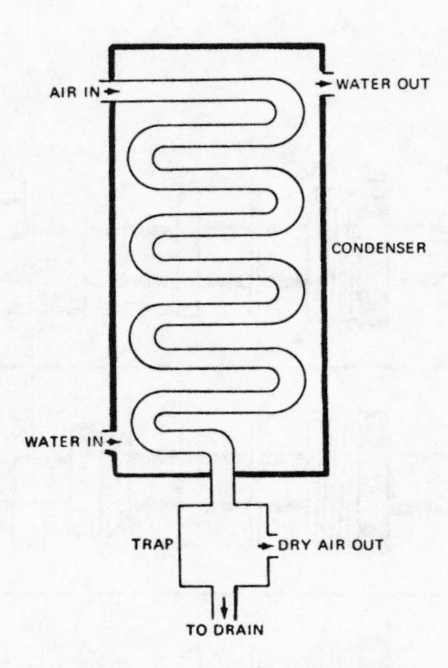

Fig. 10.11—Air dryer, after-condenser type.

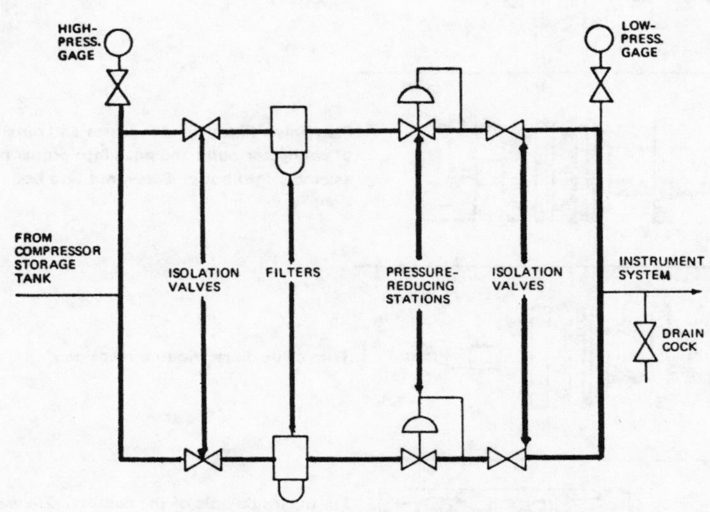

Fig. 10.12—Dual-filter regulator station. For actual installation see Fig. 10.4.

the room is filtered, the control room will stay clean as well as cool, providing a more pleasant and lower maintenance environment. Each cabinet may also be equipped with a thermometer so that the cabinet temperature can be monitored.

During set up and testing of cabinet-mounted equipment, temperature-sensitive materials, such as tempilac, may be placed in areas of high-component density and low airflow or on components that require critical temperature control to be effective. If after 8 hr of operation the temperature-sensitive materials indicate proper operating temperatures, the airflow should be turned off or blocked, and the effect of the ensuring temperature rise on the equipment operation should be noted. If any significant changes occur, an alarm annunciator should be installed to signal and warn of loss of system airflow.

Other than the electronic package located in the control room, the most critical area regarding temperature regulation is around the reactor itself. Energy dissipated in the reactor shielding material produces heat which raises the temperature of any detectors or other sensors in close proximity to the core. Those responsible for installations should be certain that the ambient temperatures of each sensor location do not exceed those recommended by the manufacturer and that any connecting cable is rated for the environment in which it must perform.

(b) Vibration Control. Every attempt should be made to mount instrumentation in vibration-free areas. When it is necessary to place instruments in high-vibration positions, it is most important that neither components, portions of the case, nor wires of any kind resonate at any of the

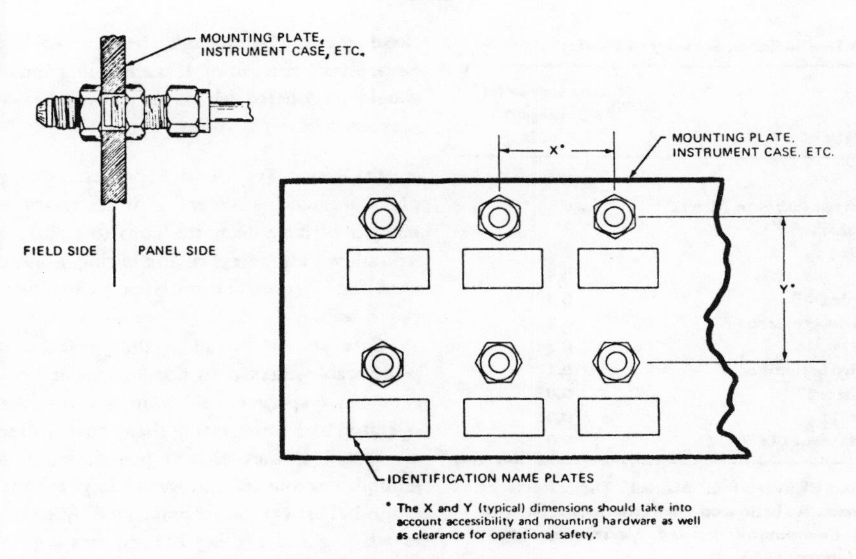

Fig. 10.13—Pneumatic terminal panel.

vibrational frequencies involved since metal fatigue will most certainly cause ultimate failure.

The easiest cure for vibrational problems is to shock mount the equipment and fasten securely all wiring harnesses by using anti-wicking tools on connector solder points. Anti-wicking tools prevent solder from flowing within stranded wire to a point beyond which the insulation has been stripped from the wire. Other approaches may also be required, including silicone rubber encapsulation of wires and connectors, special internal vibration dampening of equipment, etc. For further information on vibration control, see Defense Department Wire Specification MIL-W-5088C or MIL-W-9160D.

(c) Selection of Insulation for Radiation Environment. Wiring insulation exposed to environmental extremes of radiation should be carefully selected. It would be desirable to select wire that could withstand radiation exposures for the life of the plant (about 40 years). Extensive irradiation research programs have been conducted and numerous tables have been compiled on radiation damage to wire conductors and insulating materials.

This chapter contains several tables on radiation effects on materials. These tables are typical and may or may not agree with specific results obtained by other research organizations.

There is no substitute for experience gained in operating nuclear plants. It has been found that out-of-core detector wiring and some in-core wiring may be good for only 24 months or less. Replacement of this wiring at refueling time is considered standard operating procedure.

Table 10.1 shows the radiation stability of plastic insulating materials. Klein and Mannal concluded that only inorganic insulation materials could function within the reactor primary shield since radiation dose rates up to 10^{12} rads/hr are often experienced. The same type of insulation will be required in the containment vessel of a fast breeder reactor, where levels are expected to reach 10^5 rads/hr. Outside the primary shield but inside the containment vessel of a thermal reactor, the dose rates may range from 0.5 to 160 rads/hr, and temperatures up to 70°C may be expected.

On the basis of the foregoing assumptions and a 40-year plant lifetime, wiring inside the containment vessel may be expected to absorb 5×10^7 rads under normal conditions; a power excursion or other nuclear accident may add another 4×10^6 rads. Auxiliary structures, e.g., residual heat-removal compartments, outside the containment vessel may be expected to receive dose rates 1/100 that of objects within the containment vessel, but, in the event of an accident, these areas must be able to withstand much higher levels.

The temporary effects of radiation on elastomer-based cables include thermoluminescence, decrease in electrical resistance, and gas generation. Long-term effects include either embrittlement or softening of the insulation. Present theories tend to support the view that the cumulative radiation damage to a substance in or near a nuclear reactor depends on the total energy absorbed by the material and is not a function of the type of radiation. Accordingly, neutron damage to cables can be determined by referring to the tables for gamma-radiation damage and adjusting the total dose to account for the neutron energy. Tables 10.2 to 10.6 show degradation as a function of absorbed dose caused by gamma radiation on various parameters of a cable and for various types of cable insulations. From these findings, Blodgett and Fisher presented the data shown as Table 10.7. The table lists and rates materials that may be used successfully in various nuclear environments.

10-3.4 Installation Symbols

The designers of nuclear power plants use different symbols on drawings for the installation of instrumentation

Table 10.1—Radiation Stability of Plastics*

Material	Threshold dose for ±5% change,† 10^3 rads
Polystyrene‡	40
Phenol formaldehyde (asbestos filler)‡	40
Polyester (mineral filler)‡	4
Polyvinyl chloride‡ §	1
Polyethylene‡	0.9
Urea formaldehyde‡	0.5
Monochlorotrifluoroethylene §	0.2
Cellulose acetate §	0.2
Phenol formaldehyde (unfilled)‡	0.1
Methyl methacrylate §	0.01
Polyester (unfilled)‡	0.01
Polytetrafluoroethylene (Teflon) §	0.01

*From P. M. Klein and C. Mannal, The Effects of High-Energy Gamma Radiation on Dielectric Solids, in *AIEE Transactions on Communications and Electronics,* Part 1, Vol. 74, p. 723, January 1956.

†Based on most sensitive property, usually tensile strength.

‡Crosslinks.

§Scissions.

and electrical systems. Although there are some relevant electrical codes and standards, there still appears to be lack of uniformity throughout the industry.

Table 10.8 lists symbols typical of those currently used in the nuclear industry.

10-4 INSTALLATION OF SIGNAL AND POWER CABLES

10-4.1 Installation Hardware

(a) Conduits. Two basic types of conduit are available, aluminum and steel; other types, such as plastic, are used occasionally. In addition, plastic-coated steel and aluminum are finding wide use where corrosion is a problem. Aluminum is light in weight, free from corrosion by moisture, and easy to install, whereas steel is a far better shield against magnetic fields and has greater strength.

Where conduit is to be run through concrete (such as in biological shields), steel should be used since many concrete mixes eventually corrode aluminum, particularly with the presence of moisture. If radiation levels and conduit temperatures permit, plastic-coated steel conduits yield the best service in damp locations. Drain holes should be drilled at low points in exposed conduits to permit any accumulated moisture to escape, and unexposed conduits, such as shield penetrations, should be arranged so that moisture cannot collect in interior points. Conduit should be sized so that the installed wire cables, including allowances for expansion, fill no more than 40% of the conduit area to ensure ease of expansion and repair. Information on conduit sizes and available fittings may be found in any equipment manufacturer's catalog.

Particular attention should be paid to joints between aluminum and steel conduit, and, wherever possible, these should be avoided because of the possibility of electrolytic corrosion. If such joints must be made, they should be located where they can be easily inspected and protected from moisture.

(b) Wiring Trays and Supports. Wiring trays and supports are used wherever it is necessary to route a great number of large-diameter wires to a particular location and still allow easy access to the wiring. Solid covered trays are used for instrumentation wiring and open mesh trays for power wiring.

Care should be taken that heat buildup in enclosed power-cable trays does not lead to deterioration of insulation since power cables in enclosed trays should be operated at a lower rating than those in free air. When trays are installed, they should be bonded together to ensure ground continuity and grounding to the main building ground. This can be accomplished by several methods, such as welding and brazing the sections together, using bolted joints, or by running a ground wire along the tray sections and bonding each section to the wire with a suitable clamp (see Fig. 10-5).

10-4.2 Signal and Control Cables

(a) Power-Distribution Cables. Since power cables for instrumentation do not normally carry high currents or high voltages, a reference, such as the latest National Electrical Code (NEC), should be used to determine minimum standards of conductor type, size, etc., and installers should be aware of signal cables in the vicinity of power lines so that proper shielding measures (as shown in Fig. 10-5) can be taken. In addition to interequipment cabling, power lines in intrarack wiring should be enclosed in wireways wherever possible to improve shielding.

(b) Unshielded Control Cables. The present practice in the electrical power industry of standardizing field wires to No. 12 or No. 14 AWG with bulky insulation can cause serious installation problems if the instrument-cabinet terminal blocks are not properly sized. The design engineer and instrument-cabinet manufacturer should allow ample space in the cabinet for terminal blocks, conduits, and wireways to accommodate the large-size field wires. Single-conductor field wires with diameters of 1/4 to 5/16 in. are being used in power-station design.

Several types of single-conductor wire with small-diameter plastic insulation are durable and meet all the environmental requirements for power-station design.

Where smaller (No. 18 to No. 22 AWG) wires are used, they should normally be in the form of cables having a number of twisted pairs covered by an outer sheath, with one pair assigned to each circuit function. Avoid having several circuits tied to a single ground conductor since, if this conductor fails, a number of circuits will be put out of commission instead of only one.

It is evident that only power cables, switch commands, relay operating signals, and lines that can tolerate some cross-talk, such as communication lines, should be run in unshielded cables.

Table 10.2—Permanent Effect of Gamma Radiation on Physical Strengths of Cable Coverings*

	PVC	H.D. Poly.	SBR	C.B. CLPE	C.F. EPDM	Butyl	90°C oil base	N.F. CLPE	C.F. EPM	Silicone	PVC	Neoprene	CSPE	CPE
Tensile Strength														
Original, psi	2114	2213	1520	2045	1455	798	804	2272	872	1191	2601	2544	2113	2170
Retention after irradiation, %														
5×10^5 rads	110	96	98	122	104	96	121	102	101	76	80	104	106	112
5×10^6 rads	104	98	100	112	97	58	103	97	106	100	88	98	113	98
5×10^7 rads	79	123	82	101	93	†	98	70	119	100	61	77	124	135
1×10^8 rads	83	118	40	95	79	†	71	59	90	‡				
200% Modulus														
Original, psi	2260	2000	588	1767	1033	520	335	1260	730	859	2415	930	884	626
Retention after irradiation, %														
5×10^5 rads	94	95	106	125	100	103	121	96	116	75	81	107	116	108
5×10^6 rads	90	98	121	115	94	69	126	102	127	112	95	103	156	152
5×10^7 rads	§	§	150	§	120	†	121	108	§	98	§	160	203	§
1×10^8 rads	§	§	§	§	§	†	103	§	§	‡				
Elongation														
Original, %	260	640	460	270	470	450	870	480	300	290	250	550	560	670
Retention after irradiation, %														
5×10^5 rads	115	103	93	104	111	93	97	90	96	107	100	96	89	99
5×10^6 rads	115	103	96	96	102	87	90	96	81	90	80	93	86	63
5×10^7 rads	31		70	48	47	†	71	58	41	34	40	46	59	18
1×10^8 rads	19	2	33	37	32	†	53	25	26	‡	‡			

*From R. B. Blodgett and R. G. Fisher, Insulation and Jackets for Control and Power Cables in Thermal Reactor Nuclear Generating Stations, in *IEEE Summer Power Meeting,* Chicago, Illinois, June 1968, Institute of Electrical and Electronics Engineers, New York. †Degraded (scission). ‡Brittle. §Elongated <200%.

Note: A description of the specific wires tested is given below:

1. *PVC:* Polyvinylchloride per IPCEA S-61-402, Sec. 3.8, and UL types THW and MT, No. 4 AWG (7 strand) copper, 0.047-in. wall.

2. *H.D. Poly./PVC:* High-density polyethylene, type III, Class B, grade 3 per ASTM D1248-63T, and polyvinylchloride per IPCEA S-61-402, Sec. 3.7, and IPCEA S-19-81, Sec. 4.13.5. No. 12 AWG (7 strand) copper, 0.030-in. insulation, and 0.015-in. jacket.

3. *SBR/Neoprene:* Styrene–butadiene synthetic-rubber-based insulation per IPCEA S-19-81, Sec. 3.13, and polychloroprene-based jacket per ASTM D-752 and IPCEA S-19-81, Sec. 3.13.3, and UL type RHW. No. 14 AWG (7 strand) copper, 0.047-in. insulation, and 0.0156-in. jacket.

4. *C.B. CLPE:* Low-voltage, carbon black-filled, chemically cross-linked polyethylene per IPCEA S-66-524, Interim Standard No. 2, and UL type RHW-RHH. No. 14 AWG (7 strand) copper, 0.047-in. wall.

5. *C.F. EPDM/Neoprene:* Ozone-resisting mineral-filled EPDM-based low-voltage insulation exceeding the requirements of IPCEA S-19-81, Secs. 3.15 and 3.16, and polychloroprene-based jacket per ASTM D-752 and IPCEA S-19-81, Sec. 4.13, UL type RHH. No. 14 AWG (7 strand) copper, 0.047-in. insulation, and 0.0156-in. jacket.

6. *Butyl/Neoprene:* Ozone-resisting butyl-based insulation per IPCEA S-19-81, Secs. 3.15 and 3.16, and polychloroprene-based jacket per ASTM D-752 and IPCEA S-19-81, Sec. 4.13.3, and UL type RHW-RHH. No. 14 AWG (7 strand) copper, 0.047-in. insulation, and 0.0156-in. jacket.

7. *Oil Base/CSPE:* Ozone-resisting 90°C oil-base high-voltage insulation meeting the requirements of IPCEA S-19-81, Secs. 3.14 and 3.15, UL type RHH, and chlorosulfonated polyethylene-(CSPE) based jacket per ASTM D-752 and IPCEA S-19-81, Sec. 4.13.3, UL type RHH. No. 14 AWG (7 strand) copper, 0.047-in. insulation, and 0.0156-in. jacket.

8. *N.F. CLPE:* High-voltage, nonfilled, chemically cross-linked polyethylene (nonstaining antioxidant) per IPCEA S-66-524, Interim Standard No. 1. No. 14 AWG solid copper, 0.047-in. wall.

9. *C.F. EPM/CPE:* Ozone-resisting clay-filled EPM-bas high-voltage insulation per IPCEA S-19-81, Sec. 3.16, and UL type RHW-RHH and chlorinated polyethylene-based jacket per ASTM D-752 and IPCEA S-19-81, Sec. 4.13.3. No. 14 AWG solid copper, 0.047-in. insulation, and 0.0156-in. jacket. The insulation was discussed in IEEE paper 31 TP67-481.

10. *Silicone:* Ozone-resisting silicone rubber insulation per IPCEA S-19-81, Sec. 3.17, UL type SA. No. 14 AWG (7 strand) copper, 0.047-in. insulation, and 0.010-in. glass braid.

11. *Neoprene:* Polychloroprene-based jacket per ASTM D-752 and IPCEA S-19-81, Sec. 4.13.3, UL type RHH. No. 14 AWG solid copper, 0.047-in. wall.

12. *CSPE:* Chlorosulfonated polyethylene-based jacket per ASTM D-752 and IPCEA S-19-81, Sec. 4.13.3, UL type RHH. No. 14 AWG solid copper, 0.047-in. wall.

13. *CPE:* Chlorinated polyethylene-based jacket per ASTM D-752 and IPCEA S-19-81, Sec. 4.13.3, No. 14 AWG solid copper, 0.047-in. wall.

Table 10.3—Permanent Effect of Gamma Radiation on Dielectric Constant (k′) of Cable Coverings*†

Dose, rads	Measured after 2 hr, °C	PVC	H.D. Poly.	SBR	C.B. CLPE	C.F. EPDM	Butyl	90°C oil base	N.F. CLPE	C.F. EPM	Silicone
						k′ (S.I.C.), 40 volts/mil, 60 Hz					
None	23	4.90	2.58	3.32	3.58	3.37	4.35	3.44	2.25	3.47	3.11
	75	6.82	2.52	3.84	3.44	3.19	4.21	3.27	2.30	3.49	2.96
	90	7.32	2.51	‡	3.04	3.18	4.14	3.09	2.30	3.44	2.98
						% Change					
5×10^5	23	+3	−1	+5	−1	−4	−2	+5	+7	+8	0
	75	−4	−2	+10	−2	−4	−2	0	+3	0	−1
	90	+52	+1	‡	+4	+5	−2	+2	−4	+3	−1
5×10^6	23	+4	+39	+5	+3	−9	−20	+10	+3	+8	+29
	75	+6	+42	+6	−7	−6	0	−4	−7	+3	−8
	90	‡	+132	‡	+4	+5	0	+3	+4	+3	−8
5×10^7	23	+21	+36	+1	+3	−7	−20	+6	+3	+10	+2
	75	+41	+39	−9	−1	−8	§	+1	−1	+6	+1
	90	‡	+104	†	+9	+2	§	+10	+9	+9	0
1×10^8	23	+59	−6	+1	+2	+1	§	+7	+2	+7	+6

*See note Table 10.2.

†The high dielectric constants of the neoprene-, CSPE-, and CPE-based jacket materials were not significantly affected.

‡Loss higher than limit of bridge.

§No test, sample degraded.

Table 10.4—Permanent Effect of Gamma Radiation on d-c Resistivity of Cable Coverings*

Dose, rads	Measured after 2 hr, °C	PVC	H.D. Poly.	SBR	C.B. CLPE	C.F. EPDM	Butyl	90°C oil base	N.F. CLPE	C.F. EPM	Silicone	Neoprene	CSPE	CPE
							D-C Resistivity, 100 Teraohms-cm, 500 volts D-C							
None	23	0.15	240	2.3	70	12	76	15	141	71	0.2	10^{-3}	0.2	10^{-2}
	75	10^{-4}	25	10^{-3}	40	0.3	0.2	1.2	68	1.3	10^{-2}	10^{-4}	10^{-3}	10^{-4}
	90	10^{-4}	20	10^{-4}	37	0.3	0.1	0.1	50	1.0	10^{-3}	10^{-6}	10^{-4}	10^{-3}
							% Change							
5×10^5	23	−28	−43	+50	+33	−20	0	−1	−4	+11	+57	−31	−14	−46
	75	−90	−32	+48	+50	+32	−14	+100	−3	+10	0	−34	−32	−55
	90	−23	−90	−6	−33	−29	−51	+100	−3	0	+52	−85	−54	+198
5×10^6	23	−48	−70	+13	−59	−4	0	−1	−4	+38	0	+15	−5	0
	75	+10	−99	+40	+17	−8	−84	+66	−3	−9	+25	+4	−17	0
	90	−47	−92	0	−43	−51	−98	+90	−3	0	+15	+415	−82	+19
5×10^7	23	−67	−81	+48	−68	+53	−82	−5	−4	+25	+60	0	0	−14
	75	+100	−80	+250	−52	−34	†	+33	−4	−7	+20	+11	−17	−92
	90	+27	−99	+100	−75	−79	†	+25	−3	−40	+64	+390	−75	−85
1×10^8	23	+120	−70	+58	−7	+60	†	+35	−8	0	+60			

*See note Table 10.2.

†No test, sample degraded.

If one of these lines is terminated in a terminal strip or block, solderless crimp connectors may be used; if the line terminates in a connector, solderless or solder-type terminations may be used. When wires are terminated in a connector, covering each wire with teflon tubing, shrink tubing, or other insulation will decrease the probability of shorts.

In general, unshielded wiring is much easier to install than shielded and, if standards such as those referenced in other sections of this chapter are followed, should create no problems.

(c) **Instrument Signal Cables (Multiconductor, Shielded).** Because of cross-talk, spike induction, and other interference problems, it is important to consider the

Table 10.5—Permanent Effect of Gamma Radiation on Flame Resistance of Thin-Wall Wires in Underwriters Laboratories Flame Test*†

	PVC		H.D. Poly./ PVC		SBR/ neoprene		C.B. CLPE		C.F. EPDM/ neoprene		Butyl/ neoprene		90°C oil base/ CSPE		N.F. CLPE		C.F. EPM/ CPE		Silicone glass	
Dose, rads	0	10^8	0	10^8	0	10^8	0	10^8	0	10^8	0	10^8	0	10^8	0	10^8	0	10^8	0	10^8
Results	P	P	F	P	F	F	F	F	P	P	F	F	P	P	F	F	P	P	P	P
% flag destroyed	0	0	100	0	100	100	100	100	0	0	100	20	0	0	100	100	0	0	0	0
After burn, sec	0	0	180	0	52	60	180	100	0	0	50	80	0	0	180	180	0	0	0	0

*See note Table 10.2.
†P, pass; F, failure.

Table 10.6—Threshold (in rads) of Gamma Radiation Damage for Elastomer-Based Cable Coverings*

Property	PVC	H.D. Poly.	SBR	C.B. CLPE	C.F. EPDM	Butyl	90°C oil base	N.F. CLPE	C.F. EPM	Silicone	PVC	Neoprene	CSPE	CPE
Tensile strength	10^8	10^8	5×10^7	10^8	10^8	5×10^6	10^8	5×10^7	10^8	5×10^7	5×10^7	5×10^7	5×10^7	5×10^7
Elongation	5×10^7	5×10^6	5×10^7	5×10^7	5×10^7	5×10^6	10^8	5×10^7	5×10^7	5×10^7	5×10^7	5×10^7	5×10^7	5×10^6
Rate of oxidation	5×10^6		$>5 \times 10^7$	$>5 \times 10^7$	$>5 \times 10^7$	5×10^6	$>5 \times 10^7$	5×10^6	5×10^7	5×10^5	5×10^6	5×10^6	5×10^7	5×10^6
Dielectric loss	5×10^7	5×10^5	10^8	10^8	10^8	5×10^6	10^8	5×10^5	10^8	10^8	5×10^7	5×10^7	5×10^7	5×10^7
Electric stability	5×10^5	$>5 \times 10^7$	5×10^5	$>5 \times 10^7$	5×10^7	5×10^5	5×10^7	5×10^7	$>5 \times 10^7$	$>5 \times 10^7$	5×10^5	5×10^5	5×10^6	5×10^7
Dielectric strength	5×10^7	5×10^7	5×10^7	10^8	$>10^8$	5×10^6	10^8	$>10^8$	$>10^8$	$>10^8$	5×10^7	5×10^5	5×10^5	5×10^7
Overall threshold of damage	5×10^5	5×10^6	5×10^5	5×10^7	5×10^7	5×10^6	5×10^7	5×10^6	5×10^7	5×10^5	5×10^5	5×10^6	5×10^6	5×10^6
Highest dose still serviceable	5×10^6	5×10^7	5×10^7	10^8	10^8	5×10^6	10^8	10^8	10^8	5×10^7	5×10^6	5×10^6	5×10^7	5×10^7

*See note Table 10.2.

Table 10.7—Suggested IEEE Nuclear Environment Classification for Elastomer-Based Cable Coverings

Radiation class	Temperature Class		
	O(90°C)	A(105°C)	B(130°C)
1 (9×10^4 rads)	Silicone (see Note 1)	Silicone (see Note 1)	Silicone
2 (9×10^5 rads)	Butyl/neoprene, CSPE, CPE, and H.D. Poly.	See below	None
3 (8.8×10^8 rads)	EPDM, EPM, oil base, N.F. CLPE, and C.B. CLPE	EPDM, C.B. CLPE, and EPM	None
4 (8.8×10^9 rads)	None	None	None
5 (10^{10} rads)	None	None	None

Notes:

1. Dimethylsilicone-based insulations (IPCEA S-19-81, Par. 3.17) are suitable at their usual 130°C temperature rating only in low-radiation environments because of their sensitivity to steam and poor resistance to oxidation after irradiation. Blodgett and Fisher rate them only in classes O1, A1, and B1.

2. Carbon-black (and probably clay-filled) cross-linked polyethylenes and clay-filled EPM- or EPDM-based insulations are suitable at 105°C up to class 3 radiation levels when protected with suitable flame-resistant braids (such as the glass construction used in Blodgett and Fisher's study) or flame- and water-resistant asbestos constructions. Blodgett and Fisher rate these two materials for classes O1, O2, O3, and A1, A2, and A3.

3. Butyl and high-density polyethylenes with neoprene, CSPE, or CPE jackets or the CPE as integral-insulation jackets are suitable at their usual 90°C temperature rating only up to class 2 radiation levels. Blodgett and Fisher rate these systems only for classes O1 and O2.

4. Nonfilled cross-linked polyethylenes and oil-base insulations, when protected by a neoprene, CSPE, or COE jacket, are suitable at their usual 90°C temperature rating up to class O3. Blodgett and Fisher rate these systems for classes O1, O2, and O3.

5. SBR- and PVC-based coverings are suitable only at relatively low temperatures and radiation levels. In particular (IPCEA S-61-402, paragraphs 3.7 and 3.8), PVC's are sensitive to hot water and steam when exposed to more than 5×10^5 rads.

routing of each cable with respect to its electromagnetic environment. Mixing of low-level signals, relay commands servomotor control currents, and communication cables in the same conduit or raceway results in interference problems whether shielded cable is used or not. Good practice dictates that instrumentation cables be separated physically according to signal level and function as well as electrically by shielding etc., wherever possible. In critical circuits, such as reactor-control circuits, separate each channel's measurement and control function from all others. This will result in at least three sets of separately run conduit, color coded or identified in some way. If this policy of separation is followed along with coincidence safety logic in the control-instrumentation design as well as in wiring layout, any portion of the control system may be disconnected without causing a scram.

Isolate wiring according to function. Cables carrying high currents or voltages should be isolated from those carrying low currents or voltages, and cables carrying interference-producing signals should be isolated from those carrying direct current. A designer should use a separate conduit for all low-level (detector, thermocouple, etc.) signals, a separate conduit for high-level control signals containing shielded wires, a separate conduit for relay and contact closure leads, and a separate conduit for a-c and d-c power distribution.

If every precaution is not followed, scrams may be caused by arc welders or other noise-producing devices, such as switching d-c circuits, when placed near the detector cables.

Concerning the signal and control cables, each cable should carry only signals of the same type and, for instrumentation purposes, should have at least an overall shield (either braid or metal foil) for each group of conductors. In all instances the cable shielding should have an insulating layer over the shield to provide isolation from ground-loop currents likely to be circulating in the outer conduit. Each of the two most common types of shields (braided and foil) offers different degrees of shielding protection. The braided type of shield offers good protection for low-level signals. However, because of the effect of leakage capacity through the braided shield to ground, common-mode reflection suffers, and something better is needed for noise-free transmission of microvolt-level signals. Lapped foil shields have been developed for this purpose, and this type of solid-foil shield, plus a low-resistance drain wire, reduces the leakage capacity from about 0.1 to 0.01 pF/ft, typically. In addition, the foil shield improves shield-to-ground electrical leakage characteristics, rejection of magnetic pickup, and shield-resistance characteristics, and reduces termination problems.

Conductor pairs within the cable should be of the twisted variety since this in itself reduces interference as much as 15 dB. The use of balanced, twisted pairs is even more effective, resulting in an interference reduction of up to 80 dB. The foregoing techniques were applied in the construction of the Ballistic Missile Early Warning System (BMEWS), where many different types of cables were located in close proximity to one another. In this system inherent shielding of the cableways provided 6-dB attenua-

Table 10.8—Installation Symbols Commonly Used in the Nuclear Power Industry

Resistor

Capacitor

Battery

Alternating-current source

Thermocouple

Thermal element

Conductor and junction

2-conductor cable

Shielded 2-conductor cable

Coaxial cable

Ground

Basic contact assemblies

Electromagnetic actuator with mechanical linkage

Push-button switch

Coaxial connector

Transformer

AIR CORE IRON CORE

Fuse

Circuit breaker

Semiconductor rectifier diode

Meter

V A W

Rotating generator

GEN

Rotating motor

MOT

Winding connection 1-phase

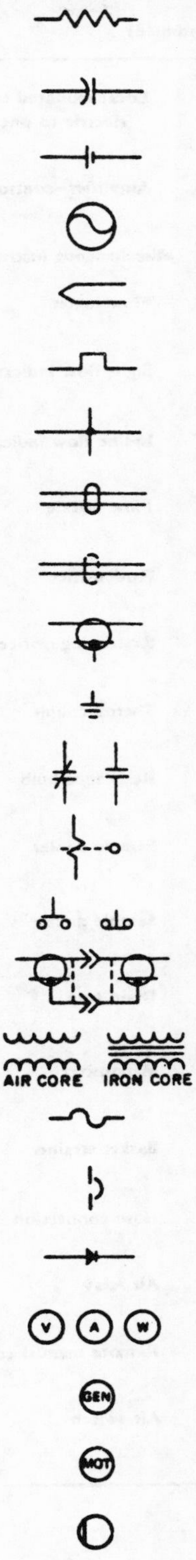

Winding connection 3-phase ungrounded

Winding connection 3-phase grounded

Piping

Primary flow line

Secondary flow line

ASME Boiler Code line

Control air line

Instrument capillary tubing

Flexible hose

Valves

Gate

Globe

Check

Stop check

Plug

Angle

Manual flow controller

Butterfly

Relief

RV

Electromatic relief

ERV

Three-way

Four-way

Throttle

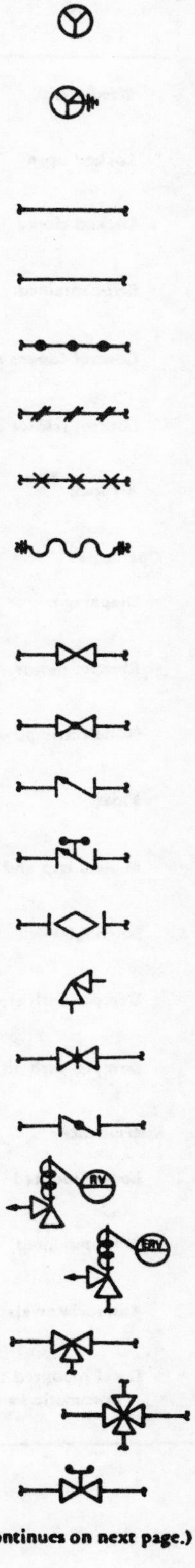

(Table continues on next page.)

Table 10.8—(Continued)

Item	Symbol label
Bleeder trip	ETV
Locked open	L.O.
Locked closed	L.C.
Self-contained	
Control (opens on air failure)	
Control (closes on air failure)	
Air lock	A.L.
Operators	
Diaphragm	
Electric motor	MOV
Nonelectric power	POV
Float	
Manual trip and reset	
Solenoid	SV
Damper with electric operator	MO
Damper with air operator	PO
Instruments	
Local mounted	PI PS
Panel mounted	PI PS
Annunciator alarm	H L XX BOARD SYMBOL
Local mounted transducer pneumatic to electric	P/E
Local mounted transducer electric to pneumatic	E/P
Amplifier–controller	AC
Miscellaneous Instruments	
Flow meter	FM
Sight flow indicator	PG
In-line flow indicator	FI
Flow nozzle	FE
Flow orifice	FE
Restricting orifice	RO
Thermocouple	TE
Resistance bulb	RTD
Sample cooler	SC
Sample nozzle	SN
Drain trap	DT
Manometer	M
Basket strainer	BS
Hose connection	
Air relay	
Remote manual control	RMC
Air switch	

tion; twisting of power and other cables provided 26 dB; and the use of balanced, twisted pairs added another 80 dB.

Because instrumentation cables must not be considered separately from the system in which they are to be used, the designer should consider terminating all cables carrying signals having frequencies greater than 10 kHz with their characteristic impedance to avoid end reflection. Termination of signal cables depends on the type of cable and the signal levels involved. In general, cables other than coaxial can be terminated using color-coded crimp connectors of the "ring" type and affixed to terminal strips. Each end of a conductor should also be marked by attaching a piece of plastic sleeving bearing the wire designation number as shown on the system interconnection diagram.

Wires within a conduit should not exceed code limits to ensure easy cable pulling. No splices should be allowed except in appropriate junction boxes [see pt. 10-4.1(a)]. The conduit "fill" should not exceed 40%, including planned additional space reserved for system changes. When cable is pulled through conduit, excessive stress should not be placed on the cables since this may result in damage to insulation in regular wiring or changes of impedance in coaxial cables. Spare conductors should be installed in each cable to permit system expansion. As a rule, running 10 to 15% more conductors than required seems to work well. This allows for additions without raising the cost excessively. However, the type of reactor installation (power, experimental, etc.) may alter this general rule.

Termination of thermocouple leads is a special case, and the manufacturer's instructions should be followed to ensure that there are no error currents produced by improper terminations. In addition, thermocouples should be kept away from cables carrying high current or voltage signal levels. Self-balancing temperature recorders respond too slowly to be affected by transients on thermocouple leads. However, electronic time-sequential multiplexing of large numbers of thermocouples into a device (such as a computer) requires that transients on the thermocouple leads be eliminated since sampling of a particular thermocouple may occur when the signal level is being influenced by a transient.

(d) Coaxial and Triaxial Cables. Coaxial and triaxial cables require care in selection and termination because signal levels are of the order of 1 mV or less. Triaxial cables provide additional low-frequency (<100 kHz) shielding attenuation of 20 to 40 dB over coaxial cables, and additional benefits, such as low leakage, may be gained by appropriately driving the inner shield as explained in pt. 10-5.5(a). Regarding installation of these cables, it is safe to use BNC type connectors of either the crimp or solder style up to 500 volts d-c. (The crimp type is popular.) Above 500 volts d-c, MHV series connectors may be used to voltages of 5000 volts d-c, except where high-frequency pulses are present. At very high pulse rates (>1 MHz), the connector impedance must match the cable, and so other types of connectors must be chosen. When cable connectors are being installed, care must be used to ensure that there are no loose ends of the shield braid to cause shorts or lower the breakdown resistance of the connector. After the cable has been assembled, the test procedure outlined in Table 10.9 is recommended.

Table 10.9—Cable-Testing Procedure

Operating voltage	Test procedure
<600 volts d-c	2 times rated voltage plus 1000 volts applied for 1 min.
>600 volts d-c	2.25 times rated voltage plus 2000 volts applied for 1 min.

Since magnetic and electric fields are responsible for most interference problems below approximately 3 MHz, low-frequency (60 Hz) pickup should be guarded against by using steel conduit around low-level coaxial and triaxial cables. The steel effectively attenuates both magnetic and electrostatic fields at all frequencies.

At termination ends of coaxial or triaxial cables, each connector should be marked with its appropriate print number as well as the number of the mating connector for the particular cable. An appropriate marking device is a plastic sleeve wrapped around the cable end and bearing the necessary information.

When triaxial cable is used, connector assembly is more critical than when BNC is used, and greater care must be taken in testing the completed cable. The tests in Table 10.9 should be applied in this case not only between center conductor and inner shield but also between the inner and the outer shield to ensure proper connector integrity. For extra protection a quantity of silicone grease may be used inside the connector to provide additional insulation and to prevent accumulation of moisture within the connector.

(e) Containment Penetrations. Penetrations for signal, control, and power cables in the reactor containment have been custom-designed. Custom-designed penetrations, in many cases, required assembly at installation and disassembly for repair. Many problems experienced with containment penetrations resulted from field-assembly conditions. Although the problems experienced with field-assembled penetrations were often similar (e.g., difficult installation, leaking seals, and poor wire termination), the variety of custom designs prevented universal solutions from being applied to similar problems.

For greater reliability and ease of installation, several manufacturers have designed and now fabricate preassembled and pretested electrical penetrations. These penetrations can be supplied with seals that are compatible with a variety of ambient environmental conditions (temperature, moisture, and nuclear-radiation level). Penetrations can be supplied with electrical conductors ranging from unshielded control and power wires to coaxial, triaxial, and other types of shielded cable. Wire terminations are available that range from pigtails and pressure or

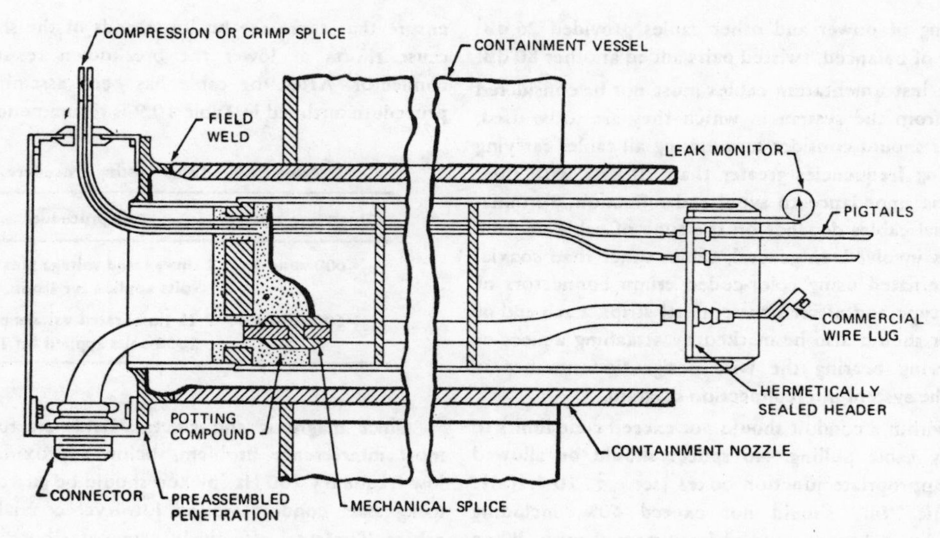

Fig. 10.14—Electrical penetration.

crimp splice tubes to special high-voltage and shielded connectors. The preassembled penetrations are tested at the factory for leak rate, conductor continuity, and insulation resistance. The assembled and tested penetrations can be equipped with a leak-monitor pressure gauge and pressurized with inert gas, thus allowing the penetration to be monitored for leaks during shipment and installation as well as during operation. The preassembled penetration can be supplied for field installation with a welding ring or a bolted flange. Figure 10.14 shows some of the features available in preassembled penetrations.

10-5 GROUNDING AND SHIELDING

10-5.1 Electrical Noise Problem

Establishing a common ground may be the goal of electrical machinery design, but it creates problems in data-measuring systems. Ground-loop currents between pieces of equipment that are gounded at separate points to a common ground introduce voltages that can affect measurements. Differences in potential between various points in a grounding system are not uncommon. These differences are caused by stray currents of any origin in the system, such as faults or transients on electric-power equipment. If low-signal-level instrumentation has multiple connections to ground, either by intention or accident, these potential differences can result in ground-loop currents.

Difficulties attributable to grounding have been experienced in several nuclear-power-reactor installations and are difficult to locate. For this reason a single-point grounding system is used in nuclear-power-plant reactor instrumentation systems. An independent grounding system, isolated from the building grounds, has been installed. The concept of an independent grounding system has advantages; however, it will not eliminate capacitive coupling or leakage resistance to ground, which also results in ground-loop currents.

The final design of instrumentation grounding depends on several factors, particularly the types of reactor instrumentation to be used and the nature of the reactor-building grounding system. Intricate reactor instrumentation systems almost always require some extensive modifications of grounding connections after the equipment is installed to obtain satisfactory operation.

10-5.2 Grounding System for Reactor Building

The grounding system for a nuclear power station must provide for: (1) instrumentation-system grounding, (2) ground connections for grounded neutral power systems, (3) a discharge path for lightning arrestors, (4) grounding of equipment frames and housings to protect equipment and personnel from dangerous electrical potentials caused by faults, and (5) communication and fire-alarm-system grounding.

Figure 10.15 shows a system using a grounding grid for a nuclear power station. A properly installed grounding grid with its associated grounding rods or grounding wells should have a total resistance across the entire grid of less than 0.2 ohm. Some nuclear power stations may use means other than a grounding grid for grounding between buildings, containment spheres, and other major systems. The essential requirement is that all major systems, subsystems, and equipment be thoroughly grounded with ample size grounding conductors and proper grounding connections.

The grounding of a reactor building should be well established. All grounding connections to stainless-steel equipment and piping should be made to stainless-steel stringers or saddles welded to the equipment with, if possible, thermite welds. Grounding connections should be accessible. Two or more grounding connections are recom-

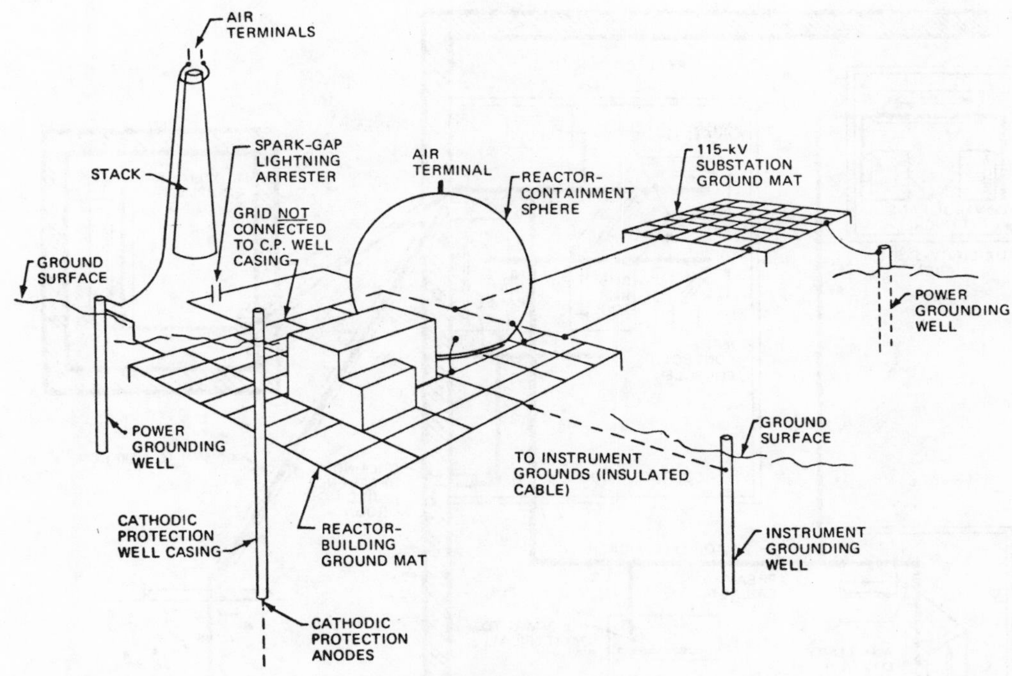

Fig. 10.15—Reactor-building grounding.

mended for large equipment. An equipment ground conductor should be included for all power circuits entering the containment. Penetrations of the independent instrumentation grounding conductors must be insulated from the containment.

10-5.3 Grounding of Electrical Switchgear and Motor Control Centers

Improper grounding of the electrical switchgear, motor control centers, and machinery has been a source of electrical noise for reactor instrumentation. A ground bus with a rating equal to the rating of the largest circuit breaker in the structure should extend throughout the largest of the switchgear assemblies. Each enclosure should be grounded directly to the ground bus. The frame of each circuit-breaker unit should be connected to the ground bus through a separate ground contact device except when the primary disconnecting devices are separated a safe distance.

All other equipment requiring ground connections should be connected to the main ground bus by a copper bar or stranded copper cable. The terminal fittings should be pressure-type solderless connections. All contact surfaces at splices should be silver-plated. The ground bus should have the same rating throughout the length of the cabinet and switchgear assemblies. Tapered ground buses should not be used. At each end of the switchgear assembly, cabinets and panels, and motor control centers, provisions should be made for connecting the ground bus to the station grounding system; these connections should consist of silvered sections of the bus.

10-5.4 Grounding of Instrument Panels and Cabinets

Grounding the instrument panels and cabinets does not normally involve the massive amounts of metal required in grounding power systems. Nevertheless, the principles are the same, and all equipment cabinets, racks, etc., must be electrically bonded together and connected to a common ground point. These connections should be either welded or brazed, particularly the connections to the ground bus from the equipment.

Each instrument-rack structure must be equipped with an electrical grounding bus. This bus is generally a 1.25-in. copper bar (Fig. 10.4) mounted in the lower section of the structure. The bus is provided with a means for connecting it to the plant grounding system. Maximum resistance measured from the grounding bus to the building ground should be less than 1 ohm. Where electrical grounding of equipment is required, the structure frame must not be used as a ground path. A conductor from the equipment to the ground bus must be provided.

So that electrical noise will not be induced in the instrument circuits, the circuit components must not use the equipment frame ground conductor as a ground. Currents in an equipment ground conductor can cause a voltage drop along the ground conductor, thus changing the point of reference for any circuit using the conductor.

Figure 10.16 shows the grounding system for a series of instrument racks and cabinets scattered throughout the reactor plant, including an instrument panel located in an adjacent building.

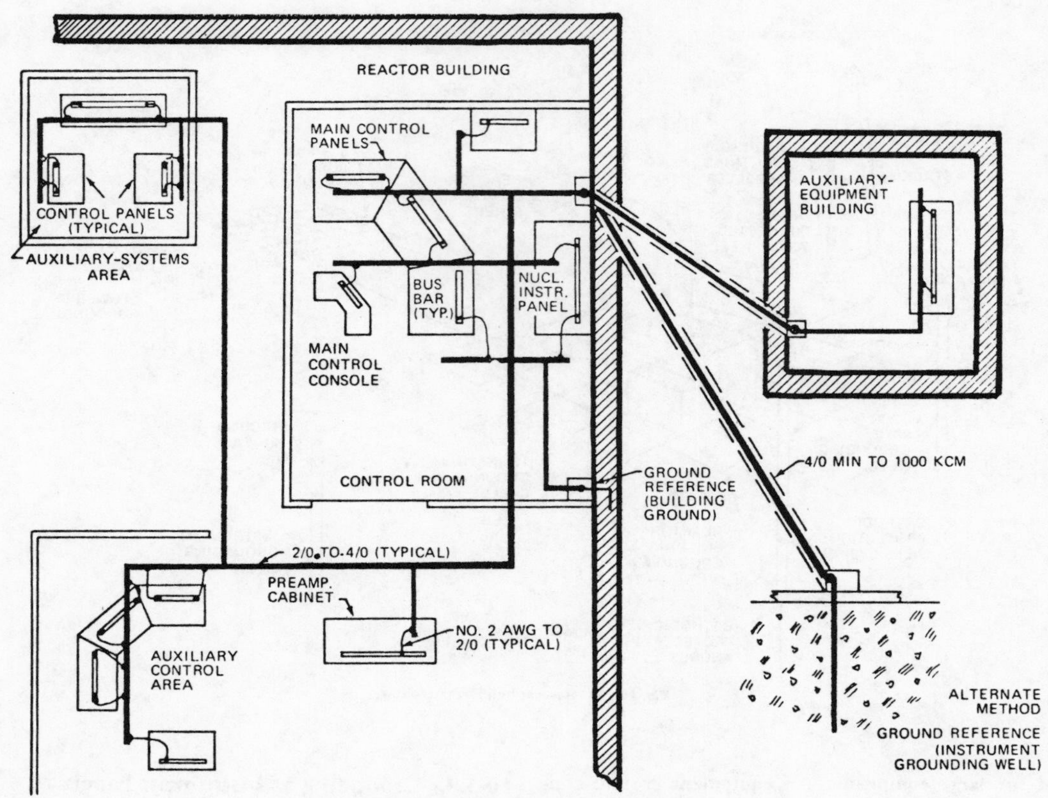

Fig. 10.16—Grounding instrument panels and cabinets.

The independent grounding system for reactor instrumentation shown in Figs. 10.15 and 10.16 terminates at an instrument grounding well and is isolated from the building grounds. The most widely accepted method employs a single-point grounding system for reactor instrumentation that is terminated at one point to the building ground. This latter method provides a ground at the amplifier cabinets in the control room; however, in some nuclear stations the grounding may terminate at the reactor near the neutron detectors.

If an independent grounding system is used, it should be entirely separate from the power grounding system. Several separate areas or zones of instrumentation, divided as to type and location, should be provided with individual grounding buses or conductors that can be connected to the independent grounding system or left floating, as operating experience dictates.

An independent ground system requires that all instrumentation be constructed with the signal grounds insulated from frames, chassis, power-supply grounds, etc. Separation of the two grounding systems involves some difficult practical problems. For example, any sensor, amplifier, or other component normally grounded to its housing requires special construction; therefore the type of grounding system used must be decided in advance.

Because of inevitable ground-loop currents over any appreciable length of building ground bus and for added reliability, relays for control circuits should always be operated with one twisted pair of wires per relay. In addition, neither of the relay control wires should be grounded except at the controlling location since to do so may introduce circulating currents in the grounded wire as well as in the ground bus.

It is good engineering practice to install suppression devices on control relays to attenuate interference from relay operation. Such devices as diodes, thyristors, or capacitor–resistor networks can be used for this purpose.

10-5.5 Grounding of Neutron-Monitoring System

The neutron-monitoring system is a vital part of both the reactor control system and the plant protection system. The system must have proper grounding and shielding. Some factors involved in determining the proper grounding and shielding methods are:

1. The length of the signal conductors between the detectors and amplifiers.

2. Methods used for internal grounding of detectors and amplifiers.

3. Methods used for grounding the electrical distribution systems of the building.

Several types of neutron sensors are widely used by the nuclear industry (see sec. 2, 3, and 4). Some variation

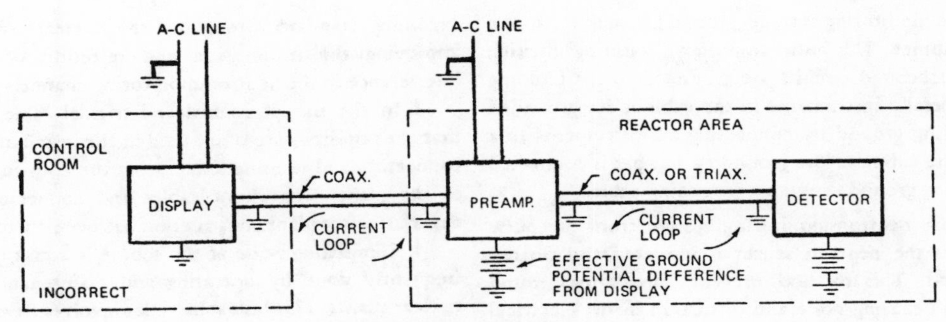

Fig. 10.17—Incorrect grounding system.

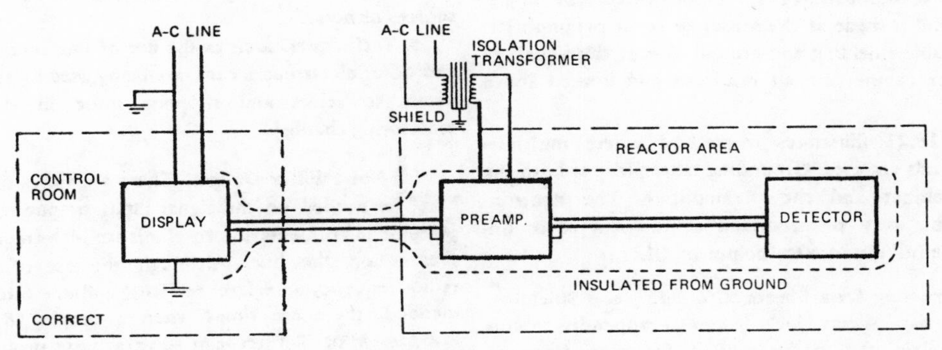

Fig. 10.18—Correct grounding system.

exists in the manufacture of nuclear-instrument circuits in regard to the method of providing internal signal grounding. Some nuclear instruments have the signal ground on the chassis, whereas others have the signal ground insulated from the chassis ground.

The following discussion concerns the grounding systems currently accepted by the nuclear industry in operating power plants.

(a) Grounding of Signal and Control Cables. The following must be kept in mind when considering the methods to be used for grounding neutron-monitoring systems:

1. There is always some potential difference between two points on the earth's surface.
2. Because a cable is connected to a ground bus, it is not necessarily a good ground.
3. Ground connections are not always noise-free.

Since it is virtually impossible to eliminate noise and induced current in ground connections, the proper procedure to follow in wiring practice consists of routing the inevitable currents around the equipment in such a manner that the signal input is not affected.

Figure 10.17 shows a typical sensor–preamplifier–count-rate-meter combination as often installed, along with some of the sources of error and interference due to potential differences generated by the system at various points. Figure 10.18 shows the same system with different grounding connections to eliminate ground loops in sensitive circuitry. If the ground loops shown in Fig. 10.17 are not removed, a "battery" voltage composed of noise is impressed on the opposite ends of the cable shield, thus effectively causing a current in the shield that may add to or modulate the desired signal. Ground loops are eliminated by isolating the system from ground except at the console. This is only one of several noise-rejection grounding techniques available. Others are electric differential input techniques and the use of balanced lines [see pt. 10-5.6(b)].

(b) Grounding and Shielding Practice. We made a survey of 10 major nuclear power plants in the United States, including 4 of the largest operating plants, according to MW(e) output in service as of January 1969. The purpose was to collect and compile data on methods used in these plants for grounding and shielding the neutron-monitoring systems. Each had experienced noise problems in the neutron-monitoring channels.

Figures 10.19 to 10.21 illustrate the grounding and shielding methods used by the nuclear power plants surveyed. Numerous modifications were made on the equipment after the plants were constructed. Such modifications as adding line filters, radio-frequency (r-f) grounds, and π filters were made to the equipment to suppress noise and interference. In a majority of the plants surveyed, a single-point ground system was used either by grounding the system at the nuclear-instrument cabinet or at the neutron detector. Other nuclear power plants provided grounding at both the neutron detector and amplifier cabinet.

Figure 10.19 illustrates a grounding method in which

the neutron-monitoring-system ground is made at the amplifier cabinet. The pulse amplifiers, counting circuits, and other associated circuits are grounded to the building ground. Generally instruments in the cabinet are grounded to the building ground by connecting all instruments to a bus bar in the cabinet. The grounding bus bar is connected to the building ground through a grounding cable.

The entire neutron-monitoring system from the pulse amplifiers to the neutron sensor is insulated and floated above ground. This method prevents circulating ground currents from causing noise and distortion in the electrical signal being transmitted to the control room.

Figure 10.20 illustrates a method where the single-system ground is made at the sensor or at the preamplifier. The signal-cable shielding and ground side of the signals in the amplifier cabinet are all insulated and floated above ground.

Figure 10.21 illustrates a method where multiple-system grounds are used. Grounding takes place at both the amplifier cabinet and the preamplifier. The neutron-detector cable may be grounded at the reactor or ungrounded the full distance to the preamplifier.

(c) Engineering Data Sheets, Grounding and Shielding. Engineering data sheets, included as an appendix to this chapter, explain the methods used for grounding the start-up channels at 10 major operating nuclear power plants in the United States. Included with the system descriptions are comments on operating problems and modifications made to the equipment to prevent noise and interference. These data sheets should be used by both the design and construction engineer in selecting the best method of grounding an instrumentation system. Information is included on how to avoid pitfalls that have been encountered in nuclear power plants and, more important, how to avoid having to make extensive modifications after the plants have been constructed.

Grounding and shielding problems in nuclear plants are often associated with the neutron-monitoring start-up channels. A number of mechanisms in a nuclear plant generate r-f and low-frequency noise, which finds its way into the neutron-monitoring start-up channels. This noise generates signals that must be cut off by a higher discriminator voltage setting, thereby materially reducing channel sensitivity. If rate amplifiers and reactor trip circuits are used with the start-up channels, inadvertent scrams result where noise increases. As the reactor power is increased and current-measuring channels take over from the pulse count-rate channels, the effect of noise is much less important.

The following conclusions can be drawn from the information contained in the engineering data sheets:

1. A single-point grounding system, grounded to the building ground at the amplifier cabinet with the signal-cable shields and neutron sensors floated above ground, appears to be most widely used by the nuclear industry. The entire system is grounded at one point.

2. The use of triaxial cable in place of coaxial cable is becoming standard throughout the nuclear industry, thus improving the suppression and rejection of noise and interference in the neutron-monitoring channels.

3. In the use of coaxial and triaxial cable, it appears that the required care is not used in the installation of cable connectors and terminations during the constructior phase of the plant. Rework of cables and connectors is often necessary after plant construction has been completed.

4. Eliminating noise at the source is a task that is quite frequently done by operating and maintenance staffs at nuclear plants. This task sometimes involves days of tedious work to isolate and eliminate the source of noise. Faulty a-c and d-c machinery, relays, switches, motor starters, etc., are sources of noise.

5. Techniques, such as the use of line filters, r-f filters, and other electronic means, are being used by some nuclear plants to reject and suppress noise in the neutron-monitoring channels.

(d) Noise Filter Design. There are times when a rapid "fix" is needed on the signal input or power input of a piece of instrumentation to eliminate high-frequency noise present on the line. Although the use of commercial radio-frequency interference (RFI) and line filters is recommended, there are times when a simple 18 dB/octave, low-pass π or T filter can solve a noise problem. Figure 10.22 shows design details for networks of this type.

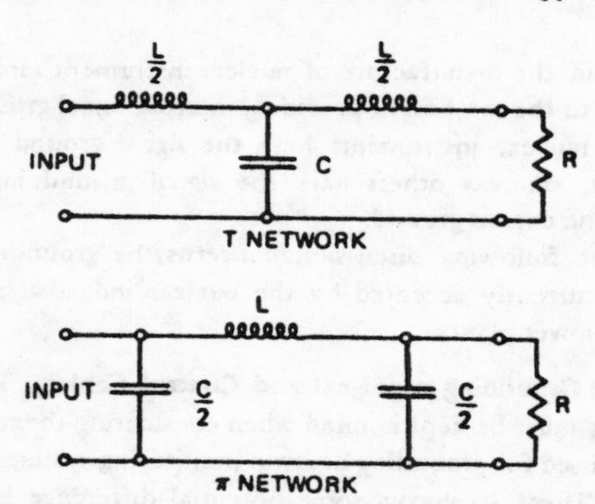

Fig. 10.22—Typical π and T networks. R, load resistance (ohms); $L = R/\pi f_c$ (henries); $C = 1/\pi f_c R$ (farads); f_c, frequency at cutoff (hertz).

When a filter of this type is installed, it should be placed in its own metal box where possible. If not possible, at least the input and output leads should be separated. Such a filter can be used either as a low-level-signal filter or as a high-level power-line filter if the components are suitably rated. These filters may be placed in power input lines to any piece of equipment in a neutron-monitoring system and may also be used in signal input leads to equipment as long as the desired input-signal frequency is below f_c. These filters are not used in pulse-amplifier inputs.

10-5.6 Process-Monitoring Instrumentation

Process-monitoring instrumentation refers to those

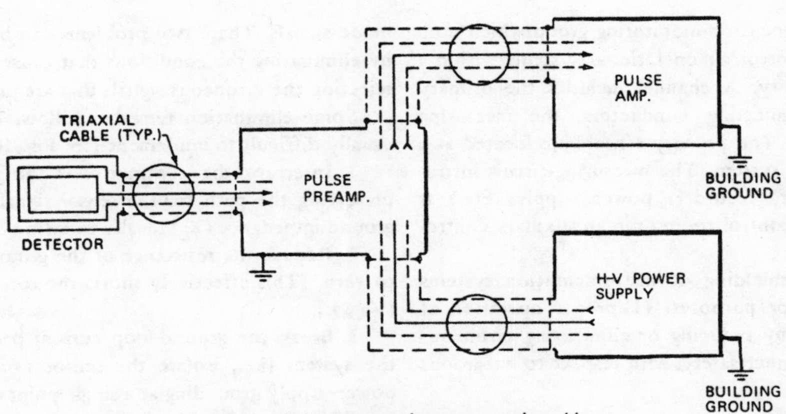

Fig. 10.19—Single-point ground system; ground at cabinet.

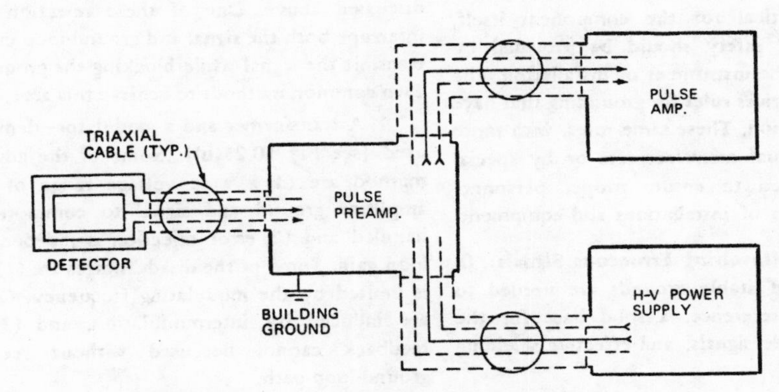

Fig. 10.20—Single-point ground system; ground at preamplifier.

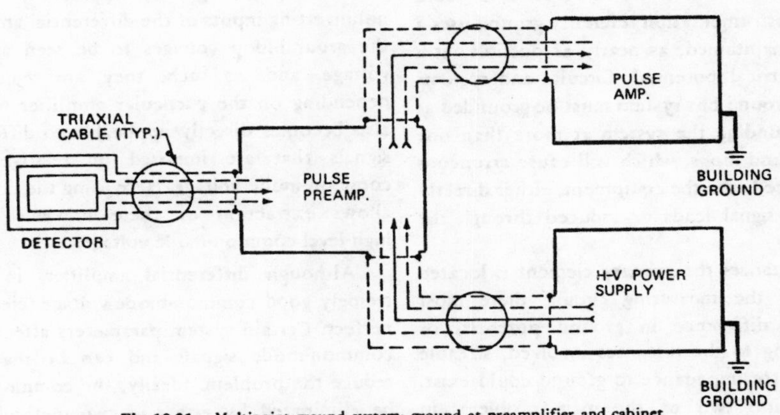

Fig. 10.21—Multipoint ground system; ground at preamplifier and cabinet.

systems outside the neutron-monitoring group which indicate, record, and control all operational systems within a nuclear reactor facility. A channel includes the primary sensor, the interconnecting conductors, the measuring circuit, and displays. The primary sensors are located at a point in the process system. The measuring-circuit instrumentation (amplifier, recorder, power supply, etc.) is located in the main control room or in an auxiliary control area.

Grounding and shielding of instrumentation systems accomplish two major purposes: (1) proper operation of the instrumentation by reducing or eliminating erroneous signals and (2) personnel safety with respect to hazardous operating parameters.

(a) Personnel Safety. Equipment grounding provides safety for personnel who might come in contact with the equipment. As far as instrumentation is concerned, two main hazards exist for which protection must be provided: electrical shock and heat. All external surfaces should be at ground potential and at a temperature less than 120°F. Where this is not practical for the component itself, protection for personnel safety should be provided by completely surrounding the instrument or installation. The National Electrical Code gives rules for grounding that have wide commercial application. These same rules, with more details added by individual manufacturers or by special applications, can be used to ensure proper personnel protection for most types of installations and equipment.

(b) System Problems (Involving Erroneous Signals). In instrument systems, good stable grounds are needed to provide a measurement reference, a solid base for the rejection of common-mode signals, and effective shielding for low-level circuits.

A stable reference for measurement can be provided positively in only one way: by referencing all measurements to a single-point ground. However, this is not possible except through the use of a floating system (i.e., all system components completely isolated). A floating system also provides a firm base for the rejection of common-mode signals. Compared to single-point grounding, the instrumentation reference ground is inferior; however, it is more generally used. The instrumentation reference ground uses a grid or bus that is maintained, as nearly as possible, at a consistent fixed electrical potential. Circuits and systems using this type of a ground bus system must be grounded at one point only. Grounding the system at more than one point will create ground loops, which will cause erroneous signals to be introduced into the equipment, either directly through one of the signal leads or induced through the shielding.

Since in most instances the primary element is located some distance from the measuring circuit, there most certainly will be a difference in ground potential or reference. Also, owing to the distances involved, sizeable differences in conductor impedance to ground could exist. These conditions cause two of the main problems in low-level measuring circuits, ground loops and common-mode signals. These two problems can be dealt with either by eliminating the conditions that cause the problem or by rejecting the erroneous signals that are produced as a result.

Some elimination remedies follow. These remedies are usually difficult to implement [see Fig. 10.23(a)].

1. Interrupt the continuity of the ground loop while preserving the path for the sensor signal [i.e., increase the ground impedance (Z_g) ideally to infinity].

2. Reduce the resistance of the ground conductor (R_g) to zero. [This effectively shorts the total ground potential (V_{AB}).]

3. Break the ground-loop current path (I_g) by floating the system (i.e., isolate the sensor or the amplifier and power-supply grounding at a single point only). Note: If the amplifier is used to feed signals to a recorder, analog-to-digital (A/D) converter, display, or other data-handling device, the path of the ground-loop current may be reestablished through these devices.

Generally, some method of rejecting the erroneous signals is more practical than the elimination remedies discussed above. One of these rejection remedies is to interrupt both the signal and ground-loop currents and then transmit the signal while blocking the ground-loop current. Two common methods to achieve this are:

1. A transformer and a modulator–demodulator can be used [see Fig. 10.23(b)]. Some of the advantages of this method are (1) a wide voltage range of both signal to amplifier ground and signal to common mode can be handled and (2) error rejection is independent of closed-loop gain. Some of the disadvantages are (1) the bandwidth is limited by the modulating frequency, (2) output errors are induced by intermodulation, and (3) error-reducing feedback cannot be used without reestablishing the ground-loop path.

2. A switched capacitor can be used [see Fig. 10.23(c)]. This method has the same advantages and disadvantages as the transformer method with the added disadvantage of poor frequency response.

A popular method of curing ground-loop problems is by using a differential amplifier. Identical fractions of the ground-loop voltage are applied to the inverting and noninverting inputs of the differential amplifier. This causes the ground-loop voltages to be seen as a common-mode voltage, and, as such, they are rejected (to a degree depending on the particular amplifier used). This method also becomes directly applicable to differential transducer signals that are imposed on relatively high levels of common-mode voltage. Here using the differential amplifier allows extraction of these low-level signals from the high-level common-mode voltages.

Although differential amplifiers in general have extremely good common-mode voltage rejection, they are not perfect. Certain system parameters affect the level of these common-mode signals and can be manipulated to help reduce the problem. Ideally, the common-mode signals can be eliminated by either of two methods [Fig. 10.23(a)]: reducing the conductor resistances of both the signal and

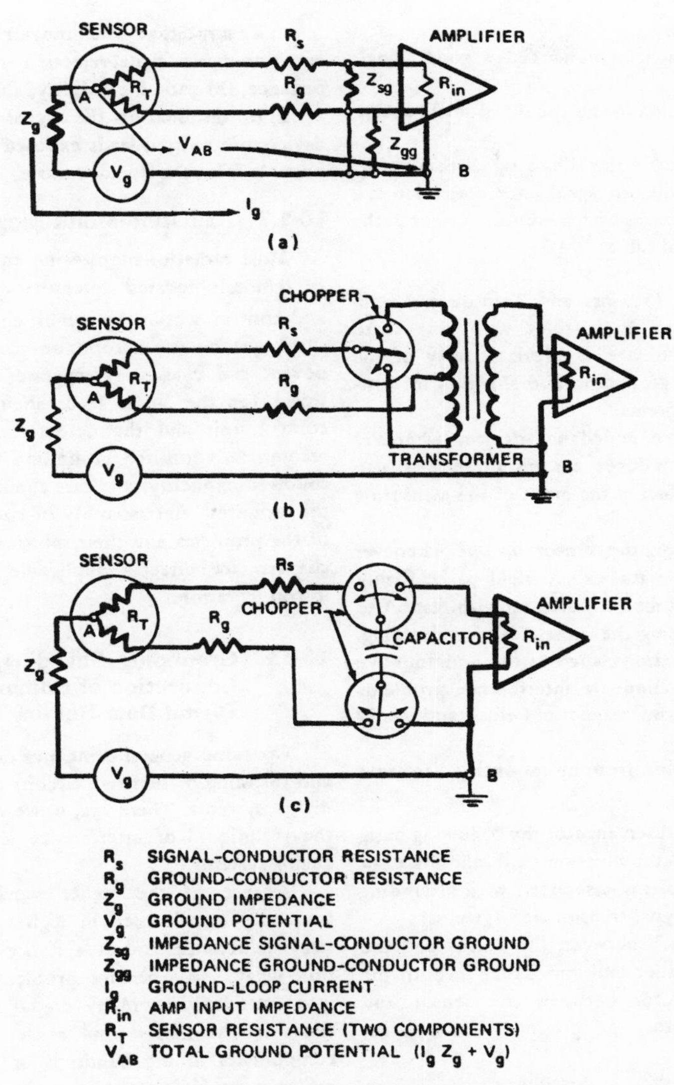

Fig. 10.23—Ground loops and common-mode signals.

ground (R_s and R_g) to zero or increasing the impedance to ground of both the signal and ground conductors (Z_{sg} and Z_{gg}) to infinity. Practically, this same result can be attained by making the ground and signal conductor resistances (R_g and R_s) equal and the signal and ground impedances to ground (Z_{sg} and Z_{gg}) equal. This can be accomplished by observing the following rules:

1. Use a "balanced line" between the sensor and the amplifier (i.e., equal resistance and impedance in both the signal and ground conductor between the sensor and amplifier and the conductor and ground). This can be done most easily by using a shielded twisted pair of conductors for transmission of the signal from the sensor to the amplifier.

2. Keep signal cables as short as possible.

3. Use a source with a center tap if possible.

Both the signal source and the cable impedance have a shunting effect on the input signal to the amplifier. Having equal impedances at each end (sensor and amplifier) idealizes maximum power transfer; however, in transferring voltage signals, this is detrimental owing to line losses caused by system impedance. Therefore, if low-impedance sensors and high-input-impedance amplifiers are selected, the amount of signal current can be kept at a minimum, thus minimizing system error due to voltage drop. For systems where higher sensor impedance is required, correspondingly higher amplifier input impedance should be used. Good practice dictates that the input impedance of the amplifier should be at least 10 times the output impedance of the sensor.

The following general rules should be observed in the installation of low-level signal systems:

1. Avoid ground loops.

2. Provide a stable signal ground and a good signal-shield ground.

3. Ground the signal circuit and the signal shield at one common point.

4. Never use a signal-cable shield as a signal conductor.

5. Ensure that the minimum signal interconnection is a uniformly twisted pair of wires with all return current paths confined to the same signal cable.

(c) Primary Elements (Sensors and Transducers). Observance of the following basic rules with respect to primary elements will eliminate or alleviate many of the problems associated with grounding and shielding of low-level-signal-transmission systems:

1. Use low signal-source impedance devices whenever possible. This not only reduces system noise but also minimizes the shunting effect at the input of the measuring circuit.

2. Use a center tap on the sensor output whenever possible. This permits the signal-cable shield to be firmly fixed and operated at a minimum potential with respect to the signal pair, thus providing the most effective shielding.

3. Use special configurations, such as the noninductive strain gage, to reduce or eliminate interference problems from electromagnetic fields, magnetic fields, and other types of induced noise.

4. Ensure proper isolation from all mounting hardware for isolated sensors.

(d) Interconnection. Observance of the following basic rules with respect to interconnection will eliminate or alleviate many of the problems associated with grounding and shielding of low-level-signal transmission systems:

1. Use a "balanced line" between the sensor and the amplifier (i.e., equal resistance and impedance in both the signal and ground conductor between the sensor and amplifier and the conductor and ground). Use twisted, shielded pair.

2. Keep signal cables as short as possible.

3. Never use splices in signal leads.

4. When using connectors (multipin): (1) use adjacent pins for signal pairs, (2) carry shield through pins adjacent to signal pins, (3) use spare pins as a shield around signal pair by grounding them together and then to the signal shield.

5. Separate low-level-signal cables and power cables by the maximum practical distance and cross them, where necessary, at right angles.

6. Isolate signal cables with conductive conduits and wireways.

7. Ensure that spare shielded conductors in signal cables are single-end grounded, with the shields grounded at the opposite end.

(e) Measuring Circuit. Observance of the following basic rules with respect to the measuring circuit will eliminate or alleviate many of the problems associated with grounding and shielding of low-level-signal transmission systems:

1. Ensure that the measuring circuit has (1) high common-mode signal-rejection ratio, (2) high input impedance, (3) good d-c stability, and (4) wide bandwidth.

2. In terminating the signal cable to the measuring device, use twisted leads exposed for as short a distance as possible from the shielded cable.

10-5.7 Radiation-Monitoring Instrumentation

Most radiation-monitoring equipment requires the use of remotely located detectors connected to the monitor and control section by a multiconductor cable. Generally, these systems use a common ground for signal reference, power, and chassis. A separate conductor as well as the shield for the signal leads should be used between the control unit and the detector unit. The manufacturer's recommendation for grounding and shielding should be followed explicitly, and care should be taken to ensure that the mounting and assembly of components is proper. Most of the problems and their solutions discussed earlier in this chapter are directly applicable to radiation-monitoring instrumentation.

10-5.8 Grounding, Shielding, and Connection of Computers and Digital Data-Holding Systems

The same general principles of isolation of signal leads and shielding of sensitive circuits apply to both analog and digital systems. There are, however, distinct differences in the techniques of interference suppression in digital and analog systems.

Because of the higher signal levels and the high-frequency signals used in digital devices, standing waves, stray inductances and capacitances, RFI, and line propagation delays may become problems if not carefully taken into consideration. Many digital control signals have rise times in the nanosecond range; so there are frequency components in the hundreds of MHz range. Connecting cables must be treated as transmission lines to pass this information from circuit to circuit.

The points discussed in the following should be considered to provide suitable interconnections between units of high-speed digital systems.

1. Cables carrying frequency components above 100 kHz should be terminated properly, i.e., both ends having an impedance match between the cable and circuit input or output. Ordinary wire has a characteristic impedance of about 150 ohms at frequencies where standing waves become a problem; so this is a good value to choose as a terminating resistor when an approximate first choice is called for. If coaxial cable is used, the terminating resistor should match the cable impedance. The output of the equipment must be able to supply the current to drive the characteristic impedance of the cable under continuous load conditions, if long-term, constant d-c signal levels are expected [Fig. 10.24(a)].

2. Another method of terminating a data cable is shown in Fig. 10.24(b). A Zener diode is used to clamp the

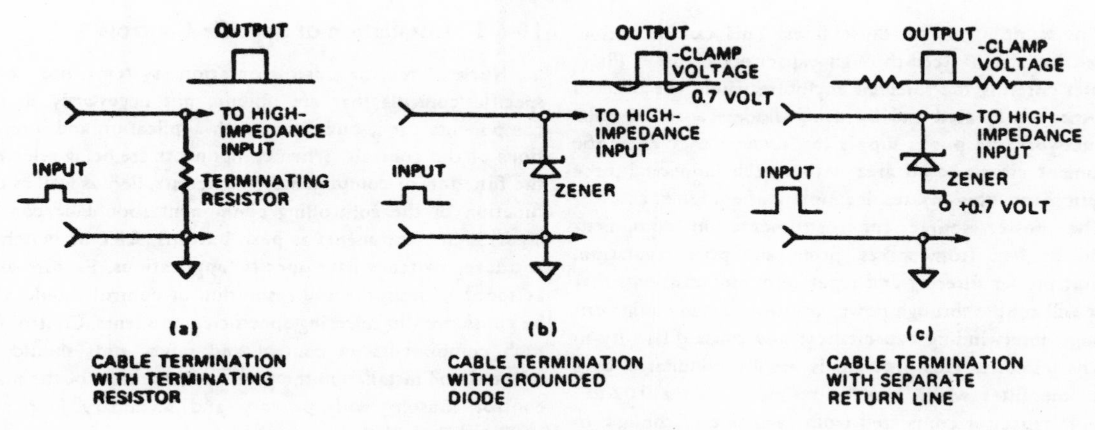

Fig. 10.24—Data-system terminations.

reflected signal to ground and limit signal excursions above the desired input signal. This circuit is useful in the control of standing waves where not enough continuous power is available to supply a low terminating impedance and a series of high-speed single-polarity pulses is to be sent along a line. Figure 10.24(c) outlines the same technique except that the grounding diode is returned to a −0.7-volt line to ensure that the reflected wave on the line is clamped to a value approaching zero.

3. The usefulness of a digital approach to instrumentation derives partly from the fact that a signal is either off or on, and all modern digital systems have varying degrees of sensitivity to noise pulses in the circuit. "Noise immunity" is usually defined as the minimum voltage difference between the higher O-voltage level and the lower I level; this typically ranges between 1 and 10 volts. Because of this, millivolt or microvolt interference levels are of no consequence; so many of the elaborate shielding and guarding techniques described in this chapter are not needed for digital systems. Inputs to pulse-sensitive devices, such as some flip-flops, should be protected by shielding or other means against spikes on the input lines; however, the rapid rise time of an induced noise spike may cause spurious triggering even though it is less than the specified d-c noise immunity level.

(a) Data-System Power and Ground Connection. In general, a small computer or other digital system may be connected to both power and ground by following standard electrical code procedures. It is important, however, that the various cabinets of the system be connected together by a low-resistance bus, such as No. 4 AWG copper wire, and the system be tied to a good earth ground. The steel frame of the building in which the computer is housed is a good ground, or, if this is not convenient, a large water pipe is also a good ground. The ground wire should be run along with the signal cables when it is used to interconnect different portions of the system. A good a-c ground for a signal return is important. As an example, the d-c resistance of a bond strap 0.002 to 0.003 in. thick, 1 in. wide, and 1 to 5 in. long would be negligible at direct current but would be about 0.1 ohm at 10 MHz and 15 ohms at 1000 MHz. This relatively large change in impedance with frequency indicates there is no substitute for a short grounding connection with a large cross section and low self-inductance.

The quality of the ground depends on the type of system since a self-contained system (one having no remote transducers or A/D converters) is more immune to noise than, for example, a system accepting thermocouple signals directly. For this reason all the interference-reduction ideas mentioned in this chapter should be applied to any system that has many small signal circuits.

10-5.9 RFI and Electromagnetic Shielding

An enclosure that has high conductivity and completely surrounds a piece of equipment forms an excellent shield against RFI radiation, provided it is grounded. A useful rule of thumb, however, is that below 2 to 3 MHz interference from one component to the next in a system is primarily electromagnetic (i.e., the coupling is by magnetic and electric fields), whereas above this frequency radiated (r-f) energy is the primary carrier of interference. This means that low-frequency shielding should be mostly composed of ferromagnetic materials to eliminate magnetic fields and high-frequency shielding should have high conductivity since magnetic fields are not significant. Often, if magnetic materials, such as steel, are used at high frequencies, the ohmic resistance they have (compared to materials such as silver) causes potential differences and subsequent electric fields to be set up in shields around sensitive circuits, thereby nullifying some of the effectiveness of the shields. In the gray area (a few megahertzs) composite shielding, such as copper-coated steel, is often used. These shields should be used wherever the adjacent equipment or components are sensitive to interference (e.g., coaxial shields applied on input leads and shields placed over any gaps in the case if either enclosed or external circuits are r-f radiation-sensitive).

Though shielding is effective in the elimination of interference, additional suppression is sometimes added by filtering inputs, outputs, and power connections to eliminate noise riding in on those lines. This can be accomplished by placing feed-through capacitors or other filters on lines entering the shielded enclosure where the affected equipment is located and by using diode–capacitor isolation networks on power-supply leads where they enter the instrument circuit-board area within each individual piece of equipment. This ensures decoupling of equipment.

The power source for instrumentation equipment should be free from spikes, jitter, and poor regulation. Without proper filtering and regulation, any transients that occur will couple through power transformers in equipment (through interwinding capacitance) and cause difficulty in sensitive circuits. This problem is usually eliminated by a power-line filter which may be no more than a 0.02-μF, 600-volt capacitor connected from each side of the line to ground.

In summary, interference can, and usually does, enter equipment wherever there is an unprotected entryway. It is up to the equipment installer to make sure that no signals other than those desired by the circuit designer enter the equipment. To accomplish this task, he must be aware of the many techniques available for suppression of interference.

10-6 INSTALLATION OF REACTOR CONTROL, DISPLAY, AND RECORDING EQUIPMENT

The control, display, and recording functions of any system constitute the major portion of the instrumentation. Man is linked to the machine by the displays and controls, and he uses the recorded data to analyze and improve operation.

10-6.1 Control–Display Relation

Proper relation of the control function to the display is the basic purpose of the system. The function of the display system is to provide information to enable the operations personnel to make decisions regarding plant and system operation. The grouping of control–display components should always reflect the use of human factors engineering. The maximum efficiency with which data can be perceived and the control action initiated indicates the effectiveness of the control–display design.

10-6.2 Installation of Hand Controls

The installation of hand controls should utilize human-factors engineering. Considerable thought must be given to aspects such as type and placement of controls as related to function. Likewise, operator qualifications and limitations should be considered. All controls should be mounted to withstand the rigors of normal use and possible abuse. Controls affecting plant or system safety should be protected from accidental actuation. Controls that are subject to extreme environments should be protected, including the marking and identification of the control.

10-6.3 Installation of Specific Controls

Nuclear reactor instrumentation systems use some specific controls that are unique, not necessarily in the components used, but rather with application and interactions of the controls. When components are being selected, the function of components being controlled as well as the function of the controlling component should be considered. Such components as push buttons, selector switches, and level switches have specific applications. Factors such as speed of response and resolution of control should also be considered in selecting specific components. Control for such components as control and safety rods should be selected and installed in the prime control areas of the main control console, with primary and secondary loop and auxiliary controls being installed according to frequency of use and importance to the system.

10-6.4 Installation of Visual Display and Recording Equipment

The location of all equipment that must be visually monitored should be given priority consideration. The physical abilities and limitations, as well as the psychological characteristics, of the operating personnel should be taken into account. Viewing distance and angle as well as illumination must be considered, along with character size, configuration, and background, to avoid eye strain, inaccurate perception, and glare. Audible display should be considered in areas where there is a possibility of failure to give the necessary visual cognizance for critical and semicritical parameters.

Compatibility with related controls and possible perceptual interaction with other displayed information are important factors in the installation of effective displays. Recording equipment also used for displaying information should adhere to the criteria for any display equipment. In addition, recording equipment must be installed so that routine operational maintenance, such as inking and chart-paper replacement, can be accomplished with maximum efficiency.

10-6.5 Installation Recommendations

The installation of an instrumentation system involves the use of a combination of practices to achieve the desired results, both from a functional and an aesthetic standpoint (see Fig. 10.25). Instruments and components should be grouped and mounted so that they are aesthetically neat and orderly as well as logically functional. Instruments with varying front-panel dimensions should be mounted so that the instrument case tops are level. Switches and pilot lights should be mounted with the center lines even. Instruments with hinged doors should be mounted so that there is no interference between adjacent instruments. In general, controls, displays, and recorders should be placed in the prime functional areas of the panels in which they are mounted. Components that require visual readout should be as close as possible to eye level. All control actuation components should be kept within easy reach (at least 18 in. above floor level).

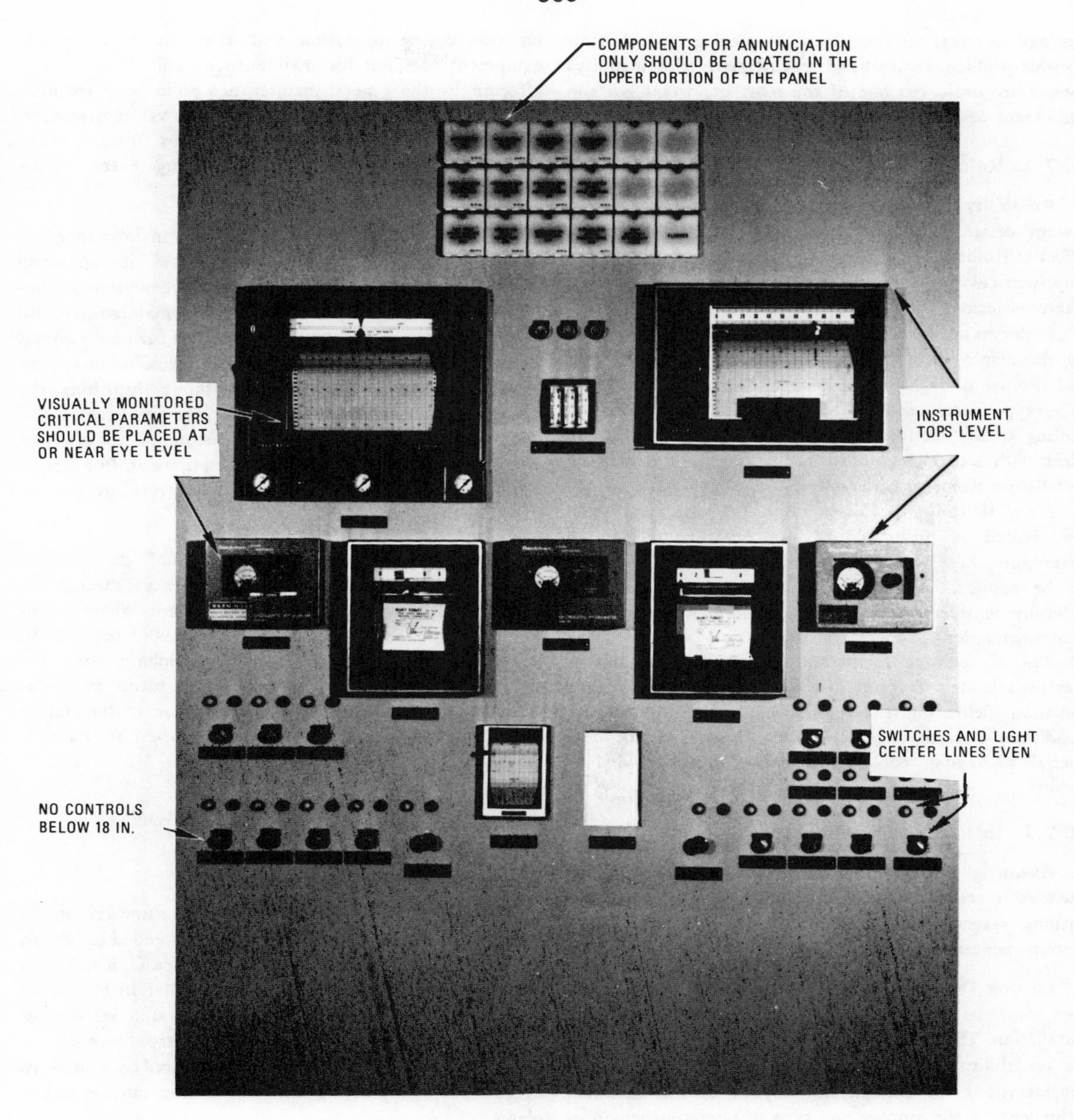
COMPONENTS FOR ANNUNCIATION ONLY SHOULD BE LOCATED IN THE UPPER PORTION OF THE PANEL
VISUALLY MONITORED CRITICAL PARAMETERS SHOULD BE PLACED AT OR NEAR EYE LEVEL
INSTRUMENT TOPS LEVEL
SWITCHES AND LIGHT CENTER LINES EVEN
NO CONTROLS BELOW 18 IN.

Fig. 10.25—Installation of layout for display and control.

10-7 INSTALLATION FOR EASE OF MAINTENANCE

10-7.1 Reliability and Maintainability

The value of any instrumentation system is largely determined by its use factor, i.e., the amount of operating time compared to the amount of downtime attributable to instrument or component failure during normal operation of the system. The use factor is a function of both the reliability and maintainability of the instrumentation system. Reliability is concerned with the *frequency of failure*, whereas maintainability is concerned with the *duration of failure* (see sec. 11).

Although in recent years gains have been made in component reliability, they have almost always been exceeded by increases in system complexity. With the certain evolution of MSI and LSI (medium-scale and large-scale integrated circuits), which will increase complexity on a component level without reducing reliability, systems reliability in the areas of total plant control and monitoring will begin to improve. Presently, the day of totally reliable, maintenance-free systems appears to be still in the future. This condition has forced the electronics industry to develop the technology of instrument and system maintainability. Although instrument manufacturers have always considered easy maintenance as a desirable feature, the complexities of today's systems have dictated renewed interest and analysis. Achieving the shortest possible problem analysis and repair time (maintainability) should be considered one of the most important goals in equipment design and installation.

10-7.2 Reliability

Reliability is primarily a function of instrument and system design. However, two aspects of installation can affect reliability: the quality of materials and of workmanship (particularly in system and component interfacing and interconnection) and the operating environment.

Problems in the quality of installation can be minimized by the employment of highly qualified, competent labor and the use of highly reliable installation hardware. Using crimp-type terminals and connectors with the proper tooling greatly reduces termination problems. Installation plans with step checkoff sheets or diagrams can improve installation efficiency and accuracy.

Many reliability problems associated with installation are created by unsuitable environmental conditions. In determining how instruments, components, and systems are to be applied, the environmental parameters must be carefully considered during installation. The manufacturer's recommendations concerning installation and operating conditions, such as temperature, humidity, cleanliness, electrostatic and magnetic fields, vibration, and nuclear radiation fields, must be given close attention. Where conditions exist that are contrary to those recommended, special packaging, cooling, etc., should be provided as required.

10-7.3 Installation for Maintainability

Assuming that all the equipment in the system is installed in accordance with the manufacturer's recommendations, system performance now becomes a function of system maintainability.

(a) Unit Design and Equipment Spacing. Several factors must be considered in the design of the system installation. The type and location of the plant determines the availability of maintenance personnel. Space and access considerations for the physical layout of the plant are influenced by the number and type of operating personnel on site during operation and also the type of test equipment required for maintenance. Unit design is also affected by the general maintenance philosophy for each instrument or system, such as repair vs. replacement, components vs. modules, and on-line vs. off-line maintenance. This, in turn, is governed to a large extent by the availability of replacement parts.

(b) Cables and Connectors. When maintenance requires removal of the equipment from its operating configuration, the ease with which the removal is accomplished is a prime factor in successful maintenance. All cables and connectors should be installed for easy removal without the use of special tools. It should always be possible to remove equipment without disturbing the operation of adjacent equipment. Likewise, cables and connectors of the operating equipment should not interfere with the removal of the defective equipment. Connectors should be indexed and coded so that incorrect interconnection is virtually impossible.

(c) Displays for Maintenance. One of the quickest ways to determine maintenance requirements is through the use of a self-annunciating or display system. Where practical, this type of failure alarm system should be used. In instrumentation systems where self-checking is used, the installation of the display unit should blend as far as possible with the operational configuration of the system and should become conspicuous only when an alarm is actuated.

10-8 SUMMARY OF INSTALLATION PRACTICES

Table 10.10 has been prepared as a summary of the material presented in this chapter. It is intended to furnish the reader with an easy reference on sound installation practices and problem areas in the installation of reactor instrumentation systems. Opinions on sound installation practices are diverse in the nuclear industry; however, the opinions expressed in Table 10.10 are shared by a majority of the people operating and maintaining nuclear power plants.

Table 10.10—Installation Practices for Reactor Instrumentation Systems

Do's	Don'ts	Do's	Don'ts
Make certain the insulation is thoroughly stripped off wire before crimping on lug	Don't leave wire strands out of lug when crimping	Use triaxial cables with a floating shield operated at a fixed potential	Don't ground the shield of a cable at both ends to avoid ground loops
Give installed connector a resistance and voltage test to assure proper operation		Use triaxial cable for low-level signals, such as from neutron detectors and ionization chambers	
Reserve one side of terminal blocks for terminating field wiring		Provide easy access to the back of instruments behind the instrument panels	Don't install instrumentation, conduit, or electrical boxes behind instruments mounted in panels
Use large-enough conductor to ensure proper grounding	Don't use panel structural member as a ground conductor	Mount all visually monitored instruments at or near eye level	Don't mount hand controls below 18 in. above the floor level
Bond all racks and chassis together and to ground	Don't use a common ground return wire for several relays	Protect critical controls from accidental actuation	
Support cables and wires at several points	Don't support cables and wires by terminations	Ground all conduits to the main building ground bus	Don't use aluminum conduit in concrete unless care is taken to prevent corrosion
Install coaxial and triaxial signal cables in metal conduits	Don't install power cables in the same conduit as signal cables	Drill drain holes in conduits at low points to allow water condensing inside to escape	Don't form bends in conduit in such a manner that water can collect in them
Keep switching command cables, such as those for relays etc., isolated from low-level signals for detectors	Don't think that all signal levels in instrumentation are the same and lump all the cables together	Use covered wiring trays for power leads, and band all tray sections together	Don't join aluminum and steel conduit to each other
Carefully inspect coaxial and triaxial cable installations and procedures to ensure work is properly done	Don't make coaxial cable out of triaxial cable by connecting both shields together	Provide adequate temperature-sensitive circuitry	Don't stack instruments in racks without regard to ventilation or temperature rise
Allow extra conductors in cables wherever practical for future expansion (10 to 15%)	Don't let cable "fill" in a conduit exceed 40% in initial installations	Be sure that radiation detectors are not overheated by neutron heating, etc.	Don't mount instrumentation in areas of vibration unless precautions are taken
Single-end ground spare shielded signal leads in a cable with the shield grounded at the opposite end	Don't ground circuits at random places or allow them to become grounded unless a ground is called for	Pull cables by hand when possible to ensure that no problems exist which will cause the cable to stick	
Provide terminal blocks with terminals adequately sized to handle the physical as well as the electrical requirements of both the interior and field wiring	Don't forget to mark each wire and cable with appropriate identification to assist in circuit tracing	Terminate high-frequency cables properly to avoid end reflections and standing waves	
Keep signal lead as short as possible	Don't use splices in signal leads	Shield all noise-sensitive circuits from electrostatic as well as magnetic fields if magnetic fields are a problem	Don't use adjacent pins in a multipin connector for signal and power circuits
Use line filters and shielded transformers wherever necessary	Don't think interference won't occur; it will	Use differential amplifiers to eliminate common-mode voltages and ground-loop problems	
Segregate loads on power lines so that motors, welders, and other machinery are not on the same line as instrumentation	Don't neglect to put interference suppressors on any sort of device that may generate interference, i.e., relays, motors, fluorescent lights, welders, and heaters	Use a sensor with a center tap where possible	
Consider radiation environment as well as temperature, moisture, etc., when choosing cables	Don't apply higher than rated voltages to coaxial and triaxial connectors	Use a balanced-line shielded twisted pair of conductors for low-level signal transmission	
		Use high-impedance measuring circuits	Don't use high-impedance sensors, if possible

Appendix

ENGINEERING DATA SHEETS ON NEUTRON-MONITORING START-UP-CHANNEL GROUNDING AND SHIELDING PRACTICES AND EXPERIENCE IN SELECTED U. S. NUCLEAR POWER PLANTS

Nuclear Power Plant A

Type of neutron detector: Fission counter

Type of signal cable: RG-71/U double shielded

Location of pulse preamplifier: Near loading-face shield at top of detector wells

Location of pulse amplifier: In amplifier cabinet at control room

Distance between pulse amplifier and preamplifier: Approximately 120 ft

Distance between neutron detector and preamplifier: Approximately 28 ft

Method of grounding: Multiple-point grounding system. Both the preamplifier and amplifier are grounded to the building ground at the point of their installation. The signal-cable shield is grounded at the preamplifier and the amplifier. The neutron detector is grounded through the signal-cable shield back at the preamplifier.

Operating problems and modifications: During the first year of operation, it was noted that the excessive noise was eliminated by adjustment of the discriminator threshold bias control on the pulse amplifier. Toward the end of the first year of operation, the noise became so excessive other methods were required to correct the problem. Examination disclosed that a No. 8 conductor was installed between the preamplifier and amplifier cabinets to establish building grounds between these units. The conductor resistance measured greater than 3 ohms. Ground circulation currents caused by an excessive pickup of r-f noise from a-c and d-c machinery throughout the reactor building produced an emf across the grounding conductor. The emf modulated the signal to the amplifier cabinet, causing excessive noise in the neutron-monitoring channels. This source of noise was eliminated by removing the No. 8 conductor and installing two No. 2/0 grounding conductors in its place.

Other shielding problems: Noise showed up in the automatic servo-control system during the first year of operation. Examination disclosed that a shield on a multiconductor shielded cable had not been terminated to ground by the construction contractor. The problem was corrected by properly terminating the shield to the reactor building ground.

Nuclear Power Plant B

Type of neutron detector: BF_3

Type of signal cable: Triaxial

Location of pulse preamplifier: At top of detector instrument well

Location of pulse amplifier: In amplifier cabinet at control room

Distance between pulse amplifier and preamplifier: 220 ft

Distance between neutron detector and preamplifier: 35 ft

Method of grounding: A single-point grounding system is used. The neutron detector is grounded to the building ground at its point of installation. The inner and outer shields of the triaxial cable are grounded at the neutron detector. One electrode of the neutron detector is connected to the inner shield. The inner shield to the preamplifier is connected to the preamplifier-signal ground. The outer conductor is insulated and floated above ground back to the amplifier cabinet. The signal ground of the preamplifier and amplifier are likewise insulated and floated above building ground, being grounded at the detector.

Operating problems and modifications: During the initial installation and preoperational testing, it was discovered that the start-up channels were excessively noisy, which prevented further operation of the plant. These problems were corrected by installing LR filters in the signal lead at the preamplifier output. Other noise problems were isolated and corrected at the source, such as faulty switches, relays, and motor starters.

Nuclear Power Plant C

Type of neutron detector: Fission counter
Type of signal cable: Triaxial
Location of pulse preamplifier: Preamplifier cabinet at reactor
Location of pulse amplifier: Control-room cabinet
Distance between pulse amplifier and preamplifier: 600 ft
Distance between neutron detector and preamplifier: 60 ft
Method of grounding: The system uses a single-point ground. The system is floated above building ground and is grounded to a special instrument ground in a 700-ft-deep well. A shield around the detector and detector thimble is connected to building ground.

Nuclear Power Plant D

Type of neutron detector: Fission counter
Type of signal cable: Coaxial
Location of pulse preamplifier: Top of counter tube at reactor
Location of pulse amplifier: Control room
Distance between pulse amplifier and preamplifier: 250 ft
Distance between neutron detector and preamplifier: 50 ft
Method of grounding: This system uses a single-point ground at the pulse amplifier. The detector and preamplifier chassis are also equipped with r-f paths to building ground at the counter tube.
Operating problems and modifications: After the system was installed, the preamplifier gain was increased to give a better signal-to-noise ratio for the signal between the preamplifier and pulse amplifier. A filter was installed in the signal output lead of the preamplifier; this filter eliminated the disturbance on the system caused by the interaction of the ground system in the pulse-amplifier cabinet and the building ground at the counter tubes. To protect against any ground loops, the shield on the high-voltage coaxial cable to the preamplifier was lifted from the preamplifier chassis and terminated into a high-valued bleeder resistance. Low-pass filters were installed on the power leads to the nuclear instrumentation to eliminate the signal induced by having high- and low-power leads running together.

Nuclear Power Plant E

Type of neutron detector: BF_3
Type of signal cable: RG-59/U coaxial
Location of pulse preamplifier: Instrument pit at edge of loading-face shield
Location of pulse amplifier: Control room
Distance between pulse amplifier and preamplifier: 200 ft
Distance between neutron detector and preamplifier: 75 ft
Method of grounding: Single-point grounding system. The detector and the preamplifier are isolated from ground. The coaxial-cable shields and all external chassis are grounded to the pulse-amplifier chassis. The pulse amplifier is grounded to the nuclear instrument panel.
Operating problems and modifications: When placed in operation, the system was plagued by noise bursts, high-level background noise, cable ringing, and electromagnetic pickup so large in magnitude that neutron pulses were not distinguished by the system. Many multiple grounds in the system were found and eliminated. The system was converted to a single grounded system by insulating the detector and preamplifier from ground and tying the system to ground at the pulse amplifier. The system was improved greatly; however, some disturbances were still present in the system. These disturbances, although not large enough to prevent operation, were annoying, and it was felt that the use of triaxial cable in the system would eliminate the disturbances altogether.

Nuclear Power Plant F

Type of neutron detector: BF_3
Type of signal cable: Triaxial
Location of pulse preamplifier: In amplifier cabinet at control room
Location of pulse amplifier: In amplifier cabinet at control room
Distance between pulse amplifier and preamplifier: 2 ft
Distance between neutron detector and preamplifier: Approximately 140 ft
Method of grounding: A single-point grounding system is used. System ground is made at the amplifier cabinet. The signal cable and neutron detector are insulated and floated above ground. The two shields in the triaxial cable are connected together to one electrode of the neutron detector. The two shields are grounded at the amplifier cabinet. The signal ground in the preamplifier and the amplifier are grounded at the amplifier cabinet. The amplifier cabinet is grounded to the building ground through grounding cable and buses.

Nuclear Power Plant G

Type of neutron detector: BF_3 and fission counters
Type of signal cable: Coaxial between detector and preamplifier; triaxial between preamplifier and amplifier
Location of pulse preamplifier: Top of neutron-detector instrument well
Location of pulse amplifier: In amplifier cabinet at control room
Distance between pulse amplifier and preamplifier: 185 ft
Distance between neutron detector and preamplifier: 25 ft
Method of grounding: A single-point grounding system is used. System ground is made at the amplifier cabinet. The signal cabling, preamplifier, and neutron detector are insulated and floated above ground, being grounded at the amplifier cabinet. The two shields in the triaxial cable are connected. Coaxial fittings are used for both triaxial and coaxial cables. The two shields are grounded at the amplifier cabinet. The signal ground in the amplifier is grounded at the amplifier cabinet. The amplifier cabinet is grounded to building ground through grounding cable and buses.
Operating problems and modifications: During the initial installation and preoperational testing, it was discovered that the start-up channels were subject to transient noise problems from several systems throughout the reactor plant. These noise problems were corrected at the source by replacing faulty relays and switches and by providing better shielding for fluorescent lighting. After several years of operation, an unusual noise problem showed up at this power reactor site. The reactor was shut down in preparation for refueling. Just prior to the start of the second fuel-loading program, r-f noise showed up on all the signal buses coming out of the containment sphere. The noise was of such magnitude that all operations were suspended. After several days of attempting to isolate the source of r-f noise, the breakers at the 2180 kv substation were opened and closed by a planned operation. The noise all but disappeared. By readjusting the discriminator threshold bias control on the pulse amplifier, the start-up channels were brought within safe operating condition again. A technical explanation has not been given as to what caused the severe r-f noise problem.

Nuclear Power Plant H

Type of neutron detector: Proportional counter
Type of signal cable: Coaxial
Location of pulse preamplifier: Source-range detector thimble junction box
Location of pulse amplifier: Nuclear instrument cabinet
Distance between pulse amplifier and preamplifier: 1210 ft
Distance between neutron detector and preamplifier: 19 ft
Method of grounding: A single-point grounding system is used. The grounding point is at the pulse-amplifier chassis. The pulse-amplifier chassis is grounded to the station grounding grid by bus bar and cable.

Operating problems and modifications: During preoperational checkout of the system, it was discovered that the majority of the coaxial connectors in the nuclear instrument system were assembled with poor workmanship. The main problems were the preparation of the cable, the soldering of the pins, and the assembly of the connectors. Many hours of rework were required to correct the problems. When the system was put in operation, the source-range channels were very noisy; a check of the system revealed that induced voltage was causing ripple on the +15-volt power supply to the preamplifier. The leads carrying the +15-volt power supply were changed from nonshielded to shielded conductors, which improved the condition but did not eliminate ripple sufficiently to not affect the operation of the source-range channels. A RL filter was designed and installed at the preamplifier in the +15-volt supply. This filter removed the ripple and corrected the noise problem.

Nuclear Power Plant I

Type of neutron detector: Fission counter

Type of signal cable: Single and double shielded coaxial

Location of pulse preamplifier: At top of detector instrument well

Location of pulse amplifier: In amplifier cabinet at control room

Distance between pulse amplifier and preamplifier: 90 ft

Distance between neutron detector and preamplifier: 24 ft

Method of grounding: Multiple-point grounding system. Both the preamplifier and the amplifier are grounded to the building ground at the point of their installation. The signal-cable shield is grounded at the preamplifier and the amplifier. The neutron detector is grounded through the signal-cable shield back at the preamplifier.

Operating problems and modifications: During the initial installation period and preoperational testing of the neutron-monitoring system, noise in the start-up channels was prohibitive. Noise was reduced by replacing the RG-8/U single-shield coaxial cable with RG-71/U double-shield coaxial cable. Other noise problems caused by the operation of relays, motor starters, and switches were eliminated at the source. The switches and motor starters for the fission-chamber carts were a source of noise causing large transients in the start-up channels. Some problems were eliminated by installing filter capacitors across switches.

Nuclear Power Plant J

Type of neutron detector: BF_3

Type of signal cable: Triaxial

Location of pulse preamplifier: In amplifier cabinet at control room

Location of pulse amplifier: In amplifier cabinet at control room

Distance between pulse amplifier and preamplifier: 2 ft

Distance between neutron detector and preamplifier: 225 ft

Method of grounding: A single-point grounding system is used. System ground is made at the amplifier cabinet. The signal cabling and neutron detector are insulated and floated above ground, being grounded at the amplifier cabinet. The inner shield of the triaxial cable is connected to one electrode of the neutron detector. The outer shield is connected to the case of the neutron detector. The two shields are grounded at the amplifier cabinet. The signal grounds in the preamplifier and amplifier are grounded at the amplifier cabinet. The amplifier cabinet is grounded to building ground through grounding cable and buses.

Operating problems and modifications: During the initial installation, it was discovered that the start-up channels were subject to transient noise problems from several systems throughout the reactor plant. The starting and stopping of cranes were a major source of noise. The noise problems were eliminated by installing capacitor filters across motor starters, relays, and switches. Some noise still remains in the system, but the level is not great enough to affect operation of the system.

SECTION 11

QUALITY, ASSURANCE AND RELIABILITY

11-1 INTRODUCTION

11-1.1 Definition of Terms

Quality assurance **comprises all those planned and systematic actions necessary to provide adequate confidence that a structure, system, or component will perform satisfactorily in service. Quality assurance includes quality control, which comprises those quality-assurance actions related to the physical characteristics of materials, structures, or systems which provide a means to control their quality to predetermined requirements.**

Reliability **is the probability that a system, channel, or component will perform a specified function under given conditions for a specified period of time without failure. For instrument channels this probability is normally a function of time because few one-shot measurements are considered in nuclear power plants.**

A product may meet all design specifications when first tested, but, if components of the product are overstressed, they will fail sooner than expected. As a result, a product that passes stringent quality-control requirements may not necessarily be reliable. This relation between the two terms should be kept in mind even though they are treated separately here.

11-1.2 Quality and Reliability Requirements in Reactor Instrumentation

In the past few years a great deal of effort has gone into the generation of definitive specifications, standards, and regulatory requirements concerning all aspects of quality for the purpose of establishing universal guidelines for all nuclear-reactor systems and components manufacturers and their vendors, as well as nuclear-reactor architect–engineers and owner/operators.

Those organizations and personnel concerned with design, construction, and operation of systems and components for domestic power plants are bound by the 18 quality-assurance criteria in Appendix B of Title 10, Part 50, in the *Code of Federal Regulations*. This document, which was promulgated by the AEC, has created an atmosphere within the nuclear energy industry of extreme concern for the quality of systems and components that are important to safety.

Another document, also issued by the AEC, but much more comprehensive, is the standard used by organizations and personnel concerned with reactor development and test facility projects. The objective of this document, RDT-F2-2T, Quality Assurance Program Requirements, "is to assure that structures, components, systems, and facilities are designed, developed, manufactured, constructed, operated, and maintained in compliance with established engineering criteria."

Certain instrumentation devices, such as in-core detectors and penetration seals, which when placed into service become an integral part of a pressure boundary, are required to meet Sec. III of the *ASME Boiler and Pressure Vessel Code*, which addresses itself to Nuclear Power Plant Components. Article NA-4000 of this section, titled Quality Assurance, "sets forth the requirements for planning, managing, and conducting quality-assurance programs for controlling the quality of work performed under this section of the Code."

In addition to the above-mentioned documents, the American National Standard Institute (ANSI), under the sponsorship of the ASME, is issuing a whole series of standards (N45.2 series) for the purpose of guiding organizations and personnel involved in design, construction, and operation of nuclear-power-plant systems and components in proper performance of the quality-related aspects of each phase of the total scope of activities.

There has never been any question about the need for a high level of quality and reliability in nuclear instrumentation, especially since it represents such a visible segment of the nuclear safety system. These documents attempt to describe the contents of an acceptable quality-assurance program, and it is up to the industry to develop such a program and still remain competitive. The remainder of this chapter endeavors to deal with this theme.

11-2 QUALITY ASSURANCE OF REACTOR INSTRUMENTATION

11-2.1 Fundamentals of Quality Assurance

Industrial quality assurance has evolved from a policing function, consisting of final inspection and test, to a defect-prevention system that begins with product conception and ends only when that product has satisfactorily fulfilled its intended function.

At some quality level the balance between cost of

failure and the cost of prevention and appraisal will be at a minimum. A successful quality system operates at or near that level. In the nuclear industry the quality level must be higher than in most industries; this tends to result in higher costs for quality assurance.

In specialized industries, such as those producing manned space vehicles and nuclear reactors, the level of quality has been set more by reliability requirements than by cost considerations. However, as competition in the nuclear industry increases and as competition between nuclear energy and other energy sources increases, quality systems in the nuclear industry will have to meet plant safety and performance requirements at reasonable cost.

(a) Modern Quality Systems. Modern quality systems prevent defects and substandard workmanship by exercising controls over design, materials, processes, and products at the appropriate place in the product cycle. All modern quality systems contain three basic elements: design control, materials control, and process and product control.

Design control consists in the preproduction efforts of Design Engineering, Manufacturing Engineering, and Quality-Control Engineering* in developing a design and production and test methods that will ensure with a high degree of confidence that a quality product can be built and sold for reasonable profit to a customer who will accept the product and remain satisfied over its expected life. Design control must be a joint effort of the three groups. Properly conducted design reviews are essential during this phase.

Materials control consists in the preproduction efforts of Quality Assurance, Materials, and Purchasing. Quality Assurance must evaluate the vendor's capability to perform and must ensure that quality requirements are included with purchase orders. In addition, incoming material must be tested or inspected on a statistically valid basis. An objective in materials control is to establish a certification program that puts the burden of quality control on the vendor rather than on Receiving Inspection. This is essential for products that will be shipped directly to a site. Materials control becomes extremely important whenever Boiler Code materials are involved. Such materials must be traceable to their original heat number regardless of the state of manufacture. This is accomplished by such techniques as color coding, electroetching, and tagging or by the use of move tickets, depending on the state of completion of the part. All materials, whether raw stock or finished goods, must be controlled by Materials, and these controls should be audited periodically to ensure that they are being followed.

Process and product control consists in the evaluation and control of manufacturing facilities (whether skilled workers or production machines) and the inspection and test of the product to ensure that the product meets all engineering specifications and quality standards. Once control of manufacturing facilities has been accomplished through planning and the development of manufacturing and inspection or test equipment, training of operators, etc., process and product control must be maintained by implementing the quality plan. This plan may require audits, automatic test, continuous sampling, or roving inspection. It will specify the points in the manufacturing cycle where certain inspection or tests, or both, are necessary, the records to be kept, control charts, calibration and maintenance schedules of critical equipment, special handling techniques, etc. Finally, there will be final inspection or tests, or both, to be performed and possible special packaging instructions and inspections to be performed at the site in many cases.

(b) Justification of the Quality System. Since about 1950 most companies have incorporated a quality system containing the basic elements. Given the talent, quality costs have been significantly reduced following the initial costs of putting the system in operation. Costs have been reduced because the production lines have produced less scrap and customers have returned fewer goods or demanded less service. The savings have more than paid for increased staff. Savings have also been realized by reducing the number of policing inspectors and testers. This reduction has been made possible by detailed planning of inspection and testing and by using automated test equipment. Cost savings have been experienced by large and small concerns, whether production-line or job-shop oriented.

Positive side effects have been experienced following the institution of a well-planned quality system: improved product design, better processes, and the development of quality mindedness in the production and engineering forces. It is especially important that such systems be set up in the nuclear instrumentation industry and that the industry be organized in such a way as to foster the philosophy of *total quality control*. The difficult, but not impossible, goal of increasing quality levels while reducing quality costs can be achieved when this happens.

(c) Organization of the Quality System. The staffing of an organization to implement and operate a modern quality system depends on the size and resources of the company. The number of engineers and test and inspection supervisors depends on three factors: volume, variety, and complexity of product. In brief, one or more quality-control engineers, process-control engineers, test-equipment engineers, foremen, planners, inspectors, and testers are necessary. In a small organization the process-control engineer can double as foreman; the quality-control engineer can do planning and specify commercially available test equipment, and inspector–testers can combine those two functions. The quality-control engineer participates in design control, writes quality plans, including inspection and test instructions for a given product, evaluates vendor's performance data, analyzes process and product measurements and product service reports, and applies the results to prevent poor-quality material and products in future purchases and production. The process-control engineer is responsible for implementing the quality-control engineer's quality plan (including control of incoming materials, processing equipment, and inspection or test equipment). He should be the leader in solving technical quality problems in the manufacturing area.

Since this chapter is intended to detail quality assurance of nuclear instrumentation systems, the foregoing summary of a quality system and its staffing must suffice. For additional information the reader is referred to Refs. 1 to 3.

(d) Special Aspects of Reactor Instrumentation Quality Control. Nuclear sensors are used to detect the presence and amount (intensity and energy) of neutrons and gammas (see sec. 2 and 3). Sensors that are located in the nuclear-reactor core are considered an integral part of the pressure vessel and therefore fall within the scope of the *ASME Boiler and Pressure Vessel Code.* As a consequence, manufacturing procedures and quality control of these sensors are stringent. Many of the processes used in the manufacture of in-core sensors are peculiar to that product (e.g., uranium and boron coating of electrodes, casting metal to ceramics, outgassing and backfilling with special mixtures, and pressures and purities of fill gases). In-process quality-control procedures rely heavily on mechanical inspection techniques, especially where mechanical tolerances are critical. Helium mass-spectrometer leak testing is important after enclosure welds are made, and insulation-resistance checks are very important wherever parts are mechanically and electrically isolated from each other. Final test and inspection depends on the end use. Actual end-use operating conditions should be simulated as closely as is economically feasible.

Nuclear-reactor readout instruments can be classified as any electronic system that accepts a signal from a nuclear sensor and converts it to usable information. Formerly, this category was normally typified by rack-mounted equipment. However, recently the trend has been to integrate various instrumentation functions directly into panels and consoles; this has necessitated drastic changes in the in-process quality control. With rack-mounted instruments, there are usually printed wire boards to be checked out either as boards or as a part of a system. Obviously the quality of work by personnel who solder components or attach leads to the boards or to instrument chassis must be very high and should be monitored at carefully chosen points in the manufacturing cycle. Quality-control problems are different from those involved in sensor manufacturing.

Nuclear instrumentation systems can be classified as a group of instruments, including sensors, that together perform a specific function, such as the reactor protection system, the neutron monitoring system, the off-gas monitoring system, the rod control system, or the area monitoring system. Usually such systems are housed in one or two panels or a panel and a console so that systems check-out can be made without having to interconnect more than two panels. These panels may house a multitude of instruments, or they may house switches, relays, and meters. The degree to which in-process check-out (continuity, insulation, etc.) can be effected is determined by the configuration of the panel.

Peripheral equipment is a catch-all term for the essential equipment involved in interfacing sensors and reactors or instruments. It includes drive mechanisms, penetration seals, etc., and must be treated individually since each piece of equipment has its own peculiar problems.

11-2.2 Design Control

(a) Early Coordination Efforts. A typical engineering organization has a development group that is responsible for the design work involved in developing new products. If the new product is tied in with a production contract, quality-assurance coordination must be factored into the design cycle as early as possible. This can be accomplished by setting up a team of four or five key personnel to review periodically the progress of the design. Normally the team would consist of representatives from Marketing, Design, Manufacturing, and Quality Assurance. In this way individuals from all the important engineering groups are keyed in to developments on new designs and are in a position to contribute from their special backgrounds. They can feed back information to their respective organizations so that timely preparations can be made for manufacturing and testing the new design.

Another approach is for the design engineer to hold one or more design reviews, depending on the complexity of the design. The participants in the design review normally include the personnel noted above plus any other interested parties. The review should be chaired by the design engineer. So that the review will be successful, all parties involved must be given all the design particulars (drawings, specifications, prototype test results, design reports, etc.) before the actual review.

(b) Design and Performance Specifications. Design and performance specifications must be forwarded to Manufacturing and Quality Assurance as soon as possible so that process development and design and construction of special manufacturing or test equipment may proceed well in advance of the release of the design to Manufacturing. For the same reason, design reviews on new products should be held at critical stages of their development, and process reviews should be held on new processes as they are being developed. The quality-control engineer should be involved in the development of new processes since it is often possible to integrate test and/or inspection equipment right into the processing equipment, thus ensuring automatic feedback to keep the process within specification limits.

(c) Process Development. The quality-control engineer is responsible for ensuring that all manufacturing processes are maintained under control throughout the manufacturing cycle. He can best do this by being well informed about all these processes and by keeping in constant contact with the manufacturing engineer. The controls he institutes must be consistent with product costs and must be compatible with the associated manufacturing equipment.

Two basic types of process control are (1) operator, or

open-loop, control, where the operator adjusts the process to keep it under control, and (2) automatic, or closed-loop, control, where the process is regulated by a feedback system. The objective of process control is to keep the process operating within predetermined operating limits.

Mechanization of measurements and process control may be accomplished in one or more stages of the manufacturing process, depending on the quality requirements placed on the product and the process itself. For example,

1. Preprocess measurement and control may be required for monitoring or controlling the materials or parts entering the process.

2. Measurement and control may be used during processing to regulate the process in response to a measured variable.

3. Postprocess measurement and control may be desirable or necessary if it is difficult or impossible to measure or control the product during the manufacturing process.

4. Features of two or more of the above techniques may be combined.

(d) Prototype Construction and Testing. During prototype construction and testing, the quality-control engineer may have an opportunity to prove out inspection equipment and in-process testing equipment and procedures. He should make every effort to have these developed in time to be used and evaluated during the phase.

Although prototype testing is normally done by the design engineer, the quality-control engineer can assist in these tests and thus gain (1) the product knowledge needed to develop a meaningful quality plan; (2) confidence in the special inspection and testing equipment that has been developed; and (3) the advance information needed to correct the quality information equipment if it does not work satisfactorily on the prototype. Prototype testing should subject the prototype to essentially every environmental extreme in which the unit is intended to operate. This may entail use of expensive equipment that may be used for qualification testing alone. Even if the test work has to be farmed out, a quality-control engineer should be involved in the prototype testing along with the design engineer so he can become familiar with the new product.

Sensor Prototype Testing. The major problem with nuclear sensors is that the operating environment of a nuclear reactor is very difficult and expensive to simulate. This is particularly the situation for in-core sensors, where meaningful tests (e.g., response, burnup rate, saturation level, and signal-to-noise ratio) require accurate simulation. Test reactors are available where, by using specially built thimbles and specially designed instrumentation, meaningful tests can be performed. These should be used whenever possible.

Circuit Prototype Testing. Circuit designs start with tests of breadboards and subassemblies. Each individual module must function within the limits of the environmental extremes for the complete assembly. The design engineer must determine in advance at what stages in the development modules and subassemblies are to be tested to environmental extremes. It may be that certain of these tests are only meaningful after the prototype instrument has been completed. A test program must be carefully planned in advance so that all necessary tests are performed in a logical sequence and testing facilities are available when needed.

System Prototype Testing. More and more systems are being standardized, and it has become essential to perform environmental tests, such as temperature rise, on entire systems. This can be accomplished by shrouding the entire operating system with a plastic hood and monitoring for hot spots with well-placed thermocouples. Appropriate functional tests are performed on prototype systems.

Peripheral Equipment Testing. Peripheral equipment is normally mechanical and may be subject to wear and fatigue failures due to a hostile environment (heat and nuclear radiation). Life tests must be run to evaluate the reliability of the assembly before proceeding with production. Weld integrity should be checked periodically in these tests.

(e) Test Specifications. The results of prototype tests help the design engineer determine the test specifications for the product. The test specification should be a formal communication to Quality Assurance which spells out the tests that Engineering believes are essential to prove out the product functionally and the limits of each test. Test conditions need to be spelled out to preclude any misunderstanding. A description of the test setup should be included.

Although Quality Assurance must have the test specification, this document is not necessarily the only criterion for the determination of test limits on the quality-control test instruction. The quality-control engineer may decide to tighten the limits set up by the test specifications if a particular manufacturing process is not as dependable as the quality-control engineer wants it to be or if the measuring capabilities of the test equipment are such that the credibility of the measurement is in doubt. For example, as a rule of thumb, the measuring equipment should be capable of reading out to at least ten times the specification limits. (This means that if a proportional counter is supposed to be capable of 10^6 counts/sec, then the count-rate meter must be capable of resolving to 10^{-7} sec or 100 nsec.) All test equipment, whether being used for engineering prototype tests or for quality-control final acceptance tests, must be periodically calibrated to standards that are traceable to the National Bureau of Standards.

(f) Engineering-Document Control. Since the test specification is an official engineering document, it must be controlled in the same manner as other engineering documents, such as engineering drawings. The control of engineering drawings, often referred to as blueprint control, must be accomplished both at the place of origin (by

Drafting or Engineering Services) and at the place of use (usually by the Production Control organization).

There are many techniques for maintaining blueprint control. These will not be described here. However, it should be pointed out that there are pitfalls that Quality Assurance personnel should be aware of. Some of these are:

1. Advanced manufacturing releases. These drawings may or may not be identical with the final release, and any planning that is based on advanced releases must always b contingent on review of the final manufacturing release.

2. Marked-up drawings. Sometimes it is necessary for the engineer to mark up a drawing prior to the issuance of a formal engineering change. There must be some system for maintaining control so that Quality Assurance can be certain the loop is closed. One technique is to maintain a log with an open entry that can only be closed when a revised drawing, containing a change identical to that of the marked-up drawing, is issued.

3. Engineering changes. These should be reviewed by the cognizant quality-control engineer before issuance to ensure that any quality planning affected by the engineering change can be revised accordingly. The only effective way this can be handled is to keep the quality-control engineer in series with the engineering change, i.e., if his signature is necessary for issuance of any change that may affect (1) health or safety; (2) functional performance, effective use, or operation; (3) interchangeability, reliability, or maintainability of the item or its repair parts; and (4) weight or appearance (where these are important factors).

(g) Process Instructions. Manufacturing process instructions are as important as engineering drawings and must be controlled; i.e., Manufacturing Engineering would be responsible for generating such documents, but Engineering and Quality-Control Engineering must have approval authority. Any changes to process instructions should be controlled in the same way as engineering changes and should have the same approvals. In this way any process-instruction changes that may affect process limits are reviewed to ensure that no product degradation results and that quality checks are appropriately changed.

(h) Quality Planning. Quality planning is developed throughout the design phase of a new product. It consists in determining all the quality checkpoints that are necessary to ensure, with a high degree of confidence, that any rational customer will be satisfied with the product for the duration of its expected lifetime and that the product satisfies all other special requirements, such as applicable standards and codes.

Quality planning must take into account software as well as hardware requirements. For example, the quality plan must specify the material certification requirements necessary for inspection of raw materials as they are received, the marking and identification of material through its machining and processing, and the in-process inspection and tests and documentation thereof, as well as the final acceptance tests and all necessary paperwork required, both internally and for customer submission. Thorough quality planning provides for each of the following:

1. Determination of control points.
2. Classification of characteristics.
3. Determination of quality levels.
4. Determination of process capabilities.
5. Determination of control procedures.
6. Appropriate record forms.
7. Disposition routines.
8. Routing and handling procedures.
9. Quality information equipment development.
10. QIE calibration.
11. QIE maintenance.
12. In-process test and inspection.
13. Final-product test, inspection, and acceptance.
14. In-process audit (both procedure and product).
15. Outgoing product audit.
16. Shipping inspection.
17. Quality data feedback.
18. Quality measurements.

A good overall quality system may provide for a general quality plan, an area quality plan, a product quality plan, a contract quality plan, and a vendor quality plan.

A general quality plan takes into account quality-control procedures that are common throughout all segments of the business and are followed regardless of what type product is being built or what manufacturing area is involved (e.g., quality information equipment calibration).

An area quality plan integrates all the individual station control plans. (A station control plan is the basic plan for each identifiable manufacturing station, such as a lathe or an electronic assembly bench. This plan is usually an integral part of the manufacturing planning and should include provision for controlling all inputs to the station including direct and indirect materials, e.g., stainless steel and cutting fluid or electronic components and solder, tooling, environment, and workmanship skills. It should also include provisions for monitoring the station continuously and checking the outgoing part or assembly as necessary.) It includes all controls and procedures that are common throughout a manufacturing area.

In an area where there is a definite flow from one station to another, such as an assembly area (as opposed to a machine shop where each article is subject to a different sequence of operations), a flow chart should be constructed to indicate the relations of the various stations to each other and to show every important manufacturing process in its proper sequence with all quality checkpoints inserted. Each manufacturing station and quality checkpoint (inspection or test) should be identified by legend and references to the applicable operating instruction or general inspection procedure.

In an area where there is no particular flow pattern, a schedule of stations should be established which describes each manufacturing operation or station and all the controls of inputs to such stations as well as specific quality checks applicable to these stations.

The area quality plan should describe the environmental conditions required in the area (such as temperature and humidity extremes and cleanliness) and the controls for maintaining such conditions. It should describe any special materials handling or in-process storage requirements peculiar to the area. And it should describe all quality-measuring tools needed for direct support of production as well as the requirements of the test and inspection equipment for the quality checkpoints in the area. Special maintenance work should be delineated for manufacturing tooling, such as stamping dies and cutting tools. Calibration cycles on test and inspection equipment in the area should be reviewed and special exception made to the general quality plan whenever there is to be a deviation from standard practice.

The quality-data feedback system should be described in the area quality plan and should include applicable quality cost data necessary for analysis, how it is to be obtained, and how it is to be fed back to Quality Control Engineering.

A quality training and awareness program is an essential part of a good quality program for each manufacturing area and should provide for both operator training and continuous upgrading and verification of quality personnel. Every quality plan must provide for an audit that ensures adherence to the quality plan in its entirety.

A product quality plan is an integrating plan that ties together the individual quality plans for each of the various assemblies and subassemblies making up the final product and includes all the controls and procedures common throughout the manufacturing cycle of that particular product. The product quality plan should reference all applicable area quality plans and workmanship standards as required.

The number of individual quality plans for assemblies and subassemblies is dictated by the complexity of the final product; however, each quality plan must contain a flow chart indicating the relation between all lower-tier parts and subassemblies. These flow charts should show each important manufacturing process with all necessary quality checkpoints. Each manufacturing station and quality checkpoint (inspection or test) should be identified by legend and referenced to the applicable method sheet or inspection or test procedure. The manufacturing-operations sheets and inspection and test procedures should be an integral part of the product quality plan.

A contract quality plan is an integrating plan that ties together all applicable area quality plans plus the product quality plan and all special customer requirements resulting from the contract. It may modify standard quality plans to the extent necessary to meet all customer requirements. The contract quality plan must contain a schedule spelling out in detail the data requirements and identifying who is responsible for them along with a schedule of target dates for each submittal.

A vendor quality plan is a plan that describes in detail the requirements of the vendor's quality system, including any and all requirements for data submittal. The vendor quality plan should also spell out the special tests or receiving inspection steps that must be taken to ensure receipt of acceptable vendor material.

(i) Process Capability Studies. In addition to making certain that the test and inspection equipment is adequate during the prototype stage, the quality-control engineer must know whether or not the production equipment is capable of meeting the engineering tolerances and, accordingly, must determine the optimum sampling plan. This information can be obtained by performing process capability studies.

"Process capability" has been defined as "that which the process is capable of producing under normal, in-control conditions." The key phrases in this definition are "the process" and "normal, in-control conditions."

The process includes the entire manufacturing process and all that enters into it, such as the raw material, the machine or equipment, the measuring device, and the skill of the operator or inspector or both. The process is a single combination of these factors. One process is with a given raw material, a given machine, a certain operator, and the like; whereas another process may be different only in the raw material used. Practically speaking, many of the processes made up by these various combinations of factors are similar in output and can be considered as one. But only those combinations which will yield the same output under the second condition, "normal and in-control conditions," can be so considered as one.

Normal and in-control conditions are those which yield parts with measurements having a predictable and normal frequency distribution compatible with the target specified. Since the process capability is a forecasted distribution of the variability for a given process, this distribution needs to be predictable not only in the spread but also in the shape. Generally, most distributions that are not normal indicate a lack of control and nonnormal conditions.

Quantitatively, the process capability is defined in terms of six standard deviation units (6σ). Within $\pm 3\sigma$ from the mean lie 99.73% of all the readings for a normal distribution. For the majority of operations, this 6σ interval includes practically all the readings and represents the capability of the process. Thus, if the process capability is less than the drawing tolerances, a certain amount of sorting and scrap will result.

The process capability study is a powerful tool. Not only can it be computed easily but also its uses are many, including providing the following information:

1. To facilitate the design of a product.
2. For acceptance of a new or reconditioned piece of equipment.
3. For scheduling work to machines.
4. For setting up a machine for a production run.
5. For establishing control limits for equipment that has a narrow process capability in comparison to the allowable tolerance band.
6. For determining the economic nominal around which to operate when the process capability exceeds the tolerance.

The following points should be kept in mind when a process capability study is being performed:

1. The study should be taken under normal conditions of operation.

2. Factors in the manufacturing process that will introduce nonrandom variations should be held constant.

3. Normally, at least 50 readings should be taken.

4. The order of the readings should be preserved.

5. The individual readings should be plotted over time.

6. The measuring devices used should normally have an accuracy of at least 10 times the tolerance spread and 8 times the capability spread.

Although computing the process capability from the range is usually the easiest and fastest method, it is also the one that is most affected by the requirement for a normal distribution. The first step in computing the process capability through ranges is to compute the ranges, R, of subgroups of the total sample of 50 pieces. If, for example, we assume a subgroup of 5, then the average range, $\overline{R}$, is the arithmetic average of each of the 10 values of R. Using the relation $\sigma = \overline{R}/d_2$ (where d_2 can be determined from the readings, see any standard text on statistical quality control[3,4]), then we find the process capability is simply $6(\overline{R}/2.326) = 2.58\overline{R}$ for subgroups of 5.

Process capability studies can produce savings by identifying losses due to inadequate processes, poor tool maintenance, unskilled operators, etc. The process capability study can help ensure optimum programming of machines and operators in making the product to specification at a minimum cost.

11-2.3 Qualification Testing

(a) Test Instruction Preparation. Once prototype testing is completed and the design specifications and drawings are released to Manufacturing, the qualification test program must be developed. The prototype testing phase establishes design feasibility and engineering specifications. On the other hand, the qualification testing phase probes whether the product really does meet the engineering specifications when manufactured under normal production conditions. Therefore, the qualification test instruction must be a comprehensive document covering all the environmental extremes considered necessary to verify product performance.

Because some of the tests on such items as in-core sensors are unusual and expensive, the qualification test instruction should be developed jointly by Quality-Control Engineering and Design Engineering. In tests of in-core sensors, where the unit being tested may become radioactive or even be destroyed during certain tests, the sequence of tests is vitally important and must be carefully considered. It is advisable to perform certain reference tests before and after each environmental test and to perform the most severe tests (such as shock and vibration) near the end of the test phase, thereby accumulating as much reliable data as possible before risking mechanical failure of the product. Tests of in-core sensors or other equipment in a neutron flux will render the tested units radioactive; these tests are normally performed last to avoid unnecessary exposure of personnel to nuclear radiation.

Prototype testing must be performed under carefully controlled conditions, taking data by automatic means if possible, with well-designed test equipment that is appropriate for the intended purpose.

(b) Pilot Production. The pilot production units must be monitored closely to ensure that quality plans are being followed and are adequate so that the cost of quality does not become exhorbitant. Naturally, building the first few production units may involve problems; nevertheless, with sufficient preplanning these problems should be minimized. It is advisable to include extra in-process inspections and tests as part of the plan for the first few production units until confidence is built up in production techniques. Photographs taken at various critical assembly stages during this stage of production have been found useful; they can be used as samples for future production.

(c) Extent of Testing and Equipment Selection. One question that never seems to be adequately answered, because of cost and scheduling difficulties, is how many units should be tested. If only one unit were to be tested, there would be little confidence in the value of the test results in determining specification limits. Assuming a relatively expensive product, three units should be the minimum to be considered for testing. However, using the results from tests on 10 units would naturally yield a higher confidence factor, and therefore using 10 units would be preferable.

The equipment to be used in qualification testing should be as accurate and reliable as is economically feasible but not necessarily the same as that to be used for production testing. Since many of the tests to be performed during the qualification testing phase may be unique, it is not unusual to have special testing equipment designed and built specifically for that purpose.

(d) Environmental and Special Functional Tests. *Nuclear Sensor Qualification: Functional Testing.* Some of the possible functional tests in qualifying nuclear sensors are:

1. Radiographic: To observe extremely tight fit-up conditions, critical welds, etc.

2. Dye penetrant: To observe critical welds for possible cracks.

3. High potential: To observe corona or high-voltage breakdown that may indicate faulty assembly or an incorrect gas fill or improper gas pressure.

4. Insulation resistance: To determine the integrity of the weld joint and the condition of the insulating surfaces.

5. Capacitance: To indicate improper assembly or an open circuit in the center conductor.

6. Mass-spectrometer leak detection: To test for seal integrity.

7. Partial-pressure gas analysis: To investigate for gas contamination.

8. Pulse-height gas analysis: To check on proper gas mixture and pressure in neutron sensors.

The choice of the test equipment to check the output of a nuclear sensor depends on what type it is, i.e., pulse-counting, d-c measuring, or mean-square-voltage (see Sec. *2, 3*, and 5). If many checks on a quantity of sensors are to be made over any length of time, it will pay to automate the test equipment and have the test data automatically recorded.

Nuclear Sensor Qualification: Environmental Testing. Environmental tests for nuclear sensor qualification include:

1. Gamma sensitivity: Up to 10^7 R/hr, depending on specification requirements.

2. Neutron sensitivity: Up to 10^{13} neutrons cm^{-2} sec^{-1}, depending on specification requirements.

3. High temperature: Other parameters would normally be tested simultaneously, such as insulation resistance or neutron sensitivity.

4. Reactor environment: This test should embody as many of the actual environmental conditions as possible.

5. Shock: This would be required for certain defense applications.

6. Vibration: The extent would be determined by end-use conditions, e.g., in-core sensors should be subjected to vibrations simulating those expected in reactor core.

7. Humidity: Most out-of-core sensors are in high-humidity environments and should therefore be tested accordingly.

8. Autoclave: This combination steam and high-pressure test for in-core sensors must be carefully controlled if used.

9. Hydrostatic: A required test for ASME Boiler Code applications, primarily for testing the integrity of weld joints and sheath.

Electronics Instrumentation Qualification: Functional Testing.

1. Power-supply input and output voltage, ripple, line, and load regulation.

2. Rise time, linearity, pulse width, waveform, and dynamic range.

3. Overall response time with simulated maximum-input cable capacitance.

4. Trip-circuit accuracy, hysteresis, and range.

5. Calibration checks to specified tolerances at all outputs.

6. Full-load test for a period of time sufficient to represent a significant fraction of the life expectancy.

Electronics Instrumentation Qualification: Environmental Testing.

1. High-temperature tests at full load.

2. Humidity tests.

3. Vibration tests.

(e) Error Correction. The qualification test program will inevitably indicate that certain design or specification changes should be made. Consequently the results should be analyzed promptly and fed back to Engineering for their use in correcting errors in, or improving, the design. If the design engineer is performing or participating in the qualification test program, this is unnecessary. This is, however, the last good place to make or recommend design changes. It may be, for example, that the instrument works satisfactorily but is very difficult to assemble, test, or maintain, thereby indicating the need for a design change.

(f) Final-Product Specifications. Final-product specifications can now be firmed up and released by Engineering. This event should trigger a review of all manufacturing and quality-control procedures as well as a review of tools, fixtures, methods, and equipment by Manufacturing and Quality-Control Engineering. Hopefully, changes will be minor and any units built during the time of qualification testing will not have to be rebuilt in any way that would necessitate repetition of the qualification tests. This decision should be made jointly by Design Engineering and Quality-Control Engineering. Should it be necessary to repeat some of the qualification tests, the extent of the testing should be considered carefully since normally a complete redesign would be required before a complete rerun of the qualification tests would be necessary.

The final-product specifications form a part of the manufacturing release from Engineering which also contains all drawings and parts lists. Materials then procures all the various parts in accordance with the specifications.

11-2.4 Purchased-Materials Quality Assurance

(a) Review and Coding of Material Requests. Vendor quality can be controlled by having Quality Engineering review all requests for production materials and parts. When the requests are reviewed, several aspects should be kept in mind: the estimated cost of the purchased item, the functional criticality of the item, and the relation of the item to the production schedule. There should be a quality plan for each purchased part. The plan should detail each parameter or characteristic to be checked (and to what statistical plan), should dictate the equipment to be used, and should describe exactly how to perform any tests or inspections other than those considered standard or routine.

Besides indicating to the vendor the relative importance of the various characteristics of the part, the classification category (see next subsection) gives the quality-control engineer a basis for determining the sample size for his receiving inspection plan. For example, he may choose to have every critical parameter or characteristic checked or verified 100% of the time, then perhaps a 0.65% acceptable quality limit (AQL) would be assigned to major characteristics, a 4.0% AQL to minor characteristics, and perhaps a one-piece sampling of characteristics that are designated incidental.

It is recommended that statistical sampling for attributes should be in accordance with MIL-STD-105.

(b) Classification of Part Characteristics. A most important step in controlling the quality of the outgoing product is to control the quality of the incoming parts and materials. The job of the vendor can be made easier (and therefore better quality assured) by classifying the various characteristics of the part or parts he is to supply. If, for example, a part has several dimensions or characteristics, some are obviously going to be more critical to the function of the final product than others. Accordingly, typical classifications used are "critical," "major," "minor," and "incidental." These are usually defined as follows:

1. A critical classification means that, should a characteristic thus classified not be within specifications, it would likely result in hazardous or unsafe conditions for individuals using, maintaining, or depending on the product or it would be likely to prevent performance of an essential function of a major end item.

2. A major classification means that, should a characteristic thus classified not be within specifications, it would be likely to reduce materially the usability of the unit or product for its intended purpose and would most likely result in a customer complaint.

3. A minor classification means that, should a characteristic thus classified not be within specifications, it would not likely reduce materially the usability of the unit or product for its intended purpose but would most likely still be found objectionable by the customer.

4. An incidental classification means that, should a characteristic thus classified not be within specifications, it would not be found objectionable except by the most critical customer (e.g., a blemish on the inside surface of an instrument housing).

(c) Test and Inspection Equipment Requirements. The test and inspection equipment needed to support a nuclear instrumentation receiving inspection area includes such items as an insulation-resistance tester (voltage variable to about 1000 volts d-c), an a-c high-potential tester (to about 3500 volts a-c), a helium mass-spectrometer leak detector (sensitive to 10^{-10} cm^3 He/sec), optical comparator (at least 14 in.), dye-penetrant test set, precision bench centers, standard volt-ohm-milliammeter, transistor and capacitor checkers as well as integrated circuit testers, and other electronic component testers.

Another possibility to be considered is source inspection, i.e., inspecting the product at the vendor's plant. This may be essential in cases where the vendor is either in trouble or is so new that he has not yet proved his ability to produce. In other cases, where the cost of the product is very high or the function of the product is critical, the quality-control engineer may consider it important to institute vendor surveillance to ensure that the vendor understands exactly what is required of him.

(d) Material Verification. Whether raw material or completed parts, positive proof of material identity must be on file, particularly in the case of materials and parts for in-core sensors where the *ASME Boiler and Pressure Vessel Code* applies. A practical way to handle this is to have the material certifications reviewed against the applicable specifications as soon as they are received. If the paperwork is correct, it still may be considered appropriate to make certain chemical spot tests to verify that the material is properly marked or to send a sample to the laboratory for chemical or spectrographic analysis. The next step is to file the "certs" by purchase-order number and have the material painted according to an identification system such as that described in Appendix A to this chapter. The more critical materials, such as stainless-steel tubing and bar stock intended for use within a reactor, should be marked along their entire length with the purchase-order number, the material identification, and heat number for ready reference at any subsequent time.

(e) Destructive Testing Procedures and Use of Laboratories. If there are any doubts concerning the validity of the certification or if the material requirements are critical, then it is sound practice to send a sample to a qualified materials-analysis laboratory for spectrographic or wet-chemistry analysis. Where heat treatment is important, a hardness test should be performed to verify surface condition. Tensile strength would have to be verified by performing a pull test on a tensile specimen.

(f) Nondestructive Testing Procedures. There are several nondestructive testing procedures available. Perhaps the least expensive is the use of a dye penetrant for locating minute cracks (e.g., in aluminum insulators and welded tubing). Ultrasonic testing is the best technique for testing large quantities of welded tubing for cracks and flaws; once the equipment is set up, tubing can be tested rapidly. Other nondestructive tests, such as radiography, are used more for in-process and final inspection than for receiving inspection.

(g) Disposition of Defective Material. Any material found defective by Receiving Inspection, either by test or inspection or as part of a defective lot, should be so identified and segregated from good or unexamined material until it can be returned to the vendor. Scheduling constraints may make it necessary to review the nature of the defect and take an alternative action, such as (1) use as is, (2) rework to drawing, (3) rework to an acceptable configuration not to drawing, (4) sort to screen out acceptable parts, (5) scrap, or (6) return to vendor.

11-2.5 Production Inspection and Test

(a) Inspection and Test Instructions. *Machined Parts.* Inspection instructions for machined parts are normally simple enough to be included as part of the production planning. When special instructions are required, a separate inspection instruction may be written and referenced on the planning sheet.

Subassemblies. Often it is necessary to ensure that subassemblies are correct before progressing to the next assembly step, particularly where the next assembly step will obscure visual access or where the next assembly operation is expensive and would be wasted should previous operations prove to be faulty. The quality engineer must assess the type of inspection or test that is needed after every production operation. He may have to design special test equipment (black boxes) or inspection fixtures, such as "go, no-go" gages. Again the complexity of the instruction determines whether it can be a part of the production planning or if a special instruction is required.

Test instructions for modules or boards containing active circuit elements should be derived from engineering test specifications. Modules or boards with only passive elements may be inspected visually and tested later as part of a complete instrument. Functional tests are normally performed on each active element-containing module or board at ambient conditions. The functional tests may include the following:

1. Zero, balance, and calibration adjustments to specification.

2. Load regulation to specified tolerances.

3. Linearity, pulse width, waveform, and dynamic range to specification.

4. Trip-circuit accuracy, hysteresis, load, and range to specification.

Electronic Assemblies. Typically nuclear electronic assemblies include rack-mounted chassis-type equipment, such as power supplies, source-range monitors (count-rate meters), intermediate-range monitors (log N current amplifiers), and power-range monitors (flux amplifiers), as well as wide-range monitors (picoammeters) with their associated logic and trip circuits. Test instructions for this equipment should be derived from engineering test specifications. For special-purpose instrumentation, test instructions may be derived from customer specifications as well as from engineering test specifications. Standard tests that may be included in the test procedures include the following:

1. Mechanical zeroing of all meters.

2. Power-supply input and output voltage, ripple, and line and load regulation checks to specified tolerances.

3. Zero, balance, and calibration adjustments to specification.

4. Rise time, linearity, pulse width, waveform, and dynamic range checks to specification.

5. Overall response time to specification with simulated maximum-input cable capacitance.

6. Trip-circuit accuracy, hysteresis, and range to specification.

7. Calibration checks to specified tolerances at all outputs.

8. A full-load run for 24 hr at maximum specified ambient temperature, followed by an operational recheck.

Nuclear Sensors and Peripheral Equipment. Test instructions for sensors should be derived from engineering test specifications and may include:

1. Gamma and neutron sensitivity checks.

2. Gamma compensation check.

3. Insulation-resistance and high-voltage (hi pot) checks.

4. Mass-spectrometer leak test.

5. Cable-resistance test at room temperature and at maximum specified temperature.

6. Pressure test at room temperature.

7. Dye-penetrant and radiographic tests as specified.

8. Continuity checks.

Systems Test Instructions. Systems test instructions may be derived from customer specifications or test specifications provided by Engineering. Standard system tests normally include the following:

1. Point-to-point wire check of all interconnections per connections diagram.

2. Tests at 1000 volts above normal control voltage and/or insulation-resistance tests at 500 volts on all power and control wiring. (Instruments, including meters and recorders, are disconnected and/or shorted during test.)

3. Functional electrical tests with interconnections to simulate field wiring, which may include:

   a. Simulated (including cable capacitance) current or pulse signals to all neutron- and gamma-monitoring channels through all ranges or decades.
   b. Externally connected loads of specified impedance.
   c. Operation as system elements of all switches, meters, relays, recorders, lamps, logic circuits, and other panel-mounted devices.
   d. Recording of specified data, such as accuracy, response times, trip points, logic operation, stability, and repeatability.

4. Functional checks on each panel-mounted process instrument, which may include the following:

   a. Rough accuracy checks at or near 10 and 90% of scale using pneumatic or electrical signals to simulate process variables.
   b. Interlocking, alarm, and trip contacts operation at panel terminal-board points.

Final Inspection of Systems. To assure the quality of the completed system, the inspector should use a checklist. A typical checklist is given in Appendix B to this chapter.

(b) Test and Inspection Equipment. *General.* Test equipment may differ for factory or field use. Factory test operations involving high production rates require automated or semiautomated devices. Care must be taken that data obtained by such devices can be verified by equipment available to the field. Factory test equipment used for development, design, and low-production-rate work should be selected from commercially available items. Equipment used for production prototype final design and tests should have specifications that can be duplicated in all significant characteristics by factory and field equipment.

Field test equipment, in addition to duplicating factory equipment characteristics, should be selected, insofar as possible, from vendors who have nationwide service capabilities (or worldwide in the case of systems sold overseas). Some required field test equipment may not be commercially available, in which case instrument-system vendors must supply special portable test equipment using normal design and production methods plus issuance of complete operation and maintenance instructions.

System manufacturers must issue formal listings of all test equipment required, including either catalog numbers or essential characteristics. If system manufacturers are responsible for field check-out, they should insist on the right to review customer-purchased test equipment for compatibility. Obviously factory training programs should use the same listed test equipment.

High-Production-Rate Test Equipment. Although a hard and fast rule cannot be made, a continuing production rate of 500 relatively complex circuit boards of one type per year usually justifies automated testing and the associated investment in design and equipment. A subsystem of interconnected assemblies comprising many circuit boards or modules of several types might justify automated tests at subsystem production rates of five per year.

Automated equipment is usually designed and built by the instrument-system manufacturer. There are also test-equipment vendors specializing in custom design and/or building of such devices. The cost of such equipment must be weighed against expected product design life. It should be recognized that the more highly automated the device, the less adaptable it is, in general, to changes in production design.

Semiautomated test equipment usually consists of specially designed devices that interface conventional signal sources, the production item under test, and conventional output readouts. The interface device accepts a circuit board, a module, or an interconnected assembly directly by mating connectors. It may also contain signal conditioners, voltage, load, or other parameter-changing elements that can, when a few switches are operated and external input signals are varied, test the production item.

Although no rigid distinction exists between semiautomated and fully automated devices, the latter would replace the manual switching and variation of input signals with electromechanical or electronic switching and stepping circuits. Manual output logging is usually replaced by a digital tape printer. The item under test is simply plugged in, a start button pushed, and the test is automatically completed with data printed out. Large interconnected assemblies can be connected by mating connectors or terminal fanning strips. One type of automated test device used for circuit-board and module testing continuously compares production items with known good standard boards or modules on a go, no-go basis and alarms when defects are found. Some testers may even localize the failed circuit element or region.

Examples of nuclear-instrument circuit boards and modules lending themselves to semiautomated or fully automated testing are d-c amplifiers, trip circuits, voltage regulators, power supplies, and generally items used with uncommon signal-conditioning boards or modules in assemblies comprising an instrument entity, such as a count-rate meter, a log N amplifier, or a mean-square-voltage neutron monitor. Base-mounted multichannel in-core power-range monitor subsystems, flux-mapping systems, and control-rod-position information systems may contain "card files" of identical circuit boards, which, along with their systems, can profitably be tested with automated equipment.

The degree of customization required for this type of test equipment precludes any detailed description here.

Field and Factory (Nonautomated) Test Equipment. Field and factory test equipment should be selected, wherever possible, from commercially available items, preferably with availability as extensive as the market to be served by the instrument manufacturer. In development work, where the circuits are not directly used in production equipment, relatively more sophisticated items can be justified; a greater variety of items can be used than would be appropriate in final design and quality-control work. Development extras include sampling oscilloscopes, spectrum analyzers, multichannel pulse-height analyzers, harmonic wave analyzers, noise generators, double-pulse generators, and similar equipment. High accuracy is desirable but not mandatory.

Test equipment used in design, quality-control, and field work need not be as sophisticated as that used in development, but it must yield accurate and reproducible results. National Bureau of Standards traceable devices to generate or measure alternating and direct voltage and current, time and frequency, resistance, inductance, capacitance, pressure, and temperature must be available, either owned or leased from a local calibration service, and used for calibration of design, quality-control, and development test equipment. Test equipment used in the field at locations where local commercial calibration service is not available must be supplemented with minimal portable standards, such as current sources, which can be periodically sent out for calibration and recertified.

As noted earlier, devices used for factory design, factory test, and field test should duplicate each other in all significant characteristics. This is especially true of equipment used for pulse or complex wave-form generation and analysis. As an example, if the designers use a 50-MHz response oscilloscope in the factory to obtain wave-form or response-time data (to be included in the instruction manual) on a fast preamplifier driven by 10-nanosecond rise-time input pulses and a 25-MHz oscilloscope in the field, the test pulses will display different wave forms and lead to unnecessary troubleshooting. Interface devices, such as terminating elements, must also be specified in detail by designers so that results can be duplicated. Considerable frustration can be experienced in reactor instrument-system check-outs because of nonduplication and failure to properly interface.

Special Test Equipment. Special test equipment denotes equipment that is required for factory or field test of production equipment but is not commercially available. Although this would include the automated test equipment previously discussed, here it shall be assumed to be portable equipment required for factory or field check-out of instruments or systems. Examples are test fixtures used to simulate control-rod-position indicators or multiple current inputs or to generate squaring circuit calibration signals. Test devices should also be provided for modules or circuit boards used in systems that do not permit bypassing during operation.

The nuclear-instrument manufacturer is responsible for

reviewing the testability of his product and offering, as manufactured and documented items, all special test equipment needed in the field to calibrate the product and demonstrate its operability. This includes jumpers for interconnecting items under test with test devices and power supplies as well as special input or output load simulators.

Specific Test Equipment for Nuclear and Process Sensors and Instruments. Nuclear sensors and channels, such as logarithmic and linear count-rate meters and logarithmic and linear current amplifiers, are used at most reactors. Process sensors and instruments for measuring temperature, pressure, level, and flow are also used. Mean-square-voltage monitors and power-averaging instruments for both in-core and out-of-core applications are gaining popularity.

The lists of equipment given here are intended to be typical rather than complete. Except where noted, required test equipment is commercially available, usually from several vendors. Overseas applications must specify proper line voltage and frequency. Some available devices may combine listed functions.

Table 11.1 lists the equipment required for testing and inspecting nuclear sensors. Table 11.2 is a similar list for eight basic nuclear instruments: log or linear count-rate meters, log N amplifiers, period meters, linear mean-square-voltage monitors, linear direct-current (often called d-c-current wide-range monitors, single-range d-c (power-range) monitors, power-averaging instruments, and process and area radiation monitors. Table 11.3 lists general-purpose equipment used in troubleshooting, instrument power-supply voltage setting, and calibration of other test devices.

Because of piping requirements, process primary sensors and associated transmitters are frequently calibrated or tested in place. This requires portable test devices. Control-room indicators, controllers, signal conditioners, recorders, and similar items lend themselves more readily to instrument shop test or calibration. A clean shop air supply must be provided for pneumatically operated instruments. Interfaces, corresponding to dummy loads for nuclear instruments, are usually far more difficult to simulate for process instruments. Static check-out of components does not always predict dynamic operation in a system with control valves and piping. Table 11.4 lists test equipment for temperature monitors, and Table 11.5 lists equipment for testing pressure and differential-pressure (flow or level) monitors. As in the listing for nuclear sensors and instruments, the listed devices are typical for the types of the instruments or sensor combinations noted. For most overseas applications, equivalent metric scales must be specified.

(c) Special Environmental Test Requirements. The original design of a nuclear instrument must be qualified in the anticipated nuclear-reactor environment, and the instrument itself may have to be subjected to that environment during production testing. An example of this is the *ASME Boiler and Pressure Vessel Code* testing of in-core detectors. All in-core instrumentation must be subjected to minimum ASME Code requirements and certified by the authorized code examiner. Obviously radiation-detection instruments must be exposed to nuclear radiations to determine whether they are working properly; nevertheless, there are practical limits of absorbed dose which should not be exceeded. Instruments that are to be subjected to high temperature should be tested at a sufficiently high temperature to ensure that they will have adequate insulation resistance when used in a nuclear reactor. The ability of the instrument to withstand low-frequency vibration, such as might be encountered during seismic disturbances, would normally be proved out during qualification testing; however, for certain sensitive equipment a vibration test may be included as a production test.

Table 11.1—Equipment Required for Testing and Inspecting Nuclear Sensors

1. *High-resistance meter* (to measure cable or detector insulation or leakage resistance): Range, 1×10^{6} to 1×10^{14} ohms full scale; switched accuracy, ±20% reading above 10% of full scale; d-c test voltages, 10, 50, 250, and 1000 volts with 10^{-5} amp limit.
2. *Low-current or voltage meter* (to measure background sensor current or cable to detector polarization current or voltage): Range (d-c), 10^{-5} to 10^{-13} amp, switched or (d-c), 1 to 100 mV, switched; input resistance, 10^{4} to 10^{11} ohms ±5%, depending on voltage or current range.
3. *Current-limited power supply* (to check detector element breakdown voltage): Range, 0 to 1500 volts d-c limited to 100 μA (or use a limiting resistor and microammeter to calculate resistor voltage drop).
4. *Test sources*
   a. *Gamma sources* to cover energy and intensity ranges specified for system. Sources of ^{137}Cs or ^{60}Co are adequate if the energy response of system is documented. Intensity must be high enough to activate the highest range (or decade) response with acceptable geometry.
   b. *Neutron sources* to provide low-range response on start-up and wide-range instruments. For some intermediate- and all single-range power-range monitors, the reactor is the test source, and initial activation and operation to full-scale ranges must be observed during start-up.

(d) In-Process Inspection. The variety of instruments and associated equipment involved in nuclear-reactor operation is so great that in this chapter it only is feasible to survey briefly the various in-process inspection and test techniques that are available and useful to the industry.

Nuclear Radiation Sensors. In-process control of coated electrodes, whether uranium or boron (see sec. 3), demands careful analytical techniques during the coating process. Proof-testing with dummy ion chambers is an excellent technique after the coating process has been completed, but, once the process has been qualified, it need only be performed on a sample basis. Two of the most important tests performed on ion chambers are the mass-spectrometer helium leak test (after practically every welding or brazing operation) and the high-voltage insulation-resistance test (after practically every important assembly operation).

Table 11.2—Equipment Required for Testing Nuclear Instruments

A. Count-Rate Meters, Log or Linear, Including Preamplifiers

1. ***Pulse generator***
   Internal repetition frequency: 5 Hz to 1 MHz continuous with rough calibration only.
   Externally synchronized repetition frequency: D-c to 2 MHz; output pulse duration, 100 nsec to 100 msec continuous.
   Output pulse rise time: <10 nsec.
   Output pulse amplitude: 0 to 30 volts peak into 1000-ohm load, continuous, negative, and positive.
2. ***Electronic counter*** (to accurately set pulse-repetition frequency)
   Range: 5 Hz to 11 MHz.
   Accuracy: 1 part in 10^6 per year (cumulative) ±1 count with minimum 10 mV rms input.
   Internal standard: 100 KHz or 1 MHz.
   Gate time: 0.1 to 10 sec.
3. ***Step attenuator*** (coaxial) (to match pulse generator to preamplifier)
   Range: 0 to 120 db in 10-db steps.
   Impedance: 50 ohms nominal.
   Power dissipation: 0.5 watt average.
4. ***Standard capacitor*** (coaxial) (to generate simulated detector charge pulses)
   Value 1, 10, or 100 pF, depending on charge range required; accuracy, ±0.1%.
5. ***Oscilloscope*** (dual-trace)
   Bandwidth: D-c to 50 MHz, above 20 mV per vertical division; D-c to 40 MHz, 5 to 20 mV per division.
   Time base: 0.1 μsec to 1 sec per horizontal division; calibrated deflection factor, 5 mV to 10 volts per division; calibrated accessories, 1:1 and 10:1 probes.
   Delay to permit viewing leading edge of triggering wave form.

B. Log N Amplifiers

1. ***Current source*** (d-c) (self-contained or calibrated resistance box)
   Range: 5×10^{-3} to 1×10^{-13} amp.
   Accuracy: ±1% at 5×10^{-3} to 1×10^{-6} amp; ±2% at 1×10^{-6} to 1×10^{-9} amp; ±3% at 1×10^{-9} to 1×10^{-11} amp; ±4% at 1×10^{-11} to 1×10^{-12} amp; and ±5% at 1×10^{-12} to 1×10^{-13} amp (accuracy may be attained with aid of voltage and temperature corrections).
   Source impedance at least a factor of 1000 greater than input impedance of device under test at test current level.
2. ***Voltage source*** (for resistance box, if used)
   Range: As required for above current range.
   Accuracy: As required for above overall accuracies.
3. ***Oscilloscope*** (to observe response times, spurious signals, etc.) See item 5 under section A of this table.

C. Period Meters (Without Self-Contained Ramp Generator)

1. ***Function generator*** (triangular wave form)
   Range: 0.01 Hz to 1 KHz, switched by decade.
   Accuracy: ± one division for 92 dial divisions (9 to 101).
   Linearity: Less than 1% over full range.
   Output: 0 to 10 volts peak to peak into 600 ohms.
2. ***10-turn potentiometer*** with calibrated dial (to reduce generator output)
   Resistance: 600 ohms.
   Linearity: <1%.
   Accuracy: ±1%.
3. ***Oscilloscope*** (see item 5 under section A of this table).
   Frequency response: ±1% at 50 Hz to 1 MHz; ±5% at 10 to 50 Hz, 1 to 10 MHz.
   Input impedance: 1 megohm shunted by 50 pF (maximum).
4. ***Oscilloscope*** (see item 5 under section A of this table).

D. Linear Mean-Square-Voltage Monitors (Including Preamplifier)

1. ***Test oscillator*** (sine wave)
   Range: 10 Hz to 10 MHz, decade switched.
   Accuracy: ±3% of dial reading.
   Output: 0.3 mV to 1 volt rms into 600 ohms.
   Attenuator: 70 db in 10-db steps over output range.
2. ***Test attenuator*** (to interface oscillator output with monitor or preamplifier)
   Special test device to provide complementary output steps that are proportional to the square root of monitor range steps. For example, if the monitor is switched to a 1-decade-less-sensitive range, the attenuator must supply a signal greater by $\sqrt{10}$ or 3.16.
3. ***True rms voltmeter***
   Range: 1 mV to 10 volts full scale, $\sqrt{10}$ range switch factor.

E. Linear Switched Direct-Current (Wide-Range) Monitors

See current and voltage source for log N amplifiers (items 1 and 2 under section B of this table) except that calibrator accuracy should be a factor of 4 better than specified monitor accuracy, range for range. This implies that some potentiometric or standard ramp or capacitor calibrator checking system, not commercially available as a single device, would be required for accuracy below about 10^{-8} amp.

F. Single-Range Direct-Current (Power-Range) Monitors

1. See current and voltage source for log N amplifiers (items 1 and 2 under section B of this table) except that calibrator accuracy should be a factor of 4 better than specified monitor accuracy. This can usually be accomplished with precision resistors and d-c current and/or voltage standards available to 0.1% or better accuracy over the current ranges involved (usually 10^{-6} to 5×10^{-3} amp).
2. ***Oscilloscope*** (see item 5 under section A of this table).

G. Power-Averaging Instruments

Test fixture (to simulate multiple sensors or flux amplifiers)
A special test device to supply constant currents or voltages to the number of channels averaged. Device must have both single-channel and averaged output self-checking ability with an accuracy a factor of 4 better than averaging instrument or must possess interfacing switches to permit external current or voltage monitoring using test devices of that accuracy.
Oscilloscope (see item 5 under section A of this table).

H. Process and Area Radiation Monitors (Typically Gamma Monitoring)

These instruments are in principle and electronic test-equipment requirements either identical to sections A, B, and E of this table or are combinations of them (many area monitors are log count-rate meters at lower radiation levels and become log current amplifiers at higher levels). In addition to the test devices described, radioactive sources to cover specified energy responses and intensities are required (see item 4a of Table 11.1).

Table 11.3—General-Purpose Equipment Used in Calibrating Test Devices, Troubleshooting, Etc.

(Some or all of the items listed below, in addition to those listed i Tables 11.1 and 11.2, are widely used for troubleshooting, fo setting instrument power-supply voltage, for calibrating test equip ment, etc.)

1. *D-c volt-ohm-ammeter* (typical electronic type)
   Voltage range: 1 mV to 1000 volts, $\sqrt{10}$ range switch factor.
   Input resistance: 10 megohms minimum.
   Current range: 1 μA to 1 amp, $\sqrt{10}$ range factor.
   Ohmmeter range: 1 ohm to 100 megohms, center scale.
   Accuracy: ±1% of full scale, all voltage ranges; ±2% of full scale, all current ranges; and ±5% at center scale, all resistance ranges.
2. *Multimeter*
   Voltage range: 1 to 5000 volts a-c or d-c at 20,000 ohms/volt d-c and 5000 ohms/volt a-c.
   Current range: 50 μA to 10 amps d-c.
   Ohmmeter range: 1 ohm to 10 megohms.
   Accuracy: ±5% of full-scale voltage and current, all ranges and ±10% of center-scale reading, all ohmmeter ranges.
3. *D-c voltage standard*
   Null voltmeter and standard voltage source, range 1 to 1000 volts full scale; decade switched, 20 mA capability as source.
   Accuracy: ±0.02% of setting or reading.
4. *A-c differential voltmeter*
   Range: 1 to 1000 volts full scale, decade or $\sqrt{10}$ range factor.
   Accuracy: ±0.15% of full scale at power-line frequency.
   (This may be used in conjunction with a regulated a-c voltage source for meter calibration.)
5. *D-c power supplies*
   Selected to match ranges of instruments' internal power supplies and also power sources if d-c powered. Supplies should have current-limiting capabilities, ±0.1% line or load regulation, and 1 mV or 0.1% of setting maximum rms noise (whichever is greater).
6. *Sinusoidal voltage regulator*
   To supply standard line voltage and specified line frequency at ±0.2% line or load regulation, 3% maximum harmonic distortion. A 1-kW minimum rating is recommended.
7. *Cables, connectors, and adapters*
   Test cables with connectors to mate any instrument used with any test device, including inter-type and tee adapters, as required.
8. *Dummy loads*
   Test loads to simulate system input and output impedances.

Table 11.4—Test Equipment for Temperature Monitoring

1. *Direct reading pyrometers* (one for each range) with 10-in. arm, assorted heads, ranges 0 to 400°F and 0 to 1200°F, and ±1% of full scale as millivolt meter.
2. *Portable potentiometers* (two at least) with assorted scales, such as 32 to 570°F (copper–constantan), 32 to 522°F (iron–constantan), 32 to 1700°F (iron–constantan), 32 to 2425°F (chromel–alumel); 0 to 14.8 mV, 0 to 53.5 mV, and 0 to 155 mV.
3. *Portable Kelvin double bridge* (for measuring system lead resistances): 8 ranges, 0.0001 to 0.0011 ohm to 2 to 22 ohms; ±0.1% accuracy; and ±0.1% repeatability.
4. *Portable Wheatstone bridge*: 4 ranges, 0.1 to 1 to 100 to 1000 ohms; ±0.1% accuracy; and ±0.1% repeatability.
5. *Decade resistance box:* 0 to 1000 ohms in 0.1-ohm steps; ±0.1% accuracy.

Table 11.5—Test Equipment for Pressure and Differential-Pressure (Flow or Level) Monitoring

1. *Pressure vacuum variator* (bellows for generating pressure to 30 psig or vacuum to −20 in. Hg): Effective volume, 12.5 in.^{3}.
2. *Dual-range deadweight tester* with pump (oil): 0 to 600 psi (5-lb increments) and 0 to 3000 psi (25-lb increments); accuracy, 0.1% at increments.
3. *Test gage* (one per range): range, −30 to 15 psi, 0 to 30 psi, 0 to 60 psi, 0 to 100 psi, 0 to 300 psi, 0 to 600 psi, 0 to 1000 psi, 0 to 1500 psi, and 0 to 2000 psi; accuracy, ±0.25% of full scale or with movable tabs capable of ±0.1% of tab point when calibrated with deadweight tester.
4. *Dial manometers* (one per range): 0 to 30 in. Hg and 0 to 60 in. Hg; accuracy, ±0.1% of full scale.
5. *Slope tube manometer*: 0.5 to 2.0 in. (Hg or water).
6. *Portable test pump* (water): 0 to 5000 psi, with test gages 0 to 160 psi, 0 to 600 psi, and 0 to 5000 psi; accuracy, ±0.25% of test gage full scale.
7. *Water-weight gage*: 20 to 3000 psi in 0.2-psi increments; accuracy, ±0.1% at increment.
8. *Differential-pressure indicator:* 0 to 200 in. water, 0.5-in. divisions; accuracy, ±0.5% of full scale.

Electronic Control and Monitoring Equipment. In-process inspection and test procedures vary according to the manufacturing techniques used. Basically, visual examination is required after every major operation or series of minor operations. (Operations include hand soldering, wire wrapping, etc.) A typical subassembly, such as a printed wire board, has all components mounted on it by one or several operators. It is then checked as a first piece by an inspector and run through a solder machine. The board is again inspected and then subjected to a performance test, after which (if all is well) the remainder of the production lot may be run. Tests from this point on may be either on a sampling basis or 100%, depending on such factors as complexity, quantity, and adequacy of succeeding tests on the next level assembly.

A common technique for verifying whether or not a board is correct is to use the first piece sample as an inspection aid against which succeeding boards are checked. The tester can be set up on the same principle: a known good board and the board under test are electronically compared by subjecting both boards to identical signals and automatically comparing outputs across a bridge circuit. Necessary adjustments can then be made by tuning for a null.

A variety of automatic and semiautomatic testing devices are available or can be designed to meet specific objectives. Such test devices can range from a simple black box to an elaborate computer arrangement that analyzes and prints out data automatically. Each situation has to be evaluated and analyzed by a competent test engineer. Computers can be effectively adapted to testing some of the more complex logic systems that are necessary in nuclear-reactor control systems.

Peripheral Equipment. Testing of peripheral equipment often presents a great challenge to the test-equipment designer because the equipment usually combines mechanical and electrical capabilities and often creates serious space problems (e.g., a drive-mechanism test in conjunction with a traversing in-core probe).

In-process inspection and test of penetration seals is extremely critical since Boiler Code requirements must be met. Records are important for such tests as leak tests under pressure. High-potential and insulation-resistance testing offer real challenges because of the safety aspect and the sheer number of combinations and permutations on seals with multiple penetrations.

(e) Serialization and Control. Control of equipment by serialization is usually a function of production control. A common technique is to maintain a notebook of consecutive serial numbers for each major subassembly. This would normally be a component that could be provided as a spare part; it would have a functional specification that could be tested and might require a data sheet completed by quality-control test.

The serial numbers can be assigned as a block when the work order is initiated. Serial numbers can be physically affixed to subassemblies in any number of ways, such as by wired-on tags, etching, screwed-on plates, and silk screening.

Test data are normally filed first by drawing number and then by serial number. Copies of data from all indentured parts are usually filed together with the top assembly in the customer project file.

Serialization of larger discrete assemblies, such as ion chambers and source-range monitors, is accomplished in the same manner as above except that it may be important to date code. One way to do this without revealing the actual date (if this is desirable) is to establish a date code such as A to L for the months January to December and A to Z for the years 1961 to 1986, and use the date code as a prefix or suffix to the regular serial number. Thus a serial number DF87432 would indicate the 87,432nd unit of a particular drawing number and that it was shipped in April 1966.

(f) Final Inspection and Test Requirements. *Radiation-Detection Devices.* Final inspection and test of radiation-detection devices is accomplished in about as many ways as there are different types of detectors. However, some general principles apply. For example, even though the best possible test for any item is to test it in its actual operating environment, this is usually impractical and often undesirable, particularly where the test causes the item to become radioactive. Therefore it is often necessary to devise substitute tests that test only certain characteristics over a limited portion of the total operating range.

Since many nuclear radiation-detection devices, such as in-core sensors, cannot be tested in their intended operating environments, the importance of in-process tests and inspections cannot be overemphasized. For example, when a uranium coating has been checked and found to be correct through in-process testing, when the various parts and components have been found to be dimensionally correct, when the final fill-gas purity and pressure have been determined to be correct, and when the continuity of the center conductor is proven, about the only item left to verify is leaktightness of the final seal. Since a high-temperature insulation-resistance test normally reveals problems of this nature, it may be this final test that gives a high degree of confidence that the unit is functionally operable.

Electronic Control and Monitoring Equipment. Final tests of electronic chassis, assemblies, and systems should simulate actual operating conditions as closely as possible. Special test equipment must be devised to load individual units with simulated inputs, e.g., pulses and ramp currents. As noted in [illegible] 11.5(a), a variety of functional electrical tests can and should be performed whenever practicable. The check list of Appendix B is also useful for inspection of systems.

Peripheral Equipment. Since final testing of peripheral equipment is usually only an extension of the in-process tests that have already been performed, it is not necessary to repeat, only to mention, that there is no substitute for a complete and thorough checklist when performing final inspection.

(g) Disposition of Rejected Units. Rejected units, regardless of whether they are small fabricated parts, large subassemblies, or completed instruments, must be properly labeled and physically separated from accepted units and the uninspected portion of the lot. The label must be distinctive and must include the drawing number of the unit, serial number of the unit, reason for rejection, and specification limits of the rejected parameter. There should also be space on the label for an inspection stamp and date.

Most manufacturing organizations use a Material Review Board (MRB), normally made up of representatives from Design Engineering, Manufacturing Engineering, and Quality- or Process-Control Engineering, to review rejected material. The objective of the material review is to determine the best disposition of the rejected material. An appropriate disposition may be to (1) scrap, (2) rework to drawing, (3) rework to MRB instructions, or (4) accept as is. A decision by the MRB should be unanimous. Obviously any decision to rework or accept must be made only after careful consideration. Sometimes special studies have to be made to determine the possible effects of accepting out-of-spec material. The personnel comprising the MRB must be experienced and knowledgeable since the board must take into account all viewpoints (safety, quality, cost, and schedule).

(h) Reporting and Disposition of Error Correction. Any well-run quality-assurance organization has a built-in feedback loop whereby errors or defects are reported and assimilated into a report that automatically exposes problem areas where attention is required. Many subtle but expensive problems are never brought to light simply because the reporting system only reports significant problems requiring immediate action or, worse yet, there is no reporting system and problems are attacked only when they threaten shipments.

A system to report defects, based on such items as total scrap and rework expense or impact on shippable sales so that reasonable priorities may be established automatically, is an excellent technique for operating the "Pareto" principle, i.e., 10% of the problems account for 90% of the excess cost. By implementing such a system, the process-control engineers or the quality-control engineers can apply their efforts where they will pay off most for the company.

11-2.6 Test- and Inspection-Equipment Engineering

Test equipment and measurement systems are generally sufficiently complex to require a special engineering group to provide measurement hardware. The requirement for test equipment is set by the quality-control engineer. The test- and inspection-equipment engineers, who are basically design oriented, take the quality-control engineer's requirements and provide the hardware that will be used by the process-control engineer in implementing the quality-control engineer's plans. The most comprehensive measurement expertise in an organization usually is in the test and inspection engineering group. Calibration to ensure correct functioning of test equipment is often a responsibility of this group as well.

Test equipment may be defined as that equipment used to measure or generate any of the various units of measure. This distinguishes between equipment that is part of the manufacturing process (and thus the responsibility of the manufacturing organization) and the test and inspection equipment. It is convenient to consider two categories of test equipment: *commercial equipment* (equipment that is commercially available and may be purchased from a vendor) and *special equipment* (equipment that is designed, usually by the test- and inspection-equipment engineering group, to solve a particular measurement problem). Since both categories involve intimate knowledge of measurement techniques (although no design skills are required in specifying commercial test equipment), both categories of test equipment are usually made the responsibility of the test- and inspection-equipment engineering group. The two categories may be further broken into mechanical and electrical equipment.

(a) Commercial Test Equipment. The availability of suitable commercial equipment should always be explored before the decision is made to design special equipment. The cost of the commercial instrument will almost without exception be less than the cost of designing a piece of special equipment. In addition to lower cost, there are a number of other advantages to using commercial equipment. Commercial equipment is generally flexible, having been designed for the broadest possible market. This flexibility reduces the chance of obsolescence when a measurement requirement changes or a new process must be measured. An additional, and often major, advantage is that, when a particular commercial instrument breaks down, a substitute is available within the plant or from the vendor.

There are many pitfalls to be avoided in purchasing commercial test equipment, and it is wise to let the responsibility rest with a group that specializes in measurement problems and test equipment. The newest manufacturer with equipment having the latest innovations is not necessarily the best choice of vendor. Unfortunately, new companies frequently drop a product line or go out of business, leaving the purchaser with an unsolvable maintenance and parts problem. The purchaser must also be wary of "specmanship," where the manufacturer carefully selects his specification wording to make his product appear better than his competitor's. An example of this might be a digital voltmeter that operates within its temperature specification on a hot day but does not meet its accuracy specification unless it is operating at the low end of its line-voltage specification and (in addition) unless it has been calibrated and adjusted during the preceding 48 hr. The final choice of manufacturer and model should be based on:

1. The ability of the equipment to perform the required task and any reasonable variations of the task.
2. The stability of the manufacturer and the proven reliability of his products.
3. The compatibility of the equipment with existing test equipment (in terms of interfacing with other equipment and maintenance and calibration).
4. Price and delivery. Price should be considered after the first three considerations since problems with any of the first three usually result in losses far exceeding any price differential in competitive products.

(b) Special Test Equipment. *Justification.* The decision to design and build special test equipment is usually made when it has been determined that there is no commercial test equipment to perform the particular measurement. Less frequently, the decision is made when commercial equipment is so general purpose and the particular measurement is so specialized that it is cheaper to design and build a special instrument than to purchase the general-purpose equipment. The most common pitfall in this latter situation is finding that, after the special test equipment has been built, a design change in the equipment being tested results in total and irreversible obsolescence of the special test equipment. Even where there is no choice but to design specialized test equipment, obsolescence through design change of the assembly under test presents a real hazard. The solution is to design all special test equipment to be as flexible as practical.

Design and Building. Special test equipment is produced in small quantities, typically only one of a kind being built. The labor costs for design and construction are the major items of expense, and material costs are not significant. This must be kept in mind. For instance, other considerations being equal, it is not profitable to devote 2 hr of design effort to avoid the use of an SCR that is $10 more expensive than another SCR. Similarly, incorporating available designs as elements in the special-equipment design can save time and money. Duplication of circuits in the unit under test is often necessary to ensure compatibility between the test equipment and the tested unit. Although the designer has greater freedom in some respects, he has some constraints that are more stringent than those imposed on the product design engineer. Accuracy of the equipment being used to measure or to generate quantities normally must be 10 times greater than the tolerance of the device under test, i.e., the test equipment itself can contribute no more than 10% to the error that is allowed for the unit under test. There are occasions when even 10% test-equipment error cannot be tolerated. There are other occasions when state-of-the-art measurements are made and a 10 : 1 accuracy ratio cannot be achieved. Although special test equipment involves construction of a single item or, at most, a limited quantity, workmanship cannot be compromised, and rugged construction is generally a must. Engineering breadboards and prototypes cannot be simply stuffed into a box and shipped out for use. Inadequate mechanical ruggedness as well as poor solder connections in breadboards and prototypes are certain to create problems in normal use. One technique, normally associated with production in quantity, can be applied to advantage in the production of special test equipment, namely, the use of printed circuit boards. These provide ruggedness and also have the short direct paths between discrete components needed in high-frequency and pulse circuits. Printed-circuit-board construction even on a single-unit basis can be as economical as a terminal-board layout. Definite cost savings can be realized if two or more units are built. Moreover, the printed circuit board ensures similar performance of the units.

The approach in designing special test equipment is similar to that in designing a commercial product. Initially the design goals are established in close cooperation with the quality-control engineer. Design alternatives are studied, and the best is selected for the particular measurement problem. In this choice a number of factors must be considered: the complexity or difficulty of the measurement, working environment, skill level of the operator, expected useful life of the test equipment, cost per measurement over the useful life, operator convenience or human engineering, ease of maintenance and calibration, cost of maintenance and calibration, accuracy, safety, and initial cost. Once the approach has been established, the design engineer develops the detailed design, including breadboards of critical circuits if the problem is electrical. Construction is best accomplished under the direction of the design engineer by people familiar with the peculiarities of test-equipment construction. The completed equipment or first item is evaluated by the design engineer and, with more-complex design problems, further evaluated by the quality-control engineer. After his acceptance the equipment is then passed through the Calibration Laboratory for formal entry into the calibration cycle. Finally, the equipment is released to the process-control engineer for application.

Documentation. The documentation problem on special equipment is as complex as that for a new product to be marketed commercially. The first documentation takes place in the engineering notebook or similar document, where various design approaches are explored and details supporting the design logic are recorded. When problems arise as the result of measurements made by test equipment, whether special or commercial, the first area questioned is the adequacy of the test equipment. This is rightfully so, and it is only sound engineering practice to have the validity of the design well documented. In many cases the normal documentation associated with construction and calibration of test equipment is not adequate for this purpose. Documentation of the design details, with one exception, must be complete and under formal control. Because of the unique nature of test equipment and the importance of decisions that are based on its performance, the designer must clearly express all the details of his design to the group constructing the equipment and the group calibrating and maintaining the equipment. The exception to this would be the assembly details that are normally associated with the new product. Since special test-equipment production usually consists of only a few items and since they are constructed under the direct supervision of the test-equipment engineer, those assembly details not essential to the accuracy of the equipment need not be documented. Calibration instructions and specifications must be provided for the Calibration Laboratory. The equipment must not only perform correctly when initially released but also must continue to perform within specification, be periodically recalibrated, and be successfully repaired when necessary, all without the direct and continuing direction of the test-equipment engineer. When more complex test-equipment designs are involved, the quality-control engineer's test or inspection instructions may not be adequate. Then a detailed instruction manual, similar to the manual for any commercial piece of test equipment, must also be provided by the test-equipment design engineer. This manual would provide the same type information customarily put into instruction manuals for commercial equipment, such as oscilloscopes, frequency counters, and spectrum analyzers.

Follow up and Feedback. Performance and continuing evaluation of special equipment is monitored in two ways. The maintenance record maintained by the Calibration Laboratory provides an accurate record of long-term performance. The process-control engineer provides rapid feedback of any critical problems. During the first months of use, he must be particularly alert to spot problems not readily apparent in an engineering evaluation. One of the common weaknesses of test equipment that is not always found in an engineering evaluation is its response to certain modes of failure in the unit under test. Under some conditions destruction of the test equipment is possible, and protection must be designed into it. The long-term performance information from the Calibration Laboratory maintenance record can be used to revise calibration recall intervals, i.e., to base them on actual performance. In addition, this record may also indicate more subtle reliability problems.

(c) The Calibration Laboratory. Both commercial and special test and inspection equipment must perform as intended by the original design engineer so that the test technician will feel confident of the results he obtains. The Calibration Laboratory ensures this performance through its various activities and provides the basis for his confidence.

The fundamental responsibility of the Calibration Laboratory is to ensure that the units of measure at a particular facility have the same dimensions as those defined and maintained at the National Bureau of Standards. There are four basic areas of responsibility:

1. Traceability of all units of measure to the National Bureau of Standards (NBS). Traceability implies both derivation of the local unit from the national unit and a known accuracy relation between the local and national units.

2. Maintenance, both preventive and corrective, must be performed to ensure continued performance of equipment within its specifications.

3. Documentation must be initially validated and thereafter maintained to ensure that the equipment can be calibrated and restored, if necessary, to the required level of performance.

4. Recall control must be established to ensure that the equipment can be relied on throughout its life to be within its specifications with a reasonable degree of confidence.

Organization and Staffing. The Calibration Laboratory can take many forms from a single person who performs all the basic functions to individual departments for each of the various tasks. In a medium- or large-scale operation, the first logical division of effort should be between the maintenance function and the standards function. Meaningful results in standards work can only be obtained when meticulous care is taken in making measurements by someone who is not only familiar with the techniques involved but also takes great pride in his work. The maintenance function also requires special talents, but not to the same degree as standards work. In either case the individual must live by the rule that the quality of work, not quantity, is of prime importance. Where there are a number of people on the Calibration Laboratory staff, there must be technically competent leadership. In small organizations this is provided by the supervisor or manager. In large organizations engineers who are specialists in the areas of electrical and mechanical measurements provide the leadership.

New Equipment Control. All new equipment, commercial or special, should pass through the Calibration Laboratory before it is released to the end user. An initial calibration is performed to ensure that the equipment is indeed within its specification. At the same time the documentation is being checked out to make certain that it is adequate for future calibration and maintenance. The performance and repair record, discussed later in this section, is initiated at this time, and the equipment is placed on a recall schedule. These steps are taken to establish a firm base for all future control of the equipment.

Calibration Standards. The traceability of any unit of measure, including a known accuracy relation to the unit as defined by the National Bureau of Standards, is the fundamental responsibility of the Calibration Laboratory. The number of intermediate steps between the end user and the NBS depends on the accuracy requirements. Traceability does not imply direct comparison of the local standards against the NBS standard. A valid traceable standard can exist even though the measurement has been passed from NBS through many intermediate laboratories provided the accuracy degeneration is correctly defined.

In each calibration laboratory some instrument or piece of equipment represents the most accurate repository at that local level for a particular unit of measure. This then becomes the primary standard for that unit in that laboratory. Several echelons of measure may still exist between this local primary standard and the end user of test equipment. In some cases the local primary standard must be used directly to calibrate test equipment; more often it is used to calibrate other standards called "secondary standards" or "working standards." Depending on the accuracy required of the local primary standard, it may be calibrated either by an independent laboratory or directly by the NBS. Every measurement made degrades the transferred value of the standard involved in that measurement to some extent. This then becomes the disadvantage of using intermediate laboratories in establishing calibration standards. In very accurate measurements, this degeneration often cannot be tolerated. Despite this basic disadvantage, all local primary standards should not, and often cannot, be calibrated directly with the NBS. Calibration at NBS is usually much more expensive than traceable calibration through an independent laboratory, and it usually involves delays that may not be tolerable where alternate equipment is not available. In addition, NBS will not work with the less-accurate standards that are often all that are required for a particular local primary standard.

One word of caution on standards of any type: no device, either mechanical or electrical, remains indefinitely stable, regardless of the ultimate source of the instability. For this reason all test equipment is placed on a calibration recall cycle. Standards, too, are subject to instability. The shiny new standard is not nearly as valuable as the old standard that has a proven record of stability. The standards of any calibration laboratory increase in value with time as their history of stability is documented.

Use of Outside Laboratories. The integrity of traceable calibration is not jeopardized in the least by using outside laboratories for calibration of either the local primary standards or the user's test equipment. In the case of standards calibration, the first consideration needs to be that the required level of accuracy for that standard can be achieved by a laboratory other than the NBS. Each intermediate step between the user and the ultimate reference at NBS degrades the transferred value of the unit. The particular outside laboratory chosen to perform the calibration must be selected with care, especially if a local primary standard is to be calibrated. There are a number of obvious things to look for: good equipment, certificates establishing traceability, neat and well-organized laboratory areas, and available and well-used reference material. These are all items that may easily be checked. More difficult to determine is the level of competence of the personnel. Regardless of the quality of the physical facilities, reliable results in precision measurements cannot be obtained without highly qualified personnel.

The choice of an outside laboratory for routine calibration of standard test equipment is not as critical, although the selection should still be made with care. The reason for using an outside laboratory instead of NBS for calibration of local primary standards is that the turn-around time is shorter than at NBS. On the other hand, the decision to use an outside laboratory for calibrating test equipment is based on economics. If the number of calibrations per year of a particular type is limited, use of an outside laboratory that already has appropriate standards and trained personnel is more practical. When the cost of performing your own calibration is being compared with the cost of having an outside laboratory perform the work, the cost of calibrating local primary standards must also be included. It is not at all uncommon for the price of a single NBS calibration to exceed the cost of the equipment being calibrated.

Frequency of Calibration. A frequency of calibration is established to give the equipment user reasonable assurance that the equipment will remain within its specifications between calibrations. The frequency depends on the type of test equipment and its application. Because the severity of operating conditions can be so widely variable, calibration intervals for a class of instruments are usually based on their history of performance. This ensures an optimum recall interval for the particular conditions of environment, use, and abuse. Initially, intervals are established by reference to handbooks and manuals that describe normal recall intervals for typical applications. Too long a calibration interval is undesirable because it diminishes the possibility of the instrument's remaining within its specifications throughout the interval. The cost of repairs per call-in usually increases in this circumstance. On the other hand, too frequent call-in has disadvantages. Calibration costs per year on the instrument are higher than they should be. In addition, the instrument is out of service more frequently than is necessary, resulting in inconvenience to the user and the need for backup equipment that would not otherwise be required.

The significance of calibration stickers should be considered. Too frequently a current calibration sticker is considered proof that the instrument is functioning correctly, and, conversely, an expired sticker is proof that the instrument is no longer within specifications. Barring errors, the sticker is only proof that the instrument was within specifications at the time of calibration. It also signifies that, if the instrument is within the calibration interval, there is a reasonable probability that it is still within specifications. If the calibration interval has expired, there is a diminished probability that the instrument remains within its specifications. With both electrical and mechanical test equipment, there is always the chance of a subtle failure that is not readily detectable. There is no substitute for intelligent use of test equipment by a user who is alert to subtle irregularities.

Call-in Techniques. It is often more difficult to break loose a piece of equipment from the user for a trip to the Calibration Laboratory than it is to repair and calibrate the instrument. Two direct approaches are generally used. When the instrument is calibrated, a sticker is put on it which gives, among other information, the date when the instrument is due for recalibration. The user himself can thus see when the equipment must be returned for recalibration, and he can schedule his use of the equipment to allow turn-in before the calibration has expired. The Calibration Laboratory keeps a record of the date the instrument is due for recalibration. Each month lists of equipment due for recalibration are circulated to alert the users so they can plan around the temporary loss of the equipment. Delinquent lists are also published by the Calibration Laboratory when equipment is not turned in as required. The attitude of the supervisor or foreman in the area using the equipment can be helpful in ensuring the timely turn-in of the equipment. Another call-in technique that can be used in some special situations is the running-time meter. Where equipment degeneration is based on running time rather than elapsed time since previous calibration, the running-time meter can be used.

Performance and Repair Records. The Calibration Laboratory must generate at least two records: the calibration sticker and the equipment-history card. The calibration sticker is placed on the instrument after calibration and contains, as a minimum, the date calibrated, the date due for recalibration, and the identification of the person who performed the calibration. Information such as equipment-use limitations and equipment accuracy may also be desirable. The equipment-history card contains all the basic data and history of the instrument. In small- to medium-size operations, this record is normally a maintenance or history record that is maintained on a one-card-per-instrument basis. In large organizations the information may be entered into a computer. The basic information that must be maintained in this record includes the description of the equipment, the identifying serial number, the name of the person who performed the last calibration, the date of the last calibration, the date recalibration is due, and the condition the equipment was in when received for repair or calibration. Besides providing the basic information for calibration recall, the history information is useful in troubleshooting, and the information on equipment condition is used to establish realistic calibration intervals.

Obsolescence Determination. Most equipment eventually reaches the point where it becomes more economical to replace it than to continue it in service. There is no problem in determining that new measurement requirements have exceeded the capabilities of existing instrumentation. More difficult to determine is when equipment should be removed from service owing to lack of use or the high cost of repair. The history card is the basic tool for making this determination. Data on the equipment condition when received and the extent of repairs necessary at each calibration can be used to decide when it is cheaper to invest in a new instrument with low maintenance cost. When this judgment is being made, maintenance costs associated with equipment abuse should be excluded since, in most cases, these costs would continue even with new equipment. Disposal of equipment for lack of use is highly dependent on an organization's specific situation. The advantages of removing the equipment from an organization's capital assets should be considered, as well as the cost of storage, equipment deterioration, and the risk of equipment obsolescence with prolonged storage.

11-2.7 Quality Control at the Reactor Site

(a) Verification of Condition on Receipt. All nuclear instruments, associated panels, sensors, wire, and coaxial cable should be inspected by Receiving Inspection for signs of damage on receipt at a reactor site. Receiving documents should then be checked to verify that the requirements of the purchase requisition have been fulfilled. The purchase requisition will contain specifications or refer to specifications that are applicable to the purchased item and will also state if vendor quality-assurance inspection and certification was a requirement before shipment.

(b) Quality Checks During Installation.* During installation, quality checks of nuclear instrumentation should be made as follows:

1. Check and ensure that coaxial connectors are installed on cables per the manufacturer's specifications. Cleanliness is very important during connector installation to maintain a high insulation resistance.

2. Check and ensure that noncoaxial connectors are installed per the manufacturer's instructions. Items to watch for are wire size, insulation removal, crimping, and pin insertion tools.

3. Insulation resistance of coaxial cables should be measured after the coaxial connectors have been installed. A rule of thumb is that the numerical value of the insulation resistance (in ohms) should be 10 or more times the reciprocal of the lowest signal current (in amperes) that the cable will carry. Where coaxial cables are used to carry a-c signals (pulses, for example), insulation-resistance values of 10^{10} ohms are more than adequate and are easy to achieve.

4. High-voltage (hi pot) tests of coaxial cables should be performed.

5. Check routing of field cables to ensure that there are no friction points where excessive wear could occur on cable or wire insulation.

6. Perform construction tests. These are functional tests performed with the equipment energized to verify that all field wiring is correctly installed. Any method for checking field wires, such as manual operation of relays and use of jumpers, is acceptable. However, care must be taken to identify and tag equipment, circuits, and systems that are to be energized and to isolate, as necessary, circuits that should not be energized.

Before installation, equipment (and even cables) may be assigned a quality-assurance number that can be used later as a quick guide to the applicable certification documents.

(c) Preoperational Check-Out Procedures. Preoperational tests are functional operating tests that are performed before putting into operation a system (e.g., neutron-monitoring, control-rod-drive, reactor protection system) and its associated instrumentation where the system is actually monitoring a process or performing a safety function. The purpose of these tests is to verify that instruments and systems function as designed and as specified in the applicable technical specifications.

Inputs to nuclear instruments that receive inputs from sensors when in actual operation may be simulated with a pulse generator, sine-wave generator, or current source as required. Trip points can be set, and the resultant functions initiated by trips (e.g., scram, rod block, and annunciation) can be checked with simulated inputs.

An acceptable preoperational check-out procedure must be detailed enough to check out every component, circuit, wire, and coaxial cable in the system covered by the procedure. It must also cover the check-out of any mechanical equipment associated with a system, e.g., in-core sensor retracting drives used with source- and intermediate-range instruments.

Field tests must also be made on in-core neutron sensors before they are put into service. The in-core sensors fall into three groups (see Chap. 3): pulse counting for source-range coverage, mean-square-voltage type for intermediate range, and direct-current type for power range. Field tests of neutron sensors are as follows:

1. Insulation-resistance tests to verify that no damage has occurred to the insulation and seals. In general, the insulation resistance should be greater than 10^{10} ohms.

2. Voltage-breakdown tests to verify that the filling gas has not escaped owing to a cracked or broken seal. Current in this test should be limited to approximately 10 μA to avoid possible damage to the insulating material.

Source tests of "dunking" chambers (fission chamber or proportional counter) that are to be used during fuel loading should be made after the fuel-loading source is placed in the reactor core. Curves of background count vs. discriminator setting must be made before the loading source is placed in the core. After the loading source has been placed in the core, but before fuel loading, discriminator curves and voltage-plateau curves should be run to determine the optimum discriminator set point and chamber operating voltage for each source-range channel. After the discriminator and voltage settings have been determined and set, a final check should be made to verify that the chambers are indeed seeing the neutron source. This final check can be made by raising and lowering the dunking chambers above and below the level of the source and verifying that the count-rate readout of the source-range channels decreases and increases accordingly. In addition, neutron pulses can be distinguished from background and gamma by monitoring the source-range instrument input signal with an oscilloscope.

Source-range in-core fission chambers should be source tested as the source-range instruments are changed from dunking chamber inputs to the permanent in-core chambers. This changeover and the tests required are made after fuel loading has been completed and the large start-up sources have been placed in the core. The same tests that were performed with the dunking chambers should be repeated. If the source-range fission chambers are retractable, positive verification that the source-range chambers are seeing neutrons can be made by retracting the chambers and noting the decrease in count rate.

The field instrument engineer is responsible for verifying that all neutron-monitoring instruments have been calibrated before fuel loading, that preoperational tests on neutron-monitoring systems have been completed satisfactorily, and that documentation exists for verification of all tests and results.

(d) Field Feedback Reporting and Analysis. The field engineer must feed back information relative to the performance of instruments and systems for which he is responsible. The information should be included in reports to the home office.

Reports of equipment failure are particularly important. To help those who must evaluate the failure, the failure report should include:

1. Catalog and serial number of failed part.
2. Description of the failed part.
3. Mode of failure.
4. Operating status at time of failure.
5. Effect on system or subsystem, if known.
6. Date of failure and approximate total operating time before failure.
7. Corrective action taken.

Field engineers' reports should be distributed to the responsible engineering groups for information and/or evaluation. For instance, if repeated failure of a particular component is observed at one or more field locations, a redesign of circuits or system may be warranted. In some systems, depending on the effect of a component failure in that system, a single failure could make it mandatory for redesign to prevent recurrence.

In situations where corrective action is initiated by a field engineer and the action involves a redesign or a deviation from approved drawings, change information should be sent to the home office immediately for Engineering approval and drawing changes before the system is put back into operation (where it is performing its intended function). Approval by telephone may be adequate in some cases when followed up in writing.

Changes initiated by Engineering and performed by the field engineer should be reported to the home office as being completed once the change has been made and the instrument or system has been retested.

Analysis of feedback from the field and determination of corrective action is the responsibility of the appropriate component of Engineering. Field Engineering is responsible for carrying out the corrective action and documenting changes.

11-2.8 Summary

A total quality system that embodies design control, materials control, process control, and product control must be implemented to attain the reliability necessary for achieving design goals relative to the appropriate level of safe and trouble-free life while still maintaining competitive costs.

The requirements of the Atomic Energy Commission and the customer fix the minimum quality standards that must be incorporated into the design of nuclear instrumentation systems. The Quality Assurance organization must ensure that these standards are upheld throughout the procurement and manufacturing cycles by establishing appropriate controls at critical points, such as receiving inspection of raw materials, parts, and subassemblies, in-process inspection and subassembly testing, final systems test, and shipping inspection. Judicious selection of the points in the manufacturing cycle where tests or inspections are to be performed as well as the selection of the correct type of quality information equipment and the generation of inspection and test instructions is the job of the quality-control engineer.

The life cycle of a particular product can be thought of in terms of distinct phases: the preproduction phase, which includes design and procurement; the production phase, which includes manufacturing, testing and packaging; and the postproduction phase, which includes shipping, customer installation, and acceptance testing and service life (particularly during the warranty period). A total quality-assurance program will ensure, with a high degree of confidence, that appropriate measures are implemented during each phase by all personnel involved with the product, from sales to customer installation and servicing. Therefore quality assurance should not be thought of as the inspection and testing operation that screens the good product from the bad; instead, it must be thought of as a company-wide program to ensure customer satisfaction with minimum cost to the company.

The quality-assurance program for a company involved in the design and manufacture of nuclear instrumentation must contain all the elements of a good total quality-assurance program. Criteria and standards promulgated by the AEC and ASME, such as Quality Assurance Criteria for Nuclear Power Plants (Appendix B to 10 CFR 50), or

Quality Assurance Program Requirements (RDT F 2-2T), or Quality Assurance (NA4000 from Sec. III of the *ASME Boiler and Pressure Vessel Code*), or Quality Assurance Program Requirements for Nuclear Power Plants (ANSI-N45.2), all describe quality-assurance programs which if properly implemented will provide an excellent QA program for anyone in the nuclear industry.

Nevertheless, it must be noted and emphasized that there is no substitute for a high degree of technical competence in the personnel involved in implementing such a system. For example, a competent quality-control engineer in this industry needs to be versatile not only in the quality-control and statistical field but also in the fields of electronics, nondestructive testing, and nuclear technology—four very specialized fields.

11-3 RELIABILITY OF REACTOR INSTRUMENTATION

11-3.1 Introduction

(a) Importance of Design Phase. Reliable instrumentation systems can only be created during the design phase. No amount of quality control, field testing, or maintenance can adequately compensate for a lack of careful planning during the conceptual design. A few designers seem to have an intuitive sense of good design, and their products enjoy a reputation for reliability. Most designers can acquire the art of producing reliable designs by adhering to some of the disciplines of reliability engineering.

A designer's most frequent failing is to be trapped into thinking only about how to design a system that will work and giving no thought as to how it might fail. Everyone is amused by the famous Rube Goldberg designs (Fig. 11.1). It is readily apparent how Rube's systems work and equally apparent how prone they are to fail. Modern reactor instrumentation systems have become so sophisticated that the designer, in concentrating his efforts on making the system work, forgets to look for ways in which it may fail. By applying the techniques of reliability analysis, a good designer should consistently turn out a product that not only works but also has a low incidence of failure.

The attitude of the designer is all important. He must have the desire to create a reliable system. He must not be lulled into the attitude that the design is adequate because the system passed a design audit or was granted an operating license. A trained reliability engineer can probe deeply and learn much about a system, but, unless he has full cooperation from the designer, the effort will fall short of complete success.

System reliability and system capability must not be confused. For example, the nameplate rating on a power supply tells the designer about the load capacity it may be expected to accommodate. Choice of a power supply with a rating in excess of the load requirements assures capability. A reliability study assumes system capability and goes on to make an assessment of the probability that the system will actually be successful in performing a given task within its capability. The two terms, capability and reliability, are related through the term "design margin," and the designer should recognize the favorable influence excess capability may have on reliability through the application of appropriate derating factors.

Effort invested in systems reliability analysis is economically attractive. The potential for cost saving lies in systematically selecting the higher reliability systems for detailed design, in choosing the simpler of two alternatives, in avoiding overdesign on portions of the system that do not contribute to reliability, in avoiding costly retrofits and in gaining a reputation for a reliable product.

(b) Reliability and Availability. The term "reliability" is frequently used in a qualitative sense to imply quality and integrity. As pointed out in Pt. 11-1 of this section, reliability is a measure of the time stability of a product's performance. This concept can be expressed in a quantitative sense in the definition of reliability approved by the IEEE:[4] "The characteristic of an item expressed by the probability that it will perform a required function under stated conditions for a stated period of time." In the general sense, this definition of reliability does not allow for failure and repair during the stated period of time. For example, consider a pressure switch monitoring the reactor pressure located where it is totally inaccessible during reactor operation. An assessment of the numerical reliability of the switch to survive one fuel cycle is a meaningful measure of its integrity.

In many other cases, inspection, test, and repair during operation are permitted, and this is, in fact, the preferred mode of operation. In this event, the most meaningful measure of integrity is called "availability" and is defined as[4] "The characteristic of an item expressed by the probability that it will be operational at a randomly selected future instant in time." Stated another way, availability is the fraction of the time the system is operational. In the long-term steady-state situation, availability is given by the equation:

$$\text{Availability} = \frac{\text{Up time}}{\text{Up time} + \text{Down time}} \qquad (11.1)$$

where the up time is approximately equal to the average time between failures and the down time is the average time to repair and restore the system to service. Since the down time is the average time to repair the system, it must include the elapsed time between the failure and its discovery, a term that may be of primary significance, especially in standby systems.

Availability, not reliability, is the term that is most applicable to the usual reactor instrumentation system, where redundancy, repair during operation, and a paramount concern for detecting and eliminating unsafe failures are prominent characteristics.

11-3.2 Preselection of a Basic System Configuration

The most effective reliability efforts are those expended during the preliminary design phase while the system configuration is being determined. During this period the reliability considerations can influence design decisions and

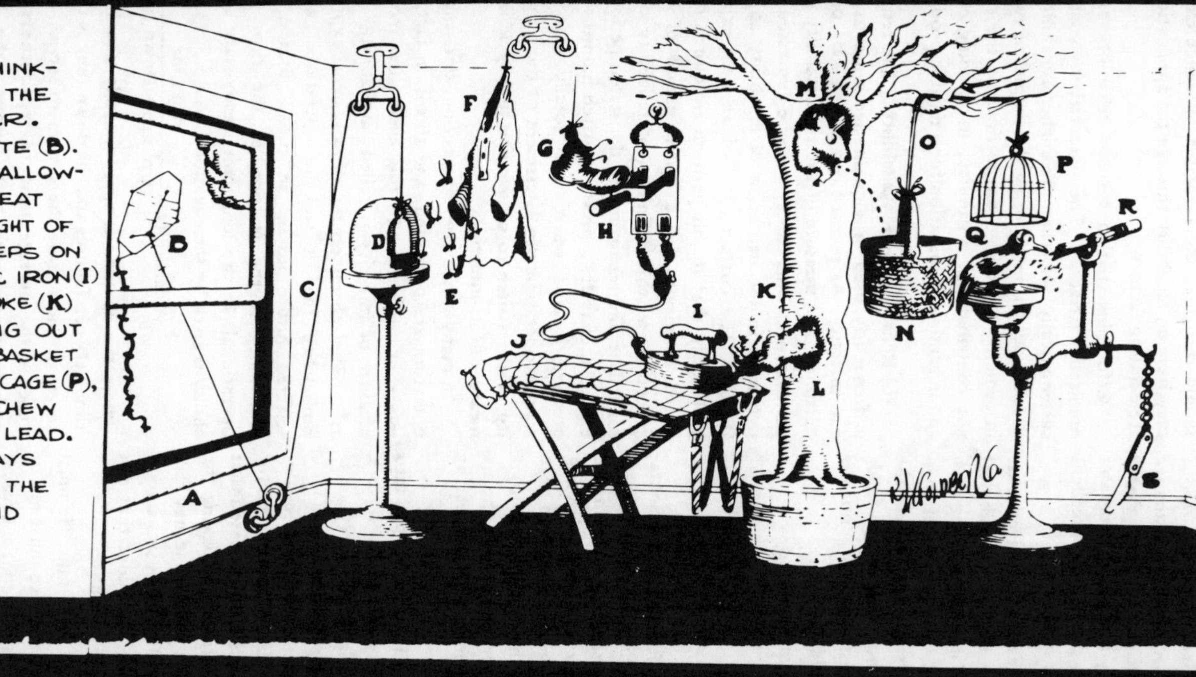

PROFESSOR BUTTS GETS HIS THINK-TANK WORKING AND EVOLVES THE SIMPLIFIED PENCIL-SHARPENER.
OPEN WINDOW (A) AND FLY KITE (B). STRING (C) LIFTS SMALL DOOR (D) ALLOWING MOTHS (E) TO ESCAPE AND EAT RED FLANNEL SHIRT (F). AS WEIGHT OF SHIRT BECOMES LESS, SHOE (G) STEPS ON SWITCH (H) WHICH HEATS ELECTRIC IRON (I) AND BURNS HOLE IN PANTS (J). SMOKE (K) ENTERS HOLE IN TREE (L) SMOKING OUT OPOSSUM (M) WHICH JUMPS INTO BASKET (N) PULLING ROPE (O) AND LIFTING CAGE (P), ALLOWING WOODPECKER (Q) TO CHEW WOOD FROM PENCIL (R) EXPOSING LEAD. EMERGENCY KNIFE (S) IS ALWAYS HANDY IN CASE OPOSSUM OR THE WOODPECKER GETS SICK AND CAN'T WORK.
A
B
C
D
E
F
G
H
I
J
K
L
M
N
O
P
Q
R
S

Fig. 11.1—Rube Goldberg's simplified pencil sharpener.

ensure that the chosen configuration is one that can be developed to meet the reliability requirements. The process is an iterative one, with the designer continually looping back and trying different system configurations until all constraints are satisfied. At this time the principal effort is being expended at the drawing board. Errors of judgment may be corrected with an eraser instead of a jackhammer.

Reliability considerations are not the only constraints imposed on the design. Other constraints include capability, size, shape, weight, cost, schedule, and customer preference, all of which must be adequately satisfied. Reliability analysis provides a disciplined framework within which the interplay of these constraints can be viewed with sharper perspective. Frequently the result is not only a more reliable system but also a better system as judged by all other applicable criteria.

In the preselection of a basic system configuration, the designer should (1) define success for the system, (2) establish adequate reliability goals, (3) propose alternate designs, and (4) evaluate the reliability potential for each design. These four tasks are discussed in the following sections.

(a) Defining "Success" for the System. The designer must know exactly what his system is expected to do. This may sound trite, but many a design has been impaired because the designer either did not fully know or perhaps had lost track of the real reason for having the system. The definition of success should include the environmental constraints in force and the length of time or the number of cycles the system is expected to endure. There may be two or more valid success definitions for the same system, each requiring a separate analysis.

For example, assume an instrumentation system associated with a set of isolation valves. There may be two definitions of success imposed, one for safety reasons and the other for operational or economic reasons:

Success #1: Given that the pipeline downstream of the isolation valve is broken (complete severance), the instrumentation system shall detect the resultant leak within 10 sec and signal the isolation valve to close.

Success #2: Given that the pipeline downstream of the isolation valve is not broken, the instrumentation shall not signal the isolation valve to close.

Operating conditions: The cable and detector environment is 120°F and 50% relative humidity prior to the break and 212°F and 100% relative humidity following the break. The instrumentation system is tested every 3 months and calibrated annually.

In every case the boundaries of the system under consideration must be explicitly defined. In the above example, it is intended that the valve itself be excluded. For this reason a transition point from the instrumentation system to the valve must be chosen so that every component or potential point of failure is certain to be included within one system or the other, but never both.

(b) Establishing Goals. The designer must have some measure of achievement for the reliability of his system. One simple and effective goal that has been used on critical systems for many years is the so-called "single-failure" criterion, namely, that the system shall fulfill its success definition in the event of failure of a single active component. Basically, this criterion has served the nuclear industry rather well in spite of some limitations. One limitation is that it is not readily adjustable to match the whole range of consequences of system failure. If the single-failure criterion were universally applied, a high-level warning instrument on a waste-water sump would need to be just as reliable as the reactor protection system, even though the consequences of failure are vastly different. In addition, the single-failure criterion does not adequately protect against multiple independent failures that are more probable than should be allowed. Despite its shortcomings, the single-failure criterion for all active components should be imposed as the minimum goal on all reactor instrumentation systems where safety and the potential for economic loss are important considerations.

The techniques of reliability analysis, properly applied, yield a numerical measure of the expected system reliability or availability. For this reason a numerical goal serves an especially useful purpose. Although such numerical goals are commonplace in the aerospace industry, they are only recently coming into use in the nuclear industry. A numerical goal can be established in any one of a number of ways.

1. Risk acceptance. Ideally the goal should be a function of the highest risk the public will accept in return for the benefits derived from nuclear power. Risk is defined as the product of the probability of failure and the consequences of that failure. The consequences may be measured on any convenient scale, such as dollars, curies of ^{131}I, and injuries. Unfortunately this concept is not very far advanced and not universally accepted. However, an examination of some of its precepts does yield some insight into the relative reliability required for various systems.

2. Grandfather systems. Even though the nuclear industry is relatively young, there are some instrumentation systems that have gained wide acceptance for a given application and enjoy a reputation for being adequately reliable. A reliability analysis of one or more of these systems will yield a numerical result that should prove useful in establishing a realistic numerical goal for new systems.

3. Industry standard goals. Industry committees concerned with the safety of nuclear plants (see sec. 12) are beginning to address themselves to the matter of goals. On the international scene, a goal of 10^{-5} probability of failure has been proposed for the reactor-protection-system scram function. The IEEE Nuclear Science Group Technical Committee on Standards (see section 14) has considered goals, but it is currently recommending that each designer set a goal to meet the particular need.[4]

(c) Proposing Alternate Designs. Before any attempt is made at a detailed design, a wide range of design alternates should be blocked out for evaluation. The instrumentation

system must be considered as an integral and essential part of the overall functional system. In other words, instrumentation systems perform an essential service for the functional systems in the plant; instrumentation does not exist for its own sake.

For example, assume that the functional system is an emergency cooling loop. The engineer designing the functional system and the instrumentation engineer must work together to propose alternates, such as the following:

1. One loop, two 100% capacity pumps per loop.
2. One loop, three 50% capacity pumps per loop.
3. Two loops, one 100% capacity pump per loop.
4. Two loops, one 100% capacity pump per loop with a crosstie.
5. Two loops, one 50% capacity pump per loop plus one 50% capacity pump shared by both loops.

The list should be made as inclusive as possible so that no worthy configuration is omitted. All proposed alternate designs should pass the capability test before being evaluated for reliability or availability. Obviously the probability for system success can vary widely, depending on the system configuration. The instrumentation systems to start and stop the pumps and open and close valves are very different for the relatively few configurations cited.

(d) Evaluating the Reliability Potential for Each Design. It is not practical to perform a detailed design on each proposed system before making a selection that is based on, among other considerations, a detailed reliability analysis. Therefore it is particularly important that the proposed designs be carefully screened to eliminate those which do not have the potential for development into a system with adequate reliability.

The foregoing may be accomplished by adhering to the following discipline:

1. Construct a simple reliability model for each proposed design. The blocks from which the models are constructed should encompass as much of the system's equipment as is reasonably possible. For example, a block called "pump" could effectively include the pump, its driving motor, coupling, and circuit breaker.

2. Use a consistent set of failure data throughout the comparative evaluations. Where failure-rate data have been reported, use them as a base, but do not hesitate to adjust them upward or downward to reflect best judgment, duty factors, or environmental conditions. Where failure-rate data do not exist, choose a value that reflects the best judgment of knowledgeable people in the field but use the same assumption consistently throughout the evaluation.

3. Reflect the expected operating conditions. If in one design certain components are exposed to environmental conditions more severe than normal, that design should be properly penalized by adjusting the failure rates upward by an appropriate K factor to reflect the higher level of imposed stress.

4. Allow each system proper credit for its compatibility with testing. In general, the unreliability of a component increases almost linearly with the interval between thorough tests. (See Fig. 11.4.) Therefore a component that is physically inaccessible for test except during a refueling shutdown should be penalized in comparison to one that is readily accessible and frequently tested.

5. Solve the models for a numerical index of reliability. If the model is really kept at its simplest level, the probabilistic solution should not be difficult. If the solution is difficult, concentrate first on simplifying the model to get an approximate solution rather than straining at the mathematics for these preliminary design evaluations.

6. Conduct sensitivity studies to identify the dominant components contributing to the unreliability of each system. This may be done by making a significant change in the assumed failure rate of a particular component and noting the change in the overall probability of system failure. Figure 11.2 shows a plot on a log–log scale of component unreliability vs. system unreliability. The reference or expected failure probab'lity for component 1 is indicated by an arrow. Note that the arrow falls on the flat portion of the curve, indicating that this particular component does not contribute significantly to system unreliability. The reference value for component 2 is on the steep part of the curve, indicating that a change in failure rate here will have a dominant effect on system unreliability. Good safety design dictates that the overall system have a sufficiently low value of unreliability. Good economic design dictates that, in general, the least expensive components should not be dominant contributors to unreliability.

7. Redesign the proposed systems, as appropriate, to minimize the areas of apparent weakness revealed by the sensitivity analysis. This becomes an iterative process, but this is where the big payoff comes, in being able to bring quickly into focus the systems with the greatest potential for detailed design consideration.

8. Reexamine the final proposals to be sure that they satisfy all other operational and physical constraints that may be appropriate.

9. Select the one or two exploratory designs that show the greatest potential for maturing through a detailed design process and for adequately satisfying all constraints, including reliability.

11-3.3 Detailed System Design for Reliability

The proposed design or designs selected from the evaluations of reliability potential are subjected to detailed design. Components of known quality are selected and applied well within their rating for the expected environment. If a new component of unknown quality is to be applied, it may be appropriate to subject it to test, particularly if the assumed best-judgment failure rate indicates (through the sensitivity study) that the component has a dominant influence on reliability.

In every possible way the designer must endeavor to emulate the model and to be certain that the boundary assumptions are satisfied. If the model assumes that the failure of one component or group of components is statistically independent of other failures, the designer should try to ensure that this is true. For example, if two

channels of instrumentation are assumed to be independent, they should be so located that only a highly improbable event could disable both. The routing of signal cables and the location of power sources must be carefully considered. Historically, localized overheating and fire have been the two most common single-event failures that can cause other failures to be interdependent. Careful judgment is required to develop a design that gives a reasonable assurance that a fire can be controlled without transgressing the independence of channels.

The designer should also recognize that interdependence can creep in by inconspicuous routes. If the required level of reliability is such that redundancy is necessary, it may be appropriate to make the redundant channel different just to increase the likelihood that an unknown deficiency or inadequacy in one channel will not be repeated in the other. Frequently this can be accomplished by functional diversity; for example, one channel can monitor temperature and another pressure, either signal containing the desired information. Where functional diversity is not possible, equipment diversity may be used to good advantage. For example, pressure can be monitored by two sets of equipment that operate on two entirely different principles. Functional diversity is to be preferred because it almost automatically includes equipment diversity.

If the model assumes that the component failure rates are constant, the designer should be sure that the maintenance and replacement practices will not allow worn-out components to remain in the system. If the model assumes that some of the components are to be tested while the plant is in operation, the designer should be sure that adequate testing facilities are provided. If the system or portions of the system perform functions in addition to a safety-related task, the designer must ensure that these additional functions do not interfere with the model of the safety-related function.

Simple, straightforward systems are easy to understand, easy to model, and tend to have high reliability. If a system is allowed to develop without the benefit of a model to emulate, the system can become complex and interwoven in such a way that modeling is extremely difficult and, intuitively, the whole system is suspect. A safe rule, then, is never design a system that cannot be reduced to a tractable reliability model.

Of course, the instrumentation system must still meet all its normal objectives of performance. The reliability discipline is simply superimposed on the usual detail design procedure. The concepts of reliable system design are not difficult nor are the associated mathematical relations. For this reason it is preferable that a designer with a good reputation for instrumentation design take on the disciplines of reliability engineering rather than interpose a reliability engineer as a series element in the design chain. A reliability engineer serves the highest purpose when used as a consultant and when he and the designer approach a problem with open minds and an honest desire to understand the system.

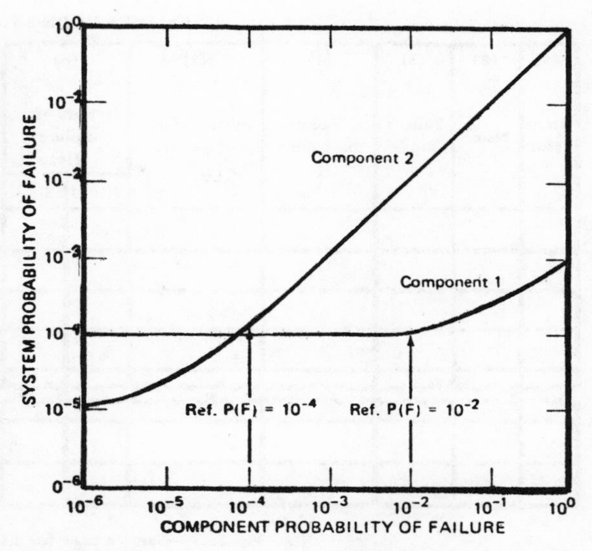

Fig. 11.2—Sensitivity study of system failure vs. component failure.

11-3.4 Evaluation of the Design

(a) The Failure Modes and Effects Analysis (FMEA). The detailed design is followed by a detailed reliability evaluation. As a minimum the system must be subjected to an FMEA. This is a subjective and nonnumerical analysis that exposes potential failure points. The FMEA identifies every component in the system by component number and name. It lists the various modes in which the particular component would fail (open, short, closed, stuck, etc.) and lists the failure mechanisms that can induce a particular failure mode. It further identifies the relation a failed component has to system failure and to the failure of other systems.

A sample page of an FMEA is shown in Fig. 11.3. There are many variations of this form, and there should be no hesitation in adapting the form to suit each analysis. The primary function of the FMEA is to provide an understanding in depth of how the system reacts to all modes of component failure. It is particularly valuable in serving as evidence of conformity with the single-failure criterion. If only a minimum effort can be budgeted on reliability analysis, it is generally best spent on an FMEA.

The FMEA form also has space to develop information on failure rates, application factors, test intervals, and repair times in preparation for the more rigorous mathematical model of reliability or availability.

(b) Detailed Reliability Model. The designer now has the information at hand to conduct a more detailed reliability or availability analysis of the system. If the designer adheres closely to the framework of the original simple reliability model, the detailed model should generally be satisfied by the same skeletal block diagram, the exception being only in the number of blocks represented. Thus the final model should be just as tractable mathematically as the exploratory model on which the design is

FAILURE MODE AND EFFECTS ANALYSIS

(1) Item No.	(2) Name	(3) Failure mode	(4) Failure mechanism	(5) System effect of failure	(6) Rate of failure effect		(7) Failure class (conse- quence)	(8) Remarks (compensating features etc.)	(9) Failure risk								
									System state			System state			System state		
					Rapid	Slow			λ	t	λt	λ	t	λt	λ	t	λt

Fig. 11.3—Sample page for Failure Mode and Effects Analysis.

based. If it is not, the designer should examine the original assumptions, particularly the one on component or channel independence, to see if they have been violated.

The main value in performing the detailed reliability analysis is in making certain that some trivial component does not make an unexpected and unwarranted contribution to unreliability. If such a discovery is made, the problem can usually be rectified by the choice of a better component, more frequent and more thorough testing, or by the judicious use of redundant components.

When it has been determined that the contribution of a given component to unreliability is trivial, it may be eliminated from the model or lumped with associated components to simplify the computation.

11-3.5 Testing for Reliability

Instrumentation systems are characterized by two kinds of signals, analog and bistable. The analog portion of the system can usually be depended on to annunciate its faults by causing either a zero output or an off-scale reading. With the proper alarms, gross failures in the analog circuit can usually be detected immediately. The bistable portion of the circuit can usually be designed so that the most probable modes of failure are to the "safe" state. However, "fail-to-safe state" design is more of a design objective than an accomplished reality, and the only way one can be sure that a bistable device will successfully change state is to test it. Accordingly the designer should address himself faithfully to the special problems of testing.

(a) Thoroughness of Testing. The test should be thorough. The objective of the test is to ensure that the system still retains its original properties. The test must reflect the conditions that will exist when the instrumentation system is called on to function. For example, a bistable trip circuit should be tested by running the analog signal up to the trip level rather than dropping the trip level down to the normal operating point. The former proves that the analog signal will not saturate before it reaches the trip level.

If possible, a complete channel should be tested end to end with the real input parameter as the variable. For example, some designers arrange to squirt hot water on a temperature sensor to drive it past the alarm point; or, in other cases, there is a provision for removing a neutron shield from a neutron sensor to give it an up-scale signal. If it is not possible, for safety or practical reasons, to vary the real input parameter, then a substitute should be used which bridges the gap. For example, if it is not safe to reduce the flow down to the trip level, then a differential pressure signal may be substituted.

Instrumentation systems that must function while the reactor is operating should be tested while the reactor is operating. If a system is repaired or maintained while the reactor is shut down, it should be tested using substitute inputs before start-up and retested after start-up in its operational mode.

(b) Testing Redundant Systems. Redundancy in a system presents special problems for testing. The objective of redundancy is to provide a system that will continue to work in spite of the failure of a component. The redundancy tends to hide the failure, and, unless the test finds the first failure, the redundant system is not appreciably better than a nonredundant system.

The maximum benefit of redundancy is realized when the product of the failure rate λ and the test interval τ is kept small compared to unity. This principle is illustrated in Fig. 11.4, where the probability of failure is plotted as a function of $\lambda\tau$ for a single and a dual (redundant) device. Note that if $\lambda\tau$ is allowed to get too large, there is little advantage in redundancy.

The ability to test should influence system design. In Fig. 11.5(a) redundancy is applied on the component level, i.e., each component is paralleled with a like component. If there is no opportunity to test or repair (e.g., in an unmanned satellite), component redundancy may have some advantages in higher reliability. However, component redundancy is difficult to test because the function is performed if either of the parallel components perform. Also, few components work compatibly in parallel without some special provisions to get them to share the load, and there are likely to be single-failure modes that fail both components.

The system redundancy shown in Fig. 11.5(b) is amenable to test because each channel can be tested independently and the first failure can be found on test. In addition, the two channels can be physically and electrically isolated from each other, and single-failure modes are minimized. System redundancy is the preferred design mode. Component redundancy should be used sparingly

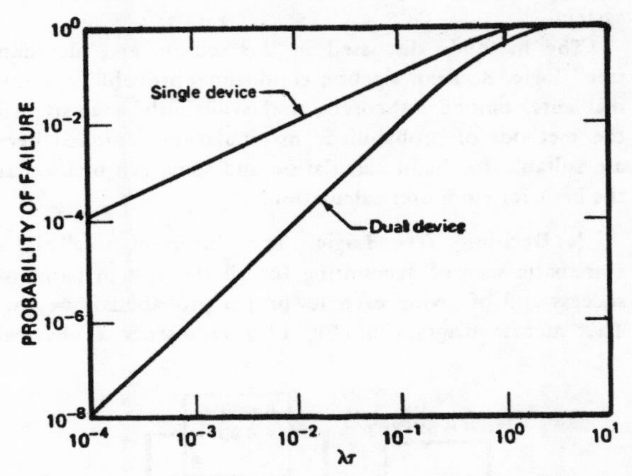

Fig. 11.4—Probability of failure of single and dual (redundant) devices as a function of λt. [From I. M. Jacobs, Safety-System Design Technology, *Nucl. Safety*, 6(9): 235 (Spring 1965).]

and only to bolster the reliability of one component in an otherwise strong chain.

(c) Staggered Tests. If there are multiple channels of instrumentation performing the same function in a redundant manner, there is an advantage to be gained in staggering the tests. For example, in a three-channel system scheduled for quarterly test, one channel should be tested each month. If the three channels comprise a one-out-of-three system, staggering the tests reduces the predicted unavailability to one-third that which would accrue for simultaneous testing. The higher the level of redundancy, the greater the benefits of staggered tests.

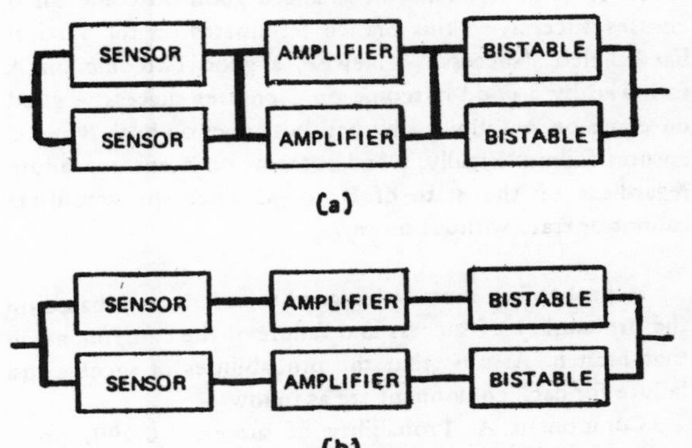

Fig. 11.5—Component (a) vs. system (b) redundancy.

Table 11.6 shows the unannounced unavailability of various logic configurations for three different schedules of testing. Note that random testing [calculated by the methods of part 11-3.6(f)] yields a result that is intermediate between simultaneous and perfectly staggered testing.

Table 11.6—Unavailability as a Function of Logic Configuration and Testing Schedule

	Unannounced unavailability		
Logic	Simultaneous testing*	Random testing†	Perfectly staggered testing*
1/2	$\frac{1}{3}\lambda^2\tau^2$	$\frac{1}{4}\lambda^2\tau^2$	$\frac{5}{24}\lambda^2\tau^2$
2/2	$\lambda\tau$	$\lambda\tau$	$\lambda\tau$
1/3	$\frac{1}{4}\lambda^3\tau^3$	$\frac{1}{8}\lambda^3\tau^3$	$\frac{1}{12}\lambda^3\tau^3$
2/3	$\lambda^2\tau^2$	$\frac{3}{4}\lambda^2\tau^2$	$\frac{2}{3}\lambda^2\tau^2$
3/3	$\frac{3}{2}\lambda\tau$	$\frac{3}{2}\lambda\tau$	$\frac{3}{2}\lambda\tau$
1/4	$\frac{1}{5}\lambda^4\tau^4$	$\frac{1}{16}\lambda^4\tau^4$	$\frac{251}{7680}\lambda^4\tau^4$
2/4	$\lambda^3\tau^3$	$\frac{1}{2}\lambda^3\tau^3$	$\frac{3}{8}\lambda^3\tau^3$
3/4	$2\lambda^2\tau^2$	$\frac{3}{2}\lambda^2\tau^2$	$\frac{11}{8}\lambda^2\tau^2$

*Derived from A. E. Green and A. J. Bourne, Safety Assessment with Reference to Automatic Protective Systems for Nuclear Reactors, British Report AHSB(S)R-117(Pt. 2).
†Derived from Sec. 11-3.6(f) of this chapter.

Another more subtle benefit of staggered testing should be noted. If all tests are run simultaneously, i.e., one test immediately following another, there is increased opportunity for human error. Suppose, for example, that the technician systematically reads the wrong scale on the calibration instrument and sets all channels to a low gain rendering them unsafe. If he proceeds directly from one test to the next, he is much more likely to repeat such an error on all channels than if the tests are spaced days apart.

(d) Provisions for Testing. The designer should anticipate the needs for testing and make appropriate provisions. If the test must be performed frequently, some built-in arrangements may be in order. If the test is simple and is performed infrequently, it may be more appropriate for the one performing the test to implement the test provisions. In any event the designer must be certain that the test can be run and that the test gear does not interfere with the ability of the system to perform its intended function.

If standard test-signal emitters are built into a channel, procedure should call for a regular cross calibration with another standard to ensure that the built-in standard remains within tolerance. If switches are built in to facilitate testing, an alarm should be initiated if the switches are not restored to "normal" before returning the channel to operation.

Observing the response of the channel to normal process variability and cross comparing with other channels monitoring the same variable is a very convincing test for the analog portions of an instrumentation channel.

Automatic testing provisions are sometimes used in reactor protection systems. These have two primary advantages:

1. The interval between tests can be reduced considerably with a corresponding reduction in system unavailability.

2. The automatic testing system can be designed to be a diagnostic help in troubleshooting.

These benefits should be balanced carefully against the disadvantages:

1. Extreme care must be exercised to ensure that a common-failure mode is not introduced in otherwise independent channels.

2. It must be demonstrated conclusively that the automatic test signal (frequently a pulse train) has the same effect on the circuit output device as a bona fide trip signal, which may build up slowly and sustain itself much longer.

3. The automatic test must encompass the whole channel, from sensor to channel output, and be thorough in discovering failures. This presents some difficulty for the tester, especially if a portion of the channel is analog in nature.

(e) Setting the Test Interval. Several factors must be considered in setting the time interval between tests. Tests spaced too far apart may allow unsafe failures to accumulate. On the other hand, tests conducted too frequently can become a burden on the plant operator if they do not truly enhance safety. The designer should consider the following factors in making the test interval a viable part of his design:

1. The tests should be made frequently enough so that the system will meet its design goal. In fact, one way to increase or decrease the availability of a system is to alter the test interval.

2. The tests should not be scheduled as "busy" work at an unrealistically short interval lest the tester lose his respect for the test and become negligent.

3. Only failure modes that are primarily time dependent need be considered in setting the test interval. For example, it is futile to test for a failure that occurs only as a result of the stress of initiating the test.

4. Wear-out due to testing should be a consideration, and, if it is necessary for the sake of safety to test so often that wear-out could be a problem, provisions must be made to monitor the component failure rates carefully and renew the components before they deteriorate to too low a level.

5. If a channel must be bypassed while it undergoes test, then the interval between tests (τ) should not be so short that the unavailability due to being bypassed is higher than the expected unavailability due to unsafe failures. For a single channel there is an optimum test interval. If the expected channel unsafe failure rate is λ and the time required to perform the test is t, the optimum test interval for highest availability is given by the expression[5]

$$\tau = \left(\frac{2t}{\lambda}\right)^{1/2} \tag{11.2}$$

In no case is it of benefit to test more often. The preceding expression does not hold for a redundant or majority logic system, there being no true optimum. However, negligible benefits are derived from testing the individual channels of a redundant system more often than is indicated by Eq. 11.2.

11-3.6 Probabilistic Manipulations

Systems are made of component parts, and the reliability of each component part is either known or can be predicted. The reliability of the overall system is what is desired. The reliability of the system can be predicted as a function of the reliabilities of its various component parts by applying the logic of success–failure events in the system.

The methods discussed in this section are "decision tree" logic, Boolean algebra, conditional probability, minimal cuts, binomial theorem, and availability analysis. All the methods of probabilistic manipulation described here are suitable for hand calculation and some can be used as the basis for computer calculation.

(a) Decision Tree Logic. The "decision tree" is a systematic way of accounting for all the system paths to success and of giving each its proper probabilistic weight. The success diagram of Fig. 11.6 represents a physical system in which two amplifiers are dependent on one power supply. Success is assured if the power supply and at least one amplifier are operating.

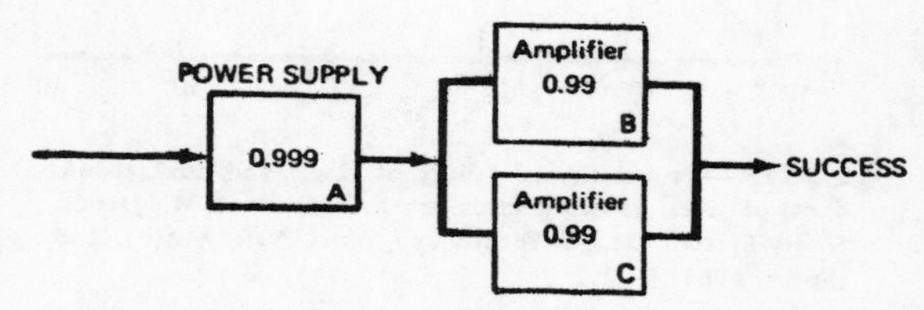

Fig. 11.6—Reliability block diagram for a series–parallel system.

The decision tree used to calculate system success is shown in Fig. 11.7. Each branch in the diagram, reading from left to right, represents a decision, go or no-go, on the success or failure of that particular component. By convention, good outcomes branch upward and bad outcomes branch downward. The components are considered in order. A good outcome on A and a good outcome on B ensures success; so this branch terminates on the vertical bar labeled "success." Likewise, a good outcome on A followed by a good outcome on C ensures success. A good outcome on A followed by bad outcomes on both B and C ensures failure. Finally, a bad outcome on A ensures failure regardless of the state of B and C since the amplifiers cannot operate without power.

A probability must be assigned to each branch based on the probability of success and failure of the components in that branch. Assume that the probabilities of success and failure for each component are as follows:

Component A: Probability of success = 0.999; probability of failure = 0.001.

Component B: Probability of success = 0.99; probability of failure = 0.01.

Component C: Probability of success = 0.99; probability of failure = 0.01.

The probability of success may be computed by tracing each success path back to the origin, taking the product of the probabilities along each path and summing the result. For success path 1, find the product of (0.99) × (0.999) = 0.98901. For success path 2, find the product of (0.99) × (0.01) × (0.999) = 0.0098901. The total probability of success is the sum of the two products, or 0.9989001.

These rules are essential:

1. The sum of the probabilities at each branch must be unity.

2. The components of the system must be considered in order until success or failure is ensured without regard to the state of any of the remaining components.

3. The various component-failure events must be statistically independent.

The least redundant component should always start the first branch and the most redundant components should be considered last to reduce the number of branches on the tree.

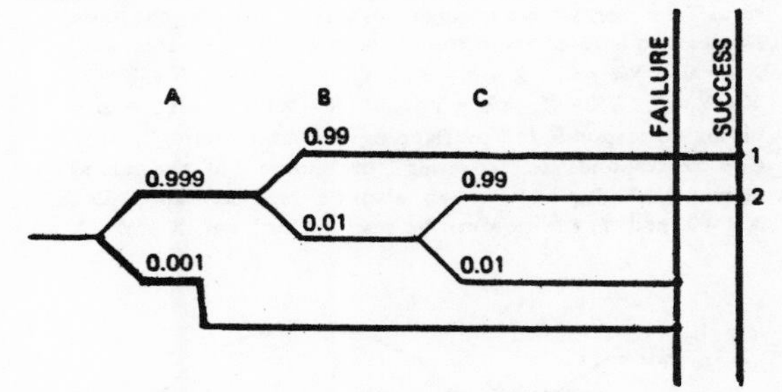

Fig. 11.7—Decision tree for predicting system success for the series–parallel system of Fig. 11.6.

The decision tree can handle events with more than two states. It can also accommodate certain physical interdependencies. Consider the success diagram of Fig. 11.8. This figure shows an amplifier driving an output transistor whose

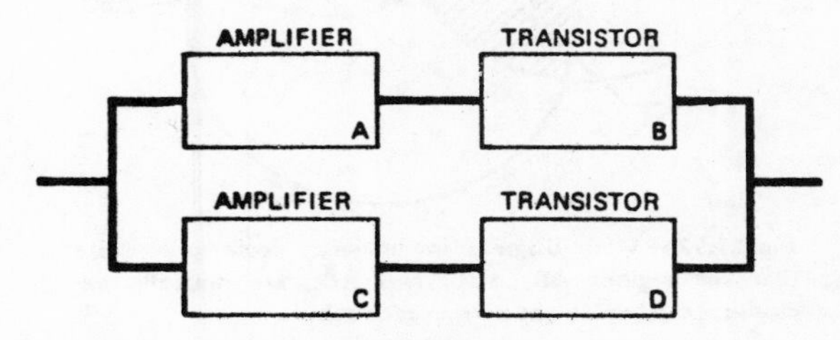

Fig. 11.8—System with multiple component failure states.

output is connected in parallel with the output transistor from another amplifier. One amplifier can fail without disturbing the other. The output transistor can be good, fail open, or fail short. If it fails short, it voids the output of the parallel output transistor and failure is certain.

The decision tree for Fig. 11.8 is shown in Fig. 11.9. Note that three branches are shown for each output transistor decision: good, open, short. Note also that any short causes failure whereas an open only leads to failure in combination with other failures. As in the previous example, a probability is assigned to each branch, summing to unity. The probability of success is the sum of the products of probabilities along each success path.

As numerical check, it is advisable to calculate the probability of system failure in the same manner and check to see that the sum of the success and failure probabilities is unity. Such a check does not guarantee against errors in the way the decision tree is branched to represent the problem. Consequently it is essential to exercise care in constructing the tree to be sure that it represents the physical system.

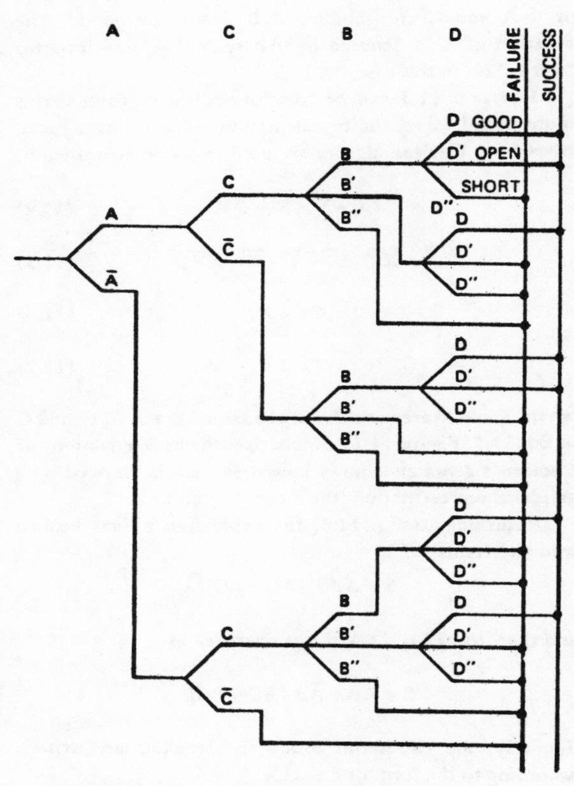

Fig. 11.9—Decision tree for the system with multiple component failure states shown in Fig. 11.8.

(b) Boolean Algebra. The techniques of using Boolean algebra as an adjunct to probabilistic calculations are documented in many sources.[6-8] For those not familiar with Boolean algebra, a simple, but often overlooked, technique will be described.

Consider a two-out-of-three, or majority, logic configuration as represented in the success diagram of Fig. 11.10. If any two of the components are good, the system is good. In Boolean algebra, the event "success" is described symbolically as

$$S = AB + AC + BC \quad (11.3)$$

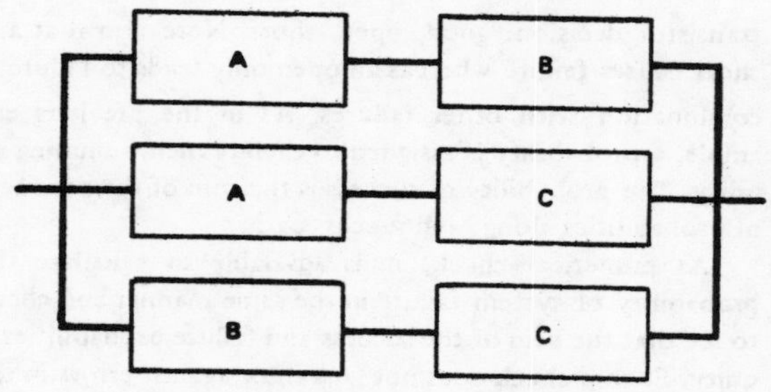

Fig. 11.10—Majority logic success model.

where AB means A *and* B and + means *or*. In words, Eq. 11.3 says, "The system is successful if A and B are good or if A and C are good or if B and C are good." The negation of A is denoted by the symbol $\overline{A}$, which means "not A" or, in this case, "A fails."

Equation 11.3 can be transformed into a form that is more useful in reliability calculations. The following relations from Boolean algebra are used in the transformation:

$$X + Y = X + \overline{X}Y \quad (11.4)$$

$$\overline{XY} = \overline{X} + \overline{Y} = \overline{X} + X\overline{Y} \quad (11.5)$$

$$XX = X \quad (11.6)$$

$$X\overline{X} = 0 \quad (11.7)$$

where X and Y are variables of the same kind as A, B, and C in Eq. 11.3. Figure 11.11 summarizes the basic equations of Boolean algebra and shows how they may be derived by a graphical representation (the Venn diagram).

Returning to Eq. 11.3, the expression is first broken into two terms:

$$S = AB + [AC + BC]$$

and then, using Eq. 11.4, this is rewritten as

$$S = AB + \overline{AB}\,[AC + BC]$$

Similarly, the two terms inside the brackets are written according to the form of Eq. 11.4;

$$S = AB + \overline{AB}\,[AC + (\overline{AC})(BC)] \quad (11.8)$$

From Eq. 11.5 the terms $\overline{AB}$ and $\overline{AC}$ can be written as $\overline{A} + A\overline{B}$ and $\overline{A} + A\overline{C}$, respectively. Substituting these into Eq. 11.8 yields

$$S = AB + (\overline{A} + A\overline{B})[AC + (\overline{A} + A\overline{C})\,BC]$$
$$= AB + \overline{A}AC + \overline{A}\overline{A}BC + \overline{A}A\overline{C}BC + A\overline{B}AC$$
$$+ A\overline{B}\overline{A}BC + A\overline{B}A\overline{C}BC$$
$$= AB + \overline{A}BC + A\overline{B}C \quad (11.9)$$

In the final step the relations (from Eqs. 11.6 and 11.7) $\overline{A}A = 0$, $\overline{A}\overline{A} = \overline{A}$, $C\overline{C} = 0$, and $AA = A$ have been used.

In Fig. 11.12, the result (Eq. 11.9) is shown graphically.

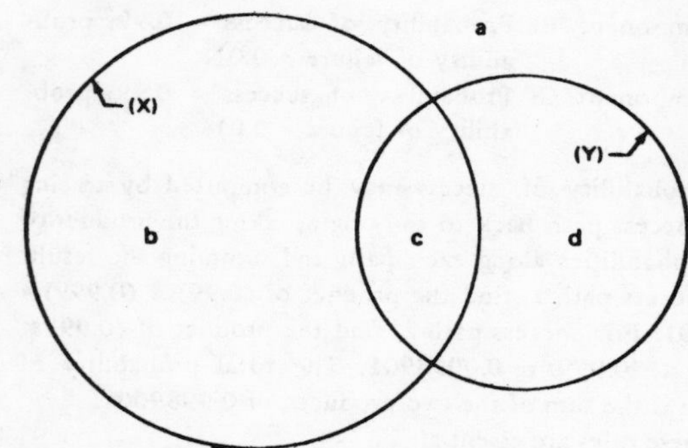

X = area inside circle (X) = areas b and c
Y = area inside circle (Y) = areas c and d
$\overline{X}$ = area outside circle (X) = areas a and d
$\overline{Y}$ = area outside circle (Y) = areas a and b
XY = overlap of circles (X) and (Y) = area c
X + Y = area covered by circles (X) and (Y) = areas b, c, and d
$X\overline{X}$ = overlap of X and $\overline{X}$ = 0 (since there is no overlap)
X + X = area covered by circle (X) and circle (X) = area covered by (X)
XX = overlap of X and X = X (since overlap is same as X)
X + $\overline{X}$ = area covered by X and $\overline{X}$ = total area (a, b, c, and d) = 1 by definition
$X\overline{Y}$ = area of overlap of X and $\overline{Y}$ = area b
$\overline{X}Y$ = area of overlap of $\overline{X}$ and Y = area d
X + $\overline{X}Y$ = area covered by X and $\overline{X}Y$ = areas b, c, and d

Fig. 11.11—Venn diagram illustrating basic Boolean equations. The diagram is a nonrigorous way to visualize the basic Boolean equations. From the above it can be seen that, e.g., $X\overline{X} = 0$, $Y\overline{Y} = 0$, $X + X = X$, $Y + Y = Y$, $X + \overline{X} = 1$, $Y + \overline{Y} = 1$, $XX = X$, $YY = Y$, and $X + \overline{X}Y = X + Y$. Multiplying corresponds to "overlapping" or "intersecting"; addition corresponds to "covering" or "union" of the added elements. Multiplication can also be read as "and," i.e., XY = X and Y; addition can be read as "or", i.e., X + Y = X

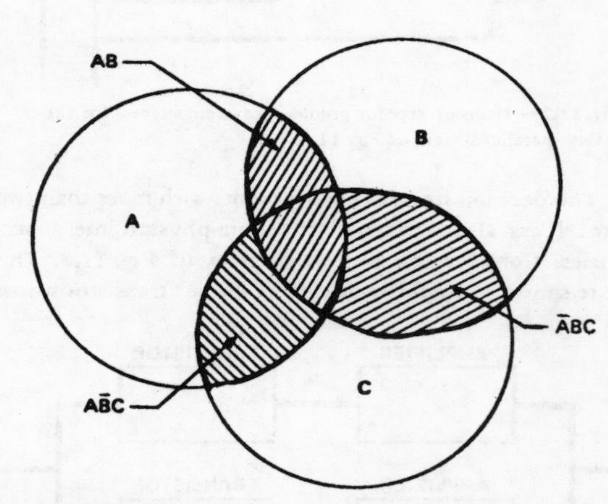

Fig. 11.12—Venn diagram for majority logic case. Note that the regions AB, $\overline{A}BC$, and $A\overline{B}C$ are mutually exclusive, i.e., they do not overlap each other.

The terms AB, $\overline{A}BC$, and $A\overline{B}C$ are seen to be mutually exclusive, i.e., the areas wherein two events intersect are all shown on the Venn diagram, but the areas representing the three terms of Eq. 11.9 do not overlap. Therefore the probability of success is simply the sum of the joint probabilities:

$$P(S) = P(AB) + P(\bar{A}BC) + P(A\bar{B}C)$$
$$= P(A)\,P(B) + P(\bar{A})\,P(B)\,P(C)$$
$$+ P(A)\,P(\bar{B})\,P(C) \qquad (11.10)$$

As a numerical example, assume $P(A) = P(B) = P(C) = 0.99$, then

$$P(S) = (0.99)(0.99) + (0.01)(0.99)(0.99)$$
$$+ (0.99)(0.01)(0.99)$$
$$= 0.999702$$

In summary, the foregoing method is as follows:

1. Write the Boolean expression that is the union of all possible success paths.
2. Separate the first term and intersect the negation of that term with the rest of the terms. Continue down to the last term.
3. Express each negated success path as the union of mutually exclusive events:

$$\overline{AB\ldots N} = \bar{A} + A\bar{B} + \ldots AB\bar{N}$$

4. Starting with the innermost enclosures, clear the expression using the Boolean relations: $A\bar{A} = 0$, $A + A = A$, and $AA = A$.
5. The probability of success is equal to the sum of the probabilities of the mutually exclusive events.

(c) Conditional Probability. Frequently the complexity of a probabilistic calculation can be reduced by the use of conditional probability. It is especially useful if a given component is repeated throughout many branches of the system success model or if the component occupies a key position in the model that makes it difficult to evaluate.

Conditional probability may be expressed as

$$P(S) = P(S|A)\,P(A) + P(S|\bar{A})\,P(\bar{A}) \qquad (11.11)$$

where $P(S)$ = the probability of system success
$P(S|A)$ = the probability of system success, given that component A is good
$P(A)$ = the probability that component A is good
$P(S|\bar{A})$ = the probability of system success, given that component A is bad
$P(\bar{A})$ = the probability that component A is bad

The method is illustrated by the solution to the reliability model resembling the bridge circuit of Fig. 11.13(a). The probability of success can be expressed in the conditional sense as

$$P(S) = P(S|E)\,P(E) + P(S|\bar{E})\,P(\bar{E}) \qquad (11.12)$$

The $P(S|E)$ may be obtained from the easily computed diagram of Fig. 11.13(b). where component E is replaced by a solid line, indicating that E is perfect. The $P(S|\bar{E})$ may be obtained from Fig. 11.13(c), where the path normally provided by component E is missing, indicating that E has failed.

This method is described fully in Ref. 9.

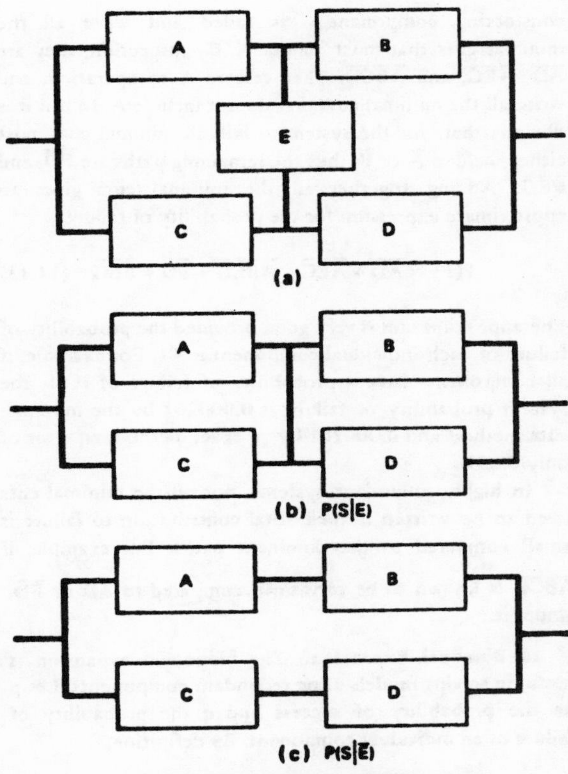

Fig. 11.13—Bridge-type reliability model.

(d) Minimal Cuts. A cut is a collection of equipments belonging to a model such that if all these equipments fail, then successful completion of the mission phase represented by that model is precluded.[10] A minimal cut is a unique set of failed equipment such that the deletion of any one piece of equipment from the cut restores the system to success.

Consider the reliability block diagram of Fig. 11.14. The hand-calculation method is as follows. Start by

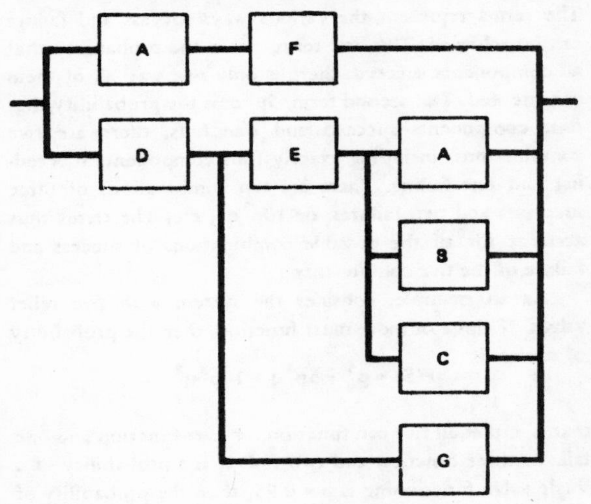

Fig. 11.14—Reliability block diagram to illustrate minimal-cuts method.

considering component A as failed and write all the minimal cuts that must include A. By inspection, they are $\overline{AD}$, $\overline{AEG}$, and $\overline{ABCG}$. Then restore A to operation, and write all the minimal cuts that do not include A. In this it is obvious that, for the system to fail, all minimal cuts must either include $\overline{A}$ or $\overline{F}$; thus the remaining paths are $\overline{FD}$, and $\overline{FEG}$. Adding together all the minimal cuts gives an approximate expression for the probability of failure:

$$P(F) \simeq \overline{AD} + \overline{AEG} + \overline{ABCG} + \overline{FD} + \overline{FEG} \quad (11.13)$$

The approximation is very good provided the probability of failure of each individual component is $\ll 1$. For example, if all components have a probability of failure of 0.01, the system probability of failure is 0.000202 by the minimal-cuts method and 0.000201 by an exact method, an error of only 0.05%.

In highly redundant systems, not all the minimal cuts need to be written if their total contribution to failure is small compared to the dominant paths. For example, if $\overline{ABCG}$ is known to be very small compared to $\overline{AD}$ or $\overline{FD}$, ignore it.

(e) Binomial Expansion. The bionomial expansion is useful in solving models using redundant components. Let p be the probability of success and q the probability of failure of an individual component. By definition,

$$p + q = 1 \quad (11.14)$$

Likewise,

$$(p + q)^n = 1$$

where n represents the level of redundancy. For example, if $n = 5$,

$$(p + q)^5 = p^5 + 5p^4q + 10p^3q^2 + 10p^2q^3 + 5pq^4 + q^5 = 1$$

The terms represent the various ways success and failure can be achieved. The first term, p^5, is the probability that all components succeed; there is only one way all of them can succeed. The second term, $5p^4q$, is the probability that four components succeed and one fails; there are five combinations, including exactly four components succeeding and one failing. There are ten combinations of three successes and two failures, or $10p^3q^2$, etc. The terms thus account for all the possible combinations of success and failure of the five components.

As an example, consider the system with five relief valves. If three or more must function, then the probability of success is

$$P(S) = p^5 + 5p^4q + 10p^3q^2$$

that is, either all five can function, or four function and one fail, or three function and two fail. If the probability of a single valve functioning is $p = 0.95$, then the probability of a single valve failing is $q = 0.05$ and the probability of system success is

$$P(S) = (0.95)^5 + 5(0.95)^4 (0.05) + 10(0.95)^3 (0.05)^2$$
$$= 0.9988$$

In general, the binomial expansion is

$$(p + q)^n = \sum_{k=0}^{n} \binom{n}{k} p^k q^{n-k} \quad (11.15)$$

where

$$\binom{n}{k} = \frac{n!}{(n-k)!\, k!}$$

The coefficients of the binomial expansion can be generated by constructing Pascal's triangle, in which each number is the sum of the two numbers above it. The procedure is shown in Fig. 11.15.

n													
0							1						
1						1		1					
2					1		2		1				
3				1		3		3		1			
4			1		4		6		4		1		
5		1		5		10		10		5		1	
6	1		6		15		20		15		6		1

Fig. 11.15—Pascal's triangle for determining the coefficients of the binomial expansion.

(f) Availability Analysis. *Series Subsystems.* If there are n subsystems in series with failure rates $\lambda_1, \lambda_2, \lambda_3, \ldots, \lambda_n$ and mean repair times of $\theta_1, \theta_2, \theta_3, \ldots, \theta_n$ and if repair is instituted on each subsystem as soon as failure occurs, then the series configuration

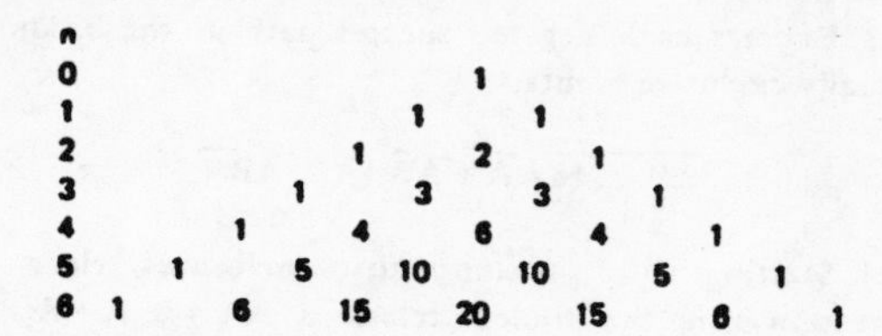

reduces to

$$\longrightarrow \boxed{\lambda_T, \theta_T} \longrightarrow$$

where

$$\lambda_T = \lambda_1 + \lambda_2 + \ldots + \lambda_n$$

and

$$\theta_T = \frac{\lambda_1\theta_1 + \lambda_2\theta_2 + \ldots + \lambda_n\theta_n}{\lambda_T}$$

The system parameters are
Mean time between failure for the system = $(1/\lambda_T)$
Mean down time for system = θ_T

Parallel Subsystems. If there are two subsystems in parallel with failure rates λ_1 and λ_2 and mean repair times θ_1 and θ_2, if either one or both subsystems in operation constitute an operating system, and if repair is instituted on each subsystem as soon as failure occurs, then the parallel configuration

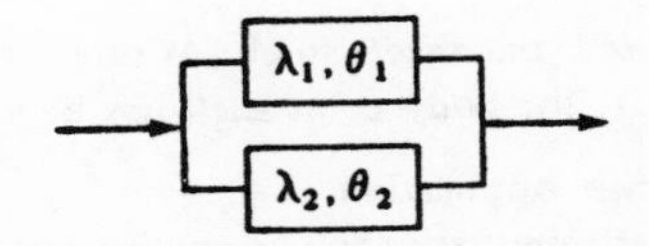

reduces to

$$\lambda_T, \theta_T$$

where

$$\lambda_T = (\lambda_1 \lambda_2)(\theta_1 + \theta_2)$$

and

$$\theta_T = \frac{\theta_1 \theta_2}{\theta_1 + \theta_2}$$

If there are n subsystems in parallel (partially redundant to each other), each having the same failure rate λ and mean repair time θ, and r out of n of these subsystems constitute an operating system, and if repair is instituted on each subsystem as soon as failure occurs, then this system

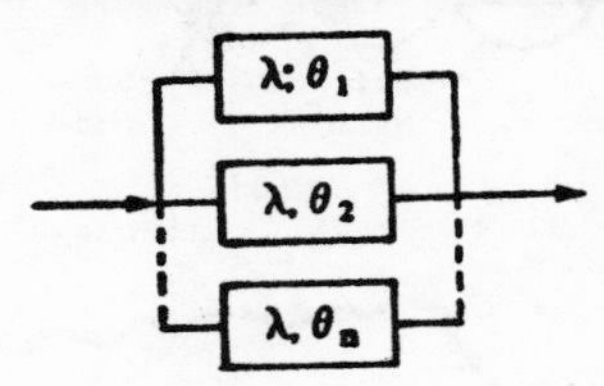

reduces to

$$\lambda_T, \theta_T$$

where

$$\lambda_T = \frac{n\binom{n-1}{r-1}}{\frac{1}{\lambda}\left(\frac{1}{\lambda\theta}\right)^{n-r}}$$

$$\binom{n-1}{r-1} = \frac{(n-1)!}{(r-1)!(n-r)!}$$

and

$$\theta_T = \frac{\theta}{n-r+1}$$

Example 1: Two-out-of-three subsystems must operate, i.e., $n = 3, r = 2$.

$$\binom{n-1}{r-1} = \frac{2!}{1!\,(2-1)!} = 2$$

$$\lambda_T = \frac{3(2)}{\frac{1}{\lambda}\left(\frac{1}{\lambda\theta}\right)} = 6\lambda^2\theta$$

$$\theta_T = \frac{\theta}{1+1} = \frac{\theta}{2}$$

$$\text{Availability} = 1 - \lambda_T\theta_T = 1 - 3(\lambda\theta)^2$$

Example 2: Three-out-of-four subsystems must operate, i.e., $n = 4, r = 3$.

$$\binom{n-1}{r-1} = \frac{3!}{2!\,(3-2)!} = 3$$

$$\lambda_T = \frac{4(3)}{\frac{1}{\lambda}\left(\frac{1}{\lambda\theta}\right)} = 12\lambda^2\theta$$

$$\theta_T = \frac{\theta}{1+1} = \frac{\theta}{2}$$

$$\text{Availability} = 1 - \lambda_T\theta_T = 1 - 6(\lambda\theta)^2$$

Note that the unavailability (= 1 − availability) of the three-out-of-four system is twice that of the two-out-of-three system.

The foregoing relations were derived from Ref. 9. In the derivations it is assumed that failure rates and repair times are exponentially distributed and that all $\lambda\theta$ products are $\ll 1$.

By these techniques any simple series-parallel availability model can be reduced to a single block with an equivalent failure rate and repair time. This technique is not applicable to a system with repeated components or components that bridge between series-parallel strings.

In many cases repair does not commence when failure occurs but rather when failure is discovered. This is particularly true of most of the failures of standby systems and of non-fail-safe failures on power-plant protection systems. In these cases the mean repair time is equal to one-half the time interval between tests plus the actual repair time.

In the event that repair starts immediately on detection of the failure at a periodic test, the following substitution in the preceding formulas will yield useful results with little error:

$$\theta^* = \frac{\tau}{2} + \theta$$

where θ^* is a new equivalent repair time including the time elapse between failure and discovery. Frequently the actual repair time is short compared to the time between tests; so the above can be approximated by $\theta^* \simeq \tau/2$.

APPENDIX A RAW-STOCK IDENTIFICATION SYSTEM

Purpose: To establish a method of identifying raw-material stock in terms of basic composition and thermal condition.

Scope: This procedure applies to raw metallic material that is to be used for production purposes in the Fabrication Department. Specifically excluded are castings and extrusions or any other material that is produced in accordance with engineering drawings and is assigned a part number distinctly different from commercial part numbers.

11-A.1 Identification System

The identification system used may be expanded to provide for materials that may be added to the listings (Table 11-A.2) by assigning striped colors, which are available within the system, or, if necessary, by using more than a single stripe.

11-A.2 System Description

All metals shall be identified by a color-coding system that uses the twelve basic colors listed in Table 11-A.1. A combination of at least two of these colors is required to identify a stocked item. The use of the colors shall be according to the system described in the following paragraphs.

(a) Body Color. One of the basic colors is assigned to each of the categories of materials given in Table 11-A.1.

Table 11-A.1—Basic Body Colors

Color	Material category
Red	Aluminum alloys
Blue	Magnesium alloys
Pink	Brasses
Yellow	Beryllium coppers
Orange	Bronzes
Green	Carbon and free-cutting steels
Light green	Spring steels
Aquamarine	Alloy steels
Brown	Stainless steels
Gray	Magnetic metals
Black	Cast irons (bars and rods)
White	Coppers

This identifies the materials as falling within a specific category. This color will be the background color or body color over which a stripe will be applied to identify the material according to the categories listed in Table 11-A.2.

(b) Body-Color Application

1. Air-dry lacquer shall be used as the body color. The colors shall match the stripe tape described below.

2. One end of bars, rods, and shapes shall be painted the body color (see Fig. 11-A.1).

3. Sheet and strip and plate stock shall be stacked with one end in the same plane and the entire end painted with the body color.

4. Coiled strip, wire, or tubing may be identified per item 3 above by a metal tag painted the body color and attached with a length of wire to the coil [Fig. 11-A.1(c)].

(c) Stripe Color. Strips of colored pressure-sensitive vinyl tape ⅛ in. wide shall be pressed across the body color painted on the ends of tags, bars, rods, or shapes. Painted stripe is used on sheet stock. Within any category of Table 11-A.1, the body color shall not be used for a stripe.

(d) Stripe Application.

1. The striping tape shall be applied across the center of the body-colored ends of bars, rods, and shapes, except in cases where the diameter is so small or the configuration such that the body color is not clearly evident after the application, then the tape shall be wrapped around the material close to the painted end.

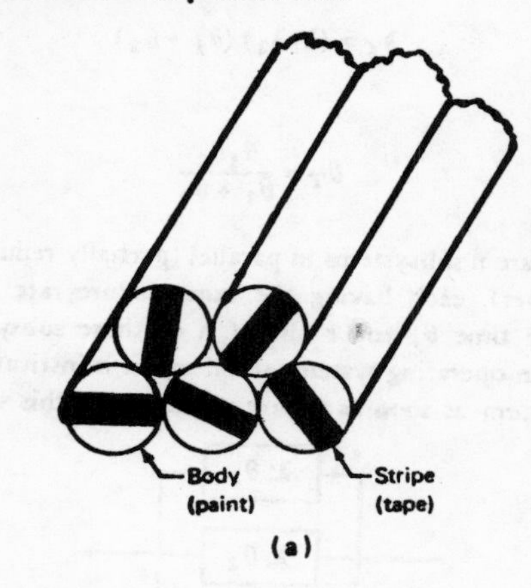

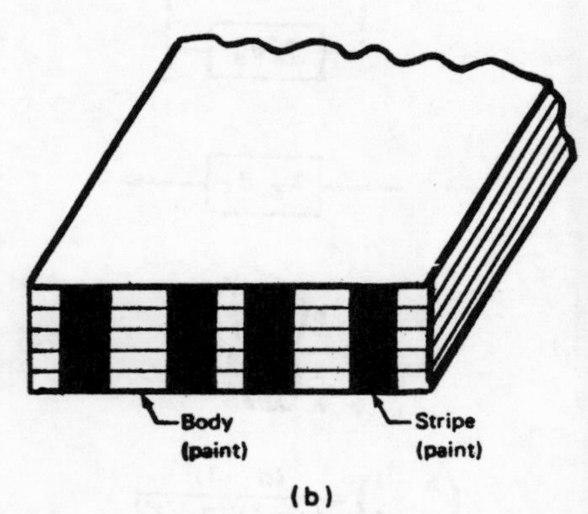

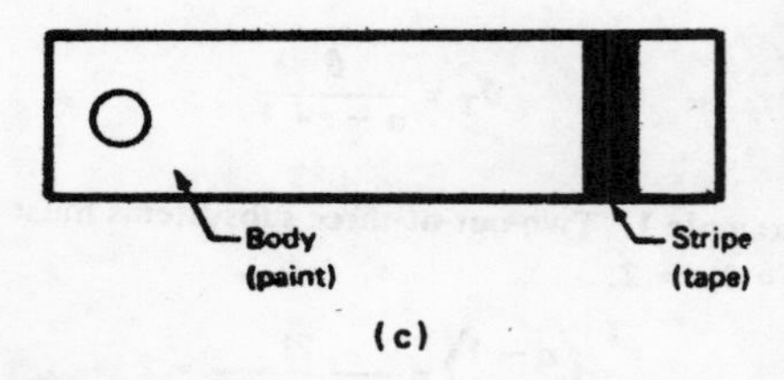

Fig. 11-A.1—Coding application. (a) Bars, rods, and shapes. (b) Sheet and strip and plate stock. (c) Coiled strip, wire, and tubing.

2. Stripes shall be painted over the body color on the ends of sheet, plate, or strip. The stripes shall be placed so that when material is cut off a sheet, the color-code identification is not lost.

3. Stripes shall be painted on coils of strip, wire, or tubing, or tape shall be applied across metal tags.

Table 11-A.2—Stripe Colors

Alloy	Stripe color	Alloy	Stripe color
Aluminum		**Bronze**	
1100-0	Blue	Tobin-SAE 73	Green
2011-T4	Black	SAE 660: QQB-691, Comp. 12	Red
2017-T4	Pink	MIL-N-994, QQB-636, SAE 63	Blue
2024-T4	Yellow	SAE type 1, class A, MIL-B-5687	Pink
3003-H14	White	**Carbon and Free-Cutting Steel**	
5050-H32	Orange	B-1113	Red
5052-0	White–orange	1018 (bar and rod)	Blue
6061-T4	Green	1010 to 1020 (sheet and strip)	White
6061-T6	Gray	MT-1015 (tube)	Yellow
6063-T5	Coppertone	**Magnetic Metal**	
7075-T6	Brown	Hi Mu 80	Red
Tool plate	Light green	Moly–perm.	Blue
Brazing sheet	White–blue	Mu–metal	Yellow
Magnesium		Conetic AA	Green
AZ31-B	Red	**Cast Iron**	
AZ31-B-O	Pink	Meehanite	Red
AZ31-B-H24	Yellow	**Stainless Steel**	
Tool plate	Orange	17-7 PH	Red–blue
Brass		301 AN	Red–yellow
QQB-613	Red	301 FH	Red–orange
QQB-626, SAE 72	Blue	301½ H	Red
WWT-791, SAE 74	Yellow	302, 304	Blue
Spring Steel		302 FH	Blue–orange
1086	Red	303S, 303SE	Pink
1095	Blue	316A	Yellow–pink
Drill Rod		321	Light green
Drill rod	Orange	410	Orange
Alloy Steel		416	Green
52100	Green	430	Gray
4140	Red	440A	Light green–yellow
4620	Blue	440C	Yellow
4130	Yellow	440F	Gray–yellow
Beryllium Copper		Carpenter #10	Blue–yellow
25A	Red	MIL-T-6845, ASTM 269	Black
25¼ H	Blue		
25½ H	Pink		
25H	Orange		
10	Green		
Copper			
OFHC	Red		
Phosphor Bronze	Yellow		

APPENDIX B QUALITY-CONTROL PANEL INSPECTION CHECKLIST

PROJECT ______________ **PANEL** ________.

INSPECTION DETAILS

1.0 WIRE
- 1.1 Correct size and type ______
- 1.2 Insulation
  - 1.2.1 Correct type ______
  - 1.2.2 Correct color ______
  - 1.2.3 Nondefective ______

2.0 TERMINALS
- 2.1 Correct type and hole size ______
- 2.2 Staking
  - 2.2.1 Full staking impression ______
  - 2.2.2 Adequate insulation grip ______
- 2.3 No damage or distortion to grip or tongue ______
- 2.4 Correct wire-insertion depth into barrel
  - 2.4.1 Wire insertion is within 1/16 in. of insulation stripback butt ______
- 2.5 Use of staking tools approved by the terminal manufacturer ______

3.0 CONNECTORS (disconnects or couplings)
- 3.1 Correct size and type ______
- 3.2 Correct clocking ______
- 3.3 Insulation sleeving over solder pots (where required) ______
- 3.4 Solder pots full of solder (no excessive overflow or peaks) ______
- 3.5 Wire insertion in solder pot is within 1/16 in. of insulation stripback butt ______
- 3.6 Connector shell mated and tight ______
- 3.7 Adapter
  - 3.7.1 Tight ______
  - 3.7.2 Lock washers under screwheads of saddle clamps ______
  - 3.7.3 Screws tightened evenly ______
  - 3.7.4 Ground return jumper tight (if used) ______
  - 3.7.5 Sufficient bushing or number of grommets under saddle straps to secure wires and relieve tension on solder pots ______
  - 3.7.6 Correct size and type of bushing ______
- 3.8 Hi-pot test of connectors (where required) ______
- 3.9 Potted connectors
  - 3.9.1 Lack of voids ______
  - 3.9.2 Lack of sponginess ______
  - 3.9.3 Correct compound type ______
  - 3.9.4 Correct cure ______
  - 3.9.5 Proper clearance ______
  - 3.9.6 Complete fill ______
- 3.10 Crimp-type pins
  - 3.10.1 Correct staking impression ______
  - 3.10.2 Correct wire-insertion depth into barrel ______
  - 3.10.3 Insulation stripback is within 1/32 in. of barrel top ______
  - 3.10.4 Use of staking tools approved by the terminal manufacturer ______
- 3.11 Pins properly seated in connector ______

4.0 CLAMPS (support)
- 4.1 Correct size (snug fit but not tight) ______
- 4.2 Clamp "lip" closed ______
- 4.3 Hardware secure ______
- 4.4 Cushion closed ______
- 4.5 Plastic clamps free of distortion and pulling ______
- 4.6 Coaxial-cable clamps are free of excessive tightness to prevent cable distortion ______

5.0 TERMINAL STRIPS
- 5.1 Correct size and type ______
- 5.2 Terminal lugs are staked back to back (if used in multiple) ______
- 5.3 No broken nodes (barriers) ______
- 5.4 Terminals are aligned in middle of nodes ______

5.5 Terminals are correctly and completely identified ____
5.6 Terminal screwhead slots free from burrs or twists ____
5.7 All wires in compression-type terminal secure and tight ____

6.0 COAXIAL CABLES
6.1 Correct type of cable and connector ____
6.2 Stripping ____
6.2.1 Inner conductor for correct length, no nicks or cut strands ____
6.2.2 Shielding and outer insulation for correct length ____
6.3 Correct assembly sequence ____
6.4 Contact pin soldered properly ____
6.5 Location of contact pin is correct after assembly ____
6.6 Hi-pot or megger test for insulation breakdown (where required) ____
6.7 Continuity check ____

7.0 SOLDER
7.1 Identity to type ____
7.2 Nonuse of acid core solder, fluxes, or paste ____
7.3 Bonding of wire to connector solder pots or eyelet by solder flow ____
7.4 No cold solder joints (frosty) ____
7.5 All wire strands in solder pot or eyelet ____
7.6 No excessive solder ____
7.7 No solder spill or splatter into receptacles, devices, components, or printed circuits ____
7.8 Excessive rosin flux removed ____
7.9 Soldered connections covered with insulation and secured (where required) ____
7.10 No overheating of terminals causing insulation scorch, barrier scorch, or component parts burn ____
7.11 Use of soldering tool with sufficient watt density to ensure acceptable solder joints ____
7.12 Equipment or hardware mounting to aid visual identification of part number or value of component after installation ____

8.0 GROUNDING
8.1 All control shafts and bushings grounded (unless otherwise specified) ____
8.2 Ground lugs are utilized in place of parts mounting facilities ____
8.3 Ground lugs mount on metal surface under screwhead ____

9.0 SPLICES (when permitted)
9.1 Inspect prior to covering or enclosing into wire bundles ____
9.2 Correct method and device used to make splice ____

10.0 PANELS, MODULES
10.1 Basic ____
10.1.1 Bonding areas clean, free of primers of paint ____
10.1.2 Equipment, devices, and operators for correct: ____
10.1.2.1 Part number ____
10.1.2.2 Rating ____
10.1.2.3 Location ____
10.1.2.4 Model ____
10.1.2.5 Type ____
10.1.2.6 Size ____
10.1.3 All parts mounted (when practical) so that values and identity are in full view ____
10.1.4 Mounting security of parts
10.1.4.1 Retaining nuts backed up with loc washers ____
10.1.4.2 Correct length of screws and bolts ____
10.1.4.3 Screwhead slots free from burrs or twists ____
10.1.4.4 Bonding areas refinished after installation of jumpers or ground straps ____
10.1.4.5 Unused mounting or terminal screws or nuts tightened ____
10.2 Wire bundles or harnesses secured
10.3 Wire bundles are routed to avoid interference with future terminations, components, moving mechanical parts, and stub-up locations ____
10.4 Slack loops are provided where needed to properly facilitate the movement of mechanical moving structures or allow access to components or sub-panel opening or removal ____
10.5 No twisting or entanglement of wire in harnesses or bundles ____
10.6 Bushings and barriers installed where required ____
10.7 Terminals correctly installed and tightened ____
10.8 Wire breakouts and wire entrances in or out of bundles are consistent. Avoid "as the crow flies" or point-to-point wire routing in favor of main wire bundle flow pattern ____
10.9 Wire terminations at terminal blocks for customer "future tie in" or external connections are consistent either to the right or left of the terminal blocks ____
10.10 All disconnect plugs, receptacles, and connectors are suitably covered to prevent entrance of foreign material and moisture ____
10.11 Correct dimensions of:
10.11.1 Chassis ____
10.11.2 Panel fronts ____
10.11.3 Mounting hole layout ____
10.11.4 Access door ____
10.11.5 Clearance requirements ____
10.12 All wire bundles, harnesses, and wire runs are suitably protected from protruding surfaces, sharp edges, and abrasive surfaces ____
10.13 Controls
10.13.1 Shafts clear panel cutout ____
10.13.2 Shafts correct length ____
10.13.3 Shafts are straight ____
10.13.4 Knobs are secured to shaft ____
10.13.5 Knobs clear front panel ____
10.13.6 Locking devices on controls (where required) ____

11.0 FINISH
11.1 Paint
11.1.1 Color ____
11.1.2 Primer ____
11.1.3 Dry (air or bake) ____
11.1.4 Touch-up blended ____
11.1.5 Free of ripples, sag, orange peel, over spray, pits, and voids ____

12.0 METERS
- 12.1 Correct range ______
- 12.2 Scale identification ______
- 12.3 Scale and index, ______
- 12.4 Size, color, mount
  - 12.4.1 Flush, semiflush, front, back ______
- 12.5 Illumination ______
- 12.6 Dial or scale background color ______
- 12.7 Dial or scale numeral color ______

13.0 INDICATING LIGHTS
- 13.1 Illumination factor ______
- 13.2 Cover lens color correct ______
- 13.3 Engraving correct (where used) ______

14.0 NAMEPLATES
- 14.1 Correct etching or attachment ______
- 14.2 Drawing number—group number ______
- 14.3 Serial number ______

15.0 WORKMANSHIP

16.0 LAYOUT IDENTIFICATION OF PARTS

17.0 IDENTIFICATION OF COMPLETED COMPONENTS

18.0 IDENTIFICATION OF PANEL FACILITIES
All controls, indicators, lights, jacks, sockets, and fuse holders marked with suitable words, phrases, or abbreviations indicating the function or use of the part ______

19.0 IDENTIFICATION OF WIRE TERMINAL POINTS

20.0 CLEANLINESS

21.0 FINAL INSPECTION
- 21.1 Compliance to procedure, approved drawings, specifications, and all applicable data ______
- 21.2 Cleanliness continuity (if not part of test) ______
- 21.3 Function or operation (where required) ______
- 21.4 Rework completed and accepted ______
- 21.5 Accessory parts, instruction sheets, or books identified ______
- 21.6 All shortages documented ______
- 21.7 Applicable test and inspection stamps ______
- 21.8 Test data complete, reviewed, and approved ______

REFERENCES

1. A. V. Feigenbaum, *Total Quality Control*, McGraw-Hill Book Company, Inc., New York, 1969.
2. J. M. Juran, *Quality Control Handbook*, McGraw-Hill Book Company, Inc., New York, 1962.
3. E. L. Grant, *Statistical Quality Control*, McGraw-Hill Book Company, Inc., New York, 1964.
4. Institute of Electrical and Electronics Engineers, IEEE-STD-352-1972, *General Principles for Reliability Analysis of Nuclear Power Generating Station Protection Systems.*
5. Benjamin Epstein and Albert Schiff, Improving Availability and Readiness of Field Equipment Through Periodic Inspection, USAEC Report UCRL-50451, Lawrence Radiation Laboratory, July 16, 1968.
6. Emanuel Parzen, *Modern Probability Theory and Its Applications*, John Wiley & Sons, Inc., New York, 1960.
7. Montgomery Phister, *Logical Design of Digital Computers*, John Wiley & Sons, Inc., New York, 1958.
8. Norman Roberts, *Mathematical Methods in Reliability Engineering*, McGraw-Hill Book Company, Inc., New York, 1964.
9. Jerome D. Braverman and I. Paul Sternberg, Reliability Modeling with Conditional Probabilistic Logic, *Proceedings of the 1966 Annual Symposium on Reliability*, sponsored by IEEE, IES, SNT, and ASQC, pp. 321-331, Institute of Electrical and Electronics Engineers.
10. Stuart A. Weisberg and John H. Schmidt, Computer Technique for Estimating System Reliability, *Proceedings of the 1966 Annual Symposium on Reliability*, sponsored by IEEE, IES, SNT, and ASQC, pp. 87-97, Institute of Electrical and Electronics Engineers.